COMPUTATIONAL METHODS FOR PLASTICITY

COMPUTATIONAL METHODS FOR PLASTICITY

THEORY AND APPLICATIONS

EA de Souza Neto

D Perić

DRJ Owen

Civil and Computational Engineering Centre, Swansea University

WILEY

A John Wiley and Sons, Ltd, Publication

This edition first published 2008

Registered office
John Wiley & Sons Ltd, The Atrium, Southern Gate, Chichester, West Sussex, PO19 8SQ, United Kingdom

For details of our global editorial offices, for customer services and for information about how to apply for permission to reuse the copyright material in this book please see our website at www.wiley.com.

Library of Congress Cataloging-in-Publication Data

Neto, Eduardo de Souza.
Computational methods for plasticity : theory and applications / Eduardo de Souza Neto, Djordje Peric, DRJ Owen.
p. cm.
Includes bibliographical references and index.
ISBN 978-0-470-69452-7 (cloth)
1. Plasticity–Mathematical models. I. Peric, Djordje. II. Owen, DRJ. III. Title.
TA418.14.N48 2008
531'.385–dc22
2008033260

A catalogue record for this book is available from the British Library.

ISBN 978-0-470-69452-7

Set in 10/12pt Times by Sunrise Setting Ltd, Torquay, UK.
Printed in Singapore by Markono.

This book is lovingly dedicated to

Deise, Patricia and André;

Mira, Nikola and Nina;

Janet, Kathryn and Lisa.

CONTENTS

PREFACE

THE purpose of this text is to describe in detail numerical techniques used in small and large strain finite element analysis of elastic and inelastic solids. Attention is focused on the derivation and description of various constitutive models – based on phenomenological hyperelasticity, elastoplasticity and elasto-viscoplasticity – together with the relevant numerical procedures and the practical issues arising in their computer implementation within a quasi-static finite element scheme. Many of the techniques discussed in the text are incorporated in the FORTRAN program, named `HYPLAS`, which accompanies this book and can be found at www.wiley.com/go/desouzaneto. This computer program has been specially written to illustrate the practical implementation of such techniques. We make no pretence that the text provides a complete account of the topics considered but rather, we see it as an attempt to present a reasonable balance of theory and numerical procedures used in the finite element simulation of the nonlinear mechanical behaviour of solids.

When we embarked on the project of writing this text, our initial idea was to produce a rather concise book – based primarily on our own research experience – whose bulk would consist of the description of numerical algorithms required for the finite element implementation of small and large strain plasticity models. As the manuscript began to take shape, it soon became clear that a book designed as such would be most appropriate to those already involved in research on computational plasticity or closely related areas, being of little use to those willing to learn computational methods in plasticity from a fundamental level. A substantial amount of background reading from other sources would be required for readers unfamiliar with topics such as basic elastoplasticity theory, tensor analysis, nonlinear continuum mechanics – particularly nonlinear kinematics – finite hyperelasticity and general dissipative constitutive theory of solids. Our initial plan was then gradually abandoned as we chose to make the text more self-contained by incorporating a considerable amount of basic theory. Also, while writing the manuscript, we decided to add more advanced (and very exciting) topics such as damage mechanics, anisotropic plasticity and the treatment of finite strain single crystal plasticity. Following this route, our task took at least three times as long to complete and the book grew to about twice the size as originally planned. There remains plenty of interesting material we would like to have included but cannot due to constraints of time and space. We are certainly far more satisfied with the text now than with its early versions, but we do not believe our final product to be optimal in any sense. We merely offer it to fill a gap in the existing literature, hoping that the reader will benefit from it in some way.

The text is arranged in three main parts. Part One presents some basic material of relevance to the subject matter of the book. It includes an overview of elementary tensor analysis, continuum mechanics and thermodynamics, the finite element method in quasi-static nonlinear solid mechanics and a brief description of the computer program `HYPLAS`. Part Two deals with small strain problems. It introduces the mathematical theory of infinitesimal plasticity as well as the relevant numerical procedures for the implementation of plasticity models within a finite element environment. Both rate-independent (elastoplastic) and rate-dependent (elasto-viscoplastic) theories are addressed and some advanced models, including anisotropic plasticity and ductile damage are also covered. Finally, in Part Three we focus on large strain problems. The theory of finite hyperelasticity is reviewed first together with details of its finite element implementation. This is followed by an introduction to large strain plasticity. Hyperelastic-based theories with multiplicative elastoplastic kinematics as well as hypoelastic-based models are discussed, together with relevant numerical procedures for their treatment. The discussion on finite plasticity and its finite element implementation culminates with a description of techniques for single crystal plasticity. Finite element techniques for large-strain near-incompressibility are also addressed.

We are indebted to many people for their direct or indirect contribution to this text. This preface would not be complete without the due acknowledgement of this fact and a record of our sincere gratitude to the following: to J.M.P. Macedo for the numerous valuable suggestions during the design of the program HYPLAS at the very early stages of this project; to R. Billardon for the many enlightening discussions on damage modelling; to R.A. Feijóo and E. Taroco for the fruitful discussions held on many occasions over a long period of time; to M. Dutko for producing some of the numerical results reported; to Y.T. Feng for helpful discussions on the arc-length method; to F.M. Andade Pires for his key contribution to the development of *F-bar*-Patch elements, for producing the related figures presented and for thoroughly reviewing early versions of the manuscript; to P.H. Saksono for his involvement in the production of isoerror maps; to A. Orlando for literally 'scanning' through key parts of the text to find inconsistencies of any kind; to L. Driemeier, W. Dettmer, M. Vaz Jr, M.C. Lobão, M. Partovi, D.C.D. Speirs, D.D. Somer, E. Saavedra, A.J.C. Molina, S. Giusti and P.J. Blanco for carefully reviewing various parts of the manuscript, spotting hard-to-find mistakes and making several important suggestions for improvement. Last, but not least, to our late colleague and friend Mike Crisfield, for the numerous illuminating and passionate discussions (often held on the beach or late in the bar) on many topics addressed in the book.

EA de Souza Neto
D Perić
DRJ Owen

Swansea

Part One

Basic concepts

1 INTRODUCTION

OVER the last four decades, the use of computational techniques based on the Finite Element Method has become a firmly established practice in the numerical solution of nonlinear solid mechanics problems both in academia and industry.

In their early days, these techniques were largely limited to infinitesimal deformation and strain problems with the main complexity arising from the nonlinear constitutive characterization of the underlying material by means of basic elastoplastic or elasto-viscoplastic theories. Applications were mostly confined to the modelling of the behaviour of solids in conventional areas of engineering and analyses were carried out on crude, user-unfriendly software that typically required highly specialized users. Since those days, this area of solid mechanics – generally known as *computational plasticity* – has experienced dramatic developments. Fuelled by the steady increase in computing power at decreasing costs together with the continuous industrial demand for accurate models of solids, the evolution of computational plasticity techniques have made possible the development of refined software packages with a considerable degree of automation that are today routinely employed by an ever-increasing number of engineers and scientists. The variety of practical problems of interest to which such techniques are currently applied with acceptable levels of predictive capability is very wide. They range from traditional engineering applications, such as stress analysis in structures, soil and rock mechanics, to the simulation of manufacturing processes such as metal forming. Also included are much less conventional applications, such as the simulation of food processing, mining operations and biological tissue behaviour. Many such problems are characterised by extremely large straining and material behaviour often described by means of rather complex constitutive equations.

1.1. Aims and scope

The main objective of this text is to describe in detail numerical techniques used in the small and large strain analysis of elastic and inelastic solids by the Finite Element Method. Particular emphasis is placed on the derivation and description of various constitutive models – based on phenomenological hyperelasticity, elastoplasticity and elasto-viscoplasticity – as well as on the relevant numerical procedures and the practical issues arising in their computer implementation. The range covered goes from basic infinitesimal isotropic to more sophisticated finite strain theories, including anisotropy. Many of the techniques discussed in the text are implemented in the FORTRAN computer program, named HYPLAS, which accompanies this book. Parts of its source code are included in the text and should help readers correlate the relevant numerical methods with their computer implementation in practice. Another important aspect to emphasise is that the performance of many of the models/techniques described in the text is documented in numerical examples. These should be of particular relevance to those involved in software research and development in computational plasticity.

In order to make the book more self-contained, we have chosen to incorporate a considerable amount of basic theory within the text. This includes some material on elementary tensor analysis, an introduction to the nonlinear mechanics and thermodynamics of continuous media and an overview of small and large strain elastoplasticity and viscoplasticity theory, finite hyperelasticity and finite element techniques in nonlinear solid mechanics. Having a sound knowledge of such topics is essential, we believe, to the clear understanding of the very problems the numerical techniques discussed in this book are meant to simulate.

We reiterate, however, that our main focus here is computational. Thus, the volume of theory and the depth at which it is presented is kept to the minimum necessary for the above task. For example, in the presentation of tensor analysis and continuum mechanics and thermodynamics, we omit most proofs for standard relations. In plasticity, viscoplasticity and hyperelasticity, we limit ourselves mainly to presenting the constitutive models together with their most relevant properties and the essential relations needed in their formulation. Issues such as material stability and the existence and uniqueness of solutions to initial boundary value problems are generally not addressed.

1.1.1. READERSHIP

This book is intended for graduate students, research engineers and scientists working in the field of computational continuum mechanics. The text requires a basic knowledge of solid mechanics – especially the theory of linear elasticity – as well as the Finite Element Method and numerical procedures for the approximate solution of ordinary differential equations. An elementary understanding of vector and tensor calculus is also very helpful. Readers wishing to follow the computer implementation of the procedures described in the text should, in addition, be familiar at a fairly basic level with the FORTRAN computer programming language. It is worth remarking here that the choice of the FORTRAN language is motivated mainly by the following:

(a) its widespread acceptance in engineering computing in general and, in particular, within the finite element community;

(b) the suitability of procedural languages for codes with relatively low level of complexity, such as `HYPLAS`. In the present case, the use of more advanced programming concepts (e.g. object-oriented programming) could add a further difficulty in the learning of the essential concepts the `HYPLAS` code is meant to convey;

(c) its relative clarity in the coding of short algorithmic procedures such as those arising typically in the implementation of elastic and inelastic material models – the main subject of this book.

1.2. Layout

In line with the above aims, the book has been divided into three parts as follows.

- *Part One: Basic concepts*. In this part we introduce concepts of fundamental relevance to the applications presented in Parts Two and Three. The following material is covered:
  - elementary tensor analysis;
  - introductory continuum mechanics and thermodynamics;
  - finite elements in quasi-static nonlinear solid mechanics;
  - a concise description of the computer program `HYPLAS`.
- *Part Two: Small strains*. Here, the theory of infinitesimal plasticity is introduced together with the relevant numerical procedures used for its implementation into a finite element environment. A relatively wide range of models is presented, including both rate-independent (elastoplastic) and rate-dependent (elasto-viscoplastic) theories. The following main topics are considered:
  - the theory of infinitesimal plasticity;
  - finite elements in infinitesimal plasticity;
  - advanced plasticity models, including anisotropy;
  - viscoplasticity;
  - elastoplasticity with damage.
- *Part Three: Large strains*. This part focuses on finite strain hyperelasticity and elastoplasticity problems. The models discussed here, as well as their computational implementation, are obviously more complex than those of Part Two. Their complexity stems partly from the finite strain kinematics. Thus, to follow Part Three, a sound knowledge of the kinematics of finite deformations discussed in Chapter 3 (in Part One) is essential. The following topics are addressed:
  - large strain isotropic hyperelasticity;
  - large strain plasticity;
  - finite element techniques for large strain incompressibility;
  - single crystal (anisotropic) finite plasticity.

The material has been organised into sixteen chapters and four appendices. These will now be briefly described. The remainder of Chapter 1 discusses the general scheme of notation adopted in the book.

Chapter 2 contains an introduction to elementary tensor analysis. In particular, the material is presented mainly in *intrinsic* (or *compact*) tensor notation – which is heavily relied upon thoughout the book.

Chapter 3 provides an introdution to the mechanics and thermodynamics of continuous media. The material presented here covers the kinematics of deformation, balance laws and constitutive theory. These topics are essential for an in-depth understanding of the theories discussed in later chapters.

Chapter 4 shows the application of the Finite Element Method to the solution of problems in quasi-static nonlinear solid mechanics. A generic dissipative constitutive model, initially presented in Chapter 3, is used as the underlying material model.

Chapter 5 describes the general structure of the program HYPLAS, where many of the techniques discussed in the book are implemented. We remark that the program description is rather concise. Further familiarisation with the program will require the reader to follow the comments in the FORTRAN source code together with the cross-referencing of the main procedures with their description in the book. This is probably more relevant to those wishing to use the HYPLAS program for research and development purposes.

Chapter 6 is devoted to the mathematical theory of infinitesimal plasticity. The main concepts associated with phenomenological time-independent plasticity are introduced here. The basic yield criteria of Tresca, von Mises, Mohr–Coulomb and Drucker–Prager are reviewed, together with the most popular plastic flow rules and hardening laws.

In Chapter 7, we introduce the essential numerical methods required in the finite element solution of initial boundary value problems with elastoplastic underlying material models. Applications of the von Mises model with both isotropic and mixed isotropic/kinematic hardening are described in detail. The most relevant subroutines of the program HYPLAS are also listed and explained in detail.

Chapter 8 focuses on the detailed description of the implementation of the basic plasticity models based on the Tresca, Mohr–Coulomb and Drucker–Prager yield criteria. Again, the relevant subroutines of HYPLAS are listed and explained in some detail.

In Chapter 9 we describe the numerical treatment of plasticity models under plane stress conditions. Different options are considered and their relative merits and limitations are discussed. Parts of source code are also included to illustrate some of the most important programming aspects. The application of the concepts introduced here to other stress-constrained states is briefly outlined at the end of the chapter.

In Chapter 10 advanced elastoplasticity models are considered. Here we describe the computational implementation of a modified Cam-Clay model for soils, a capped Drucker–Prager model for geomaterials and the Hill, Hoffman and Barlat–Lian anisotropic models for metals. The numerical techniques required for the implementation of such models are mere specialisations of the procedures already discussed in Chapters 7 and 8. However, due to the inherent complexity of the models treated in this chapter, their actual implementation is generally more intricate than those of the basic models.

Chapter 11 begins with an introduction to elasto-viscoplasticity theory within the constitutive framework for dissipative materials described in Chapter 3. The (rate-independent) plasticity theory is then obtained as a limiting case of viscoplasticity. The numerical methods for a generic viscoplastic model are described, following closely the procedures applied earlier in elastoplasticity. Application of the methodology to von Mises criterion-based viscoplastic models is described in detail.

In Chapter 12 we discuss continuum damage mechanics – the branch of Continuum Solid Mechanics devoted to the modelling of the progressive material deterioration that precedes the onset of macroscopic fracturing. Some elastoplastic damage models are reviewed and their implementation, with the relevant computational issues, is addressed in detail.

Chapter 13 introduces finite strain hyperelasticity. The basic theory is reviewed and some of the most popular isotropic models are presented. The finite element implementation of the Odgen model is discussed in detail with relevant excerpts of HYPLAS source code included. In addition, the modelling of the so-called Mullins dissipative effect by means of a hyperelastic-damage theory is addressed at the end of the chapter. This concept is closely related to those already discussed in Chapter 12 for ductile elastoplastic damage.

In Chapter 14 we introduce finite strain elastoplasticity together with the numerical procedures relevant to the finite element implementation of finite plasticity models. The main discussion is focused on hyperelastic-based finite plasticity theories with multiplicative kinematics. The finite plasticity models actually implemented in the program HYPLAS belong to this class of theories. However, for completeness, a discussion on the so-called hypoelastic-based theories is also included. The material presented is confined mostly to isotropic elastoplasticity, with anisotropy in the form of kinematic hardening added only at the end of the chapter.

Chapter 15 is concerned with the treatment of large strain incompressibility within the Finite Element Method. This issue becomes crucial in large-scale finite strain simulations where the use of low-order elements (which, without any added specific techniques, are generally inappropriate near the incompressible limit) is highly desirable. Three different approaches to tackle the problem are considered: the so-called F-*bar* method (including its more recent F-*bar*-Patch variant for simplex elements); the Enhanced Assumed Strain (EAS) technique, and the mixed u/p formulation.

Finally, in Chapter 16 we describe a general model of large-strain single-crystal plasticity together with the relevant numerical procedures for its use within a finite element environment. The implementation of a specialisation of the general model based on a planar double-slip system is described in detail. This implementation is incorporated in the program HYPLAS.

In addition to the above, four appendices are included. Appendix A is concerned with isotropic scalar- and tensor-valued functions of a symmetric tensor that are widely exploited throughout the text. It presents some important basic properties as well as formulae that can be used in practice for the computation of function values and function derivatives. Appendix B addresses the tensor exponential function. The tensor exponential is of relevance for the treatment of finite plasticity presented in Chapters 14 and 16. In Appendix C we derive the linearisation of the virtual work equation both under small and large deformations. The expressions derived here provide the basic formulae for the tangent operators required in the assembly of the tangent stiffness matrix in the finite element context. Finally, in Appendix D we describe the handling – including array storage and product operations – of second and fourth-order tensors in finite element computer programs.

1.2.1. THE USE OF BOXES

Extensive use of *boxes* has been made to summarise constitutive models and numerical algorithms in general (refer, for example, to pages 146 and 199). Boxes should be of particular use to readers interested mostly in computer implementation aspects and who wish to skip the details of derivation of the models and numerical procedures. Numerical algorithms listed in boxes are presented in the so-called *pseudo-code* format – a format that resembles the actual computer code of the procedure. For boxes describing key procedures implemented in the program HYPLAS, the name of the corresponding FORTRAN subprogram is often indicated at the top of the box (see page 221, for instance).

1.3. General scheme of notation

Throughout the text, an attempt has been made to maintain the notation as uniform as possible by assigning specific letter styles to the representation of each type of mathematical entity.

At the same time, we have tried to keep the notation in line with what is generally adopted in the present subject areas. Unfortunately, these two goals often conflict so that, in many cases, we choose to adopt the notation that is more widely accepted instead of adhering to the use of specific fonts for specific mathematical entities. Whenever such exceptions occur, their meaning should either be clear from the context or will be explicitly mentioned the first time they appear.

1.3.1. CHARACTER FONTS. GENERAL CONVENTION

The mathematical meaning associated with specific font styles is given below. Some important exceptions are also highlighted. We emphasise, however, that other exceptions, not mentioned here, may occur in the text.

- *Italic light-face letters* A, a, . . .: scalars and scalar-valued functions.
- *Italic bold-face majuscules* $\boldsymbol{B}$, $\boldsymbol{C}$, . . .: second-order tensors or tensor-valued functions. *Light-face with indices* B_{ij}, $B_{\alpha\beta}$, . . .: components of the corresponding tensors. *Important exceptions*: $\boldsymbol{A}$ (set of thermodynamical forces), $\boldsymbol{H}$ (generalised elastoplastic hardening modulus), $\boldsymbol{J}$ (generalised viscoplastic hardening constitutive function).
- *Italic bold-face minuscules* $\boldsymbol{p}$, $\boldsymbol{v}$, . . .: points, vectors and vector-valued functions. *Light-face with indices* p_i, p_α, . . .: coordinates (components) of the corresponding points (vectors).
  Important exception: $\boldsymbol{s}$ (stress tensor deviator).
- *Sans-serif (upright) bold-face letters* A, a, . . .: fourth-order tensors. *Light-face with indices* A_{ijkl}, $\mathsf{A}_{\alpha\beta\gamma\delta}$, . . .: the corresponding Cartesian components.
- *Greek light-face letters* α, β, . . ., Φ, Ψ, . . .: scalars and scalar-valued functions. *Important exception*: Ω (region of Euclidean space occupied by a generic body).
- *Greek bold-face minuscules* $\boldsymbol{\beta}$, $\boldsymbol{\sigma}$, $\boldsymbol{\tau}$, . . .: second-order tensors. *Light-face with indices* σ_{ij}, $\sigma_{\alpha\beta}$, . . .: the corresponding components.
  Important exceptions: $\boldsymbol{\alpha}$ (generic set of internal state variables), $\boldsymbol{\varphi}$ (deformation map) and $\boldsymbol{\eta}$ (when meaning virtual displacement fields).
- *Upright bold-face majuscules, minuscules and greek letters* $\mathbf{A}$, $\mathbf{a}$, $\boldsymbol{\sigma}$, . . .: finite element arrays (vectors and matrices) representing second or fourth-order tensors and general finite element operators.
- *Script majuscules* $\mathscr{A}$, $\mathscr{B}$, . . .: spaces, sets, groups, bodies.
- *German majuscules* $\mathfrak{F}$, $\mathfrak{G}$, . . .: constitutive response functionals.
- *Calligraphic majuscules* $\mathcal{X}$, $\mathcal{Y}$, . . .: generic mathematical entity (scalar, vector, tensor, field, etc.)
- *Typewriter style letters* `HYPLAS`, `SUVM`, . . .: used exclusively to denote FORTRAN procedures and variable names, instructions, etc.

1.3.2. SOME IMPORTANT CHARACTERS

The specific meaning of some important characters is listed below. We remark that some of these symbols may occasionally be used with a different connotation (which should be clear from the context).

$\boldsymbol{A}$	Generic set of thermodynamical forces
A	Finite element assembly operator (note the large font)
A	First elasticity tensor
a	Spatial elasticity tensor
$\boldsymbol{B}$	Left Cauchy–Green strain tensor
$\boldsymbol{B}^e$	Elastic left Cauchy–Green strain tensor
$\mathbf{B}$	Discrete (finite element) symmetric gradient operator (strain-displacement matrix)
$\mathscr{B}$	Generic body
$\boldsymbol{b}$	Body force
$\bar{\boldsymbol{b}}$	Reference body force
$\boldsymbol{C}$	Right Cauchy–Green strain tensor
c	Cohesion
D	Damage internal variable
$\boldsymbol{D}$	Stretching tensor
$\boldsymbol{D}^e$	Elastic stretching
$\boldsymbol{D}^p$	Plastic stretching
D	Infinitesimal consistent tangent operator
D^e	Infinitesimal elasticity tensor
D^{ep}	Infinitesimal elastoplastic consistent tangent operator
$\mathbf{D}$	Consistent tangent matrix (array representation of D)
$\mathbf{D}^e$	Elasticity matrix (array representation of D^e)
$\mathbf{D}^{ep}$	Elastoplastic consistent tangent matrix (array representation of D^{ep})
E	Young's modulus
$\boldsymbol{E}_i$	Eigenprojection of a symmetric tensor associated with the ith eigenvalue
$\mathscr{E}$	Three-dimensional Euclidean space; elastic domain
$\bar{\mathscr{E}}$	Set of plastically admissible stresses
$\boldsymbol{e}_i$	Generic base vector; unit eigenvector of a symmetric tensor associated with the ith eigenvalue

$\boldsymbol{F}$	Deformation gradient
$\boldsymbol{F}^e$	Elastic deformation gradient
$\boldsymbol{F}^p$	Plastic deformation gradient
$\mathbf{f}^{\rm ext}$	Global (finite element) external force vector
$\mathbf{f}^{\rm ext}_{(e)}$	External force vector of element e
$\mathbf{f}^{\rm int}$	Global (finite element) internal force vector
$\mathbf{f}^{\rm int}_{(e)}$	Internal force vector of element e
G	Virtual work functional; shear modulus
$\mathbf{G}$	Discrete (finite element) full gradient operator
H	Hardening modulus
$\boldsymbol{H}$	Generalised hardening modulus
I_1, I_2, I_3	Principal invariants of a tensor
I	Fourth-order identity tensor: $\mathsf{I}_{ijkl} = \delta_{ik}\delta_{jl}$
I_S	Fourth-order symmetric identity tensor: $\mathsf{I}_{ijkl} = \frac{1}{2}(\delta_{ik}\delta_{jl} + \delta_{il}\delta_{jk})$
$\mathsf{I}_{\rm d}$	Deviatoric projection tensor: $\mathsf{I}_{\rm d} \equiv \mathsf{I}_S - \frac{1}{3}\boldsymbol{I} \otimes \boldsymbol{I}$
$\boldsymbol{I}$	Second-order identity tensor
$\mathbf{I}_S$	Array representation of I_S
$\mathbf{i}$	Array representation of $\boldsymbol{I}$
J	Jacobian of the deformation map: $J \equiv \det \boldsymbol{F}$
J_2, J_3	Stress deviator invariants
$\boldsymbol{J}$	Generalised viscoplastic hardening constitutive function
K	Bulk modulus
$\mathbf{K}_T$	Global tangent stiffness matrix
$\mathbf{K}_T^{(e)}$	Tangent stiffness matrix of element e
$\mathscr{K}$	Set of kinematically admissible displacements
$\boldsymbol{L}$	Velocity gradient
$\boldsymbol{L}^e$	Elastic velocity gradient
$\boldsymbol{L}^p$	Plastic velocity gradient
$\boldsymbol{m}^\alpha$	Unit vector normal to the slip plane α of a single crystal
$\boldsymbol{N}$	Plastic flow vector
$\bar{\boldsymbol{N}}$	Unit plastic flow vector: $\bar{\boldsymbol{N}} \equiv \boldsymbol{N}/\|\boldsymbol{N}\|$
$\mathscr{O}$	The orthogonal group
$\mathscr{O}^+$	The rotation (proper orthogonal) group

$\mathbf{0}$	Zero tensor; zero array; zero generic entity
$\boldsymbol{o}$	Zero vector
$\boldsymbol{P}$	First Piola–Kirchhoff stress tensor
$\boldsymbol{p}$	Generic material point
p	Cauchy or Kirchhoff hydrostatic pressure
$\boldsymbol{Q}$	Generic orthogonal or rotation (proper orthogonal) tensor
q	von Mises (Cauchy or Kirchhoff) effective stress
$\boldsymbol{R}$	Rotation tensor obtained from the polar decomposition of $\boldsymbol{F}$
$\boldsymbol{R}^e$	Elastic rotation tensor
$\mathscr{R}$	Real set
$\mathbf{r}$	Global finite element residual (out-of-balance) force vector
s	Entropy
$\boldsymbol{s}$	Cauchy or Kirchhoff stress tensor deviator
$\boldsymbol{s}^\alpha$	Unit vector in the slip direction of slip system α of a single crystal
$\boldsymbol{t}$	Surface traction
$\bar{\boldsymbol{t}}$	Reference surface traction
$\boldsymbol{U}$	Right stretch tensor
$\boldsymbol{U}^e$	Elastic right stretch tensor
$\boldsymbol{U}^p$	Plastic right stretch tensor
$\mathscr{U}$	Space of vectors in $\mathscr{E}$
$\boldsymbol{u}$	Generic displacement vector field
$\mathbf{u}$	Global finite element nodal displacement vector
$\boldsymbol{V}$	Left stretch tensor
$\boldsymbol{V}^e$	Elastic left stretch tensor
$\boldsymbol{V}^p$	Plastic left stretch tensor
$\mathscr{V}$	Space of virtual displacements
$\boldsymbol{v}$	Generic velocity field
$\boldsymbol{W}$	Spin tensor
$\boldsymbol{W}^e$	Elastic spin tensor
$\boldsymbol{W}^p$	Plastic spin tensor
$\boldsymbol{x}$	Generic point in space
α_p	Ogden hyperelastic constants ($p = 1, \ldots, N$) for a model with N terms in the Ogden strain-energy function series
$\boldsymbol{\alpha}$	Generic set of internal state variables

$\boldsymbol{\beta}$ Back-stress tensor

$\dot{\gamma}$ Plastic multiplier

δ_{ij} Krönecker delta

$\boldsymbol{\varepsilon}$ Strain tensor; also Eulerian logarithmic strain when under large strains

$\boldsymbol{\varepsilon}^e$ Elastic strain tensor; also elastic Eulerian logarithmic strain when under large strains

$\boldsymbol{\varepsilon}^p$ Plastic strain tensor

$\boldsymbol{\mathrm{\varepsilon}}$, $\boldsymbol{\mathrm{\varepsilon}}^e$, $\boldsymbol{\mathrm{\varepsilon}}^p$ Array representation of $\boldsymbol{\varepsilon}$, $\boldsymbol{\varepsilon}^e$ and $\boldsymbol{\varepsilon}^p$, respectively

ε, ε^e, ε^p Axial total, elastic and plastic strain in one-dimensional models

$\bar{\varepsilon}^p$ Effective (or accumulated) plastic strain

$\boldsymbol{\eta}$ Virtual displacement field; relative stress tensor in kinematic hardening plasticity models

κ Isotropic hardening thermodynamical force

λ One of the Lamé constants of linear elasticity; axial stretch; load factor in proportional loading

λ^e, λ^p Elastic and plastic axial stretch

λ_i, λ_i^e, λ_i^p Total, elastic and plastic principal stretches

μ One of the Lamé constants of linear elasticity

μ_p Ogden hyperelastic constants ($p = 1, \ldots, N$) for a model with N terms in the Ogden strain-energy function series

ν Poisson ratio

Ξ Dissipation potential

$\boldsymbol{\xi}$ Isoparametric coordinates of a finite element

ρ Mass density

$\bar{\rho}$ Reference mass density

$\boldsymbol{\sigma}$ Cauchy stress tensor

σ Axial stress in one-dimensional models

σ_i Principal Cauchy stress

σ_y Yield stress (uniaxial yield stress for the conventional von Mises and Tresca models)

σ_{y0} Initial yield stress

$\boldsymbol{\mathrm{\sigma}}$ Array representation of $\boldsymbol{\sigma}$

$\boldsymbol{\tau}$ Kirchhoff stress tensor

τ_i Principal Kirchhoff stress

τ^{α}	Resolved Schmid shear stress on slip system α
$\boldsymbol{\tau}$	Array representation of $\boldsymbol{\tau}$
Φ	Yield function; damage function
φ	Deformation map; motion
Ψ	Plastic flow potential
ψ	Helmholtz free-energy per unit mass; strain-energy per unit mass
Ω	Domain of a body in the reference configuration
$\Omega^{(e)}$	Domain of finite element e in the reference configuration

1.3.3. INDICIAL NOTATION, SUBSCRIPTS AND SUPERSCRIPTS

When indicial notation is used, the following convention is adopted for subscripts:

- *Italic subscripts* i, j, k, l, . . ., as in the Cartesian components

$$u_i,\ B_{ij},\ \mathsf{a}_{ijkl},$$

  or for the basis vectors

$$\boldsymbol{e}_i,$$

  normally range over 1, 2 and 3. In a more general context (in an n-dimensional space), their range may be $1, 2, \ldots, n$.

- *Greek subscripts* α, β, γ, δ, . . .: range over 1 and 2.

- When an index appears twice in the same product, summation over the repeated index (Einstein notation) is implied unless otherwise stated. For example,

$$u_i \boldsymbol{e}_i = \sum_{i=1}^{3} u_i \boldsymbol{e}_i.$$

We remark that subscripts are *not* employed exclusively in connection with indicial notation. Different connotations are assigned to subscripts throughout the text and the actual meaning of a particular subscript should be clear from the context. For example, in the context of incremental numerical procedures, subscripts may indicate the relevant increment number. In the expression

$$\Delta\varepsilon = \varepsilon_{n+1} - \varepsilon_n,$$

the subscripts n and $n + 1$ refer to the values of ε, respectively, at the end of increments n and $n + 1$.

Superscripts

Superscripts are also used extensively throughout the text. The meaning of a particular superscript will be stated the first time it appears in the text and should be clear from the context thereafter.

1.3.4. OTHER IMPORTANT SYMBOLS AND OPERATIONS

The meanings of other important symbols and operations are listed below.

$\det(\cdot)$	Determinant of $(\cdot)$
$\text{dev}(\cdot)$	Deviator of $(\cdot)$
$\text{div}_p(\cdot)$	Material divergence of $(\cdot)$
$\text{div}_x(\cdot)$	Spatial divergence of $(\cdot)$
$\exp(\cdot)$	Exponential (including tensor exponential) of $(\cdot)$
$\ln(\cdot)$	Natural logarithm (including tensor logarithm) of $(\cdot)$
$o(\cdot)$	A term that vanishes faster than $(\cdot)$
$\text{sign}(\cdot)$	The signum function: $\text{sign}(\cdot) \equiv (\cdot)/\lvert(\cdot)\rvert$
$\text{skew}(\cdot)$	Skew-symmetric part of $(\cdot)$
$\text{sym}(\cdot)$	Symmetric part of $(\cdot)$
$\text{tr}(\cdot)$	Trace of $(\cdot)$
$\Delta(\cdot)$	Increment of $(\cdot)$. Typically, $\Delta(\cdot) = (\cdot)_{n+1} - (\cdot)_n$
$\delta(\cdot)$	Iterative increment of $(\cdot)$
$\nabla(\cdot)$	Gradient of $(\cdot)$
$\nabla_p(\cdot)$	Material gradient of $(\cdot)$
$\nabla_x(\cdot)$	Spatial gradient of $(\cdot)$
∇^s, ∇^s_p, $\nabla^s_x(\cdot)$	Corresponding symmetric gradients of $(\cdot)$
$\partial(\cdot)$	Boundary of the domain $(\cdot)$
$\partial_a(\cdot)$	Subdifferential of $(\cdot)$ with respect to a
$\dfrac{\partial}{\partial a}(\cdot)$	Derivative of $(\cdot)$ with respect to a
$\dot{(\cdot)}$	Material time derivative of $(\cdot)$
$(\cdot)^T$	The transpose of $(\cdot)$
$a \equiv b$	Means a is *defined* as b. The symbol $\equiv$ is often used to emphasise that the expression in question is a definition.
$a := b$, $a := a + b$	Assignment operation. The value of the right-hand side of the expression is *assigned* to its left-hand side. The symbol $:=$ is often used to emphasise that a given expression is an assignment operation performed by a computational algorithm.
$\mathsf{S} : \mathsf{T}$, $\mathsf{S} : \boldsymbol{T}$, $\boldsymbol{S} : \boldsymbol{T}$	Double contraction of tensors (internal product of second-order tensors)

$\boldsymbol{u} \cdot \boldsymbol{v}$, $\boldsymbol{T} \cdot \boldsymbol{u}$, $\boldsymbol{S} \cdot \boldsymbol{T}$	Single contraction of vectors and tensors. The single contraction symbol (the single dot) is usually omitted in single contractions between a tensor and a vector or between tensors; that is, $\boldsymbol{T} \cdot \boldsymbol{u}$ and $\boldsymbol{S} \cdot \boldsymbol{T}$ are normally represented simply as $\boldsymbol{T}\boldsymbol{u}$ and $\boldsymbol{S}\boldsymbol{T}$.
$\boldsymbol{u} \times \boldsymbol{v}$	Vector product
$\boldsymbol{S} \otimes \boldsymbol{T}$, $\boldsymbol{u} \otimes \boldsymbol{v}$	Tensor product of tensors or vectors
$\boldsymbol{\mathcal{X}} * \boldsymbol{\mathcal{Y}}$	The appropriate product between two generic entities, $\boldsymbol{\mathcal{X}}$ and $\boldsymbol{\mathcal{Y}}$, in a given context.
$\lvert(\cdot)\rvert$	Norm (absolute value) of the scalar $(\cdot)$
$\lVert\boldsymbol{T}\rVert$, $\lVert\boldsymbol{v}\rVert$	Euclidean norm of tensors and vectors: $\lVert\boldsymbol{T}\rVert \equiv \sqrt{\boldsymbol{T}:\boldsymbol{T}}, \quad \lVert\boldsymbol{v}\rVert \equiv \sqrt{\boldsymbol{v}\cdot\boldsymbol{v}}$
$\rightarrow$ `A`	Used in the description of arguments of FORTRAN subprograms listed in the text. An arrow pointing to the right followed by an argument name `A` means that the value of `A` at entry is used by the relevant subprogram and is not changed during its execution.
$\leftarrow$ `A`	Analogously to $\rightarrow$ above, an arrow pointing to the left followed by an argument name `A` means that the value of `A` is calculated and returned by the relevant subprogram and its value at entry is ignored.
$\leftrightarrow$ `A`	Analogously to $\rightarrow$ and $\leftarrow$ above, a double arrow followed by an argument name `A` means that the value of `A` at entry is used by the relevant subprogram and is changed during its execution.

2 ELEMENTS OF TENSOR ANALYSIS

THIS chapter introduces the notation and reviews some fundamentals of vector and tensor calculus which are extensively employed in this book. Throughout this text, preference is given to the use of *intrinsic* (or *compact*) tensor notation where no indices are used to represent mathematical entities. However, in many of the definitions introduced in this chapter, *indicial* notation is also used. This will allow readers not yet familiar with compact notation to associate compactly written entities and operations with their indicial forms, which will be expressed exclusively in terms of Cartesian coordinate systems. We note that the use of Cartesian, rather than curvilinear, coordinates for indicial representation is sufficiently general for the applications considered in this book. In the subsequent chapters, the use of indicial notation will be much less frequent. Readers who are familiar with tensor analysis and, in particular, the use of compact notation, may comfortably skip this chapter.

We remark that no proofs are given to most relations presented in this chapter. Readers interested in such proofs and a more in-depth treatment of the subject are referred to other textbooks such as Gurtin (1981).

2.1. Vectors

Let $\mathscr{E}$ be an n-dimensional Euclidean space and let $\mathscr{U}$ be the space of n-dimensional vectors associated with $\mathscr{E}$. Points of $\mathscr{E}$ and vectors $\mathscr{U}$ satisfy the basic rules of vector algebra, with which we assume the reader to be familiar.

2.1.1. INNER PRODUCT, NORM AND ORTHOGONALITY

Let

$$\boldsymbol{u} \cdot \boldsymbol{v}$$

denote the *inner product* (or *scalar product*) between two arbitrary vectors of $\mathscr{U}$. The *Euclidean norm* (or, simply, *norm*) of a vector $\boldsymbol{u}$ is defined as

$$\|\boldsymbol{u}\| = \sqrt{\boldsymbol{u} \cdot \boldsymbol{u}}, \tag{2.1}$$

and $\boldsymbol{u}$ is said to be a *unit vector* if

$$\|\boldsymbol{u}\| = 1. \tag{2.2}$$

The *zero vector*, here denoted $\boldsymbol{o}$, is the element of $\mathscr{U}$ that satisfies

$$\|\boldsymbol{o}\| = 0. \tag{2.3}$$

Computational Methods for Plasticity: Theory and Applications EA de Souza Neto, D Perić and DRJ Owen

A vector $\boldsymbol{u}$ is said to be *orthogonal* (or *perpendicular*) to a vector $\boldsymbol{v}$ if

$$\boldsymbol{u} \cdot \boldsymbol{v} = 0. \tag{2.4}$$

2.1.2. ORTHOGONAL BASES AND CARTESIAN COORDINATE FRAMES

A set $\{\boldsymbol{e}_i\} \equiv \{\boldsymbol{e}_1, \boldsymbol{e}_2, \ldots, \boldsymbol{e}_n\}$ of n mutually orthogonal vectors satisfying

$$\boldsymbol{e}_i \cdot \boldsymbol{e}_j = \delta_{ij}, \tag{2.5}$$

where

$$\delta_{ij} = \begin{cases} 1 & \text{if } i = j \\ 0 & \text{if } i \neq j \end{cases} \tag{2.6}$$

is the *Krönecker delta*, defines an *orthonormal basis* for $\mathscr{U}$.

Any vector $\boldsymbol{u} \in \mathscr{U}$ can be represented as

$$\boldsymbol{u} = u_1\, \boldsymbol{e}_1 + u_2\, \boldsymbol{e}_2 + \cdots + u_n\, \boldsymbol{e}_n = u_i\, \boldsymbol{e}_i, \tag{2.7}$$

where

$$u_i = \boldsymbol{u} \cdot \boldsymbol{e}_i, \quad i = 1, 2, \ldots, n \tag{2.8}$$

are the Cartesian components of $\boldsymbol{u}$ relative to the basis $\{\boldsymbol{e}_i\}$. Any vector of $\mathscr{U}$ is uniquely defined by its components relative to a given basis. This allows us to represent any vector $\boldsymbol{u}$ as a single column matrix, denoted $[\boldsymbol{u}]$, of components

$$[\boldsymbol{u}] = \begin{bmatrix} u_1 \\ u_2 \\ \vdots \\ u_n \end{bmatrix}. \tag{2.9}$$

An orthonormal basis, $\{\boldsymbol{e}_i\}$, together with an origin point, $\boldsymbol{x}_0 \in \mathscr{E}$, defines a *Cartesian coordinate frame*. Analogously to the representation of vectors, any point $\boldsymbol{x}$ of $\mathscr{E}$ can be represented by an array

$$[\boldsymbol{x}] = \begin{bmatrix} x_1 \\ x_2 \\ \vdots \\ x_n \end{bmatrix}, \tag{2.10}$$

of *Cartesian coordinates* of $\boldsymbol{x}$. The Cartesian coordinates $\{x_i\}$ of $\boldsymbol{x}$ are the Cartesian components of the *position vector*

$$\boldsymbol{r} = \boldsymbol{x} - \boldsymbol{x}_0, \tag{2.11}$$

of $\boldsymbol{x}$ relative to the origin $\boldsymbol{x}_0$. That is,

$$x_i = (\boldsymbol{x} - \boldsymbol{x}_0) \cdot \boldsymbol{e}_i. \tag{2.12}$$

2.2. Second-order tensors

Second-order tensors are linear transformations from $\mathscr{U}$ into $\mathscr{U}$, i.e. a second-order tensor

$$\boldsymbol{T}: \mathscr{U} \rightarrow \mathscr{U}$$

maps each vector $\boldsymbol{u}$ into a vector

$$\boldsymbol{v} = \boldsymbol{T}\boldsymbol{u}. \tag{2.13}$$

Any linear vector-valued function of a vector is a tensor. The operations of *sum* and *scalar multiplication* of tensors are defined by

$$\begin{aligned}(\boldsymbol{S}+\boldsymbol{T})\boldsymbol{u} &= \boldsymbol{S}\,\boldsymbol{u} + \boldsymbol{T}\boldsymbol{u} \\ (\alpha\,\boldsymbol{S})\boldsymbol{u} &= \alpha\,(\boldsymbol{S}\,\boldsymbol{u}),\end{aligned} \tag{2.14}$$

where $\alpha \in \mathscr{R}$. In addition, the *zero tensor*, $\boldsymbol{0}$, and the *identity tensor*, $\boldsymbol{I}$, are, respectively, the tensors that satisfy

$$\begin{aligned}\boldsymbol{0}\boldsymbol{u} &= \boldsymbol{o} \\ \boldsymbol{I}\,\boldsymbol{u} &= \boldsymbol{u}\end{aligned} \tag{2.15}$$

$\forall\, \boldsymbol{u} \in \mathscr{U}$.

The *product* of two tensors $\boldsymbol{S}$ and $\boldsymbol{T}$ is the tensor $\boldsymbol{ST}$ defined by

$$\boldsymbol{ST}\boldsymbol{u} = \boldsymbol{S}\,(\boldsymbol{T}\boldsymbol{u}). \tag{2.16}$$

In general,

$$\boldsymbol{ST} \neq \boldsymbol{TS}. \tag{2.17}$$

If $\boldsymbol{ST} = \boldsymbol{TS}$, then $\boldsymbol{S}$ and $\boldsymbol{T}$ are said to *commute*.

2.2.1. THE TRANSPOSE. SYMMETRIC AND SKEW TENSORS

The *transpose*, $\boldsymbol{T}^T$, of a tensor $\boldsymbol{T}$ is the unique tensor that satisfies

$$\boldsymbol{T}\boldsymbol{u} \cdot \boldsymbol{v} = \boldsymbol{u} \cdot \boldsymbol{T}^T\,\boldsymbol{v}, \quad \forall\, \boldsymbol{u}, \boldsymbol{v} \in \mathscr{U}. \tag{2.18}$$

If

$$\boldsymbol{T} = \boldsymbol{T}^T, \tag{2.19}$$

then $\boldsymbol{T}$ is said to be *symmetric*. If

$$\boldsymbol{T} = -\boldsymbol{T}^T, \tag{2.20}$$

then $\boldsymbol{T}$ is said to be *skew symmetric* (or, simply, *skew*).

Any tensor $\boldsymbol{T}$ can be decomposed as the sum

$$\boldsymbol{T} = \mathrm{sym}(\boldsymbol{T}) + \mathrm{skew}(\boldsymbol{T}) \tag{2.21}$$

of its *symmetric part*

$$\mathrm{sym}(\boldsymbol{T}) \equiv \tfrac{1}{2}(\boldsymbol{T} + \boldsymbol{T}^T) \tag{2.22}$$

and its *skew part*

$$\mathrm{skew}(\boldsymbol{T}) \equiv \tfrac{1}{2}(\boldsymbol{T} - \boldsymbol{T}^T). \tag{2.23}$$

Basic properties

The following basic properties involving the transpose, skew and symmetric parts of a tensor hold:

(i) $(\boldsymbol{S}+\boldsymbol{T})^T = \boldsymbol{S}^T + \boldsymbol{T}^T$.

(ii) $(\boldsymbol{S}\,\boldsymbol{T})^T = \boldsymbol{T}^T\,\boldsymbol{S}^T$.

(iii) $(\boldsymbol{T}^T)^T = \boldsymbol{T}$.

(iv) If $\boldsymbol{T}$ is symmetric, then

$$\text{skew}(\boldsymbol{T}) = \boldsymbol{0}, \quad \text{sym}(\boldsymbol{T}) = \boldsymbol{T}.$$

(v) If $\boldsymbol{T}$ is skew, then

$$\text{skew}(\boldsymbol{T}) = \boldsymbol{T}, \quad \text{sym}(\boldsymbol{T}) = \boldsymbol{0}.$$

2.2.2. TENSOR PRODUCTS

The *tensor product* of two vectors $\boldsymbol{u}$ and $\boldsymbol{v}$, denoted

$$\boldsymbol{u} \otimes \boldsymbol{v},$$

is the tensor that maps each vector $\boldsymbol{w}$ into the vector $(\boldsymbol{v}\cdot\boldsymbol{w})\boldsymbol{u}$:

$$(\boldsymbol{u}\otimes\boldsymbol{v})\,\boldsymbol{w} = (\boldsymbol{v}\cdot\boldsymbol{w})\,\boldsymbol{u}. \tag{2.24}$$

The tensor product is sometimes referred to as the *dyadic product*.

Some properties of the tensor product

The following relations hold for any vectors $\boldsymbol{s}$, $\boldsymbol{t}$, $\boldsymbol{u}$, $\boldsymbol{v}$, $\boldsymbol{w}$ and tensor $\boldsymbol{S}$:

(i) $\boldsymbol{u}\otimes(\boldsymbol{v}+\boldsymbol{w}) = \boldsymbol{u}\otimes\boldsymbol{v} + \boldsymbol{u}\otimes\boldsymbol{w}$.

(ii) $(\boldsymbol{u}\otimes\boldsymbol{v})^T = \boldsymbol{v}\otimes\boldsymbol{u}$.

(iii) $(\boldsymbol{u}\otimes\boldsymbol{v})(\boldsymbol{s}\otimes\boldsymbol{t}) = (\boldsymbol{v}\cdot\boldsymbol{s})\boldsymbol{u}\otimes\boldsymbol{t}$.

(iv) $\boldsymbol{e}_i\otimes\boldsymbol{e}_i = \boldsymbol{I}$.

(v) $\boldsymbol{S}\,(\boldsymbol{u}\otimes\boldsymbol{v}) = (\boldsymbol{S}\,\boldsymbol{u})\otimes\boldsymbol{v}$.

(vi) $(\boldsymbol{u}\otimes\boldsymbol{v})\boldsymbol{S} = (\boldsymbol{u}\otimes\boldsymbol{S}^T\boldsymbol{v})$.

2.2.3. CARTESIAN COMPONENTS AND MATRIX REPRESENTATION

Any second-order tensor $\boldsymbol{T}$ can be represented as

$$\boldsymbol{T} = T_{11}\, \boldsymbol{e}_1 \otimes \boldsymbol{e}_1 + T_{12}\, \boldsymbol{e}_1 \otimes \boldsymbol{e}_2 + \cdots + T_{nn}\, \boldsymbol{e}_n \otimes \boldsymbol{e}_n = T_{ij}\, \boldsymbol{e}_i \otimes \boldsymbol{e}_j \tag{2.25}$$

where

$$T_{ij} = \boldsymbol{e}_i \cdot \boldsymbol{T} \boldsymbol{e}_j \tag{2.26}$$

are the Cartesian components of $\boldsymbol{T}$. Note that in (2.25) no summation is implied over the index n.

Any second tensor is uniquely defined by its Cartesian components. Thus, by arranging the components T_{ij} in a matrix, we may have the following *matrix representation* for $\boldsymbol{T}$:

$$[\boldsymbol{T}] = \begin{bmatrix} T_{11} & T_{12} & \cdots & T_{1n} \\ T_{21} & T_{22} & \cdots & T_{2n} \\ \vdots & \vdots & \ddots & \vdots \\ T_{n1} & T_{n2} & \cdots & T_{nn} \end{bmatrix}. \tag{2.27}$$

For instance, the Cartesian components of the identity tensor, $\boldsymbol{I}$, read

$$I_{ij} = \delta_{ij}, \tag{2.28}$$

so that its matrix representation is given by

$$[\boldsymbol{I}] = \begin{bmatrix} 1 & 0 & \cdots & 0 \\ 0 & 1 & \cdots & 0 \\ \vdots & \vdots & \ddots & \vdots \\ 0 & 0 & \cdots & 1 \end{bmatrix}. \tag{2.29}$$

The Cartesian components of the vector $\boldsymbol{v} = \boldsymbol{T}\boldsymbol{u}$ are given by

$$v_i = [T_{jk}(\boldsymbol{e}_j \otimes \boldsymbol{e}_k)\, u_l \boldsymbol{e}_l] \cdot \boldsymbol{e}_i = T_{ij}\, u_j. \tag{2.30}$$

Thus, the array $[\boldsymbol{v}]$ of Cartesian components of $\boldsymbol{v}$ is obtained from the matrix-vector product

$$[\boldsymbol{v}] \equiv \begin{bmatrix} v_1 \\ v_2 \\ \vdots \\ v_n \end{bmatrix} = \begin{bmatrix} T_{11} & T_{12} & \cdots & T_{1n} \\ T_{21} & T_{22} & \cdots & T_{2n} \\ \vdots & \vdots & \ddots & \vdots \\ T_{n1} & T_{n2} & \cdots & T_{nn} \end{bmatrix} \begin{bmatrix} u_1 \\ u_2 \\ \vdots \\ u_n \end{bmatrix}. \tag{2.31}$$

It can be easily proved that the Cartesian components T^T_{ij} of the transpose $\boldsymbol{T}^T$ of a tensor $\boldsymbol{T}$ are given by

$$T^T_{ij} = T_{ji}. \tag{2.32}$$

Thus, $\boldsymbol{T}^T$ has the following Cartesian matrix representation

$$[\boldsymbol{T}^T] = \begin{bmatrix} T_{11} & T_{21} & \cdots & T_{n1} \\ T_{12} & T_{22} & \cdots & T_{n2} \\ \vdots & \vdots & \ddots & \vdots \\ T_{1n} & T_{2n} & \cdots & T_{nn} \end{bmatrix}. \tag{2.33}$$

2.2.4. TRACE, INNER PRODUCT AND EUCLIDEAN NORM

For any $\boldsymbol{u}, \boldsymbol{v} \in \mathscr{U}$, the *trace* of the tensor $(\boldsymbol{u} \otimes \boldsymbol{v})$ is the linear map defined as

$$\operatorname{tr}(\boldsymbol{u} \otimes \boldsymbol{v}) = \boldsymbol{u} \cdot \boldsymbol{v}. \tag{2.34}$$

For a generic tensor, $\boldsymbol{T} = T_{ij}\, \boldsymbol{e}_i \otimes \boldsymbol{e}_j$, it then follows that

$$\operatorname{tr} \boldsymbol{T} = T_{ij} \operatorname{tr}(\boldsymbol{e}_i \otimes \boldsymbol{e}_j) = T_{ij}\, \delta_{ij} = T_{ii}, \tag{2.35}$$

that is, the trace of $\boldsymbol{T}$ is the sum of the diagonal terms of the Cartesian matrix representation $[\boldsymbol{T}]$.

The *inner product*, $\boldsymbol{S} : \boldsymbol{T}$, between two tensors $\boldsymbol{S}$ and $\boldsymbol{T}$ is defined as

$$\boldsymbol{S} : \boldsymbol{T} = \operatorname{tr}(\boldsymbol{S}^T \boldsymbol{T}), \tag{2.36}$$

or, in Cartesian component form,

$$\boldsymbol{S} : \boldsymbol{T} = S_{ij}\, T_{ij}. \tag{2.37}$$

The *Euclidean norm* (or simply *norm*) of a tensor $\boldsymbol{T}$ is defined as

$$\|\boldsymbol{T}\| \equiv \sqrt{\boldsymbol{T} : \boldsymbol{T}} = \sqrt{T_{11}^2 + T_{12}^2 + \cdots + T_{nn}^2}. \tag{2.38}$$

Basic properties

The following basic properties involving the internal product of tensors hold for any tensors $\boldsymbol{R}$, $\boldsymbol{S}$, $\boldsymbol{T}$ and vectors $\boldsymbol{s}$, $\boldsymbol{t}$, $\boldsymbol{u}$, $\boldsymbol{v}$:

(i) $\boldsymbol{I} : \boldsymbol{T} = \operatorname{tr} \boldsymbol{T}$.

(ii) $\boldsymbol{R} : (\boldsymbol{S}\boldsymbol{T}) = (\boldsymbol{S}^T \boldsymbol{R}) : \boldsymbol{T} = (\boldsymbol{R}\boldsymbol{T}^T) : \boldsymbol{S}$.

(iii) $\boldsymbol{u} \cdot \boldsymbol{S}\boldsymbol{v} = \boldsymbol{S} : (\boldsymbol{u} \otimes \boldsymbol{v})$.

(iv) $(\boldsymbol{s} \otimes \boldsymbol{t}) : (\boldsymbol{u} \otimes \boldsymbol{v}) = (\boldsymbol{s} \cdot \boldsymbol{u})(\boldsymbol{t} \cdot \boldsymbol{v})$.

(v) $T_{ij} = \boldsymbol{T} : (\boldsymbol{e}_i \otimes \boldsymbol{e}_j)$.

(vi) $(\boldsymbol{u} \otimes \boldsymbol{v})_{ij} = (\boldsymbol{u} \otimes \boldsymbol{v}) : (\boldsymbol{e}_i \otimes \boldsymbol{e}_j) = u_i\, v_j$.

(vii) If $\boldsymbol{S}$ is symmetric, then $\boldsymbol{S} : \boldsymbol{T} = \boldsymbol{S} : \boldsymbol{T}^T = \boldsymbol{S} : \operatorname{sym}(\boldsymbol{T})$.

(viii) If $\boldsymbol{S}$ is skew, then $\boldsymbol{S} : \boldsymbol{T} = -\boldsymbol{S} : \boldsymbol{T}^T = \boldsymbol{S} : \operatorname{skew}(\boldsymbol{T})$.

(ix) If $\boldsymbol{S}$ is symmetric and $\boldsymbol{T}$ is skew, then $\boldsymbol{S} : \boldsymbol{T} = 0$.

2.2.5. INVERSE TENSOR. DETERMINANT

A tensor $\boldsymbol{T}$ is said to be *invertible* if its *inverse*, denoted $\boldsymbol{T}^{-1}$, satisfying

$$\boldsymbol{T}^{-1}\boldsymbol{T} = \boldsymbol{T}\boldsymbol{T}^{-1} = \boldsymbol{I} \tag{2.39}$$

exists.

The *determinant* of a tensor $\boldsymbol{T}$, denoted

$$\det \boldsymbol{T},$$

is the determinant of the matrix $[\boldsymbol{T}]$. A tensor $\boldsymbol{T}$ is invertible if and only if

$$\det \boldsymbol{T} \neq 0. \tag{2.40}$$

A tensor $\boldsymbol{T}$ is said to be *positive definite* if

$$\boldsymbol{T}\boldsymbol{u} \cdot \boldsymbol{u} > 0, \quad \forall\, \boldsymbol{u} \neq \boldsymbol{o}. \tag{2.41}$$

Any positive definite tensor is invertible.

Basic relations involving the determinant and the inverse tensor

Relation (i) below holds for any tensors $\boldsymbol{S}$ and $\boldsymbol{T}$ and relations (ii)–(iv) hold for any invertible tensors $\boldsymbol{S}$ and $\boldsymbol{T}$:

(i) $\det(\boldsymbol{S}\boldsymbol{T}) = \det \boldsymbol{S}\ \det \boldsymbol{T}$.

(ii) $\det \boldsymbol{T}^{-1} = (\det \boldsymbol{T})^{-1}$.

(iii) $(\boldsymbol{S}\boldsymbol{T})^{-1} = \boldsymbol{T}^{-1}\boldsymbol{S}^{-1}$.

(iv) $(\boldsymbol{T}^{-1})^T = (\boldsymbol{T}^T)^{-1}$.

2.2.6. ORTHOGONAL TENSORS. ROTATIONS

A tensor $\boldsymbol{Q}$ is said to be *orthogonal* if

$$\boldsymbol{Q}^T = \boldsymbol{Q}^{-1}. \tag{2.42}$$

The set of all orthogonal tensors will be denoted $\mathscr{O}$. The determinant of any orthogonal tensor equals either $+1$ or -1. An orthogonal tensor $\boldsymbol{Q}$ with

$$\det \boldsymbol{Q} = 1 \tag{2.43}$$

is called a *proper orthogonal* tensor (or a *rotation*). The set of all proper orthogonal (or rotation) tensors is the *proper orthogonal group*. It will be denoted $\mathscr{O}^+$. The product

$$\boldsymbol{Q}_1\boldsymbol{Q}_2$$

of any two orthogonal tensors $\boldsymbol{Q}_1$ and $\boldsymbol{Q}_2$ is an orthogonal tensor. If $\boldsymbol{Q}_1$ and $\boldsymbol{Q}_2$ are rotations, then the product $\boldsymbol{Q}_1\boldsymbol{Q}_2$ is also a rotation. For all vectors $\boldsymbol{u}$ and $\boldsymbol{v}$, an orthogonal tensor $\boldsymbol{Q}$ satisfies

$$\boldsymbol{Q}\boldsymbol{u} \cdot \boldsymbol{Q}\boldsymbol{v} = \boldsymbol{u} \cdot \boldsymbol{v}. \tag{2.44}$$

Rotations and changes of basis

Let $\{\boldsymbol{e}_i\}$ and $\{\boldsymbol{e}_i^*\}$ be two orthonormal bases of $\mathscr{U}$. Such bases are related by

$$\boldsymbol{e}_i^* = \boldsymbol{R}\,\boldsymbol{e}_i, \quad \text{for } i = 1, 2, \ldots, n, \tag{2.45}$$

where $\boldsymbol{R}$ is a rotation. Let $\boldsymbol{T}$ and $\boldsymbol{u}$ be, respectively, a tensor and a vector with matrix representations $[\boldsymbol{T}]$ and $[\boldsymbol{u}]$ with respect to the basis $\{\boldsymbol{e}_i\}$. The matrix representations $[\boldsymbol{T}^*]$ and $[\boldsymbol{u}^*]$ of $\boldsymbol{T}$ and $\boldsymbol{u}$ relative to the basis $\{\boldsymbol{e}_i^*\}$ are given by the following products of matrices:

$$[\boldsymbol{T}^*] = [\boldsymbol{R}]^T\,[\boldsymbol{T}]\,[\boldsymbol{R}]; \quad [\boldsymbol{u}^*] = [\boldsymbol{R}]^T\,[\boldsymbol{u}]. \tag{2.46}$$

Equivalently, in component form, we have

$$T_{ij}^* = R_{ki}\,T_{kl}\,R_{lj}, \quad u_i^* = R_{ji}\,u_j. \tag{2.47}$$

The matrix $[\boldsymbol{R}]$ is given by

$$[\boldsymbol{R}] = \begin{bmatrix} \boldsymbol{e}_1\cdot\boldsymbol{e}_1^* & \boldsymbol{e}_1\cdot\boldsymbol{e}_2^* & \cdots & \boldsymbol{e}_1\cdot\boldsymbol{e}_n^* \\ \boldsymbol{e}_2\cdot\boldsymbol{e}_1^* & \boldsymbol{e}_2\cdot\boldsymbol{e}_2^* & \cdots & \boldsymbol{e}_2\cdot\boldsymbol{e}_n^* \\ \vdots & \vdots & \ddots & \vdots \\ \boldsymbol{e}_n\cdot\boldsymbol{e}_1^* & \boldsymbol{e}_n\cdot\boldsymbol{e}_2^* & \cdots & \boldsymbol{e}_n\cdot\boldsymbol{e}_n^* \end{bmatrix}, \tag{2.48}$$

or, in component form,

$$R_{ij} = \boldsymbol{e}_i\cdot\boldsymbol{e}_j^*. \tag{2.49}$$

Example. A rotation in two dimensions

In two-dimensional space, the rotation tensor has a simple Cartesian representation. Let the tensor $\boldsymbol{R}$ be a transformation that rotates all vectors of the two-dimensional space by an (anti-clockwise positive) angle θ. The matrix representation of $\boldsymbol{R}$ reads

$$[\boldsymbol{R}] = \begin{bmatrix} \cos\theta & -\sin\theta \\ \sin\theta & \cos\theta \end{bmatrix}. \tag{2.50}$$

2.2.7. CROSS PRODUCT

Let us now restrict ourselves to the *three-dimensional* vector space. In this space, we define the *cross product* (or *vector product*) between two vectors $\boldsymbol{u}$ and $\boldsymbol{v}$ as the vector

$$\boldsymbol{w} = \boldsymbol{u}\times\boldsymbol{v}, \tag{2.51}$$

whose components are given by

$$w_i = \epsilon_{ijk}\,u_j\,v_k, \tag{2.52}$$

where ϵ_{ijk} denotes the *alternating tensor*

$$\epsilon_{ijk} = \begin{cases} +1 & \text{if } \{i, j, k\} \text{ is an even permutation of } \{1, 2, 3\} \\ -1 & \text{if } \{i, j, k\} \text{ is an odd permutation of } \{1, 2, 3\} \\ 0 & \text{if at least two indices coincide.} \end{cases} \tag{2.53}$$

Basic relations with the cross product

Some useful relations involving the cross product are listed in the following:

(i) $\boldsymbol{u} \times \boldsymbol{v} = -\boldsymbol{v} \times \boldsymbol{u}$.

(ii) $(\boldsymbol{u} \times \boldsymbol{v}) \cdot \boldsymbol{w} = (\boldsymbol{v} \times \boldsymbol{w}) \cdot \boldsymbol{u} = (\boldsymbol{w} \times \boldsymbol{u}) \cdot \boldsymbol{v}$.

(iii) $\boldsymbol{u} \times \boldsymbol{u} = \boldsymbol{o}$.

(iv) For any tensor $\boldsymbol{T}$ and a set $\{\boldsymbol{u}, \boldsymbol{v}, \boldsymbol{w}\}$ of *linearly independent* vectors,

$$\det \boldsymbol{T} = \frac{(\boldsymbol{T}\boldsymbol{u} \times \boldsymbol{T}\boldsymbol{v}) \cdot \boldsymbol{T}\boldsymbol{w}}{(\boldsymbol{u} \times \boldsymbol{v}) \cdot \boldsymbol{w}}. \tag{2.54}$$

(v) For any skew tensor $\boldsymbol{W}$, there is a unique vector $\boldsymbol{w}$, called the *axial vector* of $\boldsymbol{W}$, such that

$$\boldsymbol{W}\,\boldsymbol{u} = \boldsymbol{w} \times \boldsymbol{u}. \tag{2.55}$$

In terms of the Cartesian components of $\boldsymbol{W}$, the axial vector is expressed as

$$\boldsymbol{w} = W_{32}\,\boldsymbol{e}_1 + W_{13}\,\boldsymbol{e}_2 + W_{21}\,\boldsymbol{e}_3. \tag{2.56}$$

The matrix representation of $\boldsymbol{W}$, reads

$$[\boldsymbol{W}] = \begin{bmatrix} 0 & -w_3 & w_2 \\ w_3 & 0 & -w_1 \\ -w_2 & w_1 & 0 \end{bmatrix}, \tag{2.57}$$

where $\{w_i\}$ are the components of $\boldsymbol{w}$.

2.2.8. SPECTRAL DECOMPOSITION

Given a tensor $\boldsymbol{T}$, a non-zero vector $\boldsymbol{u}$ is said to be an *eigenvector* of $\boldsymbol{T}$ associated with the *eigenvalue* (or *principal value*) ω if

$$\boldsymbol{T}\boldsymbol{u} = \omega\,\boldsymbol{u}. \tag{2.58}$$

The space of all vectors $\boldsymbol{u}$ satisfying the above equation is called the *characteristic space* of $\boldsymbol{T}$ corresponding to ω. The following properties hold:

(i) The eigenvalues of a positive definite tensor are strictly positive.

(ii) The characteristic spaces of a symmetric tensor are mutually orthogonal.

Spectral theorem

Let $\boldsymbol{S}$ be a symmetric tensor. Then $\boldsymbol{S}$ admits the representation

$$\boldsymbol{S} = \sum_{i=1}^{n} s_i \, \boldsymbol{e}_i \otimes \boldsymbol{e}_i, \tag{2.59}$$

where $\{\boldsymbol{e}_i\}$ is an orthonormal basis for $\mathscr{U}$ consisting exclusively of eigenvectors of $\boldsymbol{S}$ and $\{s_i\}$ are the corresponding eigenvalues.† The above representation is called the *spectral decomposition* of $\boldsymbol{S}$. Relative to the basis $\{\boldsymbol{e}_i\}$, $\boldsymbol{S}$ has the following *diagonal* matrix representation

$$[\boldsymbol{S}] = \begin{bmatrix} s_1 & 0 & \cdots & 0 \\ 0 & s_2 & \cdots & 0 \\ \vdots & \vdots & \ddots & \vdots \\ 0 & 0 & \cdots & s_n \end{bmatrix}. \tag{2.60}$$

The direction of an eigenvector $\boldsymbol{e}_i$ is called a *principal axis* or *principal direction* of $\boldsymbol{S}$.

Eigenprojections

Alternatively, with $p \leq n$ defined as the number of *distinct* eigenvalues of $\boldsymbol{S}$, we may write

$$\boldsymbol{S} = \sum_{i=1}^{p} s_i \, \boldsymbol{E}_i, \tag{2.61}$$

where the symmetric tensors $\{\boldsymbol{E}_i\}$ are called the *eigenprojections* of $\boldsymbol{S}$. Each eigenprojection $\boldsymbol{E}_i$ is the orthogonal projection operator on the characteristic space of $\boldsymbol{S}$ associated with s_i. The eigenprojections have the property

$$\boldsymbol{I} = \sum_{i=1}^{p} \boldsymbol{E}_i, \tag{2.62}$$

and, if $p = n$ (no repeated eigenvalues), then

$$\boldsymbol{E}_i = \boldsymbol{e}_i \otimes \boldsymbol{e}_i \tag{2.63}$$

for $i = 1, \ldots, n$, with no summation implied on i. Also, the eigenprojections satisfy

$$\boldsymbol{E}_i : \boldsymbol{E}_j = \delta_{ij}, \quad i, j = 1, \ldots, p. \tag{2.64}$$

In closed form, $\boldsymbol{E}_i$ are given by the expression

$$\boldsymbol{E}_i = \begin{cases} \displaystyle\prod_{\substack{j=1 \\ j \neq i}}^{p} \frac{1}{s_i - s_j} \, (\boldsymbol{S} - s_j \, \boldsymbol{I}) & \text{if } p > 1 \\ \boldsymbol{I} & \text{if } p = 1. \end{cases} \tag{2.65}$$

† Note that, in the present context, s_i denotes the *eigenvalue* of tensor $\boldsymbol{S}$ associated with the unit eigenvector $\boldsymbol{e}_i$. We remark that this notation is often employed throughout this book and, whenever it is used, s_i should not be confused with the Cartesian component of a vector.

Characteristic equation. Principal invariants

Every eigenvalue s_i of a tensor $\boldsymbol{S}$ (symmetric or non-symmetric) satisfies the *characteristic equation*

$$\det(\boldsymbol{S} - s_i\,\boldsymbol{I}) = 0. \tag{2.66}$$

Two-dimensional space

In the two-dimensional space, $\det(\boldsymbol{S} - \alpha\,\boldsymbol{I})$ can be expressed as

$$\det(\boldsymbol{S} - \alpha\,\boldsymbol{I}) = \alpha^2 - \alpha\,I_1 + I_2, \tag{2.67}$$

for any $\alpha \in \mathscr{R}$, where I_1 and I_2 are the *principal invariants* of $\boldsymbol{S}$, defined as

$$\begin{aligned} I_1(\boldsymbol{S}) &\equiv \operatorname{tr}\boldsymbol{S} = S_{ii} \\ I_2(\boldsymbol{S}) &\equiv \det\boldsymbol{S} = S_{11}S_{22} - S_{12}S_{21}. \end{aligned} \tag{2.68}$$

In this case, the characteristic equation reads

$$s_i^2 - s_i\,I_1 + I_2 = 0. \tag{2.69}$$

The eigenvalues s_i are the solutions to this quadratic equation. If $\boldsymbol{S}$ is symmetric, then its principal invariants can be expressed in terms of its eigenvalues as

$$\begin{aligned} I_1 &= s_1 + s_2 \\ I_2 &= s_1 s_2. \end{aligned} \tag{2.70}$$

Three-dimensional space

In the three-dimensional space, $\det(\boldsymbol{S} - \alpha\,\boldsymbol{I})$ has the representation

$$\det(\boldsymbol{S} - \alpha\,\boldsymbol{I}) = -\alpha^3 + \alpha^2\,I_1 - \alpha\,I_2 + I_3, \tag{2.71}$$

where I_1, I_2 and I_3 are the *principal invariants* of $\boldsymbol{S}$, which are now defined by

$$\begin{aligned} I_1(\boldsymbol{S}) &\equiv \operatorname{tr}\boldsymbol{S} = S_{ii} \\ I_2(\boldsymbol{S}) &\equiv \tfrac{1}{2}[(\operatorname{tr}\boldsymbol{S})^2 - \operatorname{tr}(\boldsymbol{S}^2)] = \tfrac{1}{2}(S_{ii}S_{jj} - S_{ij}S_{ji}) \\ I_3(\boldsymbol{S}) &\equiv \det\boldsymbol{S} = \tfrac{1}{6}\epsilon_{ijk}\epsilon_{pqr}S_{ip}S_{jq}S_{kr}. \end{aligned} \tag{2.72}$$

The characteristic equation (now a *cubic* equation), whose solution is the set of eigenvalues of $\boldsymbol{S}$, reads

$$-s_i^3 + s_i^2\,I_1 - s_i\,I_2 + I_3 = 0. \tag{2.73}$$

If $\boldsymbol{S}$ is symmetric, then we have

$$\begin{aligned} I_1 &= s_1 + s_2 + s_3 \\ I_2 &= s_1 s_2 + s_2 s_3 + s_1 s_3 \\ I_3 &= s_1 s_2 s_3. \end{aligned} \tag{2.74}$$

2.2.9. POLAR DECOMPOSITION

Let $\boldsymbol{F}$ be a positive definite tensor. Then there exist symmetric positive definite tensors $\boldsymbol{U}$ and $\boldsymbol{V}$ and a rotation $\boldsymbol{R}$ such that

$$\boldsymbol{F} = \boldsymbol{R}\boldsymbol{U} = \boldsymbol{V}\boldsymbol{R}. \tag{2.75}$$

The decompositions $\boldsymbol{R}\boldsymbol{U}$ and $\boldsymbol{V}\boldsymbol{R}$ are unique and are called, respectively, the *right* and *left polar decompositions* of $\boldsymbol{F}$. The symmetric tensors $\boldsymbol{U}$ and $\boldsymbol{V}$ are given by

$$\boldsymbol{U} = \sqrt{\boldsymbol{F}^T\boldsymbol{F}}, \quad \boldsymbol{V} = \sqrt{\boldsymbol{F}\boldsymbol{F}^T}, \tag{2.76}$$

where $\sqrt{(\cdot)}$ denotes the *tensor square root* of $(\cdot)$. The square root of a symmetric tensor $\boldsymbol{S}$ is the unique tensor $\boldsymbol{T}$ that satisfies

$$\boldsymbol{T}^2 \equiv \boldsymbol{T}\boldsymbol{T} = \boldsymbol{S}. \tag{2.77}$$

With $\{s_i\}$ and $\{\boldsymbol{e}_i\}$ denoting, respectively, the eigenvalues and the basis of eigenvectors of $\boldsymbol{S}$, the spectral decomposition of its square root, $\boldsymbol{T}$, reads

$$\boldsymbol{T} = \sum_i \sqrt{s_i}\, \boldsymbol{e}_i \otimes \boldsymbol{e}_i. \tag{2.78}$$

Assertions (2.75)–(2.76) are commonly referred to as the *Polar Decomposition Theorem*. Note that the rotation $\boldsymbol{R}$ associated with an arbitrary $\boldsymbol{F}$ can be expressed as

$$\boldsymbol{R} = \boldsymbol{F}\boldsymbol{U}^{-1} = \boldsymbol{V}^{-1}\boldsymbol{F}. \tag{2.79}$$

2.3. Higher-order tensors

So far we have seen operations involving scalars, that can be considered as *zero-order* tensors, vectors, which can be classed as *first-order* tensors, and second-order tensors, which are linear operators on vectors. Linear operators of higher order, or *higher-order tensors*, are frequently employed in continuum mechanics. In this section we introduce some basic definitions and operations involving higher-order tensors.

In fact, in the definition of the cross product in Section 2.2.7, we have already made use of a higher-order tensor: the *third*-order alternating tensor whose components are defined in expression (2.53). The alternating tensor is a linear operator that maps vectors into skew symmetric tensors. It can be represented as

$$\boldsymbol{\epsilon} = \epsilon_{ijk}\, \boldsymbol{e}_i \otimes \boldsymbol{e}_j \otimes \boldsymbol{e}_k, \tag{2.80}$$

where the tensor product (now the tensor product of *three* vectors) is defined as the operator that satisfies

$$(\boldsymbol{a} \otimes \boldsymbol{b} \otimes \boldsymbol{c})\, \boldsymbol{d} = (\boldsymbol{c} \cdot \boldsymbol{d})(\boldsymbol{a} \otimes \boldsymbol{b}), \tag{2.81}$$

for arbitrary vectors $\boldsymbol{a}$, $\boldsymbol{b}$, $\boldsymbol{c}$ and $\boldsymbol{d}$. The multiplication of the alternating tensor by a vector $\boldsymbol{v}$ yields the second-order tensor

$$\boldsymbol{\epsilon}\, \boldsymbol{v} = \epsilon_{ijk} v_k\, \boldsymbol{e}_i \otimes \boldsymbol{e}_j. \tag{2.82}$$

Note that, with the above definitions, expression (2.52) can be equivalently written in compact form as

$$\boldsymbol{w} = (\boldsymbol{\epsilon}\, \boldsymbol{v})\, \boldsymbol{u}. \tag{2.83}$$

2.3.1. FOURTH-ORDER TENSORS

Fourth-order tensors are of particular relevance in continuum mechanics. A general fourth-order tensor‡ T can be represented as

$$\mathsf{T} = \mathsf{T}_{ijkl}\, \boldsymbol{e}_i \otimes \boldsymbol{e}_j \otimes \boldsymbol{e}_k \otimes \boldsymbol{e}_l. \tag{2.84}$$

Fourth-order tensors map second-order tensors into second-order tensors. They also map vectors into third-order tensors and third-order tensors into vectors.

As a direct extension of equation (2.81) we define

$$(\boldsymbol{a} \otimes \boldsymbol{b} \otimes \boldsymbol{c} \otimes \boldsymbol{d})\boldsymbol{e} = (\boldsymbol{e} \cdot \boldsymbol{d})(\boldsymbol{a} \otimes \boldsymbol{b} \otimes \boldsymbol{c}), \tag{2.85}$$

and the double contractions

$$\begin{aligned} (\boldsymbol{a} \otimes \boldsymbol{b} \otimes \boldsymbol{c} \otimes \boldsymbol{d}) : (\boldsymbol{e} \otimes \boldsymbol{f}) &= (\boldsymbol{c} \cdot \boldsymbol{e})(\boldsymbol{d} \cdot \boldsymbol{f})(\boldsymbol{a} \otimes \boldsymbol{b}), \\ (\boldsymbol{a} \otimes \boldsymbol{b} \otimes \boldsymbol{c} \otimes \boldsymbol{d}) : (\boldsymbol{e} \otimes \boldsymbol{f} \otimes \boldsymbol{g} \otimes \boldsymbol{h}) &= (\boldsymbol{c} \cdot \boldsymbol{e})(\boldsymbol{d} \cdot \boldsymbol{f})(\boldsymbol{a} \otimes \boldsymbol{b} \otimes \boldsymbol{g} \otimes \boldsymbol{h}), \end{aligned} \tag{2.86}$$

for arbitrary vectors $\boldsymbol{a}$, $\boldsymbol{b}$, $\boldsymbol{c}$, $\boldsymbol{d}$, $\boldsymbol{e}$ and $\boldsymbol{f}$. With the above definitions, the following relations are valid:

(i) $\mathsf{T}_{ijkl} = (\boldsymbol{e}_i \otimes \boldsymbol{e}_j) : \mathsf{T} : (\boldsymbol{e}_k \otimes \boldsymbol{e}_l)$.

(ii) $\mathsf{T}\,\boldsymbol{u} = \mathsf{T}_{ijkl}\, u_l\, \boldsymbol{e}_i \otimes \boldsymbol{e}_j \otimes \boldsymbol{e}_k$.

(iii) $\mathsf{T} : \boldsymbol{S} = \mathsf{T}_{ijkl}\, S_{kl}\, \boldsymbol{e}_i \otimes \boldsymbol{e}_j$.

(vi) $\boldsymbol{S} : \mathsf{T} = \mathsf{T}_{klij}\, S_{kl}\, \boldsymbol{e}_i \otimes \boldsymbol{e}_j$.

(v) $\mathsf{T} : \mathsf{S} = \mathsf{T}_{ijmn}\, \mathsf{S}_{mnkl}\, \boldsymbol{e}_i \otimes \boldsymbol{e}_j \otimes \boldsymbol{e}_k \otimes \boldsymbol{e}_l$.

Symmetry

We shall call *symmetric* any fourth-order tensor that satisfies

$$\boldsymbol{S} : \mathsf{T} : \boldsymbol{U} = (\mathsf{T} : \boldsymbol{S}) : \boldsymbol{U}, \tag{2.87}$$

for any second-order tensors $\boldsymbol{S}$ and $\boldsymbol{U}$. This definition is analogous to that of symmetric second-order tensors. The Cartesian components of symmetric fourth-order tensors satisfy the major symmetries

$$\mathsf{T}_{ijkl} = \mathsf{T}_{klij}. \tag{2.88}$$

It should be noted that other symmetries are possible in fourth-order tensors. If symmetry occurs in the last two indices, i.e. if

$$\mathsf{T}_{ijkl} = \mathsf{T}_{ijlk}, \tag{2.89}$$

the tensor has the properties

$$\mathsf{T} : \boldsymbol{S} = \mathsf{T} : \boldsymbol{S}^T, \quad \boldsymbol{S} : \mathsf{T} = (\boldsymbol{S} : \mathsf{T})^T, \tag{2.90}$$

for any $\boldsymbol{S}$. If it is symmetric in the first two indices,

$$\mathsf{T}_{ijkl} = \mathsf{T}_{jikl}, \tag{2.91}$$

then,

$$\mathsf{T} : \boldsymbol{S} = (\mathsf{T} : \boldsymbol{S})^T, \quad \boldsymbol{S} : \mathsf{T} = \boldsymbol{S}^T : \mathsf{T}. \tag{2.92}$$

‡Fourth-order tensors are represented in this text by bold face upright sans serif fonts.

Change of basis transformation

Again, let us consider the orthonormal basis $\{e_i^*\}$ defined as

$$e_i^* = \boldsymbol{R} e_i, \tag{2.93}$$

with $\boldsymbol{R}$ a rotation. The components T^*_{ijkl} of a tensor T relative to the basis defined by $\{e_i^*\}$ are given by

$$\mathsf{T}^*_{ijkl} = R_{mi} R_{nj} R_{pk} R_{ql}\, \mathsf{T}_{mnpq}, \tag{2.94}$$

where T_{mnpq} are the components relative to $\{e_i\}$.

2.3.2. GENERIC-ORDER TENSORS

Generic tensors of order m are defined as

$$\mathcal{T} = \mathcal{T}_{i_1 i_2 \cdots i_m}\, (e_{i_1} \otimes e_{i_2} \otimes \cdots \otimes e_{i_m}), \tag{2.95}$$

where, extending the previous definitions of the tensor product, we have

$$(e_{i_1} \otimes e_{i_2} \otimes \cdots \otimes e_{i_m})\boldsymbol{u} = (\boldsymbol{u} \cdot e_{i_m})(e_{i_1} \otimes e_{i_2} \otimes \cdots \otimes e_{i_{m-1}}), \tag{2.96}$$

for all $\boldsymbol{u} \in \mathscr{U}$. The definition of contraction operations is completely analogous to that seen above for fourth-order tensors.

2.4. Isotropic tensors

A tensor is said to be *isotropic* if its components are invariant under any change of basis.

2.4.1. ISOTROPIC SECOND-ORDER TENSORS

A second-order tensor, $\boldsymbol{T}$, is isotropic if

$$[\boldsymbol{T}] = [\boldsymbol{R}][\boldsymbol{T}][\boldsymbol{R}]^T \tag{2.97}$$

for any rotation $\boldsymbol{R}$. *Spherical tensors*, i.e. tensors represented as

$$\alpha \boldsymbol{I},$$

with scalar α, are the only second-order isotropic tensors.

2.4.2. ISOTROPIC FOURTH-ORDER TENSORS

A fourth-order tensor, T, is isotropic if

$$\mathsf{T}_{ijkl} = R_{mi} R_{nj} R_{pk} R_{ql}\, \mathsf{T}_{mnpq}, \tag{2.98}$$

for any rotation $\boldsymbol{R}$.

Any isotropic fourth-order tensor, U, can be represented as a linear combination of three basic isotropic tensors: I, I_T and $(\boldsymbol{I} \otimes \boldsymbol{I})$, i.e.

$$\mathsf{U} = \alpha\, \mathsf{I} + \beta\, \mathsf{I}_T + \gamma\, (\boldsymbol{I} \otimes \boldsymbol{I}), \tag{2.99}$$

where α, β and γ are scalars. The tensor I is called the *fourth-order identity*, given in component form as

$$\mathsf{I}_{ijkl} = \delta_{ik}\delta_{jl}. \tag{2.100}$$

For any second-order tensor $\boldsymbol{T}$, the fourth-order identity satisfies

$$\mathsf{I} : \boldsymbol{T} = \boldsymbol{T} : \mathsf{I} = \boldsymbol{T}, \tag{2.101}$$

and, for any fourth-order tensor, T, we have

$$\mathsf{I} : \mathsf{T} = \mathsf{T} : \mathsf{I} = \mathsf{T}. \tag{2.102}$$

The tensor I_T is the *transposition* tensor. It maps any second-order tensor onto its transpose, i.e.

$$\mathsf{I}_T : \boldsymbol{S} = \boldsymbol{S} : \mathsf{I}_T = \boldsymbol{S}^T, \tag{2.103}$$

for any $\boldsymbol{S}$. The components of I_T are

$$(\mathsf{I}_T)_{ijkl} = \delta_{il}\delta_{jk}. \tag{2.104}$$

Finally, the tensor $(\boldsymbol{I} \otimes \boldsymbol{I})$ has components

$$(\boldsymbol{I} \otimes \boldsymbol{I})_{ijkl} = \delta_{ij}\delta_{kl}. \tag{2.105}$$

When multiplying any tensor $\boldsymbol{T}$ it gives

$$(\boldsymbol{I} \otimes \boldsymbol{I}) : \boldsymbol{T} = (\operatorname{tr} \boldsymbol{T})\, \boldsymbol{I}. \tag{2.106}$$

Another important isotropic tensor that frequently appears in continuum mechanics is the tensor defined as

$$\mathsf{I}_S = \tfrac{1}{2}(\mathsf{I} + \mathsf{I}_T). \tag{2.107}$$

This tensor maps any second-order tensor into its symmetric part, i.e.

$$\mathsf{I}_S : \boldsymbol{T} = \boldsymbol{T} : \mathsf{I}_S = \operatorname{sym}(\boldsymbol{T}). \tag{2.108}$$

Obviously, for any symmetric $\boldsymbol{S}$,

$$\mathsf{I}_S : \boldsymbol{S} = \boldsymbol{S}. \tag{2.109}$$

We shall refer to this tensor as the *symmetric projection* or *symmetric identity*. Its Cartesian components are

$$(\mathsf{I}_S)_{ijkl} = \tfrac{1}{2}(\delta_{ik}\delta_{jl} + \delta_{il}\delta_{jk}). \tag{2.110}$$

2.5. Differentiation

2.5.1. THE DERIVATIVE MAP. DIRECTIONAL DERIVATIVE

Let $\mathscr{X}$ and $\mathscr{Y}$ be finite-dimensional normed vector spaces and let a function $\mathcal{Y}$ be defined as $\mathcal{Y} : \mathscr{D} \subset \mathscr{X} \rightarrow \mathscr{Y}$, where the domain $\mathscr{D}$ of $\mathcal{Y}$ is an open subset of $\mathscr{X}$.

The function $\mathcal{Y}$ is said to be *differentiable* at an argument $\mathcal{X}_0 \in \mathscr{D}$ if there exists a linear transformation $\mathrm{D}\mathcal{Y}(\mathcal{X}_0) : \mathscr{X} \rightarrow \mathscr{Y}$ such that as $\mathcal{U} \in \mathscr{X}$ approaches zero (the zero element of $\mathscr{X}$),

$$\mathcal{Y}(\mathcal{X}_0 + \mathcal{U}) = \mathcal{Y}(\mathcal{X}_0) + \mathrm{D}\mathcal{Y}(\mathcal{X}_0)\,[\mathcal{U}] + o(\mathcal{U}), \tag{2.111}$$

where

$$\mathrm{D}\mathcal{Y}(\mathcal{X}_0)\,[\mathcal{U}]$$

denotes a *linear transformation* on $\mathcal{U}$, represented in abstract notation, and

$$o(\mathcal{U})$$

is a term that approaches zero faster than $\mathcal{U}$, i.e.

$$\lim_{\mathcal{U} \rightarrow 0} \frac{\|o(\mathcal{U})\|}{\|\mathcal{U}\|} = 0. \tag{2.112}$$

If the linear map $\mathrm{D}\mathcal{Y}(\mathcal{X}_0)$ exists, it is unique and is called the *derivative* of $\mathcal{Y}$ at $\mathcal{X}_0$. The derivative satisfies

$$\mathrm{D}\mathcal{Y}(\mathcal{X}_0)\,[\mathcal{U}] = \lim_{\substack{\epsilon \rightarrow 0 \\ \epsilon \in \mathscr{R}}} \frac{1}{\epsilon}[\mathcal{Y}(\mathcal{X}_0 + \epsilon\,\mathcal{U}) - \mathcal{Y}(\mathcal{X}_0)] = \frac{\mathrm{d}}{\mathrm{d}\epsilon}\mathcal{Y}(\mathcal{X}_0 + \epsilon\,\mathcal{U})\bigg|_{\epsilon=0}, \tag{2.113}$$

for each $\mathcal{U} \in \mathscr{X}$. For a given $\mathcal{U}$, the term $\mathrm{D}\mathcal{Y}(\mathcal{X}_0)\,[\mathcal{U}]$ is called the *directional derivative* of $\mathcal{Y}$ (at $\mathcal{X}_0$) in the direction of $\mathcal{U}$.

2.5.2. LINEARISATION OF A NONLINEAR FUNCTION

The function $\mathcal{L} : \mathscr{D} \subset \mathscr{X} \rightarrow \mathscr{Y}$, defined as

$$\mathcal{L}(\mathcal{U}) \equiv \mathcal{Y}(\mathcal{X}_0) + \mathrm{D}\mathcal{Y}(\mathcal{X}_0)\,[\mathcal{U}] \tag{2.114}$$

is the called the *linearisation* of $\mathcal{Y}$ about $\mathcal{X}_0$, that is, it is the linear approximation to $\mathcal{Y}$ at $\mathcal{X}_0$. Note by observing (2.111) that the function defined in (2.114) corresponds indeed to the approximation to $\mathcal{Y}$ that ignores all higher-order (nonlinear) terms in $\mathcal{U}$.

2.5.3. THE GRADIENT

Since the derivative is a linear transformation between finite-dimensional spaces, it can be represented as

$$\mathrm{D}\mathcal{Y}(\mathcal{X}_0)\,[\mathcal{U}] = \nabla\mathcal{Y}(\mathcal{X}_0) * \mathcal{U}, \tag{2.115}$$

where $\nabla\mathcal{Y}(\mathcal{X}_0)$ is the *gradient* of $\mathcal{Y}$ at $\mathcal{X}_0$ and the symbol '$*$' denotes an *appropriate* product operation.

We remark that throughout this text we shall use the term *derivative* also as a synonym for *gradient*. We will also use the notation

$$\frac{\mathrm{d}\mathcal{Y}}{\mathrm{d}\mathcal{X}}$$

to denote the gradient and when $\mathcal{Y}$ is a function of two or more arguments, we may use the notation

$$\frac{\partial\mathcal{Y}}{\partial\mathcal{X}}$$

to emphasise that the derivative is taken with respect to $\mathcal{X}$ having the other function arguments as parameters.

Example. A scalar function

The above concepts are probably better understood by applying them to the trivial case of a scalar function of a scalar argument. Let us consider the function (Figure 2.1)

$$y(x) \equiv x^2. \tag{2.116}$$

In the present case, (2.111) reads

$$\begin{aligned} y(x_0 + u) &= y(x_0) + \mathrm{D}y(x_0)\,[u] + o(u) \\ &= x_0^2 + 2x_0 u + u^2. \end{aligned} \tag{2.117}$$

We can then identify

$$o(u) = u^2. \tag{2.118}$$

Note that, indeed, this term vanishes faster than u, i.e.

$$\lim_{u\to 0} \frac{|o(u)|}{|u|} = \lim_{u\to 0} \frac{u^2}{|u|} = 0. \tag{2.119}$$

In addition, we have

$$\mathrm{D}y(x_0)\,[u] = 2x_0 u. \tag{2.120}$$

Thus, by comparing the above expression with (2.115), the gradient of y can be promptly identified as

$$\nabla y(x_0) = 2x_0, \tag{2.121}$$

and the product '$*$' is identified with the standard product between scalars. With the above and (2.114), we find that the linearisation of y about x_0 in the present case reads

$$l(u) \equiv y(x_0) + \nabla y(x_0) u = x_0^2 + 2x_0 u. \tag{2.122}$$

2.5.4. DERIVATIVES OF FUNCTIONS OF VECTOR AND TENSOR ARGUMENTS

Let us now consider some simple illustrative examples of derivatives of functions whose arguments and/or values may be scalar, vectors or tensors.

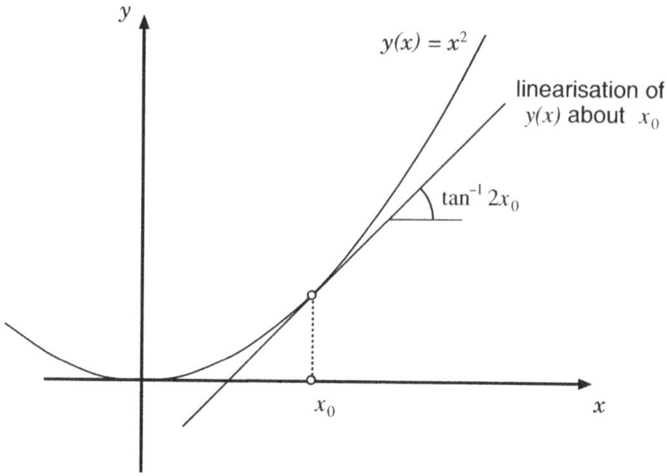

Figure 2.1. Function derivative and linearisation.

Scalar function of vector argument

We start with the scalar function of a vector argument defined by

$$y(\boldsymbol{x}) \equiv \boldsymbol{x} \cdot \boldsymbol{x} = x_i x_i. \tag{2.123}$$

In this case, we have

$$y(\boldsymbol{x} + \boldsymbol{u}) = \boldsymbol{x} \cdot \boldsymbol{x} + 2\boldsymbol{x} \cdot \boldsymbol{u} + o(\boldsymbol{u}) \tag{2.124}$$

where $o(\boldsymbol{u}) = \boldsymbol{u} \cdot \boldsymbol{u}$. It then follows that

$$\mathrm{D}y(\boldsymbol{x})\,[\boldsymbol{u}] = 2\boldsymbol{x} \cdot \boldsymbol{u}, \tag{2.125}$$

and the gradient ∇y can be identified as the vector

$$\nabla y(\boldsymbol{x}) = 2\boldsymbol{x}. \tag{2.126}$$

The generic product operation indicated in (2.115) becomes the internal product between vectors in the present case. We may alternatively use the notation

$$\frac{\mathrm{d}y}{\mathrm{d}\boldsymbol{x}} = 2\boldsymbol{x}, \tag{2.127}$$

or, in component form,

$$\left(\frac{\mathrm{d}y}{\mathrm{d}\boldsymbol{x}}\right)_i = \frac{\mathrm{d}y}{\mathrm{d}x_i} = 2x_i. \tag{2.128}$$

Scalar function of a tensor argument

Now consider the internal product between tensors

$$y(\boldsymbol{X}) \equiv \boldsymbol{X} : \boldsymbol{X} = X_{ij} X_{ij}. \tag{2.129}$$

We have

$$\begin{aligned} y(\boldsymbol{X}+\boldsymbol{U}) &= \boldsymbol{X}:\boldsymbol{X}+2\boldsymbol{X}:\boldsymbol{U}+\boldsymbol{U}:\boldsymbol{U} \\ &= y(\boldsymbol{X})+\mathrm{D}y(\boldsymbol{X})\,[\boldsymbol{U}]+o(\boldsymbol{U}). \end{aligned} \tag{2.130}$$

The gradient – a second-order tensor in this case – can be immediately identified as

$$\nabla y(\boldsymbol{X}) = 2\boldsymbol{X}, \tag{2.131}$$

or, in component form,

$$(\nabla y)_{ij} = \frac{\mathrm{d}y}{\mathrm{d}X_{ij}} = 2\,X_{ij}. \tag{2.132}$$

Tensor function of a tensor argument

We now determine the derivative of the square of a tensor

$$\boldsymbol{Y}(\boldsymbol{X}) \equiv \boldsymbol{X}^2 = \boldsymbol{X}\boldsymbol{X}. \tag{2.133}$$

This tensor-value function is equivalently expressed in component form as

$$Y_{ij} \equiv X_{ik}X_{kj}. \tag{2.134}$$

To obtain its derivative, we write

$$\begin{aligned} \boldsymbol{Y}(\boldsymbol{X}+\boldsymbol{U}) &= \boldsymbol{X}\boldsymbol{X}+\boldsymbol{X}\boldsymbol{U}+\boldsymbol{U}\boldsymbol{X}+\boldsymbol{U}\boldsymbol{U} \\ &= \boldsymbol{Y}(\boldsymbol{X})+\mathrm{D}\boldsymbol{Y}(\boldsymbol{X})\,[\boldsymbol{U}]+o(\boldsymbol{U}). \end{aligned} \tag{2.135}$$

The derivative in this case is a linear mapping between second-order tensors and can be identified as the *fourth*-order tensor $\nabla\boldsymbol{Y}(\boldsymbol{X})$ that satisfies

$$\nabla\boldsymbol{Y}(\boldsymbol{X}):\boldsymbol{U} = \mathrm{D}\boldsymbol{Y}(\boldsymbol{X})\,[\boldsymbol{U}] = \boldsymbol{X}\boldsymbol{U}+\boldsymbol{U}\boldsymbol{X} \tag{2.136}$$

for any $\boldsymbol{U}$. The Cartesian components of $\nabla\boldsymbol{Y}$ can be easily found to be given by

$$(\nabla\boldsymbol{Y}\,)_{ijkl} = X_{ik}\delta_{jl} + X_{lj}\delta_{ik}. \tag{2.137}$$

Some useful relations

We list below (without proof) some basic expressions for derivatives relevant to theoretical and computational mechanics applications:

(i) $y(\boldsymbol{x}) \equiv \|\boldsymbol{x}\|$,

$$\begin{aligned} \mathrm{D}y(\boldsymbol{x})\,[\boldsymbol{u}] &= \frac{\boldsymbol{x}}{\|\boldsymbol{x}\|}\cdot\boldsymbol{u} \\ \frac{\mathrm{d}\|\boldsymbol{x}\|}{\mathrm{d}\boldsymbol{x}} &= \frac{\boldsymbol{x}}{\|\boldsymbol{x}\|}. \end{aligned} \tag{2.138}$$

(ii) $y(\boldsymbol{X}) \equiv \|\boldsymbol{X}\|$,

$$\begin{aligned} \mathrm{D}y(\boldsymbol{X})\,[\boldsymbol{U}] &= \frac{\boldsymbol{X}}{\|\boldsymbol{X}\|} : \boldsymbol{U} \\ \frac{\mathrm{d}\|\boldsymbol{X}\|}{\mathrm{d}\boldsymbol{X}} &= \frac{\boldsymbol{X}}{\|\boldsymbol{X}\|}. \end{aligned} \tag{2.139}$$

(iii) $y(\boldsymbol{X}) \equiv \det \boldsymbol{X}$,

$$\begin{aligned} \mathrm{D}y(\boldsymbol{X})\,[\boldsymbol{U}] &= (\det \boldsymbol{X})\,\boldsymbol{X}^{-T} : \boldsymbol{U} \\ \frac{\mathrm{d}}{\mathrm{d}\boldsymbol{X}} \det \boldsymbol{X} &= (\det \boldsymbol{X})\,\boldsymbol{X}^{-T}. \end{aligned} \tag{2.140}$$

(iv) $\boldsymbol{Y}(\boldsymbol{x}) \equiv \boldsymbol{a} \otimes \boldsymbol{x}$ (with constant $\boldsymbol{a}$),

$$\begin{aligned} \mathrm{D}\boldsymbol{Y}(\boldsymbol{x})\,[\boldsymbol{u}] &= (\boldsymbol{a} \otimes \boldsymbol{I}) \cdot \boldsymbol{u} \\ \frac{\mathrm{d}}{\mathrm{d}\boldsymbol{x}}(\boldsymbol{a} \otimes \boldsymbol{x}) &= \boldsymbol{a} \otimes \boldsymbol{I}. \end{aligned} \tag{2.141}$$

(v) $\boldsymbol{Y}(\boldsymbol{x}) \equiv \boldsymbol{x} \otimes \boldsymbol{x}$,

$$\begin{aligned} \mathrm{D}\boldsymbol{Y}(\boldsymbol{x})\,[\boldsymbol{u}] &= (\boldsymbol{I} \otimes \boldsymbol{x} + \boldsymbol{x} \otimes \boldsymbol{I}) \cdot \boldsymbol{u} \\ \frac{\mathrm{d}}{\mathrm{d}\boldsymbol{x}}(\boldsymbol{x} \otimes \boldsymbol{x}) &= \boldsymbol{I} \otimes \boldsymbol{x} + \boldsymbol{x} \otimes \boldsymbol{I}. \end{aligned} \tag{2.142}$$

(vi) $\boldsymbol{Y}(\boldsymbol{X}) \equiv \boldsymbol{X}$,

$$\begin{aligned} \mathrm{D}\boldsymbol{Y}(\boldsymbol{X})\,[\boldsymbol{U}] &= \boldsymbol{U} \\ \frac{\mathrm{d}}{\mathrm{d}\boldsymbol{X}}\boldsymbol{X} &= \mathsf{I}. \end{aligned} \tag{2.143}$$

(vii) $\boldsymbol{Y}(\boldsymbol{X}) \equiv \boldsymbol{X}^{-1}$,

$$\begin{aligned} \mathrm{D}\boldsymbol{Y}(\boldsymbol{X})\,[\boldsymbol{U}] &= -\boldsymbol{X}^{-1}\boldsymbol{U}\boldsymbol{X}^{-1} = \mathsf{D} : \boldsymbol{U}; \quad \mathsf{D}_{ijkl} = -X^{-1}_{ik}X^{-1}_{lj} \\ \frac{\mathrm{d}}{\mathrm{d}\boldsymbol{X}}(\boldsymbol{X}^{-1}) &= \mathsf{D}. \end{aligned} \tag{2.144}$$

2.5.5. THE CHAIN RULE

Let the function $\boldsymbol{\mathcal{Y}} : \mathscr{D} \subset \mathscr{X} \rightarrow \mathscr{Y}$ be defined as the composition

$$\boldsymbol{\mathcal{Y}}(\boldsymbol{\mathcal{X}}) = \boldsymbol{\mathcal{W}}(\boldsymbol{\mathcal{Z}}(\boldsymbol{\mathcal{X}}))$$

of two differentiable functions

$$\boldsymbol{\mathcal{Z}} : \mathscr{D} \subset \mathscr{X} \rightarrow \mathscr{Z} \quad \text{and} \quad \boldsymbol{\mathcal{W}} : \mathscr{C} \subset \mathscr{Z} \rightarrow \mathscr{Y}.$$

The derivative of $\boldsymbol{\mathcal{Y}}$ satisfies

$$\mathrm{D}\boldsymbol{\mathcal{Y}}(\boldsymbol{\mathcal{X}})\,[\boldsymbol{\mathcal{U}}] = \mathrm{D}\boldsymbol{\mathcal{W}}(\boldsymbol{\mathcal{Z}}(\boldsymbol{\mathcal{X}}))\,[\mathrm{D}\boldsymbol{\mathcal{Z}}(\boldsymbol{\mathcal{X}})\,[\boldsymbol{\mathcal{U}}]]$$

for all $\boldsymbol{\mathcal{U}} \in \mathscr{X}$.

2.5.6. THE PRODUCT RULE

Let us consider a function $\mathcal{Y} : \mathscr{D} \subset \mathscr{X} \to \mathscr{Y}$ defined as

$$\mathcal{Y}(\mathcal{X}) = \mathcal{W}(\mathcal{X}) * \mathcal{Z}(\mathcal{X})$$

where the symbol '$*$' denotes a generic product between functions $\mathcal{W} : \mathscr{D} \subset \mathscr{X} \to \mathscr{W}$ and $\mathcal{Z} : \mathscr{D} \subset \mathscr{X} \to \mathscr{Z}$. If $\mathcal{W}$ and $\mathcal{Z}$ are vectors, for instance, this product could be the internal or tensor product between vectors. If $\mathcal{W}$ is a fourth-order tensor and $\mathcal{Z}$ is a second-order tensor, '$*$' could be the double contraction operation, and so on. If $\mathcal{W}$ and $\mathcal{Z}$ are differentiable at an argument $\mathcal{X}$, then so is $\mathcal{Y}$ and

$$\mathrm{D}\mathcal{Y}(\mathcal{X})\,[\mathcal{U}] = \mathcal{W}(\mathcal{X}) * \mathrm{D}\mathcal{Z}(\mathcal{X})\,[\mathcal{U}] + \mathrm{D}\mathcal{W}(\mathcal{X})\,[\mathcal{U}] * \mathcal{Z}(\mathcal{X}).$$

2.5.7. THE DIVERGENCE

Let $\boldsymbol{v} : \mathscr{D} \subset \mathscr{E} \to \mathscr{U}$ be a smooth vector field on $\mathscr{D}$. The *divergence* of $\boldsymbol{v}$ is the *scalar* field defined as

$$\operatorname{div} \boldsymbol{v} = \operatorname{tr}(\nabla \boldsymbol{v}) = \frac{\partial v_i}{\partial x_i}. \tag{2.145}$$

Now let $\boldsymbol{T}$ be a smooth second-order tensor field on $\mathscr{D}$. The *divergence* of $\boldsymbol{T}$ is the *vector* field, denoted div $\boldsymbol{T}$, that satisfies

$$(\operatorname{div} \boldsymbol{T}) \cdot \boldsymbol{u} = \operatorname{div}(\boldsymbol{T}^T \boldsymbol{u}), \quad \forall\, \boldsymbol{u} \in \mathscr{U}. \tag{2.146}$$

In Cartesian components, we have

$$(\operatorname{div} \boldsymbol{T})_i = \frac{\partial T_{ij}}{\partial x_j}. \tag{2.147}$$

The divergence theorem

Let $\mathscr{B} \subset \mathscr{E}$ be a closed region with piecewise smooth boundary $\partial\mathscr{B}$ and let α, $\boldsymbol{v}$ and $\boldsymbol{T}$ be smooth scalar, vector and tensor-valued fields on $\mathscr{B}$, respectively. Then,

$$\begin{aligned}
\int_{\partial\mathscr{B}} \alpha\, \boldsymbol{n}\, \mathrm{d}a &= \int_{\mathscr{B}} \nabla\alpha\, \mathrm{d}v, \\
\int_{\partial\mathscr{B}} \boldsymbol{v} \cdot \boldsymbol{n}\, \mathrm{d}a &= \int_{\mathscr{B}} \operatorname{div} \boldsymbol{v}\, \mathrm{d}v, \\
\int_{\partial\mathscr{B}} \boldsymbol{T}\boldsymbol{n}\, \mathrm{d}a &= \int_{\mathscr{B}} \operatorname{div} \boldsymbol{T}\, \mathrm{d}v,
\end{aligned} \tag{2.148}$$

where $\boldsymbol{n}$ is the field of unit outward normal vectors on $\partial\mathscr{B}$.

2.5.8. USEFUL RELATIONS INVOLVING THE GRADIENT AND THE DIVERGENCE

Let α and $\boldsymbol{T}$ be, respectively, smooth scalar and tensor fields on $\mathscr{D} \subset \mathscr{E}$ and let $\boldsymbol{u}$ and $\boldsymbol{v}$ be smooth vector fields on $\mathscr{D}$. The following useful relations hold:

(i) $\nabla(\alpha \boldsymbol{u}) = \alpha\, \nabla \boldsymbol{u} + \boldsymbol{u} \otimes \nabla\alpha$.

(ii) $\operatorname{div}(\alpha \boldsymbol{u}) = \alpha \operatorname{div} \boldsymbol{u} + \boldsymbol{u} \cdot \nabla\alpha$.

(iii) $\nabla(\boldsymbol{u} \cdot \boldsymbol{v}) = (\nabla \boldsymbol{v})^T \boldsymbol{u} + (\nabla \boldsymbol{u})^T \boldsymbol{v}$.

(iv) $\operatorname{div}(\boldsymbol{u} \otimes \boldsymbol{v}) = \boldsymbol{u} \operatorname{div} \boldsymbol{v} + (\nabla \boldsymbol{u})\boldsymbol{v}$.

(v) $\operatorname{div}(\boldsymbol{T}\boldsymbol{u}) = \boldsymbol{T}^T : \nabla \boldsymbol{u} + \boldsymbol{u} \cdot \operatorname{div} \boldsymbol{T}^T$.

(vi) $\operatorname{div}(\alpha\, \boldsymbol{T}) = \alpha \operatorname{div} \boldsymbol{T} + \boldsymbol{T} \nabla\alpha$.

The proof of the above relations is given in Gurtin (1981).

2.6. Linearisation of nonlinear problems

The linearisation of nonlinear problems plays a crucial role in theoretical and computational mechanics. An interesting presentation of the theory of linearisation is provided by Marsden and Hughes (1983) within a rather general mathematical framework and by Hughes and Pister (1978) within the context of solid mechanics problems. In theoretical continuum mechanics, the concept of linearisation is essential in the derivation of linear approximations to general nonlinear theories. In computational mechanics, on the other hand, the interest in linearisation stems mostly from the fact that numerical solutions to nonlinear problems are usually obtained by algorithms which require the solution of a sequence of linearised (or approximately linearised) problems. A typical example is the well-known Newton–Raphson iterative scheme, which is extensively exploited in the finite element framework described later in this book.

2.6.1. THE NONLINEAR PROBLEM AND ITS LINEARISED FORM

Let $\mathcal{Y} : \mathscr{D} \subset \mathscr{X} \rightarrow \mathscr{Y}$ be a generic *nonlinear* function. Our generic nonlinear problem consists in finding $\mathcal{X} \in \mathscr{D}$ such that

$$\mathcal{Y}(\mathcal{X}) = 0. \tag{2.149}$$

The *linearisation* of the problem defined by equation (2.149) about an arbitrary point $\mathcal{X}_0 \in \mathscr{D}$ at which $\mathcal{Y}$ is differentiable consists in finding $\mathcal{U} \in \mathscr{D}$ such that

$$\mathcal{L}(\mathcal{U}) \equiv \mathcal{Y}(\mathcal{X}_0) + \mathrm{D}\mathcal{Y}(\mathcal{X}_0)\, [\mathcal{U}] = 0, \tag{2.150}$$

where $\mathcal{L}(\mathcal{U})$ is the linearisation of $\mathcal{Y}$ at $\mathcal{X}_0$.

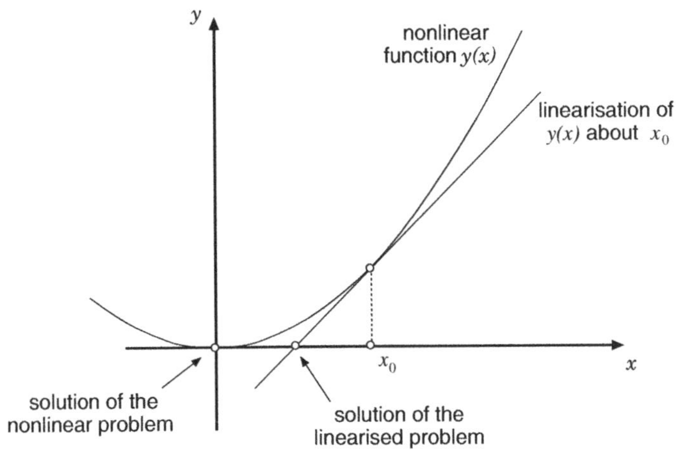

Figure 2.2. Linearisation of a nonlinear equation.

Example. The scalar case

To illustrate the above concept, let us again appeal to the case of a scalar function of a single scalar argument, $y(x) \equiv x^2$, and consider the trivial root-finding problem (refer to the graphical illustration of Figure 2.2) defined by the equation

$$y(x) = 0. \tag{2.151}$$

This is obviously a particularisation of the general nonlinear problem (2.149). In the present case, the generic linearised problem (2.150) takes the form of the following scalar equation for u at an arbitrary argument x_0:

$$l(u) \equiv y(x_0) + \nabla y(x_0)\, u = x_0^2 + 2x_0 u = 0. \tag{2.152}$$

2.6.2. LINEARISATION IN INFINITE-DIMENSIONAL FUNCTIONAL SPACES

So far, the discussion on differentiation and linearisation has been restricted to mappings between finite-dimensional vector spaces. It is important to emphasise that such concepts are equally applicable to the more general case where $\mathcal{Y}$ is a mapping between infinite-dimensional functional spaces.[§] Of particular interest in computational mechanics is the case when $\mathcal{Y}$ is a *functional*, i.e. a scalar-valued function of a function. Linearisation of problems involving functions of this type (such as the virtual work functional) are of paramount importance in the computational solution of nonlinear solid mechanics boundary value problems and will be discussed later in this book. Here, we describe a much simpler example whose purpose is only to familiarise the reader with the concept.

[§]The precise definition of such spaces falls outside the scope of the present text. The more interested reader is referred to Marsden and Hughes (1983) where the concept of linearisation is introduced in the context of mappings between Banach spaces.

A simple example

Let $\mathscr{X}$ now be a space of sufficiently smooth[¶] functions $x : \mathscr{R}^n \rightarrow \mathscr{R}$. We define a *functional* $y : \mathscr{X} \rightarrow \mathscr{R}$ as

$$y(x) \equiv \int_\Omega \sin(x(\boldsymbol{p}))\,\mathrm{d}\boldsymbol{p}, \tag{2.153}$$

where $\Omega \subset \mathscr{R}^n$ is a given integration domain. Linearisation of the above functional about a given argument (function) x_0 is the following generalisation of (2.114):

$$l(u) = y(x_0) + \mathrm{D}y(x_0)\,[u] = \int_\Omega \sin(x_0(\boldsymbol{p}))\,\mathrm{d}\boldsymbol{p} + \mathrm{D}y(x_0)\,[u], \tag{2.154}$$

where the directional derivative $\mathrm{D}y(x_0)\,[u]$ is now a linear transformation on the *function* $u \in \mathscr{X}$ and can be determined by direct generalisation of (2.113):

$$\begin{aligned}\mathrm{D}y(x_0)\,[u] &= \frac{\mathrm{d}}{\mathrm{d}\epsilon} y(x_0 + \epsilon u)\bigg|_{\epsilon=0} \\ &= \frac{\mathrm{d}}{\mathrm{d}\epsilon} \int_\Omega \sin(x_0(\boldsymbol{p}) + \epsilon\, u(\boldsymbol{p}))\,\mathrm{d}\boldsymbol{p}\bigg|_{\epsilon=0} \\ &= \int_\Omega \cos(x_0(\boldsymbol{p}))\, u(\boldsymbol{p})\,\mathrm{d}\boldsymbol{p}.\end{aligned} \tag{2.155}$$

From the above, the linearisation of the functional (2.153) at x_0 is then established as

$$l(u) = \int_\Omega \sin(x_0(\boldsymbol{p}))\,\mathrm{d}\boldsymbol{p} + \int_\Omega \cos(x_0(\boldsymbol{p}))\, u(\boldsymbol{p})\,\mathrm{d}\boldsymbol{p}. \tag{2.156}$$

[¶]To avoid a precise statement of regularity properties of functions, we frequently use the term *sufficiently smooth* in the present text, meaning that functions have a sufficient degree of regularity so that all operations in which they are involved are properly defined.

3 ELEMENTS OF CONTINUUM MECHANICS AND THERMODYNAMICS

THIS chapter reviews some basic concepts of mechanics and thermodynamics of continuous media. The definitions and notation introduced will be systematically employed throughout the subsequent chapters of this book. The material presented here is well established in the continuum mechanics literature and an effort has been made to follow the notation and nomenclature in use in standard textbooks (Billington and Tate, 1981; Bonet and Wood, 1997; Ciarlet, 1988; Gurtin, 1981; Lemaitre and Chaboche, 1990; Ogden, 1984; Spencer, 1980; Truesdell and Noll, 1965).

3.1. Kinematics of deformation

Let $\mathscr{B}$ be a *body* which occupies an open region Ω of the three-dimensional Euclidean space $\mathscr{E}$ with a regular boundary $\partial\Omega$ in its *reference* configuration. A *deformation* of $\mathscr{B}$ (Figure 3.1) is defined by a smooth one-to-one function

$$\varphi : \Omega \rightarrow \mathscr{E}$$

that maps each material particle[†] $\boldsymbol{p}$ of $\mathscr{B}$ into a point

$$\boldsymbol{x} = \varphi(\boldsymbol{p}) \tag{3.1}$$

where the particle is positioned in the *deformed* configuration of $\mathscr{B}$. The region of $\mathscr{E}$ occupied by $\mathscr{B}$ in its deformed configuration will be denoted

$$\varphi(\Omega).$$

The vector field $\boldsymbol{u}(\boldsymbol{p})$, defined by

$$\boldsymbol{u}(\boldsymbol{p}) = \varphi(\boldsymbol{p}) - \boldsymbol{p}, \tag{3.2}$$

is the *displacement* of $\boldsymbol{p}$. Thus, one may write

$$\boldsymbol{x} = \boldsymbol{p} + \boldsymbol{u}(\boldsymbol{p}). \tag{3.3}$$

[†]For convenience, material particles of $\mathscr{B}$ will be identified with their positions in the reference configuration of $\mathscr{B}$.

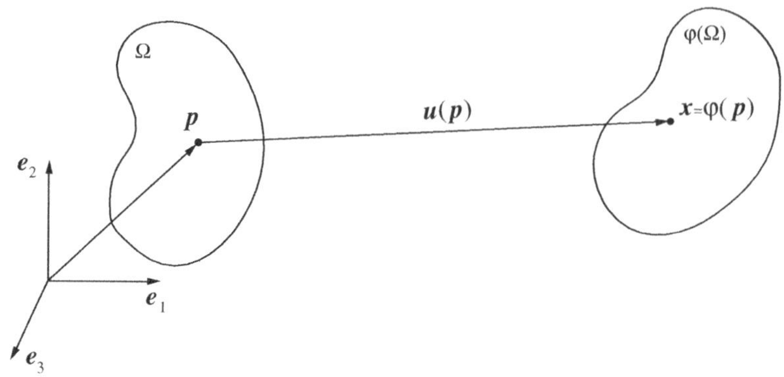

Figure 3.1. Deformation.

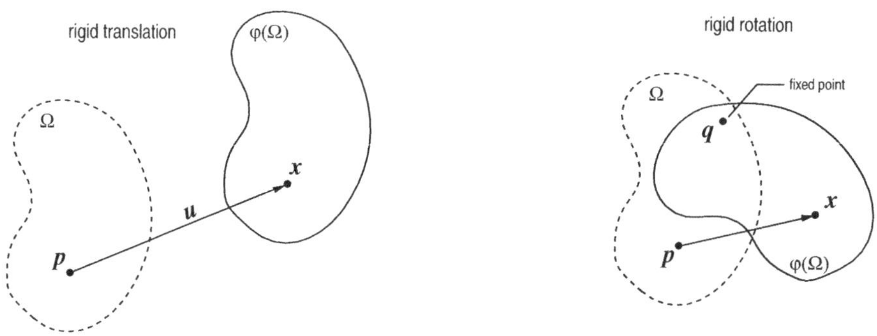

Figure 3.2. Rigid deformations.

A *rigid deformation* of $\mathscr{B}$ is a deformation that preserves the distances between all material particles of $\mathscr{B}$. A rigid deformation (Figure 3.2) can be a translation, a rotation, or a combination of a translation and a rotation. A *rigid translation* is a deformation with constant displacement vector ($\boldsymbol{u}$ independent of $\boldsymbol{p}$):

$$\varphi(\boldsymbol{p}) = \boldsymbol{p} + \boldsymbol{u}. \tag{3.4}$$

A *rigid rotation* is a deformation that can be expressed as

$$\varphi(\boldsymbol{p}) = \boldsymbol{q} + \boldsymbol{R}\,(\boldsymbol{p} - \boldsymbol{q}), \tag{3.5}$$

where $\boldsymbol{R}$ is a proper orthogonal tensor (a rotation) and $\boldsymbol{q}$ is the point about which $\mathscr{B}$ is rotated. A deformation is rigid, including translations and/or rotations, if and only if it can be expressed in the form:

$$\varphi(\boldsymbol{p}) = \varphi(\boldsymbol{q}) + \boldsymbol{R}\,(\boldsymbol{p} - \boldsymbol{q}). \tag{3.6}$$

The deformation map above represents a rigid translation with displacement $\varphi(\boldsymbol{q}) - \boldsymbol{q}$ superimposed on a rigid rotation $\boldsymbol{R}$ about point $\boldsymbol{q}$.

A time-dependent deformation of $\mathscr{B}$ is called a *motion* of $\mathscr{B}$. A motion (Figure 3.3) is defined by a function

$$\varphi : \Omega \times \mathscr{R} \rightarrow \mathscr{E},$$

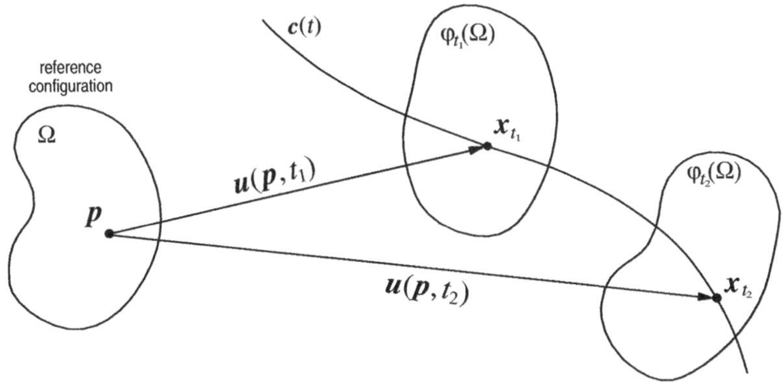

Figure 3.3. Motion.

so that for each time t, the map $\varphi(\cdot, t)$ is a deformation of $\mathscr{B}$. The deformation map at time t will be also denoted φ_t. During the motion φ, the position $\boldsymbol{x}$ of a material particle $\boldsymbol{p}$ at time t is given by

$$\boldsymbol{x} = \varphi(\boldsymbol{p}, t). \tag{3.7}$$

Similarly,

$$\varphi(\Omega, t)$$

will denote the region of $\mathscr{E}$ occupied by the body $\mathscr{B}$ at time t. In terms of the displacement field the motion is expressed as

$$\varphi(\boldsymbol{p}, t) = \boldsymbol{p} + \boldsymbol{u}(\boldsymbol{p}, t). \tag{3.8}$$

The parametric curve $\boldsymbol{c}(t)$, defined as

$$\boldsymbol{c}(t) = \varphi(\boldsymbol{p}, t) \tag{3.9}$$

for a *fixed* material point $\boldsymbol{p}$, describes the *trajectory* of $\boldsymbol{p}$ during the motion of $\mathscr{B}$.

During a motion φ, the *velocity* of a material particle $\boldsymbol{p}$ is defined by

$$\dot{\boldsymbol{x}}(\boldsymbol{p}, t) = \frac{\partial \varphi(\boldsymbol{p}, t)}{\partial t}. \tag{3.10}$$

Since at each time t the map $\varphi(\cdot, t)$ is one-to-one (and hence invertible) by assumption, material points can be expressed in terms of the place they occupy at a time t as

$$\boldsymbol{p} = \varphi^{-1}(\boldsymbol{x}, t) = \boldsymbol{x} - \boldsymbol{u}(\varphi^{-1}(\boldsymbol{x}, t), t). \tag{3.11}$$

The map φ^{-1} is called the *reference map*. Using the reference map, one may define the function

$$\boldsymbol{v}(\boldsymbol{x}, t) \equiv \dot{\boldsymbol{x}}(\varphi^{-1}(\boldsymbol{x}, t), t). \tag{3.12}$$

The field $\boldsymbol{v}$ is called the *spatial velocity* and gives the velocity of the material particle positioned at $\boldsymbol{x}$ at time t.

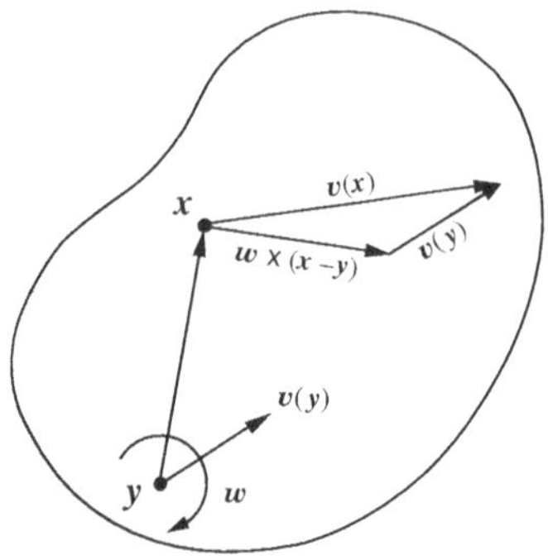

Figure 3.4. Rigid velocity.

A *rigid motion* of $\mathscr{B}$ is a motion for which, at each time t, the map $\varphi(\cdot, t)$ is a rigid deformation. A motion φ is rigid if and only if at each time t, the spatial velocity $\boldsymbol{v}$ admits the representation

$$\boldsymbol{v}(\boldsymbol{x}, t) = \boldsymbol{v}(\boldsymbol{y}, t) + \boldsymbol{W}(t)\,(\boldsymbol{x} - \boldsymbol{y}) \tag{3.13}$$

for all $\boldsymbol{x}, \boldsymbol{y} \in \varphi(\Omega, t)$, with $\boldsymbol{W}(t)$ a *skew* tensor. The velocity at $\boldsymbol{x}$ is given as the sum of a uniform velocity $\boldsymbol{v}(\boldsymbol{y}, t)$ and a superimposed rotation about the line that passes through $\boldsymbol{y}$ and is parallel to the axial vector associated to the skew tensor $\boldsymbol{W}$. By denoting $\boldsymbol{w}(t)$ the axial vector of $\boldsymbol{W}(t)$, the velocity field above can be re-written as

$$\boldsymbol{v}(\boldsymbol{x}, t) = \boldsymbol{v}(\boldsymbol{y}, t) + \boldsymbol{w}(t) \times (\boldsymbol{x} - \boldsymbol{y}), \tag{3.14}$$

which is the standard formula for the velocity field of classical rigid-body dynamics. The vector $\boldsymbol{w}(t)$ is called the *angular velocity* of the body. The rigid velocity field is schematically illustrated in Figure 3.4.

3.1.1. MATERIAL AND SPATIAL FIELDS

Both fields $\dot{\boldsymbol{x}}$ and $\boldsymbol{v}$ introduced above describe the velocity of material particles. However, $\dot{\boldsymbol{x}}$ and $\boldsymbol{v}$ have different arguments. While $\dot{\boldsymbol{x}}$ has material particle and time as arguments, the arguments of $\boldsymbol{v}$ are spatial position and time. This motivates the following definitions: Let a general time-dependent (scalar, vectorial or tensorial) field α be defined over the body $\mathscr{B}$. If the domain of α is $\Omega \times \mathscr{R}$, i.e. if the value of α is expressed as a function of material particles $\boldsymbol{p}$ (and time) then α is said to be a *material field*. On the other hand, if its domain is $\varphi_t(\Omega) \times \mathscr{R}$, then α is said to be a *spatial field*. Using (3.7), the *material description* of a spatial field $\alpha(\boldsymbol{x}, t)$ is defined by

$$\alpha_m(\boldsymbol{p}, t) = \alpha(\varphi(\boldsymbol{p}, t), t). \tag{3.15}$$

Conversely, the *spatial description* of a material field $\beta(\boldsymbol{p}, t)$ is defined by

$$\beta_s(\boldsymbol{x}, t) = \beta(\varphi^{-1}(\boldsymbol{x}, t), t). \tag{3.16}$$

It should be noted that *any* field associated with a motion of $\mathscr{B}$ can be expressed as a function of time and material particles *or* spatial position. A material (spatial) field does

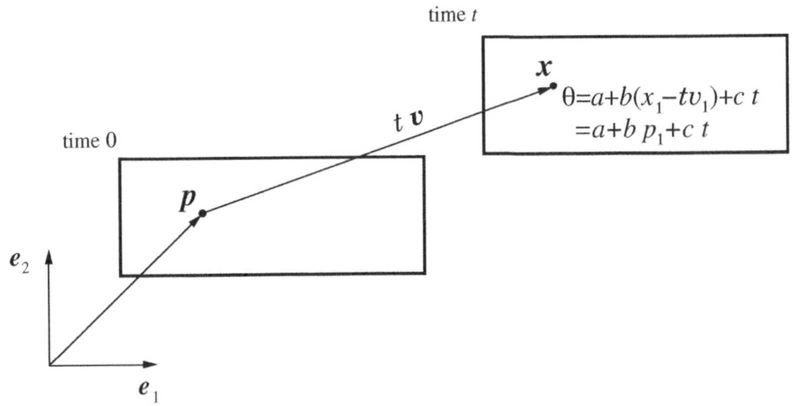

Figure 3.5. Material and spatial descriptions.

not necessarily represent a quantity physically associated with the reference (deformed) configuration of the body.

Example 3.1.1. Consider, for instance, the rectangular body of Figure 3.5 subjected to the *rigid translation*:

$$\boldsymbol{x} = \boldsymbol{\varphi}(\boldsymbol{p}, t) \equiv \boldsymbol{p} + t\,\boldsymbol{v},$$

with constant velocity $\boldsymbol{v}$. Assume that, during the motion $\boldsymbol{\varphi}$, the temperature field of the body in question is linearly distributed along its longitudinal axis and varies uniformly throughout the body at a constant rate. Taking the initial configuration (at $t = 0$) as the reference configuration (and, therefore, labelling material particles of the body with their position $\boldsymbol{p}$ at time 0), the material description of this temperature field reads

$$\theta_m(\boldsymbol{p}, t) = a + b\,p_1 + c\,t,$$

where a, b and c are constants. In view of the assumed motion $\boldsymbol{\varphi}$, the spatial description of the same field is given by

$$\theta_s(\boldsymbol{x}, t) = \theta_m(\boldsymbol{p}(\boldsymbol{x}, t), t) = a + b(x_1 - t\,v_1) + c\,t.$$

Note that, in spite of having $\boldsymbol{p}$ as one of its arguments, θ_m (as θ_s) expresses a physical quantity associated with the configuration of time t. The spatial description θ_s gives the temperature, at time t, of the material particle whose position at time t is $\boldsymbol{x}$. In experimental terms, it would be the temperature read from a thermometer held fixed in space at $\boldsymbol{x}$. The function θ_m gives the temperature, at time t, of the material particle whose position at time 0 is $\boldsymbol{p}$. It would be the temperature indicated by a thermometer attached to this material particle.

To avoid notational complexity, the subscripts m and s employed above to denote the material and spatial descriptions of general fields will not be used throughout this book unless absolutely necessary. In general, the description employed will be evident either from the context or from the argument used ($\boldsymbol{p}$ or $\boldsymbol{x}$).

3.1.2. MATERIAL AND SPATIAL GRADIENTS, DIVERGENCES AND TIME DERIVATIVES

The *material* and *spatial gradients* of a general field α, denoted respectively $\nabla_p \alpha$ and $\nabla_x \alpha$, are defined as

$$\nabla_p \alpha = \frac{\partial}{\partial \boldsymbol{p}} \alpha_m(\boldsymbol{p}, t), \quad \nabla_x \alpha = \frac{\partial}{\partial \boldsymbol{x}} \alpha_s(\boldsymbol{x}, t), \tag{3.17}$$

i.e. they are, respectively, the derivatives of α with respect to $\boldsymbol{p}$ and $\boldsymbol{x}$ holding t fixed.

Similarly, the *material* and *spatial time derivatives* of α, denoted respectively $\dot{\alpha}$ and α', are defined by

$$\dot{\alpha} = \frac{\partial}{\partial t} \alpha_m(\boldsymbol{p}, t), \quad \alpha' = \frac{\partial}{\partial t} \alpha_s(\boldsymbol{x}, t). \tag{3.18}$$

The material time derivative $\dot{\alpha}$ measures the rate of change of α at a *fixed material particle* $\boldsymbol{p}$. The spatial time derivative, on the other hand, measures the rate of change of α observed at a *fixed spatial position* $\boldsymbol{x}$. In the example of Figure 3.5, the material and spatial time derivatives of the temperature field θ are given by

$$\dot{\theta} = c, \quad \theta' = -b\, v_1 + c.$$

The material time derivative in this case corresponds to the temperature rate computed from a thermometer attached to a material particle $\boldsymbol{p}$ whilst θ' is the temperature rate observed in a thermometer held fixed in space at $\boldsymbol{x}$. Note that the extra term $-b\, v_1$ added to θ' is a contribution to the rate of change of temperature at $\boldsymbol{x}$ due to the motion of the body combined with its non-uniform distribution of temperature. This contribution vanishes if the body moves parallel to $\boldsymbol{e}_2$ ($v_1 = 0$), i.e. the direction of temperature isolines. It would also vanish if the temperature were uniform throughout the body ($b = 0$).

Analogously to (2.145) (page 37), we define the *spatial* and *material divergence* of a vector field $\boldsymbol{v}$, respectively, as

$$\operatorname{div}_p \boldsymbol{v} = \operatorname{tr}(\nabla_p \boldsymbol{v}), \quad \operatorname{div}_x \boldsymbol{v} = \operatorname{tr}(\nabla_x \boldsymbol{v}). \tag{3.19}$$

In addition (refer to (2.147)), for a tensor field $\boldsymbol{T}$, the spatial and material divergence are given, in Cartesian components, by

$$(\operatorname{div}_x \boldsymbol{T})_i = \frac{\partial T_{ij}}{\partial x_j}, \quad (\operatorname{div}_p \boldsymbol{T})_i = \frac{\partial T_{ij}}{\partial p_j}. \tag{3.20}$$

The compact definition (2.146) is also applicable to the material and spatial divergence of a tensor.

3.1.3. THE DEFORMATION GRADIENT

The *deformation gradient* of the motion φ is the second-order tensor $\boldsymbol{F}$ defined by

$$\boldsymbol{F}\,(\boldsymbol{p}, t) = \nabla_p \varphi(\boldsymbol{p}, t) = \frac{\partial \boldsymbol{x}_t}{\partial \boldsymbol{p}}. \tag{3.21}$$

In view of (3.8) it can be written as

$$\boldsymbol{F} = \boldsymbol{I} + \nabla_p \boldsymbol{u}. \tag{3.22}$$

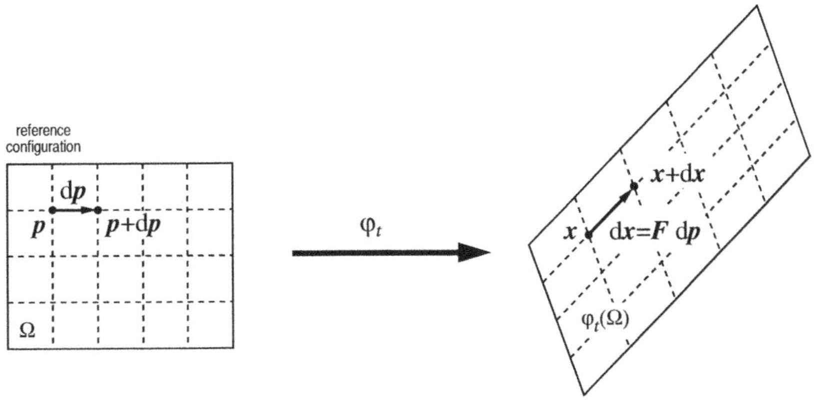

Figure 3.6. The deformation gradient.

The Cartesian components of $\boldsymbol{F}$ are given by

$$\begin{aligned} F_{ij} &= \frac{\partial x_i}{\partial p_j} \\ &= \delta_{ij} + \frac{\partial u_i}{\partial p_j}, \end{aligned} \tag{3.23}$$

where x_i denote the components of $\boldsymbol{x}_t$. In terms of the reference map (3.11), the deformation gradient may be equivalently expressed as

$$\boldsymbol{F}(\boldsymbol{x}, t) = [\nabla_x \boldsymbol{\varphi}^{-1}(\boldsymbol{x}, t)]^{-1} = [\boldsymbol{I} - \nabla_x \boldsymbol{u}]^{-1}. \tag{3.24}$$

Consider the infinitesimal material fibre $\mathrm{d}\boldsymbol{p}$ that connects two neighbouring material particles $\boldsymbol{p}$ and $\boldsymbol{p} + \mathrm{d}\boldsymbol{p}$ of a deforming body (Figure 3.6). Under the deformation $\boldsymbol{\varphi}_t$, these particles are mapped, respectively, into $\boldsymbol{x}$ and $\boldsymbol{x} + \mathrm{d}\boldsymbol{x}$. The deformation gradient is the linear operator that relates infinitesimal material fibres $\mathrm{d}\boldsymbol{p}$ with their deformed counterparts $\mathrm{d}\boldsymbol{x}$:

$$\mathrm{d}\boldsymbol{x} = \boldsymbol{F}\, \mathrm{d}\boldsymbol{p}. \tag{3.25}$$

A deformation of $\mathscr{B}$ with uniform deformation gradient ($\boldsymbol{F}$ independent of $\boldsymbol{p}$) is called a *homogeneous deformation*. A deformation is homogeneous if and only if it admits the representation

$$\boldsymbol{\varphi}(\boldsymbol{p}) = \boldsymbol{\varphi}(\boldsymbol{q}) + \boldsymbol{F}(\boldsymbol{p} - \boldsymbol{q}) \tag{3.26}$$

for all points $\boldsymbol{p}$, $\boldsymbol{q} \in \mathscr{B}$, with $\boldsymbol{F}$ a positive definite tensor. Clearly, rigid translations and rotations are homogeneous deformations.

3.1.4. VOLUME CHANGES. THE DETERMINANT OF THE DEFORMATION GRADIENT

Consider now the infinitesimal volume $\mathrm{d}v_0$ defined by the infinitesimal vectors $\mathrm{d}\boldsymbol{a}$, $\mathrm{d}\boldsymbol{b}$ and $\mathrm{d}\boldsymbol{c}$ emanating from the material particle $\boldsymbol{p}$ in the reference configuration (Figure 3.7). Trivially,

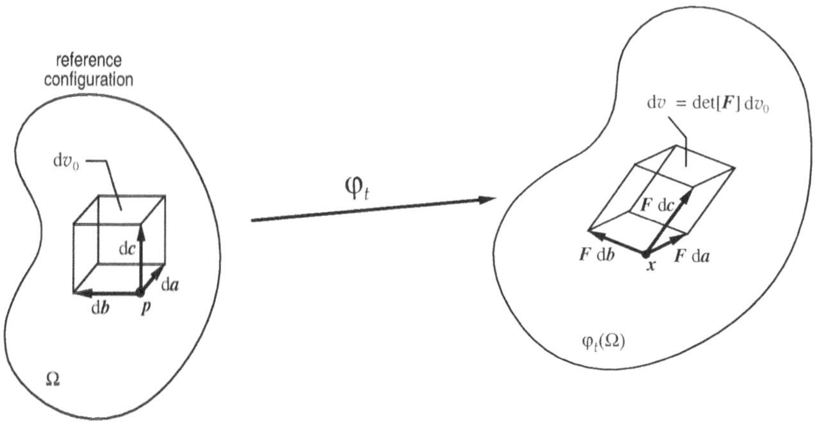

Figure 3.7. The determinant of the deformation gradient.

one has

$$dv_0 = (d\boldsymbol{a} \times d\boldsymbol{b}) \cdot d\boldsymbol{c}. \tag{3.27}$$

The deformation φ_t maps the infinitesimal vectors, respectively, into $\boldsymbol{F}\, d\boldsymbol{a}$, $\boldsymbol{F}\, d\boldsymbol{b}$ and $\boldsymbol{F}\, d\boldsymbol{c}$, so that the deformed infinitesimal volume is given by

$$dv = (\boldsymbol{F}\, d\boldsymbol{a} \times \boldsymbol{F}\, d\boldsymbol{b}) \cdot \boldsymbol{F}\, d\boldsymbol{c}. \tag{3.28}$$

By making use of identity (2.54), it follows that

$$\det \boldsymbol{F} = \frac{dv}{dv_0}, \tag{3.29}$$

i.e. the determinant of the deformation gradient represents, locally, the volume after deformation per unit reference volume (or *volume change ratio*). Throughout this book, we will adopt the following notation

$$J \equiv \det \boldsymbol{F}. \tag{3.30}$$

From (3.29) it follows that if $\det \boldsymbol{F} = 0$, then the infinitesimal volume has collapsed into a material particle. Since the body is not allowed to penetrate itself (this restriction is embodied in the assumption that the deformation map is one-to-one), this represents a physically unacceptable situation. Also note that, at the reference configuration, $\boldsymbol{F} = \boldsymbol{I}$ and, consequently, $J = 1$. Thus, a configuration with $J < 0$ cannot be reached from the reference configuration without having, at some stage, $J = 0$. Therefore, in any deformed configuration of a body, J satisfies

$$J > 0. \tag{3.31}$$

Isochoric deformations

Isochoric (or *volume-preserving*) deformations are deformations that do not produce changes in volume. A locally isochoric deformation is characterised by

$$J = 1. \tag{3.32}$$

Volumetric deformations

Volumetric deformations (i.e. *pure contractions/dilations*) are deformations consisting purely of a uniform contraction/dilation in all directions. The deformation gradient of any volumetric deformation is a spherical tensor:

$$\boldsymbol{F} = \alpha\,\boldsymbol{I}, \tag{3.33}$$

where the *scalar* α is the corresponding contraction/dilation ratio. With l_0 and l denoting, respectively, the undeformed and deformed lengths of a material fibre, for a locally volumetric deformation we have:

$$\frac{l}{l_0} = \alpha \tag{3.34}$$

in *all directions*.

3.1.5. ISOCHORIC/VOLUMETRIC SPLIT OF THE DEFORMATION GRADIENT

Any deformation can be locally decomposed as a purely volumetric deformation followed by an isochoric deformation or as an isochoric deformation followed by a pure volumetric deformation. To see this, note that the deformation gradient can always be multiplicatively split as

$$\boldsymbol{F} = \boldsymbol{F}_{\text{iso}}\,\boldsymbol{F}_{\text{v}} = \boldsymbol{F}_{\text{v}}\,\boldsymbol{F}_{\text{iso}}, \tag{3.35}$$

where

$$\boldsymbol{F}_{\text{v}} \equiv (\det \boldsymbol{F})^{\frac{1}{3}}\,\boldsymbol{I} \tag{3.36}$$

is the *volumetric* component of $\boldsymbol{F}$ and

$$\boldsymbol{F}_{\text{iso}} \equiv (\det \boldsymbol{F})^{-\frac{1}{3}}\,\boldsymbol{F} \tag{3.37}$$

is the *isochoric* (*volume-preserving* or *unimodular*) component. Note that, by construction, $\boldsymbol{F}_{\text{v}}$ corresponds indeed to a purely volumetric deformation (it has the representation (3.33)) and, since

$$\det \boldsymbol{F}_{\text{v}} = [(\det \boldsymbol{F})^{\frac{1}{3}}]^3 \det \boldsymbol{I} = \det \boldsymbol{F}, \tag{3.38}$$

$\boldsymbol{F}_{\text{v}}$ produces the same volume change as $\boldsymbol{F}$. The isochoric component in turn represents a volume preserving deformation, that is,

$$\det \boldsymbol{F}_{\text{iso}} = [(\det \boldsymbol{F})^{-\frac{1}{3}}]^3 \det \boldsymbol{F} = 1. \tag{3.39}$$

3.1.6. POLAR DECOMPOSITION. STRETCHES AND ROTATION

By applying the *polar decomposition* to the deformation gradient, one obtains:

$$\boldsymbol{F} = \boldsymbol{R}\boldsymbol{U} = \boldsymbol{V}\boldsymbol{R}, \tag{3.40}$$

where the proper orthogonal tensor $\boldsymbol{R}$ is the local *rotation tensor* and the symmetric positive definite tensors $\boldsymbol{U}$ and $\boldsymbol{V}$ are, respectively, the *right* and *left stretch tensors*. The right and left stretch tensors are related by the rotation

$$\boldsymbol{V} = \boldsymbol{R}\boldsymbol{U}\boldsymbol{R}^T. \tag{3.41}$$

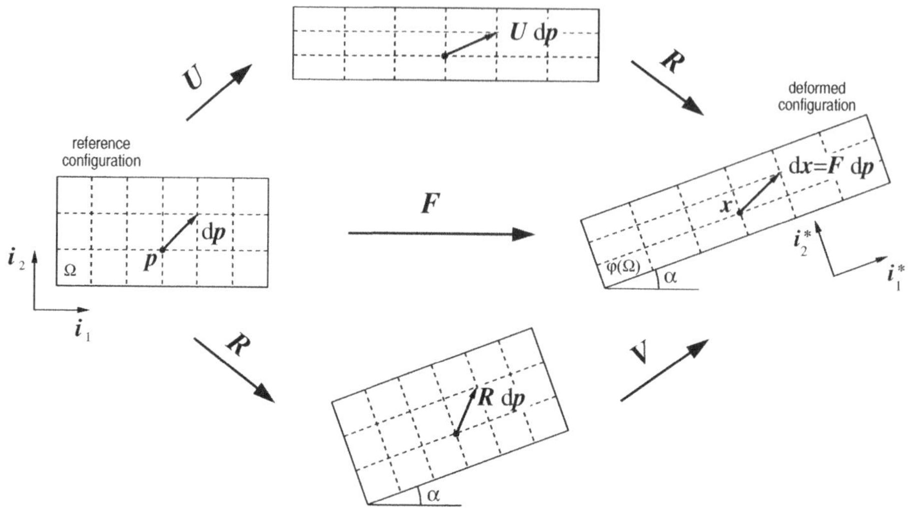

Figure 3.8. Polar decomposition of the deformation gradient. Stretches and rotation.

The stretch tensors $\boldsymbol{U}$ and $\boldsymbol{V}$ can be expressed as

$$\boldsymbol{U} = \sqrt{\boldsymbol{C}}, \quad \boldsymbol{V} = \sqrt{\boldsymbol{B}}, \tag{3.42}$$

where $\boldsymbol{C}$ and $\boldsymbol{B}$ – named, respectively, the *right* and *left Cauchy–Green strain tensors* – are defined by

$$\boldsymbol{C} = \boldsymbol{U}^2 = \boldsymbol{F}^T\boldsymbol{F}, \quad \boldsymbol{B} = \boldsymbol{V}^2 = \boldsymbol{F}\,\boldsymbol{F}^T. \tag{3.43}$$

Example 3.1.2 (A simple plane deformation). To illustrate the meaning of the polar decomposition of $\boldsymbol{F}$, a simple example consisting of a body subjected to a homogeneous deformation, i.e. with $\boldsymbol{F}$ independent of $\boldsymbol{p}$, is given in what follows. Consider the rectangular body of Figure 3.8 subjected to homogeneous stretching/compression in the directions of its longitudinal and transversal axes (respectively, the directions of $\boldsymbol{i}_1$ and $\boldsymbol{i}_2$ in the reference configuration) with a superimposed rigid rotation of angle α. With p_i and x_i denoting coordinates of $\boldsymbol{p}$ and $\boldsymbol{x}$ in the Cartesian system associated with the orthonormal basis $\{\boldsymbol{i}_1, \boldsymbol{i}_2\}$, the deformation map is defined as

$$\varphi : \begin{cases} x_1 = p_1\,\lambda_1\,\cos\alpha - p_2\,\lambda_2\,\sin\alpha \\ x_2 = p_1\,\lambda_1\,\sin\alpha + p_2\,\lambda_2\,\cos\alpha, \end{cases} \tag{3.44}$$

where the factors λ_1 and λ_2 determine how much stretching/compression occurs, respectively, along the longitudinal and transversal axes. In the basis $\{\boldsymbol{i}_1, \boldsymbol{i}_2\}$, the matrix representation of the corresponding deformation gradient is given by

$$\boldsymbol{F} = \begin{bmatrix} \lambda_1\,\cos\alpha & -\lambda_2\,\sin\alpha \\ \lambda_1\,\sin\alpha & \lambda_2\,\cos\alpha \end{bmatrix}. \tag{3.45}$$

The rotation tensor, obtained from the polar decomposition of $\boldsymbol{F}$, is represented by

$$\boldsymbol{R} = \begin{bmatrix} \cos\alpha & -\sin\alpha \\ \sin\alpha & \cos\alpha \end{bmatrix} \tag{3.46}$$

and the right and left stretch tensors by

$$\boldsymbol{U} = \begin{bmatrix} \lambda_1 & 0 \\ 0 & \lambda_2 \end{bmatrix} \tag{3.47}$$

and

$$\boldsymbol{V} = \begin{bmatrix} \lambda_1\ \cos^2\alpha + \lambda_2\ \sin^2\alpha & (\lambda_1 - \lambda_2)\ \sin\alpha\ \cos\alpha \\ (\lambda_1 - \lambda_2)\ \sin\alpha\ \cos\alpha & \lambda_1\ \sin^2\alpha + \lambda_2\ \cos^2\alpha \end{bmatrix}. \tag{3.48}$$

Insight into the meaning of the polar decomposition of the deformation gradient can be gained by focusing now on the generic infinitesimal fibre represented by $\mathrm{d}\boldsymbol{p}$ in Figure 3.8. Under deformation, $\mathrm{d}\boldsymbol{p}$ is mapped into $\mathrm{d}\boldsymbol{x} = \boldsymbol{F}\,\mathrm{d}\boldsymbol{p}$. With use of the polar decomposition of $\boldsymbol{F}$, this mapping can be split into two sequential steps. If the *right* polar decomposition $\boldsymbol{F} = \boldsymbol{R}\boldsymbol{U}$ is used, the two steps are:

1. $\mathrm{d}\boldsymbol{p} \longrightarrow \boldsymbol{U}\,\mathrm{d}\boldsymbol{p}$,
2. $\boldsymbol{U}\,\mathrm{d}\boldsymbol{p} \longrightarrow \boldsymbol{R}\,(\boldsymbol{U}\,\mathrm{d}\boldsymbol{p}) = \boldsymbol{F}\,\mathrm{d}\boldsymbol{p}$.

In the first operation, $\mathrm{d}\boldsymbol{p}$ deforms as if the body were being purely stretched (or compressed) along the directions of its longitudinal and transversal axes (which at this stage coincide with $\boldsymbol{i}_1$ and $\boldsymbol{i}_2$ respectively). The second mapping is a pure rotation (of angle α) of the deformed fibre $\boldsymbol{U}\,\mathrm{d}\boldsymbol{p}$ and corresponds to a rigid rotation of the body. If the *left* polar decomposition $\boldsymbol{F} = \boldsymbol{V}\boldsymbol{R}$ is employed instead, the sequence is reversed:

1. $\mathrm{d}\boldsymbol{p} \longrightarrow \boldsymbol{R}\,\mathrm{d}\boldsymbol{p}$,
2. $\boldsymbol{R}\,\mathrm{d}\boldsymbol{p} \longrightarrow \boldsymbol{V}(\boldsymbol{R}\,\mathrm{d}\boldsymbol{p}) = \boldsymbol{F}\,\mathrm{d}\boldsymbol{p}$.

In this case, the fibre is first rigidly rotated by an angle α. The second operation corresponds to the deformation of the fibre under pure stretching/compression of the body along its axial and transversal directions. However, due to the previous rotation, these directions coincide now with $\boldsymbol{i}_1^* = \boldsymbol{R}\,\boldsymbol{i}_1$ and $\boldsymbol{i}_2^* = \boldsymbol{R}\,\boldsymbol{i}_2$, respectively. Note that if the basis $\{\boldsymbol{i}_1^*, \boldsymbol{i}_2^*\}$ is used, the matrix representation of $\boldsymbol{V}$ reads

$$\boldsymbol{V} = \begin{bmatrix} \lambda_1 & 0 \\ 0 & \lambda_2 \end{bmatrix}, \tag{3.49}$$

so that the transformation $(\cdot) \rightarrow \boldsymbol{V}(\cdot)$ indeed corresponds to stretchings along the directions of $\boldsymbol{i}_1^*$ and $\boldsymbol{i}_2^*$.

The above example has illustrated the significance of the polar decomposition of $\boldsymbol{F}$. The discussion has been restricted to a homogeneous deformation only to ease visualisation of the stretches and rotation involved in the decomposition of the deformation gradient. It should be remarked that for a generic deformation of a body, in which $\boldsymbol{F}$ is a function of $\boldsymbol{p}$, intermediate configurations of the body corresponding to pure stretching or pure rigid rotation (such as those illustrated in Figure 3.8) do not exist in general. Nevertheless, the interpretation of $\boldsymbol{U}$ and $\boldsymbol{V}$ as pure stretchings and of $\boldsymbol{R}$ as a rigid rotation remain valid in a *local* sense. Note that for *any* deformation φ, one may write:

$$\boldsymbol{x} + \mathrm{d}\boldsymbol{x} = \varphi(\boldsymbol{p} + \mathrm{d}\boldsymbol{p}) = \boldsymbol{p} + \boldsymbol{F}(\boldsymbol{p})\, \mathrm{d}\boldsymbol{p}, \tag{3.50}$$

that is, within an infinitesimal neighbourhood of a material point $\boldsymbol{p}$, the deformation behaves like a homogeneous deformation with gradient $\boldsymbol{F}(\boldsymbol{p})$. Thus, within this infinitesimal neighbourhood of $\boldsymbol{p}$, $\boldsymbol{U}(\boldsymbol{p})$ and $\boldsymbol{V}(\boldsymbol{p})$ measure stretches from $\boldsymbol{p}$ and $\boldsymbol{R}(\boldsymbol{p})$ measures the local rigid rotation.

Spectral decomposition of the stretch tensors

Since $\boldsymbol{U}$ and $\boldsymbol{V}$ are symmetric, it follows from the *spectral theorem* that they admit the spectral decomposition

$$\boldsymbol{U} = \sum_{i=1}^{3} \lambda_i\, \boldsymbol{l}_i \otimes \boldsymbol{l}_i, \quad \boldsymbol{V} = \sum_{i=1}^{3} \lambda_i\, \boldsymbol{e}_i \otimes \boldsymbol{e}_i, \tag{3.51}$$

where the $\{\lambda_1, \lambda_2, \lambda_3\}$ are the eigenvalues of $\boldsymbol{U}$ (and $\boldsymbol{V}$) named the *principal stretches*. The vectors $\boldsymbol{l}_i$ and $\boldsymbol{e}_i$ are unit eigenvectors of $\boldsymbol{U}$ and $\boldsymbol{V}$ respectively. The triads $\{\boldsymbol{l}_1, \boldsymbol{l}_2, \boldsymbol{l}_3\}$ and $\{\boldsymbol{e}_1, \boldsymbol{e}_2, \boldsymbol{e}_3\}$ form orthonormal bases for the space $\mathscr{U}$ of vectors in $\mathscr{E}$. They are called, respectively, the *Lagrangian* and *Eulerian triads* and define the *Lagrangian* and *Eulerian principal directions*.

Substitution of (3.41) into (3.51) gives the following relationship between the eigenvectors of $\boldsymbol{V}$ and $\boldsymbol{U}$:

$$\boldsymbol{l}_i = \boldsymbol{R}\, \boldsymbol{e}_i, \tag{3.52}$$

that is, each vector $\boldsymbol{e}_i$ differs from the corresponding $\boldsymbol{l}_i$ by a rotation $\boldsymbol{R}$.

The spectral decomposition of the right and left stretch tensors implies that in *any* deformation, the local stretching from a material particle can always be expressed as a superposition of stretches along three mutually orthogonal directions. In the example discussed above, illustrated by Figure 3.8, $\{\lambda_1, \lambda_2\}$ are the principal stretches and the Lagrangian and Eulerian bases are, respectively, $\{\boldsymbol{i}_1, \boldsymbol{i}_2\}$ and $\{\boldsymbol{i}_1^*, \boldsymbol{i}_2^*\}$.

3.1.7. STRAIN MEASURES

In the above section, we have seen that in a local sense, i.e. within an infinitesimal neighbourhood of a generic material particle $\boldsymbol{p}$, pure rotations can be distinguished from pure stretching by means of the polar decomposition of the deformation gradient. Under the action of pure rotations, the distances between particles within this neighbourhood remain fixed. When the distances between material particles are identical to their values in the reference configuration, we say that the region surrounding $\boldsymbol{p}$ is *unstrained*. In this case, the difference between the deformed neighbourhood of $\boldsymbol{p}$ and its reference configuration is a rigid deformation. Pure stretching, on the other hand, characterised by $\boldsymbol{U}$ or $\boldsymbol{V}$, changes the distance between material particles. Under stretching, we say that the region surrounding $\boldsymbol{p}$ is *strained*. To quantify straining, i.e. to evaluate how much $\boldsymbol{U}$ (or $\boldsymbol{V}$) departs from $\boldsymbol{I}$ (a rigid deformation), some kind of *strain measure* needs to be defined.

Let us consider, again, the generic material fibre represented by the infinitesimal vector $\mathrm{d}\boldsymbol{p}$ that emanates from $\boldsymbol{p}$ (Figure 3.8 serves as an illustration). The deformation maps $\mathrm{d}\boldsymbol{p}$ into $\mathrm{d}\boldsymbol{x} = \boldsymbol{F}\,\mathrm{d}\boldsymbol{p}$. Thus, the square of the deformed length of the material fibre in question reads

$$\|\mathrm{d}\boldsymbol{x}\|^2 = \boldsymbol{F}\,\mathrm{d}\boldsymbol{p} \cdot \boldsymbol{F}\,\mathrm{d}\boldsymbol{p} = \boldsymbol{C}\,\mathrm{d}\boldsymbol{p} \cdot \mathrm{d}\boldsymbol{p} = (\boldsymbol{I} + 2\,\boldsymbol{E}^{(2)})\,\mathrm{d}\boldsymbol{p} \cdot \mathrm{d}\boldsymbol{p}, \tag{3.53}$$

where $\boldsymbol{C} = \boldsymbol{F}^T\,\boldsymbol{F} = \boldsymbol{U}^2$ is the right Cauchy–Green tensor and the *strain measure* $\boldsymbol{E}^{(2)}$ (the meaning of the superscript will be made clear below) is the so-called *Green–Lagrange* strain tensor *defined* as

$$\begin{aligned}\boldsymbol{E}^{(2)} &= \tfrac{1}{2}(\boldsymbol{C} - \boldsymbol{I})\\ &= \tfrac{1}{2}[\nabla_p \boldsymbol{u} + (\nabla_p \boldsymbol{u})^T + (\nabla_p \boldsymbol{u})^T \nabla_p \boldsymbol{u}].\end{aligned} \tag{3.54}$$

No straining occurs, that is, the size of any infinitesimal material fibre emanating from $\boldsymbol{p}$ remains constant ($\|\mathrm{d}\boldsymbol{x}\| = \|\mathrm{d}\boldsymbol{p}\|,\ \ \forall\ \mathrm{d}\boldsymbol{p}$), if and only if $\boldsymbol{E}^{(2)} = \boldsymbol{0}$. This condition is equivalent to $\boldsymbol{C} = \boldsymbol{U} = \boldsymbol{I}$, implying that $\boldsymbol{F}$ is an orthogonal tensor and the deformation is rigid (pure translation and/or rotation) in the neighbourhood of $\boldsymbol{p}$. From the definition of $\boldsymbol{E}^{(2)}$, its eigenvectors coincide with the Lagrangian triad so that it can be expressed as

$$\boldsymbol{E}^{(2)} = \sum_{i=1}^{3} \tfrac{1}{2}(\lambda_i^2 - 1)\,\boldsymbol{l}_i \otimes \boldsymbol{l}_i, \tag{3.55}$$

and, since it measures strains along the principal *Lagrangian* directions, it is called a *Lagrangian strain measure*.

It must be emphasised that the Green–Lagrange strain measure is *defined* by expression (3.54). It is by no means the unique way of quantifying straining. In fact, the definition of a strain measure is somewhat arbitrary and a specific choice is usually dictated by mathematical and physical convenience. An important family of *Lagrangian strain tensors*, i.e. strain measures based on the Lagrangian triad, is defined by Seth (1964), Hill (1978) and Ogden (1984)

$$\boldsymbol{E}^{(m)} = \begin{cases} \dfrac{1}{m}(\boldsymbol{U}^m - \boldsymbol{I}) & m \neq 0 \\ \ln[\boldsymbol{U}] & m = 0 \end{cases} \tag{3.56}$$

where m is a real number and $\ln[\,\cdot\,]$ denotes the *tensor logarithm* of $[\,\cdot\,]$. Equivalently, in terms of its spectral decomposition, (3.56) may be rephrased as

$$\boldsymbol{E}^{(m)} = \sum_{i=1}^{3} f(\lambda_i)\,\boldsymbol{l}_i \otimes \boldsymbol{l}_i, \tag{3.57}$$

where

$$f(\lambda_i) = \begin{cases} \dfrac{1}{m}(\lambda_i^m - 1) & m \neq 0 \\ \ln \lambda_i & m = 0. \end{cases} \tag{3.58}$$

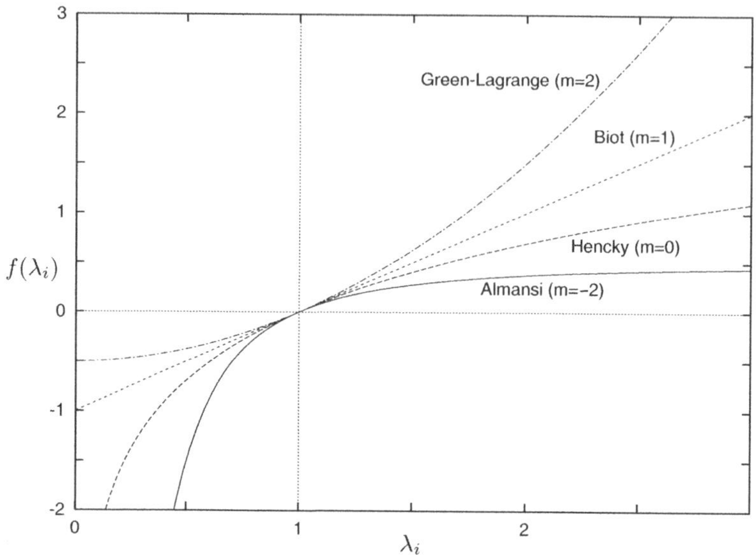

Figure 3.9. Strain measures. Principal strain as a function of the principal stretch for various strain measures.

The Green–Lagrange strain tensor, $\boldsymbol{E}^{(2)}$, is a particular member of this family (with $m = 2$). Other commonly used members of this family are the Biot ($m = 1$), Hencky ($m = 0$) and Almansi ($m = -2$) strain tensors. Note that for any m, the associated strain tensor vanishes if and only if the deformation gradient represents, locally, a rigid deformation, i.e.

$$\boldsymbol{E}^{(m)} = \boldsymbol{0} \quad \Longleftrightarrow \quad \boldsymbol{U} = \boldsymbol{I} \quad \Longleftrightarrow \quad \boldsymbol{F} = \boldsymbol{R}. \tag{3.59}$$

To illustrate the relationship between the stretch and strain tensors, the principal strain for various strain measures is plotted in Figure 3.9 as a function of the corresponding principal stretch.

Analogously to the strain measures discussed above, it is also possible to define tensors that measure strain along the principal *Eulerian* directions or, simply, *Eulerian strain tensors*. Based on the *left* stretch tensor, the Eulerian counterpart of the Lagrangian family of strain measures above is defined by

$$\boldsymbol{\varepsilon}^{(m)} = \begin{cases} \dfrac{1}{m}(\boldsymbol{V}^m - \boldsymbol{I}) & m \neq 0 \\ \ln[\boldsymbol{V}] & m = 0, \end{cases} \tag{3.60}$$

or, using the Eulerian triad,

$$\boldsymbol{\varepsilon}^{(m)} = \sum_{i=1}^{3} f(\lambda_i)\, \boldsymbol{e}_i \otimes \boldsymbol{e}_i. \tag{3.61}$$

Lagrangian and Eulerian strain tensors are related by

$$\boldsymbol{\varepsilon}^{(m)} = \boldsymbol{R}\, \boldsymbol{E}^{(m)} \boldsymbol{R}^T, \tag{3.62}$$

that is, they differ by the local rotation $\boldsymbol{R}$.

3.1.8. THE VELOCITY GRADIENT. RATE OF DEFORMATION AND SPIN

The spatial field $\boldsymbol{L}$, defined as

$$\boldsymbol{L} = \nabla_x \boldsymbol{v}, \tag{3.63}$$

is named the *velocity gradient*. Equivalently, with application of the chain rule one has

$$\boldsymbol{L} = \frac{\partial}{\partial t}\left(\frac{\partial \boldsymbol{\varphi}}{\partial \boldsymbol{p}}\right)\frac{\partial \boldsymbol{p}}{\partial \boldsymbol{x}} = \dot{\boldsymbol{F}}\,\boldsymbol{F}^{-1}. \tag{3.64}$$

Two important tensors are obtained by splitting $\boldsymbol{L}$ into its symmetric and skew parts. Namely, the *rate of deformation* tensor (also referred to as the *stretching* tensor), $\boldsymbol{D}$, and the *spin* tensor, $\boldsymbol{W}$, are defined by

$$\boldsymbol{D} = \text{sym}(\boldsymbol{L}), \quad \boldsymbol{W} = \text{skew}(\boldsymbol{L}). \tag{3.65}$$

To gain insight into the physical meaning of the tensors $\boldsymbol{D}$ and $\boldsymbol{W}$, it is convenient to consider a body undergoing a motion with uniform (independent of $\boldsymbol{x}$) velocity gradient. For such a motion the velocity field reads

$$\boldsymbol{v}(\boldsymbol{x}, t) = \boldsymbol{v}(\boldsymbol{y}, t) + \boldsymbol{L}(t)\,(\boldsymbol{x} - \boldsymbol{y}). \tag{3.66}$$

If the decomposition of $\boldsymbol{L}$ into its symmetric and skew parts is introduced, the velocity field can be split as

$$\boldsymbol{v}(\boldsymbol{x}, t) = \boldsymbol{v}^R(\boldsymbol{x}, t) + \boldsymbol{v}^S(\boldsymbol{x}, t), \tag{3.67}$$

where the following definitions have been used:

$$\begin{aligned} \boldsymbol{v}^R(\boldsymbol{x}, t) &= \boldsymbol{v}(\boldsymbol{y}, t) + \boldsymbol{W}(t)\,(\boldsymbol{x} - \boldsymbol{y}), \\ \boldsymbol{v}^S(\boldsymbol{x}, t) &= \boldsymbol{D}(t)\,(\boldsymbol{x} - \boldsymbol{y}). \end{aligned} \tag{3.68}$$

By recalling expression (3.13), the velocity $\boldsymbol{v}^R$, associated with the spin tensor $\boldsymbol{W}$, can be immediately identified as a *rigid* velocity. The only contribution to straining is then provided by the term $\boldsymbol{v}^S$, associated with the rate of deformation tensor. Note that, due to its symmetry, $\boldsymbol{D}$ admits the representation

$$\boldsymbol{D} = \sum_{i=1}^{3} d_i\, \boldsymbol{e}_i \otimes \boldsymbol{e}_i, \tag{3.69}$$

with d_i and $\{\boldsymbol{e}_i\}$, respectively, the eigenvalues and an orthonormal basis of eigenvectors of $\boldsymbol{D}$. With the spectral representation above, the velocity field $\boldsymbol{v}^S$ can be decomposed as a sum of three linearly independent velocities of the form:

$$d_i\,(\boldsymbol{e}_i \otimes \boldsymbol{e}_i)\,(\boldsymbol{x} - \boldsymbol{y}),$$

with no summation implied on i, so that the components of $\boldsymbol{v}^S$ relative to the basis $\{\boldsymbol{e}_1, \boldsymbol{e}_2, \boldsymbol{e}_3\}$ are given by

$$v_i^S = d_i\,(x_i - y_i), \tag{3.70}$$

again with no summation implied, where x_i and y_i denote the coordinates of points $\boldsymbol{x}$ and $\boldsymbol{y}$ in a Cartesian system associated to $\{\boldsymbol{e}_1, \boldsymbol{e}_2, \boldsymbol{e}_3\}$. As schematically illustrated in Figure 3.10,

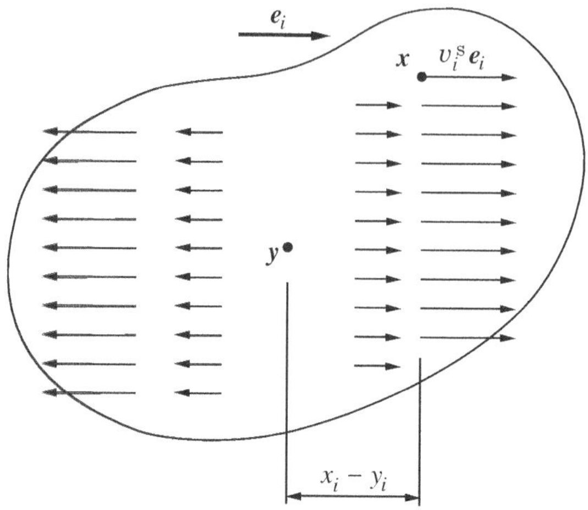

Figure 3.10. Straining velocity field.

each v_i^S corresponds to a velocity field that purely stretches the body in the direction of $\boldsymbol{e}_i$, with the plane perpendicular to $\boldsymbol{e}_i$ that passes through $\boldsymbol{y}$ fixed. Thus, the rate of deformation tensor corresponds indeed to a pure stretching of the body.

If a general motion (in which $\boldsymbol{L}$ is not necessarily uniform) is considered, the above decomposition of the velocity field into the sum of a rigid velocity and a straining velocity remains valid in the *local* sense. In this case, consider a point $\boldsymbol{x}$ and a point $\boldsymbol{x} + \mathrm{d}\boldsymbol{x}$ lying within an infinitesimal neighbourhood of $\boldsymbol{x}$. The velocity field within this infinitesimal neighbourhood of $\boldsymbol{x}$ is given by

$$\boldsymbol{v}(\boldsymbol{x} + \mathrm{d}\boldsymbol{x}, t) = \boldsymbol{v}(\boldsymbol{x}, t) + \boldsymbol{L}(\boldsymbol{x}, t)\, \mathrm{d}\boldsymbol{x}, \tag{3.71}$$

so that, in any motion, the velocity field can be *locally* decomposed as a sum of a rigid velocity

$$\boldsymbol{v}(\boldsymbol{x}, t) + \boldsymbol{W}(\boldsymbol{x}, t)\, \mathrm{d}\boldsymbol{x},$$

associated with the spin tensor $\boldsymbol{W}$, and a straining velocity

$$\boldsymbol{D}(\boldsymbol{x}, t)\, \mathrm{d}\boldsymbol{x},$$

associated exclusively to the rate of deformation tensor $\boldsymbol{D}$.

3.1.9. RATE OF VOLUME CHANGE

The *rate of volume change*, $\dot{J}$, is related to the rate of deformation tensor through the expression

$$\dot{J} = J \operatorname{tr} \boldsymbol{D}. \tag{3.72}$$

To derive this expression, we first apply the chain rule to obtain

$$\dot{J} \equiv (\det \boldsymbol{F})^{\dot{}} = \frac{\partial(\det \boldsymbol{F})}{\partial \boldsymbol{F}} : \dot{\boldsymbol{F}} = J\, \boldsymbol{F}^{-T} : \dot{\boldsymbol{F}}, \tag{3.73}$$

where we have made use of relation (2.140) (page 36) for the derivative of the determinant. This, together with definition (2.36) (page 22) of the trace of a tensor and the fact that the skew symmetry of $\boldsymbol{W}$ implies

$$\text{tr}\,\boldsymbol{L} = \text{tr}\,\boldsymbol{D}, \tag{3.74}$$

leading to (3.72).

Also note that from the definition (2.145) of the divergence of a vector field we have

$$\text{tr}\,\boldsymbol{D} = \text{div}_x\,\boldsymbol{v}, \tag{3.75}$$

so that the rate of volume change can be equivalently expressed as

$$\dot{J} = J\,\text{div}_x\,\boldsymbol{v}. \tag{3.76}$$

3.2. Infinitesimal deformations

Small or infinitesimal deformations are deformations with sufficiently small displacement gradient, $\nabla_p \boldsymbol{u}$. For such deformations, the description of kinematics can be substantially simplified.

3.2.1. THE INFINITESIMAL STRAIN TENSOR

Recall definition (3.43) of the Cauchy–Green tensors. In terms of the displacement gradient, one has

$$\begin{aligned} \boldsymbol{C} &= \boldsymbol{I} + \nabla_p \boldsymbol{u} + (\nabla_p \boldsymbol{u})^T + (\nabla_p \boldsymbol{u})^T\, \nabla_p \boldsymbol{u}, \\ \boldsymbol{B} &= \boldsymbol{I} + \nabla_p \boldsymbol{u} + (\nabla_p \boldsymbol{u})^T + \nabla_p \boldsymbol{u}\, (\nabla_p \boldsymbol{u})^T. \end{aligned} \tag{3.77}$$

If the displacement gradient is sufficiently small, the second-order terms in $\nabla_p \boldsymbol{u}$ of the expressions above can be neglected so that, under small deformations, the following approximation can be made

$$\boldsymbol{C} \approx \boldsymbol{B} \approx \boldsymbol{I} + \nabla_p \boldsymbol{u} + (\nabla_p \boldsymbol{u})^T. \tag{3.78}$$

From the above expression and the definitions of the Green–Lagrange strain tensor $\boldsymbol{E}^{(2)}$ and its Eulerian counterpart $\boldsymbol{\varepsilon}^{(2)}$, it follows that, to the same order of approximation,

$$\boldsymbol{E}^{(2)} \approx \boldsymbol{\varepsilon}^{(2)} \approx \tfrac{1}{2}[\nabla_p \boldsymbol{u} + (\nabla_p \boldsymbol{u})^T]. \tag{3.79}$$

This motivates the definition of the *infinitesimal strain tensor* to measure strains under small deformations

$$\boldsymbol{\varepsilon} \equiv \nabla_p^s \boldsymbol{u}, \tag{3.80}$$

where we have introduced the notation

$$\nabla^s(\cdot) = \text{sym}[\nabla(\cdot)] = \tfrac{1}{2}[\nabla(\cdot) + \nabla(\cdot)^T], \tag{3.81}$$

for the *symmetric gradient* of a vector field. It is worth pointing out here that $\boldsymbol{\varepsilon}$ is a *linear* functional of $\boldsymbol{u}$. This fact greatly simplifies the description of small deformations.

In fact, it can be easily shown that not only $\boldsymbol{E}^{(2)}$ and $\boldsymbol{\varepsilon}^{(2)}$ but *all* Lagrangian and Eulerian strain measures defined by expressions (3.56) and (3.60) have the same small deformation limit, i.e. for any m and to within an error of second order in $\nabla_p \boldsymbol{u}$, one has

$$\boldsymbol{\varepsilon}^{(m)} \approx \boldsymbol{E}^{(m)} \approx \boldsymbol{\varepsilon}. \tag{3.82}$$

3.2.2. INFINITESIMAL RIGID DEFORMATIONS

In terms of the infinitesimal strain tensor, the square of the deformed length of a generic material fibre $\mathrm{d}\boldsymbol{p}$ (recall the text preceding expression (3.53)) reads

$$\|\mathrm{d}\boldsymbol{x}\|^2 = (\boldsymbol{I} + 2\,\boldsymbol{\varepsilon})\,\mathrm{d}\boldsymbol{p}\cdot\mathrm{d}\boldsymbol{p} + o(\nabla_p\boldsymbol{u}) \tag{3.83}$$

with $o(\nabla_p\boldsymbol{u})$ a term of second order in $\nabla_p\boldsymbol{u}$. It is clear from this expression that, to within an error of $o(\nabla_p\boldsymbol{u})$, only the symmetric part $\boldsymbol{\varepsilon}$ of $\nabla_p\boldsymbol{u}$ is associated with local straining. The skew part of $\nabla_p\boldsymbol{u}$ produces no straining and is associated exclusively with local infinitesimal rigid rotations. For a pure local infinitesimal rigid rotation ($\|\mathrm{d}\boldsymbol{x}\| = \|\mathrm{d}\boldsymbol{p}\|,\ \ \forall\ \ \|\mathrm{d}\boldsymbol{p}\|$) the tensor $\boldsymbol{\varepsilon}$ vanishes or, equivalently, $\nabla_p\boldsymbol{u}$ is skew.

For a body under an arbitrary homogeneous deformation ($\nabla_p\boldsymbol{u}$ independent of $\boldsymbol{p}$), the displacement field can be written as

$$\boldsymbol{u}(\boldsymbol{p}) = \boldsymbol{u}(\boldsymbol{q}) + \nabla_p\boldsymbol{u}\,(\boldsymbol{p}-\boldsymbol{q}), \tag{3.84}$$

for all points $\boldsymbol{p}$ and $\boldsymbol{q}$. For infinitesimal rigid deformations and within an approximation of second order in the displacement gradient, $\nabla_p\boldsymbol{u}$ is skew and the field $\boldsymbol{u}$ can be written as

$$\boldsymbol{u}(\boldsymbol{p}) = \boldsymbol{u}(\boldsymbol{q}) + \boldsymbol{A}\,(\boldsymbol{p}-\boldsymbol{q}), \tag{3.85}$$

for all points $\boldsymbol{p}$ and $\boldsymbol{q}$ with $\boldsymbol{A} \equiv \nabla_p\boldsymbol{u}$ a skew tensor. Alternatively, with $\boldsymbol{a}$ denoting the axial vector of $\boldsymbol{A}$, $\boldsymbol{u}$ can be expressed as

$$\boldsymbol{u}(\boldsymbol{p}) = \boldsymbol{u}(\boldsymbol{q}) + \boldsymbol{a}\times(\boldsymbol{p}-\boldsymbol{q}). \tag{3.86}$$

Any displacement that admits the representation (3.85)–(3.86) is called an *infinitesimal rigid displacement* field. Note that infinitesimal rigid displacements have the same representation as rigid velocity fields (see expressions (3.13) and (3.14)).

3.2.3. INFINITESIMAL ISOCHORIC AND VOLUMETRIC DEFORMATIONS

Analogously to the isochoric/volumetric split of the deformation gradient in the finite strain context (refer to Section 3.1.5), the infinitesimal strain tensor $\boldsymbol{\varepsilon}$ can also be split into a purely volumetric and a volume-preserving contribution. The isochoric/volumetric split of the infinitesimal strain tensor is *additive* (in contrast to the *multiplicative* split of the deformation gradient in the finite strain theory) and reads

$$\boldsymbol{\varepsilon} = \boldsymbol{\varepsilon}_{\mathrm{d}} + \boldsymbol{\varepsilon}_{\mathrm{v}}, \tag{3.87}$$

where

$$\boldsymbol{\varepsilon}_{\mathrm{d}} \equiv \boldsymbol{\varepsilon} - \boldsymbol{\varepsilon}_{\mathrm{v}} \tag{3.88}$$

is the isochoric component, known as the *strain deviator* or *deviatoric strain*, which measures pure infinitesimal distortions. The tensor

$$\boldsymbol{\varepsilon}_{\mathrm{v}} \equiv \tfrac{1}{3}\,\varepsilon_{\mathrm{v}}\,\boldsymbol{I} \tag{3.89}$$

is the infinitesimal *volumetric strain tensor*. The scalar invariant of ε, defined as

$$\varepsilon_{\rm v} \equiv I_1(\varepsilon) = \mathrm{tr}\,\varepsilon = \mathrm{tr}\,\nabla^s \boldsymbol{u} = \mathrm{tr}\,\nabla \boldsymbol{u} \tag{3.90}$$

is named the infinitesimal *volumetric strain*. An infinitesimal deformation is volume-preserving if and only if

$$\varepsilon_{\rm v} = 0. \tag{3.91}$$

The tensors $\varepsilon_{\rm d}$ and $\varepsilon_{\rm v}$ can be equivalently written in terms of linear operations on ε as

$$\varepsilon_{\rm d} = [\mathsf{I}_S - \tfrac{1}{3}\,\boldsymbol{I}\otimes\boldsymbol{I}] : \varepsilon, \quad \varepsilon_{\rm v} = \tfrac{1}{3}\,(\boldsymbol{I}\otimes\boldsymbol{I}) : \varepsilon. \tag{3.92}$$

It should be noted that the strain deviator is a *traceless* tensor, i.e.

$$\mathrm{tr}\,\varepsilon_{\rm d} = 0. \tag{3.93}$$

The fourth-order tensor defined as

$$\mathsf{I}_{\rm d} \equiv \mathsf{I}_S - \tfrac{1}{3}\,\boldsymbol{I}\otimes\boldsymbol{I}, \tag{3.94}$$

is referred to as the *deviatoric projection tensor*. It projects second-order symmetric tensors into the *deviatoric subspace*, i.e. into the space of traceless tensors. Throughout this book we shall often use the alternative notation

$$\mathrm{dev}(\boldsymbol{S})$$

to represent the deviator of a symmetric tensor $\boldsymbol{S}$, i.e.

$$\mathrm{dev}(\boldsymbol{S}) \equiv \mathsf{I}_{\rm d} : \boldsymbol{S}. \tag{3.95}$$

From finite to infinitesimal isochoric and volumetric strains

Analogously to Section 3.2.1, where the infinitesimal strain tensor is derived from the finite strain theory, the above isochoric/volumetric split can also be obtained from its finite deformation counterpart by neglecting higher order terms in $\nabla_p \boldsymbol{u}$.

To show this, let us consider the Green–Lagrange strain tensor, $\boldsymbol{E}^{(2)}$. Following the isochoric/volumetric split of the deformation gradient given by (3.35), we define the corresponding isochoric and volumetric Green–Lagrange strains

$$\boldsymbol{E}^{(2)}_{\rm iso} \equiv \tfrac{1}{2}(\boldsymbol{C}_{\rm iso} - \boldsymbol{I}); \quad \boldsymbol{E}^{(2)}_{\rm v} \equiv \tfrac{1}{2}(\boldsymbol{C}_{\rm v} - \boldsymbol{I}), \tag{3.96}$$

where

$$\boldsymbol{C}_{\rm iso} \equiv \boldsymbol{F}_{\rm iso}^T\,\boldsymbol{F}_{\rm iso} = (\det\,\boldsymbol{F})^{-\frac{2}{3}}\,\boldsymbol{F}^T\,\boldsymbol{F} = (\det\,\boldsymbol{F})^{-\frac{2}{3}}\,\boldsymbol{C} \tag{3.97}$$

and

$$\boldsymbol{C}_{\rm v} \equiv \boldsymbol{F}_{\rm v}^T\,\boldsymbol{F}_{\rm v} = (\det\,\boldsymbol{F})^{\frac{2}{3}}\,\boldsymbol{I}. \tag{3.98}$$

Now we proceed to show that, under small strain conditions (small $\nabla_p \boldsymbol{u}$), the volumetric Green–Lagrange strain defined above leads to definition (3.90). From $(3.96)_2$ and (3.98), we have

$$\boldsymbol{E}^{(2)}_{\rm v} = \tfrac{1}{2}\,[(\det\,\boldsymbol{F})^{\frac{2}{3}} - 1]\,\boldsymbol{I}. \tag{3.99}$$

From the standard concepts of differentiation discussed in Section 2.5 together with the definition $\boldsymbol{F} = \boldsymbol{I} + \nabla_p \boldsymbol{u}$ and the expression given in (iii) of page 36 for the derivative of the determinant, we find that

$$\begin{aligned} \det \boldsymbol{F} &= \det(\boldsymbol{I} + \nabla_p \boldsymbol{u}) \\ &= \det \boldsymbol{I} + (\det \boldsymbol{I})\, \boldsymbol{I} : \nabla_p \boldsymbol{u} + o(\nabla_p \boldsymbol{u}) \\ &= 1 + \operatorname{tr} \nabla_p \boldsymbol{u} + o(\nabla_p \boldsymbol{u}) \end{aligned} \tag{3.100}$$

and

$$(\det \boldsymbol{F})^{\frac{2}{3}} = 1 + \tfrac{2}{3} \operatorname{tr} \nabla_p \boldsymbol{u} + o(\nabla_p \boldsymbol{u}). \tag{3.101}$$

With the substitution of the above expression into (3.99), we then obtain

$$\boldsymbol{E}_{\mathrm{v}}^{(2)} = \varepsilon_{\mathrm{v}} + \boldsymbol{o}(\nabla_p \boldsymbol{u}). \tag{3.102}$$

Thus, if higher-order terms are neglected, we have the following approximation

$$\boldsymbol{E}_{\mathrm{v}}^{(2)} \approx \varepsilon_{\mathrm{v}}. \tag{3.103}$$

Following a completely analogous procedure with the isochoric Green–Lagrange strain, we obtain

$$\begin{aligned} \boldsymbol{E}_{\mathrm{iso}}^{(2)} &= \tfrac{1}{2}[(\det \boldsymbol{F})^{-\frac{2}{3}}(\boldsymbol{I} + \nabla_p^T \boldsymbol{u} + \nabla_p \boldsymbol{u} + \nabla_p^T \boldsymbol{u}\, \nabla_p \boldsymbol{u}) - \boldsymbol{I}] \\ &= \tfrac{1}{2}\{[1 - \tfrac{2}{3} \operatorname{tr} \nabla_p \boldsymbol{u} + o(\nabla_p \boldsymbol{u})][\boldsymbol{I} + \nabla_p^T \boldsymbol{u} + \nabla_p \boldsymbol{u} + o(\nabla_p \boldsymbol{u})] - \boldsymbol{I}\} \\ &= \boldsymbol{\varepsilon} - \tfrac{1}{3}(\operatorname{tr} \nabla_p \boldsymbol{u})\boldsymbol{I} + o(\nabla_p \boldsymbol{u}) \\ &= \boldsymbol{\varepsilon}_{\mathrm{d}} + \boldsymbol{o}(\nabla_p \boldsymbol{u}). \end{aligned} \tag{3.104}$$

Thus, to within second-order terms in $\nabla_p \boldsymbol{u}$, we have

$$\boldsymbol{E}_{\mathrm{iso}}^{(2)} \approx \boldsymbol{\varepsilon}_{\mathrm{d}}. \tag{3.105}$$

The infinitesimal limits above are valid for all Lagrangian and Eulerian finite strain measures defined by expressions (3.56) and (3.60).

3.3. Forces. Stress Measures

The previous sections of this chapter have been limited to the mathematical description of the kinematics of deformation. In particular, concepts such as the deformation gradient, rotations and the different strain measures used to quantify internal straining are of utmost importance in the formulation of the mechanical and thermodynamical theory of continua. It should be noted that, thus far, no reference has been made to *forces* and how they are transferred within continuum bodies.

The forces associated with the mechanical description of a body can be classed into three categories:‡

‡Stress *couples* could also be considered but these are outside the scope of this book and fall within the realm of the so-called *polar* continuum theories (Cosserat and Cosserat, 1909; Toupin, 1962; Truesdell and Noll, 1965).

1. *Boundary forces*. Forces applied to the boundary of the body such as those resulting from contact with another body. The dimension of boundary forces is force per unit area.
2. *Body forces*. Forces exerted on the interior of the body. Gravitational and magnetic forces are typical examples of such forces. The dimension of body forces is force per unit mass (or volume).
3. *Internal interactions between adjacent parts of a body*. The dimension of such interactions is force per unit area.

Internal interaction forces arise from the action of one part of the body upon an adjacent part and are transmitted across the surface that separate them. Boundary forces represent interactions between the exterior and the interior of a body and, as internal interactions, are transmitted across a surface (the boundary of the body in this case). Thus, boundary forces and interactions between distinct parts of a body are forces of essentially the same type and will be collectively called *surface forces*. To describe surface forces mathematically, the concept of *stress* as well as the different ways of quantifying it are introduced in this section.

3.3.1. CAUCHY'S AXIOM. THE CAUCHY STRESS VECTOR

Crucial to the description of surface forces is *Cauchy's axiom* stated in what follows. Consider a body $\mathscr{B}$ in an arbitrarily deformed configuration (Figure 3.11). Let $\mathscr{S}$ be an oriented surface of $\mathscr{B}$ with unit normal vector $\boldsymbol{n}$ at a point $\boldsymbol{x}$. Cauchy's axiom states that 'At $\boldsymbol{x}$, the surface force, i.e. the force per unit area, exerted across $\mathscr{S}$ by the material on the side of $\mathscr{S}$ into which $\boldsymbol{n}$ is pointing upon the material on the other side of $\mathscr{S}$ depends on $\mathscr{S}$ only through its normal $\boldsymbol{n}$'. This means that identical forces are transmitted across *any* surfaces with normal $\boldsymbol{n}$ at $\boldsymbol{x}$ (such as surfaces $\mathscr{S}$ and $\mathscr{T}$ in Figure 3.11). This force (per unit area) is called the *Cauchy stress vector* and will be denoted

$$\boldsymbol{t}(\boldsymbol{n}),$$

with dependence on $\boldsymbol{x}$ and time omitted for notational convenience. If $\mathscr{S}$ belongs to the boundary of $\mathscr{B}$, then the Cauchy stress vector represents the contact force exerted by the surrounding environment on $\mathscr{B}$.

3.3.2. THE AXIOM OF MOMENTUM BALANCE

Let $\mathscr{B}$ now be subjected to a system of surface forces, $\boldsymbol{t}(\boldsymbol{x}, \boldsymbol{n})$, and body forces, $\boldsymbol{b}(\boldsymbol{x})$. The spatial field $\boldsymbol{b}(\boldsymbol{x})$ represents force per unit mass acting on the interior of $\mathscr{B}$. The *axiom of momentum balance* asserts that 'For any part $\mathscr{P}$ of the deformed configuration of $\mathscr{B}$, with boundary $\mathscr{S}$, the *balance of linear momentum*,

$$\int_{\mathscr{S}} \boldsymbol{t}(\boldsymbol{n})\,\mathrm{d}a + \int_{\mathscr{P}} \rho\,\boldsymbol{b}\,\mathrm{d}v = \int_{\mathscr{P}} \rho\,\dot{\boldsymbol{v}}\,\mathrm{d}v \tag{3.106}$$

and the *balance of angular momentum*,

$$\int_{\mathscr{S}} \boldsymbol{x} \times \boldsymbol{t}(\boldsymbol{n})\,\mathrm{d}a + \int_{\mathscr{P}} \boldsymbol{x} \times \rho\,\boldsymbol{b}\,\mathrm{d}v = \int_{\mathscr{P}} \boldsymbol{x} \times \rho\,\dot{\boldsymbol{v}}\,\mathrm{d}v \tag{3.107}$$

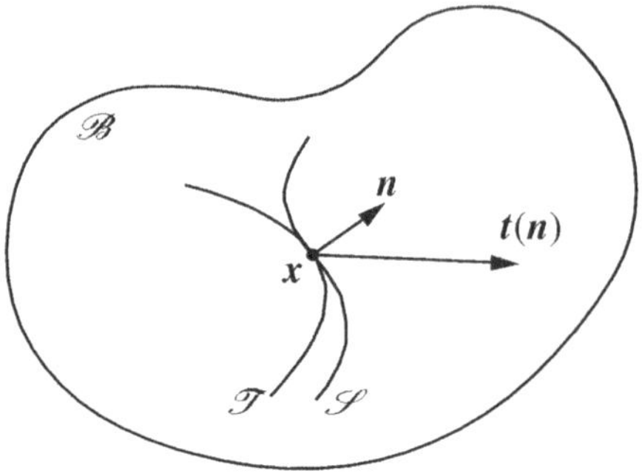

Figure 3.11. Surface forces.

are satisfied, with $\rho = \rho(\boldsymbol{x})$ denoting the *mass density* field, i.e. the mass per unit volume in the deformed configuration of $\mathscr{B}$'. The right-hand sides of (3.106) and (3.107) contain the *inertia* terms, with $\dot{\boldsymbol{v}} = \ddot{\boldsymbol{u}}$ denoting the acceleration field of $\mathscr{B}$.

3.3.3. THE CAUCHY STRESS TENSOR

One of the most fundamental results in continuum mechanics is *Cauchy's theorem* which establishes that, as a consequence of the axiom of momentum balance, the dependence of the surface force $\boldsymbol{t}$ upon the normal $\boldsymbol{n}$ is *linear*, i.e. there exists (recall Section 2.2, starting page 19) a *second-order tensor* field $\boldsymbol{\sigma}(\boldsymbol{x})$ such that the Cauchy stress vector (see Figure 3.12) is given by

$$\boldsymbol{t}(\boldsymbol{x}, \boldsymbol{n}) = \boldsymbol{\sigma}(\boldsymbol{x})\, \boldsymbol{n}. \tag{3.108}$$

Further, $\boldsymbol{\sigma}$ is *symmetric*,[§]

$$\boldsymbol{\sigma} = \boldsymbol{\sigma}^T. \tag{3.109}$$

The tensor $\boldsymbol{\sigma}$ is called the *Cauchy stress tensor* and is often referred to as the *true stress tensor* or, simply, *stress tensor*. Formal proofs to Cauchy's theorem can be found, among others, in Wang and Truesdell (1973), Gurtin (1972, 1981), Gurtin and Martins (1976), Marsden and Hughes (1983) and Ciarlet (1988).

At this point, it should be emphasised that, in real life bodies, forces are actually transferred by atomic interactions which are clearly *discrete* quantities. The continuum mathematical representation of such interactions by means of a stress tensor is meaningful only in an average sense and is valid only for a sufficiently large volume of material. This observation applies equally to quantities such as strain measures or any other continuum fields associated with the body. The smallest volume of material for which the continuum representation makes sense is called the *representative volume element*.

Cauchy stress components

Using an orthonormal basis $\{\boldsymbol{e}_1, \boldsymbol{e}_2, \boldsymbol{e}_3\}$, the Cauchy stress tensor is represented as

$$\boldsymbol{\sigma} = \sigma_{ij}\, \boldsymbol{e}_i \otimes \boldsymbol{e}_j, \tag{3.110}$$

[§]The symmetry of $\boldsymbol{\sigma}$ is a result of the balance of angular momentum.

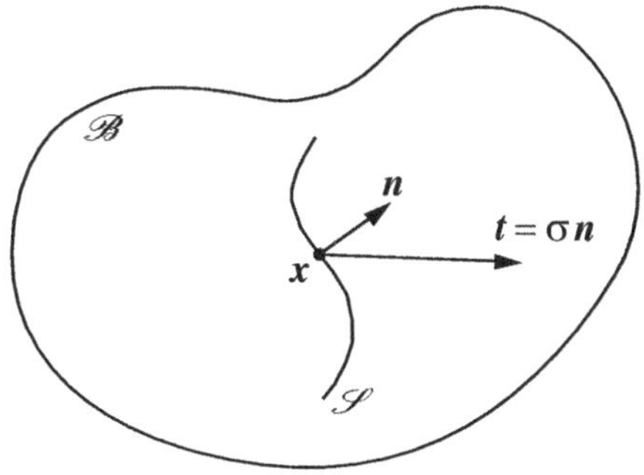

Figure 3.12. The Cauchy stress.

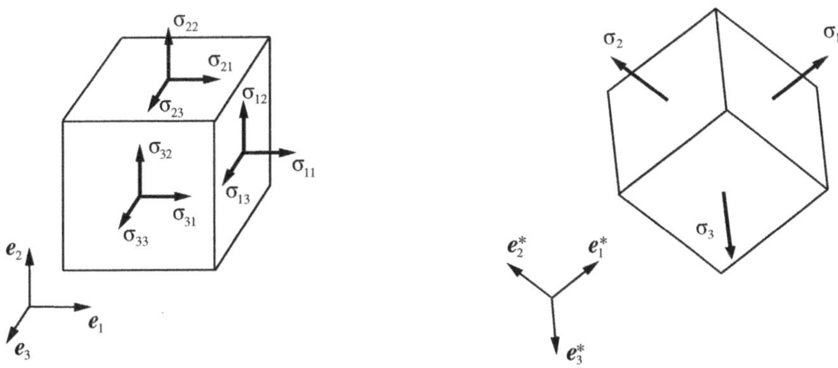

Figure 3.13. Cauchy stress tensor components and principal Cauchy stresses.

with summation on repeated indices implied and the components σ_{ij} given by

$$\sigma_{ij} = (\boldsymbol{\sigma}\, \boldsymbol{e}_i) \cdot \boldsymbol{e}_j. \tag{3.111}$$

From (3.108), it follows that the vector $\boldsymbol{\sigma}\, \boldsymbol{e}_i$ is the force per unit area exerted across a surface whose unit normal vector is $\boldsymbol{e}_i$ at the point of interest. The component σ_{ij} of the Cauchy stress tensor is the magnitude of the projection of $\boldsymbol{\sigma}\, \boldsymbol{e}_i$ in the direction of $\boldsymbol{e}_j$. The schematic representation of such projections is illustrated in Figure 3.13 where an infinitesimal cube with faces normal to the base vectors $\boldsymbol{e}_1$, $\boldsymbol{e}_2$ and $\boldsymbol{e}_3$ is considered. The components σ_{11}, σ_{22} and σ_{33} represent the tractions *normal* to the faces of the infinitesimal cube whereas the remaining components, σ_{12}, σ_{13}, σ_{21}, σ_{23}, σ_{31} and σ_{32} are the *shear* tractions acting parallel to the faces.

Principal Cauchy stresses

Due to its symmetry, the Cauchy stress tensor admits the spectral representation

$$\boldsymbol{\sigma} = \sum_{i=1}^{3} \sigma_i\, \boldsymbol{e}_i^* \otimes \boldsymbol{e}_i^*, \tag{3.112}$$

that is, there exists an orthonormal basis $\{\boldsymbol{e}_1^*, \boldsymbol{e}_2^*, \boldsymbol{e}_3^*\}$, for which all shear components of the Cauchy stress tensor vanish and only the normal components may be non-zero. The normal components, σ_i, are the eigenvalues of $\boldsymbol{\sigma}$ and are called the *principal Cauchy stresses*. The directions defined by the basis $\{\boldsymbol{e}_1^*, \boldsymbol{e}_2^*, \boldsymbol{e}_3^*\}$ are named the *principal stress directions*. The schematic representation of the forces acting on the faces of the infinitesimal cube oriented according to the principal stress directions is shown in Figure 3.13. The forces are exclusively normal to the faces of this cube. Note that, analogously to the representation of the stress tensor in terms of principal stresses, the spectral decomposition has been used in Section 3.1 to represent the stretch tensors $\boldsymbol{U}$ and $\boldsymbol{V}$ in terms of principal stretches (see expression (3.51)).

Deviatoric and hydrostatic stresses

It is often convenient, particularly for the purpose of constitutive modelling, to split the stress tensor $\boldsymbol{\sigma}$ into the sum of a *spherical* and a *traceless* component

$$\boldsymbol{\sigma} = \boldsymbol{s} + p\,\boldsymbol{I}, \tag{3.113}$$

where the invariant

$$p \equiv \tfrac{1}{3}\,I_1(\boldsymbol{\sigma}) = \tfrac{1}{3}\,\mathrm{tr}\,\boldsymbol{\sigma} \tag{3.114}$$

is the *hydrostatic pressure* (also referred to as *hydrostatic stress*, *mean stress* or *mean normal pressure*), and

$$\boldsymbol{s} \equiv \boldsymbol{\sigma} - p\,\boldsymbol{I} = \mathsf{I}_{\mathrm{d}} : \boldsymbol{\sigma}, \tag{3.115}$$

with I_d defined by (3.94), is a traceless tensor named the *deviatoric stress* or *stress deviator*. The tensor

$$p\,\boldsymbol{I} = \tfrac{1}{3}(\boldsymbol{I} \otimes \boldsymbol{I}) : \boldsymbol{\sigma} \tag{3.116}$$

is called the *spherical stress tensor*. The above decomposition is analogous to the isochoric/volumetric split of the infinitesimal strain tensor discussed in Section 3.2.3.

3.3.4. THE FIRST PIOLA–KIRCHHOFF STRESS

The traction vector $\boldsymbol{t}$ of expression (3.108) measures the force exerted across a material surface per unit *deformed* area. Crucial to the definition of the first Piola–Kirchhoff stress is the counterpart $\bar{\boldsymbol{t}}$ of $\boldsymbol{t}$ that measures, at the point of interest, the force that acts across any surface whose normal is $\boldsymbol{n}$ in the deformed configuration per unit *reference* area. With $\mathrm{d}a$ denoting an infinitesimal element of area of a surface normal to $\boldsymbol{n}$ in the deformed configuration and with $\mathrm{d}a_0$ being its undeformed counterpart, $\bar{\boldsymbol{t}}$ is expressed as (Figure 3.14)

$$\bar{\boldsymbol{t}} = \frac{\mathrm{d}a}{\mathrm{d}a_0}\,\boldsymbol{t} = \frac{\mathrm{d}a}{\mathrm{d}a_0}\,\boldsymbol{\sigma}\,\boldsymbol{n}. \tag{3.117}$$

Consider the surface $\mathscr{S}$ in the *reference* configuration of $\mathscr{B}$ (Figure 3.14). Let $\mathrm{d}\boldsymbol{p}_1$ and $\mathrm{d}\boldsymbol{p}_2$ be infinitesimal (linearly independent) vectors tangent to $\mathscr{S}$ at the material point $\boldsymbol{p}$ and let $\mathrm{d}a_0$ be the area element generated by $\mathrm{d}\boldsymbol{p}_1$ and $\mathrm{d}\boldsymbol{p}_2$. With $\boldsymbol{m}$ denoting the unit normal to $\mathscr{S}$ at $\boldsymbol{p}$, one has

$$\boldsymbol{m}\,\mathrm{d}a_0 = \mathrm{d}\boldsymbol{p}_1 \times \mathrm{d}\boldsymbol{p}_2. \tag{3.118}$$

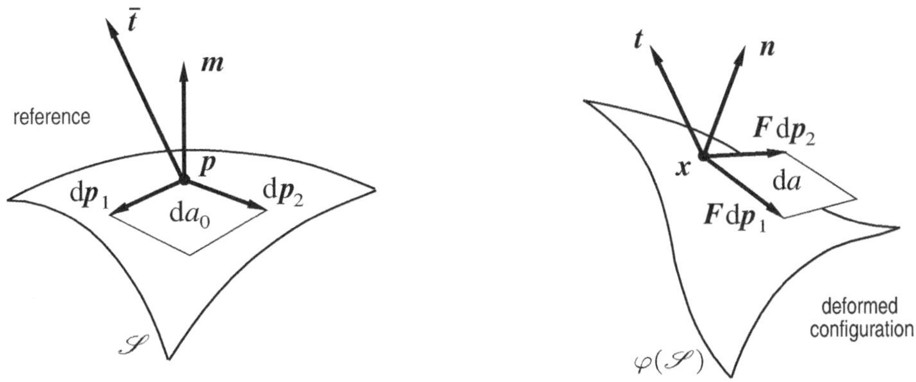

Figure 3.14. The first Piola–Kirchhoff stress tensor.

Under deformation, the tangent vectors $\mathrm{d}\boldsymbol{p}_1$ and $\mathrm{d}\boldsymbol{p}_2$ are mapped, respectively, into $\boldsymbol{F}\,\mathrm{d}\boldsymbol{p}_1$ and $\boldsymbol{F}\,\mathrm{d}\boldsymbol{p}_2$ so that the unit normal to the deformed configuration of $\mathscr{S}$ reads

$$\boldsymbol{n}\,\mathrm{d}a = \boldsymbol{F}\,\mathrm{d}\boldsymbol{p}_1 \times \boldsymbol{F}\,\mathrm{d}\boldsymbol{p}_2, \tag{3.119}$$

where $\mathrm{d}a$ is the corresponding deformed area element. Pre-multiplication of both sides of the expression above by $\boldsymbol{F}^T$ together with use of the identity

$$\boldsymbol{S}\boldsymbol{u} \times \boldsymbol{S}\boldsymbol{v} = (\det \boldsymbol{S})\,\boldsymbol{S}^{-T}\,(\boldsymbol{u} \times \boldsymbol{v}), \tag{3.120}$$

valid for any invertible tensor $\boldsymbol{S}$ and vectors $\boldsymbol{u}$ and $\boldsymbol{v}$, leads to

$$\boldsymbol{F}^T\boldsymbol{n}\,\mathrm{d}a = J\,\mathrm{d}\boldsymbol{p}_1 \times \mathrm{d}\boldsymbol{p}_2 = J\,\boldsymbol{m}\,\mathrm{d}a_0, \tag{3.121}$$

where $J \equiv \det \boldsymbol{F}$. This is equivalent to

$$\frac{\mathrm{d}a}{\mathrm{d}a_0}\,\boldsymbol{n} = J\,\boldsymbol{F}^{-T}\boldsymbol{m}. \tag{3.122}$$

Finally, with substitution of the expression above into (3.117), $\bar{\boldsymbol{t}}$ may be written in terms of the reference unit normal $\boldsymbol{m}$ as

$$\bar{\boldsymbol{t}} = J\,\boldsymbol{\sigma}\,\boldsymbol{F}^{-T}\,\boldsymbol{m}. \tag{3.123}$$

This last expression motivates the following definition

$$\boldsymbol{P} \equiv J\,\boldsymbol{\sigma}\,\boldsymbol{F}^{-T}, \tag{3.124}$$

so that the force transmitted across $\mathscr{S}$ measured per unit *reference* area reads

$$\bar{\boldsymbol{t}} = \boldsymbol{P}\,\boldsymbol{m}. \tag{3.125}$$

The tensor $\boldsymbol{P}$ is called the *first Piola–Kirchhoff stress* and is often referred to as the *Piola–Kirchhoff stress* or *nominal stress*.¶ The vector $\bar{\boldsymbol{t}}$ is obtained by applying the first Piola–Kirchhoff stress to the unit vector $\boldsymbol{m}$, normal to the *reference* configuration of $\mathscr{S}$ at the point of interest. Note that in contrast to the Cauchy stress, $\boldsymbol{P}$ is generally *unsymmetric*.

¶Some authors (Billington and Tate, 1981; Nemat-Nasser, 1999) define the nominal stress as the transpose of the first Piola–Kirchhoff stress tensor.

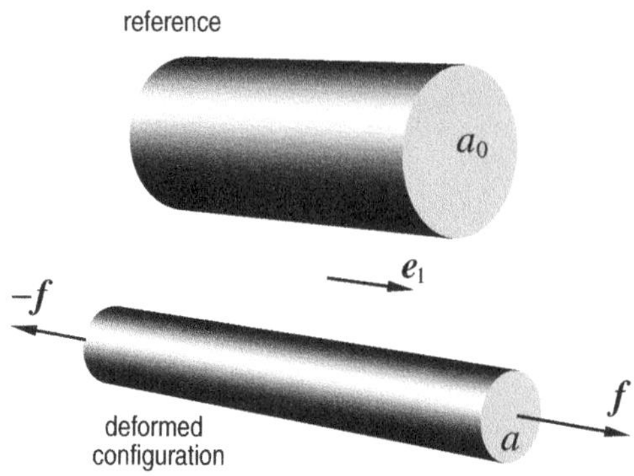

Figure 3.15. The first Piola–Kirchhoff stress. Example.

Example 3.3.1 (The Piola–Kirchhoff stress). Consider a cylindrical bar (Figure 3.15) with cross-sectional area a_0 in its initial configuration (taken as reference). During a uniaxial experiment this bar is stretched along its longitudinal axis (direction of $\boldsymbol{e}_1$) with a simultaneous reduction of its cross section. Assume that the final deformed configuration of the bar corresponds to a state of homogeneous deformation with cross-sectional area a. Furthermore, assume that the bar is subjected to a state of *uniaxial* stress, with constant $\boldsymbol{\sigma}$ given by

$$\boldsymbol{\sigma} = \sigma_{11}\, \boldsymbol{e}_1 \otimes \boldsymbol{e}_1.$$

Let $\boldsymbol{f} = f\, \boldsymbol{e}_1$ be the total force applied to the deformed configuration of the bar (by the experimental equipment). Under the assumption of uniform stress distribution in the cross-section of the bar, force balance requires that the Cauchy stress component σ_{11} be given by

$$\sigma_{11} = \frac{f}{a}.$$

In practice, the force $\boldsymbol{f}$ (and not the stress component) is what can actually be measured in an experiment. Thus, after $\boldsymbol{f}$ is measured, the Cauchy stress σ_{11} is determined according to the expression above. If instead of a, the *reference* cross-sectional area a_0 is used, then the *first Piola–Kirchhoff* or *nominal* stress component is determined

$$P_{11} = \frac{f}{a_0}.$$

It is obvious that, in this case, the corresponding tractions $\boldsymbol{t}$ and $\bar{\boldsymbol{t}}$, respectively per unit deformed and reference area, are simply

$$\boldsymbol{t} = \sigma_{11}\, \boldsymbol{e}_1 = \frac{1}{a}\, \boldsymbol{f}, \quad \bar{\boldsymbol{t}} = P_{11}\, \boldsymbol{e}_1 = \frac{1}{a_0}\, \boldsymbol{f}.$$

3.3.5. THE SECOND PIOLA–KIRCHHOFF STRESS

The *Second Piola–Kirchhoff* stress tensor, denoted $\boldsymbol{S}$, is the tensor defined as

$$\boldsymbol{S} \equiv J\boldsymbol{F}^{-1}\boldsymbol{\sigma}\boldsymbol{F}^{-T}. \tag{3.126}$$

Note that from this definition, we have

$$S^T = J F^{-1} \sigma^T F^{-T}, \tag{3.127}$$

so that the symmetry of σ implies that S is symmetric.

3.3.6. THE KIRCHHOFF STRESS

Another important measure of stress is the *Kirchhoff stress tensor*, τ, defined by

$$\tau \equiv J\,\sigma. \tag{3.128}$$

Due to the symmetry of σ, the Kirchhoff stress is symmetric. Its spectral representation reads

$$\tau = \sum_{i=1}^{3} \tau_i\, e_i^* \otimes e_i^*, \tag{3.129}$$

where the *principal Kirchhoff stresses*, τ_i, are related to the principal Cauchy stresses, σ_i, by

$$\tau_i = J\,\sigma_i. \tag{3.130}$$

Later in this book, frequent reference to the principal Kirchhoff stresses will be made in the formulation of various constitutive models.

3.4. Fundamental laws of thermodynamics

In order to state the fundamental laws of thermodynamics, it is necessary to introduce the scalar fields θ, e, s and r defined over $\mathscr{B}$ which represent, respectively, the *temperature*, *specific internal energy*, *specific entropy* and the *density of heat production*. In addition, b and q will denote the vector fields corresponding, respectively, to the *body force* (force per unit volume in the deformed configuration) and *heat flux*.

3.4.1. CONSERVATION OF MASS

The postulate of conservation of mass requires that

$$\dot{\rho} + \rho\, \mathrm{div}_x\, \dot{u} = 0. \tag{3.131}$$

3.4.2. MOMENTUM BALANCE

In terms of the Cauchy stress tensor, whose existence has been established in Section 3.3.3, the balance of momentum for $\mathscr{B}$ can be expressed by the following partial differential equation with boundary condition:‖

$$\begin{aligned} \mathrm{div}_x\, \sigma + b &= \rho\, \ddot{u} \quad &&\text{in } \varphi(\Omega) \\ t &= \sigma\, n &&\text{in } \varphi(\partial\Omega), \end{aligned} \tag{3.132}$$

‖Equations (3.132) are also a result of Cauchy's theorem, alluded to in page 62.

where $\boldsymbol{n}$ is the outward unit vector normal to the deformed boundary $\varphi(\partial\Omega)$ of $\mathscr{B}$ and $\boldsymbol{t}$ is the applied boundary traction vector field on $\varphi(\partial\Omega)$. Equations (3.132) are often referred to as the *strong*, *local* or *point-wise* form of equilibrium. Equation $(3.132)_1$ is known as *Cauchy's equation of motion*.

The above momentum balance equations are formulated in the *spatial* (deformed) configuration. Equivalently, they may be expressed in the *reference* (or material) configuration of $\mathscr{B}$ in terms of the first Piola–Kirchhoff stress tensor as

$$\begin{aligned} \operatorname{div}_p \boldsymbol{P} + \bar{\boldsymbol{b}} = \bar{\rho}\,\ddot{\boldsymbol{u}} \quad & \text{in } \Omega \\ \bar{\boldsymbol{t}} = \boldsymbol{P}\,\boldsymbol{m} \quad & \text{in } \partial\Omega, \end{aligned} \tag{3.133}$$

where

$$\bar{\boldsymbol{b}} = J\,\boldsymbol{b} \tag{3.134}$$

is the *reference body force*, i.e. the body force measured per unit volume in the reference configuration,

$$\bar{\rho} = J\,\rho, \tag{3.135}$$

is the *reference density* (mass per unit volume in the reference configuration), $\bar{\boldsymbol{t}}$ is the *reference boundary traction* (boundary force per unit reference area) and $\boldsymbol{m}$ is the outward normal to the boundary of $\mathscr{B}$ in its reference configuration.

3.4.3. THE FIRST PRINCIPLE

The first principle of thermodynamics postulates the *conservation of energy*. Before stating this principle, it is convenient to introduce the product

$$\boldsymbol{\sigma} : \boldsymbol{D},$$

which represents the *stress power* per unit volume in the deformed configuration of a body. The first principle of thermodynamics is mathematically expressed by the equation

$$\rho\,\dot{e} = \boldsymbol{\sigma} : \boldsymbol{D} + \rho\,r - \operatorname{div}_x \boldsymbol{q}. \tag{3.136}$$

In words, the rate of internal energy per unit deformed volume must equal the sum of the stress power and heat production per unit deformed volume minus the spatial divergence of the heat flux.

3.4.4. THE SECOND PRINCIPLE

The second principle of thermodynamics postulates the *irreversibility of entropy production*. It is expressed by means of the inequality

$$\rho\,\dot{s} + \operatorname{div}_x \left[\frac{\boldsymbol{q}}{\theta} \right] - \frac{\rho\,r}{\theta} \geq 0. \tag{3.137}$$

3.4.5. THE CLAUSIUS–DUHEM INEQUALITY

By combination of the first and second principles stated above, one easily obtains the fundamental inequality

$$\rho\,\dot{s} + \mathrm{div}_x\left[\frac{\boldsymbol{q}}{\theta}\right] - \frac{1}{\theta}(\rho\,\dot{e} - \boldsymbol{\sigma} : \boldsymbol{D} + \mathrm{div}_x\,\boldsymbol{q}) \geq 0. \tag{3.138}$$

The introduction of the *specific free energy* ψ (also known as the *Helmholtz free energy per unit mass*), defined by

$$\psi = e - \theta\,s, \tag{3.139}$$

along with the identity

$$\mathrm{div}_x\left[\frac{\boldsymbol{q}}{\theta}\right] = \frac{1}{\theta}\,\mathrm{div}_x\,\boldsymbol{q} - \frac{1}{\theta^2}\,\boldsymbol{q}\cdot\nabla_x\theta, \tag{3.140}$$

into the above fundamental inequality results in the *Clausius–Duhem inequality*

$$\boldsymbol{\sigma} : \boldsymbol{D} - \rho(\dot{\psi} + s\,\dot{\theta}) - \frac{1}{\theta}\,\boldsymbol{q}\cdot\boldsymbol{g} \geq 0, \tag{3.141}$$

where we have defined $\boldsymbol{g} = \nabla_x\theta$. The left-hand side of (3.141) represents the dissipation per unit *deformed* volume. Equivalently, by making use of (3.135), the Clausius–Duhem inequality can be expressed in terms of dissipation per unit *reference* volume as

$$\boldsymbol{\tau} : \boldsymbol{D} - \bar{\rho}(\dot{\psi} + s\,\dot{\theta}) - \frac{J}{\theta}\,\boldsymbol{q}\cdot\boldsymbol{g} \geq 0. \tag{3.142}$$

3.5. Constitutive theory

The balance principles presented so far are valid for any continuum body, regardless of the material of which the body is made. In order to distinguish between different types of material, a *constitutive model* must be introduced. In this section, we review the principles that form the basis of the constitutive theories discussed in later chapters of this book. We start by stating, in Section 3.5.1, three fundamental axioms that define a rather general class of constitutive models of continua. The use of internal variables to formulate constitutive models of dissipative materials is then addressed in Section 3.5.2. We remark that all dissipative constitutive models discussed in Parts Two and Three of this book are based on the internal variable approach. Section 3.5.4 summarises a generic *purely mechanical* internal variable model. The discussion on constitutive theory ends in Section 3.5.5 where the fundamental constitutive initial value problems are stated.

3.5.1. CONSTITUTIVE AXIOMS

In the present context, the axioms stated in this section must be satisfied for any constitutive model. Before going further, it is convenient to introduce the definitions of *thermokinetic* and *calorodynamic* processes (Truesdell, 1969). A *thermokinetic process* of $\mathscr{B}$ is a pair of fields

$$\boldsymbol{\varphi}(\boldsymbol{p}, t) \quad \text{and} \quad \theta(\boldsymbol{p}, t).$$

A set

$$\{\boldsymbol{\sigma}(\boldsymbol{p}, t),\ e(\boldsymbol{p}, t),\ s(\boldsymbol{p}, t),\ r(\boldsymbol{p}, t),\ \boldsymbol{b}(\boldsymbol{p}, t),\ \boldsymbol{q}(\boldsymbol{p}, t)\}$$

of fields satisfying the balance of momentum, the first and the second principles of thermodynamics is called a *calorodynamic process* of $\mathscr{B}$.

Thermodynamic determinism

The basic axiom underlying the constitutive theory discussed here is the *principle of thermodynamically compatible determinism* (Truesdell, 1969). It postulates that 'the history of the thermokinetic process to which a neighbourhood of a point $\boldsymbol{p}$ of $\mathscr{B}$ has been subjected determines a calorodynamic process for $\mathscr{B}$ at $\boldsymbol{p}$'. In particular, we shall be concerned with so-called *simple materials*, for which the *local* history (history at point $\boldsymbol{p}$ only) of $\boldsymbol{F}$, θ and $\boldsymbol{g}$ suffices to determine the history of the thermokinetic process for constitutive purposes. In this case, regarding the body force $\boldsymbol{b}$ and heat supply r as delivered, respectively, by the linear momentum balance $(3.132)_1$ and conservation of energy (3.136) and introducing the specific free energy (3.139), the principle of thermodynamic determinism implies the existence of constitutive functionals $\mathfrak{F}$, $\mathfrak{G}$, $\mathfrak{H}$ and $\mathfrak{I}$ of the histories of $\boldsymbol{F}$, θ and $\boldsymbol{g}$ such that, for a point $\boldsymbol{p}$,

$$
\begin{aligned}
\boldsymbol{\sigma}(t) &= \mathfrak{F}\,(\boldsymbol{F}^{\,t}, \theta^{t}, \boldsymbol{g}^{t}) \\
\psi(t) &= \mathfrak{G}(\boldsymbol{F}^{\,t}, \theta^{t}, \boldsymbol{g}^{t}) \\
s(t) &= \mathfrak{H}(\boldsymbol{F}^{\,t}, \theta^{t}, \boldsymbol{g}^{t}) \\
\boldsymbol{q}(t) &= \mathfrak{I}\,(\boldsymbol{F}^{\,t}, \theta^{t}, \boldsymbol{g}^{t})
\end{aligned}
\tag{3.143}
$$

and the Clausius–Duhem inequality (3.141) holds for every thermokinetic process of $\mathscr{B}$. The dependence on $\boldsymbol{p}$ is understood on both sides of (3.143) and $(\cdot)^t$ on the right-hand sides denotes the *history* of $(\cdot)$ at $\boldsymbol{p}$ up to time t.

Material objectivity

Another fundamental axiom of the constitutive theory is the *principle of material objectivity* (or *frame invariance*). It states that 'the material response is independent of the observer'. The motion φ^* is related to the motion φ by a change in observer if it can be expressed as

$$
\varphi^*(\boldsymbol{p}, t) = \boldsymbol{y}(t) + \boldsymbol{Q}(t)\,[\varphi(\boldsymbol{p}, t) - \boldsymbol{x}_0], \tag{3.144}
$$

where $\boldsymbol{y}(t)$ is a point in space, $\boldsymbol{Q}(t)$ is a rotation and $\varphi(\boldsymbol{p}, t) - \boldsymbol{x}_0$ is the position vector of $\varphi(\boldsymbol{p}, t)$ relative to an arbitrary origin $\boldsymbol{x}_0$. This relation corresponds to a rigid relative movement between the different observers and the deformation gradient corresponding to φ^* is given by

$$
\boldsymbol{F}^* = \boldsymbol{Q}\,\boldsymbol{F}. \tag{3.145}
$$

Scalar fields (such as θ, ψ and s) are unaffected by a change in observer but the Cauchy stress $\boldsymbol{\sigma}(t)$, heat flux $\boldsymbol{q}(t)$ and the temperature gradient $\boldsymbol{g}(t)$ transform according to the rules

$$
\begin{aligned}
\boldsymbol{\sigma} &\longrightarrow \boldsymbol{\sigma}^* = \boldsymbol{Q}\,\boldsymbol{\sigma}\,\boldsymbol{Q}^T \\
\boldsymbol{q} &\longrightarrow \boldsymbol{q}^* = \boldsymbol{Q}\,\boldsymbol{q} \\
\boldsymbol{g} &\longrightarrow \boldsymbol{g}^* = \boldsymbol{Q}\,\boldsymbol{g}.
\end{aligned}
\tag{3.146}
$$

The principle of material objectivity places restrictions on the constitutive functionals (3.143). Formally, it requires that $\mathfrak{F}$, $\mathfrak{G}$, $\mathfrak{H}$ and $\mathfrak{I}$ satisfy

$$\begin{aligned} \boldsymbol{\sigma}^*(t) &= \mathfrak{F}\,(\boldsymbol{F}^{\,t*}, \theta^t, \boldsymbol{g}^{t*}) \\ \psi(t) &= \mathfrak{G}(\boldsymbol{F}^{\,t*}, \theta^t, \boldsymbol{g}^{t*}) \\ s(t) &= \mathfrak{H}(\boldsymbol{F}^{\,t*}, \theta^t, \boldsymbol{g}^{t*}) \\ \boldsymbol{q}^*(t) &= \mathfrak{I}\,(\boldsymbol{F}^{\,t*}, \theta^t, \boldsymbol{g}^{t*}) \end{aligned} \tag{3.147}$$

for any transformation of the form (3.145, 3.146).

Material symmetry

The *symmetry group* of a material is the set of density preserving changes of reference configuration under which the material response functionals $\mathfrak{F}$, $\mathfrak{G}$, $\mathfrak{H}$ and $\mathfrak{I}$ are not affected. The symmetry group of a *solid* material is a subset of the proper orthogonal group $\mathscr{O}^+$, that is, a set of *rotations*. Thus, the symmetry group of a solid material is the set of rotations of the reference configuration under which the response functionals remain unchanged. This concept is expressed mathematically as follows. A subgroup $\mathscr{S}$ of $\mathscr{O}^+$ is said to be the symmetry group of the material defined by the constitutive functionals $\mathfrak{F}$, $\mathfrak{G}$, $\mathfrak{H}$ and $\mathfrak{I}$ if the relations

$$\begin{aligned} \mathfrak{F}\,(\boldsymbol{F}^{\,t}, \theta^t, \boldsymbol{g}^t) &= \mathfrak{F}\,([\boldsymbol{F}\,\boldsymbol{Q}]^t, \theta^t, \boldsymbol{g}^t) \\ \mathfrak{G}\,(\boldsymbol{F}^{\,t}, \theta^t, \boldsymbol{g}^t) &= \mathfrak{G}([\boldsymbol{F}\,\boldsymbol{Q}]^t, \theta^t, \boldsymbol{g}^t) \\ \mathfrak{H}\,(\boldsymbol{F}^{\,t}, \theta^t, \boldsymbol{g}^t) &= \mathfrak{H}([\boldsymbol{F}\,\boldsymbol{Q}]^t, \theta^t, \boldsymbol{g}^t) \\ \mathfrak{I}\,(\boldsymbol{F}^{\,t}, \theta^t, \boldsymbol{g}^t) &= \mathfrak{I}\,([\boldsymbol{F}\,\boldsymbol{Q}]^t, \theta^t, \boldsymbol{g}^t) \end{aligned} \tag{3.148}$$

hold for any time-*independent* rotation $\boldsymbol{Q} \in \mathscr{S}$. A solid is said to be *isotropic*** if its symmetry group is the entire proper orthogonal group. In the development of any constitutive model, the constitutive functionals must comply with the restrictions imposed by the symmetries of the material in question.

3.5.2. THERMODYNAMICS WITH INTERNAL VARIABLES

The constitutive equations (3.143) written in terms of functionals of the history of $\boldsymbol{F}$, θ and $\boldsymbol{g}$, in that format, are far too general to have practical utility in modelling real materials undergoing real thermodynamical process. This is especially true if one has in mind the experimental identification of the constitutive functionals and the solution of the boundary value problems of practical interest. Therefore, it is imperative that simplifying assumptions are added to the general forms of the constitutive relations stated above.

An effective alternative to the general description based on history functionals is the adoption of the so-called *thermodynamics with internal variables*. The starting point of the thermodynamics with internal variables is the hypothesis that at any instant of a thermodynamical process the thermodynamic state (defined by $\boldsymbol{\sigma}$, ψ, s and $\boldsymbol{q}$) at a given

**We remark that most constitutive models discussed in this book are isotropic.

point $\boldsymbol{p}$ can be completely determined by the knowledge of a finite number of *state variables*. The thermodynamic state depends only on the *instantaneous* value of the state variables and not on their past history.

Mathematically, state variable models can be seen as particular instances of the general history functional-based constitutive theory. The relationship between the two approaches is discussed in detail by Kestin and Bataille (1977) and Bataille and Kestin (1979). In general terms, state variable models can be obtained from the general history functional-based description by re-defining the history of the thermokinetic process in terms of a finite number of parameters (the state variables).

The state variables

For the applications with which we are mostly concerned, it will be convenient to assume that at any time t, the thermodynamic state at a point is determined by the following set of *state variables*:

$$\{\boldsymbol{F},\ \theta,\ \boldsymbol{g},\ \boldsymbol{\alpha}\},$$

where $\boldsymbol{F}$, θ and $\boldsymbol{g}$ are the *instantaneous* values of deformation gradient, temperature and the temperature gradient and

$$\boldsymbol{\alpha}=\{\alpha_k\} \tag{3.149}$$

is a set of *internal variables* containing, in general, entities of scalar, vectorial and tensorial nature associated with dissipative mechanisms.

Thermodynamic potential. Stress constitutive equation

Following the above hypothesis, the specific free energy is assumed to have the form[††]

$$\psi=\psi(\boldsymbol{F},\ \theta,\ \boldsymbol{\alpha}), \tag{3.150}$$

so that its rate of change is given by

$$\dot{\psi}=\frac{\partial\psi}{\partial\boldsymbol{F}}:\dot{\boldsymbol{F}}+\frac{\partial\psi}{\partial\theta}\,\dot{\theta}+\frac{\partial\psi}{\partial\alpha_k}\,\dot{\alpha}_k, \tag{3.151}$$

where summation over k is implied. In this case, using the connection

$$\boldsymbol{\sigma}:\boldsymbol{D}=\boldsymbol{\sigma}\,\boldsymbol{F}^{-T}:\dot{\boldsymbol{F}}, \tag{3.152}$$

for the stress power, one obtains for the Clausius–Duhem inequality

$$\left(\boldsymbol{\sigma}\,\boldsymbol{F}^{-T}-\rho\,\frac{\partial\psi}{\partial\boldsymbol{F}}\right):\dot{\boldsymbol{F}}-\rho\left(s+\frac{\partial\psi}{\partial\theta}\right)\dot{\theta}-\rho\,\frac{\partial\psi}{\partial\alpha_k}\,\dot{\alpha}_k-\frac{1}{\theta}\,\boldsymbol{q}\cdot\boldsymbol{g}\geq 0. \tag{3.153}$$

Equivalently, in terms of power per unit *reference* volume, we have

$$\left(\boldsymbol{P}-\bar{\rho}\,\frac{\partial\psi}{\partial\boldsymbol{F}}\right):\dot{\boldsymbol{F}}-\bar{\rho}\left(s+\frac{\partial\psi}{\partial\theta}\right)\dot{\theta}-\bar{\rho}\,\frac{\partial\psi}{\partial\alpha_k}\,\dot{\alpha}_k-\frac{J}{\theta}\,\boldsymbol{q}\cdot\boldsymbol{g}\geq 0. \tag{3.154}$$

[††]The dependence of ψ on the temperature gradient is disregarded as it contradicts the second principle of thermodynamics (Coleman and Gurtin, 1967).

Expression (3.154) is obtained from (3.153), by simply using relation (3.135).

The principle of thermodynamic determinism requires that the constitutive equations must be such that the above inequality holds for any thermokinetic process. Thus, (3.154) must remain valid for any pair of functions $\{\dot{\boldsymbol{F}}(t), \dot{\theta}(t)\}$. This implies the constitutive equations

$$\boldsymbol{P} = \bar{\rho}\, \frac{\partial \psi}{\partial \boldsymbol{F}}, \qquad s = -\frac{\partial \psi}{\partial \theta}, \tag{3.155}$$

for the first Piola–Kirchhoff stress and entropy. Equation $(3.155)_1$ is equivalent to the following constitutive relations for the Cauchy and Kirchoff stress tensors:

$$\boldsymbol{\sigma} = \frac{1}{J}\, \bar{\rho}\, \frac{\partial \psi}{\partial \boldsymbol{F}}\, \boldsymbol{F}^T, \qquad \boldsymbol{\tau} = \bar{\rho}\, \frac{\partial \psi}{\partial \boldsymbol{F}}\, \boldsymbol{F}^T. \tag{3.156}$$

Thermodynamical forces

For each internal variable α_k of the set $\boldsymbol{\alpha}$, we define the *conjugate thermodynamical force*

$$A_k \equiv \bar{\rho}\, \frac{\partial \psi}{\partial \alpha_k}. \tag{3.157}$$

With this definition and the identities (3.155), the Clausius–Duhem inequality can be rewritten as

$$-A_k * \dot{\alpha}_k - \frac{J}{\theta}\, \boldsymbol{q} \cdot \boldsymbol{g} \geq 0, \tag{3.158}$$

where we recall that the symbol '$*$' denotes the appropriate product operation between A_k and $\dot{\alpha}_k$. In what follows, we will adopt for convenience the notation

$$\boldsymbol{A} \equiv \{A_k\} \tag{3.159}$$

for the set of thermodynamical forces, so that (3.158) can be expressed in a more compact form as

$$-\boldsymbol{A} * \dot{\boldsymbol{\alpha}} - \frac{J}{\theta}\, \boldsymbol{q} \cdot \boldsymbol{g} \geq 0. \tag{3.160}$$

Dissipation. Evolution of the internal variables

In order to completely characterise a constitutive model, complementary laws associated with the dissipative mechanisms are required. Namely, constitutive equations for the flux variables $\frac{1}{\theta}\boldsymbol{q}$ and $\dot{\boldsymbol{\alpha}}$ must be postulated. In the general case, we assume that the flux variables are given functions of the state variables. The following constitutive equations are then postulated

$$\begin{aligned} \dot{\boldsymbol{\alpha}} &= f(\boldsymbol{F}, \theta, \boldsymbol{g}, \boldsymbol{\alpha}) \\ \frac{1}{\theta}\boldsymbol{q} &= h(\boldsymbol{F}, \theta, \boldsymbol{g}, \boldsymbol{\alpha}). \end{aligned} \tag{3.161}$$

Recalling the principle of thermodynamic determinism, the Clausius–Duhem inequality, now expressed by (3.158), must hold for any process. This requirement places restrictions on the possible forms of the general constitutive functions f and h in (3.161) (the reader is referred to Coleman and Gurtin, 1967; Truesdell, 1969, for further details on this issue). It is also important to mention that when internal variables of vectorial or tensorial nature are present, it is frequently convenient to re-formulate $(3.161)_1$ in terms of so-called *objective rates* rather than the standard material time derivative of $\boldsymbol{\alpha}$. Objective rates are insensitive to rigid-body motions and may be essential in the definition of a *frame invariant* evolution law for variables representing physical states associated with material directions. Objective rates are discussed in Section 14.10 (starting page 615) in the context of the hypoelastic-based formulation of plasticity models.

Dissipation potential. Normal dissipativity

An effective way of ensuring that (3.158) is satisfied consists in postulating the existence of a scalar-valued *dissipation potential* of the form

$$\Xi = \Xi(\boldsymbol{A}, \boldsymbol{g}; \boldsymbol{F}, \theta, \boldsymbol{\alpha}), \tag{3.162}$$

where the state variables $\boldsymbol{F}$, θ and $\boldsymbol{\alpha}$ appear as parameters. The potential Ξ is assumed convex with respect to each A_k and $\boldsymbol{g}$, non-negative and zero valued at the origin, $\{\boldsymbol{A}, \boldsymbol{g}\} = \{\boldsymbol{0}, \boldsymbol{0}\}$. In addition, the hypothesis of *normal dissipativity* is introduced, i.e. the flux variables are assumed to be determined by the laws

$$\dot{\alpha}_k = -\frac{\partial \Xi}{\partial A_k}, \quad \frac{1}{\theta}\boldsymbol{q} = -\frac{\partial \Xi}{\partial \boldsymbol{g}}. \tag{3.163}$$

A constitutive model defined by (3.150), (3.155) and (3.163) satisfies *a priori* the dissipation inequality. It should be noted, however, that the constitutive description by means of convex potentials as described above is *not* a consequence of thermodynamics but, rather, a convenient tool for formulating constitutive equations without violating thermodynamics. Examples of constitutive models supported by experimental evidence which do not admit representation by means of dissipation potentials are discussed by Onat and Leckie (1988).

3.5.3. PHENOMENOLOGICAL AND MICROMECHANICAL APPROACHES

The success of a constitutive model intended to describe the behaviour of a particular material depends critically on the choice of an appropriate set of internal variables. Since no plausible model will be general enough to describe the response of a material under all processes, we should have in mind that the choice of internal variables must be guided not only by the specific material in question but, equally importantly, by the processes (i.e. the range of thermokinetic processes defined by strain and temperature histories as well as the time span) under which the model is meant to describe the behaviour of the material. The importance of considering the possible thermokinetic processes when devising a constitutive model can be clearly illustrated, for instance, by considering a simple steel bar. When subjected to a sufficiently small axial strain at room temperature, the bar exhibits a behaviour that can be accurately modelled by linear elasticity theory (generalised Hooke's law). If strains become larger, however, linear elasticity may no longer capture the observed response satisfactorily. In this case, a plasticity theory may be more appropriate. With further increase in complexity of the strain history (by introducing, say, cyclic extension), other phenomena such as internal damaging and possibly fracturing may take place and more refined constitutive models, incorporating a larger number of state variables, will be required. Due account of the possible temperature histories and time span to be considered is also fundamental. At higher temperatures, the long-term behaviour of the steel bar subjected to even a very small strain, may no longer be accurately modelled by the linear elasticity theory. In this case the introduction of time-dependent effects (creep/relaxation) may be essential to produce an acceptable model. In an extreme situation, if the temperature rises above melting point, the bar will cease to be a solid. Under such circumstances, a fluid mechanics theory will be needed to describe the behaviour of the material.

In general, due to the difficulty involved in the identification of the underlying dissipative mechanisms, the choice of the appropriate set of internal variables is somewhat subtle and tends to be biased by the preferences and background of the investigator. In simplistic terms, we may say that constitutive modelling by means of internal variables relies either on a micromechanical or on a phenomenological approach. The micromechanical approach involves the determination of mechanisms and related variables at the atomic, molecular or crystalline levels. In general, these variables are discrete quantities and their continuum (macroscopic) counterparts can be defined by means of homogenisation techniques. The phenomenological approach, on the other hand, is based on the study of the response of the *representative volume element*, i.e. the element of matter large enough to be regarded as a homogeneous continuum. The internal variables in this case will be directly associated with the dissipative behaviour observed at the *macroscopic* level in terms of continuum quantities (such as strain, temperature, etc.). Despite the macroscopic nature of theories derived on the basis of the phenomenological methodology, it should be expected that 'good' phenomenological internal variables will be somehow related to the underlying microscopic dissipation mechanisms.

The phenomenological approach to irreversible thermodynamics has been particularly successful in the field of solid mechanics. Numerous well-established models of solids, such as classical isotropic elastoplasticity and viscoplasticity, discussed in Parts Two and Three of this book, have been developed on a purely phenomenological basis providing evidence of how powerful such an approach to irreversible thermodynamics can be when the major concern is the description of the essentially macroscopic behaviour. In some instances, however, the inclusion of microscopic information becomes essential and a purely phenomenological methodology is unlikely to describe the behaviour of the material with sufficient accuracy. One such case is illustrated in Chapter 16, where a microscopically-based continuum model of ductile metallic crystals is described.

3.5.4. THE PURELY MECHANICAL THEORY

Thermal effects are ignored in the constitutive theories addressed in Parts Two and Three of this book. It is, therefore, convenient at this point to summarise the general internal variable-based constitutive equations in the purely mechanical case. By removing the thermally-related terms of the above theory, we end up with the following set of mechanical constitutive equations:

$$\begin{cases} \psi = \psi(\boldsymbol{F}, \boldsymbol{\alpha}) \\[2ex] \boldsymbol{P} = \bar{\rho}\,\dfrac{\partial \psi}{\partial \boldsymbol{F}} \\[2ex] \dot{\boldsymbol{\alpha}} = f(\boldsymbol{F}, \boldsymbol{\alpha}). \end{cases} \tag{3.164}$$

The infinitesimal strain case

In *the infinitesimal strain case*, the infinitesimal strain tensor, ε, replaces the deformation gradient and the stress tensor $\boldsymbol{\sigma}$ of the infinitesimal theory replaces the first Piola–Kirchhoff stress. We then have the general constitutive law

$$\begin{cases} \psi = \psi(\varepsilon, \boldsymbol{\alpha}) \\ \boldsymbol{\sigma} = \bar{\rho}\, \dfrac{\partial \psi}{\partial \varepsilon} \\ \dot{\boldsymbol{\alpha}} = f(\varepsilon, \boldsymbol{\alpha}). \end{cases} \tag{3.165}$$

3.5.5. THE CONSTITUTIVE INITIAL VALUE PROBLEM

Our basic constitutive problem is defined as follows: '*Given the history of the deformation gradient (and the history of temperature and temperature gradient, if thermal effects are considered), find the free-energy and stress (plus entropy and heat flux, in the thermomechanical case) according to the constitutive law conceptually expressed by (3.143)*'. If the internal variable approach is adopted in the formulation of the constitutive equations, the generic constitutive problem reduces to the following fundamental mechanical *initial value problem*.

Problem 3.1 (The mechanical constitutive initial value problem). *Given the initial values of the internal variables $\boldsymbol{\alpha}(t_0)$ and the history of the deformation gradient,*

$$\boldsymbol{F}(t), \quad t \in [t_0, T],$$

find the functions $\boldsymbol{P}(t)$ and $\boldsymbol{\alpha}(t)$, for the first Piola–Kirchhoff stress and the set of internal variables, such that the constitutive equations

$$\begin{cases} \boldsymbol{P}(t) = \bar{\rho}\, \dfrac{\partial \psi}{\partial \boldsymbol{F}}\bigg|_t \\ \dot{\boldsymbol{\alpha}}(t) = f(\boldsymbol{F}(t), \boldsymbol{\alpha}(t)) \end{cases} \tag{3.166}$$

are satisfied for every $t \in [t_0, T]$.

In the infinitesimal case, $\boldsymbol{P}$ and $\boldsymbol{F}$ are replaced with $\boldsymbol{\sigma}$ and ε, respectively, in the above initial value problem. For completeness, the infinitesimal constitutive initial value problem is stated in the following.

Problem 3.2 (The infinitesimal constitutive initial value problem). *Given the initial values of the internal variables $\boldsymbol{\alpha}(t_0)$ and the history of the infinitesimal strain tensor,*

$$\varepsilon(t), \quad t \in [t_0, T],$$

find the functions $\boldsymbol{\sigma}(t)$ and $\boldsymbol{\alpha}(t)$, for the stress tensor and the set of internal variables, such that the constitutive equations

$$\begin{cases} \boldsymbol{\sigma}(t) = \bar{\rho}\, \dfrac{\partial \psi}{\partial \varepsilon}\bigg|_t \\ \dot{\boldsymbol{\alpha}}(t) = f(\varepsilon(t), \boldsymbol{\alpha}(t)) \end{cases} \tag{3.167}$$

are satisfied for every $t \in [t_0, T]$.

3.6. Weak equilibrium. The principle of virtual work

The *strong* (point-wise, local or differential) forms of the momentum balance have been stated in Section 3.4 by expressions (3.132) and (3.133). In this section, we state the momentum balance equations in their corresponding *weak* (global or integral) forms. The weak equilibrium statement – the *Principle of Virtual Work* – is fundamental to the definition of the basic initial boundary value problem stated in Section 3.7 and, as we shall see in Chapter 4, is the starting point of displacement-based finite element procedures for the analysis of solids.

Again, let us consider the body $\mathscr{B}$ which occupies the region $\Omega \subset \mathscr{E}$ with boundary $\partial\Omega$ in its reference configuration subjected to body forces in its interior and surface tractions on its boundary. In its deformed configuration, $\mathscr{B}$ occupies the region $\varphi(\Omega)$ with boundary $\varphi(\partial\Omega)$ defined through the deformation map φ.

3.6.1. THE SPATIAL VERSION

The *spatial* version of the principle of virtual work states that the body $\mathscr{B}$ is in *equilibrium* if, and only if, its Cauchy stress field, $\boldsymbol{\sigma}$, satisfies the equation

$$\int_{\varphi(\Omega)} [\boldsymbol{\sigma} : \nabla_x \boldsymbol{\eta} - (\boldsymbol{b} - \rho\, \ddot{\boldsymbol{u}}) \cdot \boldsymbol{\eta}]\, \mathrm{d}v - \int_{\varphi(\partial\Omega)} \boldsymbol{t} \cdot \boldsymbol{\eta}\, \mathrm{d}a = 0, \quad \forall\, \boldsymbol{\eta} \in \mathscr{V}, \tag{3.168}$$

where $\boldsymbol{b}$ and $\boldsymbol{t}$ are the body force per unit deformed volume and boundary traction per unit deformed area and $\mathscr{V}$ is the *space of virtual displacements* of $\mathscr{B}$, i.e. the space of sufficiently regular arbitrary displacements

$$\boldsymbol{\eta} : \varphi(\Omega) \to \mathscr{U}.$$

Equivalence between strong and weak equilibrium statements

When the stress field $\boldsymbol{\sigma}$ is sufficiently smooth, the virtual work equation is equivalent to the strong momentum balance equations (3.132). To show this, let us start by assuming that the field $\boldsymbol{\sigma}$ is sufficiently regular so that we can use the identity (v) of Section 2.5.8 (page 38) to obtain

$$\boldsymbol{\sigma} : \nabla_x \boldsymbol{\eta} = \mathrm{div}_x(\boldsymbol{\sigma}\, \boldsymbol{\eta}) - (\mathrm{div}_x \boldsymbol{\sigma}) \cdot \boldsymbol{\eta}. \tag{3.169}$$

In obtaining the above identity we have used the fact that $\boldsymbol{\sigma}$ is symmetric. Next, by substituting the above expression into (3.168), it follows that

$$\int_{\varphi(\Omega)} [\mathrm{div}_x(\boldsymbol{\sigma}\, \boldsymbol{\eta}) - (\mathrm{div}_x\, \boldsymbol{\sigma} + \boldsymbol{b} - \rho \ddot{\boldsymbol{u}}) \cdot \boldsymbol{\eta}]\, \mathrm{d}v - \int_{\varphi(\partial\Omega)} \boldsymbol{t} \cdot \boldsymbol{\eta}\, \mathrm{d}a = 0, \quad \forall\, \boldsymbol{\eta} \in \mathscr{V}. \tag{3.170}$$

We now concentrate on the first term within the square brackets of the above equation. The divergence theorem (expression $(2.148)_2$, page 37) implies the following identity

$$\int_{\varphi(\Omega)} \mathrm{div}_x(\boldsymbol{\sigma}\, \boldsymbol{\eta})\, \mathrm{d}v = \int_{\varphi(\partial\Omega)} \boldsymbol{\sigma}\, \boldsymbol{\eta} \cdot \boldsymbol{n}\, \mathrm{d}a. \tag{3.171}$$

By taking into account the symmetry of $\boldsymbol{\sigma}$, which implies $\boldsymbol{\sigma}\boldsymbol{\eta} \cdot \boldsymbol{n} = \boldsymbol{\sigma}\boldsymbol{n} \cdot \boldsymbol{\eta}$, together with the above identity, equation (3.170) can be rewritten in the equivalent form

$$\int_{\varphi(\Omega)} (\mathrm{div}_x\, \boldsymbol{\sigma} + \boldsymbol{b} - \rho \ddot{\boldsymbol{u}}) \cdot \boldsymbol{\eta}\, \mathrm{d}v + \int_{\varphi(\partial\Omega)} (\boldsymbol{t} - \boldsymbol{\sigma}\, \boldsymbol{n}) \cdot \boldsymbol{\eta}\, \mathrm{d}a = 0, \quad \forall\, \boldsymbol{\eta} \in \mathscr{V}. \tag{3.172}$$

Finally, since this equation holds for all virtual displacement fields $\boldsymbol{\eta}$, then it follows from the fundamental theorem of variational calculus (refer, for instance, to Gurtin 1972; Oden 1979 or Reddy 1998) that each bracketed term of the above equation must vanish pointwise within their respective domains, i.e. we recover the strong equilibrium equations (3.132).

Conversely, the strong form yields the weak form of equilibrium. This can be shown in a relatively straightforward manner by applying a weighted residual method to the strong form together with use of the divergence theorem.

3.6.2. THE MATERIAL VERSION

The virtual work equation can be equivalently expressed in the *reference* configuration of $\mathscr{B}$. The corresponding *material* (or *reference*) version of the Principle of Virtual Work states that $\mathscr{B}$ is in equilibrium if and only if its first Piola–Kirchhoff stress field, $\boldsymbol{P}$, satisfies

$$\int_{\Omega} [\boldsymbol{P} : \nabla_p \boldsymbol{\eta} - (\bar{\boldsymbol{b}} - \bar{\rho}\ddot{\boldsymbol{u}}) \cdot \boldsymbol{\eta}] \, \mathrm{d}v - \int_{\partial\Omega} \bar{\boldsymbol{t}} \cdot \boldsymbol{\eta} \, \mathrm{d}a = 0, \quad \forall \, \boldsymbol{\eta} \in \mathscr{V}, \tag{3.173}$$

where $\bar{\boldsymbol{b}}$ and $\bar{\boldsymbol{t}}$ are, respectively, the body force per unit reference volume and the surface traction per unit reference area and $\bar{\rho}$ is the mass density in the reference configuration. The space of virtual displacements, $\mathscr{V}$, is accordingly defined as the space of sufficiently regular arbitrary displacement fields

$$\boldsymbol{\eta} : \Omega \rightarrow \mathscr{U}.$$

The material version of the virtual work equation is obtained by introducing, in its spatial counterpart, the identities

$$\boldsymbol{\sigma} = \frac{1}{J} \boldsymbol{P} \, \boldsymbol{F}^T; \quad \nabla_x \boldsymbol{a} = \nabla_p \boldsymbol{a} \, \boldsymbol{F}^{-1}, \tag{3.174}$$

where the second expression holds for a generic vector field $\boldsymbol{a}$, and making use of the standard relation (Gurtin, 1981)

$$\int_{\varphi(\Omega)} a(\boldsymbol{x}) \, \mathrm{d}v = \int_{\Omega} J(\boldsymbol{p}) \, a(\boldsymbol{\varphi}(\boldsymbol{p})) \, \mathrm{d}v, \tag{3.175}$$

valid for any scalar field a.

The proof of equivalence between (3.173) and the strong form (3.133) under conditions of sufficient regularity is then analogous to that given for the spatial version discussed in Section 3.6.1 above.

3.6.3. THE INFINITESIMAL CASE

Under infinitesimal deformations, reference and deformed configurations coincide and the virtual work equation reads simply

$$\int_{\Omega} [\boldsymbol{\sigma} : \nabla \boldsymbol{\eta} - (\boldsymbol{b} - \rho\ddot{\boldsymbol{u}}) \cdot \boldsymbol{\eta}] \, dv - \int_{\partial\Omega} \boldsymbol{t} \cdot \boldsymbol{\eta} \, \mathrm{d}a = 0, \quad \forall \, \boldsymbol{\eta} \in \mathscr{V}. \tag{3.176}$$

3.7. The quasi-static initial boundary value problem

Having defined the generic constitutive initial value problems in Section 3.5 and the weak equilibrium statements in Section 3.6, we are now in a position to state the weak form of fundamental initial boundary value problems, whose numerical solution by the finite element method is the main subject of the subsequent chapters of this book. The problems formulated here are restricted to *quasi-static* conditions, where inertia effects are ignored. This is the case on which the numerical methods discussed in this book are focused.

3.7.1. FINITE DEFORMATIONS

Let the body $\mathscr{B}$ (Figure 3.16) be subjected to a prescribed history of body forces

$$\boldsymbol{b}(t), \quad t \in [t_0, T]$$

in its interior. In the above, dependence of $\boldsymbol{b}$ on $\boldsymbol{x}$ is implicitly assumed. In addition, the following boundary conditions are imposed.

(i) *The natural boundary condition.* The history of the surface traction

$$\boldsymbol{t}(t), \quad t \in [t_0, T],$$

with dependence on $\boldsymbol{x}$ implied, is prescribed over the portion of the boundary of $\mathscr{B}$ that occupies the region $\partial\Omega_t$ in its reference configuration.

(ii) *The essential boundary condition.* The *motion* is a prescribed function on the part of the boundary of $\mathscr{B}$ that occupies the region $\partial\Omega_u$ in the reference configuration

$$\bar{\varphi}(\boldsymbol{p}, t) = \boldsymbol{p} + \bar{\boldsymbol{u}}(\boldsymbol{p}, t) \quad t \in [t_0, T],\ \boldsymbol{p} \in \partial\Omega_u,$$

where $\bar{\boldsymbol{u}}$ is the corresponding prescribed boundary displacement field. For simplicity, it is assumed here that $\partial\Omega_u \bigcap \partial\Omega_t = \emptyset$. We define the set of *kinematically admissible displacements* of $\mathscr{B}$ as the set of all sufficiently regular displacement functions that satisfy the kinematic constraint (the essential boundary condition)

$$\mathscr{K} = \{\boldsymbol{u} : \Omega \times \mathscr{R} \to \mathscr{U} \mid \boldsymbol{u}(\boldsymbol{p}, t) = \bar{\boldsymbol{u}}(\boldsymbol{p}, t),\ t \in [t_0, T],\ \boldsymbol{p} \in \partial\Omega_u\}. \tag{3.177}$$

The body $\mathscr{B}$ is assumed to be made from a generic material modelled by the internal variable-based constitutive equations associated with Problem 3.1 (page 76) and the internal variable field, $\boldsymbol{\alpha}$, is known at the initial time t_0, i.e.

$$\boldsymbol{\alpha}(\boldsymbol{p}, t_0) = \boldsymbol{\alpha}_0(\boldsymbol{p}). \tag{3.178}$$

The fundamental quasi-static initial boundary value problem is stated in its spatial version in the following.

Problem 3.3 (The spatial quasi-static initial boundary value problem). *Find a kinematically admissible displacement function, $\boldsymbol{u} \in \mathscr{K}$, such that, for all $t \in [t_0, T]$, the virtual work*

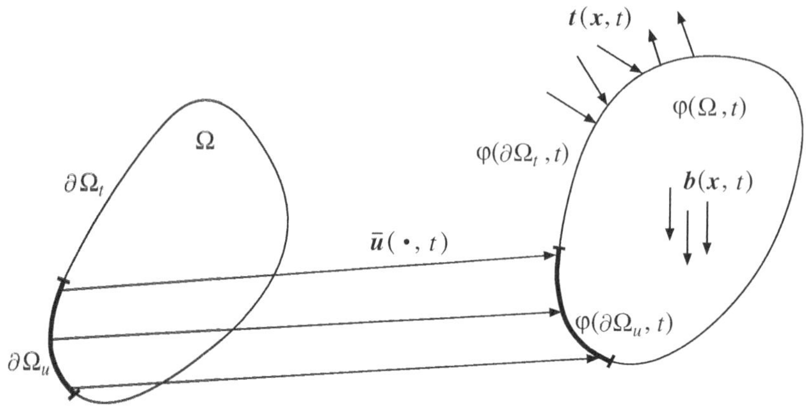

Figure 3.16. The initial boundary value problem. Schematic illustration.

equation is satisfied

$$\int_{\varphi(\Omega,t)} [\boldsymbol{\sigma}(t) : \nabla_x \boldsymbol{\eta} - \boldsymbol{b}(t) \cdot \boldsymbol{\eta}] \, \mathrm{d}v - \int_{\varphi(\partial\Omega_t,t)} \boldsymbol{t}(t) \cdot \boldsymbol{\eta} \, \mathrm{d}a = 0, \quad \forall \, \boldsymbol{\eta} \in \mathscr{V}_t. \tag{3.179}$$

The space of virtual displacements at time t is defined by

$$\mathscr{V}_t = \{\boldsymbol{\eta} : \varphi(\Omega, t) \rightarrow \mathscr{U} \mid \boldsymbol{\eta} = \boldsymbol{0} \text{ on } \varphi(\partial\Omega_u, t)\} \tag{3.180}$$

and, at each point of $\mathscr{B}$, the Cauchy stress is given by

$$\boldsymbol{\sigma}(t) = \boldsymbol{P}(t)\boldsymbol{F}(t)^T / J(t), \tag{3.181}$$

where $\boldsymbol{P}(t)$ is the solution of constitutive initial value Problem 3.1 (page 76) with prescribed deformation gradient

$$\boldsymbol{F}(t) = \nabla_p \boldsymbol{\varphi}(\boldsymbol{p}, t) = \boldsymbol{I} + \nabla_p \boldsymbol{u}(\boldsymbol{p}, t). \tag{3.182}$$

The problem can be equivalently formulated in the reference configuration of $\mathscr{B}$ in terms of the material version of the principle of virtual work (3.173). For completeness, we state the material version of the fundamental initial boundary value problem in the following.

Problem 3.4 (The material quasi-static initial boundary value problem). *Find a kinematically admissible displacement function, $\boldsymbol{u} \in \mathscr{K}$, such that, for all $t \in [t_0, T]$,*

$$\int_{\Omega} [\boldsymbol{P}(t) : \nabla_p \boldsymbol{\eta} - \bar{\boldsymbol{b}}(t) \cdot \boldsymbol{\eta}] \, \mathrm{d}v - \int_{\partial\Omega_t} \bar{\boldsymbol{t}}(t) \cdot \boldsymbol{\eta} \, \mathrm{d}a = 0, \quad \forall \, \boldsymbol{\eta} \in \mathscr{V}, \tag{3.183}$$

where

$$\mathscr{V} = \{\boldsymbol{\eta} : \Omega \rightarrow \mathscr{U} \mid \boldsymbol{\eta} = \boldsymbol{0} \text{ on } \partial\Omega_u\} \tag{3.184}$$

and the Piola–Kirchhoff stress, $\boldsymbol{P}(t)$, is the solution of initial value Problem 3.1 with prescribed deformation gradient (3.182).

3.7.2. THE INFINITESIMAL PROBLEM

Under infinitesimal deformations, the quasi-static initial boundary value problem is based on the weak form (3.176). It is stated in the following.

Problem 3.5 (The infinitesimal quasi-static initial boundary value problem). *Find a kinematically admissible displacement, $\boldsymbol{u} \in \mathscr{K}$, such that, for $t \in [t_0, T]$,*

$$\int_{\Omega} [\boldsymbol{\sigma}(t) : \nabla \boldsymbol{\eta} - \boldsymbol{b}(t) \cdot \boldsymbol{\eta}] \, \mathrm{d}v - \int_{\partial \Omega_t} \boldsymbol{t}(t) \cdot \boldsymbol{\eta} \, \mathrm{d}a = 0, \quad \forall \, \boldsymbol{\eta} \in \mathscr{V}, \tag{3.185}$$

where

$$\mathscr{V} = \{\boldsymbol{\eta} : \Omega \rightarrow \mathscr{U} \mid \boldsymbol{\eta} = \mathbf{0} \text{ on } \partial \Omega_u\} \tag{3.186}$$

and, at each point $\mathbf{p}$, $\boldsymbol{\sigma}(t)$ is the solution of the constitutive initial value Problem 3.2 (page 76) with prescribed strain

$$\boldsymbol{\varepsilon}(t) = \nabla^s \boldsymbol{u}(\boldsymbol{p}, t). \tag{3.187}$$

4 THE FINITE ELEMENT METHOD IN QUASI-STATIC NONLINEAR SOLID MECHANICS

IN this chapter we present a summary of the application of the Finite Element Method to quasi-static nonlinear solid mechanics. Our aim here is to review the main techniques involved in the finite element solution of the quasi-static initial boundary value problems stated in Section 3.7. No attempt is made to provide a comprehensive review of the subject. For a detailed account on the Finite Element Method, and related techniques we refer to standard textbooks (Bathe, 1996; Belytschko *et al.*, 2000; Hughes, 1987; Wriggers, 2001; Zienkiewicz and Taylor, 2000). The material presented here covers the basic procedures that form the backbone of the computer implementation provided in the finite element program `HYPLAS` and should be helpful to readers who wish to learn more about the code in Chapter 5. Throughout the chapter, reference is frequently made to subroutines of `HYPLAS` involved in the computational implementation of the procedures described.

At this point, it is probably worth emphasising that two major numerical approximations (Figure 4.1) are necessary in the finite element solution of the generic initial boundary value problems stated in Section 3.7:

1. *A time discretisation of the underlying constitutive initial value problem.* A numerical integration scheme is introduced to solve the initial value problem defined by the constitutive equations of the model that relate stresses to the history of deformations. The general constitutive initial value problems are those stated in Section 3.5.5. Upon introduction of the numerical integration scheme, the original time-continuum constitutive equations are transformed into *incremental* (or *time-discrete*) counterparts.

2. *A finite element discretisation.* This comprises a standard finite element approximation of the virtual work statement where the domain of the body and the associated functional sets are replaced with finite-dimensional counterparts generated by finite element interpolation functions.

With the introduction of the above approximations, the original initial boundary value problem is reduced to a set of *incremental (generally nonlinear) algebraic finite element equations* to be solved at each time station of the considered time interval. In addition to items 1 and 2 above, the present chapter addresses the solution of the associated algebraic system, with particular emphasis on the quadratically convergent Newton–Raphson algorithm. The *arc-length* technique, which becomes crucial in the solution of problems involving structural instability, is also described.

Computational Methods for Plasticity: Theory and Applications EA de Souza Neto, D Perić and DRJ Owen

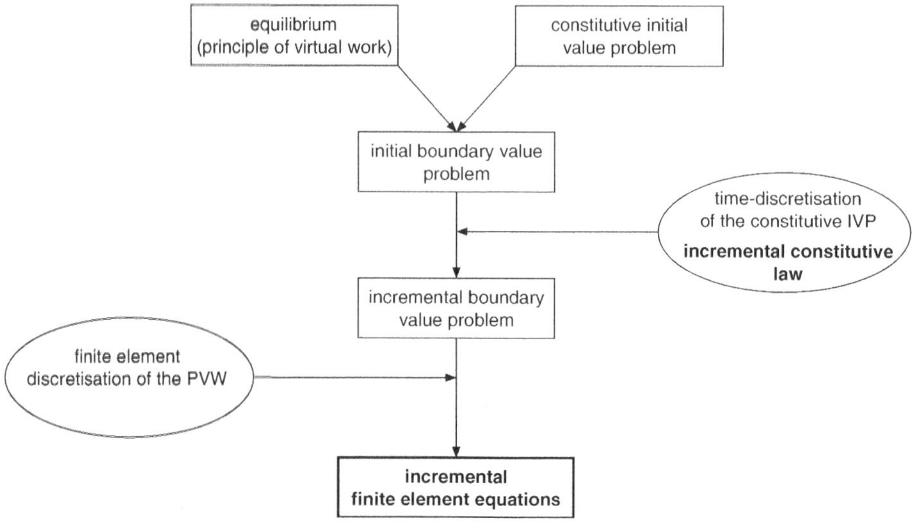

Figure 4.1. Numerical approximations. Reducing the initial boundary value problem to a set of incremental finite element equations.

We start in Section 4.1 with a brief description of displacement-based finite element methods for small strain problems. Most of the basic finite element matrices and arrays are introduced here. At this stage, the underlying material is assumed elastic so that no time discretisation of the constitutive equations is required and attention can be focused on the finite element discretisation alone. The treatment of path-dependent (inelastic) materials, still within the small strain context, is introduced in Section 4.2. Here, the numerical integration of the constitutive initial value problem is briefly discussed and a generic incremental form of the original equation is introduced. Associated nonlinear solution schemes, particularly the Newton–Raphson scheme, are also discussed. In Section 4.3 we move on to the realm of finite deformations and strains and describe a general finite element strategy based on the spatial configuration (spatial version of the virtual work equation). The consistent linearisation of the field equations, required by the Newton–Raphson scheme for the solution of the nonlinear discretised equation system, is also discussed. Finally, in Section 4.4, we review the *arc-length* technique.

Before proceeding, we would like to draw the attention of the reader to the notation adopted. Finite element arrays (matrices, vectors) are represented *exclusively* with *upright* bold-faced Roman and Greek fonts:

$$\mathbf{B}, \quad \mathbf{u}, \quad \boldsymbol{\sigma}, \ldots$$

so that they can be distinguished from standard vector and second-order tensor quantities that are represented with *italic* bold-faced fonts.

4.1. Displacement-based finite elements

Let us start by considering the infinitesimal deformation case. We also assume that the underlying material is *elastic*; that is, the stress tensor is a (possibly nonlinear) function of the strain tensor only (without dependence on internal variables). Since the strain tensor is in turn a *functional* of the displacement field, we have the following constitutive equation

$$\boldsymbol{\sigma} = \boldsymbol{\sigma}(\boldsymbol{\varepsilon}(\boldsymbol{u})) = \boldsymbol{\sigma}(\nabla^s \boldsymbol{u}). \tag{4.1}$$

With the above constitutive equation, the infinitesimal problem of Section 3.7.2 reduces to finding a kinematically admissible displacement field $\boldsymbol{u} \in \mathscr{K}$ such that

$$\int_\Omega [\boldsymbol{\sigma}(\nabla^s \boldsymbol{u}) : \nabla^s \boldsymbol{\eta} - \boldsymbol{b} \cdot \boldsymbol{\eta}] \, \mathrm{d}v - \int_{\partial\Omega_t} \boldsymbol{t} \cdot \boldsymbol{\eta} \, \mathrm{d}a = 0, \quad \forall \, \boldsymbol{\eta} \in \mathscr{V} \tag{4.2}$$

holds.

4.1.1. FINITE ELEMENT INTERPOLATION

The finite element method for numerical solution of the above problem consists of replacing the functional sets $\mathscr{K}$ and $\mathscr{V}$ with discrete subsets ${}^h\mathscr{K}$ and ${}^h\mathscr{V}$ generated by a finite element discretisation h of the domain Ω.

Let the generic finite element e be defined by n_{node} *nodes* with one *shape function* (or *interpolation function*) $N_i^{(e)}(\boldsymbol{x})$ associated with each node i whose coordinate is $\boldsymbol{x}^i$. Figure 4.2 illustrates a three-noded triangular element with linear shape functions. Each shape function $N_i^{(e)}(\boldsymbol{x})$ is defined so that its value is unity at node i, i.e.

$$N_i^{(e)}(\boldsymbol{x}^i) = 1, \tag{4.3}$$

and zero at any other node of the element:

$$N_i^{(e)}(\boldsymbol{x}^j) = 0, \quad \text{for } j \neq i. \tag{4.4}$$

Now let $a(\boldsymbol{x})$ be a generic field defined over the domain Ω^e of the element. The finite element interpolation ${}^h a$ of the field a within element e is obtained as

$${}^h a(\boldsymbol{x}) \equiv \sum_{i=1}^{n_{\text{node}}} a^i \, N_i^{(e)}(\boldsymbol{x}), \tag{4.5}$$

where a^i is the value of a at node i:

$$a^i \equiv a(\boldsymbol{x}^i). \tag{4.6}$$

Now let $a(\boldsymbol{x})$ be defined over the entire domain Ω. In the finite element interpolation of a we discretise Ω with a *mesh* of n_{elem} finite elements. Within each element, a is interpolated as described in the above. The interpolated function, now defined over the approximate domain

$${}^h\Omega = \bigcup_{e=1}^{n_{\text{elem}}} \Omega^{(e)}, \tag{4.7}$$

is given by

$${}^h a(\boldsymbol{x}) = \sum_{i=1}^{n_{\text{poin}}} a^i \, N_i^g(\boldsymbol{x}), \tag{4.8}$$

where N_i^g is a piecewise polynomial function – the *global shape function* – associated with the *global* node i and n_{poin} is the total number of nodal points in the finite element mesh. A typical case, where a plane domain has been discretised by a mesh of triangular finite elements, is illustrated in Figure 4.3.

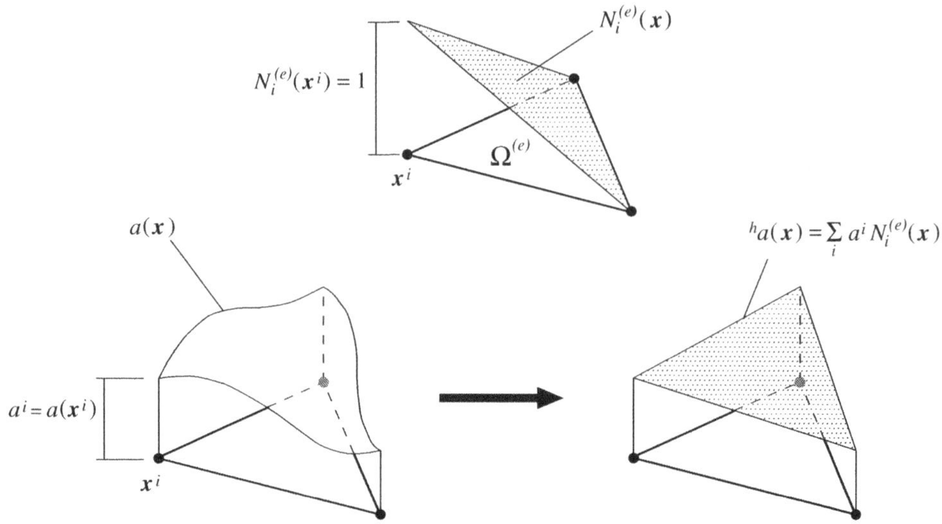

Figure 4.2. Finite element interpolation. The element shape function.

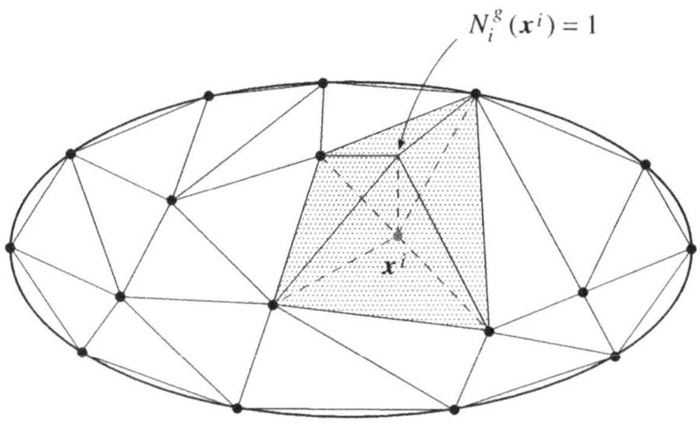

Figure 4.3. Finite element interpolation. The global shape function.

4.1.2. THE DISCRETISED VIRTUAL WORK

With the introduction of the above interpolation procedure we generate the finite-dimensional sets

$$ {}^h\mathscr{K} \equiv \left\{ {}^h\boldsymbol{u}(\boldsymbol{x}) = \sum_{i=1}^{n_{\rm poin}} \boldsymbol{u}^i\, N_i^g(\boldsymbol{x}) \mid \boldsymbol{u}^i = \bar{\boldsymbol{u}}(\boldsymbol{x}^i) \text{ if } \boldsymbol{x}^i \in \partial\Omega_u \right\} \tag{4.9} $$

and

$$ {}^h\mathscr{V} \equiv \left\{ {}^h\boldsymbol{\eta}(\boldsymbol{x}) = \sum_{i=1}^{n_{\rm poin}} \boldsymbol{\eta}^i\, N_i^g(\boldsymbol{x}) \mid \boldsymbol{\eta}^i = \boldsymbol{0} \text{ if } \boldsymbol{x}^i \in \partial\Omega_u \right\}. \tag{4.10} $$

The finite element approximation to the continuum variational equation (4.2) is then obtained by replacing the functional sets $\mathscr{K}$ and $\mathscr{V}$ with the above defined finite-dimensional subsets.

Matrix notation

To derive the discretised form of the virtual work equation it is convenient to introduce the standard matrix notation that follows. Let n_{dim} be the number of spatial dimensions. Firstly, we define the *global interpolation matrix*:

$$\mathbf{N}^g(\boldsymbol{x}) = [\text{diag}[N_1^g(\boldsymbol{x})] \quad \text{diag}[N_2^g(\boldsymbol{x})] \quad \cdots \quad \text{diag}[N_{n_{\text{poin}}}^g(\boldsymbol{x})]], \tag{4.11}$$

where $\text{diag}[N_i^g]$ denotes the $n_{\text{dim}} \times n_{\text{dim}}$ diagonal matrix defined as

$$\text{diag}[N_i^g] = \begin{bmatrix} N_i^g & 0 & \cdots & 0 \\ 0 & N_i^g & \cdots & 0 \\ & & \ddots & \\ 0 & 0 & \cdots & N_i^g \end{bmatrix}. \tag{4.12}$$

We also define the *global vector of nodal displacements*:

$$\mathbf{u} = [u_1^1, \ldots, u_{n_{\text{dim}}}^1, \ldots\ldots, u_1^{n_{\text{poin}}}, \ldots, u_{n_{\text{dim}}}^{n_{\text{poin}}}]^T, \tag{4.13}$$

where the generic element u_i^j is the i-component of the displacement vector at the global node j. Any element ${}^h\boldsymbol{u} \in {}^h\mathscr{K}$ can be represented as

$${}^h\boldsymbol{u} = \mathbf{N}^g\, \mathbf{u}, \tag{4.14}$$

where the components of $\mathbf{u}$ corresponding to nodal points on $\partial\Omega_u$ satisfy the prescribed kinematic constraints and dependence on $\boldsymbol{x}$ has been omitted for notational convenience. Analogously, any virtual displacement ${}^h\boldsymbol{\eta} \in {}^h\mathscr{V}$ has the representation

$${}^h\boldsymbol{\eta} = \mathbf{N}^g\, \boldsymbol{\eta}, \tag{4.15}$$

where

$$\boldsymbol{\eta} = [\eta_1^1, \ldots, \eta_{n_{\text{dim}}}^1, \ldots\ldots, \eta_1^{n_{\text{poin}}}, \ldots, \eta_{n_{\text{dim}}}^{n_{\text{poin}}}]^T \tag{4.16}$$

is the vector of virtual nodal displacements. In accordance with the definition of ${}^h\mathscr{V}$, the components of $\boldsymbol{\eta}$ associated with nodal points belonging to $\partial\Omega_u$ must vanish.

It is also convenient to introduce the global *discrete symmetric gradient operator* (or *strain-displacement matrix*) which, in two dimensions (plane stress and plane strain problems), has the format

$$\mathbf{B}^g = \begin{bmatrix} N_{1,1}^g & 0 & N_{2,1}^g & 0 & \cdots & N_{n_{\text{poin}},1}^g & 0 \\ 0 & N_{1,2}^g & 0 & N_{2,2}^g & \cdots & 0 & N_{n_{\text{poin}},2}^g \\ N_{1,2}^g & N_{1,1}^g & N_{2,2}^g & N_{2,1}^g & \cdots & N_{n_{\text{poin}},2}^g & N_{n_{\text{poin}},1}^g \end{bmatrix}, \tag{4.17}$$

where

$$(\cdot)_{i,j} = \frac{\partial(\cdot)_i}{\partial x_j}. \tag{4.18}$$

For a generic vector field $\boldsymbol{v} = \mathbf{N}^g\ \mathbf{v}$, we have

$$\mathbf{B}^g\ \mathbf{v} = \begin{bmatrix} v_{1,1} \\ v_{2,2} \\ v_{1,2} + v_{2,1} \end{bmatrix} = \begin{bmatrix} (\nabla^s \mathbf{v})_{11} \\ (\nabla^s \mathbf{v})_{22} \\ 2\,(\nabla^s \mathbf{v})_{12} \end{bmatrix}; \tag{4.19}$$

that is, the multiplication of $\mathbf{B}^g$ by a global vector of nodal displacements gives the array of engineering strains ordered according to the convention discussed in Appendix D.

Finally, the array of stress components in plane strain/stress problems is defined as†

$$\boldsymbol{\sigma} = [\sigma_{11}, \sigma_{22}, \sigma_{12}]^T. \tag{4.20}$$

In axisymmetric analyses, $\boldsymbol{\sigma}$ is defined by

$$\boldsymbol{\sigma} = [\sigma_{11}, \sigma_{22}, \sigma_{12}, \sigma_{33}]^T, \tag{4.21}$$

where the hoop direction is associated with the index 3. In three-dimensional problems, we have

$$\boldsymbol{\sigma} = [\sigma_{11}, \sigma_{22}, \sigma_{33}, \sigma_{12}, \sigma_{23}, \sigma_{13}]^T. \tag{4.22}$$

The discrete boundary value problem

With the above notation at hand, the replacement of $\mathscr{K}$ and $\mathscr{V}$ respectively with ${}^h\mathscr{K}$ and ${}^h\mathscr{V}$ in (4.2) gives the following discretised virtual work expression

$$\int_{^h\Omega} [\boldsymbol{\sigma}^T \mathbf{B}^g\ \boldsymbol{\eta} - \boldsymbol{b} \cdot \mathbf{N}^g\ \boldsymbol{\eta}]\ \mathrm{d}v - \int_{\partial^h\Omega_t} \boldsymbol{t} \cdot \mathbf{N}^g\ \boldsymbol{\eta}\ \mathrm{d}a = 0, \quad \forall\, \boldsymbol{\eta} \in {}^h\mathscr{V} \tag{4.23}$$

which can be conveniently rearranged as

$$\left\{ \int_{^h\Omega} [(\mathbf{B}^g)^T\ \boldsymbol{\sigma} - (\mathbf{N}^g)^T\ \boldsymbol{b}]\ \mathrm{d}v - \int_{\partial^h\Omega_t} (\mathbf{N}^g)^T\ \boldsymbol{t}\ \mathrm{d}a \right\}^T \boldsymbol{\eta} = 0, \quad \forall\, \boldsymbol{\eta} \in {}^h\mathscr{V}. \tag{4.24}$$

Since the above equation is satisfied for all vectors $\boldsymbol{\eta}$ of the form (4.16), the term within the curly brackets on the left-hand side must vanish. Also, recall that the stress tensor components depend on the strain components which, in turn, depend on the displacement field defined by $\mathbf{u}$. The finite element *discrete boundary value problem* is then formulated as follows. Find the *global vector of nodal displacements*, $\mathbf{u}$, such that

$$\mathbf{f}^{\,\mathrm{int}}(\mathbf{u}) - \mathbf{f}^{\,\mathrm{ext}} = \mathbf{0}, \tag{4.25}$$

where $\mathbf{f}^{\,\mathrm{int}}$ and $\mathbf{f}^{\,\mathrm{ext}}$ are, respectively, the *internal* and *external global force vectors*

$$\begin{aligned} \mathbf{f}^{\,\mathrm{int}} &= \int_{^h\Omega} (\mathbf{B}^g)^T \boldsymbol{\sigma}\ \mathrm{d}v \\ \mathbf{f}^{\,\mathrm{ext}} &= \int_{^h\Omega} (\mathbf{N}^g)^T \boldsymbol{b}\ \mathrm{d}v + \int_{\partial^h\Omega_t} (\mathbf{N}^g)^T \boldsymbol{t}\ \mathrm{d}a. \end{aligned} \tag{4.26}$$

†The computational array representation of tensors is described in more detail in Appendix D. The reader unfamiliar with this notation is advised to read Appendix D before proceeding.

The element force vectors

In actual finite element computations, the above force vectors are obtained as the assemblies

$$\begin{aligned}\mathbf{f}^{\text{int}} &= \mathop{\mathsf{A}}_{e=1}^{n_{\text{elem}}} (\mathbf{f}^{\text{int}}_{(e)})\\ \mathbf{f}^{\text{ext}} &= \mathop{\mathsf{A}}_{e=1}^{n_{\text{elem}}} (\mathbf{f}^{\text{ext}}_{(e)})\end{aligned} \tag{4.27}$$

of the *element vectors*

$$\begin{aligned}\mathbf{f}^{\text{int}}_{(e)} &= \int_{\Omega^{(e)}} \mathbf{B}^T \boldsymbol{\sigma}\, \mathrm{d}v\\ \mathbf{f}^{\text{ext}}_{(e)} &= \int_{\Omega^{(e)}} \mathbf{N}^T \boldsymbol{b}\, \mathrm{d}v + \int_{\partial\Omega^{(e)}_t} \mathbf{N}^T \boldsymbol{t}\, \mathrm{d}a.\end{aligned} \tag{4.28}$$

The *finite element assembly operator*, A, implies that each component of the global force associated with a particular global node is obtained as the sum of the corresponding contributions from the element force vectors of all elements that share that global node. For a generic element e, the interpolation matrix $\mathbf{N}$ is given by

$$\mathbf{N} = [\text{diag}[N^{(e)}_1(\boldsymbol{x})] \quad \text{diag}[N^{(e)}_2(\boldsymbol{x})] \quad \cdots \quad \text{diag}[N^{(e)}_{n_{\text{node}}}(\boldsymbol{x})]], \tag{4.29}$$

and matrix $\mathbf{B}$ (in plane stress/strain analyses) has the standard format

$$\mathbf{B} = \begin{bmatrix} N^{(e)}_{1,1} & 0 & N^{(e)}_{2,1} & 0 & \cdots & N^{(e)}_{n_{\text{node}},1} & 0\\ 0 & N^{(e)}_{1,2} & 0 & N^{(e)}_{2,2} & \cdots & 0 & N^{(e)}_{n_{\text{node}},2}\\ N^{(e)}_{1,2} & N^{(e)}_{1,1} & N^{(e)}_{2,2} & N^{(e)}_{2,1} & \cdots & N^{(e)}_{n_{\text{node}},2} & N^{(e)}_{n_{\text{node}},1}\end{bmatrix}. \tag{4.30}$$

The multiplication of matrix $\mathbf{B}$ by an elemental vector of nodal displacements produces the array of engineering strains within the element.

Numerical integration. Gaussian quadratures

As a general rule, the exact integrals of (4.28) are replaced with *Gaussian quadratures* for numerical evaluation of the element force vectors. Let Γ be a standard integration domain. A Gaussian quadrature approximation with n_{gaus} Gauss points to the integral of a generic function f over Γ reads

$$\int_\Gamma f(\boldsymbol{\xi})\, \mathrm{d}\boldsymbol{\xi} \approx \sum_{i=1}^{n_{\text{gaus}}} w_i\, f(\boldsymbol{\xi}_i), \tag{4.31}$$

where $\boldsymbol{\xi}_i$ $(i = 1, \ldots, n_{\text{gaus}})$ are the *positions* (*coordinates*) of the Gauss points in the standard domain Γ and w_i $(i = 1, \ldots, n_{\text{gaus}})$ are the corresponding *weights*.

Now let g be a generic function defined over the element domain, $\Omega^{(e)}$, and let $\boldsymbol{x} : \Gamma \rightarrow \Omega^{(e)}$ map the standard domain Γ onto $\Omega^{(e)}$. By a straightforward transformation of variables the Gaussian quadrature for approximation of the integral of g over $\Omega^{(e)}$ is obtained as

$$\int_{\Omega^{(e)}} g(\boldsymbol{x})\,\mathrm{d}\boldsymbol{x} = \int_{\Gamma} g(\boldsymbol{x}(\boldsymbol{\xi}))\,j(\boldsymbol{\xi})\,\mathrm{d}\boldsymbol{\xi} \approx \sum_{i=1}^{n_{\text{gaus}}} w_i\,g_i\,j_i, \tag{4.32}$$

where

$$j(\boldsymbol{\xi}) = \det[\partial\boldsymbol{x}/\partial\boldsymbol{\xi}] \tag{4.33}$$

is the determinant of the *Jacobian* of the transformation $\boldsymbol{x}$ and we have defined

$$j_i = j(\boldsymbol{\xi}_i) \tag{4.34}$$

and

$$g_i = g(\boldsymbol{x}(\boldsymbol{\xi}_i)). \tag{4.35}$$

Similarly to (4.32), we define the Gaussian quadrature to approximate integrals over the *boundary*, $\partial\Omega^{(e)}$, of an element. With $\partial\Gamma$ denoting the standard integration interval and the function g defined over $\partial\Omega^{(e)}$ we have

$$\int_{\partial\Omega^{(e)}} g(\boldsymbol{x}^b)\,\mathrm{d}\boldsymbol{x}^b = \int_{\partial\Gamma} g(\boldsymbol{x}^b(\boldsymbol{\xi}^b))\,j^b(\boldsymbol{\xi}^b)\,\mathrm{d}\boldsymbol{\xi}^b \approx \sum_{i=1}^{n_{\text{gausb}}} w_i^b\,g_i\,j_i^b, \tag{4.36}$$

where w_i^b are the weights at the corresponding boundary integration points $\boldsymbol{\xi}_i^b$ and j_i^b denotes the Jacobian of the boundary transformation $\boldsymbol{x}^b : \partial\Gamma \to \partial\Omega^{(e)}$.

By applying Gaussian quadratures with n_{gausp} and n_{gausb} points for integrals, respectively, over the element domain and the relevant portion of its boundary ($\partial\Omega_t^{(e)}$), the element arrays $\mathbf{f}_{(e)}^{\text{int}}$ and $\mathbf{f}_{(e)}^{\text{ext}}$ are evaluated as

$$\begin{aligned} \mathbf{f}_{(e)}^{\text{int}} &= \sum_{i=1}^{n_{\text{gausp}}} w_i\,\mathbf{B}_i^T\boldsymbol{\sigma}_i\,j_i \\ \mathbf{f}_{(e)}^{\text{ext}} &= \sum_{i=1}^{n_{\text{gausp}}} w_i\,\mathbf{N}_i^T\,\boldsymbol{b}_i\,j_i + \sum_{i=1}^{n_{\text{gausb}}} w_i^b\,\mathbf{N}_i^T\,\boldsymbol{t}_i\,j_i^b. \end{aligned} \tag{4.37}$$

4.1.3. SOME TYPICAL ISOPARAMETRIC ELEMENTS

For completeness, some commonly used two-dimensional isoparametric elements, which are coded in program HYPLAS, are reviewed in this section.

Linear triangle

We start by considering the simplest two-dimensional element – the three-noded linear triangle. This element is named TRI_3 in program HYPLAS. Figures 4.2 and 4.3 have already illustrated element and global shape functions associated with the linear triangle. Let the element be defined over the standard domain Γ (Figure 4.4). The element shape functions are

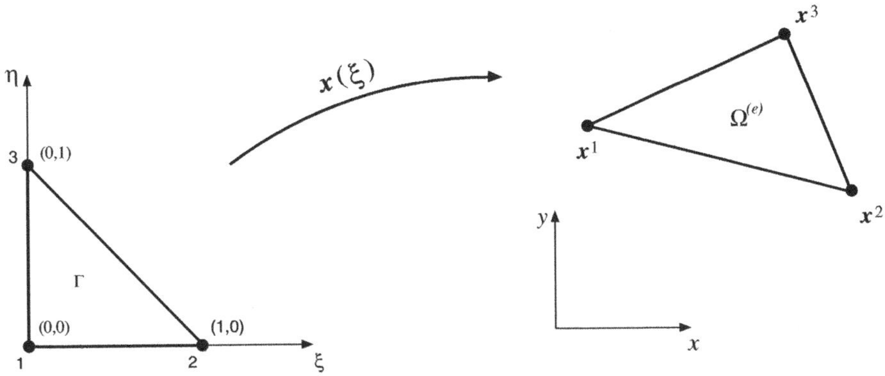

Figure 4.4. Linear triangle.

defined as

$$\begin{aligned} N_1(\boldsymbol{\xi}) &= N_1(\xi,\eta) = 1-\xi-\eta \\ N_2(\boldsymbol{\xi}) &= N_2(\xi,\eta) = \xi \\ N_3(\boldsymbol{\xi}) &= N_3(\xi,\eta) = \eta. \end{aligned} \tag{4.38}$$

The function $\boldsymbol{x}(\boldsymbol{\xi})$ that maps Γ onto $\Omega^{(e)}$ is given by

$$\boldsymbol{x}(\boldsymbol{\xi}) = \sum_{i=1}^{3} N_i(\boldsymbol{\xi})\, \boldsymbol{x}^i, \tag{4.39}$$

or, in component form,

$$\begin{aligned} x(\xi,\eta) &= \sum_{i=1}^{3} N_i(\xi,\eta)\, x^i \\ y(\xi,\eta) &= \sum_{i=1}^{3} N_i(\xi,\eta)\, y^i. \end{aligned} \tag{4.40}$$

In HYPLAS, the shape functions (as well as their derivatives) for the three-node triangle are coded in subroutine SFT3.

This element is typically used with a one-point Gauss quadrature. The corresponding Gauss-point position and weight are

$$\boldsymbol{\xi}^1 = \left(\tfrac{1}{3}, \tfrac{1}{3}\right), \quad w_1 = \tfrac{1}{2}. \tag{4.41}$$

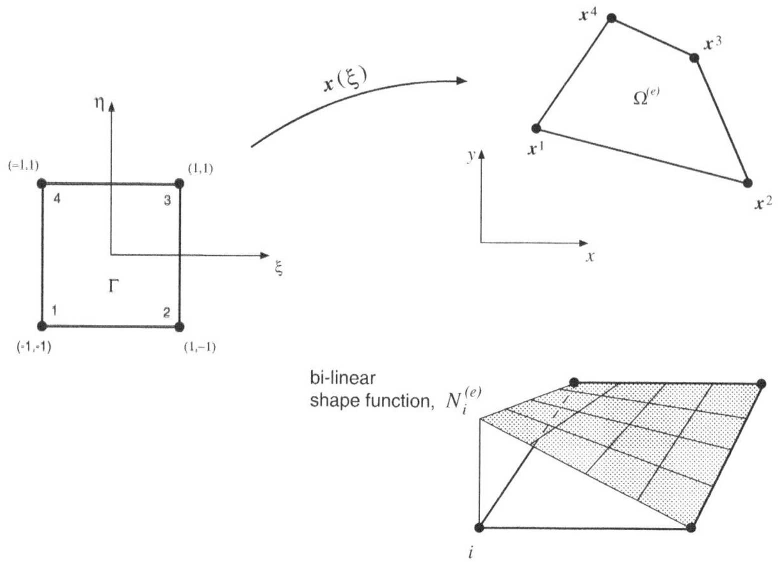

Figure 4.5. Bi-linear quadrilateral.

Bi-linear quadrilateral

The bi-linear quadrilateral (named QUAD_4 in HYPLAS) has the following shape functions:

$$
\begin{aligned}
N_1(\boldsymbol{\xi}) &= \tfrac{1}{4}(1 - \xi - \eta + \xi\eta) \\
N_2(\boldsymbol{\xi}) &= \tfrac{1}{4}(1 - \xi + \eta - \xi\eta) \\
N_3(\boldsymbol{\xi}) &= \tfrac{1}{4}(1 + \xi + \eta + \xi\eta) \\
N_4(\boldsymbol{\xi}) &= \tfrac{1}{4}(1 + \xi - \eta - \xi\eta).
\end{aligned}
\tag{4.42}
$$

The above shape functions are computed in subroutine SFQ4 of HYPLAS. The element is illustrated in Figure 4.5. The four-point (2×2) Gauss quadrature is usually adopted for this element. The corresponding sampling point positions and weights can be found in subroutine GAUS2D.

Eight-noded quadrilateral

Another important isoparametric element, which is widely used in infinitesimal elastoplastic analysis, is the eight-noded isoparametric quadrilateral (element QUAD_8 of HYPLAS). The corresponding shape functions will not be listed here. We refer to subroutine SFQ8 where the shape functions and derivatives for this element are computed.

In elastoplastic analysis under plane strain and axisymmetric conditions and isochoric plastic flow, this element is almost always used with a four-Gauss point (reduced) numerical integration rule. The reduced integration is particularly helpful in avoiding the phenomenon of volumetric ‘locking’ observed when low-order elements are used under near incompressibility conditions (volumetric locking is discussed in Chapter 15). Under plane stress, the nine-point (3×3) Gauss quadrature is recommended.

4.1.4. EXAMPLE. LINEAR ELASTICITY

So far we have left the stress tensor, whose components are grouped in the stress array $\boldsymbol{\sigma}$, as a generic (possibly nonlinear) function of the strain tensor, ε. A simple example of application of the Finite Element Method is given in isotropic *linear elasticity*, where the stress tensor is a linear function of the strain tensor:

$$\boldsymbol{\sigma} = \mathsf{D}^e : \boldsymbol{\varepsilon}, \tag{4.43}$$

and the fourth-order isotropic *elasticity tensor* has the classical general format

$$\mathsf{D}^e = 2G\,\mathsf{I}_S + A\,(K - \tfrac{2}{3}G)\,\boldsymbol{I} \otimes \boldsymbol{I}, \tag{4.44}$$

where G and K are, respectively, the shear and bulk moduli, $\boldsymbol{I}$ is the second-order identity and I_S is the fourth-order symmetric identity tensor. In plane strain, axisymmetric and three-dimensional analyses, the constant A is given by

$$A = 1, \tag{4.45}$$

whereas, in plane stress,

$$A = \frac{2G}{K + \frac{4}{3}G}. \tag{4.46}$$

Matrix notation

Let us now define the *array of engineering strains*.‡ Analogously to the array $\boldsymbol{\sigma}$ of stress components, we define for plane strain/stress,

$$\boldsymbol{\varepsilon} = [\varepsilon_{11}, \varepsilon_{22}, 2\varepsilon_{12}]^T. \tag{4.47}$$

Under axisymmetric conditions, we have

$$\boldsymbol{\varepsilon} = [\varepsilon_{11}, \varepsilon_{22}, 2\varepsilon_{12}, \varepsilon_{33}]^T, \tag{4.48}$$

and, for three-dimensional analyses,

$$\boldsymbol{\varepsilon} = [\varepsilon_{11}, \varepsilon_{22}, \varepsilon_{33}, 2\varepsilon_{12}, 2\varepsilon_{23}, 2\varepsilon_{13}]^T. \tag{4.49}$$

The elastic law can then be written equivalently in terms of the arrays $\boldsymbol{\sigma}$ and $\boldsymbol{\varepsilon}$ as

$$\boldsymbol{\sigma} = \mathbf{D}^e\, \boldsymbol{\varepsilon}, \tag{4.50}$$

where the *elasticity matrix*, $\mathbf{D}^e$, has the general form

$$\mathbf{D}^e = 2G\,\mathbf{I}_S + A\,(K - \tfrac{2}{3}G)\,\mathbf{i}\,\mathbf{i}^T, \tag{4.51}$$

where $\mathbf{I}_S$ is the array of components of I_S (refer to Section D.2, starting on page 761) and $\mathbf{i}$ is the array representation of the second-order identity. For example, in axisymmetric problems we have

$$\mathbf{i} = [1 \quad 1 \quad 0 \quad 1]^T. \tag{4.52}$$

‡Refer to Appendix D for further details on the computational array representation of tensors in the finite element context.

The finite element equilibrium equation

From the definition of the strain-displacement matrix, it follows that

$$\boldsymbol{\varepsilon} = \mathbf{B}^g \, \mathbf{u}, \tag{4.53}$$

so that, in view of the linear elastic constitutive law for $\boldsymbol{\sigma}$, we have

$$\boldsymbol{\sigma} = \mathbf{D}^e \, \mathbf{B}^g \, \mathbf{u}. \tag{4.54}$$

With the substitution of the above relation into $(4.26)_1$, the global internal force vector reduces to the following linear function of $\mathbf{u}$:

$$\mathbf{f}^{\,\text{int}}(\mathbf{u}) = \left[\int_{{}^h\Omega} (\mathbf{B}^g)^T \, \mathbf{D}^e \, \mathbf{B}^g \, \mathrm{d}v \right] \mathbf{u}, \tag{4.55}$$

or, equivalently,

$$\mathbf{f}^{\,\text{int}}(\mathbf{u}) = \mathbf{K}\,\mathbf{u}, \tag{4.56}$$

where $\mathbf{K}$ is the *global stiffness matrix* assembled from the *element stiffnesses*

$$\mathbf{K} = \mathop{\mathsf{A}}_{e=1}^{n_{\text{elem}}} \mathbf{K}^{(e)}, \tag{4.57}$$

with

$$\mathbf{K}^{(e)} = \int_{\Omega^{(e)}} \mathbf{B}^T \, \mathbf{D}^e \, \mathbf{B} \, \mathrm{d}v. \tag{4.58}$$

Finally, with (4.56), the discrete boundary value problem defined by the equilibrium equation (4.25) is reduced to the solution of the following linear system of algebraic equations for the global nodal displacement vector $\mathbf{u}$:

$$\mathbf{K}\,\mathbf{u} = \mathbf{f}^{\,\text{ext}}. \tag{4.59}$$

In program `HYPLAS`, the solution of the above system is undertaken by the classical *frontal* method (Hinton and Owen, 1977; Irons, 1970). The frontal linear equation solver is coded in subroutine `FRONT`.

4.2. Path-dependent materials. The incremental finite element procedure

Let us now assume that the constitutive equations of the underlying material model are *path-dependent*; that is, the stress tensor is no longer a function of the instantaneous value of the infinitesimal strain only. It is now dependent on the *history* of strains to which the solid has been subjected. The stress tensor now is the solution of a constitutive initial value problem whose general form is stated in Problem 3.2 (page 76). Examples of path-dependent materials are the general elastoplastic and elasto-viscoplastic models that will be dealt with in Parts Two and Three of this book.

4.2.1. THE INCREMENTAL CONSTITUTIVE FUNCTION

Given a generic path-dependent model, the solution of the constitutive initial value problem for a given set of initial conditions is usually not known for complex strain paths $\varepsilon(t)$. Therefore, the use of an appropriate numerical algorithm for integration of the rate constitutive equations is an essential requirement in the finite element simulation of problems involving such models. The choice of a particular technique for integration of a constitutive law will be dependent on the characteristics of the model considered. In general, algorithms for integration of rate constitutive equations are obtained by adopting some kind of time (or pseudo-time) discretisation along with some hypothesis on the deformation path between adjacent time stations. Within the context of the purely mechanical theory, considering the time increment $[t_n, t_{n+1}]$ and given the set $\boldsymbol{\alpha}_n$ of internal variables at t_n, the strain tensor $\boldsymbol{\varepsilon}_{n+1}$ at time t_{n+1} must determine the stress $\boldsymbol{\sigma}_{n+1}$ uniquely through the integration algorithm. This requirement may be regarded as the numerical counterpart of the principle of thermodynamic determinism stated in Section 3.5.1. Such an algorithm defines an (approximate) *incremental constitutive function*, $\hat{\boldsymbol{\sigma}}$, for the stress tensor:

$$\boldsymbol{\sigma}_{n+1} = \hat{\boldsymbol{\sigma}}\,(\boldsymbol{\alpha}_n, \boldsymbol{\varepsilon}_{n+1}), \tag{4.60}$$

whose outcome, $\boldsymbol{\sigma}_{n+1}$, is expected to converge to the exact solution of the actual evolution problem as the strain increments are reduced. The numerical constitutive law is nonlinear in general and is path-*independent* within one increment; that is, within each increment $\boldsymbol{\sigma}_{n+1}$ is a function of $\boldsymbol{\varepsilon}_{n+1}$ alone (note that the argument $\boldsymbol{\alpha}_n$ is constant within $[t_n, t_{n+1}]$), analogous to a nonlinear elastic law. The integration algorithm also defines a similar incremental constitutive function for the internal variables of the model:

$$\boldsymbol{\alpha}_{n+1} = \hat{\boldsymbol{\alpha}}\,(\boldsymbol{\alpha}_n, \boldsymbol{\varepsilon}_{n+1}). \tag{4.61}$$

In the context of elastoplasticity, procedures such as the elastic predictor/return mapping algorithms, thoroughly discussed in Part Two of this book, provide concrete examples of numerical integration schemes for path-dependent constitutive laws.

4.2.2. THE INCREMENTAL BOUNDARY VALUE PROBLEM

Having defined the above generic incremental constitutive law, we can state the *incremental* (or *time-discrete*) version of the initial boundary value problem of Section 3.7.2 as follows.

Problem 4.1 (The infinitesimal incremental boundary value problem). *Given the set* $\boldsymbol{\alpha}_n$ *of internal variables at time* t_n*, find a displacement field* $\boldsymbol{u}_{n+1} \in \mathscr{K}_{n+1}$ *such that*

$$\int_{\Omega} [\hat{\boldsymbol{\sigma}}(\boldsymbol{\alpha}_n, \nabla^s \boldsymbol{u}_{n+1}) : \nabla^s \boldsymbol{\eta} - \boldsymbol{b}_{n+1} \cdot \boldsymbol{\eta}]\, \mathrm{d}v - \int_{\partial\Omega_t} \boldsymbol{t}_{n+1} \cdot \boldsymbol{\eta}\, \mathrm{d}a = 0 \tag{4.62}$$

for any $\boldsymbol{\eta} \in \mathscr{V}$*, where* $\mathbf{b}_{n+1}$ *and* $\mathbf{t}_{n+1}$ *are the body force and surface traction fields prescribed at time station* t_{n+1}*. The set* $\mathscr{K}_{n+1}$ *is defined as*

$$\mathscr{K}_{n+1} = \{\boldsymbol{u} : \Omega \rightarrow \mathscr{U} \mid \boldsymbol{u} = \bar{\boldsymbol{u}}_{n+1} \text{ on } \partial\Omega_u\}, \tag{4.63}$$

where $\bar{\boldsymbol{u}}_{n+1}$ *is the prescribed boundary displacement at* t_{n+1}.

4.2.3. THE NONLINEAR INCREMENTAL FINITE ELEMENT EQUATION

After a standard finite element discretisation of (4.62) the problem is reduced to the following. Find the nodal displacement vector $\mathbf{u}_{n+1}$ at time t_{n+1} such that the incremental finite element equilibrium equation

$$\mathbf{r}(\mathbf{u}_{n+1}) \equiv \mathbf{f}^{\text{int}}(\mathbf{u}_{n+1}) - \mathbf{f}^{\text{ext}}_{n+1} = \mathbf{0}, \tag{4.64}$$

is satisfied, where $\mathbf{f}^{\text{int}}(\mathbf{u}_{n+1})$ and $\mathbf{f}^{\text{ext}}_{n+1}$ are assembled from the element vectors

$$\begin{aligned}\mathbf{f}^{\text{int}}_{(e)} &= \int_{\Omega^{(e)}} \mathbf{B}^T\, \hat{\sigma}(\alpha_n, \varepsilon(\mathbf{u}_{n+1}))\, \mathrm{d}v\\ \mathbf{f}^{\text{ext}}_{(e)} &= \int_{\Omega^{(e)}} \mathbf{N}^T\, \boldsymbol{b}_{n+1}\, \mathrm{d}v + \int_{\partial\Omega^{(e)}_t} \mathbf{N}^T\, \boldsymbol{t}_{n+1}\, \mathrm{d}a.\end{aligned} \tag{4.65}$$

Equation (4.64) is generally nonlinear. The source of its nonlinearity is the nonlinearity of the incremental constitutive function that takes part in the definition of the element internal force vector above.

The incremental finite element scheme is summarised in Box 4.1 where the particular case of *proportional loading* (implemented in HYPLAS) is considered. Proportional loading is characterised by body force and surface traction fields given, at an arbitrary instant t_{n+1}, by

$$\begin{aligned}\boldsymbol{b}_{n+1} &= \lambda_{n+1}\, \tilde{\boldsymbol{b}}\\ \boldsymbol{t}_{n+1} &= \lambda_{n+1}\, \tilde{\boldsymbol{t}},\end{aligned} \tag{4.66}$$

where λ_{n+1} is the prescribed *load factor* at t_{n+1} and $\tilde{\boldsymbol{b}}$ and $\tilde{\boldsymbol{t}}$ are prescribed *constant* (in time) fields. In this case, the global external force vector reduces to

$$\mathbf{f}^{\text{ext}}_{n+1} = \lambda_{n+1}\, \bar{\mathbf{f}}^{\text{ext}}, \tag{4.67}$$

where $\bar{\mathbf{f}}^{\text{ext}}$ is computed only once at the beginning of the incremental procedure as the assembly of element vectors

$$\bar{\mathbf{f}}^{\text{ext}}_{(e)} = \int_{\Omega^{(e)}} \mathbf{N}^T\, \tilde{\boldsymbol{b}}\, \mathrm{d}v + \int_{\partial\Omega^{(e)}_t} \mathbf{N}^T\, \tilde{\boldsymbol{t}}\, \mathrm{d}a. \tag{4.68}$$

4.2.4. NONLINEAR SOLUTION. THE NEWTON–RAPHSON SCHEME

The Newton–Raphson algorithm is particularly attractive for the solution of the nonlinear incremental equation (4.64). Due to its quadratic rates of asymptotic convergence, this method tends to produce relatively robust and efficient incremental nonlinear finite element schemes.

Each iteration of the Newton–Raphson scheme comprises the solution of the linearised version of the discretised incremental equilibrium equation (4.64)[§] – or, equivalently, the discrete version of the linearised virtual work equation (C.12) (page 754). The finite element

[§]The reader is referred to Section 2.6 (from page 38) for details on linearisation of nonlinear problems.

Box 4.1. The incremental nonlinear finite element scheme.

(i) Assemble global external force, $\bar{\mathbf{f}}^{\text{ext}}$ and set up the proportional loading curve, $\lambda(t)$

(ii) Initialise increment counter, $i := 1$

(iii) Set load factor $\lambda_i := \lambda(t_i)$ at the end of the current interval $[t_{i-1}, t_i]$

(iv) Solve the nonlinear equation

$$\mathbf{f}^{\text{int}}(\mathbf{u}_i) - \lambda_i \bar{\mathbf{f}}^{\text{ext}} = \mathbf{0}$$

for $\mathbf{u}_i$ and obtain updated stresses and state variables, $\{\boldsymbol{\alpha}_i, \boldsymbol{\sigma}_i\}$

(v) IF $i < n_{\text{incr}}$ (prescribed number of increments) THEN

$i := i + 1$ and GOTO (iii)

(vi) EXIT

discretisation of (C.12) reads

$$\left\{\int_{^h\Omega} (\mathbf{B}^g)^T \mathbf{D}\, \mathbf{B}^g \,\mathrm{d}v\right\} \delta\mathbf{u} \cdot \boldsymbol{\eta} = -\left\{\int_{^h\Omega} [(\mathbf{B}^g)^T\, \boldsymbol{\sigma} - (\mathbf{N}^g)^T\, \boldsymbol{b}] \,\mathrm{d}v \right.$$
$$\left. - \int_{\partial^h\Omega_t} (\mathbf{N}^g)^T\, \boldsymbol{t} \,\mathrm{d}a\right\} \cdot \boldsymbol{\eta}, \quad \forall \boldsymbol{\eta} \in {}^h\mathscr{V}. \tag{4.69}$$

By the same argument leading to (4.25), the above gives

$$\left\{\int_{^h\Omega} (\mathbf{B}^g)^T \mathbf{D}\, \mathbf{B}^g \,\mathrm{d}v\right\} \delta\mathbf{u} = -\left\{\int_{^h\Omega} [(\mathbf{B}^g)^T\, \boldsymbol{\sigma} - (\mathbf{N}^g)^T\, \boldsymbol{b}] \,\mathrm{d}v \right.$$
$$\left. - \int_{\partial^h\Omega_t} (\mathbf{N}^g)^T\, \boldsymbol{t} \,\mathrm{d}a\right\}. \tag{4.70}$$

At a state defined by the global displacement vector $\mathbf{u}_{n+1}^{(k-1)}$, the typical iteration (k) of the Newton–Raphson scheme consists of solving the linear system of equations

$$\mathbf{K}_T\, \delta\mathbf{u}^{(k)} = -\mathbf{r}^{(k-1)}, \tag{4.71}$$

for $\delta\mathbf{u}^{(k)}$, where we have defined the *residual* (or *out-of-balance force*) vector

$$\mathbf{r}^{(k-1)} \equiv \mathbf{f}^{\text{int}}(\mathbf{u}_{n+1}^{(k-1)}) - \mathbf{f}_{n+1}^{\text{ext}}, \tag{4.72}$$

and $\mathbf{K}_T$ is the *global tangent stiffness matrix*:

$$\mathbf{K}_T \equiv \int_{^h\Omega} (\mathbf{B}^g)^T \mathbf{D}\, \mathbf{B}^g \,\mathrm{d}v = \left.\frac{\partial \mathbf{r}}{\partial \mathbf{u}_{n+1}}\right|_{\mathbf{u}_{n+1}^{(k-1)}}. \tag{4.73}$$

With the solution $\delta\mathbf{u}^{(k)}$ of the linear system (4.71) at hand, we apply the Newton correction to the global displacement

$$\mathbf{u}_{n+1}^{(k)} = \mathbf{u}_{n+1}^{(k-1)} + \delta\mathbf{u}^{(k)}, \tag{4.74}$$

or, in terms of displacement *increments*,

$$\mathbf{u}_{n+1}^{(k)} = \mathbf{u}_n + \Delta\mathbf{u}^{(k)}, \tag{4.75}$$

where $\Delta\mathbf{u}^{(k)}$ is the *incremental displacement vector*:

$$\Delta\mathbf{u}^{(k)} = \Delta\mathbf{u}^{(k-1)} + \delta\mathbf{u}^{(k)}. \tag{4.76}$$

The method is schematically illustrated in Figure 4.6. The Newton–Raphson iterations are repeated until after some iteration (m), the following *convergence criterion* is satisfied:

$$\frac{|\mathbf{r}^{(m)}|}{|\mathbf{f}_{n+1}^{\,\mathrm{ext}}|} \leq \epsilon_{\mathrm{tol}}, \tag{4.77}$$

where ϵ_{tol} is a sufficiently small specified *equilibrium convergence tolerance*. The corresponding displacement vector, $\mathbf{u}_{n+1}^{(m)}$, is then accepted as sufficiently close to the solution of (4.64)

$$\mathbf{u}_{n+1} := \mathbf{u}_{n+1}^{(m)}. \tag{4.78}$$

To start up the Newton–Raphson iterations, we need an initial guess, $\mathbf{u}_{n+1}^{(0)}$. The initial guess is usually taken as the converged (equilibrium) displacement vector at the end of the previous increment,

$$\mathbf{u}_{n+1}^{(0)} = \mathbf{u}_n \quad \text{or} \quad \Delta\mathbf{u}_{n+1}^{(0)} = \mathbf{0}. \tag{4.79}$$

The overall Newton–Raphson algorithm for solution of the nonlinear finite element equations is summarised in Box 4.2 in pseudo-code format. The procedure is implemented in program HYPLAS. To help familiarise the reader with the program, the main subroutines of HYPLAS associated with some of the items listed in Box 4.2 are named within square brackets.

4.2.5. THE CONSISTENT TANGENT MODULUS

The global tangent stiffness defined by (4.73) is the assembly of the *element tangent stiffness* matrices

$$\mathbf{K}_T^{(e)} = \int_{\Omega^{(e)}} \mathbf{B}^T\, \mathbf{D}\, \mathbf{B}\, \mathrm{d}v, \tag{4.80}$$

where $\mathbf{D}$ is the *consistent tangent matrix* – the matrix form of the fourth-order *consistent tangent operator*

$$\mathsf{D} \equiv \left.\frac{\partial\hat{\boldsymbol{\sigma}}}{\partial\boldsymbol{\varepsilon}_{n+1}}\right|_{\boldsymbol{\varepsilon}_{n+1}^{(k-1)}}. \tag{4.81}$$

This tensor possesses the symmetries

$$\mathsf{D}_{ijkl} = \mathsf{D}_{jikl} = \mathsf{D}_{ijlk}. \tag{4.82}$$

The consistent tangent operator is the derivative of the *incremental* constitutive function $\hat{\boldsymbol{\sigma}}$. This *generally implicit* function is typically defined by some numerical algorithm for integration of the rate constitutive equations of the model. The full (exact) linearisation of the finite element equations in the context of path-dependent materials, including the exact

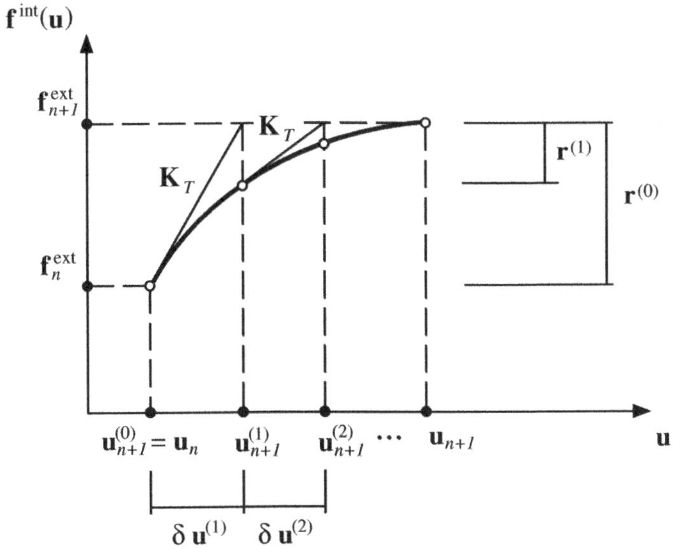

Figure 4.6. The Newton–Raphson algorithm for the incremental finite element equilibrium equation.

linearisation of the incremental constitutive function, has been first addressed by Nagtegaal (1982). In particular, this author has emphasised the need for linearisation of the incremental stress updating procedure, rather than to appeal to the rate stress–strain tangential relation (commonly used at the time), in order to achieve quadratic rates of asymptotic convergence in the iterative solution of the finite element equilibrium equations. This concept was later formalised by Simo and Taylor (1985) who coined the term *consistent* tangent to refer to the tangent operator *consistent* with the relevant numerical algorithm for integration of the path-dependent rate constitutive equations. The derivation of consistent tangent operators is thoroughly discussed in Part Two of this book, in the context of elastoplasticity and elastoviscoplasticity. We remark that the full Newton–Raphson scheme, which relies on consistent tangent operators, is available for all material models implemented in `HYPLAS`.

4.2.6. ALTERNATIVE NONLINEAR SOLUTION SCHEMES

Other iterative methods can be used to solve the nonlinear incremental finite element equations. A straightforward alternative to the above full Newton–Raphson scheme are the so-called *modified Newton methods*. These methods consist of replacing the tangential stiffness $\mathbf{K}_T$, which in the standard method is updated in every iteration, by a constant counterpart. Traditional approaches include (Owen and Hinton (1980) and Zienkiewicz and Taylor (2000))

(i) the use of the *initial* tangent stiffness throughout all increments;

(ii) the use of a constant stiffness *within each increment*. The stiffness is updated at the beginning of every increment;

Box 4.2. The Newton–Raphson scheme for solution of the incremental nonlinear finite element equation (infinitesimal strains).

(i) $k := 0$. Set initial guess and residual

$$\mathbf{u}_{n+1}^{(0)} := \mathbf{u}_n; \qquad \mathbf{r} := \mathbf{f}^{\text{int}}(\mathbf{u}_n) - \lambda_{n+1}\bar{\mathbf{f}}^{\text{ext}}$$

(ii) Compute consistent tangent matrices [MATICT]

$$\mathbf{D} := \partial\hat{\sigma}/\partial\varepsilon_{n+1}$$

(iii) Assemble element tangent stiffness matrices [ELEIST, STSTD2]

$$\mathbf{K}_T^{(e)} := \sum_{i=1}^{n_{\text{gausp}}} w_i \, j_i \, \mathbf{B}_i^T \, \mathbf{D}_i \, \mathbf{B}_i$$

(iv) $k := k + 1$. Assemble global stiffness and solve for $\delta\mathbf{u}^{(k)}$ [FRONT]

$$\mathbf{K}_T \, \delta\mathbf{u}^{(k)} = -\mathbf{r}^{(k-1)}$$

(v) Apply Newton correction to displacements [UPCONF]

$$\mathbf{u}_{n+1}^{(k)} := \mathbf{u}_{n+1}^{(k-1)} + \delta\mathbf{u}^{(k)}$$

(vi) Update strains [IFSTD2]

$$\varepsilon_{n+1}^{(k)} := \mathbf{B}\,\mathbf{u}_{n+1}^{(k)}$$

(vii) Use constitutive integration algorithm to update stresses and other state variables [MATISU]

$$\boldsymbol{\sigma}_{n+1}^{(k)} := \hat{\boldsymbol{\sigma}}(\boldsymbol{\alpha}_n, \boldsymbol{\varepsilon}_{n+1}^{(k)}); \qquad \boldsymbol{\alpha}_{n+1}^{(k)} := \hat{\boldsymbol{\alpha}}(\boldsymbol{\alpha}_n, \boldsymbol{\varepsilon}_{n+1}^{(k)})$$

(viii) Compute element internal force vectors [INTFOR, IFSTD2]

$$\mathbf{f}_{(e)}^{\text{int}} := \sum_{i=1}^{n_{\text{gausp}}} w_i \, j_i \, \mathbf{B}_i^T \, \boldsymbol{\sigma}_{n+1}^{(k)}\Big|_i$$

(ix) Assemble global internal force vector and update residual [CONVER]

$$\mathbf{r} := \mathbf{f}^{\text{int}} - \lambda_{n+1}\bar{\mathbf{f}}^{\text{ext}}$$

(x) Check for convergence [CONVER]

IF $\|\mathbf{r}\|/\|\mathbf{f}^{\text{ext}}\| \leq \epsilon_{\text{tol}}$ THEN set $(\cdot)_{n+1} := (\cdot)_{n+1}^{(k)}$ and EXIT

ELSE GOTO (ii)

(iii) a variant of method (ii) where the stiffness matrix is updated after a certain number of iterations within each increment.

For such schemes, the stiffness matrix does not need to be computed or factorised whenever it is reused. Some procedures of this type are available in the HYPLAS program. The convergence rates of such methods are far slower than the quadratic rates of the classical Newton algorithm and numerical experience shows that much faster solutions are obtained generally with the full Newton–Raphson scheme.

Other alternatives include the *quasi-Newton methods*. Such methods, whose roots are in the theory of optimisation (Haftka *et al.*, 1990), require neither the direct computation of the tangent stiffness nor its inversion (linear system solution). Instead, at a new iteration, an updated approximation of the inverse of the stiffness matrix is obtained which depends on the residual and displacement vectors at the end of the previous two iterations and the inverse stiffness approximation employed in the previous iteration of the algorithm. The *Broyden–Fletcher–Goldfarb–Shanno* algorithm (BFGS) is a particularly popular quasi-Newton scheme. Its application in the finite element context has been introduced by Matthies and Strang (1979). At the beginning of an iteration (k), the approximate inverse stiffness of the BFGS scheme is updated by the formula

$$\mathbf{K}_{(k)}^{-1} = \mathbf{K}_{(k-1)}^{-1} - \frac{\mathbf{v}\mathbf{w}^T\mathbf{K}_{(k-1)}^{-1}}{\mathbf{v}^T\mathbf{w}} - \frac{\mathbf{K}_{(k-1)}^{-1}\mathbf{w}\mathbf{v}^T}{\mathbf{v}^T\mathbf{w}} + \left(1 + \frac{\mathbf{w}^T\mathbf{K}_{(k-1)}^{-1}\mathbf{w}}{\mathbf{v}^T\mathbf{w}}\right)\frac{\mathbf{v}\mathbf{w}^T}{\mathbf{v}^T\mathbf{w}}, \tag{4.83}$$

where

$$\mathbf{v} = \mathbf{u}^{(k-1)} - \mathbf{u}^{(k-2)}, \quad \mathbf{w} = \mathbf{r}^{(k-1)} - \mathbf{r}^{(k-2)}. \tag{4.84}$$

These methods can be useful, for instance, when the derivation of the tangent operator (4.81) is difficult – a situation that may arise in dealing with complex material models/integration algorithms. Again, the rates of convergence are much slower than those produced by the standard Newton algorithm.

4.2.7. NON-INCREMENTAL PROCEDURES FOR PATH-DEPENDENT MATERIALS

The framework discussed above for the solution of quasi-static solid mechanics problems with path-dependent material models relies crucially upon the use of numerical integration procedures whereby the numerical solution of the constitutive initial value problem is undertaken in sequential steps in time. Such procedures generate approximate (incremental) constitutive functions for the stress tensor with general form (4.60), whose substitution into the generic initial boundary value problem of Section 3.7.2 (page 81) results in the incremental version stated in Problem 4.1 (page 95). The numerical solution of the initial boundary value problem in this case is obtained by splitting the considered time interval, say $[t_0, T]$, into a number of sufficiently small time steps and then solving sequentially, for each step, the corresponding incremental boundary value problem whose unknown is the displacement field over the domain Ω at the end of the time step (instant t_{n+1} in Problem 4.1).

An interesting alternative to the above general incremental procedure deserves special mention: the so-called *Large Time Increment Method*, or *LATIN* Method for short, originally proposed by Ladevèze (1984, 1989, 1999) is a *non-incremental* iterative procedure relying on a radically different approach to the treatment of the problem. Its computational implementation is considerably different from that of the incremental approach explored and explained in detail throughout this book. The LATIN method falls outside the scope of the present text. The interested reader is referred to Ladevèze (1999) for further details. Essentially, this method generates a sequence of *histories* $\{\boldsymbol{v}^i, \boldsymbol{\sigma}^i\}$ of velocity and stress fields over $\Omega \times [t_0, T]$ – note that the *entire* time domain considered is covered by each function of the sequence – that converges to a pair $\{\boldsymbol{v}^*, \boldsymbol{\sigma}^*\}$ whose stress satisfies equilibrium (over the entire time domain $[t_0, T]$) and is related to the strain history resulting from $\boldsymbol{v}^*$ through the constitutive equations of the model. Each iteration i of the LATIN method produces a new velocity history, $\boldsymbol{v}^i$, of the form

$$\boldsymbol{v}^i(\boldsymbol{x},t) = \boldsymbol{v}^{i-1}(\boldsymbol{x},t) + \sum_{k=1}^{p} g_k(t)\, \boldsymbol{w}_k(\boldsymbol{x}), \tag{4.85}$$

where g_k are suitably chosen scalar-valued basis functions defined over $[t_0, T]$ and $\boldsymbol{w}_k$ are vector-valued functions defined over Ω. The functions $\boldsymbol{w}_k$ are generated by conventional finite element interpolation and their corresponding nodal values are unknowns at each iteration of the method. Analogous approximations are used to describe the sequence of strain rate and stress rate histories. The overall accuracy of the history approximations depends crucially upon the choice of the total number, p, of terms used to describe the iterative history differences. In contrast to the conventional incremental procedure, the single increment of the LATIN Method comprising the interval $[t_0, T]$ may contain several cycles of a cyclic load. This is illustrated, for instance, by Boisse *et al.* (1990, 1989) in elastoplastic applications. Further applications are found in Boucard *et al.* (1997), Cognard and Ladèveze (1993), Cognard *et al.* (1999), Ladevèze and Perego (2000) and Dureisseix *et al.* (2003).

4.3. Large strain formulation

We now focus our attention on the use of the Finite Element Method in the solution of solid mechanics problems involving *finite* deformations and strains. The starting point here is the finite strain version of the Principle of Virtual Work and the corresponding initial boundary value problem stated at the end of Chapter 3.

4.3.1. THE INCREMENTAL CONSTITUTIVE FUNCTION

At the outset, we shall assume that the underlying material model is *path-dependent*. Thus, the comments made in Section 4.2.1 about the need for a numerical integration scheme to update the stresses and other state variables of the model apply equally to the present case. Analogously to (4.60) the corresponding algorithmic incremental constitutive function is defined in the following general format

$$\boldsymbol{\sigma}_{n+1} = \hat{\boldsymbol{\sigma}}\,(\boldsymbol{\alpha}_n, \boldsymbol{F}_{n+1}), \tag{4.86}$$

where $\boldsymbol{F}_{n+1}$ is the *deformation gradient* prescribed at the end of the standard interval $[t_n, t_{n+1}]$ and $\boldsymbol{\sigma}$ is now the *Cauchy* stress tensor; that is, given the set $\boldsymbol{\alpha}_n$ of internal variables, the prescribed deformation gradient $\boldsymbol{F}_{n+1}$ will determine the Cauchy stress tensor uniquely through the incremental constitutive function $\hat{\boldsymbol{\sigma}}$ defined by means of some algorithm for numerical integration of the constitutive equations of the model. An equivalent incremental function can be defined for the *Kirchhoff* stress

$$\boldsymbol{\tau}_{n+1} = \hat{\boldsymbol{\tau}}\,(\boldsymbol{\alpha}_n, \boldsymbol{F}_{n+1}) = J_{n+1}\, \hat{\boldsymbol{\sigma}}\,(\boldsymbol{\alpha}_n, \boldsymbol{F}_{n+1}), \tag{4.87}$$

where $J_{n+1} = \det[\boldsymbol{F}_{n+1}]$.

It is important to note that path-*independent* (finite elasticity) laws can also be written in incremental form as particular cases of the above functions. In such cases, no internal variables are needed and $\boldsymbol{\sigma}_{n+1}$ (or $\boldsymbol{\tau}_{n+1}$) can usually be computed by simple function evaluation (rather than by some numerical integration algorithm). Thus, the framework presented in this section is equally applicable to finite elasticity models. Finite strain path-dependent models (including constitutive integration algorithms) as well as hyperelastic theories are discussed in Part Three of this book.

4.3.2. THE INCREMENTAL BOUNDARY VALUE PROBLEM

The finite strain incremental boundary value problem is obtained by introducing the above time-discrete constitutive law into the original time-continuum initial boundary value problem of Section 3.7.1 (starting on page 79). The problem – here considered in its *spatial* version – is stated in the following.

Problem 4.2 (The finite strain incremental boundary value problem). *Given the field* $\boldsymbol{\alpha}_n$ *at time* t_n *and given the body forces and surface traction fields at* t_{n+1}*, find a kinematically admissible configuration* $\varphi_{n+1}(\Omega) \in \mathscr{K}_{n+1}$ *such that the virtual work equation*

$$\int_{\varphi_{n+1}(\Omega)} [\hat{\boldsymbol{\sigma}}(\boldsymbol{\alpha}_n, \boldsymbol{F}_{n+1}) : \nabla_x^s \boldsymbol{\eta} - \boldsymbol{b}_{n+1} \cdot \boldsymbol{\eta}] \, \mathrm{d}v - \int_{\varphi_{n+1}(\partial\Omega_t)} \boldsymbol{t}_{n+1} \cdot \boldsymbol{\eta} \, \mathrm{d}a = 0, \tag{4.88}$$

is satisfied for any $\boldsymbol{\eta} \in \mathscr{V}$*, where* φ_{n+1} *is the deformation map at* t_{n+1}

$$\boldsymbol{x}_{n+1} = \varphi_{n+1}(\boldsymbol{p}) = \boldsymbol{p} + \boldsymbol{u}_{n+1}(\boldsymbol{p}), \tag{4.89}$$

and

$$\boldsymbol{F}_{n+1} = \nabla_p \varphi_{n+1} = \boldsymbol{I} + \nabla_p \boldsymbol{u}_{n+1}. \tag{4.90}$$

4.3.3. THE FINITE ELEMENT EQUILIBRIUM EQUATION

The finite element discretisation of the above equation is completely analogous to the discretisation of its infinitesimal counterpart. The discrete problem consists in finding a kinematically admissible global displacement vector $\mathbf{u}_{n+1}$ that satisfies the standard incremental equilibrium equation

$$\mathbf{r}(\mathbf{u}_{n+1}) \equiv \mathbf{f}^{\,\mathrm{int}}(\mathbf{u}_{n+1}) - \mathbf{f}^{\,\mathrm{ext}}_{n+1} = \mathbf{0}, \tag{4.91}$$

where, now, the internal and external force vectors are defined as

$$\begin{aligned} \mathbf{f}^{\,\mathrm{int}}_{(e)} &= \int_{\varphi_{n+1}(\Omega^{(e)})} \mathbf{B}^T \, \hat{\boldsymbol{\sigma}}(\boldsymbol{\alpha}_n, \boldsymbol{F}(\mathbf{u}_{n+1})) \, \mathrm{d}v \\ \mathbf{f}^{\,\mathrm{ext}}_{(e)} &= \int_{\varphi_{n+1}(\Omega^{(e)})} \mathbf{N}^T \, \boldsymbol{b}_{n+1} \, \mathrm{d}v + \int_{\varphi_{n+1}(\partial\Omega_t^{(e)})} \mathbf{N}^T \, \boldsymbol{t}_{n+1} \, da. \end{aligned} \tag{4.92}$$

The $\mathbf{B}$-matrix above is the *spatial discrete symmetric gradient operator*. It has the same format as that of its small strain counterpart (such as, for example, that given by (4.30) for the plane stress and plane strain cases), but its shape function derivatives are *spatial* derivatives; that is, derivatives with respect to the *spatial* coordinates of the finite element mesh (at the deformed configuration defined by $\mathbf{u}_{n+1}$).

4.3.4. LINEARISATION. THE CONSISTENT SPATIAL TANGENT MODULUS

Analogously to the infinitesimal deformations case of Section 4.2.4, the linearised version of (4.91), to be used in the Newton–Raphson algorithm for solution of the large deformation incremental equilibrium equation, can be obtained by applying a finite element discretisation to the linearised form (C.32) (page 757) of the virtual work equation (4.88).

Starting from (C.32) and following the same arguments as those resulting in (4.70) we obtain the discrete form of the linearised virtual work equation for large strain problems in spatial description:¶

$$\left\{\int_{\varphi({}^h\Omega)} (\mathbf{G}^g)^T \mathbf{a}\, \mathbf{G}^g \,\mathrm{d}v\right\}\delta\mathbf{u} = -\left\{\int_{\varphi({}^h\Omega)} [(\mathbf{B}^g)^T\, \boldsymbol{\sigma} - (\mathbf{N}^g)^T\, \boldsymbol{b}]\,\mathrm{d}v - \int_{\varphi(\partial^h\Omega)} (\mathbf{N}^g)^T\, \boldsymbol{t}\,\mathrm{d}a\right\}, \tag{4.93}$$

where $\mathbf{a}$ is the matrix of components of the spatial tangent modulus (C.31) ordered according to the convention described in Section D.2.1 (starting page 763). The matrix $\mathbf{G}^g$ is the *global discrete spatial gradient operator*. In plane strain and plane stress analyses, for example, it is given by

$$\mathbf{G}^g = \begin{bmatrix} N^g_{1,1} & 0 & N^g_{2,1} & 0 & \cdots & N^g_{n_{\text{poin}},1} & 0 \\ 0 & N^g_{1,1} & 0 & N^g_{2,1} & \cdots & 0 & N^g_{n_{\text{poin}},1} \\ N^g_{1,2} & 0 & N^g_{2,2} & 0 & \cdots & N^g_{n_{\text{poin}},2} & 0 \\ 0 & N^g_{1,2} & 0 & N^g_{2,2} & \cdots & 0 & N^g_{n_{\text{poin}},2} \end{bmatrix}. \tag{4.94}$$

Its multiplication by, say, the global vector $\delta\mathbf{u}$ gives the array of components of the spatial gradient of $\delta\mathbf{u}$ ordered according to the convention of Section D.2.1.

Similarly to the infinitesimal strain case, the generic Newton–Raphson iteration (k) here requires the solution of the standard linear system for $\delta\mathbf{u}$:

$$\mathbf{K}_T\, \delta\mathbf{u}^{(k)} = -\mathbf{r}^{(k-1)}, \tag{4.95}$$

where the global tangent stiffness matrix $\mathbf{K}_T$ corresponds to the bracketed term on the left-hand side of (4.93). This matrix is constructed in practice by assembling the element tangent stiffness matrices defined as

$$\mathbf{K}_T^{(e)} = \int_{\varphi^{(k)}_{n+1}(\Omega^{(e)})} \mathbf{G}^T\, \mathbf{a}\, \mathbf{G}\,\mathrm{d}v, \tag{4.96}$$

where $\mathbf{G}$ is the *element* discrete spatial gradient operator which, in plane stress/strain analyses, has the format

$$\mathbf{G} = \begin{bmatrix} N^{(e)}_{1,1} & 0 & N^{(e)}_{2,1} & 0 & \cdots & N^{(e)}_{n_{\text{node}},1} & 0 \\ 0 & N^{(e)}_{1,1} & 0 & N^{(e)}_{2,1} & \cdots & 0 & N^{(e)}_{n_{\text{node}},1} \\ N^{(e)}_{1,2} & 0 & N^{(e)}_{2,2} & 0 & \cdots & N^{(e)}_{n_{\text{node}},2} & 0 \\ 0 & N^{(e)}_{1,2} & 0 & N^{(e)}_{2,2} & \cdots & 0 & N^{(e)}_{n_{\text{node}},2} \end{bmatrix}. \tag{4.97}$$

¶We remark that this expression is valid when the external loads $\boldsymbol{b}$ and $\boldsymbol{t}$ are configuration-*independent*, i.e. they do not depend on the displacement $\boldsymbol{u}$. Its extension to account for configuration-dependent loads is briefly addressed in Section 4.3.6.

Box 4.3. The Newton–Raphson scheme for the *large strain* incremental nonlinear finite element equation. Modifications to Box 4.2.

(ii) Compute consistent spatial tangent modulus matrix [MATICT]

$$\mathsf{a}_{ijkl} := \frac{1}{J} \frac{\partial \hat{\tau}_{ij}}{\partial F_{km}} F_{lm} - \sigma_{il}\, \delta_{jk}$$

(iii) Assemble element tangent stiffness matrices [ELEIST, STSTD2]

$$\mathbf{K}_T^{(e)} := \sum_{i=1}^{n_{\text{gausp}}} w_i\, j_i\, \mathbf{G}_i^T\, \mathbf{a}_i\, \mathbf{G}_i$$

(vi) Update deformation gradient [IFSTD2]

$$\boldsymbol{F}_{n+1}^{(k)} := (\boldsymbol{I} - \nabla_x \boldsymbol{u}_{n+1}^{(k)})^{-1}$$

(vii) Use constitutive integration algorithm to update stresses and other state variables [MATISU]

$$\boldsymbol{\sigma}_{n+1}^{(k)} := \hat{\boldsymbol{\sigma}}(\boldsymbol{\alpha}_n, \boldsymbol{F}_{n+1}^{(k)}); \qquad \boldsymbol{\alpha}_{n+1}^{(k)} := \hat{\boldsymbol{\alpha}}(\boldsymbol{\alpha}_n, \boldsymbol{F}_{n+1}^{(k)})$$

The consistent spatial tangent modulus

Let us recall here the expression for the Cartesian components of the *consistent spatial tangent modulus* derived in Section C.2.2:

$$\mathsf{a}_{ijkl} = \frac{1}{J} \frac{\partial \tau_{ij}}{\partial F_{km}} F_{lm} - \sigma_{il}\, \delta_{jk}. \tag{4.98}$$

On the right-hand side of the above expression, the only terms that depend on the constitutive equations of the model are the derivatives of the Kirchhoff stress tensor components. In the general case considered here, the Kirchhoff stress tensor is the outcome of the *algorithmic* incremental constitutive function (4.87). As in the infinitesimal case, the incremental function is usually *implicit* and is defined by the particular algorithm used to integrate the constitutive equations of the model. Thus, the derivatives $\partial \tau_{ij}/\partial F_{km}$ are the components of the derivative of this generally implicit incremental constitutive function, that is,

$$\frac{\partial \tau_{ij}}{\partial F_{km}} = \left[\frac{\partial \hat{\boldsymbol{\tau}}}{\partial \boldsymbol{F}_{n+1}} \right]_{ijkm}. \tag{4.99}$$

The overall Newton procedure in the finite strain case is completely analogous to that shown in Box 4.2 for the infinitesimal theory. The only necessary change is the replacement of some of the items of Box 4.2 with the finite strain counterparts listed in Box 4.3.

Examples of path-dependent material implementations where $\boldsymbol{\tau}$ is defined by an implicit algorithmic function are given in Part Three of this book. The derivation of explicit expressions for the above implicit function derivative may become quite intricate but involve nothing more than standard concepts of linearisation. Finite elasticity (also addressed in Part Three) is a particular instance where $\boldsymbol{\tau}_{n+1}$ is obtained by direct function evaluation. In such a case, $\hat{\boldsymbol{\tau}}$ is an *explicit* function of the deformation gradient and its derivative can, as a general rule, be derived in a relatively straightforward manner.

4.3.5. MATERIAL AND GEOMETRIC STIFFNESSES

In finite element computations, the tangent stiffness (4.96) is frequently split into the so-called *material* and *geometric* stiffnesses

$$\mathbf{K}_T^{(e)} = \mathbf{K}_M^{(e)} + \mathbf{K}_G^{(e)}, \tag{4.100}$$

defined by

$$\begin{aligned} \mathbf{K}_M^{(e)} &= \int_{\varphi_{n+1}^{(k)}(\Omega^{(e)})} \mathbf{B}^T \, \mathbf{c} \, \mathbf{B} \, \mathrm{d}v \\ \mathbf{K}_G^{(e)} &= \int_{\varphi_{n+1}^{(k)}(\Omega^{(e)})} \mathbf{G}^T \, \mathbf{S} \, \mathbf{G} \, \mathrm{d}v, \end{aligned} \tag{4.101}$$

where $\mathbf{c}$ is the array representation of the fourth-order tensor

$$\mathsf{c}_{ijkl} = \mathsf{a}_{ijkl} - \sigma_{jl} \, \delta_{ik}, \tag{4.102}$$

and $\mathbf{S}$ is the array representation of the tensor

$$\mathsf{S}_{ijkl} = \sigma_{jl} \, \delta_{ik}. \tag{4.103}$$

As the infinitesimal tangent operator (4.81), the tensor c has the symmetries

$$\mathsf{c}_{ijkl} = \mathsf{c}_{jikl} = \mathsf{c}_{ijlk}. \tag{4.104}$$

The material stiffness matrix $(4.101)_1$ has the same format as that of the infinitesimal stiffness matrix (4.80) except that, in the finite strain case, the matrix form of the tensor defined by (4.102) replaces the matrix form of the infinitesimal consistent tangent operator. In computational terms, this allows the subroutines for computation of the infinitesimal element tangent stiffness matrix to be reused to compute the material stiffness of the finite strain case. If this approach is adopted, the full tangent stiffness under finite strains can be obtained by simply adding the geometrical term $(4.101)_2$ to the material stiffness. We remark, however, that in program `HYPLAS` the form (4.96) is adopted instead; that is, under finite strains, we firstly compute the matrix form of tensor a and then obtain the element stiffness by performing a single computation with (4.96).

4.3.6. CONFIGURATION-DEPENDENT LOADS. THE LOAD-STIFFNESS MATRIX

If the external load depends on the configuration of the body, that is, if

$$\mathbf{f}_{n+1}^{\text{ext}} = \mathbf{f}^{\text{ext}}(\mathbf{u}_{n+1}), \tag{4.105}$$

then the tangent stiffness, that takes part in (4.95), is replaced by

$$\mathbf{K}_T + \mathbf{K}_L,$$

where the added contribution, $\mathbf{K}_L$, is the so-called *load-correction* or *load-stiffness* matrix. The load stiffness is the derivative of the external force vector with respect to the displacement vector:

$$\mathbf{K}_L = \left.\frac{\partial \mathbf{f}^{\text{ext}}}{\partial \mathbf{u}_{n+1}}\right|_{\mathbf{u}_{n+1}^{(k-1)}}. \tag{4.106}$$

The consideration of configuration-dependence of external loads becomes important in situations where substantial changes in direction and/or intensity of loads may result from the deformation of the loaded body. A typical example is given by pressure loading of rubber membranes (see some of the numerical examples of Section 13.6). The explicit derivation of the load stiffness will not be addressed in this book. The reader is referred to Schweizerhof and Ramm (1984) for details of derivation (see also Argyris and Symeonidis, 1981).

4.4. Unstable equilibrium. The arc-length method

Many finite deformation problems of practical interest are characterised by the existence of *unstable* equilibrium configurations. Common examples appear in *buckling* instability and problems involving *snap-through* or *snap-back* phenomena. Let us consider the case of proportional loading. An *equilibrium path* is formed by the set of all load factor-displacement pairs

$$\{\lambda, \boldsymbol{u}\}$$

which satisfy equilibrium. Typical equilibrium paths associated with snap-through and snap-back behaviour are illustrated in Figure 4.7. In such situations, the standard *load-controlled* finite element scheme of the previous section, where the displacement vector is found for a *prescribed* load factor, cannot be used in general once a limit point (points A of Figure 4.7) is reached. To trace an equilibrium path beyond limit points we need to resort to *continuation techniques*, of which the *arc-length* method appears to be the most popular (Crisfield, 1981, 1983, 1991, 1997; de Souza Neto and Feng, 1999; Feng *et al.*, 1997, 1995, 1996; Ramm, 1981; Riks, 1972, 1979; Wempner, 1971). A description of the arc-length method is given in this section. The method is implemented in program HYPLAS. Readers who wish to skip the description of this technique are referred directly to Box 4.4 (page 109), where the combined Newton–Raphson/cylindrical arc-length scheme provided in HYPLAS is summarised in pseudo-code format. The names of the main routines involved in the computational implementation of the most relevant procedures of Box 4.4 are shown within square brackets.

4.4.1. THE ARC-LENGTH METHOD

Let us consider, again, the standard interval $[t_n, t_{n+1}]$ and let

$$\Delta\lambda \equiv \lambda_{n+1} - \lambda_n \tag{4.107}$$

be the corresponding *incremental load factor*. To derive the arc-length method, we allow $\Delta\lambda$ to become a *variable* and redefine the residual equation (4.91) as

$$\mathbf{r}(\mathbf{u}_{n+1}, \Delta\lambda) \equiv \mathbf{f}^{\text{int}}(\mathbf{u}_{n+1}) - (\lambda_n + \Delta\lambda)\,\bar{\mathbf{f}}^{\text{ext}} = \mathbf{0}. \tag{4.108}$$

The arc-length method consists of adding an extra constraint to the above augmented residual equation so as to limit the 'length' of the incremental solution.

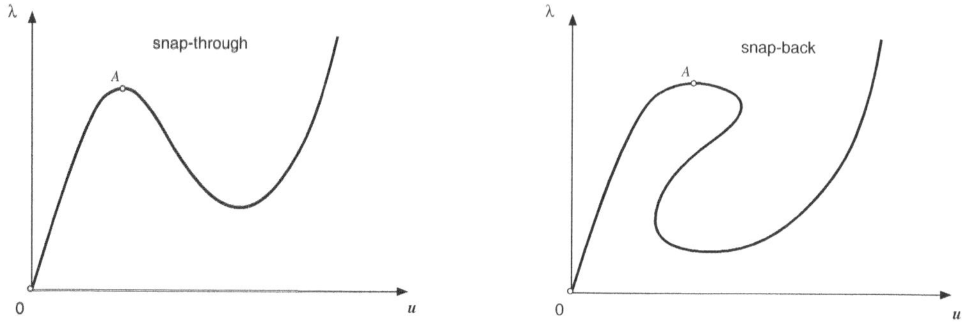

Figure 4.7. Unstable equilibrium. Snap-through and snap-back behaviour.

The spherical arc-length method

For the so-called *spherical* arc-length method, the constraint equation has the general format

$$\Delta\mathbf{u}^T\Delta\mathbf{u} + \Delta\lambda^2\,\psi^2\,\bar{\mathbf{f}}^{\text{ext}\,T}\,\bar{\mathbf{f}}^{\text{ext}} - l^2 = 0, \tag{4.109}$$

where $\Delta\mathbf{u}$ and $\Delta\lambda$ are converged incremental quantities, l is a prescribed incremental solution length and ψ is a prescribed scaling parameter.

The cylindrical arc-length method

For the more widely used *cylindrical* arc-length method (implemented in HYPLAS), the scaling parameter ψ is set to zero and the constraint equation reads simply

$$\Delta\mathbf{u}^T\Delta\mathbf{u} = l^2. \tag{4.110}$$

In this case, the constraint equation requires that the Euclidean norm of the converged incremental displacement be l, i.e. the equilibrium solution at the end of the increment lies at an intersection between the solution path and a ball of radius l in the space of nodal displacements (a *cylinder* in the λ-$\mathbf{u}$ space) centred at the equilibrium configuration $\mathbf{u}_n$ of the beginning of the increment. A graphical representation is shown in Figure 4.8 which illustrates a system with two degrees of freedom. The possible intersections are denoted A and B.

4.4.2. THE COMBINED NEWTON–RAPHSON/ARC-LENGTH PROCEDURE

In summary, having specified the required l, an equilibrium solution is obtained by solving, for both $\mathbf{u}_{n+1}$ and $\Delta\lambda$, the residual equation (4.108) in conjunction with one of the above 'length' constraint equations ((4.109) or (4.110)). The standard Newton–Raphson algorithm for iterative solution of the augmented arc-length system of nonlinear equations is derived by simply linearising (4.108) together with the relevant arc-length constraint. For the *cylindrical* version of the arc-length method, the linearised system to be solved for $\delta\mathbf{u}^{(k)}$ and $\delta\lambda^{(k)}$ at

Box 4.4. The combined Newton–Raphson/arc-length scheme.

(i) Initialise iteration counter, $k := 0$, and set initial guess for displacement and incremental load factor

$$\mathbf{u}_{n+1}^{(0)} := \mathbf{u}_n; \qquad \lambda_{n+1}^{(0)} := \lambda_n; \qquad \mathbf{r}^{(0)} := \mathbf{f}^{\,\mathrm{int}}(\mathbf{u}_n) - \lambda_n \bar{\mathbf{f}}^{\,\mathrm{ext}}$$

(ii) Assemble stiffness, $\mathbf{K}_T := \mathbf{K}_T(\mathbf{u}_{n+1}^{(k)})$ [FRONT, ELEIST]

(iii) Set $k := k + 1$. Solve the linear systems for $\delta\mathbf{u}^*$ and for the *tangential solution*, $\delta\bar{\mathbf{u}}$ [subroutine FRONT]

$$\mathbf{K}_T\, \delta\mathbf{u}^* = -\mathbf{r}^{(k-1)}; \qquad \mathbf{K}_T\, \delta\bar{\mathbf{u}} = -\bar{\mathbf{f}}^{\,\mathrm{ext}}$$

(iv) Find iterative load factor $\delta\lambda^{(k)}$ [ARCLEN]

(a) If $k = 1$ (predictor solution) then compute

$$\delta\lambda^{(1)} := \mathrm{sign}(\Delta\mathbf{u}_n^T\, \delta\bar{\mathbf{u}}) \frac{l}{\sqrt{\delta\bar{\mathbf{u}}^T \delta\bar{\mathbf{u}}}}$$

(b) If $k \neq 1$ then solve the cylindrical arc-length constraint equation

$$a\, \delta\lambda^{(k)^2} + b\, \delta\lambda^{(k)} + c = 0$$

with coefficients defined by (4.117) and choose root $\delta\lambda^{(k)}$ according to (4.118)

(v) Apply correction to incremental load factor [ARCLEN]

$$\lambda_{n+1}^{(k)} := \lambda_{n+1}^{(k-1)} + \delta\lambda^{(k)}$$

(vi) Compute iterative displacement [ARCLEN]

$$\delta\mathbf{u}^{(k)} := \delta\mathbf{u}^* + \delta\lambda^{(k)}\, \delta\bar{\mathbf{u}}$$

(vii) Correction to total and incremental displacements [UPCONF]

$$\mathbf{u}_{n+1}^{(k)} := \mathbf{u}_{n+1}^{(k-1)} + \delta\mathbf{u}^{(k)}; \qquad \Delta\mathbf{u}_{n+1}^{(k)} := \Delta\mathbf{u}_{n+1}^{(k-1)} + \delta\mathbf{u}^{(k)}$$

(viii) Update residual, $\mathbf{r}^{(k)} := \mathbf{f}^{\,\mathrm{int}}(\mathbf{u}_{n+1}^{(k)}) - \lambda_{n+1}^{(k)} \bar{\mathbf{f}}^{\,\mathrm{ext}}$ [INTFOR, CONVER]

(ix) Check for convergence [CONVER]

IF $\|\mathbf{r}^{(k)}\|/\|\mathbf{f}^{\,\mathrm{ext}}\| \leq \epsilon_{\mathrm{tol}}$ THEN set $(\cdot)_{n+1} := (\cdot)_{n+1}^{(k)}$ and EXIT

ELSE GOTO (ii)

the kth Newton iteration reads

$$\begin{bmatrix} \mathbf{K}_T(\mathbf{u}^{(k-1)}) & -\bar{\mathbf{f}}^{\mathrm{ext}} \\ 2\Delta\mathbf{u}^{(k-1)^T} & 0 \end{bmatrix} \begin{Bmatrix} \delta\mathbf{u}^{(k)} \\ \delta\lambda^{(k)} \end{Bmatrix} = -\begin{Bmatrix} \mathbf{r}(\mathbf{u}^{(k-1)}, \Delta\lambda^{(k-1)}) \\ \Delta\mathbf{u}^{(k-1)^T} \Delta\mathbf{u}^{(k-1)} - l^2 \end{Bmatrix}, \tag{4.111}$$

where the subscripts $n + 1$ have been dropped for notational convenience.

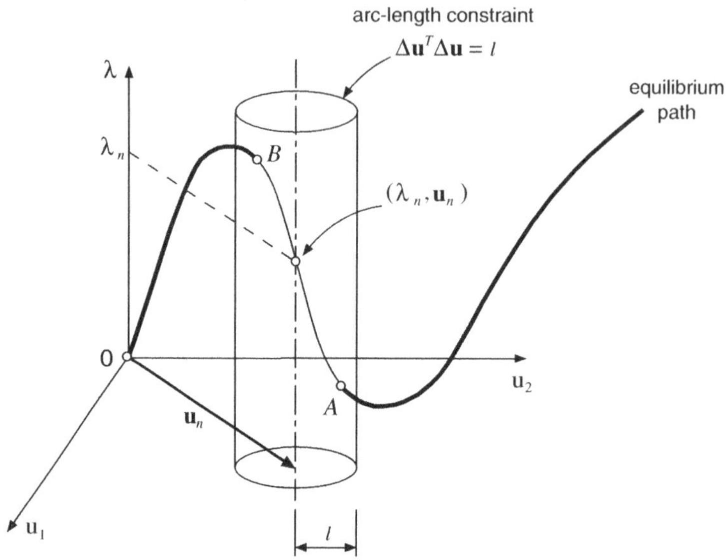

Figure 4.8. The cylindrical arc-length method.

The non-consistent scheme

In practice, instead of solving the above coupled system (whose coefficients matrix is non-banded), it is more convenient to adopt what is sometimes referred to as the *non-consistent* scheme, where the original augmented system to be solved at each iteration is replaced by the following equations:

$$\begin{aligned} &\left[\mathbf{K}_T(\mathbf{u}^{(k-1)}) \quad -\bar{\mathbf{f}}^{\text{ext}}\right] \begin{Bmatrix} \delta\mathbf{u}^{(k)} \\ \delta\lambda^{(k)} \end{Bmatrix} = -\mathbf{r}(\mathbf{u}^{(k-1)}, \Delta\lambda^{(k-1)}) \\ &\Delta\mathbf{u}^{(k)^T}\Delta\mathbf{u}^{(k)} = l^2. \end{aligned} \tag{4.112}$$

In the iterations (4.111), the arc-length constraint is guaranteed to hold only at the converged solution whereas, in the non-consistent scheme, the arc-length constraint is enforced at *every* iteration. Both iterative schemes lead to identical converged equilibrium solutions $\{\Delta\lambda, \mathbf{u}\}$. From $(4.112)_1$, we have

$$\delta\mathbf{u}^{(k)} = \delta\mathbf{u}^* + \delta\lambda\, \delta\bar{\mathbf{u}}, \tag{4.113}$$

where $\delta\mathbf{u}^*$ is the iterative displacement stemming from the Newton–Raphson algorithm for the standard load controlled scheme:

$$\delta\mathbf{u}^* \equiv -\mathbf{K}_T^{-1}\, \mathbf{r}^{(k-1)}, \tag{4.114}$$

and $\delta\bar{\mathbf{u}}$ is the so-called *tangential solution* defined by

$$\delta\bar{\mathbf{u}} \equiv \mathbf{K}_T^{-1}\, \bar{\mathbf{f}}^{\text{ext}}. \tag{4.115}$$

Substitution of (4.113) into the relation $\Delta\mathbf{u}^{(k)} = \Delta\mathbf{u}^{(k-1)} + \delta\mathbf{u}^{(k)}$, followed by the enforcement of the constraint $(4.112)_2$, results in the quadratic equation for the iterative load factor $\delta\lambda^{(k)}$

$$a\,\delta\lambda^{(k)^2} + b\,\delta\lambda^{(k)} + c = 0, \tag{4.116}$$

with coefficients

$$\begin{aligned} a &= \delta\bar{\mathbf{u}}^T\delta\bar{\mathbf{u}} \\ b &= 2(\Delta\mathbf{u}^{(k-1)} + \delta\mathbf{u}^*)^T\delta\bar{\mathbf{u}} \\ c &= (\Delta\mathbf{u}^{(k-1)} + \delta\mathbf{u}^*)^T(\Delta\mathbf{u}^{(k-1)} + \delta\mathbf{u}^*) - l^2. \end{aligned} \tag{4.117}$$

The choice of the appropriate root

The iterative load factor is normally chosen as the solution to the quadratic equation that yields the minimum angle between $\Delta\mathbf{u}^{(k-1)}$ and $\Delta\mathbf{u}^{(k)}$ (Crisfield, 1991), i.e. $\delta\lambda^{(k)}$ is the solution of (4.116) which gives the maximum product $\Delta\mathbf{u}^{(k)\,T}\Delta\mathbf{u}^{(k-1)}$:

$$\delta\lambda^{(k)} = \arg\Big[\max_{\delta\hat{\lambda}\,|\,a\,\delta\hat{\lambda}^2+b\,\delta\hat{\lambda}+c=0}\{(\Delta\mathbf{u}^{(k-1)} + \delta\mathbf{u}^* + \delta\hat{\lambda}\,\delta\bar{\mathbf{u}})^T\Delta\mathbf{u}^{(k-1)}\}\Big]. \tag{4.118}$$

The incremental load factor, $\Delta\lambda^{(k)}$, is updated according to

$$\Delta\lambda^{(k)} = \Delta\lambda^{(k-1)} + \delta\lambda^{(k)}. \tag{4.119}$$

4.4.3. THE PREDICTOR SOLUTION

When $k = 1$ (the predictor solution), the above criterion cannot be used for determination of the appropriate root $\delta\gamma^{(k)}$ as the initial guess, $\Delta\mathbf{u}^{(0)} = \mathbf{0}$, does not contain information about the path being currently followed by the iterative procedure. The two possible values of $\delta\lambda^{(1)}$ for the predictor solution are

$$\delta\lambda^{(1)} = \pm\frac{l}{\sqrt{\delta\bar{\mathbf{u}}^T\delta\bar{\mathbf{u}}}}, \tag{4.120}$$

and the success of the path-following technique depends crucially on the choice of the appropriate sign for the iterative load factor. If the wrong choice is made, the predictor solution will 'track back' on the current path (Crisfield, 1991). Many procedures are currently used to predict the continuation direction, i.e. to choose the sign of $\delta\lambda^{(1)}$ that carries on tracing the current solution path. Some criteria are listed below:

(a) *Stiffness determinant.* Follow the sign of the stiffness determinant $|\mathbf{K}_T(\mathbf{u}^{(0)})|$:

$$\text{sign}(\delta\lambda^{(1)}) = \text{sign}(|\mathbf{K}_T(\mathbf{u}^{(0)})|); \tag{4.121}$$

(b) *Incremental work.* Follow the sign of the predictor work increment:

$$\text{sign}(\delta\lambda^{(1)}) = \text{sign}(\delta\bar{\mathbf{u}}^T\,\bar{\mathbf{f}}^{\text{ext}}); \tag{4.122}$$

(c) *Secant path* (Feng *et al.*, 1995, 1996). The sign of $\delta\lambda^{(1)}$ is determined as

$$\text{sign}(\delta\lambda^{(1)}) = \text{sign}(\Delta\mathbf{u}_n^T\,\delta\bar{\mathbf{u}}). \tag{4.123}$$

Criteria (a) and (c) are implemented in program HYPLAS. Procedure (a) is widely used in commercial finite element codes and works well in the absence of bifurcations. In the presence of bifurcations, however, it is known not to be appropriate and fails in most cases. As pointed out by Crisfield (1991), its ill-conditioned behaviour stems from the fact that the sign of $|\mathbf{K}_T|$ changes either when a limit point or when a bifurcation point is passed. In this case, the predictor cannot distinguish between these two quite different situations, unless further (usually computationally expensive) analyses are undertaken. In the presence of a bifurcation, instead of following the current equilibrium path, the solution will oscillate about the bifurcation point. This property is mathematically proved by Feng *et al.* (1997). Procedure (b), on the other hand, is 'blind' to bifurcations and can continue to trace an equilibrium path after passing a bifurcation point. However, this criterion proves ineffective in the descending branch of the load-deflection curve in 'snap-back' problems, where the predicted positive 'slope' will provoke a 'back tracking' load increase. One important feature shared by the criteria (a) and (b) is the fact that they rely exclusively on information relative to the *current* equilibrium point (at the beginning of the increment). The decision on the sign of $\delta\lambda$ is made without considering the *history* of the currently traced equilibrium path. In situations such as the ones pointed out above, this may result in wrong direction prediction. In contrast, a key point concerning criterion (c) is the fact that $\Delta\mathbf{u}_n$ carries with it information about the *history* of the current equilibrium path. The importance of this fact is established in the discussion that follows, where, in particular, we show by means of geometric arguments that the secant predictor can easily overcome the problems associated with criteria (a) and (b).

Secant path. Geometric interpretation

Firstly, let us concentrate on the meaning of the tangential solution, $\delta\bar{\mathbf{u}}$. As schematically illustrated in Figure 4.9, $\delta\bar{\mathbf{u}}$ is a vector tangent to the equilibrium path in the space of displacements. In view of its definition (4.115), it points in the direction of the *positive gradient* of λ. Thus, $\delta\bar{\mathbf{u}}$ provides information on the direction (starting from the current configuration) along which the load factor increases – direction associated with the choice of positive $\delta\lambda$. Note that $\delta\bar{\mathbf{u}}$ is not defined at load-reversing points (cf. points S_1 and S_2 of Figure 4.9) where $\mathbf{K}_T$ is singular. The incremental displacement $\Delta\mathbf{u}_n$, on the other hand, is secant to the equilibrium path and, when sufficiently small, gives a good approximation to the solution curve within the interval $[t_{n-1}, t_n]$. In other words, $\Delta\mathbf{u}_n$ approximates the *history* of the current equilibrium path within $[t_{n-1}, t_n]$. A schematic representation is shown in Figure 4.10. Clearly, when sufficiently small, $\Delta\mathbf{u}_n$ indicates the 'forward' direction of the current solution path at $\mathbf{u}_n$. In order to keep tracing the current path without returning to previously obtained points, this criterion requires that the incremental displacement of the predictor solution, $\delta\lambda\delta\bar{\mathbf{u}}$, be in the 'forward' direction, that is,

$$\delta\lambda\, \Delta\mathbf{u}_n^T \delta\bar{\mathbf{u}} > 0. \tag{4.124}$$

The above inequality is equivalent to (4.123). In essence, a positive product, $\Delta\mathbf{u}_n^T\delta\bar{\mathbf{u}}$, indicates that the load factor is currently increasing so that the positive sign should be chosen for the predictor $\delta\lambda$ to carry on following the present equilibrium path (Figure 4.10(a)). Similarly, if $\Delta\mathbf{u}_n^T\delta\bar{\mathbf{u}}$ is negative, the load factor is decreasing and the negative sign should be chosen (Figure 4.10(b)).

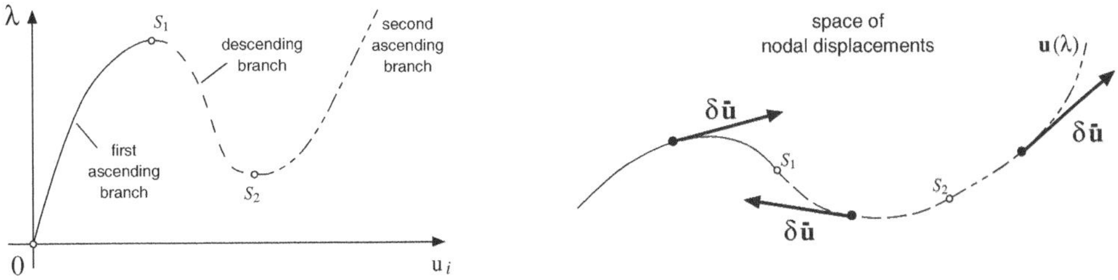

Figure 4.9. The tangential solution, $\delta\bar{\mathbf{u}}$. (Reproduced with permission from On the determination of the path direction for arc-length methods in the presence of bifurcations and 'snap-backs', EA de Souza Neto and YT Feng, *Computer Methods in Applied Mechanics and Engineering*, Vol 179, Issue 12 © 1999 Elsevier Science S.A.)

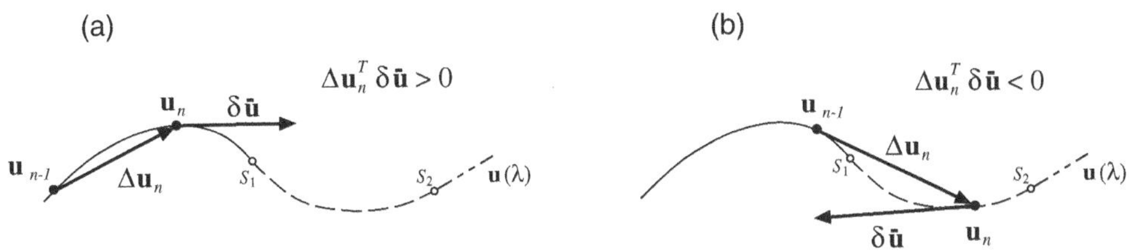

Figure 4.10. The secant path direction prediction criterion. (a) Load factor is currently increasing: choose $\delta\lambda > 0$; (b) load factor is currently decreasing: choose $\delta\lambda < 0$. (Reproduced with permission from On the determination of the path direction for arc-length methods in the presence of bifurcations and 'snap-backs', EA de Souza Neto and YT Feng, *Computer Methods in Applied Mechanics and Engineering*, Vol 179, Issue 12 © 1999 Elsevier Science S.A.)

It is important to emphasise that the secant path criterion is insensitive to the presence of bifurcations and, provided that $\Delta\mathbf{u}_n$ is sufficiently small, does not show the basic deficiency associated with the criterion based on the sign of $|\mathbf{K}_T|$. Also, in the descending branch of 'snap back' curves, the secant path predictor should indicate that the load is decreasing (negative $\delta\lambda$), overcoming the problem associated with the predictor incremental work criterion. Numerical experiments demonstrating the effectiveness of the secant path criterion in situations where criteria (a) and (b) fail are shown by de Souza Neto and Feng (1999).

Step-size limitation

The need for *sufficiently small* $\Delta\mathbf{u}_n$ has been stressed above as a condition for producing reliable indications of the 'forward' direction. With regard to this aspect, it is important to recall that a natural limitation on the maximum size of $\Delta\mathbf{u}_n$ is already imposed by the Newton–Raphson algorithm. If the convergence radius of the Newton–Raphson scheme is smaller than the maximum size of $\Delta\mathbf{u}_n$ that produces an accurate direction prediction, then no further increment size restrictions are introduced by the use of the secant path predictor criterion. In fact, numerical experience shows that the maximum increment size is usually dictated by the Newton–Raphson algorithm rather than by the direction predictor criterion.

5 OVERVIEW OF THE PROGRAM STRUCTURE

IN Chapter 4, we have described a general strategy for the finite element simulation of linear and nonlinear solid mechanics problems, including generic path-dependent constitutive models and finite deformations and strains. Most procedures described there are incorporated in the standard version of the program HYPLAS that accompanies this book. Throughout their description, whenever appropriate, reference has been made to the subroutines of HYPLAS where some of the most relevant computational operations associated with such numerical procedures are carried out.

In the present chapter we provide a more thorough description of the HYPLAS program, with emphasis on its structure. The depth at which details of the code are described here is only what the authors believe to be sufficient to help readers find their way through HYPLAS. It is by no means intended to be a comprehensive guide to the code. The present chapter is particularly relevant for researchers who wish to understand and/or modify the basic code by including new procedures, material models, finite elements, etc. Those who are only interested in the theoretical and numerical aspects of nonlinear solid mechanics may skip this chapter and move on to Part Two of the book.

5.1. Introduction

HYPLAS is a finite element code for implicit small and large strain analysis of hyperelastic and elastoplastic solids in plane stress, plane strain and axisymmetric states.

5.1.1. OBJECTIVES

The main purpose of HYPLAS is to illustrate the computational implementation of numerical procedures described throughout this book. The general framework adopted is that described in Chapter 4. It comprises:

(i) a general displacement-based incremental finite element procedure;

(ii) iterative schemes (e.g. Newton–Raphson) for the solution of nonlinear incremental finite element equations;

(iii) an arc-length scheme for problems involving structural instability.

Within this framework, numerous (elastic and elastoplastic) material models have been incorporated. The theory and numerical methods underlying the implementation of each

Computational Methods for Plasticity: Theory and Applications EA de Souza Neto, D Perić and DRJ Owen

material model are thoroughly discussed in Parts Two and Three of this book. Part Two focuses on the infinitesimal theory and Part Three on the finite strain range. The reader will find there a detailed description of the associated procedures including, in many cases, the pseudo-code and the corresponding FORTRAN source code of the relevant routines.

5.1.2. REMARKS ON PROGRAM STRUCTURE

In order to treat a wider range of material models and finite element types within the above general framework, the structure of HYPLAS has been designed so as to have element type-specific and material model-specific procedures as more or less self-contained modules which remain confined to relatively small parts of the program. Element and material model-related data are also self-contained to some extent.

Due to this modularity, the code of, say, a constitutive integration algorithm for a particular material model can be easily identified with the corresponding description provided in this book. In addition, the adopted program structure makes the incorporation of new material models and finite element types relatively straightforward tasks (once the associated numerical procedures have been coded by the developer). This feature is particularly relevant for researchers who wish to use HYPLAS for the development, implementation and testing of new constitutive models/algorithms as well as finite elements.

Clarity of coding

When developing a computer code, programmers are frequently faced with the dilemma: clarity versus efficiency. Finite element codes are no exception. We do not mean that clarity of coding and computational efficiency are opposing concepts but, as a general rule, excessive optimisation of computer operations tends to result in cumbersome, difficult-to-follow source codes.

In view of the educational purposes of HYPLAS, the programming philosophy adopted in its development has been biased towards code clarity and modularity. This is particularly true for the numerical procedures whose theory is described in this text.

Cross-referencing. Comment lines

The reader will also find that numerous comments have been added to the source code. Frequently, the comments indicate where in the book the procedure being carried out is described. Whenever convenient, the routines of HYPLAS involved in the computational implementation of procedures described in the book are also indicated in the text. This should allow a relatively easy cross-referencing between theory and corresponding computer implementation.

5.1.3. PORTABILITY

The code of HYPLAS is entirely written in (almost) standard FORTRAN 77 programming language (ANSI, 1978). The existing exceptions to the FORTRAN 77 standards are known to be accepted by most currently available compilers.

5.2. The main program

The main program of HYPLAS can be divided into three basic parts:

1. *Data input and initialisation.* The data input and initialisation phase is carried out at the very beginning of the execution of the program. At this stage all data defining the problem to be analysed are read from the relevant files and all necessary arrays are initialised.

2. *The incremental finite element procedure.* This is the main body of the program, where the numerical procedures discussed in Chapter 4 are implemented. Essentially, it consists of a main loop over load increments with a nested loop over equilibrium iterations. The procedure has been conceptually described in Boxes 4.1–4.4 of the previous chapter.

3. *Output of converged results.* This comprises all output operations necessary to print converged finite element solutions into the results file and/or dump an image of the database into a restart output file. Output routines are called from inside the main loop over load increments.

These components are described in more detail in the following sections.

5.3. Data input and initialisation

The first operations carried out as the execution of HYPLAS is initiated, comprise the opening of files accessed during the execution, reading of input data and initialising of most global arrays. Input data can be read:

(a) either from an *input data file* only,

(b) or from an *input restart file* and an *input data file*.

The input data file is an ASCII format file which, in case (a), contains all information about the problem to be analysed. Under option (b), most of the problem definition is read from an *input restart file*. This is a binary file generated during a previous execution of HYPLAS. It contains an image of the complete database at the time of its creation. In this case (case (b)), the finite element analysis is *restarted* from where the restart file was created. The only information read from the relevant (ASCII) input data file, is the data related to the definition of the proportional load incrementation programme. The flowchart of the procedure is shown in Figure 5.1, where the main routines involved are indicated.

5.3.1. THE GLOBAL DATABASE

The global database of HYPLAS stores most arrays and program control parameters required in the finite element solution process. The corresponding data can be classed into two basic categories:

(i) *Problem-defining (fixed) data.* These are data that do not change during the solution process such as initial nodal coordinates, topology of the finite element mesh (nodal connectivities), material and element properties, kinematic constraints and applied external loads as well as various control parameters required in the finite element analysis.

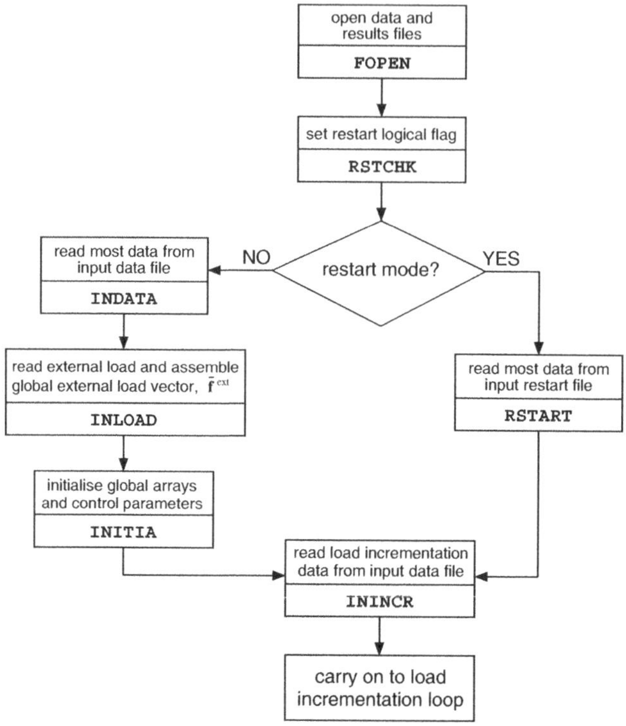

Figure 5.1. Data input and initialisation. Flowchart.

(ii) *Solution-related (variable) data*. Data that change during the solution process. Typical solution-related data are nodal displacements, current nodal coordinates (changes in large deformation analysis only), stresses and state variables in general at Gauss points, arrays storing element internal forces, etc.

The global database is defined in the *include* file **GLBDBASE.INC**. The reader is referred to the comment lines of this file for a description of the data stored in the arrays and variables of the global database. These will be further explained in this chapter only when convenient for a better understanding of the program. All arrays of the database have fixed dimension and are grouped in a number of `COMMON` blocks. File **GLBDBASE.INC** is included in all routines that require data stored in any of the `COMMON` blocks of the global database. The array-dimensioning parameters defining the dimensions of the global database arrays are set in the include files **MAXDIM.INC**, **MATERIAL.INC** and **ELEMENTS.INC** (refer to their comment lines). File **MAXDIM.INC** sets the parameters that define the maximum admissible problem size that can be analysed. Files **MATERIAL.INC** and **ELEMENTS.INC** set, respectively, maximum array dimensions required by the currently implemented material models and element types. These are discussed further in Sections 5.6 and 5.7.

Variables and arrays (or array components) containing data of type (i) are set once and for all in subroutines `INDATA`, `INLOAD` and `ININCR`, according to the input data file. If in restart mode, the data otherwise read and set in `INDATA` and `INLOAD` are retrieved from the input restart file by subroutine `RSTART`. In the data input and initialisation phase of `HYPLAS`, data of type (ii) are initialised in subroutine INITIA. If in restart mode, current values at the last equilibrium (converged) solution are retrieved instead from the input restart file in RSTART.

5.3.2. MAIN PROBLEM-DEFINING DATA. SUBROUTINE INDATA

This routine reads, from the input data file, the geometry of the problem, mesh topology, kinematic constraints, material properties, element properties as well as most program control parameters including: analysis type (plane stress, strain or axisymmetric) flag, nonlinear solution algorithm, etc.

INDATA also carries out a number of checks to ensure the validity of the given data. If unacceptable data is detected (such as undefined or multiply defined nodal numbers, coordinates, etc.), then an error message is printed (by routine ERRPRT) and the execution of the program is aborted.

5.3.3. EXTERNAL LOADING. SUBROUTINE INLOAD

Subroutine INLOAD reads the input data defining the external loading and assembles the global external force vector

$$\bar{\mathbf{f}}^{\rm ext}$$

(refer to expressions (4.67) and (4.68)). The components of $\bar{\mathbf{f}}^{\rm ext}$ are stored in the global array

RLOAD.

The loading types implemented in INLOAD are

- *Point load.* Forces defined directly at nodal points. In this case, the corresponding prescribed values are assigned to their appropriate position in array RLOAD.
- *Gravity load.* Assembled from the element load vectors

$$\rho\, g \int_{\Omega^{(e)}} \mathbf{N}^T \boldsymbol{g}\, \mathrm{d}v \approx \rho\, g \sum_{i=1}^{n_{\rm gausp}} w_i\, j(\boldsymbol{\xi}_i)\, \mathbf{N}(\boldsymbol{\xi}_i)^T\, \boldsymbol{g}, \tag{5.1}$$

  where the input data ρ, g and $\boldsymbol{g}$ are, respectively, the material density, the gravity acceleration and the unit vector defining the direction of the gravity acceleration.
- *Distributed load at element boundaries (edges in two dimensions).* Assembly of element vectors

$$\int_{\partial\Omega_t^{(e)}} \mathbf{N}_b^T\, \boldsymbol{t}\, \mathrm{d}a \approx \sum_{i=1}^{n_{\rm gausb}} w_i\, j(\xi_i)\, \mathbf{N}_b(\xi_i)^T\, \boldsymbol{t}, \tag{5.2}$$

  where $\partial\Omega_t^{(e)}$ is the loaded edge of the boundary of element e, $\mathbf{N}_b$ is the *boundary* shape function associated with the loaded boundary and $n_{\rm gausb}$ is the number of Gauss points used for numerical integration over the element boundary. Array $\boldsymbol{t}$ is the array of nodal pressure vectors of the loaded edge. The above numerical integration is carried out for each loaded edge of the element.

Some checks are also made in INLOAD to detect incorrect data.

5.3.4. INITIALISATION OF VARIABLE DATA. SUBROUTINE INITIA

This routine initialises some arrays and problem control variables. Arrays such as displacements (incremental, iterative and total) are initialised to zero. Stresses and other state variables (most arrays of COMMON block STATE), which are defined at the Gauss point level, are initialised by material-specific routines called from subroutine MATISW. The convenience of using material-specific routines for state variable initialisation is discussed in Section 5.7.

5.4. The load incrementation loop. Overview

Let us now focus on the core of the finite element analysis code: the load incrementation loop. The main load incrementation loop carries out the proportional loading programme either with *prescribed* (fixed) load increments (where the incremental load factors are prescribed in the input data file) or via an arc-length method. In the latter case, the proportional load increments are determined along the process according to the *cylindrical* arc-length procedure described in the previous chapter.

Within each step of the loop over load increments, an iterative procedure (typically the Newton–Raphson algorithm) is carried out to solve the nonlinear equilibrium problem. The pseudo-code of the overall scheme is given in Boxes 4.1–4.3 for the fixed increments option. The inner (equilibrium iteration) loop for the combined Newton–Raphson/arc-length scheme is summarised in Box 4.4.

5.4.1. FIXED INCREMENTS OPTION

A flowchart illustrating the main steps of the load increment loop under the fixed increments option is shown in Figure 5.2. The main routines called from the HYPLAS main program are indicated.

Under this option, the incremental load factors for each increment are fixed, as defined in the input data file. If the increment cutting procedure (Section 5.4.3) is activated, the current incremental load factor is split into two equally sized sub-increments. In the present implementation, the loop over load increments is ended either if the load programme is successfully completed (the last prescribed increment converges) or if the number of successive increment cuts causes the sub-increment stack array DFSUB to become full.

5.4.2. ARC-LENGTH CONTROL OPTION

Under *arc-length* control, the flow of the main program is a small variation of that followed with prescribed load increments. The corresponding flowchart is shown in Figure 5.3. The essential difference is that the proportional load increment factor here is recalculated (routine ARCLEN) in each pass of the iteration loop whereas, with fixed increments, the external load is incremented (in INCREM) before the iteration loop starts. Also, the arc-length method requires the solution of the linear system (in FRONT) for *two* right-hand sides (refer to item (iii) of Box 4.4).

In the present arc-length implementation, the arc length, l, is changed at the end of every increment (after convergence) according to the number of iterations required to achieve convergence within the prescribed tolerance. The idea is to allow the arc length to increase when convergence is 'easy' and to reduce it if convergence is 'difficult', trying to obtain

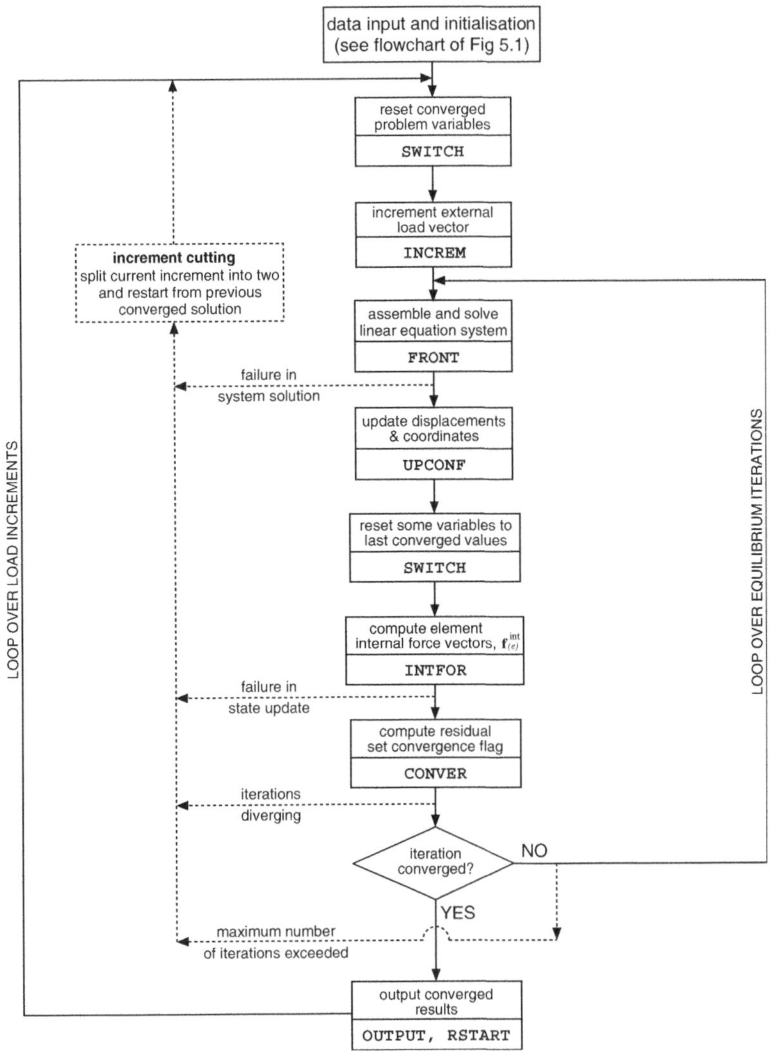

Figure 5.2. The load incrementation loop. Fixed increments option.

convergence in a *desired* number of iterations, n_{itdes} (prescribed in the input data file). With n_{itact} and l_{curr} denoting, respectively, the number of iterations for convergence and the arc length used in the current step, the new length l_{next} to be used in the next step is set (in subroutine LENGTH) as

$$l_{\text{next}} = l_{\text{curr}} \frac{n_{\text{itdes}}}{n_{\text{itact}}}. \tag{5.3}$$

The new length is not allowed to be larger than a maximum value $l_{\max}$.

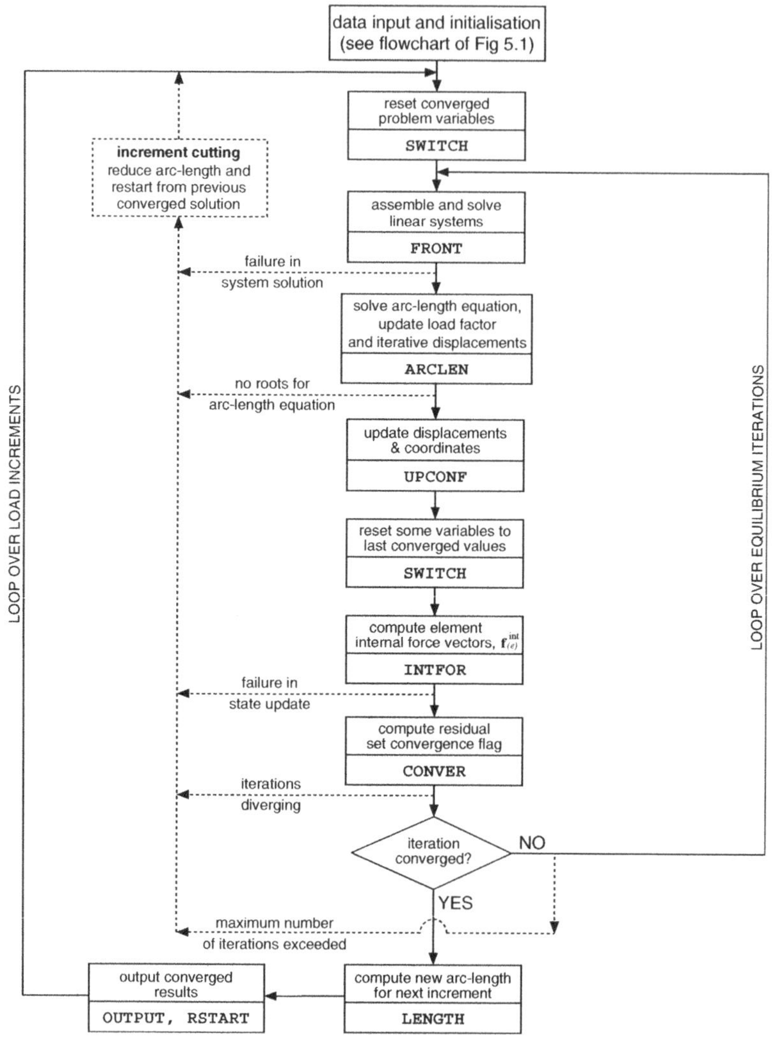

Figure 5.3. The load incrementation loop. Arc-length control.

Start-up length.

Obviously, at the beginning of the first iteration of the very first load increment, the above length adjustment cannot be applied. Since it is not normally easy to imagine the magnitude of l that would be reasonable for a particular problem, the start-up length, l_0, is calculated (in ARCLEN) by specifying (in the input data file) an initial load increment factor, $\Delta\lambda_0$, and setting

$$l_0 = \sqrt{\Delta\lambda_0^2 \, \delta\bar{\mathbf{u}}^T \bar{\mathbf{u}}} \tag{5.4}$$

where $\bar{\mathbf{u}}$ is the initial *tangential solution* (see item (iii) of Box 4.4). The maximum length $l_{\rm max}$ is also set at this stage as

$$l_{\rm max} = p\, l_0, \tag{5.5}$$

where p is the *maximum arc-length parameter* (defined in the input data file).

Predictor solution.

The present implementation allows the choice of the sign of the *predictor solution* (refer to Section 4.4.3) to be made by two criteria: the *stiffness determinant sign* or the *secant path* predictor. The predictor solution is computed in subroutine `ARCLEN`.

Stopping the load incrementation loop.

Under arc-length control, the iteration loop is broken and the program stopped if a maximum prescribed load factor has been exceeded or if a maximum prescribed number of successful (converged) load steps has been completed.

5.4.3. AUTOMATIC INCREMENT CUTTING

An automatic *increment cutting* facility is available in `HYPLAS` (see dashed lines in the flowcharts of Figures 5.2 and 5.3). This facility proves extremely useful in general nonlinear analysis. It frequently occurs that a converged equilibrium solution cannot be attained in the equilibrium iterations. Whenever that happens, the increment cutting procedure is activated and the current step is restarted (from the last converged solution) with reduced increment size. The causes of failure to converge are

- The load increment is too large. The total prescribed load (with fixed increments option) is beyond the limit load of the structure or the initial guess, $\mathbf{u}_{n+1}^{(0)}$ (refer to Box 4.2), falls outside the convergence radius of the iterative solution method. In this case the residual norm will either diverge or not converge within a reasonable number of iterations. A possible alternative (not available in `HYPLAS`) to increase the convergence radius is the incorporation of *line-searches* (Crisfield, 1991) within the equilibrium iterations.

- The algorithm for numerical integration of path-dependent constitutive equations (the *state update procedure*) fails to produce a solution at a Gauss point. This is normally caused by excessively large strain increments. Highly nonlinear material models can be particularly prone to this type of ill behaviour and some more complex algorithms may fail even for relatively small increments. Certain measures can be taken to make constitutive algorithms more robust and avoid the problem. This issue is addressed in Part Two of this book.

- The linear equation system cannot be solved in `FRONT` due to a zero pivot in the stiffness matrix. This may occur for material models whose tangent operator may become singular under certain states of stress.

- *For arc-length control only*. The arc-length constraint equation (4.116), solved in routine ARCLEN, may have no real roots. This problem is also generally overcome by reducing the increment size (the *arc length*, l, in this case) and restarting the current step with the reduced value of l.

5.4.4. THE LINEAR SOLVER. SUBROUTINE FRONT

The linear system solution is carried out in subroutine FRONT which uses the classical *frontal* method (Hinton and Owen, 1977; Irons, 1970). In the frontal method, the linear system is solved by Gauss elimination. The equations are assembled, with the appropriate contribution of individual elements to the global stiffness matrix and load vector, and eliminated *simultaneously*. Details of the technique are described elsewhere (Hinton and Owen, 1977) and will not be discussed here.

In the present implementation, the equation system is solved for up to two right-hand sides. Note that the solution for two right-hand sides is required by the arc-length method. In this case, the stiffness matrix is reduced only once. The number of right-hand sides for which the solution is required is set in the local variable NRHS. Another local control flag used by this routine is the integer MODE. If MODE=1, the standard system

$$\mathbf{K}_T\,\delta\mathbf{u} = -\mathbf{r} \tag{5.6}$$

is solved for $\delta\mathbf{u}$. If MODE=2, the *tangential solution* of the arc-length method, $\delta\bar{\mathbf{u}}$, is obtained instead by solving

$$\mathbf{K}_T\,\delta\bar{\mathbf{u}} = -\bar{\mathbf{f}}^{\text{ext}}. \tag{5.7}$$

If MODE=3, then both solutions are obtained. The variable KRESL controls whether or not a new stiffness is computed. If KRESL=1, then a new stiffness is required. When the standard Newton–Raphson algorithm is used, KRESL is always set to 1. If KRESL=2, then the stiffness of the previous iteration is to be reused in the current iteration and only the load term will be reduced (this situation will occur, for instance, when the initial stiffness method is used).

The global arrays of HYPLAS used exclusively by the frontal solver are not included in the global database (file **GLBDBASE.INC**). They are grouped in the COMMON block named FRONTA.

5.4.5. INTERNAL FORCE CALCULATION. SUBROUTINE INTFOR

Subroutine INTFOR computes the element internal force vector of all elements of the mesh. For each element, it calls the internal force evaluation routine of the corresponding *element class*. Stresses and all other relevant variables defined at the Gauss point level are updated (at a lower layer of code) by material-specific state updating procedures during internal force computation. Further details on the structure of HYPLAS regarding the evaluation of internal forces and general state updating are given in Section 5.5.

5.4.6. SWITCHING DATA. SUBROUTINE SWITCH

Some arrays of the global database store data related to the current iteration (generally not an equilibrium state) as well as values obtained in the last converged (equilibrium) solution.

This applies to all arrays of COMMON block STATE (see comments in file **GLBDBASE.INC**). Switching between current and last converged data is managed by subroutine SWITCH as follows (refer to source code). Before a material-specific state updating procedure is called (this is required during the calculation of the internal force vector in the element class-specific routines called from ELEIIF), the previous converged values are assigned to the positions corresponding to the current values. These are passed on to the state-updating procedure which returns (in the same position) the updated (or current) values. When the iterative procedure converges, the converged values need updating. The current values (which have converged at this point) are then assigned to the corresponding array positions that store converged data. Appropriate switching/resetting is also needed when increment cutting is activated. Most arrays of COMMON block STATE are switched in material-specific routines (see Section 5.7). In large deformation analysis, array COORD (in COMMON block MESH) of nodal coordinates also needs switching between current and last converged values. Other arrays, such as the total, incremental and iterative nodal displacement vectors also require switching/resetting. The type of switching operation carried out by SWITCH depends on the integer argument MODE.

5.4.7. OUTPUT OF CONVERGED RESULTS. SUBROUTINES OUTPUT AND RSTART

Once the equilibrium iteration loop has converged, the results are, if required, printed in the results file. Printed results can be nodal displacements, reactions, state variables at Gauss points and (extrapolated) state variables at nodes. The operations of output to results file are performed by subroutine OUTPUT. At this point, an image of the complete database of the program can also be dumped into a (binary) restart file to allow HYPLAS to be restarted later from that point. Output to the restart file is carried out by routine RSTART.

Output control, i.e. when and what to output, is made with the output control flags stored in arrays NOUTP and NOUTPV.

5.5. Material and element modularity

In the above, we have provided an overview of the main program with a summary description of the input phase and the main load incrementation loop. The basic steps executed are, to a large extent, independent of finite element types and material models adopted. In designing the structure of HYPLAS, this feature has been exploited to allow a relatively high degree of *modularity* of finite element types and constitutive models/algorithms. Brief reference to the modularity of elements and material models in HYPLAS has been made in Section 5.1.2. This will be addressed here in more detail.

The basic idea consists of confining element-specific and material-specific operations to localised areas of the code and avoiding interference with general procedures that are not material- or element-related. The concept is illustrated in the following example.

5.5.1. EXAMPLE. MODULARISATION OF INTERNAL FORCE COMPUTATION

An example where clear distinction can be made between element-specific operations, material-specific operations and the 'rest' of the program occurs in the computation of the element internal force vector. This is illustrated in the call tree of Figure 5.4. When

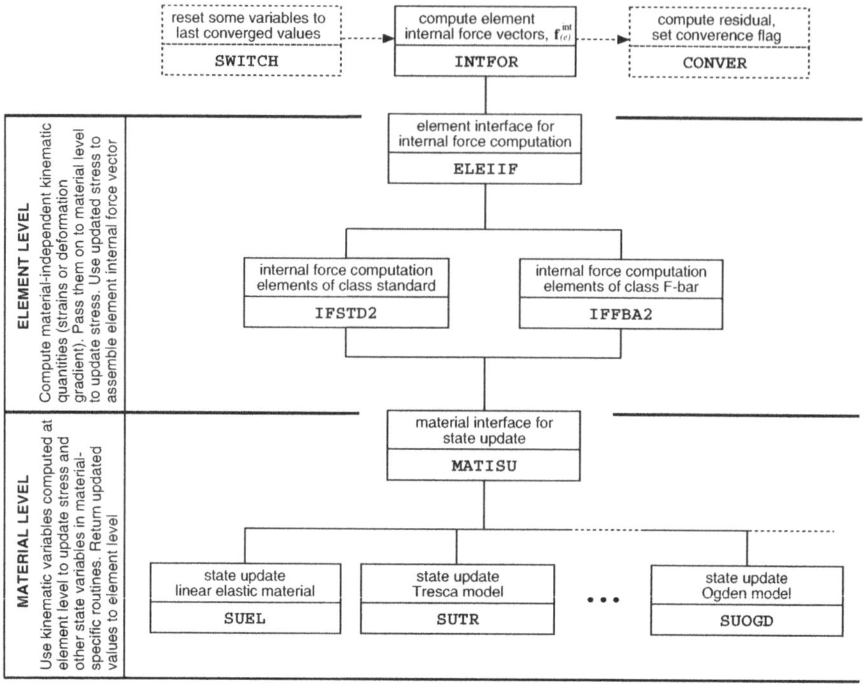

Figure 5.4. Element and material modularity in internal force computation.

element internal forces need to be computed within the equilibrium iteration loop, the main program calls subroutine INTFOR (refer to the flowcharts of Figures 5.2 and 5.3). INTFOR loops over all elements of the mesh and calls the element interface routine for internal force calculation ELEIIF (see call tree of 5.4). Before ELEIIF is called, only operations related to the general incremental procedure, with no distinction between different types of finite element or material, are carried out.

Material and element levels

At this point, it is worth recalling how the element internal force vector is computed:

$$\mathbf{f}_{(e)}^{\text{int}} := \sum_{i=1}^{n_{\text{gausp}}} w_i \, j(\boldsymbol{\xi}_i) \, \mathbf{B}(\boldsymbol{\xi}_i)^T \boldsymbol{\sigma}|_{\xi_i}. \tag{5.8}$$

The Gaussian quadrature weights, w_i, the Gauss points positions, $\boldsymbol{\xi}_i$ and the way in which the Jacobian, j, and the strain-displacement matrix, $\mathbf{B}$, are computed depend only on the particular element being used. More precisely, they depend on the element *class* to which the adopted element belongs.† The calculation of these quantities is independent of the underlying material model. The stress array, on the other hand, is the outcome of a state update procedure which depends exclusively on the adopted material model and the

†The definition of *element class* is given in Section 5.6.

algorithm used to integrate its constitutive equations (if the material is path-dependent). In the general path-dependent case, under infinitesimal strains, the state update algorithm defines an incremental constitutive function

$$\boldsymbol{\sigma} = \hat{\boldsymbol{\sigma}}(\boldsymbol{\alpha}_n, \boldsymbol{\varepsilon}_{n+1}) \tag{5.9}$$

within the typical interval $[t_n, t_{n+1}]$. In the above, $\boldsymbol{\alpha}_n$ is the set of internal variables at the converged state at t_n and $\boldsymbol{\varepsilon}_{n+1}$ is the given strain at the end of the interval. The incremental constitutive function can be expressed in the equivalent form[‡]

$$\boldsymbol{\sigma} = \hat{\boldsymbol{\sigma}}(\boldsymbol{\alpha}_n, \Delta\boldsymbol{\varepsilon}), \tag{5.10}$$

where the function arguments are the *incremental* strains,

$$\Delta\boldsymbol{\varepsilon} \equiv \boldsymbol{\varepsilon}_{n+1} - \boldsymbol{\varepsilon}_n = \nabla^s(\Delta\boldsymbol{u}), \tag{5.11}$$

and the set $\boldsymbol{\alpha}_n$ which now contains the converged internal state variables at time station t_n and the strain $\boldsymbol{\varepsilon}_n$.[§] In HYPLAS, the incremental strain is the actual argument passed into the state update interface in the infinitesimal case. In finite strain analysis, the corresponding argument is the incremental deformation gradient,

$$\boldsymbol{F}_\Delta \equiv \boldsymbol{I} + \nabla_n(\Delta\boldsymbol{u}) = [\boldsymbol{I} - \nabla_{n+1}(\Delta\boldsymbol{u})]^{-1}, \tag{5.12}$$

and the incremental constitutive function is represented as

$$\boldsymbol{\sigma} = \hat{\boldsymbol{\sigma}}(\boldsymbol{\alpha}_n, \boldsymbol{F}_\Delta). \tag{5.13}$$

Based on the above comments, the computation of the internal force vector can be split into two well-defined levels:

1. *The element level.* Here, all element-related quantities (w_i, j, $\mathbf{B}$) are computed together with the essential kinematic variable $\Delta\boldsymbol{\varepsilon}$ ($\boldsymbol{F}_\Delta$ in large strains). The incremental strain (or deformation gradient) is passed on to the material level which returns the updated stress. With the updated stress at hand, the internal force vector is assembled according to (5.8).

2. *The material level.* This is the lowest layer of code. It receives the incremental strain (or deformation gradient) from the element level, retrieves the relevant converged state variables $\boldsymbol{\alpha}_n$ stored in appropriate arrays and updates the stress by using the material model-specific subroutine that defines the incremental function (5.10). The updated stress is returned to the element level.

[‡]Note that, formally, the function on the right-hand side of (5.9) differs from that on the right-hand side of (5.10) in that it has different arguments. Throughout the text, however, we shall employ the symbol $\hat{\boldsymbol{\sigma}}$ to denote algorithmic (incremental) constitutive functions for the stress tensor in general, regardless of their list of arguments.

[§]As we shall see later, in computations with elastoplastic/viscoplastic models, it will be convenient to have the *elastic* strain of time station t_n, rather than the strain $\boldsymbol{\varepsilon}_n$, as part of $\boldsymbol{\alpha}_n$.

HYPLAS ***implementation***

In HYPLAS, the *element level* starts in subroutine ELEIIF. This routine identifies the *element class* and calls the appropriate element class-specific internal force computation routine. The element class-specific routine (e.g. IFSTD2 for standard two-dimensional isoparametric elements), in turn, calls the material interface for state update, MATISU. The material level starts here. MATISU identifies the material model/algorithm in question and calls the corresponding material-specific state updating procedure (e.g. routine SUTR for the Tresca elastoplastic model).

As we shall see in the following sections, the modularity concept discussed here is applied not only to the internal force computation, but also to the evaluation of the element tangent stiffness and some input/output operations. It makes the program particularly suitable for the incorporation of new elements and material models and/or modification of the existing ones.

5.6. Elements. Implementation and management

Finite elements in HYPLAS are grouped into element *classes*. One element class may contain several element *types*. Within the program, each element class and type is identified by a unique enumeration parameter set in the include file **ELEMENTS.INC**. This file defines our element database. It also sets maximum dimensioning parameters for some arrays of the global database whose size depends on properties of the available elements.

We define as an element class (this definition is by no means strict) a family of elements whose main operations for computation of internal force vector and tangent stiffness matrix follow the same steps. Thus, each element class has one subroutine for computation of the internal force vector and one subroutine for stiffness evaluation. These routines are used by all element types of the class. What differentiates element types within the same class are only trivial characteristics such as number of nodes, Gaussian integration rule, shape functions and so on.

All two-dimensional isoparametric elements of HYPLAS, for instance, form one class. The internal force computation for this class is carried out in routine IFSTD2 (see call tree of Figure 5.4). The stiffness is computed in STSTD2.

5.6.1. ELEMENT PROPERTIES. ELEMENT ROUTINES FOR DATA INPUT

The essential data defining one element type within a class is stored in COMMON block ELEMEN (refer to file **GLBDBASE.INC**). The data is stored in the arrays

IELPRP and RELPRP,

that contain, respectively, integer and real element properties. Typical element properties are listed below:

- *Integer element properties*. Number of nodes, number of degrees of freedom, number of Gauss points, number of boundaries, node ordering on boundaries, etc.
- *Real element properties*. Gauss quadrature weights and positions for domain and boundary integration, extrapolation matrices to extrapolate Gauss point values to nodes (used for output purposes only), etc.

Each group of elements in the structure (as defined in the input data file) is assigned one element type. The properties of the element type assigned to an element group IGRUP are stored in the columns

IELPRP(,IGRUP) and RELPRP(,IGRUP),

of the element properties arrays. These data are used by the corresponding element class-specific subroutines for computation of internal force and tangent stiffness matrices. Element properties are also used in the evaluation of the external load vector carried out in INLOAD.

Element type-specific data input routines

Element properties are assigned to the arrays of COMMON block ELEMEN during the input phase of the program (refer to source code of subroutine INDATA). To read the relevant data from the input data file and assign all necessary properties to IELPRP and RELPRP, each element type uses its own routine. Such subroutines are named following the convention:

RSxxxx.

For instance, subroutines RST3 and RSQ4 read and set data, respectively, for the isoparametric three-noded triangle and four-noded bi-linear quadrilateral. These routines are called only once during the program execution.

5.6.2. ELEMENT INTERFACES. INTERNAL FORCE AND STIFFNESS COMPUTATION

The modularisation of elements in the internal force computation has been discussed in Section 5.5.1. Subroutine ELEIIF in this case is the *element interface* routine that controls the evaluation of the element internal force vector. The basic task of the element interface here is to identify the class of the element whose internal force is required and call the appropriate class-specific internal force evaluation routine. For standard isoparametric two-dimensional elements, the class-specific routine is IFSTD2. For F-bar elements (discussed in Chapter 15), the corresponding class-specific routine is IFFBA2.

The evaluation of the element tangent stiffness matrix is also carried out in a modular structure completely analogous to that shown in Figure 5.4. The corresponding call tree is shown in Figure 5.5. At the element level, we have the interface routine ELEIST (called from FRONT) whose task is to identify element classes and call the class-specific routine for computation of the element tangent stiffness. The class-specific routines for standard and F-bar elements are STSTD2 and STFBA2, respectively.

5.6.3. IMPLEMENTING A NEW FINITE ELEMENT

In this section, we provide a summary of the basic path to be followed in order to include a new finite element into HYPLAS.

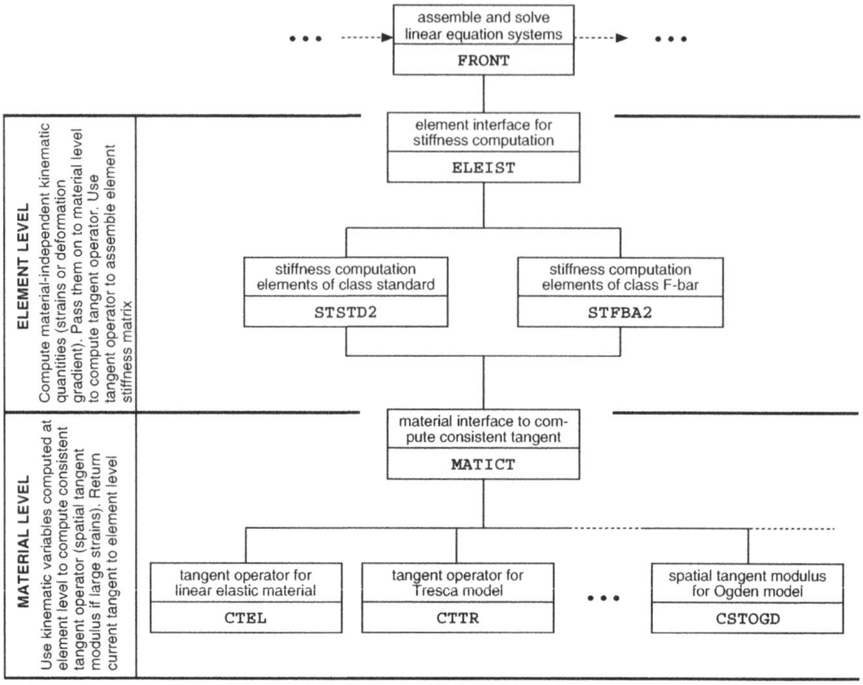

Figure 5.5. Consistent tangent computation. Modular structure.

A new element of an existing class

Assume that we want to include a new element type belonging to an existing element class. In this case, basic class-related internal force and stiffness computation routines already exist and do not need to be changed. The implementation of the new element requires coding of the following new element type-specific procedures:

1. The element data input/setting routine (`RSxxxx`).

2. The routine for evaluation of shape functions and shape function derivatives (`SFxxxx`). These routines have not been discussed above. They are called by `SHPFUN` which is used by element classes available in the program. Subroutine `SHPFUN` is the interface for shape function/shape function derivatives computation. It identifies the element *type* and calls the corresponding element type-specific shape function/derivative routine (refer to the source code of subroutine `SFT3` or `SFQ4`). New classes of element that do not use this structure will not require shape function routines coded in this way.

In addition, a new element-type identification parameter has to be added to the include file **ELEMENTS.INC**. For the interested reader a good exercise could be to include, say, the nine-noded Lagrangian isoparametric quadrilateral.

A new element class

Obviously, this will require more work than to include an element of an existing class. In addition to the above procedures, we need to code new element class-specific stiffness and internal force evaluation routines. This is the most demanding part. Also, a new element-*class* identification parameter has to be added to file **ELEMENTS.INC**, and calls to the new routines have to be included in the element interfaces `ELEIIF` and `ELEIST`.

5.7. Material models: implementation and management

Material models (or *types*) are also grouped into *classes*. Material types and classes are identified by identification parameters defined in the include file **MATERIAL.INC**. This file is the *material database* of `HYPLAS`. Grouping materials into classes is not absolutely necessary for modularity but becomes convenient particularly in the finite strain regime when, for instance, a number of models may require identical transformation of the basic kinematic variable ($\boldsymbol{F}_{\Delta}$) before the essential state updating procedure can be applied. A typical example is the class of isotropic elastoplasticity models implemented in the program (which also includes the linear elastic model). The identification parameter of this class is `HYPEPL` (see file **MATERIAL.INC**). Under finite strains, all materials of this class are logarithmic strain-based extensions of the corresponding infinitesimal models (the linear elastic model, for instance, turns into the Hencky material). The Hencky material is discussed in Chapter 13 and the finite elastoplasticity models are described in Chapter 14. In this case, the actual stress updating requires the computation of a logarithmic strain measure which is obtained for all models by performing identical operations on the basic kinematic variable, $\boldsymbol{F}_{\Delta}$. It then turns out conveniently – it avoids excessive repetition of code – to group all such models into a single material *class* and to perform the common extra kinematic operations before the actual material *type*-related procedure (which is unique for each model) starts.

5.7.1. MATERIAL PROPERTIES. MATERIAL-SPECIFIC DATA INPUT

Integer and real material properties are stored, respectively, in the arrays

`IPROPS` and `RPROPS`,

of `COMMON` block `MATERL` (see include file **GLBDBASE.INC**). The data stored in these arrays include all parameters required by the continuum constitutive model and, if required, may also contain other parameters related to the specific numerical algorithm adopted to integrate the constitutive equations of the model. Examples of integer and real material properties are:

- *Real material properties.* Usual properties such as Young's modulus, Poisson's ratio, parameters defining hardening curves of elastoplastic materials, etc.

- *Integer material properties.* Number of points defining a piecewise linear hardening curve for elastoplastic materials, number of terms in the series defining an Ogden-type finite strain hyperelasticity strain energy function, etc.

The situation here is analogous to the storage of element properties discussed in Section 5.6.1. Each group of elements defined by the user in the input data file is assigned one material type with one set of material properties. The material properties of a generic group IGRUP are stored in the columns

IPROPS(,IGRUP) and RPROPS(,IGRUP),

of the material properties arrays.

Material-specific data input routines

In the program, material-specific data are read from the input data file in material-specific subroutines; that is, each material type has its own self-contained subprogram that reads all material model/algorithm-related data and stores them in arrays IPROPS and RPROPS. The routines that execute this task are named following the convention:

RDxxxx.

For example, all data for the von Mises piecewise linear isotropic hardening model are read and set in subroutine RDVM. For the Ogden hyperelastic model, the corresponding routine is RDOGD.

The material interface for data input

Calls to all material-specific input data routines are controlled by the material interface routine

MATIRD

(**MAT**erial **I**nterface routine for **R**eading and setting material-specific **D**ata). The interface routine identifies the material type in question and calls the appropriate input routine.

5.7.2. STATE VARIABLES AND OTHER GAUSS POINT QUANTITIES. MATERIAL-SPECIFIC STATE UPDATING ROUTINES

All variables that define the *state* of the material at the Gauss point level are stored in the arrays of COMMON block STATE (defined in file **GLBDBASE.INC**). These include the stress components (stored in array STRSG), general real state variables (stored in RSTAVA), the Gauss point thickness (array THKGP) as well as the arrays RALGVA and LALGVA which store, respectively, real and logical *algorithmic* variables.

Array RSTAVA may store any state variable (other than the stress components) needed by the state updating procedure and/or required for output purposes. Typical real state variables are internal variables of the model, strain components, etc. The algorithmic variables stored in RALGVA and LALGVA are variables related to the constitutive integration algorithm employed to update the state of the material. Simple examples of real algorithmic variables are the incremental plastic multipliers associated with elastoplastic algorithms. A logical algorithmic variable can be, for instance, a flag telling us whether the material point is in elastic or elastoplastic regime. The Gauss point thickness stored in THKGP is a variable only in large strain analysis under plane stress.

The arrays of `COMMON` block `STATE` store both *current* and *last converged* values of the corresponding variables. The array positions having the last index equal to 1, such as

```
STRSG( , ,1)
```

for the Gauss point stresses, store the current values. Positions with the last index equal to 2, such as

```
STRSG( , ,2),
```

contain the last equilibrium converged values.

Material-specific state updating procedure

Here, we described in more detail the structure of the *material level* layer of Figure 5.4. Let us recall that, within the overall incremental finite element scheme, each material model implementation is defined by incremental constitutive functions, $\hat{\boldsymbol{\sigma}}$ and $\hat{\boldsymbol{\alpha}}$, such that the state $\{\boldsymbol{\sigma}_{n+1}, \boldsymbol{\alpha}_{n+1}\}$ at a Gauss point is symbolically represented as

$$\begin{aligned} \boldsymbol{\sigma}_{n+1} &= \hat{\boldsymbol{\sigma}}(\boldsymbol{\alpha}_n, \Delta\boldsymbol{\varepsilon}) \\ \boldsymbol{\alpha}_{n+1} &= \hat{\boldsymbol{\alpha}}(\boldsymbol{\alpha}_n, \Delta\boldsymbol{\varepsilon}). \end{aligned} \tag{5.14}$$

In the finite strain case, the above functions are defined in terms of the incremental deformation gradient, $\boldsymbol{F}_\Delta$, rather than $\Delta\boldsymbol{\varepsilon}$. For path-dependent constitutive models, the format of the incremental functions depends not only on the material model in question but also on the algorithm used to integrate its constitutive equations.

In the `HYPLAS` program, each material model implementation has a material-specific *state-updating procedure* that defines a pair of functions $\{\hat{\boldsymbol{\sigma}}, \hat{\boldsymbol{\alpha}}\}$. The state-updating procedures (subroutines) are named according to the rule

```
SUxxxx.
```

For the von Mises elastoplastic model, for instance, we have the subroutines `SUVM` (general case) and `SUVMPS` (plane stress algorithm).

The material interface for state updating

The management of calls to material-specific state-updating subroutines is carried out by the material interface routine

```
MATISU
```

(**MAT**erial **I**nterface for **S**tate-**U**pdating routine calls). This routine identifies the material *class* and *type*, performs extra class-specific kinematic operations (if necessary) and then calls the material *type*-specific routine `SUxxxx`.

5.7.3. MATERIAL-SPECIFIC SWITCHING/INITIALISING ROUTINES

The nature of the variables stored in the arrays of `COMMON` block `STATE` depends on the material model and algorithm in question. To keep the modularity of material models, it is thus convenient to perform the operations of initialisation of variables as well as switching between current and previous converged values in material-specific routines. The switching/initialising subroutines are named as

SWxxxx.

For the von Mises and Ogden models we have, respectively, the routines SWVM and SWOGD.

The material interface for switching/initialisation

All calls to material-specific switching/initialisation routines are made from the single material interface routine

MATISW

(**MAT**erial **I**nterface for **SW**itching/initialisation of data).

5.7.4. MATERIAL-SPECIFIC TANGENT COMPUTATION ROUTINES

The call tree for computation of material-related tangent operators is illustrated in Figure 5.5 (material level). In line with the material modularity discussed in the above, the computation of the consistent tangent operators (4.81) and (4.98), required in the assemblage of element tangent stiffness matrices, is also carried out in material-specific routines. The name convention adopted for such routines is

CTxxxx

for computation of *infinitesimal* consistent tangent operators (defined by (4.81)), and

CSTxxx

for consistent spatial tangent moduli (defined by (4.98)) needed in finite strain analyses. In this case, the material-specific routine for the von Mises model is named CTVM. The Ogden hyperelastic model uses subroutine CSTOGD.

The material interface for tangent computation

A material interface routine is also used to call the tangent operator computation subroutines. It identifies the material and calls the appropriate tangent computation routine. The **MAT**erial **I**nterface for **C**onsistent **T**angent computation is named

MATICT.

5.7.5. MATERIAL-SPECIFIC RESULTS OUTPUT ROUTINES

Finally, we describe the last material-related operation: the output of material-specific data to a results file. Whenever required, all variables other than the stress components and principal stresses (which are common to all material models) are written to the results file by material-specific routines.

The reason for adoption of material-specific routines in the output phase is simple: variables that are important for certain material models may be meaningless for other models.

For instance, the von Mises effective stress is important for metals but may be of no relevance for concrete failure models. Thus, the use of material-specific routines at this stage is convenient in order to print out only relevant results and keep, at the same time, a clean and modular structure. These routines are named in `HYPLAS` according to the convention

`ORxxxx.`

The output routine for the von Mises model, for instance, is named `ORVM`.

Material interface for results output

Again, a material interface routine is used to control calls to material-specific routines. The **MAT**erial **I**nterface to **O**utput **R**esults is the subroutine

`MATIOR.`

5.7.6. IMPLEMENTING A NEW MATERIAL MODEL

Within the modular structure described above, the incorporation of a new material model/algorithm into `HYPLAS`, requires the following basic steps:

1. *Add new entries to the materials database.* Add a new material-type identification parameter to the materials database (include file **MATERIAL.INC**). A new material class identification parameter needs to be added too, only if the new material belongs to a newly defined class. The maximum dimensioning parameters of file **MATERIAL.INC** have to be increased accordingly if the new material model requires more storage space.

2. *Write new material-specific routines.* The developer needs to code *five* material-specific routines. The crucial material-defining routines are

   2.1 The state updating procedure, `SUxxxx`. For path-dependent materials this is a numerical algorithm for integration of the corresponding constitutive equations.

   2.2 The routine for computation of the associated consistent tangent operator, `CTxxxx` (`CSTxxx` in finite strains).

   2.3 The routine for switching and initialising material-related variables, `SWxxxx`.

   Most of the programming work is concentrated in items 2.1 and 2.2. In addition, the following auxiliary routines are required:

   2.4 The routine for material data input `RDxxxx`.

   2.5 The routine for output of material-related results `ORxxxx`.

3. *Add new calls to the material interface routines.* Finally, calls to the above routines have to be added to the corresponding material interfaces: `MATISU`, `MATICT`, `MATISW`, `MATIRD` and `MATIOR`.

Note that changes to the existing code are needed only in the material database file **MATERIAL.INC** and in the above material interface routines.

Part Two

Small strains

6 THE MATHEMATICAL THEORY OF PLASTICITY

THE mathematical theory of plasticity provides a general framework for the continuum constitutive description of the behaviour of an important class of materials. Basically, the theory of plasticity is concerned with solids that, after being subjected to a loading programme, may sustain permanent (or *plastic*) deformations when completely unloaded. In particular, this theory is restricted to the description of materials (and conditions) for which the permanent deformations do *not* depend on the rate of application of loads and is often referred to as *rate-independent* plasticity. Materials whose behaviour can be adequately described by the theory of plasticity are called *plastic* (or *rate-independent plastic*) materials. A large number of engineering materials, such as metals, concrete, rocks, clays and soils in general, may be modelled as plastic under a wide range of circumstances of practical interest. The origins of the theory of plasticity can be traced back to the middle of the nineteenth century and, following the substantial development that took place, particularly in the first half of the twentieth century, this theory is today established on sound mathematical foundations and is regarded as one of the most successful phenomenological constitutive models of solid materials.

The present chapter reviews the mathematical theory of plasticity. The theory presented here is restricted to infinitesimal deformations and provides the basis for the numerical simulation of the behaviour of elastoplastic solids to be discussed in Chapter 7. We remark that only the most important concepts and mathematical expressions are reviewed. Attention is focused on the description of mathematical models of elastoplastic materials and, in particular, issues such as limit analysis and slip-line field theory are not addressed. For a more comprehensive treatment of the theory of plasticity, the reader is referred to Hill (1950), Prager (1959), Lubliner (1990) and Jirásek and Bažant (2002). A more mathematically oriented approach to the subject is presented by Halphen and Nguyen (1975), Duvaut and Lions (1976), Matthies (1979), Suquet (1981) and Han and Reddy (1999).

This chapter is organised as follows. In Section 6.1, aspects of the phenomenological behaviour of materials classed as plastic are discussed and the main properties are pointed out in the analysis of a simple uniaxial tension experiment. The discussion is followed, in Section 6.2, by the formulation of a mathematical model of the uniaxial experiment. The uniaxial model, though simple, embodies all the essential concepts of the mathematical theory of plasticity and provides the foundation for the general multidimensional model established in Section 6.3. The remainder of the chapter focuses on the detailed description of the plasticity models most commonly used in engineering analysis: the models of Tresca, von Mises, Mohr–Coulomb and Drucker–Prager. The corresponding yield criteria are

Computational Methods for Plasticity: Theory and Applications EA de Souza Neto, D Perić and DRJ Owen

described in Section 6.4. Plastic flow rules and hardening laws are addressed, respectively, in Sections 6.5 and 6.6.

6.1. Phenomenological aspects

In spite of their qualitatively distinct mechanical responses, materials as contrasting as metals and soils share some important features of their phenomenological behaviour that make them amenable to modelling by means of the theory of plasticity. To illustrate such common features, a uniaxial tension experiment with a metallic bar is discussed in what follows.

Typically, uniaxial tension tests with ductile metals produce stress–strain curves of the type shown in Figure 6.1. In the schematic diagram of Figure 6.1, where the axial stress, σ, is plotted against the axial strain, ε, a load programme has been considered in which the bar is initially subjected to a monotonic increase in axial stress from zero to a prescribed value, σ_0. The bar is then unloaded back to an unstressed state and subsequently reloaded to a higher stress level σ_1. The stress–strain curve follows the path $O_0Y_0Z_0O_1Y_1Z_1$ shown. In this path, the initial line segment O_0Y_0 is virtually straight and, if the bar is unloaded from point Y_0 (or before it is reached), it returns to the original unstressed state O_0. Thus, in segment O_0Y_0 the behaviour of the material is regarded as *linear elastic*. Beyond Y_0, the slope of the stress–strain curve changes dramatically and if the stress (or strain) loading is reversed at, say, point Z_0, the bar returns to an unstressed state via path Z_0O_1. The new unstressed state, O_1, differs from the initial unstressed state, O_0, in that a permanent change in the shape of the bar is observed. This shape change is represented in the graph by the permanent (or *plastic*) axial strain ε^p. Monotonic reloading of the bar to a stress level σ_1 will follow the path $O_1Y_1Z_1$. Similarly to the initial *elastic* segment O_0Y_0, the portion O_1Y_1 is also virtually straight and unloading from Y_1 (or before Y_1 is reached) will bring the stress–strain state back to the unstressed configuration O_1, with no further plastic straining of the bar. Therefore, the behaviour of the material in the segment O_1Y_1 may also be regarded as linear elastic. Here, it is important to emphasise that, even though some discrepancy between unloading and reloading curves (such as lines Z_0O_1 and O_1Y_1) is observed in typical experiments, the actual difference between them is in fact much smaller than that shown in the diagram of Figure 6.1. Again, loading beyond an *elastic limit* (point Y_1 in this case) will cause further increase in plastic deformation.

Some important phenomenological properties can be identified in the above described uniaxial test. They are enumerated below:

1. The existence of an *elastic domain*, i.e. a range of stresses within which the behaviour of the material can be considered as purely elastic, without evolution of permanent (plastic) strains. The elastic domain is delimited by the so-called *yield stress*. In Figure 6.1, segments O_0Y_0 and O_1Y_1 define the elastic domain at two different states. The associated yield stresses correspond to points Y_0 and Y_1.

2. If the material is further loaded at the yield stress, then *plastic yielding* (or *plastic flow*), i.e. evolution of plastic strains, takes place.

3. Accompanying the evolution of the plastic strain, an *evolution of the yield stress* itself is also observed (note that the yield stresses corresponding to points Y_0 and Y_1 are different). This phenomenon is known as *hardening*.

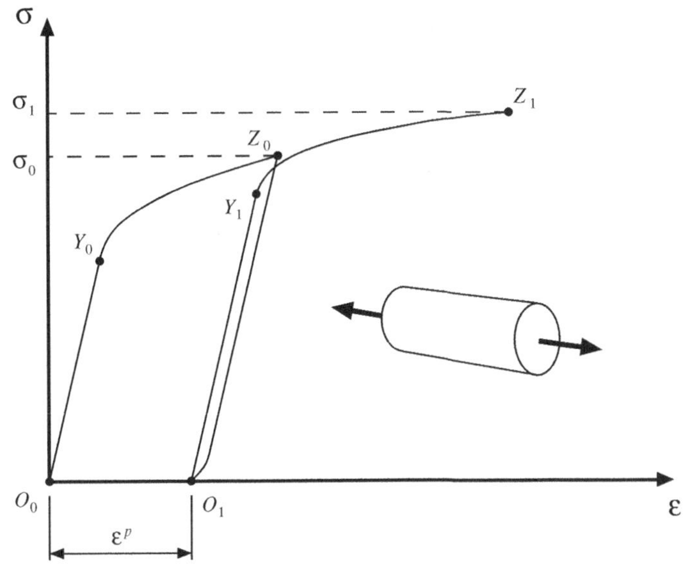

Figure 6.1. Uniaxial tension experiment with ductile metals.

It is emphasised that the above properties can be observed not only in metals but also in a wide variety of materials such as concrete, rocks, soils and many others. Obviously, the microscopic mechanisms that give rise to these common phenomenological characteristics can be completely distinct for different types of material. It is also important to note that, according to the type of material, different experimental procedures may be required for the verification of such properties. For instance, in materials such as soils, which typically cannot resist tensile stresses, uniaxial tension tests do not make physical sense. In this case, experiments such as *triaxial shear tests*, in which the sides of the specimen are subjected to a confining hydrostatic pressure prior to the application of longitudinal compression, are more appropriate.

The object of the mathematical theory of plasticity is to provide continuum constitutive models capable of describing (qualitatively and quantitavely) with sufficient accuracy the phenomenological behaviour of materials that possess the characteristics discussed in the above.

6.2. One-dimensional constitutive model

A simple mathematical model of the uniaxial experiment discussed in the previous section is formulated in what follows. In spite of its simplicity the one-dimensional constitutive model contains all the essential features that form the basis of the mathematical theory of plasticity.

At the outset, the original stress–strain curve of Figure 6.1, that resulted from the loading programme described in the previous section, is approximated by the idealised version shown in Figure 6.2. The assumptions involved in the approximation are summarised in the following. Firstly, the difference between unloading and reloading curves (segments Z_0O_1 and O_1Y_1 of Figure 6.1) is ignored and points Z_0 and Y_1, that correspond respectively to

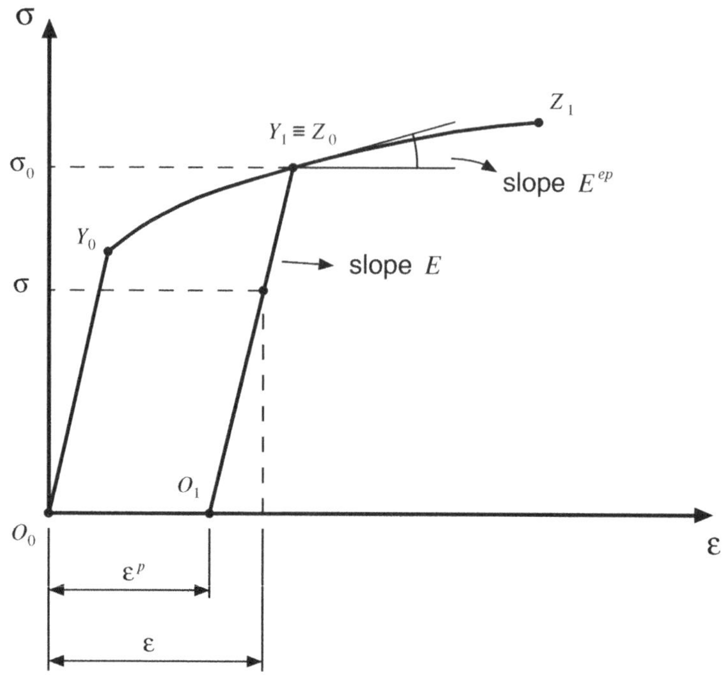

Figure 6.2. Uniaxial tension experiment. Mathematical model.

the beginning of unloading and the onset of plastic yielding upon subsequent reloading, are assumed to coincide. The transition between the elastic region and the elastoplastic regime is now clearly marked by a non-smooth change of slope (points Y_0 and Y_1). During plastic yielding, the stress–strain curve always follows the path defined by $O_0Y_0Y_1Z_1$. This path is normally referred to as the *virgin curve* and is obtained by a continuous monotonic loading from the initial unstressed state O_0.

Under the above assumptions, after being monotonically loaded from the initial unstressed state to the stress level σ_0, the behaviour of the bar between states O_1 and Y_1 is considered to be linear elastic, with constant plastic strain, ε^p, and yield limit, σ_0. Thus, within the segment O_1Y_1, the uniaxial stress corresponding to a configuration with *total* strain ε is given by

$$\sigma = E\,(\varepsilon - \varepsilon^p), \tag{6.1}$$

where E denotes the Young's modulus of the material of the bar. Note that the difference between the total strain and the current plastic strain, $\varepsilon - \varepsilon^p$, is *fully reversible*; that is, upon complete unloading of the bar, $\varepsilon - \varepsilon^p$ is fully recovered without further evolution of plastic strains. This motivates the additive decomposition of the axial strain described in the following section.

6.2.1. ELASTOPLASTIC DECOMPOSITION OF THE AXIAL STRAIN

One of the chief hypotheses underlying the small strain theory of plasticity is the decomposition of the *total strain*, ε, into the sum of an *elastic* (or reversible) component, ε^e, and a *plastic* (or permanent) component, ε^p,

$$\varepsilon = \varepsilon^e + \varepsilon^p, \tag{6.2}$$

where the *elastic strain* has been defined as

$$\varepsilon^e = \varepsilon - \varepsilon^p. \tag{6.3}$$

6.2.2. THE ELASTIC UNIAXIAL CONSTITUTIVE LAW

Following the above definition of the elastic axial strain, the constitutive law for the axial stress can be expressed as

$$\sigma = E\,\varepsilon^e. \tag{6.4}$$

The next step in the definition of the uniaxial constitutive model is to derive formulae that express mathematically the fundamental phenomenological properties enumerated in Section 6.1. The items 1 and 2 of Section 6.1 are associated with the formulation of a *yield criterion* and a *plastic flow rule*, whereas item 3 requires the formulation of a *hardening law*. These are described in the following.

6.2.3. THE YIELD FUNCTION AND THE YIELD CRITERION

The existence of an elastic domain delimited by a yield stress has been pointed out in item 1 of Section 6.1. With the introduction of a *yield function*, Φ, of the form

$$\Phi(\sigma, \sigma_y) = |\sigma| - \sigma_y, \tag{6.5}$$

the *elastic domain* at a state with uniaxial yield stress σ_y can be defined in the one-dimensional plasticity model as the set

$$\mathscr{E} = \{\sigma \mid \Phi(\sigma, \sigma_y) < 0\}, \tag{6.6}$$

or, equivalently, the elastic domain is the set of stresses σ that satisfy

$$|\sigma| < \sigma_y. \tag{6.7}$$

Generalising the results of the uniaxial *tension* test discussed, it has been assumed in the above that the yield stress in compression is identical to that in tension. The corresponding idealised elastic domain is illustrated in Figure 6.3.

It should be noted that, at any stage, no stress level is allowed above the current yield stress, i.e. *plastically admissible stresses* lie either in the elastic domain or on its boundary (the yield limit). Thus, any admissible stress must satisfy the restriction

$$\Phi(\sigma, \sigma_y) \leq 0. \tag{6.8}$$

For stress levels within the elastic domain, only elastic straining may occur, whereas on its boundary (at the yield stress), either elastic unloading or plastic yielding (or plastic loading) takes place. This *yield criterion* can be expressed by

$$\begin{aligned} &\text{If } \Phi(\sigma, \sigma_y) < 0 \implies \dot{\varepsilon}^p = 0, \\ &\text{If } \Phi(\sigma, \sigma_y) = 0 \implies \begin{cases} \dot{\varepsilon}^p = 0 & \text{for elastic unloading,} \\ \dot{\varepsilon}^p \neq 0 & \text{for plastic loading.} \end{cases} \end{aligned} \tag{6.9}$$

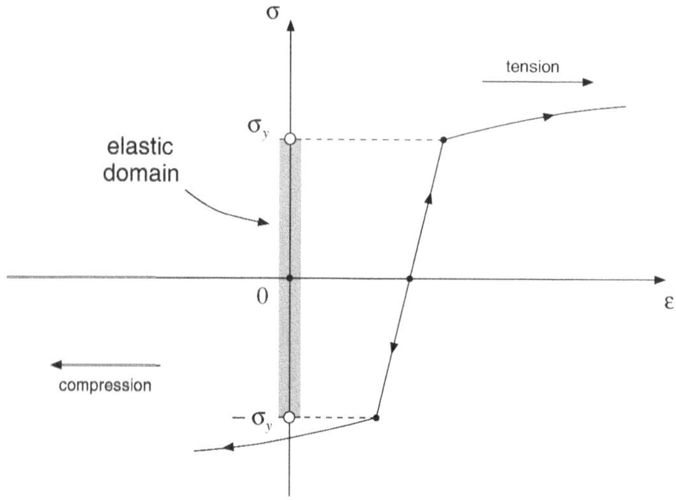

Figure 6.3. Uniaxial model. Elastic domain.

6.2.4. THE PLASTIC FLOW RULE. LOADING/UNLOADING CONDITIONS

Expressions (6.9) above have defined a criterion for plastic yielding, i.e. they have set the conditions under which plastic straining may occur. By noting in Figure 6.3 that, upon plastic loading, the plastic strain rate $\dot{\varepsilon}^p$ is positive (stretching) under tension (positive σ) and negative (compressive) under compression (negative σ), the *plastic flow rule* for the uniaxial model can be formally established as

$$\dot{\varepsilon}^p = \dot{\gamma}\ \mathrm{sign}(\sigma), \tag{6.10}$$

where sign is the *signum* function defined as

$$\mathrm{sign}(a) = \begin{cases} +1 & \text{if } a \geq 0 \\ -1 & \text{if } a < 0 \end{cases} \tag{6.11}$$

for any scalar a and the scalar $\dot{\gamma}$ is termed the *plastic multiplier*. The plastic multiplier is *non-negative*,

$$\dot{\gamma} \geq 0, \tag{6.12}$$

and satisfies the *complementarity condition*

$$\Phi\,\dot{\gamma} = 0. \tag{6.13}$$

The constitutive equations (6.10) to (6.13) imply that, as stated in the yield criterion (6.9), the plastic strain rate vanishes within the elastic domain, i.e.

$$\Phi < 0 \implies \dot{\gamma} = 0 \implies \dot{\varepsilon}^p = 0, \tag{6.14}$$

and plastic flow ($\dot{\varepsilon}^p \neq 0$) may occur only when the stress level σ coincides with the current yield stress

$$|\sigma| = \sigma_y \implies \Phi = 0 \implies \dot{\gamma} \geq 0. \tag{6.15}$$

Expressions (6.8), (6.12) and (6.13) define the so-called *loading/unloading conditions* of the elasticplastic model; that is, the constraints

$$\Phi \leq 0, \quad \dot{\gamma} \geq 0, \quad \dot{\gamma}\Phi = 0, \tag{6.16}$$

establish when plastic flow may occur.

6.2.5. THE HARDENING LAW

Finally, the complete characterisation of the uniaxial model is achieved with the introduction of the *hardening law*. As remarked in item 3 of Section 6.1, an evolution of the yield stress accompanies the evolution of the plastic strain. This phenomenon, known as *hardening*, can be incorporated into the uniaxial model simply by assuming that, in the definition (6.5) of Φ, the yield stress σ_y is a given function

$$\sigma_y = \sigma_y(\bar{\varepsilon}^p) \tag{6.17}$$

of the *accumulated* axial plastic strain, $\bar{\varepsilon}^p$. The accumulated axial plastic strain is defined as

$$\bar{\varepsilon}^p \equiv \int_0^t |\dot{\varepsilon}^p| \, \mathrm{d}t, \tag{6.18}$$

thus ensuring that both tensile and compressive plastic straining contribute to $\bar{\varepsilon}^p$. Clearly, in a monotonic tensile test we have

$$\bar{\varepsilon}^p = \varepsilon^p, \tag{6.19}$$

whereas in a monotonic compressive uniaxial test,

$$\bar{\varepsilon}^p = -\varepsilon^p. \tag{6.20}$$

The curve defined by the hardening function $\sigma_y(\bar{\varepsilon}^p)$ is usually referred to as the *hardening curve* (Figure 6.4).

From the definition of $\bar{\varepsilon}^p$, it follows that its evolution law is given by

$$\dot{\bar{\varepsilon}}^p = |\dot{\varepsilon}^p|, \tag{6.21}$$

which, in view of the plastic flow rule, is equivalent to

$$\dot{\bar{\varepsilon}}^p = \dot{\gamma}. \tag{6.22}$$

6.2.6. SUMMARY OF THE MODEL

The overall one-dimensional plasticity model is defined by the constitutive equations (6.2), (6.4), (6.5), (6.10), (6.16), (6.17), and (6.22). The model is summarised in Box 6.1.

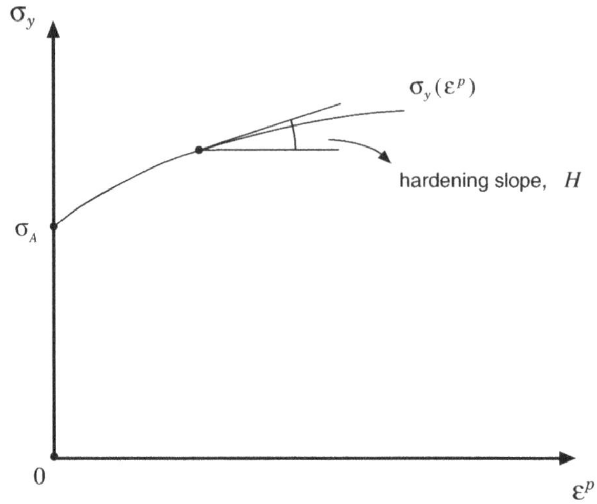

Figure 6.4. One-dimensional model. Hardening curve.

Box 6.1. One-dimensional elastoplastic constitutive model.

1. Elastoplastic split of the axial strain

$$\varepsilon = \varepsilon^e + \varepsilon^p$$

2. Uniaxial elastic law

$$\sigma = E\,\varepsilon^e$$

3. Yield function

$$\Phi(\sigma, \sigma_y) = |\sigma| - \sigma_y$$

4. Plastic flow rule

$$\dot{\varepsilon}^p = \dot{\gamma}\ \mathrm{sign}(\sigma)$$

5. Hardening law

$$\sigma_y = \sigma_y(\bar{\varepsilon}^p)$$

$$\dot{\bar{\varepsilon}}^p = \dot{\gamma}$$

6. Loading/unloading criterion

$$\Phi \leq 0, \quad \dot{\gamma} \geq 0, \quad \dot{\gamma}\Phi = 0$$

6.2.7. DETERMINATION OF THE PLASTIC MULTIPLIER

So far, in the uniaxial plasticity model introduced above, the *plastic multiplier*, $\dot{\gamma}$, was left indeterminate during plastic yielding. Indeed, expressions (6.12) and (6.13) just tell us that $\dot{\gamma}$ vanishes during elastic straining but may assume *any* non-negative value during plastic flow. In order to eliminate this indetermination, it should be noted firstly that, *during plastic flow*, the value of the yield function remains constant

$$\Phi = 0, \tag{6.23}$$

as the absolute value of the current stress always coincides with the current yield stress. Therefore, the following additional complementarity condition may be established:

$$\dot{\Phi}\,\dot{\gamma} = 0 \tag{6.24}$$

which implies that the rate of Φ vanishes whenever plastic yielding occurs ($\dot{\gamma} \neq 0$),

$$\dot{\Phi} = 0, \tag{6.25}$$

and, during elastic straining, ($\dot{\gamma} = 0$), $\dot{\Phi}$ may assume any value. Equation (6.25) is called the *consistency condition*. By taking the time derivative of the yield function (6.5), one obtains

$$\dot{\Phi} = \operatorname{sign}(\sigma)\,\dot{\sigma} - H\,\dot{\bar{\varepsilon}}^p, \tag{6.26}$$

where H is called the *hardening modulus*, or *hardening slope*, and is defined as (refer to Figure 6.4)

$$H = H(\bar{\varepsilon}^p) = \frac{\mathrm{d}\sigma_y}{\mathrm{d}\bar{\varepsilon}^p}. \tag{6.27}$$

Under plastic yielding, equation (6.25) holds so that one has the following expression for the stress rate

$$\operatorname{sign}(\sigma)\,\dot{\sigma} = H\,\dot{\bar{\varepsilon}}^p. \tag{6.28}$$

From the elastic law, it follows that

$$\dot{\sigma} = E(\dot{\varepsilon} - \dot{\varepsilon}^p). \tag{6.29}$$

Finally, by combining the above expression with (6.22), (6.28) and (6.10), the plastic multiplier, $\dot{\gamma}$, is *uniquely* determined during plastic yielding as

$$\dot{\gamma} = \frac{E}{H+E}\operatorname{sign}(\sigma)\,\dot{\varepsilon} = \frac{E}{H+E}\,|\dot{\varepsilon}|. \tag{6.30}$$

6.2.8. THE ELASTOPLASTIC TANGENT MODULUS

Let us now return to the stress–strain curve of Figure 6.2. Plastic flow at a generic yield limit produces the following tangent relation between strain and stress

$$\dot{\sigma} = E^{ep}\,\dot{\varepsilon}, \tag{6.31}$$

where E^{ep} is called the *elastoplastic tangent modulus*. By combining expressions (6.31), (6.29), the flow rule (6.10) and (6.30) the following expression is obtained for the elastoplastic tangent modulus

$$E^{ep} = \frac{E\,H}{E+H}. \tag{6.32}$$

Equivalently, the hardening modulus, H, can be expressed in terms of the elastic modulus and the elastoplastic modulus as

$$H = \frac{E^{ep}}{1 - E^{ep}/E}. \tag{6.33}$$

6.3. General elastoplastic constitutive model

A mathematical model of a uniaxial tension experiment with a ductile metal has been described in the previous section. As already mentioned, the one-dimensional equations contain all basic components of a general elastoplastic constitutive model:

- the elastoplastic strain decomposition;
- an elastic law;
- a yield criterion, stated with the use of a yield function;
- a plastic flow rule defining the evolution of the plastic strain; and
- a hardening law, characterising the evolution of the yield limit.

The generalisation of these concepts for application in two- and three-dimensional situations is described in this section.

6.3.1. ADDITIVE DECOMPOSITION OF THE STRAIN TENSOR

Following the decomposition of the uniaxial strain given in the previous section, the corresponding generalisation is obtained by splitting the strain *tensor*, $\boldsymbol{\varepsilon}$, into the sum of an elastic component, $\boldsymbol{\varepsilon}^e$, and a plastic component, $\boldsymbol{\varepsilon}^p$; that is,

$$\boldsymbol{\varepsilon} = \boldsymbol{\varepsilon}^e + \boldsymbol{\varepsilon}^p. \tag{6.34}$$

The tensors $\boldsymbol{\varepsilon}^e$ and $\boldsymbol{\varepsilon}^p$ are known, respectively, as the *elastic strain tensor* and the *plastic strain tensor*. The corresponding rate form of the additive split reads

$$\dot{\boldsymbol{\varepsilon}} = \dot{\boldsymbol{\varepsilon}}^e + \dot{\boldsymbol{\varepsilon}}^p. \tag{6.35}$$

Note that (6.35) together with the given initial condition

$$\boldsymbol{\varepsilon}(t_0) = \boldsymbol{\varepsilon}^e(t_0) + \boldsymbol{\varepsilon}^p(t_0) \tag{6.36}$$

at a (pseudo-)time t_0 is equivalent to (6.34).

6.3.2. THE FREE ENERGY POTENTIAL AND THE ELASTIC LAW

The formulation of general dissipative models of solids within the framework of thermodynamics with an internal variable has been addressed in Section 3.5 of Chapter 3. Recall that the free energy potential plays a crucial role in the derivation of the model and provides the constitutive law for stress. The starting point of the theories of plasticity treated in this book is the assumption that the free energy, ψ, is a function

$$\psi(\boldsymbol{\varepsilon}, \boldsymbol{\varepsilon}^p, \boldsymbol{\alpha}),$$

of the total strain, the plastic strain (taken as an internal variable) and a set $\boldsymbol{\alpha}$ of internal variables associated with the phenomenon of hardening. It is usual to assume that the free energy can be split as

$$\begin{aligned}\psi(\varepsilon, \varepsilon^p, \alpha) &= \psi^e(\varepsilon - \varepsilon^p) + \psi^p(\alpha) \\ &= \psi^e(\varepsilon^e) + \psi^p(\alpha)\end{aligned} \tag{6.37}$$

into a sum of an elastic contribution, ψ^e, whose dependence upon strains and internal variables appears only through the elastic strain, and a contribution due to hardening, ψ^p.

Following the above expression for the free energy, the Clausius–Duhem inequality reads

$$\left(\sigma - \bar{\rho}\,\frac{\partial\psi^e}{\partial\varepsilon^e}\right) : \dot{\varepsilon}^e + \sigma : \dot{\varepsilon}^p - A * \dot{\alpha} \geq 0, \tag{6.38}$$

where

$$A \equiv \bar{\rho}\,\partial\psi^p/\partial\alpha \tag{6.39}$$

is the *hardening thermodynamical force* and we note that $-\sigma$ is the thermodynamical force associated with the plastic strain while the symbol $*$ indicates the appropriate product between A and $\dot{\alpha}$. The above inequality implies a general elastic law of the form

$$\sigma = \bar{\rho}\,\frac{\partial\psi^e}{\partial\varepsilon^e}, \tag{6.40}$$

so that the requirement of non-negative dissipation can be reduced to

$$\Upsilon^p(\sigma, A; \dot{\varepsilon}^p, \dot{\alpha}) \geq 0, \tag{6.41}$$

where the function Υ^p, defined by

$$\Upsilon^p(\sigma, A; \dot{\varepsilon}^p, \dot{\alpha}) \equiv \sigma : \dot{\varepsilon}^p - A * \dot{\alpha}, \tag{6.42}$$

is called the *plastic dissipation function*.

This chapter is focused on materials whose elastic behaviour is *linear* (as in the uniaxial model of the previous section) and isotropic. In this case, the elastic contribution to the free energy is given by

$$\begin{aligned}\bar{\rho}\,\psi^e(\varepsilon^e) &= \tfrac{1}{2}\,\varepsilon^e : \mathsf{D}^e : \varepsilon^e \\ &= G\,\varepsilon^e_d : \varepsilon^e_d + \tfrac{1}{2}\,K\,(\varepsilon^e_v)^2\end{aligned} \tag{6.43}$$

where D^e is the standard isotropic elasticity tensor and G and K are, respectively the shear and bulk moduli. The tensor ε^e_d is the deviatoric component of the elastic strain and $\varepsilon^e_v \equiv \mathrm{tr}[\varepsilon^e]$ is the volumetric elastic strain. Thus, the general counterpart of uniaxial elastic law (6.4) is given by

$$\begin{aligned}\sigma &= \mathsf{D}^e : \varepsilon^e \\ &= 2G\,\varepsilon^e_d + K\,\varepsilon^e_v\,I.\end{aligned} \tag{6.44}$$

6.3.3. THE YIELD CRITERION AND THE YIELD SURFACE

Recall that in the uniaxial yield criterion it was established that plastic flow may occur when the uniaxial stress attains a critical value. This principle could be expressed by means of a yield function which is negative when only elastic deformations are possible and reaches zero when plastic flow is imminent. Extension of this concept to the three-dimensional case is obtained by stating that plastic flow may occur only when

$$\Phi(\boldsymbol{\sigma}, \boldsymbol{A}) = 0, \tag{6.45}$$

where the scalar yield function, Φ, is now a function of the stress *tensor* and a set $\boldsymbol{A}$ of hardening thermodynamical forces. Analogously to the uniaxial case, a yield function defines the *elastic domain* as the set

$$\mathscr{E} = \{\boldsymbol{\sigma} \mid \Phi(\boldsymbol{\sigma}, \boldsymbol{A}) < 0\} \tag{6.46}$$

of stresses for which plastic yielding is not possible. Any stress lying in the elastic domain or on its boundary is said to be *plastically admissible*. We then define the *set of plastically admissible stresses* (or *plastically admissible domain*) as

$$\bar{\mathscr{E}} = \{\boldsymbol{\sigma} \mid \Phi(\boldsymbol{\sigma}, \boldsymbol{A}) \leq 0\}. \tag{6.47}$$

The yield locus, i.e. the set of stresses for which plastic yielding may occur, is the boundary of the elastic domain, where $\Phi(\boldsymbol{\sigma}, \boldsymbol{A}) = 0$. The yield locus in this case is represented by a hypersurface in the space of stresses. This hypersurface is termed the *yield surface* and is defined as

$$\mathscr{Y} = \{\boldsymbol{\sigma} \mid \Phi(\boldsymbol{\sigma}, \boldsymbol{A}) = 0\}. \tag{6.48}$$

6.3.4. PLASTIC FLOW RULE AND HARDENING LAW

The complete characterisation of the general plasticity model requires the definition of the evolution laws for the internal variables, i.e. the variables associated with the dissipative phenomena. In the present case, the internal variables are the plastic strain tensor and the set $\boldsymbol{\alpha}$ of hardening variables. The following plastic flow rule and hardening law are then postulated

$$\dot{\boldsymbol{\varepsilon}}^p = \dot{\gamma}\, \boldsymbol{N} \tag{6.49}$$

$$\dot{\boldsymbol{\alpha}} = \dot{\gamma}\, \boldsymbol{H}, \tag{6.50}$$

where the tensor

$$\boldsymbol{N} = \boldsymbol{N}(\boldsymbol{\sigma}, \boldsymbol{A}) \tag{6.51}$$

is termed the *flow vector* and the function

$$\boldsymbol{H} = \boldsymbol{H}(\boldsymbol{\sigma}, \boldsymbol{A}) \tag{6.52}$$

is the *generalised hardening modulus* which defines the evolution of the hardening variables. The evolution equations (6.49) and (6.50) are complemented by the loading/unloading conditions

$$\Phi \leq 0, \quad \dot{\gamma} \geq 0, \quad \Phi\dot{\gamma} = 0, \tag{6.53}$$

that define when evolution of plastic strains and internal variables ($\dot{\gamma} \neq 0$) may occur.

For convenience, the general plasticity model resulting from the above equations is listed in Box 6.2.

Box 6.2. A general elastoplastic constitutive model.

1. Additive decomposition of the strain tensor

$$\boldsymbol{\varepsilon} = \boldsymbol{\varepsilon}^e + \boldsymbol{\varepsilon}^p$$

or

$$\dot{\boldsymbol{\varepsilon}} = \dot{\boldsymbol{\varepsilon}}^e + \dot{\boldsymbol{\varepsilon}}^p, \quad \boldsymbol{\varepsilon}(t_0) = \boldsymbol{\varepsilon}^e(t_0) + \boldsymbol{\varepsilon}^p(t_0)$$

2. Free-energy function

$$\psi = \psi(\boldsymbol{\varepsilon}^e, \boldsymbol{\alpha})$$

where $\boldsymbol{\alpha}$ is a set of hardening internal variables

3. Constitutive equation for $\boldsymbol{\sigma}$ and hardening thermodynamic forces $\boldsymbol{A}$

$$\boldsymbol{\sigma} = \bar{\rho}\,\frac{\partial \psi}{\partial \boldsymbol{\varepsilon}^e}, \quad \boldsymbol{A} = \bar{\rho}\,\frac{\partial \psi}{\partial \boldsymbol{\alpha}}$$

4. Yield function

$$\Phi = \Phi(\boldsymbol{\sigma}, \boldsymbol{A})$$

5. Plastic flow rule and hardening law

$$\dot{\boldsymbol{\varepsilon}}^p = \dot{\gamma}\,\boldsymbol{N}(\boldsymbol{\sigma}, \boldsymbol{A})$$

$$\dot{\boldsymbol{\alpha}} = \dot{\gamma}\,\boldsymbol{H}(\boldsymbol{\sigma}, \boldsymbol{A})$$

6. Loading/unloading criterion

$$\Phi \leq 0, \quad \dot{\gamma} \geq 0, \quad \dot{\gamma}\Phi = 0$$

6.3.5. FLOW RULES DERIVED FROM A FLOW POTENTIAL

In the formulation of multidimensional plasticity models, it is often convenient to define the flow rule (and possibly the hardening law) in terms of a *flow* (or *plastic*) *potential*. The starting point of such an approach is to postulate the existence of a flow potential with general form

$$\Psi = \Psi(\boldsymbol{\sigma}, \boldsymbol{A}) \tag{6.54}$$

from which the flow vector, $\boldsymbol{N}$, is obtained as

$$\boldsymbol{N} \equiv \frac{\partial \Psi}{\partial \boldsymbol{\sigma}}. \tag{6.55}$$

If the hardening law is assumed to be derived from the same potential, then we have in addition

$$\boldsymbol{H} \equiv -\frac{\partial \Psi}{\partial \boldsymbol{A}}. \tag{6.56}$$

When such an approach is adopted, the plastic potential, Ψ, is required to be a non-negative convex function of both $\boldsymbol{\sigma}$ and $\boldsymbol{A}$ and zero-valued at the origin,

$$\Psi(\mathbf{0}, \mathbf{0}) = 0. \tag{6.57}$$

These restrictions ensure that the dissipation inequality (6.41) is satisfied *a priori* by the evolution equations (6.49) and (6.50).

Associative flow rule

As we shall see later, many plasticity models, particularly for ductile metals, have their yield function, Φ, as a flow potential, i.e.

$$\Psi \equiv \Phi. \tag{6.58}$$

Such models are called *associative* (or *associated*) plasticity models. The issue of associativity will be further discussed in Section 6.5.1.

6.3.6. THE PLASTIC MULTIPLIER

Here we extend to the multidimensional case the procedure for the determination of the plastic multiplier, $\dot{\gamma}$, described in Section 6.2.7 for the one-dimensional plasticity model. Following the same arguments employed in Section 6.2.7, the starting point in the determination of $\dot{\gamma}$ is the consideration of the additional complementarity equation

$$\dot{\Phi}\,\dot{\gamma} = 0, \tag{6.59}$$

which implies the *consistency condition*

$$\dot{\Phi} = 0 \tag{6.60}$$

under plastic yielding (when $\dot{\gamma} \neq 0$). By differentiating the yield function with respect to time, we obtain

$$\dot{\Phi} = \frac{\partial \Phi}{\partial \boldsymbol{\sigma}} : \dot{\boldsymbol{\sigma}} + \frac{\partial \Phi}{\partial \boldsymbol{A}} * \dot{\boldsymbol{A}}. \tag{6.61}$$

By taking into account the additive split of the strain tensor, the elastic law and the plastic flow rule (6.49), we promptly find the obvious rate form

$$\dot{\boldsymbol{\sigma}} = \mathsf{D}^e : (\dot{\boldsymbol{\varepsilon}} - \dot{\boldsymbol{\varepsilon}}^p) = \mathsf{D}^e : (\dot{\boldsymbol{\varepsilon}} - \dot{\gamma}\,\boldsymbol{N}). \tag{6.62}$$

This, together with the definition of $\boldsymbol{A}$ in terms of the free-energy potential (refer to expression (6.39)) and the evolution law (6.50), allow us to write (6.61) equivalently as

$$\begin{aligned}\dot{\Phi} &= \frac{\partial \Phi}{\partial \boldsymbol{\sigma}} : \mathsf{D}^e : (\dot{\boldsymbol{\varepsilon}} - \dot{\boldsymbol{\varepsilon}}^p) + \frac{\partial \Phi}{\partial \boldsymbol{A}} * \bar{\rho}\frac{\partial^2 \psi^p}{\partial \boldsymbol{\alpha}^2} * \dot{\boldsymbol{\alpha}}.\\ &= \frac{\partial \Phi}{\partial \boldsymbol{\sigma}} : \mathsf{D}^e : (\dot{\boldsymbol{\varepsilon}} - \dot{\gamma}\,\boldsymbol{N}) + \dot{\gamma}\,\frac{\partial \Phi}{\partial \boldsymbol{A}} * \bar{\rho}\frac{\partial^2 \psi^p}{\partial \boldsymbol{\alpha}^2} * \boldsymbol{H}.\end{aligned} \tag{6.63}$$

Finally, the above expression and the consistency condition (6.60) lead to the following closed formula for the plastic multiplier

$$\dot{\gamma} = \frac{\partial \Phi / \partial \boldsymbol{\sigma} : \mathsf{D}^e : \dot{\boldsymbol{\varepsilon}}}{\partial \Phi / \partial \boldsymbol{\sigma} : \mathsf{D}^e : \boldsymbol{N} - \partial \Phi / \partial \boldsymbol{A} * \bar{\rho}\partial^2 \psi^p / \partial \boldsymbol{\alpha}^2 * \boldsymbol{H}}. \tag{6.64}$$

6.3.7. RELATION TO THE GENERAL CONTINUUM CONSTITUTIVE THEORY

At this point, we should emphasise that the general rate-independent plasticity model described above can under some conditions be shown to be a particular instance of the general constitutive theory postulated in Section 3.5.2, starting page 71. The link between the two theories can be clearly demonstrated when rate-independent plasticity is obtained as a limit case of rate-dependent plasticity (or viscoplasticity).

However, since the theory of elasto-viscoplasticity is introduced only in Chapter 11, we find it convenient to carry on focusing on rate-independent plasticity and postpone the demonstration until that chapter. Those wishing to see now the link between rate-independent plasticity and the general constitutive theory are referred to Section 11.4.3, starting on page 452. We remark, though, that the concept of subdifferential, introduced below in Section 6.3.9, is fundamental to the demonstration. Readers not yet familiar with this concept are advised to read through Section 6.3.9 before moving to Section 11.4.3.

6.3.8. RATE FORM AND THE ELASTOPLASTIC TANGENT OPERATOR

In the elastic regime, the rate constitutive equation for stress reads simply

$$\dot{\boldsymbol{\sigma}} = \mathsf{D}^e : \dot{\boldsymbol{\varepsilon}}. \tag{6.65}$$

Under plastic flow, the corresponding rate relation can be obtained by introducing expression (6.64) into (6.62). The rate equation reads

$$\dot{\boldsymbol{\sigma}} = \mathsf{D}^{ep} : \dot{\boldsymbol{\varepsilon}}, \tag{6.66}$$

where D^{ep} is the *elastoplastic tangent modulus* given by

$$\mathsf{D}^{ep} = \mathsf{D}^e - \frac{(\mathsf{D}^e : \boldsymbol{N}) \otimes (\mathsf{D}^e : \partial\Phi/\partial\boldsymbol{\sigma})}{\partial\Phi/\partial\boldsymbol{\sigma} : \mathsf{D}^e : \boldsymbol{N} - \partial\Phi/\partial\boldsymbol{A} * \bar{\rho}\partial^2\psi^p/\partial\boldsymbol{\alpha}^2 * \boldsymbol{H}}. \tag{6.67}$$

In obtaining the above expression, we have made use of the fact that the symmetry (refer to equation (2.87), page 29) of the elasticity tensor implies

$$\partial\Phi/\partial\boldsymbol{\sigma} : \mathsf{D}^e : \dot{\boldsymbol{\varepsilon}} = \mathsf{D}^e : \partial\Phi/\partial\boldsymbol{\sigma} : \dot{\boldsymbol{\varepsilon}}. \tag{6.68}$$

The fourth-order tensor D^{ep} is the multidimensional generalisation of the scalar modulus E^{ep} associated with the slope of the uniaxial stress–strain curve under plastic flow. In the computational plasticity literature, D^{ep} is frequently referred to as the *continuum* elastoplastic tangent operator.

Remark 6.1 (The symmetry of D^{ep}). Note that if the plastic flow rule is *associative*, i.e. if $\boldsymbol{N} \equiv \partial\Phi/\partial\boldsymbol{\sigma}$, then the continuum elastoplastic tangent operator is *symmetric*. For models with non-associative plastic flow, D^{ep} is generally unsymmetric.

6.3.9. NON-SMOOTH POTENTIALS AND THE SUBDIFFERENTIAL

It should be noted that expressions (6.55) and (6.56) only make sense if the potential Ψ is differentiable. When that happens, the flow vector, $\boldsymbol{N}$, can be interpreted as a vector

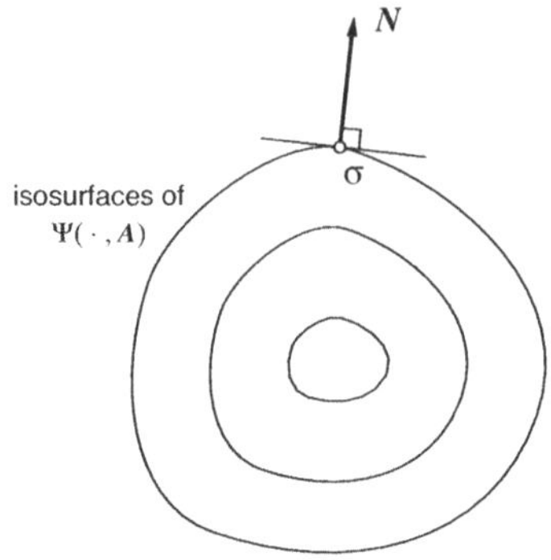

Figure 6.5. The flow vector. Smooth potential.

normal to the iso-surfaces of function Ψ in the space of stresses (with fixed $\boldsymbol{A}$). A schematic representation of $\boldsymbol{N}$ in this case is shown in Figure 6.5. The generalised modulus, $\boldsymbol{H}$, can be interpreted in a completely analogous way.

The requirement of differentiability of the flow potential is, however, too restrictive and many practical plasticity models are based on the use of a non-differentiable Ψ. Specific examples are given later in this chapter. For a more comprehensive account of such theories the reader is referred to Duvaut and Lions (1976), Eve *et al.* (1990) and Han and Reddy (1999). In such cases, the function Ψ is called a *pseudo-potential* or *generalised potential* and the formulation of the evolution laws for the internal variables can be dealt with by introducing the concept of *subdifferential sets*, which generalises the classical definition of derivative.[†]

Subgradients and the subdifferential

Let us consider a *scalar* function $y : \mathscr{R}^n \rightarrow \mathscr{R}$. The *subdifferential* of y at a point $\bar{\boldsymbol{x}}$ is the set

$$\partial y(\bar{\boldsymbol{x}}) = \{s \in \mathscr{R}^n \mid y(\boldsymbol{x}) - y(\bar{\boldsymbol{x}}) \geq \boldsymbol{s} \cdot (\boldsymbol{x} - \bar{\boldsymbol{x}}),\ \forall\, \boldsymbol{x} \in \mathscr{R}^n\}. \tag{6.69}$$

If the set ∂y is not empty at $\bar{\boldsymbol{x}}$, the function y is said to be *subdifferentiable* at $\bar{\boldsymbol{x}}$. The elements of ∂y are called *subgradients* of y. If the function y is *differentiable*, then the subdifferential contains a *unique* subgradient which coincides with the derivative of y,

$$\partial y = \left\{ \frac{\mathrm{d}y}{\mathrm{d}\boldsymbol{x}} \right\}. \tag{6.70}$$

A schematic illustration of the concept of subdifferential is shown in Figure 6.6 for $n = 1$. In this case, when y is subdifferentiable (but not necessarily differentiable) at a point $\bar{x}$, the subdifferential at that point is composed of all slopes s lying between the slopes on the right and left of $\bar{x}$ (the two one-sided derivatives of y at $\bar{x}$).

[†]The concept of subdifferential sets is exploited extensively in convex analysis. The reader is referred to Rockafellar (1970), Part V, for a detailed account of the theory of subdifferentiable functions.

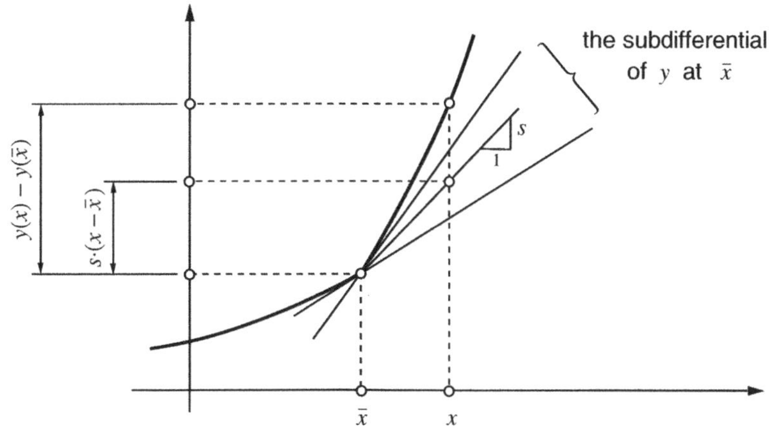

Figure 6.6. The subdifferential of a convex function.

Plastic flow with subdifferentiable flow potentials

Assume now that the (pseudo-) potential Ψ is a subdifferentiable function of $\boldsymbol{\sigma}$ and $\boldsymbol{A}$. At points where Ψ is non-differentiable in $\boldsymbol{\sigma}$, the isosurfaces of Ψ in the space of stresses contain a singularity (corner) where the normal direction is not uniquely defined. A typical situation is schematically illustrated in Figure 6.7 where two distinct normals, $\boldsymbol{N}_1$ and $\boldsymbol{N}_2$, are assumed to exist. In this case, the subdifferential of Ψ with respect to $\boldsymbol{\sigma}$, denoted $\partial_{\boldsymbol{\sigma}}\Psi$, is the set of vectors contained in the cone defined by all linear combinations (with positive coefficients) of $\boldsymbol{N}_1$ and $\boldsymbol{N}_2$. The generalisation of the plastic flow rule (6.49) is obtained by replacing expression (6.55) for the flow vector with

$$\boldsymbol{N} \in \partial_{\boldsymbol{\sigma}}\Psi, \tag{6.71}$$

i.e. the flow vector $\boldsymbol{N}$ is now assumed to be a *subgradient* of Ψ. Analogously, the evolution law (6.50) for $\boldsymbol{\alpha}$ can be generalised with the replacement of the definition (6.56) by

$$\boldsymbol{H} \in -\partial_{\boldsymbol{A}}\Psi. \tag{6.72}$$

At this point, it should be remarked that differentiability of Ψ with respect to the stress tensor is violated for some very basic plasticity models, such as the Tresca, Mohr–Coulomb and Drucker–Prager theories to be seen later. Therefore, the concepts of subgradient and subdifferential sets introduced above are important in the formulation of evolution laws for $\boldsymbol{\varepsilon}^p$.

An alternative definition of the plastic flow rule with non-smooth potentials, which incorporates a wide class of models, is obtained as follows. Firstly assume that a *finite* number, n, of distinct normals ($\boldsymbol{N}_1, \boldsymbol{N}_2, \ldots, \boldsymbol{N}_n$) is defined at a generic singular point of an isosurface of Ψ. In this case, any subgradient of Ψ can be written as a linear combination

$$c_1\boldsymbol{N}_1 + c_2\boldsymbol{N}_2 + \cdots + c_n\boldsymbol{N}_n,$$

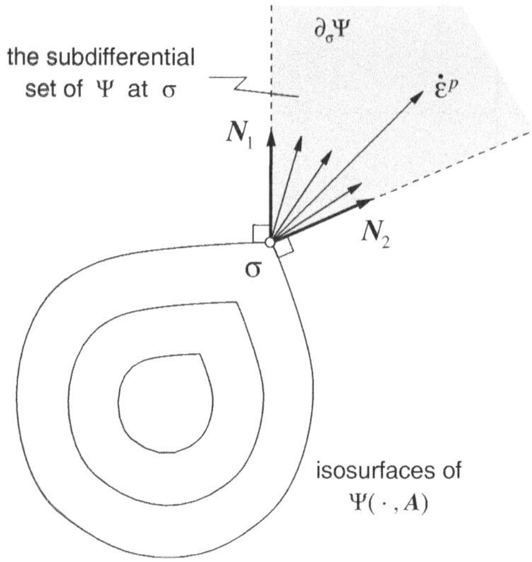

Figure 6.7. The flow vector. Non-smooth potential.

with *non-negative* coefficients $c_1, c_2, \ldots, c_n$.‡ Based on this observation, the flow rule (6.49) can be generalised as

$$\dot{\varepsilon}^p = \sum_{i=1}^{n} \dot{\gamma}_i \, \boldsymbol{N}_i, \tag{6.73}$$

with all n plastic multipliers required to be non-negative

$$\dot{\gamma}_i \geq 0, \quad i = 1, \ldots, n. \tag{6.74}$$

The generalisation of the plastic flow law, in this format, was originally proposed by Koiter (1953).

Multisurface models

The above concepts are particularly useful in defining evolution laws for *multisurface* plasticity models. In a generic multisurface model, the elastic domain is bound by a *set* of n surfaces in the space of stresses which intersect in a non-smooth fashion. In this case, n yield functions (Φ_i, $i = 1, \ldots, n$) are defined so that each bounding surface is given by an equation

$$\Phi_i(\boldsymbol{\sigma}, \boldsymbol{A}) = 0. \tag{6.75}$$

The elastic domain in this case reads

$$\mathscr{E} = \{\boldsymbol{\sigma} \mid \Phi_i(\boldsymbol{\sigma}, \boldsymbol{A}) < 0, \; i = 1, \ldots, n\}, \tag{6.76}$$

and the yield surface, i.e. the boundary of $\mathscr{E}$, is the set of all stresses such that $\Phi_i(\boldsymbol{\sigma}, \boldsymbol{A}) = 0$ for at least one i and $\Phi_j(\boldsymbol{\sigma}, \boldsymbol{A}) \leq 0$ for all other indices $j \neq i$.

‡It should be emphasised that this representation is not valid for certain types of singularity where the corresponding subdifferential set cannot be generated by a finite number of vectors.

Assuming associativity ($\Psi \equiv \Phi$), the situation discussed previously, where the subgradient of the flow potential is a linear combination of a finite number of normals, is recovered. Thus, the plastic flow rule can be written in the general form (6.73) with the normals being defined here as

$$\boldsymbol{N}_i = \frac{\partial \Phi_i}{\partial \boldsymbol{\sigma}}. \tag{6.77}$$

In the present case, the standard loading/unloading criterion (6.53) is replaced by the generalisation

$$\Phi_i \leq 0, \quad \dot{\gamma}_i \geq 0, \quad \Phi_i \dot{\gamma}_i = 0, \tag{6.78}$$

which must hold for each $i = 1, \ldots, n$. Note that summation on repeated indices is not implied in the above law.

6.4. Classical yield criteria

The general constitutive model for elastoplastic materials has been established in the previous section. There, the yield criterion has been stated in its general form, without reference to any particular criteria. In this section, some of the most common yield criteria used in engineering practice are described in detail; namely, the criteria of *Tresca*, *von Mises*, *Mohr–Coulomb* and *Drucker–Prager*.

6.4.1. THE TRESCA YIELD CRITERION

This criterion was proposed by Tresca (1868) to describe plastic yielding in metals. The Tresca yield criterion assumes that *plastic yielding begins when the maximum shear stress reaches a critical value.*

Recall the spectral representation of the stress tensor,

$$\boldsymbol{\sigma} = \sum_{i=1}^{3} \sigma_i \, \boldsymbol{e}_i \otimes \boldsymbol{e}_i, \tag{6.79}$$

where σ_i are the *principal stresses* and $\boldsymbol{e}_i$ the associated unit eigenvectors, and let $\sigma_{\max}$ and $\sigma_{\min}$ be, respectively, the maximum and minimum principal stresses

$$\begin{aligned} \sigma_{\max} &= \max(\sigma_1, \sigma_2, \sigma_3); \\ \sigma_{\min} &= \min(\sigma_1, \sigma_2, \sigma_3). \end{aligned} \tag{6.80}$$

The *maximum shear stress*, $\tau_{\max}$, is given by

$$\tau_{\max} = \tfrac{1}{2}(\sigma_{\max} - \sigma_{\min}). \tag{6.81}$$

According to the Tresca criterion, the onset of plastic yielding is defined by the condition

$$\tfrac{1}{2}(\sigma_{\max} - \sigma_{\min}) = \tau_y(\alpha), \tag{6.82}$$

where τ_y is the *shear yield stress*, here assumed to be a function of a *hardening internal variable*, α, to be defined later. The shear yield stress is the yield limit under a state of pure shear.

In view of (6.82), the yield function associated with the Tresca yield criterion can be represented as

$$\Phi(\boldsymbol{\sigma}) = \tfrac{1}{2}(\sigma_{\max} - \sigma_{\min}) - \tau_y(\alpha), \tag{6.83}$$

with the onset of yielding characterised by $\Phi = 0$. Alternatively, the Tresca yield function may be defined as

$$\Phi(\boldsymbol{\sigma}) = (\sigma_{\max} - \sigma_{\min}) - \sigma_y(\alpha), \tag{6.84}$$

where σ_y is the *uniaxial* yield stress

$$\sigma_y = 2\,\tau_y, \tag{6.85}$$

that is, it is the stress level at which plastic yielding begins under *uniaxial* stress conditions. That σ_y is indeed the uniaxial yield stress for the Tresca theory can be established by noting that, when plastic yielding begins under uniaxial stress conditions, we have

$$\sigma_{\max} = \sigma_y, \quad \sigma_{\min} = 0. \tag{6.86}$$

The substitution of the above into (6.82) gives (6.85). The elastic domain for the Tresca criterion can be defined as

$$\mathscr{E} = \{\boldsymbol{\sigma} \mid \Phi(\boldsymbol{\sigma}, \sigma_y) < 0\}. \tag{6.87}$$

Pressure-insensitivity

Due to its definition exclusively in terms of *shear* stress, the Tresca criterion is *pressure insensitive*, that is, the hydrostatic pressure component,

$$p \equiv \tfrac{1}{3}\,\mathrm{tr}[\boldsymbol{\sigma}] = \tfrac{1}{3}\,(\sigma_1 + \sigma_2 + \sigma_3), \tag{6.88}$$

of the stress tensor does *not* affect yielding. Indeed, note that the superposition of an arbitrary pressure, p^*, on the stress tensor does not affect the value of the Tresca yield function

$$\Phi(\boldsymbol{\sigma} + p^*\boldsymbol{I}) = \Phi(\boldsymbol{\sigma}). \tag{6.89}$$

We remark that the von Mises criterion described in Section 6.4.2 below is also pressure-insensitive. This property is particularly relevant in the modelling of metals as, for these materials, the influence of the hydrostatic stress on yielding is usually negligible in practice.

Isotropy

One very important aspect of the Tresca criterion is its *isotropy* (a property shared by the von Mises, Mohr–Coulomb and Drucker–Prager criteria described in the following sections).

Note that, since Φ in (6.83) or (6.84) is defined as a function of the principal stresses, the Tresca yield function is an *isotropic* function of the stress tensor (refer to Section A.1, page 731, for the definition of isotropic scalar functions of a symmetric tensor), i.e. it satisfies

$$\Phi(\boldsymbol{\sigma}) = \Phi(\boldsymbol{Q}\boldsymbol{\sigma}\boldsymbol{Q}^T) \tag{6.90}$$

for all rotations $\boldsymbol{Q}$; that is, rotations of the state of stress do not affect the value of the yield function.

At this point, it is convenient to introduce the following definition: *A plastic yield criterion is said to be isotropic if it is defined in terms of an isotropic yield function of the stress tensor.*

Graphical representation

Since any isotropic scalar function of a symmetric tensor can be described as a function of the principal values of its argument, it follows that any iso-surface (i.e. any subset of the function domain with fixed function value) of such functions can be graphically represented as a surface in the space of principal values of the argument. This allows, in particular, the yield surface (refer to expression (6.48), page 150) of any isotropic yield criterion to be represented in a particularly simple and useful format as a three-dimensional surface in the space of principal stresses.

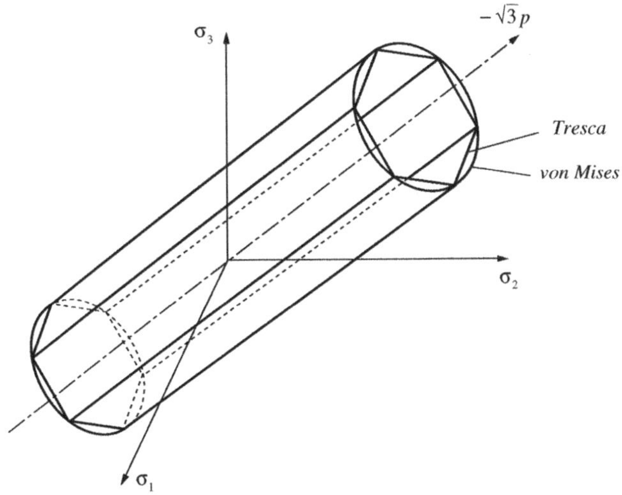

Figure 6.8. The Tresca and von Mises yield surfaces in principal stress space.

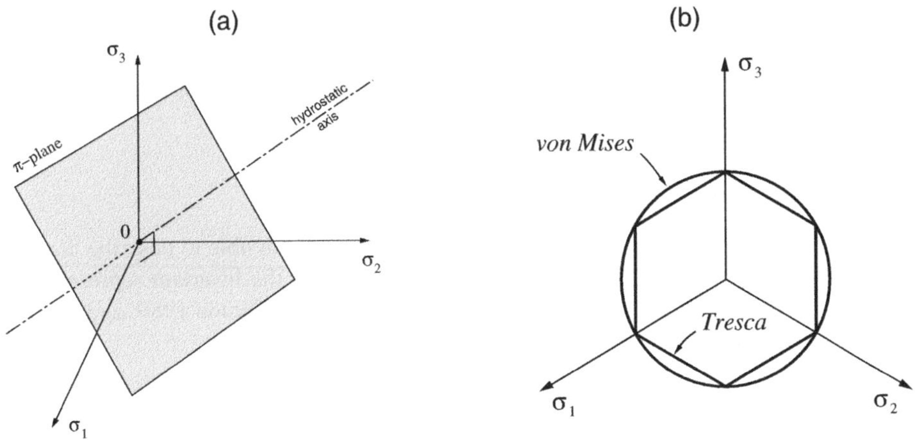

Figure 6.9. (a) The π-plane in principal stress space and, (b) the π-plane representation of the Tresca and von Mises yield surfaces.

In principal stress space, the Tresca yield surface, i.e. the set of stresses for which $\Phi = 0$, is graphically represented by the surface of an infinite hexagonal prism with axis coinciding with the hydrostatic line (also known as the space diagonal), defined by $\sigma_1 = \sigma_2 = \sigma_3$. This is illustrated in Figure 6.8. The elastic domain (for which $\Phi < 0$) corresponds to the interior of the prism. Due to the assumed insensitivity to pressure, a further simplification in the representation of the yield surface is possible in this case. The Tresca yield surface may be represented, without loss of generality, by its projection on the subspace of stresses with zero hydrostatic pressure component ($\sigma_1 + \sigma_2 + \sigma_3 = 0$). This subspace is called the *deviatoric plane*, also referred to as the *π-plane*. It is graphically illustrated in Figure 6.9(a). Figure 6.9(b) shows the π-plane projection of the Tresca yield surface.

Multisurface representation

Equivalently to the above representation, the Tresca yield criterion can be expressed by means of the following six yield functions

$$\begin{aligned}
\Phi_1(\boldsymbol{\sigma}, \sigma_y) &= \sigma_1 - \sigma_3 - \sigma_y \\
\Phi_2(\boldsymbol{\sigma}, \sigma_y) &= \sigma_2 - \sigma_3 - \sigma_y \\
\Phi_3(\boldsymbol{\sigma}, \sigma_y) &= \sigma_2 - \sigma_1 - \sigma_y \\
\Phi_4(\boldsymbol{\sigma}, \sigma_y) &= \sigma_3 - \sigma_1 - \sigma_y \\
\Phi_5(\boldsymbol{\sigma}, \sigma_y) &= \sigma_3 - \sigma_2 - \sigma_y \\
\Phi_6(\boldsymbol{\sigma}, \sigma_y) &= \sigma_1 - \sigma_2 - \sigma_y,
\end{aligned} \tag{6.91}$$

so that, for fixed σ_y, the equation

$$\Phi_i(\boldsymbol{\sigma}, \sigma_y) = 0 \tag{6.92}$$

corresponds to a *plane* in the space of principal stresses for each $i = 1, \ldots, 6$ (Figure 6.10).

In the multisurface representation, the elastic domain for a given σ_y can be defined as

$$\mathscr{E} = \{\boldsymbol{\sigma} \mid \Phi_i(\boldsymbol{\sigma}, \sigma_y) < 0,\ i = 1, \ldots, 6\}. \tag{6.93}$$

Definitions (6.87) and (6.93) are completely equivalent. The yield surface – the boundary of $\mathscr{E}$ – is defined in this case as the set of stresses for which $\Phi_i(\boldsymbol{\sigma}, \sigma_y) = 0$ for at least one i with $\Phi_j(\boldsymbol{\sigma}, \sigma_y) \leq 0$ for $j \neq i$.

Invariant representation

Alternatively to the representations discussed above, it is also possible to describe the yield locus of the Tresca criterion in terms of *stress invariants*. In the invariant representation, proposed by Nayak and Zienkiewicz (1972) (see also Owen and Hinton 1980, and Crisfield 1997), the yield function assumes the format

$$\Phi = 2\sqrt{J_2}\cos\theta - \sigma_y, \tag{6.94}$$

where $J_2 = J_2(\boldsymbol{s})$ is the invariant of the stress deviator, $\boldsymbol{s}$, defined by

$$J_2 \equiv -I_2(\boldsymbol{s}) = \tfrac{1}{2}\,\mathrm{tr}[\boldsymbol{s}^2] = \tfrac{1}{2}\,\boldsymbol{s} : \boldsymbol{s} = \tfrac{1}{2}\,\|\boldsymbol{s}\|^2. \tag{6.95}$$

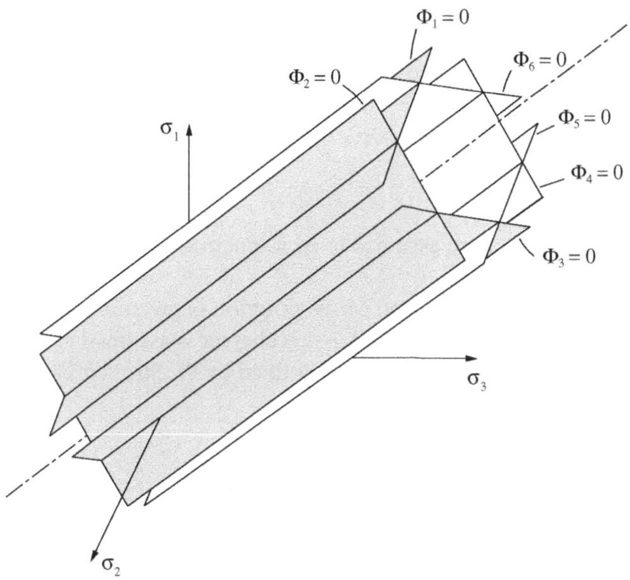

Figure 6.10. The Tresca criterion. Multisurface representation in principal stress space.

Recall that the stress deviator is given by

$$\boldsymbol{s} \equiv \boldsymbol{\sigma} - \tfrac{1}{3}(\mathrm{tr}\boldsymbol{\sigma})\boldsymbol{I}. \tag{6.96}$$

The *Lode angle*, θ, is a function of the deviatoric stress defined as

$$\theta \equiv \tfrac{1}{3}\sin^{-1}\left(\frac{-3\sqrt{3}\,J_3}{2J_2^{\frac{3}{2}}}\right), \tag{6.97}$$

where J_3 is the third principal invariant of stress deviator[§]

$$J_3 \equiv I_3(\boldsymbol{s}) \equiv \det \boldsymbol{s} = \tfrac{1}{3}\,\mathrm{tr}(\boldsymbol{s})^3. \tag{6.98}$$

The Lode angle is the angle, on the deviatoric plane, between $\boldsymbol{s}$ and the nearest pure shear line (a pure shear line is graphically represented in Figure 6.11). It satisfies

$$-\frac{\pi}{6} \leq \theta \leq \frac{\pi}{6}. \tag{6.99}$$

Despite being used often in computational plasticity, the above invariant representation results in rather cumbersome algorithms for integration of the evolution equations of the Tresca model. This is essentially due to the high degree of nonlinearity introduced by the trigonometric function involved in the definition of the Lode angle. The multisurface representation, on the other hand, is found by the authors to provide an optimal parametrisation of the Tresca surface which results in a simpler numerical algorithm and will be adopted in the computational implementation of the model addressed in Chapter 8.

[§]The equivalence between the two rightmost terms in (6.98) is established by summing the characteristic equation (2.73) (page 27) for $i = 1, 2, 3$ and taking into account the fact that $I_1(\boldsymbol{s}) = 0$ ($\boldsymbol{s}$ is a traceless tensor) and that $\mathrm{tr}(\boldsymbol{S})^3 = \sum_i s_i^3$ for any symmetric tensor $\boldsymbol{S}$.

6.4.2. THE VON MISES YIELD CRITERION

Also appropriate to describe plastic yielding in metals, this criterion was proposed by von Mises (1913). According to the von Mises criterion, *plastic yielding begins when the J_2 stress deviator invariant reaches a critical value*. This condition is mathematically represented by the equation

$$J_2 = R(\alpha), \tag{6.100}$$

where R is the critical value, here assumed to be a function of a hardening internal variable, α, to be defined later.

The physical interpretation of the von Mises criterion is given in the following. Since the elastic behaviour of the materials described in this chapter is assumed to be linear elastic, the stored elastic strain-energy at the generic state defined by the stress σ can be decomposed as the sum

$$\psi^e = \psi^e_d + \psi^e_v, \tag{6.101}$$

of a *distortional* contribution,

$$\bar{\rho}\, \psi^e_d = \frac{1}{2\,G}\, \boldsymbol{s} : \boldsymbol{s} = \frac{1}{G}\, J_2, \tag{6.102}$$

and a *volumetric* contribution,

$$\bar{\rho}\, \psi^e_v = \frac{1}{K}\, p^2, \tag{6.103}$$

where G and K are, respectively, the shear and bulk modulus. In view of (6.102), the von Mises criterion is equivalent to stating that *plastic yielding begins when the distortional elastic strain-energy reaches a critical value*. The corresponding critical value of the distortional energy is

$$\frac{1}{G}\, R.$$

It should be noted that, as in the Tresca criterion, the pressure component of the stress tensor does not take part in the definition of the von Mises criterion and only the deviatoric stress can influence plastic yielding. Thus, the von Mises criterion is also *pressure-insensitive*.

In a state of *pure shear*, i.e. a state with stress tensor

$$[\boldsymbol{\sigma}] = \begin{bmatrix} 0 & \tau & 0 \\ \tau & 0 & 0 \\ 0 & 0 & 0 \end{bmatrix}, \tag{6.104}$$

we have, $\boldsymbol{s} = \boldsymbol{\sigma}$ and

$$J_2 = \tau^2. \tag{6.105}$$

Thus, a yield function for the von Mises criterion can be defined as

$$\Phi(\boldsymbol{\sigma}) = \sqrt{J_2(\boldsymbol{s}(\boldsymbol{\sigma}))} - \tau_y, \tag{6.106}$$

where $\tau_y \equiv \sqrt{R}$ is the *shear* yield stress. Let us now consider a state of *uniaxial stress*:

$$[\boldsymbol{\sigma}] = \begin{bmatrix} \sigma & 0 & 0 \\ 0 & 0 & 0 \\ 0 & 0 & 0 \end{bmatrix}. \tag{6.107}$$

In this case, we have

$$[\boldsymbol{s}] = \begin{bmatrix} \frac{2}{3}\sigma & 0 & 0 \\ 0 & -\frac{1}{3}\sigma & 0 \\ 0 & 0 & -\frac{1}{3}\sigma \end{bmatrix} \tag{6.108}$$

and

$$J_2 = \tfrac{1}{3}\sigma^2. \tag{6.109}$$

The above expression for the J_2-invariant suggests the following alternative definition of the von Mises yield function:

$$\Phi(\boldsymbol{\sigma}) = q(\boldsymbol{\sigma}) - \sigma_y, \tag{6.110}$$

where $\sigma_y \equiv \sqrt{3R}$ is the *uniaxial* yield stress and

$$q(\boldsymbol{\sigma}) \equiv \sqrt{3\, J_2(\boldsymbol{s}(\boldsymbol{\sigma}))} \tag{6.111}$$

is termed the *von Mises effective* or *equivalent stress*. The uniaxial and shear yield stresses for the von Mises criterion are related by

$$\sigma_y = \sqrt{3}\,\tau_y. \tag{6.112}$$

Note that this relation differs from that of the Tresca criterion given by (6.85). Obviously, due to its definition in terms of an invariant of the stress tensor, the von Mises yield function is an isotropic function of $\boldsymbol{\sigma}$.

The von Mises and Tresca criteria may be set to agree with one another in either uniaxial stress or pure shear states. If they are set by using the yield functions (6.84) and (6.110) so that both predict the same uniaxial yield stress σ_y, then, under pure shear, the von Mises criterion will predict a yield stress $2/\sqrt{3}$ (≈ 1.155) times that given by the Tresca criterion. On the other hand, if both criteria are made to coincide under pure shear (expressions (6.83) and (6.106) with the same τ_y), then, in uniaxial stress states, the von Mises criterion will predict yielding at a stress level $\sqrt{3}/2$ (≈ 0.866) times the level predicted by Tresca's law.

The yield surface ($\Phi = 0$) associated with the von Mises criterion is represented, in the space of principal stresses, by the surface of an infinite circular cylinder, the axis of which coincides with the hydrostatic axis. The von Mises surface is illustrated in Figure 6.8 where it has been set to match the Tresca surface (shown in the same figure) under uniaxial stress. The corresponding π-plane representation is shown in Figure 6.9(b). The von Mises circle intersects the vertices of the Tresca hexagon. The yield surfaces for the von Mises and Tresca criteria set to coincide in shear is shown in Figure 6.11. In this case, the von Mises circle is tangent to the sides of the Tresca hexagon. It is remarked that, for many metals, experimentally determined yield surfaces fall between the von Mises and Tresca surfaces. A more general model, which includes both the Tresca and the von Mises yield surfaces as particular cases (and, in addition, allows for anisotropy of the yield surface), is described in Section 10.3.4 (starting page 427).

6.4.3. THE MOHR–COULOMB YIELD CRITERION

The criteria presented so far are pressure-insensitive and adequate to describe metals. For materials such as soils, rocks and concrete, whose behaviour is generally characterised by

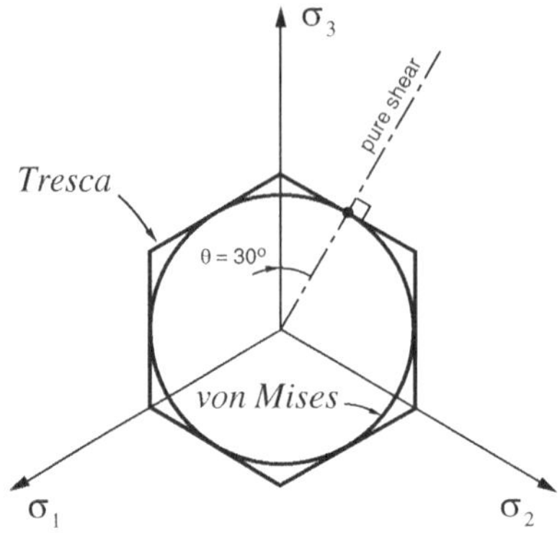

Figure 6.11. Yield surfaces for the Tresca and von Mises criteria coinciding in pure shear.

a strong dependence of the yield limit on the hydrostatic pressure, appropriate description of plastic yielding requires the introduction of *pressure-sensitivity*. A classical example of a pressure-sensitive law is given by the Mohr–Coulomb yield criterion described in the following.

The Mohr–Coulomb criterion is based on the assumption that the phenomenon of macroscopic plastic yielding is, essentially, the result of frictional sliding between material particles. Generalising Coulomb's friction law, this criterion states that *plastic yielding begins when, on a plane in the body, the shearing stress, τ, and the normal stress, σ_n, reach the critical combination*

$$\tau = c - \sigma_n \tan\phi, \tag{6.113}$$

where c is the cohesion and ϕ is the angle of internal friction or frictional angle. In the above, the normal stress, σ_n, was assumed tensile positive.

The yield locus of the Mohr–Coulomb criterion is the set of all stress states such that there exists a plane in which (6.113) holds. The Mohr–Coulomb yield locus can be easily visualised in the Mohr plane representation shown in Figure 6.12. It is the set of all stresses whose largest Mohr circle, i.e. the circle associated with the maximum and minimum principal stresses ($\sigma_{\max}$ and $\sigma_{\min}$, respectively), is tangent to the critical line defined by $\tau = c - \sigma_n \tan\phi$. The elastic domain for the Mohr–Coulomb law is the set of stresses whose all three Mohr circles are below the critical line. From Figure 6.12, the yield condition (6.113) is found to be equivalent to the following form in terms of principal stresses

$$\frac{\sigma_{\max} - \sigma_{\min}}{2} \cos\phi = c - \left(\frac{\sigma_{\max} + \sigma_{\min}}{2} + \frac{\sigma_{\max} - \sigma_{\min}}{2} \sin\phi \right) \tan\phi, \tag{6.114}$$

which, rearranged, gives

$$(\sigma_{\max} - \sigma_{\min}) + (\sigma_{\max} + \sigma_{\min}) \sin\phi = 2\,c\, \cos\phi. \tag{6.115}$$

In view of (6.115), a yield function expressed in terms of the principal stresses can be immediately defined for the Mohr–Coulomb criterion as

$$\Phi(\boldsymbol{\sigma}, c) = (\sigma_{\max} - \sigma_{\min}) + (\sigma_{\max} + \sigma_{\min})\, \sin\phi - 2\,c\, \cos\phi. \tag{6.116}$$

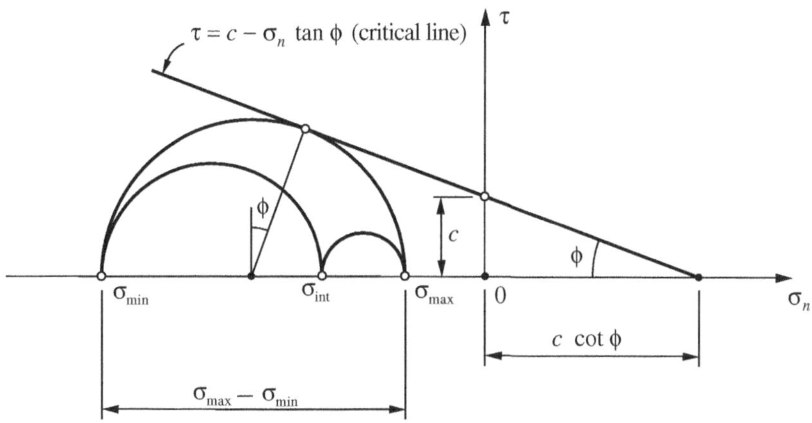

Figure 6.12. The Mohr–Coulomb criterion. Mohr plane representation.

Due to its definition in terms of principal stresses, this yield function is an isotropic function of $\boldsymbol{\sigma}$. The corresponding yield surface ($\Phi = 0$) is a hexagonal pyramid aligned with the hydrostatic axis and whose apex is located at

$$p = c \cot \phi \tag{6.117}$$

on the tensile side of the hydrostatic axis. The Mohr–Coulomb surface is illustrated in Figure 6.13. Its pyramidal shape, as opposed to the prismatic shape of the Tresca surface, is a consequence of the *pressure-sensitivity* of the Mohr–Coulomb criterion. It should be noted, however, that both criteria coincide in the absence of internal friction, i.e. when $\phi = 0$. As no stress state is allowed on the outside of the yield surface, the apex of the pyramid (point A in the figure) defines the limit of resistance of the material to tensile pressures. Limited strength under tensile pressure is a typical characteristic of materials such as concrete, rock and soils, to which the Mohr–Coulomb criterion is most applicable.

Multisurface representation

Analogously to the multisurface representation of the Tresca criterion, the Mohr–Coulomb criterion can also be expressed by means of six functions:

$$\begin{aligned}
\Phi_1(\boldsymbol{\sigma}, c) &= \sigma_1 - \sigma_3 + (\sigma_1 + \sigma_3) \sin \phi - 2\,c \cos \phi \\
\Phi_2(\boldsymbol{\sigma}, c) &= \sigma_2 - \sigma_3 + (\sigma_2 + \sigma_3) \sin \phi - 2\,c \cos \phi \\
\Phi_3(\boldsymbol{\sigma}, c) &= \sigma_2 - \sigma_1 + (\sigma_2 + \sigma_1) \sin \phi - 2\,c \cos \phi \\
\Phi_4(\boldsymbol{\sigma}, c) &= \sigma_3 - \sigma_1 + (\sigma_3 + \sigma_1) \sin \phi - 2\,c \cos \phi \\
\Phi_5(\boldsymbol{\sigma}, c) &= \sigma_3 - \sigma_2 + (\sigma_3 + \sigma_2) \sin \phi - 2\,c \cos \phi \\
\Phi_6(\boldsymbol{\sigma}, c) &= \sigma_1 - \sigma_2 + (\sigma_1 + \sigma_2) \sin \phi - 2\,c \cos \phi,
\end{aligned} \tag{6.118}$$

whose roots, $\Phi_i(\boldsymbol{\sigma}, c) = 0$ (for fixed c), define six planes in the principal stress space. Each plane contains one face of the Mohr–Coulomb pyramid represented in Figure 6.13.

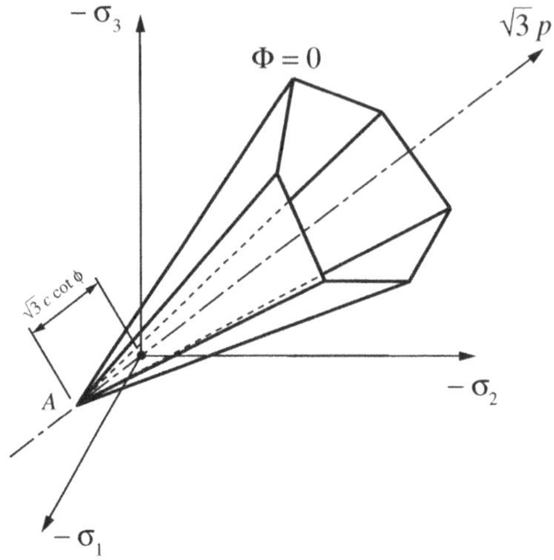

Figure 6.13. The Mohr–Coulomb yield surface in principal stress space.

The definition of the elastic domain and the yield surface in the multisurface representation is completely analogous to that of the Tresca criterion.

Invariant representation

Analogously to the invariant representation (6.94) of the Tresca criterion, the Mohr–Coulomb yield function can be expressed as (Owen and Hinton 1980, and Crisfield 1997):

$$\Phi = \left(\cos\theta - \frac{1}{\sqrt{3}}\sin\theta\,\sin\phi\right)\sqrt{J_2(\boldsymbol{s})} + p(\boldsymbol{\sigma})\,\sin\phi - c\,\cos\phi, \tag{6.119}$$

where the Lode angle, θ, is defined in (6.97). As for the Tresca model, in spite of its frequent use in computational plasticity, the invariant representation of the Mohr–Coulomb surface renders more complex numerical algorithms so that the multisurface representation is preferred in the computational implementation of the model described in Chapter 8.

6.4.4. THE DRUCKER–PRAGER YIELD CRITERION

This criterion has been proposed by Drucker and Prager (1952) as a smooth approximation to the Mohr–Coulomb law. It consists of a modification of the von Mises criterion in which an extra term is included to introduce pressure-sensitivity. The Drucker–Prager criterion states that *plastic yielding begins when the* J_2 *invariant of the deviatoric stress and the hydrostatic stress, p, reach a critical combination.* The onset of plastic yielding occurs when the equation

$$\sqrt{J_2(\boldsymbol{s})} + \eta\, p = \bar{c}, \tag{6.120}$$

is satisfied, where η and $\bar{c}$ are material parameters. Represented in the principal stress space, the yield locus of this criterion is a circular cone whose axis is the hydrostatic line. For $\eta = 0$, the von Mises cylinder is recovered. The Drucker–Prager cone is illustrated in Figure 6.14.

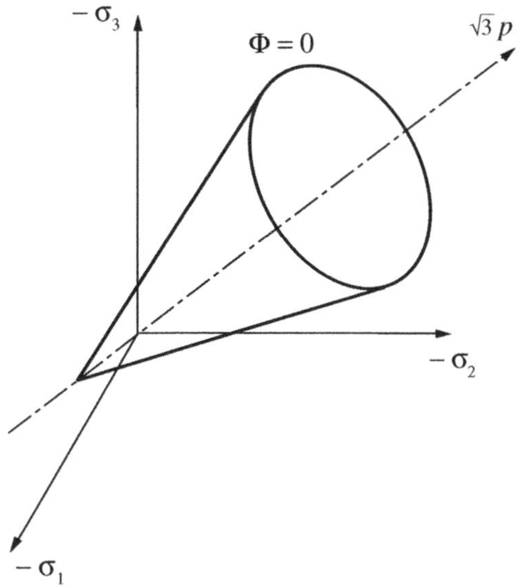

Figure 6.14. The Drucker–Prager yield surface in principal stress space.

In order to approximate the Mohr–Coulomb yield surface, it is convenient to define the Drucker–Prager yield function as

$$\Phi(\boldsymbol{\sigma}, c) = \sqrt{J_2(\boldsymbol{s}(\boldsymbol{\sigma}))} + \eta\, p(\boldsymbol{\sigma}) - \xi\, c, \tag{6.121}$$

where c is the *cohesion* and the parameters η and ξ are chosen according to the required approximation to the Mohr–Coulomb criterion. Note that the isotropy of the Mohr–Coulomb yield function follows from the fact that it is defined in terms of invariants of the stress tensor ($J_2(\boldsymbol{s})$ and p). Two of the most common approximations used are obtained by making the yield surfaces of the Drucker–Prager and Mohr–Coulomb criteria coincident either at the outer or inner edges of the Mohr–Coulomb surface. Coincidence at the *outer* edges is obtained when

$$\eta = \frac{6\,\sin\phi}{\sqrt{3}\,(3 - \sin\phi)}, \qquad \xi = \frac{6\,\cos\phi}{\sqrt{3}\,(3 - \sin\phi)}, \tag{6.122}$$

whereas, coincidence at the *inner* edges is given by the choice

$$\eta = \frac{6\,\sin\phi}{\sqrt{3}\,(3 + \sin\phi)}, \qquad \xi = \frac{6\,\cos\phi}{\sqrt{3}\,(3 + \sin\phi)}. \tag{6.123}$$

The outer and inner cones are known, respectively, as the *compression cone* and the *extension cone*. The inner cone matches the Mohr–Coulomb criterion in uniaxial tension and biaxial compression. The outer edge approximation matches the Mohr–Coulomb surface in uniaxial compression and biaxial tension. The π-plane section of both surfaces is shown in Figure 6.15. Another popular Drucker–Prager approximation to the Mohr–Coulomb criterion is obtained by forcing both criteria to predict identical collapse loads under *plane strain* conditions. In this case (the reader is referred to Section 4.7 of Chen and Mizuno (1990) for derivation) the constants η and ξ read

$$\eta = \frac{3\ \tan\phi}{\sqrt{9 + 12\ \tan^2\phi}}, \qquad \xi = \frac{3}{\sqrt{9 + 12\ \tan^2\phi}}. \tag{6.124}$$

For all three sets of parameters above, the apex of the approximating Drucker–Prager cone coincides with the apex of the corresponding Mohr–Coulomb yield surface. It should be emphasised here that any of the above approximations to the Mohr–Coulomb criterion can give a poor description of the material behaviour for certain states of stress. Thus, according to the dominant stress state in a particular problem to be analysed, other approximations may be more appropriate. For instance, under plane stress, which can be assumed in the analysis of structures such as concrete walls, it may be convenient to use an approximation such that both criteria match under, say, uniaxial tensile and uniaxial compressive stress states. For the Mohr–Coulomb criterion to fit a given uniaxial tensile strength, f'_t, and a given uniaxial compressive strength, f'_c, the parameters ϕ and c have to be chosen as

$$\phi = \sin^{-1}\left(\frac{f'_c - f'_t}{f'_c + f'_t}\right), \qquad c = \frac{f'_c\, f'_t}{f'_c - f'_t}\ \tan\phi. \tag{6.125}$$

The corresponding Drucker–Prager cone (Figure 6.16) that predicts the same uniaxial failure loads is obtained by setting

$$\eta = \frac{3\ \sin\phi}{\sqrt{3}}, \qquad \xi = \frac{2\cos\phi}{\sqrt{3}}. \tag{6.126}$$

Its apex no longer coincides with the apex of the original Mohr–Coulomb pyramid. For problems where the failure mechanism is indeed dominated by uniaxial tension/compression, the above approximation should produce reasonable results. However, if for a particular problem, failure occurs under biaxial *compression* instead (with stresses near point f'_{bc} of Figure 6.16), then the above approximation will largely overestimate the limit load, particularly for high ratios f'_c/f'_t which are typical for concrete. Under such a condition, a different approximation (such as the inner cone that matches point f'_{bc}) needs to be adopted to produce sensible predictions. Another useful approximation for plane stress, where the Drucker–Prager cone coincides with the Mohr–Coulomb surface in biaxial tension (point f'_{bt}) and biaxial compression (point f'_{bc}), is obtained by setting

$$\eta = \frac{3\ \sin\phi}{2\sqrt{3}}, \qquad \xi = \frac{2\cos\phi}{\sqrt{3}}. \tag{6.127}$$

Drucker–Prager approximations to the Mohr–Coulomb criterion are thoroughly discussed by Chen (1982), Chen and Mizuno (1990) and Zienkiewicz *et al.* (1978).

6.5. Plastic flow rules

6.5.1. ASSOCIATIVE AND NON-ASSOCIATIVE PLASTICITY

It has already been said that a plasticity model is classed as *associative* if the yield function, Φ, is taken as the flow potential, i.e.

$$\Psi = \Phi. \tag{6.128}$$

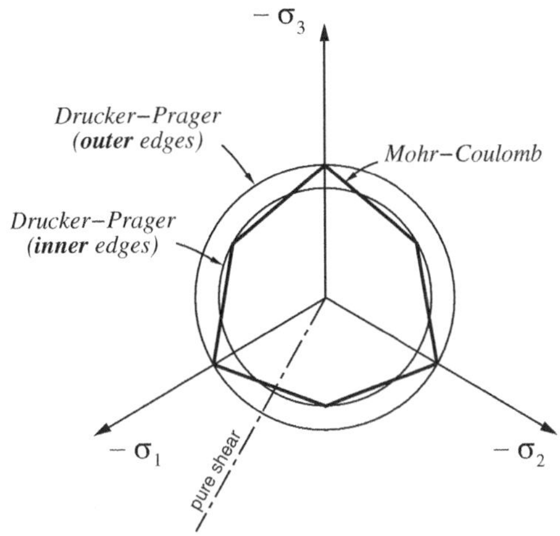

Figure 6.15. The π-plane section of the Mohr–Coulomb surface and the Drucker–Prager approximations.

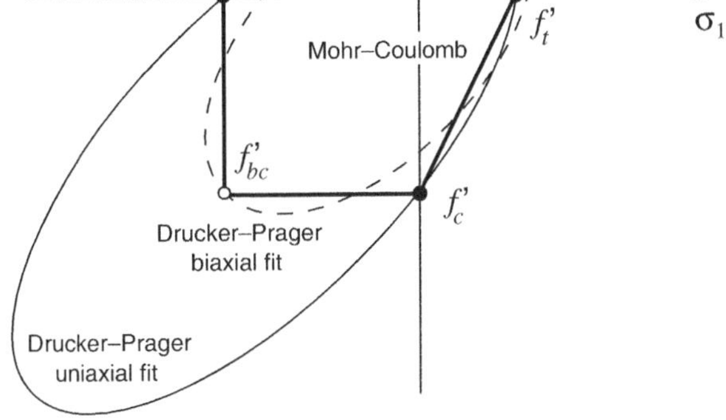

Figure 6.16. Plane stress. Drucker–Prager approximation matching the Mohr–Coulomb surface in uniaxial tension and uniaxial compression.

Any other choice of flow potential characterises a *non-associative* (or *non-associated*) plasticity model.

In associative models, the evolution equations for the plastic strain and hardening variables are given by

$$\dot{\varepsilon}^p = \dot{\gamma}\,\frac{\partial \Phi}{\partial \boldsymbol{\sigma}}, \tag{6.129}$$

and

$$\dot{\boldsymbol{\alpha}} = -\dot{\gamma}\,\frac{\partial \Phi}{\partial \boldsymbol{A}}. \tag{6.130}$$

Associativity implies that the plastic strain rate is a tensor *normal to the yield surface* in the space of stresses. In the generalised case of non-smooth yield surfaces, the flow vector is a subgradient of the yield function, i.e. we have

$$\dot{\varepsilon}^p = \dot{\gamma}\boldsymbol{N}; \quad \boldsymbol{N} \in \partial_\sigma \Phi. \tag{6.131}$$

In non-associative models, the plastic strain rate is not normal to the yield surface in general.

6.5.2. ASSOCIATIVE LAWS AND THE PRINCIPLE OF MAXIMUM PLASTIC DISSIPATION

It can be shown that the associative laws are a consequence of the *principle of maximum plastic dissipation*. Before stating the principle of maximum plastic dissipation, recall that for a state defined by a hardening force $\boldsymbol{A}$, the admissible stress states are those that satisfy $\Phi(\boldsymbol{\sigma}, \boldsymbol{A}) \leq 0$. Thus, it makes sense to define

$$\mathscr{A} = \{(\boldsymbol{\sigma}, \boldsymbol{A}) \mid \Phi(\boldsymbol{\sigma}, \boldsymbol{A}) \leq 0\} \tag{6.132}$$

as the set of all admissible pairs (combinations) of stress and hardening force. The principle of maximum dissipation postulates that *among all admissible pairs* $(\boldsymbol{\sigma}^*, \boldsymbol{A}^*) \in \mathscr{A}$, *the actual state* $(\boldsymbol{\sigma}, \boldsymbol{A})$ *maximises the dissipation function* (6.42) *for a given plastic strain rate,* $\dot{\varepsilon}^p$, *and rate* $\dot{\boldsymbol{\alpha}}$ *of hardening internal variables*. The principle of maximum plastic dissipation requires that, for given $(\dot{\varepsilon}^p, \dot{\boldsymbol{\alpha}})$,

$$\Upsilon^p(\boldsymbol{\sigma}, \boldsymbol{A}; \dot{\varepsilon}^p, \dot{\boldsymbol{\alpha}}) \geq \Upsilon^p(\boldsymbol{\sigma}^*, \boldsymbol{A}^*; \dot{\varepsilon}^p, \dot{\boldsymbol{\alpha}}), \ \forall\, (\boldsymbol{\sigma}^*, \boldsymbol{A}^*) \in \mathscr{A}. \tag{6.133}$$

In other words, the actual state $(\boldsymbol{\sigma}, \boldsymbol{A})$ of stress and hardening force is a solution to the following constrained optimisation problem:

$$\begin{aligned} &\text{maximise } \Upsilon^p(\boldsymbol{\sigma}^*, \boldsymbol{A}^*, \dot{\varepsilon}^p, \dot{\boldsymbol{\alpha}}) \\ &\text{subject to } \Phi(\boldsymbol{\sigma}^*, \boldsymbol{A}^*) \leq 0. \end{aligned} \tag{6.134}$$

The *Kuhn–Tucker optimality conditions* (Luenberger, 1973, Chapter 10) for this optimisation problem are precisely the associative plastic flow rule (6.129), the associative hardening rule (6.130) and the loading/unloading conditions

$$\Phi(\boldsymbol{\sigma}, \boldsymbol{A}) \leq 0, \quad \dot{\gamma} \geq 0, \quad \Phi(\boldsymbol{\sigma}, \boldsymbol{A})\dot{\gamma} = 0. \tag{6.135}$$

Remark 6.2. The postulate of maximum plastic dissipation, and the corresponding associative laws, are *not* universal. Based on physical considerations, maximum dissipation has been shown to hold in crystal plasticity and is particularly successful when applied to the description of metals. Nevertheless, for many materials, particularly soils and granular materials in general, associative laws frequently do not correspond to experimental evidence. In such cases, the maximum dissipation postulate is clearly not applicable and the use of non-associative laws is essential.

6.5.3. CLASSICAL FLOW RULES

The Prandtl–Reuss equations

The Prandtl–Reuss plasticity law is the flow rule obtained by taking the von Mises yield function (6.110) as the flow potential. The corresponding flow vector is given by

$$\boldsymbol{N} \equiv \frac{\partial \Phi}{\partial \boldsymbol{\sigma}} = \frac{\partial}{\partial \boldsymbol{\sigma}}[\sqrt{3\, J_2(\boldsymbol{s})}] = \sqrt{\tfrac{3}{2}}\, \frac{\boldsymbol{s}}{\|\boldsymbol{s}\|}, \tag{6.136}$$

and the flow rule results in

$$\dot{\boldsymbol{\varepsilon}}^p = \dot{\gamma}\, \sqrt{\tfrac{3}{2}}\, \frac{\boldsymbol{s}}{\|\boldsymbol{s}\|}. \tag{6.137}$$

Here, it should be noted that the Prandtl–Reuss flow vector is the derivative of an isotropic scalar function of a symmetric tensor – the von Mises yield function. Thus (refer to Section A.1.2, page 732, where the derivative of isotropic functions of this type is discussed), $\boldsymbol{N}$ and $\boldsymbol{\sigma}$ are coaxial, i.e. the principal directions of $\boldsymbol{N}$ coincide with those of $\boldsymbol{\sigma}$. Due to the pressure-insensitivity of the von Mises yield function, the plastic flow vector is *deviatoric*. The Prandtl–Reuss flow vector is a tensor parallel to the deviatoric projection $\boldsymbol{s}$ of the stress tensor. Its principal stress representation is depicted in Figure 6.17. The Prandtl–Reuss rule is usually employed in conjunction with the von Mises criterion and the resulting plasticity model is referred to as the von Mises associative model or, simply, the von Mises model.

Associative Tresca

The associative Tresca flow rule utilises the yield function (6.84) as the flow potential. Since Φ here is also an isotropic function of $\boldsymbol{\sigma}$, the rate of plastic strain has the same principal directions as $\boldsymbol{\sigma}$. The Tresca yield function is differentiable when the three principal stresses are distinct ($\sigma_1 \neq \sigma_2 \neq \sigma_3$) and non-differentiable when two principal stresses coincide (at the edges of the Tresca hexagonal prism). Hence, the Tresca associative plastic flow rule is generally expressed as

$$\dot{\boldsymbol{\varepsilon}}^p = \dot{\gamma} \boldsymbol{N}, \tag{6.138}$$

where $\boldsymbol{N}$ is a subgradient of the Tresca function

$$\boldsymbol{N} \in \partial_{\sigma} \Phi. \tag{6.139}$$

Its multisurface-based representation reads

$$\dot{\boldsymbol{\varepsilon}}^p = \sum_{i=1}^{6} \dot{\gamma}^i \boldsymbol{N}^i = \sum_{i=1}^{6} \dot{\gamma}^i \frac{\partial \Phi_i}{\partial \boldsymbol{\sigma}}, \tag{6.140}$$

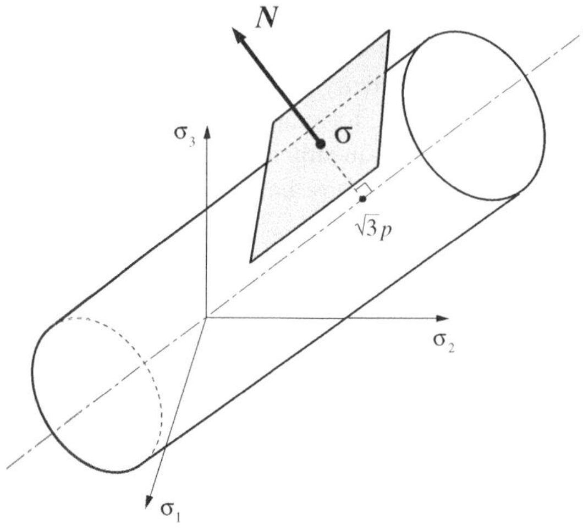

Figure 6.17. The Prandtl–Reuss flow vector.

with the yield functions Φ_i defined by (6.91). Each vector $\boldsymbol{N}^i$ is normal to the plane defined by $\Phi_i = 0$.

The above flow rule can be alternatively expressed as follows. Firstly assume, without loss of generality, that the principal stresses are ordered as $\sigma_1 \geq \sigma_2 \geq \sigma_3$, so that the discussion can be concentrated on the sextant of the π-plane illustrated in Figure 6.18. Three different possibilities have to be considered in this sextant:

(a) yielding at a stress state on the *side* (main plane) of the Tresca hexagon ($\Phi_1 = 0$, $\Phi_2 < 0$ and $\Phi_6 < 0$);

(b) yielding from the *right corner*, R ($\Phi_1 = 0$, $\Phi_6 = 0$ and $\Phi_2 < 0$); and

(c) Yielding from the *left corner*, L ($\Phi_1 = 0$, $\Phi_2 = 0$ and $\Phi_6 < 0$).

When the stress is on the side of the hexagon, only one multiplier may be non-zero and the plastic flow rule reads

$$\dot{\boldsymbol{\varepsilon}}^p = \dot{\gamma}\, \boldsymbol{N}^a, \tag{6.141}$$

where the flow vector is the normal to the plane $\Phi_1 = 0$, given by

$$\begin{aligned} \boldsymbol{N}^a \equiv \boldsymbol{N}^1 &= \frac{\partial \Phi_1}{\partial \boldsymbol{\sigma}} = \frac{\partial}{\partial \boldsymbol{\sigma}}(\sigma_1 - \sigma_3) \\ &= \boldsymbol{e}_1 \otimes \boldsymbol{e}_1 - \boldsymbol{e}_3 \otimes \boldsymbol{e}_3, \end{aligned} \tag{6.142}$$

with $\boldsymbol{e}_i$ denoting the eigenvector of $\boldsymbol{\sigma}$ associated with the principal stress σ_i. In deriving the last right-hand side of (6.142), use has been made of the expression (A.27) of page 736 for the derivative of an eigenvalue of a symmetric tensor.

At the right and left corners of the hexagon, where two planes intersect, two multipliers may be non-zero. Thus, the plastic flow equation is

$$\dot{\boldsymbol{\varepsilon}}^p = \dot{\gamma}^a\, \boldsymbol{N}^a + \dot{\gamma}^b\, \boldsymbol{N}^b. \tag{6.143}$$

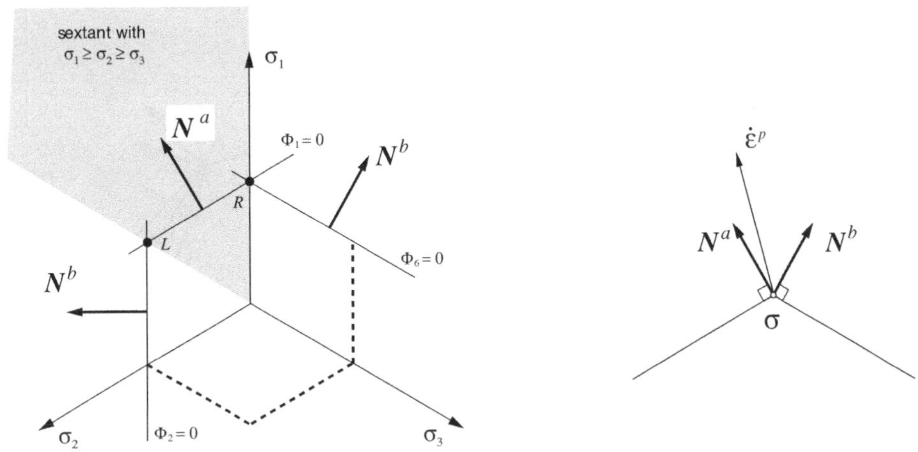

Figure 6.18. The associative Tresca flow rule.

The vector $\boldsymbol{N}^a$ is the normal to the plane $\Phi_1 = 0$, already defined. In the *right corner* (repeated minimum principal stress), the second vector, $\boldsymbol{N}^b$, is normal to the plane $\Phi_6 = 0$ and is obtained analogously to (6.142) as

$$\boldsymbol{N}^b \equiv \boldsymbol{N}^6 = \boldsymbol{e}_1 \otimes \boldsymbol{e}_1 - \boldsymbol{e}_2 \otimes \boldsymbol{e}_2. \tag{6.144}$$

In the *left corner* (repeated maximum principal stress), $\boldsymbol{N}^b$, is normal to the plane $\Phi_2 = 0$,

$$\boldsymbol{N}^b \equiv \boldsymbol{N}^2 = \boldsymbol{e}_2 \otimes \boldsymbol{e}_2 - \boldsymbol{e}_3 \otimes \boldsymbol{e}_3. \tag{6.145}$$

It should be noted that, as for the Prandtl–Reuss rule, the plastic flow predicted by the associative Tresca law is *volume-preserving*. Indeed, note that, in the above, we have trivially

$$\text{tr}\, \boldsymbol{N}^a = \text{tr}\, \boldsymbol{N}^b = 0. \tag{6.146}$$

This is due to the pressure-insensitivity of the Tresca yield function.

Associative and non-associative Mohr–Coulomb

In the associative Mohr–Coulomb law, the Mohr–Coulomb yield function (6.116) is adopted as the flow potential. Its multisurface representation is based on the yield functions (6.118). The flow rule, which requires consideration of the intersections between the yield surfaces, is derived in a manner analogous to the Tresca law described above. However, it should be noted that in addition to the edge singularities, the present surface has an extra singularity in its apex (Figure 6.13). Plastic yielding may then take place from a face, from an edge or from the apex of the Mohr–Coulomb pyramid.

Again, in the derivation of the flow rules at faces and edges, it is convenient to assume that the principal stresses are ordered as $\sigma_1 \geq \sigma_2 \geq \sigma_3$ so that, without loss of generality, the analysis can be reduced to a single sextant of a cross-section of the Mohr–Coulomb pyramid as illustrated in Figure 6.19. The situation is identical to Tresca's (Figure 6.18) except that

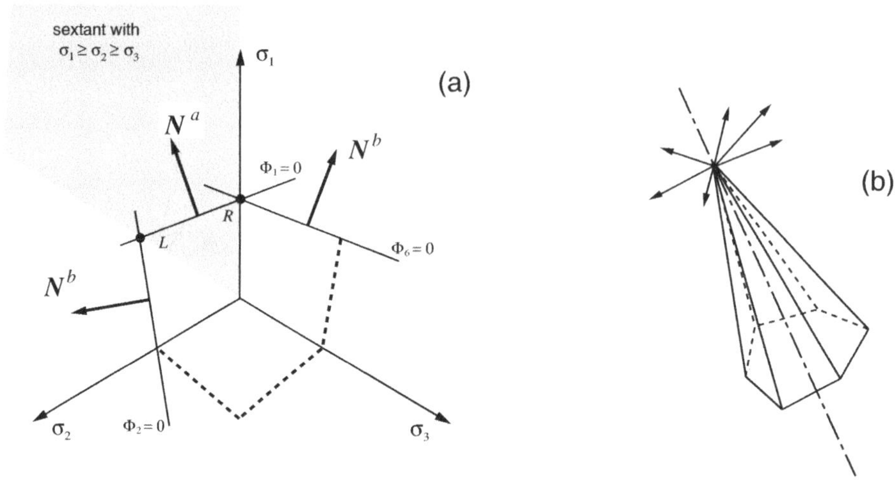

Figure 6.19. The Mohr–Coulomb flow rule; (a) faces and edges, and (b) apex.

the normal vectors $\boldsymbol{N}^a$ and $\boldsymbol{N}^b$ are no longer deviatoric, i.e. they have a non-zero component along the hydrostatic axis (the vectors shown in Figure 6.19 are deviatoric projections of the actual normals). For plastic yielding from the face, the flow rule is given by

$$\dot{\varepsilon}^p = \dot{\gamma}\, \boldsymbol{N}^a, \tag{6.147}$$

where $\boldsymbol{N}^a$ is normal to the plane $\Phi_1 = 0$,

$$\begin{aligned} \boldsymbol{N}^a &= \frac{\partial \Phi_1}{\partial \boldsymbol{\sigma}} = \frac{\partial}{\partial \boldsymbol{\sigma}}[\sigma_1 - \sigma_3 + (\sigma_1 + \sigma_3)\sin\phi] \\ &= (1 + \sin\phi)\boldsymbol{e}_1 \otimes \boldsymbol{e}_1 - (1 - \sin\phi)\boldsymbol{e}_3 \otimes \boldsymbol{e}_3. \end{aligned} \tag{6.148}$$

At the corners, the above flow rule is replaced by

$$\dot{\varepsilon}^p = \dot{\gamma}^a\, \boldsymbol{N}^a + \dot{\gamma}^b\, \boldsymbol{N}^b. \tag{6.149}$$

At the *right* (extension) corner, R, the second vector, $\boldsymbol{N}^b$, is normal to the plane $\Phi_6 = 0$ and is given by

$$\boldsymbol{N}^b = (1 + \sin\phi)\, \boldsymbol{e}_1 \otimes \boldsymbol{e}_1 - (1 - \sin\phi)\, \boldsymbol{e}_2 \otimes \boldsymbol{e}_2, \tag{6.150}$$

whereas, at the *left* (compression) corner, L, the tensor $\boldsymbol{N}^b$ is normal to the plane $\Phi_2 = 0$,

$$\boldsymbol{N}^b = (1 + \sin\phi)\, \boldsymbol{e}_2 \otimes \boldsymbol{e}_2 - (1 - \sin\phi)\, \boldsymbol{e}_3 \otimes \boldsymbol{e}_3. \tag{6.151}$$

At the apex of the Mohr–Coulomb surface, all six planes intersect and, therefore, six normals are defined and up to six plastic multipliers may be non-zero. This situation is schematically illustrated in Figure 6.19(b). The plastic strain rate tensor lies within the pyramid defined by the six normals:

$$\dot{\varepsilon}^p = \sum_{i=1}^{6} \dot{\gamma}^i\, \boldsymbol{N}^i. \tag{6.152}$$

It is important to note that, due to the pressure sensitivity of the Mohr–Coulomb criterion, the associative Mohr–Coulomb rule predicts a non-zero *volumetric* plastic straining. This is in contrast to the Prandtl–Reuss and associative Tresca laws. The volumetric component of the plastic strain rate in the associative Mohr–Coulomb law can be obtained by expanding (6.152) in principal stress space taking into account the definitions of $\boldsymbol{N}^i$. This gives

$$\begin{bmatrix} \dot{\varepsilon}_1^p \\ \dot{\varepsilon}_2^p \\ \dot{\varepsilon}_3^p \end{bmatrix} = \begin{bmatrix} \alpha & 0 & \beta & \beta & 0 & \alpha \\ 0 & \alpha & \alpha & 0 & \beta & \beta \\ \beta & \beta & 0 & \alpha & \alpha & 0 \end{bmatrix} \begin{bmatrix} \dot{\gamma}^1 \\ \dot{\gamma}^2 \\ \dot{\gamma}^3 \\ \dot{\gamma}^4 \\ \dot{\gamma}^5 \\ \dot{\gamma}^6 \end{bmatrix}, \tag{6.153}$$

where

$$\alpha \equiv 1 + \sin\phi, \quad \beta \equiv -1 + \sin\phi. \tag{6.154}$$

The above trivially yields

$$\dot{\varepsilon}_v^p \equiv \dot{\varepsilon}_1^p + \dot{\varepsilon}_2^p + \dot{\varepsilon}_3^p = 2\,\sin\phi \sum_{i=1}^{6} \dot{\gamma}^i. \tag{6.155}$$

As all $\dot{\gamma}^i$'s are non-negative, the volumetric plastic strain rate is positive and, therefore, *dilatant*. The phenomenon of dilatancy during plastic flow is observed for many materials, particularly geomaterials. However, the dilatancy predicted by the associative Mohr–Coulomb law is often excessive. To overcome this problem, it is necessary to use a *non-associated* flow rule in conjunction with the Mohr–Coulomb criterion. The non-associated Mohr–Coulomb law adopts, as flow potential, a Mohr–Coulomb yield function with the frictional angle ϕ replaced by a different (smaller) angle ψ. The angle ψ is called the *dilatancy angle* and the amount of dilation predicted is proportional to its sine. Note that for $\psi = 0$, the plastic flow becomes purely deviatoric and the flow rule reduces to the associative Tresca law.

Associative and non-associative Drucker–Prager

The associative Drucker–Prager model employs as flow potential the yield function defined by (6.121). To derive the corresponding flow rule, one should note first that the Drucker–Prager function is singular at the apex of the yield surface and is smooth anywhere else. Thus, two situations need to be considered:

(a) plastic yielding at (smooth portion of) the cone surface; and

(b) plastic yielding at the apex.

At the cone surface, where the Drucker–Prager yield function is differentiable, the flow vector is obtained by simply differentiating (6.121) which gives (Figure 6.20(a))

$$\boldsymbol{N} = \frac{1}{2\sqrt{J_2(\boldsymbol{s})}}\,\boldsymbol{s} + \frac{\eta}{3}\,\boldsymbol{I}, \tag{6.156}$$

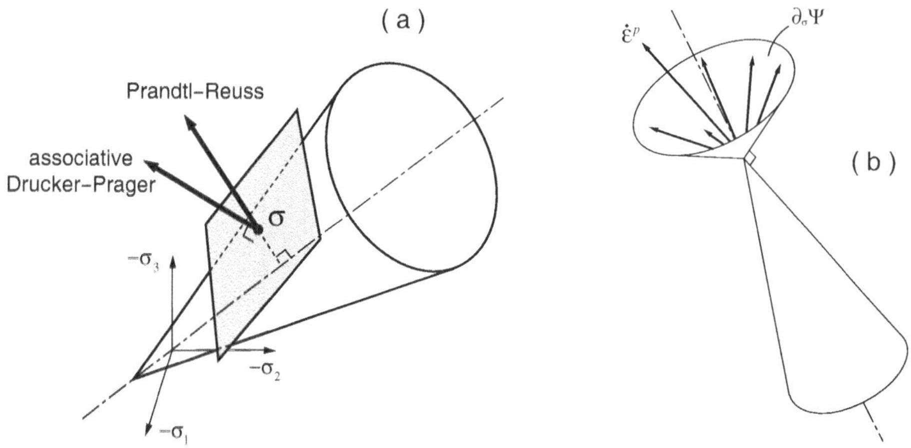

Figure 6.20. The Drucker–Prager flow vector; (a) cone surface, and (b) apex.

where η is given by $(6.122)_1$, $(6.123)_1$ or $(6.124)_1$, according to the chosen approximation to the Mohr–Coulomb surface. The flow rule is then

$$\dot{\varepsilon}^p = \dot{\gamma}\, \boldsymbol{N}. \tag{6.157}$$

The deviatoric/volumetric decomposition of the Drucker–Prager flow vector gives

$$\boldsymbol{N}_d = \frac{1}{2\sqrt{J_2(\boldsymbol{s})}}\, \boldsymbol{s}, \qquad N_v = \eta. \tag{6.158}$$

At the apex singularity, the flow vector is an element of the subdifferential of the yield function (6.121):

$$\boldsymbol{N} \in \partial_\sigma \Phi. \tag{6.159}$$

It lies within the complementary cone to the Drucker–Prager yield surface, i.e. the cone whose wall is normal to the Drucker–Prager cone illustrated in Figure 6.20(b). From standard properties of subdifferentials (Rockafellar, 1970; Rockafellar and Wets, 1998) it can be established that the deviatoric/volumetric split of $\boldsymbol{N}$ in this case is given by

$$\boldsymbol{N}_d \in \partial_\sigma \Phi_d, \qquad N_v = \eta, \tag{6.160}$$

where $\Phi_d \equiv \sqrt{J_2(\boldsymbol{s})}$. Expressions (6.157), (6.158) and (6.160) result in the following rate of (dilatant) volumetric plastic strain for the associative Drucker–Prager flow rule:

$$\dot{\varepsilon}^p_v = \dot{\gamma}\, \eta. \tag{6.161}$$

This expression is analogous to (6.155).

Similarly to the associative Mohr–Coulomb flow rule, the often excessive dilatancy predicted by the associated rule in the present case is avoided by using a non-associated law. The non-associative Drucker–Prager law is obtained by adopting, as the flow potential, a Drucker–Prager yield function with the frictional angle ϕ replaced by a dilatancy angle

$\psi < \phi$; that is, we define

$$\Psi(\boldsymbol{\sigma}, c) = \sqrt{J_2(\boldsymbol{s}(\boldsymbol{\sigma}))} + \bar{\eta}\, p, \tag{6.162}$$

where $\bar{\eta}$ is obtained by replacing ϕ with ψ in the definition of η given by $(6.122)_1$, $(6.123)_1$ or $(6.124)_1$. In other words,

$$\bar{\eta} = \frac{6\ \sin \psi}{\sqrt{3}\,(3 - \sin \psi)}, \tag{6.163}$$

when the outer cone approximation to the Mohr–Coulomb criterion is employed. When the inner cone approximation is used,

$$\bar{\eta} = \frac{6\ \sin \psi}{\sqrt{3}\,(3 + \sin \psi)}, \tag{6.164}$$

whereas, for the plane strain match,

$$\bar{\eta} = \frac{3\ \tan \psi}{\sqrt{9 + 12\ \tan^2 \psi}}. \tag{6.165}$$

The non-associated Drucker–Prager flow vector differs from its associated counterpart only in the volumetric component which, for the non-associated case, reads

$$N_v = \bar{\eta}. \tag{6.166}$$

If the dilatancy angle of the non-associative potential is chosen as $\psi = 0$, then the volumetric component, N_v, vanishes and the flow rule reduces to the Prandtl–Reuss law that predicts volume-preserving plastic flow (refer to Figure 6.20(a)).

6.6. Hardening laws

The phenomenon of hardening has been identified in the uniaxial experiment described in Section 6.1. Essentially, hardening is characterised by a dependence of yield stress level upon the history of plastic straining to which the body has been subjected. In the uniaxial model, formulated in Section 6.2, this phenomenon has been incorporated by allowing the uniaxial yield stress to vary (as a function of the axial accumulated plastic strain) during plastic flow. In the two- and three-dimensional situations, hardening is represented by changes in the hardening thermodynamical force, $\boldsymbol{A}$, during plastic yielding. These changes may, in general, affect the size, shape and orientation of the yield surface, defined by $\Phi(\boldsymbol{\sigma}, \boldsymbol{A}) = 0$.

6.6.1. PERFECT PLASTICITY

A material model is said to be *perfectly plastic* if *no hardening* is allowed, that is, the yield stress level does *not* depend in any way on the degree of plastification. In this case, the yield surface remains fixed regardless of any deformation process the material may experience and, in a uniaxial test, the elastoplastic modulus, E^{ep}, vanishes. In the von Mises, Tresca, Drucker–Prager and Mohr–Coulomb models described above, perfect plasticity corresponds to a *constant* uniaxial yield stress, σ_y (or constant cohesion, c). Figure 6.21 shows the stress–strain curve of a typical uniaxial cyclic (tension–compression) test with a perfectly plastic

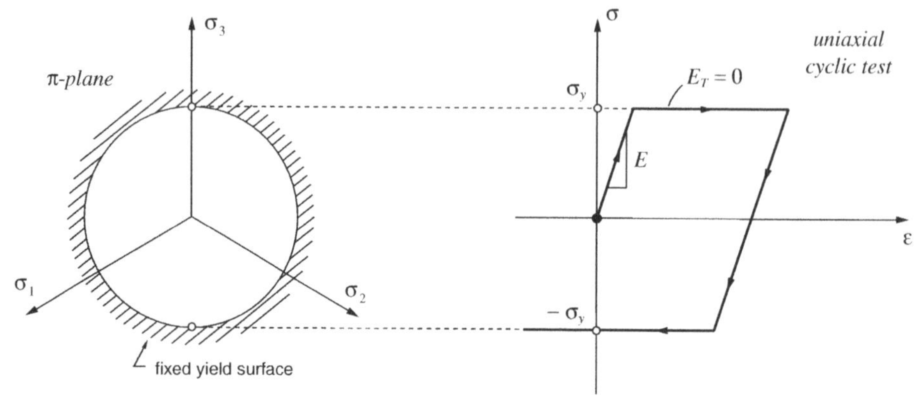

Figure 6.21. Perfect plasticity. Uniaxial test and π-plane representation.

von Mises model along with the corresponding π-plane representation of the yield surface. Perfectly plastic models are particularly suitable for the analysis of the stability of structures and soils and are widely employed in engineering practice for the determination of limit loads and safety factors.

6.6.2. ISOTROPIC HARDENING

A plasticity model is said to be *isotropic hardening* if the evolution of the yield surface is such that, at any state of hardening, it corresponds to a uniform (isotropic) expansion of the initial yield surface, without translation. The uniaxial model described in Section 6.2 is a typical example of an isotropic hardening model. For that model, the elastic domain expands equally in tension and compression during plastic flow. For a multiaxial plasticity model with a von Mises yield surface, isotropic hardening corresponds to the increase in radius of the von Mises cylinder in principal stress space. This, together with a typical stress–strain curve for a uniaxial cyclic test for an isotropic hardening von Mises model is illustrated in Figure 6.22.

The choice of a suitable set (denoted α in Section 6.3) of hardening internal variables must be obviously dependent on the specific characteristics of the material considered. In metal plasticity, for instance, the hardening internal variable is intrinsically connected with the density of dislocations in the crystallographic microstructure that causes an isotropic increase in resistance to plastic flow. In the constitutive description of isotropic hardening, the set α normally contains a single *scalar* variable, which determines the size of the yield surface. Two approaches, *strain hardening* and *work hardening*, are particularly popular in the treatment of isotropic hardening and are suitable for modelling the behaviour of a wide range of materials. These are described below.

Strain hardening

In this case the hardening internal state variable is some suitably chosen scalar measure of *strain*. A typical example is the von Mises *effective plastic strain*, also referred to as the

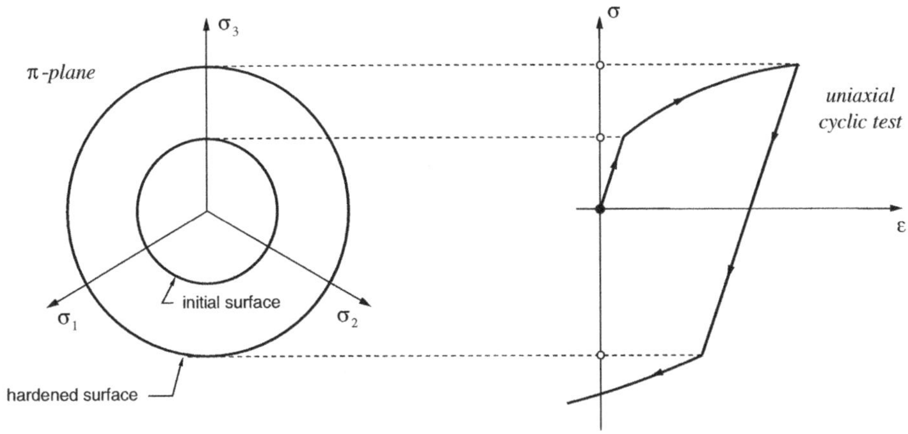

Figure 6.22. Isotropic hardening. Uniaxial test and π-plane representation.

von Mises *equivalent* or *accumulated plastic strain*, defined as

$$\bar{\varepsilon}^p \equiv \int_0^t \sqrt{\tfrac{2}{3}\,\dot{\varepsilon}^p : \dot{\varepsilon}^p}\,\mathrm{d}t = \int_0^t \sqrt{\tfrac{2}{3}}\,\|\dot{\varepsilon}^p\|\,\mathrm{d}t. \tag{6.167}$$

The above definition generalises the accumulated axial plastic strain (6.18) (page 145) of the one-dimensional model to the multiaxially strained case. Its rate evolution equation reads

$$\dot{\bar{\varepsilon}}^p = \sqrt{\tfrac{2}{3}\,\dot{\varepsilon}^p : \dot{\varepsilon}^p} = \sqrt{\tfrac{2}{3}}\,\|\dot{\varepsilon}^p\|, \tag{6.168}$$

or, equivalently, in view of the Prandtl–Reuss flow equation (6.137),

$$\dot{\bar{\varepsilon}}^p = \dot{\gamma}. \tag{6.169}$$

Accordingly, a von Mises isotropic *strain-hardening* model is obtained by letting the uniaxial yield stress be a function of the accumulated plastic strain:

$$\sigma_y = \sigma_y(\bar{\varepsilon}^p). \tag{6.170}$$

This function defines the *strain-hardening curve* (or *strain-hardening function*) that can be obtained, for instance, from a uniaxial tensile test.

Behaviour under uniaxial stress conditions

Under uniaxial stress conditions the von Mises model with isotropic strain hardening reproduces the behaviour of the one-dimensional plasticity model discussed in Section 6.2 and summarised in Box 6.1 (page 146). This is demonstrated in the following. Let us assume that both models share the same Young's modulus, E, and hardening function $\sigma_y = \sigma_y(\bar{\varepsilon}^p)$. Clearly, the two models have identical uniaxial elastic behaviour and initial yield stress. Hence, we only need to show next that their behaviour under plastic yielding is also identical.

Under a uniaxial stress state with axial stress σ and axial stress rate $\dot{\sigma}$ in the direction of the base vector $\boldsymbol{e}_1$, the matrix representations of the stress tensor and the stress rate tensor in the three-dimensional model are given by

$$[\boldsymbol{\sigma}] = \sigma \begin{bmatrix} 1 & 0 & 0 \\ 0 & 0 & 0 \\ 0 & 0 & 0 \end{bmatrix}, \quad [\dot{\boldsymbol{\sigma}}] = \dot{\sigma} \begin{bmatrix} 1 & 0 & 0 \\ 0 & 0 & 0 \\ 0 & 0 & 0 \end{bmatrix}. \tag{6.171}$$

The corresponding stress deviator reads

$$[\boldsymbol{s}] = \tfrac{2}{3}\sigma \begin{bmatrix} 1 & 0 & 0 \\ 0 & -\frac{1}{2} & 0 \\ 0 & 0 & -\frac{1}{2} \end{bmatrix}. \tag{6.172}$$

In this case, the Prandtl–Reuss flow equation (6.137) gives

$$[\dot{\boldsymbol{\varepsilon}}^p] = \dot{\varepsilon}^p \begin{bmatrix} 1 & 0 & 0 \\ 0 & -\frac{1}{2} & 0 \\ 0 & 0 & -\frac{1}{2} \end{bmatrix}, \tag{6.173}$$

where

$$\dot{\varepsilon}^p = \dot{\gamma}\,\mathrm{sign}(\sigma) \tag{6.174}$$

is the axial plastic strain rate. Note that the above expression coincides with the one-dimensional plastic flow rule (6.10). Now, we recall the consistency condition (6.60), which must be satisfied under plastic flow. In the present case, by taking the derivatives of the von Mises yield function (6.110), with σ_y defined by (6.170), we obtain

$$\dot{\Phi} = \boldsymbol{N} : \dot{\boldsymbol{\sigma}} - H\,\dot{\bar{\varepsilon}}^p = 0, \tag{6.175}$$

where $\boldsymbol{N} \equiv \partial\Phi/\partial\boldsymbol{\sigma}$ is the Prandtl–Reuss flow vector (6.136) and $H = H(\bar{\varepsilon}^p)$ is the hardening modulus defined in (6.27). To conclude the demonstration, we combine (6.175) with (6.136), $(6.171)_2$ and (6.172) to recover (6.28) and, then, following the same arguments as in the one-dimensional case we find that, under uniaxial stress conditions, the isotropic strain hardening von Mises model predicts the tangential axial stress–strain relation

$$\dot{\sigma} = \frac{EH}{E+H}\,\dot{\varepsilon}, \tag{6.176}$$

which is identical to equation (6.31) of the one-dimensional model.

Work hardening

In work-hardening models, the variable defining the state of hardening is the dissipated *plastic work*,¶ w^p, defined by

$$w^p \equiv \int_0^t \boldsymbol{\sigma} : \dot{\boldsymbol{\varepsilon}}^p \,\mathrm{d}t. \tag{6.177}$$

¶The term *work hardening* is adopted by many authors as a synonym for the phenomenon of hardening in general. Materials that harden, i.e. materials whose yield stress level depends on the history of strains, are frequently referred to as *work-hardening materials*. In this text, however, the term *work hardening* is reserved for plasticity models in which the dissipated plastic work is taken as the state variable associated with hardening.

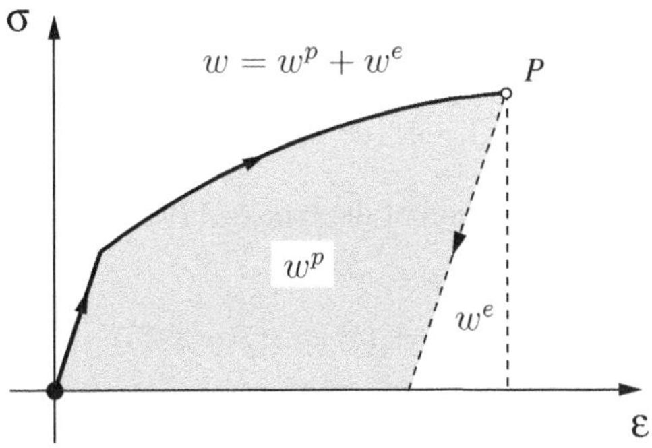

Figure 6.23. The plastic work.

In a uniaxial test, for instance (Figure 6.23), the total work w necessary to deform the material up to point P is given by the total area under the corresponding stress–strain curve. Part of this work, w^e, is stored in the form of elastic energy and is fully recovered upon elastic unloading. The remaining (shaded) area, w^p, is the *plastic work*. It corresponds to the energy dissipated by the plastic mechanisms and cannot be recovered. From the definition of w^p, its evolution equation is given by

$$\dot{w}^p = \boldsymbol{\sigma} : \dot{\boldsymbol{\varepsilon}}^p. \tag{6.178}$$

An isotropic *work-hardening* von Mises model is obtained by postulating

$$\sigma_y = \sigma_y(w^p). \tag{6.179}$$

This defines the *work-hardening curve* (or *work-hardening function*).

Equivalence between strain and work hardening

Under some circumstances, the strain-hardening and work-hardening descriptions are equivalent. This is shown in the following for the von Mises model with associative flow rule (6.137).

The substitution of (6.137) into (6.178), together with the identity $\sqrt{3/2}\|\boldsymbol{s}\| = \sigma_y$ valid for the von Mises model under plastic flow, gives

$$\dot{w}^p = \sigma_y \, \dot{\bar{\varepsilon}}^p, \tag{6.180}$$

or, equivalently,

$$\frac{\mathrm{d}w^p}{\mathrm{d}\bar{\varepsilon}^p} = \sigma_y. \tag{6.181}$$

As σ_y is strictly positive ($\sigma_y > 0$), the above differential relation implies that the mapping between w^p and $\bar{\varepsilon}^p$ is one-to-one and, therefore, invertible so that

$$w^p = w^p(\bar{\varepsilon}^p) \tag{6.182}$$

and

$$\bar{\varepsilon}^p = \bar{\varepsilon}^p(w^p). \tag{6.183}$$

This allows any given strain-hardening function of the type (6.170) to be expressed as an equivalent work-hardening function,

$$\sigma_y(\bar{\varepsilon}^p) = \tilde{\sigma}_y(w^p) \equiv \sigma_y(\bar{\varepsilon}^p(w^p)), \tag{6.184}$$

and any given work-hardening function of the type (6.179) to be expressed as an equivalent strain-hardening function,

$$\sigma_y(w^p) = \hat{\sigma}_y(\bar{\varepsilon}^p) \equiv \sigma_y(w^p(\bar{\varepsilon}^p)). \tag{6.185}$$

Expressions (6.184) and (6.185) establish the equivalence between the strain and work-hardening descriptions for the von Mises model with associative flow rule.

Linear and nonlinear hardening

A model is said to be *linear hardening* if the strain-hardening function (6.170) is linear, i.e. if it can be expressed as

$$\sigma_y(\bar{\varepsilon}^p) = \sigma_{y_0} + H\bar{\varepsilon}^p, \tag{6.186}$$

with *constant* σ_{y_0} and H. The constant σ_{y_0} is the *initial yield stress*, i.e. the uniaxial yield stress at the initial (*virgin*) state of the material, and H is called the *linear isotropic hardening modulus*. Any other hardening model is said to be *nonlinear hardening*. Note that perfect plasticity (defined in Section 6.6.1) is obtained if we set $H = 0$ in (6.186).

It should also be noted that a linear *work*-hardening function corresponds in general to an equivalent *nonlinear* strain-hardening function (i.e. a nonlinear hardening model). This can be easily established by observing that (6.181) defines a nonlinear relation between w^p and $\bar{\varepsilon}^p$ if σ_y is not a constant.

6.6.3. THERMODYNAMICAL ASPECTS. ASSOCIATIVE ISOTROPIC HARDENING

Within the formalism of thermodynamics with internal variables, the above isotropic strain-hardening law corresponds to the assumption that the plastic contribution, ψ^p, to the free energy (recall expression (6.37), page 149) is a function of a single scalar argument – the accumulated plastic strain. That is, the set $\boldsymbol{\alpha}$ of hardening variables is defined as

$$\boldsymbol{\alpha} \equiv \{\bar{\varepsilon}^p\} \tag{6.187}$$

and

$$\psi^p = \psi^p(\bar{\varepsilon}^p). \tag{6.188}$$

The set of hardening thermodynamic forces in this case specialises as

$$\boldsymbol{A} \equiv \{\kappa\}, \tag{6.189}$$

where the scalar thermodynamic force, κ, associated to isotropic hardening is defined by

$$\kappa \equiv \bar{\rho}\frac{\partial \psi^p}{\partial \bar{\varepsilon}^p} = \kappa(\bar{\varepsilon}^p). \tag{6.190}$$

The hardening curve is postulated in terms of κ as

$$\sigma_y(\bar{\varepsilon}^p) \equiv \sigma_{y_0} + \kappa(\bar{\varepsilon}^p). \tag{6.191}$$

If the state of hardening is defined in terms of cohesion (or shear yield stress), c (or τ_y) replaces σ_y in (6.191). Note that the hardening modulus H, initially defined in (6.27), represents the rate of change of the hardening thermodynamic force with respect to the hardening internal variable, i.e.

$$H(\bar{\varepsilon}^p) \equiv \frac{\partial \sigma_y}{\partial \bar{\varepsilon}^p} = \frac{\partial \kappa}{\partial \bar{\varepsilon}^p}. \tag{6.192}$$

For the strain-hardening von Mises model the evolution law (6.168) and (6.169) for the internal variable, $\bar{\varepsilon}^p$, follows from the hypothesis of *associativity*, that relies on the choice of the yield function as the plastic potential. The associative evolution equation for $\bar{\varepsilon}^p$ in this case is a specialisation of (6.130); that is, we have

$$\dot{\bar{\varepsilon}}^p = \dot{\gamma}\boldsymbol{H} = \dot{\gamma}. \tag{6.193}$$

The associative generalised modulus $\boldsymbol{H}$ is given by

$$\boldsymbol{H} = -\frac{\partial \Phi}{\partial \boldsymbol{A}} \equiv -\frac{\partial \Phi}{\partial \kappa} = 1, \tag{6.194}$$

where Φ is the von Mises yield function (6.110). A hardening law defined by means of the associativity hypothesis is called an *associative hardening* law. Any other hardening rule is said to be *non-associative*.

Multisurface models with associative hardening

Analogously to the associative plastic flow rule definition (6.73), (6.77) and (6.78), associative hardening for multisurface plasticity models can be defined by postulating the following generic evolution equation for the accumulated plastic strain:

$$\dot{\bar{\varepsilon}}^p = -\sum_{i=1}^{n} \dot{\gamma}^i \frac{\partial \Phi_i}{\partial \kappa}. \tag{6.195}$$

Note that, here, the accumulated plastic strain, $\bar{\varepsilon}^p$, is being *defined* by evolution equation (6.195). Its actual physical meaning depends on the specific format of the functions Φ_i and is generally different from that of (6.167) adopted for the von Mises model.

A simple example of associative isotropic hardening law of the type (6.195) is obtained for the Tresca model. Here, we refer to the plastic flow equations (6.141) and (6.143), defined respectively on the side (smooth portion) and corner of the Tresca yield surface. The corresponding associative evolution equations that define the accumulated plastic strain $\bar{\varepsilon}^p$ are

$$\dot{\bar{\varepsilon}}^p = -\dot{\gamma}\frac{\partial \Phi_1}{\partial \kappa} = \dot{\gamma} \tag{6.196}$$

and

$$\dot{\bar{\varepsilon}}^p = -\dot{\gamma}^a \frac{\partial \Phi_1}{\partial \kappa} - \dot{\gamma}^b \frac{\partial \Phi_6}{\partial \kappa} = \dot{\gamma}^a + \dot{\gamma}^b, \tag{6.197}$$

respectively, where functions Φ_1 and Φ_6 are defined by (6.91) with σ_y related to κ through (6.191).

For Mohr–Coulomb plasticity, one of the possibilities in defining a hardening law is to assume the *cohesion*, c, that takes part of the yield function (6.116) or (6.121) to be a function of the hardening internal variable:

$$c = c(\bar{\varepsilon}^p). \tag{6.198}$$

This type of hardening description is often used in practice in the modelling of soils – for which cohesion is a fundamental strength parameter. This assumption will be adopted in the computer implementation of Mohr–Coulomb and Drucker–Prager models described in Chapter 8. If hardening associativity is also assumed, then similarly to (6.191) we define

$$c(\bar{\varepsilon}^p) = c_0 + \kappa(\bar{\varepsilon}^p), \tag{6.199}$$

and the internal variable $\bar{\varepsilon}^p$ – the accumulated plastic strain for associative Mohr–Coulomb hardening – is defined by the corresponding particularisation of general evolution law (6.195). This gives the general expression

$$\dot{\bar{\varepsilon}}^p = 2\cos\phi \sum_{i=1}^{6} \dot{\gamma}^i. \tag{6.200}$$

When flow takes place at the smooth portion of a Mohr–Coulomb pyramid face, this is reduced to

$$\dot{\bar{\varepsilon}}^p = 2\cos\phi\,\dot{\gamma}. \tag{6.201}$$

At the corners (refer to the plastic flow equation (6.149)), we have

$$\dot{\bar{\varepsilon}}^p = 2\cos\phi\,(\dot{\gamma}^a + \dot{\gamma}^b). \tag{6.202}$$

Note that if it is insisted to adopt the von Mises accumulated plastic strain rate definition (6.167) in conjunction, say, with the Tresca model with associative plastic flow, (6.141) to (6.143), the evolution equation for $\bar{\varepsilon}^p$ will result in

$$\dot{\bar{\varepsilon}}^p = \sqrt{\tfrac{2}{3}\,\dot{\varepsilon}^p : \dot{\varepsilon}^p} = \tfrac{2}{\sqrt{3}}\,\dot{\gamma} \tag{6.203}$$

for flow from the smooth portions of the Tresca surface, and

$$\dot{\bar{\varepsilon}}^p = \sqrt{\tfrac{2}{3}\,\dot{\varepsilon}^p : \dot{\varepsilon}^p} = \tfrac{2}{\sqrt{3}}\,\sqrt{(\dot{\gamma}^a)^2 + \dot{\gamma}^a\dot{\gamma}^b + (\dot{\gamma}^b)^2}, \tag{6.204}$$

for flow from a corner. In this case, the isotropic hardening law is *non-associative* in spite of the associativity of the plastic flow rule.

Drucker–Prager associative hardening

Associative hardening for Drucker–Prager plasticity is obtained by combining assumption (6.199) and the yield function definition (6.121) with the general associative evolution law (6.130) for the hardening internal variable. The accumulated plastic strain in this case is then defined by the evolution equation

$$\dot{\bar{\varepsilon}}^p = -\dot{\gamma}\,\frac{\partial\Phi}{\partial\kappa} = \dot{\gamma}\,\xi. \tag{6.205}$$

Other hardening models

Further refinements to capture hardening behaviour more accurately can be incorporated in Mohr–Coulomb based plasticity models by assuming, in addition, the frictional angle to be a function, for example, of the accumulated plastic strain:

$$\phi = \phi(\bar{\varepsilon}^p). \tag{6.206}$$

For Drucker–Prager-based models, the above would correspond to having

$$\eta = \eta(\bar{\varepsilon}^p), \quad \xi = \xi(\bar{\varepsilon}^p). \tag{6.207}$$

The direction of plastic flow is generally affected by the history of plastic straining in materials such as soils and rocks. This phenomenon can be accounted for in non-associative flow Mohr–Coulomb type models by letting the dilatancy angle, ψ, be a function of the hardening internal variable. For Drucker–Prager-based models, this can be obtained by having the parameter $\bar{\eta}$ as a function of the hardening variable.

6.6.4. KINEMATIC HARDENING. THE BAUSCHINGER EFFECT

When the yield surfaces preserve their shape and size but *translate* in the stress space as a rigid body, *kinematic hardening* is said to take place. It is frequently observed in experiments that, after being loaded (and hardened) in one direction, many materials show a decreased resistance to plastic yielding in the opposite direction (Lemaitre and Chaboche, 1990). This phenomenon is known as the *Bauschinger effect* and can be modelled with the introduction of kinematic hardening. A number of constitutive models have been proposed to describe elastoplastic behaviour under cyclic loading conditions (Lemaitre and Chaboche, 1990; Mróz, 1967; Skrzypek, 1993). The typical result of a uniaxial cyclic test showing the Bauschinger effect is illustrated in Figure 6.24. The evolution of a kinematically hardening von Mises-type yield surface (in the deviatoric plane) used to model the phenomenon is shown alongside. The yield function for the kinematically hardening model is given by

$$\Phi(\boldsymbol{\sigma}, \boldsymbol{\beta}) = \sqrt{3\, J_2(\boldsymbol{\eta}(\boldsymbol{\sigma}, \boldsymbol{\beta}))} - \sigma_y, \tag{6.208}$$

where

$$\boldsymbol{\eta}(\boldsymbol{\sigma}, \boldsymbol{\beta}) \equiv \boldsymbol{s}(\boldsymbol{\sigma}) - \boldsymbol{\beta} \tag{6.209}$$

is the *relative stress* tensor, defined as the difference between the stress deviator and the symmetric deviatoric (stress-like) tensor, $\boldsymbol{\beta}$, known as the *back-stress* tensor. Note that, by definition, the relative stress is *deviatoric*. The back-stress tensor is the thermodynamical force associated with kinematic hardening and represents the translation (Figure 6.24) of the yield surface in the space of stresses. The *constant* σ_y in (6.208) defines the radius of the yield surface. When $\boldsymbol{\beta} = \boldsymbol{0}$, we have $\boldsymbol{\eta} = \boldsymbol{s}$ and the yield surface defined by $\Phi = 0$ is the isotropic von Mises yield surface with uniaxial yield stress σ_y.

It is important to observe that, unlike the isotropically hardening von Mises model, the yield function Φ defined by (6.208) is *not* an isotropic function of the stress tensor for kinematically hardened states ($\boldsymbol{\beta} \neq \boldsymbol{0}$). The function (6.208) is an isotropic function of the *relative stress*, $\boldsymbol{\eta}$. Analogously to expression (6.208), it is possible to introduce kinematic hardening in other plasticity models simply by replacing $\boldsymbol{\sigma}$ with a relative stress measure, defined as the difference $\boldsymbol{\sigma} - \boldsymbol{\beta}$, in the definition of the corresponding yield function.

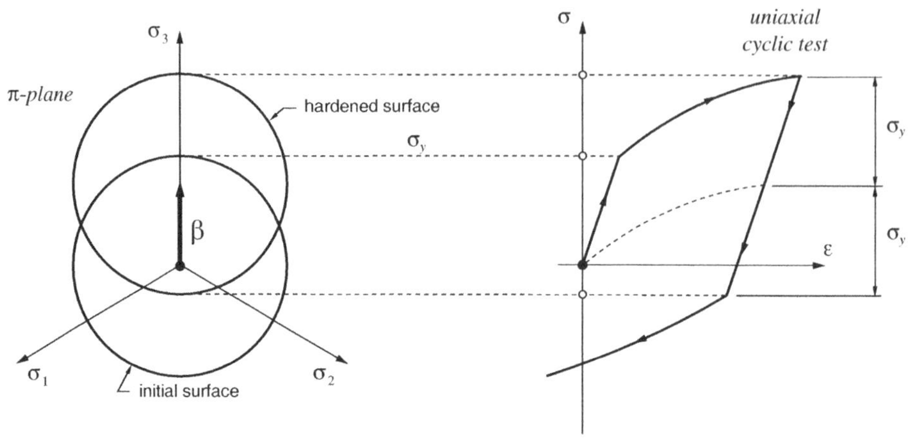

Figure 6.24. Kinematic hardening and the *Bauschinger effect*. Uniaxial test and π-plane representation. Loading in one direction results in decreased resistance to plastic yielding in the opposite direction.

Plastic flow rule with kinematic hardening

The von Mises model with kinematic hardening is used in conjunction with an *associative* flow rule. The flow vector in this case reads

$$\boldsymbol{N} \equiv \frac{\partial \Phi}{\partial \boldsymbol{\sigma}} = \sqrt{\tfrac{3}{2}}\,\frac{\boldsymbol{\eta}}{\|\boldsymbol{\eta}\|} \tag{6.210}$$

and we have the following plastic strain rate equation:

$$\dot{\boldsymbol{\varepsilon}}^p = \dot{\gamma}\,\boldsymbol{N} = \dot{\gamma}\sqrt{\tfrac{3}{2}}\,\frac{\boldsymbol{\eta}}{\|\boldsymbol{\eta}\|}. \tag{6.211}$$

This rule extends the Prandtl–Reuss equation to account for kinematic hardening. Note that the plastic flow is in the direction of the (deviatoric) relative stress, $\boldsymbol{\eta}$, and coincides with the Prandtl–Reuss equation if $\boldsymbol{\beta} = \boldsymbol{0}$.

Prager's linear kinematic hardening

To complete the definition of the kinematic hardening plasticity model, evolution equations for $\boldsymbol{\beta}$ are required. One of the most commonly used laws is *Prager's linear kinematic hardening rule*, where the rate evolution equation for $\boldsymbol{\beta}$ is given by

$$\dot{\boldsymbol{\beta}} = \tfrac{2}{3} H\, \dot{\boldsymbol{\varepsilon}}^p = \dot{\gamma}\sqrt{\tfrac{2}{3}}\, H \frac{\boldsymbol{\eta}}{\|\boldsymbol{\eta}\|}. \tag{6.212}$$

The material constant H is the *linear kinematic hardening modulus*.

Behaviour under monotonic uniaxial stress loading

For monotonic loading under uniaxial stress conditions, the stress–strain behaviour of the model defined by equations (6.208), with constant $\sigma_y = \sigma_{y_0}$, (6.211) and (6.212) and initial state of hardening defined by $\boldsymbol{\beta} = \mathbf{0}$ is identical to the behaviour of the purely isotropic hardening von Mises model with linear hardening curve (6.186) and initial state of hardening $\bar{\varepsilon}^p = 0$. It is assumed in this statement that both models share the same Young's modulus, E. Under the above conditions, it is clear that both models have the same elastic behaviour and uniaxial yield stress, σ_{y_0}. To show that their plastic behaviour also coincides, let us consider again a uniaxial test with loading in the direction of the base vector $\boldsymbol{e}_1$. In this case, the stress, stress rate and stress deviator tensors have the matrix representations given in (6.171) and (6.172). Now note that the integration of the rate equation (6.212) with initial condition $\boldsymbol{\beta} = \mathbf{0}$ (i.e. $\boldsymbol{\eta} = \boldsymbol{s}$) and $\boldsymbol{s}$ as in (6.172) gives a back-stress tensor of the form

$$[\boldsymbol{\beta}] = \beta \begin{bmatrix} 1 & 0 & 0 \\ 0 & -\frac{1}{2} & 0 \\ 0 & 0 & -\frac{1}{2} \end{bmatrix}, \tag{6.213}$$

where β is the axial back-stress component. With the above, we obtain for the relative stress tensor

$$[\boldsymbol{\eta}] = \eta \begin{bmatrix} 1 & 0 & 0 \\ 0 & -\frac{1}{2} & 0 \\ 0 & 0 & -\frac{1}{2} \end{bmatrix}, \tag{6.214}$$

where

$$\eta = \tfrac{2}{3}\sigma - \beta \tag{6.215}$$

is the axial relative stress. From (6.212) and (6.214) we obtain

$$[\dot{\boldsymbol{\beta}}] = \tfrac{2}{3} H \dot{\varepsilon}^p \begin{bmatrix} 1 & 0 & 0 \\ 0 & -\frac{1}{2} & 0 \\ 0 & 0 & -\frac{1}{2} \end{bmatrix}, \tag{6.216}$$

where $\dot{\varepsilon}^p$ is the axial plastic strain rate given by

$$\dot{\varepsilon}^p = \dot{\gamma}\, \text{sign}(\eta). \tag{6.217}$$

Now, by recalling (6.60) and specialising (6.61) for the present case we have that, under plastic yielding, the following consistency condition must be satisfied:

$$\dot{\Phi} = \frac{\partial \Phi}{\partial \boldsymbol{\sigma}} : \dot{\boldsymbol{\sigma}} + \frac{\partial \Phi}{\partial \boldsymbol{\beta}} : \dot{\boldsymbol{\beta}} = 0. \tag{6.218}$$

After some straightforward tensor algebra, taking into account $(6.171)_2$ and the above expressions for $\dot{\boldsymbol{\beta}}$, $\boldsymbol{\beta}$, the definition of $\boldsymbol{\eta}$, and the identity

$$\frac{\partial \Phi}{\partial \boldsymbol{\beta}} = -\frac{\partial \Phi}{\partial \boldsymbol{\sigma}} = -\sqrt{\frac{3}{2}} \frac{\boldsymbol{\eta}}{\|\boldsymbol{\eta}\|}, \tag{6.219}$$

equation (6.218) yields

$$\dot{\sigma} = H \dot{\varepsilon}^p. \tag{6.220}$$

Then, with the introduction of the elastoplastic split of the axial strain rate, together with the equation

$$\dot{\sigma} = E \dot{\varepsilon}^e, \tag{6.221}$$

of the linear elastic model under uniaxial stress conditions, into (6.220), we obtain

$$\dot{\sigma} = \frac{EH}{E+H}\,\dot{\varepsilon}, \tag{6.222}$$

which coincides with the stress rate equation (6.176) of the von Mises isotropic strain-hardening model with constant H. To complete the demonstration, let us assume that the uniaxial loading is monotonic, i.e. we have either $\dot{\varepsilon} > 0$ or $\dot{\varepsilon} < 0$ throughout the entire loading process. In this case, the integration of (6.222) having the initial yield stress (σ_{y_0} for both models) as the initial condition produces the same stress–strain curve as the isotropic model.

Armstrong–Frederick hardening

A refinement upon the linear kinematic hardening law proposed by Armstrong and Frederick (1966) is obtained by introducing an extra term in the above expression (refer to Lemaitre and Chaboche (1990), Chapter 5, or Jirásek and Bažant (2002), Chapter 20, for details) with the evolution of $\boldsymbol{\beta}$ given by

$$\begin{aligned}\dot{\boldsymbol{\beta}} &= \frac{2}{3}\,H\,\dot{\boldsymbol{\varepsilon}}^p - \dot{\gamma}\,b\,\boldsymbol{\beta} \\ &= \dot{\gamma}\left(\frac{2}{3}H\,\frac{\partial\Phi}{\partial\boldsymbol{\sigma}} - b\,\boldsymbol{\beta}\right),\end{aligned} \tag{6.223}$$

where b is a material constant. The extra term $-\dot{\gamma}\,b\,\boldsymbol{\beta}$ introduces the effect of saturation in the kinematic hardening rule. In the case of the von Mises criterion, the saturation corresponds to a maximum limit value for the norm of $\boldsymbol{\beta}$, at which the material behaves as perfectly plastic.

Nonlinear extension to Prager's rule

Another possible improvement upon Prager's linear kinematic hardening rule is the introduction of nonlinearity by replacing the constant kinematic hardening modulus, H, of (6.212) with a generic function of the accumulated plastic strain, $\bar{\varepsilon}^p$,

$$\dot{\boldsymbol{\beta}} = \frac{2}{3}H(\bar{\varepsilon}^p)\,\dot{\boldsymbol{\varepsilon}}^p = \dot{\gamma}\,\frac{2}{3}H(\bar{\varepsilon}^p)\,\frac{\partial\Phi}{\partial\boldsymbol{\sigma}}. \tag{6.224}$$

In this case, a scalar function

$$\bar{\beta} \equiv \bar{\beta}(\bar{\varepsilon}^p), \tag{6.225}$$

such that

$$H(\bar{\varepsilon}^p) = \frac{\mathrm{d}\bar{\beta}}{\mathrm{d}\bar{\varepsilon}^p}, \tag{6.226}$$

defines the *kinematic hardening curve*. This curve can be obtained from simple uniaxial tests in a manner analogous to the determination of the hardening curve for the purely isotropic hardening model.

Thermodynamical aspects of kinematic hardening

From the thermodynamical viewpoint, the above kinematic hardening laws follow from the assumption that the plastic contribution, ψ^p, to the free energy is a function of a second-order tensor-valued internal variable, $\boldsymbol{X}$,

$$\psi^p = \psi^p(\boldsymbol{X}). \tag{6.227}$$

The variable $\boldsymbol{X}$ is related to self-equilibrated residual stresses that remain after elastic unloading. These stresses may increase or decrease resistance to plastic slip according to the direction considered. The kinematic hardening thermodynamical force – the back-stress tensor, $\boldsymbol{\beta}$ – is then defined as the derivative

$$\boldsymbol{\beta} \equiv \frac{\partial \psi^p}{\partial \boldsymbol{X}}. \tag{6.228}$$

For the Armstrong–Frederick kinematic hardening law (6.223), for instance, we have

$$\psi^p(\boldsymbol{X}) = \frac{a}{2}\, \boldsymbol{X} : \boldsymbol{X}, \tag{6.229}$$

where the material constant a has been defined as

$$a \equiv \tfrac{2}{3}\, H. \tag{6.230}$$

The back-stress tensor (6.228) is then a scalar multiple of $\boldsymbol{X}$, given by

$$\boldsymbol{\beta} = a\, \boldsymbol{X}. \tag{6.231}$$

The evolution law for the internal variable $\boldsymbol{X}$ is obtained by postulating a flow potential

$$\Psi \equiv \Phi + \frac{b}{2a} \boldsymbol{\beta} : \boldsymbol{\beta}, \tag{6.232}$$

and assuming normal dissipativity

$$\dot{\boldsymbol{X}} \equiv -\dot{\gamma}\, \frac{\partial \Psi}{\partial \boldsymbol{\beta}} = -\dot{\gamma} \left(\frac{\partial \Phi}{\partial \boldsymbol{\beta}} + \frac{b}{a}\, \boldsymbol{\beta} \right). \tag{6.233}$$

Obviously (since $\Psi \neq \Phi$), this evolution law is *non-associative*. The equivalence between the above equation and (6.223) can be established by taking into account (6.231) and the fact that, since Φ is obtained from a non-kinematic hardening yield function by replacing the argument $\boldsymbol{\sigma}$ with $\boldsymbol{\sigma} - \boldsymbol{\beta}$, we have

$$\frac{\partial \Phi}{\partial \boldsymbol{\beta}} = -\frac{\partial \Phi}{\partial \boldsymbol{\sigma}}. \tag{6.234}$$

6.6.5. MIXED ISOTROPIC/KINEMATIC HARDENING

Rather than purely isotropic or purely kinematic hardening, real-life materials show in general a combination of both; that is, under plastic straining, the yield surface expands/shrinks and translates simultaneously in stress space. Thus, more realistic plasticity models can be obtained by combining the above laws for isotropic and kinematic hardening.

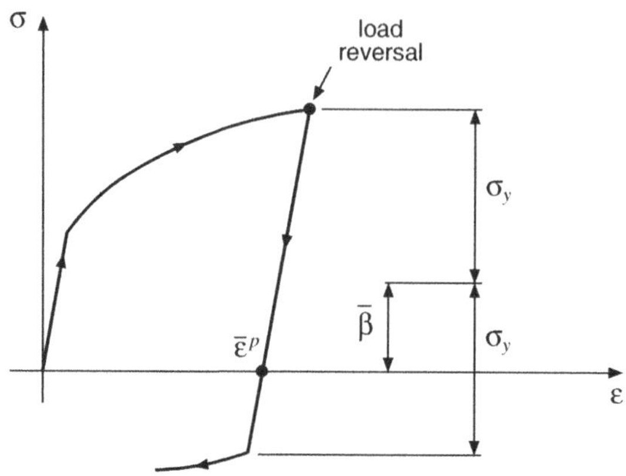

Figure 6.25. Mixed hardening. Uniaxial test with load reversal.

For example, a relatively simple von Mises-based model with mixed isotropic/kinematic hardening can be devised by adopting the yield function (6.208) and allowing σ_y to be a function of $\bar{\varepsilon}^p$. If the nonlinear rule defined by (6.224) and (6.225) is adopted, the hardening behaviour of the model is determined by the curves

$$\sigma_y = \sigma_y(\bar{\varepsilon}^p), \quad \bar{\beta} = \bar{\beta}(\bar{\varepsilon}^p), \tag{6.235}$$

which can be obtained from relatively simple uniaxial tests with load reversal (see schematic illustration of Figure 6.25). At each point $\bar{\varepsilon}^p$, the *kinematic hardening stress*, $\bar{\beta}$, is the kinematic contribution to overall hardening.

A more refined mixed hardening model can be devised by coupling the Armstrong–Frederick law (6.223) with the von Mises-type yield function (6.208) where σ_y, as in $(6.235)_1$, is a function of the accumulated plastic strain. A model including mixed hardening of this type is discussed in Section 12.3 (starting on page 478) in the context of damage mechanics.

7 FINITE ELEMENTS IN SMALL-STRAIN PLASTICITY PROBLEMS

IN the previous chapter, the mathematical theory of plasticity has been reviewed. A general small-strain elastoplastic constitutive model has been established within the formalism of thermodynamics with internal variables and the most popular theories, namely, the von Mises, Tresca, Mohr–Coulomb and Drucker–Prager models, have been described in detail.

Obviously, due to the mathematical complexity of such constitutive theories, an exact solution to boundary value problems of practical engineering interest can only be obtained under very simplified conditions. The existing analytical solutions are normally restricted to perfectly plastic models and are used for the determination of limit loads and steady plastic flow of bodies with simple geometries (Chakrabarty, 1987; Hill, 1950; Lubliner, 1990; Prager, 1959; Skrzypek, 1993). The analysis of the behaviour of elastoplastic structures and soils under more realistic conditions requires the adoption of an adequate numerical framework capable of producing approximate solutions within reasonable accuracy. As pointed out in Chapter 4, the approximate solution to such problems is addressed in this book within the context of the Finite Element Method. In fact, the Finite Element Method is by far the most commonly adopted procedure for the solution of elastoplastic problems. Since the first reported applications of finite elements in plasticity in the mid-1960s, a substantial development of the related numerical techniques has occurred. Today, the Finite Element Method is regarded as the most powerful and reliable tool for the analysis of solid mechanics problems involving elastoplastic materials and is adopted by the vast majority of commercial software packages for elastoplastic stress analysis.

This chapter describes in detail the numerical/computational procedures necessary for the implicit finite element solution of small strain plasticity problems within the framework of Chapter 4. For the sake of generality, the methodologies presented in this chapter are initially derived taking the general plasticity model introduced in Chapter 6 (summarised in Box 6.2, page 151) as the underlying constitutive model. Practical application of the theory and procedures introduced, including a complete description of the algorithms and corresponding FORTRAN subroutines of the `HYPLAS` program, is then made to the particular case of the von Mises model with nonlinear isotropic hardening. The choice of this model is motivated here by the simplicity of its computational implementation. A set of numerical examples is also presented. Further application of the theory is made at the end of the chapter to a mixed isotropic/kinematic hardening version of the von Mises model. This model is also included in the `HYPLAS` program. Application to the Tresca, Mohr–Coulomb and Drucker–Prager models is left for Chapter 8.

7.1. Preliminary implementation aspects

At this point, it is convenient to recall that, within the modular structure of the HYPLAS program (refer to Section 5.7, from page 131), each material model implementation is defined by a set of five material-specific routines. Each one of these routines is called by a corresponding material interface subroutine that is common to all models. There are two most fundamental material-specific operations:

1. The *state update procedure* which, in the case of elastoplastic materials, requires the formulation of a scheme for numerical integration of the rate elastoplastic evolution equations. Within a (pseudo-) time increment $[t_n, t_{n+1}]$, the state update procedure gives the stresses $\boldsymbol{\sigma}_{n+1}$ and the internal variables $\boldsymbol{\alpha}_{n+1}$ at the end of the increment as a function of the internal variables $\boldsymbol{\alpha}_n$ at the beginning of the increment and the strains $\boldsymbol{\varepsilon}_{n+1}$ at the end of the increment:

$$\boldsymbol{\sigma}_{n+1} = \hat{\boldsymbol{\sigma}}(\boldsymbol{\alpha}_n, \boldsymbol{\varepsilon}_{n+1}), \tag{7.1}$$

$$\boldsymbol{\alpha}_{n+1} = \hat{\boldsymbol{\alpha}}(\boldsymbol{\alpha}_n, \boldsymbol{\varepsilon}_{n+1}). \tag{7.2}$$

   The incremental constitutive functions $\hat{\boldsymbol{\sigma}}$ and $\hat{\boldsymbol{\alpha}}$ are defined by the integration algorithm adopted and the stress delivered by $\hat{\boldsymbol{\sigma}}$ is used to assemble the element internal force vector

$$\mathbf{f}_e^{\text{int}} = \sum_{i=1}^{n_{\text{gausp}}} j_i \, w_i \, \mathbf{B}_i^T \, \sigma_{n+1}|_i. \tag{7.3}$$

2. The computation of the associated *consistent tangent modulus*

$$\mathsf{D} \equiv \frac{\partial \hat{\boldsymbol{\sigma}}}{\partial \boldsymbol{\varepsilon}_{n+1}}, \tag{7.4}$$

   to be used whenever the evaluation of a new tangent stiffness matrix is required by the selected iterative scheme for solution of the nonlinear finite element equilibrium equations. The element tangent stiffness matrix is computed as

$$\mathbf{K}_T^{(e)} = \sum_{i=1}^{n_{\text{gausp}}} w_i \, j_i \, \mathbf{B}_i^T \, \mathbf{D}_i \, \mathbf{B}_i. \tag{7.5}$$

These are the primary procedures that effectively define a particular constitutive model/algorithm in the program.

This chapter will focus precisely on the two topics listed above, specialised to the case of elastoplastic materials. Following the nomenclature established in Chapter 5 (refer to the call trees of Figures 5.4 and 5.5 respectively on pages 126 and 130), the computational implementation of the state update algorithms and tangent moduli addressed in the present chapter appears at the lowest layer of the *material level* in the computation of the element internal force vector and tangent stiffness matrix.

7.2. General numerical integration algorithm for elastoplastic constitutive equations

The importance of the state-updating procedure within the overall finite element scheme has been stressed in Chapter 4. In the case of path-dependent materials, such as elastoplastic materials, the updating scheme usually requires the formulation of a numerical algorithm for integration of the corresponding rate constitutive equations. This requirement stems from the fact that analytical solutions to the initial value problem defined by the elastoplastic equations are generally not known for complex strain paths. An important point that one should bear in mind regarding the formulation of state-updating procedures is that the accuracy of the overall finite element scheme depends crucially on the accuracy of the particular numerical algorithm adopted. This section describes a numerical procedure for integration of the general elastoplastic model of Section 6.3. The strategy presented here is later specialised and applied to the von Mises model in Section 7.3. Specialisation to the other basic plasticity models of Chapter 6, i.e. the Tresca, Mohr–Coulomb and Drucker–Prager models, is presented in Chapter 8 and its plane stress implementation is addressed in Chapter 9. Further applications of the algorithms described in the present chapter are made in Chapter 10, in the context of advanced plasticity models, and in Chapter 12, where the numerical implementation of damage mechanics models is discussed.

7.2.1. THE ELASTOPLASTIC CONSTITUTIVE INITIAL VALUE PROBLEM

Consider a point $\boldsymbol{p}$ of a body $\mathscr{B}$ with constitutive behaviour described by the general elastoplastic model of Box 6.2 (page 151). Assume that at a given (pseudo-)time t_0 the elastic strain, $\boldsymbol{\varepsilon}^e(t_0)$, the plastic strain tensor, $\boldsymbol{\varepsilon}^p(t_0)$, and all elements of the set $\boldsymbol{\alpha}(t_0)$ of hardening internal variables are known at point $\boldsymbol{p}$. Furthermore, let the motion of $\mathscr{B}$ be prescribed between t_0 and a subsequent instant, T. Clearly, the prescribed motion defines the *history* of the strain tensor, $\boldsymbol{\varepsilon}(t)$, at the material point of interest between instants t_0 and T. The basic elastoplastic constitutive initial value problem at point $\boldsymbol{p}$ is stated in the following.

Problem 7.1 (The elastoplastic constitutive initial value problem). *Given the initial values $\boldsymbol{\varepsilon}^e(t_0)$ and $\boldsymbol{\alpha}(t_0)$ and given the history of the strain tensor, $\boldsymbol{\varepsilon}(t)$, $t \in [t_0, T]$, find the functions $\boldsymbol{\varepsilon}^e(t)$, $\boldsymbol{\alpha}(t)$ and $\dot{\gamma}(t)$ for the elastic strain, hardening internal variables set and plastic multiplier that satisfy the reduced general elastoplastic constitutive equations*

$$\begin{aligned} \dot{\boldsymbol{\varepsilon}}^e(t) &= \dot{\boldsymbol{\varepsilon}}(t) - \dot{\gamma}(t)\, \boldsymbol{N}(\boldsymbol{\sigma}(t), \boldsymbol{A}(t)) \\ \dot{\boldsymbol{\alpha}}(t) &= \dot{\gamma}(t)\, \boldsymbol{H}(\boldsymbol{\sigma}(t), \boldsymbol{A}(t)) \end{aligned} \tag{7.6}$$

$$\dot{\gamma}(t) \geq 0, \quad \Phi(\boldsymbol{\sigma}(t), \boldsymbol{A}(t)) \leq 0, \quad \dot{\gamma}(t)\, \Phi(\boldsymbol{\sigma}(t), \boldsymbol{A}(t)) = 0 \tag{7.7}$$

for each instant $t \in [t_0, T]$, with

$$\boldsymbol{\sigma}(t) = \bar{\rho}\, \frac{\partial \psi}{\partial \boldsymbol{\varepsilon}^e}\bigg|_t, \quad \boldsymbol{A}(t) = \bar{\rho}\, \frac{\partial \psi}{\partial \boldsymbol{\alpha}}\bigg|_t. \tag{7.8}$$

Remark 7.1. We refer to the system of differential equations (7.6) as *reduced* in that it is obtained from the model of Box 6.2 by incorporating the plastic flow equation into the additive strain rate decomposition. In this way, the plastic strain does not appear explicitly in the system and the only unknowns are the elastic strain, the set of hardening internal variables and the plastic multiplier. Note that with the solution $\boldsymbol{\varepsilon}^e(t)$ of Problem 7.1 at hand, the history of the plastic strain tensor is obtained from the trivial relation

$$\boldsymbol{\varepsilon}^p(t) = \boldsymbol{\varepsilon}(t) - \boldsymbol{\varepsilon}^e(t), \tag{7.9}$$

so that the history of all variables involved in the definition of the elastoplastic model of Box 6.2 is determined.

As already mentioned, exact solutions to Problem 7.1, when yield functions and flow rules such as the ones described in the previous chapter are adopted, may only be obtained for very simple prescribed strain histories. Even in such cases, the derivation of the analytical solutions is normally cumbersome. For complex deformation paths, which are more likely to occur in realistic engineering problems, analytical solutions are not available in general and the adoption of a numerical technique to find an approximate solution becomes absolutely essential. A general framework for the numerical solution of the constitutive initial value problem of elastoplasticity is described below.

7.2.2. EULER DISCRETISATION: THE INCREMENTAL CONSTITUTIVE PROBLEM

The starting point here is the adoption of an *Euler scheme* to discretise equations (7.6) and (7.7). For simplicity we shall choose to adopt a *backward* (or fully implicit) Euler scheme.† Accordingly, for integration within a generic (pseudo-)time interval $[t_n, t_{n+1}]$, we replace all rate quantities in (7.7)–(7.6) with corresponding incremental values within the considered interval and the functions $\boldsymbol{N}$, $\boldsymbol{H}$ and Φ with their values at the end of the interval, t_{n+1}. The resulting discrete version of Problem 7.1 is stated in the following.

Problem 7.2 (The incremental elastoplastic constitutive problem). *Given the values $\boldsymbol{\varepsilon}^e_n$ and $\boldsymbol{\alpha}_n$, of the elastic strain and internal variables set at the beginning of the pseudo-time interval $[t_n, t_{n+1}]$, and given the prescribed incremental strain $\Delta\boldsymbol{\varepsilon}$ for this interval, solve the following system of algebraic equations*

$$\begin{aligned} \boldsymbol{\varepsilon}^e_{n+1} &= \boldsymbol{\varepsilon}^e_n + \Delta\boldsymbol{\varepsilon} - \Delta\gamma\, \boldsymbol{N}(\boldsymbol{\sigma}_{n+1}, \boldsymbol{A}_{n+1}) \\ \boldsymbol{\alpha}_{n+1} &= \boldsymbol{\alpha}_n + \Delta\gamma\, \boldsymbol{H}(\boldsymbol{\sigma}_{n+1}, \boldsymbol{A}_{n+1}) \end{aligned} \tag{7.10}$$

for the unknowns $\boldsymbol{\varepsilon}^e_{n+1}$, $\boldsymbol{\alpha}_{n+1}$ and $\Delta\gamma$, subjected to the constraints

$$\Delta\gamma \geq 0, \quad \Phi(\boldsymbol{\sigma}_{n+1}, \boldsymbol{A}_{n+1}) \leq 0, \quad \Delta\gamma\, \Phi(\boldsymbol{\sigma}_{n+1}, \boldsymbol{A}_{n+1}) = 0, \tag{7.11}$$

where

$$\boldsymbol{\sigma}_{n+1} = \bar{\rho}\, \frac{\partial\psi}{\partial\boldsymbol{\varepsilon}^e}\bigg|_{n+1}, \qquad \boldsymbol{A}_{n+1} = \bar{\rho}\, \frac{\partial\psi}{\partial\boldsymbol{\alpha}}\bigg|_{n+1}. \tag{7.12}$$

In the above, we have adopted the obvious notation

$$\Delta(\cdot) \equiv (\cdot)_{n+1} - (\cdot)_n, \tag{7.13}$$

†Other Euler-based schemes can be equally adopted. These will be discussed later in Section 7.2.7.

with $(\cdot)_n$ and $(\cdot)_{n+1}$ denoting the value of $(\cdot)$ respectively at t_n and t_{n+1}. The increment $\Delta\gamma$ will be called the *incremental plastic multiplier*. Note that once the solution ε^e_{n+1} has been obtained, the plastic strain at t_{n+1} can be calculated as

$$\varepsilon^p_{n+1} = \varepsilon^p_n + \Delta\varepsilon - \Delta\varepsilon^e, \tag{7.14}$$

so that all variables of the model are known at the end of the interval $[t_n, t_{n+1}]$.

Solution of the incremental problem

Due to the presence of the discrete complementarity condition (7.11), the solution of the incremental elastoplastic problem (7.10)–(7.12) does not follow directly the conventional procedure for standard initial value problems (i.e. initial value problems without equations of the type (7.7)). Nevertheless, as we shall see, the solution scheme in the present case remains rather simple with the discrete complementarity condition giving rise to a two-step algorithm derived in the following.

Firstly, note that $(7.11)_1$ allows only for the two (mutually exclusive) possibilities enumerated below:

1. Null incremental plastic multiplier,

$$\Delta\gamma = 0. \tag{7.15}$$

In this case there is no plastic flow or evolution of internal variables within the considered interval $[t_n, t_{n+1}]$, i.e. the step is purely elastic. The constraint $(7.11)_3$ is automatically satisfied, ε^e_{n+1} and $\boldsymbol{\alpha}_{n+1}$ are given by

$$\begin{aligned} \varepsilon^e_{n+1} &= \varepsilon^e_n + \Delta\varepsilon \\ \boldsymbol{\alpha}_{n+1} &= \boldsymbol{\alpha}_n \end{aligned} \tag{7.16}$$

and, in addition, the constraint

$$\Phi(\boldsymbol{\sigma}_{n+1}, \boldsymbol{A}_{n+1}) \leq 0, \tag{7.17}$$

must hold, where $\boldsymbol{\sigma}_{n+1}$ and $\boldsymbol{A}_{n+1}$ are functions of ε^e_{n+1} and $\boldsymbol{A}_{n+1}$ defined through the potential relations (7.12).

2. Strictly positive plastic multiplier,

$$\Delta\gamma > 0. \tag{7.18}$$

In this case, ε^e_{n+1}, $\boldsymbol{\alpha}_{n+1}$ and $\Delta\gamma$ satisfy

$$\begin{aligned} \varepsilon^e_{n+1} &= \varepsilon^e_n + \Delta\varepsilon - \Delta\gamma\, \boldsymbol{N}(\boldsymbol{\sigma}_{n+1}, \boldsymbol{A}_{n+1}) \\ \boldsymbol{\alpha}_{n+1} &= \boldsymbol{\alpha}_n + \Delta\gamma\, \boldsymbol{H}(\boldsymbol{\sigma}_{n+1}, \boldsymbol{A}_{n+1}) \end{aligned} \tag{7.19}$$

and $(7.11)_2$ combined with $(7.11)_3$ result in the constraint

$$\Phi(\boldsymbol{\sigma}_{n+1}, \boldsymbol{A}_{n+1}) = 0. \tag{7.20}$$

In summary, we have just seen that any solution to Problem 7.2 is either given by (7.16), in which case it must satisfy constraint (7.17), or it is a solution of the algebraic system of equations (7.19)–(7.20), subjected to the constraint (7.18); that is, only one of two possible sets of equations provides a solution to Problem 7.2. It remains now to devise a procedure whereby the correct solution can be chosen. This is next described in detail.

7.2.3. THE ELASTIC PREDICTOR/PLASTIC CORRECTOR ALGORITHM

The nature of the above problem motivates the establishment of a (conceptually very simple) two-step algorithm in which the two possible sets of equations are employed sequentially and the final solution is selected as the only valid one. The strategy adopted is the following:

(a) *The Elastic Trial Step.*
Firstly, we assume that the first of the above two situations ($\Delta\gamma = 0$) occurs; that is, we assume that the step $[t_n, t_{n+1}]$ is *elastic*. The solution given by (7.16), which is not necessarily the actual solution to Problem 7.2, will be called the *elastic trial solution*, and will be denoted

$$\begin{aligned} \boldsymbol{\varepsilon}^{e\,\text{trial}}_{n+1} &= \boldsymbol{\varepsilon}^e_n + \Delta\boldsymbol{\varepsilon} \\ \boldsymbol{\alpha}^{\text{trial}}_{n+1} &= \boldsymbol{\alpha}_n. \end{aligned} \tag{7.21}$$

The corresponding stress and hardening force will be called the *elastic trial stress* and *elastic trial hardening force*, given by

$$\boldsymbol{\sigma}^{\text{trial}}_{n+1} = \bar{\rho}\,\frac{\partial\psi}{\partial\boldsymbol{\varepsilon}^e}\bigg|^{\text{trial}}_{n+1}, \qquad \boldsymbol{A}^{\text{trial}}_{n+1} = \bar{\rho}\,\frac{\partial\psi}{\partial\boldsymbol{\alpha}}\bigg|^{\text{trial}}_{n+1}. \tag{7.22}$$

The above variables are collectively called the *elastic trial state*. Now note that, to be the actual solution, the elastic trial state has, in addition, to satisfy (7.17). We then proceed as follows. If

$$\Phi^{\text{trial}} \equiv \Phi(\boldsymbol{\sigma}^{\text{trial}}_{n+1}, \boldsymbol{A}^{\text{trial}}_{n+1}) \leq 0, \tag{7.23}$$

that is, if the *elastic trial state lies within the elastic domain or on the yield surface*, it is accepted as a solution to Problem 7.2. In this case, we update

$$(\cdot)_{n+1} := (\cdot)^{\text{trial}}_{n+1} \tag{7.24}$$

and the algorithm is terminated. Otherwise, the elastic trial state is not plastically admissible and a solution to Problem 7.2 must be obtained from the *plastic corrector step* described below.

(b) *The Plastic Corrector Step (or Return-Mapping Algorithm).*
The only option left now is to solve the system (7.19)–(7.20) of algebraic equations subject to the constraint (7.18). Using the elastic trial state definition above, we rewrite the algebraic system equivalently as

$$\begin{aligned} \boldsymbol{\varepsilon}^e_{n+1} &= \boldsymbol{\varepsilon}^{e\,\text{trial}}_{n+1} - \Delta\gamma\,\boldsymbol{N}(\boldsymbol{\sigma}_{n+1}, \boldsymbol{A}_{n+1}) \\ \boldsymbol{\alpha}_{n+1} &= \boldsymbol{\alpha}^{\text{trial}}_{n+1} + \Delta\gamma\,\boldsymbol{H}(\boldsymbol{\sigma}_{n+1}, \boldsymbol{A}_{n+1}) \\ \Phi(\boldsymbol{\sigma}_{n+1}, &\boldsymbol{A}_{n+1}) = 0, \end{aligned} \tag{7.25}$$

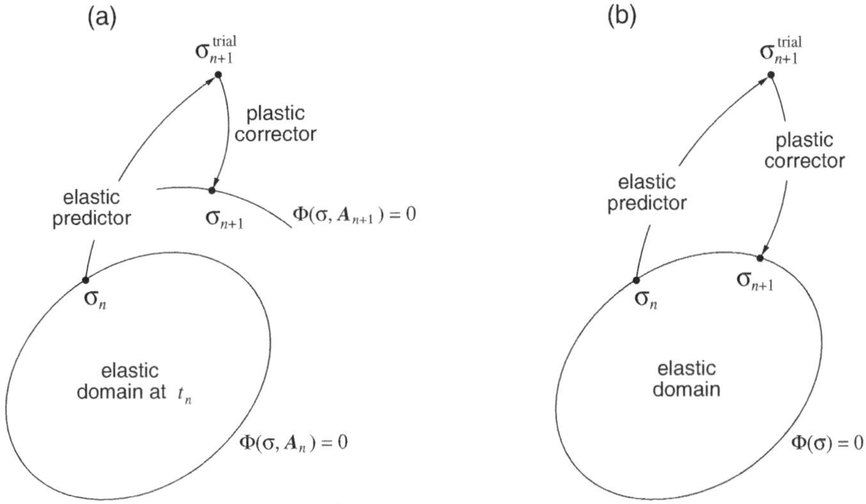

Figure 7.1. General return mapping schemes. Geometric interpretation: (a) hardening plasticity; and (b) perfect plasticity.

which are, of course, complemented with the potential relations (7.12). The plastic corrector stage of the algorithm then consists in finding a solution ε^e_{n+1}, α_{n+1} and $\Delta\gamma$ for (7.25) that satisfies

$$\Delta\gamma > 0. \tag{7.26}$$

Remark 7.2. The procedure of item (b) above possesses an appealing geometric interpretation as illustrated in Figure 7.1. Consider the yield surface at the elastic trial state. The elastic trial stress, $\boldsymbol{\sigma}^{\text{trial}}_{n+1}$, in this case lies outside the plastically admissible domain (i.e. neither in the elastic domain nor on the yield surface). Upon solution of the algebraic system (7.25), equation $(7.25)_3$, which is commonly referred to as the *plastic consistency equation*, ensures that the stress, $\boldsymbol{\sigma}_{n+1}$, at the end of the interval $[t_n, t_{n+1}]$ lies on the updated yield surface; that is, the elastic trial stress *returns* to the yield surface so that plastic consistency is re-established in the updated state. In the case of perfect plasticity, $\boldsymbol{\sigma}^{\text{trial}}_{n+1}$ returns to a *fixed* surface. Due to this interpretation the procedure of item (b) is referred to as the *return mapping algorithm* and (7.25) are called the *return mapping equations*. The first algorithm of this type appears to have been the *radial return method* proposed in the pioneering work of Wilkins (1964).

Summary of the overall procedure

In summary, the application of an Euler difference scheme to find an approximate solution of the constitutive initial value problem of elastoplasticity – Problem 7.1 – has resulted in a numerical algorithm that involves two steps: the *elastic predictor*, in which the evolution problem is solved as if the material were purely elastic within the interval considered, followed by the *return mapping*, which accounts for plastic flow and enforces plastic admissibility. The return mapping procedure is executed only if the elastic trial state violates plastic admissibility. The schematic diagram of Figure 7.2 shows the main steps in the

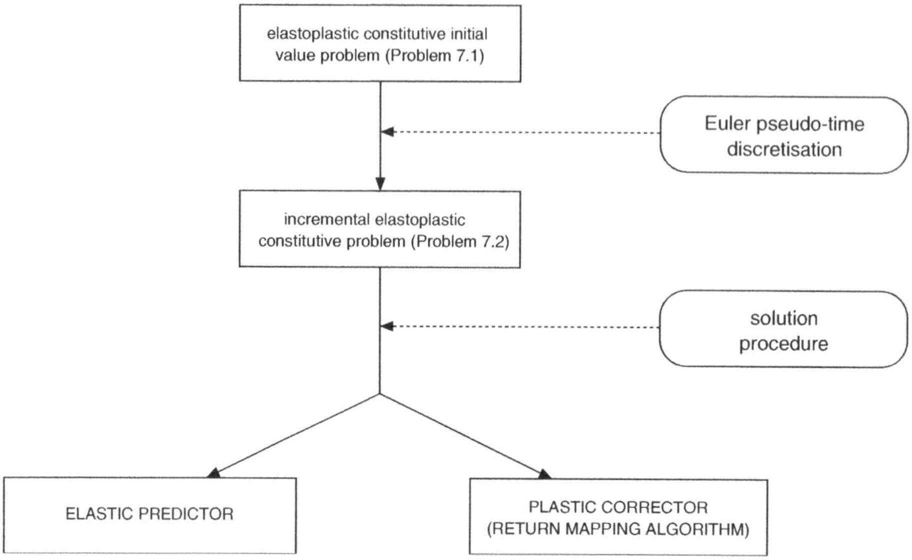

Figure 7.2. From the initial value problem of elastoplasticity to the elastic predictor/return-mapping integration algorithm. Schematic diagram.

derivation of the overall integration algorithm. The algorithm described above has been obtained by adopting, in particular, the backward Euler scheme to discretise elastoplastic constitutive equations and is, therefore, termed the *backward, fully implicit* or simply *implicit* elastic predictor/return mapping scheme. This algorithm is conveniently summarised in Box 7.1 in pseudo-code format. We remark that different discretisation schemes may be used instead, each one resulting in a different return mapping algorithm, but all having the same elastic predictor step. Alternatives to the backward Euler-based algorithm will be discussed in Section 7.2.7.

7.2.4. SOLUTION OF THE RETURN-MAPPING EQUATIONS

Let us now focus on the solution of the return mapping equations. It should be noted that the algebraic system (7.25) is generally nonlinear and, in addition, has to be solved subjected to the constraint (7.26). The procedure commonly adopted in practice is quite simple. Firstly, the algebraic system (7.25) is solved on its own, i.e. without regard for the constraint equation (7.26), by some iterative procedure. If the found solution satisfies (7.26), then it is accepted as a solution to Problem 7.2. Note that if no solution exists with strictly positive incremental plastic multiplier, then Problem 7.2 does not have a solution.

As far as the iterative procedure for the solution of the return-mapping equations is concerned, the standard Newton–Raphson scheme is often an optimal choice and will be adopted *exclusively* throughout this book (and in the elastoplastic implementations of the HYPLAS program). This choice is motivated mainly by the quadratic rates of convergence achieved by this method which, as a general rule, results in very computationally efficient return mapping procedures. Alternative techniques, such as quasi-Newton procedures in general, could be used instead. The main argument in favour of these methods is normally based on

Box 7.1. Fully implicit elastic predictor/return-mapping algorithm for numerical integration of general elastoplastic constitutive equations.

(i) Elastic predictor. Given $\Delta\boldsymbol{\varepsilon}$ and the state variables at t_n, evaluate the *elastic trial state*

$$\boldsymbol{\varepsilon}^{e\,\text{trial}}_{n+1} = \boldsymbol{\varepsilon}^e_n + \Delta\boldsymbol{\varepsilon}$$

$$\boldsymbol{\alpha}^{\text{trial}}_{n+1} = \boldsymbol{\alpha}_n$$

$$\boldsymbol{\sigma}^{\text{trial}}_{n+1} = \bar{\rho}\,\frac{\partial\psi}{\partial\boldsymbol{\varepsilon}^e}\bigg|^{\text{trial}}_{n+1}, \qquad \boldsymbol{A}^{\text{trial}}_{n+1} = \bar{\rho}\,\frac{\partial\psi}{\partial\boldsymbol{\alpha}}\bigg|^{\text{trial}}_{n+1}$$

(ii) Check plastic admissibility

$$\text{IF } \Phi(\boldsymbol{\sigma}^{\text{trial}}_{n+1}, \boldsymbol{A}^{\text{trial}}_{n+1}) \leq 0$$

$$\text{THEN set } \;(\cdot)_{n+1} = (\cdot)^{\text{trial}}_{n+1} \;\text{ and EXIT}$$

(iii) Return mapping. Solve the system

$$\begin{Bmatrix} \boldsymbol{\varepsilon}^e_{n+1} - \boldsymbol{\varepsilon}^{e\,\text{trial}}_{n+1} + \Delta\gamma\,\boldsymbol{N}_{n+1} \\ \boldsymbol{\alpha}_{n+1} - \boldsymbol{\alpha}^{\text{trial}}_{n+1} - \Delta\gamma\,\boldsymbol{H}_{n+1} \\ \Phi(\boldsymbol{\sigma}_{n+1}, \boldsymbol{A}_{n+1}) \end{Bmatrix} = \begin{Bmatrix} \mathbf{0} \\ \mathbf{0} \\ 0 \end{Bmatrix}$$

for $\boldsymbol{\varepsilon}^e_{n+1}$, $\boldsymbol{\alpha}_{n+1}$ and $\Delta\gamma$, with

$$\boldsymbol{\sigma}_{n+1} = \bar{\rho}\,\frac{\partial\psi}{\partial\boldsymbol{\varepsilon}^e}\bigg|_{n+1}, \qquad \boldsymbol{A}_{n+1} = \bar{\rho}\,\frac{\partial\psi}{\partial\boldsymbol{\alpha}}\bigg|_{n+1}$$

(iv) EXIT

the fact that, in contrast to the Newton–Raphson algorithm, they do not require the exact *gradients* of the associated residual functions (defined by the left-hand side of the equations of item (iii) of Box 7.1) and, for complex material models, the derivation and computational implementation of residual gradients may prove a tedious exercise if performed manually. However, this argument is substantially weakened by considering that currently available symbolic manipulation software packages, such as *MATHEMATICA*® (Wolfram, 1991), can handle the closed-form calculation of derivatives quite easily, making the derivation and computational implementation of residual gradients a relatively straightforward task, even for complex plasticity models. Added to this is the fact that the relative ease of quasi-Newton procedures comes at the expense of lower rates of convergence and, generally, less efficiency in the return-mapping procedure (and poorer performance of the overall finite element scheme). Also, as will be seen later, the return-mapping residual derivatives are needed to compute the *consistent tangent operator*, used to assemble the tangent stiffness matrix. Thus, if the full Newton–Raphson scheme is adopted to solve the *global* finite element equilibrium equations, the return mapping residual derivatives will have to be computed anyway.

Remark 7.3. One aspect of the return-mapping scheme deserves particular attention. It might happen that the corresponding nonlinear system of algebraic equations has a solution with

$\Delta\gamma < 0$. Since $\Delta\gamma$ is required to be positive, such a solution is physically meaningless and must be disregarded if the Newton–Raphson algorithm converges to it. Fortunately, numerical experience shows that such a situation is not usual and, in particular, does not occur for the basic plasticity models implemented in this chapter and in Chapter 8 for the normal range of material parameters. It may occur, however, for complex constitutive models. In such cases, a strategy must be developed to ensure that the converged incremental plastic multiplier is positive. This issue is discussed in Chapter 12 in the context of damage mechanics plasticity models. Another important consideration is that, for some plasticity models, the degree of nonlinearity of the return-mapping equations is so high that the convergence radius of the Newton scheme becomes substantially reduced. In such cases, procedures such as *line-searches* (as suggested by Dutko *et al.* 1993 and described in Chapter 10 for the Barlat–Lian anisotropic plasticity model) can be incorporated into the standard Newton–Raphson algorithm in order to expand its radius of convergence. The use of *improved initial guesses* in the Newton–Raphson iterations may also prove effective in ensuring convergence for highly nonlinear models (refer to Chapter 12). Another alternative can be the use of a *sub-stepping* scheme (Huerta *et al.*, 1999). This approach consists of dividing the total strain increment $\Delta\varepsilon$ into a number m of *sub-increments*

$$\Delta\varepsilon = \Delta\varepsilon_1 + \Delta\varepsilon_2 + \cdots + \Delta\varepsilon_m,$$

and then applying the integration algorithm sequentially m times to reach the approximate state at t_{n+1}. Each strain sub-increment $\Delta\varepsilon_i$, $i = 1, \ldots, m$, has to be sufficiently small to ensure convergence of the Newton–Raphson iterations. More recently, Armero and Pérez-Foguet (2002) and Pérez-Foguet and Armero (2002) have explored the closest point projection interpretation (Section 7.2.5 below) and the associated variational structure of the return-mapping equations to devise globally convergent root-finding algorithms (i.e. algorithms that converge regardless of the initial guess) based on a combination of the Newton–Raphson scheme and constrained line-search procedures.

7.2.5. CLOSEST POINT PROJECTION INTERPRETATION

For materials with linear elastic response, i.e. materials for which the stress is given by

$$\boldsymbol{\sigma} = \mathsf{D}^e : \boldsymbol{\varepsilon}^e, \tag{7.27}$$

with *constant* elasticity tensor, D^e, equation $(7.25)_1$ can be rewritten equivalently in terms of stresses as

$$\boldsymbol{\sigma}_{n+1} = \boldsymbol{\sigma}^{\text{trial}}_{n+1} - \Delta\gamma\, \mathsf{D}^e : \boldsymbol{N}_{n+1}. \tag{7.28}$$

In this case, the updated stress, $\boldsymbol{\sigma}_{n+1}$, obtained by the implicit return mapping (Figure 7.3) is the projection of the trial stress $\boldsymbol{\sigma}^{\text{trial}}_{n+1}$ onto the updated yield surface along the direction of the tensor

$$\mathsf{D}^e : \boldsymbol{N}_{n+1}.$$

For *perfectly plastic* materials with *associative flow rule*, the implicit return mapping can be interpreted as a *closest point projection* of the trial stress onto the set

$$\mathscr{A} = \{\boldsymbol{\sigma} \mid \Phi(\boldsymbol{\sigma}) \leq 0\} \tag{7.29}$$

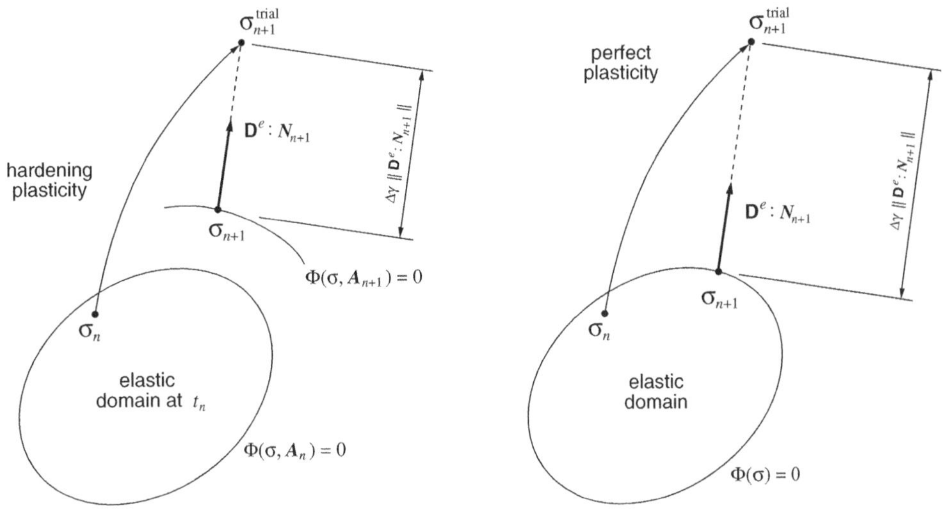

Figure 7.3. The fully implicit return mapping. Geometric interpretation for materials with linear elastic response.

of plastically admissible stresses. With the *energy norm* defined by

$$\|\boldsymbol{\sigma}\|_{\mathsf{D}^e} \equiv \sqrt{\boldsymbol{\sigma} : [\mathsf{D}^e]^{-1} : \boldsymbol{\sigma}}, \tag{7.30}$$

and the associated measure of *distance* between two generic stress states given by

$$d(\boldsymbol{\sigma}_a, \boldsymbol{\sigma}_b) \equiv \|\boldsymbol{\sigma}_a - \boldsymbol{\sigma}_b\|_{\mathsf{D}^e}, \tag{7.31}$$

the updated stress is the admissible stress that lies *closest* to the elastic trial stress, i.e.

$$\boldsymbol{\sigma}_{n+1} = \arg\{ \min_{\boldsymbol{\sigma}\in\mathscr{A}} [d(\boldsymbol{\sigma}, \boldsymbol{\sigma}_{n+1}^{\text{trial}})] \}. \tag{7.32}$$

The interpretation of the implicit return mapping as a closest point projection of the trial stress remains valid for linearly hardening materials provided that a suitable definition of distance in the space of stress *and* hardening forces is introduced (see Simo *et al.* (1988b) for details).

7.2.6. ALTERNATIVE JUSTIFICATION: OPERATOR SPLIT METHOD

An interesting alternative justification for the elastic predictor/return mapping scheme for elastoplasticity is provided by Simo and Hughes (1987). These authors arrive at the elastic predictor/return mapping procedure by exploiting the additive decomposition of the total strain rate of the elastoplastic constitutive initial value problem within the context of product formula or operator split numerical algorithms (the reader is referred to Chorin *et al.* (1978) for further details on operator split algorithms).

7.2.7. OTHER ELASTIC PREDICTOR/RETURN-MAPPING SCHEMES

It has been mentioned earlier that procedures other than the backward Euler difference scheme may be used to discretise the elastoplastic constitutive initial value problem (Problem 7.1). Obviously, the accuracy and stability (the concepts of accuracy and stability are reviewed in Section 7.2.10) of the overall elastic predictor/return mapping algorithm will depend on the particular strategy adopted. Two important families of algorithms for elastoplasticity, which incorporate the backward (or fully implicit) Euler approach as a particular case, can be derived by adopting generalised versions of the classical *trapezoidal* and *midpoint* rules in the discretisation of Problem 7.1. These procedures have been proposed by Ortiz and Popov (1985). Another popular scheme is the so-called *cutting plane* algorithm proposed by Simo and Ortiz (1985) (see also Ortiz and Simo 1986). These procedures are described below. It should be emphasised that in all cases discussed here the use of a different discretisation rule will affect *only* the return mapping part of the overall integration algorithm (item (iii) of Box 7.1). The elastic predictor stage, as previously described, remains unchanged.

The generalised trapezoidal return mapping

A generalisation of the classical trapezoidal rule and application to the numerical approximation of the elastoplastic constitutive problem results in a family of return mapping algorithms whose general expression for the associated algebraic system of nonlinear equations is given by

$$\begin{aligned} \boldsymbol{\varepsilon}^e_{n+1} &= \boldsymbol{\varepsilon}^{e\,\text{trial}}_{n+1} - \Delta\gamma\,[(1-\theta)\boldsymbol{N}_n + \theta\boldsymbol{N}_{n+1}] \\ \boldsymbol{\alpha}_{n+1} &= \boldsymbol{\alpha}_n + \Delta\gamma\,[(1-\theta)\boldsymbol{H}_n + \theta\boldsymbol{H}_{n+1}] \\ \Phi(\boldsymbol{\sigma}_{n+1}, \boldsymbol{A}_{n+1}) &= 0, \end{aligned} \tag{7.33}$$

where, of course, it is implicitly understood that constraint (7.26) is satisfied. Individual members of this family of algorithms are defined by the prescribed parameter θ, which must lie within the interval

$$0 \le \theta \le 1. \tag{7.34}$$

The incremental flow vector, in the present case given by

$$(1-\theta)\boldsymbol{N}_n + \theta\boldsymbol{N}_{n+1},$$

is a linear combination of the flow vectors at times t_n and t_{n+1}. The corresponding discrete plastic flow equation is

$$\boldsymbol{\varepsilon}^p_{n+1} = \boldsymbol{\varepsilon}^p_n + \Delta\gamma\,[(1-\theta)\boldsymbol{N}_n + \theta\boldsymbol{N}_{n+1}]. \tag{7.35}$$

The geometric interpretation of the generalised trapezoidal return algorithm in the space of stresses is shown in Figure 7.4 for materials with linear elastic response. In this case, expression $(7.33)_1$ may be equivalently written in terms of stresses as

$$\boldsymbol{\sigma}_{n+1} = \boldsymbol{\sigma}^{\text{trial}}_{n+1} - \Delta\gamma\,\mathsf{D}^e : [(1-\theta)\boldsymbol{N}_n + \theta\boldsymbol{N}_{n+1}], \tag{7.36}$$

so that the updated stress can be interpreted as the projection of the trial stress onto the updated yield surface along the direction

$$\mathsf{D}^e : [(1-\theta)\boldsymbol{N}_n + \theta\boldsymbol{N}_{n+1}].$$

It is important to note that the previously described fully implicit return mapping is recovered by choosing $\theta = 1$ in the generalised trapezoidal rule. The choice $\theta = 0$ corresponds to a *fully explicit* return mapping.

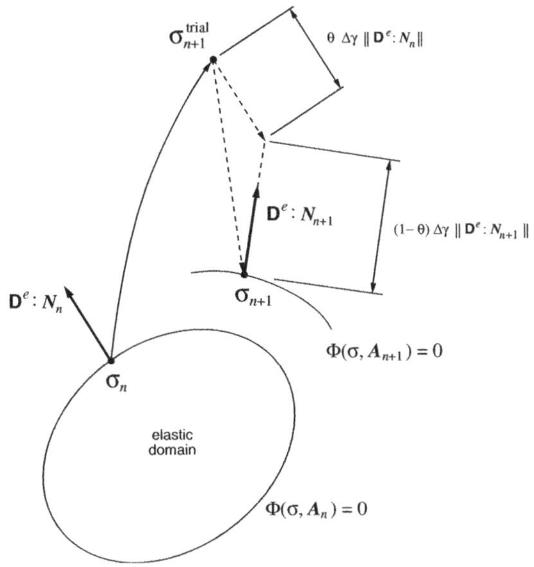

Figure 7.4. The generalised trapezoidal return mapping. Geometric interpretation for materials with linear elastic response.

The generalised midpoint return mapping

A generalisation of the midpoint rule for integration of the elastoplastic constitutive initial value problem gives the following return mapping equations

$$
\begin{aligned}
\varepsilon^e_{n+1} &= \varepsilon^{e\ \text{trial}}_{n+1} - \Delta\gamma\, \boldsymbol{N}_{n+\theta} \\
\boldsymbol{\alpha}_{n+1} &= \boldsymbol{\alpha}_n + \Delta\gamma\, \boldsymbol{H}_{n+\theta} \\
\Phi(\boldsymbol{\sigma}_{n+1}, \boldsymbol{A}_{n+1}) &= 0,
\end{aligned}
\tag{7.37}
$$

where

$$
\begin{aligned}
\boldsymbol{N}_{n+\theta} &= \boldsymbol{N}(\boldsymbol{\sigma}_{n+\theta}, \boldsymbol{A}_{n+\theta}) \\
\boldsymbol{H}_{n+\theta} &= \boldsymbol{H}(\boldsymbol{\sigma}_{n+\theta}, \boldsymbol{A}_{n+\theta}),
\end{aligned}
\tag{7.38}
$$

with the *generalised midpoint state* defined by the variables

$$
\begin{aligned}
\boldsymbol{\sigma}_{n+\theta} &= (1-\theta)\boldsymbol{\sigma}_{n+1} + \theta\, \boldsymbol{\sigma}_n \\
\boldsymbol{A}_{n+\theta} &= (1-\theta)\boldsymbol{A}_n + \theta\, \boldsymbol{A}_{n+1}.
\end{aligned}
\tag{7.39}
$$

The parameter θ is, again, a prescribed constant within the interval $0 \le \theta \le 1$. In the present case, the discrete plastic flow rule reads

$$
\varepsilon^p_{n+1} = \varepsilon^p_n + \Delta\gamma\, \boldsymbol{N}_{n+\theta}. \tag{7.40}
$$

Note that the generalised midpoint rule coincides with the generalised trapezoidal rule (7.33) for $\theta = 1$ and $\theta = 0$, which correspond, respectively, to the fully implicit and fully explicit

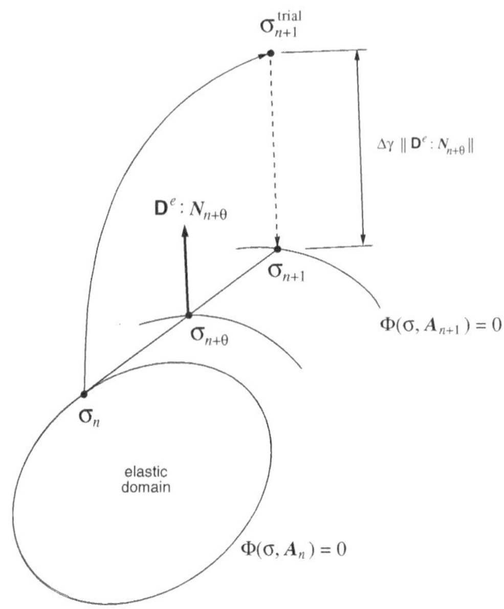

Figure 7.5. The generalised midpoint return mapping. Geometric interpretation for materials with linear elastic response.

return mappings. The geometric interpretation of the generalised midpoint return algorithm is given in Figure 7.5 for materials with linear elastic response. The updated stress, in this case, is given by

$$\boldsymbol{\sigma}_{n+1} = \boldsymbol{\sigma}^{\text{trial}}_{n+1} - \Delta\gamma\, \mathsf{D}^e : \boldsymbol{N}_{n+\theta}, \tag{7.41}$$

and corresponds to the projection of the elastic trial stress, $\boldsymbol{\sigma}^{\text{trial}}_{n+1}$, onto the updated yield surface at t_{n+1} along the direction of

$$\mathsf{D}^e : \boldsymbol{N}_{n+\theta}.$$

A possible variation of the above generalised midpoint rule is obtained by replacing the consistency condition $(7.37)_3$ with the alternative

$$\Phi(\boldsymbol{\sigma}_{n+\theta}, \boldsymbol{A}_{n+\theta}) = 0. \tag{7.42}$$

In this case, plastic consistency is enforced upon the *generalised midpoint*, rather than the updated, state. This version of the generalised return mapping is discussed in detail by Simo and Govindjee (1991) who highlight, in particular, the fact that in this case the symmetry of the associated consistent tangent operators[‡] is ensured for fully associative models – a property not generally preserved by the family of algorithms based on $(7.37)_3$, as noted by Ortiz and Martin (1989).

Remark 7.4. As for the fully implicit return mapping, both generalised trapezoidal and midpoint return-mapping algorithms, defined respectively by (7.33) and (7.37), require the

[‡]Consistent tangent operators for return-mapping schemes will be discussed in Section 7.4. In particular, a discussion on the possible symmetry of such operators is provided in Section 7.4.6.

solution of a (generally nonlinear) system of equations for any prescribed parameter θ. As before, an efficient scheme is normally obtained by adopting the quadratically convergent Newton–Raphson method. It is interesting to note that, for the *fully explicit* member of these families of algorithms ($\theta = 0$), the return-mapping equations can be substantially simplified. In this case, all terms, with the exception of $\Delta\gamma$, on the right-hand side of (7.37) are known values at t_n. Therefore, the original system variables, $\boldsymbol{\varepsilon}^e_{n+1}$ and $\boldsymbol{\alpha}_{n+1}$ become functions solely of the incremental multiplier $\Delta\gamma$ and so does $\boldsymbol{\sigma}_{n+1}$ and $\boldsymbol{A}_{n+1}$. The return-mapping equations in this case can then be reduced to a single *scalar* (generally nonlinear) equation of the form

$$\tilde{\Phi}(\Delta\gamma) \equiv \Phi(\boldsymbol{\sigma}_{n+1}(\Delta\gamma), \boldsymbol{A}_{n+1}(\Delta\gamma)) = 0, \tag{7.43}$$

for *any* elastoplastic model. For certain material models, simplifications of this type are also possible for $\theta \neq 0$ and are crucially important for the computational efficiency of the associated return-mapping procedure. This idea is exploited in Section 7.3, where the derivation of implicit return mapping for the von Mises model is described in detail.

The cutting-plane algorithm

Let us start by observing that the return-mapping equations $(7.25)_{1,2}$ are a backward Euler approximation to the initial value problem defined by the differential equation

$$\begin{aligned} \dot{\boldsymbol{\varepsilon}}^e &= -\dot{\gamma}\boldsymbol{N}(\boldsymbol{\sigma}, \boldsymbol{A}) \\ \dot{\boldsymbol{\alpha}} &= \dot{\gamma}\boldsymbol{H}(\boldsymbol{\sigma}, \boldsymbol{A}), \end{aligned} \tag{7.44}$$

with initial condition

$$\boldsymbol{\varepsilon}^e = \boldsymbol{\varepsilon}^{e\ \mathrm{trial}}_{n+1}, \quad \boldsymbol{\alpha}^e = \boldsymbol{\alpha}^{e\ \mathrm{trial}}_{n+1}. \tag{7.45}$$

The implicit return-mapping scheme consists in solving the above problem, subject to the consistency condition $(7.25)_3$, by means of the backward Euler scheme. The cutting-plane method, which we shall describe below, is an alternative scheme for the numerical solution of the return-mapping problem.

The first step to derive the cutting-plane algorithm, is to recast the potential relation (7.8) into the equivalent rate form

$$\begin{aligned} \dot{\boldsymbol{\sigma}} &= \mathsf{D}^e(\boldsymbol{\varepsilon}^e, \boldsymbol{\alpha}) : \dot{\boldsymbol{\varepsilon}}^e + \mathsf{E}(\boldsymbol{\varepsilon}^e, \boldsymbol{\alpha}) * \dot{\boldsymbol{\alpha}} \\ \dot{\boldsymbol{A}} &= \mathsf{F}(\boldsymbol{\varepsilon}^e, \boldsymbol{\alpha}) * \dot{\boldsymbol{\varepsilon}}^e + \mathsf{G}(\boldsymbol{\varepsilon}^e, \boldsymbol{\alpha}) * \dot{\boldsymbol{\alpha}}, \end{aligned} \tag{7.46}$$

where $*$ denotes the product of the appropriate type. The above rate constitutive equations for $\boldsymbol{\sigma}$ and $\boldsymbol{A}$ are obtained by a straightforward application of the chain rule to (7.8). The fourth-order tensor $\mathsf{D}^e(\boldsymbol{\varepsilon}^e, \boldsymbol{\alpha})$ in $(7.46)_1$ is the elastic modulus

$$\mathsf{D}^e(\boldsymbol{\varepsilon}^e, \boldsymbol{\alpha}) = \bar{\rho}\,\frac{\partial^2\psi}{\partial\boldsymbol{\varepsilon}^{e\,2}}, \tag{7.47}$$

which, in general, is assumed to be a function of $\boldsymbol{\varepsilon}^e$ and $\boldsymbol{\alpha}$. The operator $\mathsf{E}(\boldsymbol{\varepsilon}^e, \boldsymbol{\alpha})$ denotes the tangent modulus associated with the hardening response, defined by

$$\mathsf{E}(\boldsymbol{\varepsilon}^e, \boldsymbol{\alpha}) = \bar{\rho}\,\frac{\partial^2\psi}{\partial\boldsymbol{\varepsilon}^e\partial\boldsymbol{\alpha}}. \tag{7.48}$$

In addition, the tangential operators F and G are defined as

$$\mathsf{F}(\varepsilon^e, \boldsymbol{\alpha}) = \bar{\rho}\, \frac{\partial^2 \psi}{\partial \boldsymbol{\alpha} \partial \varepsilon^e} \tag{7.49}$$

and

$$\mathsf{G}(\varepsilon^e, \boldsymbol{\alpha}) = \bar{\rho}\, \frac{\partial^2 \psi}{\partial \boldsymbol{\alpha}^2}. \tag{7.50}$$

It should be noted that D^e is *constant* for materials whose elastic behaviour is linear. Also, under the assumption (6.37) (page 149) of decoupling between elasticity and hardening, the tangent operators E and F vanish and so does the second term on the right-hand side of $(7.46)_1$ and the first term on the right-hand side of $(7.46)_2$. However, for models in which elasticity is coupled with dissipative phenomena, such as Lemaitre's damage theory discussed in Chapter 12, these terms do not vanish in general.

By combining (7.44) with (7.46) and making use of complementary potential relations that give

$$\varepsilon^e = \varepsilon^e(\boldsymbol{\sigma}, \boldsymbol{A}), \quad \boldsymbol{\alpha} = \boldsymbol{\alpha}(\boldsymbol{\sigma}, \boldsymbol{A}), \tag{7.51}$$

we redefine the initial value problem (7.44)–(7.45) equivalently in terms of stress and hardening force as

$$\begin{aligned} \dot{\boldsymbol{\sigma}} &= \dot{\gamma}[-\mathsf{D}^e(\boldsymbol{\sigma}, \boldsymbol{A}) : \boldsymbol{N}(\boldsymbol{\sigma}, \boldsymbol{A}) + \mathsf{E}(\boldsymbol{\sigma}, \boldsymbol{A}) * \boldsymbol{H}(\boldsymbol{\sigma}, \boldsymbol{A})] \\ \dot{\boldsymbol{A}} &= \dot{\gamma}[-\mathsf{F}(\boldsymbol{\sigma}, \boldsymbol{A}) * \boldsymbol{N}(\boldsymbol{\sigma}, \boldsymbol{A}) + \mathsf{G}(\boldsymbol{\sigma}, \boldsymbol{A}) * \boldsymbol{H}(\boldsymbol{\sigma}, \boldsymbol{A})], \end{aligned} \tag{7.52}$$

with the obvious initial condition

$$\boldsymbol{\sigma} = \boldsymbol{\sigma}^{\text{trial}}_{n+1}, \quad \boldsymbol{A} = \boldsymbol{A}^{\text{trial}}_{n+1}. \tag{7.53}$$

In summary, the return-mapping problem now comprises the initial value problem (7.52)–(7.53) in conjunction with the plastic consistency constraint

$$\Phi(\boldsymbol{\sigma}_{n+1}, \boldsymbol{A}_{n+1}) = 0. \tag{7.54}$$

Basically, the cutting plane return-mapping algorithm is an iterative procedure for numerical solution of the return-mapping problem whereby the linear approximation to the plastic consistency equation is solved at each iteration. In a typical cutting-plane iteration, (k), the plastic consistency equation is linearised[§] about the current (known) state, $\{\boldsymbol{\sigma}^{(k)}_{n+1}, \boldsymbol{A}^{(k)}_{n+1}\}$, of the stress and hardening force. A new state,

$$\{\boldsymbol{\sigma}^{(k+1)}_{n+1}, \boldsymbol{A}^{(k+1)}_{n+1}\},$$

is then obtained as the solution of the linearised equation

$$\begin{aligned} \Phi(\boldsymbol{\sigma}^{(k)}_{n+1}, \boldsymbol{A}^{(k)}_{n+1}) + \bar{\boldsymbol{N}}^{(k)}_{n+1} : [\boldsymbol{\sigma}^{(k+1)}_{n+1} - \boldsymbol{\sigma}^{(k)}_{n+1}] \\ + \bar{\boldsymbol{H}}^{(k)}_{n+1} * [\boldsymbol{A}^{(k+1)}_{n+1} - \boldsymbol{A}^{(k)}_{n+1}] = 0, \end{aligned} \tag{7.55}$$

[§]Refer to Section 2.6, (from page 38) for the definition of the linearisation of a nonlinear problem.

where $\bar{\boldsymbol{N}}^{(k)}_{n+1}$ and $\bar{\boldsymbol{H}}^{(k)}_{n+1}$ have been defined as

$$\bar{\boldsymbol{N}}^{(k)}_{n+1} = \left.\frac{\partial \Phi}{\partial \boldsymbol{\sigma}}\right|_{\boldsymbol{\sigma}^{(k)}_{n+1}, \boldsymbol{A}^{(k)}_{n+1}}, \qquad \bar{\boldsymbol{H}}^{(k)}_{n+1} = \left.\frac{\partial \Phi}{\partial \boldsymbol{A}}\right|_{\boldsymbol{\sigma}^{(k)}_{n+1}, \boldsymbol{A}^{(k)}_{n+1}}. \tag{7.56}$$

Note that for fully associative plasticity ($\Psi \equiv \Phi$), $\bar{\boldsymbol{N}}$ and $\bar{\boldsymbol{H}}$ coincide, respectively, with the flow vector, $\boldsymbol{N}$, and $\boldsymbol{H}$.

To solve (7.55) we proceed as follows. We discretise the rate constitutive equations (7.52) using a *forward* (or *explicit*) Euler scheme. This gives the formula

$$\begin{aligned} \boldsymbol{\sigma}^{(k+1)}_{n+1} - \boldsymbol{\sigma}^{(k)}_{n+1} &= \Delta\gamma \left[-\mathsf{D}^{e\,(k)}_{n+1} : \boldsymbol{N}^{(k)}_{n+1} + \mathsf{E}^{(k)}_{n+1} * \boldsymbol{H}^{(k)}_{n+1} \right] \\ \boldsymbol{A}^{(k+1)}_{n+1} - \boldsymbol{A}^{(k)}_{n+1} &= \Delta\gamma \left[-\mathsf{F}^{(k)}_{n+1} * \boldsymbol{N}^{(k)}_{n+1} + \mathsf{G}^{(k)}_{n+1} * \boldsymbol{H}^{(k)}_{n+1} \right]. \end{aligned} \tag{7.57}$$

With the substitution of the above equations into (7.55), the following expression is obtained for $\Delta\gamma$ in *closed form*

$$\begin{aligned} \Delta\gamma = \Phi^{(k)}_{n+1} \Big/ \Big\{ \bar{\boldsymbol{N}}^{(k)}_{n+1} &: \left[\mathsf{D}^{e\,(k)}_{n+1} : \boldsymbol{N}^{(k)}_{n+1} - \mathsf{E}^{(k)}_{n+1} * \boldsymbol{H}^{(k)}_{n+1} \right] \\ &+ \bar{\boldsymbol{H}}^{(k)}_{n+1} * \left[\mathsf{F}^{(k)}_{n+1} * \boldsymbol{N}^{(k)}_{n+1} - \mathsf{G}^{(k)}_{n+1} * \boldsymbol{H}^{(k)}_{n+1} \right] \Big\}, \end{aligned} \tag{7.58}$$

and the new state $\{\boldsymbol{\sigma}^{(k+1)}_{n+1}, \boldsymbol{A}^{(k+1)}_{n+1}\}$ is computed by substituting the value obtained for $\Delta\gamma$ into (7.57).

Starting (with $k = 0$) from the initial condition of the return-mapping problem

$$\{\boldsymbol{\sigma}^{(0)}_{n+1}, \boldsymbol{A}^{(0)}_{n+1}\} = \{\boldsymbol{\sigma}^{\text{trial}}_{n+1}, \boldsymbol{A}^{\text{trial}}_{n+1}\}, \tag{7.59}$$

the repeated application of the above iteration generates a sequence of states

$$\{\boldsymbol{\sigma}^{(k)}_{n+1}, \boldsymbol{A}^{(k)}_{n+1}\}, \quad k = 0, 1, 2, \dots .$$

The cutting-plane iterations continue until the value of the yield function is sufficiently close to zero, i.e. the iterative process is interrupted at an iteration (k) if the state $\{\boldsymbol{\sigma}^{(k)}_{n+1}, \boldsymbol{A}^{(k)}_{n+1}\}$ satisfies the convergence criterion

$$\Phi(\boldsymbol{\sigma}^{(k)}_{n+1}, \boldsymbol{A}^{(k)}_{n+1}) \leq \epsilon_{\text{tol}}, \tag{7.60}$$

where ϵ_{tol} is a prescribed convergence tolerance. The cutting-plane return-mapping algorithm is summarised in Box 7.2 in pseudo-code format.

Remark 7.5. The geometric interpretation of the cutting-plane algorithm is illustrated in Figure 7.6. The interpretation here remains valid for materials with a *nonlinear* elastic response. In each iteration, the new stress, $\{\boldsymbol{\sigma}^{(k+1)}_{n+1}, \boldsymbol{A}^{(k+1)}_{n+1}\}$, is obtained by projecting the current stress, $\{\boldsymbol{\sigma}^{(k)}_{n+1}, \boldsymbol{A}^{(k)}_{n+1}\}$, onto a *cutting plane* defined by the linearised consistency condition¶. The projection is made along the direction of the tensor

$$\mathsf{D}^{e\,(k)}_{n+1} : \boldsymbol{N}^{(k)}_{n+1} - \mathsf{E}^{(k)}_{n+1} * \boldsymbol{H}^{(k)}_{n+1}.$$

¶Note that the linearised consistency condition (7.55) defines a *hyperplane* in the space of stresses and hardening forces.

Box 7.2. Cutting-plane return-mapping algorithm for general elastoplastic models.

(i) Given the *elastic trial state*, set $k = 0$ and

$$\boldsymbol{\sigma}_{n+1}^{(0)} = \boldsymbol{\sigma}_{n+1}^{\text{trial}}, \quad \boldsymbol{A}_{n+1}^{(0)} = \boldsymbol{A}_{n+1}^{\text{trial}}$$

(ii) Perform cutting-plane iteration

$$\Delta\gamma = \Phi_{n+1}^{(k)} \Big/ \Big\{ \bar{\boldsymbol{N}}_{n+1}^{(k)} : \Big[\mathsf{D}_{n+1}^{e\,(k)} : \boldsymbol{N}_{n+1}^{(k)} - \mathsf{E}_{n+1}^{(k)} * \boldsymbol{H}_{n+1}^{(k)} \Big] + \bar{\boldsymbol{H}}_{n+1}^{(k)} * \Big[\mathsf{F}_{n+1}^{(k)} * \boldsymbol{N}_{n+1}^{(k)} - \mathsf{G}_{n+1}^{(k)} * \boldsymbol{H}_{n+1}^{(k)} \Big] \Big\}$$

$$\boldsymbol{\sigma}_{n+1}^{(k+1)} = \boldsymbol{\sigma}_{n+1}^{(k)} - \Delta\gamma \Big[\mathsf{D}_{n+1}^{e\,(k)} : \boldsymbol{N}_{n+1}^{(k)} - \mathsf{E}_{n+1}^{(k)} * \boldsymbol{H}_{n+1}^{(k)} \Big]$$

$$\boldsymbol{A}_{n+1}^{(k+1)} = \boldsymbol{A}_{n+1}^{(k)} - \Delta\gamma \Big[\mathsf{F}_{n+1}^{(k)} * \boldsymbol{N}_{n+1}^{(k)} - \mathsf{G}_{n+1}^{(k)} * \boldsymbol{H}_{n+1}^{(k)} \Big]$$

(iii) Check convergence

IF $\Phi(\boldsymbol{\sigma}_{n+1}^{(k)}, \boldsymbol{A}_{n+1}^{(k)}) \leq \epsilon_{\text{tol}}$ THEN update

$$\boldsymbol{\sigma}_{n+1} = \boldsymbol{\sigma}_{n+1}^{(k)}; \qquad \boldsymbol{A}_{n+1} = \boldsymbol{A}_{n+1}^{(k)};$$

$$\boldsymbol{\varepsilon}_{n+1}^{e} = \boldsymbol{\varepsilon}^{e}(\boldsymbol{\sigma}_{n+1}^{(k)}, \boldsymbol{A}_{n+1}^{(k)}); \qquad \boldsymbol{\alpha}_{n+1} = \boldsymbol{\alpha}(\boldsymbol{\sigma}_{n+1}^{(k)}, \boldsymbol{A}_{n+1}^{(k)});$$

$$\boldsymbol{\varepsilon}_{n+1}^{p} = \boldsymbol{\varepsilon}_{n+1} - \boldsymbol{\varepsilon}_{n+1}^{e}; \qquad \text{and EXIT}$$

ELSE

set $k := k + 1$ and GO TO (ii)

In the limit of the iterative process, plastic consistency is restored and the cutting plane is tangent to the actual yield surface.

Remark 7.6. The iterations of the cutting-plane algorithm converge to the solution $\Phi = 0$ at *quadratic* rates. Quadratic rates of convergence are achieved here despite the fact that only relatively simple function evaluations are performed in each iteration. This is in contrast to the previously described return-mapping schemes for which the computation of (generally more complex) residual derivatives is required in order to achieve quadratic rates of convergence. This makes the cutting-plane algorithm particularly attractive for more complex plasticity models. However, it is important to emphasise that the incremental stress–strain function

$$\boldsymbol{\sigma}_{n+1} = \hat{\boldsymbol{\sigma}}(\boldsymbol{\alpha}_n, \boldsymbol{\varepsilon}_{n+1}), \tag{7.61}$$

delivered by the cutting-plane algorithm, is *not* amenable to linearisation and derivation of a consistent tangent modulus[||]

$$\mathsf{D} = \frac{\partial \hat{\boldsymbol{\sigma}}}{\partial \boldsymbol{\varepsilon}_{n+1}}. \tag{7.62}$$

[||] The notion of a consistent tangent modulus for elastoplasticity numerical integration algorithms is introduced in Section 7.4, to which readers unfamiliar with the concept are referred.

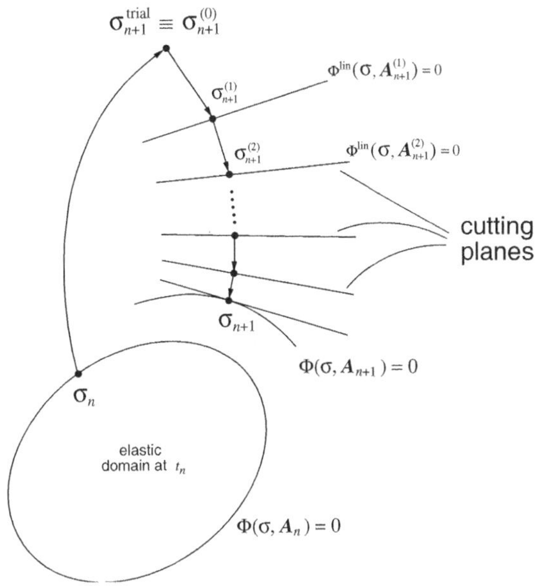

Figure 7.6. The cutting-plane return mapping. Geometric interpretation.

Within the implicit finite element framework of Chapter 4 (see Section 4.2.5, page 98), the lack of a consistent tangent modulus represents a serious limitation for it does *not* allow the use of the (quadratically convergent) Newton–Raphson algorithm in the solution of the (global) finite element equilibrium equations. However, in explicit codes, which do not require the solution of a global system of equilibrium equations, the use of cutting-plane algorithms could be an attractive option.

7.2.8. PLASTICITY AND DIFFERENTIAL-ALGEBRAIC EQUATIONS

Under continuous plastic loading, the system of equations that characterises the elastoplastic constitutive problem (Problem 7.1) reduces to

$$\begin{aligned} \dot{\varepsilon}^e(t) &= \dot{\varepsilon}(t) - \dot{\gamma}(t)\, \boldsymbol{N}(\boldsymbol{\sigma}(t), \boldsymbol{A}(t)) \\ \dot{\alpha}(t) &= \dot{\gamma}(t)\, \boldsymbol{H}(\boldsymbol{\sigma}(t), \boldsymbol{A}(t)) \\ \Phi(\boldsymbol{\sigma}(t), \boldsymbol{A}(t)) &= 0, \end{aligned} \tag{7.63}$$

where the potential relations (7.8) can be used to express $\boldsymbol{\sigma}$ and $\boldsymbol{A}$ as explicit functions of ε^e and α. Formally, such a system is classed as a system of *differential-algebraic equations* (DAE). These generally contain ordinary differential equations (equations $(7.63)_{1,2}$ in the present case) complemented by a constraint in the form of algebraic equations (equation $(7.63)_3$).

A detailed account on numerical methods for DAEs is provided in the textbook by Ascher and Petzold (1998) (see also Gear (1971) for an early reference on the subject). The identification of plasticity equations with DAEs allows the use of numerical methods devised for this type of equation, as well as the corresponding methods of analysis, in the treatment of plasticity problems. By exploiting these ideas, Papadopoulos and Taylor (1994) proposed a two-step backward difference formula of second-order accuracy for von Mises type plasticity. Results from DAE theory were used by these authors to prove the stability and accuracy order of their method (here, the concepts of accuracy and stability are briefly reviewed below in Section 7.2.10). The use of such tools in plasticity is also discussed by Simo (1998).

7.2.9. ALTERNATIVE MATHEMATICAL PROGRAMMING-BASED ALGORITHMS

The class of elastic predictor/return-mapping algorithms described thus far is by no means the only possible approach to the formulation of incremental elastoplastic constitutive algorithms. An important class of numerical procedures developed on the basis of mathematical programming concepts (Bazaraa and Shetty, 1979; Luenberger, 1973) has been proposed by a number of authors. For details on this class of methods, we refer to the work of Maier (1970), Martin *et al.* (1987), Feijóo and Zouain (1988), Zouain *et al.* (1988, 1992), Caddemi and Martin (1991), Reddy and Martin (1991), Comi *et al.* (1991) and Romano *et al.* (1993). By exploiting the potential structure that characterises many elastoplasticity models and concepts of convex analysis, some of these methods reduce the complete initial boundary value problem to a mathematical programming problem that can be solved by means of standard numerical algorithms. For other methods of this class, only the elastoplastic constitutive evolution problem at the Gauss point level is reduced to a mathematical programming problem.

The effectiveness of such methodologies has been demonstrated in the numerical solution of a wide range of initial boundary value problems. To the authors' knowledge, however, it appears that the relative efficiency of mathematical programming-based algorithms (as compared to elastic predictor/plastic corrector approaches) has not been assessed so far.

One interesting aspect of formulations such as the one proposed by Feijóo and Zouain (1988) is the fact that singularities (corners) of multisurface models (e.g. Tresca) do not require special treatment and are naturally accommodated within the mathematical programming environment. In contrast, even though singular models can be efficiently integrated within the elastic predictor/return-mapping framework (as we shall see in Chapter 8), extra complexity is inevitably introduced in the presence of non-smooth corners on the yield surface.

7.2.10. ACCURACY AND STABILITY CONSIDERATIONS

Accuracy and stability are crucially important aspects of numerical algorithms for the solution of initial value problems in general. In the present context, the accuracy and stability of the adopted numerical scheme for integration of the elastoplastic equations at the Gauss point level is directly associated with the effectiveness and reliability of the overall finite element solution scheme. Generally speaking, 'accurate' finite element solutions to initial boundary value problems involving elastoplastic materials should not be expected if 'inaccurate' state-update procedures are used at the Gauss point level. In this section, accuracy and stability properties of the return-mapping schemes described above are discussed. It is emphasised that no attempt is made here to derive (usually complex) mathematical proofs of the accuracy and stability of elastoplastic integration algorithms. Rather, the purpose of this section is to provide the reader with some results that can be useful in helping to decide which algorithm to adopt when considering the computational implementation of a particular plasticity model.

The concepts of accuracy order, finite step accuracy and stability

Before proceeding, it is convenient at this point to give a more precise definition of what accuracy and stability mean in the discussion that follows. The term *accuracy* will be used here in two different contexts:

(a) With a mathematically precise definition, the *accuracy order* of an algorithm is a measure of how accurately the discretised equations approximate their differential counterparts within an *infinitesimal* vicinity of the initial conditions. Let us consider a typical (well-posed) initial value problem defined by an ordinary differential equation

$$\dot{x}(t) = f(x, t) \tag{7.64}$$

over the generic domain

$$t \in [t_0, t_f], \tag{7.65}$$

subjected to the initial condition

$$x(t_0) = x_0, \tag{7.66}$$

with the superimposed dot denoting differentiation with respect to the independent variable t. A numerical method for approximate solution of (7.64)–(7.66) is said to be *first-order accurate*, or to have *consistency of order one*) (Gear, 1971), if the update formula for the approximate value x_{n+1} of the unknown function x at the end of the generic step $[t_n, t_{n+1} = t_n + \Delta t] \in [t_0, t_f]$ satisfies

$$\left. \frac{\mathrm{d}}{\mathrm{d}\Delta t}\, x_{n+1} \right|_{\Delta t=0} = \dot{x}(t_n). \tag{7.67}$$

The *numerical error*, i.e. the maximum difference between exact and numerical solutions (*excluding machine round-off errors*) within $(t_0, t_f]$, produced by first-order accurate algorithms is proportional to the step size, Δt, as $\Delta t \to 0$. If, in addition to first-order accuracy, the algorithm satisfies

$$\left. \frac{\mathrm{d}^2}{\mathrm{d}\Delta t^2}\, x_{n+1} \right|_{\Delta t=0} = \ddot{x}(t_n), \tag{7.68}$$

then it is said to be *second-order accurate* (or to have *consistency of order two*). Second-order accurate algorithms produce numerical errors that are proportional to the *square* of the step size, Δt^2, as $\Delta t \to 0$. In summary, the accuracy order of an algorithm provides information about its behaviour as the step size Δt (strain increment size in the case of elastoplasticity) approaches zero. It gives no indication, however, as to the actual numerical errors incurred under steps of finite size; that is, for a given finite step size Δt, the choice of a method of higher accuracy order rather than one of lower order does not necessarily produce a more accurate numerical solution. This aspect of numerical methods for ordinary differential equations is discussed, for instance, by Gear (1971). Here, we shall address this issue in the following.

(b) With a more practical (and less precise) definition, the term *finite step accuracy* will be adopted here to describe the actual accuracy of particular algorithms under *finite* step sizes. In the context of plasticity, finite step accuracy may be measured by means of numerical experiments in which the state-update algorithm is used to integrate the elastoplastic equations under a wide range of initial conditions and strain increment sizes and directions. Finite step accuracy measurements can give important information on the practical limitations of integration algorithms, especially with regard to the permissible size of strain increments for which the error remains within reasonable bounds.

The property of *stability* of an algorithm for numerical solution of the typical problem (7.64)–(7.66) relates to how perturbations to the initial condition propagate throughout the stepping procedure. Generally speaking, an algorithm is said to be *stable* if the variations in the numerical solution that result from a perturbation of the initial conditions are bounded within the domain $[t_0, t_f]$. The reader is referred, for instance, to Ascher and Petzold (1998) for various definitions of stability in this context. In the case of perfectly plastic models (Ortiz and Popov, 1985), an elastic predictor/plastic corrector scheme (under plastic yielding) is said to be *stable* if given two arbitrary distinct initial stress states at t_n, $\boldsymbol{\sigma}_n$ and $\boldsymbol{\sigma}^*_n$, the corresponding updated values at t_{n+1} are bounded by

$$d(\boldsymbol{\sigma}_{n+1}, \boldsymbol{\sigma}^*_{n+1}) \leq d(\boldsymbol{\sigma}_n, \boldsymbol{\sigma}^*_n) \tag{7.69}$$

where d is a measure of distance on the yield surface defined such that for any pair $\{\boldsymbol{\sigma}_a, \boldsymbol{\sigma}_b\}$ of stresses on the yield surface, we have

$$d(\boldsymbol{\sigma}_a, \boldsymbol{\sigma}_b) \equiv \inf_{\lambda} \int_{\lambda} \|\boldsymbol{\sigma}'(s)\| \, \mathrm{d}s, \tag{7.70}$$

with λ denoting arbitrary paths on the yield surface connecting $\boldsymbol{\sigma}_a$ and $\boldsymbol{\sigma}_b$. For unstable algorithms, on the other hand, perturbations to the initial conditions propagate in an unbounded manner so that small changes in increment size, or even machine round-off errors, may produce dramatic changes in the numerical results which completely invalidate the solution.

The property of stability together with first-order accuracy (or consistency of order one) of the numerical method, are necessary and sufficient conditions for the numerical solution to converge to the exact solution as the increment size tends to zero. An algorithm that satisfies the first-order accuracy and stability requirements is said to be *convergent*. If an algorithm is not stable or not (at least) first-order accurate, then it does not converge to the exact solution.

Remarks on the accuracy order and stability of elastoplastic algorithms

A detailed study of the accuracy and stability characteristics of the elastic predictor/return-mapping algorithms based on the generalised trapezoidal and midpoint rules has been carried out by Ortiz and Popov (1985). Their analysis has concentrated on perfectly plastic materials with linear elastic response and possible non-associative flow rule. In spite of the restriction to perfect plasticity and linear elastic law, their analysis provides a very interesting insight into the properties of the trapezoidal and midpoint algorithms described in Section 7.2.7 which allow for general elastic and hardening laws. It is remarked that consideration of general nonlinear elastic behaviour and hardening response introduces substantial mathematical complexity into an analysis of this type and results for such general conditions do not seem to be currently available.

Essentially, it has been shown by Ortiz and Popov (1985) that *all* members of the families of integration algorithms based on the generalised midpoint and trapezoidal rules are (at least) *first-order accurate*. Recall that the integration algorithm based on the *fully implicit* return mapping, described in Section 7.2.3, is a particular (common) member of both families (for $\theta = 1$) and, therefore, is also first-order accurate. In addition, *second-order accuracy* is attained for the choice $\theta = \frac{1}{2}$ in both trapezoidal and midpoint algorithms. In summary, as far as accuracy order is concerned, all trapezoidal and midpoint algorithms satisfy the necessary condition of first-order accuracy for convergence.

The next step in the analysis of these families of algorithms is the assessment of their stability properties. The analysis carried out by Ortiz and Popov (1985) shows that the stability of the generalised trapezoidal algorithms depends strongly on the shape of the yield surface. For surfaces with constant curvature in the space of stresses, such as the von Mises surface, unconditional stability is obtained for $\frac{1}{2} \leq \theta \leq 1$. With increasing curvature, the interval for which the generalised trapezoidal algorithms remain unconditionally stable narrows around $\theta = 1$ (the fully implicit scheme). In the limit, for surfaces with infinite curvature, such as the Tresca and Mohr–Coulomb surfaces which have sharp corners, the only unconditionally stable algorithm is the fully implicit. For the algorithms based on the generalised midpoint rule, on the other hand, stability does not depend on the shape of the yield surface. These algorithms are unconditionally stable for $\theta \geq \frac{1}{2}$.

It is worth mentioning that a similar analysis of accuracy and stability has been carried out by Simo and Govindjee (1991) for the version of the generalised midpoint return-mapping scheme of Section 7.2.7 characterised by the midpoint plastic consistency equation (7.42). The analysis was focused on materials with an *associative* hardening and flow rules and, again, restricted to the case of linear elastic behaviour. Their results are similar to those of Ortiz and Popov and show that the midpoint algorithms based on the midpoint consistency condition are second-order accurate for $\theta = \frac{1}{2}$ and first-order accurate otherwise. In addition, as for the midpoint schemes based on the fully implicit consistency $(7.37)_3$, their algorithms are stable for $\theta \geq \frac{1}{2}$.

Attention is now focused on the elastoplastic algorithm based on the cutting-plane return mapping. Since the sequence of states generated by the cutting-plane return mapping is obtained by fully linearising the plastic corrector evolution problem (7.44,7.45), it follows that, as the Backward Euler-based scheme (which also approximates this problem numerically), the cutting-plane procedure is naturally first-order accurate. The stability of the cutting-plane algorithm, however, is only conditional. It is pointed out by Simo and Ortiz (1985) that the elastoplastic algorithm based on the cutting-plane return mapping is unconditionally stable for materials with *associative* flow rule.

Finite step accuracy aspects: the iso-error maps

As mentioned previously, the accuracy order of an algorithm gives information about how well the discretised equations approximate the original differential equations in an infinitesimal neighbourhood of the initial conditions and, therefore, should be expected to predict its behaviour only for relatively *small* increments. Within the context of the fully implicit Finite Element Method described in Chapter 4 and adopted in the HYPLAS program, it is desirable that the state-update procedures employed at the Gauss point level be sufficiently accurate for strain increments as large as possible in order to ensure that the global finite element solution remains within reasonable bounds of accuracy for large load increments. From this point of view, the extension of the accuracy order study to the assessment of the accuracy of elastoplastic integration algorithms under finite steps becomes crucial.

Systematic finite step accuracy analyses of elastoplastic algorithms have been carried out firstly by Krieg and Krieg (1977) who investigated the behaviour of procedures for integration of the von Mises perfectly plastic model. The fully implicit algorithm, which for this particular model is termed the *radial return method*, was among the procedures assessed by these authors. Their assessment was based on the use of *iso-error maps*. It is remarked that, since the original application by Krieg and Krieg (1977), iso-error maps have proved very effective and are currently accepted as the most reliable (if not the only) tool for the assessment of the finite step accuracy of integration algorithms for elastoplasticity (de Souza Neto *et al.*, 1994a; Dutko *et al.*, 1993; Fuschi *et al.*, 1992; Ortiz and Popov, 1985; Simo and Taylor, 1986).

In order to generate a typical iso-error map, consider an arbitrary stress state at a point P on the yield surface as shown in Figure 7.7(a). Here, the perfectly plastic associative von Mises model is adopted for illustration, with the yield surface of Figure 7.7(a) represented in the deviatoric plane. From this point a sequence of strain increments is applied corresponding to specified normalised *elastic trial* stress increments of the form

$$\Delta\boldsymbol{\sigma}^{\text{trial}} = \frac{\Delta\sigma_T}{q}\,\boldsymbol{T} + \frac{\Delta\sigma_N}{q}\,\boldsymbol{N}, \tag{7.71}$$

where $\boldsymbol{N}$ and $\boldsymbol{T}$ are, respectively, the unit (in Euclidean norm) normal and tangent vectors to the yield surface and q is the von Mises equivalent stress at P. As a result of the numerical integration algorithm, an approximated stress $\boldsymbol{\sigma}^{\text{num}}$ is computed for each increment. Calling $\boldsymbol{\sigma}^{\text{exact}}$ the exact solution of the stress integration problem, the *error* associated with each increment is defined as

$$ERROR = \frac{\sqrt{(\boldsymbol{\sigma}^{\text{exact}} - \boldsymbol{\sigma}^{\text{num}}) : (\boldsymbol{\sigma}^{\text{exact}} - \boldsymbol{\sigma}^{\text{num}})}}{\sqrt{\boldsymbol{\sigma}^{\text{exact}} : \boldsymbol{\sigma}^{\text{exact}}}}. \tag{7.72}$$

By varying the prescribed increment sizes $\Delta\sigma_{\text{T}}$ and $\Delta\sigma_{\text{N}}$, respectively associated with the tangential and normal directions to the yield surface, an error field is obtained. The contour plot of this error field is the *iso-error map* for P. The iso-error map corresponding to the fully implicit scheme for integration of the perfectly plastic von Mises model under general stress state is shown in Figure 7.7(b). For this particular model, the starting point P is immaterial. It is important to note that, as analytical solutions are generally not available, $\boldsymbol{\sigma}^{\text{exact}}$ is taken as the numerical solution obtained by dividing each strain increment into a sufficiently large number of subincrements**. To give a general idea of reasonable sub-increment sizes, it is worth mentioning that one thousand sub-increments have been used to produce the 'exact' solutions for the map of Figure 7.7(b). It should be emphasised that the particular increment directions and error measure employed above are suitable for the von Mises perfectly plastic

**Note that this assumption is valid if the algorithm under study is *convergent*, i.e. first-order accurate and stable, for the particular plasticity model considered.

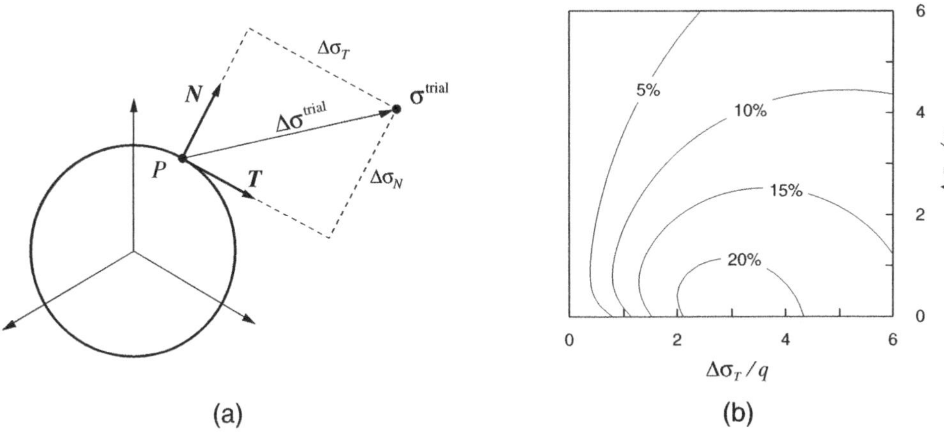

Figure 7.7. Iso-error map: (a) typical increment directions; and (b) a typical iso-error map.

model, but may not be so for different models. In the case of hardening materials, for instance, it might be useful to include the error associated with the hardening internal variable in the definition of the error measure.

Having described the concept of iso-error maps, let us now return to the discussion regarding the properties of the integration algorithms of Sections 7.2.3 and 7.2.7 concentrating on their finite step accuracy characteristics. A finite step accuracy analysis of the generalised trapezoidal and midpoint rule integration algorithms, based on iso-error maps, has been carried out by Ortiz and Popov (1985) in the same paper in which they presented the analysis of accuracy and stability of these algorithms. The iso-error maps discussed by these authors were restricted to the perfectly plastic von Mises model – a model for which the generalised trapezoidal and midpoint algorithms coincide. Essentially, these authors have found that, for small increments, the best accuracy is obtained with $\theta = \frac{1}{2}$. This is in obvious agreement with the (infinitesimal) accuracy analysis that established second-order accuracy for this particular choice of θ. With increasing increment size, however, the performance of the second-order algorithm deteriorates. For reasonably sized increments, likely to occur in real problems, a choice of θ between 0.7 and 0.8 gives the best accuracy. Also, the fully implicit algorithm ($\theta = 1$), which will be adopted in the numerical implementation described in the following sections of this chapter, gives better accuracy than the second-order algorithm. It is worth remarking that completely analogous results have been found by Fuschi *et al.* (1992) for the variant of the midpoint rule based algorithm proposed by Simo and Govindjee (1991).

7.3. Application: integration algorithm for the isotropically hardening von Mises model

As far as the computer implementation of elastic predictor/return-mapping state-update procedures is concerned, the von Mises model is the simplest type discussed in this book. The specialisation of the *fully implicit* algorithm of Box 7.1 (page 199) to the von Mises model, featuring the standard associative law and linear elastic behaviour and including general nonlinear isotropic strain hardening, is described in this section. It is remarked that the *implicit* algorithm is exclusively adopted throughout this book (and in the HYPLAS program) for integration of elastoplastic models. Its particularisation for the Tresca, Mohr–Coulomb and Drucker–Prager models is described later, in Chapter 8.

The choice of the fully implicit algorithm here and in the remainder of this book is essentially motivated by:

(i) its (unconditional) stability which, in conjunction with its first-order accuracy, ensures that the resulting algorithm is convergent;

(ii) its generally 'good' finite step accuracy;

(iii) its suitability for the derivation of associated consistent tangent operators – a property (not shared by the cutting-plane algorithm) that is absolutely essential for its use in conjunction with a (quadratically convergent) global Newton–Raphson procedure; and

(iv) its relatively simple computational implementation, as compared to other members of the families of generalised trapezoidal and midpoint algorithms discussed in Section 7.2.7.

7.3.1. THE IMPLEMENTED MODEL

The von Mises yield criterion as well as the corresponding associative flow rule and possible hardening laws have been thoroughly discussed in Chapter 6. Before going further, it is convenient at this point to list the basic equations used in the present implementation of the von Mises model. Essentially, the model comprises:

1. A linear elastic law

$$\boldsymbol{\sigma} = \mathsf{D}^e : \boldsymbol{\varepsilon}^e, \tag{7.73}$$

where D^e is the standard isotropic elasticity tensor.

2. A yield function of the form

$$\Phi(\boldsymbol{\sigma}, \sigma_y) = \sqrt{3\, J_2(\boldsymbol{s}(\boldsymbol{\sigma}))} - \sigma_y, \tag{7.74}$$

where

$$\sigma_y = \sigma_y(\bar{\varepsilon}^p) \tag{7.75}$$

is the uniaxial yield stress – a function of the accumulated plastic strain, $\bar{\varepsilon}^p$.

3. A standard associative flow rule

$$\dot{\boldsymbol{\varepsilon}}^p = \dot{\gamma}\, \boldsymbol{N} = \dot{\gamma}\, \frac{\partial \Phi}{\partial \boldsymbol{\sigma}}, \tag{7.76}$$

with the (Prandtl–Reuss) flow vector, $\boldsymbol{N}$, explicitly given by

$$\boldsymbol{N} \equiv \frac{\partial \Phi}{\partial \boldsymbol{\sigma}} = \sqrt{\frac{3}{2}}\, \frac{\boldsymbol{s}}{\|\boldsymbol{s}\|}. \tag{7.77}$$

4. An associative hardening rule, with the evolution equation for the hardening internal variable given by

$$\dot{\bar{\varepsilon}}^p = \sqrt{\tfrac{2}{3}}\, \|\dot{\boldsymbol{\varepsilon}}^p\| = \dot{\gamma}. \tag{7.78}$$

7.3.2. THE IMPLICIT ELASTIC PREDICTOR/RETURN-MAPPING SCHEME

Given the increment of strain

$$\Delta\varepsilon = \varepsilon_{n+1} - \varepsilon_n, \tag{7.79}$$

corresponding to a typical (pseudo-) time increment $[t_n, t_{n+1}]$, and given the state variables $\{\varepsilon^e_n, \bar{\varepsilon}^p_n\}$ at t_n, the elastic trial strain and trial accumulated plastic strain are given by

$$\begin{aligned} \varepsilon^{e\ \mathrm{trial}}_{n+1} &= \varepsilon^e_n + \Delta\varepsilon \\ \bar{\varepsilon}^{p\ \mathrm{trial}}_{n+1} &= \bar{\varepsilon}^p_n. \end{aligned} \tag{7.80}$$

The corresponding trial stress is computed as

$$\boldsymbol{\sigma}^{\mathrm{trial}}_{n+1} = \mathsf{D}^e : \varepsilon^{e\ \mathrm{trial}}_{n+1}, \tag{7.81}$$

or, equivalently, by applying the hydrostatic/deviatoric decomposition

$$\boldsymbol{s}^{\mathrm{trial}}_{n+1} = 2G\, \varepsilon^{e\ \mathrm{trial}}_{\mathrm{d}\ n+1}, \quad p^{\mathrm{trial}}_{n+1} = K\, \varepsilon^{e\ \mathrm{trial}}_{\mathrm{v}\ n+1}, \tag{7.82}$$

where s and p denote, respectively, the deviatoric and hydrostatic stresses, G and K are, respectively, the shear and bulk moduli and the subscripts d and v in the elastic trial strain denote, respectively, the deviatoric and volumetric components. The trial yield stress is simply

$$\sigma^{\mathrm{trial}}_{y\ n+1} = \sigma_y(\bar{\varepsilon}^p_n) = \sigma_{y\,n}. \tag{7.83}$$

Having computed the elastic trial state, the next step in the algorithm is to check whether $\boldsymbol{\sigma}^{\mathrm{trial}}_{n+1}$ lies inside or outside of the trial yield surface:

- If $\boldsymbol{\sigma}^{\mathrm{trial}}_{n+1}$ lies inside of the trial yield surface, i.e. if

  $$\Phi(\boldsymbol{\sigma}^{\mathrm{trial}}_{n+1}, \sigma_{y\,n}) \le 0,$$

  then the process within the interval $[t_n, t_{n+1}]$ is purely elastic and the elastic trial state itself is the solution to the integration problem. In this case,

  $$\begin{aligned} \varepsilon^e_{n+1} &= \varepsilon^{e\ \mathrm{trial}}_{n+1} \\ \boldsymbol{\sigma}_{n+1} &= \boldsymbol{\sigma}^{\mathrm{trial}}_{n+1} \\ \bar{\varepsilon}^p_{n+1} &= \bar{\varepsilon}^{p\ \mathrm{trial}}_{n+1} = \bar{\varepsilon}^p_n \\ \sigma_{y\ n+1} &= \sigma^{\mathrm{trial}}_{y\ n+1} = \sigma_{y\,n} \end{aligned} \tag{7.84}$$

  is updated.

- Otherwise, the process is elastoplastic within the interval $[t_n, t_{n+1}]$ and the *return-mapping* procedure described below has to be applied.

Recall that the general implicit return-mapping procedure corresponds to solving the system (7.25) of nonlinear equations. In the present case, direct specialisation of the general fully implicit return-mapping equations (7.25) to the von Mises model gives the following set of nonlinear equations:

$$\begin{aligned}
\boldsymbol{\varepsilon}^e_{n+1} &= \boldsymbol{\varepsilon}^{e\ \text{trial}}_{n+1} - \Delta\gamma \sqrt{\frac{3}{2}}\,\frac{\boldsymbol{s}_{n+1}}{\|\boldsymbol{s}_{n+1}\|}\\
\bar{\varepsilon}^p_{n+1} &= \bar{\varepsilon}^p_n + \Delta\gamma \\
\sqrt{3\,J_2(\boldsymbol{s}_{n+1})} &- \sigma_y(\bar{\varepsilon}^p_{n+1}) = 0,
\end{aligned} \tag{7.85}$$

which has to be solved for $\boldsymbol{\varepsilon}^e_{n+1}$, $\bar{\varepsilon}^p_{n+1}$ and $\Delta\gamma$ and where

$$\boldsymbol{s}_{n+1} = \boldsymbol{s}_{n+1}(\boldsymbol{\varepsilon}^e_{n+1}) = 2G\,\text{dev}[\boldsymbol{\varepsilon}^e_{n+1}]. \tag{7.86}$$

After the solution of the above system, the plastic strain tensor can be updated according to the following formula:

$$\boldsymbol{\varepsilon}^p_{n+1} = \boldsymbol{\varepsilon}^p_n + \Delta\gamma \sqrt{\frac{3}{2}}\,\frac{\boldsymbol{s}_{n+1}}{\|\boldsymbol{s}_{n+1}\|}. \tag{7.87}$$

Single-equation return mapping

It is remarked here that the above system can be substantially simplified. In fact, as shall be seen in what follows, the return mapping for the von Mises model can be reduced to a *single* nonlinear equation having the incremental plastic multiplier $\Delta\gamma$ as the unknown. It should be emphasised that this reduction in the number of equations is of extreme importance in order to make the state-update procedure more computationally efficient and, of course, to improve the performance of the overall finite element scheme. Upon simplification of the system (7.85), it should be noted firstly that the von Mises flow vector is *purely deviatoric* so that the deviatoric/volumetric split of $(7.85)_1$ gives

$$\begin{aligned}
\varepsilon^e_{v\,n+1} &= \varepsilon^{e\ \text{trial}}_{v\,n+1}\\
\boldsymbol{\varepsilon}^e_{d\,n+1} &= \boldsymbol{\varepsilon}^{e\ \text{trial}}_{d\,n+1} - \Delta\gamma \sqrt{\frac{3}{2}}\,\frac{\boldsymbol{s}_{n+1}}{\|\boldsymbol{s}_{n+1}\|}.
\end{aligned} \tag{7.88}$$

Equivalently, in terms of stresses, we have

$$\begin{aligned}
p_{n+1} &= p^{\text{trial}}_{n+1}\\
\boldsymbol{s}_{n+1} &= \boldsymbol{s}^{\text{trial}}_{n+1} - \Delta\gamma\, 2G \sqrt{\frac{3}{2}}\,\frac{\boldsymbol{s}_{n+1}}{\|\boldsymbol{s}_{n+1}\|},
\end{aligned} \tag{7.89}$$

that is, the return mapping affects *only* the deviatoric stress component. The hydrostatic stress, p_{n+1}, has the value computed in the elastic predictor stage and can, therefore, be eliminated from the system of equations. Further simplification follows by noting that by rearranging the deviatoric stress update formula $(7.89)_2$ we obtain

$$\left(1 + \sqrt{\frac{3}{2}}\,\frac{\Delta\gamma\, 2G}{\|\boldsymbol{s}_{n+1}\|}\right)\boldsymbol{s}_{n+1} = \boldsymbol{s}^{\text{trial}}_{n+1}$$

i.e. the trial and updated deviatoric stresses are *co-linear*. This implies that

$$\frac{\boldsymbol{s}_{n+1}}{\|\boldsymbol{s}_{n+1}\|} = \frac{\boldsymbol{s}^{\text{trial}}_{n+1}}{\|\boldsymbol{s}^{\text{trial}}_{n+1}\|},$$

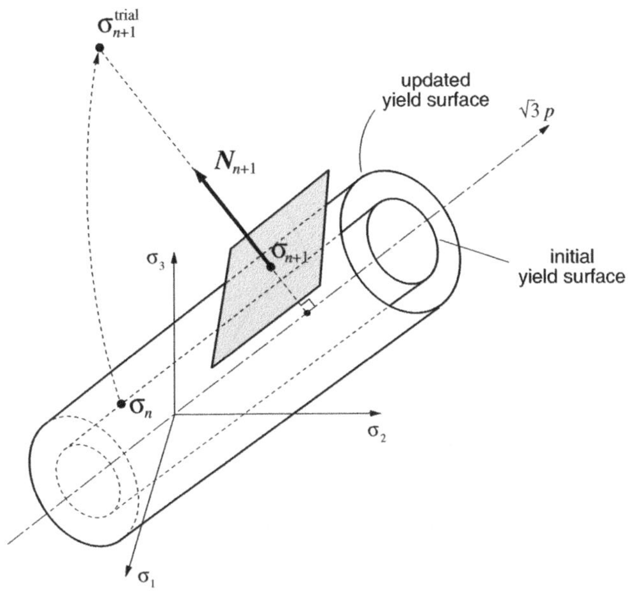

Figure 7.8. The implicit elastic predictor/return-mapping scheme for the von Mises model. Geometric interpretation in principal stress space.

so that the flow vectors at the trial and updated states coincide. Substitution of the above identity into $(7.89)_2$ leads to the following simpler update formula for the deviatoric stress:

$$\begin{aligned} \boldsymbol{s}_{n+1} &= \left(1 - \sqrt{\frac{3}{2}}\frac{\Delta\gamma\, 2G}{\|\boldsymbol{s}_{n+1}^{\text{trial}}\|}\right)\boldsymbol{s}_{n+1}^{\text{trial}} \\ &= \left(1 - \frac{\Delta\gamma\, 3G}{q_{n+1}^{\text{trial}}}\right)\boldsymbol{s}_{n+1}^{\text{trial}}, \end{aligned} \tag{7.90}$$

where $q_{n+1}^{\text{trial}} \equiv \sqrt{3\, J_2(\boldsymbol{s}_{n+1}^{\text{trial}})}$ is the elastic trial von Mises effective stress. Note that, since $\boldsymbol{s}_{n+1}^{\text{trial}}$ is a constant tensor in the return mapping, the deviatoric stress, $\boldsymbol{s}_{n+1}$, is a (linear) function of $\Delta\gamma$ only in the above update formula. Expression (7.90) implies that, in the fully implicit algorithm for the von Mises model, the updated deviatoric stress is obtained by scaling down the trial deviatoric stress by the factor $1 - \Delta\gamma\, 3G/q_{n+1}^{\text{trial}}$. The geometric representations of this update formula in the principal stress space and deviatoric plane are illustrated, respectively, in Figures 7.8 and 7.9.

Finally, with substitution of (7.90) and $(7.85)_2$ into the plastic consistency condition $(7.85)_3$, the system (7.85) of equations of the return mapping for the von Mises model is reduced to the following *scalar* (generally nonlinear) equation having the incremental plastic multiplier, $\Delta\gamma$, as the only unknown:

$$\tilde{\Phi}(\Delta\gamma) \equiv q_{n+1}^{\text{trial}} - 3G\,\Delta\gamma - \sigma_y(\bar{\varepsilon}_n^p + \Delta\gamma) = 0. \tag{7.91}$$

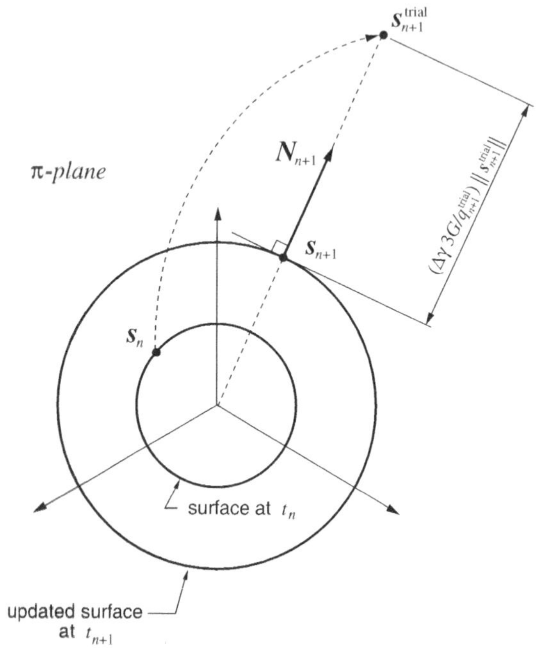

Figure 7.9. The implicit elastic predictor/return-mapping scheme for the von Mises model. Geometric interpretation in the deviatoric plane.

The above equation is then solved by the Newton–Raphson method and, with its solution $\Delta\gamma$ at hand, the state variables are updated as follows:

$$
\begin{aligned}
\boldsymbol{s}_{n+1} &= \left(1 - \frac{\Delta\gamma\, 3G}{q_{n+1}^{\text{trial}}}\right) \boldsymbol{s}_{n+1}^{\text{trial}} \\
\boldsymbol{\sigma}_{n+1} &= \boldsymbol{s}_{n+1} + p_{n+1}^{\text{trial}}\, \boldsymbol{I} \\
\boldsymbol{\varepsilon}_{n+1}^{e} &= [\mathsf{D}^e]^{-1} : \boldsymbol{\sigma}_{n+1} = \frac{1}{2G}\, \boldsymbol{s}_{n+1} + \frac{1}{3} \varepsilon_{\text{v}\, n+1}^{e\ \text{trial}} \\
\bar{\varepsilon}_{n+1}^{p} &= \bar{\varepsilon}_{n}^{p} + \Delta\gamma.
\end{aligned}
\tag{7.92}
$$

If required, the plastic strain tensor is updated by means of (7.87).

The overall elastic predictor/return-mapping algorithm for the von Mises model is summarised in Boxes 7.3 and 7.4. The procedure shown in these boxes is implemented in subroutine `SUVM` of the `HYPLAS` program. This routine is described in detail in Section 7.3.5.

7.3.3. THE INCREMENTAL CONSTITUTIVE FUNCTION FOR THE STRESS

From the update formulae $(7.92)_{1,2}$ and relations (7.81,7.82) we establish after simple manipulations that the updated stress tensor, $\boldsymbol{\sigma}_{n+1}$, can be expressed in terms of $\bar{\varepsilon}_n^p$ and

Box 7.3. Fully implicit elastic predictor/return-mapping algorithm for the von Mises model with nonlinear isotropic hardening.

HYPLAS procedure:	SUVM

(i) Elastic predictor. Given $\Delta\boldsymbol{\varepsilon}$ and the state variables at t_n, evaluate the *elastic trial state*

$$\boldsymbol{\varepsilon}^{e\ \text{trial}}_{n+1} := \boldsymbol{\varepsilon}^e_n + \Delta\boldsymbol{\varepsilon}$$

$$\bar{\varepsilon}^{p\ \text{trial}}_{n+1} := \bar{\varepsilon}^p_n$$

$$p^{\text{trial}}_{n+1} := K\,\varepsilon^{e\ \text{trial}}_{\text{v}\ n+1};\qquad \boldsymbol{s}^{\text{trial}}_{n+1} := 2G\,\boldsymbol{\varepsilon}^{e\ \text{trial}}_{\text{d}\ n+1}$$

$$q^{\text{trial}}_{n+1} := \sqrt{\tfrac{3}{2}\,\boldsymbol{s}^{\text{trial}}_{n+1} : \boldsymbol{s}^{\text{trial}}_{n+1}}$$

(ii) Check plastic admissibility

$$\text{IF } q^{\text{trial}}_{n+1} - \sigma_y(\bar{\varepsilon}^{p\ \text{trial}}_{n+1}) \le 0$$

$$\text{THEN set}\quad (\cdot)_{n+1} := (\cdot)^{\text{trial}}_{n+1}\quad \text{and EXIT}$$

(iii) Return mapping. Solve the equation

$$\tilde{\Phi}(\Delta\gamma) \equiv q^{\text{trial}}_{n+1} - 3G\,\Delta\gamma - \sigma_y(\bar{\varepsilon}^p_n + \Delta\gamma) = 0$$

for $\Delta\gamma$ using the Newton–Raphson method – GOTO Box 7.4 – and update the state variables

$$p_{n+1} := p^{\text{trial}}_{n+1};\qquad \boldsymbol{s}_{n+1} := \left(1 - \frac{\Delta\gamma\,3G}{q^{\text{trial}}_{n+1}}\right)\boldsymbol{s}^{\text{trial}}_{n+1}$$

$$\boldsymbol{\sigma}_{n+1} := \boldsymbol{s}_{n+1} + p_{n+1}\,\boldsymbol{I}$$

$$\boldsymbol{\varepsilon}^e_{n+1} = \frac{1}{2G}\,\boldsymbol{s}_{n+1} + \frac{1}{3}\,\varepsilon^{e\ \text{trial}}_{\text{v}\ n+1}\,\boldsymbol{I}$$

$$\bar{\varepsilon}^p_{n+1} := \bar{\varepsilon}^p_n + \Delta\gamma$$

(iv) EXIT

$\varepsilon^{e\ \text{trial}}_{n+1}$ by means of the following *incremental constitutive function*:

$$\boldsymbol{\sigma}_{n+1} = \bar{\boldsymbol{\sigma}}_{n+1}(\bar{\varepsilon}^p_n, \boldsymbol{\varepsilon}^{e\ \text{trial}}_{n+1}) \equiv \left[\mathbf{D}^e - \hat{H}(\Phi^{\text{trial}})\,\frac{\Delta\gamma\,6G^2}{q^{\text{trial}}_{n+1}}\,\mathbf{I}_{\text{d}}\right] : \boldsymbol{\varepsilon}^{e\ \text{trial}}_{n+1}, \tag{7.93}$$

where $\hat{H}$ is the *Heaviside step function* defined as

$$\hat{H}(a) \equiv \begin{cases} 1 & \text{if } a > 0 \\ 0 & \text{if } a \le 0 \end{cases},\qquad \text{for any scalar } a, \tag{7.94}$$

$\mathbf{I}_d$ is the deviatoric projection tensor defined by (3.94) (page 59),

$$\begin{aligned} q^{\text{trial}}_{n+1} &= \sqrt{\tfrac{3}{2}}\|\boldsymbol{s}^{\text{trial}}_{n+1}\| = 2G\sqrt{\tfrac{3}{2}}\|\boldsymbol{\varepsilon}^{e\ \text{trial}}_{\text{d}\ n+1}\| \\ &= q^{\text{trial}}_{n+1}(\boldsymbol{\varepsilon}^{e\ \text{trial}}_{n+1}) \equiv 2G\sqrt{\tfrac{3}{2}}\,\|\,\mathbf{I}_{\text{d}} : \boldsymbol{\varepsilon}^{e\ \text{trial}}_{n+1}\|, \end{aligned} \tag{7.95}$$

Box 7.4. The Newton–Raphson algorithm for solution of the return-mapping equation of the von Mises model.

HYPLAS procedure:	SUVM

(i) Initialise iteration counter, $k := 0$, set initial guess for $\Delta\gamma$

$$\Delta\gamma^{(0)} := 0$$

and corresponding residual (yield function value)

$$\tilde{\Phi} := q_{n+1}^{\text{trial}} - \sigma_y(\bar{\varepsilon}_n^p)$$

(ii) Perform Newton–Raphson iteration

$$H := \left.\frac{\mathrm{d}\sigma_y}{\mathrm{d}\bar{\varepsilon}^p}\right|_{\bar{\varepsilon}_n^p+\Delta\gamma} \quad \text{(hardening slope)}$$

$$d := \frac{\mathrm{d}\tilde{\Phi}}{\mathrm{d}\Delta\gamma} = -3G - H \quad \text{(residual derivative)}$$

$$\Delta\gamma := \Delta\gamma - \frac{\tilde{\Phi}}{d} \quad \text{(new guess for } \Delta\gamma)$$

(iii) Check for convergence

$$\tilde{\Phi} := q_{n+1}^{\text{trial}} - 3G\,\Delta\gamma - \sigma_y(\bar{\varepsilon}_n^p + \Delta\gamma)$$

$$\text{IF} \quad |\tilde{\Phi}| \leq \epsilon_{\text{tol}} \quad \text{THEN} \quad \text{RETURN to Box 7.3}$$

(iv) GOTO (ii)

Φ^{trial} is the value of the yield function at the elastic trial state:

$$\Phi^{\text{trial}} = \Phi^{\text{trial}}(\bar{\varepsilon}_n^p, \boldsymbol{\varepsilon}_{n+1}^{e\ \text{trial}}) \equiv q_{n+1}^{\text{trial}}(\boldsymbol{\varepsilon}_{n+1}^{e\ \text{trial}}) - \sigma_y(\bar{\varepsilon}_n^p), \tag{7.96}$$

and

$$\Delta\gamma = \Delta\gamma(\bar{\varepsilon}_n^p, \boldsymbol{\varepsilon}_{n+1}^{e\ \text{trial}}) \tag{7.97}$$

is the *implicit* function of $\boldsymbol{\varepsilon}_{n+1}^{e\ \text{trial}}$ and $\bar{\varepsilon}_n^p$ defined as the solution of the consistency equation (7.91).

Clearly, (7.93) defines $\boldsymbol{\sigma}_{n+1}$ as an *implicit* function of the elastic trial strain and $\bar{\varepsilon}_n^p$. Equivalently, since $\boldsymbol{\varepsilon}_{n+1}^{e\ \text{trial}} = \boldsymbol{\varepsilon}_{n+1} - \boldsymbol{\varepsilon}_n^p$, we may write

$$\boldsymbol{\sigma}_{n+1} = \hat{\boldsymbol{\sigma}}_{n+1}(\bar{\varepsilon}_n^p, \boldsymbol{\varepsilon}_n^p, \boldsymbol{\varepsilon}_{n+1}) \equiv \bar{\boldsymbol{\sigma}}_{n+1}(\bar{\varepsilon}_n^p, \boldsymbol{\varepsilon}_{n+1} - \boldsymbol{\varepsilon}_n^p). \tag{7.98}$$

For a given state at t_n, the functions (7.93) and (7.98) express the updated stress as implicit functions, respectively, of the elastic trial stress and the total elastic strain at t_{n+1}.

Remark 7.7. The use of incremental algorithmic constitutive functions has been first alluded to in this book in Chapter 4 in the formulation of incremental boundary value problems involving path-dependent materials. The reader is referred to expression (4.60) (page 95), and the text surrounding it for details. The function $\hat{\boldsymbol{\sigma}}$ defined in (7.98, 7.93) is the particularisation of the generic incremental function (4.60) for the von Mises model with isotropic strain hardening (for which $\boldsymbol{\alpha}_n = \{\bar{\varepsilon}^p_n, \boldsymbol{\varepsilon}^p_n\}$) integrated numerically by the fully implicit elastic predictor/return-mapping algorithm.

7.3.4. LINEAR ISOTROPIC HARDENING AND PERFECT PLASTICITY: THE CLOSED-FORM RETURN MAPPING

It should be noted that the only source of nonlinearity in the von Mises return-mapping equation (7.91) is the *hardening curve*, defined by the given function $\sigma_y = \sigma_y(\bar{\varepsilon}^p)$. For *linear hardening* materials, this function is linear and is expressed by

$$\sigma_y(\bar{\varepsilon}^p) = \sigma_0 + H\,\bar{\varepsilon}^p, \tag{7.99}$$

where σ_0 is the initial yield stress of the virgin material and H is the (constant) hardening modulus. In such cases, (7.91) reads

$$\tilde{\Phi}(\Delta\gamma) \equiv q^{\text{trial}}_{n+1} - 3G\,\Delta\gamma - [\sigma_0 + (\bar{\varepsilon}^p_n + \Delta\gamma)\,H] = 0 \tag{7.100}$$

and the incremental plastic multiplier can be obtained in *closed form* as

$$\Delta\gamma = \frac{\Phi^{\text{trial}}}{3G + H}. \tag{7.101}$$

Thus, for linearly hardening von Mises materials, the above closed expression replaces the Newton–Raphson algorithm of Box 7.4 in the return-mapping procedure.

In the case of *perfect plasticity* ($H = 0$), the expression for $\Delta\gamma$ reads

$$\Delta\gamma = \frac{\Phi^{\text{trial}}}{3G}. \tag{7.102}$$

The geometric interpretation of the fully implicit algorithm for the perfectly plastic von Mises model is illustrated in Figure 7.10. In this case, the updated stress is simply the projection of the elastic trial stress onto the fixed yield surface along its *radial* direction. It is the *closest point* projection of the trial stress onto the yield surface.

d

The explicit incremental constitutive function

Under the assumption of linear hardening, we substitute the explicit formula (7.101) for $\Delta\gamma$ into (7.93) and obtain, after a straightforward manipulation, the following incremental constitutive function for the updated stress:

$$\begin{aligned}\boldsymbol{\sigma}_{n+1} &= \bar{\boldsymbol{\sigma}}_{n+1}(\boldsymbol{\varepsilon}^{e\,\text{trial}}_{n+1}, \bar{\varepsilon}^p_n)\\ &\equiv \left[\mathsf{D}^e - \hat{H}(\Phi^{\text{trial}})\,\frac{6G^2}{3G+H}\left(1 - \frac{\sigma_y(\bar{\varepsilon}^p_n)}{q^{\text{trial}}_{n+1}}\right)\mathsf{I}_d\right] : \boldsymbol{\varepsilon}^{e\,\text{trial}}_{n+1}.\end{aligned} \tag{7.103}$$

In contrast to the general case (7.93), $\bar{\boldsymbol{\sigma}}_{n+1}$ in the above definition is an *explicit* function. We remark, however, that explicit incremental constitutive functions are obtainable in the context of implicit integration algorithms only under very special circumstances (such as linear hardening in the present case). For more realistic models, such functions are usually implicit.

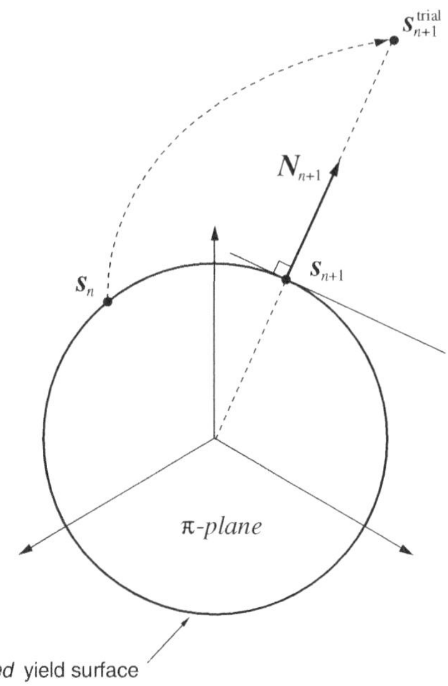

Figure 7.10. The perfectly plastic von Mises model. Geometric interpretation of the implicit return-mapping scheme as the *closest point projection algorithm.*

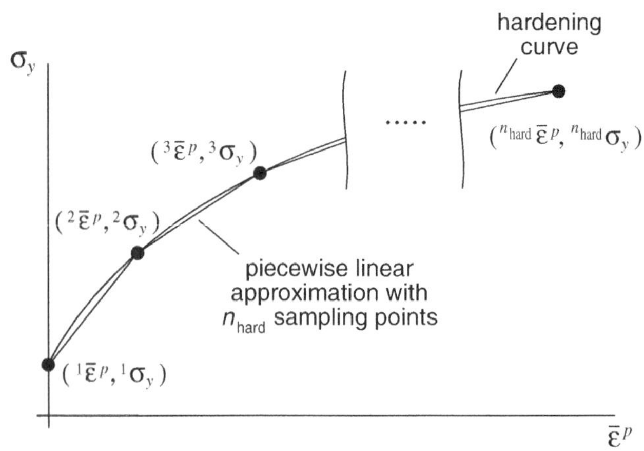

Figure 7.11. Piecewise linear hardening.

7.3.5. SUBROUTINE SUVM

The FORTRAN implementation of the fully implicit elastic predictor/return-mapping algorithm for the von Mises model is described in detail in this section. This procedure, summarised in Boxes 7.3 and 7.4, is carried out in subroutine `SUVM` (**S**tate **U**pdate procedure for the **V**on **M**ises model). In the `HYPLAS` program, subroutine `SUVM` is called by the material interface `MATISU` during the computation of the finite element internal force vector.

In the present implementation, the isotropic hardening curve defined by

$$\sigma_y(\bar{\varepsilon}^p)$$

has been assumed to be *piecewise linear*. This means that any (arbitrarily nonlinear) hardening curve can be adequately approximated by using a sufficiently large number of sampling pairs

$$\{\bar{\varepsilon}^p, \sigma_y\}$$

with linear interpolation between adjacent pairs. A piecewise linear approximation, with $n_{\rm hard}$ sampling points, to a generic hardening curve is illustrated in Figure 7.11. In fact, actual experimental data for hardening curves are normally obtained as a set of points $(\sigma_y, \bar{\varepsilon}^p)$. Thus, the present implementation of the von Mises model allows for the experimental data to be used directly in the definition of the hardening curve.

The FORTRAN source code of SUVM is listed below.

```
      SUBROUTINE SUVM
     1(   DGAMA      ,IPROPS     ,LALGVA     ,NTYPE      ,RPROPS     ,
     2    RSTAVA     ,STRAT      ,STRES      )
      IMPLICIT DOUBLE PRECISION (A-H,O-Z)
      PARAMETER(IPHARD=4  ,MSTRE=4)
      LOGICAL IFPLAS, LALGVA(2), SUFAIL
      DIMENSION
     1    IPROPS(*)           ,RPROPS(*)           ,RSTAVA(MSTRE+1)    ,
     2    STRAT(MSTRE)        ,STRES(MSTRE)
      DIMENSION
     1    EET(MSTRE)
      DATA
     1    R0   ,RP5  ,R1   ,R2   ,R3   ,TOL   /
     2    0.0D0,0.5D0,1.0D0,2.0D0,3.0D0,1.D-06/
      DATA MXITER / 50 /
C***********************************************************************
C STATE UPDATE PROCEDURE FOR THE VON MISES ELASTO-PLASTIC MATERIAL MODEL
C WITH NON-LINEAR (PIECEWISE LINEAR) ISOTROPIC HARDENING:
C IMPLICIT ELASTIC PREDICTOR/RETURN MAPPING ALGORITHM (BOXES 7.3-4).
C PLANE STRAIN AND AXISYMMETRIC IMPLEMENTATIONS.
C***********************************************************************
C Stop program if neither plane strain nor axisymmetric state
      IF(NTYPE.NE.2.AND.NTYPE.NE.3)CALL ERRPRT('EI0013')
C Initialise some algorithmic and internal variables
      DGAMA=R0
      IFPLAS=.FALSE.
      SUFAIL=.FALSE.
      EPBARN=RSTAVA(MSTRE+1)
C Set some material properties
      YOUNG=RPROPS(2)
      POISS=RPROPS(3)
      NHARD=IPROPS(3)
C Shear and bulk moduli and other necessary constants
      GMODU=YOUNG/(R2*(R1+POISS))
      BULK=YOUNG/(R3*(R1-R2*POISS))
      R2G=R2*GMODU
      R3G=R3*GMODU
C Elastic predictor: Compute elastic trial state
C ----------------------------------------------
C Volumetric strain and pressure stress
      EEV=STRAT(1)+STRAT(2)+STRAT(4)
```

```
      P=BULK*EEV
C Elastic trial deviatoric strain
      EEVD3=EEV/R3
      EET(1)=STRAT(1)-EEVD3
      EET(2)=STRAT(2)-EEVD3
      EET(4)=STRAT(4)-EEVD3
C Convert engineering shear component into physical component
      EET(3)=STRAT(3)/R2
C Compute trial effective stress and uniaxial yield stress
      VARJ2T=R2G*R2G*(EET(3)*EET(3)+RP5*(EET(1)*EET(1)+
     1                     EET(2)*EET(2)+EET(4)*EET(4)))
      QTRIAL=SQRT(R3*VARJ2T)
      SIGMAY=PLFUN(EPBARN,NHARD,RPROPS(IPHARD))
C Check for plastic admissibility
C -------------------------------
      PHI=QTRIAL-SIGMAY
      IF(PHI/SIGMAY.GT.TOL)THEN
C Plastic step: Apply return mapping - use Newton-Raphson algorithm
C               to solve the return mapping equation (Box 7.4)
C -------------------------------------------------------------------
        IFPLAS=.TRUE.
        EPBAR=EPBARN
        DO 10 NRITER=1,MXITER
C Compute residual derivative
          DENOM=-R3G-DPLFUN(EPBAR,NHARD,RPROPS(IPHARD))
C Compute Newton-Raphson increment and update variable DGAMA
          DDGAMA=-PHI/DENOM
          DGAMA=DGAMA+DDGAMA
C Compute new residual
          EPBAR=EPBAR+DDGAMA
          SIGMAY=PLFUN(EPBAR,NHARD,RPROPS(IPHARD))
          PHI=QTRIAL-R3G*DGAMA-SIGMAY
C Check convergence
          RESNOR=ABS(PHI/SIGMAY)
          IF(RESNOR.LE.TOL)THEN
C update accumulated plastic strain
            RSTAVA(MSTRE+1)=EPBAR
C update stress components
            FACTOR=R2G*(R1-R3G*DGAMA/QTRIAL)
            STRES(1)=FACTOR*EET(1)+P
            STRES(2)=FACTOR*EET(2)+P
            STRES(3)=FACTOR*EET(3)
            STRES(4)=FACTOR*EET(4)+P
C compute converged elastic (engineering) strain components
            FACTOR=FACTOR/R2G
            RSTAVA(1)=FACTOR*EET(1)+EEVD3
            RSTAVA(2)=FACTOR*EET(2)+EEVD3
            RSTAVA(3)=FACTOR*EET(3)*R2
            RSTAVA(4)=FACTOR*EET(4)+EEVD3
            GOTO 999
          ENDIF
   10   CONTINUE
C reset failure flag and print warning message if the algorithm fails
        SUFAIL=.TRUE.
        CALL ERRPRT('WE0004')
      ELSE
C Elastic step: Update stress using linear elastic law
C ---------------------------------------------------
        STRES(1)=R2G*EET(1)+P
        STRES(2)=R2G*EET(2)+P
```

```
            STRES(3)=R2G*EET(3)
            STRES(4)=R2G*EET(4)+P
C elastic engineering strain
            RSTAVA(1)=STRAT(1)
            RSTAVA(2)=STRAT(2)
            RSTAVA(3)=STRAT(3)
            RSTAVA(4)=STRAT(4)
          ENDIF
  999 CONTINUE
C Update some algorithmic variables before exit
      LALGVA(1)=IFPLAS
      LALGVA(2)=SUFAIL
      RETURN
      END
```

The arguments of SUVM

$\leftarrow$ DGAMA $[\Delta\gamma]$. The incremental plastic multiplier. If the increment is elastic, it returns as 0. Otherwise, it is obtained as the solution of the return-mapping equation (7.91).

$\rightarrow$ IPROPS. Array of integer material properties. IPROPS(3) contains the number of sampling points for the piecewise linear hardening curve, NHARD, and is the only integer material property required by the present subroutine. IPROPS(3) is set in subroutine RDVM during the input phase of HYPLAS.

$\leftarrow$ LALGVA. Array of logical algorithmic flags or variables. For the von Mises model it contains the plastic yielding flag, IFPLAS, and the state update failure flag SUFAIL. If the step is elastic, the plastic yielding flag returns as .FALSE., otherwise its return value is .TRUE.. The state update failure flag returns as .TRUE. only if the Newton scheme of the return-mapping algorithm fails to converge. In this case, the state variables are *not updated* in SUVM and a warning message is sent (by calling subroutine ERRPRT) to the results file and standard output. When the state update procedure fails (for any material model), increment cutting is activated in the main program; that is, the calculations for the current load increment are aborted and restarted with a smaller increment.

$\rightarrow$ NTYPE. Stress state type flag.

$\rightarrow$ RPROPS. Array of real material properties. It contains the elastic properties (Young's modulus, E, and Poisson's ratio, ν) and the pairs $\{{}^{i}\bar{\varepsilon}^{p}, {}^{i}\sigma_{y}\}$, $i = 1, \ldots, n_{\text{hard}}$, of sampling points along the hardening curve (see illustration of Figure 7.11):

$$RPROPS = [E, \nu, {}^{1}\bar{\varepsilon}^{p}, {}^{1}\sigma_{y}, {}^{2}\bar{\varepsilon}^{p}, {}^{2}\sigma_{y}, \ldots\ldots, {}^{n_{\text{hard}}}\bar{\varepsilon}^{p}, {}^{n_{\text{hard}}}\sigma_{y}].$$

Array RPROPS is set in subroutine RDVM during the input phase of HYPLAS.

$\leftrightarrow$ RSTAVA $[\varepsilon^{e}, \bar{\varepsilon}^{p}]$. Array of state variables other than the stress components. For the present material model implementation, this array contains the elastic strain (returned as the updated value ε^{e}_{n+1}) and the equivalent plastic strain (last converged solution value $\bar{\varepsilon}^{p}_{n}$ on entry, updated value $\bar{\varepsilon}^{p}_{n+1}$ on exit). The last converged elastic strain, ε^{e}_{n}, is used in subroutine MATISU to compute the elastic trial strain before the present subroutine is called.

→ STRAT [$\varepsilon^{e\ \text{trial}}_{n+1}$]. Array containing the elastic trial (engineering) strains. It is computed as shown in item (i) of Box 7.3 in subroutine MATISU. The calculation of array STRAT in MATISU is common to all elastoplastic material models of the class whose identification parameter is HYPEPL.

← STRES [$\boldsymbol{\sigma}_{n+1}$]. Array containing the updated stress tensor components.

Some local variables and arrays of SUVM

- EET [$\varepsilon^{e\ \text{trial}}_{\text{d}\ n+1}$]. Array of elastic trial deviatoric strain components.
- EEV [$\varepsilon^{e\ \text{trial}}_{\text{v}\ n+1}$]. Elastic trial volumetric strain. Equals the actual elastic volumetric strain for the present model.
- MXITER. Maximum number of iterations allowed in the Newton–Raphson procedure for solution of the return-mapping equation.
- NRITER. Iteration counter for the Newton–Raphson algorithm.
- TOL [ϵ_{tol}]. Convergence tolerance for the Newton–Raphson algorithm.

Most remaining local variables of SUVM have been named so as to resemble the corresponding notation of Section 7.3.2. Thus, the operations carried out in SUVM can be easily identified with those indicated in Boxes 7.3 and 7.4.

Some function calls from SUVM

- DPLFUN. This function computes the derivative of a piecewise linear function. Used in SUVM to compute the *slope of the hardening curve*, H.
- PLFUN. Piecewise linear function defined by a set of pairs $(x, f(x))$. Used in SUVM as the *hardening function* $\sigma_y(\bar{\varepsilon}^p)$. Recall that the hardening curve is defined here by the NHARD pairs of sampling points stored in array RPROPS.

7.4. The consistent tangent modulus

Within the description of the incremental finite element framework of Chapter 4, reference has been made in Section 4.2.5 (page 98) to the use of consistent tangent operators:

$$\mathsf{D} \equiv \frac{\partial \boldsymbol{\sigma}_{n+1}}{\partial \boldsymbol{\varepsilon}_{n+1}}, \tag{7.104}$$

in the computation of the element tangent stiffness.

Elastic materials

In the case of purely elastic materials, the stress is an *explicit* function of the strain tensor

$$\boldsymbol{\sigma} = \boldsymbol{\sigma}(\boldsymbol{\varepsilon}) = \bar{\rho}\,\frac{\partial\psi(\boldsymbol{\varepsilon})}{\partial\boldsymbol{\varepsilon}}, \tag{7.105}$$

so that the tangent modulus can be explicitly derived as

$$\mathsf{D} = \mathsf{D}^e \equiv \bar{\rho}\,\frac{\partial^2\psi(\boldsymbol{\varepsilon})}{\partial\boldsymbol{\varepsilon}^2}. \tag{7.106}$$

For linear elastic materials, the consistent tangent modulus is the standard (constant) elasticity tensor

$$\mathsf{D} = 2G\,\mathsf{I}_d + K\,\boldsymbol{I}\otimes\boldsymbol{I}. \tag{7.107}$$

7.4.1. CONSISTENT TANGENT OPERATORS IN ELASTOPLASTICITY

As shown in Section 7.2, elastoplastic materials require in general some kind of numerical integration algorithm to update the stress tensor. Basically, given the known internal variable set $\boldsymbol{\alpha}_n$ and the new prescribed total strain $\boldsymbol{\varepsilon}_{n+1}$ as input, each of the general integration procedures described in Section 7.2 will deliver the updated stress $\boldsymbol{\sigma}_{n+1}$ as the result of the application of a particular numerical algorithm (see diagram of Figure 7.12). This defines an *algorithmic* incremental constitutive function, $\hat{\boldsymbol{\sigma}}$, for the stress tensor with general form

$$\boldsymbol{\sigma}_{n+1} = \hat{\boldsymbol{\sigma}}(\boldsymbol{\alpha}_n, \boldsymbol{\varepsilon}_{n+1}). \tag{7.108}$$

Algorithmic functions of this type have been first referred to in Section 4.2.1 (page 95), in the formulation of incremental boundary value problems with general path-dependent material models. Specific examples of incremental constitutive functions have been obtained earlier in this chapter for the fully implicit elastic predictor/return-mapping implementation of the von Mises model with isotropic strain hardening. Expressions (7.93, 7.98) show the corresponding (implicit) incremental constitutive function for the model with nonlinear hardening and expression (7.103) shows the particular (explicit) format obtained under linear hardening.

Within a load increment $[t_n, t_{n+1}]$, the internal variable set $\boldsymbol{\alpha}_n$ given as argument of $\hat{\boldsymbol{\sigma}}$ is *fixed*. Only the guesses for the total strain, $\boldsymbol{\varepsilon}_{n+1}$ – associated with the guesses for the displacement field, $\boldsymbol{u}_{n+1}$ – change during the global Newton–Raphson equilibrium iterations (refer to Section 4.2, from page 94 for details on the global Newton–Raphson procedure). In other words, within each global load increment, the stress $\boldsymbol{\sigma}_{n+1}$ delivered by the integration algorithm is a function of the total strain tensor *only*. This function – $\hat{\boldsymbol{\sigma}}(\boldsymbol{\alpha}_n, \boldsymbol{\varepsilon}_{n+1})$ with fixed $\boldsymbol{\alpha}_n$ – defines a path-*independent* stress/strain relation within the interval $[t_n, t_{n+1}]$, equivalent to a (nonlinear) elastic law. The consistent tangent modulus in this case is precisely the derivative of this equivalent nonlinear elastic law:

$$\mathsf{D} \equiv \frac{\mathrm{d}\boldsymbol{\sigma}_{n+1}}{\mathrm{d}\boldsymbol{\varepsilon}_{n+1}} = \left.\frac{\partial\hat{\boldsymbol{\sigma}}}{\partial\boldsymbol{\varepsilon}_{n+1}}\right|_{\boldsymbol{\alpha}_n}, \tag{7.109}$$

i.e. it is the derivative of the algorithmic function $\hat{\boldsymbol{\sigma}}$ with respect to $\boldsymbol{\varepsilon}_{n+1}$ with $\boldsymbol{\alpha}_n$ held constant.

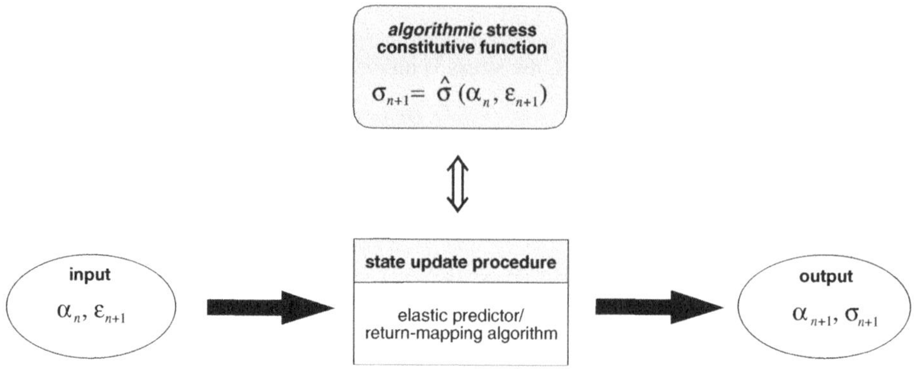

Figure 7.12. The algorithmic constitutive function for the stress tensor.

At this point, it is worth remarking that, in the context of the multiplicative finite strain plasticity framework discussed in Chapter 14, no measure of *total* nonlinear strain analogous to ε_{n+1} is used (or needed) in the definition of elastoplastic constitutive models. An *elastic trial* strain measure (analogous to the infinitesimal tensor $\varepsilon^{e\ \text{trial}}_{n+1}$), however, does appear naturally in the formulation of the corresponding elastic predictor/return-mapping schemes. Clearly, since $\varepsilon^{e\ \text{trial}}_{n+1} = \varepsilon_{n+1} - \varepsilon^p_n$, the incremental constitutive function for $\boldsymbol{\sigma}_{n+1}$ can always be expressed equivalently as[††]

$$\boldsymbol{\sigma}_{n+1} = \bar{\boldsymbol{\sigma}}(\boldsymbol{\alpha}_n, \varepsilon^{e\ \text{trial}}_{n+1}) \equiv \hat{\boldsymbol{\sigma}}(\boldsymbol{\alpha}_n, \varepsilon_{n+1}), \tag{7.110}$$

in terms of the elastic trial strain and the internal variable set at t_n, and we also have the trivial identity

$$\mathsf{D} = \frac{\partial \hat{\boldsymbol{\sigma}}}{\partial \varepsilon_{n+1}} = \frac{\partial \bar{\boldsymbol{\sigma}}}{\partial \varepsilon^{e\ \text{trial}}_{n+1}} \tag{7.111}$$

for the consistent tangent operator. To make the material presented here formally valid also for the large-strain case addressed in Chapter 14 (where a total strain tensor is not defined), we shall adopt in what follows the rightmost term of (7.111) as the definition of consistent tangent operator.[‡‡]

The elastic and elastoplastic tangents

Before going into further details, it is worth remarking at this point that the algorithmic function $\hat{\boldsymbol{\sigma}}$ is *non-differentiable* in general. This is clearly seen by noting the presence of the Heaviside step function – a non-differentiable function – in definition (7.93), (7.98) of the incremental constitutive law for the implicitly integrated von Mises model with nonlinear isotropic strain hardening. For states lying within the elastic domain, i.e. states corresponding

[††] This obvious equivalence has already been used in (7.98).

[‡‡] Also note that, in order to maximise the number of routines shared by small and large-strain isotropic plasticity models in the HYPLAS program, we have conveniently chosen the *elastic trial strain*, $\varepsilon^{e\ \text{trial}}_{n+1}$, rather than the total strain ε_{n+1}, as the actual input argument to the elastoplastic integration procedures for models of the present type (see routine SUVM, listed on page 227, for instance).

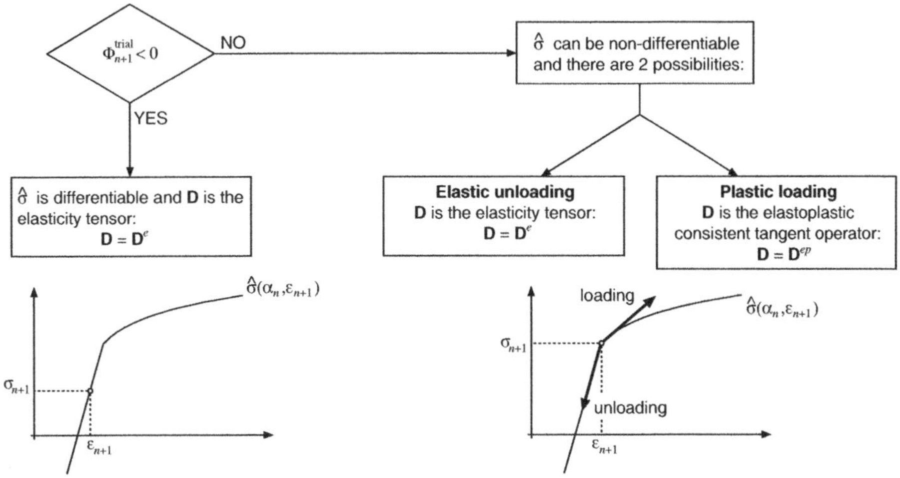

Figure 7.13. The tangent moduli consistent with elastic predictor/return-mapping integration algorithms.

to $\Phi^{\rm trial} < 0$ in (7.93), (7.98), any infinitesimal change of total strain can only be elastic, with the stress $\boldsymbol{\sigma}_{n+1}$ evolving along the (smooth) elastic curve (see graphical representation of the uniaxial stress case in Figure 7.13). In this case the function $\hat{\boldsymbol{\sigma}}$ is differentiable. At states with $\Phi^{\rm trial} > 0$, the function $\hat{\boldsymbol{\sigma}}$ is also differentiable if the hardening curve is smooth. Infinitesimal changes of $\boldsymbol{\varepsilon}_{n+1}$ will move the stress along the elastoplastic part of the incremental stress–strain curve. However, at states with $\Phi^{\rm trial} = 0$ in (7.93, 7.98) – where the Heaviside step function is non-differentiable – either elastic unloading or plastic straining may occur in the incremental constitutive law. The incremental constitutive function is obviously *non-differentiable* in this case. The tangent modulus D is not uniquely defined and two tangent stress–strain relations exist: an *elastic* tangent relation, defined for elastic unloading, and an *elastoplastic* tangent relation, defined for plastic loading. Consider the one-dimensional case illustrated in Figure 7.13. Even though $\hat{\boldsymbol{\sigma}}$ is non-differentiable its two *one-sided derivatives* – the *elastic* and the *elastoplastic* tangents – are well defined. In the multidimensional case, these are generalised respectively as the elastic tangent modulus, D^e, and the elastoplastic consistent tangent modulus, D^{ep}. The elastic tangent is associated with the elastic predictor procedure whereas the elastoplastic tangent is related to the plastic corrector (return-mapping) procedure. Clearly, when assembling the tangent stiffness matrix required by the Newton–Raphson iterative procedure for the global incremental equilibrium problem, the appropriate choice of tangent operator must be made.

The elastic tangent

If the stress is inside the elastic domain ($\Phi^{\rm trial} < 0$) or if it is on the yield surface ($\Phi^{\rm trial} = 0$) and elastic unloading is assumed to occur, the tangent modulus D consistent with any of the integration algorithms previously discussed is simply given by

$$\mathsf{D} = \mathsf{D}^e \equiv \bar{\rho}\, \frac{\partial^2 \psi}{\partial \boldsymbol{\varepsilon}^{e\,2}}, \tag{7.112}$$

i.e. it is the standard *elastic* modulus. Note that, in this case, the stress $\boldsymbol{\sigma}_{n+1}$ is the outcome of the *elastic predictor*, which employs only the elastic constitutive law. This procedure is common to all algorithms described in Section 7.2.

For elastoplastic materials whose elastic response is linear, such as the von Mises model and all other models described in Chapter 6, the *elastic consistent tangent* is the standard elasticity operator (7.107).

The elastoplastic tangent: the derivative of an implicit function

If the stress is on the yield surface, i.e. $\Phi^{\text{trial}} > 0$ or $\Phi^{\text{trial}} = 0$ and it is assumed that further plastic loading is going to occur, then the tangent operator is called the *elastoplastic consistent tangent* and is denoted D^{ep}. It is important to emphasise here that elastoplastic consistent tangent operators cannot be derived for the cutting-plane algorithm (refer to Remark 7.6, page 208) so that the discussion that follows is restricted to the families of elastic predictor/return-mapping procedures based on the generalised midpoint and trapezoidal algorithms.

Crucial to the derivation of the elastoplastic consistent tangent moduli is the observation that under plastic yielding, the outcome $\boldsymbol{\sigma}_{n+1}$ of any member of the families of generalised midpoint and trapezoidal integration algorithms is the solution of a nonlinear system of algebraic equations in the plastic corrector (return-mapping) procedure. The system solved in the return mapping depends on the particular algorithm adopted. In this case, it is clear that $\boldsymbol{\sigma}_{n+1}$ is defined *implicitly* through the corresponding nonlinear system. Note, for example, that the term $\Delta\gamma$ in (7.93) is an implicit function of $\boldsymbol{\varepsilon}^{e\ \text{trial}}_{n+1}$ (or $\boldsymbol{\varepsilon}_{n+1}$) defined as the solution of an algebraic nonlinear equation. In other words, the updated stress $\boldsymbol{\sigma}_{n+1}$ is an *implicit function* of the elastic trial strain $\boldsymbol{\varepsilon}^{e\ \text{trial}}_{n+1}$ in this case. Thus, the consistent tangent operator

$$\mathsf{D}^{ep} = \frac{\partial \hat{\boldsymbol{\sigma}}}{\partial \boldsymbol{\varepsilon}^{e\ \text{trial}}_{n+1}}$$

is simply the *derivative of the implicit function defined by the return-mapping equations* and is derived by following the standard procedure for differentiation of implicit functions.

As an illustration of the above concepts, the elastoplastic tangent consistent with the fully implicit algorithm for the von Mises model is derived below.

7.4.2. THE ELASTOPLASTIC CONSISTENT TANGENT FOR THE VON MISES MODEL WITH ISOTROPIC HARDENING

The implicit elastic predictor/return-mapping algorithm for the von Mises model has been described in detail in Section 7.3. There, it was remarked that from the computational point of view the implementation of the von Mises model is the simplest described in this book. In this section, the elastoplastic tangent operator consistent with the von Mises implicit return mapping is derived step by step. The idea is to use this algorithm as an example to provide the reader with a clear picture of the procedure for the derivation of elastoplastic consistent tangent operators. The simplicity of this particular return-mapping scheme avoids the complications associated with more complex models/algorithms. The derivation of elastoplastic tangent operators consistent with the implicit return mapping for *general* plasticity models is addressed later, in Section 7.4.4. The application of the generic procedure to the isotropically hardening von Mises model, which offers an alternative route to the derivation presented below, is described in Section 7.4.5.

The incremental algorithmic constitutive function for the implicitly integrated von Mises model with nonlinear isotropic strain hardening is given by (7.93). Under plastic flow, i.e. when the return-mapping procedure is used, the update formula for $\boldsymbol{\sigma}_{n+1}$ reads

$$\boldsymbol{\sigma}_{n+1} = \left[\mathsf{D}^e - \frac{\Delta\gamma\, 6G^2}{q_{n+1}^{\text{trial}}}\,\mathsf{I}_\text{d}\right] : \boldsymbol{\varepsilon}_{n+1}^{e\ \text{trial}}, \tag{7.113}$$

where $\Delta\gamma$ is the solution of the return-mapping equation of the algorithm (Box 7.3),

$$\tilde{\Phi}(\Delta\gamma) \equiv q_{n+1}^{\text{trial}} - 3G\,\Delta\gamma - \sigma_y(\bar{\varepsilon}_n^p + \Delta\gamma) = 0. \tag{7.114}$$

In the above, the elastic trial von Mises effective stress, q_{n+1}^{trial}, is the function of the elastic trial strain defined by (7.95). The elastoplastic consistent tangent modulus for the present model/algorithm combination is obtained by differentiating (7.113).

A straightforward application of tensor differentiation rules to (7.113) gives

$$\begin{aligned}\frac{\partial \boldsymbol{\sigma}_{n+1}}{\partial \boldsymbol{\varepsilon}_{n+1}^{e\ \text{trial}}} &= \mathsf{D}^e - \frac{\Delta\gamma\, 6G^2}{q_{n+1}^{\text{trial}}}\,\mathsf{I}_\text{d} - \frac{6G^2}{q_{n+1}^{\text{trial}}}\,\boldsymbol{\varepsilon}_{\text{d}\,n+1}^{e\ \text{trial}} \otimes \frac{\partial \Delta\gamma}{\partial \boldsymbol{\varepsilon}_{n+1}^{e\ \text{trial}}}\\ &\quad + \frac{\Delta\gamma\, 6G^2}{(q_{n+1}^{\text{trial}})^2}\,\boldsymbol{\varepsilon}_{\text{d}\,n+1}^{e\ \text{trial}} \otimes \frac{\partial q_{n+1}^{\text{trial}}}{\partial \boldsymbol{\varepsilon}_{n+1}^{e\ \text{trial}}}.\end{aligned} \tag{7.115}$$

From (7.95) and relation (2.139) (page 36), for the tensor norm derivative, we obtain

$$\frac{\partial q_{n+1}^{\text{trial}}}{\partial \boldsymbol{\varepsilon}_{n+1}^{e\ \text{trial}}} = 2G\sqrt{\tfrac{3}{2}}\,\bar{\boldsymbol{N}}_{n+1}, \tag{7.116}$$

where we have conveniently defined the *unit* flow vector

$$\bar{\boldsymbol{N}}_{n+1} \equiv \sqrt{\tfrac{2}{3}}\,\boldsymbol{N}_{n+1} = \frac{\boldsymbol{s}_{n+1}^{\text{trial}}}{\|\boldsymbol{s}_{n+1}^{\text{trial}}\|} = \frac{\boldsymbol{\varepsilon}_{\text{d}\,n+1}^{e\ \text{trial}}}{\|\boldsymbol{\varepsilon}_{\text{d}\,n+1}^{e\ \text{trial}}\|} \tag{7.117}$$

and we have made use of the trivial identity: $\boldsymbol{\varepsilon}_{\text{d}\,n+1}^{e\ \text{trial}} : \mathsf{I}_\text{d} = \boldsymbol{\varepsilon}_{\text{d}\,n+1}^{e\ \text{trial}}$, when applying the chain rule. Further, the differentiation of the implicit equation (7.114) for $\Delta\gamma$, taking (7.116) into account gives

$$\begin{aligned}\frac{\partial \Delta\gamma}{\partial \boldsymbol{\varepsilon}_{n+1}^{e\ \text{trial}}} &= \frac{1}{3G+H}\,\frac{\partial q_{n+1}^{\text{trial}}}{\partial \boldsymbol{\varepsilon}_{n+1}^{e\ \text{trial}}}\\ &= \frac{2G}{3G+H}\sqrt{\tfrac{3}{2}}\,\bar{\boldsymbol{N}}_{n+1},\end{aligned} \tag{7.118}$$

where H is the slope of the hardening curve:

$$H \equiv \left.\frac{\mathrm{d}\sigma_y}{\mathrm{d}\bar{\varepsilon}^p}\right|_{\bar{\varepsilon}_n^p + \Delta\gamma}. \tag{7.119}$$

Finally, by substituting (7.116) and (7.118) into (7.115), we obtain, after a straightforward manipulation making use of (7.95) and definition (7.117), the following expression for the elastoplastic tangent operator consistent with the implicit return-mapping scheme for the isotropically hardening von Mises model:

$$\begin{aligned}\mathsf{D}^{ep} &= \mathsf{D}^e - \frac{\Delta\gamma\, 6G^2}{q^{\text{trial}}_{n+1}}\,\mathsf{I}_{\text{d}} + 6G^2\bigg(\frac{\Delta\gamma}{q^{\text{trial}}_{n+1}} - \frac{1}{3G+H}\bigg)\,\bar{\boldsymbol{N}}_{n+1}\otimes\bar{\boldsymbol{N}}_{n+1}\\ &= 2G\bigg(1-\frac{\Delta\gamma\, 3G}{q^{\text{trial}}_{n+1}}\bigg)\,\mathsf{I}_{\text{d}}\\ &\quad + 6G^2\bigg(\frac{\Delta\gamma}{q^{\text{trial}}_{n+1}} - \frac{1}{3G+H}\bigg)\,\bar{\boldsymbol{N}}_{n+1}\otimes\bar{\boldsymbol{N}}_{n+1} + K\,\boldsymbol{I}\otimes\boldsymbol{I}.\end{aligned} \tag{7.120}$$

It should be noted that the operator D^{ep} in the present case, i.e. for this particular model and numerical integration algorithm is *symmetric*. The symmetry of consistent elastoplastic tangent operators will be further commented upon in Section 7.4.6. In the `HYPLAS` program, the above tangent operator is computed in subroutine `CTVM`. Its implementation is described in detail in Section 7.4.3.

Remark 7.8. Within the global (equilibrium) Newton–Raphson scheme, the value of $\Delta\gamma$, q^{trial} and H, as well as the incremental unit flow vector, $\bar{\boldsymbol{N}}_{n+1}$, that take part in (7.120) are those obtained for the Gauss point of interest in the return-mapping procedure of the *previous global iteration*. For the first iteration of any global load increment, $\Delta\gamma$ is zero.

The continuum tangent operator

The concept of tangent operators in plasticity has been initially discussed in Sections 6.2.8 and 6.3.8 (from pages 147 and 153, respectively) in the time-continuum setting. In Section 6.3.8, the corresponding elastoplastic *continuum* tangent operator has been derived for the generic multi-dimensional plasticity model. Its closed form is given by expression (6.67). Let us now particularise this formula for the von Mises model with isotropic strain hardening. Firstly we consider (6.187)–(6.192). In this case we have

$$\bar{\rho}\frac{\partial^2\psi^p}{\partial\boldsymbol{\alpha}^2} = \bar{\rho}\frac{\partial^2\psi^p}{\partial\bar{\varepsilon}^{p\,2}} = \frac{\partial\kappa}{\partial\bar{\varepsilon}^p} = H. \tag{7.121}$$

With the above, together with (6.194) and the associative flow vector definition (6.136) for the von Mises model, we find that expression (6.67) particularises in the following format:

$$\mathsf{D}^{ep}_{\text{c}} = \mathsf{D}^e - \frac{(\mathsf{D}^e:\boldsymbol{N})\otimes(\mathsf{D}^e:\boldsymbol{N})}{\boldsymbol{N}:\mathsf{D}^e:\boldsymbol{N}+H}, \tag{7.122}$$

where we have used the subscript 'c' to emphasise that the above operator is the *continuum* tangent modulus. With D^e defined by (7.107), and taking into consideration the fact that for the von Mises model $\boldsymbol{N}$ is a deviatoric tensor, we have

$$\mathsf{D}^e:\boldsymbol{N} = 2G\,\boldsymbol{N}. \tag{7.123}$$

In addition, using (6.136), it follows that

$$\boldsymbol{N}:\mathsf{D}^e:\boldsymbol{N} = 3G. \tag{7.124}$$

By introducing these results into (7.122), we obtain the following explicit expression for the continuum tangent operator for the von Mises model with isotropic strain hardening:

$$\mathbf{D}^{ep}_{\mathrm{c}} = \mathbf{D}^{e} - \frac{6G^2}{3G+H}\,\bar{\boldsymbol{N}} \otimes \bar{\boldsymbol{N}}, \tag{7.125}$$

where $\bar{\boldsymbol{N}}$ is the unit flow vector at the current state.

Remark 7.9. The difference between the elastoplastic consistent tangent operator and its continuum counterpart above lies only in the terms that contain $\Delta\gamma$ in expression (7.120). Note that we may write

$$\mathbf{D}^{ep} = \mathbf{D}^{ep}_{\mathrm{c}} - \frac{\Delta\gamma\, 6G^2}{q^{\mathrm{trial}}_{n+1}}\,[\,\mathbf{I}_d - \bar{\boldsymbol{N}}_{n+1} \otimes \bar{\boldsymbol{N}}_{n+1}]. \tag{7.126}$$

If $\Delta\gamma$ is set to zero (as in the first iteration of any load increment), the continuum tangent is recovered. This fact (Ortiz and Martin, 1989) is a mere consequence of the consistency of the numerical method (backward Euler-based in the present case) adopted in the discretisation of the time-continuum elastoplasticity equations. For large steps, when the value of $\Delta\gamma$ is large, the difference between the continuum and the consistent operator can be substantial. In such cases, the use of the continuum tangent in the assemblage of the stiffness matrix results in a dramatic degradation of the convergence rate of the global iterative procedure. Clearly, if the continuum tangent is used in conjunction with the return-mapping scheme, then the global iterative procedure is *not* the Newton–Raphson algorithm. In this case, the global iterations are a form of approximation to the Newton–Raphson scheme. Early implicit elastoplastic implementations (Owen and Hinton, 1980) relied exclusively on the use of the continuum tangent operator. The use of the consistent tangent operator in this context was introduced by Simo and Taylor (1985).

7.4.3. SUBROUTINE CTVM

This section describes subroutine CTVM (**C**onsistent **T**angent operator for the **V**on **M**ises model) in detail. In the HYPLAS program, this routine is called by the material interface MATICT at each Gauss point during the evaluation of the element tangent stiffness matrix. It computes either the elastic tangent, $\mathbf{D}^e$, given by (7.107), or the elastoplastic tangent operator, $\mathbf{D}^{ep}$, given by expression (7.120). The elastoplastic operator is consistent with the implicit elastic predictor/return-mapping algorithm for the von Mises model with nonlinear (piecewise linear) isotropic hardening. The algorithm is implemented in subroutine SUVM.

The FORTRAN source code of CTVM is listed below.

```
      SUBROUTINE CTVM
     1(   DGAMA      ,DMATX      ,EPFLAG     ,IPROPS     ,NTYPE      ,
     2    RPROPS     ,RSTAVA     ,STRES      )
      IMPLICIT DOUBLE PRECISION (A-H,O-Z)
      PARAMETER(IPHARD=4   ,MSTRE=4)
      LOGICAL EPFLAG
      DIMENSION
     1    DMATX(MSTRE,MSTRE),IPROPS(*)           ,RPROPS(*)           ,
     2    RSTAVA(MSTRE+1)    ,STRES(MSTRE)
      DIMENSION
```

```
     1     DEVPRJ(MSTRE,MSTRE),FOID(MSTRE,MSTRE)   ,S(MSTRE)          ,
     2     SOID(MSTRE)
      DATA
     1     FOID(1,1),FOID(1,2),FOID(1,3),FOID(1,4)/
     2     1.0D0    ,0.0D0    ,0.0D0    ,0.0D0    /
     3     FOID(2,1),FOID(2,2),FOID(2,3),FOID(2,4)/
     4     0.0D0    ,1.0D0    ,0.0D0    ,0.0D0    /
     5     FOID(3,1),FOID(3,2),FOID(3,3),FOID(3,4)/
     6     0.0D0    ,0.0D0    ,0.5D0    ,0.0D0    /
     7     FOID(4,1),FOID(4,2),FOID(4,3),FOID(4,4)/
     8     0.0D0    ,0.0D0    ,0.0D0    ,1.0D0    /
      DATA
     1     SOID(1)  ,SOID(2)  ,SOID(3)  ,SOID(4)  /
     2     1.0D0    ,1.0D0    ,0.0D0    ,1.0D0    /
      DATA
     1     R1   ,R2   ,R3   ,R6   /
     2     1.0D0,2.0D0,3.0D0,6.0D0/
C***********************************************************************
C COMPUTATION OF THE CONSISTENT TANGENT MODULUS FOR VON MISES TYPE
C ELASTO-PLASTIC MATERIAL WITH PIECE-WISE LINEAR ISOTROPIC HARDENING.
C PLANE STRAIN AND AXISYMMETRIC IMPLEMENTATIONS.
C***********************************************************************
C Stops program if neither plane strain nor axisymmetric state
      IF(NTYPE.NE.2.AND.NTYPE.NE.3)CALL ERRPRT('EI0030')
C Current accumulated plastic strain
      EPBAR=RSTAVA(MSTRE+1)
C Set material properties
      YOUNG=RPROPS(2)
      POISS=RPROPS(3)
      NHARD=IPROPS(3)
C Shear and bulk moduli
      GMODU=YOUNG/(R2*(R1+POISS))
      BULK=YOUNG/(R3*(R1-R2*POISS))
      R2G=R2*GMODU
      R1D3=R1/R3
C Set deviatoric projection tensor
      IF(NTYPE.EQ.2)THEN
        NSTRE=3
      ELSEIF(NTYPE.EQ.3)THEN
        NSTRE=4
      ENDIF
      DO 20 I=1,NSTRE
        DO 10 J=1,NSTRE
          DEVPRJ(I,J)=FOID(I,J)-SOID(I)*SOID(J)*R1D3
   10   CONTINUE
   20 CONTINUE
      IF(EPFLAG)THEN
C Compute elastoplastic consistent tangent
C -----------------------------------------
        R3G=R3*GMODU
        ROO3D2=SQRT(R3/R2)
C Hydrostatic pressure
        P=(STRES(1)+STRES(2)+STRES(4))*R1D3
C Deviatoric stress components
        S(1)=STRES(1)-P
        S(2)=STRES(2)-P
        S(3)=STRES(3)
        S(4)=STRES(4)-P
C Recover last elastic trial von Mises effective stress
        SNORM=SQRT(S(1)*S(1)+S(2)*S(2)+R2*S(3)*S(3)+S(4)*S(4))
        Q=ROO3D2*SNORM
```

```
        QTRIAL=Q+R3G*DGAMA
C Assemble elastoplastic tangent (upper triangle only)
        AFACT=R2G*(R1-R3G*DGAMA/QTRIAL)
        BFACT=R6*GMODU*GMODU*(DGAMA/QTRIAL-
     1         R1/(R3G+DPLFUN(EPBAR,NHARD,RPROPS(IPHARD))))/
     2         (SNORM*SNORM)
        DO 40 I=1,NSTRE
          DO 30 J=I,NSTRE
            DMATX(I,J)=AFACT*DEVPRJ(I,J)+BFACT*S(I)*S(J)+
     1                 BULK*SOID(I)*SOID(J)
   30     CONTINUE
   40   CONTINUE
      ELSE
C Compute elasticity matrix (upper triangle only)
C -----------------------------------------------
        DO 60 I=1,NSTRE
          DO 50 J=I,NSTRE
            DMATX(I,J)=R2G*DEVPRJ(I,J)+BULK*SOID(I)*SOID(J)
   50     CONTINUE
   60   CONTINUE
      ENDIF
C Assemble lower triangle
C -----------------------
      DO 80 J=1,NSTRE-1
        DO 70 I=J+1,NSTRE
          DMATX(I,J)=DMATX(J,I)
   70   CONTINUE
   80 CONTINUE
      RETURN
      END
```

The arguments of CTVM

→ DGAMA [$\Delta\gamma$]. The incremental plastic multiplier obtained from the return mapping of SUVM in the previous global (equilibrium) iteration. It is set to zero (before being passed into CTVM) for the first iteration of every load increment.

← DMATX [either $\mathbf{D}^e$ or $\mathbf{D}^{ep}$]. Tangent operator (stored in matrix form) consistent with the implicit elastic predictor/return-mapping algorithm for the von Mises model. The implicit algorithm has been described in Section 7.3.2. The associated elastoplastic consistent tangent, $\mathbf{D}^{ep}$, was derived in Section 7.4.2 above and is given by expression (7.120).

→ EPFLAG. Elastoplastic tangent logical flag. If .TRUE., DMATX returns as the elastoplastic consistent tangent operator, $\mathbf{D}^{ep}$. If .FALSE., DMATX returns as the elastic matrix, $\mathbf{D}^e$. The value of EPFLAG is set in subroutine MATICT. The procedure for setting EPFLAG in MATICT is common to all elastoplastic models of the material class HYPEPL.

→ IPROPS. Array of integer material properties (see description on page 227).

→ NTYPE. Stress state type flag. NTYPE=2 for plane strain and NTYPE=3 for axisymmetric problems.

→ RPROPS. Array of real material properties (see description on page 227).

→ RSTAVA [ε^e, $\bar{\varepsilon}^p$]. Array of updated state variables other than the stress components (output of SUVM) (see page 227 for complete description).

→ STRES [$\boldsymbol{\sigma}_{n+1}$]. Array containing the updated stress tensor components (output of SUVM).

Some local arrays of CTVM

- DEVPRJ [I_d]. Fourth-order deviatoric projection tensor stored in array form. Refer to Appendix D for rules of array conversion.
- FOID [I_S]. Fourth-order symmetric identity tensor stored in array form according to the convention shown in (D.16) (page 762).
- SOID [$\boldsymbol{I}$]. Second-order identity tensor stored in array form.

The names of most local variables and arrays of CTVM follow closely the notation of Section 7.4.2, where the elastoplastic consistent tangent operator has been derived.

Function calls from CTVM

- DPLFUN. Called to compute the slope, H, of the piecewise linear hardening curve (see details on page 228).

7.4.4. THE GENERAL ELASTOPLASTIC CONSISTENT TANGENT OPERATOR FOR IMPLICIT RETURN MAPPINGS

A symbolic expression for the elastoplastic tangent consistent with the implicit return-mapping algorithm for the general plasticity model is derived in this section. The general implicit algorithm for plasticity was derived in Section 7.2.3. The corresponding return-mapping equations, summarised in Box 7.1 [item (iii)], are repeated below for convenience:

$$\left\{\begin{matrix} \boldsymbol{\varepsilon}^e_{n+1} - \boldsymbol{\varepsilon}^{e\,\text{trial}}_{n+1} + \Delta\gamma\, \boldsymbol{N}_{n+1} \\ \boldsymbol{\alpha}_{n+1} - \boldsymbol{\alpha}_n - \Delta\gamma\, \boldsymbol{H}_{n+1} \\ \Phi(\boldsymbol{\sigma}_{n+1}, \boldsymbol{A}_{n+1}) \end{matrix}\right\} = \left\{\begin{matrix} \mathbf{0} \\ \mathbf{0} \\ 0 \end{matrix}\right\}.$$

The basic unknowns of the above system of algebraic equations are: the updated elastic strain, $\boldsymbol{\varepsilon}^e_{n+1}$; the updated set of internal variables, $\boldsymbol{\alpha}_{n+1}$; and the incremental plastic multiplier, $\Delta\gamma$. The corresponding updated stress tensor delivered by the above return mapping is evaluated from the standard potential form

$$\boldsymbol{\sigma}_{n+1} = \bar{\rho}\, \frac{\partial \psi}{\partial \boldsymbol{\varepsilon}^e}\bigg|_{n+1}.$$

Clearly, by changing the elastic trial strain, $\boldsymbol{\varepsilon}^{e\,\text{trial}}_{n+1}$, that takes part in the return map equations, the solution $\{\boldsymbol{\varepsilon}^e_{n+1}, \boldsymbol{\alpha}_{n+1}, \Delta\gamma\}$ will change accordingly and so will the updated stress $\boldsymbol{\sigma}_{n+1}$. In this way, the algebraic system of equations define an implicit function for the stress tensor at t_{n+1}. As discussed in the preceding sections, the elastoplastic consistent tangent operator

$$\mathsf{D}^{ep} \equiv \frac{\partial \boldsymbol{\sigma}_{n+1}}{\partial \boldsymbol{\varepsilon}^{e\ \mathrm{trial}}_{n+1}},$$

i.e. the derivative of the implicit function for stress, gives the linear tangent relationship between $\boldsymbol{\varepsilon}^{e\ \mathrm{trial}}_{n+1}$ and $\boldsymbol{\sigma}_{n+1}$. Thus, the first step in its derivation is to *linearise* the return-mapping equations (having the elastic trial strain – the system input – also as a variable). Straightforward differentiation of the general return-mapping equations yields the following linearised form:

$$\begin{Bmatrix} \mathrm{d}\boldsymbol{\varepsilon}^e + \Delta\gamma \dfrac{\partial \boldsymbol{N}}{\partial \boldsymbol{\sigma}} : \mathrm{d}\boldsymbol{\sigma} + \Delta\gamma \dfrac{\partial \boldsymbol{N}}{\partial \boldsymbol{A}} * \mathrm{d}\boldsymbol{A} + \mathrm{d}\Delta\gamma\, \boldsymbol{N} \\[2ex] \mathrm{d}\boldsymbol{\alpha} - \Delta\gamma \dfrac{\partial \boldsymbol{H}}{\partial \boldsymbol{\sigma}} * \mathrm{d}\boldsymbol{\sigma} - \Delta\gamma \dfrac{\partial \boldsymbol{H}}{\partial \boldsymbol{A}} * \mathrm{d}\boldsymbol{A} - \mathrm{d}\Delta\gamma\, \boldsymbol{H} \\[2ex] \dfrac{\partial \Phi}{\partial \boldsymbol{\sigma}} : \mathrm{d}\boldsymbol{\sigma} + \dfrac{\partial \Phi}{\partial \boldsymbol{A}} * \mathrm{d}\boldsymbol{A} \end{Bmatrix} = \begin{Bmatrix} \mathrm{d}\boldsymbol{\varepsilon}^{e\ \mathrm{trial}} \\[2ex] \boldsymbol{0} \\[2ex] 0 \end{Bmatrix}, \tag{7.127}$$

where the subscripts $n+1$ have been suppressed for convenience.

Recall that, for the differentials $\mathrm{d}\boldsymbol{\sigma}$ and $\mathrm{d}\boldsymbol{A}$, we have the following relations (refer to the rate elastoplasticity equations (7.52) on page 206):

$$\begin{aligned} \mathrm{d}\boldsymbol{\sigma} &= \mathsf{D}^e : \mathrm{d}\boldsymbol{\varepsilon}^e + \mathsf{E} * \mathrm{d}\boldsymbol{\alpha} \\ \mathrm{d}\boldsymbol{A} &= \mathsf{F} * \mathrm{d}\boldsymbol{\varepsilon}^e + \mathsf{G} * \mathrm{d}\boldsymbol{\alpha}. \end{aligned} \tag{7.128}$$

The linear operators E, F and G are defined in (7.48)–(7.50). Inversion of this relation gives the following general expression

$$\begin{aligned} \mathrm{d}\boldsymbol{\varepsilon}^e &= \mathsf{C} : \mathrm{d}\boldsymbol{\sigma} + \mathsf{B} * \mathrm{d}\boldsymbol{A} \\ \mathrm{d}\boldsymbol{\alpha} &= \mathsf{A} * \mathrm{d}\boldsymbol{\sigma} + \mathsf{J} * \mathrm{d}\boldsymbol{A}, \end{aligned} \tag{7.129}$$

where C, B, A and J are suitably defined linear operators. Note that, under the classical assumption of decoupling between elasticity and plastic hardening:

$$\psi(\boldsymbol{\varepsilon}^e, \boldsymbol{\alpha}) = \psi^e(\boldsymbol{\varepsilon}^e) + \psi^p(\boldsymbol{\alpha}), \tag{7.130}$$

the tangent moduli E and F vanish and so do B and A. In this case we end up with the following relation:

$$\begin{aligned} \mathrm{d}\boldsymbol{\varepsilon}^e &= \mathsf{C} : \mathrm{d}\boldsymbol{\sigma}\,; \\ \mathrm{d}\boldsymbol{\alpha} &= \mathsf{J} * \mathrm{d}\boldsymbol{A}, \end{aligned} \tag{7.131}$$

where $\mathsf{C} = [\mathsf{D}^e]^{-1}$ and $\mathsf{J} = \mathsf{G}^{-1}$.

By substituting (7.129) into (7.127), the linearised return-mapping equations can be rewritten in the following symbolic matrix representation:†

$$\begin{bmatrix} \mathsf{C} + \Delta\gamma \dfrac{\partial \boldsymbol{N}}{\partial \boldsymbol{\sigma}} & \mathsf{B} + \Delta\gamma \dfrac{\partial \boldsymbol{N}}{\partial \boldsymbol{A}} & \boldsymbol{N} \\ \mathsf{A} - \Delta\gamma \dfrac{\partial \boldsymbol{H}}{\partial \boldsymbol{\sigma}} & \mathsf{J} - \Delta\gamma \dfrac{\partial \boldsymbol{H}}{\partial \boldsymbol{A}} & -\boldsymbol{H} \\ \dfrac{\partial \Phi}{\partial \boldsymbol{\sigma}} & \dfrac{\partial \Phi}{\partial \boldsymbol{A}} & 0 \end{bmatrix} \begin{bmatrix} \mathrm{d}\boldsymbol{\sigma} \\ \mathrm{d}\boldsymbol{A} \\ \mathrm{d}\Delta\gamma \end{bmatrix} = \begin{bmatrix} \mathrm{d}\boldsymbol{\varepsilon}^{e\ \text{trial}} \\ \mathbf{0} \\ 0 \end{bmatrix}. \tag{7.132}$$

Finally, by inverting the above linear relation, we obtain

$$\begin{bmatrix} \mathrm{d}\boldsymbol{\sigma} \\ \mathrm{d}\boldsymbol{A} \\ \mathrm{d}\Delta\gamma \end{bmatrix} = \begin{bmatrix} \mathsf{D}_{11} & \mathsf{D}_{12} & \boldsymbol{D}_{13} \\ \mathsf{D}_{21} & \mathsf{D}_{22} & \mathsf{D}_{23} \\ \boldsymbol{D}_{31} & \mathsf{D}_{32} & \boldsymbol{D}_{33} \end{bmatrix} \begin{bmatrix} \mathrm{d}\boldsymbol{\varepsilon}^{e\ \text{trial}} \\ \mathbf{0} \\ 0 \end{bmatrix}. \tag{7.133}$$

From the above, we obtain the generalised tangent operators consistent with the implicit return-mapping algorithm. The generalised operators are: The *elastoplastic consistent tangent modulus*,

$$\mathsf{D}_{11} = \frac{\mathrm{d}\boldsymbol{\sigma}_{n+1}}{\mathrm{d}\boldsymbol{\varepsilon}^{e\ \text{trial}}_{n+1}} \equiv \mathsf{D}^{ep}. \tag{7.134}$$

The operator

$$\mathsf{D}_{21} = \frac{\mathrm{d}\boldsymbol{A}_{n+1}}{\mathrm{d}\boldsymbol{\varepsilon}^{e\ \text{trial}}_{n+1}}, \tag{7.135}$$

which gives the tangent relation between increments of elastic trial strain and thermodynamical force set $\boldsymbol{A}_{n+1}$; and

$$\boldsymbol{D}_{31} = \frac{\mathrm{d}\Delta\gamma}{\mathrm{d}\boldsymbol{\varepsilon}^{e\ \text{trial}}_{n+1}}, \tag{7.136}$$

which is the tangent operator (a second-order tensor in this case) relating increments of elastic trial strain and incremental plastic multiplier. Note that the dimension and order of the linear operator D_{21} depends on the definition of set $\boldsymbol{A}$. As far as finite element computations are concerned, only the elastoplastic consistent tangent operator is of relevance.

7.4.5. ILLUSTRATION: THE VON MISES MODEL WITH ISOTROPIC HARDENING

For the sake of generality, no assumptions have been made about the nature of the sets $\boldsymbol{\alpha}$, of internal state variables, and $\boldsymbol{A}$, of associated forces, in the above symbolic derivation.

†Note that each element of the symbolic matrix of expression (7.132) is a generic tensor (or set containing scalars, tensors, etc.) of appropriate order. The product operations between the elements of the matrix and the elements of the 'vector' containing the differentials $\mathrm{d}\boldsymbol{\sigma}$, etc. are the corresponding appropriate products. This symbolic matrix notation will be used frequently in this book to represent linearised systems.

As discussed in Chapter 6, these sets may contain as many variables as necessary for the appropriate description of the material behaviour. Depending on the particular model, each element of these sets may be a scalar, vector or tensor of any order. With such a high degree of generality, it may not be easy to have a clear picture of what operations are involved in the above derivation. In order to provide the reader with a better view of the derivation of the elastoplastic consistent tangent operator, the general procedure described above is here particularised for the implicit return mapping for the von Mises model with isotropic strain hardening.

Firstly, we recall the discussion around expressions (6.187–6.194) (page 182) for the von Mises isotropically hardening model. We then have

$$\frac{\partial \boldsymbol{H}}{\partial \boldsymbol{A}} \equiv \frac{\partial \boldsymbol{H}}{\partial \kappa} = 0; \qquad \frac{\partial \boldsymbol{H}}{\partial \boldsymbol{\sigma}} = \boldsymbol{0}. \tag{7.137}$$

In addition, as elasticity and plastic hardening are decoupled in the von Mises model, the operators B and A vanish, as pointed out in the discussion following expression (7.129). The remaining (non-vanishing) operators are

$$\mathsf{C} = [\mathsf{D}^e]^{-1} \tag{7.138}$$

and

$$\mathsf{J} = \mathsf{G}^{-1} = \left(\bar{\rho}\frac{\partial^2 \psi}{\partial \bar{\varepsilon}^{p\,2}}\right)^{-1} = \left(\frac{\partial \sigma_y}{\partial \bar{\varepsilon}^p}\right)^{-1} = \frac{1}{H}, \tag{7.139}$$

where H is the hardening modulus (slope of the hardening curve). From the definition (6.136) of the associative flow vector for the von Mises model, it also follows that

$$\frac{\partial \boldsymbol{N}}{\partial \boldsymbol{A}} \equiv \frac{\partial \boldsymbol{N}}{\partial \kappa} = \boldsymbol{0}. \tag{7.140}$$

Substitution of the above expressions and (6.194) into (7.132) results in the following linearised form:

$$\begin{bmatrix} [\mathsf{D}^e]^{-1} + \Delta\gamma \dfrac{\partial \boldsymbol{N}}{\partial \boldsymbol{\sigma}} & 0 & \boldsymbol{N} \\ 0 & \dfrac{1}{H} & -1 \\ \boldsymbol{N} & -1 & 0 \end{bmatrix} \begin{bmatrix} \mathrm{d}\boldsymbol{\sigma} \\ \mathrm{d}\sigma_y \\ \mathrm{d}\Delta\gamma \end{bmatrix} = \begin{bmatrix} \mathrm{d}\boldsymbol{\varepsilon}^{e\ \mathrm{trial}} \\ 0 \\ 0 \end{bmatrix}, \tag{7.141}$$

where by taking (6.136) into account, the flow vector derivative is easily obtained as

$$\frac{\partial \boldsymbol{N}}{\partial \boldsymbol{\sigma}} = \frac{\partial^2 \Phi}{\partial \boldsymbol{\sigma}^2} = \frac{3}{2q}(\mathsf{I}_{\mathrm{d}} - \bar{\boldsymbol{N}} \otimes \bar{\boldsymbol{N}}), \tag{7.142}$$

where $\bar{\boldsymbol{N}}$ is the unit flow vector (7.117).

The tangential relation (7.141) can be inverted trivially as follows. We start by observing that the first row of (7.141) is equivalent to

$$\mathrm{d}\boldsymbol{\sigma} = \mathsf{P} : (\mathrm{d}\boldsymbol{\varepsilon}^{e\ \mathrm{trial}} - \mathrm{d}\Delta\gamma\, \boldsymbol{N}), \tag{7.143}$$

where

$$\mathsf{P} \equiv \left(\mathsf{I}_S + \Delta\gamma\, \mathsf{D}^e : \frac{\partial \boldsymbol{N}}{\partial \boldsymbol{\sigma}} \right)^{-1} : \mathsf{D}^e. \tag{7.144}$$

The second row of (7.141), in turn, gives

$$\mathrm{d}\sigma_y = H\, \mathrm{d}\Delta\gamma. \tag{7.145}$$

This, combined with the third row and (7.143), results in

$$\begin{aligned} \mathrm{d}\Delta\gamma &= \frac{1}{H}\, \boldsymbol{N} : \mathrm{d}\boldsymbol{\sigma} \\ &= \frac{1}{H}(\boldsymbol{N} : \mathsf{P} : \mathrm{d}\boldsymbol{\varepsilon}^{e\ \mathrm{trial}} - \mathrm{d}\Delta\gamma\, \boldsymbol{N} : \mathsf{P} : \boldsymbol{N}), \end{aligned} \tag{7.146}$$

or, after rearranging,

$$\mathrm{d}\Delta\gamma = \frac{1}{\boldsymbol{N} : \mathsf{P} : \boldsymbol{N} + H}\, \boldsymbol{N} : \mathsf{P} : \mathrm{d}\boldsymbol{\varepsilon}^{e\ \mathrm{trial}}. \tag{7.147}$$

Finally, by substituting (7.147) into (7.143), we obtain the following formula for the elastoplastic consistent tangent operator

$$\begin{aligned} \mathsf{D}^{ep} &= \mathsf{P} - \frac{1}{\boldsymbol{N} : \mathsf{P} : \boldsymbol{N} + H}\, (\mathsf{P} : \boldsymbol{N}) \otimes (\mathsf{P} : \boldsymbol{N}) \\ &= \mathsf{P} - \frac{1}{\bar{\boldsymbol{N}} : \mathsf{P} : \bar{\boldsymbol{N}} + \frac{2}{3}H}\, (\mathsf{P} : \bar{\boldsymbol{N}}) \otimes (\mathsf{P} : \bar{\boldsymbol{N}}). \end{aligned} \tag{7.148}$$

The above expression for D^{ep} is an equivalent representation to (7.120) and could be used as an alternative in the computational implementation of the model. Also note the similarity between (7.148) and the continuum tangent operator (7.122). As expected (refer to Remark 7.9, page 235), the above formula recovers (7.122) if $\Delta\gamma = 0$.

Alternative formula

Yet another representation, alternative to (7.148), is sometimes used in computational applications. The alternative formula is obtained simply by replacing the first right-hand side of (7.146) into the first row of (7.141) and then inverting the resulting differential relation. The corresponding expression for the elastoplastic consistent tangent reads

$$\mathsf{D}^{ep} = \left\{ [\mathsf{D}^e]^{-1} + \Delta\gamma\, \frac{\partial \boldsymbol{N}}{\partial \boldsymbol{\sigma}} + \frac{1}{H}\, \boldsymbol{N} \otimes \boldsymbol{N} \right\}^{-1}, \tag{7.149}$$

or, in view of (7.142), and making use of the unit flow vector definition,

$$\mathsf{D}^{ep} = \left\{ [\mathsf{D}^e]^{-1} + \frac{3\Delta\gamma}{2q}\, \mathsf{I}_{\mathrm{d}} + \frac{3}{2}\left(\frac{1}{H} - \frac{\Delta\gamma}{q} \right) \bar{\boldsymbol{N}} \otimes \bar{\boldsymbol{N}} \right\}^{-1}. \tag{7.150}$$

The use of an analogous formula is reported, for instance, by Dutko *et al.* (1993) in the context of anisotropic plasticity.

Remark 7.10. The presence of the hardening modulus in the denominator in the above formula does not allow its use in practical computations if $H = 0$ (perfect plasticity).

7.4.6. TANGENT OPERATOR SYMMETRY: INCREMENTAL POTENTIALS

It has been emphasised in Chapter 6 (refer to Remark 6.1, page 153) that the time-continuum elastoplastic tangent modulus with general expression (6.67) is symmetric for models with associative plastic flow rule. This property is not transferred in general to its time-discrete (or consistent) counterpart, whose general symbolic expression was derived in Section 7.4.4. An interesting discussion on this topic is provided by Ortiz and Martin (1989). These authors show, for instance, that for the family of generalised midpoint algorithms based on the return-mapping equations (7.37), operator symmetry is preserved only for the fully implicit member ($\theta = 0$), when applied to perfectly plastic associative flow models or, in the presence of hardening, to *fully associative* models (i.e. models with associative flow *and* hardening rules). If the plastic flow rule is associative but the hardening rule does not derive from the associative relation (6.130), then, even though the continuum tangent is symmetric, its implicit consistent counterpart is not. Note that the von Mises model with isotropic hardening discussed in the preceding sections is fully associative so that its elastoplastic tangent consistent with the fully implicit algorithm given by the equivalent formulae (7.120), (7.148) or (7.150), is symmetric. In a further discussion on this issue, Simo and Govindjee (1991) show that for the variant of the family of midpoint algorithms with consistency equation (7.42), the symmetry of the corresponding consistent tangents is guaranteed for any $\theta \in [0, 1]$, again only for perfectly plastic associative flow models or, in the presence of hardening, for *fully associative* models. For models with associative flow but non-associative hardening, the consistent tangents for this family of algorithms are generally unsymmetric.

Incremental potentials

The symmetry of the consistent tangent operator implies that at any state $\{\boldsymbol{\varepsilon}_n, \boldsymbol{\varepsilon}_n^p, \boldsymbol{\alpha}_n\}$ there exists an *incremental potential*,

$$\psi_n = \tilde{\psi}_n(\boldsymbol{\varepsilon}_{n+1}) = \bar{\psi}_n(\boldsymbol{\varepsilon}^{e\ \mathrm{trial}}_{n+1}) \tag{7.151}$$

analogous to the free-energy function, ψ, such that the updated stress, $\boldsymbol{\sigma}_{n+1}$, delivered by the integration algorithm is given by

$$\boldsymbol{\sigma}_{n+1} = \bar{\rho}\,\frac{\partial \tilde{\psi}_n}{\partial \boldsymbol{\varepsilon}_{n+1}} = \bar{\rho}\,\frac{\partial \bar{\psi}_n}{\partial \boldsymbol{\varepsilon}^{e\ \mathrm{trial}}_{n+1}}. \tag{7.152}$$

Analogously to (7.106), the corresponding (symmetric) consistent tangent operator in such cases obeys the potential relations

$$\mathsf{D}^{ep} \equiv \frac{\partial \boldsymbol{\sigma}_{n+1}}{\partial \boldsymbol{\varepsilon}_{n+1}} = \bar{\rho}\,\frac{\partial^2 \tilde{\psi}_n}{\partial \boldsymbol{\varepsilon}^2_{n+1}} = \frac{\partial \boldsymbol{\sigma}_{n+1}}{\partial \boldsymbol{\varepsilon}^{e\ \mathrm{trial}}_{n+1}} = \bar{\rho}\,\frac{\partial^2 \bar{\psi}_n}{\partial \boldsymbol{\varepsilon}^{e\ \mathrm{trial}\,2}_{n+1}}. \tag{7.153}$$

Remark 7.11. For elastoplastic model/algorithm combinations that preserve the consistent tangent operator symmetry, the incremental equilibrium boundary value problem is endowed with a potential structure analogous to that of hyperelasticity boundary value problems. This allows methods conventionally used in hyperelasticity to be adopted in the study of elasto-plastic incremental boundary value problems. Also, many important properties associated with the regularity and stability of the solution of (time-continuum) elastoplasticity problems depend crucially on the symmetry of the continuum tangent operator. When symmetry is preserved in the consistent tangent operator, many such properties can be transferred to the discretised problem. These issues fall outside the scope of this book. Interested readers are referred to Martin (1975) and Rice (1976) for further details.

Example: the von Mises model with linear strain hardening

Let us now focus on the fully implicit elastic predictor/return-mapping scheme for the von Mises model with linear isotropic strain hardening. This particular combination of elastoplasticity model and constitutive integration algorithm has the following incremental (pseudo-) potential:

$$
\begin{aligned}
\bar{\rho}\psi_n = {} & \frac{1}{2}\,(\mathsf{D}^e : \boldsymbol{\varepsilon}^{e\ \mathrm{trial}}_{n+1}) : \boldsymbol{\varepsilon}^{e\ \mathrm{trial}}_{n+1} \\
& - \hat{H}(\Phi^{\mathrm{trial}}_{n+1})\,\frac{6G^2}{3G+H}\left(\frac{1}{2}\|\boldsymbol{\varepsilon}^{e\ \mathrm{trial}}_{d\ n+1}\|^2 - \frac{\sigma_y}{2G\sqrt{3/2}}\|\boldsymbol{\varepsilon}^{e\ \mathrm{trial}}_{d\ n+1}\|\right).
\end{aligned}
\tag{7.154}
$$

That the incremental constitutive function (7.103) is indeed obtained from the potential relation (7.152) with the above potential can be established by a straightforward application of tensor differentiation rules taking expression (2.139) (page 36), into account together with the identity $\boldsymbol{\varepsilon}^{e\ \mathrm{trial}}_{\mathrm{d}\ n+1} = \mathsf{I}_{\mathrm{d}} : \boldsymbol{\varepsilon}^{e\ \mathrm{trial}}_{n+1}$.

7.5. Numerical examples with the von Mises model

So far, this chapter has presented a general framework for the numerical treatment of elastoplasticity within the context of implicit incremental finite element procedures. Firstly, a general theory of elastic predictor/return-mapping integration algorithms for elastoplasticity has been presented followed by its specialisation (including a detailed description of the numerical implementation) to the fully implicit scheme for the von Mises model with nonlinear isotropic hardening. Subsequently, the concept of consistent tangent operators in elastoplasticity has been introduced and also specialised, with a complete description of the tangent operator consistent with the implicit algorithm, for the same model.

The practical application of these numerical procedures is illustrated in this section by a comprehensive set of benchmark numerical examples. The examples presented here are restricted to the von Mises model with isotropic hardening, whose details of numerical implementation have been given in the previous sections. All finite element analyses have been carried out in the HYPLAS program. Most results are compared with existing analytical solutions. In all examples, the full Newton–Raphson algorithm, resulting from the use of the consistent tangent operator described in Sections 7.4.2 and 7.4.3, has been adopted.

7.5.1. INTERNALLY PRESSURISED CYLINDER

This example considers the simulation of the behaviour of a long metallic thick-walled cylinder subjected to internal pressure. The geometry of the problem, the material properties and the adopted finite element mesh is illustrated in Figure 7.14. The standard eight-noded quadrilateral element, with four (2×2) Gauss integration points is adopted. The analysis

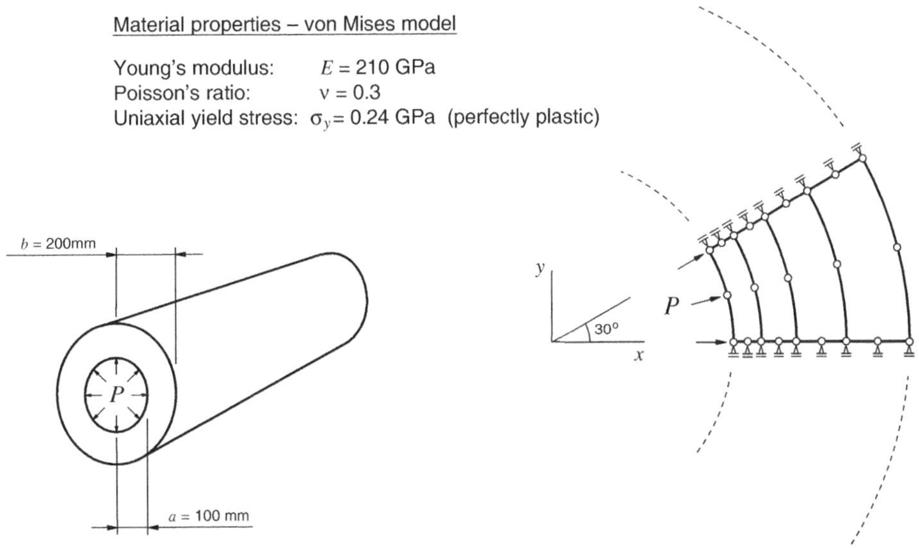

Figure 7.14. Internally pressurised cylinder. Geometry, material properties and finite element mesh.

is carried out assuming plane strain conditions. Due to symmetry, only a 30° segment of the whole cylinder cross-section is discretised with the appropriate symmetry displacement constraints imposed on the edge nodes. The pressure, P, prescribed on the inner surface, is increased gradually until a collapse (limit) load is reached. For the present problem (see Figure 7.15), plastic yielding starts at the inner surface (with radial coordinate $r = a$) and develops gradually, in the form of a cylindrical plastic front (with radius c), toward the outer face of the cylinder ($r = b$). Collapse occurs when the plastic front reaches the outer face ($c = b$) and the entire cylinder becomes plastified. At the limit load, the cylinder can expand indefinitely without further increase in the applied pressure. A closed-form solution to this problem has been derived by Hill (1950). It relates the applied pressure to the radius c of the plastic front by means of the expression

$$\frac{P}{Y} = \ln\left(\frac{c}{a}\right) + \frac{1}{2}\left(1 - \frac{c^2}{b^2}\right), \tag{7.155}$$

where, for the von Mises model, $Y = \sigma_y/\sqrt{3}$. Plastic yielding begins when $c = a$, which corresponds to the yielding pressure

$$P_0 = \frac{Y}{2}\left(1 - \frac{a^2}{b^2}\right). \tag{7.156}$$

Before plastic yielding starts ($P < P_0$), the radial displacement, u_b, of the outer surface is a linear function of P, given by

$$u_b = \frac{2Pb}{E(b^2/a^2 - 1)}(1 - \nu^2), \quad P < P_0. \tag{7.157}$$

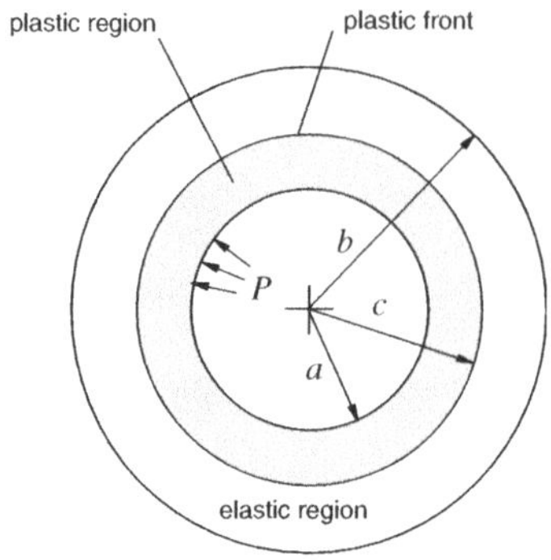

Figure 7.15. Internally pressurised cylinder. Partly plastified cross-section.

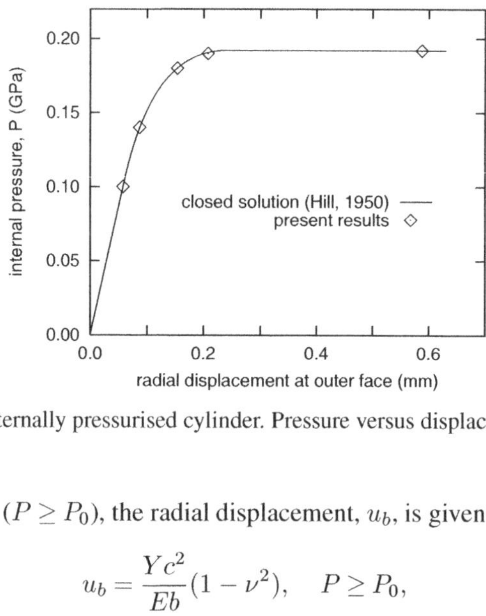

Figure 7.16. Internally pressurised cylinder. Pressure versus displacement diagram.

Under plastic yielding ($P \geq P_0$), the radial displacement, u_b, is given by the formula

$$u_b = \frac{Yc^2}{Eb}(1 - \nu^2), \quad P \geq P_0, \tag{7.158}$$

where c can be evaluated as an implicit function of P through (7.155). A diagram showing the applied pressure versus the radial displacement at the outer face of the cylinder obtained in the finite element simulation is plotted in Figure 7.16 together with Hill's closed-form solution. The high accuracy of the finite element results is clear. The limit load is reached when $c = b$. It then follows from (7.155) that the limit pressure is

$$P_{\text{lim}} = \frac{2\sigma_y}{\sqrt{3}} \ln(b/a), \tag{7.159}$$

which for the present dimensions and material parameters gives

$$P_{\text{lim}} \approx 0.19209\text{GPa}. \tag{7.160}$$

In the finite element solution, the limit load is assumed to have been reached when equilibrium can no longer be obtained (global iterations do not converge) with a reasonably small

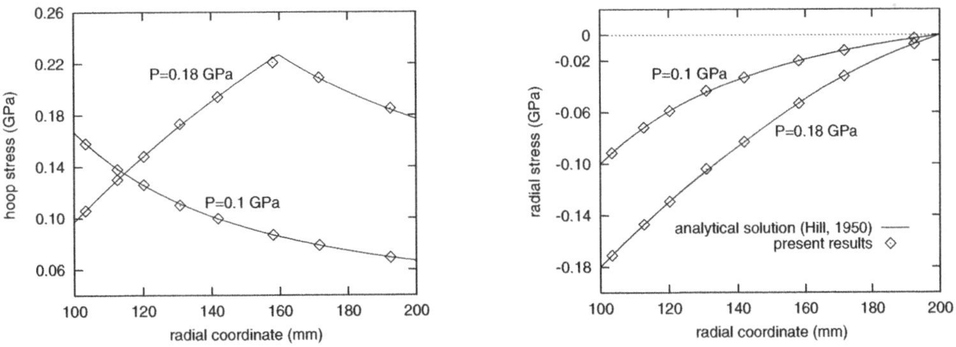

Figure 7.17. Internally pressurised cylinder. Hoop and radial stress distributions at different levels of applied internal pressure. Finite element results are computed at Gauss integration points.

load increment. The limit pressure obtained in the present simulation is about 0.19209 GPa, which is virtually identical to Hill's closed-form solution. The hoop and radial stresses obtained at the Gauss points are plotted in Figure 7.17 together with Hill's solution for $P = 0.1$ GPa and $P = 0.18$ GPa. In the plastic region, where the radial coordinate r satisfies $a \le r \le c$, the radial and hoop stresses obtained by Hill are given, respectively, by

$$\sigma_r = Y\left[-\frac{1}{2} - \ln\left(\frac{c}{r}\right) + \frac{c^2}{2b^2}\right], \quad \sigma_\theta = Y\left[\frac{1}{2} - \ln\left(\frac{c}{r}\right) + \frac{c^2}{2b^2}\right]. \tag{7.161}$$

In the elastic region, $c \le r \le b$, the stresses are

$$\sigma_r = -\frac{Yc^2}{2b^2}\left(\frac{b^2}{r^2} - 1\right), \quad \sigma_\theta = \frac{Yc^2}{2b^2}\left(\frac{b^2}{r^2} + 1\right). \tag{7.162}$$

The finite element results at the Gauss points are also in very close agreement with Hill's solution. It is worth mentioning that at the pressure level $P = 0.1$ GPa the entire cylinder is still elastic. At the level $P = 0.18$ GPa the plastic front has already progressed considerably and $c \approx 159.79$. The transition between the elastic and plastic zones is clearly marked by the drastic change in the slope of the curve for the hoop stress shown in Figure 7.17.

7.5.2. INTERNALLY PRESSURISED SPHERICAL SHELL

An axisymmetric finite element analysis of an internally pressurised thick-walled spherical shell is carried out in this example. The geometry of the problem is illustrated in Figure 7.18. The material parameters and the finite element mesh adopted are identical to those of Example 7.5.1. As in the previous example, plastic yielding starts at the inner surface and propagates through the shell as the internal pressure is increased. Collapse takes place when the (spherical) plastic front reaches the outer boundary and the entire shell is under plastic flow. An analytical solution to this problem has also been derived by Hill (1950). Adopting the notation of Example 7.5.1, the analytical radial displacement of the outer face of the shell under plastic yielding ($P \ge P_0$) is

$$u_b = \frac{\sigma_y\, c^3}{Eb^2}(1 - \nu), \quad P \ge P_0, \tag{7.163}$$

where the radius of the plastic front, c, is associated with the applied pressure, P, through the expression

$$P = 2\sigma_y \ln\left(\frac{c}{a}\right) + \frac{2\sigma_y}{3}\left(1 - \frac{c^3}{b^3}\right), \tag{7.164}$$

and the yielding pressure, P_0, is

$$P_0 = \frac{2\sigma_y}{3}\left(1 - \frac{a^3}{b^3}\right). \tag{7.165}$$

When the whole shell is elastic, i.e. when $P < P_0$, the displacement u_b is given by

$$u_b = \frac{3Pb}{2E(b^3/a^3 - 1)}(1 - \nu), \quad P < P_0. \tag{7.166}$$

The displacement u_b obtained in the finite element solution is plotted against the applied pressure in Figure 7.18. The analytical curve is also plotted for comparison. The analytical limit load, corresponding to $c = b$, is obtained as

$$P_{\lim} = 2\sigma_y \ln(b/a), \tag{7.167}$$

which, for the present problem gives

$$P_{\lim} \approx 0.33271 \text{ GPa}. \tag{7.168}$$

Again, in the numerical solution, it is assumed that the limit load has been achieved when convergence for sufficiently small load increments cannot be obtained. The limit pressure obtained in the present analysis is approximately 0.33769 GPa, which is nearly identical to the exact value. The hoop and radial stresses obtained at the Gauss points in the finite element simulation are plotted in Figure 7.19 together with the corresponding analytical solution. The agreement between numerical and exact results is excellent. The analytical expressions for the hoop and radial stresses are

$$\sigma_r = -2Y\left[\ln\left(\frac{c}{r}\right) + \frac{1}{3}\left(1 - \frac{c^3}{2b^3}\right)\right], \quad \sigma_\theta = 2Y\left[\frac{1}{2} - \ln\left(\frac{c}{r}\right) - \frac{1}{3}\left(1 - \frac{c^3}{2b^3}\right)\right], \tag{7.169}$$

in the plastic region ($a \le r \le c$). In the elastic region ($c \le r \le b$), the stresses are

$$\sigma_r = -\frac{2\sigma_y c^3}{3b^3}\left(\frac{b^3}{r^3} - 1\right), \quad \sigma_\theta = \frac{2\sigma_y c^3}{3b^3}\left(\frac{b^3}{2r^3} + 1\right). \tag{7.170}$$

Residual stresses

After complete unloading from a partly plastic state, the shell will be subjected to a field of (self-equilibrated) *residual stresses*. The residual stress field produces an increase in the yield pressure so that pre-loading can be applied to strengthen the shell. For a shell that has been monotonically loaded to a pressure level P and then completely unloaded without the occurrence of reverse plastic flow (i.e. the unloading process is purely elastic), the analytical

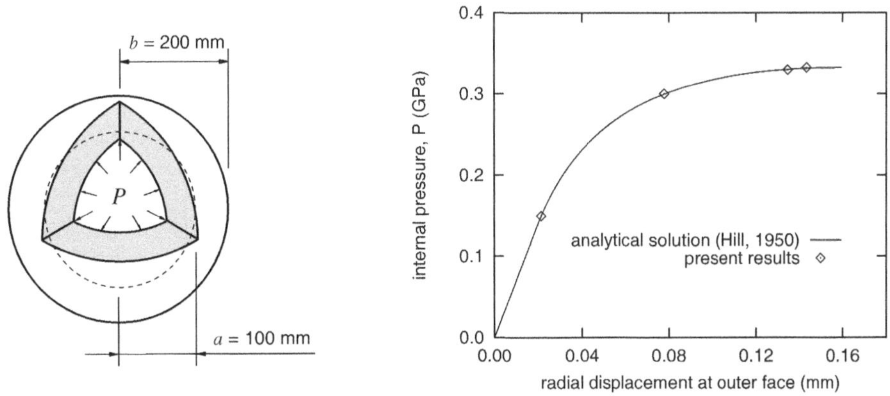

Figure 7.18. Internally pressurised spherical shell. Geometry and pressure versus displacement curve.

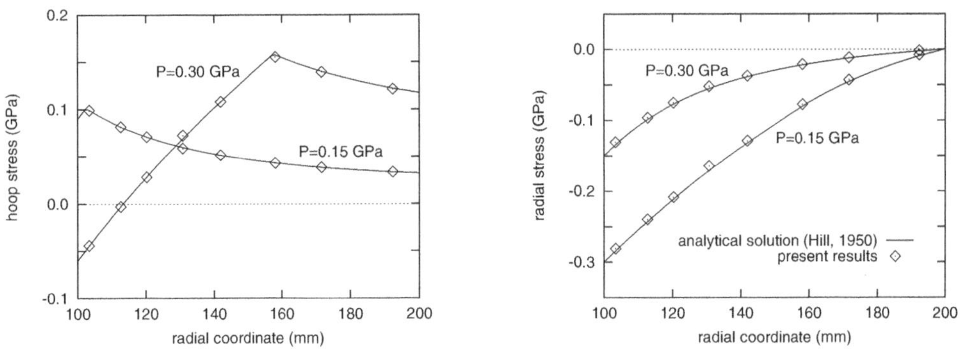

Figure 7.19. Internally pressurised spherical shell. Hoop and radial stress distributions at different levels of applied internal pressure. Finite element results are computed at Gauss integration points.

distribution of residual hoop and radial stresses along the sphere radius are given by

$$\left.\begin{aligned}\sigma_r^R &= -\frac{2\sigma_y}{3}\left(\frac{c^3}{a^3}-\frac{P}{P_0}\right)\left(\frac{a^3}{r^3}-\frac{a^3}{b^3}\right)\\ \sigma_\theta^R &= \frac{2\sigma_y}{3}\left(\frac{c^3}{a^3}-\frac{P}{P_0}\right)\left(\frac{a^3}{2r^3}+\frac{a^3}{b^3}\right)\end{aligned}\right\}, \quad \text{for } c \le r \le b, \tag{7.171}$$

and

$$\left.\begin{aligned}\sigma_r^R &= -\frac{2\sigma_y}{3}\left[\frac{P}{P_0}\left(1-\frac{a^3}{r^3}\right)-3\ln\left(\frac{r}{a}\right)\right]\\ \sigma_\theta^R &= -\frac{2\sigma_y}{3}\left[\frac{3}{2}-\frac{P}{P_0}\left(1+\frac{a^3}{2r^3}\right)+3\ln\left(\frac{r}{a}\right)\right]\end{aligned}\right\}, \quad \text{for } a \le r \le c. \tag{7.172}$$

In the above expressions, c is the radius of the plastic region at the maximum pressure level attained before unloading. The maximum pressure that can be applied without the occurrence

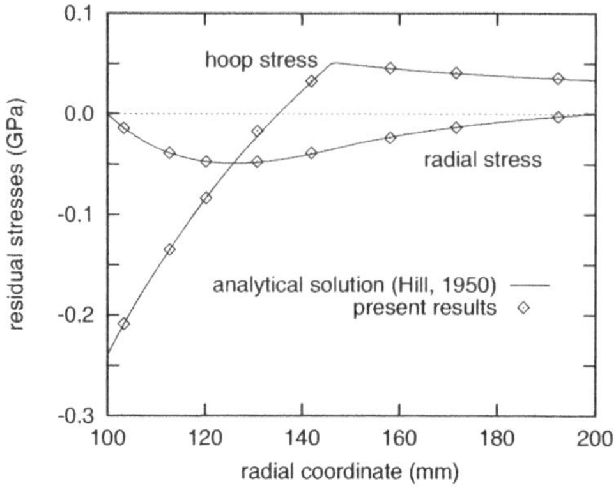

Figure 7.20. Internally pressurised spherical shell. Residual hoop and radial stress distributions resulting from pre-loading to $P = 0.28$ GPa.

of reverse plastic flow during unloading is $P = 2P_0$, which for the present dimensions and material constants is

$$P = 0.28\ \text{GPa}. \tag{7.173}$$

For thinner shells, with $b/a < 1.701$, the maximum pressure that can be applied in pre-loading is $P = P_{\text{lim}}$. Here, the finite element simulation of a pre-loading operation (with maximum load $P = 0.28$ GPa) has also been carried out. The resulting residual stresses are shown in Figure 7.20 where the above analytical solution is also plotted for comparison. Again, the high accuracy of the finite element solution is evident.

7.5.3. UNIFORMLY LOADED CIRCULAR PLATE

A simply supported uniformly loaded circular plate is analysed in this example. The plate, of radius R and thickness h, is simply supported on its edge and is subjected to a uniform pressure P on its top surface. The dimensions of the problem, the material parameters and the finite element mesh adopted are shown in Figure 7.21. The plate is discretised by ten eight-noded axisymmetric quadrilaterals with four Gauss integration points. The elements are distributed in two layers across the thickness. As in the previous two examples, the material model adopted is elastic-perfectly plastic with a von Mises yield criterion and the purpose of the analysis is to determine the limit load for the plate. The limit load for the present problem (to any desired accuracy) can be obtained by using methods of limit analysis combined with the finite difference method. The procedure is described by Skrzypek (1993) and the corresponding limit load is

$$P_{\text{lim}} \approx \frac{6.52\, M_y}{R^2}, \tag{7.174}$$

where

$$M_y \equiv \sigma_y h^2/4 \tag{7.175}$$

is the *yield bending moment*. For the present dimensions and yield stress, we have

$$P_{\text{lim}} \approx 260.8. \tag{7.176}$$

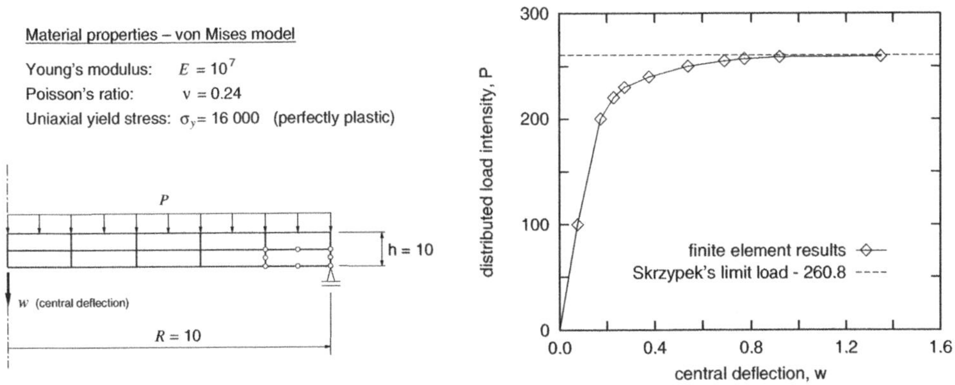

Figure 7.21. Uniformly loaded circular plate. Geometry and load versus deflection diagram.

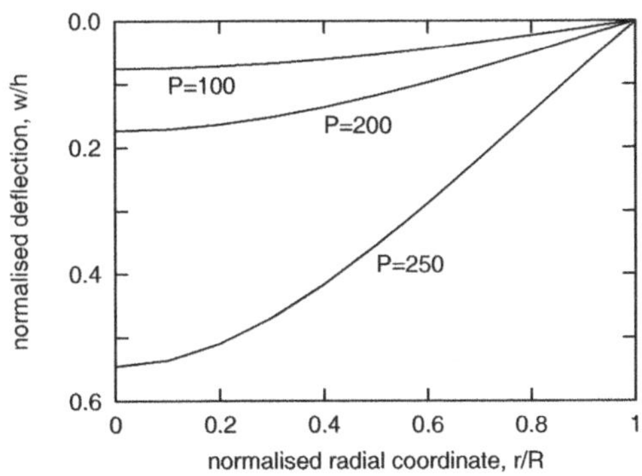

Figure 7.22. Uniformly loaded circular plate. Deflection profiles at different levels of load obtained in the finite element analysis.

In the finite element analysis, starting from the unloaded state, the pressure P is increased gradually, in ten increments, until collapse occurs. The deflection at the centre of the plate obtained in the numerical simulation is plotted against the applied pressure in the diagram of Figure 7.21. The accuracy of the finite element procedure in capturing the collapse is evident. The limit load obtained in the finite element analysis (taken as the load above which equilibrium iterations can no longer converge for a sufficiently small load increment) is

$$P_{\text{lim}}^{\text{fe}} \approx 259.8. \tag{7.177}$$

The relative error is about 0.4%. Note that such a high accuracy has been obtained despite the use of a relatively coarse mesh. Deflection profiles obtained at different stages of the loading process are shown in Figure 7.22. In the present problem, plastic yielding starts at the top and bottom surfaces of the plate around its centre and propagates toward the neutral plane with increasing load until collapse occurs. The propagation of the plastic zones is illustrated in Figure 7.23. The plastic and elastic regions, represented respectively by the shaded and white areas, are shown at different stages of the loading process. At $P = 100$ the plate is still purely elastic. At $P = 200$ plastic yield has already taken place at the bottom and top surfaces

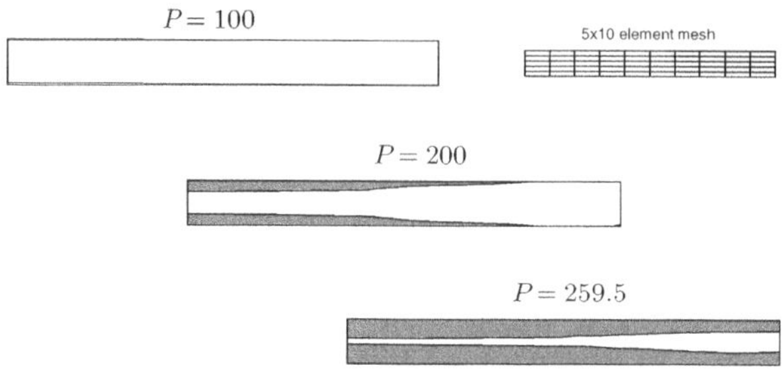

Figure 7.23. Uniformly loaded circular plate. Evolution of the plastic region.

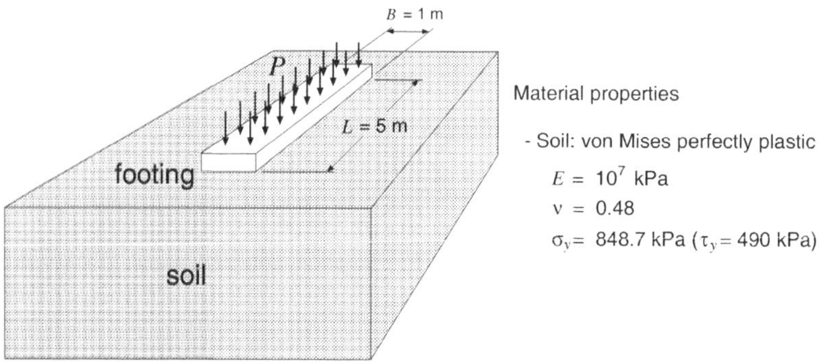

Figure 7.24. Strip footing. Problem definition.

around the plate centre and at $P = 259.5$ only a narrow layer surrounding the neutral plane remains elastic and collapse is imminent. In order to obtain a more accurate definition of the plastic zones, a finer mesh of 5×10 elements has been used. It is remarked that the limit load obtained with the finer mesh is virtually identical to the one obtained with the previous coarser mesh.

7.5.4. STRIP-FOOTING COLLAPSE

This example shows the application of the finite element method to the determination of the bearing capacity (limit load) of a strip footing. The problem consists of a long rectangular footing, of width $B = 1$ m and length $L = 5$ m, lying on soil (assumed as an infinite medium). The geometry of the problem is illustrated in Figure 7.24. The footing is subjected to a vertical pressure P and the purpose of the present analysis is to determine the collapse pressure $P_{\rm lim}$. The soil is assumed to be weightless and is modelled as a von Mises perfectly plastic material with the material constants shown in Figure 7.24. Due to the long length of the footing, the present problem is solved by assuming a plane strain state. The adopted finite element model is shown in Figure 7.25. Due to obvious symmetry, only one half of the cross-section is discretised. A mesh of 135 eight-noded quadrilaterals (with four-point quadrature) is used with a total of 446 nodes. In order to emulate the infinite medium assumption, the finite element mesh discretises a sufficiently large domain of soil with a depth and half-width of 5 m. The footing is assumed to be rigid and smooth (no friction at the footing/soil

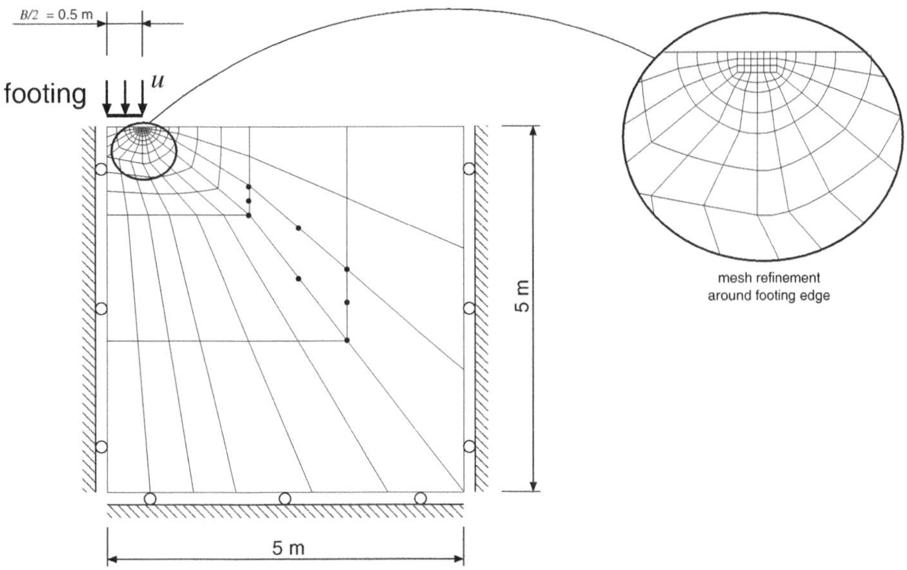

Figure 7.25. Strip footing. Finite element model.

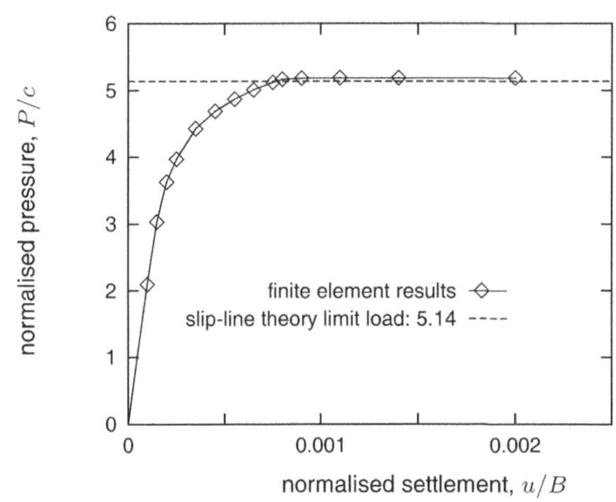

Figure 7.26. Strip footing. Load-displacement curve.

interface). This corresponds to prescribing the vertical displacement (the *settlement*), u, of the nodes under the footing and allowing their horizontal displacement to be unconstrained. A total displacement of $u = 0.002$ m is applied in 14 increments. The corresponding (average) pressure P supported by the footing is computed as the total reaction on the footing divided by the width B. Solutions to this problem, based on slip-line theory, have been derived by Prandtl and Hill (Hill, 1950). They give the following limit pressure:

$$P_{\text{lim}} = (2 + \pi)c \approx 5.14\,c \approx 2.97\,\sigma_y, \tag{7.178}$$

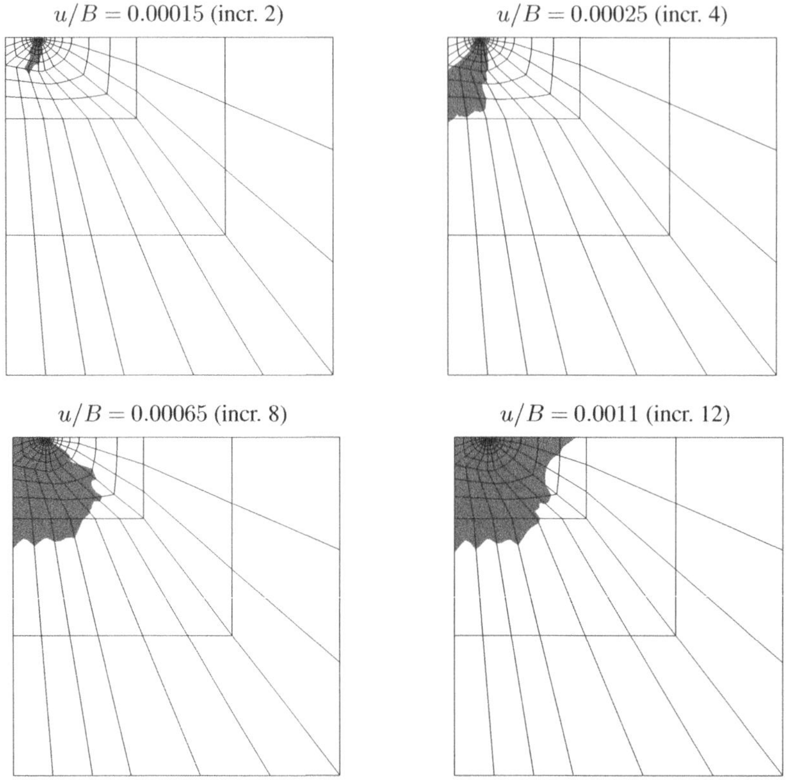

Figure 7.27. Strip footing. Evolution of the plastic zone.

where the cohesion or shear strength, c, for the von Mises model is given by $c = \sigma_y/\sqrt{3}$. The normalised average pressure obtained in the finite element simulation is plotted in Figure 7.26 versus the normalised settlement. The limit pressure obtained is in excellent agreement with the slip-line solution, with a relative error of approximately 0.9%. It is remarked that the accuracy can be increased by refining the finite element mesh. The evolution of the plastic zone during the process of loading is shown in Figure 7.27. The plastic zones correspond to the shaded area in the plot. At the 12th increment, the collapse load has already been reached. The incremental nodal displacement field corresponding to increment 12 is shown in Figure 7.28. The sizes of the nodal displacement vectors plotted in Figure 7.28 have been greatly exaggerated and are not to the same scale as the underlying finite element mesh. At the collapse load, the nodes located sufficiently far from the footing are virtually fixed and only the nodes surrounding the footing move. Note that the mechanism depicted in Figure 7.28 is in agreement with Hill's slip-line solution (Hill, 1950) which, as pointed out by Hill, is valid when the footing is assumed to be smooth (unconstrained horizontal displacements under the footing). It is important to emphasise here that, except under undrained conditions, the plastic flow in soils is strongly affected by the hydrostatic stress. Thus, pressure insensitive plasticity models, such as the von Mises material adopted in the present example, give in general a rather poor representation of the actual behaviour. The simulation of the collapse of

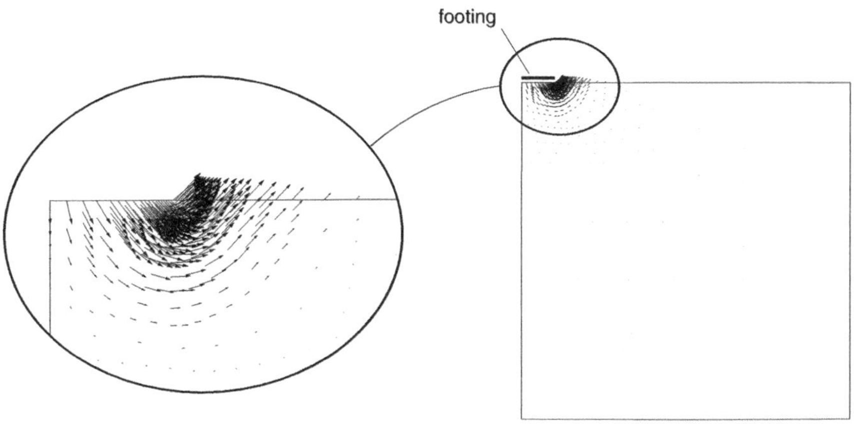

Figure 7.28. Strip footing. Incremental nodal displacement field at collapse (increment 12).

the strip footing using more appropriate pressure-sensitive plasticity models is carried out in Chapter 8.

7.5.5. DOUBLE-NOTCHED TENSILE SPECIMEN

In this example, the plane strain analysis of a deep double-edge-notched tensile specimen is carried out. As in all examples of this chapter, the von Mises elastoplastic model is adopted. Here, two situations are considered:

1. The material is assumed *elastic-perfectly plastic*.
2. *Linear isotropic hardening* is assumed.

Under the hypothesis of perfect plasticity, the present problem is simply the tensile version of the strip footing problem described above and is often adopted as a benchmark test for the convergence properties of finite elements in incompressible plasticity (Nagtegaal *et al.*, 1974; Simo and Rifai, 1990). The geometry, material properties and the finite element mesh used are shown in Figure 7.29. The specimen, of width $w = 10$ and length $l = 30$, contains two deep notches and its two halves are connected only by a ligament of width $b = 1$. Only one symmetric quarter of the specimen, discretised by the 153-element mesh shown in Figure 7.29, is used in the finite element computations. Again, the standard eight-noded quadrilateral element with four Gauss integration points is adopted. The total number of nodes is 506. The specimen is stretched by prescribing the vertical displacement u on the top nodes of the mesh. A total displacement $u = 0.17$ is applied incrementally. With R denoting the total reaction on the restrained edge, the *net axial stress*, $\bar{\sigma}$, on the ligament is given by

$$\bar{\sigma} = \frac{R}{b}. \tag{7.179}$$

In terms of the net stress on the ligament, the limit load obtained by Prandtl for the present problem (under the hypothesis of perfect plasticity) is given by

$$\bar{\sigma}_{\text{lim}} = \frac{(2+\pi)}{\sqrt{3}}\, \sigma_y \approx 2.97\, \sigma_y. \tag{7.180}$$

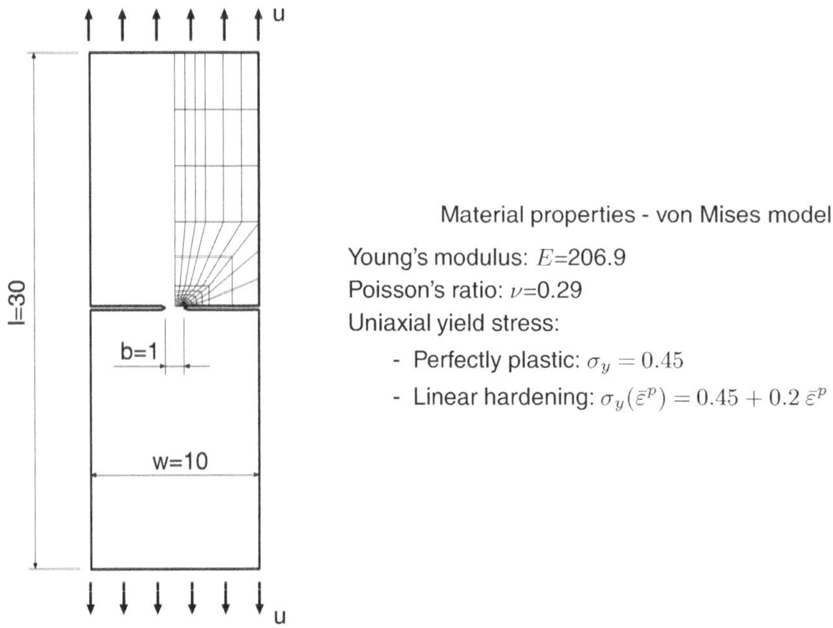

Figure 7.29. Double-notched tensile specimen. Geometry, material properties and finite element mesh.

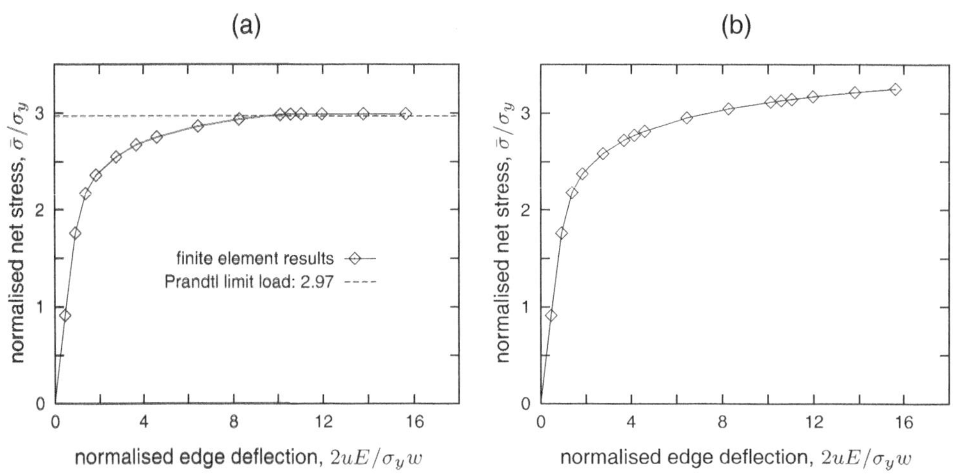

Figure 7.30. Double-notched specimen. Load-deflection curves: (a) perfect plasticity; and (b) linear hardening.

In Figure 7.30(a), the normalised axial net stress, $\bar{\sigma}/\sigma_y$, obtained in the perfectly plastic finite element analysis is plotted against the normalised deflection of the top edge, $2uE/\sigma_y w$. The limit load obtained with the present mesh is only approximately 0.8% higher than the Prandtl solution. The load-deflection curve obtained for the linearly hardening model is shown in Figure 7.30(b). In this case, no limit load exists.

7.6. Further application: the von Mises model with nonlinear mixed hardening

Having used the simple set-up of the von Mises model with isotropic hardening, we hope to have provided the reader with a clear picture of the basic components of the finite element implementation of plasticity models. In this section we move one step forward and apply the same concepts to derive an implicit integration algorithm (together with its associated consistent tangent operator) for a version of the von Mises model that combines general nonlinear isotropic and kinematic hardening. The mixed hardening rule considered here is the one described in Section 6.6.5 (page 189).

It is worth remarking that consideration of kinematic hardening may become crucial in applications involving *cyclic loads* – situations where the *Bauschinger effect* may not be disregarded without significant loss of accuracy. The algorithm/tangent operator derived here are also incorporated into `HYPLAS`. As we shall see, the structure of these routines is very similar to those of the isotropic hardening-only counterparts. The computational implementation of these procedures is a straightforward extension of routines `SUVM` (for the integration algorithm) and `CTVM` (for the consistent tangent). The corresponding routines for the mixed hardening model are named, respectively, `SUVMMX` and `CTVMX`. Their source code is not included in the text.

Readers wishing to skip the details of the present derivations are referred directly to Box 7.5 (page 260) for the integration algorithm and formula (7.213) for the corresponding elastoplastic consistent tangent modulus.

7.6.1. THE MIXED HARDENING MODEL: SUMMARY

Let us start by summarising the equations of the model to be used in what follows. In addition to the standard linear elastic law, we have:

1. The kinematically hardening von Mises yield function

$$\begin{aligned}\Phi(\boldsymbol{\sigma}, \boldsymbol{\beta}, \sigma_y) &\equiv \sqrt{3\, J_2(\boldsymbol{s}(\boldsymbol{\sigma}) - \boldsymbol{\beta})} - \sigma_y \\ &= \sqrt{\tfrac{3}{2}}\, \|\boldsymbol{\eta}\| - \sigma_y \end{aligned} \tag{7.181}$$

   where $\boldsymbol{\beta}$ is the backstress tensor which defines the translation of the centre of the (kinematically hardened) von Mises circle in the deviatoric plane, $\boldsymbol{\eta}$ is the *relative stress*

$$\boldsymbol{\eta} \equiv \boldsymbol{s} - \boldsymbol{\beta}, \tag{7.182}$$

   and σ_y defines now only the radius of the yield surface and not necessarily the uniaxial yield stress as in the isotropic hardening-only model.

2. Associative law for the plastic flow:

$$\dot{\boldsymbol{\varepsilon}}^p = \dot{\gamma}\, \frac{\partial \Phi}{\partial \boldsymbol{\sigma}} = \dot{\gamma} \sqrt{\frac{3}{2}}\, \frac{\boldsymbol{\eta}}{\|\boldsymbol{\eta}\|}. \tag{7.183}$$

3. General isotropic strain hardening defined by the isotropic hardening curve

$$\sigma_y = \sigma_y(\bar{\varepsilon}^p). \tag{7.184}$$

4. Nonlinear kinematic hardening defined by the following evolution law for the back-stress:

$$\dot{\boldsymbol{\beta}} = \frac{2}{3}\, H^{\mathrm{k}}(\bar{\varepsilon}^p)\, \dot{\boldsymbol{\varepsilon}}^p = \dot{\gamma}\, \sqrt{\frac{2}{3}}\, H^{\mathrm{k}}(\bar{\varepsilon}^p)\, \frac{\boldsymbol{\eta}}{\|\boldsymbol{\eta}\|}, \tag{7.185}$$

with

$$H^{\mathrm{k}}(\bar{\varepsilon}^p) \equiv \frac{\mathrm{d}\bar{\beta}}{\mathrm{d}\bar{\varepsilon}^p} \tag{7.186}$$

denoting the slope of the given kinematic hardening curve:

$$\bar{\beta} = \bar{\beta}(\bar{\varepsilon}^p). \tag{7.187}$$

Recall that the *kinematic hardening stress*, $\bar{\beta}$, is the kinematic contribution to overall hardening that can be obtained from uniaxial tests with load reversal.

7.6.2. THE IMPLICIT RETURN-MAPPING SCHEME

Particularisation of the general implicit return-mapping equations (7.25) to the present case, where $\boldsymbol{\alpha} \equiv \{\bar{\varepsilon}^p, \boldsymbol{\beta}\}$, gives

$$\begin{aligned}
&\boldsymbol{\varepsilon}^e_{n+1} = \boldsymbol{\varepsilon}^{e\ \mathrm{trial}}_{n+1} - \Delta\gamma\, \sqrt{\frac{3}{2}}\, \frac{\boldsymbol{\eta}_{n+1}}{\|\boldsymbol{\eta}_{n+1}\|} \\
&\boldsymbol{\beta}_{n+1} = \boldsymbol{\beta}_n + \Delta\gamma\, \sqrt{\frac{2}{3}}\, H^{\mathrm{k}}(\bar{\varepsilon}^p_{n+1}) \frac{\boldsymbol{\eta}_{n+1}}{\|\boldsymbol{\eta}_{n+1}\|} \\
&\bar{\varepsilon}^p_{n+1} = \bar{\varepsilon}^p_n + \Delta\gamma \\
&\sqrt{\frac{3}{2}}\, \|\boldsymbol{\eta}_{n+1}\| - \sigma_y(\bar{\varepsilon}^p_{n+1}) = 0.
\end{aligned} \tag{7.188}$$

As in the isotropic hardening-only case, the above system can be reduced to the solution of a *single* scalar equation for the plastic multiplier $\Delta\gamma$. To achieve this, firstly note that equation $(7.188)_1$, together with the elastic law, gives the following incremental update for the stress deviator

$$\boldsymbol{s}_{n+1} = \boldsymbol{s}^{\mathrm{trial}}_{n+1} - 2G\, \Delta\gamma\, \sqrt{\frac{3}{2}}\, \frac{\boldsymbol{\eta}_{n+1}}{\|\boldsymbol{\eta}_{n+1}\|}. \tag{7.189}$$

By subtracting $(7.188)_2$ from (7.189), we obtain

$$\boldsymbol{\eta}_{n+1} = \boldsymbol{\eta}^{\mathrm{trial}}_{n+1} - \Delta\gamma\, \sqrt{\frac{3}{2}} \left[2G + \frac{2}{3} H^{\mathrm{k}}(\bar{\varepsilon}^p_{n+1})\right] \frac{\boldsymbol{\eta}_{n+1}}{\|\boldsymbol{\eta}_{n+1}\|}, \tag{7.190}$$

where we have defined the *elastic trial relative stress* as

$$\boldsymbol{\eta}^{\mathrm{trial}}_{n+1} \equiv \boldsymbol{s}^{\mathrm{trial}}_{n+1} - \boldsymbol{\beta}_n. \tag{7.191}$$

Note that (7.190) implies that $\boldsymbol{\eta}_{n+1}$ and $\boldsymbol{\eta}^{\mathrm{trial}}_{n+1}$ are *colinear*, i.e. $\boldsymbol{\eta}_{n+1}$ is a scalar multiple of $\boldsymbol{\eta}^{\mathrm{trial}}_{n+1}$. This gives the identity

$$\frac{\boldsymbol{\eta}_{n+1}}{\|\boldsymbol{\eta}_{n+1}\|} = \frac{\boldsymbol{\eta}^{\mathrm{trial}}_{n+1}}{\|\boldsymbol{\eta}^{\mathrm{trial}}_{n+1}\|}, \tag{7.192}$$

which, substituted into (7.190), renders the simpler update formula for the relative stress

$$\boldsymbol{\eta}_{n+1} = \left\{1 - \frac{\Delta\gamma}{\bar{q}^{\text{trial}}_{n+1}}[3G + H^{\text{k}}(\bar{\varepsilon}^p_{n+1})]\right\}\boldsymbol{\eta}^{\text{trial}}_{n+1}, \tag{7.193}$$

where $\bar{q}^{\text{trial}}_{n+1}$ is the elastic trial *relative effective stress*, defined by

$$\bar{q}^{\text{trial}}_{n+1} \equiv \sqrt{\tfrac{3}{2}}\,\|\boldsymbol{\eta}^{\text{trial}}_{n+1}\|. \tag{7.194}$$

Then, by substituting (7.193), together with $(7.188)_3$, into $(7.188)_4$, the return mapping reduces to the following generally nonlinear scalar equation for $\Delta\gamma$:

$$\tilde{\Phi}(\Delta\gamma) \equiv \bar{q}^{\text{trial}}_{n+1} - \Delta\gamma[3G + H^{\text{k}}(\bar{\varepsilon}^p_n + \Delta\gamma)] - \sigma_y(\bar{\varepsilon}^p_n + \Delta\gamma) = 0. \tag{7.195}$$

Note that, also consistently with the backward Euler approximation, we have

$$\Delta\gamma\, H^{\text{k}}(\bar{\varepsilon}^p_{n+1}) \approx \bar{\beta}_{n+1} - \bar{\beta}_n, \tag{7.196}$$

where $\bar{\beta}_k \equiv \bar{\beta}(\bar{\varepsilon}^p_k)$. Since the curve $\bar{\beta}(\bar{\varepsilon}^p)$ (in the form of discrete sampling points) rather than the slope $H^{\text{k}}(\bar{\varepsilon}^p)$ is what is normally available from uniaxial experiments, it is convenient to use the above approximation and work only with the kinematic hardening stress in the return-mapping equations. With the adoption of this approach, we replace $(7.188)_2$ and (7.193), respectively, with

$$\boldsymbol{\beta}_{n+1} = \boldsymbol{\beta}_n + \sqrt{\frac{2}{3}}\,(\bar{\beta}_{n+1} - \bar{\beta}_n)\,\frac{\boldsymbol{\eta}_{n+1}}{\|\boldsymbol{\eta}_{n+1}\|} \tag{7.197}$$

and

$$\boldsymbol{\eta}_{n+1} = \left(1 - \frac{3G\Delta\gamma + \bar{\beta}_{n+1} - \bar{\beta}_n}{\bar{q}^{\text{trial}}_{n+1}}\right)\boldsymbol{\eta}^{\text{trial}}_{n+1}. \tag{7.198}$$

Correspondingly, the return equation (7.195) is replaced by

$$\tilde{\Phi}(\Delta\gamma) \equiv \bar{q}^{\text{trial}}_{n+1} - 3G\Delta\gamma - \bar{\beta}(\bar{\varepsilon}^p_n + \Delta\gamma) + \bar{\beta}_n - \sigma_y(\bar{\varepsilon}^p_n + \Delta\gamma) = 0. \tag{7.199}$$

The mth Newton–Raphson iterative correction to $\Delta\gamma$ in the solution of the above equation reads

$$\Delta\gamma^{(m)} := \Delta\gamma^{(m-1)} - \frac{\tilde{\Phi}(\Delta\gamma^{(m-1)})}{d}, \tag{7.200}$$

where

$$d = -3G - H^{\text{k}}(\bar{\varepsilon}^p_n + \Delta\gamma^{(m-1)}) - H^{\text{i}}(\bar{\varepsilon}^p_n + \Delta\gamma^{(m-1)}) \tag{7.201}$$

and $H^{\text{i}} \equiv \sigma'_y$ is the slope of the isotropic hardening curve.

The overall integration algorithm is listed in Box 7.5 in pseudo-code format.

Box 7.5. Implicit elastic predictor/return-mapping algorithm for the von Mises model with mixed nonlinear hardening.

HYPLAS procedure:	SUVMMX

(i) Elastic predictor. Given $\Delta\boldsymbol{\varepsilon}$ and the state variables at t_n, evaluate the *elastic trial state*

$$\boldsymbol{\varepsilon}^{e\ \mathrm{trial}}_{n+1} := \boldsymbol{\varepsilon}^{e}_{n} + \Delta\boldsymbol{\varepsilon}$$

$$\bar{\varepsilon}^{p\ \mathrm{trial}}_{n+1} := \bar{\varepsilon}^{p}_{n}; \qquad \boldsymbol{\beta}^{\mathrm{trial}}_{n+1} := \boldsymbol{\beta}_n$$

$$p^{\mathrm{trial}}_{n+1} := K\varepsilon^{e\ \mathrm{trial}}_{\mathrm{v}\ n+1}; \qquad \boldsymbol{s}^{\mathrm{trial}}_{n+1} := 2G\,\boldsymbol{\varepsilon}^{e\ \mathrm{trial}}_{\mathrm{d}\ n+1}$$

$$\boldsymbol{\eta}^{\mathrm{trial}}_{n+1} := \boldsymbol{s}^{\mathrm{trial}}_{n+1} - \boldsymbol{\beta}_n; \qquad \bar{q}^{\mathrm{trial}}_{n+1} := \sqrt{\tfrac{3}{2}}\,\|\boldsymbol{\eta}^{\mathrm{trial}}_{n+1}\|$$

(ii) Check plastic admissibility

$$\text{IF } \bar{q}^{\mathrm{trial}}_{n+1} - \sigma_y(\bar{\varepsilon}^{p\ \mathrm{trial}}_{n+1}) \le 0$$

$$\text{THEN set} \quad (\cdot)_{n+1} := (\cdot)^{\mathrm{trial}}_{n+1} \quad \text{and EXIT}$$

(iii) Return mapping. Solve the equation for $\Delta\gamma$

$$\bar{q}^{\mathrm{trial}}_{n+1} - 3G\Delta\gamma - \bar{\beta}(\bar{\varepsilon}^{p}_{n} + \Delta\gamma) + \bar{\beta}_n - \sigma_y(\bar{\varepsilon}^{p}_{n} + \Delta\gamma) = 0$$

where $\bar{\beta}_n = \bar{\beta}(\bar{\varepsilon}^p_n)$ and update

$$\bar{\varepsilon}^{p}_{n+1} := \bar{\varepsilon}^{p}_{n} + \Delta\gamma; \qquad \bar{\beta}_{n+1} := \bar{\beta}(\bar{\varepsilon}^{p}_{n+1})$$

$$\boldsymbol{\beta}_{n+1} := \boldsymbol{\beta}_n + \sqrt{\frac{2}{3}}(\bar{\beta}_{n+1} - \bar{\beta}_n)\frac{\boldsymbol{\eta}^{\mathrm{trial}}_{n+1}}{\|\boldsymbol{\eta}^{\mathrm{trial}}_{n+1}\|}$$

$$p_{n+1} := p^{\mathrm{trial}}_{n+1}; \qquad \boldsymbol{s}_{n+1} := \boldsymbol{s}^{\mathrm{trial}}_{n+1} - 2G\Delta\gamma\sqrt{\frac{3}{2}}\frac{\boldsymbol{\eta}^{\mathrm{trial}}_{n+1}}{\|\boldsymbol{\eta}^{\mathrm{trial}}_{n+1}\|}$$

$$\boldsymbol{\sigma}_{n+1} := \boldsymbol{s}_{n+1} + p_{n+1}\,\boldsymbol{I}; \qquad \boldsymbol{\varepsilon}^{e}_{n+1} = \frac{1}{2G}\,\boldsymbol{s}_{n+1} + \frac{1}{3}\,\varepsilon^{e\ \mathrm{trial}}_{\mathrm{v}\ n+1}\,\boldsymbol{I}$$

(iv) EXIT

7.6.3. THE INCREMENTAL CONSTITUTIVE FUNCTION

By means of straightforward tensor manipulations, we can easily establish that the incremental constitutive function, analogous to (7.93), for the present stress updating procedure reads

$$\begin{aligned}\boldsymbol{\sigma}_{n+1} &= \bar{\boldsymbol{\sigma}}_{n+1}(\bar{\varepsilon}^p_n, \boldsymbol{\beta}_n, \boldsymbol{\varepsilon}^{e\ \mathrm{trial}}_{n+1}) \\ &\equiv \left[\mathsf{D}^e - \hat{H}(\Phi^{\mathrm{trial}})\,\frac{\Delta\gamma\,6G^2}{\bar{q}^{\mathrm{trial}}_{n+1}}\,\mathsf{I}_{\mathrm{d}}\right] : \boldsymbol{\varepsilon}^{e\ \mathrm{trial}}_{n+1} + \hat{H}(\Phi^{\mathrm{trial}})\,\frac{\Delta\gamma\,3G}{\bar{q}^{\mathrm{trial}}_{n+1}}\,\boldsymbol{\beta}_n,\end{aligned} \tag{7.202}$$

where $\hat{H}$ is the Heaviside step function (7.94),

$$\bar{q}^{\,\text{trial}}_{n+1} = \sqrt{\tfrac{3}{2}}\|\boldsymbol{s}^{\text{trial}}_{n+1} - \boldsymbol{\beta}_n\| = \sqrt{\tfrac{3}{2}}\|2G\boldsymbol{\varepsilon}^{e\ \text{trial}}_{\text{d}\ n+1} - \boldsymbol{\beta}_n\|$$
$$= \bar{q}^{\,\text{trial}}_{n+1}(\boldsymbol{\beta}_n, \boldsymbol{\varepsilon}^{e\ \text{trial}}_{n+1}) \equiv \sqrt{\tfrac{3}{2}}\,\|2G\,\mathsf{I}_{\text{d}} : \boldsymbol{\varepsilon}^{e\ \text{trial}}_{n+1} - \boldsymbol{\beta}_n\|, \tag{7.203}$$

Φ^{trial} is the function

$$\Phi^{\text{trial}} = \Phi^{\text{trial}}(\bar{\varepsilon}^p_n, \boldsymbol{\beta}_n, \boldsymbol{\varepsilon}^{e\ \text{trial}}_{n+1}) \equiv \bar{q}^{\,\text{trial}}_{n+1}(\boldsymbol{\beta}_n, \boldsymbol{\varepsilon}^{e\ \text{trial}}_{n+1}) - \sigma_y(\bar{\varepsilon}^p_n), \tag{7.204}$$

and

$$\Delta\gamma = \Delta\gamma(\bar{\varepsilon}^p_n, \boldsymbol{\beta}_n, \boldsymbol{\varepsilon}^{e\ \text{trial}}_{n+1}) \tag{7.205}$$

is the implicit function of $\boldsymbol{\varepsilon}^{e\ \text{trial}}_{n+1}$ and the internal variables $\{\bar{\varepsilon}^p_n, \boldsymbol{\beta}_n\}$ defined by the consistency equation (7.199).

7.6.4. LINEAR HARDENING: CLOSED-FORM RETURN MAPPING

If the kinematic and isotropic hardening are linear, i.e., if

$$\sigma_y(\bar{\varepsilon}^p) = \sigma_{y_0} + H^{\text{i}}\,\bar{\varepsilon}^p, \quad \bar{\beta}(\bar{\varepsilon}^p) = H^{\text{k}}\,\bar{\varepsilon}^p, \tag{7.206}$$

with constant σ_{y_0}, H^{i} and H^{k}, then the present return mapping equation has the following closed form solution:

$$\Delta\gamma = \frac{\Phi^{\text{trial}}}{3G + H^{\text{k}} + H^{\text{i}}}. \tag{7.207}$$

7.6.5. COMPUTATIONAL IMPLEMENTATION ASPECTS

As anticipated at the beginning of Section 7.6, the algorithm of Box 7.5 is a straightforward extension of the algorithm for the isotropic hardening-only model listed in Boxes 7.3 and 7.4. The present mixed hardening algorithm is implemented in subroutine `SUVMMX` (**S**tate **U**pdate procedure for the **V**on **M**ises model with nonlinear **M**i**X**ed hardening). This routine is a simple modification of routine `SUVM` (listed in Section 7.3.5). The main modifications comprise the redefinition of the return-mapping equation and the incorporation of new state variables – the components of the *backstress tensor*, $\boldsymbol{\beta}$. These components are stored in array `RSTAVA` of real-state variables. Also note that, here, *two* hardening curves (rather than one in the isotropic hardening-only case) are required: one for σ_y (already implemented in `SUVM`) and a second one for $\bar{\beta}$. As for the isotropic hardening curve, the kinematic hardening curve,

$$\bar{\beta}(\bar{\varepsilon}^p),$$

is assumed to be piecewise linear, defined by the same number of (user supplied) sampling pairs as the isotropic hardening curve

$$\{\bar{\varepsilon}^p, \bar{\beta}\}.$$

The extra properties are read and stored in array `RPROPS` (real material properties) during the data input phase of `HYPLAS`.

7.6.6. THE ELASTOPLASTIC CONSISTENT TANGENT

Under plastic flow ($\Phi^{\text{trial}} > 0$), the algorithmic incremental constitutive function (7.202) gives

$$\boldsymbol{\sigma}_{n+1} = \left[\mathsf{D}^e - \frac{\Delta\gamma\, 6G^2}{\bar{q}^{\text{trial}}_{n+1}}\,\mathsf{I}_{\text{d}}\right] : \boldsymbol{\varepsilon}^{e\ \text{trial}}_{n+1} + \frac{\Delta\gamma\, 3G}{\bar{q}^{\text{trial}}_{n+1}}\,\boldsymbol{\beta}_n. \tag{7.208}$$

The elastoplastic consistent tangent operator for the present case is obtained by differentiating the above relation. The derivation follows closely that presented in Section 7.4.2 (from page 232) for the isotropic hardening-only model.

The differentiation of (7.208) gives

$$\begin{aligned}\frac{\partial\boldsymbol{\sigma}_{n+1}}{\partial\boldsymbol{\varepsilon}^{e\ \text{trial}}_{n+1}} = \mathsf{D}^e &- \frac{\Delta\gamma\, 6G^2}{\bar{q}^{\text{trial}}_{n+1}}\,\mathsf{I}_{\text{d}} - \frac{3G}{\bar{q}^{\text{trial}}_{n+1}}\,\boldsymbol{\eta}^{\text{trial}}_{n+1}\otimes\frac{\partial\Delta\gamma}{\partial\boldsymbol{\varepsilon}^{e\ \text{trial}}_{n+1}}\\ &+ \frac{\Delta\gamma\, 3G}{(\bar{q}^{\text{trial}}_{n+1})^2}\,\boldsymbol{\eta}^{\text{trial}}_{n+1}\otimes\frac{\partial\bar{q}^{\text{trial}}_{n+1}}{\partial\boldsymbol{\varepsilon}^{e\ \text{trial}}_{n+1}},\end{aligned} \tag{7.209}$$

where we have made use of the connection $\boldsymbol{\eta}^{\text{trial}}_{n+1} = 2G\,\mathsf{I}_{\text{d}} : \boldsymbol{\varepsilon}^{e\ \text{trial}}_{n+1} - \boldsymbol{\beta}_n$. The derivative of the trial relative effective stress is obtained analogously to (7.116). The corresponding expression in the present case has the same format as (7.116), i.e.

$$\frac{\partial\bar{q}^{\text{trial}}_{n+1}}{\partial\boldsymbol{\varepsilon}^{e\ \text{trial}}_{n+1}} = 2G\sqrt{\frac{2}{3}}\,\bar{\boldsymbol{N}}_{n+1}, \tag{7.210}$$

with the unit flow vector here defined as

$$\bar{\boldsymbol{N}}_{n+1} \equiv \frac{\boldsymbol{\eta}^{\text{trial}}_{n+1}}{\|\boldsymbol{\eta}^{\text{trial}}_{n+1}\|}. \tag{7.211}$$

The incremental plastic multiplier derivative is obtained by differentiating the return-mapping equation (7.199) with respect to $\boldsymbol{\varepsilon}^{e\ \text{trial}}_{n+1}$. This gives

$$\begin{aligned}\frac{\partial\Delta\gamma}{\partial\boldsymbol{\varepsilon}^{e\ \text{trial}}_{n+1}} &= \frac{1}{3G + H^{\text{k}} + H^{\text{i}}}\,\frac{\partial\bar{q}^{\text{trial}}_{n+1}}{\partial\boldsymbol{\varepsilon}^{e\ \text{trial}}_{n+1}}\\ &= \frac{2G}{3G + H^{\text{k}} + H^{\text{i}}}\sqrt{\frac{3}{2}}\,\bar{\boldsymbol{N}}_{n+1}.\end{aligned} \tag{7.212}$$

Then, with the substitution of (7.210) and (7.212) into (7.209) we obtain, after a straightforward algebra taking (7.194) and (7.211) into account, the following closed-form expression for the elastoplastic consistent tangent operator:

$$\begin{aligned}\mathsf{D}^{ep} &= \mathsf{D}^e - \frac{\Delta\gamma\, 6G^2}{\bar{q}^{\text{trial}}_{n+1}}\,\mathsf{I}_{\text{d}} + 6G^2\left(\frac{\Delta\gamma}{\bar{q}^{\text{trial}}_{n+1}} - \frac{1}{3G + H^{\text{k}} + H^{\text{i}}}\right)\bar{\boldsymbol{N}}_{n+1}\otimes\bar{\boldsymbol{N}}_{n+1}\\ &= 2G\left(1 - \frac{\Delta\gamma\, 3G}{\bar{q}^{\text{trial}}_{n+1}}\right)\mathsf{I}_{\text{d}}\\ &\quad + 6G^2\left(\frac{\Delta\gamma}{\bar{q}^{\text{trial}}_{n+1}} - \frac{1}{3G + H^{\text{k}} + H^{\text{i}}}\right)\bar{\boldsymbol{N}}_{n+1}\otimes\bar{\boldsymbol{N}}_{n+1} + K\,\boldsymbol{I}\otimes\boldsymbol{I}.\end{aligned} \tag{7.213}$$

Implementation

The above tangent operator is computed in subroutine `CTVMMX` (**C**onsistent **T**angent operator for the **V**on Mises model with nonlinear **M**i**X**ed hardening) of `HYPLAS`. As the reader may verify `CTVMMX` is a straightforward variation of routine `CTVM` (refer to Section 7.4.3, from page 235) coded for the isotropic hardening-only model. The basic difference is that, in `CTVMMX`, the incremental flow vector and the scalar factors multiplying the fourth-order tensors are redefined according to the above formulae.

8 COMPUTATIONS WITH OTHER BASIC PLASTICITY MODELS

IN Chapter 7, some general schemes for numerical integration of elastoplastic constitutive equations have been reviewed together with related issues such as the computation of consistent tangent operators and error analysis. To illustrate such concepts, the von Mises model – the simplest of the models discussed in this book – has been used as an example and the corresponding implicit elastic predictor/return-mapping algorithm, together with the associated consistent tangent, have been derived in detail for the isotropic and mixed hardening cases. The complete FORTRAN implementation of the related computational procedures within the finite element environment of program HYPLAS has also been presented for the isotropic hardening version of the model.

In the present chapter, the complete computational implementation of other basic plasticity models discussed in Chapter 6, including a step-by-step derivation and FORTRAN coding of the necessary procedures, is described in detail. Namely, the models discussed here are

- the Tresca model,
- the Mohr–Coulomb model, and
- the Drucker–Prager model,

all featuring linear elastic behaviour and generally a nonlinear isotropic hardening law. The associative plastic flow rule is adopted for the Tresca model whereas, for the Mohr–Coulomb and Drucker–Prager models, the generally non-associated laws discussed in Section 6.5.3 of Chapter 6 are considered. The algorithms for integration of the corresponding elastoplastic constitutive equations derived here are specialisations of the elastic predictor/return mapping scheme based on the *fully implicit* discretisation of the plastic corrector equations, discussed in Section 7.2.3. The associated consistent tangent moduli are also derived in detail and an error analysis based on iso-error maps is presented for each model considered.

We remark that the only new concept introduced in this chapter concerns the computational treatment of singularities (corners) in the yield surface. Note that all three models addressed here feature such singularities where, in particular, the Tresca and Mohr–Coulomb criteria admit a multisurface representation with corresponding multivector flow rules.

The main sections of this chapter are three: Sections 8.1, 8.2 and 8.3. These sections describe, respectively, the implementation of the Tresca, Mohr–Coulomb and Drucker–Prager models. At the beginning of each of the main sections, a table has been added indicating the location of flowcharts, pseudo-code and FORTRAN source code of the relevant computational procedures as well as iso-error maps for the particular material model of interest. These summarise the most important results and the tables should be particularly helpful to readers who wish to skip the details of derivation and concentrate only on the practical aspects of computational implementation. Also at the beginning of each main section, a summary of the constitutive equations of the model concerned is presented. This layout makes the main sections self-contained to a certain extent and has been chosen in order to avoid readers having to refer back to Chapter 6, where a more detailed description of the constitutive equations adopted here is given.

Applications of the numerical procedures derived in the main sections are illustrated in Section 8.4 by means a comprehensive set of benchmarking numerical examples.

8.1. The Tresca model

This section describes the specialisation of the implicit algorithm of Section 7.2.3 to the Tresca model with general nonlinear isotropic strain hardening. The corresponding consistent tangent operator is also derived and an accuracy assessment of the integration algorithm, by means of iso-error maps, is presented. For the reader who is interested only in the practical implementation aspects of the material presented in this section, the location of the main results is summarised below.

integration algorithm	flowchart	Figure 8.3
	pseudo-code	Boxes 8.1–8.3
	FORTRAN code	subroutine SUTR (Section 8.1.2)
iso-error maps		Figures 8.6 and 8.7
consistent tangent – FORTRAN code –		subroutine CTTR (Section 8.1.5)

The model

All ingredients of the Tresca plasticity model implemented here have been fully described in Chapter 6. In the following we summarise for convenience only its main equations.

The multisurface representation of the Tresca model is defined by means of the six yield functions, $\Phi_1, \ldots, \Phi_6$, given in (6.91) (page 160). The corresponding associative plastic flow rule is defined by the rate equation

$$\dot{\varepsilon}^p = \sum_{i=1}^{6} \dot{\gamma}^i \, \boldsymbol{N}^i, \tag{8.1}$$

where

$$\boldsymbol{N}^i \equiv \frac{\partial \Phi_i}{\partial \boldsymbol{\sigma}}, \tag{8.2}$$

together with the loading/unloading conditions

$$\Phi_i \leq 0, \quad \dot{\gamma}^i \geq 0, \quad \Phi_i \dot{\gamma}^i = 0 \tag{8.3}$$

with no summation on the repeated index.

Isotropic strain hardening is incorporated, as in the von Mises model implementation discussed in Chapter 7, by assuming the uniaxial yield stress σ_y that takes part in the yield functions Φ_i to be a given (generally nonlinear) function of the accumulated plastic strain:

$$\sigma_y = \sigma_y(\bar{\varepsilon}^p). \tag{8.4}$$

With the principal stresses ordered, without loss of generality, as $\sigma_1 \geq \sigma_2 \geq \sigma_3$ (refer to Figure 6.18, page 173), the model can be completely defined with reference to a single sextant of the principal stress space. In this case, only three possibilities are identified for the associative plastic flow:

1. Plastic flow from the flat portion (main plane), where the Tresca yielding function is differentiable. In this case,
$$\dot{\boldsymbol{\varepsilon}}^p = \dot{\gamma} \boldsymbol{N}^a, \tag{8.5}$$
where the flow vector $\boldsymbol{N}^a$ is the normal to the corresponding plane defined by $\Phi_1 = 0$:
$$\boldsymbol{N}^a \equiv \boldsymbol{N}^1 = \boldsymbol{e}_1 \otimes \boldsymbol{e}_1 - \boldsymbol{e}_3 \otimes \boldsymbol{e}_3. \tag{8.6}$$

2. Plastic flow from the *right corner*, R. The rate of plastic strain in this case is a linear combination with non-negative coefficients of the normals to the two intersecting planes at this point. The flow rule reads
$$\dot{\boldsymbol{\varepsilon}}^p = \dot{\gamma}^a \boldsymbol{N}^a + \dot{\gamma}^b \boldsymbol{N}^b, \tag{8.7}$$
where $\boldsymbol{N}^a$ is the normal to the main plane, defined by (8.6), and $\boldsymbol{N}^b$ is the normal to the plane on the right of the main plane defined by $\Phi_6 = 0$:
$$\boldsymbol{N}^b \equiv \boldsymbol{N}^6 = \boldsymbol{e}_1 \otimes \boldsymbol{e}_1 - \boldsymbol{e}_2 \otimes \boldsymbol{e}_2. \tag{8.8}$$

3. Plastic flow from the *left corner*, L. The flow rule, in this case, is the same as the one above, but with $\boldsymbol{N}^b$ being the normal to the plane on the left of the main plane (defined by $\Phi_2 = 0$):
$$\boldsymbol{N}^b \equiv \boldsymbol{N}^2 = \boldsymbol{e}_2 \otimes \boldsymbol{e}_2 - \boldsymbol{e}_3 \otimes \boldsymbol{e}_3. \tag{8.9}$$

Under the assumption of associative hardening (for further details, refer to the text surrounding equation (6.196), page 183), the accumulated plastic strain for this model is defined by means of its evolution equations, which are given respectively by

$$\dot{\bar{\varepsilon}}^p = \dot{\gamma} \tag{8.10}$$

for plastic flow from the main plane (item 1 above), and

$$\dot{\bar{\varepsilon}}^p = \dot{\gamma}^a + \dot{\gamma}^b \tag{8.11}$$

for plastic flow from the right and left corners (items 2 and 3, respectively).

8.1.1. THE IMPLICIT INTEGRATION ALGORITHM IN PRINCIPAL STRESSES

The derivation of the implicit elastic predictor/plastic corrector integration algorithm for the Tresca model follows much the same path as the derivation of algorithm for the von Mises model, with an extra degree of complexity introduced due to the existence of singularities (corners) on the yield surface. Early numerical implementations of singular-surface models have resorted to the smoothing of corners leading to singularity-free formulations (Nayak and Zienkiewicz, 1972; Owen and Hinton, 1980). In such approaches, however, numerical instabilities frequently arise in connection with the high curvature of the smoothed corners. More recently, formulations based on the multisurface-based description of the plastic flow have been shown to avoid such instabilities providing an effective solution to the integration of elastoplastic models with non-smooth yield surfaces (Crisfield, 1987; de Borst, 1987; Pankaj and Bićanić, 1989; Pramono and Willam, 1989; Simo *et al.*, 1988b). An alternative approach based on a semi-analytical integration algorithm is described by Sloan and Booker (1992). It is worth remarking that in most of the computational plasticity literature – especially in the derivation of return-mapping-type schemes – the numerical treatment of the Tresca model is based on the representation of the Tresca yield function in terms of *stress invariants* (refer to expression (6.94), page 160). The main drawback of such a representation lies in the fact that the Tresca yield function is a highly nonlinear function of the stress invariants and its use results in rather intricate expressions for the corresponding integration algorithm and consistent tangent operator. The algorithm described in what follows was proposed by Perić and de Souza Neto (1999) and differs from the invariant-based procedures in that it relies entirely on the *principal stress* representation of the Tresca criterion. The principal stress-based algorithm results in a simpler computational implementation of this particular model. For instance, note that the Tresca surface can be represented by means of a set of *linear* functions of the principal stresses.

The elastic predictor stage here is identical to the one in the von Mises model implementation so that, given the incremental strain $\Delta\varepsilon$, and the state variables at t_n, the elastic trial state is computed as in (7.80)–(7.83).

The next step is to verify whether the trial state violates the plastic consistency constraint. At this point, the algorithm for the Tresca model becomes substantially different from the previously described von Mises algorithm. Due to the principal stress approach adopted here, the *spectral decomposition* of the elastic trial stress is required prior to the consistency check. The principal elastic trial stresses, σ_i^{trial} $(i = 1, 2, 3)$, and the corresponding eigenvectors, $\boldsymbol{e}_i$, are computed in closed form by means of the formulae described in Appendix A. With the principal trial stresses ordered as $\sigma_1^{\text{trial}} \geq \sigma_2^{\text{trial}} \geq \sigma_3^{\text{trial}}$ (for notational convenience, the subscript $n+1$ is omitted from the trial and updated principal stresses), the consistency check follows as:

- If $\Phi^{\text{trial}} \equiv \sigma_1^{\text{trial}} - \sigma_3^{\text{trial}} - \sigma_y(\bar{\varepsilon}_n^p) \leq \epsilon_{\text{tol}}$, then the process is elastic and

$$(\cdot)_{n+1} := (\cdot)_{n+1}^{\text{trial}}. \tag{8.12}$$

- Otherwise, the updated state at t_{n+1} is obtained by the return-mapping procedure described in what follows.

It is emphasised here that, as in the algorithm for the von Mises model, substantial simplification is possible when the general implicit return mapping equations (7.25) are particularized for the Tresca model. To derive the simplified equations, consider the implicit update formula for the elastic strain obtained as a result of the discretisation of (8.1):

$$\boldsymbol{\varepsilon}^e_{n+1} = \boldsymbol{\varepsilon}^{e\ \text{trial}}_{n+1} - \sum_{i=1}^{6} \Delta\gamma^i \boldsymbol{N}^i_{n+1}. \tag{8.13}$$

Due to the isotropic linear elastic law and the fact that the Tresca flow vector is purely deviatoric, this expression is equivalent, as in the von Mises model, to independent update formulae for the hydrostatic and the deviatoric stress:

$$\begin{aligned} p_{n+1} &= p^{\text{trial}}_{n+1} \\ \boldsymbol{s}_{n+1} &= \boldsymbol{s}^{\text{trial}}_{n+1} - 2G \sum_{i=1}^{6} \Delta\gamma^i \boldsymbol{N}^i_{n+1}. \end{aligned} \tag{8.14}$$

Note that by definition of the flow vectors, $\boldsymbol{N}^i_{n+1}$ and $\boldsymbol{s}_{n+1}$ share the same principal directions. Thus, expression $(8.14)_2$ implies that these principal directions are also shared by $\boldsymbol{s}^{\text{trial}}_{n+1}$ so that the update formula for the deviatoric stress can be equivalently expressed in terms of principal stresses as†

$$s_j = s^{\text{trial}}_j - 2G \sum_{i=1}^{6} \Delta\gamma^i N^i_j, \quad (j = 1, 2, 3), \tag{8.15}$$

where N^i_j denotes the eigenvalues of $\boldsymbol{N}^i_{n+1}$.

The three possible return mappings

Crucial to the derivation of the return-mapping algorithm is the observation that the plastic flow rule (8.5)–(8.9) has three different explicit forms. The actual form to be used in discretised form to update the stress depends on the location of the (unknown) updated stress on the (unknown) updated yield surface. The discretisation of (8.5)–(8.9) results in the three following possibilities for the return-mapping algorithm:

1. The updated stress lies on the *main plane*. In this case, there is only one possible non-zero multiplier and the flow vector is given by (8.6). This substituted in (8.15) gives the following update formula for the principal deviatoric stresses:

$$\begin{aligned} s_1 &= s^{\text{trial}}_1 - 2G\,\Delta\gamma \\ s_2 &= s^{\text{trial}}_2 \\ s_3 &= s^{\text{trial}}_3 + 2G\,\Delta\gamma. \end{aligned} \tag{8.16}$$

   On the main plane, the update formula for the accumulated plastic strain is given by the implicit discrete counterpart of (8.10):

$$\bar{\varepsilon}^p_{n+1} = \bar{\varepsilon}^p_n + \Delta\gamma. \tag{8.17}$$

†The update formula in terms of principal deviatoric stresses is equally valid for the von Mises model. However, for the von Mises model, the return-mapping algorithm (described in the previous section) turns out to be more efficient when the standard formulation in terms of the deviatoric stress tensor components (rather than its eigenvalues) is adopted.

The return mapping, in the present case, is obtained with the introduction of (8.16) and (8.17) into the discrete consistency condition $\Phi_1(\boldsymbol{\sigma}_{n+1}, \sigma_y(\bar{\varepsilon}^p_{n+1})) = 0$. This results in the following *scalar* (generally nonlinear) equation in $\Delta\gamma$:

$$\tilde{\Phi}(\Delta\gamma) \equiv s_1^{\text{trial}} - s_3^{\text{trial}} - 4G\,\Delta\gamma - \sigma_y(\bar{\varepsilon}^p_n + \Delta\gamma) = 0. \tag{8.18}$$

2. The updated stress lies on the *right* corner. Here, two plastic multipliers may be non-zero and the incremental plastic strain is obtained from the discretisation of (8.7) as

$$\Delta\boldsymbol{\varepsilon}^p = \Delta\gamma^a\,\boldsymbol{N}^a_{n+1} + \Delta\gamma^b\,\boldsymbol{N}^b_{n+1}, \tag{8.19}$$

where $\boldsymbol{N}^a$ and $\boldsymbol{N}^b$ denote, respectively, the normals to the main plane and to the plane on its right side given by (8.6) and (8.8). This results in the following explicit expression for the update of the principal deviatoric stresses:

$$\begin{aligned} s_1 &= s_1^{\text{trial}} - 2G\,(\Delta\gamma^a + \Delta\gamma^b) \\ s_2 &= s_2^{\text{trial}} + 2G\,\Delta\gamma^b \\ s_3 &= s_3^{\text{trial}} + 2G\,\Delta\gamma^a. \end{aligned} \tag{8.20}$$

The incremental law for $\bar{\varepsilon}^p$ in this case is the discrete counterpart of (8.11):

$$\bar{\varepsilon}^p_{n+1} = \bar{\varepsilon}^p_n + \Delta\gamma^a + \Delta\gamma^b. \tag{8.21}$$

Analogously to the one-vector return mapping of item 1, the *two-vector* return-mapping equation is obtained here by introducing (8.20) and (8.21) into the discrete consistency condition resulting from (8.3):

$$\Phi^a_{n+1} \equiv \Phi_1(\boldsymbol{\sigma}_{n+1}, \sigma_y(\bar{\varepsilon}^p_{n+1})) = 0, \quad \Phi^b_{n+1} \equiv \Phi_6(\boldsymbol{\sigma}_{n+1}, \sigma_y(\bar{\varepsilon}^p_{n+1})) = 0. \tag{8.22}$$

This yields the following set of two algebraic equations for $\Delta\gamma^a$ and $\Delta\gamma^b$:

$$\begin{aligned} \tilde{\Phi}^a(\Delta\gamma^a, \Delta\gamma^b) &\equiv s_1^{\text{trial}} - s_3^{\text{trial}} - 2G(2\Delta\gamma^a + \Delta\gamma^b) - \tilde{\sigma}_y(\Delta\gamma^a, \Delta\gamma^b) = 0 \\ \tilde{\Phi}^b(\Delta\gamma^a, \Delta\gamma^b) &\equiv s_1^{\text{trial}} - s_2^{\text{trial}} - 2G(\Delta\gamma^a + 2\Delta\gamma^b) - \tilde{\sigma}_y(\Delta\gamma^a, \Delta\gamma^b) = 0, \end{aligned} \tag{8.23}$$

where $\tilde{\sigma}_y$ has been defined as

$$\tilde{\sigma}_y(\Delta\gamma^a, \Delta\gamma^b) \equiv \sigma_y(\bar{\varepsilon}^p_n + \Delta\gamma^a + \Delta\gamma^b). \tag{8.24}$$

3. The updated stress lies on the *left* corner. The situation now is completely analogous to item 2 above with the difference that, here, $\boldsymbol{N}^b$ is the normal to the plane on the *left* of the main plane, given by (8.9). The deviatoric principal stress update formula in the present case reads

$$\begin{aligned} s_1 &= s_1^{\text{trial}} - 2G\,\Delta\gamma^a \\ s_2 &= s_2^{\text{trial}} - 2G\,\Delta\gamma^b \\ s_3 &= s_3^{\text{trial}} + 2G\,(\Delta\gamma^a + \Delta\gamma^b). \end{aligned} \tag{8.25}$$

The update formula for $\bar{\varepsilon}^p$ is the same as for the return map to the right corner and the final equations to be solved in the return-mapping algorithm are

$$\tilde{\Phi}^a(\Delta\gamma^a, \Delta\gamma^b) \equiv s_1^{\text{trial}} - s_3^{\text{trial}} - 2G(2\Delta\gamma^a + \Delta\gamma^b) - \tilde{\sigma}_y(\Delta\gamma^a, \Delta\gamma^b) = 0$$
$$\tilde{\Phi}^b(\Delta\gamma^a, \Delta\gamma^b) \equiv s_2^{\text{trial}} - s_3^{\text{trial}} - 2G(\Delta\gamma^a + 2\Delta\gamma^b) - \tilde{\sigma}_y(\Delta\gamma^a, \Delta\gamma^b) = 0. \tag{8.26}$$

The last of the above equations enforces

$$\Phi_2(\boldsymbol{\sigma}_{n+1}, \sigma_y(\bar{\varepsilon}^p_{n+1})) = 0. \tag{8.27}$$

In summary, it has been shown above that the essential return-mapping algorithm for the Tresca model, formulated in principal deviatoric stresses, may have three different explicit forms. The particular form to be employed depends on the position of the updated stress on the yield surface. In any case, similarly to the von Mises return mapping, an equation (or system of two equations for the corners) is solved firstly for the plastic multiplier $\Delta\gamma$ ($\Delta\gamma^a$ and $\Delta\gamma^b$ for the corners). With the plastic multiplier(s) at hand, the principal deviatoric stresses, s_i, and the accumulated plastic strain, $\bar{\varepsilon}^p$, are updated by the formulae corresponding to the particular case considered (main plane, right or left corner). The updated stress tensor components, required in the finite element computations, are then obtained simply by assembling

$$\boldsymbol{\sigma}_{n+1} := \sum_{i=1}^{3} (s_i + p_{n+1})\, \boldsymbol{e}_i \otimes \boldsymbol{e}_i. \tag{8.28}$$

Selection of the appropriate return mapping

Having described the possible return procedures, the question now is: as the final location of the updated stress on the (updated) yield surface is not known in advance, how can one decide which return mapping to apply in the actual computational implementation of the model? In the present case, a rather simple and effective algorithm to select the appropriate return mapping can be derived based on the geometric characteristics of the Tresca surface. It is remarked that the selection algorithm, described below, ensures that the resulting updated state rigorously satisfies the general implicit return-mapping equations (7.25).

Firstly, consider the application of the return algorithm to the *main plane* (item 1 above) for an arbitrary trial stress. The geometric interpretation is illustrated in Figure 8.1. In this case, the updated deviatoric stress, $\boldsymbol{s}_{n+1}$, lies necessarily on the plane

$$s_1 - s_3 - \sigma_y(\bar{\varepsilon}^p_{n+1}) = 0. \tag{8.29}$$

Upon application of the main plane return algorithm, the final stress can end up either inside or outside the main sextant:

$$s_1 \geq s_2 \geq s_3, \tag{8.30}$$

represented by the shaded areas in Figure 8.1. If the updated principal stresses satisfy the above relation, then they are in the main sextant (Figure 8.1(a)), and, clearly, satisfy the consistency condition, i.e. they lie on the updated yield surface, and the result of the return mapping to the main plane is valid. If the updated stress falls outside the main sextant

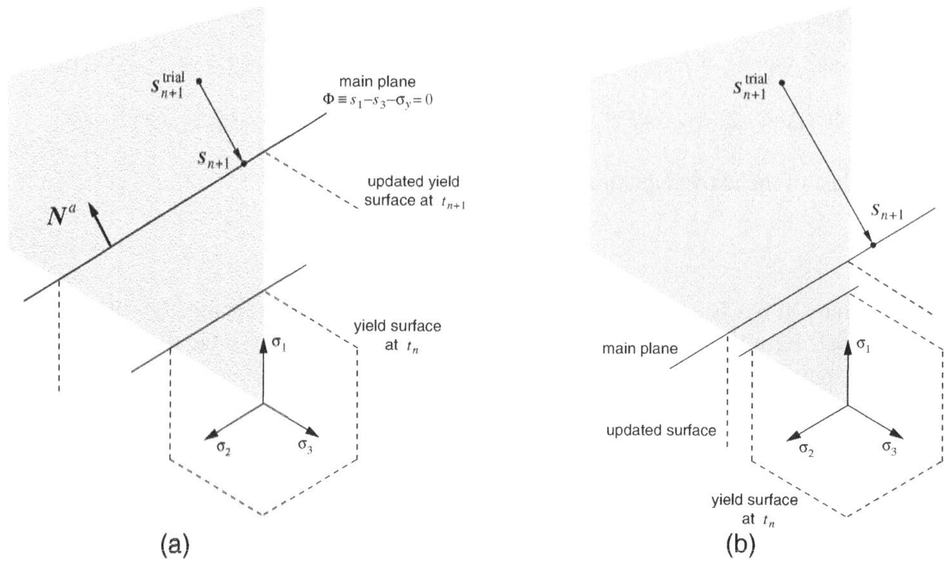

Figure 8.1. The implicit algorithm for the Tresca model in principal stresses: (a) valid return to main plane; and (b) invalid return to main plane – converged stress outside the original sextant. (Reproduced with permission from A new computational model for Tresca plasticity at finite strains with an optimal parametrization in the principal space, D Perić and EA de Souza Neto, *Computer Methods in Applied Mechanics and Engineering*, Vol 171 © 1999 Elsevier Science S.A.)

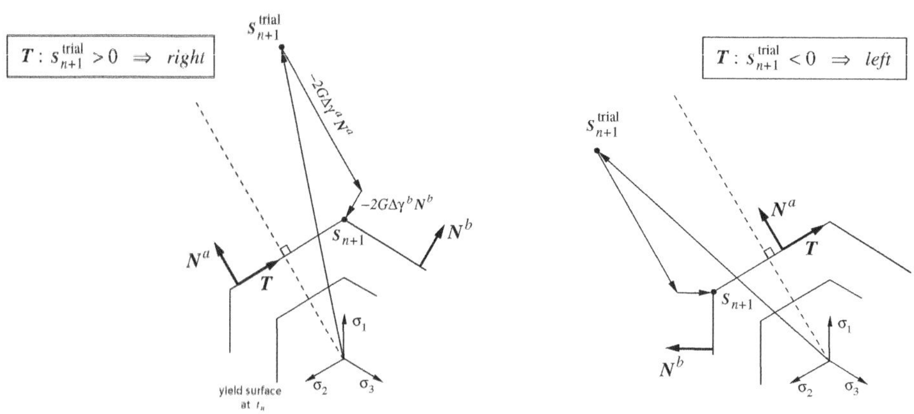

Figure 8.2. The implicit algorithm for the Tresca model in principal stresses. Selection of the appropriate return mapping to corner. (Reproduced with permission from A new computational model for Tresca plasticity at finite strains with an optimal parametrization in the principal space, D Perić and EA de Souza Neto, *Computer Methods in Applied Mechanics and Engineering*, Vol 171 © 1999 Elsevier Science S.A.)

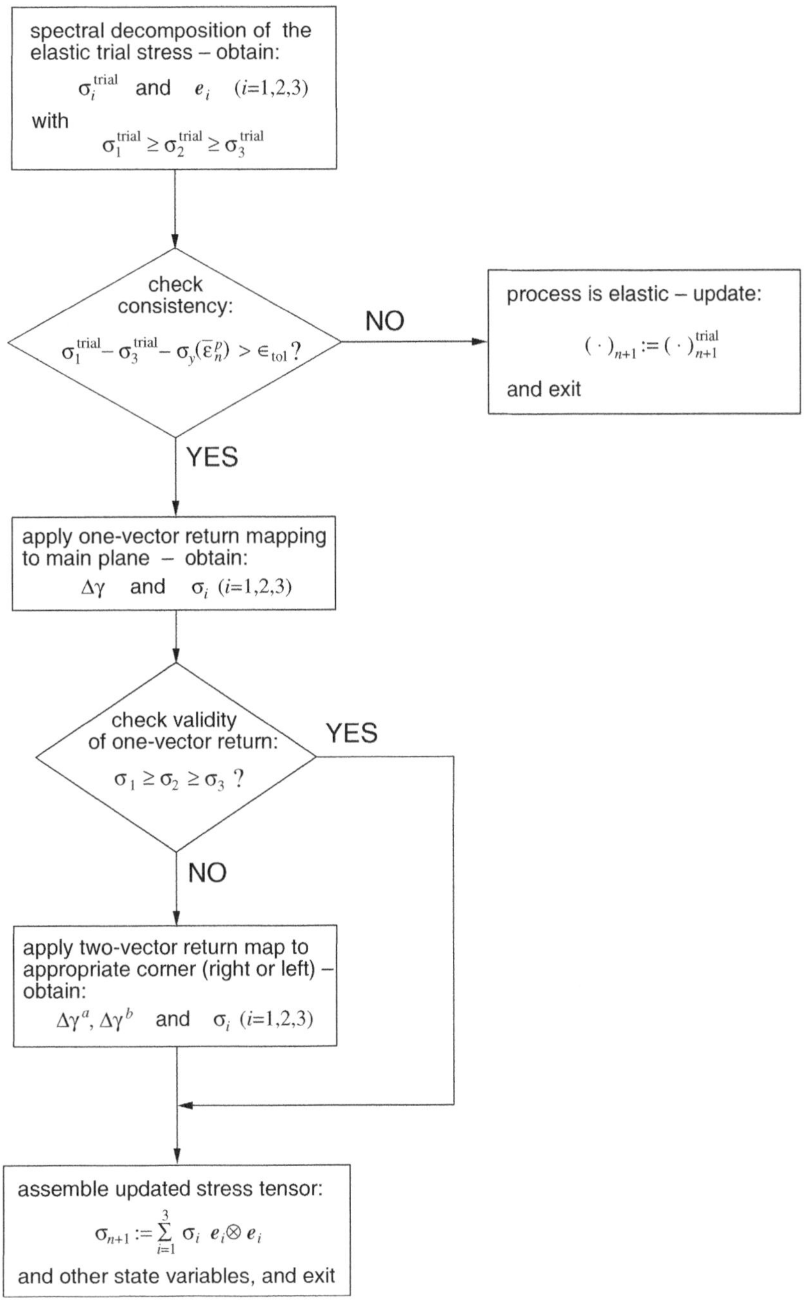

Figure 8.3. Flowchart of the integration algorithm for the Tresca model in principal stresses. Procedure implemented in subroutine SUTR of program HYPLAS.

(Figure 8.1(b)), i.e. if relation (8.30) is not satisfied, then, in spite of being on the main plane, the updated stress lies outside the updated elastic domain and, therefore, violates the consistency condition. In this case, the result of the return to the main plane is obviously *not* valid and the correct algorithm to be applied must be either the return to the left or to right corner, described, respectively, in items 2 and 3 above. The appropriate algorithm to be applied (right or left) can be easily determined by, again, considering the geometric properties of the Tresca yield surface: if $\boldsymbol{s}_{n+1}^{\text{trial}}$ lies on the right side of the line that passes through the origin of stresses and is orthogonal to the main plane (the pure shear line), represented by the dotted line in Figure 8.2, then it can be easily visualized that $\boldsymbol{s}_{n+1}^{\text{trial}}$ could only return to the left corner if $\Delta\gamma^b$ were negative, which is unacceptable. In this case, the only possible return is to the right corner. Using the same argument, one concludes that if the trial deviatoric stress is on the left side of the dotted line, then the appropriate return is to the left corner. It is very simple to determine, in actual computations, whether $\boldsymbol{s}_{n+1}^{\text{trial}}$ lies on the right or left of the dotted line. Let $\boldsymbol{T}$ be a tangent vector to the main plane as illustrated in Figure 8.2. Clearly, $\boldsymbol{T}$ has the same eigenvectors as $\boldsymbol{s}_{n+1}^{\text{trial}}$ and its eigenvalues may be chosen as

$$T_1 = 1, \quad T_2 = -2, \quad T_3 = 1. \tag{8.31}$$

If the scalar product

$$\boldsymbol{T} : \boldsymbol{s}_{n+1}^{\text{trial}} = \sum_{i=1}^{3} T_i \, s_i^{\text{trial}} = s_1^{\text{trial}} + s_3^{\text{trial}} - 2\, s_2^{\text{trial}}, \tag{8.32}$$

between $\boldsymbol{T}$ and $\boldsymbol{s}_{n+1}^{\text{trial}}$, is positive, then $\boldsymbol{s}_{n+1}^{\text{trial}}$ is on the right side of the dotted line and the return mapping to the right corner is applied. If the product is negative, then the trial stress is on the left side and the return mapping to the left corner is applied.

With the appropriate choice of return mappings based on the above, the implicit elastic predictor/return-mapping algorithm for the Tresca model is completely defined. The flowchart of the procedure is illustrated in Figure 8.3. The algorithm, implemented in subroutine SUTR (**S**tate **U**pdate procedure for the **TR**esca model), is described in detail in Boxes 8.1–8.3. Box 8.1 shows the main algorithm and Boxes 8.2 and 8.3 show, respectively, the Newton–Raphson scheme for solution of the return mapping to the main plane and the two-vector return mappings to the corners.

Box 8.1. Implicit elastic predictor/return-mapping algorithm for the Tresca model.

HYPLAS procedure:	SUTR

(i) Elastic predictor. Given $\Delta\boldsymbol{\varepsilon}$ and the state variables at t_n, evaluate the *elastic trial state*

$$\boldsymbol{\varepsilon}^{e\,\text{trial}}_{n+1} := \boldsymbol{\varepsilon}^{e}_{n} + \Delta\boldsymbol{\varepsilon}; \quad \bar{\varepsilon}^{p\,\text{trial}}_{n+1} := \bar{\varepsilon}^{p}_{n}$$

$$p^{\text{trial}}_{n+1} := K\,\varepsilon^{e\,\text{trial}}_{\text{v}\,n+1}; \quad \boldsymbol{s}^{\text{trial}}_{n+1} := 2G\,\boldsymbol{\varepsilon}^{e\,\text{trial}}_{\text{d}\,n+1}$$

(ii) Spectral decomposition of $\boldsymbol{s}^{\text{trial}}$ (routine SPDEC2). Compute

$$s^{\text{trial}}_1 \geq s^{\text{trial}}_2 \geq s^{\text{trial}}_3 \quad \text{and} \quad \boldsymbol{e}_i\ (i = 1, 2, 3)$$

(iii) Check plastic admissibility

$$\text{IF } s^{\text{trial}}_1 - s^{\text{trial}}_3 - \sigma_y(\bar{\varepsilon}^{p\,\text{trial}}_{n+1}) \leq 0$$

$$\text{THEN set } \quad (\cdot)_{n+1} := (\cdot)^{\text{trial}}_{n+1} \quad \text{and EXIT}$$

(iv) Return mapping

(iv.a) Return to main plane – GOTO Box 8.2

(iv.b) Check validity of main plane return

IF $s_1 \geq s_2 \geq s_3$ THEN return is valid – GOTO (v)

(iv.c) Return to corner

IF $s^{\text{trial}}_1 + s^{\text{trial}}_3 - 2\,s^{\text{trial}}_2 > 0$

THEN apply return to **right** corner – GOTO Box 8.3

ELSE apply return to **left** corner – GOTO Box 8.3

(v) Assemble updated stress

$$p_{n+1} := p^{\text{trial}}_{n+1}$$

$$\boldsymbol{\sigma}_{n+1} := \textstyle\sum_{i=1}^{3} (s_i + p_{n+1})\,\boldsymbol{e}_i \otimes \boldsymbol{e}_i$$

and update elastic strain

$$\boldsymbol{\varepsilon}^{e}_{n+1} := \frac{1}{2G}\,\boldsymbol{s}_{n+1} + \frac{1}{3}\,\varepsilon^{e\;trial}_{\text{v}\;n+1}\,\boldsymbol{I}$$

(vi) EXIT

Box 8.2. The Tresca model. One-vector return mapping to main plane.

HYPLAS procedure:	SUTR

(i) Set initial guess for $\Delta\gamma$

$$\Delta\gamma := 0$$

and corresponding residual (yield function value)

$$\tilde{\Phi} := s_1^{\text{trial}} - s_3^{\text{trial}} - \sigma_y(\bar{\varepsilon}_n^p)$$

(ii) Perform Newton–Raphson iteration

$$H := \left.\frac{\mathrm{d}\sigma_y}{\mathrm{d}\bar{\varepsilon}^p}\right|_{\bar{\varepsilon}_n^p + \Delta\gamma} \qquad \text{(hardening slope)}$$

$$d := \frac{\mathrm{d}\tilde{\Phi}}{\mathrm{d}\Delta\gamma} = -4G - H \qquad \text{(residual derivative)}$$

$$\Delta\gamma := \Delta\gamma - \frac{\tilde{\Phi}}{d} \qquad \text{(new guess for } \Delta\gamma)$$

(iii) Check for convergence

$$\tilde{\Phi} := s_1^{\text{trial}} - s_3^{\text{trial}} - 4G\,\Delta\gamma - \sigma_y(\bar{\varepsilon}_n^p + \Delta\gamma)$$

IF $|\tilde{\Phi}| \leq \epsilon_{\text{tol}}$ THEN update

$$s_1 := s_1^{\text{trial}} - 2G\,\Delta\gamma$$

$$s_2 := s_2^{\text{trial}}$$

$$s_3 := s_3^{\text{trial}} + 2G\,\Delta\gamma$$

$$\bar{\varepsilon}_{n+1}^p := \bar{\varepsilon}_n^p + \Delta\gamma$$

and RETURN to Box 8.1

(iv) GOTO (ii)

Box 8.3. The Tresca model. Two-vector return mappings to corners.

HYPLAS procedure:	SUTR

(i) Set initial guess for $\Delta\gamma^a$ and $\Delta\gamma^b$

$$\Delta\gamma^a := 0 \qquad \Delta\gamma^b := 0$$

and corresponding residual

$$\begin{bmatrix} \tilde{\Phi}^a \\ \tilde{\Phi}^b \end{bmatrix} := \begin{bmatrix} \bar{s}^a - \sigma_y(\bar{\varepsilon}^p_n) \\ \bar{s}^b - \sigma_y(\bar{\varepsilon}^p_n) \end{bmatrix}$$

where

$$\bar{s}^a = s_1^{\text{trial}} - s_3^{\text{trial}}, \qquad \bar{s}^b = \begin{cases} s_1^{\text{trial}} - s_2^{\text{trial}}, & \text{for } \textbf{right} \text{ corner} \\ s_2^{\text{trial}} - s_3^{\text{trial}}, & \text{for } \textbf{left} \text{ corner} \end{cases}$$

(ii) Perform Newton–Raphson iteration

$$\overline{\Delta\gamma} := \Delta\gamma^a + \Delta\gamma^b$$

$$\bar{\varepsilon}^p_{n+1} := \bar{\varepsilon}^p_n + \overline{\Delta\gamma} \quad \text{(update } \bar{\varepsilon}^p\text{)}$$

$$H := \left.\frac{\mathrm{d}\sigma_y}{\mathrm{d}\bar{\varepsilon}^p}\right|_{\bar{\varepsilon}^p_{n+1}} \qquad \text{(hardening slope)}$$

residual derivative:

$$\mathbf{d} := \begin{bmatrix} \dfrac{\mathrm{d}\tilde{\Phi}^a}{\mathrm{d}\Delta\gamma^a} & \dfrac{\mathrm{d}\tilde{\Phi}^a}{\mathrm{d}\Delta\gamma^b} \\ \dfrac{\mathrm{d}\tilde{\Phi}^b}{\mathrm{d}\Delta\gamma^a} & \dfrac{\mathrm{d}\tilde{\Phi}^b}{\mathrm{d}\Delta\gamma^b} \end{bmatrix} = \begin{bmatrix} -4G - H & -2G - H \\ -2G - H & -4G - H \end{bmatrix}$$

new guess for $\Delta\gamma^a$ and $\Delta\gamma^b$:

$$\begin{bmatrix} \Delta\gamma^a \\ \Delta\gamma^b \end{bmatrix} := \begin{bmatrix} \Delta\gamma^a \\ \Delta\gamma^b \end{bmatrix} - \mathbf{d}^{-1} \begin{bmatrix} \tilde{\Phi}^a \\ \tilde{\Phi}^b \end{bmatrix}$$

continued on page 278

Box 8.3 (contd. from page 277). The Tresca model. Two-vector return mappings to corners (implemented in subroutine SUTR).

(iii) Check for convergence

$$\begin{bmatrix} \tilde{\Phi}^a \\ \tilde{\Phi}^b \end{bmatrix} := \begin{bmatrix} \bar{s}^a - 2G(2\Delta\gamma^a + \Delta\gamma^b) - \sigma_y(\bar{\varepsilon}^p_{n+1}) \\ \bar{s}^b - 2G(\Delta\gamma^a + 2\Delta\gamma^b) - \sigma_y(\bar{\varepsilon}^p_{n+1}) \end{bmatrix}$$

IF $|\tilde{\Phi}^a| + |\tilde{\Phi}^b| \leq \epsilon_{\text{tol}}$ THEN update

$$\left.\begin{aligned} s_1 &:= s_1^{\text{trial}} - 2G(\Delta\gamma^a + \Delta\gamma^b) \\ s_2 &:= s_2^{\text{trial}} + 2G\Delta\gamma^b \\ s_3 &:= s_3^{\text{trial}} + 2G\Delta\gamma^a \end{aligned}\right\} \text{for **right** corner}$$

$$\left.\begin{aligned} s_1 &:= s_1^{\text{trial}} - 2G\Delta\gamma^a \\ s_2 &:= s_2^{\text{trial}} - 2G\Delta\gamma^b \\ s_3 &:= s_3^{\text{trial}} + 2G(\Delta\gamma^a + \Delta\gamma^b) \end{aligned}\right\} \text{for **left** corner}$$

and RETURN to Box 8.1

(iv) GOTO (ii)

8.1.2. SUBROUTINE SUTR

As in the von Mises state update procedure implemented in subroutine SUVM (Section 7.3.5, starting page 224), *piecewise linear isotropic hardening* is assumed in the present implementation of the implicit integration algorithm for the Tresca model. This hardening rule is adopted due to its flexibility in fitting experimental data. The reader is referred to Section 7.3.5 for details of implementation of the piecewise linear hardening law.

The FORTRAN source code of SUTR is listed below.

```
      SUBROUTINE SUTR
     1(   DGAM       ,IPROPS     ,LALGVA     ,NTYPE      ,RPROPS     ,
     2    RSTAVA     ,STRAT      ,STRES      )
      IMPLICIT DOUBLE PRECISION (A-H,O-Z)
      PARAMETER(IPHARD=4  ,MSTRE=4)
C Arguments
      LOGICAL
     1    LALGVA(4)
      DIMENSION
     1    DGAM(2)              ,IPROPS(*)            ,RPROPS(*)            ,
     2    RSTAVA(MSTRE+1)      ,STRAT(MSTRE)         ,STRES(MSTRE)
C Local arrays and variables
      LOGICAL
     1    DUMMY, IFPLAS, RIGHT, SUFAIL, TWOVEC
      DIMENSION
     1    EIGPRJ(MSTRE,2)      ,PSTRS(3)             ,STREST(3)

      DATA
     1    R0   ,R1   ,R2   ,R3   ,R4   ,SMALL ,TOL   /
     2    0.0D0,1.0D0,2.0D0,3.0D0,4.0D0,1.D-10,1.D-10/
      DATA MXITER / 50 /
C***********************************************************************
C STRESS UPDATE PROCEDURE FOR TRESCA TYPE ELASTO-PLASTIC MATERIAL WITH
C PIECE-WISE LINEAR ISOTROPIC HARDENING:
C IMPLICIT ELASTIC PREDICTOR/RETURN MAPPING ALGORITHM (Boxes 8.1-3).
C PLANE STRAIN AND AXISYMMETRIC IMPLEMENTATIONS.
C***********************************************************************
C Stops program if neither plane strain nor axisymmetric state
      IF(NTYPE.NE.2.AND.NTYPE.NE.3)CALL ERRPRT('EI0029')
C Initialize some algorithmic and internal variables
      DGAMA=R0
      DGAMB=R0
      IFPLAS=.FALSE.
      SUFAIL=.FALSE.
      EPBARN=RSTAVA(MSTRE+1)
      EPBAR=EPBARN
C Set some material properties
      YOUNG=RPROPS(2)
      POISS=RPROPS(3)
      NHARD=IPROPS(3)
C Set some constants
      GMODU=YOUNG/(R2*(R1+POISS))
      BULK=YOUNG/(R3*(R1-R2*POISS))
      R2G=R2*GMODU
      R4G=R4*GMODU
      R1D3=R1/R3
C Compute elastic trial state
C ---------------------------
C Volumetric strain and pressure stress
      EEV=STRAT(1)+STRAT(2)+STRAT(4)
      P=BULK*EEV
```

```
C Spectral decomposition of the elastic trial deviatoric stress
      EEVD3=EEV*R1D3
      STREST(1)=R2G*(STRAT(1)-EEVD3)
      STREST(2)=R2G*(STRAT(2)-EEVD3)
      STREST(3)=GMODU*STRAT(3)
      CALL SPDEC2(EIGPRJ,PSTRS,DUMMY,STREST)
      PSTRS(3)=R2G*(STRAT(4)-EEVD3)
C Identify maximum (PSTRS1) and minimum (PSTRS3) principal stresses
      II=1
      JJ=1
      PSTRS1=PSTRS(II)
      PSTRS3=PSTRS(JJ)
      DO 10 I=2,3
        IF(PSTRS(I).GE.PSTRS1)THEN
          II=I
          PSTRS1=PSTRS(II)
        ENDIF
        IF(PSTRS(I).LT.PSTRS3)THEN
          JJ=I
          PSTRS3=PSTRS(JJ)
        ENDIF
   10 CONTINUE
      IF(II.NE.1.AND.JJ.NE.1)MM=1
      IF(II.NE.2.AND.JJ.NE.2)MM=2
      IF(II.NE.3.AND.JJ.NE.3)MM=3
      PSTRS2=PSTRS(MM)
C Compute trial yield function and check for plastic consistency
C -----------------------------------------------------------------
      SHMAXT=PSTRS1-PSTRS3
      SIGMAY=PLFUN(EPBARN,NHARD,RPROPS(IPHARD))
      PHIA=SHMAXT-SIGMAY
      IF(PHIA/SIGMAY.GT.TOL)THEN
C Plastic step: Apply return mapping
C ===================================
        IFPLAS=.TRUE.
C identify possible two-vector return: right or left of main plane
        SCAPRD=PSTRS1+PSTRS3-PSTRS2*R2
        IF(SCAPRD.GE.R0)THEN
          RIGHT=.TRUE.
        ELSE
          RIGHT=.FALSE.
        ENDIF
C Apply one-vector return mapping first (return to main plane)
C ---------------------------------------------------------------
        TWOVEC=.FALSE.
C Start Newton-Raphson iterations
        DO 20 NRITER=1,MXITER
C Compute residual derivative
          DENOM=-R4G-DPLFUN(EPBAR,NHARD,RPROPS(IPHARD))
C Compute Newton-Raphson increment and update variable DGAMA
          DDGAMA=-PHIA/DENOM
          DGAMA=DGAMA+DDGAMA
C Compute new residual
          EPBAR=EPBARN+DGAMA
          SIGMAY=PLFUN(EPBAR,NHARD,RPROPS(IPHARD))
          SHMAX=SHMAXT-R4G*DGAMA
          PHIA=SHMAX-SIGMAY
C Check convergence
          RESNOR=ABS(PHIA/SIGMAY)
          IF(RESNOR.LE.TOL)THEN
C Check validity of one-vector return
            S1=PSTRS1-R2G*DGAMA
```

```
          S2=PSTRS2
          S3=PSTRS3+R2G*DGAMA
          DELTA=DMAX1(ABS(S1),ABS(S2),ABS(S3))*SMALL
          IF(S1+DELTA.GE.S2.AND.S2+DELTA.GE.S3)THEN
C converged stress is in the same sextant as trial stress -> 1-vector
C return is valid. Update EPBAR and principal deviatoric stresses
            RSTAVA(MSTRE+1)=EPBAR
            PSTRS1=S1
            PSTRS3=S3
            GOTO 50
          ELSE
C 1-vector return is not valid - go to two-vector procedure
            GOTO 30
          ENDIF
        ENDIF
   20 CONTINUE
C failure of stress update procedure
      SUFAIL=.TRUE.
      CALL ERRPRT('WE0001')
      GOTO 999
   30 CONTINUE
C Apply two-vector return mapping (return to corner - right or left)
C -------------------------------------------------------------------
      TWOVEC=.TRUE.
      DGAMA=R0
      DGABAR=R1
      EPBAR=EPBARN
      SIGMAY=PLFUN(EPBARN,NHARD,RPROPS(IPHARD))
      SHMXTA=PSTRS1-PSTRS3
      IF(RIGHT)THEN
        SHMXTB=PSTRS1-PSTRS2
      ELSE
        SHMXTB=PSTRS2-PSTRS3
      ENDIF
      PHIA=SHMXTA-SIGMAY
      PHIB=SHMXTB-SIGMAY
C Start Newton-Raphson iterations
      DO 40 NRITER=1,MXITER
C Compute residual derivative
        HSLOPE=DPLFUN(EPBAR,NHARD,RPROPS(IPHARD))
        DRVAA=-R4G-HSLOPE
        DRVAB=-R2G-HSLOPE
        DRVBA=-R2G-HSLOPE
        DRVBB=-R4G-HSLOPE
C Compute Newton-Raphson increment and update variables DGAMA and DGAMB
        R1DDET=R1/(DRVAA*DRVBB-DRVAB*DRVBA)
        DDGAMA=(-DRVBB*PHIA+DRVAB*PHIB)*R1DDET
        DDGAMB=(DRVBA*PHIA-DRVAA*PHIB)*R1DDET
        DGAMA=DGAMA+DDGAMA
        DGAMB=DGAMB+DDGAMB
C Compute new residual
        DGABAR=DGAMA+DGAMB
        EPBAR=EPBARN+DGABAR
        SIGMAY=PLFUN(EPBAR,NHARD,RPROPS(IPHARD))
        PHIA=SHMXTA-R2G*(R2*DGAMA+DGAMB)-SIGMAY
        PHIB=SHMXTB-R2G*(DGAMA+R2*DGAMB)-SIGMAY
C Check convergence
        RESNOR=(ABS(PHIA)+ABS(PHIB))/SIGMAY
        IF(RESNOR.LE.TOL)THEN
C Update EPBAR and principal deviatoric stresses
          RSTAVA(MSTRE+1)=EPBAR
          IF(RIGHT)THEN
            PSTRS1=PSTRS1-R2G*(DGAMA+DGAMB)
            PSTRS3=PSTRS3+R2G*DGAMA
```

```
                PSTRS2=PSTRS2+R2G*DGAMB
              ELSE
                PSTRS1=PSTRS1-R2G*DGAMA
                PSTRS3=PSTRS3+R2G*(DGAMA+DGAMB)
                PSTRS2=PSTRS2-R2G*DGAMB
              ENDIF
              GOTO 50
            ENDIF
   40   CONTINUE
C failure of stress update procedure
        SUFAIL=.TRUE.
        CALL ERRPRT('WE0001')
        GOTO 999
   50   CONTINUE
C update stress components
C ------------------------
        PSTRS(II)=PSTRS1
        PSTRS(JJ)=PSTRS3
        PSTRS(MM)=PSTRS2
        STRES(1)=PSTRS(1)*EIGPRJ(1,1)+PSTRS(2)*EIGPRJ(1,2)+P
        STRES(2)=PSTRS(1)*EIGPRJ(2,1)+PSTRS(2)*EIGPRJ(2,2)+P
        STRES(3)=PSTRS(1)*EIGPRJ(3,1)+PSTRS(2)*EIGPRJ(3,2)
        STRES(4)=PSTRS(3)+P
C and elastic engineering strain
        RSTAVA(1)=(STRES(1)-P)/R2G+EEVD3
        RSTAVA(2)=(STRES(2)-P)/R2G+EEVD3
        RSTAVA(3)=STRES(3)/GMODU
        RSTAVA(4)=PSTRS(3)/R2G+EEVD3
      ELSE
C Elastic step: update stress using linear elastic law
C =====================================================
        STRES(1)=STREST(1)+P
        STRES(2)=STREST(2)+P
        STRES(3)=STREST(3)
        STRES(4)=PSTRS(3)+P
C elastic engineering strain
        RSTAVA(1)=STRAT(1)
        RSTAVA(2)=STRAT(2)
        RSTAVA(3)=STRAT(3)
        RSTAVA(4)=STRAT(4)
      ENDIF
  999 CONTINUE
C Update algorithmic variables before exit
C ========================================
      DGAM(1)=DGAMA
      DGAM(2)=DGAMB
      LALGVA(1)=IFPLAS
      LALGVA(2)=SUFAIL
      LALGVA(3)=TWOVEC
      LALGVA(4)=RIGHT
      RETURN
      END
```

The arguments of SUTR

The arguments of this subroutine are identical to those of SUVM for the von Mises model implementation (see list starting on page 227), except for the following:

← DGAM [$\Delta\gamma^a$, $\Delta\gamma^b$]. This argument now is an array that stores up to two incremental plastic multipliers. If the increment is elastic, both multipliers are set to 0. Otherwise, the multiplier(s) are obtained as the solution of one of the three possible return mapping equation sets (main plane, right or left corner).

← LALGVA. In addition to the plastic yielding flag IFPLAS and the state update failure flag SUFAIL used also in SUVM, the array LALGVA of logical algorithnmic flags now stores the two-vector return flag TWOVEC and the right corner return flag RIGHT. The two-vector return flag is set to .TRUE. if one of the two possible two-vector return mappings (right or left corner) is used. It is set to .FALSE. otherwise. The right corner return flag is set to .TRUE. if the possible two-vector return is to the right corner to .FALSE. if to the left corner. The flags TWOVEC and RIGHT are required by subroutine CTTR to decide which elastoplastic tangent operator to compute (consistent with main plane, right or left corner return mapping).

Also, IPROPS(3) and array RPROPS are, for the present material model implementation, set in subroutine RDTR during the data input phase of HYPLAS.

Some local variables and arrays of SUTR

- EIGPRJ [$\boldsymbol{e}_i \otimes \boldsymbol{e}_i$, $i = 1, 2$]. Matrix containing the components of the in-plane eigen-projection tensors of the elastic trial stress.
- PSTRS [$\sigma_i^{\rm trial}$ or σ_i]. Array of principal trial stresses or principal updated stresses.
- STREST [$\boldsymbol{\sigma}_{n+1}^{\rm trial}$]. Array of components of the trial stress tensor.
- TOL [$\epsilon_{\rm tol}$]. Convergence tolerance for the Newton–Raphson algorithm used to solve the return-mapping equations.

Function calls from SUTR

- DPLFUN. Used to evaluate the slope of the piecewise linear hardening curve.
- ERRPRT. Called to send warning message to results file and standard output in case of failure of the return-mapping algorithm. Corresponding message is in file **ERROR.RUN**.
- PLFUN. Used in SUTR as the *piecewise linear* hardening function $\sigma_y(\bar{\varepsilon}^p)$.
- SPDEC2. Called to perform the spectral decomposition of $\boldsymbol{\sigma}_{n+1}^{\rm trial}$.

8.1.3. FINITE STEP ACCURACY: ISO-ERROR MAPS

The use of iso-error maps to assess the accuracy of numerical integration algorithms for elastoplasticity under realistic finite steps has been discussed in Section 7.2.10. This procedure is adopted here to assess the accuracy of the implicit elastic predictor/return-mapping algorithm for the Tresca model described above. Recall that for the von Mises model (given as example in Section 7.2.10) the starting point for the map is immaterial due to the

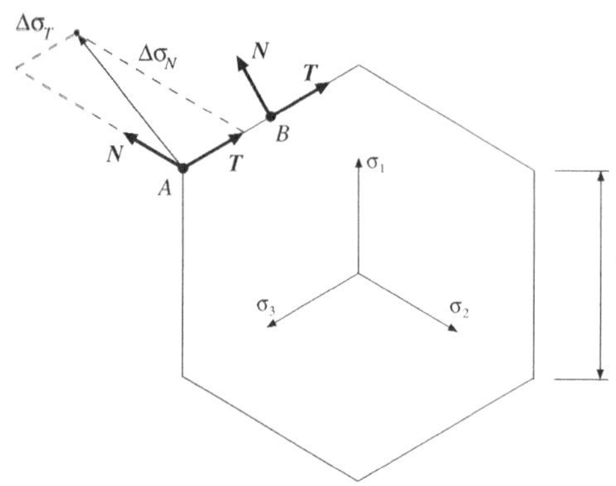

Figure 8.4. Iso-error maps for the implicit algorithm for the Tresca model. Increment directions.

symmetry of the yield surface. For the Tresca model, however, the starting point has to be taken into consideration. Here, two starting points, A and B, lying on the deviatoric plane are considered. These points are shown in Figure 8.4. Point A is on the singularity of the Tresca hexagon and point B corresponds to a state of pure shear. The increments $\Delta\boldsymbol{\sigma}^{\text{trial}}$ are of the form

$$\Delta\boldsymbol{\sigma}^{\text{trial}} = \frac{\Delta\sigma_N}{r}\boldsymbol{N} + \frac{\Delta\sigma_T}{r}\boldsymbol{T}, \tag{8.33}$$

where $\boldsymbol{N}$ and $\boldsymbol{T}$ are unit (in Euclidean norm) tensors defining the increment directions and the scaling factor,

$$r \equiv \sqrt{\tfrac{2}{3}}\,\sigma_y, \tag{8.34}$$

is the size of the sides of the Tresca hexagon (same as the radius of the von Mises cylinder matching the Tresca prism on the edges). Note that, at the singularity (point A), the tensors $\boldsymbol{N}$ and $\boldsymbol{T}$ are not orthogonal to each other. Two cases are considered in the iso-error maps constructed in this section:

- perfect plasticity;
- nonlinear (piecewise linear) isotropic hardening.

In both cases, the error plotted is the (Euclidean) norm of the relative difference between the numerical and 'exact' updated stresses. With isotropic hardening, the hardening curve adopted is shown in Figure 8.5. A total of 16 sampling points have been used to define the curve.

The error maps obtained for the perfectly plastic case are shown in Figure 8.6. Note that for both starting points, there is a large area within which the integration error vanishes. This is obviously a very desirable feature. However, within a narrow band of increments the relative error can be as high as 40% for the increment range considered in the present assessment. The reason for such relatively high errors can be easily explained by graphically performing the projection of $\boldsymbol{\sigma}^{\text{trial}}$ onto the (fixed) yield surface and comparing the resulting stress with the (exact) one obtained when the increment $\Delta\boldsymbol{\sigma}^{\text{trial}}$ is divided into two substeps of suitably

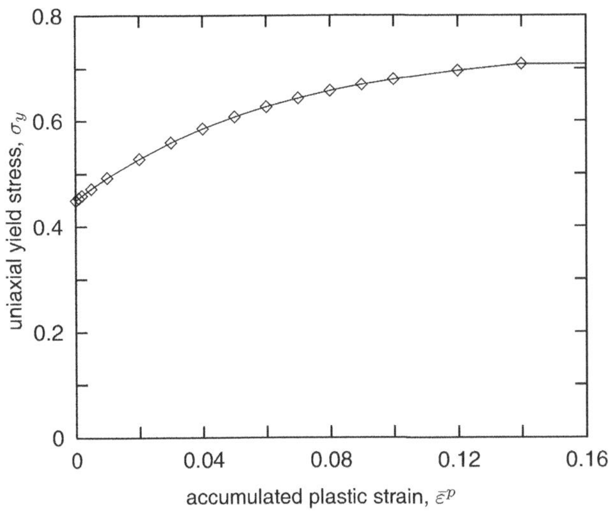

Figure 8.5. Iso-error maps for the Tresca model. Isotropic hardening curve.

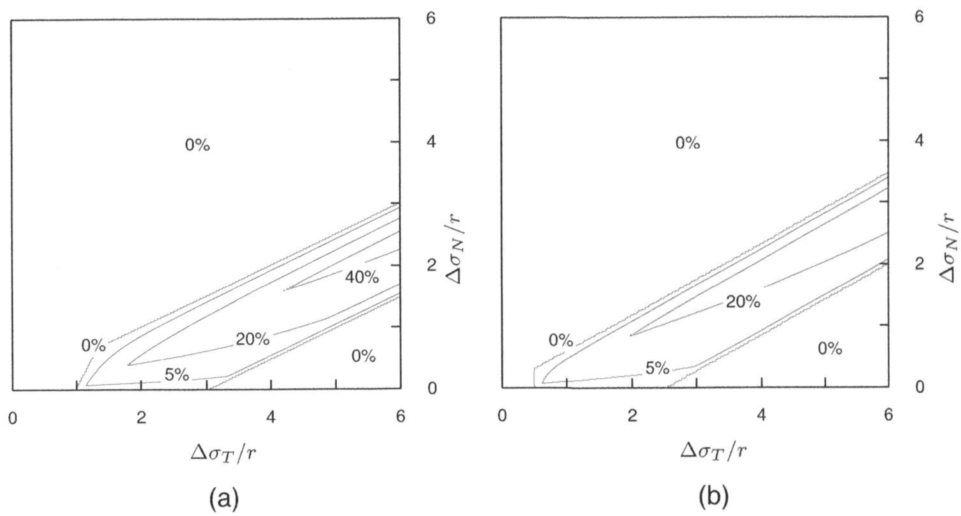

Figure 8.6. Iso-error maps. *Perfectly plastic* Tresca model: (a) point A; and (b) point B.

chosen sizes. In this case (perfect plasticity), the integration error is reduced to zero for any increment if the substepping procedure is included in the return-mapping algorithm. The error maps obtained with isotropic hardening are shown in Figure 8.7. Again, the areas in which the error is reasonably small are quite large, but errors are high within a narrow band of increments. A comparison of the maps obtained with and without hardening suggests that, for practical purposes, the accuracy expected from the implicit algorithm for the Tresca model is not influenced by hardening.

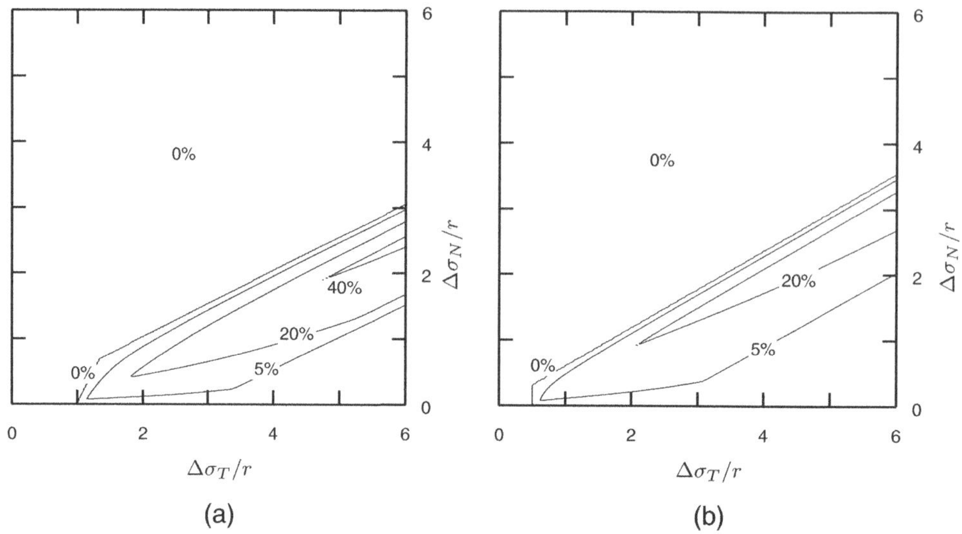

Figure 8.7. Iso-error maps. Tresca model *with hardening*: (a) point A; and (b) point B.

8.1.4. THE CONSISTENT TANGENT OPERATOR FOR THE TRESCA MODEL

The derivation of tangent moduli consistent with elastic predictor/return-mapping schemes for plasticity models has been thoroughly explained in Section 7.4. There, the tangent operator consistent with the implicit algorithm for the von Mises model – the simplest of all elastoplastic algorithms discussed in this book – has been derived in detail, illustrating the application of the concept of consistent tangent operators. In the present section, this concept is applied in the derivation of the tangent modulus consistent with the implicit algorithm for the Tresca model.

Recall that the elastic tangent (the elasticity tensor) is the same for all basic elastoplastic models:

$$\mathsf{D}^e = 2G\,[\mathsf{I}_S - \tfrac{1}{3}\boldsymbol{I}\otimes\boldsymbol{I}] + K\,\boldsymbol{I}\otimes\boldsymbol{I}, \tag{8.35}$$

so that, in what follows, attention will be focused on the derivation of the elastoplastic tangent modulus – the elastoplastic tangent – consistent with the return mapping for the Tresca model.

To start with, recall from Section 7.4.1 that the elastoplastic tangent modulus associated with a particular return-mapping scheme is the derivative

$$\mathsf{D}^{ep} \equiv \frac{\mathrm{d}\boldsymbol{\sigma}_{n+1}}{\mathrm{d}\boldsymbol{\varepsilon}^{e\ \mathrm{trial}}_{n+1}} \tag{8.36}$$

where $\boldsymbol{\sigma}_{n+1}$ is the outcome of the implicit function defined by the corresponding return-mapping equations. For the return mapping for the Tresca model presented in Section 8.1.1, the implicit constitutive function for the stress tensor is defined by the procedures of items (ii), (iv) and (v) of Box 8.1. It is important to note that, for the Tresca model, there are *three possible sets* of equations to be solved:

(i) return to main plane;

(ii) return to right corner; and

(iii) return to left corner,

each one defining a different implicit function for $\boldsymbol{\sigma}_{n+1}$. In this case, the tangent operator used to assemble the stiffness matrix will be the derivative of one of the three possible implicit functions. The particular derivative adopted is chosen as the one consistent with the previous application of the return-mapping algorithm for the Gauss point in question. In other words, if the return to the main plane was used in the previous equilibrium iteration of the current load increment (or the last iteration of the previous converged load increment if the present iteration is the first of the current increment) the tangent modulus will be the derivative of the function associated with the return mapping to the main plane. Similarly, if the return to the right (left) corner was used in previous stress update, the tangent modulus will be consistent with the implicit function defined by the right (left) corner return-mapping equations. The tangent moduli consistent with the three possible sets of equations are derived below.

The elastoplastic tangent: the derivative of an isotropic tensor function

Before deriving the elastoplastic tangent moduli for the Tresca model it is crucial to observe that the corresponding return mappings are effectively carried out in *principal stresses*. Thus, the Tresca return-mapping equations define implicit functions for the principal stresses, σ_i, at t_{n+1} of the form

$$\sigma_i = \tilde{\sigma}_i(\varepsilon_1^{e\ \mathrm{trial}}, \varepsilon_2^{e\ \mathrm{trial}}, \varepsilon_3^{e\ \mathrm{trial}}), \quad i = 1, 2, 3, \tag{8.37}$$

where the eigenprojection tensors, $\boldsymbol{e}_i \otimes \boldsymbol{e}_i$, of $\boldsymbol{\sigma}_{n+1}$ coincide with those of $\boldsymbol{\varepsilon}_{n+1}^{e\ \mathrm{trial}}$. In addition, due to the isotropy of the model, the algorithmic constitutive function for the stress tensor is *isotropic*.

In summary, the integration algorithm for the Tresca model defines the stress tensor, $\boldsymbol{\sigma}_{n+1}$, as an isotropic tensor function of a single tensor[‡] – the elastic trial strain, $\boldsymbol{\varepsilon}_{n+1}^{e\ \mathrm{trial}}$. This function, defined in (8.37), can be identified as a particular case of the class of tensor-valued functions of a single tensor discussed in Appendix A. Its derivative – the elastoplastic consistent tangent operator – can be computed by means of the general procedure presented in Sections A.3 and A.4. The procedure for computation of derivatives of this type of function is implemented in subroutine DGISO2. In order to compute D^{ep}, the main arguments required by this routine are

- the principal stresses, σ_i, and the corresponding unit eigenvectors;
- the principal elastic trial strains, $\varepsilon_i^{e\ \mathrm{trial}}$; and
- the partial derivatives, $\partial\sigma_i/\partial\varepsilon_j^{e\ \mathrm{trial}}$, of the principal stresses with respect to the principal elastic trial strains associated with the algorithmic functions (8.37).

With these values at hand, the tangent operator D^{ep} is assembled as described in Box A.3. The principal stresses, the elastic trial strains and the associated eigenvectors are computed in a trivial manner. The principal stresses derivatives, $\partial\sigma_i/\partial\varepsilon_j^{e\ \mathrm{trial}}$, are the result of the consistent linearisation of the (principal stress-based) return-mapping algorithm. Their explicit forms, derived below, depend on the particular algorithm used (main plane, right or left corner).

[‡]This property is also valid for the von Mises integration algorithm discussed in Chapter 7 as well as for the other algorithms discussed later in the present chapter.

Principal stress derivatives for the one-vector return

Since the return mapping for the Tresca model affects only the deviatoric component of the stress tensor, it is convenient to derive first the derivatives

$$\frac{\partial s_i}{\partial \varepsilon^{e\ \mathrm{trial}}_{\mathrm{d}\,j}},$$

of the principal deviatoric stresses with respect to the principal deviatoric elastic trial strains.

When the return to the main plane is applied, the principal deviatoric stresses are updated according to expressions (8.16–8.18) (item (iii) of Box 8.2). By using the elastic deviatoric relation, $s_i^{\mathrm{trial}} = 2G\,\varepsilon^{e\ \mathrm{trial}}_{\mathrm{d}\,i}$, and differentiating the principal stress update formula (8.16), we obtain

$$\begin{aligned} \mathrm{d}s_1 &= 2G(\mathrm{d}\varepsilon^{e\ \mathrm{trial}}_{\mathrm{d}\,1} - \mathrm{d}\Delta\gamma) \\ \mathrm{d}s_2 &= 2G\mathrm{d}\varepsilon^{e\ \mathrm{trial}}_{\mathrm{d}\,2} \\ \mathrm{d}s_3 &= 2G(\mathrm{d}\varepsilon^{e\ \mathrm{trial}}_{\mathrm{d}\,3} + \mathrm{d}\Delta\gamma). \end{aligned} \tag{8.38}$$

With the differentiation of the return-mapping equation (8.18) the linearised form of the consistency condition is obtained as

$$\mathrm{d}\tilde{\Phi} = 2G(\mathrm{d}\varepsilon^{e\ \mathrm{trial}}_{\mathrm{d}\,1} - \mathrm{d}\varepsilon^{e\ \mathrm{trial}}_{\mathrm{d}\,3} - 2\,\mathrm{d}\Delta\gamma) - H\,\mathrm{d}\Delta\gamma = 0, \tag{8.39}$$

where $H \equiv \mathrm{d}\sigma_y/\mathrm{d}\bar{\varepsilon}^p$ is the slope of the hardening curve. From the above expression, we obtain

$$\mathrm{d}\Delta\gamma = \frac{2G}{4G+H}(\mathrm{d}\varepsilon^{e\ \mathrm{trial}}_{\mathrm{d}\,1} - \mathrm{d}\varepsilon^{e\ \mathrm{trial}}_{\mathrm{d}\,3}). \tag{8.40}$$

With the substitution of (8.40) into (8.38) it follows that the partial derivatives $\partial s_i/\partial \varepsilon^{e\ \mathrm{trial}}_{\mathrm{d}\,j}$, arranged in matrix format, are given by

$$\left[\frac{\partial s_i}{\partial \varepsilon^{e\ \mathrm{trial}}_{\mathrm{d}\,j}}\right] = \begin{bmatrix} 2G(1-f) & 0 & 2Gf \\ 0 & 2G & 0 \\ 2Gf & 0 & 2G(1-f) \end{bmatrix} \tag{8.41}$$

where the factor f has been defined as

$$f \equiv \frac{2G}{4G+H}. \tag{8.42}$$

The complete derivatives.

So far, only the expressions for the *deviatoric* principal stress derivatives have been derived. In order to obtain the complete derivatives $\partial\sigma_i/\partial\varepsilon_j^{e\ \mathrm{trial}}$, required in the assemblage of the elastoplastic tangent (these derivatives must be passed as arguments to the general routine

DGISO2), note that for the present algorithm we have

$$\tilde{\sigma}_i(\varepsilon_1^{e\,\text{trial}}, \varepsilon_2^{e\,\text{trial}}, \varepsilon_3^{e\,\text{trial}}) = \tilde{s}_i(\varepsilon_{\text{d}\,1}^{e\,\text{trial}}, \varepsilon_{\text{d}\,2}^{e\,\text{trial}}, \varepsilon_{\text{d}\,3}^{e\,\text{trial}}) + p(\varepsilon_{\text{v}}^{e\,\text{trial}}), \tag{8.43}$$

i.e. the principal deviatoric stresses are functions solely of the principal deviatoric elastic trial strains and the hydrostatic stress is a function of the volumetric elastic trial strain only. The principal deviatoric trial strains are given by

$$\varepsilon_{\text{d}\,i}^{e\,\text{trial}} = \varepsilon_i^{e\,\text{trial}} - \tfrac{1}{3}\,\varepsilon_{\text{v}}^{e\,\text{trial}}, \tag{8.44}$$

where

$$\varepsilon_{\text{v}}^{e\,\text{trial}} = \varepsilon_1^{e\,\text{trial}} + \varepsilon_2^{e\,\text{trial}} + \varepsilon_3^{e\,\text{trial}}. \tag{8.45}$$

Finally, the substitution of these expressions into (8.43) followed by a straightforward application of the chain rule leads to the following expression for the principal stress derivatives:

$$\frac{\partial \sigma_i}{\partial \varepsilon_j^{e\,\text{trial}}} = \frac{\partial s_i}{\partial \varepsilon_{\text{d}\,k}^{e\,\text{trial}}}(\delta_{kj} - \tfrac{1}{3}) + K, \tag{8.46}$$

with summation on repeated indices implied. Here, K is the bulk modulus and δ_{kj} denotes the Krönecker delta.

Principal stress derivatives for the right corner return

Now consider the return mapping to the *right* corner of the Tresca hexagon. In this case, the principal deviatoric stresses are updated by (8.20) where the incremental plastic multipliers are the solution to the return-mapping equations (8.23) (see also item (iii) of Box 8.3). Differentiation of the stress update formulae followed by the use of the deviatoric elastic law gives

$$\begin{aligned} \text{d}s_1 &= 2G(\text{d}\varepsilon_{\text{d}\,1}^{e\,\text{trial}} - \text{d}\Delta\gamma^a - \text{d}\Delta\gamma^b) \\ \text{d}s_2 &= 2G(\text{d}\varepsilon_{\text{d}\,2}^{e\,\text{trial}} + \text{d}\Delta\gamma^b) \\ \text{d}s_3 &= 2G(\text{d}\varepsilon_{\text{d}\,3}^{e\,\text{trial}} + \text{d}\Delta\gamma^a). \end{aligned} \tag{8.47}$$

The differentials of the incremental plastic multipliers are obtained by linearising the consistency condition (return-mapping equations) as follows. Differentiation of (8.23) and use of the elastic law yields the linearised equation

$$\begin{bmatrix} \text{d}\Phi^a \\ \text{d}\Phi^b \end{bmatrix} = \mathbf{d} \begin{bmatrix} \text{d}\Delta\gamma^a \\ \text{d}\Delta\gamma^b \end{bmatrix} + 2G \begin{bmatrix} \text{d}\varepsilon_{\text{d}\,1}^{e\,\text{trial}} - \text{d}\varepsilon_{\text{d}\,3}^{e\,\text{trial}} \\ \text{d}\varepsilon_{\text{d}\,1}^{e\,\text{trial}} - \text{d}\varepsilon_{\text{d}\,2}^{e\,\text{trial}} \end{bmatrix} = \begin{bmatrix} 0 \\ 0 \end{bmatrix}, \tag{8.48}$$

where the matrix $\mathbf{d}$, also used in the Newton–Raphson algorithm for solution of the return-mapping equation in Box 8.3, is defined as

$$\mathbf{d} = \begin{bmatrix} d_{aa} & d_{ab} \\ d_{ba} & d_{bb} \end{bmatrix}, \tag{8.49}$$

with

$$\begin{aligned} d_{aa} &= -4G - H, \\ d_{ab} &= -2G - H, \\ d_{ba} &= -2G - H, \\ d_{bb} &= -4G - H. \end{aligned} \tag{8.50}$$

Solution of the linearised equation for $\mathrm{d}\Delta\gamma^a$ and $\mathrm{d}\Delta\gamma^b$ gives

$$\begin{bmatrix} \mathrm{d}\Delta\gamma^a \\ \mathrm{d}\Delta\gamma^b \end{bmatrix} = -\frac{2G}{\det \mathbf{d}} \begin{bmatrix} d_{bb} & -d_{ab} \\ -d_{ba} & d_{aa} \end{bmatrix} \begin{bmatrix} \mathrm{d}\varepsilon^{e\ \text{trial}}_{\text{d}\,1} - \mathrm{d}\varepsilon^{e\ \text{trial}}_{\text{d}\,3} \\ \mathrm{d}\varepsilon^{e\ \text{trial}}_{\text{d}\,1} - \mathrm{d}\varepsilon^{e\ \text{trial}}_{\text{d}\,2} \end{bmatrix}, \tag{8.51}$$

which substituted into (8.47) leads, after a straightforward manipulation, to the expressions for the derivatives of the principal deviatoric stresses consistent with the return mapping to the right corner. Arranged in matrix format, the resulting derivatives are

$$\left[\frac{\partial s_i}{\partial \varepsilon^{e\ \text{trial}}_{\text{d}\,j}}\right] = \begin{bmatrix} 2G\left(1 - \dfrac{8G^2}{\det \mathbf{d}}\right) & \dfrac{4G^2}{\det \mathbf{d}}(d_{ab} - d_{aa}) & \dfrac{4G^2}{\det \mathbf{d}}(d_{ba} - d_{bb}) \\ \dfrac{8G^3}{\det \mathbf{d}} & 2G\left(1 + \dfrac{2G\, d_{aa}}{\det \mathbf{d}}\right) & -\dfrac{4G^2}{\det \mathbf{d}} d_{ba} \\ \dfrac{8G^3}{\det \mathbf{d}} & -\dfrac{4G^2}{\det \mathbf{d}} d_{ab} & 2G\left(1 + \dfrac{2G\, d_{bb}}{\det \mathbf{d}}\right) \end{bmatrix}. \tag{8.52}$$

With the above deviatoric principal stresses derivatives at hand, the principal stress derivatives, $\partial\sigma_i/\partial\varepsilon^{e\ \text{trial}}_j$, are obtained by applying formula (8.46).

Derivatives consistent with the left corner return

The principal stress derivatives consistent with the return mapping to the *left* corner are obtained by following exactly the same steps as in the above derivation for the right corner return. The final expression for the deviatoric principal stress derivatives is given by

$$\left[\frac{\partial s_i}{\partial \varepsilon^{e\ \text{trial}}_{\text{d}\,j}}\right] = \begin{bmatrix} 2G\left(1 + \dfrac{2G\, d_{bb}}{\det \mathbf{d}}\right) & -\dfrac{4G^2}{\det \mathbf{d}} d_{ab} & \dfrac{8G^3}{\det \mathbf{d}} \\ -\dfrac{4G^2}{\det \mathbf{d}} d_{ba} & 2G\left(1 + \dfrac{2G\, d_{aa}}{\det \mathbf{d}}\right) & \dfrac{8G^3}{\det \mathbf{d}} \\ \dfrac{4G^2}{\det \mathbf{d}}(d_{ba} - d_{bb}) & \dfrac{4G^2}{\det \mathbf{d}}(d_{ab} - d_{aa}) & 2G\left(1 - \dfrac{8G^2}{\det \mathbf{d}}\right) \end{bmatrix} \tag{8.53}$$

where, again, the derivatives $\partial\sigma_i/\partial\varepsilon^{e\ \text{trial}}_j$ are obtained by applying (8.46).

8.1.5. SUBROUTINE CTTR

The computation of the (either elastic or elastoplastic) tangent operator consistent with the Tresca integration algorithm is implemented in subroutine CTTR (**C**onsistent **T**angent operator for the **TR**esca model) of program HYPLAS. The integration algorithm is coded in subroutine SUTR. The elastic tangent is evaluated here in a trivial manner. The computation of the *elastoplastic* consistent tangent comprises the following basic steps:

1. Firstly, the principal stresses derivatives consistent with the appropriate return mapping (main plane, right or left corner) are evaluated according to the expressions derived in the above.

2. The tangent operator, $\mathbf{D}^{ep}$, (in matrix form) is then assembled in subroutine DGISO2 (**D**erivative of **G**eneral **ISO**tropic tensor functions of a single tensor in **2**-D and axisymmetric conditions). This routine is called by CTTR which passes the principal stress derivatives (as well as other necessary variables) as arguments.

The FORTRAN source code of CTTR is listed below.

```
      SUBROUTINE CTTR
     1(   DMATX      ,EPFLAG     ,IPROPS     ,LALGVA     ,NTYPE      ,
     2    RPROPS     ,RSTAVA     ,STRAT      ,STRES      )
      IMPLICIT DOUBLE PRECISION (A-H,O-Z)
      PARAMETER(IPHARD=4  ,MDIM=3,  MSTRE=4)
      LOGICAL EPFLAG, LALGVA(4), OUTOFP, RIGHT, REPEAT, TWOVEC
      DIMENSION
     1    DMATX(MSTRE,MSTRE) ,IPROPS(*)          ,RPROPS(*)          ,
     2    RSTAVA(MSTRE+1)    ,STRAT(*)           ,STRES(*)
      DIMENSION
     1    DPSTRS(MDIM,MDIM)  ,DPSTRE(MDIM,MDIM)  ,EIGPRJ(MSTRE,2)    ,
     2    FOID(MSTRE,MSTRE)  ,PSTRS(MDIM)        ,PSTRA(MDIM)        ,
     3    SOID(MSTRE)        ,STRAC(MSTRE)
      DATA
     1    FOID(1,1),FOID(1,2),FOID(1,3),FOID(1,4)/
     2    1.0D0    ,0.0D0    ,0.0D0    ,0.0D0    /
     3    FOID(2,1),FOID(2,2),FOID(2,3),FOID(2,4)/
     4    0.0D0    ,1.0D0    ,0.0D0    ,0.0D0    /
     5    FOID(3,1),FOID(3,2),FOID(3,3),FOID(3,4)/
     6    0.0D0    ,0.0D0    ,0.5D0    ,0.0D0    /
     7    FOID(4,1),FOID(4,2),FOID(4,3),FOID(4,4)/
     8    0.0D0    ,0.0D0    ,0.0D0    ,1.0D0    /
      DATA
     1    SOID(1)  ,SOID(2)  ,SOID(3)  ,SOID(4)  /
     2    1.0D0    ,1.0D0    ,0.0D0    ,1.0D0    /
      DATA
     1    R0   ,RP5  ,R1   ,R2   ,R3   ,R4   /
     2    0.0D0,0.5D0,1.0D0,2.0D0,3.0D0,4.0D0/
C***********************************************************************
C COMPUTATION OF CONSISTENT TANGENT MODULUS FOR TRESCA TYPE
C ELASTO-PLASTIC MATERIAL WITH PIECE-WISE LINEAR ISOTROPIC HARDENING.
C PLANE STRAIN AND AXISYMMETRIC IMPLEMENTATIONS.
C***********************************************************************
C Stops program if neither plane strain nor axisymmetric state
      IF(NTYPE.NE.2.AND.NTYPE.NE.3)CALL ERRPRT('EI0028')
C Current accumulated plastic strain
      EPBAR=RSTAVA(MSTRE+1)
```

```
C Set material properties
      YOUNG=RPROPS(2)
      POISS=RPROPS(3)
      NHARD=IPROPS(3)
C Set needed algorithmic variables
      TWOVEC=LALGVA(3)
      RIGHT=LALGVA(4)
C Set some constants
      GMODU=YOUNG/(R2*(R1+POISS))
      BULK=YOUNG/(R3*(R1-R2*POISS))
      R2G=R2*GMODU
      R4G=R4*GMODU
      R1D3=R1/R3
      R2D3=R2*R1D3
      IF(EPFLAG)THEN
C Compute elastoplastic consistent tangent
C ---------------------------------------
C Spectral decomposition of the elastic trial strain
        STRAC(1)=STRAT(1)
        STRAC(2)=STRAT(2)
        STRAC(3)=STRAT(3)*RP5
        CALL SPDEC2(EIGPRJ,PSTRA,REPEAT,STRAC)
        PSTRA(3)=STRAT(4)
C and current total stress
        PSTRS(1)=STRES(1)*EIGPRJ(1,1)+STRES(2)*EIGPRJ(2,1)+
     1                             R2*STRES(3)*EIGPRJ(3,1)
        PSTRS(2)=STRES(1)*EIGPRJ(1,2)+STRES(2)*EIGPRJ(2,2)+
     1                             R2*STRES(3)*EIGPRJ(3,2)
        PSTRS(3)=STRES(4)
C Identify directions of maximum and minimum principal trial stresses
        II=1
        JJ=1
        PSTMAX=PSTRA(II)
        PSTMIN=PSTRA(JJ)
        DO 10 I=2,3
          IF(PSTRA(I).GE.PSTMAX)THEN
            II=I
            PSTMAX=PSTRA(II)
          ENDIF
          IF(PSTRA(I).LT.PSTMIN)THEN
            JJ=I
            PSTMIN=PSTRA(JJ)
          ENDIF
   10   CONTINUE
        IF(II.NE.1.AND.JJ.NE.1)MM=1
        IF(II.NE.2.AND.JJ.NE.2)MM=2
        IF(II.NE.3.AND.JJ.NE.3)MM=3
        IF(TWOVEC)THEN
C Tangent consistent with two-vector return algorithm
          HSLOPE=DPLFUN(EPBAR,NHARD,RPROPS(IPHARD))
          DAA=R4G+HSLOPE
          DAB=R2G+HSLOPE
          DBA=R2G+HSLOPE
          DBB=R4G+HSLOPE
          DET=DAA*DBB-DAB*DBA
          R2GDD=R2G/DET
          R4G2DD=R2G*R2GDD
          IF(RIGHT)THEN
C ...returned to right corner
            DPSTRS(II,II)=R2G*(R1-R2GDD*R4G)
            DPSTRS(II,MM)=R4G2DD*(DAA-DAB)
```

```
            DPSTRS(II,JJ)=R4G2DD*(DBB-DBA)
            DPSTRS(MM,II)=R4G2DD*R2G
            DPSTRS(MM,MM)=R2G*(R1-R2GDD*DAA)
            DPSTRS(MM,JJ)=R4G2DD*DBA
            DPSTRS(JJ,II)=R4G2DD*R2G
            DPSTRS(JJ,MM)=R4G2DD*DAB
            DPSTRS(JJ,JJ)=R2G*(R1-R2GDD*DBB)
          ELSE
C ...returned to left corner
            DPSTRS(II,II)=R2G*(R1-R2GDD*DBB)
            DPSTRS(II,MM)=R4G2DD*DAB
            DPSTRS(II,JJ)=R4G2DD*R2G
            DPSTRS(MM,II)=R4G2DD*DBA
            DPSTRS(MM,MM)=R2G*(R1-R2GDD*DAA)
            DPSTRS(MM,JJ)=R4G2DD*R2G
            DPSTRS(JJ,II)=R4G2DD*(DBB-DBA)
            DPSTRS(JJ,MM)=R4G2DD*(DAA-DAB)
            DPSTRS(JJ,JJ)=R2G*(R1-R2GDD*R4G)
          ENDIF
        ELSE
C Tangent consistent with one-vector return algorithm
          FACTOR=R2G/(R4G+DPLFUN(EPBAR,NHARD,RPROPS(IPHARD)))
          DPSTRS(II,II)=R2G*(R1-FACTOR)
          DPSTRS(II,MM)=R0
          DPSTRS(II,JJ)=R2G*FACTOR
          DPSTRS(MM,II)=DPSTRS(II,MM)
          DPSTRS(MM,MM)=R2G
          DPSTRS(MM,JJ)=R0
          DPSTRS(JJ,II)=DPSTRS(II,JJ)
          DPSTRS(JJ,MM)=DPSTRS(MM,JJ)
          DPSTRS(JJ,JJ)=DPSTRS(II,II)
        ENDIF
        DPSTRE(1,1)=+DPSTRS(1,1)*R2D3-DPSTRS(1,2)*R1D3-DPSTRS(1,3)*R1D3+
     1                                                              BULK
        DPSTRE(2,1)=+DPSTRS(2,1)*R2D3-DPSTRS(2,2)*R1D3-DPSTRS(2,3)*R1D3+
     1                                                              BULK
        DPSTRE(3,1)=+DPSTRS(3,1)*R2D3-DPSTRS(3,2)*R1D3-DPSTRS(3,3)*R1D3+
     1                                                              BULK
        DPSTRE(1,2)=-DPSTRS(1,1)*R1D3+DPSTRS(1,2)*R2D3-DPSTRS(1,3)*R1D3+
     1                                                              BULK
        DPSTRE(2,2)=-DPSTRS(2,1)*R1D3+DPSTRS(2,2)*R2D3-DPSTRS(2,3)*R1D3+
     1                                                              BULK
        DPSTRE(3,2)=-DPSTRS(3,1)*R1D3+DPSTRS(3,2)*R2D3-DPSTRS(3,3)*R1D3+
     1                                                              BULK
        DPSTRE(1,3)=-DPSTRS(1,1)*R1D3-DPSTRS(1,2)*R1D3+DPSTRS(1,3)*R2D3+
     1                                                              BULK
        DPSTRE(2,3)=-DPSTRS(2,1)*R1D3-DPSTRS(2,2)*R1D3+DPSTRS(2,3)*R2D3+
     1                                                              BULK
        DPSTRE(3,3)=-DPSTRS(3,1)*R1D3-DPSTRS(3,2)*R1D3+DPSTRS(3,3)*R2D3+
     1                                                              BULK
        IF(NTYPE.EQ.2)THEN
          OUTOFP=.FALSE.
        ELSEIF(NTYPE.EQ.3)THEN
          OUTOFP=.TRUE.
        ENDIF
        CALL DGISO2
     1(   DPSTRE     ,DMATX       ,EIGPRJ      ,PSTRA       ,PSTRS       ,
     2    OUTOFP     ,REPEAT      )
      ELSE
C Compute elasticity matrix
```

```
C -------------------------
          IF(NTYPE.EQ.2)THEN
            NSTRE=3
          ELSEIF(NTYPE.EQ.3)THEN
            NSTRE=4
          ENDIF
C
          FACTOR=BULK-R2G*R1D3
          DO 50 I=1,NSTRE
            DO 40 J=I,NSTRE
              DMATX(I,J)=R2G*FOID(I,J)+FACTOR*SOID(I)*SOID(J)
   40       CONTINUE
   50     CONTINUE
          DO 70 J=1,NSTRE-1
            DO 60 I=J+1,NSTRE
              DMATX(I,J)=DMATX(J,I)
   60       CONTINUE
   70     CONTINUE
        ENDIF
        RETURN
        END
```

The arguments of CTTR

In addition to the arguments of SUTR (see list on page 282), whose output values are taken as input values for (and are not changed by) CTTR, the present routine requires the following:

← DMATX [either $\mathbf{D}^e$ or $\mathbf{D}^{ep}$]. As in the von Mises implementation (refer to page 237).

→ EPFLAG. Elastoplastic tangent logical flag (refer to page 237 for a description).

The elements of the array LALGVA of logical algorithmic flags, set in subroutine SUTR, indicate which elastoplastic tangent is to be computed here (consistent with main plane, right or left return algorithm).

Some local arrays of CTTR

- DPSTRE [$\partial\sigma_i/\partial\varepsilon_j^{e\ \mathrm{trial}}$]. Matrix of derivatives of the principal stresses.
- DPSTRS [$\partial s_i/\partial\varepsilon_{\mathrm{d}\,j}^{e\ \mathrm{trial}}$]. Matrix of derivatives of the principal deviatoric stresses.
- EIGPRJ [$\boldsymbol{e}_i \otimes \boldsymbol{e}_i$, i=1,2]. Matrix containing the components of the in-plane eigenprojection tensors of the elastic trial strain.
- FOID [$\mathbf{I}_S$]. Fourth-order identity tensor stored in array form according to (D.16), page 762.
- SOID [$\boldsymbol{I}$]. Second-order identity tensor stored in array form.
- STRAT [$\varepsilon_i^{e\ \mathrm{trial}}$]. Array containing the principal elastic trial strains.

The names of most local variables defined in CTTR resemble resemble the notation of Section 8.1.4.

Important function calls from `CTTR`

- `DGISO2`. Called to assemble the elastoplastic consistent tangent operator as a particular case of the general derivatives of isotropic tensor function discussed in Section A.3 of Appendix A.
- `SPDEC2`. Called to perform the spectral decomposition of the elastic trial strain.

8.2. The Mohr–Coulomb model

This section is devoted to the computational implementation of a classical (generally non-associative) Mohr–Coulomb model with nonlinear isotropic strain hardening. For quick reference, the following table shows where the main results of this section can be found.

integration algorithm	flowchart	Figure 8.10
	pseudo-code	Boxes 8.4–8.7
	FORTRAN code	subroutine `SUMC` (Section 8.2.2)
iso-error maps		Figure 8.12
consistent tangent – FORTRAN code –		subroutine `CTMC` (Section 8.2.5)

As far as yield surface singularities are concerned, the algorithm described here is entirely based on the ideas that underly the numerical integration scheme for the Tresca model of the previous section. Here, however, an extra return-mapping procedure – the return to the *apex* of the Mohr–Coulomb yield surface – is required in the numerical integration scheme.

The implemented Mohr–Coulomb model

The Mohr–Coulomb yield criterion as well as the corresponding flow rules and possible hardening laws have been thoroughly discussed in Chapter 6. In what follows, we summarised the actual equations adopted in the present computational implementation of the Mohr–Coulomb model.

The multisurface description of the Mohr–Coulomb plastic flow rule has the representation (8.1)–(8.3) in the associative case with the yield functions Φ_i defined by (6.118) (page 165). In the generally non-associative case, whose implementation is described here, the flow vectors are re-defined as

$$\boldsymbol{N}^i \equiv \frac{\partial \Psi_i}{\partial \boldsymbol{\sigma}}, \tag{8.54}$$

where each Ψ_i has the same format as the corresponding Φ_i, except that the frictional angle ϕ is replaced with the dilatancy angle $\psi \leq \phi$ (assumed constant) in the definition of Ψ_i.

With the principal stresses ordered as $\sigma_1 \geq \sigma_2 \geq \sigma_3$, the flow rule may be formulated within the sextant of the principal stress space depicted in Figure 6.19(a) (page 174). There are four distinct possibilities for the plastic flow definition.

1. Plastic flow from the smooth (flat) portion of the main plane, where only one multiplier may be non-zero. The generally non-associative flow vector in this case is reduced to

$$\boldsymbol{N}^a \equiv \boldsymbol{N}^1 = (1 + \sin\psi)\boldsymbol{e}_1 \otimes \boldsymbol{e}_1 - (1 - \sin\psi)\boldsymbol{e}_3 \otimes \boldsymbol{e}_3, \tag{8.55}$$

   and

$$\dot{\boldsymbol{\varepsilon}}^p = \dot{\gamma}\boldsymbol{N}^a. \tag{8.56}$$

2. Plastic flow from the *right edge*. Here, only two plastic multipliers can be non-zero and the flow rule is explicitly defined as

$$\dot{\boldsymbol{\varepsilon}}^p = \dot{\gamma}^a\boldsymbol{N}^a + \dot{\gamma}^b\boldsymbol{N}^b, \tag{8.57}$$

   where $\boldsymbol{N}^a$ is the same as in the above, $\boldsymbol{N}^b$ is the flow vector of the plane on the right of the main plane

$$\boldsymbol{N}^b \equiv \boldsymbol{N}^6 = (1 + \sin\psi)\boldsymbol{e}_1 \otimes \boldsymbol{e}_1 - (1 - \sin\psi)\boldsymbol{e}_2 \otimes \boldsymbol{e}_2. \tag{8.58}$$

3. Plastic flow from the *left edge*. The plastic flow is defined as above but with $\boldsymbol{N}^b$ being the flow vector of the plane on the left of the main plane

$$\boldsymbol{N}^b \equiv \boldsymbol{N}^2 = (1 + \sin\psi)\boldsymbol{e}_2 \otimes \boldsymbol{e}_2 - (1 - \sin\psi)\boldsymbol{e}_3 \otimes \boldsymbol{e}_3. \tag{8.59}$$

4. Plastic flow from the *apex* of the Mohr–Coulomb pyramid. In this case, up to six multipliers may be non-zero, as the plastic rate tensor lies within the pyramid formed by the six vectors of the Mohr–Coulomb model (refer to Figure 6.19.(b) which illustrates the associative flow case). The plastic strain rate equation in this case is left in its general format

$$\dot{\boldsymbol{\varepsilon}}^p = \sum_{i=1}^{6} \dot{\gamma}^i\boldsymbol{N}^i. \tag{8.60}$$

In the present implementation, isotropic strain hardening is included by letting the cohesion c that takes part in the definition of the yield functions Φ_i be a generic nonlinear function of the accumulated plastic strain:

$$c = c(\bar{\varepsilon}^p). \tag{8.61}$$

Hardening associativity is assumed. The general equation for the evolution of $\bar{\varepsilon}^p$ in this case is given by (6.200) on page 184. For flow from the main plane as discussed above, only one multiplier may be non-zero and (6.200) is reduced to

$$\dot{\bar{\varepsilon}}^p = 2\cos\phi\,\dot{\gamma}. \tag{8.62}$$

For plastic flow from an edge of the Mohr–Coulomb pyramid, the evolution of $\bar{\varepsilon}^p$ reads

$$\dot{\bar{\varepsilon}}^p = 2\cos\phi\,(\dot{\gamma}^a + \dot{\gamma}^b). \tag{8.63}$$

At the apex, the general equation (6.200) applies. In this case, the following general relation

$$\dot{\bar{\varepsilon}}^p = \frac{\cos\phi}{\sin\psi}\,\dot{\varepsilon}^p_v, \tag{8.64}$$

between the rate of accumulated plastic strain and the volumetric plastic strain rate, will be crucial in the computational implementation of the model. This relation can be obtained by combining the counterpart of (6.155) (page 175) under non-associative plastic flow:

$$\dot{\varepsilon}^p_{\rm v} = 2\,\sin\psi\,\sum_{i=1}^{6}\dot{\gamma}^i, \tag{8.65}$$

and (6.200).

8.2.1. INTEGRATION ALGORITHM FOR THE MOHR–COULOMB MODEL

As for the algorithm for the Tresca model, the essential stress-updating is carried out here in the principal stress space. Hence, after the computation of the elastic trial state, the spectral decomposition of the trial stress, $\boldsymbol{\sigma}^{\rm trial}$, is performed. With the principal trial stresses arranged as $\sigma_1^{\rm trial} \geq \sigma_2^{\rm trial} \geq \sigma_3^{\rm trial}$, the plastic consistency check proceeds as follows:

- If

  $$\Phi^{\rm trial} \equiv \sigma_1^{\rm trial} - \sigma_3^{\rm trial} + (\sigma_1^{\rm trial} + \sigma_3^{\rm trial})\,\sin\phi - 2\,c(\bar{\varepsilon}^p_n)\,\cos\phi \leq 0,$$

  then the step is elastic and all state variables are updated as

  $$(\cdot)_{n+1} := (\cdot)^{\rm trial}_{n+1}.$$

- Otherwise, a return-mapping procedure is carried out.

Recall that the general return-mapping update formula for the stress tensor is given by

$$\boldsymbol{\sigma}_{n+1} = \boldsymbol{\sigma}^{\rm trial}_{n+1} - \mathsf{D}^e : \Delta\boldsymbol{\varepsilon}^p. \tag{8.66}$$

With the discretisation of the non-associated Mohr–Coulomb flow rule (8.60), this gives

$$\boldsymbol{\sigma}_{n+1} = \boldsymbol{\sigma}^{\rm trial}_{n+1} - \mathsf{D}^e : \sum_{i=1}^{6}\Delta\gamma^i\boldsymbol{N}^i_{n+1}. \tag{8.67}$$

Due to the isotropy of the model, the stress update formula can be written equivalently in terms of principal stresses as

$$\sigma_j = \sigma_j^{\rm trial} - \sum_{i=1}^{6}\Delta\gamma^i\,(2G\,[\boldsymbol{N}^i_{\rm d}]_j - K\,N^i_{\rm v}), \tag{8.68}$$

for $j = 1, 2, 3$. In the above formula, $N^i_{\rm v} \equiv {\rm tr}[\boldsymbol{N}^i]$ is the volumetric component of the flow vector $\boldsymbol{N}^i$ at the updated state and $[\boldsymbol{N}^i_{\rm d}]_j$ denotes the $j^{\rm th}$ eigenvalue of its deviatoric projection.

The four possible return mappings

Following the four possible descriptions of the Mohr–Coulomb flow rule, the corresponding backward Euler-based return-mapping algorithm has four distinct explicit forms, which depend on the location of the updated stress, $\boldsymbol{\sigma}_{n+1}$, on the yield surface. The equations for each of the four possible return mappings are derived below:

1. The updated stress lies on the *main plane* (smooth portion). The flow vector in this case is defined by (8.55). Introduction of this expression in (8.68) gives the following principal stress updating formulae:

$$
\begin{aligned}
\sigma_1 &= \sigma_1^{\text{trial}} - \Delta\gamma\,[2G(1 + \tfrac{1}{3}\sin\psi) + 2K\ \sin\psi] \\
\sigma_2 &= \sigma_2^{\text{trial}} + \Delta\gamma\,(\tfrac{4}{3}G - 2K)\sin\psi \\
\sigma_3 &= \sigma_3^{\text{trial}} + \Delta\gamma\,[2G(1 - \tfrac{1}{3}\sin\psi) - 2K\ \sin\psi].
\end{aligned}
\tag{8.69}
$$

   The corresponding increment of accumulated plastic strain reads

$$\Delta\bar{\varepsilon}^p = 2\cos\phi\,\Delta\gamma. \tag{8.70}$$

   The incremental plastic multiplier, $\Delta\gamma$, is obtained by solving the return-mapping equation

$$
\begin{aligned}
\tilde{\Phi}(\Delta\gamma) \equiv (\sigma_1^{\text{trial}} - \sigma_3^{\text{trial}}) + (\sigma_1^{\text{trial}} + \sigma_3^{\text{trial}})\sin\phi \\
- 2\,c(\bar{\varepsilon}_n^p + \Delta\bar{\varepsilon}^p)\cos\phi - a\Delta\gamma = 0,
\end{aligned}
\tag{8.71}
$$

   where $\Delta\bar{\varepsilon}^p$ is the (linear) function of $\Delta\gamma$ defined by (8.70) and a is the constant

$$a = 4G(1 + \tfrac{1}{3}\sin\phi\sin\psi) + 4K\sin\phi\sin\psi. \tag{8.72}$$

   Equation (8.71) is obtained by introducing the update formulae (8.69) into the main plane equation $\Phi_1 = 0$.

2. The updated stress lies on the *right edge* of the pyramid. The incremental plastic strain, obtained from the backward Euler discretisation of (8.57), is in this case given by the two-vector formula

$$\Delta\boldsymbol{\varepsilon}^p = \Delta\gamma^a\boldsymbol{N}^a + \Delta\gamma^b\boldsymbol{N}^b. \tag{8.73}$$

   The corresponding principal stress update formulae follow from the substitution of the above expression into (8.68):

$$
\begin{aligned}
\sigma_1 &= \sigma_1^{\text{trial}} - [2G(1 + \tfrac{1}{3}\sin\psi) + 2K\ \sin\psi](\Delta\gamma^a + \Delta\gamma^b) \\
\sigma_2 &= \sigma_2^{\text{trial}} + (\tfrac{4}{3}G - 2K)\sin\psi\,\Delta\gamma^a + [2G(1 - \tfrac{1}{3}\sin\psi) - 2K\ \sin\psi]\Delta\gamma^b \\
\sigma_3 &= \sigma_3^{\text{trial}} + [2G(1 - \tfrac{1}{3}\sin\psi) - 2K\ \sin\psi]\Delta\gamma^a + (\tfrac{4}{3}G - 2K)\sin\psi\,\Delta\gamma^b.
\end{aligned}
\tag{8.74}
$$

   The increment of accumulated plastic strain is the discrete form of (8.63):

$$\Delta\bar{\varepsilon}^p = 2\cos\phi\,(\Delta\gamma^a + \Delta\gamma^b). \tag{8.75}$$

At the right edge, the updated principal stresses are such that the equations of the main plane, $\Phi_1 = 0$, and the plane on its right, $\Phi_6 = 0$, are simultaneously satisfied. This condition gives the following set of two return-mapping equations to be solved for $\Delta\gamma^a$ and $\Delta\gamma^b$:

$$\begin{aligned}\tilde{\Phi}^a(\Delta\gamma^a, \Delta\gamma^b) &\equiv \sigma_1^{\text{trial}} - \sigma_3^{\text{trial}} + (\sigma_1^{\text{trial}} + \sigma_3^{\text{trial}}) \sin\phi \\ &\quad - 2\cos\phi\, c(\bar{\varepsilon}_n^p + \Delta\bar{\varepsilon}^p) - a\,\Delta\gamma^a - b\,\Delta\gamma^b = 0 \\ \tilde{\Phi}^b(\Delta\gamma^a, \Delta\gamma^b) &\equiv \sigma_1^{\text{trial}} - \sigma_2^{\text{trial}} + (\sigma_1^{\text{trial}} + \sigma_2^{\text{trial}}) \sin\phi \\ &\quad - 2\cos\phi\, c(\bar{\varepsilon}_n^p + \Delta\bar{\varepsilon}^p) - b\,\Delta\gamma^a - a\,\Delta\gamma^b = 0.\end{aligned} \tag{8.76}$$

The incremental accumulated plastic strain, $\Delta\bar{\varepsilon}^p$, appearing in the above equations is the linear function of $\Delta\gamma^a$ and $\Delta\gamma^b$ previously defined and the constant b is defined by

$$b = 2G(1 + \sin\phi + \sin\psi - \tfrac{1}{3}\sin\phi\sin\psi) + 4K\sin\phi\sin\psi. \tag{8.77}$$

3. The updated stress lies on the *left edge*. The return mapping to the left edge is completely analogous to the return to the right edge derived in the above. The essential difference is that the tensor $\boldsymbol{N}^b$ is now defined by (8.59). The resulting principal stress updating formulae are

$$\begin{aligned}\sigma_1 &= \sigma_1^{\text{trial}} - [2G(1 + \tfrac{1}{3}\sin\psi) + 2K\ \sin\psi]\Delta\gamma^a + (\tfrac{4}{3}G - 2K)\sin\psi\,\Delta\gamma^b \\ \sigma_2 &= \sigma_2^{\text{trial}} + (\tfrac{4}{3}G - 2K)\sin\psi\,\Delta\gamma^a - [2G(1 + \tfrac{1}{3}\sin\psi) + 2K\ \sin\psi]\Delta\gamma^b \\ \sigma_3 &= \sigma_3^{\text{trial}} + [2G(1 - \tfrac{1}{3}\sin\psi) - 2K\ \sin\psi](\Delta\gamma^a + \Delta\gamma^b),\end{aligned} \tag{8.78}$$

and the corresponding increment of accumulated plastic strain is also given by (8.75). The incremental plastic multipliers are the solution of the following return-mapping equations:

$$\begin{aligned}\tilde{\Phi}^a(\Delta\gamma^a, \Delta\gamma^b) &\equiv \sigma_1^{\text{trial}} - \sigma_3^{\text{trial}} + (\sigma_1^{\text{trial}} + \sigma_3^{\text{trial}}) \sin\phi \\ &\quad - 2\cos\phi\, c(\bar{\varepsilon}_n^p + \Delta\bar{\varepsilon}^p) - a\,\Delta\gamma^a - b\,\Delta\gamma^b = 0 \\ \tilde{\Phi}^b(\Delta\gamma^a, \Delta\gamma^b) &\equiv \sigma_2^{\text{trial}} - \sigma_3^{\text{trial}} + (\sigma_2^{\text{trial}} + \sigma_3^{\text{trial}}) \sin\phi \\ &\quad - 2\cos\phi\, c(\bar{\varepsilon}_n^p + \Delta\bar{\varepsilon}^p) - b\,\Delta\gamma^a - a\,\Delta\gamma^b = 0\end{aligned} \tag{8.79}$$

where the constant b has been redefined as

$$b = 2G(1 - \sin\phi - \sin\psi - \tfrac{1}{3}\sin\phi\sin\psi) + 4K\sin\phi\sin\psi. \tag{8.80}$$

The above equations represent the intersection of the main plane, $\Phi_1 = 0$, with the plane on its left, $\Phi_2 = 0$.

4. The updated stress lies on the *apex*. The apex of the Mohr–Coulomb pyramid is the point along the hydrostatic axis for which

$$p = c\cot\phi. \tag{8.81}$$

Now recall the general hydrostatic pressure-updating equation for a return mapping with an underlying linear description of the elastic behaviour:

$$p_{n+1} = p_{n+1}^{\text{trial}} - K \, \Delta\varepsilon_{\text{v}}^p. \tag{8.82}$$

The updated hydrostatic stress in the present case must lie on the apex of the updated Mohr–Coulomb pyramid. This condition is enforced by introducing the update formula (8.82), together with the hardening curve and the discretised evolution equation for $\bar{\varepsilon}^p$, into the apex equation (8.81). We then have (Figure 8.8)

$$c(\bar{\varepsilon}_n^p + \Delta\bar{\varepsilon}^p) \cot\phi - p_{n+1}^{\text{trial}} + K \, \Delta\varepsilon_{\text{v}}^p = 0. \tag{8.83}$$

Now consider the discrete version of (8.64):

$$\Delta\bar{\varepsilon}^p = \alpha \Delta\varepsilon_{\text{v}}^p; \qquad \alpha \equiv \frac{\cos\phi}{\sin\psi}. \tag{8.84}$$

The substitution of this expression into (8.83) yields the final apex return-mapping equation:

$$c(\bar{\varepsilon}_n^p + \alpha\Delta\varepsilon_{\text{v}}^p) \cot\phi - p_{n+1}^{\text{trial}} + K \, \Delta\varepsilon_{\text{v}}^p = 0, \tag{8.85}$$

for the unknown $\Delta\varepsilon_{\text{v}}^p$. Once the solution is found, we update

$$\begin{aligned} \bar{\varepsilon}_{n+1}^p &:= \bar{\varepsilon}_n^p + \alpha\Delta\varepsilon_v^p, \\ \boldsymbol{\sigma}_{n+1} &:= p_{n+1} \, \boldsymbol{I}, \end{aligned} \tag{8.86}$$

with p_{n+1} given by (8.82). Note that under zero dilatancy ($\psi = 0$), the return to apex does not make sense as the purely deviatoric flow vector (combination of deviatoric flow vectors) does not produce volumetric plastic flow. In this case, $\alpha \to \infty$.

Remark 8.1. Under the assumption of linear hardening, with $c = c_0 + H\bar{\varepsilon}^p$, all four sets of return-mapping equations are linear and, therefore, can be solved in closed form.

Remark 8.2. When $\phi = \psi = 0$ all equations of the one-vector return mapping to the main plane and the two-vector return mappings to the right and left edge reduce to those of the Tresca model implementation, derived in Section 8.1.1.

Strategy for selection of the appropriate return mapping

The strategy for selection of the appropriate return mapping to be used is a direct extension of the procedure adopted in the return algorithm for the Tresca model, whose description starts on page 271. As in the Tresca algorithm, the present strategy is simple and ensures that the resulting updated state rigorously satisfies the general implicit return-mapping equations (7.25), regardless of the prescribed hardening/softening curve. In essence, the strategy comprises the following steps.

1. Firstly apply the return mapping to the main plane.

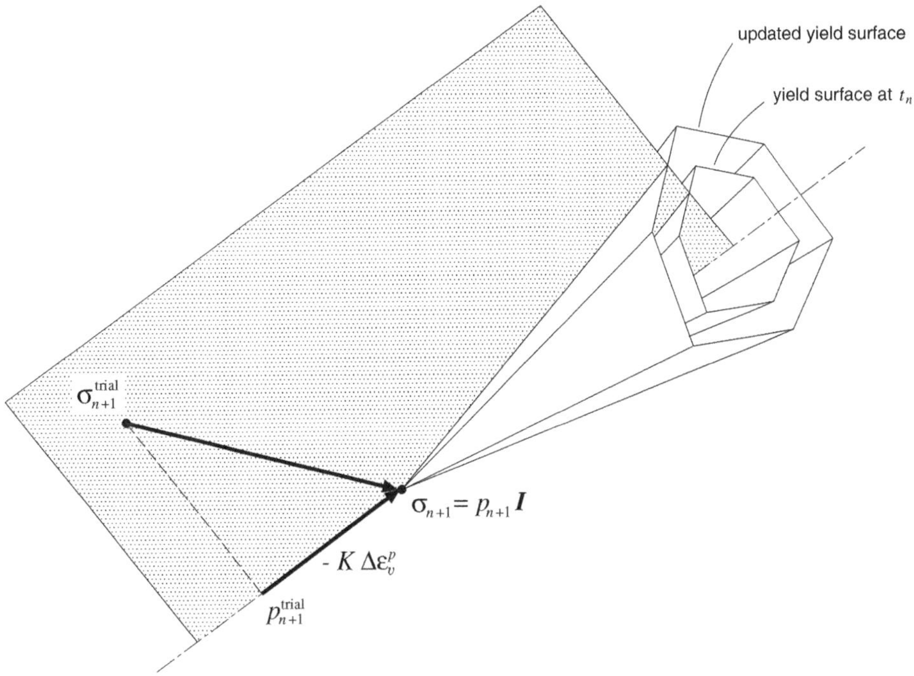

Figure 8.8. Mohr–Coulomb model. Return mapping to apex.

2. Check validity: if the updated principal stresses remain in the same sextant as the trial stresses, i.e. if

$$\sigma_1 \geq \sigma_2 \geq \sigma_3, \tag{8.87}$$

then the return to the main plane is valid and the corresponding updated state is accepted.

3. Otherwise, try either the return mapping to the right or to the left edge (the procedure to select which one to try at this stage is described later).

4. Check validity: if the updated principal stresses remain in the same sextant as the trial stress, then the return mapping carried out in item 3 is valid and the updated state is accepted.

5. Otherwise, apply the return algorithm to the apex. The result obtained now is necessarily valid and does not require further checks.

The reader should note that any *returned* stress that fails to pass the validity check of items 2 and 4 lies necessarily outside the updated elastic domain (which is fixed in the perfectly plastic case) and is, therefore, not admissible. This can be easily verified by simple geometrical considerations analogous to those of the Tresca model given in Figure 8.1.

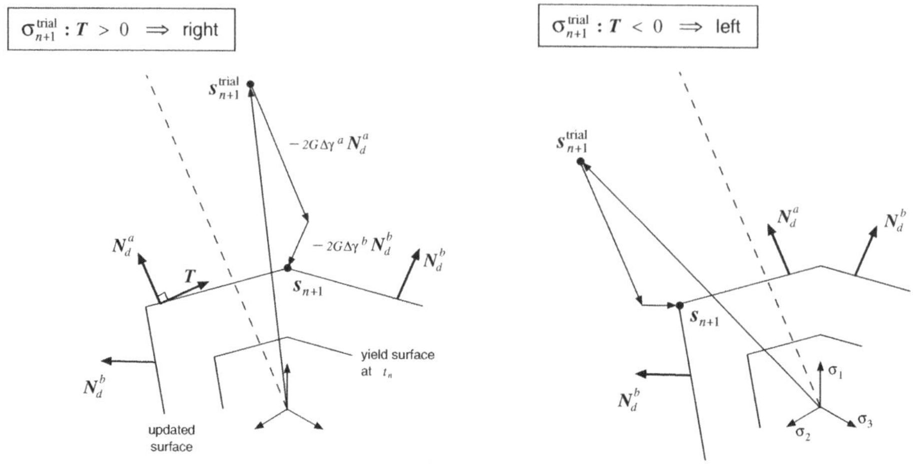

Figure 8.9. Mohr–Coulomb model. Selection of appropriate return mapping to edge.

Selection of right or left edge return.

The determination of which return mapping is to be tried in item 3 above is based purely on geometrical arguments and is also analogous to the procedure schematically illustrated in Figure 8.2 for the Tresca model. Firstly, it should be noted that at any edge of the Mohr–Coulomb pyramid we have

$$s_i = s_j$$

for at least one pair $\{s_i, s_j\}$ with $i \neq j$. According to the present convention, $s_2 = s_3$ corresponds to a point on the right edge and $s_1 = s_2$ to a point on the left edge. On the deviatoric plane, these points correspond to corners of the Mohr–Coulomb pyramid cross section (irregular hexagon). Consider now the *deviatoric* part to the two-vector edge return algorithms (Figure 8.9):

$$\boldsymbol{s}_{n+1} = \boldsymbol{s}^{\rm trial}_{n+1} - 2G(\Delta\gamma^a\, \boldsymbol{N}^a_{\rm d} + \Delta\gamma^b\, \boldsymbol{N}^b_{\rm d}), \tag{8.88}$$

where $\boldsymbol{N}^a_{\rm d}$ and $\boldsymbol{N}^b_{\rm d}$ are the deviatoric components of $\boldsymbol{N}^a$ and $\boldsymbol{N}^b$, respectively. Clearly, $\boldsymbol{N}^b$ in the above is normal to the right (left) plane if $\boldsymbol{s}_{n+1}$ lies on the right (left) edge. The dotted line of Figure 8.9 is the deviatoric projection of the plane parallel to $\boldsymbol{N}^a$ that contains the hydrostatic line. If $\boldsymbol{\sigma}^{\rm trial}_{n+1}$ lies on the right side of this plane ($\boldsymbol{s}^{\rm trial}_{n+1}$ lies on the right of the dotted line) the only possible way to return to the left edge consistently with the above formula is to have negative $\Delta\gamma^b$ – an unacceptable solution. In this case, if $\boldsymbol{\sigma}^{\rm trial}_{n+1}$ returns to an edge, it can only be the right edge. Similarly, if $\boldsymbol{s}^{\rm trial}_{n+1}$ lies on the left of the dotted line, the only possibly valid edge return algorithm is the return to the left edge.

Based on the above considerations, the procedure for selection of the appropriate edge return is extremely simple. Let $\boldsymbol{T}$ be a deviatoric tensor orthogonal to the deviatoric projection $\boldsymbol{N}^a_{\rm d}$, pointing to the right. Here $\boldsymbol{T}$ is chosen with the following eigenvalues:

$$T_1 = 1 - \sin\psi, \quad T_2 = -2, \quad T_3 = 1 + \sin\psi. \tag{8.89}$$

Its eigenvectors coincide with those of $\boldsymbol{\sigma}_{n+1}^{\text{trial}}$. To select the appropriate edge return, the scalar product

$$\begin{aligned} S &\equiv \boldsymbol{T} : \boldsymbol{\sigma}_{n+1}^{\text{trial}} \\ &= T_1\, \sigma_1^{\text{trial}} + T_2\, \sigma_2^{\text{trial}} + T_3\, \sigma_3^{\text{trial}} \\ &= (1 - \sin \psi)\, \sigma_1^{\text{trial}} - 2\, \sigma_2^{\text{trial}} + (1 + \sin \psi)\, \sigma_3^{\text{trial}}, \end{aligned} \tag{8.90}$$

is computed first. If $S > 0$ the possible edge return is to the right. Otherwise, it is the return mapping to the left edge.

The overall integration algorithm

Having described the four possible return mappings together with a simple and robust strategy to select the valid return mapping, the implicit elastic predictor/return-mapping algorithm for the Mohr–Coulomb model is completely defined.

The flowchart of the overall integration algorithm is shown in Figure 8.10. The corresponding pseudo-code is conveniently summarised in Boxes 8.4–8.7. The procedure is implemented in subroutine `SUMC` (**S**tate **U**pdate procedure for the **M**ohr–**C**oulomb model) of program `HYPLAS`. Its FORTRAN code and details of its computational implementation are explained in Section 8.2.2 below.

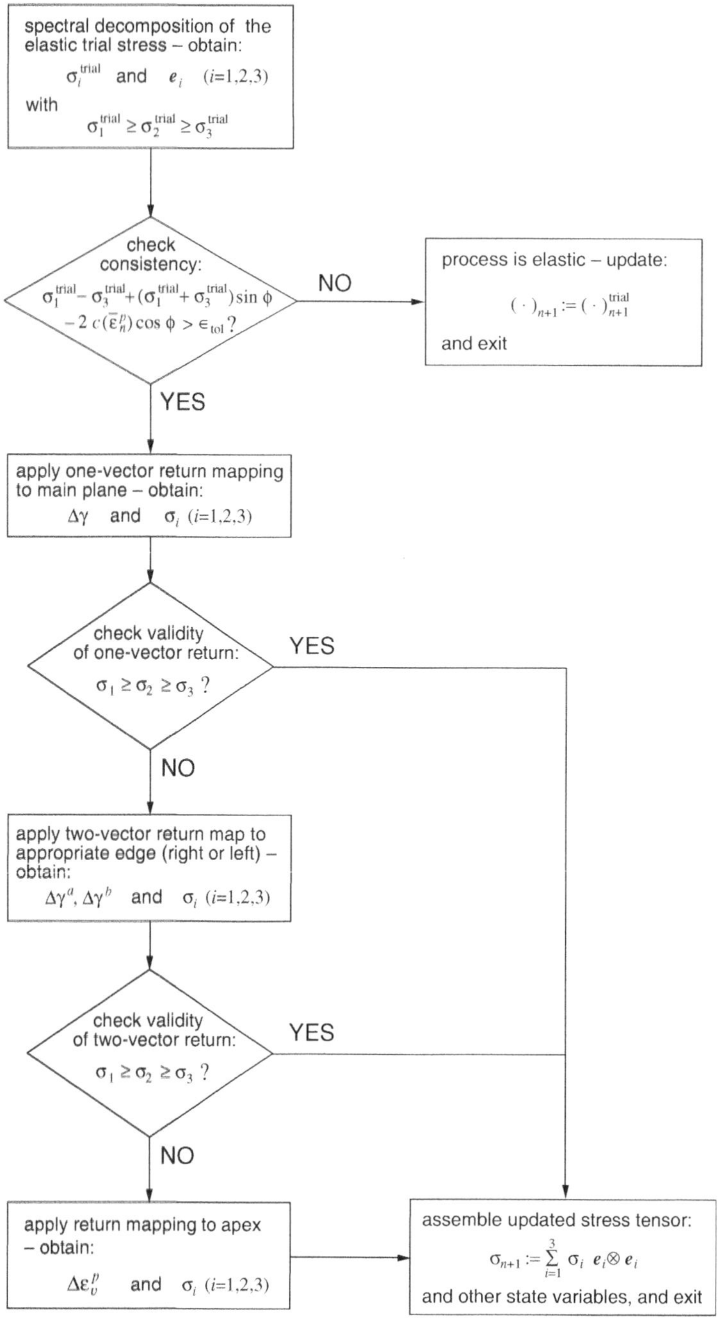

Figure 8.10. Flowchart of the integration algorithm for the Mohr–Coulomb model in principal stresses. Procedure implemented in subroutine SUMC of program HYPLAS.

Box 8.4. Implicit elastic predictor/return mapping algorithm for the Mohr–Coulomb model.

HYPLAS procedure:	SUMC

(i) Elastic predictor. Given $\Delta\boldsymbol{\varepsilon}$ and the state variables at t_n, evaluate the *elastic trial state*

$$\boldsymbol{\varepsilon}^{e\,\text{trial}}_{n+1} := \boldsymbol{\varepsilon}^e_n + \Delta\boldsymbol{\varepsilon}; \qquad \bar{\varepsilon}^{p\,\text{trial}}_{n+1} := \bar{\varepsilon}^p_n$$

$$\boldsymbol{\sigma}^{\text{trial}}_{n+1} := 2G\,\boldsymbol{\varepsilon}^{e\,\text{trial}}_{\text{d}\,n+1} + K\,\varepsilon^{e\,\text{trial}}_{\text{v}\,n+1}\,\boldsymbol{I}$$

(ii) Spectral decomposition of σ^{trial} (routine SPDEC2). Compute

$$\sigma_1^{\text{trial}} \geq \sigma_2^{\text{trial}} \geq \sigma_3^{\text{trial}} \quad \text{and} \quad \boldsymbol{e}_i \quad (i=1,2,3)$$

(iii) Check plastic admissibility

IF $\sigma_1^{\text{trial}} - \sigma_3^{\text{trial}} + (\sigma_1^{\text{trial}} + \sigma_3^{\text{trial}})\sin\phi - 2\,c(\bar{\varepsilon}^{p\,\text{trial}}_{n+1})\cos\phi \leq 0$

THEN set $(\cdot)_{n+1} := (\cdot)^{\text{trial}}_{n+1}$ and EXIT

(iv) Return mapping

(iv.a) Return to main plane – GOTO Box 8.5

(iv.b) Check validity of main plane return

IF $\sigma_1 \geq \sigma_2 \geq \sigma_3$ THEN return is valid – GOTO (v)

(iv.c) Return to edge

IF $(1-\sin\psi)\sigma_1^{\text{trial}} - 2\sigma_3^{\text{trial}} + (1+\sin\psi)\sigma_2^{\text{trial}} > 0$

THEN apply return to **right** edge – GOTO Box 8.6

ELSE apply return to **left** edge – GOTO Box 8.6

(iv.d) Check validity of edge return

IF $\sigma_1 \geq \sigma_2 \geq \sigma_3$ THEN return is valid – GOTO (v)

(iv.e) Return to apex – GOTO Box 8.7

(v) Assemble updated stress tensor

$$\boldsymbol{\sigma}_{n+1} := \sum_{i=1}^{3} \sigma_i\,\boldsymbol{e}_i \otimes \boldsymbol{e}_i$$

and update elastic strain

$$\boldsymbol{\varepsilon}^e_{n+1} := \frac{1}{2G}\,\boldsymbol{s}_{n+1} + \frac{p_{n+1}}{3K}\,\boldsymbol{I}$$

(vi) EXIT

Box 8.5. The Mohr–Coulomb model. One-vector return mapping to main plane.

HYPLAS procedure:	SUMC

(i) Set initial guess for $\Delta\gamma$

$$\Delta\gamma := 0; \quad \bar{\varepsilon}^p_{n+1} := \bar{\varepsilon}^p_n$$

and corresponding residual (yield function value)

$$\tilde{\Phi} := \sigma_1^{\text{trial}} - \sigma_3^{\text{trial}} + (\sigma_1^{\text{trial}} + \sigma_3^{\text{trial}}) \sin\phi - 2\, c(\bar{\varepsilon}^p_n) \cos\phi$$

(ii) Perform Newton–Raphson iteration for $\Delta\gamma$

$$H := \left.\frac{\mathrm{d}c}{\mathrm{d}\bar{\varepsilon}^p}\right|_{\bar{\varepsilon}^p_{n+1}} \qquad \text{(hardening slope)}$$

$$d := \frac{\mathrm{d}\tilde{\Phi}}{\mathrm{d}\Delta\gamma} = -4G(1 + \tfrac{1}{3}\sin\psi\sin\phi) - 4K\ \sin\psi\sin\phi - 4H\cos^2\phi \qquad \text{(residual derivative)}$$

$$\Delta\gamma := \Delta\gamma - \tilde{\Phi}/d \qquad \text{(update } \Delta\gamma\text{)}$$

(iii) Check for convergence

$$\bar{\varepsilon}^p_{n+1} := \bar{\varepsilon}^p_n + 2\cos\phi\,\Delta\gamma$$

$$\begin{aligned}\tilde{\Phi} := {} & \sigma_1^{\text{trial}} - \sigma_3^{\text{trial}} + (\sigma_1^{\text{trial}} + \sigma_3^{\text{trial}})\sin\phi \\ & - [4G(1 + \tfrac{1}{3}\sin\psi\sin\phi) + 4K\ \sin\phi\sin\psi]\Delta\gamma \\ & - 2\, c(\bar{\varepsilon}^p_{n+1})\cos\phi\end{aligned}$$

IF $|\tilde{\Phi}| \leq \epsilon_{\text{tol}}$ THEN update

$$\sigma_1 := \sigma_1^{\text{trial}} - [2G(1 + \tfrac{1}{3}\sin\psi) + 2K\ \sin\psi]\Delta\gamma$$

$$\sigma_2 := \sigma_2^{\text{trial}} + (\tfrac{4}{3}G - 2K)\sin\psi\,\Delta\gamma$$

$$\sigma_3 := \sigma_3^{\text{trial}} + [2G(1 - \tfrac{1}{3}\sin\psi) - 2K\ \sin\psi]\Delta\gamma$$

and RETURN to Box 8.4

(iv) GOTO (ii)

Box 8.6. The Mohr–Coulomb model. Two-vector return mappings to edges.

HYPLAS procedure:	SUMC

(i) Set initial guess for $\Delta\gamma^a$ and $\Delta\gamma^b$

$$\Delta\gamma^a := 0, \quad \Delta\gamma^b := 0; \quad \bar{\varepsilon}^p_{n+1} := \bar{\varepsilon}^p_n$$

and corresponding residual

$$\begin{bmatrix} \tilde{\Phi}^a \\ \tilde{\Phi}^b \end{bmatrix} := \begin{bmatrix} \bar{\sigma}^a - 2\,c(\bar{\varepsilon}^p_n)\cos\phi \\ \bar{\sigma}^b - 2\,c(\bar{\varepsilon}^p_n)\cos\phi \end{bmatrix}$$

where

$$\bar{\sigma}^a = \sigma_1^{\rm trial} - \sigma_3^{\rm trial} + (\sigma_1^{\rm trial} + \sigma_3^{\rm trial})\sin\phi$$

$$\bar{\sigma}^b = \begin{cases} \sigma_1^{\rm trial} - \sigma_2^{\rm trial} + (\sigma_1^{\rm trial} + \sigma_2^{\rm trial})\sin\phi & \textbf{right}\text{ edge} \\ \sigma_2^{\rm trial} - \sigma_3^{\rm trial} + (\sigma_2^{\rm trial} + \sigma_3^{\rm trial})\sin\phi & \textbf{left}\text{ edge} \end{cases}$$

(ii) Perform Newton–Raphson iteration for $\Delta\gamma^a$ and $\Delta\gamma^b$

$$H := \left.\frac{\mathrm{d}c}{\mathrm{d}\bar{\varepsilon}^p}\right|_{\bar{\varepsilon}^p_{n+1}}$$

$$a := 4G(1 + \tfrac{1}{3}\sin\phi\sin\psi) + 4K\sin\phi\sin\psi$$

$$b := \begin{cases} 2G(1 + \sin\phi + \sin\psi - \frac{1}{3}\sin\phi\sin\psi) + 4K\sin\phi\sin\psi & \textbf{right}\text{ edge} \\ 2G(1 - \sin\phi - \sin\psi - \frac{1}{3}\sin\phi\sin\psi) + 4K\sin\phi\sin\psi & \textbf{left}\text{ edge} \end{cases}$$

residual derivative matrix:

$$\mathbf{d} := \begin{bmatrix} \dfrac{\partial\tilde{\Phi}^a}{\partial\Delta\gamma^a} & \dfrac{\partial\tilde{\Phi}^a}{\partial\Delta\gamma^b} \\ \dfrac{\partial\tilde{\Phi}^b}{\partial\Delta\gamma^a} & \dfrac{\partial\tilde{\Phi}^b}{\partial\Delta\gamma^b} \end{bmatrix} = \begin{bmatrix} -a - 4H\cos^2\phi & -b - 4H\cos^2\phi \\ -b - 4H\cos^2\phi & -a - 4H\cos^2\phi \end{bmatrix}$$

new guess for $\Delta\gamma^a$ and $\Delta\gamma^b$:

$$\begin{bmatrix} \Delta\gamma^a \\ \Delta\gamma^b \end{bmatrix} := \begin{bmatrix} \Delta\gamma^a \\ \Delta\gamma^b \end{bmatrix} - \mathbf{d}^{-1}\begin{bmatrix} \tilde{\Phi}^a \\ \tilde{\Phi}^b \end{bmatrix}$$

contd on page 308

Box 8.6 (contd. from page 307). The Mohr–Coulomb model. Two-vector return mappings to edges (implemented in subroutine SUMC)

(iii) Compute new residual and check for convergence

$$\bar{\varepsilon}^p_{n+1} := \bar{\varepsilon}^p_n + 2\cos\phi\,(\Delta\gamma^a + \Delta\gamma^b)$$

$$\begin{bmatrix}\tilde{\Phi}^a \\ \tilde{\Phi}^b\end{bmatrix} := \begin{bmatrix}\bar{\sigma}^a - a\,\Delta\gamma^a - b\,\Delta\gamma^b - 2\,c(\bar{\varepsilon}^p_{n+1})\cos\phi \\ \bar{\sigma}^b - b\,\Delta\gamma^a - a\,\Delta\gamma^b - 2\,c(\bar{\varepsilon}^p_{n+1})\cos\phi\end{bmatrix}$$

IF $|\tilde{\Phi}^a| + |\tilde{\Phi}^b| \le \epsilon_{\rm tol}$ THEN update

for **right** edge:

$$\sigma_1 := \sigma_1^{\rm trial} - [2G(1 + \tfrac{1}{3}\sin\psi) + 2K\sin\psi](\Delta\gamma^a + \Delta\gamma^b)$$
$$\sigma_2 := \sigma_2^{\rm trial} + (\tfrac{4G}{3} - 2K)\sin\psi\,\Delta\gamma^a + [2G(1 - \tfrac{1}{3}\sin\psi) - 2K\sin\psi]\Delta\gamma^b$$
$$\sigma_3 := \sigma_3^{\rm trial} + [2G(1 - \tfrac{1}{3}\sin\psi) - 2K\sin\psi]\Delta\gamma^a + (\tfrac{4G}{3} - 2K)\sin\psi\,\Delta\gamma^b$$

for **left** edge:

$$\sigma_1 := \sigma_1^{\rm trial} - [2G(1 + \tfrac{1}{3}\sin\psi) + 2K\sin\psi]\Delta\gamma^b + (\tfrac{4G}{3} - 2K)\sin\psi\,\Delta\gamma^b$$
$$\sigma_2 := \sigma_2^{\rm trial} + (\tfrac{4G}{3} - 2K)\sin\psi\,\Delta\gamma^a - [2G(1 + \tfrac{1}{3}\sin\psi) + 2K\sin\psi]\Delta\gamma^b$$
$$\sigma_3 := \sigma_3^{\rm trial} + [2G(1 - \tfrac{1}{3}\sin\psi) - 2K\sin\psi](\Delta\gamma^a + \Delta\gamma^b)$$

and RETURN to Box 8.4

(iv) GOTO (ii)

Box 8.7. The Mohr–Coulomb model. Return mapping to apex.

HYPLAS procedure:	SUMC

(i) Set initial guess for $\Delta\varepsilon_v^p$

$$\Delta\varepsilon_v^p := 0; \quad \bar{\varepsilon}_{n+1}^p := \bar{\varepsilon}_n^p$$

and corresponding residual (refer to equation (8.85))

$$r := c(\bar{\varepsilon}_n^p)\ \cot\phi - p_{n+1}^{\text{trial}}$$

(ii) Perform Newton–Raphson iteration

$$H := \left.\frac{\mathrm{d}c}{\mathrm{d}\bar{\varepsilon}^p}\right|_{\bar{\varepsilon}_{n+1}^p} \qquad \text{(hardening slope)}$$

$$d := \frac{H\cos\phi\cot\phi}{\sin\psi} + K \qquad \text{(residual derivative)}$$

$$\Delta\varepsilon_{\text{v}}^p := \Delta\varepsilon_{\text{v}}^p - r/d \qquad \text{(update } \Delta\varepsilon_{\text{v}}^p\text{)}$$

(iii) Compute new residual and check for convergence

$$\bar{\varepsilon}_{n+1}^p := \bar{\varepsilon}_n^p + \tfrac{\cos\phi}{\sin\psi}\Delta\varepsilon_{\text{v}}^p$$

$$p_{n+1} := p_{n+1}^{\text{trial}} - K\,\Delta\varepsilon_{\text{v}}^p$$

$$r := c(\bar{\varepsilon}_{n+1}^p)\cot\phi - p_{n+1}$$

IF $|r| \le \epsilon_{\text{tol}}$ THEN update

$$\sigma_1 := \sigma_2 := \sigma_3 := p_{n+1}$$

and RETURN to Box 8.4

(iv) GOTO (ii)

8.2.2. SUBROUTINE SUMC

The present implementation of the implicit elastic predictor/return-mapping algorithm for the Mohr–Coulomb model incorporates a *piecewise linear isotropic hardening curve* in order to approximate general nonlinear hardening. As in the algorithms for the von Mises and Tresca models (subroutines SUVM and SUTR, respectively), the hardening curve is here defined by a user-specified number of sampling points. For the hardening law adopted in the present implementation, each sampling point i is a pair

$$\{{}^{i}\bar{\varepsilon}^{p}, {}^{i}c\}$$

defined by the accumulated plastic strain and the corresponding cohesion. It is pointed out that perfect plasticity – the most common assumption accompanying the use of the Mohr–Coulomb model – is obtained by simply defining two sampling points with identical cohesion. The FORTRAN source code of subroutine SUMC is listed below.

```
      SUBROUTINE SUMC
     1(   DGAM       ,IPROPS     ,LALGVA     ,NTYPE      ,RPROPS     ,
     2    RSTAVA     ,STRAT      ,STRES      )
      IMPLICIT DOUBLE PRECISION (A-H,O-Z)
      PARAMETER(IPHARD=7   ,MSTRE=4)
C Arguments
      LOGICAL
     1    LALGVA(5)
      DIMENSION
     1    DGAM(2)             ,IPROPS(*)           ,RPROPS(*)           ,
     2    RSTAVA(MSTRE+1)     ,STRAT(MSTRE)        ,STRES(MSTRE)
C Local variables and arrays
      LOGICAL
     1    APEX, DUMMY, EDGE, IFPLAS, RIGHT, SUFAIL
      DIMENSION
     1    EIGPRJ(MSTRE,2)     ,PSTRS(3)            ,STREST(3)
      DATA
     1    R0    ,R1    ,R2    ,R3    ,R4    ,SMALL ,TOL   /
     2    0.0D0,1.0D0,2.0D0,3.0D0,4.0D0,1.D-06,1.D-10/
      DATA MXITER / 50 /
C***********************************************************************
C STATE UPDATE PROCEDURE FOR MOHR-COULOMB TYPE ELASTO-PLASTIC MATERIAL
C WITH ASSOCIATIVE/NON-ASSOCIATIVE FLOW RULE AND PIECE-WISE LINEAR
C ISOTROPIC HARDENING:
C IMPLICIT ELASTIC PREDICTOR/RETURN MAPPING ALGORITHM (BOXES 8.4-7).
C PLANE STRAIN AND AXISYMMETRIC IMPLMENTATIONS.
C***********************************************************************
C Stops program if neither plane strain nor plane stress state
      IF(NTYPE.NE.2.AND.NTYPE.NE.3)CALL ERRPRT('EI0027')
C Initialize some algorithmic and internal variables
      DGAMA=R0
      DGAMB=R0
      IFPLAS=.FALSE.
      SUFAIL=.FALSE.
      EDGE=.FALSE.
      APEX=.FALSE.
      EPBARN=RSTAVA(MSTRE+1)
      EPBAR=EPBARN
C Set some material properties
      YOUNG=RPROPS(2)
      POISS=RPROPS(3)
      SINPHI=RPROPS(4)
```

```
      COSPHI=RPROPS(5)
      SINPSI=RPROPS(6)
      NHARD=IPROPS(3)
C Set some constants
      GMODU=YOUNG/(R2*(R1+POISS))
      BULK=YOUNG/(R3*(R1-R2*POISS))
      R2G=R2*GMODU
      R4G=R4*GMODU
      R2BULK=R2*BULK
      R2CPHI=R2*COSPHI
      R1D3=R1/R3
C Compute elastic trial state
C --------------------------
C Elastic trial volumetric strain and pressure stress
      EETV=STRAT(1)+STRAT(2)+STRAT(4)
      PT=BULK*EETV
C Spectral decomposition of the elastic trial stress
      EETVD3=EETV*R1D3
      STREST(1)=R2G*(STRAT(1)-EETVD3)+PT
      STREST(2)=R2G*(STRAT(2)-EETVD3)+PT
      STREST(3)=GMODU*STRAT(3)
      CALL SPDEC2(EIGPRJ,PSTRS,DUMMY,STREST)
      PSTRS(3)=R2G*(STRAT(4)-EETVD3)+PT
C Identify maximum (PSTRS1) and minimum (PSTRS3) principal stresses
      II=1
      JJ=1
      PSTRS1=PSTRS(II)
      PSTRS3=PSTRS(JJ)
      DO 10 I=2,3
        IF(PSTRS(I).GE.PSTRS1)THEN
          II=I
          PSTRS1=PSTRS(II)
        ENDIF
        IF(PSTRS(I).LT.PSTRS3)THEN
          JJ=I
          PSTRS3=PSTRS(JJ)
        ENDIF
   10 CONTINUE
      IF(II.NE.1.AND.JJ.NE.1)MM=1
      IF(II.NE.2.AND.JJ.NE.2)MM=2
      IF(II.NE.3.AND.JJ.NE.3)MM=3
      PSTRS2=PSTRS(MM)
C Compute trial yield function and check for plastic consistency
C ---------------------------------------------------------------
      COHE=PLFUN(EPBARN,NHARD,RPROPS(IPHARD))
      SMCT=PSTRS1-PSTRS3+(PSTRS1+PSTRS3)*SINPHI
      PHIA=SMCT-R2CPHI*COHE
      RES=PHIA
      IF(COHE.NE.R0)RES=RES/ABS(COHE)
      IF(RES.GT.TOL)THEN
C Plastic step: Apply return mapping
C ==================================
        IFPLAS=.TRUE.
C identify possible edge return: either right or left of main plane
        SCAPRD=PSTRS1*(R1-SINPSI)+PSTRS2*(-R2)+PSTRS3*(R1+SINPSI)
        IF(SCAPRD.GE.R0)THEN
          RIGHT=.TRUE.
        ELSE
          RIGHT=.FALSE.
        ENDIF
C Apply one-vector return mapping first (return to MAIN PLANE)
C ------------------------------------------------------------
        SPHSPS=SINPHI*SINPSI
```

```
          CONSTA=R4G*(R1+R1D3*SPHSPS)+R4*BULK*SPHSPS
          R4C2PH=R2CPHI*R2CPHI
C Start Newton-Raphson iterations for DGAMA
          DO 20 NRITER=1,MXITER
C Compute residual derivative
            DENOM=-CONSTA-R4C2PH*DPLFUN(EPBAR,NHARD,RPROPS(IPHARD))
C Compute Newton-Raphson increment and update variable DGAMA
            DDGAMA=-PHIA/DENOM
            DGAMA=DGAMA+DDGAMA
C Compute new residual
            EPBAR=EPBARN+R2CPHI*DGAMA
            COHE=PLFUN(EPBAR,NHARD,RPROPS(IPHARD))
            PHIA=SMCT-CONSTA*DGAMA-R2CPHI*COHE
C Check convergence
            RESNOR=ABS(PHIA)
            IF(SMCT.NE.R0)RESNOR=RESNOR/ABS(SMCT)
            IF(RESNOR.LE.TOL)THEN
C Check validity of 1-vector return (check sextant of converged stress)
              S1=PSTRS1-(R2G*(R1+R1D3*SINPSI)+R2BULK*SINPSI)*DGAMA
              S2=PSTRS2+(R4G*R1D3-R2BULK)*SINPSI*DGAMA
              S3=PSTRS3+(R2G*(R1-R1D3*SINPSI)-R2BULK*SINPSI)*DGAMA
              DELTA=DMAX1(ABS(S1),ABS(S2),ABS(S3))*SMALL
              IF(S1+DELTA.GE.S2.AND.S2+DELTA.GE.S3)THEN
C converged stress is in the same sextant as trial stress -> 1-vector
C return is valid.
                P=(S1+S2+S3)*R1D3
                GOTO 70
              ELSE
C converged stress is not in the same sextant -> 1-vector result is
C not valid. Go to two-vector return map to edge
                GOTO 30
              ENDIF
            ENDIF
   20     CONTINUE
C failure of stress update procedure
          SUFAIL=.TRUE.
          CALL ERRPRT('WE0003')
          GOTO 999
   30     CONTINUE
C Apply two-vector return mapping to appropriate EDGE
C --------------------------------------------------
          DGAMA=R0
          EPBAR=EPBARN
          COHE=PLFUN(EPBARN,NHARD,RPROPS(IPHARD))
          SMCTA=PSTRS1-PSTRS3+(PSTRS1+PSTRS3)*SINPHI
          IF(RIGHT)THEN
            SMCTB=PSTRS1-PSTRS2+(PSTRS1+PSTRS2)*SINPHI
          ELSE
            SMCTB=PSTRS2-PSTRS3+(PSTRS2+PSTRS3)*SINPHI
          ENDIF
          PHIA=SMCTA-R2CPHI*COHE
          PHIB=SMCTB-R2CPHI*COHE
          IF(RIGHT)THEN
            CONSTB=R2G*(R1+SINPHI+SINPSI-R1D3*SPHSPS)+R4*BULK*SPHSPS
          ELSE
            CONSTB=R2G*(R1-SINPHI-SINPSI-R1D3*SPHSPS)+R4*BULK*SPHSPS
          ENDIF
C Start Newton-Raphson iterations for DGAMA and DGAMB
          DO 40 NRITER=1,MXITER
C Compute residual derivative matrix
            FACTA=R4C2PH*DPLFUN(EPBAR,NHARD,RPROPS(IPHARD))
            DRVAA=-CONSTA-FACTA
```

```
            DRVAB=-CONSTB-FACTA
            DRVBA=-CONSTB-FACTA
            DRVBB=-CONSTA-FACTA
C Compute Newton-Raphson increment and update variables DGAMA and DGAMB
            R1DDET=R1/(DRVAA*DRVBB-DRVAB*DRVBA)
            DDGAMA=(-DRVBB*PHIA+DRVAB*PHIB)*R1DDET
            DDGAMB=(DRVBA*PHIA-DRVAA*PHIB)*R1DDET
            DGAMA=DGAMA+DDGAMA
            DGAMB=DGAMB+DDGAMB
C Compute new residual
            EPBAR=EPBARN+R2CPHI*(DGAMA+DGAMB)
            COHE=PLFUN(EPBAR,NHARD,RPROPS(IPHARD))
            PHIA=SMCTA-CONSTA*DGAMA-CONSTB*DGAMB-R2CPHI*COHE
            PHIB=SMCTB-CONSTB*DGAMA-CONSTA*DGAMB-R2CPHI*COHE
C Check convergence
            RESNOR=(ABS(PHIA)+ABS(PHIB))
            FACTOR=(ABS(SMCTA)+ABS(SMCTB))
            IF(FACTOR.NE.R0)RESNOR=RESNOR/FACTOR
            IF(RESNOR.LE.TOL)THEN
C Check validity of 2-vector return to edge
              AUX1=R2G*(R1+R1D3*SINPSI)+R2BULK*SINPSI
              AUX2=(R4G*R1D3-R2BULK)*SINPSI
              AUX3=R2G*(R1-R1D3*SINPSI)-R2BULK*SINPSI
              IF(RIGHT)THEN
                S1=PSTRS1-AUX1*(DGAMA+DGAMB)
                S2=PSTRS2+AUX2*DGAMA+AUX3*DGAMB
                S3=PSTRS3+AUX3*DGAMA+AUX2*DGAMB
              ELSE
                S1=PSTRS1-AUX1*DGAMA+AUX2*DGAMB
                S2=PSTRS2+AUX2*DGAMA-AUX1*DGAMB
                S3=PSTRS3+AUX3*(DGAMA+DGAMB)
              ENDIF
              DELTA=DMAX1(ABS(S1),ABS(S2),ABS(S3))*SMALL
              IF(S1+DELTA.GE.S2.AND.S2+DELTA.GE.S3)THEN
C converged stress is in the same sextant as trial stress -> 2-vector
C return to edge is valid.
                EDGE=.TRUE.
                P=(S1+S2+S3)*R1D3
                GOTO 70
              ELSE
C converged stress is not in the same sextant -> 2-vector return to edge
C is not valid. Go to two-vector return map to APEX
                GOTO 50
              ENDIF
            ENDIF
   40   CONTINUE
C failure of stress update procedure
        SUFAIL=.TRUE.
        CALL ERRPRT('WE0003')
        GOTO 999
   50   CONTINUE
C Apply multi-vector return mapping to APEX
C ---------------------------------------
C Check conditions for which return to apex does not make sense
        IF(SINPHI.EQ.R0)CALL ERRPRT('EE0009')
        IF(SINPSI.EQ.R0)CALL ERRPRT('EE0010')
C Set initial guess for volumetric plastic strain increment DEPV
        DEPV=R0
        EPBAR=EPBARN
        COHE=PLFUN(EPBAR,NHARD,RPROPS(IPHARD))
        COTPHI=COSPHI/SINPHI
        RES=COTPHI*COHE-PT
C Newton-Raphson iterations for DEPV
```

```
        DO 60 NRITER=1,MXITER
          DENOM=COSPHI*COTPHI/SINPSI*DPLFUN(EPBAR,NHARD,RPROPS(IPHARD))+
     1          BULK
          DDEPV=-RES/DENOM
          DEPV=DEPV+DDEPV
          EPBAR=EPBARN+COSPHI/SINPSI*DEPV
          COHE=PLFUN(EPBAR,NHARD,RPROPS(IPHARD))
          P=PT-BULK*DEPV
          RES=COTPHI*COHE-P
C check for convergence
          RESNOR=ABS(RES)
          IF(PT.NE.R0)RESNOR=RESNOR/ABS(PT)
          IF(RESNOR.LE.TOL)THEN
            APEX=.TRUE.
            DGAMA=DEPV
            DGAMB=R0
C update principal stresses
            S1=P
            S2=P
            S3=P
            GOTO 70
          ENDIF
   60   CONTINUE
        SUFAIL=.TRUE.
        CALL ERRPRT('WE0003')
        GOTO 999
   70   CONTINUE
C update internal variable EPBAR  and stress components
C -----------------------------------------------------
        RSTAVA(MSTRE+1)=EPBAR
        PSTRS(II)=S1
        PSTRS(JJ)=S3
        PSTRS(MM)=S2
        STRES(1)=PSTRS(1)*EIGPRJ(1,1)+PSTRS(2)*EIGPRJ(1,2)
        STRES(2)=PSTRS(1)*EIGPRJ(2,1)+PSTRS(2)*EIGPRJ(2,2)
        STRES(3)=PSTRS(1)*EIGPRJ(3,1)+PSTRS(2)*EIGPRJ(3,2)
        STRES(4)=PSTRS(3)
C and elastic engineering strain
        EEVD3=P/BULK*R1D3
        RSTAVA(1)=(STRES(1)-P)/R2G+EEVD3
        RSTAVA(2)=(STRES(2)-P)/R2G+EEVD3
        RSTAVA(3)=STRES(3)/GMODU
        RSTAVA(4)=(STRES(4)-P)/R2G+EEVD3
      ELSE
C Elastic step: update stress using linear elastic law
C ====================================================
        STRES(1)=STREST(1)
        STRES(2)=STREST(2)
        STRES(3)=STREST(3)
        STRES(4)=PSTRS(3)
C elastic engineering strain
        RSTAVA(1)=STRAT(1)
        RSTAVA(2)=STRAT(2)
        RSTAVA(3)=STRAT(3)
        RSTAVA(4)=STRAT(4)
      ENDIF
  999 CONTINUE
C Update algorithmic variables before exit
C ========================================
      DGAM(1)=DGAMA
      DGAM(2)=DGAMB
      LALGVA(1)=IFPLAS
```

```
      LALGVA(2)=SUFAIL
      LALGVA(3)=EDGE
      LALGVA(4)=RIGHT
      LALGVA(5)=APEX
      RETURN
      END
```

The arguments of SUMC

This subroutine has the same list of arguments as SUTR (for the Tresca model implementation – see page 282). Note, however, that here IPROPS(3) is set in subroutine RDMC. Array RPROPS of real material properties stores the following in the present case:

$$\text{RPROPS} = [E, \nu, \sin\phi, \cos\phi, \sin\psi, {}^{1}\bar{\varepsilon}^{p}, {}^{1}c, {}^{2}\bar{\varepsilon}^{p}, {}^{2}c, \ldots\ldots, {}^{n_{\rm hard}}\bar{\varepsilon}^{p}, {}^{n_{\rm hard}}c],$$

where ${}^{i}c$ denote the cohesion at the different points supplied along the (piecewise linear) isotropic hardening curve. The array LALGVA of logical algorithmic values also differs from that of the Tresca model implementation. Here, we have:

← LALGVA. For the Mohr–Coulomb model this array contains the plastic yielding flag, IFPLAS, the state update failure flag SUFAIL, the edge return flag EDGE, the right edge return flag RIGHT and the apex return flag APEX. Flags IFPLAS and SUFAIL are set in the same way as in the other material model implementations already described. The edge return flag is set to .TRUE. if one of the two possible edge return mappings (right or left edge) is used. It is set to .FALSE. otherwise. The right edge return flag is set to .TRUE. if the possible edge return is to the right edge and is set to .FALSE. otherwise. The flag APEX is set to .TRUE. only if the return mapping to the apex is applied. The flags EDGE, RIGHT and APEX are required by subroutine CTMC to decide which elastoplastic tangent operator to compute (consistent with main plane, right, left edge or apex return mapping).

Local variables and arrays and function calls

Subroutine SUMC calls the same functions as and has a very similar structure to its counterpart SUTR for the Tresca model. Refer to page 283 for a description of important local variables and arrays.

8.2.3. ACCURACY: ISO-ERROR MAPS

The finite step accuracy of the integration algorithm coded in SUMC is assessed in this section by means of iso-error maps. The maps constructed here are restricted to the perfectly plastic Mohr–Coulomb model and associative flow rule with $\phi = 20^{\circ}$. Three starting points lying on the deviatoric plane are considered. These are shown in Figure 8.11 together with the corresponding increment directions. The increment direction vectors $\boldsymbol{T}$ on all three points are deviatoric. The vector $\boldsymbol{N}$ at point A is normal to the Mohr–Coulomb surface. At points B and C, $\boldsymbol{N}$ is chosen as the mean normal between the corresponding adjacent planes. Note that, in view of the pressure sensitivity of the Mohr–Coulomb model, a more comprehensive accuracy assessment would have to include increments in other directions with more significant *hydrostatic* components. For instance, iso-error maps could be plotted for points B and C

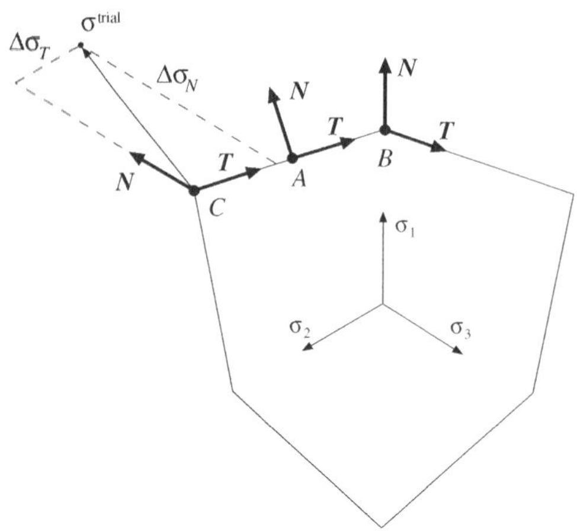

Figure 8.11. Iso-error maps for the Mohr–Coulomb model. Starting points and increment directions.

having the vector $\boldsymbol{T}$ parallel to the corresponding edge of the Mohr–Coulomb pyramid. It is remarked, however, that the results presented in this section give a very good insight into the accuracy properties of the present algorithm.

The maps obtained are shown in Figure 8.12. Similarly to the integration algorithm for the Tresca model (Section 8.1.3), large areas in which the integration error vanishes are observed in the present case for all starting points. Within some narrow bands, however, the integration error can be high. Recall that, in the absence of hardening, the present integration algorithm simply projects the trial stress onto the yield surface along a suitable direction. The reason for the observed relatively high errors can be verified by performing the projections graphically in a single step and comparing with the result obtained when the increment is divided into substeps.

8.2.4. CONSISTENT TANGENT OPERATOR FOR THE MOHR–COULOMB MODEL

The next step for the incorporation of the Mohr–Coulomb model into program `HYPLAS` is the derivation of the tangent operator consistent with the integration algorithm of Section 8.2.1 (coded in `SUMC`). In what follows, only the derivation of the *elastoplastic* tangent is addressed.

Firstly, it is remarked that the consistent elastoplastic tangent for the Mohr–Coulomb model is derived here following the same basic ideas that underly the computation of D^{ep} in the Tresca model implementation; that is, the tangent modulus in the present case is also computed as the derivative of an isotropic tensor function of a single tensor defined in terms of principal values (refer to Section 8.1.4 for details of this approach).

Essentially, in order to compute the elastoplastic tangent operator, we need only to derive expressions for the partial derivatives

$$\frac{\partial \tilde{\sigma}_i}{\partial \varepsilon_j^{e\;\text{trial}}}, \quad i, j = 1, 2, 3$$

where $\tilde{\sigma}_i$ are the implicit functions for the principal updated stresses

$$\sigma_i = \tilde{\sigma}_i(\varepsilon_1^{e\;\text{trial}}, \varepsilon_2^{e\;\text{trial}}, \varepsilon_3^{e\;\text{trial}}), \tag{8.91}$$

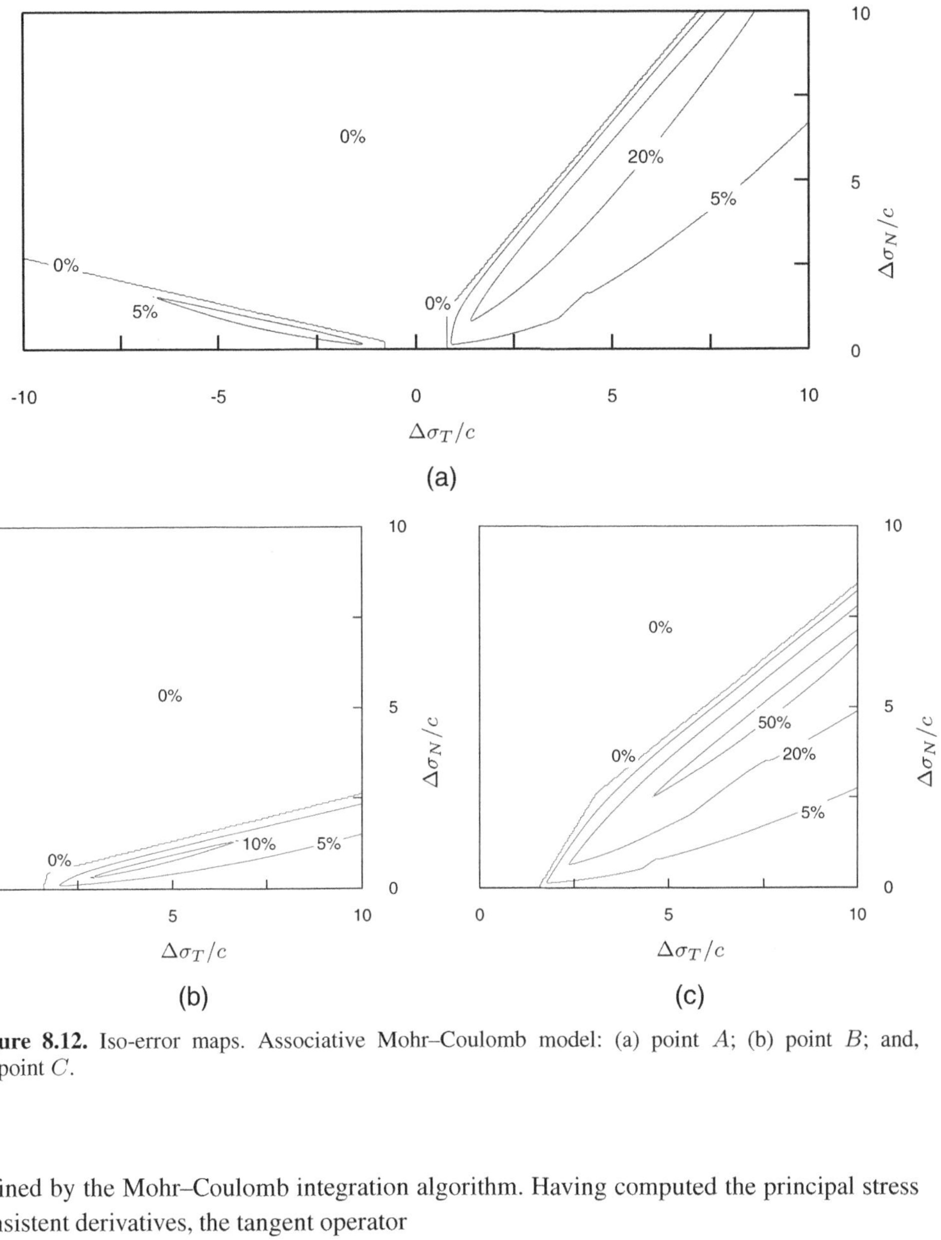

Figure 8.12. Iso-error maps. Associative Mohr–Coulomb model: (a) point A; (b) point B; and, (c) point C.

defined by the Mohr–Coulomb integration algorithm. Having computed the principal stress consistent derivatives, the tangent operator

$$\mathbf{D}^{ep} \equiv \frac{\mathrm{d}\boldsymbol{\sigma}_{n+1}}{\mathrm{d}\boldsymbol{\varepsilon}^{e\ \mathrm{trial}}_{n+1}} \tag{8.92}$$

can then be assembled as described in Section A.3 of Appendix A (in the program, the matrix form of $\mathbf{D}^{ep}$ is assembled in subroutine DGISO2).

Recall that the present model admits four distinct return-mapping sets of equations: return to main plane, right, left edge and apex.[§] For each one there is a different explicit form for the principal stress derivatives and for D^{ep}. In the finite element program, the actual form adopted in the assemblage of the tangent stiffness matrix is the one consistent with the previous application of the return-mapping procedure at the Gauss point in question. The derivation of the principal stresses derivatives consistent with each of the four possible return mappings is addressed in what follows.

Principal stress derivatives for the main plane return

The first step in the derivation is to differentiate the main plane return principal stress update formulae (8.69). This gives

$$\begin{aligned}
\mathrm{d}\sigma_1 &= \mathrm{d}\sigma_1^{\text{trial}} - [2G(1+\tfrac{1}{3}\sin\psi) + 2K\sin\psi]\,\mathrm{d}\Delta\gamma \\
\mathrm{d}\sigma_2 &= \mathrm{d}\sigma_2^{\text{trial}} + \left(\frac{4G}{3} - 2K\right)\sin\psi\,\mathrm{d}\Delta\gamma \\
\mathrm{d}\sigma_3 &= \mathrm{d}\sigma_3^{\text{trial}} + [2G(1-\tfrac{1}{3}\sin\psi) - 2K\sin\psi]\,\mathrm{d}\Delta\gamma.
\end{aligned} \tag{8.93}$$

The differential of $\Delta\gamma$ is derived by enforcing the linearised consistency condition

$$\mathrm{d}\tilde{\Phi} = \mathrm{d}\sigma_1^{\text{trial}} - \mathrm{d}\sigma_3^{\text{trial}} + (\mathrm{d}\sigma_1^{\text{trial}} + \mathrm{d}\sigma_3^{\text{trial}})\sin\phi - (4H\cos^2\phi + a)\,\mathrm{d}\Delta\gamma = 0, \tag{8.94}$$

which is obtained by differentiating (8.71). In this last expression, H is the isotropic hardening modulus (slope of the hardening curve):

$$H \equiv \left.\frac{\mathrm{d}c}{\mathrm{d}\bar{\varepsilon}^p}\right|_{\bar{\varepsilon}^p_{n+1}}, \tag{8.95}$$

and the constant a is defined by (8.72). Combination of the two differential forms above, together with the use of the linear elastic relation

$$\boldsymbol{\sigma}_{n+1}^{\text{trial}} = \mathsf{D}^e : \boldsymbol{\varepsilon}_{n+1}^{e\ \text{trial}}, \tag{8.96}$$

renders the final expressions for the derivatives of the principal stresses. The resulting expressions – not explicitly written in the present text – are implemented in subroutine `CTMC` and can be easily identified from the FORTRAN source code listed in Section 8.2.5. In `CTMC`, the computed principal stress derivatives are stored in matrix `DPSTRS`.

Principal stress derivatives for the edge return mappings

To obtain the principal stress derivatives consistent with the right edge return, we need first to differentiate the update formulae (8.74) and then substitute the differentials $\mathrm{d}\Delta\gamma^a$ and $\mathrm{d}\Delta\gamma^b$ obtained by linearising the consistency equations (8.76). The derivation is lengthier than the previous one and will not be shown here. The final results can be easily identified in subroutine `CTMC`. The same comments apply to the derivation of the derivatives for the left edge return mapping, whose main steps involve the differentiation of the expressions given in (8.78) and (8.79).

[§] The situation here is completely analogous to the Tresca model implementation where three return map equation sets exist.

Principal stress derivatives for the apex return

Differentiation of the apex return stress update formulae (8.82) and $(8.86)_2$ yields

$$\begin{aligned} \mathrm{d}\sigma_i &= \mathrm{d}p^{\mathrm{trial}}_{n+1} - K\, \mathrm{d}\Delta\varepsilon^p_{\mathrm{v}} \\ &= K\, (\mathrm{d}\varepsilon^{\mathrm{trial}}_1 + \mathrm{d}\varepsilon^{\mathrm{trial}}_2 + \mathrm{d}\varepsilon^{\mathrm{trial}}_3 - \mathrm{d}\Delta\varepsilon^p_{\mathrm{v}}) \end{aligned} \tag{8.97}$$

for $i = 1, 2, 3$. By differentiating (8.85), the following linearised consistency condition is obtained:

$$\frac{H \cos\phi \cot\phi}{\sin\psi}\, \mathrm{d}\Delta\varepsilon^p_{\mathrm{v}} - K\, (\mathrm{d}\varepsilon^{e\ \mathrm{trial}}_1 + \mathrm{d}\varepsilon^{e\ \mathrm{trial}}_2 + \mathrm{d}\varepsilon^{e\ \mathrm{trial}}_3 - \mathrm{d}\Delta\varepsilon^p_{\mathrm{v}}) = 0, \tag{8.98}$$

which rearranged gives

$$\mathrm{d}\Delta\varepsilon^p_{\mathrm{v}} = \frac{K}{K + \frac{\cos\phi\cot\phi}{\sin\psi} H} (\mathrm{d}\varepsilon^{\mathrm{trial}}_1 + \mathrm{d}\varepsilon^{\mathrm{trial}}_2 + \mathrm{d}\varepsilon^{\mathrm{trial}}_3). \tag{8.99}$$

The substitution of this expression into (8.97) yields the principal stress derivatives consistent with the apex return mapping:

$$\frac{\partial\sigma_i}{\partial\varepsilon^{e\ \mathrm{trial}}_j} = K\left(1 - \frac{K}{K + \frac{\cos\phi\cot\phi}{\sin\psi} H}\right), \tag{8.100}$$

for $i, j = 1, 2, 3$.

Remark 8.3. In the absence of hardening ($H = 0$) all principal stress derivatives consistent with the apex return vanish. Clearly, as in this case the apex remains *fixed* in stress space, we have trivially $\mathbf{D}^{ep} = \mathbf{0}$.

8.2.5. SUBROUTINE CTMC

The computation of the tangent moduli consistent with the integration algorithm for the Mohr–Coulomb model is implemented in subroutine CTMC (**C**onsistent **T**angent for the **M**ohr–**C**oulomb model). This routine is consistent with the algorithm coded in SUMC. It returns either the elastic tangent or the elastoplastic tangent depending on the entry value of the logical flag EPFLAG (one of its arguments). When the elastoplastic tangent is required, consistency with the appropriate return algorithm (main plane, right, left edge or apex) is defined by the list of logical algorithmic variables passed as arguments in array LALGVA whose values are set in routine SUMC. The source code of CTMC is listed in the following.

```
      SUBROUTINE CTMC
     1(   DMATX      ,EPFLAG     ,IPROPS     ,LALGVA     ,NTYPE      ,
     2    RPROPS     ,RSTAVA     ,STRAT      ,STRES      )
      IMPLICIT DOUBLE PRECISION (A-H,O-Z)
      PARAMETER(IPHARD=7  ,MDIM=3,  MSTRE=4)
C Arguments
      LOGICAL EPFLAG  ,LALGVA(5)
      DIMENSION
     1    DMATX(MSTRE,MSTRE) ,IPROPS(*)           ,RPROPS(*)           ,
```

```
     2     RSTAVA(MSTRE+1)     ,STRAT(MSTRE)        ,STRES(MSTRE)
C Local arrays and variables
      LOGICAL APEX ,EDGE ,OUTOFP ,RIGHT ,REPEAT
      DIMENSION
     1     DPSTRS(MDIM,MDIM)   ,EIGPRJ(MSTRE,2)     ,FOID(MSTRE,MSTRE)   ,
     2     PSTRS(MDIM)         ,PSTRA(MDIM)         ,SOID(MSTRE)         ,
     3     STRAC(MSTRE)
      DATA
     1     FOID(1,1),FOID(1,2),FOID(1,3),FOID(1,4)/
     2     1.0D0    ,0.0D0    ,0.0D0    ,0.0D0    /
     3     FOID(2,1),FOID(2,2),FOID(2,3),FOID(2,4)/
     4     0.0D0    ,1.0D0    ,0.0D0    ,0.0D0    /
     5     FOID(3,1),FOID(3,2),FOID(3,3),FOID(3,4)/
     6     0.0D0    ,0.0D0    ,0.5D0    ,0.0D0    /
     7     FOID(4,1),FOID(4,2),FOID(4,3),FOID(4,4)/
     8     0.0D0    ,0.0D0    ,0.0D0    ,1.0D0    /
      DATA
     1     SOID(1)  ,SOID(2)  ,SOID(3)  ,SOID(4)  /
     2     1.0D0    ,1.0D0    ,0.0D0    ,1.0D0    /
      DATA
     1     RP5  ,R1   ,R2   ,R3   ,R4   /
     2     0.5D0,1.0D0,2.0D0,3.0D0,4.0D0/
C***********************************************************************
C COMPUTATION OF CONSISTENT TANGENT MODULUS FOR MOHR-COULOMB TYPE
C ELASTO-PLASTIC MATERIAL WITH ASSOCIATIVE/NON-ASSOCIATIVE FLOW RULE AND
C PIECE-WISE LINEAR ISOTROPIC HARDENING.
C PLANE STRAIN AND AXISYMMETRIC IMPLEMENTATIONS.
C***********************************************************************
C Stops program if neither plane strain nor axisymmetric state
      IF(NTYPE.NE.2.AND.NTYPE.NE.3)CALL ERRPRT('EI0026')
C Current accumulated plastic strain
      EPBAR=RSTAVA(MSTRE+1)
C Set material properties
      YOUNG=RPROPS(2)
      POISS=RPROPS(3)
      SINPHI=RPROPS(4)
      COSPHI=RPROPS(5)
      SINPSI=RPROPS(6)
      NHARD=IPROPS(3)
C Set needed algorithmic variables
      EDGE=LALGVA(3)
      RIGHT=LALGVA(4)
      APEX=LALGVA(5)
C Set some constants
      GMODU=YOUNG/(R2*(R1+POISS))
      BULK=YOUNG/(R3*(R1-R2*POISS))
      R2G=R2*GMODU
      R4G=R4*GMODU
      R2BULK=R2*BULK
      R2CPHI=R2*COSPHI
      R4C2PH=R2CPHI*R2CPHI
      R1D3=R1/R3
      R2D3=R2*R1D3
      R2GD3=R2G*R1D3
      R4GD3=R4G*R1D3
      IF(EPFLAG)THEN
C Compute elastoplastic consistent tangent
C ---------------------------------------
C Spectral decomposition of the elastic trial strain
        STRAC(1)=STRAT(1)
        STRAC(2)=STRAT(2)
```

```
      STRAC(3)=STRAT(3)*RP5
      CALL SPDEC2(EIGPRJ,PSTRA,REPEAT,STRAC)
      PSTRA(3)=STRAT(4)
C and current total stress
      PSTRS(1)=STRES(1)*EIGPRJ(1,1)+STRES(2)*EIGPRJ(2,1)+
     1                           R2*STRES(3)*EIGPRJ(3,1)
      PSTRS(2)=STRES(1)*EIGPRJ(1,2)+STRES(2)*EIGPRJ(2,2)+
     1                           R2*STRES(3)*EIGPRJ(3,2)
      PSTRS(3)=STRES(4)
C Identify directions of maximum and minimum principal trial stresses
      II=1
      JJ=1
      PSTMAX=PSTRA(II)
      PSTMIN=PSTRA(JJ)
      DO 10 I=2,3
        IF(PSTRA(I).GE.PSTMAX)THEN
          II=I
          PSTMAX=PSTRA(II)
        ENDIF
        IF(PSTRA(I).LT.PSTMIN)THEN
          JJ=I
          PSTMIN=PSTRA(JJ)
        ENDIF
   10 CONTINUE
      IF(II.NE.1.AND.JJ.NE.1)MM=1
      IF(II.NE.2.AND.JJ.NE.2)MM=2
      IF(II.NE.3.AND.JJ.NE.3)MM=3
      IF(EDGE)THEN
C Tangent consistent with 2-vector return to edge
        SPHSPS=SINPHI*SINPSI
        CONSTA=R4G*(R1+R1D3*SPHSPS)+R4*BULK*SPHSPS
        IF(RIGHT)THEN
          CONSTB=R2G*(R1+SINPHI+SINPSI-R1D3*SPHSPS)+R4*BULK*SPHSPS
        ELSE
          CONSTB=R2G*(R1-SINPHI-SINPSI-R1D3*SPHSPS)+R4*BULK*SPHSPS
        ENDIF
        FACTA=R4C2PH*DPLFUN(EPBAR,NHARD,RPROPS(IPHARD))
        DRVAA=-CONSTA-FACTA
        DRVAB=-CONSTB-FACTA
        DRVBA=-CONSTB-FACTA
        DRVBB=-CONSTA-FACTA
        AUX1=R2G*(R1+R1D3*SINPSI)+R2BULK*SINPSI
        AUX2=(R4GD3-R2BULK)*SINPSI
        AUX3=R2G*(R1-R1D3*SINPSI)-R2BULK*SINPSI
        R1DDET=R1/(DRVAA*DRVBB-DRVAB*DRVBA)
        IF(RIGHT)THEN
C ...returned to right edge
          DPSTRS(II,II)=BULK+R4GD3+AUX1*(-DRVAB+DRVBB+DRVAA-DRVBA)*
     1                  (R2G+(R2BULK+R2GD3)*SINPHI)*R1DDET
          DPSTRS(II,MM)=BULK-R2GD3+AUX1*(R2G*(DRVAB-DRVAA)+
     1                  ((-DRVAB+DRVBB+DRVAA-DRVBA)*(R2BULK+R2GD3)+
     2                  (DRVBA-DRVBB)*R2G)*SINPHI)*R1DDET
          DPSTRS(II,JJ)=BULK-R2GD3+AUX1*(R2G*(DRVBA-DRVBB)+
     1                  ((-DRVAB+DRVBB+DRVAA-DRVBA)*(R2BULK+R2GD3)+
     2                  (DRVAB-DRVAA)*R2G)*SINPHI)*R1DDET
          DPSTRS(MM,II)=BULK-R2GD3+(AUX2*(DRVAB-DRVBB)+AUX3*(DRVBA-
     1                  DRVAA))*(R2G+(R2BULK+R2GD3)*SINPHI)*R1DDET
          DPSTRS(MM,MM)=BULK+R4GD3+(AUX2*((R2BULK*(DRVAB-DRVBB)+
     1                  (DRVAB*R2GD3+DRVBB*R4GD3))*SINPHI-DRVAB*R2G)+
     2                  AUX3*(DRVAA*R2G+(R2BULK*(DRVBA-DRVAA)-
     3                  (DRVAA*R2GD3+DRVBA*R4GD3))*SINPHI))*R1DDET
```

```
            DPSTRS(MM,JJ)=BULK-R2GD3+(AUX2*((R2BULK*(DRVAB-DRVBB)-
     1                   (DRVBB*R2GD3+DRVAB*R4GD3))*SINPHI+DRVBB*R2G)+
     2                   AUX3*((R2BULK*(DRVBA-DRVAA)+(DRVAA*R4GD3+
     3                   DRVBA*R2GD3))*SINPHI-DRVBA*R2G))*R1DDET
            DPSTRS(JJ,II)=BULK-R2GD3+((AUX2*(DRVBA-DRVAA)+AUX3*(DRVAB-
     1                   DRVBB))*((R2BULK+R2GD3)*SINPHI+R2G))*R1DDET
            DPSTRS(JJ,MM)=BULK-R2GD3+(AUX2*(((R2BULK*(DRVBA-DRVAA)-
     1                   (DRVBA*R4GD3+DRVAA*R2GD3))*SINPHI)+DRVAA*R2G)+
     2                   AUX3*(((R2BULK*(DRVAB-DRVBB)+(DRVAB*R2GD3+
     3                   DRVBB*R4GD3))*SINPHI)-DRVAB*R2G))*R1DDET
            DPSTRS(JJ,JJ)=BULK+R4GD3+(AUX2*(((R2BULK*(DRVBA-DRVAA)+
     1                   (DRVAA*R4GD3+DRVBA*R2GD3))*SINPHI)-DRVBA*R2G)+
     2                   AUX3*(((R2BULK*(DRVAB-DRVBB)-(DRVAB*R4GD3+
     3                   DRVBB*R2GD3))*SINPHI)+DRVBB*R2G))*R1DDET
          ELSE
C ...returned to left edge
            DPSTRS(II,II)=BULK+R4GD3+(AUX1*(((R2BULK*(DRVBB-DRVAB)+
     1                   (DRVAB*R4GD3+DRVBB*R2GD3))*SINPHI)+DRVBB*R2G)+
     2                   AUX2*(((R2BULK*(DRVBA-DRVAA)+(DRVAA*R4GD3+
     3                   DRVBA*R2GD3))*SINPHI)+DRVBA*R2G))*R1DDET
            DPSTRS(II,MM)=BULK-R2GD3+(AUX1*(((R2BULK*(DRVBB-DRVAB)-
     1                   (DRVAB*R2GD3+DRVBB*R4GD3))*SINPHI)-DRVAB*R2G)+
     2                   AUX2*(((R2BULK*(DRVBA-DRVAA)-(DRVAA*R2GD3+
     3                   DRVBA*R4GD3))*SINPHI)-DRVAA*R2G))*R1DDET
            DPSTRS(II,JJ)=BULK-R2GD3+((AUX1*(DRVBB-DRVAB)+AUX2*(DRVBA-
     1                   DRVAA))*(((R2BULK+R2GD3)*SINPHI)-R2G))*R1DDET
            DPSTRS(MM,II)=BULK-R2GD3+(AUX1*(((R2BULK*(DRVAA-DRVBA)-
     1                   (DRVAA*R4GD3+DRVBA*R2GD3))*SINPHI)-DRVBA*R2G)+
     2                   AUX2*(((R2BULK*(DRVAB-DRVBB)-(DRVAB*R4GD3+
     3                   DRVBB*R2GD3))*SINPHI)-DRVBB*R2G))*R1DDET
            DPSTRS(MM,MM)=BULK+R4GD3+(AUX1*(((R2BULK*(DRVAA-DRVBA)+
     1                   (DRVAA*R2GD3+DRVBA*R4GD3))*SINPHI)+DRVAA*R2G)+
     2                   AUX2*(((R2BULK*(DRVAB-DRVBB)+(DRVAB*R2GD3+
     3                   DRVBB*R4GD3))*SINPHI)+DRVAB*R2G))*R1DDET
            DPSTRS(MM,JJ)=BULK-R2GD3+((AUX1*(DRVAA-DRVBA)+AUX2*(DRVAB-
     1                   DRVBB))*(((R2BULK+R2GD3)*SINPHI)-R2G))*R1DDET
            DPSTRS(JJ,II)=BULK-R2GD3+(AUX3*(((R2BULK*(DRVAB-DRVBB-DRVAA+
     1                   DRVBA)+(DRVAA-DRVAB)*R4GD3+(DRVBA-DRVBB)*
     2                   R2GD3)*SINPHI)+(DRVBA-DRVBB)*R2G))*R1DDET
            DPSTRS(JJ,MM)=BULK-R2GD3+(AUX3*(((R2BULK*(DRVAB-DRVBB-DRVAA+
     1                   DRVBA)+(DRVAB-DRVAA)*R2GD3+(DRVBB-DRVBA)*
     2                   R4GD3)*SINPHI)+(DRVAB-DRVAA)*R2G))*R1DDET
            DPSTRS(JJ,JJ)=BULK+R4GD3+(AUX3*(DRVAB-DRVBB-DRVAA+DRVBA)*
     1                   (((R2BULK+R2GD3)*SINPHI)-R2G))*R1DDET
          ENDIF
        ELSEIF(APEX)THEN
C Tangent consistent with multi-vector return to apex
          COTPHI=COSPHI/SINPHI
          DSIDEJ=BULK*(R1-(BULK/(BULK+
     1          DPLFUN(EPBAR,NHARD,RPROPS(IPHARD))*COTPHI*COSPHI/
     2          SINPSI)))
          DPSTRS(II,II)=DSIDEJ
          DPSTRS(II,MM)=DSIDEJ
          DPSTRS(II,JJ)=DSIDEJ
          DPSTRS(MM,II)=DSIDEJ
          DPSTRS(MM,MM)=DSIDEJ
          DPSTRS(MM,JJ)=DSIDEJ
          DPSTRS(JJ,II)=DSIDEJ
          DPSTRS(JJ,MM)=DSIDEJ
```

```
            DPSTRS(JJ,JJ)=DSIDEJ
          ELSE
C Tangent consistent with 1-vector return to main active plane
            SPHSPS=SINPHI*SINPSI
            CONSTA=R4G*(R1+R1D3*SPHSPS)+R4*BULK*SPHSPS
            DENOM=-CONSTA-R4C2PH*DPLFUN(EPBAR,NHARD,RPROPS(IPHARD))
            B1=(R2G*(R1+R1D3*SINPSI)+R2BULK*SINPSI)/DENOM
            B2=(R4G*R1D3-R2BULK)*SINPSI/DENOM
            B3=(R2G*(R1-R1D3*SINPSI)-R2BULK*SINPSI)/DENOM
            DPSTRS(II,II)=R2G*(R2D3+B1*(R1+R1D3*SINPHI))+
     1                    BULK*(R1+R2*B1*SINPHI)
            DPSTRS(II,MM)=R1D3*(R3*BULK-R2G)*(R1+R2*B1*SINPHI)
            DPSTRS(II,JJ)=R2G*(-R1D3-B1*(R1-R1D3*SINPHI))+
     1                    BULK*(R1+R2*B1*SINPHI)
            DPSTRS(MM,II)=R2G*(-R1D3-B2*(R1+R1D3*SINPHI))+
     1                    BULK*(R1-R2*B2*SINPHI)
            DPSTRS(MM,MM)=R4G*R1D3*(R1+B2*SINPHI)+BULK*(R1-R2*B2*SINPHI)
            DPSTRS(MM,JJ)=R2G*(-R1D3+B2*(R1-R1D3*SINPHI))+
     1                    BULK*(R1-R2*B2*SINPHI)
            DPSTRS(JJ,II)=R2G*(-R1D3-B3*(R1+R1D3*SINPHI))+
     1                    BULK*(R1-R2*B3*SINPHI)
            DPSTRS(JJ,MM)=R1D3*(R3*BULK-R2G)*(R1-R2*B3*SINPHI)
            DPSTRS(JJ,JJ)=R2G*(R2D3+B3*(R1-R1D3*SINPHI))+
     1                    BULK*(R1-R2*B3*SINPHI)
          ENDIF
C
          IF(NTYPE.EQ.2)THEN
            OUTOFP=.FALSE.
          ELSEIF(NTYPE.EQ.3)THEN
            OUTOFP=.TRUE.
          ENDIF
          CALL DGISO2
     1(   DPSTRS     ,DMATX      ,EIGPRJ     ,PSTRA      ,PSTRS      ,
     2    OUTOFP     ,REPEAT     )
        ELSE
C Compute elasticity matrix
C -------------------------
          IF(NTYPE.EQ.2)THEN
            NSTRE=3
          ELSEIF(NTYPE.EQ.3)THEN
            NSTRE=4
          ENDIF
          FACTOR=BULK-R2G*R1D3
          DO 50 I=1,NSTRE
            DO 40 J=I,NSTRE
              DMATX(I,J)=R2G*FOID(I,J)+FACTOR*SOID(I)*SOID(J)
   40       CONTINUE
   50     CONTINUE
          DO 70 J=1,NSTRE-1
            DO 60 I=J+1,NSTRE
              DMATX(I,J)=DMATX(J,I)
   60       CONTINUE
   70     CONTINUE
        ENDIF
        RETURN
        END
```

The arguments of `CTMC`

The arguments here are those of `SUMC` (exept for the array `DGAM`) plus `DMATX` and `EPFLAG` which carry, respectively, the tangent operator and the elastoplastic flag that indicates whether the elastic or elastoplastic operator is to be computed in `CTMC`. The output values of the arguments of `SUMC` are taken as input values by `CTMC`.

Function calls from `CTMC`

Function calls here are the same as from subroutine `CTTR` (for the Tresca model implementation). Refer to page 295.

8.3. The Drucker–Prager model

We now move on to the last model implementation discussed in this chapter. Here we focus on a Drucker–Prager model with piecewise linear isotropic strain hardening. The most important results of this section can be found as indicated in the table below.

integration algorithm	flowchart	Figure 8.15
	pseudo-code	Boxes 8.8–8.10
	FORTRAN code	subroutine `SUDP` (Section 8.3.2)
iso-error map		Figure 8.16
consistent tangent – FORTRAN code –		subroutine `CTDP` (Section 8.3.5)

Following the format of the previous sections, a brief summary of the constitutive equations adopted in the present implementation is presented below.

The implemented Drucker–Prager constitutive equations

The Drucker–Prager yield surface is defined by means of the yield function

$$\Phi(\boldsymbol{\sigma}, c) = \sqrt{J_2(\boldsymbol{s}(\boldsymbol{\sigma}))} + \eta\, p(\boldsymbol{\sigma}) - \xi\, c, \tag{8.101}$$

where

$$J_2 = \tfrac{1}{2}\boldsymbol{s} : \boldsymbol{s}; \quad \boldsymbol{s} = \boldsymbol{\sigma} - p(\boldsymbol{\sigma})\, \boldsymbol{I}, \tag{8.102}$$

$p = 1/3\,\mathrm{tr}[\boldsymbol{\sigma}]$ is the hydrostatic pressure and c is the cohesion. The constants η and ξ are chosen according to the required approximation to the Mohr–Coulomb criterion. Their formulae for outer edge, inner edge, plane strain, uniaxial tension/compression and biaxial tension/compression matching are given, respectively, by expressions (6.122), (6.123) and (6.124), (6.126) and (6.127).

The generally non-associative flow rule is adopted in the present implementation of the Drucker–Prager model. The corresponding flow potential is

$$\Psi(\boldsymbol{\sigma}, c) = \sqrt{J_2(\boldsymbol{s}(\boldsymbol{\sigma}))} + \bar{\eta}\, p(\boldsymbol{\sigma}), \tag{8.103}$$

where the constant $\bar{\eta}$ depends on the dilatancy angle, ψ, and according to the adopted Drucker–Prager approximation to the Mohr–Coulomb law may be given by (6.163), (6.164) or (6.165) respectively for the outer, inner edge and plane strain match. The flow rule reads

$$\dot{\boldsymbol{\varepsilon}}^p = \dot{\gamma}\, \boldsymbol{N}, \tag{8.104}$$

where, on the smooth portion of the yield surface, the flow vector is given by

$$\boldsymbol{N} = \frac{\partial \Psi}{\partial \boldsymbol{\sigma}} = \frac{1}{2\sqrt{J_2(\boldsymbol{s})}}\, \boldsymbol{s} + \frac{\bar{\eta}}{3} \boldsymbol{I}. \tag{8.105}$$

At the cone apex, where the potential is singular, the flow vector is a subgradient of Ψ; that is, $\boldsymbol{N}$ is a vector contained in the complementary cone (refer to the schematic illustration of Figure 6.20(b), page 176).

As for the Mohr–Coulomb implementation of the previous section, associative isotropic strain hardening is adopted here. The hardening law is defined by (8.61), together with the associative evolution equation for the *accumulated plastic strain* derived for Drucker–Prager plasticity in Section 6.6.3:

$$\dot{\bar{\varepsilon}}^p = \dot{\gamma}\, \xi. \tag{8.106}$$

Analogously to the Mohr–Coulomb implementation, the Drucker–Prager counterpart of relation (8.64):

$$\dot{\bar{\varepsilon}}^p = \frac{\xi}{\bar{\eta}}\, \dot{\varepsilon}^p_{\mathrm{v}}, \tag{8.107}$$

will be needed in the derivation of the return-mapping equation for the apex of the Drucker–Prager cone. This relation is obtained by combining the non-associative version of (6.161) with the evolution law (8.106).

8.3.1. INTEGRATION ALGORITHM FOR THE DRUCKER–PRAGER MODEL

The integration algorithm for the Drucker–Prager model is simpler than its counterparts presented earlier in this chapter for the Tresca and Mohr–Coulomb models. This relative simplicity stems from two main features of the present material model. Firstly, only *one* singularity exists in the Drucker–Prager yield surface – its *apex*. This is in contrast with the two and three possible singularities identified, respectively, in the implementation of the Tresca and Mohr–Coulomb models. Secondly, the Drucker–Prager yield surface, as well as the flow vector field resulting from the generally non-associative flow potential, are fully symmetric about the hydrostatic axis.

Consider the general return-mapping update formula for the stress tensor for materials with linear elastic law:

$$\boldsymbol{\sigma}_{n+1} = \boldsymbol{\sigma}^{\mathrm{trial}}_{n+1} - \Delta\gamma\, \mathsf{D}^e : \boldsymbol{N}_{n+1}, \tag{8.108}$$

where $-\Delta\gamma\, \mathsf{D}^e : \boldsymbol{N}_{n+1}$ is the *return vector*. As a consequence of the symmetry about the pressure axis (see Figure 6.20, page 176), whenever the above formula is applied (regardless of whether $\boldsymbol{\sigma}_{n+1}$ lies on the smooth portion of the cone or on its apex), the return vector is always parallel to the plane that contains $\boldsymbol{\sigma}_{n+1}^{\text{trial}}$ and the hydrostatic axis (see Figure 8.14, page 329). Thus, without loss of generality, the return-mapping algorithm can be completely formulated in such a plane of the stress space. The corresponding algorithms are derived in what follows. Note that, since the definition of the flow vector $\boldsymbol{N}_{n+1}$ in the smooth portion of the cone differs from that at the apex singularity, two possible explicit forms exist for the return-mapping algorithm. These are treated separately below.

Return to the smooth portion of the cone

On the smooth portion of the cone, the flow vector is defined by (8.105). The corresponding increment of plastic strain then reads

$$\Delta\boldsymbol{\varepsilon}^p = \Delta\gamma\,\boldsymbol{N}_{n+1} = \Delta\gamma\left(\frac{1}{2\sqrt{J_2(\boldsymbol{s})}}\,\boldsymbol{s}_{n+1} + \frac{\bar{\eta}}{3}\,\boldsymbol{I}\right). \tag{8.109}$$

The corresponding stress update formula is

$$\begin{aligned}\boldsymbol{\sigma}_{n+1} &= \boldsymbol{\sigma}_{n+1}^{\text{trial}} - \Delta\gamma[2G\,(\boldsymbol{N}_{\text{d}})_{n+1} + K\,(\boldsymbol{N}_{\text{v}})_{n+1}]\\ &= \boldsymbol{\sigma}_{n+1}^{\text{trial}} - \Delta\gamma\left(\frac{G}{\sqrt{J_2(\boldsymbol{s})}}\,\boldsymbol{s}_{n+1} + \frac{K\,\bar{\eta}}{3}\,\boldsymbol{I}\right).\end{aligned} \tag{8.110}$$

Simplification of the above expression can be obtained by noting that, due to the definition of J_2, the following identity holds

$$\frac{\boldsymbol{s}_{n+1}}{J_2(\boldsymbol{s}_{n+1})} = \frac{\boldsymbol{s}_{n+1}^{\text{trial}}}{J_2(\boldsymbol{s}_{n+1}^{\text{trial}})}. \tag{8.111}$$

Its substitution into the stress update formula gives

$$\boldsymbol{\sigma}_{n+1} = \boldsymbol{\sigma}_{n+1}^{\text{trial}} - \Delta\gamma\left(\frac{G}{\sqrt{J_2(\boldsymbol{s}^{\text{trial}})}}\,\boldsymbol{s}_{n+1}^{\text{trial}} + \frac{K\,\bar{\eta}}{3}\,\boldsymbol{I}\right), \tag{8.112}$$

which, equivalently, in terms of deviatoric and hydrostatic components reads

$$\begin{aligned}\boldsymbol{s}_{n+1} &= \left(1 - \frac{G\,\Delta\gamma}{\sqrt{J_2(\boldsymbol{s}_{n+1}^{\text{trial}})}}\right)\boldsymbol{s}_{n+1}^{\text{trial}}\\ p_{n+1} &= p_{n+1}^{\text{trial}} - K\,\bar{\eta}\,\Delta\gamma.\end{aligned} \tag{8.113}$$

It should be noted that the deviatoric updated stress, $\boldsymbol{s}_{n+1}$, is obtained by simply scaling down the trial deviatoric stress, $\boldsymbol{s}_{n+1}^{\text{trial}}$, by a factor that depends on $\Delta\gamma$. Recall that in the von Mises implicit return mapping derived in Chapter 7 the updated deviatoric stress is also obtained by scaling its elastic trial counterpart.

The consistency condition in the present case is given by

$$\Phi_{n+1} \equiv \sqrt{J_2(\boldsymbol{s}_{n+1})} + \eta\,p_{n+1} - \xi\,c(\bar{\varepsilon}_{n+1}^p) = 0, \tag{8.114}$$

where the update accumulated plastic strain is obtained from the discrete version of (8.106):

$$\bar{\varepsilon}^p_{n+1} = \bar{\varepsilon}^p_n + \Delta\bar{\varepsilon}^p \tag{8.115}$$

with

$$\Delta\bar{\varepsilon}^p = \xi\,\Delta\gamma. \tag{8.116}$$

The substitution of the above expression and (8.113) into the consistency condition results in the following (generally nonlinear) equation for $\Delta\gamma$:

$$\begin{aligned}\tilde{\Phi}(\Delta\gamma) \equiv \sqrt{J_2(\boldsymbol{s}^{\text{trial}}_{n+1})} - G\,\Delta\gamma + \eta\,(p^{\text{trial}}_{n+1} - K\,\bar{\eta}\,\Delta\gamma)\\ - \xi\, c(\bar{\varepsilon}^p_n + \xi\,\Delta\gamma) = 0.\end{aligned} \tag{8.117}$$

After solution of the above equation, the stress is updated by (8.113).

Return to the apex

At the apex, the return vector must be contained in the complementary cone schematically illustrated in Figure 6.20(b) of page 176. The derivation of the return-mapping procedure in this case is completely analogous to that of the Mohr–Coulomb apex return described in item 4 of page 299. The consistency condition (8.114) is reduced to

$$c(\bar{\varepsilon}^p_n + \Delta\bar{\varepsilon}^p)\,\frac{\xi}{\bar{\eta}} - p^{\text{trial}}_{n+1} + K\,\Delta\varepsilon^p_{\text{v}} = 0. \tag{8.118}$$

Further, with the introduction of the discretised form of (8.107) in the above equation, we obtain the final return-mapping equation for the Drucker–Prager apex:

$$r(\Delta\varepsilon^p_{\text{v}}) \equiv c(\bar{\varepsilon}^p_n + \alpha\Delta\varepsilon^p_{\text{v}})\,\beta - p^{\text{trial}}_{n+1} + K\,\Delta\varepsilon^p_{\text{v}} = 0, \tag{8.119}$$

where

$$\alpha \equiv \frac{\xi}{\eta}, \quad \beta \equiv \frac{\xi}{\bar{\eta}}. \tag{8.120}$$

This equation is analogous to the return-mapping equation (8.85) of the Mohr–Coulomb implementation. Its geometrical representation is shown in Figure 8.13. After the solution of (8.119) for $\Delta\varepsilon^p_{\text{v}}$, we update

$$\begin{aligned}\bar{\varepsilon}^p_{n+1} &:= \bar{\varepsilon}^p_n + \alpha\Delta\varepsilon^p_{\text{v}},\\ \boldsymbol{\sigma}_{n+1} &:= (p^{\text{trial}}_{n+1} - K\,\Delta\varepsilon^p_{\text{v}})\,\boldsymbol{I}.\end{aligned} \tag{8.121}$$

Note that for non-dilatant flow ($\bar{\eta} = 0$), the return to apex does not make sense in the present context and the comments made immediately after expression (8.86) for the Mohr–Coulomb implementation apply.

Remark 8.4. For *perfectly plastic* materials, c is constant and, for *linearly hardening* models, the hardening function reads $c(\bar{\varepsilon}^p) = c_0 + H\,\bar{\varepsilon}^p$, where H denotes the hardening modulus. In such cases, the return-mapping equations (8.117) and (8.119) respectively for the smooth portion and apex of the Drucker–Prager cone are *linear* and $\Delta\gamma$ is computed in closed form.

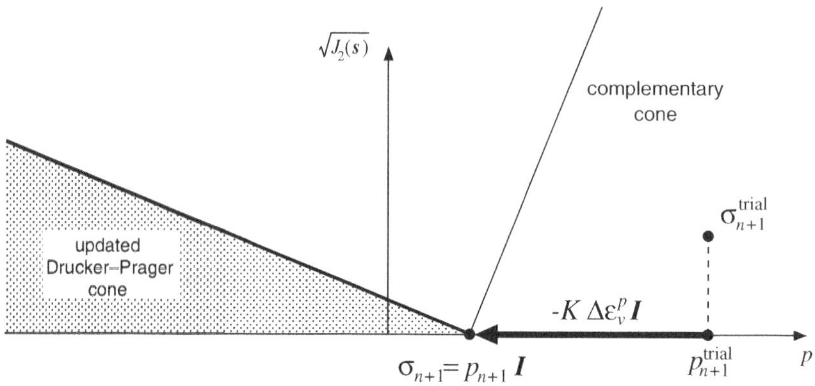

Figure 8.13. Drucker–Prager model. Return mapping to apex.

Selection of the appropriate return mapping

Having derived the two possible forms of return mapping for the Drucker–Prager model, a strategy for selection of the appropriate one is necessary for the complete definition of the integration algorithm. Recall that the purpose of the selection procedure is to ensure that the validated return mapping – in the present case, either the return to the smooth wall or the return to the apex – rigorously satisfies the general implicit return-mapping equations (7.25), derived in Chapter 7.

The selection strategy for the Drucker–Prager model is simple. There are only two possible return mappings. Thus, if the application of one of them generates a contradiction, then the other one must be valid. A possible selection strategy is summarised in the following steps (refer to Figure 8.14 for a graphical representation):

- Firstly, apply the return algorithm to the smooth part of the Drucker–Prager cone. From the corresponding deviatoric stress update formula $(8.113)_1$, it follows that

$$\sqrt{J_2(\boldsymbol{s}_{n+1})} = \sqrt{J_2(\boldsymbol{s}_{n+1}^{\text{trial}})} - G\,\Delta\gamma. \tag{8.122}$$

  If, after determination of $\Delta\gamma$ from the related consistency condition, the following is satisfied

$$a^{\text{cone}} \equiv \sqrt{J_2(\boldsymbol{s}_{n+1}^{\text{trial}})} - G\,\Delta\gamma \geq 0, \tag{8.123}$$

  then the returned stress indeed lies on the Drucker–Prager cone and the return mapping is validated.

- Otherwise, the returned stress lies outside the updated elastic domain and is not admissible. In this case, the return mapping to the apex must be applied. The results obtained by the apex return are then necessarily valid.

The above selection procedure ensures the consistency of the selected return algorithm regardless of any prescribed hardening/softening law.

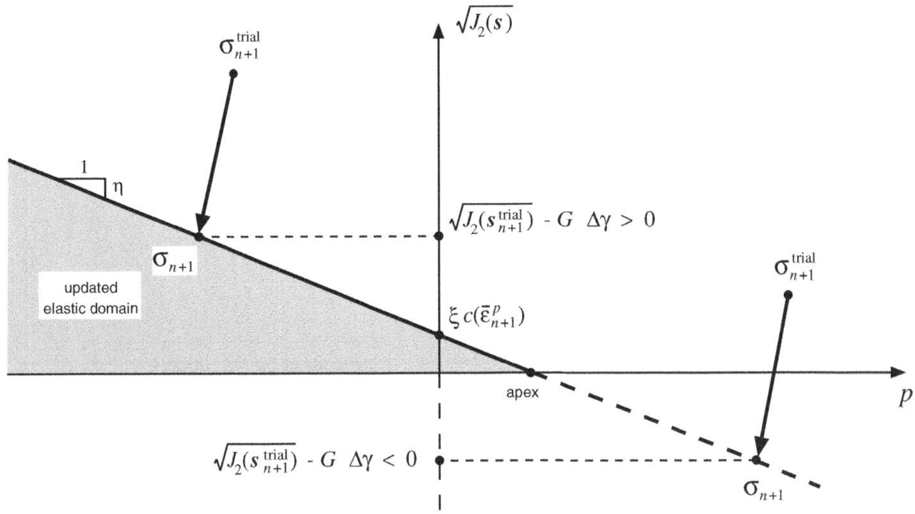

Figure 8.14. Drucker–Prager model. Selection of appropriate return mapping.

The overall integration algorithm

Essentially, the overall integration algorithm for the Drucker–Prager model comprises the elastic predictor stage – which is common to all models addressed in this chapter – followed by the two return mappings described above, with the associated selection strategy. A flowchart of the procedure is shown in Figure 8.15. The corresponding pseudo-code is summarised in Boxes 8.8–8.10. The integration algorithm for the Drucker–Prager model has been implemented in subroutine SUDP of program HYPLAS and is explained in detail below.

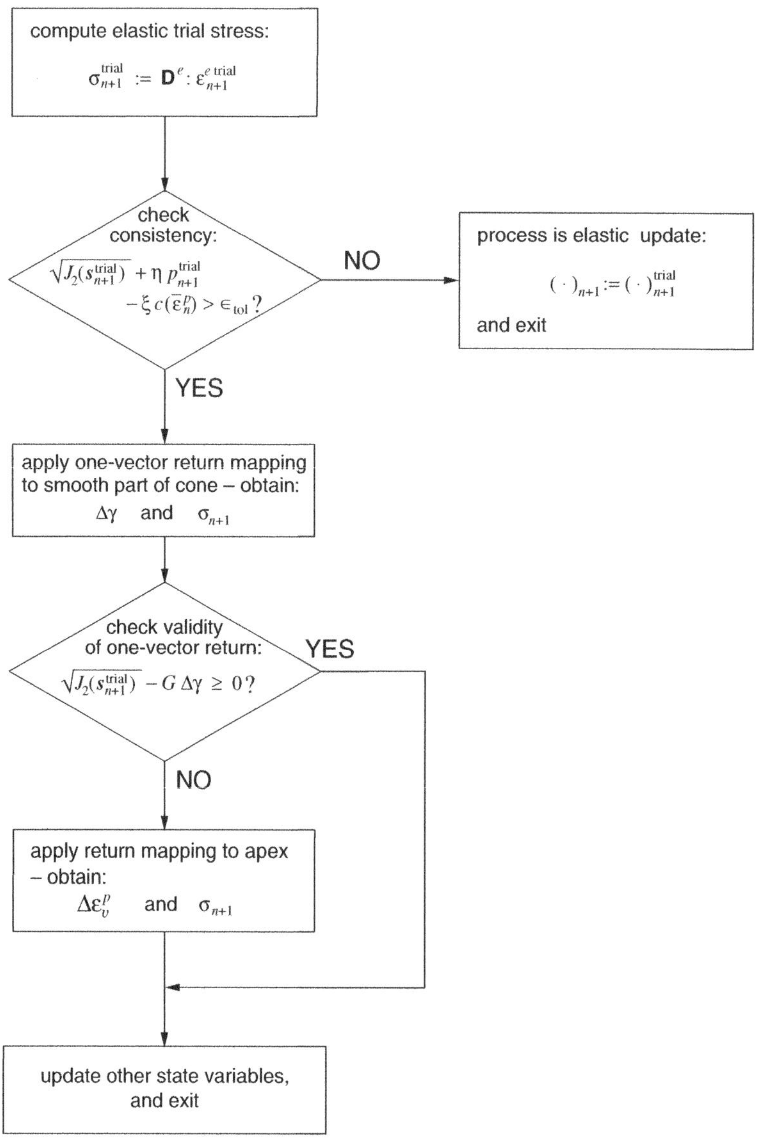

Figure 8.15. Flowchart of the implicit elastic predictor/return-mapping scheme for the Drucker–Prager model. Procedure implemented in subroutine `SUDP` of program `HYPLAS`.

Box 8.8. Implicit elastic predictor/return-mapping algorithm for the Drucker–Prager model.

HYPLAS procedure:	SUDP

(i) Elastic predictor. Given $\Delta\varepsilon$ and the state variables at t_n, evaluate the *elastic trial state*

$$\varepsilon^{e\,\text{trial}}_{n+1} := \varepsilon^e_n + \Delta\varepsilon \qquad \bar{\varepsilon}^{p\,\text{trial}}_{n+1} := \bar{\varepsilon}^p_n$$
$$s^{\text{trial}}_{n+1} := 2G\,\varepsilon^{e\,\text{trial}}_{\text{d}\,n+1} \qquad p^{\text{trial}}_{n+1} := K\,\varepsilon^{e\,\text{trial}}_{\text{v}\,n+1}$$

(ii) Check plastic admissibility

IF $\sqrt{J_2(s^{\text{trial}}_{n+1})} + \eta\,p^{\text{trial}}_{n+1} - \xi\,c(\bar{\varepsilon}^{p\,\text{trial}}_{n+1}) \leq 0$

THEN set $(\cdot)_{n+1} := (\cdot)^{\text{trial}}_{n+1}$ and EXIT

(iii) Return mapping

(a) Return to smooth portion of cone – GOTO Box 8.9

(b) Check validity

IF $\sqrt{J_2(s^{\text{trial}}_{n+1})} - G\,\Delta\gamma \geq 0$

THEN return is valid – GOTO (iv)

(c) Return to apex – GOTO Box 8.10

(iv) Update elastic strain

$$\varepsilon^e_{n+1} := \frac{1}{2G}\,s_{n+1} + \frac{p_{n+1}}{3K}\,I$$

(v) EXIT

Box 8.9. The Drucker–Prager model. Return mapping to smooth portion of cone.

HYPLAS procedure:	SUDP

(i) Set initial guess for $\Delta\gamma$

$$\Delta\gamma := 0, \quad \bar{\varepsilon}^p_{n+1} := \bar{\varepsilon}^p_n$$

and corresponding residual (yield function value)

$$\tilde{\Phi} := \sqrt{J_2(\boldsymbol{s}^{\text{trial}}_{n+1})} + \eta\, p^{\text{trial}}_{n+1} - \xi\, c(\bar{\varepsilon}^p_n)$$

(ii) Perform Newton–Raphson iteration for $\Delta\gamma$

$$H := \left.\frac{\mathrm{d}c}{\mathrm{d}\bar{\varepsilon}^p}\right|_{\bar{\varepsilon}^p_{n+1}} \qquad \text{(hardening slope)}$$

$$d := \frac{\mathrm{d}\tilde{\Phi}}{\mathrm{d}\Delta\gamma} = -G - K\bar{\eta}\eta - \xi^2 H \qquad \text{(residual derivative)}$$

$$\Delta\gamma := \Delta\gamma - \tilde{\Phi}/d \qquad \text{(new guess for } \Delta\gamma\text{)}$$

(iii) Check convergence

$$\bar{\varepsilon}^p_{n+1} := \bar{\varepsilon}^p_n + \xi\,\Delta\gamma$$

$$\tilde{\Phi} := \sqrt{J_2(\boldsymbol{s}^{\text{trial}}_{n+1})} - G\Delta\gamma + \eta\,(p^{\text{trial}} - K\,\bar{\eta}\,\Delta\gamma) - \xi\, c(\bar{\varepsilon}^p_{n+1})$$

IF $|\tilde{\Phi}| \le \epsilon_{\text{tol}}$ THEN update

$$\boldsymbol{s}_{n+1} := \left(1 - \frac{G\,\Delta\gamma}{\sqrt{J_2(\boldsymbol{s}^{\text{trial}}_{n+1})}}\right)\boldsymbol{s}^{\text{trial}}_{n+1}$$

$$p_{n+1} := p^{\text{trial}}_{n+1} - K\,\bar{\eta}\,\Delta\gamma$$

and RETURN to Box 8.8

(iv) GOTO (ii)

Box 8.10. The Drucker–Prager model. Return mapping to apex.

HYPLAS procedure:	SUDP

(i) Set initial guess for $\Delta\varepsilon^p_{\rm v}$

$$\Delta\varepsilon^p_{\rm v} := 0, \quad \bar{\varepsilon}^p_{n+1} := \bar{\varepsilon}^p_n$$

and corresponding residual (refer to equation (8.119))

$$r := c(\bar{\varepsilon}^p_n)\,\beta - p^{\rm trial}_{n+1}$$

(ii) Perform Newton–Raphson iteration for $\Delta\varepsilon^p_{\rm v}$

$$H := \left.\frac{{\rm d}c}{{\rm d}\bar{\varepsilon}^p}\right|_{\bar{\varepsilon}^p_{n+1}} \qquad \text{(hardening slope)}$$

$$d := \alpha\beta H + K \qquad \text{(residual derivative)}$$

$$\Delta\varepsilon^p_{\rm v} := \Delta\varepsilon^p_{\rm v} - r/d \quad (\text{update } \Delta\varepsilon^p_{\rm v})$$

(iii) Compute new residual and check convergence

$$\bar{\varepsilon}^p_{n+1} := \bar{\varepsilon}^p_n + \alpha\Delta\varepsilon^p_{\rm v}$$

$$p_{n+1} := p^{\rm trial}_{n+1} - K\,\Delta\varepsilon^p_{\rm v}$$

$$r := \beta\, c(\bar{\varepsilon}^p_{n+1}) - p_{n+1}$$

IF $|r| \leq \epsilon_{\rm tol}$ THEN update

$$\boldsymbol{\sigma}_{n+1} := p_{n+1}\,\boldsymbol{I}$$

and RETURN to Box 8.4

(iv) GOTO (ii)

8.3.2. SUBROUTINE SUDP

The above integration algorithm has been implemented in subroutine SUDP (**S**tate **U**pdate procedure for the **D**rucker–**P**rager model) of program HYPLAS. As in the Mohr–Coulomb implementation of subroutine SUMC (Section 8.2.2), a general piecewise isotropic hardening law is assumed with the (user-supplied) hardening curve defined by a set of pairs:

$$\{{}^{i}\bar{\varepsilon}^{p}, {}^{i}c\}.$$

The source code of SUDP is listed below.

```
      SUBROUTINE SUDP
     1(   DGAM       ,IPROPS      ,LALGVA      ,NTYPE       ,RPROPS      ,
     2    RSTAVA     ,STRAT       ,STRES       )
      IMPLICIT DOUBLE PRECISION (A-H,O-Z)
      PARAMETER(IPHARD=7  ,MSTRE=4)
      LOGICAL APEX, IFPLAS, LALGVA(3), SUFAIL
      DIMENSION
     1    IPROPS(*)           ,RPROPS(*)           ,RSTAVA(MSTRE+1)     ,
     2    STRAT(MSTRE)        ,STRES(MSTRE)
      DIMENSION
     1    STRIAL(MSTRE)
      DATA
     1    R0   ,RP5  ,R1   ,R2   ,R3   ,TOL   /
     2    0.0D0,0.5D0,1.0D0,2.0D0,3.0D0,1.D-08/
      DATA MAXRT / 50 /
C***********************************************************************
C STRESS UPDATE PROCEDURE FOR DRUCKER PRAGER TYPE ELASTO-PLASTIC
C MATERIAL WITH ASSOCIATIVE/NON-ASSOCIATIVE FLOW RULE AND PIECE_WISE
C LINEAR ISOTROPIC HARDENING:
C IMPLICIT ELASTIC PREDICTOR/RETURN MAPPING ALGORITHM (Boxes 8.8-10)
C***********************************************************************
C Stops program if neither plane strain nor axisymmetric
      IF(NTYPE.NE.2.AND.NTYPE.NE.3)CALL ERRPRT('EI0016')
C Initialize some algorithmic and internal variables
      DGAMA=R0
      IFPLAS=.FALSE.
      SUFAIL=.FALSE.
      EPBARN=RSTAVA(MSTRE+1)
      EPBAR=EPBARN
C Set some material properties
      YOUNG=RPROPS(2)
      POISS=RPROPS(3)
      ETA=RPROPS(4)
      XI=RPROPS(5)
      ETABAR=RPROPS(6)
      NHARD=IPROPS(3)
C and some constants
      GMODU=YOUNG/(R2*(R1+POISS))
      BULK=YOUNG/(R3*(R1-R2*POISS))
      R2G=R2*GMODU
      R1D3=R1/R3
C Compute elastic trial state
C ---------------------------
C Elastic trial volumetric strain and pressure stress
      EETV=STRAT(1)+STRAT(2)+STRAT(4)
      PT=BULK*EETV
C Elastic trial deviatoric stress
      EEVD3=EETV*R1D3
```

```
      STRIAL(1)=R2G*(STRAT(1)-EEVD3)
      STRIAL(2)=R2G*(STRAT(2)-EEVD3)
      STRIAL(4)=R2G*(STRAT(4)-EEVD3)
C shear component
      STRIAL(3)=R2G*(STRAT(3)*RP5)
C Compute elastic trial stress J2 invariant and cohesion
      VARJ2T=STRIAL(3)*STRIAL(3)+RP5*(STRIAL(1)*STRIAL(1)+
     1       STRIAL(2)*STRIAL(2)+STRIAL(4)*STRIAL(4))
      COHE=PLFUN(EPBARN,NHARD,RPROPS(IPHARD))
C Check for plastic consistency
C ----------------------------
      SQRJ2T=SQRT(VARJ2T)
      PHI=SQRJ2T+ETA*PT-XI*COHE
      RES=PHI
      IF(COHE.NE.R0)RES=RES/ABS(COHE)
      IF(RES.GT.TOL)THEN
C Plastic step: Use return mapping
C ================================
        IFPLAS=.TRUE.
        APEX=.FALSE.
C Apply return mapping to smooth portion of cone - Box 8.9
C -------------------------------------------------------
        DO 20 IPTER1=1,MAXRT
C Compute residual derivative
          DENOM=-GMODU-BULK*ETABAR*ETA-
     1          XI*XI*DPLFUN(EPBAR,NHARD,RPROPS(IPHARD))
C Compute Newton-Raphson increment and update variable DGAMA
          DDGAMA=-PHI/DENOM
          DGAMA=DGAMA+DDGAMA
C Compute new residual
          EPBAR=EPBARN+XI*DGAMA
          COHE=PLFUN(EPBAR,NHARD,RPROPS(IPHARD))
          SQRJ2=SQRJ2T-GMODU*DGAMA
          P=PT-BULK*ETABAR*DGAMA
          PHI=SQRJ2+ETA*P-XI*COHE
C Check convergence
          RESNOR=ABS(PHI)
          IF(COHE.NE.R0)RESNOR=RESNOR/ABS(COHE)
          IF(RESNOR.LE.TOL)THEN
C Check validity of return to smooth portion
            IF(SQRJ2.GE.R0)THEN
C results are valid, update stress components and other variables
              IF(SQRJ2T.EQ.R0)THEN
                FACTOR=R0
              ELSE
                FACTOR=R1-GMODU*DGAMA/SQRJ2T
              ENDIF
              GOTO 50
            ELSE
C smooth wall return not valid - go to apex return procedure
              GOTO 30
            ENDIF
          ENDIF
   20   CONTINUE
C failure of stress update procedure
        SUFAIL=.TRUE.
        CALL ERRPRT('WE0002')
        GOTO 999
   30   CONTINUE
C Apply return mapping to APEX - Box 8.10
C ---------------------------------------
C perform checks and set some variables
```

```
        APEX=.TRUE.
        IF(ETA.EQ.R0)CALL ERRPRT('EE0011')
        IF(ETABAR.EQ.R0)CALL ERRPRT('EE0012')
        ALPHA=XI/ETABAR
        BETA=XI/ETA
C Set initial guess for unknown DEPV and start iterations
        DEPV=R0
        EPBAR=EPBARN
        COHE=PLFUN(EPBAR,NHARD,RPROPS(IPHARD))
        RES=BETA*COHE-PT
        DO 40 IPTER2=1,MAXRT
          DENOM=ALPHA*BETA*DPLFUN(EPBAR,NHARD,RPROPS(IPHARD))+BULK
C Compute Newton-Raphson increment and update variable DEPV
          DDEPV=-RES/DENOM
          DEPV=DEPV+DDEPV
C Compute new residual
          EPBAR=EPBARN+ALPHA*DEPV
          COHE=PLFUN(EPBAR,NHARD,RPROPS(IPHARD))
          P=PT-BULK*DEPV
          RES=BETA*COHE-P
C Check convergence
          RESNOR=ABS(RES)
          IF(COHE.NE.R0)RESNOR=RESNOR/ABS(COHE)
          IF(RESNOR.LE.TOL)THEN
C update stress components and other variables
            DGAMA=DEPV/ETABAR
            FACTOR=R0
            GOTO 50
          ENDIF
  40    CONTINUE
C failure of stress update procedure
        SUFAIL=.TRUE.
        CALL ERRPRT('WE0002')
        GOTO 999
C Store converged stress components and other state variables
C -------------------------------------------------------------
  50    CONTINUE
        STRES(1)=FACTOR*STRIAL(1)+P
        STRES(2)=FACTOR*STRIAL(2)+P
        STRES(3)=FACTOR*STRIAL(3)
        STRES(4)=FACTOR*STRIAL(4)+P
C update EPBAR
        RSTAVA(MSTRE+1)=EPBAR
C compute converged elastic (engineering) strain components
        FACTOR=FACTOR/R2G
        EEVD3=P/(BULK*R3)
        RSTAVA(1)=FACTOR*STRIAL(1)+EEVD3
        RSTAVA(2)=FACTOR*STRIAL(2)+EEVD3
        RSTAVA(3)=FACTOR*STRIAL(3)*R2
        RSTAVA(4)=FACTOR*STRIAL(4)+EEVD3
      ELSE
C Elastic step: update stress using linear elastic law
C =====================================================
        STRES(1)=STRIAL(1)+PT
        STRES(2)=STRIAL(2)+PT
        STRES(3)=STRIAL(3)
        STRES(4)=STRIAL(4)+PT
C elastic engineering strain
        RSTAVA(1)=STRAT(1)
        RSTAVA(2)=STRAT(2)
        RSTAVA(3)=STRAT(3)
```

```
         RSTAVA(4)=STRAT(4)
       ENDIF
   999 CONTINUE
C Update some algorithmic variables before exit
C =============================================
       LALGVA(1)=IFPLAS
       LALGVA(2)=SUFAIL
       LALGVA(3)=APEX
       DGAM=DGAMA
       RETURN
       END
```

The arguments of SUDP

The arguments of SUDP are the same as those of SUMC (refer to list on page 315) for the Mohr–Coulomb model implementation. Note, however, that in the present case IPROPS(3) is set in subroutine RDDP. Also, for the Drucker–Prager model, array RPROPS stores the following:

$$\mathtt{RPROPS} = [E, \nu, \eta, \xi, \bar{\eta}, {}^{1}\bar{\varepsilon}^{p}, {}^{1}c, {}^{2}\bar{\varepsilon}^{p}, {}^{2}c, \ldots\ldots, {}^{n_{\rm hard}}\bar{\varepsilon}^{p}, {}^{n_{\rm hard}}c],$$

and array LALGVA is defined as

- ← LALGVA. For the Drucker–Prager model this array contains the plastic yielding flag, IFPLAS, the state update failure flag SUFAIL and the apex return flag APEX. Flags IFPLAS and SUFAIL are set in the same way as in the previous implementations discussed. The apex return flag is set to .TRUE. if the return mapping to the apex is selected. It is set to .FALSE. otherwise. The logical flag APEX is required by subroutine CTDP to decide which elastoplastic tangent operator to compute (consistent with the return to either the smooth part or apex of the Drucker–Prager cone).

8.3.3. ISO-ERROR MAP

An iso-error map for the integration algorithm described above for the Drucker–Prager model is presented in this section. The analysis is restricted to the perfectly plastic case with associative flow rule. The Drucker–Prager parameters adopted in the test are chosen so as to match the outer edges of the Mohr–Coulomb criterion with frictional angle $\phi = 20^{\circ}$. Due to the symmetry of the Drucker–Prager yield surface about the hydrostatic axis, the error maps obtained from all points within the same cross-section of the Drucker–Prager cone are identical. Here the error map is constructed having a generic point of the deviatoric plane as the starting point. The unit tensors $\boldsymbol{N}$ and $\boldsymbol{T}$ defining the increment directions are, respectively, normal and tangent to the yield surface. The tangent tensor here is deviatoric. The resulting error map is plotted in Figure 8.16. The increment normalising factor q is the von Mises effective stress at the starting point. Due to the pressure-sensitivity of the model (as for the Mohr–Coulomb case), a more comprehensive accuracy assessment would require increments in other directions, i.e. with *non-deviatoric* tangents to the yield surface as well.

8.3.4. CONSISTENT TANGENT OPERATOR FOR THE DRUCKER–PRAGER MODEL

Again, the elastic tangent is the standard elasticity matrix so that the derivation below is focused only on the *elastoplastic* tangents associated with the integration algorithm described

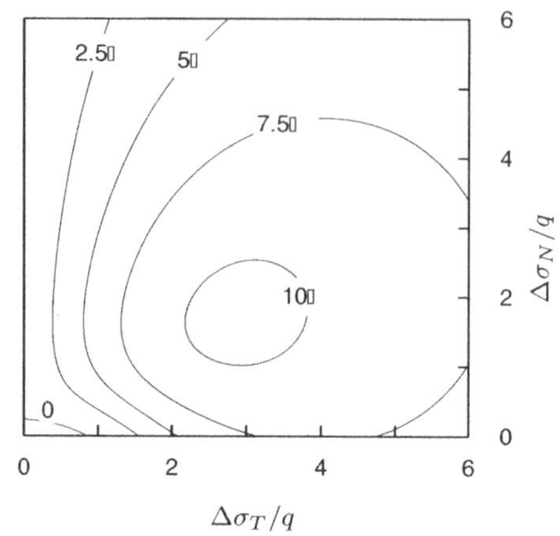

Figure 8.16. Iso-error map. Drucker–Prager model with associative flow.

above for the Drucker–Prager model. For this model, there are two possible elastoplastic tangents: one consistent with the return to the smooth portion of the cone and the other consistent with the return to the apex. Following the procedure adopted in the implementation of the Tresca and Mohr–Coulomb models where, respectively, three and four elastoplastic explicit forms of tangent exist, the actual elastoplastic tangents used to assemble the stiffness matrix are the ones consistent with the last application of the return-mapping procedure at each Gauss point.

Tangent consistent with the smooth portion return

Firstly, let us consider the *deviatoric* stress update formula of the return mapping to the smooth part of the cone:

$$\boldsymbol{s}_{n+1} = \left(1 - \frac{G\,\Delta\gamma}{\sqrt{J_2(\boldsymbol{s}^{\text{trial}})}}\right)\boldsymbol{s}^{\text{trial}}_{n+1} = 2G\left(1 - \frac{\Delta\gamma}{\sqrt{2}\,\|\boldsymbol{\varepsilon}^{e\ \text{trial}}_{\text{d}\ n+1}\|}\right)\boldsymbol{\varepsilon}^{e\ \text{trial}}_{\text{d}\ n+1}. \tag{8.124}$$

Straightforward differentiation of the above expression yields

$$\begin{aligned} \mathrm{d}\boldsymbol{s}_{n+1} = 2G\bigg[&\left(1 - \frac{\Delta\gamma}{\sqrt{2}\,\|\boldsymbol{\varepsilon}^{e\ \text{trial}}_{\text{d}\ n+1}\|}\right)\mathrm{d}\boldsymbol{\varepsilon}^{e\ \text{trial}}_{\text{d}\ n+1} \\ &+ \frac{\Delta\gamma}{\sqrt{2}\,\|\boldsymbol{\varepsilon}^{e\ \text{trial}}_{\text{d}\ n+1}\|}\,\boldsymbol{D}\otimes\boldsymbol{D} : \mathrm{d}\boldsymbol{\varepsilon}^{e\ \text{trial}}_{\text{d}\ n+1} - \frac{1}{\sqrt{2}}\,\mathrm{d}\Delta\gamma\,\boldsymbol{D}\bigg], \end{aligned} \tag{8.125}$$

where the second-order tensor $\boldsymbol{D}$ is the unit tensor parallel to $\boldsymbol{\varepsilon}^{e\ \text{trial}}_{\text{d}\ n+1}$:

$$\boldsymbol{D} \equiv \frac{\boldsymbol{\varepsilon}^{e\ \text{trial}}_{\text{d}\ n+1}}{\|\boldsymbol{\varepsilon}^{e\ \text{trial}}_{\text{d}\ n+1}\|}. \tag{8.126}$$

Similarly, the differentiation of the hydrostatic pressure update formula

$$p_{n+1} = p^{\text{trial}}_{n+1} - K\bar{\eta}\,\Delta\gamma = K(\varepsilon^{e\ \text{trial}}_{\text{v}\ n+1} - \bar{\eta}\,\Delta\gamma), \tag{8.127}$$

gives

$$\mathrm{d}p_{n+1} = K\,(\mathrm{d}\varepsilon^{e\,\mathrm{trial}}_{\mathrm{v}\,n+1} - \bar{\eta}\,\mathrm{d}\Delta\gamma). \tag{8.128}$$

The expression relating $\mathrm{d}\Delta\gamma$ and the differentials of trial strain is obtained by linearising the consistency condition (8.117). The linearisation of (8.117), in conjunction with the use of the elastic relation, gives the following equation:

$$\begin{aligned}\mathrm{d}\tilde{\Phi} &= \sqrt{2}\,G\,\boldsymbol{D} : \mathrm{d}\boldsymbol{\varepsilon}^{e\,\mathrm{trial}}_{\mathrm{d}\,n+1} + K\eta\,\mathrm{d}\varepsilon^{e\,\mathrm{trial}}_{\mathrm{v}\,n+1} \\ &\quad - (G + K\eta\bar{\eta} + \xi^2 H)\,\mathrm{d}\Delta\gamma = 0,\end{aligned} \tag{8.129}$$

which results in the identity

$$\mathrm{d}\Delta\gamma = \frac{1}{G + K\eta\bar{\eta} + \xi^2 H}(\sqrt{2}\,G\,\boldsymbol{D} : \mathrm{d}\boldsymbol{\varepsilon}^{e\,\mathrm{trial}}_{\mathrm{d}\,n+1} + K\eta\,\mathrm{d}\varepsilon^{e\,\mathrm{trial}}_{\mathrm{v}\,n+1}). \tag{8.130}$$

Finally, with the substitution of (8.130) into (8.125) and (8.128), and use of the identity

$$\mathsf{D}^{ep} \equiv \frac{\mathrm{d}\boldsymbol{\sigma}_{n+1}}{\mathrm{d}\boldsymbol{\varepsilon}^{e\,\mathrm{trial}}_{n+1}} = \frac{\mathrm{d}\boldsymbol{s}_{n+1}}{\mathrm{d}\boldsymbol{\varepsilon}^{e\,\mathrm{trial}}_{n+1}} + \boldsymbol{I} \otimes \frac{\mathrm{d}p_{n+1}}{\mathrm{d}\boldsymbol{\varepsilon}^{e\,\mathrm{trial}}_{n+1}}, \tag{8.131}$$

the explicit expression for the elastoplastic tangent consistent with the one-vector return is obtained after some straightforward manipulations as

$$\begin{aligned}\mathsf{D}^{ep} &= 2G\left(1 - \frac{\Delta\gamma}{\sqrt{2}\|\boldsymbol{\varepsilon}^{e\,\mathrm{trial}}_{\mathrm{d}\,n+1}\|}\right)\mathsf{I}_{\mathrm{d}} + 2G\left(\frac{\Delta\gamma}{\sqrt{2}\|\boldsymbol{\varepsilon}^{e\,\mathrm{trial}}_{\mathrm{d}\,n+1}\|} - GA\right)\boldsymbol{D} \otimes \boldsymbol{D} \\ &\quad - \sqrt{2}\,GAK\,(\eta\,\boldsymbol{D} \otimes \boldsymbol{I} + \bar{\eta}\,\boldsymbol{I} \otimes \boldsymbol{D}) + K\,(1 - K\eta\bar{\eta}A)\,\boldsymbol{I} \otimes \boldsymbol{I},\end{aligned} \tag{8.132}$$

where I_d is the deviatoric projection tensor and A is defined as

$$A = \frac{1}{G + K\eta\bar{\eta} + \xi^2 H}. \tag{8.133}$$

Tangent consistent with the return mapping to the apex

Since the stress deviator, $\boldsymbol{s}_{n+1}$, vanishes at the apex, the differential of the updated stress consistent with the application of the apex return is simply

$$\mathrm{d}\boldsymbol{\sigma}_{n+1} = \mathrm{d}p_{n+1}\,\boldsymbol{I}. \tag{8.134}$$

Accordingly, the associated elastoplastic tangent modulus is given by

$$\mathsf{D}^{ep} \equiv \frac{\mathrm{d}\boldsymbol{\sigma}_{n+1}}{\mathrm{d}\boldsymbol{\varepsilon}^{e\,\mathrm{trial}}_{n+1}} = \boldsymbol{I} \otimes \frac{\mathrm{d}p_{n+1}}{\mathrm{d}\boldsymbol{\varepsilon}^{e\,\mathrm{trial}}_{n+1}}. \tag{8.135}$$

Now, recall the hydrostatic pressure update formula:

$$p_{n+1} = p^{\mathrm{trial}}_{n+1} - K\,\Delta\varepsilon^{p}_{\mathrm{v}}. \tag{8.136}$$

Its differentiation gives

$$\mathrm{d}p_{n+1} = K\,\boldsymbol{I} : \mathrm{d}\boldsymbol{\varepsilon}^{e\,\mathrm{trial}}_{n+1} - K\,\mathrm{d}\Delta\varepsilon^{p}_{\mathrm{v}}. \tag{8.137}$$

The explicit expression for the $\mathrm{d}\Delta\varepsilon^p_{\mathrm{v}}$ in terms of the elastic trial strain differential is obtained by differentiating the residual equation (8.119). This gives

$$\mathrm{d}r = (K + \alpha\beta H)\,\mathrm{d}\Delta\varepsilon^p_{\mathrm{v}} - K\,\boldsymbol{I} : \mathrm{d}\varepsilon^{e\ \mathrm{trial}}_{n+1} = 0 \tag{8.138}$$

and yields the expression

$$\mathrm{d}\Delta\varepsilon^p_{\mathrm{v}} = \left(\frac{K}{K + \alpha\beta H}\right)\boldsymbol{I} : \mathrm{d}\varepsilon^{e\ \mathrm{trial}}_{n+1}. \tag{8.139}$$

Finally, with the substitution of the above formula into (8.137) and use of (8.135) we obtain the expression for the tangent operator consistent with the apex return mapping:

$$\mathsf{D}^{ep} = K\left(1 - \frac{K}{K + \alpha\beta H}\right)\boldsymbol{I} \otimes \boldsymbol{I}. \tag{8.140}$$

This formula is in complete analogy with that of the apex return for the Mohr–Coulomb model, given by (8.100) in terms of principal stress derivatives.

Remark 8.5. The comments made in Remark 8.3 are equally valid for the Drucker–Prager model. Note that, in the absence of hardening ($H = 0$) the above tangent operator vanishes.

8.3.5. SUBROUTINE CTDP

The computation of the tangent modulus consistent with the implicit elastic predictor/return-mapping scheme for the Drucker–Prager model is performed in subroutine CTDP (**C**onsistent **T**angent modulus for the **D**rucker–**P**rager model). The implicit algorithm with which the present tangent computation is consistent is coded in subroutine SUDP. This routine returns either the standard elastic tangent or the consistent elastoplastic tangent operator depending on the entry value of the logical argument EPFLAG. The elastoplastic tangents computed in CTDP are those given by expressions (8.132) and (8.140). If the elastoplastic tangent is requested, the decision on whether to compute the tangent consistent with the return mapping to the smooth part of the cone or the return mapping to the apex is made based on the entry value of the logical algorithmic variable APEX stored in array LALGVA. The actual tangent computed is consistent with the last application of the return mapping (in the Gauss integration point of interest) carried out in SUDP. The source code of CTDP is listed below.

```
      SUBROUTINE CTDP
     1(   DGAM       ,DMATX      ,EPFLAG     ,IPROPS     ,LALGVA     ,
     2    NTYPE      ,RPROPS     ,RSTAVA     ,STRAT      )
      IMPLICIT DOUBLE PRECISION (A-H,O-Z)
      PARAMETER(IPHARD=7   ,MSTRE=4)
      LOGICAL APEX, EPFLAG, LALGVA(3)
      DIMENSION
     1    DMATX(MSTRE,MSTRE),IPROPS(*)             ,RPROPS(*)            ,
     2    RSTAVA(MSTRE+1)    ,STRAT(MSTRE)
      DIMENSION
     1    EETD(MSTRE)         ,FOID(MSTRE,MSTRE)   ,SOID(MSTRE)          ,
     2    UNIDEV(MSTRE)
      DATA
     1    FOID(1,1),FOID(1,2),FOID(1,3),FOID(1,4)/
```

```
     2    1.0D0     ,0.0D0     ,0.0D0     ,0.0D0     /
     3    FOID(2,1),FOID(2,2),FOID(2,3),FOID(2,4)/
     4    0.0D0     ,1.0D0     ,0.0D0     ,0.0D0     /
     5    FOID(3,1),FOID(3,2),FOID(3,3),FOID(3,4)/
     6    0.0D0     ,0.0D0     ,0.5D0     ,0.0D0     /
     7    FOID(4,1),FOID(4,2),FOID(4,3),FOID(4,4)/
     8    0.0D0     ,0.0D0     ,0.0D0     ,1.0D0     /
      DATA
     1    SOID(1)  ,SOID(2)  ,SOID(3)  ,SOID(4)  /
     2    1.0D0     ,1.0D0     ,0.0D0     ,1.0D0     /
      DATA
     1    R0   ,R1   ,RP5  ,R2   ,R3   /
     2    0.0D0,1.0D0,0.5D0,2.0D0,3.0D0/
C***********************************************************************
C COMPUTATION OF CONSISTENT TANGENT MODULUS FOR DRUCKER-PRAGER TYPE
C ELASTO-PLASTIC MATERIAL WITH ASSOCIATIVE/NON-ASSOCIATIVE FLOW RULE AND
C PIECE-WISE LINEAR ISOTROPIC HARDENING
C***********************************************************************
      IF(NTYPE.EQ.2)THEN
        NSTRE=3
      ELSEIF(NTYPE.EQ.3)THEN
        NSTRE=4
      ELSE
        CALL ERRPRT('EI0017')
      ENDIF
C Retrieve accumulated plastic strain, DGAMA and APEX algorithm flag
      EPBAR=RSTAVA(MSTRE+1)
      DGAMA=DGAM
      APEX=LALGVA(3)
C Set some material properties
      YOUNG=RPROPS(2)
      POISS=RPROPS(3)
      ETA=RPROPS(4)
      XI=RPROPS(5)
      ETABAR=RPROPS(6)
      NHARD=IPROPS(3)
C and some constants
      GMODU=YOUNG/(R2*(R1+POISS))
      BULK=YOUNG/(R3*(R1-R2*POISS))
      R2G=R2*GMODU
      R1D3=R1/R3
      ROOT2=SQRT(R2)
C
      IF(EPFLAG)THEN
C Compute elastoplastic consistent tangent
C ========================================
C Hardening slope
        HSLOPE=DPLFUN(EPBAR,NHARD,RPROPS(IPHARD))
        IF(APEX)THEN
C Elastoplastic tangent consistent with apex return
C -------------------------------------------------
          ALPHA=XI/ETABAR
          BETA=XI/ETA
          AFACT=BULK*(R1-BULK/(BULK+ALPHA*BETA*HSLOPE))
          DO 20 I=1,NSTRE
            DO 10 J=1,NSTRE
              DMATX(I,J)=AFACT*SOID(I)*SOID(J)
   10       CONTINUE
   20     CONTINUE
        ELSE
C Elastoplastic tangent consistent with smooth cone wall return
```

```
C ----------------------------------------------------------------
C Elastic trial deviatoric (physical) strain
            EEVD3=(STRAT(1)+STRAT(2)+STRAT(4))*R1D3
            EETD(1)=STRAT(1)-EEVD3
            EETD(2)=STRAT(2)-EEVD3
            EETD(3)=STRAT(3)*RP5
            EETD(4)=STRAT(4)-EEVD3
            ETDNOR=SQRT(EETD(1)*EETD(1)+EETD(2)*EETD(2)+
     1               R2*EETD(3)*EETD(3)+EETD(4)*EETD(4))
C Unit deviatoric flow vector
            IF(ETDNOR.NE.R0)THEN
              EDNINV=R1/ETDNOR
            ELSE
              EDNINV=R0
            ENDIF
            DO 30 I=1,NSTRE
              UNIDEV(I)=EETD(I)*EDNINV
   30       CONTINUE
C Assemble tangent
            AUX=R1/(GMODU+BULK*ETA*ETABAR+XI*XI*HSLOPE)
            AFACT=R2G*(R1-DGAMA/(ROOT2*ETDNOR))
            AFACD3=AFACT*R1D3
            BFACT=R2G*(DGAMA/(ROOT2*ETDNOR)-GMODU*AUX)
            CFACT=-ROOT2*GMODU*BULK*AUX
            DFACT=BULK*(R1-BULK*ETA*ETABAR*AUX)
            DO 50 I=1,NSTRE
              DO 40 J=1,NSTRE
                DMATX(I,J)=AFACT*FOID(I,J)+BFACT*UNIDEV(I)*UNIDEV(J)+
     1                     CFACT*(ETA*UNIDEV(I)*SOID(J)+
     2                            ETABAR*SOID(I)*UNIDEV(J))+
     3                     (DFACT-AFACD3)*SOID(I)*SOID(J)
   40         CONTINUE
   50       CONTINUE
          ENDIF
        ELSE
C Compute elasticity matrix
C =========================
          FACTOR=BULK-R2G*R1D3
          DO 70 I=1,NSTRE
            DO 60 J=I,NSTRE
              DMATX(I,J)=R2G*FOID(I,J)+FACTOR*SOID(I)*SOID(J)
   60       CONTINUE
   70     CONTINUE
          DO 90 J=1,NSTRE-1
            DO 80 I=J+1,NSTRE
              DMATX(I,J)=DMATX(J,I)
   80       CONTINUE
   90     CONTINUE
        ENDIF
        RETURN
        END
```

The arguments of CTDP

The situation here is completely analogous to that of subroutine CTMC (refer to page 324), with state update related arguments coming in the present case from subroutine SUDP.

Some local variables and arrays of `CTDP`

- `ETDNOR` [$\|\varepsilon^{e\ \mathrm{trial}}_{n+1}\|$]. Norm of the deviatoric component of the elastic trial strain tensor.
- `UNIDEV` [$\boldsymbol{D}$]. Array of components of the unit tensor parallel to the deviatoric elastic trial strain, $\boldsymbol{D} \equiv \varepsilon^{e\ \mathrm{trial}}_{\mathrm{d}\ n+1}/\|\varepsilon^{e\ \mathrm{trial}}_{\mathrm{d}\ n+1}\|$.

8.4. Examples

This section presents a set of benchmark numerical examples involving all material model implementations described in the previous sections of this chapter. The examples presented comprise metal plasticity (with the Tresca model) as well as some typical soil mechanics applications with pressure-sensitive models (Mohr–Coulomb and Drucker–Prager). It is noted that all results presented here have been obtained with the standard version of program `HYPLAS` that accompanies this book. Whenever available, theoretical solutions are presented and compared with the results obtained in order to illustrate the accuracy of the numerical procedures described in the preceding sections. Without exception, the full Newton–Raphson algorithm has been adopted in the iterative solution of the global finite element equilibrium equations.

8.4.1. BENDING OF A V-NOTCHED TRESCA BAR

In this example, the collapse of a wide rectangular metal bar containing a deep 90° V-shaped notch and subjected to pure bending is analysed. The bar is wide enough so that the analysis can be carried out under the assumption of plane strain conditions. The material is modelled as elastic-perfectly plastic with the Tresca yield criterion. The geometry and material parameters are shown in Figure 8.17. An analytical upper bound to the limit load for this problem has been obtained by Green (1953) based on the theory of slip-line fields. The reader is referred to Chakrabarty (1987) for details of the solution. For a notch angle of 90° the associated upper bound moment per unit width of the bar is

$$M_{\mathrm{u}} \approx 0.623\, c\, a^2 \tag{8.141}$$

where a is the thickness of the bar in the neck and the cohesion or shear strength, c, for the Tresca model is given as $c = \sigma_y/2$. For the present material parameters and geometry, it gives

$$M_{\mathrm{u}} \approx 1.869\ \mathrm{Nm} \tag{8.142}$$

per millimetre width of the bar. The finite element model adopted in the present analysis is shown in Figure 8.18. For symmetry reasons, only half of the bar is discretised with symmetry kinematical constraints imposed on the nodes across the neck. A mesh of 312 eight-noded elements (with 2×2 Gauss quadrature) is adopted with a total of 1001 nodes. The moment M is applied by means of two opposite nodal forces of equal intensity F prescribed on the nodes indicated in Figure 8.18. Due to the distance between the neck and the points where the forces are applied, a condition close to pure bending is obtained near the neck. A total number of 13 load increments is applied in the finite element analysis. The vertical deflection obtained for the nodes where the forces are applied is plotted in Figure 8.19 against the normalised

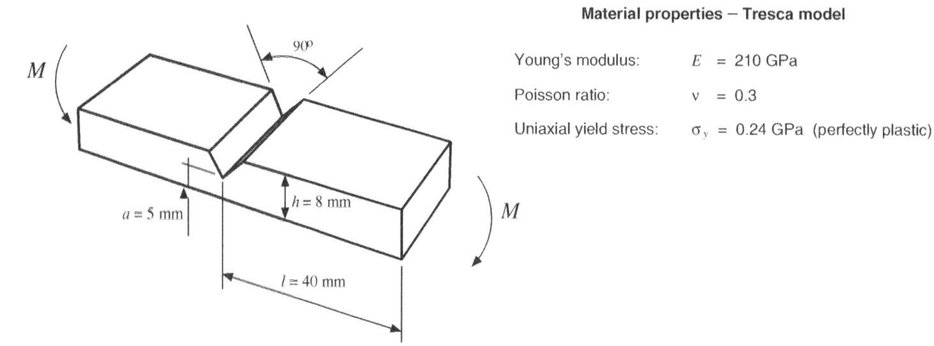

Figure 8.17. Bending of a V-notched bar. Problem definition. (Reproduced with permission from A new computational model for Tresca plasticity at finite strains with an optimal parametrization in the principal space, D Perić and EA de Souza Neto, *Computer Methods in Applied Mechanics and Engineering*, Vol 171 © 1999 Elsevier Science S.A.)

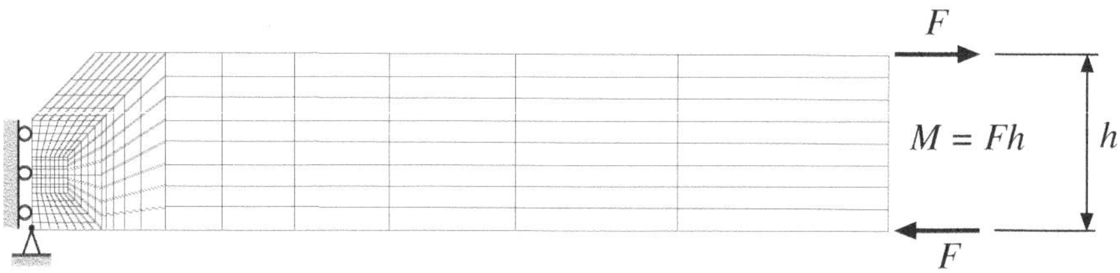

Figure 8.18. Bending of a V-notched bar. Finite element model. (Reproduced with permission from A new computational model for Tresca plasticity at finite strains with an optimal parametrization in the principal space, D Perić and EA de Souza Neto, *Computer Methods in Applied Mechanics and Engineering*, Vol 171 © 1999 Elsevier Science S.A.)

bending moment per unit width, M/ca^2. The limit load determined by the present analysis is

$$\frac{M^{\rm fe}_{\rm lim}}{c\,a^2} \approx 0.636. \tag{8.143}$$

This result is in very close agreement (about 2% higher) with Green's upper bound. The slip-line field proposed by Green (1953) for the present dimensions is schematically illustrated in Figure 8.20(a). Alongside, in Figure 8.20(b), the contour plot of the incremental accumulated plastic strain, $\Delta\bar{\varepsilon}^p$, corresponding to load increment 12 of the finite element computation is shown for comparison. In this increment the bar is effectively collapsing, i.e. a small increment of applied load results in extremely large incremental deflections. The contour plot of $\Delta\bar{\varepsilon}^p$ shows the area where the plastic process is concentrated and clearly illustrates the collapse mechanism. Note that the collapse mechanism predicted by the finite element simulation is in agreement with Green's slip-line field.

8.4.2. END-LOADED TAPERED CANTILEVER

The collapse of a wide end-loaded tapered cantilever is analysed in this example. The geometry of the cantilever as well as the mesh adopted in the finite element analysis are

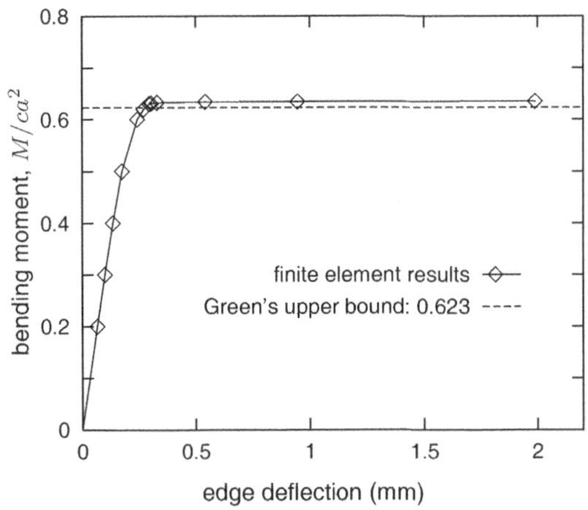

Figure 8.19. Bending of a V-notched bar. Moment-deflection diagram. (Reproduced with permission from A new computational model for Tresca plasticity at finite strains with an optimal parametrization in the principal space, D Perić and EA de Souza Neto, *Computer Methods in Applied Mechanics and Engineering*, Vol 171 © 1999 Elsevier Science S.A.)

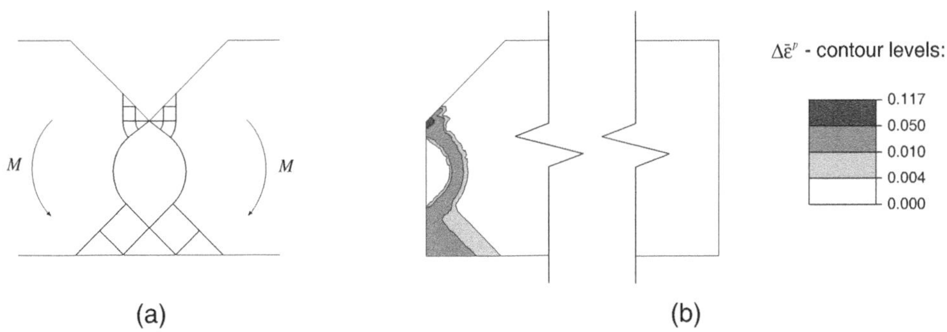

Figure 8.20. Bending of a V-notched bar: (a) slip-line field of the upper bound solution by Green (1953); (b) finite element result under collapse (increment 12): incremental accumulated plastic strain, $\Delta\bar{\varepsilon}^p$. (Reproduced with permission from A new computational model for Tresca plasticity at finite strains with an optimal parametrization in the principal space, D Perić and EA de Souza Neto, *Computer Methods in Applied Mechanics and Engineering*, Vol 171 © 1999 Elsevier Science S.A.)

illustrated in Figure 8.21. The width of the cantilever is assumed sufficiently large so that a plane strain analysis is carried out. The cantilever is firmly supported (clamped) on one edge and the applied load is a uniformly distributed shearing traction, of intensity S, acting on the opposite edge. The cantilever is assumed to be made of a Tresca elastic-perfectly plastic material with the same constants as in the previous example. Upper bound limit loads for tapered cantilevers have been obtained analytically by Green (1954) by means of slip-line field theory.¶ For the present geometry, the normalised upper bound for the shear traction

¶Green's limit loads are *exact* for taper angles $\theta \leq 75°$. His solution for $\theta > 75°$, such as in the present example, are only upper bounds. Nevertheless, due to experimental confirmation of the proposed slip-line fields, Green argues that it is unlikely that his upper-bound solutions will be above the exact theoretical limit load.

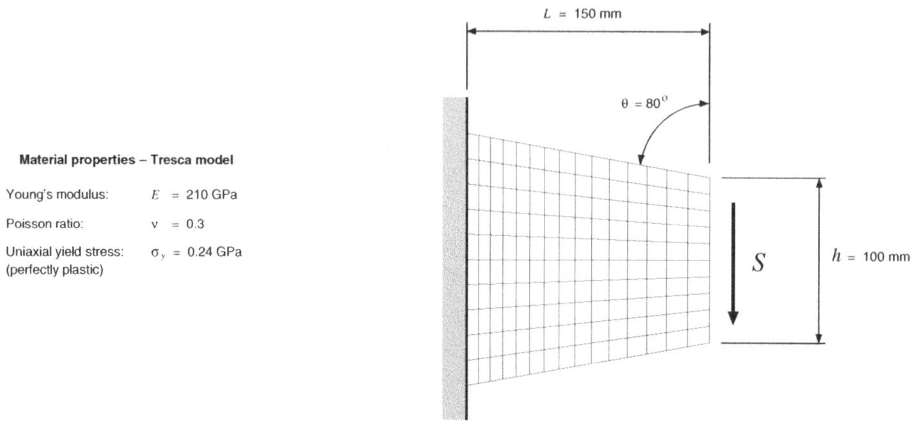

Figure 8.21. End-loaded tapered cantilever. Problem definition and finite element mesh.

found by Green is

$$\frac{S_{\rm u}}{c} \approx 0.775 \tag{8.144}$$

where c is the shear strength given, for the Tresca model, by $c = \sigma_y/2$. In the finite element analysis, the load is applied incrementally until collapse occurs. A diagram showing the vertical deflection, u, of the mid-node of the free edge versus the applied force, S, obtained in the finite element analysis is plotted in Figure 8.22. The corresponding collapse load (i.e. the load above which convergence of equilibrium iterations cannot be attained for sufficiently small increments) is

$$S^{\rm fe}_{\rm lim} \approx 0.786, \tag{8.145}$$

which is about 1.4% above Green's upper bound. The present limit load has been reached in eight increments. The slip-line field proposed by Green (1954) is schematically illustrated in Figure 8.23(a). The associated collapse mechanism is of the plastic hinge type. At collapse, the portion to the right of the slip-line field rotates by sliding over the circular arc BC. Figure 8.23(b) shows the incremental nodal displacements obtained in the last increment (increment 8), when the cantilever is effectively collapsing. It can be seen that the finite element solution has captured Green's collapse mechanism quite accurately. In this plot, the sizes of the vectors have been largely exaggerated and do not correspond to the norms of the actual nodal displacements. The increment of accumulated plastic strain, $\Delta\bar{\varepsilon}^p$, obtained in the same load increment is illustrated in Figure 8.23(c). The contour plot shows that, at collapse, the plastic process predicted by the finite element analysis is confined to an area which is in very close agreement with Green's slip-line field.

8.4.3. STRIP-FOOTING COLLAPSE

The plane strain analysis of a strip footing has been carried out in detail in Example 7.5.4 (page 252) of Chapter 7, where the suitability of the finite element framework of program HYPLAS for the determination of collapse loads has been made evident. However, the analysis performed in that example was restricted to the use of the von Mises material model. In the

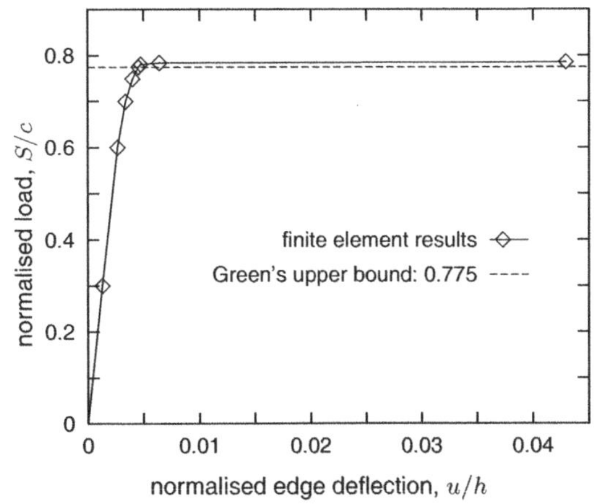

Figure 8.22. End-loaded tapered cantilever. Force-deflection diagram.

present example, the same strip-footing problem is analysed but the soil is now modelled as a Tresca, Mohr–Coulomb and Drucker–Prager material. The objective is to illustrate the performance of the algorithms described earlier in the present chapter for these models. The adopted common material parameters are

Young's modulus	$E = 10^7$ kPa
Poisson's ratio	$\nu = 0.48$
Cohesion	$c = 490$ kPa

The dimensions, boundary conditions and the finite element mesh adopted are identical to those of Example 7.5.4 (see Figure 7.24). The soil is assumed to be weightless. Note that due to symmetry, only half of the problem is discretised. The mesh contains 135 eight-noded quadrilaterals (four Gauss point quadrature) and a total of 446 nodes. The footing is assumed to be rigid and perfectly smooth (no friction). Thus, the loading consists of the prescribed vertical displacement, u, of the nodes under the footing as shown in Figure 7.24. The horizontal displacement of these nodes is unconstrained.

Tresca

Firstly, the analysis is carried out using the Tresca model. The corresponding uniaxial yield stress (required by the input data procedure for the Tresca material in HYPLAS) is $\sigma_y = 2c = 980$ kPa. A final settlement $u/B = 0.002$ is applied in 14 increments. The resulting load-settlement curve is plotted in Figure 8.24. This curve and the corresponding limit load

$$\frac{P_{\lim}^{\text{fe}}}{c} \approx 5.19, \tag{8.146}$$

are practically identical to those obtained with the von Mises model in Example 7.5.4 (plotted in Figure 7.26, page 253), where the same material parameters have been used. The

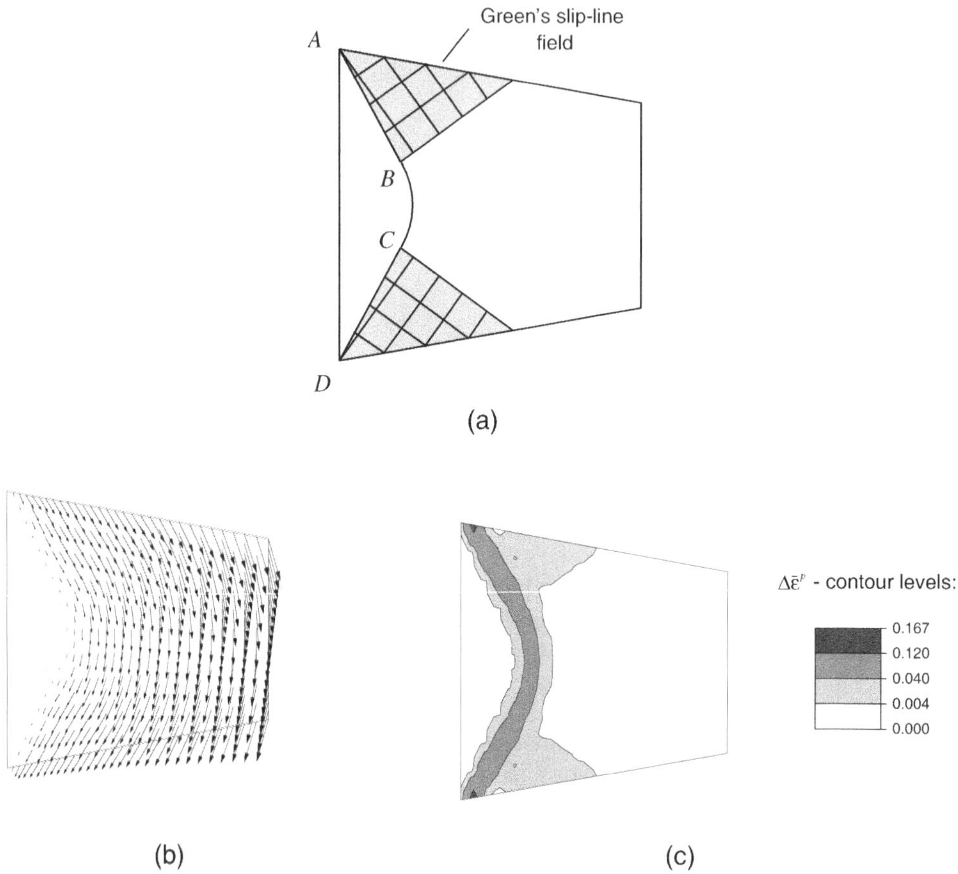

Figure 8.23. End-loaded tapered cantilever. *Analytical solution*: (a) slip-line field proposed by Green (1954); *finite element results under collapse (increment 8)*: (b) incremental nodal displacements; (c) incremental accumulated plastic strain, $\Delta\bar{\varepsilon}^p$.

numerically determined limit pressure is less than 1% above the slip-line theory solution

$$\frac{P_{\text{lim}}}{c} \approx 5.14. \tag{8.147}$$

The failure mechanism captured in the finite element analysis with the Tresca model is in agreement with that proposed by Hill (1950) for perfectly smooth footings (unconstrained horizontal displacements under the footing). This mechanism, as obtained in the simulation with the von Mises model carried out in the previous chapter, is illustrated in Figure 7.28.

Mohr–Coulomb

The analysis of the strip-footing problem with the Mohr–Coulomb material model is carried out assuming a frictional angle

$$\phi = 20^\circ. \tag{8.148}$$

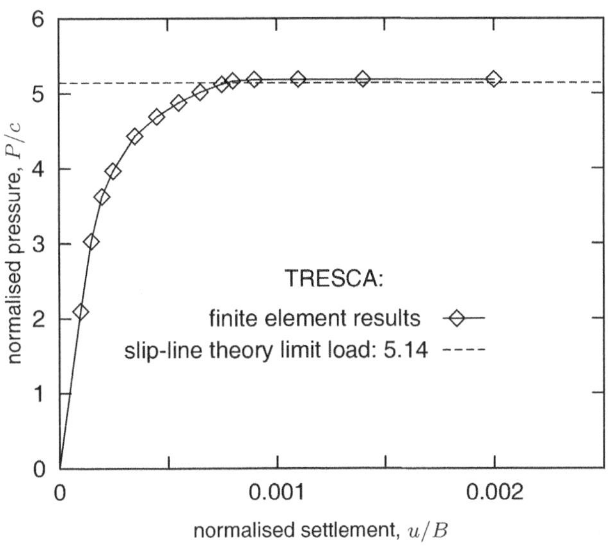

Figure 8.24. Strip footing with Tresca model. Load–displacement curve.

Two flow rules are considered:

- associative ($\phi = \psi = 20^{\rm o}$);
- non-associative with dilatancy angle $\psi = 10^{\rm o}$.

The prescribed final settlement $u/B = 0.0025$ is applied in six and seven increments, respectively, for the associative and non-associative laws. The corresponding load–settlement curves obtained in the analyses are plotted in Figure 8.25. The limit loads obtained with the associative and non-associative law are virtually identical

$$\frac{P^{\rm fe}_{\rm lim}}{c} \approx 15.0. \tag{8.149}$$

This value is in excellent agreement (about 1.4% higher) with Prandtl's solution

$$\frac{P_{\rm lim}}{c} \approx 14.8. \tag{8.150}$$

Drucker–Prager

Three of the Drucker–Prager approximations to the Mohr–Coulomb criterion discussed in Section 6.4.4 (and implemented in program HYPLAS) are used here in the analysis of the footing:

- the *outer* (or extension) cone;
- the *inner* (or compression) cone; and

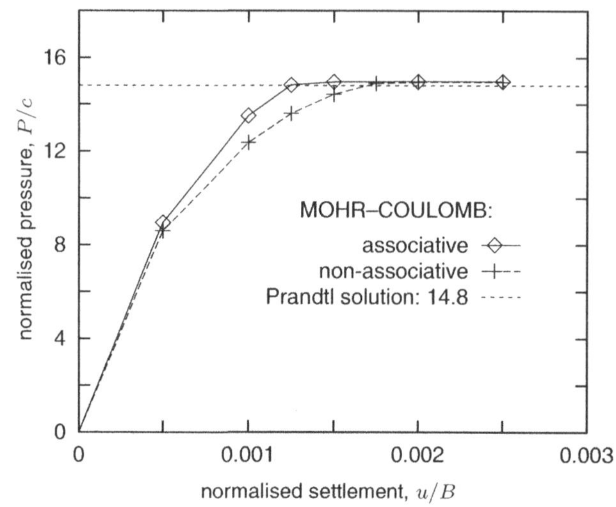

Figure 8.25. Strip footing with Mohr–Coulomb model. Load-displacement curves.

- the *plane strain match* to the Mohr–Coulomb model.

Only the associative law ($\phi = \psi = 20^o$) is considered. For all approximations, a total settlement $u/B = 0.004$ is imposed gradually in 10 increments. The load-settlement diagrams obtained are shown in Figure 8.26. The limit load predicted by the outer cone approximation is well above Prandtl's solution. The normalised limit pressure in this case is approximately 35.7 (about 140% above Prandtl's solution). The normalised limit pressure predicted by the inner cone is around 17.3 (about 17% above Prandtl's solution). For the plane strain match on the other hand, the results are in excellent agreement with Prandtl's solution. In this case, the predicted limit pressure is

$$P_{\rm lim}^{\rm fe}/c \approx 15.0, \tag{8.151}$$

which is practically identical to the limit loads obtained with the Mohr–Coulomb model.

8.4.4. CIRCULAR-FOOTING COLLAPSE

This section considers the axisymmetric analogue of the footing collapse problem described in the previous section. It consists of a circular footing of diameter $B = 1$ m (assumed rigid and smooth) lying on soil. The geometry is shown in Figure 8.27. The mesh, loading and boundary conditions are the same as in the previous example (these are illustrated in Figure 7.24, page 252). The load consists of the prescribed vertical displacements (settlement) of the nodes under the foot. In the present analysis, a total settlement $u/B = 0.002$ is reached in nine increments. The corresponding load-displacement diagram obtained is depicted in Figure 8.28. The collapse load

$$P_{\rm lim}^{\rm fe}/c \approx 20.5, \tag{8.152}$$

is effectively attained in seven increments. A complete solution to the circular (rigid smooth) footing problem has been obtained by Cox *et al.* (1961). The corresponding limit loads are

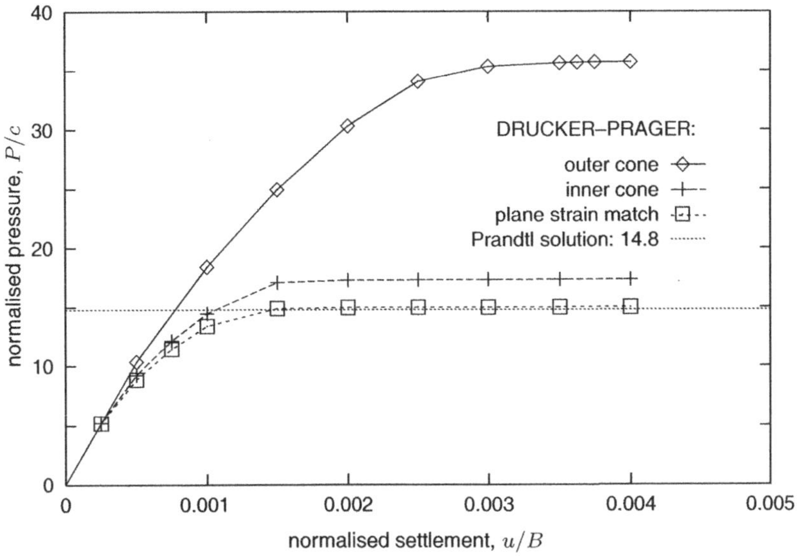

Figure 8.26. Strip footing with Drucker–Prager model. Load-displacement curves.

determined (to any desired accuracy) by the finite difference method (the limit loads for several frictional angles are tabulated in Chen, 1975). For the present frictional angle ($\phi = 20^o$) the limit average pressure on the footing is

$$P_{\mathrm{lim}}/c \approx 20.1. \tag{8.153}$$

The corresponding slip-line net is schematically illustrated in Figure 8.27. The limit load obtained in the finite element analysis is in close agreement with (8.153). The relative error is less than 2%. The incremental nodal displacement vectors near the footing obtained in increment 8 (where the collapse load has effectively been reached) are plotted in Figure 8.29. The vector sizes have been amplified to ease visualisation. The failure mechanism captured in the finite element analysis is in agreement with the slip-line field of Cox *et al.* (1961).

8.4.5. SLOPE STABILITY

In this example, the plane strain analysis of an inclined earth embankment subjected to self-weight is carried out. The objective is to assess the safety of the slope illustrated in Figure 8.30. The soil is modelled as a (perfectly plastic) Mohr–Coulomb material with the constants shown in Figure 8.30.

Limit analysis solution

The limit analysis of slopes under gravity load is described by Chen (1975). If the slope angle, β, is greater than the internal friction angle, ϕ, as in the present problem, the slope will collapse when the ratio

$$N = h\,\frac{\gamma}{c} \tag{8.154}$$

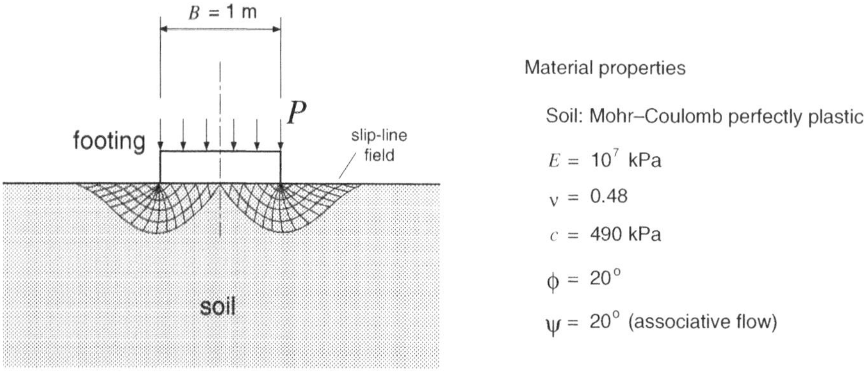

Figure 8.27. Circular footing. Geometry.

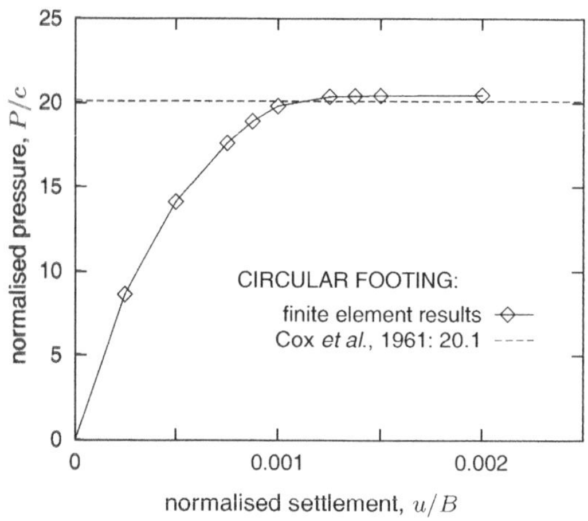

Figure 8.28. Circular footing. Load-displacement curve.

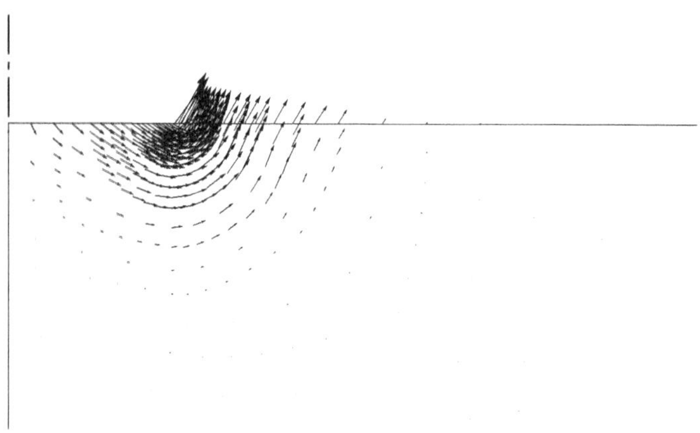

Figure 8.29. Circular footing. Incremental nodal displacement vectors at increment 8.

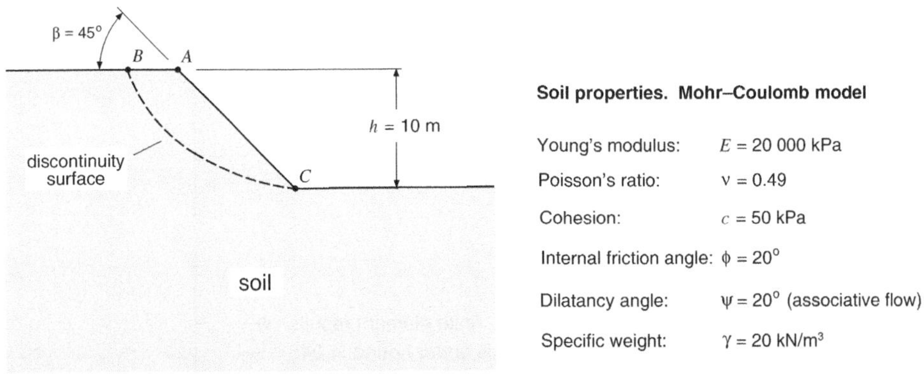

Figure 8.30. Slope stability. Geometry and material properties.

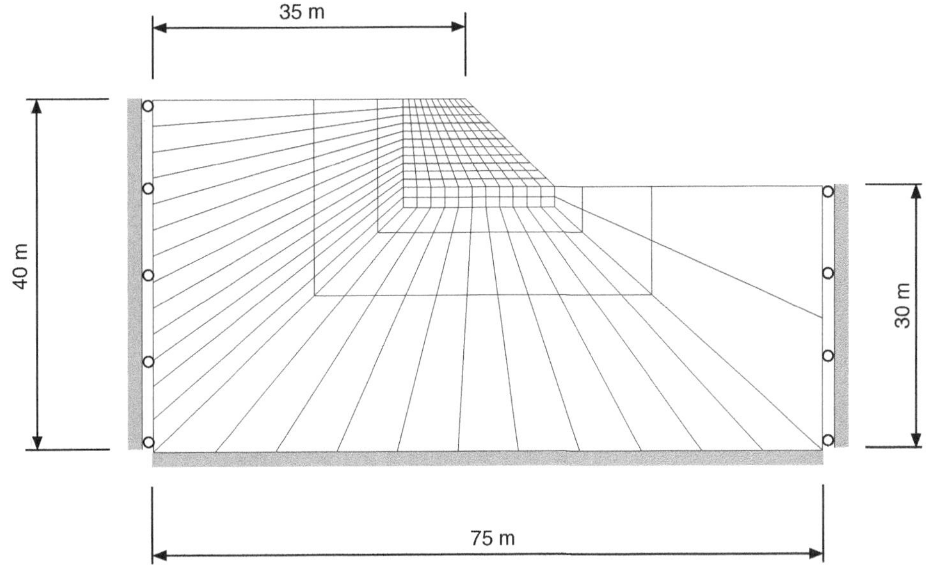

Figure 8.31. Slope stability. Finite element mesh.

reaches a critical limit. Here, h denotes the slope height, c is the cohesion and γ is the specific weight of the soil

$$\gamma = \rho\, g, \tag{8.155}$$

with ρ being the mass density and g the gravitational acceleration. For the present geometry and material constants, the ratio N is

$$N = 4. \tag{8.156}$$

The critical value of N, denoted N_s, is called the *stability factor*. The stability factor tabulated by Chen (1975) for the angles $\beta = 45^\circ$ and $\phi = 20^\circ$, adopted in the present example, is

$$N_s \approx 16.18.$$

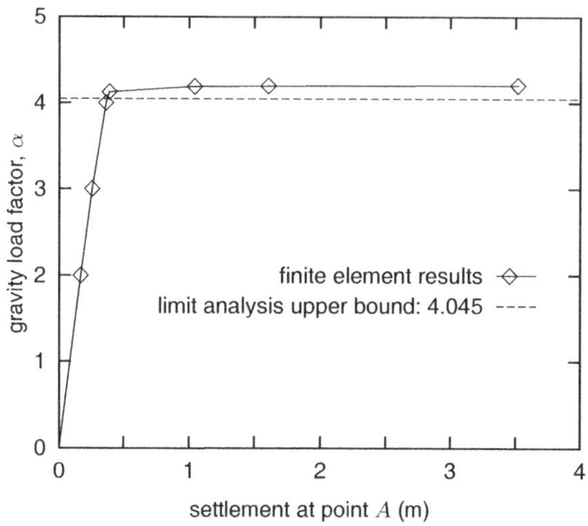

Figure 8.32. Slope stability. Load-settlement diagram.

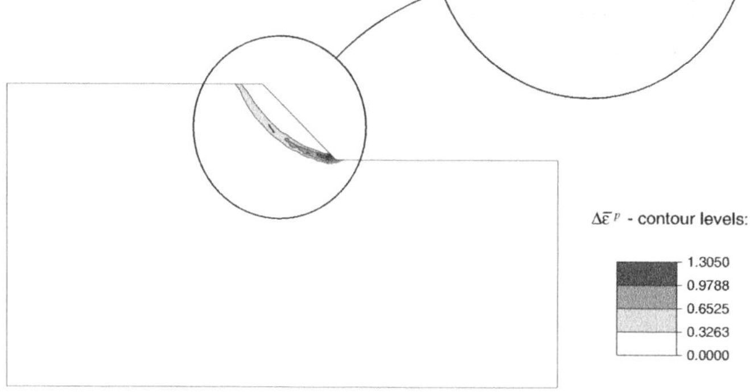

Figure 8.33. Slope stability. Increment of accumulated plastic strain and displacement at collapse (increment 6).

This value corresponds to a collapse mechanism (Figure 8.30) in which the portion ABC of the embankment slides along a logarithmic spiral discontinuity line (the dotted line) that passes through the toe (point C). A safety factor based on limit analysis can be defined as $\alpha_{\text{lim}} = N_s/N$. For the present dimensions and material properties it gives

$$\alpha_{\text{lim}} \approx 4.045. \tag{8.157}$$

Finite element analysis

The determination of the safety factor by means of the Finite Element Method with program `HYPLAS` is carried out by using the gravity load option. With

$$\bar{g} = \alpha\, g \tag{8.158}$$

denoting the applied gravity acceleration, the loading process starts from $\alpha = 0$ (completely unloaded). The load factor, α, is then increased gradually until collapse occurs. The value of α at collapse is the safety factor.

The adopted mesh, shown in Figure 8.31, consists of 221 eight-noded quadrilaterals (2×2 integration quadrature) with a total of 718 nodes. The discretised domain is made large enough to avoid interference of the collapse mechanism with the boundaries of the finite element mesh. In the finite element analysis, a total of seven load increments are applied. The settlement at the top corner of the slope (point A of Figure 8.30) is plotted in Figure 8.32 against the load factor α. The limit value of α, i.e. the safety factor, found in the incremental finite element analysis is

$$\alpha^{\text{fe}}_{\text{lim}} \approx 4.2. \tag{8.159}$$

It is about 3.8% above the limit analysis solution. The contour plot of incremental accumulated plastic strain, $\Delta\bar{\varepsilon}^p$, as well as the incremental nodal displacements obtained at increment 6 are depicted in Figure 8.33. The slope is effectively collapsing during this increment. The failure mechanism captured by the finite element simulation is in good agreement with the logarithmic spiral slide line referred to above.

9 PLANE STRESS PLASTICITY

IN the numerical solution of elastoplastic initial value problems addressed in the previous two chapters, attention has been focused on generic three-dimensional states of stress and strain – conditions under which the elastoplastic models were formulated in Chapter 6. In spite of the generality of the algorithmic formulations presented in Chapters 7 and 8, the actual computational implementation of the elastic predictor/return-mapping algorithms shown in the corresponding subroutines of program HYPLAS has been restricted to states of *plane strain* and *axisymmetric* deformation. From the theoretical point of view, the classes of problem with such constraints on the possible *strains* are those in which certain components of the strain tensor vanish during the given history of strains in the general elastoplastic initial value problem – stated in Problem 7.1, page 193. Clearly, having formulated the generic three-dimensional integration algorithms (as we did in Chapters 7 and 8), particularisation for states of constrained strain (used in the HYPLAS subroutines presented) is trivially obtained simply by eliminating the relevant *strain* components from the formulation. Note that, if the corresponding *stress* component is generally non-vanishing (such as the normal out-of-plane stress in plane strain problems), then it will be determined naturally as an outcome of the numerical integration algorithm.

The conditions addressed in the present chapter are quite different from those discussed in Chapters 7 and 8. Here, we are concerned with *plane stress* problems in which certain components of *stress*, rather than strain, are constrained to be zero. In spite of its triviality within the realm of linear elasticity, the treatment of the plane stress constraint in elastoplastic problems requires further consideration and justifies a chapter of its own. The main difficulty lies in the fact that as some components of the stress tensor are now *prescribed*, the original statement of the elastoplastic initial value problem – Problem 7.1 (page 193), where *all* stress components are *unknown* – no longer applies and needs to be reformulated. As a result, the three-dimensional numerical integration algorithms of Chapters 7 and 8, which have been developed specifically to solve Problem 7.1, cannot be used in general without modifications.

9.1. The basic plane stress plasticity problem

The plane stress assumption is typically introduced in the analysis of bodies in which one of the dimensions – the thickness – is much smaller than the others and is subjected to loads that generate dominant stresses perpendicular to the thickness direction (Figure 9.1). Typical situations of engineering interest where the plane stress assumption is applicable are found in the analysis of plates subjected to in-plane loads and thin membranes with negligible stiffness in bending and transverse shear. With indices 1 and 2 associated with the in-plane directions and index 3 corresponding to the normal (transverse or thickness) direction, a generic plane

Computational Methods for Plasticity: Theory and Applications EA de Souza Neto, D Perić and DRJ Owen

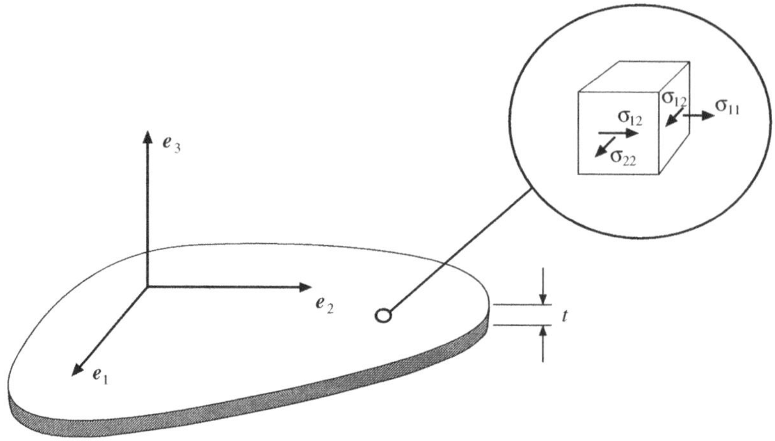

Figure 9.1. The plane stress state.

stress state is defined by a stress tensor

$$[\boldsymbol{\sigma}] = \begin{bmatrix} \sigma_{11} & \sigma_{12} & 0 \\ \sigma_{12} & \sigma_{22} & 0 \\ 0 & 0 & 0 \end{bmatrix}. \tag{9.1}$$

The corresponding subspace of plane stress states can be defined as

$$\mathscr{S}_{\rm pl} \equiv \{\boldsymbol{\sigma} \mid \sigma_{13} = \sigma_{23} = \sigma_{33} = 0\}. \tag{9.2}$$

9.1.1. PLANE STRESS LINEAR ELASTICITY

Before going further, let us now review the simple case of linear isotropic plane stress elasticity. The relationship between the non-zero stress components and the in-plane strains is given by the well-known expression

$$\begin{Bmatrix} \sigma_{11} \\ \sigma_{22} \\ \sigma_{12} \end{Bmatrix} = \frac{E}{1-\nu^2} \begin{bmatrix} 1 & \nu & 0 \\ \nu & 1 & 0 \\ 0 & 0 & \dfrac{1-\nu}{2} \end{bmatrix} \begin{Bmatrix} \varepsilon_{11} \\ \varepsilon_{22} \\ 2\varepsilon_{12} \end{Bmatrix}, \tag{9.3}$$

which, equivalently, in terms of the shear and bulk moduli reads

$$\begin{Bmatrix} \sigma_{11} \\ \sigma_{22} \\ \sigma_{12} \end{Bmatrix} = 2G \begin{bmatrix} 1+\alpha & \alpha & 0 \\ \alpha & 1+\alpha & 0 \\ 0 & 0 & \frac{1}{2} \end{bmatrix} \begin{Bmatrix} \varepsilon_{11} \\ \varepsilon_{22} \\ 2\varepsilon_{12} \end{Bmatrix}, \tag{9.4}$$

where

$$\alpha = \frac{3K-2G}{3K+4G}. \tag{9.5}$$

This relationship is derived directly from the three-dimensional elastic law as follows. Firstly, in the three-dimensional law

$$\begin{Bmatrix} \sigma_{11} \\ \sigma_{22} \\ \sigma_{33} \\ \sigma_{12} \\ \sigma_{23} \\ \sigma_{13} \end{Bmatrix} = \begin{bmatrix} K+\frac{4}{3}G & K-\frac{2}{3}G & K-\frac{2}{3}G & 0 & 0 & 0 \\ K-\frac{2}{3}G & K+\frac{4}{3}G & K-\frac{2}{3}G & 0 & 0 & 0 \\ K-\frac{2}{3}G & K-\frac{2}{3}G & K+\frac{4}{3}G & 0 & 0 & 0 \\ 0 & 0 & 0 & G & 0 & 0 \\ 0 & 0 & 0 & 0 & G & 0 \\ 0 & 0 & 0 & 0 & 0 & G \end{bmatrix} \begin{Bmatrix} \varepsilon_{11} \\ \varepsilon_{22} \\ \varepsilon_{33} \\ 2\varepsilon_{12} \\ 2\varepsilon_{23} \\ 2\varepsilon_{13} \end{Bmatrix}, \tag{9.6}$$

we prescribe the *in-plane strain* components, ε_{11}, ε_{22} and ε_{12} together with the (zero) *out-of-plane stress* components, σ_{33}, σ_{23} and σ_{13}. This gives the system

$$\begin{Bmatrix} \sigma_{11}^* \\ \sigma_{22}^* \\ 0 \\ \sigma_{12}^* \\ 0 \\ 0 \end{Bmatrix} = \begin{bmatrix} K+\frac{4}{3}G & K-\frac{2}{3}G & K-\frac{2}{3}G & 0 & 0 & 0 \\ K-\frac{2}{3}G & K+\frac{4}{3}G & K-\frac{2}{3}G & 0 & 0 & 0 \\ K-\frac{2}{3}G & K-\frac{2}{3}G & K+\frac{4}{3}G & 0 & 0 & 0 \\ 0 & 0 & 0 & G & 0 & 0 \\ 0 & 0 & 0 & 0 & G & 0 \\ 0 & 0 & 0 & 0 & 0 & G \end{bmatrix} \begin{Bmatrix} \varepsilon_{11} \\ \varepsilon_{22} \\ \varepsilon_{33}^* \\ 2\varepsilon_{12} \\ 2\varepsilon_{23}^* \\ 2\varepsilon_{13}^* \end{Bmatrix}, \tag{9.7}$$

where the star superscript has been used to point out the *unknown* quantities. Solution of the above system yields the plane stress elastic law together with the well-known expressions for the out-of-plane strains

$$\begin{gathered} \varepsilon_{13} = 0, \quad \varepsilon_{23} = 0, \\ \varepsilon_{33} = -\frac{3K-2G}{3K+4G}(\varepsilon_{11}+\varepsilon_{22}) = -\frac{\nu}{1-\nu}(\varepsilon_{11}+\varepsilon_{22}). \end{gathered} \tag{9.8}$$

In summary, the constitutive equations of plane stress elasticity are obtained by prescribing the in-plane strains and enforcing the plane stress constraint on the three-dimensional law, and then obtaining, as a result, the in-plane stresses and out-of-plane strains.

9.1.2. THE CONSTRAINED ELASTOPLASTIC INITIAL VALUE PROBLEM

The concept of constraining the original three-dimensional constitutive equation in order to obtain its plane stress counterpart can be extended to the elastoplastic case. The underlying idea is the same: we prescribe only the in-plane strains in the three-dimensional equation and use the plane stress constraint as an additional condition so as to determine the in-plane stresses and out-of-plane strains. The original three-dimensional elastoplastic initial value problem (Problem 7.1, page 193) is accordingly redefined in constrained form as follows.

Problem 9.1 (The plane stress elastoplastic initial value problem). *Given the initial values* $\boldsymbol{\varepsilon}^e(t_0)$ *and* $\boldsymbol{\alpha}(t_0)$, *of the elastic strain tensor and internal variable set, and given the history of the* in-plane *components of the strain tensor*

$$\{\varepsilon_{11}(t), \varepsilon_{22}(t), \varepsilon_{12}(t)\}, \quad t \in [t_0, T],$$

find the functions $\boldsymbol{\varepsilon}^e(t)$, $\boldsymbol{\alpha}(t)$ *and* $\dot{\gamma}(t)$, *for the elastic strain, internal variable set and plastic multiplier that satisfy the reduced general elastoplastic three-dimensional constitutive equations*

$$\begin{aligned} \dot{\boldsymbol{\varepsilon}}(t) &= \dot{\boldsymbol{\varepsilon}}^e(t) - \dot{\gamma}(t)\, \boldsymbol{N}(\boldsymbol{\sigma}(t), \boldsymbol{A}(t)) \\ \dot{\boldsymbol{\alpha}}(t) &= \dot{\gamma}(t)\, \boldsymbol{H}(\boldsymbol{\sigma}(t), \boldsymbol{A}(t)) \end{aligned} \tag{9.9}$$

and

$$\dot{\gamma}(t) \geq 0, \quad \Phi(\boldsymbol{\sigma}(t), \boldsymbol{A}(t)) \leq 0, \quad \dot{\gamma}(t)\, \Phi(\boldsymbol{\sigma}(t), \boldsymbol{A}(t)) = 0, \tag{9.10}$$

with

$$\boldsymbol{\sigma}(t) = \bar{\rho}\, \frac{\partial \psi}{\partial \boldsymbol{\varepsilon}^e}, \quad \boldsymbol{A}(t) = \bar{\rho}\, \frac{\partial \psi}{\partial \boldsymbol{\alpha}}, \tag{9.11}$$

and, in addition, *satisfy the plane stress constraint*

$$\boldsymbol{\sigma}(t) \in \mathscr{S}_{\mathrm{pl}} \tag{9.12}$$

for each instant $t \in [t_0, T]$.

9.1.3. PROCEDURES FOR PLANE STRESS PLASTICITY

In order to find a numerical approximation to the solution of the above plane stress elastoplastic evolution problem, some modification is required in the elastic predictor/return-mapping algorithms discussed in Chapters 7 and 8. Issues related to the *explicit* implementation of plane stress models are addressed by Marques (1984). For the *implicit* scheme, to which the present chapter is restricted, three main general approaches may be adopted to deal with the problem (these have been reviewed and compared by Millard 1995):

(a) direct inclusion of the plane stress constraint into the three-dimensional elastic predictor and plastic corrector algorithm equations (7.21), (7.22) and (7.25) (refer to page 196) applied at the *Gauss point* level. This approach can also be implemented by means of a nested Newton return-mapping iteration for plane stress enforcement. The nested iteration procedure is implemented in program HYPLAS for the Drucker–Prager model;

(b) use of the standard three-dimensional return mapping at the Gauss point level with the plane stress condition added as a structural constraint at the *global* structural level;

(c) use of plane stress-projected constitutive equations where, similarly to the procedure of item (a), the plane stress constraint is enforced at the *Gauss point* level. This strategy has been implemented in program HYPLAS for the von Mises isotropically hardening model.

Remark 9.1. Approaches (a), (b) and (c) are *equivalent* in that, provided the same numerical scheme is used in the discretisation of the elastoplastic evolution equations, the three methodologies produce *identical* incremental plane stress constitutive equations. The essential difference between distinct approaches in this case lies in the way that the plane stress constraint (9.12) imposed upon the three-dimensional model is enforced.

Having the general three-dimensional plasticity model of Chapter 6 and the implicit elastic predictor/return-mapping scheme of Section 7.2.3 as the underlying model/algorithm, approaches (a) and (b) are discussed below in Sections 9.2, and 9.3, respectively. In Section 9.2, an outline of the FORTRAN code of the von Mises model implementation with nested iterations is shown. The third strategy is described in Section 9.4 where the von Mises model is used as an example for completeness. The complete description of the computational implementation, including the FORTRAN code, is also given in Section 9.4.

9.2. Plane stress constraint at the Gauss point level

Approach (a) is a direct extension of the elastic predictor/return-mapping algorithms discussed in Chapter 7 in order to solve the plane stress elastoplastic evolution problem stated in Problem 9.1 above. This procedure appears to have been suggested firstly by Aravas (1987) in the context of pressure-dependent plasticity. For the fully implicit algorithm, it consists of simply adding the extra equations

$$\begin{aligned} \sigma_{13}(\boldsymbol{\varepsilon}^e_{n+1}, \boldsymbol{\alpha}_{n+1}) &= 0 \\ \sigma_{23}(\boldsymbol{\varepsilon}^e_{n+1}, \boldsymbol{\alpha}_{n+1}) &= 0 \\ \sigma_{33}(\boldsymbol{\varepsilon}^e_{n+1}, \boldsymbol{\alpha}_{n+1}) &= 0 \end{aligned} \tag{9.13}$$

to the general system of return-mapping equations (7.25) (page 196) and letting the out-of-plane components of the elastic trial strain tensor become *unknowns* of the new system. After performing a *plane stress-constrained* elastic trial state evaluation,[†] plastic consistency is checked in the usual way. If plastic admissibility is violated, i.e. if

$$\Phi(\boldsymbol{\sigma}^{\text{trial}}_{n+1}, \boldsymbol{A}_n) > 0, \tag{9.14}$$

then the return-mapping procedure has to be applied. The return mapping now requires the solution of the *augmented* algebraic system

$$\begin{aligned} \boldsymbol{\varepsilon}^e_{n+1} &= \boldsymbol{\varepsilon}^{e\ \text{trial}}_{n+1} - \Delta\gamma\, \boldsymbol{N}_{n+1} \\ \boldsymbol{\alpha}_{n+1} &= \boldsymbol{\alpha}_n + \Delta\gamma\, \boldsymbol{H}_{n+1} \\ \Phi(\boldsymbol{\sigma}_{n+1}, \boldsymbol{A}_{n+1}) &= 0 \\ \sigma_{13}(\boldsymbol{\varepsilon}^e_{n+1}, \boldsymbol{\alpha}_{n+1}) &= 0 \\ \sigma_{23}(\boldsymbol{\varepsilon}^e_{n+1}, \boldsymbol{\alpha}_{n+1}) &= 0 \\ \sigma_{33}(\boldsymbol{\varepsilon}^e_{n+1}, \boldsymbol{\alpha}_{n+1}) &= 0 \end{aligned} \tag{9.15}$$

for the variables $\boldsymbol{\varepsilon}^e_{n+1}$, $\boldsymbol{\alpha}_{n+1}$, $\Delta\gamma$ *and* the elastic trial components

$$\varepsilon^{e\ \text{trial}}_{13}, \quad \varepsilon^{e\ \text{trial}}_{23}, \quad \varepsilon^{e\ \text{trial}}_{33}.$$

As in the standard non-constrained case, the solution of the above system can be undertaken by the Newton–Raphson algorithm.

[†]Note that if the elastic behaviour is nonlinear, the plane stress-constrained elastic equations may not be obtained as trivially as in the linear case reviewed in Section 9.1.

9.2.1. IMPLEMENTATION ASPECTS

The above procedure can be implemented even for very complex plasticity models. It should be noted that the number of unknowns of the system is increased (as compared to the standard three-dimensional algorithm) which makes this approach computationally more expensive. It should be emphasised, however, that, as for the three-dimensional case, considerable simplification of the return-mapping algebraic system may be achieved by exploiting particular properties of the material model in question. We note that such simplifications have reduced the return mapping for the von Mises model to a *single* nonlinear equation (Section 7.3, page 215). Under elastoplastic isotropy, for instance, the plane stress condition implies that

$$\varepsilon_{13} = \varepsilon^e_{13} = \varepsilon^p_{13} = 0, \quad \varepsilon_{23} = \varepsilon^e_{23} = \varepsilon^p_{23} = 0, \tag{9.16}$$

which ensures that σ_{13} and σ_{23} vanish. In this case the only extra equation to be added to the general three-dimensional return mapping is[‡]

$$\sigma_{33}(\boldsymbol{\varepsilon}^e_{n+1}, \boldsymbol{\alpha}_{n+1}) = 0, \tag{9.17}$$

with the associated extra unknown $\varepsilon^{e\ \text{trial}}_{33}$.

9.2.2. PLANE STRESS ENFORCEMENT WITH NESTED ITERATIONS

It is also possible to deal with the plane stress problem by using the original three-dimensional algorithm (without modification) within a Newton–Raphson loop for the enforcement of the plane stress constraint at the Gauss point level. This approach, introduced by Dodds (1987) within the implicit implementation of the von Mises model, is described in the following. For simplicity, let us consider the isotropic case where stress and strain transverse shear components vanish identically as discussed above. In order to describe the procedure, it is convenient to employ the matrix notation

$$\boldsymbol{\sigma} = [\sigma_{11}\ \sigma_{22}\ \sigma_{12}]^T, \quad \boldsymbol{\varepsilon} = [\varepsilon_{11}\ \varepsilon_{22}\ 2\varepsilon_{12}]^T. \tag{9.18}$$

During a typical equilibrium iteration, the in-plane displacements are prescribed and so is the in-plane strain array $\boldsymbol{\varepsilon}_{n+1}$. Instead of giving $\boldsymbol{\varepsilon}_{n+1}$ (or $\boldsymbol{\varepsilon}^{e\ \text{trial}}_{n+1}$) as the input of an augmented algebraic system (as described in the above), we proceed as follows. Firstly, we define some initial guess for the unknown $(\varepsilon^e_{33})^{\text{trial}}_{n+1}$. One possible guess can be the previously (equilibrium) converged out-of-plane elastic strain, i.e. we can set

$$\varepsilon^{e\ \text{trial}}_{33} := (\varepsilon^e_{33})_n. \tag{9.19}$$

Here the subscript $n+1$ has been dropped for notational convenience. Next, we use the augmented strain array

$$[\boldsymbol{\varepsilon}^{e\ \text{trial}^T}\ \varepsilon^{e\ \text{trial}}_{33}]^T$$

as the input of the integration algorithm for the *axisymmetric* case.[§] Note that an out-of-plane strain component must be prescribed for the axisymmetric algorithm.

[‡]The original algorithm proposed by Aravas (1987) was derived under such isotropy conditions, having Gurson's void growth model (described in Section 12.5, from page 496) as the underlying material.

[§]Axisymmetric implementations of the general integration algorithm have been included in the corresponding routines for all material models discussed in Chapters 7 and 8.

After application of the axisymmetric elastic predictor/return-mapping procedure, the corresponding routine will return the augmented stress array

$$[\boldsymbol{\sigma}_{n+1}^T \; \sigma_{33}]^T.$$

If $\sigma_{33} = 0$ (or, in computational terms, $|\sigma_{33}| \leq \epsilon_{\mathrm{tol}}$) then the guess $\varepsilon_{33}^{e\,\mathrm{trial}}$ indeed solves the plane stress problem, and the solution obtained by the axisymmetric algorithm is the one we are looking for. Otherwise, we apply a Newton–Raphson correction to obtain another guess

$$\varepsilon_{33}^{e\,\mathrm{trial}} := \varepsilon_{33}^{e\,\mathrm{trial}} - \frac{\sigma_{33}}{\mathrm{D}_{22}}, \tag{9.20}$$

where D_{22} is the component of the *axisymmetric* consistent tangent matrix

$$\begin{bmatrix} \mathrm{d}\sigma_{11} \\ \mathrm{d}\sigma_{22} \\ \mathrm{d}\sigma_{12} \\ \hline \mathrm{d}\sigma_{33} \end{bmatrix} = \left[\begin{array}{c|c} \mathbf{D}_{11} & \mathbf{D}_{12} \\ \hline \mathbf{D}_{21} & \mathrm{D}_{22} \end{array}\right] \begin{bmatrix} \mathrm{d}\varepsilon_{11}^{e\,\mathrm{trial}} \\ \mathrm{d}\varepsilon_{22}^{e\,\mathrm{trial}} \\ 2\mathrm{d}\varepsilon_{12}^{e\,\mathrm{trial}} \\ \hline \mathrm{d}\varepsilon_{33}^{e\,\mathrm{trial}} \end{bmatrix}, \tag{9.21}$$

that relates the transverse stress and strain components.¶ We repeat this process until we find the elastic trial thickness strain $\varepsilon_{33}^{e\,\mathrm{trial}}$ that, together with the in-plane strain kinematically prescribed by the global equilibrium iteration, results in zero (or sufficiently small) transverse stress upon application of the axisymmetric algorithm. The resulting overall algorithm is shown in Box 9.1 in pseudo-code format. An outline of its specialisation to the von Mises model is presented in the next section. Its complete computational implementation for the Drucker–Prager model is included in program HYPLAS. The associated subroutine (not listed in this text) is named SUDPPN (**S**tate **U**pdate procedure for the **D**rucker–**P**rager model in **P**lane stress with **N**ested iterations).

Remark 9.2. The above methodology involves two nested Newton–Raphson iteration loops: one (outer) global equilibrium loop having a nested (inner) iteration loop to enforce the plane stress constraint with $\varepsilon_{33}^{e\,\mathrm{trial}}$ as the unknown. Thus, for each global equilibrium loop, a number of iterations will be required in each Gauss point to ensure that the final stress is plane (to within a prescribed numerical tolerance). Since the axisymmetric integration algorithm is applied in each iteration of the inner loop, the present procedure is computationally expensive. It is important to remember, however, that the cost of the calculations carried out at the Gauss point level increases approximately linearly with the problem size, whereas the cost of the solution of the global linearised problem increases at a much higher nonlinear rate. In other words, the relative cost of the Gauss point computations decreases considerably as the problem size increases and the present procedure can be used efficiently in large problems.

Remark 9.3. The solution obtained by the nested iteration is identical to that obtained by the previously discussed strategy based on the augmented return-mapping system. Its main advantage over the augmented system procedure lies in its simplicity. Essentially, for *any*

¶The implementation of the axisymmetric consistent tangent for the material models of Chapters 7 and 8 is also included in the corresponding routines of HYPLAS.

Box 9.1. Plane stress constraint at the Gauss point level with nested iterations.

(i) Set initial guess for the elastic trial thickness strain

$$\varepsilon_{33}^{e\ \mathrm{trial}} := (\varepsilon_{33}^{e})_n$$

The in-plane elastic trial strain $\boldsymbol{\varepsilon}_{n+1}^{e\ trial}$ is prescribed

(ii) Apply the axisymmetric integration algorithm and obtain the augmented stress array

$$[\boldsymbol{\sigma}^T\ \sigma_{33}]^T$$

(iii) Check plane stress convergence

$$\text{IF } |\sigma_{33}| < \epsilon_{\mathrm{tol}} \text{ THEN EXIT}$$

(iv) Compute component D_{22} of the axisymmetric consistent tangent matrix (see expression (9.21))

(v) Apply Newton–Raphson correction to the thickness trial strain

$$\varepsilon_{33}^{e\ \mathrm{trial}} := \varepsilon_{33}^{e\ \mathrm{trial}} - \frac{\sigma_{33}}{\mathrm{D}_{22}}$$

(vi) GOTO (ii)

material model, the nested iteration approach allows a straightforward adaptation of existing general procedures (which include the three-dimensional or axisymmetric implementations) to cope with the plane stress problem, without the need for modification of the original integration algorithms.

Remark 9.4. The present strategy is compatible with material models other than elasto-plastic, such as generic nonlinear elastic materials. Note that for a linear elastic model, for instance, the nested iteration loop produces the solution that satisfies the standard plane equilibrium together with the trivial plane stress linear elastic law (9.3, 9.4). This is what happens before plastic yielding in analyses involving linear elasticplastic materials such as the von Mises model.

9.2.3. PLANE STRESS VON MISES WITH NESTED ITERATIONS

To illustrate the above concepts, the von Mises model is used as an example and a possible FORTRAN implementation is outlined in this section. Basically the implementation of the von Mises model within the present methodology requires the use of the generic elastic predictor/return-mapping algorithm together with the corresponding consistent tangent calculation procedure. The generic (plane strain/axisymmetric) procedures are included in program `HYPLAS`. The corresponding subroutines are, respectively, `SUVM`, described in Section 7.3.5 (page 224), and `CTVM`, whose complete description is given in Section 7.4.3 (page 235). Following the steps of Box 9.1, an outline of the FORTRAN implementation of the von Mises model in plane stress is given by

```
C Newton-Raphson iteration loop for plane stress enforcement (Box 9.1)
C -----------------------------------------------------------------------
C Set initial guess for elastic trial thickness strain. Use previously
C converged elastic thickness strain.
      E33TRL=RSTAVA(4)
C Start N-R loop
      DO 20 ITER=1,MXITER
C Set state variables to values at beginning of increment
        DO 10 I=1,MSTRE+1
          RSTAUX(I)=RSTAVA(I)
   10   CONTINUE
C Use axisymmetric integration algorithm to compute stresses
        STRAT(4)=E33TRL
        CALL SUVM
     1(   DGAMA      ,IPROPS     ,LALGVA     ,3          ,RPROPS     ,
     2    RSTAUX     ,STRAT      ,STRES      )
        IFPLAS=LALGVA(1)
        SUFAIL=LALGVA(2)
        IF(SUFAIL)GOTO 999
C Check plane stress convergence
        IF(ABS(STRES(4)).LE.TOL)THEN
C...and break N-R loop in case of convergence
          GOTO 30
        ENDIF
C Compute axisymmetric consistent tangent components
        EPFLAG=IFPLAS
        CALL CTVM
     1(   DGAMA      ,DMATX      ,EPFLAG     ,IPROPS     ,3          ,
     2    RPROPS     ,RSTAUX     ,STRES      )
C Apply Newton-Raphson correction to normal elastic trial strain
        D22=DMATX(4,4)
        E33TRL=E33TRL-STRES(4)/D22
   20 CONTINUE
      WRITE(*,*)'Plane stress enforcement loop failed to converge'
      SUFAIL=.TRUE.
      LALGVA(2)=SUFAIL
      GOTO 999
   30 CONTINUE
C Set state variables to current updated values
      DO 40 I=1,MSTRE+1
        RSTAVA(I)=RSTAUX(I)
   40 CONTINUE
  999 CONTINUE
      RETURN
```

The incorporation of the procedure into HYPLAS is a straightforward programming exercise. In the above, we have created the local array RSTAUX to store the state variables temporarily. The actual array of state variables, RSTAVA, is updated only if the plane stress enforcement loop converges.

It should be emphasised that no effort has been made to optimise the FORTRAN code suggested above. The purpose of the above code is solely to illustrate the simplicity of the nested iteration approach in the plane stress implementation of models whose generic implementation already exists. Note that when CTVM is called, *all* components of the axisymmetric consistent tangent matrix are computed. The procedure can be made more efficient if a new subroutine that computes *only* the scalar D_{22} is used instead. This is the only component of the consistent tangent required in the plane stress enforcement loop.

9.2.4. THE CONSISTENT TANGENT FOR THE NESTED ITERATION PROCEDURE

To obtain the tangent operator consistent with the above nested iteration algorithm, we first differentiate the *residual equation*

$$\sigma_{33} = 0$$

of the plane stress enforcement loop. This, together with (9.21) gives

$$\mathrm{d}\sigma_{33} = \mathbf{D}_{21}\, \mathrm{d}\boldsymbol{\varepsilon}^{e\ \mathrm{trial}}_{n+1} + \mathrm{D}_{22}\, \mathrm{d}\varepsilon^{e\ \mathrm{trial}}_{33} = 0, \tag{9.22}$$

which renders

$$\mathrm{d}\varepsilon^{e\ \mathrm{trial}}_{33} = \frac{-1}{\mathrm{D}_{22}}\, \mathbf{D}_{21}\, \mathrm{d}\varepsilon^{e\ \mathrm{trial}}. \tag{9.23}$$

Substitution of the above expression into (9.21) results in the following consistent tangent relation between the in-plane stress and strain components

$$\frac{\mathrm{d}\boldsymbol{\sigma}_{n+1}}{\mathrm{d}\boldsymbol{\varepsilon}^{e\ \mathrm{trial}}_{n+1}} = \mathbf{D}_{11} - \frac{1}{\mathrm{D}_{22}}\, \mathbf{D}_{12}\mathbf{D}_{21}. \tag{9.24}$$

9.2.5. CONSISTENT TANGENT COMPUTATION FOR THE VON MISES MODEL

For completeness, we outline below the implementation of the above expressions for the von Mises model. Using the (existing) generic consistent tangent calculation routine for the von Mises model, `CTVM` (Section 7.4.3, page 235), the corresponding plane stress consistent tangent matrix can be computed by the following FORTRAN code:

```
C Compute the axisymmetric consistent tangent matrix
      CALL CTVM
     1(   DGAMA      ,DMATX      ,EPFLAG     ,IPROPS     ,3          ,
     2    RPROPS     ,RSTAVA     ,STRES      )
C Decompose into submatrices
      D12(1)=DMATX(1,4)
      D12(2)=DMATX(2,4)
      D12(3)=DMATX(3,4)
      D21(1)=DMATX(4,1)
      D21(2)=DMATX(4,2)
      D21(3)=DMATX(4,3)
      D22=DMATX(4,4)
C Assemble plane stress consistent tangent matrix
      DO 20 I=1,3
        DO 10 J=1,3
          DMATX(I,J)=DMATX(I,J)-D12(I)*D21(J)/D22
   10   CONTINUE
   20 CONTINUE
```

Its incorporation into `HYPLAS` can be left as an exercise. We remark that the full implementation of a tangent matrix consistent with the nested iteration approach is already available in `HYPLAS` for the Drucker–Prager model. The associated FORTRAN subroutine (not listed here) is named `CTDPPN` (**C**onsistent **T**angent for the **D**rucker–**P**rager model in **P**lane stress with **N**ested iterations).

9.3. Plane stress constraint at the structural level

One interesting alternative to the above procedure consists in enforcing the plane stress constraint at the *structural* level, rather than in the constitutive integration algorithm applied at the Gauss point level. This approach has been proposed by de Borst (1991). Its main advantage is the fact that, as the above described nested iteration approach, it does *not* require the modification of the three-dimensional elastic predictor/return mapping algorithm. The basic idea of the method is to use the standard three-dimensional algorithm (without modification) at the Gauss points and to introduce the plane stress constraint in the *global* finite element equations. At the Gauss point level, the plane stress constraint is satisfied only at converged (equilibrium) configurations. The method is described in the following.

Firstly, let us write the general finite element equilibrium residual equation for plane problems (plane stress/strain). At the generic (pseudo-) time station, t_{n+1}, we have

$$\mathbf{r} \equiv \int_\Omega \mathbf{B}^T \boldsymbol{\sigma}_{n+1} \, \mathrm{d}\Omega - \mathbf{f}^{\text{ext}}_{n+1} = \mathbf{0}, \tag{9.25}$$

where $\mathbf{B}$ is the *in-plane* symmetric gradient operator (with only the three rows associated with the in-plane strain components ε_{11}, ε_{22} and ε_{12}), $\mathbf{f}^{\text{ext}}_{n+1}$ is the applied external force vector at t_{n+1}, which contains only in-plane forces, and $\boldsymbol{\sigma}$ is the array of *in-plane* stress components

$$\boldsymbol{\sigma}_{n+1} \equiv \begin{bmatrix} \hat{\sigma}_{11}(\boldsymbol{\varepsilon}^{e\,\text{trial}}_{n+1}, \varepsilon^{e\,\text{trial}}_{13}, \varepsilon^{e\,\text{trial}}_{23}, \varepsilon^{e\,\text{trial}}_{33}) \\ \hat{\sigma}_{12}(\boldsymbol{\varepsilon}^{e\,\text{trial}}_{n+1}, \varepsilon^{e\,\text{trial}}_{13}, \varepsilon^{e\,\text{trial}}_{23}, \varepsilon^{e\,\text{trial}}_{33}) \\ \hat{\sigma}_{22}(\boldsymbol{\varepsilon}^{e\,\text{trial}}_{n+1}, \varepsilon^{e\,\text{trial}}_{13}, \varepsilon^{e\,\text{trial}}_{23}, \varepsilon^{e\,\text{trial}}_{33}) \end{bmatrix}, \tag{9.26}$$

where

$$\boldsymbol{\varepsilon}^{e\,\text{trial}}_{n+1} \equiv \boldsymbol{\varepsilon}^e_n + \mathbf{B}\,\Delta\mathbf{u} = [\,\varepsilon^{e\,\text{trial}}_{11} \; \varepsilon^{e\,\text{trial}}_{22} \; \varepsilon^{e\,\text{trial}}_{12}\,]^T, \tag{9.27}$$

with $\Delta\mathbf{u}$ denoting the array of incremental nodal displacements.

$$\Delta\mathbf{u} = \mathbf{u}_{n+1} - \mathbf{u}_n. \tag{9.28}$$

Here and in what follows, $\hat{\sigma}_{ij}$ (with the superimposed hat) denote the algorithmic constitutive functionals for the stress components defined by the *three-dimensional* elastoplastic integration algorithm.[‖] Note that in the plane *strain* case, the three-dimensional algorithm is used with zero prescribed out-of-plane strains.

9.3.1. THE METHOD

Again, for simplicity, we shall restrict ourselves in what follows to the more usual situation of elastoplastic isotropy where the transverse shear components of stress and (total, elastic, elastic trial, and plastic) strain vanish identically

$$\varepsilon_{13} = \varepsilon_{23} = 0, \quad \sigma_{13} = \sigma_{23} = 0.$$

In this case, the dependence of the in-plane stress components on the transverse shear strain components is removed from (9.26). Under such a condition, the present strategy consists

‖For notational convenience, the functional dependence of the stress components on the state variables at t_n is not explicitly indicated.

simply in adding the zero transverse stress constraint to the finite element equilibrium equation (9.25). The resulting residual equations are accordingly redefined as

$$\begin{cases} \displaystyle\int_\Omega \mathbf{B}^T \boldsymbol{\sigma}_{n+1} \, \mathrm{d}\Omega - \mathbf{f}^{\text{ext}}_{n+1} = \mathbf{0} \\ \hat{\sigma}_{33}(\boldsymbol{\varepsilon}^{e\ \text{trial}}_{n+1}, \varepsilon^{e\ \text{trial}}_{33}) = 0 \qquad \text{in each Gauss point,} \end{cases} \tag{9.29}$$

where the unknowns are the nodal incremental displacements $\Delta\mathbf{u}$ (which enter the equations through the dependence of $\boldsymbol{\varepsilon}^{e\ \text{trial}}_{n+1}$ upon it) *and* the thickness strain $\varepsilon^{e\ \text{trial}}_{33}$ in each Gauss point. In the above, $\hat{\sigma}_{33}$ is the algorithmic constitutive function for the transverse stress component resulting from the three-dimensional integration algorithm.

9.3.2. THE IMPLEMENTATION

In the finite element procedure, both the equilibrium and plane stress conditions are solved *simultaneously*. The Newton–Raphson scheme for solution of the new residual equation is obtained by linearising (9.29). The linearised version of (9.29), where the unknowns are the iterative changes of nodal displacement and thickness strain (or elastic trial strain) at Gauss points, reads

$$\begin{cases} \displaystyle\mathop{\mathsf{A}}_{e=1}^{n_{\text{elem}}} \sum_{i=1}^{n_{\text{gausp}}} w^{(i)} \mathbf{B}^{(i)T} \{\mathbf{D}^{(i)}_{11} \mathbf{B}^{(i)} \delta\mathbf{u}_{(e)} + \mathbf{D}^{(i)}_{12} \delta\varepsilon^{(i)}_{33}\} = -\mathbf{r} \\ \mathbf{D}^{(i)}_{21} \mathbf{B}^{(i)} \delta\mathbf{u}_{(e)} + \mathrm{D}^{(i)}_{22} \delta\varepsilon^{(i)}_{33} = -\sigma^{(i)}_{33} \qquad \text{for each g.p. } i \text{ of each elem. } e \end{cases} \tag{9.30}$$

where $\mathbf{D}_{11}$, $\mathbf{D}_{12}$, $\mathbf{D}_{21}$ and D_{22} are submatrices of the matrix form of the consistent tangent operator associated with the *axisymmetric* integration algorithm (see expression (9.21)).

The linearised residual equation can be simplified as follows. Firstly, equation $(9.30)_2$ can be solved explicitly for $\delta\varepsilon_{33}$. This gives

$$\delta\varepsilon_{33} = -\frac{1}{\mathrm{D}_{22}}[\mathbf{D}_{21}\, \mathbf{B}\, \delta\mathbf{u}_{(e)} + \sigma_{33}] = -\frac{1}{\mathrm{D}_{22}}[\mathbf{D}_{21} \delta\boldsymbol{\varepsilon} + \sigma_{33}], \tag{9.31}$$

where the superscript i has been omitted. With the substitution of this result into $(9.30)_1$ we obtain the linearised equation for the iterative nodal displacement array

$$\mathbf{K}^*\, \delta\mathbf{u} = -\mathbf{r}^*, \tag{9.32}$$

where the *modified tangent stiffness matrix*, $\mathbf{K}^*$, is defined as

$$\mathbf{K}^* \equiv \int_\Omega \mathbf{B}^T \left[\mathbf{D}_{11} - \frac{1}{\mathrm{D}_{22}} \mathbf{D}_{12}\, \mathbf{D}_{21}\right] \mathbf{B}\, \mathrm{d}\Omega, \tag{9.33}$$

and the *modified residual vector* is defined by

$$\begin{aligned} \mathbf{r}^* &\equiv \mathbf{r} - \int_\Omega \mathbf{B}^T \left[\frac{\sigma_{33}}{\mathrm{D}_{22}} \mathbf{D}_{12}\right] \mathrm{d}\Omega \\ &= \int_\Omega \mathbf{B}^T \left[\boldsymbol{\sigma} - \frac{\sigma_{33}}{\mathrm{D}_{22}} \mathbf{D}_{12}\right] \mathrm{d}\Omega - \mathbf{f}^{\text{ext}}. \end{aligned} \tag{9.34}$$

Box 9.2. Plane stress constraint at the structural level. The modified global equilibrium iteration loop.

(i) Compute modified tangent stiffness

$$\mathbf{K}^* := \int_\Omega \mathbf{B}^T \left[\mathbf{D}_{11} - \frac{1}{\mathrm{D}_{22}} \mathbf{D}_{12}\, \mathbf{D}_{21} \right] \mathbf{B}\, \mathrm{d}\Omega$$

(ii) Apply Newton–Raphson correction to nodal displacements

$$\delta\mathbf{u} := -[\mathbf{K}^*]^{-1}\, \mathbf{r}^*; \quad \mathbf{u} := \mathbf{u} + \delta\mathbf{u}$$

(iii) Update in-plane elastic trial strain

$$\delta\boldsymbol{\varepsilon} := \mathbf{B}\, \delta\mathbf{u}_{(e)}; \quad \boldsymbol{\varepsilon}^{e\ \mathrm{trial}}_{n+1} := \boldsymbol{\varepsilon}^{e\ \mathrm{trial}}_{n+1} + \delta\boldsymbol{\varepsilon}$$

(iv) Apply Newton–Raphson correction to thickness elastic trial strain

$$\delta\varepsilon_{33} := -\frac{1}{\mathrm{D}_{22}} [\mathbf{D}_{21} \delta\boldsymbol{\varepsilon} + \sigma_{33}]$$
$$\varepsilon^{e\ \mathrm{trial}}_{33} := \varepsilon^{e\ \mathrm{trial}}_{33} + \delta\varepsilon_{33}$$

(v) Update stresses

$$\boldsymbol{\sigma}_{n+1} := \hat{\boldsymbol{\sigma}}(\boldsymbol{\varepsilon}^{e\ \mathrm{trial}}_{n+1}, \varepsilon^{e\ \mathrm{trial}}_{33})$$
$$\sigma_{33} := \hat{\sigma}_{33}(\boldsymbol{\varepsilon}^{e\ \mathrm{trial}}_{n+1}, \varepsilon^{e\ \mathrm{trial}}_{33})$$

using the *three-dimensional* elastoplastic algorithm

(vi) Compute modified residual vector

$$\mathbf{r}^* := \int_\Omega \mathbf{B}^T \left[\boldsymbol{\sigma}_{n+1} - \frac{\sigma_{33}}{\mathrm{D}_{22}} \mathbf{D}_{12} \right] \mathrm{d}\Omega$$

(vii) Modified convergence check

IF $\|\mathbf{r}^*\| < \epsilon_{\mathrm{tol}}$ AND $|\sigma_{33}| < \epsilon_{\mathrm{tol}}$ for all Gauss points of the structure, THEN EXIT

ELSE GOTO (i)

In summary, the present procedure requires only a relatively simple modification of the global equilibrium Newton–Raphson iteration loop. The main modification consists in replacing the standard linearised equilibrium equation with (9.32) and using, after computation of the iterative displacement $\delta\mathbf{u}$, the explicit expression (9.31) to determine the corresponding iterative change of thickness strain (or thickness elastic trial strain), $\delta\varepsilon_{33}$. The overall algorithm is summarised in Box 9.2, where the modified Newton–Raphson global equilibrium iteration loop is shown in standard pseudo-code format.

Remark 9.5. It is clear from the above that the global Newton–Raphson procedure now searches for both nodal displacements and Gauss point transverse strains that satisfy equilibrium and the plane stress constraint simultaneously. Thus, in general, the plane stress condition will be attained only at (converged) equilibrium configurations. This is at variance with the procedures presented previously, in Section 9.2, where the plane stress constraint is always enforced at each Gauss point, whether or not the global equilibrium convergence criterion is satisfied.

Remark 9.6. The enforcement of the plane stress constraint at the structural level is compatible only with the *implicit* finite element scheme – the only scheme discussed in this book. Unlike the procedures discussed in Section 9.2 and the use of plane stress-projected evolution equations (addressed in Section 9.4), the present strategy cannot be used in conjunction with *explicit transient dynamic* finite element schemes. Explicit schemes do not incorporate equilibrium iterations.

9.4. Plane stress-projected plasticity models

Finally, we address the methodology of item (c) on page 360: the use of *plane stress-projected* plasticity models. Before proceeding, it is convenient to define precisely the meaning of *plane stress-projected* model. A plane stress-projected plasticity model is defined by a set of evolution equations that

(a) involve *only in-plane* stress and strain components and, at the same time;

(b) are equivalent to the three-dimensional model with the added plane stress constraint defined in Problem 9.1.

In this case, the *in-plane* stress/strain components are the *primary* variables which have to be determined directly by solving the rate evolution equations. The out-of-plane strains are *dependent* variables which can be calculated '*a posteriori*' as functions of the in-plane variables that solve the equations of the plane stress-projected model. The situation here is completely analogous to plane stress linear elasticity where expressions (9.3) or (9.4) define the *plane stress-projected* elastic constitutive equation and the out-of-plane strains are given in closed form by (9.8).

Remark 9.7. The use of plane-stress projected equations leads in general to more efficient computational procedures. This is due to the use of a reduced set of equations involving only the relevant in-plane components. However, plane stress-projected equations can only be formulated if the underlying three-dimensional model is sufficiently simple to allow closed-form relationships between in-plane and out-of-plane variables to be derived from the plane stress constraint,

$$\sigma_{13} = \sigma_{23} = \sigma_{33} = 0.$$

For complex models, it may not be easy or feasible to derive such relations. In such cases, we recommend the use of the methods discussed in Sections 9.2 and 9.3.

The methodology for derivation and implementation of plane stress-projected elastoplastic models is made clear in the next section where the plane stress-projected von Mises model, incorporating general nonlinear isotropic hardening, is derived in detail together with the corresponding integration algorithm and consistent tangent operator. Plane stress-projected equations for the von Mises model have been originally derived and used in a computational context by Simo and Taylor (1986) and Jetteur (1986). The model adopted by Simo and Taylor includes combined isotropic and Prager's linear kinematic hardening whereas the derivation presented by Jetteur is restricted to purely isotropic hardening. Another application of the projected equations concept is reported by Ramm and Matzenmiller (1987) in the context of large deformation shell analysis. The equations and algorithms derived by these authors are analogous to those presented here but account for transverse shear stresses and strains, which become important in the analysis of thick shells.

9.4.1. THE PLANE STRESS-PROJECTED VON MISES MODEL

The reader who wishes to skip the details of the derivation of the plane stress-projected von Mises model is referred directly to Box 9.3 (page 374) where the corresponding final equations are summarised. The (three-dimensional) von Mises model with isotropic strain hardening is defined by the following set of equations

$$\begin{aligned}
&\dot{\boldsymbol{\varepsilon}} = \dot{\boldsymbol{\varepsilon}}^e + \dot{\boldsymbol{\varepsilon}}^p \\
&\boldsymbol{\sigma} = \mathsf{D}^e : \boldsymbol{\varepsilon}^e \\
&\Phi = \sqrt{3 J_2(\boldsymbol{s})} - \sigma_y(\bar{\varepsilon}^p) \\
&\dot{\boldsymbol{\varepsilon}}^p = \dot{\gamma}\, \frac{\partial \Phi}{\partial \boldsymbol{\sigma}} = \dot{\gamma} \sqrt{\frac{3}{2}} \frac{\boldsymbol{s}}{\|\boldsymbol{s}\|} \\
&\dot{\bar{\varepsilon}}^p = \dot{\gamma} \\
&\dot{\gamma} \geq 0, \quad \Phi \leq 0, \quad \dot{\gamma}\, \Phi = 0.
\end{aligned} \tag{9.35}$$

Under plane stress, the above equations are complemented with the constraints

$$\sigma_{13}(\boldsymbol{\varepsilon}^e) = 0, \quad \sigma_{23}(\boldsymbol{\varepsilon}^e) = 0, \quad \sigma_{33}(\boldsymbol{\varepsilon}^e) = 0. \tag{9.36}$$

The plane stress-projected equations

To obtain the plane stress-projected von Mises equations, we firstly need to derive explicit expressions relating the elastic and plastic out-of-plane strain components to the in-plane components. We start by observing that, as the elastic behaviour is linear and isotropic, (9.36) is equivalent to the well-known relations

$$\varepsilon^e_{13} = 0, \quad \varepsilon^e_{23} = 0, \quad \varepsilon^e_{33} = -\frac{\nu}{1-\nu}(\varepsilon^e_{11} + \varepsilon^e_{22}). \tag{9.37}$$

In addition, the plastic incompressibility that follows from $(9.35)_4$ gives

$$\varepsilon^p_{33} = -(\varepsilon^p_{11} + \varepsilon^p_{22}). \tag{9.38}$$

As $s_{13} = s_{23} = 0$ under a plane stress state, it also follows from $(9.35)_4$ that

$$\varepsilon^p_{13} = \varepsilon^p_{23} = 0. \tag{9.39}$$

The above relations completely define the out-of-plane strain components as functions of the in-plane values. Thus, under plane stress, the history of the out-of-plane strain components is automatically prescribed as

$$
\begin{aligned}
\varepsilon_{13}(t) &= \varepsilon^e_{13}(t) + \varepsilon^p_{13}(t) \equiv 0 \\
\varepsilon_{23}(t) &= \varepsilon^e_{23}(t) + \varepsilon^p_{23}(t) \equiv 0 \\
\varepsilon_{33}(t) &= -\left[\frac{\nu}{1-\nu}(\varepsilon^e_{11}(t) + \varepsilon^e_{22}(t)) + \varepsilon^p_{11}(t) + \varepsilon^p_{22}(t)\right].
\end{aligned} \tag{9.40}
$$

Finally, with the history of the out-of-plane (elastic and plastic) strain components completely prescribed, we can now eliminate them from the original set of equations (9.35). This results in the *plane stress-projected* set of equations which, in component form, is given by

$$
\begin{cases}
\dot{\varepsilon}_{\alpha\beta} = \dot{\varepsilon}^e_{\alpha\beta} + \dot{\varepsilon}^p_{\alpha\beta} \\
\sigma_{\alpha\beta} = \bar{\mathsf{D}}^e_{\alpha\beta\gamma\delta}\, \varepsilon^e_{\gamma\delta} \\
\Phi = \sqrt{\tfrac{3}{2}[\sigma_{\alpha\beta}\,\sigma_{\alpha\beta} - \tfrac{1}{3}(\sigma_{\alpha\alpha})^2]} - \sigma_y(\bar{\varepsilon}^p) \\
\dot{\varepsilon}^p_{\alpha\beta} = \dot{\gamma}\sqrt{\dfrac{3}{2}}\,\dfrac{s_{\alpha\beta}}{\sqrt{\sigma_{\gamma\delta}\,\sigma_{\gamma\delta} - \frac{1}{3}(\sigma_{\gamma\gamma})^2}} \\
\dot{\bar{\varepsilon}}^p = \dot{\gamma} \\
\dot{\gamma} \geq 0, \quad \Phi \leq 0, \quad \dot{\gamma}\,\Phi = 0,
\end{cases} \tag{9.41}
$$

where summation on repeated indices is implied with Greek subscripts ranging between 1 and 2. In equation $(9.41)_2$, $\bar{\mathsf{D}}^e_{\alpha\beta\gamma\delta}$ denote the components of the plane stress elasticity tensor – the fourth-order counterpart of the matrix taking part in expressions (9.3, 9.4). In deriving $(9.41)_{3,4}$ we have made use of the identity

$$
\|\boldsymbol{s}\| = \sqrt{\sigma_{\alpha\beta}\,\sigma_{\alpha\beta} - \tfrac{1}{3}(\sigma_{\alpha\alpha})^2}, \tag{9.42}
$$

which holds under plane stress condition. This identity is obtained by introducing the plane stress constraint into the definition

$$
\|\boldsymbol{s}\| \equiv \sqrt{\boldsymbol{s}:\boldsymbol{s}}; \quad \boldsymbol{s} \equiv \boldsymbol{\sigma} - \tfrac{1}{3}\mathrm{tr}[\boldsymbol{\sigma}]\,\boldsymbol{I}. \tag{9.43}
$$

Matrix notation

A conveniently compact representation of the plane stress-projected von Mises equations can be obtained by making use of the matrix notation

$$
\begin{aligned}
&\boldsymbol{\sigma} = [\sigma_{11}\ \sigma_{22}\ \sigma_{12}]^T, \quad \mathbf{s} = [s_{11}\ s_{22}\ s_{12}]^T \\
&\boldsymbol{\varepsilon} = [\varepsilon_{11}\ \varepsilon_{22}\ 2\varepsilon_{12}]^T, \quad \boldsymbol{\varepsilon}^e = [\varepsilon^e_{11}\ \varepsilon^e_{22}\ 2\varepsilon^e_{12}]^T \\
&\boldsymbol{\varepsilon}^p = [\varepsilon^p_{11}\ \varepsilon^p_{22}\ 2\varepsilon^p_{12}]^T.
\end{aligned} \tag{9.44}
$$

Note that upright bold-face characters are used to denote the matrix representation. From the definition of the deviatoric stress tensor, the array s of plane deviator components can be expressed as

$$\mathbf{s} = \mathbf{P}\boldsymbol{\sigma}, \tag{9.45}$$

where $\mathbf{P}$ is the matrix defined by

$$\mathbf{P} \equiv \frac{1}{3}\begin{bmatrix} 2 & -1 & 0 \\ -1 & 2 & 0 \\ 0 & 0 & 6 \end{bmatrix}. \tag{9.46}$$

With the above matrix/vector notation at hand, the plane stress-projected von Mises equations (9.41) can be equivalently written in compact form as

$$\begin{cases} \dot{\boldsymbol{\varepsilon}} = \dot{\boldsymbol{\varepsilon}}^e + \dot{\boldsymbol{\varepsilon}}^p \\ \boldsymbol{\sigma} = \mathbf{D}^e\ \boldsymbol{\varepsilon}^e \\ \Phi = \sqrt{\tfrac{3}{2}\boldsymbol{\sigma}^T\mathbf{P}\boldsymbol{\sigma}} - \sigma_y(\bar{\varepsilon}^p) \\ \dot{\boldsymbol{\varepsilon}}^p = \dot{\gamma}\dfrac{\partial\Phi}{\partial\boldsymbol{\sigma}} = \dot{\gamma}\sqrt{\dfrac{3}{2}}\dfrac{\mathbf{P}\boldsymbol{\sigma}}{\sqrt{\boldsymbol{\sigma}^T\mathbf{P}\boldsymbol{\sigma}}} \\ \dot{\bar{\varepsilon}}^p = \dot{\gamma} \\ \dot{\gamma} \geq 0, \quad \Phi \leq 0, \quad \dot{\gamma}\,\Phi = 0, \end{cases} \tag{9.47}$$

where $\mathbf{D}^e$ is the plane stress elasticity matrix. For the derivation of the integration algorithm to be described in the next section, it is more convenient to use the squared form of the yield function which, in the present case, can be handled more easily. Then, in the set (9.47) of the evolution equations, we replace $(9.47)_3$, $(9.47)_4$ and $(9.47)_5$, respectively, with

$$\begin{aligned} \Phi &= \tfrac{1}{2}\boldsymbol{\sigma}^T\mathbf{P}\boldsymbol{\sigma} - \tfrac{1}{3}\sigma_y^2(\bar{\varepsilon}^p) \\ \dot{\boldsymbol{\varepsilon}}^p &= \dot{\gamma}\frac{\partial\Phi}{\partial\boldsymbol{\sigma}} = \dot{\gamma}\,\mathbf{P}\boldsymbol{\sigma} \\ \dot{\bar{\varepsilon}}^p &= \dot{\gamma}\sqrt{\tfrac{2}{3}\boldsymbol{\sigma}^T\mathbf{P}\boldsymbol{\sigma}}. \end{aligned} \tag{9.48}$$

The complete set of equations of the plane stress-projected von Mises model with isotropic hardening is conveniently grouped in Box 9.3. The corresponding implicit elastic predictor/return-mapping algorithm is derived subsequently.

9.4.2. THE PLANE STRESS-PROJECTED INTEGRATION ALGORITHM

The implicit integration algorithm for the plane stress-projected von Mises model is completely analogous to that derived in Section 7.3.2 (from page 217). Within the framework of general elastic predictor/return-mapping schemes discussed in Chapter 7, and having $[t_n, t_{n+1}]$ as the underlying (pseudo-)time interval, we start by computing the elastic

Box 9.3. Plane stress-projected von Mises model with isotropic hardening.

1. Elastoplastic strain split

$$\dot{\boldsymbol{\varepsilon}} = \dot{\boldsymbol{\varepsilon}}^e + \dot{\boldsymbol{\varepsilon}}^p$$

$$\boldsymbol{\varepsilon} \equiv [\varepsilon_{11}\ \varepsilon_{22}\ 2\varepsilon_{12}]^T, \quad \boldsymbol{\varepsilon}^e \equiv [\varepsilon^e_{11}\ \varepsilon^e_{22}\ 2\varepsilon^e_{12}]^T, \quad \boldsymbol{\varepsilon}^p \equiv [\varepsilon^p_{11}\ \varepsilon^p_{22}\ 2\varepsilon^p_{12}]^T$$

2. Elastic law

$$\boldsymbol{\sigma} = \mathbf{D}^e\ \boldsymbol{\varepsilon}^e$$

$$\boldsymbol{\sigma} \equiv [\sigma_{11}\ \sigma_{22}\ \sigma_{12}]^T, \quad \mathbf{D}^e \equiv \frac{E}{1-\nu^2}\begin{bmatrix} 1 & \nu & 0 \\ \nu & 1 & 0 \\ 0 & 0 & \dfrac{1-\nu}{2} \end{bmatrix}$$

3. Yield function definition

$$\Phi = \tfrac{1}{2}\boldsymbol{\sigma}^T\mathbf{P}\boldsymbol{\sigma} - \tfrac{1}{3}\sigma_y^2(\bar{\varepsilon}^p); \quad \mathbf{P} \equiv \tfrac{1}{3}\begin{bmatrix} 2 & -1 & 0 \\ -1 & 2 & 0 \\ 0 & 0 & 6 \end{bmatrix}$$

4. Plastic flow rule

$$\dot{\boldsymbol{\varepsilon}}^p = \dot{\gamma}\frac{\partial \Phi}{\partial \boldsymbol{\sigma}} = \dot{\gamma}\,\mathbf{P}\boldsymbol{\sigma}$$

5. Hardening variable evolution

$$\dot{\bar{\varepsilon}}^p = \dot{\gamma}\sqrt{\tfrac{2}{3}\boldsymbol{\sigma}^T\mathbf{P}\boldsymbol{\sigma}}$$

6. Loading/unloading criterion

$$\dot{\gamma} \geq 0, \quad \Phi \leq 0, \quad \dot{\gamma}\,\Phi = 0$$

predictor state

$$\begin{aligned} \boldsymbol{\varepsilon}^{e\ \mathrm{trial}}_{n+1} &= \boldsymbol{\varepsilon}^e_n + \Delta\boldsymbol{\varepsilon} \\ \boldsymbol{\sigma}^{\mathrm{trial}}_{n+1} &= \mathbf{D}^e\ \boldsymbol{\varepsilon}^{e\ \mathrm{trial}}_{n+1} \\ \bar{\varepsilon}^{p\ \mathrm{trial}}_{n+1} &= \bar{\varepsilon}^p_n, \end{aligned} \tag{9.49}$$

where $\Delta\boldsymbol{\varepsilon}$ is the given (in-plane) strain increment associated with the interval $[t_n, t_{n+1}]$

$$\Delta\boldsymbol{\varepsilon} = \boldsymbol{\varepsilon}^e_{n+1} - \boldsymbol{\varepsilon}^e_n. \tag{9.50}$$

The next step is to check for plastic admissibility of the elastic trial state. We then compute

$$\Phi^{\mathrm{trial}} = \tfrac{1}{2}(\boldsymbol{\sigma}^{\mathrm{trial}}_{n+1})^T\mathbf{P}\boldsymbol{\sigma}^{\mathrm{trial}}_{n+1} - \tfrac{1}{3}\sigma_y^2(\bar{\varepsilon}^{p\ \mathrm{trial}}_{n+1}). \tag{9.51}$$

If the elastic trial state is admissible, i.e. if

$$\Phi^{\mathrm{trial}} \leq 0, \tag{9.52}$$

then the process is elastic within $[t_n, t_{n+1}]$ and we update

$$(\cdot)_{n+1} = (\cdot)^{\text{trial}}_{n+1}. \tag{9.53}$$

Otherwise, the return-mapping algorithm needs to be applied as described in the following.

The plane stress-projected return mapping

The implicit return mapping in the present case consists in solving the following system of algebraic equations

$$\begin{aligned}
\boldsymbol{\varepsilon}^e_{n+1} &= \boldsymbol{\varepsilon}^{e\,\text{trial}}_{n+1} - \Delta\gamma\,\mathbf{P}\boldsymbol{\sigma}_{n+1} \\
\bar{\varepsilon}^p_{n+1} &= \bar{\varepsilon}^p_n + \Delta\gamma\sqrt{\tfrac{2}{3}(\boldsymbol{\sigma}_{n+1})^T\mathbf{P}\boldsymbol{\sigma}_{n+1}} \\
\tfrac{1}{2}\boldsymbol{\sigma}^T_{n+1}\mathbf{P}\boldsymbol{\sigma}_{n+1} &- \tfrac{1}{3}\sigma^2_y(\bar{\varepsilon}^p_{n+1}) = 0,
\end{aligned} \tag{9.54}$$

for $\boldsymbol{\varepsilon}^e_{n+1}$, $\bar{\varepsilon}^p_{n+1}$ and $\Delta\gamma$ where $\boldsymbol{\sigma}_{n+1}$ is a function of $\boldsymbol{\varepsilon}^e_{n+1}$ defined by the elastic law.

As in the three-dimensional case discussed in Section 7.3.2 (page 217), it is also possible to reduce the number of unknowns of the plane stress-projected return-mapping equations for the von Mises model. The five-variable return-mapping system (9.54) can be reduced to a *single* scalar nonlinear equation having the incremental plastic multiplier as the unknown. To this end, we substitute $(9.54)_2$ into $(9.54)_3$ and rearrange $(9.54)_1$ using the inverse elastic law. The original return-mapping system reduces to

$$\begin{aligned}
&\boldsymbol{\sigma}_{n+1} = [\mathbf{C} + \Delta\gamma\,\mathbf{P}]^{-1}\mathbf{C}\,\boldsymbol{\sigma}^{\text{trial}}_{n+1} \\
&\tfrac{1}{2}\boldsymbol{\sigma}^T_{n+1}\mathbf{P}\boldsymbol{\sigma}_{n+1} - \tfrac{1}{3}\sigma^2_y\Big(\bar{\varepsilon}^p_n + \Delta\gamma\sqrt{\tfrac{2}{3}(\boldsymbol{\sigma}_{n+1})^T\mathbf{P}\boldsymbol{\sigma}_{n+1}}\,\Big) = 0,
\end{aligned} \tag{9.55}$$

where the unknowns now are the stress array, $\boldsymbol{\sigma}_{n+1}$, and the plastic multiplier, $\Delta\gamma$, and $\mathbf{C}$ is the inverse of the elastic matrix

$$\mathbf{C} \equiv (\mathbf{D}^e)^{-1}. \tag{9.56}$$

Finally, by substituting $(9.55)_1$ into the consistency condition $(9.55)_2$, the return mapping for the plane stress von Mises model is reduced to the following *scalar* nonlinear equation having $\Delta\gamma$ as the only unknown

$$\tilde{\Phi}(\Delta\gamma) \equiv \tfrac{1}{2}\xi(\Delta\gamma) - \tfrac{1}{3}\sigma^2_y\Big(\bar{\varepsilon}^p_n + \Delta\gamma\sqrt{\tfrac{2}{3}\xi(\Delta\gamma)}\,\Big) = 0. \tag{9.57}$$

Here we have conveniently defined

$$\xi(\Delta\gamma) \equiv (\boldsymbol{\sigma}^{\text{trial}}_{n+1})^T\mathbf{A}^T(\Delta\gamma)\,\mathbf{P}\,\mathbf{A}(\Delta\gamma)\boldsymbol{\sigma}^{\text{trial}}_{n+1}, \tag{9.58}$$

with

$$\mathbf{A}(\Delta\gamma) \equiv [\mathbf{C} + \Delta\gamma\,\mathbf{P}]^{-1}\mathbf{C}. \tag{9.59}$$

Thus, the return mapping for the plane stress von Mises model is carried out as follows. Firstly, we solve the consistency equation (9.57) using the Newton–Raphson algorithm. With the solution $\Delta\gamma$ at hand, we update the other variables

$$\begin{aligned}\boldsymbol{\sigma}_{n+1} &= \mathbf{A}(\Delta\gamma)\,\boldsymbol{\sigma}^{\text{trial}}_{n+1}\\ \boldsymbol{\varepsilon}^{e}_{n+1} &= \mathbf{C}\,\boldsymbol{\sigma}_{n+1}\\ \bar{\varepsilon}^{p}_{n+1} &= \bar{\varepsilon}^{p}_{n} + \Delta\gamma\sqrt{\tfrac{2}{3}\xi(\Delta\gamma)}.\end{aligned} \tag{9.60}$$

If required, the in-plane plastic strains are updated by

$$\boldsymbol{\varepsilon}^{p}_{n+1} = \boldsymbol{\varepsilon}^{p}_{n} + \Delta\gamma\,\mathbf{P}\boldsymbol{\sigma}_{n+1}. \tag{9.61}$$

Remark 9.8. The plane stress-projected return-mapping equation (9.57) is *always nonlinear* in $\Delta\gamma$, regardless of the prescribed hardening function σ_y. This is in contrast with its three-dimensional counterpart (see equation (7.91), page 219, and the comments made at the beginning of Section 7.3.4, page 223) whose only source of nonlinearity is the hardening curve and which becomes linear for perfectly plastic and linearly hardening models.

The plane stress-projected elastic predictor/return-mapping algorithm for the von Mises model with isotropic hardening is summarised in Boxes 9.4 and 9.5 in pseudo-code format. In program `HYPLAS`, the procedure is implemented in subroutine `SUVMPS` (**S**tate **U**pdate procedure for the **V**on **M**ises model in **P**lane **S**tress), whose FORTRAN code is shown in Section 9.4.3. The solution of the plane stress-projected consistency equation (9.57) is undertaken by the Newton–Raphson algorithm. The Newton–Raphson procedure is shown in Box 9.5. Its implementation follows the simplification obtained below by exploiting the properties of matrices $\mathbf{P}$ and $\mathbf{D}^e$.

Newton–Raphson return-mapping solution

The reader might have realised already that the Newton–Raphson algorithm for solution of the consistency equation (9.57) is potentially cumbersome. This is because the scalar function ξ defined in (9.58) depends on a matrix $\mathbf{A}(\Delta\gamma)$ which, in turn, is a function whose definition (9.59) involves the sum, inversion and multiplication of other matrices. Fortunately, in the present case where the elastic behaviour is *isotropic*, the explicit expression for (9.57) can be significantly simplified. The key point to be observed is the fact that matrices $\mathbf{P}$ and $\mathbf{D}^e$ share the same eigenvectors so that they both have diagonal representation on the same basis (as do $\mathbf{C}$ and $\mathbf{A}$). By applying the orthogonal transformation

$$\mathbf{Q} = \begin{bmatrix} \dfrac{1}{\sqrt{2}} & \dfrac{1}{\sqrt{2}} & 0\\ -\dfrac{1}{\sqrt{2}} & \dfrac{1}{\sqrt{2}} & 0\\ 0 & 0 & 1\end{bmatrix} \tag{9.62}$$

Box 9.4. Implicit elastic predictor/return-mapping algorithm. Plane stress-projected von Mises model with nonlinear isotropic hardening.

HYPLAS procedure:	SUVMPS

(i) Elastic predictor. Given the in-plane incremental strains, $\Delta\varepsilon$, and the state variables at t_n, evaluate the *elastic trial state*

$$\boldsymbol{\varepsilon}^{e\,\text{trial}}_{n+1} := \boldsymbol{\varepsilon}^{e}_{n} + \Delta\boldsymbol{\varepsilon}$$

$$\bar{\varepsilon}^{p\,\text{trial}}_{n+1} := \bar{\varepsilon}^{p}_{n}$$

$$\boldsymbol{\sigma}^{\text{trial}}_{n+1} := \mathbf{D}^{e}\,\boldsymbol{\varepsilon}^{e\,\text{trial}}_{n+1}$$

(ii) Check plastic admissibility

$$a_1 := (\sigma^{\text{trial}}_{11} + \sigma^{\text{trial}}_{22})^2; \quad a_2 := (\sigma^{\text{trial}}_{22} - \sigma^{\text{trial}}_{11})^2; \quad a_3 := (\sigma^{\text{trial}}_{12})^2$$

$$\xi^{\text{trial}} := \tfrac{1}{6}a_1 + \tfrac{1}{2}a_2 + 2a_3$$

$$\text{IF} \quad \Phi^{\text{trial}} := \tfrac{1}{2}\xi^{\text{trial}} - \tfrac{1}{3}\sigma_y^2(\bar{\varepsilon}^{p\,\text{trial}}_{n+1}) \le 0$$

$$\text{THEN set} \quad (\cdot)_{n+1} := (\cdot)^{\text{trial}}_{n+1} \quad \text{and EXIT}$$

(iii) Return mapping. Solve the nonlinear equation

$$\tilde{\Phi}(\Delta\gamma) = 0$$

for $\Delta\gamma$ using the Newton–Raphson method – GOTO Box 9.5 – and update the state variables

$$\boldsymbol{\sigma}_{n+1} := \mathbf{A}(\Delta\gamma)\,\boldsymbol{\sigma}^{\text{trial}}_{n+1}$$

$$\boldsymbol{\varepsilon}^{e}_{n+1} := \mathbf{C}\,\boldsymbol{\sigma}_{n+1}$$

$$\bar{\varepsilon}^{p}_{n+1} := \bar{\varepsilon}^{p}_{n} + \Delta\gamma\sqrt{\tfrac{2}{3}\xi(\Delta\gamma)}$$

where $\mathbf{C} \equiv [\mathbf{D}^e]^{-1}$ and matrix $\mathbf{A}(\Delta\gamma)$ is given in (9.69)

(iv) EXIT

to $\mathbf{P}$ and $\mathbf{D}^e$, we obtain the diagonal representations

$$\mathbf{P}^* \equiv \mathbf{Q}\mathbf{P}\mathbf{Q}^T = \begin{bmatrix} \frac{1}{3} & 0 & 0 \\ 0 & 1 & 0 \\ 0 & 0 & 2 \end{bmatrix} \tag{9.63}$$

and

$$\mathbf{D}^{e*} \equiv \mathbf{Q}\mathbf{D}^e\mathbf{Q}^T = \begin{bmatrix} \dfrac{E}{1-\nu} & 0 & 0 \\ 0 & 2G & 0 \\ 0 & 0 & G \end{bmatrix}. \tag{9.64}$$

On the same basis, the matrix $\mathbf{A}(\Delta\gamma)$ has the diagonal representation

$$\mathbf{A}^*(\Delta\gamma) \equiv [\mathbf{C}^* + \Delta\gamma\mathbf{P}^*]^{-1}\mathbf{C}^*$$
$$= \begin{bmatrix} \dfrac{3(1-\nu)}{3(1-\nu)+E\Delta\gamma} & 0 & 0 \\ 0 & \dfrac{1}{1+2G\,\Delta\gamma} & 0 \\ 0 & 0 & \dfrac{1}{1+2G\,\Delta\gamma} \end{bmatrix}, \tag{9.65}$$

where $\mathbf{C}^* = [\mathbf{D}^{e*}]^{-1}$. The corresponding representation of the elastic trial stress array reads

$$\boldsymbol{\sigma}_{n+1}^{\text{trial}\,*} \equiv \mathbf{Q}\,\boldsymbol{\sigma}_{n+1}^{\text{trial}} = \begin{bmatrix} \frac{1}{\sqrt{2}}(\sigma_{11}^{\text{trial}} + \sigma_{22}^{\text{trial}}) \\ \frac{1}{\sqrt{2}}(\sigma_{22}^{\text{trial}} - \sigma_{11}^{\text{trial}}) \\ \sigma_{12}^{\text{trial}} \end{bmatrix}. \tag{9.66}$$

With the above transformed variables, expression (9.58) can be written in the simpler form

$$\begin{aligned}\xi(\Delta\gamma) &\equiv (\boldsymbol{\sigma}_{n+1}^{\text{trial}})^T\,\mathbf{A}^T(\Delta\gamma)\,\mathbf{P}\,\mathbf{A}(\Delta\gamma)\boldsymbol{\sigma}_{n+1}^{\text{trial}} \\ &= (\boldsymbol{\sigma}_{n+1}^{\text{trial}\,*})^T[\mathbf{A}^*(\Delta\gamma)]^2\,\mathbf{P}^*\,\boldsymbol{\sigma}_{n+1}^{\text{trial}\,*} \\ &= \frac{(\sigma_{11}^{\text{trial}} + \sigma_{22}^{\text{trial}})^2}{6\big[1+\frac{E\,\Delta\gamma}{3(1-\nu)}\big]^2} + \frac{\frac{1}{2}(\sigma_{22}^{\text{trial}} - \sigma_{11}^{\text{trial}})^2 + 2(\sigma_{12}^{\text{trial}})^2}{(1+2G\,\Delta\gamma)^2}.\end{aligned} \tag{9.67}$$

The pseudo-code of the Newton–Raphson procedure for solution of the return-mapping equation (9.57), using the above expression for ξ, is shown in Box 9.5.

Note that matrix $\mathbf{A}(\Delta\gamma)$, which takes part in the stress update formula $(9.60)_1$, can be written as

$$\mathbf{A}(\Delta\gamma) = \mathbf{Q}^T\,\mathbf{A}^*(\Delta\gamma)\,\mathbf{Q}. \tag{9.68}$$

Thus, in view of (9.62) and (9.65), it can be expressed in the simple explicit form

$$\mathbf{A}(\Delta\gamma) = \begin{bmatrix} \frac{1}{2}(A_{11}^* + A_{22}^*) & \frac{1}{2}(A_{11}^* - A_{22}^*) & 0 \\ \frac{1}{2}(A_{11}^* - A_{22}^*) & \frac{1}{2}(A_{11}^* + A_{22}^*) & 0 \\ 0 & 0 & A_{33}^* \end{bmatrix} \tag{9.69}$$

with

$$A_{11}^* = \frac{3(1-\nu)}{3(1-\nu)+E\,\Delta\gamma}, \quad A_{22}^* = \frac{1}{1+2G\,\Delta\gamma}, \quad A_{33}^* = A_{22}^*. \tag{9.70}$$

9.4.3. SUBROUTINE SUVMPS

The above plane stress-projected elastic predictor/return-mapping algorithm is implemented in subroutine SUVMPS of HYPLAS. The list of arguments of SUVMPS is identical to that of its general (plane strain/axisymmetric) counterpart described in Section 7.3.5 (page 224). We refer to that section for the description of the arguments. Note that the same approach to

Box 9.5. Newton–Raphson algorithm for solution of the return-mapping equation of the plane stress-projected von Mises model.

HYPLAS procedure:	SUVMPS

(i) Set initial guess for $\Delta\gamma$

$$\Delta\gamma := 0$$

and corresponding residual, $\tilde{\Phi}$

$$\xi := \tfrac{1}{6}(\sigma_{11}^{\rm trial} + \sigma_{22}^{\rm trial})^2 + \tfrac{1}{2}(\sigma_{22}^{\rm trial} - \sigma_{11}^{\rm trial})^2 + 2(\sigma_{12}^{\rm trial})^2$$

$$\tilde{\Phi} := \tfrac{1}{2}\xi - \tfrac{1}{3}\sigma_y^2(\bar{\varepsilon}_n^p)$$

(ii) Perform Newton–Raphson iteration

$$H := \left.\frac{\mathrm{d}\sigma_y}{\mathrm{d}\bar{\varepsilon}^p}\right|_{\bar{\varepsilon}_n^p + \Delta\gamma\sqrt{2\xi/3}} \qquad \text{(hardening slope)}$$

$$\xi' := -\frac{(\sigma_{11}^{\rm trial} + \sigma_{22}^{\rm trial})^2}{9\big[1 + \frac{E\,\Delta\gamma}{3(1-\nu)}\big]^3}\,\frac{E}{(1-\nu)} - 2G\frac{(\sigma_{22}^{\rm trial} - \sigma_{11}^{\rm trial})^2 + 4(\sigma_{12}^{\rm trial})^2}{(1+2G\,\Delta\gamma)^3}$$

$$\bar{H} := 2\,\sigma_y\bigg(\bar{\varepsilon}_n^p + \Delta\gamma\sqrt{\frac{2}{3}\xi}\bigg)\,H\sqrt{\frac{2}{3}}\bigg(\sqrt{\xi} + \frac{\Delta\gamma\,\xi'}{2\sqrt{\xi}}\bigg)$$

$$\tilde{\Phi}' := \tfrac{1}{2}\xi' - \tfrac{1}{3}\bar{H} \qquad \text{(residual derivative)}$$

$$\Delta\gamma := \Delta\gamma - \frac{\tilde{\Phi}}{\tilde{\Phi}'} \qquad \text{(new guess for } \Delta\gamma)$$

(iii) Check for convergence

$$\xi := \frac{(\sigma_{11}^{\rm trial} + \sigma_{22}^{\rm trial})^2}{6\big[1 + \frac{E\,\Delta\gamma}{3(1-\nu)}\big]^2} + \frac{\frac{1}{2}(\sigma_{22}^{\rm trial} - \sigma_{11}^{\rm trial})^2 + 2(\sigma_{12}^{\rm trial})^2}{(1+2G\,\Delta\gamma)^2}$$

$$\tilde{\Phi} := \tfrac{1}{2}\xi - \tfrac{1}{3}\sigma_y^2\bigg(\bar{\varepsilon}_n^p + \Delta\gamma\sqrt{\tfrac{2}{3}\xi}\bigg)$$

IF $|\tilde{\Phi}| \le \epsilon_{\rm tol}$ THEN RETURN to Box 9.4

(iv) GOTO (ii)

the description of nonlinear isotropic hardening, which uses a piecewise linear hardening curve, is adopted here. Again, the symbolic names of most variables resemble those of the corresponding name given in the text. This should allow readers easily to correlate the FORTRAN instructions with the corresponding expressions of Boxes 9.4 and 9.5, where the plane stress integration algorithm is summarised. The complete list of the FORTRAN source code of SUVMPS is given below.

```
      SUBROUTINE SUVMPS
     1(   DGAMA      ,IPROPS     ,LALGVA     ,NTYPE      ,RPROPS     ,
     2    RSTAVA     ,STRAT      ,STRES      )
      IMPLICIT DOUBLE PRECISION (A-H,O-Z)
      PARAMETER( IPHARD=4  ,MSTRE=4  ,NSTRE=3 )
      LOGICAL IFPLAS, LALGVA(2), SUFAIL
      DIMENSION
     1    IPROPS(*)          ,RPROPS(*)          ,RSTAVA(MSTRE+1)    ,
     2    STRAT(MSTRE)       ,STRES(MSTRE)
      DIMENSION
     1    EET(MSTRE)         ,STREST(NSTRE)
      DATA
     1    R0   ,RP5  ,R1   ,R2   ,R3   ,R4   ,R6   ,TOL   /
     2    0.0D0,0.5D0,1.0D0,2.0D0,3.0D0,4.0D0,6.0D0,1.D-08/
      DATA MXITER / 50 /
C***********************************************************************
C STATE UPDATE PROCEDURE FOR THE VON MISES ELASTO-PLASTIC MODEL WITH
C NON-LINEAR (PIECEWISE LINEAR) ISOTROPIC HARDENING IN PLANE STRESS:
C IMPLICIT PLANE STRESS-PROJECTED ELASTIC PREDICTOR/RETURN MAPPING
C ALGORITHM (BOXES 9.4-5).
C***********************************************************************
C Stop program if not plane stress
      IF(NTYPE.NE.1)CALL ERRPRT('EI0031')
C Initialise some algorithmic and internal variables
      DGAMA=R0
      IFPLAS=.FALSE.
      SUFAIL=.FALSE.
C...set previously (equilibrium) converged accumulated plastic strain
      EPBARN=RSTAVA(MSTRE+1)
C Set some material properties
      YOUNG=RPROPS(2)
      POISS=RPROPS(3)
      NHARD=IPROPS(3)
C Shear and bulk moduli and other necessary constants
      GMODU=YOUNG/(R2*(R1+POISS))
      BULK=YOUNG/(R3*(R1-R2*POISS))
      R2G=R2*GMODU
      R4G=R4*GMODU
      R1D3=R1/R3
      R1D6=R1/R6
      R2D3=R2*R1D3
      SQR2D3=SQRT(R2D3)
      R4GD3=R4G*R1D3
C Elastic predictor: Compute elastic trial state
C ----------------------------------------------
C Volumetric strain
      FACTOR=R2G/(BULK+R4GD3)
      EEV=(STRAT(1)+STRAT(2))*FACTOR
C Elastic trial deviatoric strain
      EEVD3=EEV/R3
      EET(1)=STRAT(1)-EEVD3
      EET(2)=STRAT(2)-EEVD3
C Convert engineering shear component into physical component
      EET(3)=STRAT(3)*RP5
C Elastic trial stress components
      PT=BULK*EEV
      STREST(1)=R2G*EET(1)+PT
      STREST(2)=R2G*EET(2)+PT
      STREST(3)=R2G*EET(3)
C Compute yield function value at trial state
      A1=(STREST(1)+STREST(2))*(STREST(1)+STREST(2))
```

```
      A2=(STREST(2)-STREST(1))*(STREST(2)-STREST(1))
      A3=STREST(3)*STREST(3)
      XI=R1D6*A1+RP5*A2+R2*A3
      SIGMAY=PLFUN(EPBARN,NHARD,RPROPS(IPHARD))
C...yield function
      PHI=RP5*XI-R1D3*SIGMAY*SIGMAY
C Check for plastic admissibility
C ------------------------------
      IF(PHI/SIGMAY.GT.TOL)THEN
C Plastic step: Apply return mapping - use Newton-Raphson algorithm
C               to solve the plane stress-projected return mapping
C               equation for the plastic multiplier (Box 9.5)
C -----------------------------------------------------------------
        IFPLAS=.TRUE.
        EPBAR=EPBARN
        SQRTXI=SQRT(XI)
        B1=R1
        B2=R1
        FMODU=YOUNG/(R3*(R1-POISS))
        DO 10 NRITER=1,MXITER
C Compute residual derivative
          HSLOPE=DPLFUN(EPBAR,NHARD,RPROPS(IPHARD))
          DXI=-A1*FMODU/(R3*B1*B1*B1)-R2G*(A2+R4*A3)/(B2*B2*B2)
          HBAR=R2*SIGMAY*HSLOPE*SQR2D3*(SQRTXI+DGAMA*DXI/(R2*SQRTXI))
          DPHI=RP5*DXI-R1D3*HBAR
C Compute Newton-Raphson increment and update equation variable DGAMA
          DGAMA=DGAMA-PHI/DPHI
C Compute new residual (yield function value)
          B1=R1+FMODU*DGAMA
          B2=R1+R2G*DGAMA
          XI=R1D6*A1/(B1*B1)+(RP5*A2+R2*A3)/(B2*B2)
          SQRTXI=SQRT(XI)
          EPBAR=EPBARN+DGAMA*SQR2D3*SQRTXI
          SIGMAY=PLFUN(EPBAR,NHARD,RPROPS(IPHARD))
          PHI=RP5*XI-R1D3*SIGMAY*SIGMAY
C Check for convergence
          RESNOR=ABS(PHI/SIGMAY)
          IF(RESNOR.LE.TOL)THEN
C update accumulated plastic strain
            RSTAVA(MSTRE+1)=EPBAR
C update stress components:   sigma := A sigma^trial
            ASTAR1=R3*(R1-POISS)/(R3*(R1-POISS)+YOUNG*DGAMA)
            ASTAR2=R1/(R1+R2G*DGAMA)
            A11=RP5*(ASTAR1+ASTAR2)
            A22=A11
            A12=RP5*(ASTAR1-ASTAR2)
            A21=A12
            A33=ASTAR2
            STRES(1)=A11*STREST(1)+A12*STREST(2)
            STRES(2)=A21*STREST(1)+A22*STREST(2)
            STRES(3)=A33*STREST(3)
C compute corresponding elastic (engineering) strain components
            FACTG=R1/R2G
            P=R1D3*(STRES(1)+STRES(2))
            EEV=P/BULK
            EEVD3=R1D3*EEV
            RSTAVA(1)=FACTG*(R2D3*STRES(1)-R1D3*STRES(2))+EEVD3
            RSTAVA(2)=FACTG*(R2D3*STRES(2)-R1D3*STRES(1))+EEVD3
            RSTAVA(3)=FACTG*STRES(3)*R2
            RSTAVA(4)=-POISS/(R1-POISS)*(RSTAVA(1)+RSTAVA(2))
            GOTO 999
```

```
          ENDIF
   10   CONTINUE
C reset failure flag and print warning message if N-R algorithm fails
        SUFAIL=.TRUE.
        CALL ERRPRT('WE0013')
      ELSE
C Elastic step: Update stress using linear elastic law
C ----------------------------------------------------
        STRES(1)=STREST(1)
        STRES(2)=STREST(2)
        STRES(3)=STREST(3)
C elastic engineering strain
        RSTAVA(1)=STRAT(1)
        RSTAVA(2)=STRAT(2)
        RSTAVA(3)=STRAT(3)
        RSTAVA(4)=-POISS/(R1-POISS)*(STRAT(1)+STRAT(2))
      ENDIF
  999 CONTINUE
C Update some algorithmic variables before exit
      LALGVA(1)=IFPLAS
      LALGVA(2)=SUFAIL
      RETURN
      END
```

9.4.4. THE ELASTOPLASTIC CONSISTENT TANGENT OPERATOR

With the matrix notation adopted throughout the preceding sections, the elastoplastic consistent tangent matrix is defined as

$$\mathbf{D}^{ep} \equiv \frac{\mathrm{d}\boldsymbol{\sigma}_{n+1}}{\mathrm{d}\boldsymbol{\varepsilon}_{n+1}} = \frac{\mathrm{d}\boldsymbol{\sigma}_{n+1}}{\mathrm{d}\boldsymbol{\varepsilon}^{\mathrm{trial}}_{n+1}}, \tag{9.71}$$

where $\boldsymbol{\sigma}_{n+1}$ is the outcome of the plane stress-projected return-mapping algorithm. The explicit expression for $\mathbf{D}^{ep}$ is derived analogously to its three-dimensional counterpart discussed in Section 7.4.2 (page 232). We start by differentiating $(9.54)_1$ which, together with the elastic law, gives

$$\mathrm{d}\boldsymbol{\sigma}_{n+1} = \mathbf{E}[\mathrm{d}\boldsymbol{\varepsilon}^{\mathrm{trial}}_{n+1} - \mathrm{d}\Delta\gamma\, \mathbf{P}\boldsymbol{\sigma}_{n+1}], \tag{9.72}$$

where we have defined

$$\mathbf{E} \equiv [\mathbf{C} + \Delta\gamma \mathbf{P}]^{-1}. \tag{9.73}$$

Differentiation of the plastic consistency (9.57) yields the equation

$$\begin{aligned} \mathrm{d}\tilde{\Phi} &= \tfrac{1}{2}\,\mathrm{d}\xi - \tfrac{2}{3}\,\sigma_y\, H \sqrt{\frac{2}{3}}\left(\mathrm{d}\Delta\gamma\,\sqrt{\xi} + \frac{\Delta\gamma}{2\sqrt{\xi}}\,\mathrm{d}\xi\right) \\ &= \tfrac{1}{2}\,\mathrm{d}\xi - \tfrac{2}{3}\,H\,(\xi\,\mathrm{d}\Delta\gamma + \tfrac{1}{2}\Delta\gamma\,\mathrm{d}\xi) = 0. \end{aligned} \tag{9.74}$$

In the above, we have used the identity $\sigma_y = \sqrt{\frac{3}{2}\xi}$, which holds under plastic flow. From the above equation, it immediately follows that

$$\mathrm{d}\Delta\gamma = \frac{3}{4H\xi}(1 - \tfrac{2}{3}\,H\,\Delta\gamma)\,\mathrm{d}\xi. \tag{9.75}$$

By using the elastic law and the above definition of matrix $\mathbf{E}$, the scalar ξ defined by (9.58) can be equivalently written as

$$\xi = (\boldsymbol{\varepsilon}^{\text{trial}}_{n+1})^T \mathbf{E}\,\mathbf{P}\,\mathbf{E}\;\boldsymbol{\varepsilon}^{\text{trial}}_{n+1}. \tag{9.76}$$

Straightforward differentiation of the above expression gives:**

$$\begin{aligned} \mathrm{d}\xi &= 2(\boldsymbol{\varepsilon}^{\text{trial}}_{n+1})^T\;\mathbf{E}\,\mathbf{P}\,\mathbf{E}\,\mathrm{d}\boldsymbol{\varepsilon}^{\text{trial}}_{n+1} + 2(\boldsymbol{\varepsilon}^{\text{trial}}_{n+1})^T\;\mathrm{d}\mathbf{E}\,\mathbf{P}\,\mathbf{E}\;\boldsymbol{\varepsilon}^{\text{trial}}_{n+1} \\ &= 2(\boldsymbol{\sigma}^T_{n+1}\mathbf{P}\,\mathbf{E}\,\mathrm{d}\boldsymbol{\varepsilon}^{\text{trial}}_{n+1} - \boldsymbol{\sigma}^T_{n+1}\mathbf{P}\,\mathbf{E}\,\mathbf{P}\;\boldsymbol{\sigma}_{n+1}\,\mathrm{d}\Delta\gamma). \end{aligned} \tag{9.77}$$

Substitution of the above formula into (9.75) followed by the substitution of the resulting expression into (9.72) and straightforward algebraic manipulations gives

$$\mathrm{d}\boldsymbol{\sigma}_{n+1} = [\mathbf{E} - \alpha\,(\mathbf{E}\mathbf{P}\boldsymbol{\sigma}_{n+1}) \otimes (\mathbf{E}\mathbf{P}\boldsymbol{\sigma}_{n+1})]\;\mathrm{d}\boldsymbol{\varepsilon}^{\text{trial}}_{n+1}, \tag{9.78}$$

where the scalar α is defined as

$$\alpha = \frac{1}{\boldsymbol{\sigma}^T_{n+1}\mathbf{P}\mathbf{E}\mathbf{P}\boldsymbol{\sigma}_{n+1} + \frac{2\xi H}{3-2H\,\Delta\gamma}}. \tag{9.79}$$

The explicit expression for the matrix form of the elastoplastic tangent operator consistent with the von Mises plane stress-projected return mapping is then given by

$$\mathbf{D}^{ep} = \mathbf{E} - \alpha\,(\mathbf{E}\mathbf{P}\boldsymbol{\sigma}_{n+1}) \otimes (\mathbf{E}\mathbf{P}\boldsymbol{\sigma}_{n+1}). \tag{9.80}$$

The step-by-step procedure for evaluation of the above elastoplastic tangent is summarised in Box 9.6 in pseudo-code format. Its computation in program `HYPLAS` is carried out in subroutine `CTVMPS` (**C**onsistent **T**angent operator for the **V**on **M**ises model in **P**lane **S**tress). This subroutine is described in the next section.

9.4.5. SUBROUTINE `CTVMPS`

The FORTRAN implementation of the tangent operator consistent with the implicit plane stress-projected integration algorithm for the von Mises model is described in this section. As in the standard implementation (plane strain/axisymmetric) of routine `CTVM` (see Section 7.4.3, page 235), the present routine returns either the elastic or the elastoplastic tangent operator. The elastic tangent is the standard plane stress linear elasticity matrix. The elastoplastic tangent is that given by expression (9.80) and its evaluation follows the flow of Box 9.6. The argument list of `CTVMPS` is identical to that of its general counterpart `CTVM`. The reader is therefore referred to Section 7.4.3 for a complete description of the arguments of `CTVMPS`.

**In deriving (9.77) we have made use of the standard relation for the differential of the inverse of a matrix

$$\mathrm{d}(\mathbf{M}^{-1}) = -\mathbf{M}^{-1}\,\mathrm{d}\mathbf{M}\,\mathbf{M}^{-1},$$

valid for any invertible $\mathbf{M}$. This relation, together with the use of the chain rule in (9.73), yields the identity

$$\mathrm{d}\mathbf{E} = -\mathbf{E}\,\mathbf{P}\,\mathbf{E}\,\mathrm{d}\Delta\gamma.$$

Box 9.6. Computation of the elastoplastic tangent operator for the plane stress-projected von Mises model.

HYPLAS procedure:	CTVMPS

(i) Given $\boldsymbol{\sigma}_{n+1}$, $\bar{\varepsilon}^p_{n+1}$ and $\Delta\gamma$ (outcome of the return mapping of Box 9.4), compute

$$\xi := \boldsymbol{\sigma}^T_{n+1}\mathbf{P}\boldsymbol{\sigma}_{n+1}$$

$$H := \left.\frac{\mathrm{d}\sigma_y}{\mathrm{d}\bar{\varepsilon}^p}\right|_{\bar{\varepsilon}^p_{n+1}} \qquad \text{(hardening slope)}$$

$$\mathbf{E} := [\mathbf{C} + \Delta\gamma\,\mathbf{P}]^{-1}$$

$$\mathbf{n} := \mathbf{E}\mathbf{P}\boldsymbol{\sigma}_{n+1}$$

$$\alpha := \frac{1}{\boldsymbol{\sigma}^T_{n+1}\mathbf{P}\mathbf{n} + \frac{2\xi H}{3-2H\,\Delta\gamma}}$$

(ii) Assemble the elastoplastic tangent

$$\mathbf{D}^{ep} := \mathbf{E} - \alpha\,\mathbf{n}\otimes\mathbf{n}$$

Note that analogously to matrix $\mathbf{A}$ expressed in (9.69), matrix $\mathbf{E}$ appearing in the procedure of Box 9.6 can be represented as

$$\mathbf{E}(\Delta\gamma) = \begin{bmatrix} \frac{1}{2}(E^*_{11} + E^*_{22}) & \frac{1}{2}(E^*_{11} - E^*_{22}) & 0 \\ \frac{1}{2}(E^*_{11} - E^*_{22}) & \frac{1}{2}(E^*_{11} + E^*_{22}) & 0 \\ 0 & 0 & E^*_{33} \end{bmatrix} \tag{9.81}$$

with

$$E^*_{11} = \frac{3E}{3(1-\nu) + E\,\Delta\gamma}, \qquad E^*_{22} = \frac{2G}{1 + 2G\,\Delta\gamma}, \qquad E^*_{33} = \frac{E^*_{22}}{2}. \tag{9.82}$$

A similar expression is obtained for the product $\mathbf{EP}$. This representation is exploited in subroutine CTVMPS, whose FORTRAN source code is listed below.

```
      SUBROUTINE CTVMPS
     1(   DGAMA      ,DMATX      ,EPFLAG      ,IPROPS     ,NTYPE      ,
     2    RPROPS     ,RSTAVA     ,STRES       )
      IMPLICIT DOUBLE PRECISION (A-H,O-Z)
      PARAMETER(IPHARD=4   ,MSTRE=4)
      LOGICAL EPFLAG
C Array arguments
      DIMENSION
     1    DMATX(MSTRE,MSTRE),IPROPS(*)            ,RPROPS(*)          ,
     2    RSTAVA(MSTRE+1)    ,STRES(MSTRE)
C Local arrays
      DIMENSION
     1    FOID(MSTRE,MSTRE)  ,SOID(MSTRE)         ,VECN(3)
```

```
      DATA
     1    FOID(1,1),FOID(1,2),FOID(1,3),FOID(1,4)/
     2    1.0D0    ,0.0D0    ,0.0D0    ,0.0D0    /
     3    FOID(2,1),FOID(2,2),FOID(2,3),FOID(2,4)/
     4    0.0D0    ,1.0D0    ,0.0D0    ,0.0D0    /
     5    FOID(3,1),FOID(3,2),FOID(3,3),FOID(3,4)/
     6    0.0D0    ,0.0D0    ,0.5D0    ,0.0D0    /
     7    FOID(4,1),FOID(4,2),FOID(4,3),FOID(4,4)/
     8    0.0D0    ,0.0D0    ,0.0D0    ,1.0D0    /
      DATA
     1    SOID(1)  ,SOID(2)  ,SOID(3)  ,SOID(4)  /
     2    1.0D0    ,1.0D0    ,0.0D0    ,1.0D0    /
      DATA
     1    RP5  ,R1   ,R2   ,R3   ,R4   /
     2    0.5D0,1.0D0,2.0D0,3.0D0,4.0D0/
C***********************************************************************
C COMPUTATION OF THE CONSISTENT TANGENT MODULUS FOR VON MISES TYPE
C ELASTO-PLASTIC MATERIAL WITH PIECE-WISE LINEAR ISOTROPIC HARDENING.
C PLANE STRESS IMPLEMENTATION ONLY.
C***********************************************************************
C Stops program if not plane stress
      IF(NTYPE.NE.1)CALL ERRPRT('EI0032')
C Current accumulated plastic strain
      EPBAR=RSTAVA(MSTRE+1)
C Set material properties
      YOUNG=RPROPS(2)
      POISS=RPROPS(3)
      NHARD=IPROPS(3)
C Shear and bulk moduli
      GMODU=YOUNG/(R2*(R1+POISS))
      BULK=YOUNG/(R3*(R1-R2*POISS))
      R2G=R2*GMODU
      R1D3=R1/R3
      R2D3=R2*R1D3
      IF(EPFLAG)THEN
C Compute elastoplastic consistent tangent (Box 9.6)
C ==================================================
C Item (i):
C ---------
C Compute XI
        XI=R2D3*(STRES(1)*STRES(1)+STRES(2)*STRES(2)-STRES(1)*STRES(2))+
     1     R2*STRES(3)*STRES(3)
C Hardening slope
        HSLOPE=DPLFUN(EPBAR,NHARD,RPROPS(IPHARD))
C Matrix E components
        ESTAR1=R3*YOUNG/(R3*(R1-POISS)+YOUNG*DGAMA)
        ESTAR2=R2G/(R1+R2G*DGAMA)
        ESTAR3=GMODU/(R1+R2G*DGAMA)
        E11=RP5*(ESTAR1+ESTAR2)
        E22=E11
        E12=RP5*(ESTAR1-ESTAR2)
        E33=ESTAR3
C Components of the matrix product EP
        EPSTA1=R1D3*ESTAR1
        EPSTA2=ESTAR2
        EPSTA3=EPSTA2
        EP11=RP5*(EPSTA1+EPSTA2)
        EP22=EP11
        EP12=RP5*(EPSTA1-EPSTA2)
        EP21=EP12
        EP33=EPSTA3
```

```
C Vector n
        VECN(1)=EP11*STRES(1)+EP12*STRES(2)
        VECN(2)=EP21*STRES(1)+EP22*STRES(2)
        VECN(3)=EP33*STRES(3)
C Scalar alpha
        DENOM1=STRES(1)*(R2D3*VECN(1)-R1D3*VECN(2))+
     1         STRES(2)*(R2D3*VECN(2)-R1D3*VECN(1))+
     2         STRES(3)*R2*VECN(3)
        DENOM2=R2*XI*HSLOPE/(R3-R2*HSLOPE*DGAMA)
        ALPHA=R1/(DENOM1+DENOM2)
C Item (ii): Assemble elasto-plastic tangent
C -----------------------------------------
        DMATX(1,1)=E11-ALPHA*VECN(1)*VECN(1)
        DMATX(1,2)=E12-ALPHA*VECN(1)*VECN(2)
        DMATX(1,3)=-ALPHA*VECN(1)*VECN(3)
        DMATX(2,1)=DMATX(1,2)
        DMATX(2,2)=E22-ALPHA*VECN(2)*VECN(2)
        DMATX(2,3)=-ALPHA*VECN(2)*VECN(3)
        DMATX(3,1)=DMATX(1,3)
        DMATX(3,2)=DMATX(2,3)
        DMATX(3,3)=E33-ALPHA*VECN(3)*VECN(3)
      ELSE
C Compute plane stress elasticity matrix
C ======================================
        NSTRE=3
        R4GD3=R4*GMODU/R3
        FACTOR=(BULK-R2G/R3)*(R2G/(BULK+R4GD3))
        DO 20 I=1,NSTRE
          DO 10 J=I,NSTRE
            DMATX(I,J)=R2G*FOID(I,J)+FACTOR*SOID(I)*SOID(J)
   10     CONTINUE
   20   CONTINUE
C lower triangle
        DO 40 J=1,NSTRE-1
          DO 30 I=J+1,NSTRE
            DMATX(I,J)=DMATX(J,I)
   30     CONTINUE
   40   CONTINUE
      ENDIF
      RETURN
      END
```

9.5. Numerical examples

In this section, we present a selected set of benchmarking numerical examples involving plane stress plasticity. The results have been obtained with program HYPLAS with the full Newton–Raphson scheme selected for the equilibrium iterations. In all examples, the equilibrium convergence tolerance for the relative residual norm was set to $10^{-7}\%$. It is important to remark that such a small convergence tolerance is used essentially to emphasise the quadratic rates of convergence obtained by adopting the tangent modulus consistent with the relevant integration algorithm. The numerical results obtained in the first two examples is compared with existing analytical solutions.

9.5.1. COLLAPSE OF AN END-LOADED CANTILEVER

In this example, we study the problem of the plastic collapse of an end-loaded cantilever beam with a rectangular cross-section. The geometry, boundary conditions, finite element model and material properties are illustrated in Figure 9.2. A perfectly plastic von Mises material model in plane stress is assumed. The plane stress implementation of the von Mises model, which follows the plane-stress projected approach, has been fully described in Sections 9.4.2 and 9.4.4. The cantilever is modelled with 2×50 eight-noded quadrilaterals (element type QUAD_8 of HYPLAS) with 2×2-point (reduced) Gauss quadrature. The analytical limit load for this problem is given by

$$F_{\text{lim}} = \frac{\sigma_y \, bh^2}{4L},$$

where b, h are, respectively, the breadth and width of the cross-section and L is the length of the cantilever. The above solution is derived in detail in section 4.2.2 of Lubliner (1990). For the present geometry and material properties, the analytical limit load is

$$F_{\text{lim}} = 30\,\text{KN}.$$

The numerical solution is obtained here by applying six load increments to the loaded node shown in Figure 9.2. The resulting vertical deflection of the loaded node is plotted against the applied force for each increment in Figure 9.3. The load level at the end of the last increment is $F = 30.3594$ KN. At this point, the cantilever is effectively collapsing and equilibrium can no longer be found for reasonably sized load increments. It should be noted that the numerically obtained collapse load is in excellent agreement with the theoretical limit. The relative error is approximately 1.2%. Finally, to emphasise the quadratic rates of equilibrium convergence attained as a result of the consistent linearisation of the algorithmic constitutive rule, the evolution of the residuals (out-of-balance forces) during the global Newton–Raphson iterations is shown in Table 9.1 for increments 2, 4 and 6. Note that for increment 6 (the last load increment) the cantilever is effectively collapsing (see the load-deflection diagram of Figure 9.3). At that stage, substantial diagonal decay (with consequent round-off errors) is found in the linearised finite element system of equations requiring a larger number of iterations for convergence.

9.5.2. INFINITE PLATE WITH A CIRCULAR HOLE

The problem considered in this numerical example consists of a large flat plate containing a circular hole which is expanded by the action of a gradually increasing uniform radial pressure applied on its edge. The plate is made from a von Mises perfectly plastic material. The analytical solution to this problem, where the plate is considered to be an infinite medium in plane stress state, is found in Section 5.7 of Chakrabarty (1987). Here, the problem is solved by using the mesh shown in Figure 9.4. In the finite element model, the plate has a large radius (as compared to the radius of the hole) in order to emulate the infinite planar medium. Due to the symmetry of the problem around the centre of the hole only a 10° slice of the geometry is discretised. A total number of 11 eight-noded quadrilaterals (element type QUAD_8 of HYPLAS) with five-point quadrature are used (the quadrature points are indicated by crosses within a typical element in Figure 9.4). Under plane stress conditions, eight-noded quadrilaterals are usually employed in conjunction with the nine-point Gaussian quadrature.

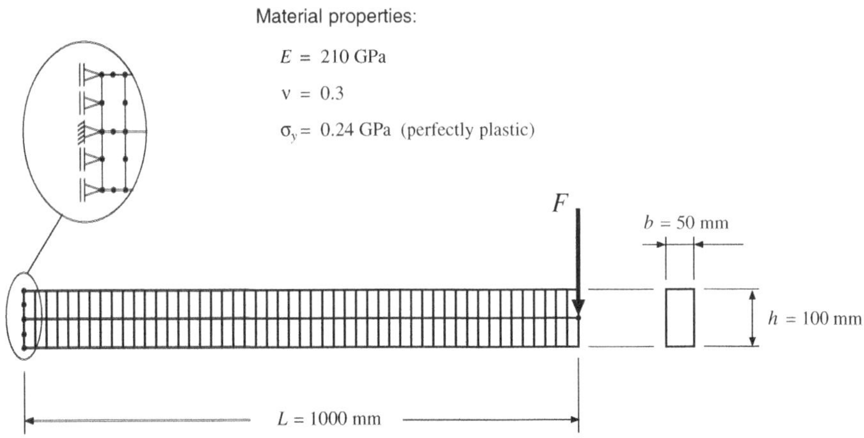

Figure 9.2. End-loaded cantilever. Problem definition and finite element model.

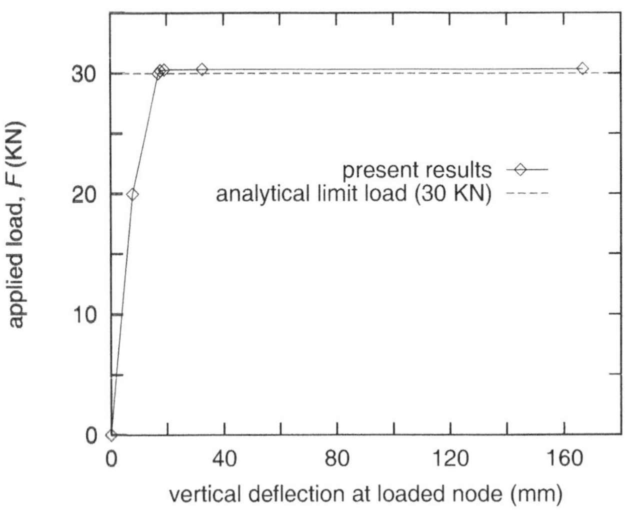

Figure 9.3. End-loaded cantilever. Load-deflection diagram.

As opposed to plane strain applications, underintegration is not required here, as locking is not an issue in plane stress problems. The choice of the five-point integration rule in the present example, is informed by the fact that it produces, with less computational effort, results virtually identical to those obtained by the nine-point Gauss quadrature. In the finite element simulation, the pressure p is increased gradually until a limit load is attained. Seven load increments are used to reach $p = 0.2853$ – the numerical pressure limit at which global convergence cannot be obtained for any significant load increment. The resulting normalised radial expansion – u/a, where u is the radial displacement of the mid-side node of the loaded edge – is plotted in Figure 9.5 against the normalised applied pressure, p/σ_y. The analytical

Table 9.1. End-loaded cantilever. Convergence table.

	Normalised residuals (%)		
Iteration number	Increment 2 $\Delta F = 10$ KN	Increment 4 $\Delta F = 0.0625$ KN	Increment 6 $\Delta F = 0.015625$ KN
1	$0.895831\ E+01$	$0.464825\ E+00$	$0.205624\ E+01$
2	$0.380886\ E+01$	$0.634491\ E+00$	$0.248066\ E+01$
3	$0.629249\ E-01$	$0.295108\ E-01$	$0.128691\ E+01$
4	$0.486441\ E-04$	$0.139026\ E-03$	$0.533184\ E+00$
5	$0.526293\ E-10$	$0.414654\ E-08$	$0.717926\ E-01$
6			$0.104585\ E-02$
7			$0.162250\ E-06$
8			$0.549382\ E-09$

normalised limit pressure is

$$\frac{p_{\text{lim}}}{\sigma_y} = \frac{2}{\sqrt{3}} \approx 1.155.$$

The limit pressure obtained in the simulation is about 3% higher than the analytical value. The hoop and radial stress distribution along the radial direction obtained at different stages of loading is shown in Figure 9.6. The numerical results shown on the graph are those obtained at the central integration point of each of the first five elements on the left side of the mesh. The corresponding analytical solution is also plotted for comparison. Similarly to the problem studied in Section 7.5.1 (page 244), plastification here starts at the edge of the hole and progresses as a cylindrical plastic front whose radius c increases as the load is increased. In the present problem, when the above limit pressure is reached, the radius of the plastic zone has the limit value

$$\frac{c_{\text{lim}}}{a} = \sqrt{e^{\pi/\sqrt{3}} \cos\frac{\pi}{3}} \approx 1.751.$$

In the elastic region – where the radial coordinate r satisfies $r \geq c$ – the radial and hoop stresses are given, respectively, by

$$\sigma_r = -\frac{\sigma_y\, c^2}{\sqrt{3}\, r^2} \quad \text{and} \quad \sigma_\theta = \frac{\sigma_y\, c^2}{\sqrt{3}\, r^2}.$$

Within the plastic region of the plate – where $a \geq r \geq c$ – the analytical stress distributions are given by

$$\sigma_r = -\frac{2\sigma_y}{\sqrt{3}} \sin\left(\frac{\pi}{6} + \phi\right) \quad \text{and} \quad \sigma_\theta = \frac{2\sigma_y}{\sqrt{3}} \sin\left(\frac{\pi}{6} - \phi\right),$$

where, for a given $c \leq c_{\text{lim}}$, ϕ is the implicit function of r defined by

$$\frac{c^2}{r^2} = e^{\sqrt{3}\phi} \cos\phi.$$

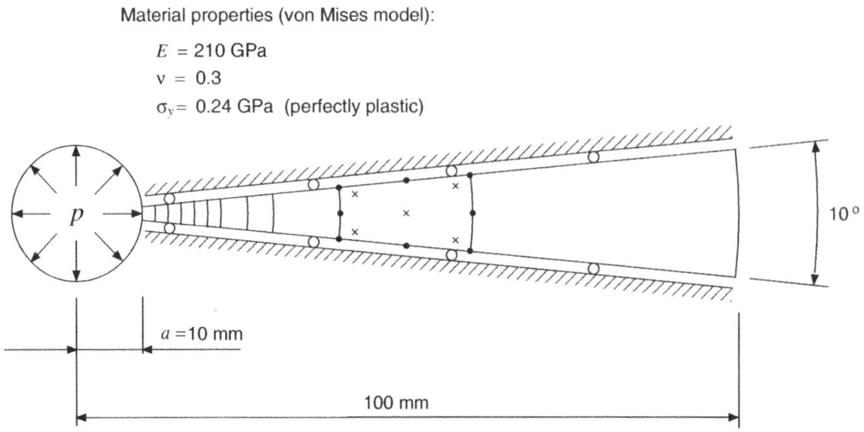

Figure 9.4. Infinite plate with a hole. Problem definition and finite element model.

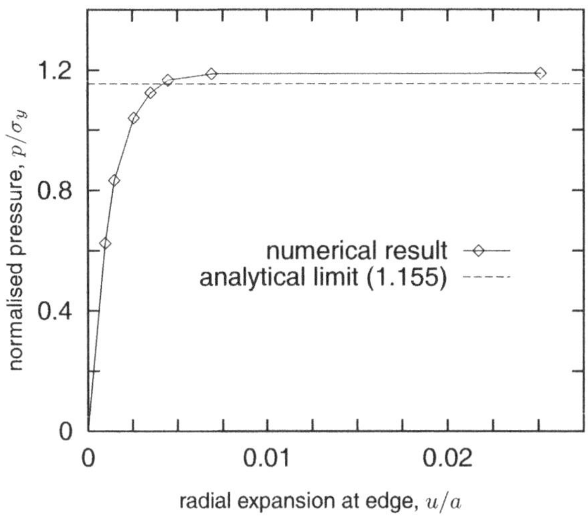

Figure 9.5. Infinite plate with a hole. Pressure-expansion diagram.

9.5.3. STRETCHING OF A PERFORATED RECTANGULAR PLATE

This example presents the numerical simulation of a thin perforated plate under plane stress subjected to stretching along its longitudinal axis. This classical example is frequently used as a benchmark for the plane stress implementation of plasticity models (Fuschi *et al.*, 1992; Simo and Taylor, 1986). A linearly hardening von Mises law is adopted. The geometry, boundary conditions and material properties are shown in Figure 9.7. The finite element mesh used contains 576 three-noded constant strain triangles (element type `TRI_3` of `HYPLAS`) with a total number of 325 nodes. For symmetry reasons, only one quarter of the complete domain is discretised with the appropriate kinematic constraints being imposed along the symmetry edges. A total vertical displacement $u = 0.14$ mm is imposed on the constrained edge in seven

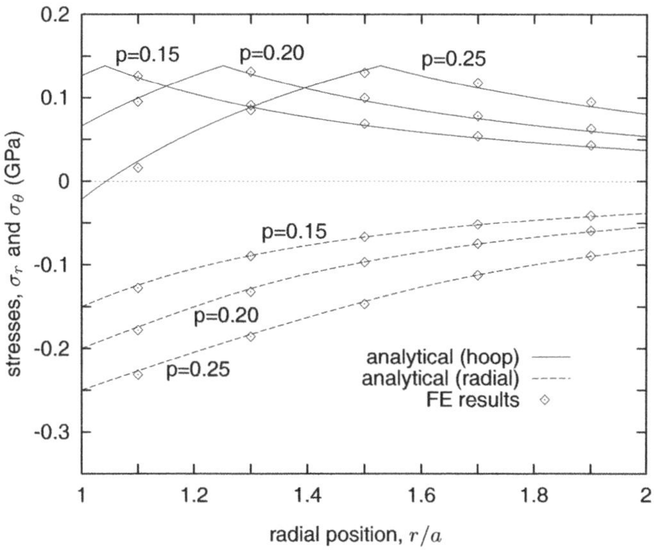

Figure 9.6. Infinite plate with a hole. Stress distribution in the radial direction (r denotes the radial coordinate). Analytical and finite element results for different applied pressures (indicated pressures levels in GPa).

increments $\Delta u = 0.02$ mm. The top edge nodes (except that on the symmetry line) are free to move horizontally. The reaction force obtained over the complete edge is plotted in Figure 9.8 versus the prescribed displacement. In this problem, plastic yielding starts at the intersection between the bottom symmetry line and the edge of the hole and spreads in an oblique front until the entire cross-section becomes plastic. The evolution of the plastic front is depicted in Figure 9.9, which shows contour plots of accumulated plastic strain at different stages of the loading process.

9.5.4. UNIFORM LOADING OF A CONCRETE SHEAR WALL

A plain concrete shear wall subjected to uniformly distributed load on its upper edge is analysed in this example. The geometry, boundary conditions and the finite element mesh adopted are shown in Figure 9.10. The mesh used to discretise the symmetric half of the problem contains 12×24 equally sized eight-noded quadrilaterals with four-point Gauss quadrature. The given material properties for the concrete (also shown in Figure 9.10) are its uniaxial compression and tensile strengths, denoted f_c' and f_t', respectively. Here, the concrete is modelled as an associative perfectly plastic Drucker–Prager material in plane stress. We remark that, in contrast to the plane stress-projected approach of the von Mises implementation used in the previous examples, the HYPLAS implementation of the plane stress Drucker–Prager model follows the nested iteration methodology described in Section 9.2.2. Two different sets of material parameters are adopted with the plane stress section of the Drucker–Prager surface matching:

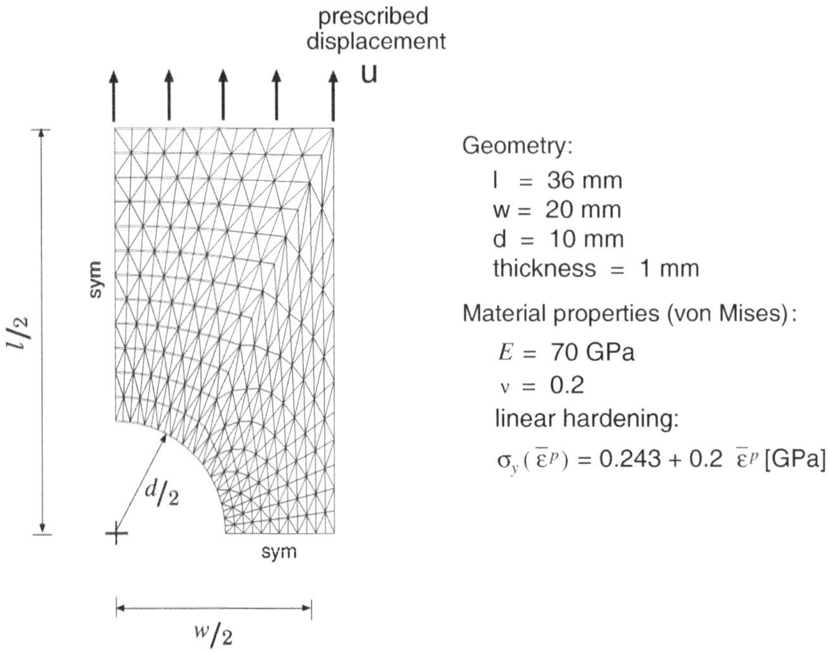

Figure 9.7. Stretching of a perforated rectangular plate. Geometry, boundary conditions and finite element mesh.

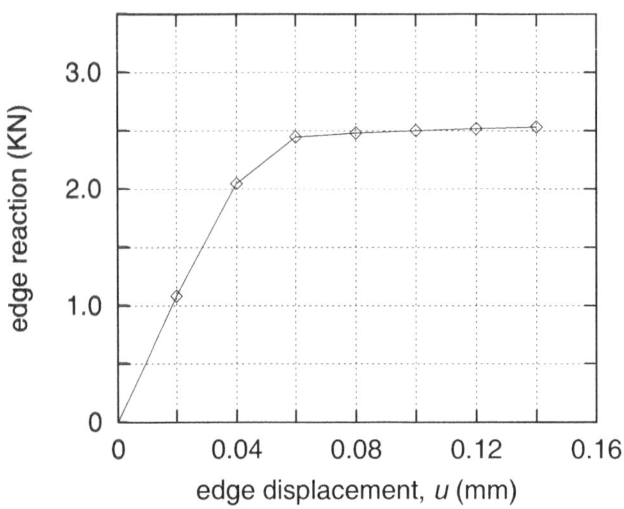

Figure 9.8. Stretching of a perforated rectangular plate. Reaction–deflection diagram.

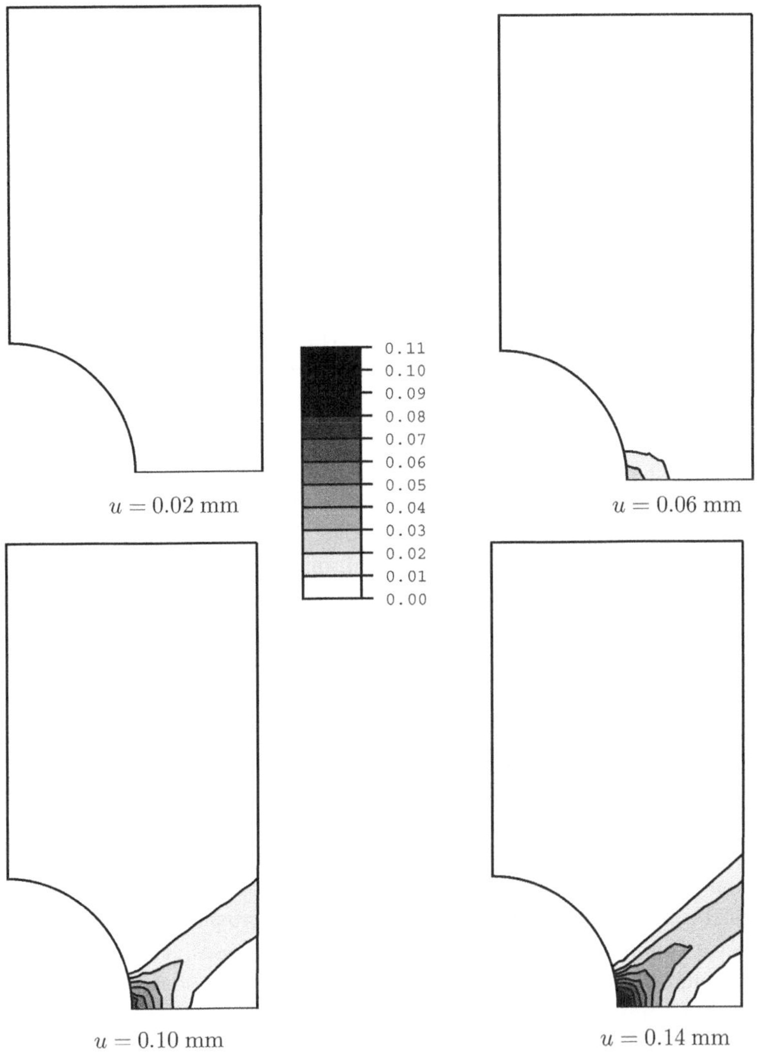

Figure 9.9. Stretching of a perforated rectangular plate. Contour plots of accumulated plastic strain, $\bar{\varepsilon}^p$.

(a) the given uniaxial data;

(b) the biaxial tension and compression limits of the Mohr–Coulomb surface that matches the given uniaxial data.

A schematic representation of the corresponding yield surfaces is given in Figure 6.16 (page 169). The frictional angle, ϕ, and the cohesion, c, for the above are given by (refer to expressions (6.125–6.127))

$$\phi = 56.57^{\circ}, \quad c = 4.58 \text{ MPa}.$$

Table 9.2. Shear wall. Convergence table.

	Normalised residuals (%)		
Iteration number	Increment 2	Increment 6	Increment 9
1	$0.102911\ E+02$	$0.101283\ E+02$	$0.567046\ E+01$
2	$0.495802\ E+00$	$0.557241\ E+01$	$0.134854\ E+01$
3	$0.822378\ E-03$	$0.198868\ E-00$	$0.904595\ E-02$
4	$0.447672\ E-08$	$0.653716\ E-04$	$0.282426\ E-05$
5		$0.541467\ E-10$	$0.132435\ E-10$

The Drucker–Prager parameters η and ξ for cases (a) and (b) are automatically set to the correct values by selecting the flag for the appropriate Drucker–Prager match of the Mohr–Coulomb surface in the input data file for HYPLAS. In the numerical simulation, the load is applied incrementally, with arc-length control, until collapse occurs. A total of 16 and 10 increments are used, respectively, in cases (a) and (b) to reach a numerical limit load. To emphasise the quadratic (equilibrium) convergence rates resulting from the consistent linearisation of the nested iteration-based integration algorithm, Table 9.2 shows the evolution of the normalised residuals (out-of-balance forces) during typical global Newton iterations at sample load increments. The applied load versus vertical deflection of the midpoint of the lower edge (point A of Figure 9.10) is shown in Figure 9.11. This example illustrates the huge discrepancy that may exist between results obtained with different types of Drucker–Prager approximations to the Mohr–Coulomb criterion. The limit load obtained with the uniaxial match is about nine times that obtained with the biaxial match. The numerical limit load in the uniaxial match case is 3045 KN/m whereas with the biaxial match the limit load is 335.5 KN/m. In the present example, failure occurs by crushing of the material near the support under a combination of shear and compressive stresses. The failure mechanism is illustrated in Figure 9.12 where the direction of the incremental nodal displacements at failure are plotted for the biaxial match case. As pointed out in Section 6.4.4, limit loads can be significantly overestimated when the uniaxial match is used under such conditions, becoming worse towards the biaxial compression state. This is particularly true in the present case where the ratio f_c'/f_t' is high. The biaxial match on the other hand produces a conservative prediction. Other predictions can be obtained by using the outer or inner edge Drucker–Prager approximations to the Mohr–Coulomb surface (refer to Section 6.4.4, from page 166) corresponding to the given uniaxial data. In these cases (not shown on the load-deflection graphs) the predicted limit loads are, respectively, 2539 KN/m and 276 KN/m. The most conservative one is that produced by the inner match approximation.

It should be emphasised that, in practice, shear walls are made from *reinforced* rather than plain concrete. Thus, the present example is not meant to be a practical engineering application. Instead, it serves to illustrate the performance of the (nested iteration) plane stress Drucker–Prager model. Within the finite element environment of HYPLAS, effects of reinforcement could be accounted for by adding steel elements (the reinforcement) on top of an appropriately designed background plain concrete mesh.

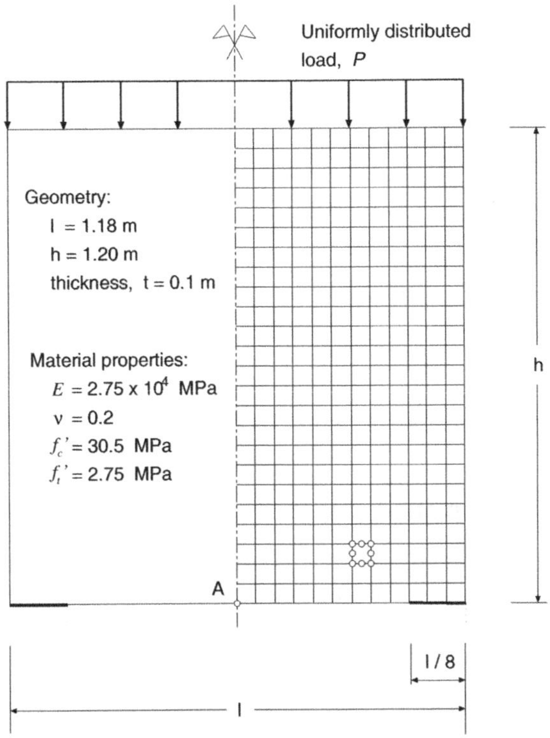

Figure 9.10. Shear wall. Geometry and finite element model.

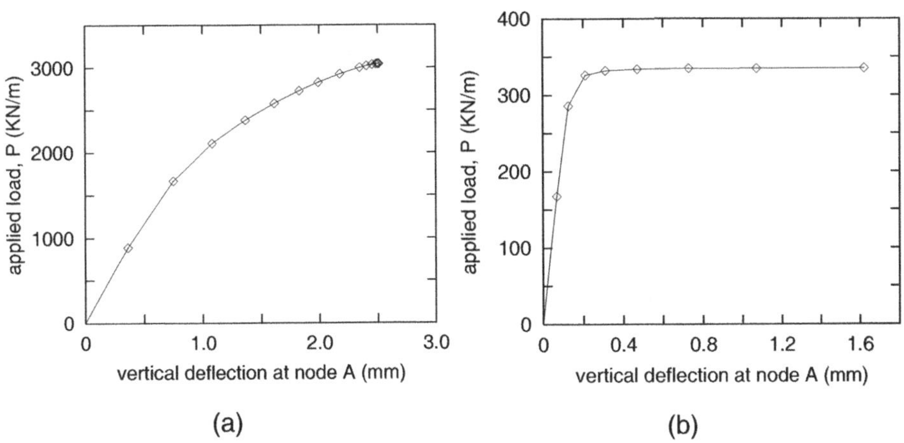

Figure 9.11. Shear wall. Load versus vertical deflection at point A: (a) uniaxial match; (b) biaxial match.

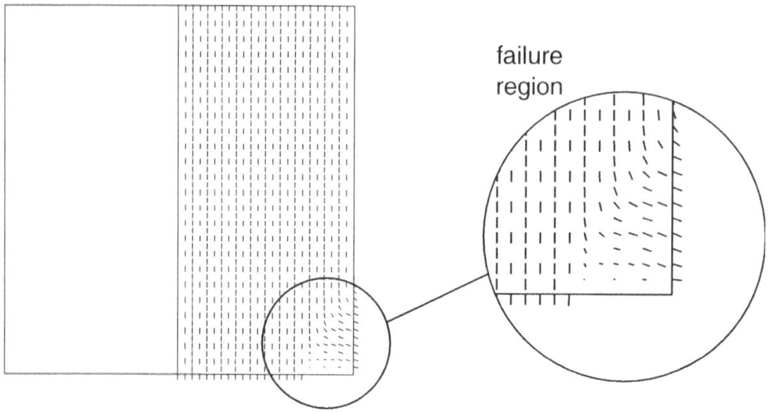

Figure 9.12. Shear wall. Incremental displacements at failure.

9.6. Other stress-constrained states

Plane stress is certainly not the only stress-constrained state of practical interest in engineering analysis and design. States of stress arising typically in structures such as beams, plates and shells can, under many practical conditions, be approximated by constraining appropriate components of the stress tensor. Within the context of linear elasticity, the classical Bernoulli and Timoshenko beam theories as well as plates and shells theories such as the Mindlin–Reissner and Kirchhoff–Love models are clear examples where certain components of the stress tensor are assumed to vanish. The extension of such theories to the elastoplastic range can be derived by means of the concepts reviewed in the previous sections of this chapter. To illustrate further applications of those concepts, we derive in the following a set of stress-constrained elastoplastic constitutive equations suitable for implementation with three-dimensional Timoshenko beam finite elements. The equations derived here are completely analogous to the plane-stress projected equations discussed earlier. In the present case we will end up with a *beam state-projected* set of evolution equations for von Mises plasticity. The corresponding integration algorithm and consistent tangent operator, which have the same format as those of the plane stress-projected model, are also derived.

9.6.1. A THREE-DIMENSIONAL VON MISES TIMOSHENKO BEAM

Let us consider here a three-dimensional beam with plane cross-sections $\mathscr{C}$ orthogonal to the centroidal axis (Figure 9.13). Following the standard kinematic hypotheses that characterise two-dimensional Timoshenko beams, it is assumed that upon a generic three-dimensional deformation the cross-sections remain plane. This assumption results in the following representation for the strain components ε_{13}, ε_{23} and ε_{33} at any point with local coordinates (η_1, η_2, ζ)

$$
\begin{aligned}
\varepsilon_{13} &= \tfrac{1}{2}[\gamma_1(\zeta) - \eta_2\, \kappa_3(\zeta) - \omega_2(\zeta)] \\
\varepsilon_{23} &= \tfrac{1}{2}[\gamma_2(\zeta) + \eta_1\, \kappa_3(\zeta) + \omega_1(\zeta)] \qquad (9.83) \\
\varepsilon_{33} &= \gamma_3(\zeta) + \eta_2\, \kappa_1(\zeta) - \eta_1\, \kappa_2(\zeta)
\end{aligned}
$$

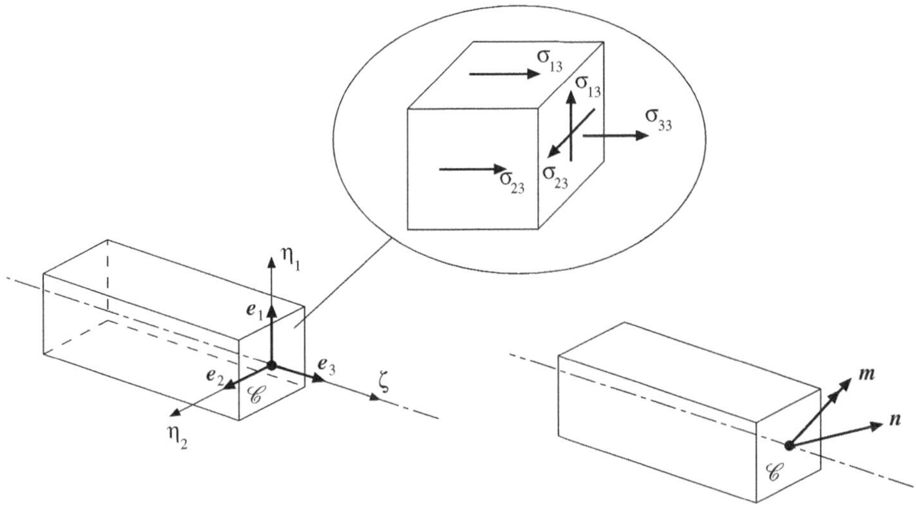

Figure 9.13. Three-dimensional Timoshenko beam state.

where γ_i and κ_i are the components of the *generalised strain arrays*

$$\boldsymbol{\gamma} \equiv \frac{\partial \bar{\boldsymbol{u}}}{\partial \zeta}, \qquad \boldsymbol{\kappa} \equiv \frac{\partial \boldsymbol{\omega}}{\partial \zeta} \tag{9.84}$$

with $\bar{\boldsymbol{u}}(\zeta)$ and $\boldsymbol{\omega}(\zeta)$ denoting, respectively, the *displacement of the centroidal axis* and the three-dimensional *rotation* of the cross-section. The components γ_1 and γ_2 are associated with transverse shear and γ_3 is the longitudinal strain at the centroidal axis. Components κ_1 and κ_2 correspond to the beam curvature deformation whereas κ_3 is the torsional strain.

The constraint on the stress tensor

In addition, a three-dimensional Timoshenko beam stress state is characterised by the following constraints on the stress tensor components

$$\sigma_{11} = \sigma_{22} = \sigma_{12} = 0 \tag{9.85}$$

so that a generic stress state is given by

$$\boldsymbol{\sigma} = \begin{bmatrix} 0 & 0 & \sigma_{13} \\ 0 & 0 & \sigma_{23} \\ \sigma_{13} & \sigma_{23} & \sigma_{33} \end{bmatrix} \tag{9.86}$$

with σ_{13}, σ_{23} and σ_{33} being the only possible non-vanishing components.

Resultant forces

The *generalised forces* associated with the generalised strain vectors $\boldsymbol{\gamma}$ and $\boldsymbol{\kappa}$ are, respectively, the *resultant force*, denoted $\boldsymbol{n}$, and the *resultant moment*, denoted $\boldsymbol{m}$. These are obtained by integration of the non-vanishing stress components over the beam cross-section. The resultant force has the local components

$$\begin{aligned} n_1 &= \int_{\mathscr{C}} \sigma_{13}\, \mathrm{d}A \\ n_2 &= \int_{\mathscr{C}} \sigma_{23}\, \mathrm{d}A \\ n_3 &= \int_{\mathscr{C}} \sigma_{33}\, \mathrm{d}A \end{aligned} \tag{9.87}$$

whereas the resultant moment is defined by

$$\begin{aligned} m_1 &= \int_{\mathscr{C}} \eta_2\, \sigma_{33}\, \mathrm{d}A \\ m_2 &= -\int_{\mathscr{C}} \eta_1\, \sigma_{33}\, \mathrm{d}A \\ m_3 &= \int_{\mathscr{C}} (\eta_1\, \sigma_{23} - \eta_2\, \sigma_{13})\, \mathrm{d}A. \end{aligned} \tag{9.88}$$

Resultant forces/generalised strains constitutive relation

In the present context, the constitutive relationship between the history of the generalised strains and the resultant forces is obtained as follows. Given the history of $\boldsymbol{\gamma}$ and $\boldsymbol{\kappa}$ (directly obtained from the history of generalised displacements, $\bar{\boldsymbol{u}}$ and $\boldsymbol{\omega}$) the history of the strain components ε_{13}, ε_{23} and ε_{33} is automatically prescribed by (9.83). With the prescribed strain components, we need to integrate the elastoplastic evolution equations so as to produce the corresponding stress $\boldsymbol{\sigma}$. Having done so, the resultant forces $\boldsymbol{n}$ and $\boldsymbol{m}$ are then obtained by integration of the stresses over the cross-section according to their definition. At the element level, the cross-section integration can be performed numerically by means of, say, a Gauss quadrature. Clearly, the only step that depends on the material model is the integration of the elastoplastic evolution problem. In the present case only the history of strain components ε_{13}, ε_{23} and ε_{33} is prescribed and the constraint (9.85) has to be satisfied. The dependent variables are the strain components ε_{11}, ε_{22} and ε_{12} which can be calculated *a posteriori*. In what follows, we concentrate on the definition of the von Mises elastoplastic evolution problem subjected to the present stress constraint.

The beam state-projected von Mises model

Analogously to what has been done in the derivation of the plane stress-projected von Mises model, it is convenient here to introduce the matrix notation

$$\boldsymbol{\sigma} \equiv \begin{Bmatrix} \sigma_{13} \\ \sigma_{23} \\ \sigma_{33} \end{Bmatrix}, \quad \mathbf{s} \equiv \begin{Bmatrix} 2s_{13} \\ 2s_{23} \\ s_{33} \end{Bmatrix} = \begin{Bmatrix} 2\sigma_{13} \\ 2\sigma_{23} \\ \frac{2}{3}\sigma_{33} \end{Bmatrix}, \tag{9.89}$$

where s_{ij} denote components of the deviatoric stress tensor. Note that the generally non-vanishing components s_{11} and s_{22}, missing in the above representation of the stress deviator, are *dependent* on s_{33}

$$s_{11} = s_{22} = -\tfrac{1}{2}\, s_{33} = -\tfrac{1}{3}\sigma_{33}. \tag{9.90}$$

The stress and stress deviator arrays are related by

$$\mathbf{s} = \mathbf{P}\,\boldsymbol{\sigma}, \tag{9.91}$$

where the projection matrix $\mathbf{P}$ is here given by

$$\mathbf{P} \equiv \begin{bmatrix} 2 & 0 & 0 \\ 0 & 2 & 0 \\ 0 & 0 & \frac{2}{3} \end{bmatrix}. \tag{9.92}$$

In order to formulate the constrained evolution equations, it is also convenient to define the following notation for the independent (engineering) strain components

$$\boldsymbol{\varepsilon} \equiv \begin{Bmatrix} 2\varepsilon_{13} \\ 2\varepsilon_{23} \\ \varepsilon_{33} \end{Bmatrix}, \quad \boldsymbol{\varepsilon}^e \equiv \begin{Bmatrix} 2\varepsilon^e_{13} \\ 2\varepsilon^e_{23} \\ \varepsilon^e_{33} \end{Bmatrix}, \quad \boldsymbol{\varepsilon}^p \equiv \begin{Bmatrix} 2\varepsilon^p_{13} \\ 2\varepsilon^p_{23} \\ \varepsilon^p_{33} \end{Bmatrix}. \tag{9.93}$$

The yield function.

With the above notation, the *projected* version of the von Mises yield function can be defined as

$$\Phi(\boldsymbol{\sigma}, \bar{\varepsilon}^p) \equiv \sqrt{\tfrac{3}{2}\,\boldsymbol{\sigma}^T\,\mathbf{P}\,\boldsymbol{\sigma}} - \sigma_y(\bar{\varepsilon}^p). \tag{9.94}$$

The projected yield function has values identical to those of the three-dimensional von Mises function for stress states satisfying the three-dimensional Timoshenko beam constraint (9.86) and has only the relevant non-vanishing stress components as arguments.

The beam state-projected von Mises equations.

Finally, the beam state-projected von Mises evolution equations are defined as follows. Given the history of the total independent strain components, $\varepsilon_{13}(t)$, $\varepsilon_{23}(t)$, and $\varepsilon_{33}(t)$ – the history of the strain arrays $\boldsymbol{\varepsilon}$ – the basic initial value problem now consists of finding the functions $\boldsymbol{\varepsilon}^p(t)$, $\boldsymbol{\varepsilon}^e(t)$, $\boldsymbol{\sigma}(t)$ and $\bar{\varepsilon}^p(t)$ such that the following constrained plasticity equations are satisfied

$$\begin{aligned} \dot{\boldsymbol{\varepsilon}}^e &= \dot{\boldsymbol{\varepsilon}} - \dot{\boldsymbol{\varepsilon}}^p \\ \boldsymbol{\sigma} &= \mathbf{D}^e\,\boldsymbol{\varepsilon}^e \\ \dot{\boldsymbol{\varepsilon}}^p &= \dot{\gamma}\,\frac{\partial\Phi}{\partial\boldsymbol{\sigma}} = \dot{\gamma}\,\sqrt{\tfrac{3}{2}}\,\frac{\mathbf{P}\boldsymbol{\sigma}}{\sqrt{\boldsymbol{\sigma}^T\mathbf{P}\boldsymbol{\sigma}}} \\ \dot{\bar{\varepsilon}}^p &= \dot{\gamma} \end{aligned} \tag{9.95}$$

where

$$\Phi(\boldsymbol{\sigma}, \bar{\varepsilon}^p) \ge 0, \quad \dot{\gamma} \ge 0, \quad \dot{\gamma}\,\Phi(\boldsymbol{\sigma}, \bar{\varepsilon}^p) > 0. \tag{9.96}$$

The elastic matrix, $\mathbf{D}^e$, for the three-dimensional Timoshenko beam is given simply by

$$\mathbf{D}^e = \begin{bmatrix} G & 0 & 0 \\ 0 & G & 0 \\ 0 & 0 & E \end{bmatrix}. \tag{9.97}$$

The above set of evolution equations is analogous to that of the plane stress-projected von Mises model of Box 9.3. Under stress states that satisfy the constraint (9.86), the rates $\dot{\varepsilon}^p_{13}$, $\dot{\varepsilon}^p_{23}$ and $\dot{\varepsilon}^p_{33}$ resulting from the original three-dimensional elastoplastic equations are identical to the rates resulting from the above constrained equations. The main difference between the two formulations is the fact that, in the three-dimensional case, the history of the total strain components, ε_{11}, ε_{22} and ε_{12}, is prescribed in the underlying initial value problem, whereas, in the constrained case, these components are dependent variables whose evolution is a consequence of the constraint on the stress state.

Evolution of the dependent strain components.

Having determined the history of the independent strains by integrating the constrained elastoplastic evolution equations, the history of the *dependent* strain components, ε_{11}, ε_{12} and ε_{22}, is obtained as follows. As $\sigma_{12} = 0$ and, consequently, $\dot{\varepsilon}^p_{12} = 0$ for constrained stress states, the elastoplastic additive strain decomposition together with the elastic relation, $\sigma_{12} = 2G\,\varepsilon^e_{12}$, implies that

$$\varepsilon_{12} = \varepsilon^e_{12} = \varepsilon^p_{12} = 0. \tag{9.98}$$

The three-dimensional Timoshenko beam constraint on stresses also imply that

$$\varepsilon^e_{11} = \varepsilon^e_{22} = \frac{-K}{2(G+K)}\,\varepsilon^e_{33}. \tag{9.99}$$

From the three-dimensional von Mises equations, it follows that, for constrained stress states, we have $\dot{\varepsilon}^p_{11} = \dot{\varepsilon}^p_{22} = -\frac{1}{2}\dot{\varepsilon}^p_{33}$, so that the components ε^p_{11} and ε^p_{22} are promptly found to be given by

$$\varepsilon^p_{11} = \varepsilon^p_{22} = -\tfrac{1}{2}\varepsilon^p_{33}. \tag{9.100}$$

Finally, the total strain components ε_{11} and ε_{22} are obtained from the additive elastoplastic split of the total strain tensor

$$\varepsilon_{11} = \varepsilon_{22} = \varepsilon^e_{11} + \varepsilon^p_{11}. \tag{9.101}$$

9.6.2. THE BEAM STATE-PROJECTED INTEGRATION ALGORITHM

The derivation of the implicit integration algorithm for the beam state-projected von Mises model follows the same steps as the derivation of the plane stress-projected algorithm described in Section 9.4.2. The final return-mapping equation for the plastic multiplier can be cast in the same format as equations (9.57, 9.58), with matrix $\mathbf{P}$ redefined by (9.92) and matrix $\mathbf{A}(\Delta\gamma)$ now given by the simpler diagonal form

$$\mathbf{A}(\Delta\gamma) = \begin{bmatrix} \dfrac{1}{1+2G\,\Delta\gamma} & 0 & 0 \\ 0 & \dfrac{1}{1+2G\,\Delta\gamma} & 0 \\ 0 & 0 & \dfrac{1}{1+\frac{2}{3}E\,\Delta\gamma} \end{bmatrix}. \tag{9.102}$$

With the above matrix, the function $\xi(\Delta\gamma)$ defined in (9.58) has the following explicit expression

$$\xi(\Delta\gamma) \equiv \frac{2[(\sigma_{13}^{\text{trial}})^2 + (\sigma_{23}^{\text{trial}})^2]}{(1 + 2G\,\Delta\gamma)^2} + \frac{2(\sigma_{33}^{\text{trial}})^2}{3(1 + \frac{2}{3}E\,\Delta\gamma)^2}. \tag{9.103}$$

The overall elastic predictor/return-mapping algorithm is identical to that shown in Boxes 9.4 and 9.5 except that $\mathbf{A}$ and ξ (as well as $\mathbf{D}^{\mathrm{e}}$ and the stress and strain arrays) have now the above new definitions and

$$\xi^{\text{trial}} := 2[(\sigma_{13}^{\text{trial}})^2 + (\sigma_{23}^{\text{trial}})^2] + \tfrac{2}{3}(\sigma_{33}^{\text{trial}})^2. \tag{9.104}$$

In the Newton algorithm of Box 9.5, the computation of the derivative ξ' is replaced by

$$\xi' := \frac{-8G[(\sigma_{13}^{\text{trial}})^2 + (\sigma_{23}^{\text{trial}})^2]}{(1 + 2G\,\Delta\gamma)^3} - \frac{8E\,(\sigma_{33}^{\text{trial}})^2}{9(1 + \frac{2}{3}E\,\Delta\gamma)^3}. \tag{9.105}$$

The elastoplastic consistent tangent modulus.

As in the return mapping described above, the derivation of the corresponding elastoplastic consistent tangent, $\mathbf{D}^{ep}$, is analogous to that of the plane stress-projected model addressed in Section 9.4.4. The final expression has the same format as (9.80) where the matrix $\mathbf{E}$ is here redefined as

$$\mathbf{E}(\Delta\gamma) = \begin{bmatrix} \dfrac{G}{1 + 2G\,\Delta\gamma} & 0 & 0 \\ 0 & \dfrac{G}{1 + 2G\,\Delta\gamma} & 0 \\ 0 & 0 & \dfrac{E}{1 + \frac{2}{3}E\,\Delta\gamma} \end{bmatrix}. \tag{9.106}$$

With the appropriate redefinition of the relevant terms, the elastoplastic consistent tangent can be computed as in Box 9.6.

10 ADVANCED PLASTICITY MODELS

THIS chapter is devoted to more advanced plasticity models. Recall that, in Chapters 7–9, all examples of numerical implementation of plasticity models have been limited to basic theories. Here, we move one step further and apply the same underlying concepts to more advanced models. Sections 10.1 and 10.2 describe, respectively, the treatment of a modified Cam-Clay model and a capped Drucker–Prager model – both mainly applicable to the description of geomaterials. Section 10.3 introduces the modelling of anisotropic plasticity. In this context, the Hill model (Hill, 1950), a model based on the Hoffman anisotropic criterion (Hoffman, 1967) – here referred to as the Hoffman model – and the Barlat–Lian model for sheet metals (Barlat and Lian, 1989) are discussed in detail, together with the computational treatment of the Hoffman and Barlat–Lian models with isotropic strain hardening.

The new theoretical concepts introduced in this chapter are, essentially, the modelling of plastic compaction – a phenomenon of particular relevance in the behaviour of geomaterials – and the modelling of plastic anisotropy. Note, however, that the computational implementation of elastoplastic models possessing such new features reduces to mere specialisations of the general framework discussed in the preceding chapters.

10.1. A modified Cam-Clay model for soils

The modified Cam-Clay model was originally proposed by Roscoe and Burland (1968) to model the plastic behaviour of soils. Models of this type are frequently used in the finite element modelling of soil mechanics problems. These models are characterised by plastic compressibility, hardening (softening) associated with compressive (dilatant) plastic flow and a possibly nonlinear elasticity law to model the hydrostatic pressure/volumetric elastic strain relation (refer, for instance, to Naylor *et al.* 1981, and Muir Wood 1990). In many cases, when dealing with soil consolidation problems, Cam-Clay-type models can be used within a multiphase environment with coupling between solid behaviour and flow through porous media. Such problems are outside the scope of this book and we shall describe in this section a version of the modified Cam-Clay plasticity model featuring a linear elasticity law and a standard volumetric plastic strain-dependent isotropic hardening rule. In soil mechanics applications, the use of a linear elastic law can be justified whenever the considered range of hydrostatic pressures is sufficiently narrow.

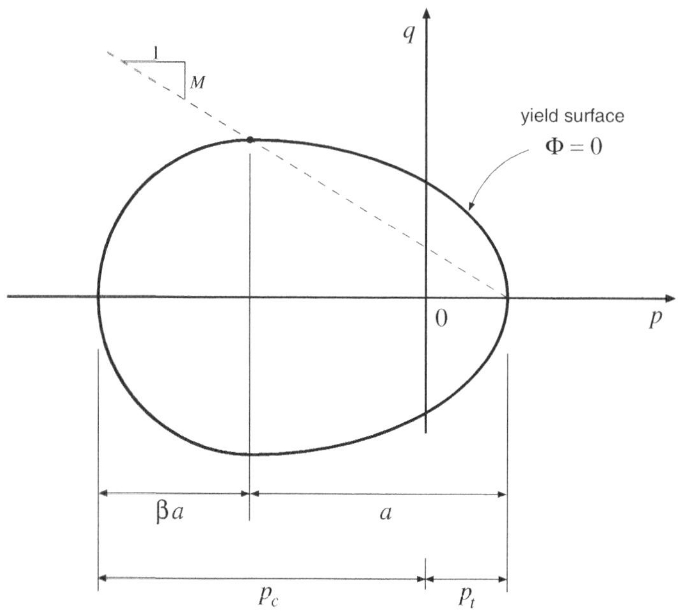

Figure 10.1. Modified Cam-Clay model. Yield surface.

10.1.1. THE MODEL

Let p and q denote, respectively, the hydrostatic stress and the von Mises effective stress. The elastic domain of the modified Cam-Clay model is delimited by an elliptic yield surface in the p–q space. Here, we will focus on a version of the modified Cam-Clay model whose yield surface (refer to Figure 10.1) is defined by a yield function of the form

$$\Phi(\boldsymbol{\sigma}, a) = \frac{1}{b^2}\,[p(\boldsymbol{\sigma}) - p_t + a]^2 + \left[\frac{q(\boldsymbol{\sigma})}{M}\right]^2 - a^2, \tag{10.1}$$

where the constant M is the ratio between the two radii of the Cam-Clay ellipse, a is the radius of the ellipse along the pressure axis and p_t is the tensile yield hydrostatic stress. The parameter b takes the values

$$b = \begin{cases} 1 & \text{if } p \geq p_t - a \\ \beta & \text{if } p < p_t - a, \end{cases} \tag{10.2}$$

where β is a material constant. This parameter modifies the radius of the second half of the ellipse on the compressive side of the hydrostatic axis. If $\beta = 1$, the yield locus becomes an ellipse with radii a and Ma, respectively, along directions p and q. Note that, for any $\beta > 0$, the two horizontal halves of the surface intersect in a smooth fashion at $(p, q) = (p_t - a, Ma)$. The dashed line in Figure 10.1 is named the *critical state line*. The parameter M is the slope of the critical state line. The portion of the p–q plane on the right of the intersection between the critical state line and the Cam-Clay ellipse is named the *supercritical* region and the portion on its left is the *subcritical* region.

Flow rule

The plastic flow equation can be defined by postulating associativity, which gives

$$\dot{\boldsymbol{\varepsilon}}^p = \dot{\gamma}\,\frac{\partial\Phi}{\partial\boldsymbol{\sigma}} = \dot{\gamma}\,\boldsymbol{N} = \dot{\gamma}\,(\boldsymbol{N}_{\rm d} + N_{\rm v}\boldsymbol{I}), \tag{10.3}$$

where we have conveniently split the flow vector, $\boldsymbol{N} = \partial\Phi/\partial\boldsymbol{\sigma}$, into deviatoric and volumetric components, respectively denoted $\boldsymbol{N}_{\rm d}$ and $N_{\rm v}$:

$$\boldsymbol{N}_{\rm d} = \frac{3}{M^2}\,\boldsymbol{s}; \quad N_{\rm v} = \frac{2}{b^2}\,(p - p_t + a). \tag{10.4}$$

The deviatoric and volumetric plastic strain rates then read

$$\dot{\boldsymbol{\varepsilon}}^p_{\rm d} = \dot{\gamma}\,\boldsymbol{N}_{\rm d}; \quad \dot{\varepsilon}^p_{\rm v} = \dot{\gamma}\,N_{\rm v}. \tag{10.5}$$

The plastic flow defined by the above associative rule is compressive[†] ($\dot{\varepsilon}^p_{\rm v} < 0$) if $p < p_t - a$ (subcritical states), dilatant ($\dot{\varepsilon}^p_{\rm v} > 0$) if $p > p_t - a$ (supercritical states) and isochoric ($\dot{\varepsilon}^p_{\rm v} = 0$) at $p = p_t - a$ (critical state).

Hardening law

A simple way of incorporating hardening into the model consists in letting the yield surface parameter a be a function of a hardening internal variable,[‡] α. For many plastically compressible materials, and for soils in particular, the state of hardening is largely dependent upon the volumetric plastic strain (or plastic compaction),

$$\varepsilon^p_{\rm v} \equiv \mathrm{tr}[\boldsymbol{\varepsilon}^p]. \tag{10.6}$$

Because in actual applications, soils will be predominantly subjected to *compressive* strains, the hardening variable, α, will be chosen as the *compressive-positive* volumetric plastic strain:

$$\alpha \equiv -\varepsilon^p_{\rm v} \tag{10.7}$$

and the hardening behaviour will be defined by means of the experimentally determined hardening function

$$a = a(\alpha). \tag{10.8}$$

Equivalently, in terms of the *compressive yield pressure* (or *compaction pressure*), p_c (also assumed compressive-positive) hardening can be defined by

$$p_c(\alpha) \equiv (1+\beta)\,a(\alpha) - p_t. \tag{10.9}$$

Under this assumption, the evolution of the volumetric plastic strain will change the size of the modified Cam-Clay yield surface while maintaining its shape. Hardening in this case is characterised by a change in compressive yield pressure, having the intersection of the yield surface with the p-axis at $p = p_t$ (the tensile yield pressure) fixed.

[†]Consistently with the continuum mechanics sign convention adopted throughout this book, compressive volumetric strains and hydrostatic pressures have a negative sign here. It should be noted that the opposite sign convention is commonly adopted in soil mechanics texts, where compressive volumetric strains and hydrostatic pressures are assumed positive.

[‡]In a more general setting, other yield surface parameters such as M or β could also be functions of hardening variables.

Remark 10.1 (Hardening behaviour). The hardening function $a(\alpha)$ (or $p_c(\alpha)$) is usually expected to be monotonically increasing in its argument, that is, a increases with increasing plastic compression. In this case, following the comments on the volumetric plastic strain rate made immediately after expression (10.5), hardening (expansion of the elastic domain) occurs only in the subcritical region, i.e. under hydrostatic pressures $p < p_t - a$. For pressures $p > p_t - a$ (supercritical states) plastic flow will cause *softening* (contraction of the elastic domain) and at $p = p_t - a$ (the critical state) the model behaves as perfectly plastic (fixed yield surface).

10.1.2. COMPUTATIONAL IMPLEMENTATION

Here we shall adopt a linear elasticity law to model the reversible behaviour of the modified Cam-Clay model. We remark, however, that it is sometimes convenient to adopt nonlinear elasticity laws with models of this type. The implementation of Cam-Clay-type models with nonlinear elastic laws is described, for instance, by Simo and Meschke (1993) and Owen *et al.* (1998).

The return-mapping algorithm

The elastic trial state computation follows the usual format:

$$s^{\text{trial}} = 2G\,\varepsilon_{\text{d}}^{e\,\text{trial}}; \qquad p^{\text{trial}} = K\,\varepsilon_{\text{v}}^{e\,\text{trial}}. \tag{10.10}$$

The return-mapping algorithm is derived as follows. Firstly, by observing the deviatoric/volumetric split (10.4) of the plastic flow and the elastic relations, the following stress update equations consistent with the implicit discretisation of the plastic flow rule are obtained:

$$\begin{cases} s_{n+1} = s^{\text{trial}} - 2G\Delta\gamma\,\dfrac{3}{M^2}\,s_{n+1}; \\[2ex] p_{n+1} = p^{\text{trial}} - K\Delta\gamma\,\dfrac{2}{b^2}\,[p_{n+1} - p_t + a(\alpha_{n+1})], \end{cases} \tag{10.11}$$

where the updated hardening variable is given by the equation

$$\alpha_{n+1} = \alpha_n - \Delta\gamma\,\frac{2}{b^2}\,[p_{n+1} - p_t + a(\alpha_{n+1})]. \tag{10.12}$$

In addition, the discrete consistency equation here reads

$$\Phi_{n+1} = \frac{1}{b^2}\,[p_{n+1} - p_t + a(\alpha_{n+1})]^2 + \left(\frac{q_{n+1}}{M}\right)^2 - [a(\alpha_{n+1})]^2 = 0, \tag{10.13}$$

where q_{n+1} is the von Mises effective stress associated with s_{n+1}.

Particularisation of the general return-mapping equations (7.25) (page 196) for the present model results in the system of equations comprising (10.11)–(10.13), having α_{n+1}, p_{n+1}, the linearly independent components of s_{n+1} and the incremental plastic multiplier, $\Delta\gamma$, as the unknowns.

Reduced equation system

In order to obtain an equivalent reduced system of equations for the return mapping, we start by noting that $(10.11)_1$ implies that for a given $\boldsymbol{s}^{\text{trial}}$ the updated stress deviator, $\boldsymbol{s}_{n+1}$, is a function of $\Delta\gamma$ only; that is, after a simple rearrangement in $(10.11)_1$ we obtain

$$\boldsymbol{s}_{n+1} = \boldsymbol{s}(\Delta\gamma) \equiv \frac{M^2}{M^2 + 6G\,\Delta\gamma}\,\boldsymbol{s}^{\text{trial}}, \tag{10.14}$$

and, similarly, for q_{n+1},

$$q_{n+1} = q(\Delta\gamma) \equiv \frac{M^2}{M^2 + 6G\,\Delta\gamma}\,q^{\text{trial}}. \tag{10.15}$$

In addition, by combining $(10.11)_2$ and (10.12) we can write

$$p_{n+1} = p(\alpha_{n+1}) \equiv p^{\text{trial}} + K\,(\alpha_{n+1} - \alpha_n). \tag{10.16}$$

Finally, the substitution of (10.15) and (10.16) in (10.12) and (10.13) leads to the following reduced system of two scalar equations with unknowns $\Delta\gamma$ and α_{n+1}:

$$\begin{Bmatrix} R_1 \\ \\ R_2 \end{Bmatrix} \equiv \begin{Bmatrix} \dfrac{1}{b^2}\,[p(\alpha) - p_t + a(\alpha)]^2 + \left[\dfrac{q(\Delta\gamma)}{M}\right]^2 - [a(\alpha)]^2 \\ \\ \alpha - \alpha_n + \Delta\gamma\,\dfrac{2}{b^2}[p(\alpha) - p_t + a(\alpha)] \end{Bmatrix} = \begin{Bmatrix} 0 \\ \\ 0 \end{Bmatrix}, \tag{10.17}$$

where the subscript $n+1$ of α_{n+1} has been omitted for notational convenience. Once the solution to the above return-mapping equations is obtained, the stress-updating procedure is completed with the computation of $\boldsymbol{s}_{n+1}$ by means of (10.14), followed by the calculation of the updated stress tensor through the standard relation

$$\boldsymbol{\sigma}_{n+1} = \boldsymbol{s}_{n+1} + p_{n+1}\,\boldsymbol{I}. \tag{10.18}$$

It is worth commenting here that system (10.17) could in principle be further reduced to a single scalar equation having α_{n+1} as the only unknown by trivially solving the second equation of (10.17) for $\Delta\gamma$:

$$\Delta\gamma = \Delta\gamma(\alpha_{n+1}) \equiv \frac{b^2(\alpha_n - \alpha_{n+1})}{2[p(\alpha_{n+1}) - p_t + a(\alpha_{n+1})]}, \tag{10.19}$$

and then replacing this result into the first equation. It should be noted, however, that the above expression cannot be used in practice at $p = p_t - a$.

Newton–Raphson solution

The solution of (10.17) can be undertaken as usual by the Newton–Raphson algorithm. For completeness, an explicit expression for the linear system to be solved for the iterative corrections $\delta\Delta\gamma^{(k)}$ and $\delta\alpha^{(k)}$ at the typical k^{th} Newton iteration is shown below:

$$\begin{bmatrix} \dfrac{-12\,G}{M^2+6G\Delta\gamma}\left(\dfrac{q}{M}\right)^2 & \dfrac{2\bar{p}}{b^2}(K+H)-2aH \\ \dfrac{2\bar{p}}{b^2} & 1+\dfrac{2\Delta\gamma}{b^2}(K+H) \end{bmatrix} \begin{Bmatrix} \delta\Delta\gamma^{(k)} \\ \delta\alpha^{(k)} \end{Bmatrix} = \begin{Bmatrix} R_1^{(k-1)} \\ R_2^{(k-1)} \end{Bmatrix}, \tag{10.20}$$

where H is the hardening modulus (the slope of the hardening curve):

$$H = H(\alpha) \equiv \frac{\mathrm{d}a}{\mathrm{d}\alpha}, \tag{10.21}$$

the scalar $\bar{p}$ is defined as

$$\bar{p} = \bar{p}(\alpha) \equiv p(\alpha) - p_t + a(\alpha), \tag{10.22}$$

and all terms of the derivative matrix on the right-hand side are evaluated at $(\Delta\gamma, \alpha) = (\Delta\gamma^{(k-1)}, \alpha_{n+1}^{(k-1)})$.

The elastoplastic consistent tangent

To derive the elastoplastic consistent tangent operator for the above implementation of the modified Cam-Clay model, we resort to the general procedure of Section 7.4.4 (from page 238), which was also applied in Section 7.4.5 (page 240) as an alternative to the derivation of the elastoplastic consistent tangent for the implicit integration of the von Mises model. Accordingly, we start by differentiating the full system of return-mapping equations which, in the present case, comprises (10.11)–(10.13) together with the trivial update formula (10.18). The differentiation, in which the elastic trial strain tensor is also treated as a variable, yields the tangential relationship

$$\begin{bmatrix} [\mathsf{D}^e]^{-1} + \Delta\gamma\,\partial\boldsymbol{N}/\partial\boldsymbol{\sigma} & 0 & \boldsymbol{N} \\ \Delta\gamma\,\partial N_{\mathrm{v}}/\partial\boldsymbol{\sigma} & 1+\dfrac{2H\Delta\gamma}{b^2} & N_{\mathrm{v}} \\ \boldsymbol{N} & \partial\Phi/\partial\alpha & 0 \end{bmatrix} \begin{Bmatrix} \mathrm{d}\boldsymbol{\sigma}_{n+1} \\ \mathrm{d}\alpha_{n+1} \\ \mathrm{d}\Delta\gamma \end{Bmatrix} = \begin{Bmatrix} \mathrm{d}\boldsymbol{\varepsilon}^{e\ \mathrm{trial}} \\ 0 \\ 0 \end{Bmatrix}, \tag{10.23}$$

where, following the deviatoric/volumetric split of (10.3), we write

$$\frac{\partial \boldsymbol{N}}{\partial \boldsymbol{\sigma}} = \frac{\partial \boldsymbol{N}_{\mathrm{d}}}{\partial \boldsymbol{\sigma}} + \boldsymbol{I} \otimes \frac{\partial N_{\mathrm{v}}}{\partial \boldsymbol{\sigma}} \tag{10.24}$$

with

$$\frac{\partial \boldsymbol{N}_{\mathrm{d}}}{\partial \boldsymbol{\sigma}} = \frac{3}{M^2}\,\mathsf{I}_{\mathrm{d}}; \qquad \frac{\partial N_{\mathrm{v}}}{\partial \boldsymbol{\sigma}} = \frac{2}{3b^2}\,\boldsymbol{I}. \tag{10.25}$$

In addition, from (10.13) we obtain

$$\frac{\partial \Phi}{\partial \alpha} = \frac{2}{b^2}\,(p - p_t + a)H - 2aH = (N_{\mathrm{v}} - 2a)H. \tag{10.26}$$

Here we should note that tangential relation (10.23) has a similar format to that of (7.141) which was obtained for the implicit implementation of the isotropically hardening von Mises model. Accordingly, its inversion follows completely analogous steps to those leading to (7.148) and yields the following formula for the elastoplastic consistent tangent operator:

$$\begin{aligned} \mathsf{D}^{ep} &\equiv \frac{\partial \boldsymbol{\sigma}_{n+1}}{\partial \boldsymbol{\varepsilon}^{e\ \mathrm{trial}}} \\ &= \mathsf{P} - \frac{1}{\tilde{\boldsymbol{N}} : \mathsf{P} : \boldsymbol{N} + (N_{\mathrm{v}} - 2a) N_{\mathrm{v}} H} (\mathsf{P} : \boldsymbol{N}) \otimes (\mathsf{P} : \tilde{\boldsymbol{N}}), \end{aligned} \tag{10.27}$$

where

$$\mathsf{P} \equiv \left(\mathsf{I}_S + \Delta\gamma\, \mathsf{D}^e : \frac{\partial \boldsymbol{N}}{\partial \boldsymbol{\sigma}} \right)^{-1} : \mathsf{D}^e \tag{10.28}$$

with flow vector derivative given by (10.24), and

$$\begin{aligned} \tilde{\boldsymbol{N}} &\equiv -\Delta\gamma (N_{\mathrm{v}} - 2a) H \frac{\partial N_{\mathrm{v}}}{\partial \boldsymbol{\sigma}} + \left(1 + \frac{2H\Delta\gamma}{b^2}\right) \boldsymbol{N} \\ &= \frac{-2\Delta\gamma (N_{\mathrm{v}} - 2a) H}{3b^2} \boldsymbol{I} + \left(1 + \frac{2H\Delta\gamma}{b^2}\right) \boldsymbol{N}. \end{aligned} \tag{10.29}$$

Since in general $\tilde{\boldsymbol{N}} \neq \boldsymbol{N}$, the elastoplastic tangent operator (10.27) is generally *unsymmetric*. Its non-symmetry is a consequence of the non-associativity of the adopted hardening law. The reader is referred to the discussion at the beginning of Section 7.4.6, from page 243. Also note that in the absence of hardening, that is, if $H = 0$, $\tilde{\boldsymbol{N}}$ coincides with $\boldsymbol{N}$ and symmetry is recovered.

10.2. A capped Drucker–Prager model for geomaterials

The *standard* Drucker–Prager model has been described in detail in Chapter 6 and its computational implementation has been addressed in Section 8.3 (from page 324). For that model (as well as for the Mohr–Coulomb model), the application of an arbitrary compressive hydrostatic pressure alone does not cause plastic flow. Plastic flow under compressive hydrostatic pressure may only be triggered with the superposition of shearing stresses. For many geomaterials, however, compressive plastic flow (compaction) under (possibly pure) compressive hydrostatic stresses can be an important feature of the overall behaviour. In circumstances where plastic compaction is relevant, the standard Drucker–Prager or Mohr–Coulomb plasticity models are not able to capture the actual material behaviour.

The modified Cam-Clay model discussed in the previous section already incorporates the possibility of plastic compaction. Another alternative for modelling such a phenomenon consists in bounding the standard Drucker–Prager yield surface with a *cap* on the compressive side of the hydrostatic axis. A particularly simple choice is the adoption of an *elliptical* cap as the compressive bounding surface. The p–q plane representation of the yield surface so obtained is illustrated in Figure 10.2. The equations of the resulting multisurface (two-surface) model, followed by an outline of its numerical treatment, is presented below. The implementation of a Cap-type model similar to that described here is presented by Simo *et al.* (1988a).

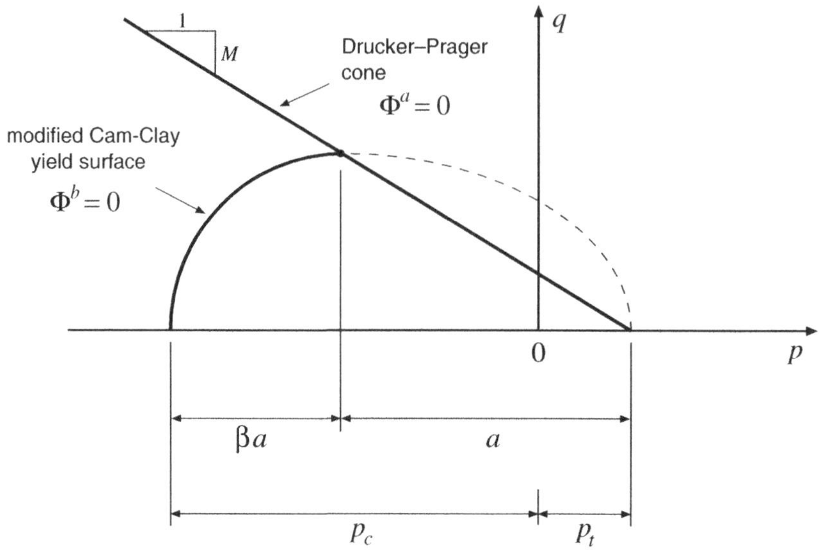

Figure 10.2. Capped Drucker–Prager model. Yield surface.

10.2.1. CAPPED DRUCKER–PRAGER MODEL

The yield functions of the model are

$$\begin{aligned}\Phi^a(\boldsymbol{\sigma}) &\equiv \sqrt{J_2(\boldsymbol{s}(\boldsymbol{\sigma}))} + \eta\,[p(\boldsymbol{\sigma}) - p_t];\\ \Phi^b(\boldsymbol{\sigma}, a) &= \frac{1}{\beta^2}\,[p(\boldsymbol{\sigma}) - p_t + a]^2 + \left[\frac{q(\boldsymbol{\sigma})}{M}\right]^2 - a^2,\end{aligned} \tag{10.30}$$

where Φ^a is an equivalent representation of the Drucker–Prager yield function obtained from its original expression (6.121) (page 167) by using the trivial relation

$$c = \frac{\eta}{\xi}\,p_t, \tag{10.31}$$

and $\Phi^b = 0$ defines the elliptical cap – the modified Cam-Clay yield surface for subcritical states discussed in the previous section of this chapter. In $(10.30)_1$,

$$\eta = M/\sqrt{3}\,. \tag{10.32}$$

For a fixed a, the set of plastically admissible stress states is defined by

$$\mathscr{A} = \{\boldsymbol{\sigma} \mid \Phi^a(\boldsymbol{\sigma}) \le 0,\ \Phi^b(\boldsymbol{\sigma}, a) \le 0\}. \tag{10.33}$$

Plastic flow rule

On the cone, the standard generally non-associative Drucker–Prager flow rule is adopted. On the smooth part of the cone, the plastic strain rate equation is given by (6.157)

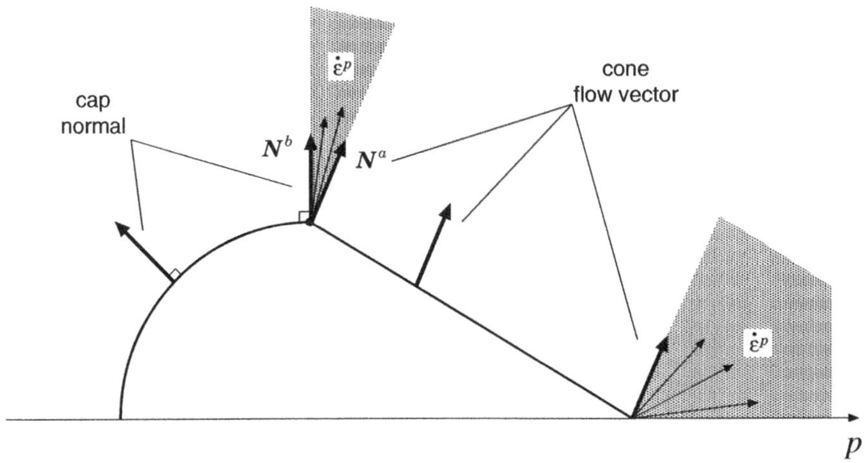

Figure 10.3. Capped Drucker–Prager model. Flow vectors.

(page 176), $(6.158)_1$ and (6.166). At the cone apex, the flow vector has deviatoric and volumetric components $(6.160)_1$ and (6.166), respectively. On the cap, the (associative) flow rule is that of the modified Cam-Clay model, given by (10.3) and (10.4).

Following the general representation of associative plastic flow rules for multisurface plasticity models given by (6.73) (page 156), (6.77) and (6.78), the plastic strain rate at the intersection between the Drucker–Prager cone and the elliptical cap reads

$$\dot{\varepsilon}^p = \dot{\gamma}^a \, \boldsymbol{N}^a + \dot{\gamma}^b \, \boldsymbol{N}^b, \tag{10.34}$$

where $\boldsymbol{N}^a$ and $\boldsymbol{N}^b$ are, respectively, the non-associative Drucker–Prager flow vector at the smooth portion of the cone and the associative modified Cam-Clay flow vector referred to above. The flow vector $\boldsymbol{N}^a$ has the standard form

$$\boldsymbol{N}^a = \frac{1}{2\sqrt{J_2(\boldsymbol{s})}} \, \boldsymbol{s} + \frac{\bar{\eta}}{3} \boldsymbol{I}. \tag{10.35}$$

At the cap/cone intersection, where $p = p_t - a$, the normal to the cap with general expression (10.4) is a deviatoric tensor:

$$\boldsymbol{N}^b = \boldsymbol{N}^b_{\rm d} = \frac{3}{M^2} \, \boldsymbol{s}. \tag{10.36}$$

A general p-q plane illustration of the plastic flow directions for the resulting model is given in Figure 10.3.

Hardening

As for the previously discussed modified Cam-Clay model (refer to expressions (10.9) and (10.8)) isotropic strain hardening is incorporated by letting the compaction pressure p_c (or, equivalently, the yield surface parameter a) be a function of the compressive-positive *volumetric plastic strain*, α, defined in (10.7). The Drucker–Prager cone remains fixed and is independent of the hardening variable.

As p_c increases (decreases), the cap moves in the compression (tensile) direction along the hydrostatic line, expanding (shrinking) the elastic domain.

10.2.2. THE IMPLICIT INTEGRATION ALGORITHM

The integration algorithm for the present model is a combination of the elastic predictor/return-mapping schemes derived in Sections 8.3.1 (from page 325) and 10.1.2 for the Drucker–Prager and modified Cam-Clay models, respectively. The two-surface model admits four possible return mappings:

1. return to the smooth portion of the Drucker–Prager cone;
2. return to the cone apex;
3. return to the elliptic cap;
4. return to the cone/cap intersection.

The return mappings to the smooth portion and apex of the Drucker–Prager cone are a particular instance of the procedure described in Section 8.3 (from page 324), with the added simplicity of perfect plasticity. In this case (refer to Remark 8.4, page 327), the return-mapping equations can be solved in closed form. The return mapping to the cap surface, in turn, is that of the modified Cam-Clay model discussed in Section 10.1.2.

The plastic corrector algorithm for the cone/cap intersection is a two-vector return mapping where the incremental plastic strain is given by the discrete form of (10.34)–(10.36):

$$\Delta \varepsilon^p = \Delta\gamma^a \, \boldsymbol{N}^a + \Delta\gamma^b \, \boldsymbol{N}^b. \tag{10.37}$$

Its volumetric component reads

$$\varepsilon^p_{\rm v} = \bar{\eta} \, \Delta\gamma^a. \tag{10.38}$$

so that the update formula for the hardening internal variable at the cone/cap intersection is given by

$$\alpha_{n+1} = \alpha_n - \bar{\eta} \, \Delta\gamma^a. \tag{10.39}$$

The return-mapping equation for the cone/cap intersection is derived by introducing (10.37) into the general stress update formula

$$\begin{aligned} \boldsymbol{s}_{n+1} &= \boldsymbol{s}^{\rm trial}_{n+1} - 2G\Delta\varepsilon^p_{\rm d}, \\ p_{n+1} &= p^{\rm trial}_{n+1} - K\Delta\varepsilon^p_{\rm v}, \end{aligned} \tag{10.40}$$

and then substituting the resulting expression, together with (10.39) into the cone/cap intersection consistency equation

$$\begin{aligned} \Phi^a_{n+1} &\equiv \sqrt{J_2(\boldsymbol{s}_{n+1})} + \eta[p_{n+1} - p_t] = 0 \\ \Phi^b_{n+1} &\equiv \frac{1}{\beta^2}\,[p_{n+1} - p_t + a(\alpha_{n+1})]^2 + \left[\frac{q(\boldsymbol{s}_{n+1})}{M}\right]^2 - [a(\alpha_{n+1})]^2 = 0. \end{aligned} \tag{10.41}$$

The derivation of the final equations for the unknown multipliers $\Delta\gamma^a$ and $\Delta\gamma^b$ is straightforward and will be left as an exercise for the interested reader.

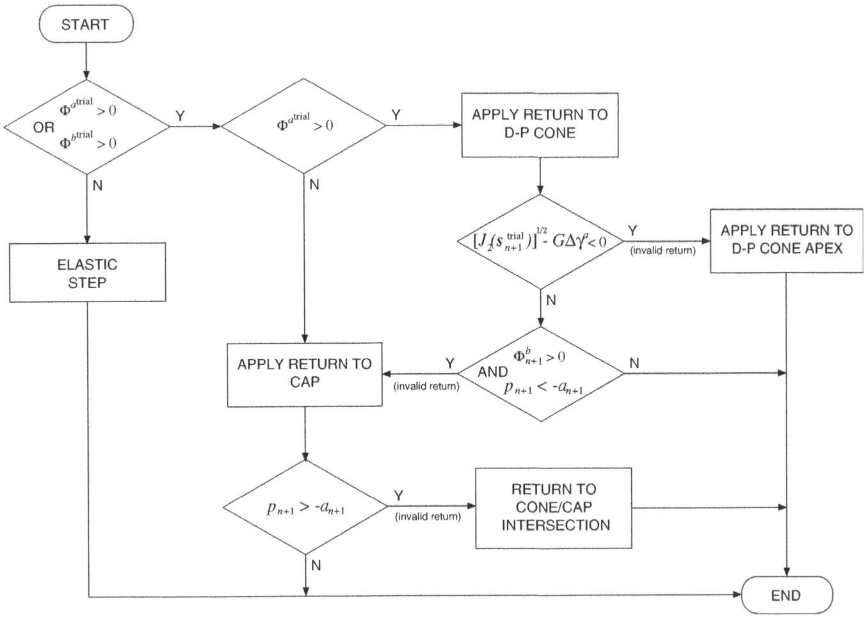

Figure 10.4. Capped Drucker–Prager model. Algorithm for selection of the correct return-mapping procedure.

Selection of the appropriate return mapping

As discussed in detail in Chapter 8, when dealing with multivector implementations of multisurface plasticity models, an appropriate algorithm has to be devised to select, for a given elastic trial state, the return mapping that fully satisfies plastic consistency. The selection algorithm in the present case is an extension of that of the conventional Drucker–Prager model represented in the flowchart of Figure 8.15 (page 330). Its validity can also be established on the basis of geometrical arguments. This will be left as an exercise for the reader. The summary of a possible selection algorithm is shown in the flowchart of Figure 10.4.

10.2.3. THE ELASTOPLASTIC CONSISTENT TANGENT OPERATOR

The elastoplastic consistent tangent will be dependent on the particular return mapping considered, as four distinct return procedures are possible for the present model. The situation here is an extension of that studied in Chapter 8 for the standard Drucker–Prager model. The basic rule for selection of the elastoplastic tangent to be used is the same: the tangent operator must be consistent with the last return-mapping procedure used in the material point (Gauss point of the finite element mesh) considered; that is, if the return to the main Drucker–Prager cone was used in the last stress update for that point, then the current elastoplastic tangent will be consistent with the Drucker–Prager cone return. If the cone–cap intersection was the last return mapping used, then the current tangent will be consistent with the cone–cap intersection return, and so on.

The tangent operator consistent with the Drucker–Prager cone and cone apex return algorithms are those derived in Section 8.3.4 (from page 337) which, for the present model, must be particularised for the perfectly plastic case. Note that, in this case, at the cone apex, we have

$$\mathsf{D}^{ep} = \mathbf{0}. \tag{10.42}$$

For the cap return, the tangent operator is that of (10.27) with $b = \beta$ in (10.29). The derivation of an explicit form for the tangent for the cone/cap intersection return follows an analogous procedure to those of the two-vector return mappings discussed in Chapter 8 and will be left as an exercise for the reader.

10.3. Anisotropic plasticity: the Hill, Hoffman and Barlat–Lian models

The elastoplasticity theories discussed so far in this book have been limited by the assumption of elastoplastic *isotropy*; that is, elastic properties (Young's modulus and Poisson's ratio) as well as plastic properties such as yield stress and hardening behaviour have been assumed independent of material orientation. However, in many circumstances of practical interest, material behaviour is truly *anisotropic* with substantial discrepancy among phenomenological properties observed in different material directions.

An important instance of anisotropic behaviour arises in polycrystalline metals. Such metals are aggregates of single crystals whose individual plastic behaviour is highly anisotropic (physical aspects and the numerical treatment of single crystal plasticity are addressed in Chapter 16). Polycrystalline aggregates with sufficiently random crystal orientation distribution present an effectively isotropic macroscopic behaviour and can be modelled by conventional isotropic plasticity theories. However, when such materials undergo manufacturing processes characterised by extreme straining along preferential directions (rolling is a typical example), the development of texturing (reorientation of crystals according to the direction of straining) results in a final crystallographic arrangement whose macroscopic behaviour is clearly anisotropic. This property is known as *strain-induced* plastic anisotropy and is often present in formed metal components. Another class of materials whose anisotropy can be of particular relevance are composites in general. Many such materials (fibre-reinforced polymers are typical examples) are designed so as to produce optimal strength/performance along predefined directions.

In such cases, the isotropy assumption may lead to a poor representation of the actual behaviour and the formulation and use of appropriate anisotropic plasticity models becomes crucial to ensure reasonable accuracy in finite element predictions. In this section, we address the modelling and computational implementation of anisotropic phenomenological plasticity models. Attention is focused on the orthotropic models proposed by Hill (1948, 1950), Hoffman (1967) and Barlat and Lian (1989).

10.3.1. THE HILL ORTHOTROPIC MODEL

The Hill criterion has been introduced as an orthotropic extension of the standard von Mises criterion in order to model the anisotropy often found in formed steel. With σ_{ij} denoting the stress tensor components on an orthonormal basis $\{e_1, e_2, e_3\}$ whose vectors coincide with the principal axes of plastic orthotropy, the yield function associated with the Hill criterion can be cast in the following form:

$$\Phi(\boldsymbol{\sigma}, \bar{\sigma}) = F_1\,(\sigma_{11} - \sigma_{22})^2 + F_2\,(\sigma_{22} - \sigma_{33})^2 + F_3\,(\sigma_{33} - \sigma_{11})^2 + F_4\,\sigma_{12}^2 + F_5\,\sigma_{23}^2 + F_6\,\sigma_{13}^2 - \bar{\sigma}^2, \tag{10.43}$$

where $\bar{\sigma}$ is the *relative yield stress* (a non-dimensional scalar) which defines the size (state of hardening) of the yield surface in the six-dimensional space of stress components. The constants F_1, F_2 and F_3 are defined as

$$\begin{aligned} F_1 &= \frac{1}{2}\left(\frac{1}{(\sigma_{11}^0)^2} + \frac{1}{(\sigma_{22}^0)^2} - \frac{1}{(\sigma_{33}^0)^2}\right); \\ F_2 &= \frac{1}{2}\left(\frac{-1}{(\sigma_{11}^0)^2} + \frac{1}{(\sigma_{22}^0)^2} + \frac{1}{(\sigma_{33}^0)^2}\right); \\ F_3 &= \frac{1}{2}\left(\frac{1}{(\sigma_{11}^0)^2} - \frac{1}{(\sigma_{22}^0)^2} + \frac{1}{(\sigma_{33}^0)^2}\right), \end{aligned} \tag{10.44}$$

where σ_{11}^0, σ_{22}^0 and σ_{33}^0 are, respectively, the (generally distinct) uniaxial yield stresses in the directions of $\boldsymbol{e}_1$, $\boldsymbol{e}_2$ and $\boldsymbol{e}_3$ when $\bar{\sigma} = 1$ (here assumed to be the *initial* state of the material). The constants F_4, F_5 and F_6 are defined as

$$F_4 = \frac{1}{(\sigma_{12}^0)^2}; \quad F_5 = \frac{1}{(\sigma_{23}^0)^2}; \quad F_6 = \frac{1}{(\sigma_{13}^0)^2}, \tag{10.45}$$

where σ_{ij}^0 are the three (generally distinct) initial (when $\bar{\sigma} = 1$) yield stresses in pure shear along the corresponding planes orthogonal to the principal directions of orthotropy. For a generic state of hardening (i.e. not necessarily unity $\bar{\sigma}$), the yield stress under a stress state with a single non-zero stress component σ_{ij} is

$$\sigma_{ij}^y = \bar{\sigma}\,\sigma_{ij}^0. \tag{10.46}$$

Indeed, note that under such a stress state it follows from (10.43) that

$$\Phi(\boldsymbol{\sigma}, \bar{\sigma}) = 0 \Longleftrightarrow \sigma_{ij} = \bar{\sigma}\,\sigma_{ij}^0. \tag{10.47}$$

Pressure insensitivity

One important feature of the Hill criterion is the fact that, as for the von Mises and Tresca criteria, it is *pressure insensitive*. This allows us to express its yield function in terms of the components of the stress deviator, $\boldsymbol{s}$, only. By recalling the trivial identities

$$\sigma_{ii} - \sigma_{jj} = s_{ii} - s_{jj}, \tag{10.48}$$

with no summation on repeated indices, and

$$\sigma_{ij} = s_{ij}, \quad \text{for } i \neq j, \tag{10.49}$$

the Hill yield function can be equivalently written as

$$\Phi(\boldsymbol{\sigma}, \bar{\sigma}) = F_1\,(s_{11} - s_{22})^2 + F_2\,(s_{22} - s_{33})^2 + F_3\,(s_{11} - s_{33})^2 + F_4\,s_{12}^2 + F_5\,s_{23}^2 + F_6\,s_{13}^2 - \bar{\sigma}^2. \tag{10.50}$$

Von Mises criterion as a particular case

Another important feature of the Hill criterion is that it recovers the standard von Mises criterion if

$$\sigma_{11}^0 = \sigma_{22}^0 = \sigma_{33}^0 = \sqrt{3}\,\sigma_{12}^0 = \sqrt{3}\,\sigma_{23}^0 = \sqrt{3}\,\sigma_{13}^0. \tag{10.51}$$

To see this, we assume that the above relation holds and set $\sigma_{11}^0 = \sigma_y$ and $\bar{\sigma} = 1$ in (10.50). After a straightforward algebraic manipulation, where we make use of the trivial relation

$$s_{11} + s_{22} + s_{33} = 0, \tag{10.52}$$

we obtain

$$\begin{aligned}\Phi(\boldsymbol{\sigma}, \bar{\sigma}) &= \frac{3}{2\sigma_y^2}[s_{11}^2 + s_{22}^2 + s_{33}^2 + 2(s_{12}^2 + s_{23}^2 + s_{13}^2)] - 1 \\ &= \frac{3}{\sigma_y^2}\, J_2(\boldsymbol{s}) - 1.\end{aligned} \tag{10.53}$$

The corresponding yield surface is that of the von Mises model with uniaxial yield stress σ_y:

$$\sqrt{3\, J_2(\boldsymbol{s})} = \sigma_y. \tag{10.54}$$

Geometric representation

Unlike isotropic models, whose yield functions can always be expressed in terms of principal stresses and the corresponding yield surfaces can be visualised in the three-dimensional principal stress space, yield surfaces for anisotropic models cannot be easily represented graphically. Anisotropic yield surfaces are truly six-dimensional hypersurfaces in the space of stress components. Nevertheless, visualisation of *projections* of such hypersurfaces on two- or three-dimensional subsets of the six-dimensional stress space is possible and can provide a very good insight into the properties of the anisotropic model. Two- and three-dimensional projections are obtained by fixing, respectively, four and three stress components and then plotting the corresponding yield locus on the subset of the two or three free stress components. In general, the two- or three-dimensional yield locus will change if the fixed stress components are changed. Thus, an appropriate visual study of anisotropic yield criteria requires plotting such projections with different combinations of fixed stress values.

In order to give the reader some visual insight into properties of the Hill criterion, we will plot here the corresponding yield surfaces under states of plane stress. Before proceeding to the graphical representation, it is worth remarking that at any given state of shear stress (defined by the components σ_{12}, σ_{23} and σ_{13} relative to the axes of material orthotropy), the yield surface defined by the Hill criterion (with material constants within the 'usable' range of the Hill model) in the three-dimensional σ_{11}-σ_{22}-σ_{33} space is a (generally elliptic) cylinder whose axis is the hydrostatic line ($\sigma_{11} = \sigma_{22} = \sigma_{33}$). The size of the cylinder cross-section (and its intersection with the plane stress space) depends upon the state of shear stress. For plane stress states (on plane $\{\boldsymbol{e}_1, \boldsymbol{e}_2\}$), the Hill yield surface is defined by

$$(F_1 + F_3)\,\sigma_{11}^2 + (F_1 + F_2)\,\sigma_{22}^2 - 2F_1\,\sigma_{11}\sigma_{22} + F_4\,\sigma_{12}^2 - \bar{\sigma}^2 = 0. \tag{10.55}$$

In order to study the effect of the material constants (the yield stresses) on the yield surface, we conveniently assume the following relations:

$$\sigma_{22}^0 = a_{22}\,\sigma_{11}^0; \quad \sigma_{33}^0 = a_{33}\,\sigma_{11}^0;$$
$$\sigma_{12}^0 = \frac{a_{12}}{\sqrt{3}}\,\sigma_{11}^0; \quad \sigma_{23}^0 = \frac{a_{23}}{\sqrt{3}}\,\sigma_{11}^0; \quad \sigma_{13}^0 = \frac{a_{13}}{\sqrt{3}}\,\sigma_{11}^0, \tag{10.56}$$

where a_{ij} are parameters defining all initial yield stresses of the model as a function of σ_{11}^0. Note that the von Mises surface, with uniaxial yield stress σ_{11}^0 is recovered if we set

$$a_{22} = a_{33} = a_{12} = a_{23} = a_{13} = 1. \tag{10.57}$$

Firstly, let us focus on the effect of the variation of direct yield stresses on the Hill yield surface. This effect is illustrated in Figures 10.5 and 10.6. In Figure 10.5, the Hill yield surface on the σ_{11}-σ_{22} space is plotted for different values of direct yield stress σ_{22}^0 (i.e. different values of a_{22}), assuming no shear stresses and

$$a_{33} = 1. \tag{10.58}$$

In this case (absence of shear stresses), the parameters a_{12}, a_{23} and a_{13} have no effect on the Hill yield function. For $a_{22} > 0.5$, the Hill surface is an ellipse intersecting the horizontal axis at ± 1 and the vertical axis at $\pm a_{22}$. The surface with $a_{22} = 1$ corresponds to that of the von Mises model. If a_{22} increases (see surface plotted for $a_{22} = 1.5$), the von Mises ellipse is stretched in the σ_{22} direction. If a_{22} decreases (refer to the surface with $a_{22} = 0.7$), the ellipse is compressed in the vertical direction but substantially stretched along its longer radius. In the limit, when $a_{22} = 0.5$ (one direct yield stress is half of the other two), the original ellipse degenerates into two parallel straight lines (two hyperplanes in the six-dimensional stress space). For $a_{22} < 0.5$ (see surface with $a_{22} = 0.3$), the yield locus produced by Hill's function is a set of two hyperbolic (non-convex) surfaces in stress space. Clearly, in such cases, the elastic domain becomes unbounded in some directions and does not correspond to physical behaviour. Thus, the Hill criterion is to be used only within certain limits of yield strength variation among the orthotropy directions. At this point, it is important to emphasise that the Hill criterion was originally proposed to model anisotropy of *formed steel* components. For such materials, the maximum variation of yield strength between the different orthotropy directions is typically less than about 5 to 10%. Under such conditions, the Hill criterion can provide reasonable approximations to the actual yield surfaces. In the next representation of the Hill surface shown in Figure 10.6, we illustrate the effect of the variation of the direct yield stress σ_{33}^0 (variation of a_{33}) in the absence of shear stresses and with $\sigma_{22}^0 = \sigma_{11}^0$ ($a_{22} = 1$). For $a_{33} > 0.5$, the Hill surface is always an ellipse intersecting the horizontal and vertical axis at ± 1. As σ_{33}^0 increases (see surface with $a_{33} = 1.5$), the ellipse stretches in the biaxial state direction. When σ_{33}^0 decreases (see surfaces with $a_{33} = 0.7$ and 0.6) the ellipse is compressed in the biaxial state direction and stretched in a pure shear ($\sigma_{11} = -\sigma_{22}$) direction. When $a_{33} = 0.5$ (one direct yield stress is half of the other two), the ellipse degenerates into two straight lines and becomes a hyperbola for $a_{33} < 0.5$. Finally, in Figure 10.7 we illustrate the effect of shear stresses on the Hill yield surface. Essentially, any increase in shear stress will shrink the yield surface isotropically (this effect is also present in the von Mises criterion and can only be seen if, as opposed to the usual representation in principal stress space, the surface is represented in the space of direct stresses along *fixed* axes). Obviously, if $\sigma_{12} = \bar{\sigma}\sigma_{12}^0$ (the shear stress has reached its yield limit) the σ_{11}-σ_{22} space surface degenerates to a point at $(0, 0)$.

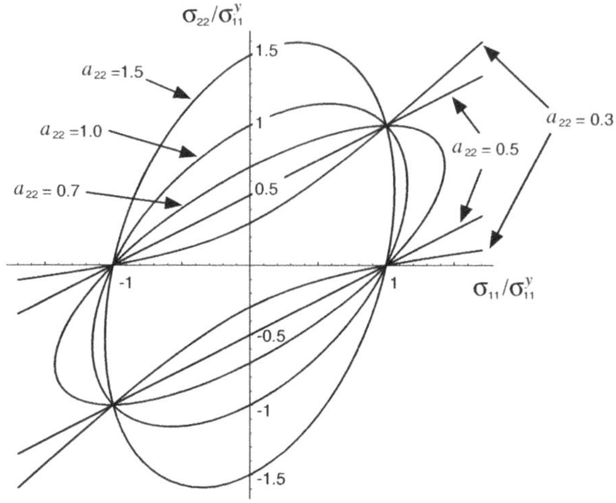

Figure 10.5. The Hill orthotropic criterion. Effect of variation of direct yield stress σ_{22}^0 on the yield surface. Surface plot for various values of a_{22} with no shear stresses and $a_{33} = 1$ ($\sigma_{33}^0 = \sigma_{11}^0$).

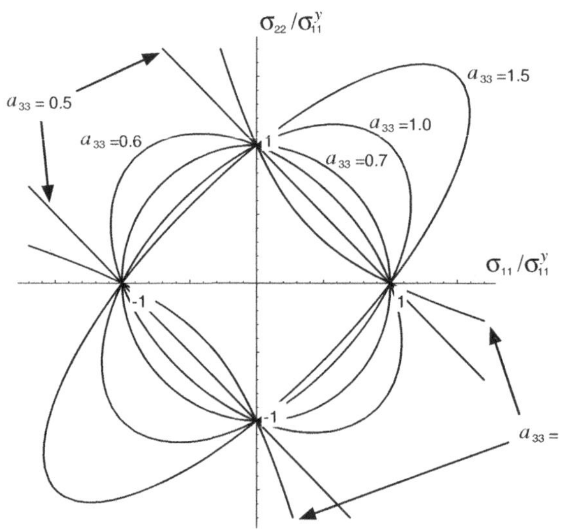

Figure 10.6. The Hill orthotropic criterion. Effect of variation of direct yield stress σ_{33}^0 (the transversal yield stress) on the yield surface. Surface plot for various values of a_{33} with no shear stresses and $a_{22} = 1$ ($\sigma_{22}^0 = \sigma_{11}^0$).

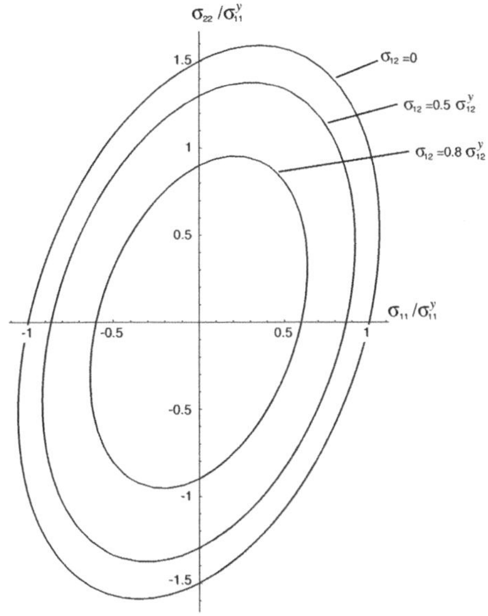

Figure 10.7. The Hill orthotropic criterion. Effect of shear stress on the yield surface. Surface plot for various values of σ_{12} with $\sigma_{23} = \sigma_{13} = 0$, $a_{22} = 1.5$ and $a_{12} = a_{33} = 1$.

Hardening law

Analogously to the standard strain-hardening von Mises model thoroughly discussed in Chapters 6, 7 and 9, *isotropic* strain hardening can be easily incorporated into the Hill model by assuming the relative yield stress $\bar{\sigma}$ in (10.43) to be a given function of the von Mises accumulated plastic strain, $\bar{\varepsilon}^p$; that is, we postulate

$$\bar{\sigma} = \bar{\sigma}(\bar{\varepsilon}^p). \tag{10.59}$$

The material in this case is assumed to remain orthotropic with constant axes of orthotropy. In addition, in view of (10.46), all six yield stresses of the Hill criterion will change in strict proportion (isotropically) as the accumulated plastic strain increases.

Remark 10.2. Since general straining of an initially orthotropic material is usually expected to change the yield stresses in different directions by different amounts, or even lead to loss of orthotropy, the model resulting from the above assumptions provides only a first approximation to the phenomenon of hardening. We remark, however, that the phenomenological modelling of hardening in plastically anisotropic materials is a complex issue which remains open at present. The approximation provided by simple laws of the above type can be very useful in the finite element analysis of plastically orthotropic materials and, undoubtedly, represents a substantial gain in predictive capability (as compared to the use of isotropic plasticity models) in situations where plastic anisotropy is an important feature of the material behaviour.

Hill's associative flow rule

The flow rule adopted by Hill (1950) in conjunction with the above orthotropic criterion is associative; that is, the plastic flow is given by the standard equation

$$\dot{\varepsilon}^p = \dot{\gamma}\,\frac{\partial \Phi}{\partial \boldsymbol{\sigma}}. \tag{10.60}$$

Here, compact tensorial representation is not particularly convenient and the plastic flow rule is best presented in its equivalent component-wise format

$$\dot{\varepsilon}^p_{ij} = \dot{\gamma}\,\frac{\partial \Phi}{\partial \sigma_{ij}}. \tag{10.61}$$

Before presenting the final expressions, it is important to note that the shear stress contributions to (10.43) take into account the trivial identity

$$\sigma_{ij} = \sigma_{ji}, \quad i \neq j. \tag{10.62}$$

The actual contributions to Φ from the independent shear components, σ_{ij} $(i \neq j)$, of the stress tensor read

$$\tfrac{1}{2}[F_4(\sigma_{12}^2 + \sigma_{21}^2) + F_5(\sigma_{23}^2 + \sigma_{32}^2) + F_6(\sigma_{13}^2 + \sigma_{31}^2)]. \tag{10.63}$$

With the above considerations, Hill's associative flow rule is found to be given by

$$\begin{aligned}
\dot{\varepsilon}^p_{11} &= \dot{\gamma}\, 2[F_1(\sigma_{11} - \sigma_{22}) + F_3(\sigma_{11} - \sigma_{33})],\\
\dot{\varepsilon}^p_{22} &= \dot{\gamma}\, 2[F_1(\sigma_{22} - \sigma_{11}) + F_2(\sigma_{22} - \sigma_{33})],\\
\dot{\varepsilon}^p_{33} &= \dot{\gamma}\, 2[F_2(\sigma_{33} - \sigma_{22}) + F_3(\sigma_{33} - \sigma_{11})],\\
\dot{\varepsilon}^p_{12} = \dot{\gamma}\, F_4\, \sigma_{12}, &\quad \dot{\varepsilon}^p_{23} = \dot{\gamma}\, F_5\, \sigma_{23}, \quad \dot{\varepsilon}^p_{13} = \dot{\gamma}\, F_6\, \sigma_{13}.
\end{aligned} \tag{10.64}$$

As expected, owing to the pressure-insensitivity of the Hill criterion, the above associative law is volume-preserving. Indeed, (10.64) implies

$$\operatorname{tr} \dot{\varepsilon}^p \equiv \dot{\varepsilon}^p_{11} + \dot{\varepsilon}^p_{22} + \dot{\varepsilon}^p_{33} = 0. \tag{10.65}$$

10.3.2. TENSION–COMPRESSION DISTINCTION: THE HOFFMAN MODEL

For many materials, a marked difference is observed between yield stress levels in tension and compression (the Bauschinger effect). This phenomenon is particularly noticeable in some composite materials and is also commonly observed in worked metals. In order to model such effects in orthotropic materials, Hoffman (1967) proposed an extension to the Hill criterion described by the following yield function:[§]

$$\begin{aligned}
\Phi(\boldsymbol{\sigma}, \bar{\sigma}) = {} & C_1\,(\sigma_{11} - \sigma_{22})^2 + C_2\,(\sigma_{22} - \sigma_{33})^2 + C_3\,(\sigma_{33} - \sigma_{11})^2\\
& + C_4\,\sigma_{12}^2 + C_5\,\sigma_{23}^2 + C_6\,\sigma_{13}^2 + C_7\,\sigma_{11} + C_8\,\sigma_{22} + C_9\,\sigma_{33} - \bar{\sigma}^2,
\end{aligned} \tag{10.66}$$

[§]This function was in fact proposed by Hoffman (1967) originally to define a *fracture* criterion for brittle materials, with no reference to plastic flow modelling. What we refer to as the Hoffman model here, is an extension of the Hill elastoplastic model with yield criterion and plastic flow rule based on Hoffman's function (10.66).

where $\bar{\sigma}$ is the non-dimensional relative yield stress (analogous to that of the Hill criterion) and $C_1, C_2, \ldots, C_9$ are material constants defined as

$$C_1 = \frac{1}{2}\left(\frac{1}{\sigma^t_{11}\sigma^c_{11}} + \frac{1}{\sigma^t_{22}\sigma^c_{22}} - \frac{1}{\sigma^t_{33}\sigma^c_{33}}\right);$$
$$C_2 = \frac{1}{2}\left(\frac{-1}{\sigma^t_{11}\sigma^c_{11}} + \frac{1}{\sigma^t_{22}\sigma^c_{22}} + \frac{1}{\sigma^t_{33}\sigma^c_{33}}\right); \quad (10.67)$$
$$C_3 = \frac{1}{2}\left(\frac{1}{\sigma^t_{11}\sigma^c_{11}} - \frac{1}{\sigma^t_{22}\sigma^c_{22}} + \frac{1}{\sigma^t_{33}\sigma^c_{33}}\right),$$

with σ^t_{ii} and σ^c_{ii} (no summation on repeated indices) denoting the initial (i.e. when $\bar{\sigma} = 1$) direct yield stresses along the orthotropy direction i, respectively in tension and compression, and

$$C_4 = \frac{1}{(\sigma^0_{12})^2}; \quad C_5 = \frac{1}{(\sigma^0_{23})^2}; \quad C_6 = \frac{1}{(\sigma^0_{13})^2}, \quad (10.68)$$

and

$$C_7 = \frac{\sigma^c_{11} - \sigma^t_{11}}{\sigma^c_{11}\sigma^t_{11}}; \quad C_8 = \frac{\sigma^c_{22} - \sigma^t_{22}}{\sigma^c_{22}\sigma^t_{22}}; \quad C_9 = \frac{\sigma^c_{33} - \sigma^t_{33}}{\sigma^c_{33}\sigma^t_{33}}. \quad (10.69)$$

The constants σ^0_{12}, σ^0_{23} and σ^0_{13} have the same meaning as in the Hill criterion, i.e. they denote the initial yield stresses in states of pure shear on the planes of orthotropy.

Remark 10.3 (Hill criterion as a particular case). If for each principal direction of orthotropy the direct yield stress in tension coincides with the direct yield stress in compression, i.e. if we set

$$\sigma^c_{11} = \sigma^t_{11} = \sigma^0_{11}; \quad \sigma^c_{22} = \sigma^t_{22} = \sigma^0_{22}; \quad \sigma^c_{33} = \sigma^t_{33} = \sigma^0_{33}, \quad (10.70)$$

then $C_7 = C_8 = C_9 = 0$ and the constants $C_1, \ldots, C_6$ coincide with the constants $F_1, \ldots, F_6$ of (10.43). In this case, the Hill criterion is recovered. However, it should be noted that, in general, the Hoffman criterion is (unlike the Hill criterion) *pressure-sensitive*. Indeed, under a state of pure hydrostatic pressure ($\sigma_{11} = \sigma_{22} = \sigma_{33} = p$), the contribution of the linear terms on the direct stresses to Φ in (10.66) will be

$$(C_7 + C_8 + C_9)p. \quad (10.71)$$

This contribution – a linear function of the hydrostatic pressure – is identical to that of the Drucker–Prager criterion (refer to expression (6.121), page 167). The constraint

$$C_7 + C_8 + C_9 = 0 \quad (10.72)$$

is the necessary and sufficient condition for the Hoffman criterion to be pressure-insensitive.

In its general form (within the range of applicability of the criterion), the Hoffman yield surface is an elliptic cone in the σ_{11}-σ_{22}-σ_{33} space. The cone intersects the σ_{11}, σ_{22} and σ_{33} axes at the corresponding prescribed values of (tensile and compressive) direct yield stresses. A graphical representation of the Hoffman criterion is shown in Figure 10.8. Again, only the intersection of the yield surface with the σ_{11}-σ_{22} plane, in the absence of shear stresses, is

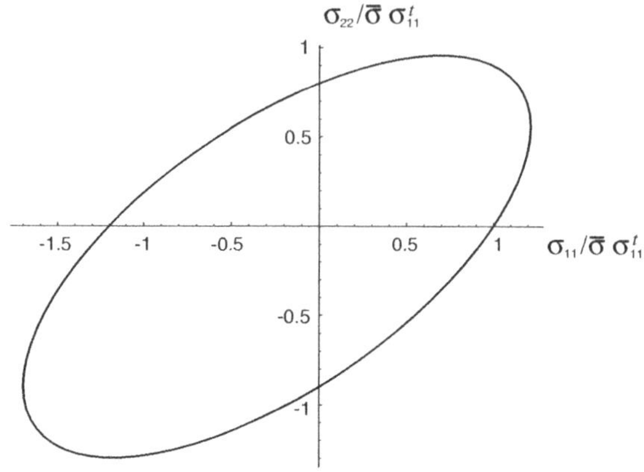

Figure 10.8. The Hoffman orthotropic criterion.

plotted. For illustration purposes only, the following relations have been chosen:

$$\sigma_{11}^c = 1.2\,\sigma_{11}^t; \quad \sigma_{22}^c = 0.9\,\sigma_{11}^t; \quad \sigma_{22}^t = 0.8\,\sigma_{11}^t;$$
$$\sigma_{33}^c = \sigma_{11}^c; \quad \sigma_{33}^t = \sigma_{11}^t. \tag{10.73}$$

As for the Hill criterion (refer to Figure 10.7), the Hoffman surface shrinks isotropically with an increase in shear stresses. The effect of (isotropic) hardening can also be incorporated into the Hoffman criterion by assuming $\bar{\sigma}$ to be a function of the accumulated plastic strain.

Again, it is important to observe that (as for the Hill criterion) the Hoffman model can only be used within certain limits of yield strength variation among the directions of orthotropy. Excessive variations of direct (tensile or compressive) yield stress can cause the originally elliptic surface to degenerate into a hyperbola that does not model the actual behaviour of solids.

Flow rule

Here, an associative flow rule is also postulated. Analogously to (10.64), the associative component-wise plastic flow equations for the Hoffman model are given by

$$\begin{aligned}
\dot{\varepsilon}_{11}^p &= \dot{\gamma}\,[C_7 + 2C_1(\sigma_{11}-\sigma_{22}) + 2C_3(\sigma_{11}-\sigma_{33})],\\
\dot{\varepsilon}_{22}^p &= \dot{\gamma}\,[C_8 + 2C_1(\sigma_{22}-\sigma_{11}) + 2C_2(\sigma_{22}-\sigma_{33})],\\
\dot{\varepsilon}_{33}^p &= \dot{\gamma}\,[C_9 + 2C_2(\sigma_{33}-\sigma_{22}) + 2C_3(\sigma_{33}-\sigma_{11})],\\
\dot{\varepsilon}_{12}^p = \dot{\gamma}\,C_4\,\sigma_{12}, &\quad \dot{\varepsilon}_{23}^p = \dot{\gamma}\,C_5\,\sigma_{23}, \quad \dot{\varepsilon}_{13}^p = \dot{\gamma}\,C_6\,\sigma_{13}.
\end{aligned} \tag{10.74}$$

Remark 10.4 (Non-isochoric plastic flow). As a result of the pressure-sensitivity of the Hoffman criterion, the above associative flow rule is *not volume-preserving* in general. The volumetric plastic strain rate for the present model is

$$\operatorname{tr}\dot{\boldsymbol{\varepsilon}}^p \equiv \dot{\varepsilon}_{11}^p + \dot{\varepsilon}_{22}^p + \dot{\varepsilon}_{33}^p = \dot{\gamma}\,(C_7 + C_8 + C_9), \tag{10.75}$$

and the plastic flow is isochoric if and only if constraint (10.72) holds.

10.3.3. IMPLEMENTATION OF THE HOFFMAN MODEL

The computational implementation of the Hoffman model is described in what follows. For simplicity, the discussion will be limited to the plane strain and axisymmetric cases. The extension to the full three-dimensional situation is trivial. Early implementations of the Hill and Hoffman anisotropic models are described, respectively, by de Borst and Feenstra (1990) and Schellekens and de Borst (1990). A plane stress implementation of the Hoffman model in the context of explicit dynamics finite element analysis is given by Koh *et al.* (1995).

The orthotropic elastic law

As plastically orthotropic composites usually also present a marked elastic orthotropy, we shall assume here an *orthotropic elasticity* law. Clearly, in this case it makes sense to consider an elastic law whose planes of symmetry coincide with the planes of symmetry of the Hoffman criterion. Using the finite element array notation, the stress–elastic strain relation then reads

$$\boldsymbol{\sigma} = \mathbf{D}\,\boldsymbol{\varepsilon}^e, \tag{10.76}$$

where

$$\boldsymbol{\sigma}^e = [\sigma_{11}\ \ \sigma_{22}\ \ \sigma_{12}\ \ \sigma_{33}]^T; \qquad \boldsymbol{\varepsilon}^e = [\varepsilon^e_{11}\ \ \varepsilon^e_{22}\ \ 2\varepsilon^e_{12}\ \ \varepsilon^e_{33}]^T, \tag{10.77}$$

and $\mathbf{D}$ denotes the *orthotropic elasticity matrix* given by

$$\mathbf{D} = \begin{bmatrix} D_{11} & D_{12} & 0 & D_{13} \\ D_{12} & D_{22} & 0 & D_{23} \\ 0 & 0 & D_{44} & 0 \\ D_{13} & D_{23} & 0 & D_{33} \end{bmatrix}, \tag{10.78}$$

with

$$D_{ii} = \frac{E_i(1-\nu_{jk}\nu_{kj})}{(1-\nu_{ki}\nu_{ik})(1-\nu_{jk}\nu_{kj}) - (\nu_{ij}+\nu_{ik}\nu_{kj})(\nu_{ji}+\nu_{jk}\nu_{ki})} \tag{10.79}$$

for $i = 1, 2, 3$ (no summation on the repeated index) and (i, j, k) denoting cyclic permutations of $(1, 2, 3)$,

$$D_{ij} = \frac{\nu_{ji}+\nu_{jk}\nu_{ki}}{1-\nu_{jk}\nu_{kj}}\, D_{ii} \tag{10.80}$$

for $i \neq j$ and, again, no summation on the repeated index, and

$$D_{44} = G_{12}. \tag{10.81}$$

The total number of independent elastic constants is nine (in the full three-dimensional case). These are: the Young's moduli, E_1, E_2 and E_3, associated with directions 1, 2 and 3 of orthotropy respectively; the shear moduli, G_{12}, G_{23} and G_{31} associated, respectively, with planes 12, 23 and 31 and satisfying

$$G_{ij} = G_{ji}; \tag{10.82}$$

and three Poisson's ratios, ν_{12}, ν_{23} and ν_{31}, where ν_{ij} is defined as the ratio between the contraction in direction j and extension in direction i under a uniaxial stress state along i.

The following relation applies:

$$\nu_{ji} = \nu_{ij}\,\frac{E_j}{E_i}. \tag{10.83}$$

In the plane strain/axisymmetric case, with the elasticity operator represented in (10.78), the transverse shear moduli G_{23} and G_{31} are not required.

Elastic trial state

The computation of the elastic trial state proceeds as usual. That is, we compute

$$\boldsymbol{\sigma}^{\text{trial}} = \mathbf{D}\,\boldsymbol{\varepsilon}^{e\ \text{trial}}, \tag{10.84}$$

where, clearly, the $\mathbf{D}$ matrix here is the orthotropic operator (10.78). The trial accumulated plastic strain (isotropic hardening state variable) is

$$\bar{\varepsilon}^{p\ \text{trial}} = \bar{\varepsilon}^p_n. \tag{10.85}$$

If the trial state is outside the elastic domain defined by the yield function (10.66), the return mapping described below is applied.

Return-mapping algorithm

For computer implementation purposes, it is convenient to write the Hoffman yield function (10.66) in the following equivalent form in terms of the array of stress components:

$$\Phi(\boldsymbol{\sigma}, \bar{\sigma}) = \tfrac{1}{2}\,\boldsymbol{\sigma}^T\mathbf{P}\boldsymbol{\sigma} + \mathbf{q}^T\boldsymbol{\sigma} - \bar{\sigma}^2, \tag{10.86}$$

where

$$\mathbf{P} = 2\begin{bmatrix} C_1+C_3 & -C_1 & 0 & -C_3 \\ -C_1 & C_2+C_1 & 0 & -C_2 \\ 0 & 0 & C_4 & 0 \\ -C_3 & -C_2 & 0 & C_3+C_2 \end{bmatrix} \tag{10.87}$$

and

$$\mathbf{q} = [\, C_7 \quad C_8 \quad 0 \quad C_9 \,]^T. \tag{10.88}$$

With the above notation, the flow rule (10.74) can be equivalently expressed in terms of the engineering plastic strain rate array as

$$\dot{\boldsymbol{\varepsilon}}^p = \dot{\gamma}\,(\mathbf{P}\boldsymbol{\sigma} + \mathbf{q}). \tag{10.89}$$

The rate of accumulated plastic strain, in turn, can be represented as

$$\begin{aligned} \dot{\bar{\varepsilon}}^p &\equiv \sqrt{\tfrac{2}{3}\,\dot{\varepsilon}^p : \dot{\varepsilon}^p} = \dot{\gamma}\,\sqrt{\tfrac{2}{3}\,(\dot{\boldsymbol{\varepsilon}}^p)^T\,\mathbf{Z}\,\dot{\boldsymbol{\varepsilon}}^p} \\ &= \dot{\gamma}\,\sqrt{\tfrac{2}{3}\,(\mathbf{P}\boldsymbol{\sigma}+\mathbf{q})^T\,\mathbf{Z}\,(\mathbf{P}\boldsymbol{\sigma}+\mathbf{q})}, \end{aligned} \tag{10.90}$$

where we have used the definition

$$\mathbf{Z} = \begin{bmatrix} 1 & 0 & 0 & 0 \\ 0 & 1 & 0 & 0 \\ 0 & 0 & \frac{1}{2} & 0 \\ 0 & 0 & 0 & 1 \end{bmatrix}. \tag{10.91}$$

With the above relations at hand, a direct particularisation of the general return-mapping equations (7.25) (page 196) is obtained as

$$\left\{ \begin{array}{c} \boldsymbol{\varepsilon}^e_{n+1} - \boldsymbol{\varepsilon}^{e\ \mathrm{trial}} + \Delta\gamma\,(\mathbf{P}\boldsymbol{\sigma}_{n+1} + \mathbf{q}) \\ \bar{\varepsilon}^p_{n+1} - \bar{\varepsilon}^p_n - \Delta\gamma[\frac{2}{3}\,(\mathbf{P}\boldsymbol{\sigma}_{n+1} + \mathbf{q})^T\,\mathbf{Z}\,(\mathbf{P}\boldsymbol{\sigma}_{n+1} + \mathbf{q})]^{\frac{1}{2}} \\ \frac{1}{2}\,\boldsymbol{\sigma}^T_{n+1}\mathbf{P}\boldsymbol{\sigma}_{n+1} + \mathbf{q}^T\boldsymbol{\sigma}_{n+1} - [\bar{\sigma}(\bar{\varepsilon}^p_{n+1})]^2 \end{array} \right\} = \left\{ \begin{array}{c} \mathbf{0} \\ 0 \\ 0 \end{array} \right\}. \tag{10.92}$$

Single-equation return mapping

Analogously to the implementation of the von Mises model, the return mapping here can also be reduced to the solution of a single equation for the plastic multiplier, $\Delta\gamma$. To obtain the reduced equation, we start by noting that, by using the linear elastic law, the first equation of system (10.92) can be expressed, after a straightforward manipulation, as

$$\boldsymbol{\sigma}_{n+1} = \boldsymbol{\sigma}(\Delta\gamma) \equiv (\mathbf{I} + \Delta\gamma\,\mathbf{DP})^{-1}(\boldsymbol{\sigma}^{\mathrm{trial}} - \Delta\gamma\,\mathbf{Dq}); \tag{10.93}$$

that is, the updated stress, $\boldsymbol{\sigma}_{n+1}$ is a function of $\Delta\gamma$, exclusively. Further, introduction of the above function in the second equation of (10.92) gives

$$\bar{\varepsilon}^p_{n+1} = \bar{\varepsilon}^p(\Delta\gamma) \equiv \bar{\varepsilon}^p_n + \Delta\gamma\{\tfrac{2}{3}\,[\mathbf{P}\boldsymbol{\sigma}(\Delta\gamma) + \mathbf{q}]^T\,\mathbf{Z}\,[\mathbf{P}\boldsymbol{\sigma}(\Delta\gamma) + \mathbf{q}]\}^{\frac{1}{2}}. \tag{10.94}$$

Finally, with the substitution of the last two expressions in the third equation of (10.92) – the discretised plastic consistency – the original return-mapping system is reduced to the following nonlinear scalar equation for $\Delta\gamma$:

$$\tilde{\Phi}(\Delta\gamma) \equiv \tfrac{1}{2}\,[\boldsymbol{\sigma}(\Delta\gamma)]^T\mathbf{P}\boldsymbol{\sigma}(\Delta\gamma) + \mathbf{q}^T\boldsymbol{\sigma}(\Delta\gamma) - [\bar{\sigma}(\bar{\varepsilon}^p(\Delta\gamma))]^2 = 0. \tag{10.95}$$

In summary, the stress-updating procedure for the Hoffman model comprises the solution of (10.95), followed by the update of stress and accumulated plastic strain according to (10.93) and (10.94), respectively.

Newton–Raphson solution

In the Newton–Raphson iterative scheme to solve (10.95), the kth guess for the solution $\Delta\gamma$ is obtained as

$$\Delta\gamma^{(k)} = \Delta\gamma^{(k-1)} + \delta\Delta\gamma^{(k)}, \tag{10.96}$$

where

$$\delta\Delta\gamma^{(k)} = -\left.\frac{\tilde{\Phi}(\Delta\gamma)}{\mathrm{d}\tilde{\Phi}/\mathrm{d}\Delta\gamma}\right|^{(k-1)}. \tag{10.97}$$

The derivative of $\tilde{\Phi}$ is given by

$$\frac{\mathrm{d}\tilde{\Phi}}{\mathrm{d}\Delta\gamma} = [\mathbf{P}\boldsymbol{\sigma}(\Delta\gamma) + \mathbf{q}]^T \frac{\mathrm{d}\boldsymbol{\sigma}}{\mathrm{d}\Delta\gamma} - 2\,\bar{\sigma}(\Delta\gamma)\,\bar{H}\,\frac{\mathrm{d}\bar{\varepsilon}^p}{\mathrm{d}\Delta\gamma}, \tag{10.98}$$

where $\bar{H} \equiv \mathrm{d}\bar{\sigma}/\mathrm{d}\bar{\varepsilon}^p$ is the slope of the (non-dimensional) hardening curve. To obtain an explicit expression for the derivative of function $\boldsymbol{\sigma}$ defined by (10.93), we need to use formula (vii) of page 36. A straightforward manipulation then gives

$$\frac{\mathrm{d}\boldsymbol{\sigma}}{\mathrm{d}\Delta\gamma} = -(\mathbf{I} + \Delta\gamma\,\mathbf{DP})^{-1}\mathbf{D}\,[\mathbf{P}\,\boldsymbol{\sigma}(\Delta\gamma) + \mathbf{q}]. \tag{10.99}$$

With the above expressions at hand, the derivation of a complete explicit formula for the derivative of $\tilde{\Phi}$ is a rather simple exercise which we shall leave for the interested reader.

The elastoplastic consistent tangent

As in the modified Cam-Clay implementation presented earlier in this chapter (refer to the text preceding equation (10.23), page 408), to derive the elastoplastic consistent tangent operator for the above implementation of the Hoffman model, we shall follow the general procedure of Section 7.4.4 (from page 238). Here, we start by differentiating the system (10.92), which together with the elastic law, gives the following linearised system in finite element array notation:

$$\begin{bmatrix} \mathbf{D}^{-1} + \Delta\gamma\mathbf{P} & 0 & \mathbf{N} \\ -\dfrac{2\Delta\gamma}{3\eta}\mathbf{N}^T\mathbf{ZP} & 1 & -\eta \\ \mathbf{N}^T & -2\,\bar{\sigma}\bar{H} & 0 \end{bmatrix} \begin{Bmatrix} \mathrm{d}\boldsymbol{\sigma}_{n+1} \\ \mathrm{d}\bar{\varepsilon}^p_{n+1} \\ \mathrm{d}\Delta\gamma \end{Bmatrix} = \begin{Bmatrix} \mathrm{d}\boldsymbol{\varepsilon}^{e\ \mathrm{trial}} \\ 0 \\ 0 \end{Bmatrix}, \tag{10.100}$$

where $\bar{\sigma} = \bar{\sigma}(\bar{\varepsilon}^p_{n+1})$ and $\bar{H}$ is the non-dimensional hardening modulus

$$\bar{H} = \frac{\mathrm{d}\bar{\sigma}}{\mathrm{d}\bar{\varepsilon}^p}, \tag{10.101}$$

also evaluated at $\bar{\varepsilon}^p_{n+1}$. The flow vector $\mathbf{N}$ is given by

$$\mathbf{N} = \mathbf{P}\boldsymbol{\sigma}_{n+1} + \mathbf{q}, \tag{10.102}$$

and the scalar η is defined as

$$\eta = \sqrt{\tfrac{2}{3}\mathbf{N}^T\mathbf{Z}\,\mathbf{N}}. \tag{10.103}$$

Finally, with the inversion of (10.100) we obtain

$$\mathbf{D}^{ep} \equiv \frac{\mathrm{d}\boldsymbol{\sigma}_{n+1}}{\mathrm{d}\boldsymbol{\varepsilon}^{e\ \mathrm{trial}}} = \left[\mathbf{D}^{-1} + \Delta\gamma\,\mathbf{P} + \frac{1}{2\bar{\sigma}\bar{H}\eta}\,\mathbf{NN}^T - \frac{2\Delta\gamma}{3\eta^2}\,\mathbf{N}^T\mathbf{NZP}\right]^{-1}. \tag{10.104}$$

The above operator is clearly unsymmetric. Its non-symmetry is a consequence of the fact that the adopted hardening rule is non-associative (despite the associativity of the plastic flow rule). Refer to the comments made in Section 7.4.6, page 243. Note that the above formula is analogous to that given by (7.149) for the von Mises isotropically hardening model. Its use, in this format, is restricted strictly to hardening models ($\bar{H} \neq 0$) owing to the presence of the hardening modulus in the denominator of the third summand of the term within square brackets. For use under the assumption of perfect plasticity ($H = 0$), an alternative representation can be obtained following completely analogous steps to those leading to expression (7.148). This will be left as an exercise for the interested reader.

10.3.4. THE BARLAT–LIAN MODEL FOR SHEET METALS

The Barlat–Lian criterion (Barlat and Lian, 1989) has been proposed to model the behaviour of orthotropic metallic sheets (typically rolled materials) under plane stress. Unlike the other plasticity models discussed in this book, whose equations are firstly formulated in the six-dimensional stress space and corresponding plane stress versions are subsequently addressed within the framework of Chapter 9, the Barlat–Lian criterion was originally defined in plane stress format. The corresponding yield function, written at the outset exclusively in terms of in-plane components of the stress tensor, reads

$$\Phi(\boldsymbol{\sigma}, \sigma_{11}^y) = f(\boldsymbol{\sigma}) - 2\,(\sigma_{11}^y)^M, \tag{10.105}$$

where

$$f(\boldsymbol{\sigma}) \equiv a|K_1 + K_2|^M + a|K_1 - K_2|^M + (2 - a)|2\,K_2|^M, \tag{10.106}$$

with

$$K_1 = \frac{\sigma_{11} + h\,\sigma_{22}}{2}; \quad K_2 = \sqrt{\left(\frac{\sigma_{11} - h\,\sigma_{22}}{2}\right)^2 + b^2\,\sigma_{12}^2}, \tag{10.107}$$

where M, a, b and h are material constants and σ_{11}^y is the uniaxial yield stress in the principal orthotropy direction 1. The yield function (10.105) is convex (Barlat and Lian, 1989) if

$$M > 1, \quad a, b, h > 0, \quad a < 2. \tag{10.108}$$

To give the reader an idea of realistic values for these constants, the following have been determined by Lege *et al.* (1989) for an aluminium alloy:

$$M = 8, \quad a = 1.24, \quad b = 1.02, \quad h = 1.15. \tag{10.109}$$

The constant M defines the curvature of the yield surface. This can be seen in Figure 10.9 where the projection of the Barlat–Lian yield surface on the σ_{11}-σ_{22} plane is shown (in the absence of shear stresses) for different values of M and $a = h = 1$ (note that the value of b is immaterial in the absence of shear). We remark that the criterion becomes isotropic when $a = b = h = 1$. Under such a condition, the standard von Mises locus is recovered when $M = 2$ and the Tresca yield locus is recovered for $M = 1$ and $M \to \infty$. The parameter h relates the uniaxial yield strength, σ_{22}^y, in the principal orthotropy direction 2 with that of direction 1; that is,

$$\sigma_{22}^y = \frac{\sigma_{11}^y}{h}. \tag{10.110}$$

The effect of the choice of h on the Barlat–Lian yield surface is illustrated in Figure 10.10(a). The effect of constant a is shown in Figure 10.10(b). Similarly to the previously discussed

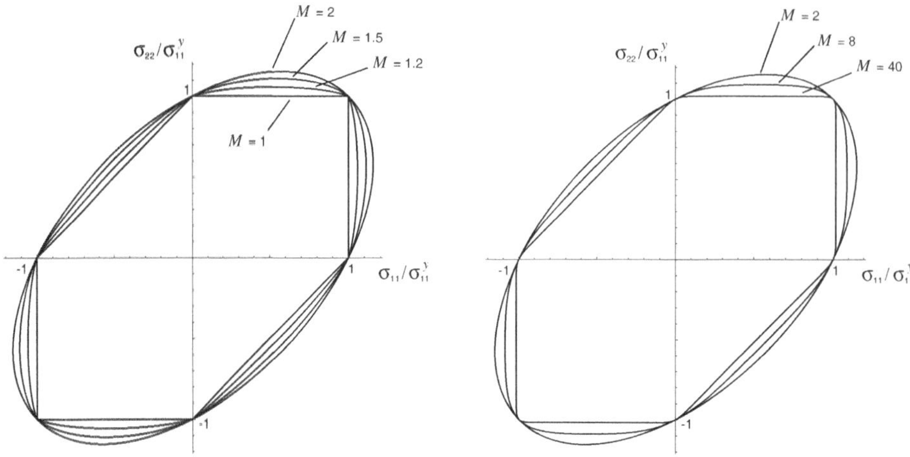

Figure 10.9. The Barlat–Lian criterion. Yield surfaces on σ_{11}-σ_{22} plane (in the absence of shear stress) for various values of M with $a = h = 1$.

Hill and Hoffman criteria, the presence of shear stresses will shrink the Barlat–Lian yield surface. This effect is illustrated in Figure 10.11, where we plot yield surfaces obtained with $M = 8$, $a = b = 1$ and $h = 2/3$ at different levels of shear stress. We remark that (not shown in Figure 10.11) an increase (decrease) in constant b will increase (decrease) the rate at which the surface shrinks with increasing shear stress.

For the purposes of computational implementation of the model, it turns out to be more convenient to describe the yield criterion for the Barlat–Lian model equivalently in terms of the following alternative definition of the yield function:

$$\Phi(\boldsymbol{\sigma}, \sigma_{11}^y) = g(\boldsymbol{\sigma}) - \sigma_{11}^y, \tag{10.111}$$

where

$$g(\boldsymbol{\sigma}) \equiv [\tfrac{1}{2} f(\boldsymbol{\sigma})]^{1/M}. \tag{10.112}$$

Note that the yield function now has dimension of stress. In the original definition (10.105), the yield function has dimension of stress to the power M, which may produce computationally intractable numbers for large values of material constant M.

Hardening law

Analogously to the Hill and Hoffman criteria, isotropic hardening can be incorporated into the present model by letting σ_{11}^y be a prescribed function of a scalar strain-like hardening internal variable, α:

$$\sigma_{11}^y = \sigma_{11}^y(\alpha). \tag{10.113}$$

Here we take the yield function (10.111) as the plastic potential and assume the evolution of the isotropic hardening variable to be governed by the standard relation

$$\dot{\alpha} = -\dot{\gamma} \frac{\partial \Phi}{\partial \sigma_{11}^y} = \dot{\gamma}. \tag{10.114}$$

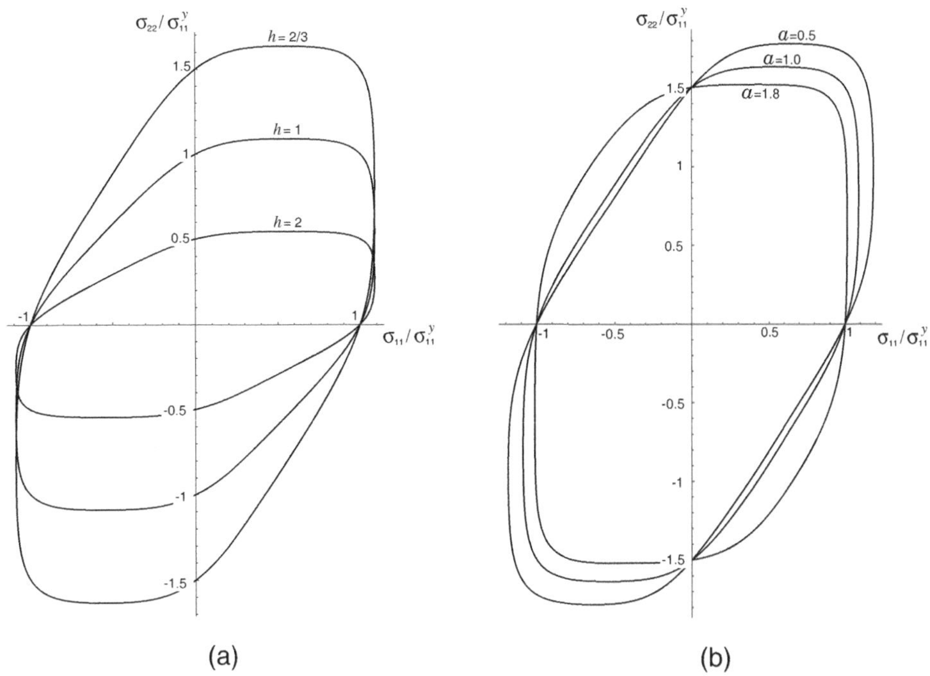

Figure 10.10. The Barlat–Lian criterion. Yield surfaces in the absence of shear for $M = 8$: (a) effect of parameter h with $a = 1$; (b) effect of parameter a with $h = 2/3$.

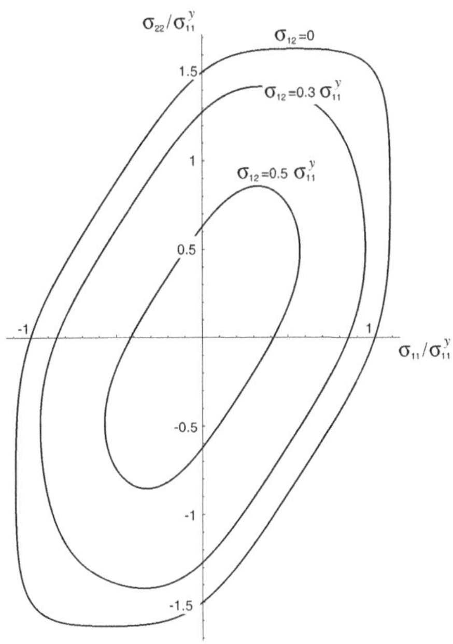

Figure 10.11. The Barlat–Lian criterion. Effect of shear stress on the yield surface for $M = 8$, $a = 1$, $h = 2/3$ and $b = 1$.

Flow rule

The Barlat–Lian yield function (10.111) is represented analogously to the plane stress-projected von Mises yield function given in $(9.41)_3$, page 372. The *in-plane* associative plastic strain rate is given simply by

$$\dot{\boldsymbol{\varepsilon}}^p = \dot{\gamma}\, \boldsymbol{N}, \tag{10.115}$$

where

$$\boldsymbol{N} \equiv \frac{\partial \Phi}{\partial \boldsymbol{\sigma}} = \frac{\partial g}{\partial \boldsymbol{\sigma}} = \frac{1}{2M}\left(\frac{f}{2}\right)^{(1-M)/M} \frac{\partial f}{\partial \boldsymbol{\sigma}}. \tag{10.116}$$

Equivalently, in component form, we have

$$\dot{\varepsilon}^p_{\alpha\beta} = \dot{\gamma}\, \frac{1}{2M}\left(\frac{f}{2}\right)^{(1-M)/M} \frac{\partial f}{\partial \sigma_{\alpha\beta}}, \qquad \alpha, \beta = 1, 2. \tag{10.117}$$

The corresponding explicit formulae for the derivatives of f are

$$\begin{aligned}\frac{\partial f}{\partial \sigma_{11}} = \frac{M}{2}\bigg\{ & a(K_1 - K_2)\,|K_1 - K_2|^{M-2}\left(1 - \frac{\sigma_{11} - h\sigma_{22}}{2K_2}\right) \\ & + a(K_1 + K_2)\,|K_1 + K_2|^{M-2}\left(1 - \frac{\sigma_{11} - h\sigma_{22}}{2K_2}\right) \\ & + 2^M(2 - a)K_2^{M-1}\,\frac{\sigma_{11} - h\sigma_{22}}{2K_2}\bigg\},\end{aligned} \tag{10.118}$$

$$\begin{aligned}\frac{\partial f}{\partial \sigma_{22}} = \frac{Mh}{2}\bigg\{ & a(K_1 - K_2)\,|K_1 + K_2|^{M-2}\left(1 - \frac{\sigma_{11} - h\sigma_{22}}{2K_2}\right) \\ & + a(K_1 + K_2)\,|K_1 - K_2|^{M-2}\left(1 - \frac{\sigma_{11} - h\sigma_{22}}{2K_2}\right) \\ & - 2^M(2 - a)K_2^{M-1}\,\frac{\sigma_{11} - h\sigma_{22}}{2K_2}\bigg\},\end{aligned} \tag{10.119}$$

$$\begin{aligned}\frac{\partial f}{\partial \sigma_{12}} = \frac{Mb^2\sigma_{12}}{2K_2}\{ & a(K_1 + K_2)\,|K_1 + K_2|^{M-2} \\ & - a(K_1 - K_2)\,|K_1 - K_2|^{M-2} + 2(2 - a)(2K_2)^{M-1}\}.\end{aligned} \tag{10.120}$$

Note that, in deriving (10.120), we have taken into account the tensorial nature of the shear stress contribution in (10.105) (refer to the comments surrounding expression (10.63)). The thickness plastic strain rate is obtained by imposing plastic incompressibility and the transversal plastic shear strain rates are assumed to vanish. Thus, the complete set of plastic flow equations comprises (10.117)–(10.120) together with the out-of plane rates

$$\dot{\varepsilon}^p_{33} = -\dot{\varepsilon}^p_{11} - \dot{\varepsilon}^p_{22}, \quad \dot{\varepsilon}^p_{13} = \dot{\varepsilon}^p_{23} = 0. \tag{10.121}$$

Finally, to complete the definition of the model, we assume the material to be elastically isotropic. Then, following the procedure applied to the plane stress-projected von Mises model, all out-of-plane strain components can be recovered *a posteriori* as functions of the in-plane elastic and plastic strains according to relations (9.37)–(9.40) (refer to page 371).

10.3.5. IMPLEMENTATION OF THE BARLAT–LIAN MODEL

The implicit elastic predictor/plastic corrector computational implementation of the (isotropically strain hardening) Barlat–Lian model, including the corresponding consistent tangent operator, was originally proposed by Dutko *et al.* (1993). The implementation is a particular case of the general methodology adopted throughout this book but, as we shall see, requires the inclusion of an extra procedure – a *line-search* algorithm – to expand the radius of convergence of the Newton–Raphson scheme used to solve the return mapping equations (refer to Remark 7.3, page 199, for a discussion on the possible need for strategies of this nature). The corresponding algorithm and associated consistent tangent operator are described in the following.

The return-mapping equations

The elastic predictor stage of the algorithm in the present case is clearly identical to that of the plane stress-projected von Mises model as described in Subsection 9.4.2 (refer to Box 9.4, page 377) and, therefore, requires no further consideration.

The return mapping is a specialisation of (7.25) (page 196), requiring the solution of the algebraic system of equations

$$\left\{\begin{matrix} \boldsymbol{\varepsilon}^e_{n+1} - \boldsymbol{\varepsilon}^{e\ \mathrm{trial}} + \Delta\gamma\, \boldsymbol{N}_{n+1} \\ \alpha_{n+1} - \alpha_n - \Delta\gamma \\ \Phi(\boldsymbol{\sigma}(\varepsilon^e_{n+1}), \kappa(\alpha_{n+1})) \end{matrix}\right\} = \left\{\begin{matrix} \mathbf{0} \\ 0 \\ 0 \end{matrix}\right\} \tag{10.122}$$

for the unknowns ε^e_{n+1}, α_{n+1} and $\Delta\gamma$, where $\boldsymbol{N}_{n+1}$ is the flow vector (10.116) evaluated at the updated state. We remark that only the *in-plane* components of the elastic strain take part in the above equation. For convenience in the description of the return-mapping algorithm we have adopted, here and in what follows, the notation

$$\kappa \equiv \sigma^y_{11}. \tag{10.123}$$

In the derivation of the return mapping presented below, we shall assume linear hardening, i.e.

$$\kappa = \kappa_0 + H\, \alpha, \tag{10.124}$$

where H is the hardening modulus. Then, in the discrete setting, the evolution of κ reads

$$\kappa_{n+1} = \kappa_n + H\, \Delta\gamma. \tag{10.125}$$

With this relation and taking into account the linear elasticity law, the return-mapping equation set can be written as

$$\bar{\mathbf{R}} \equiv \left\{\begin{matrix} [\mathsf{D}^e]^{-1} : (\boldsymbol{\sigma}_{n+1} - \boldsymbol{\sigma}^{\mathrm{trial}}) + \Delta\gamma\, \boldsymbol{N}_{n+1} \\ H^{-1}\,(\kappa_{n+1} - \kappa_n) - \Delta\gamma \\ \Phi(\boldsymbol{\sigma}_{n+1}, \kappa_{n+1}) \end{matrix}\right\} = \left\{\begin{matrix} \mathbf{0} \\ 0 \\ 0 \end{matrix}\right\}, \tag{10.126}$$

where D^e here denotes the isotropic (plane stress) elasticity operator and the unknowns are $\boldsymbol{\sigma}_{n+1}$, κ_{n+1} and $\Delta\gamma$.

Note that, unlike previously described implementations such as those of the von Mises and even the orthotropic Hoffman model, the return-mapping equation system in the present case cannot be reduced to a single scalar equation.

Newton–Raphson iteration

As usual, the solution of equation (10.126) is undertaken by the Newton–Raphson iterative scheme. To do this, it is convenient to split the residual vector $\bar{\mathbf{R}}$ as

$$\bar{\mathbf{R}} = \begin{Bmatrix} \mathbf{R} \\ \Phi \end{Bmatrix}; \quad \mathbf{R} \equiv \begin{Bmatrix} [\mathsf{D}^e]^{-1} : (\boldsymbol{\sigma}_{n+1} - \boldsymbol{\sigma}^{\text{trial}}) + \Delta\gamma\, \boldsymbol{N}_{n+1} \\ H^{-1}\,(\kappa_{n+1} - \kappa_n) - \Delta\gamma \end{Bmatrix}. \tag{10.127}$$

Accordingly, the typical k^{th} Newton iteration comprises the solution of the linear system

$$\begin{bmatrix} [\mathsf{D}^e]^{-1} + \Delta\gamma\, \partial \boldsymbol{N}/\partial\boldsymbol{\sigma} & 0 & \boldsymbol{N} \\ 0 & 1/H & -1 \\ \boldsymbol{N} & -1 & 0 \end{bmatrix}^{(k-1)} \begin{Bmatrix} \delta\boldsymbol{\sigma}^{(k)} \\ \delta\kappa^{(k)} \\ \delta\Delta\gamma^{(k)} \end{Bmatrix} = - \begin{Bmatrix} \mathbf{R}^{(k-1)} \\ \\ \Phi^{(k-1)} \end{Bmatrix} \tag{10.128}$$

for the iterative increments of $\boldsymbol{\sigma}_{n+1}$, κ_{n+1} and $\Delta\gamma$. Following (10.116), the flow vector derivative can be expressed as

$$\frac{\partial \boldsymbol{N}}{\partial \boldsymbol{\sigma}} = \frac{\partial^2 \Phi}{\partial \boldsymbol{\sigma}} = \frac{1}{2M}\left(\frac{f}{2}\right)^{\frac{1-M}{M}} \frac{\partial^2 f}{\partial \boldsymbol{\sigma}^2} + \frac{1-M}{4M^2}\left(\frac{f}{2}\right)^{\frac{1-2M}{M}} \frac{\partial f}{\partial \boldsymbol{\sigma}} \otimes \frac{\partial f}{\partial \boldsymbol{\sigma}}. \tag{10.129}$$

Straightforward manipulation of system (10.128) gives the following solution (iterative increments) in symbolic form:

$$\delta\Delta\gamma^{(k)} = \frac{\Phi^{(k-1)} - [\boldsymbol{N}^{(k-1)} \quad -1\,]\, \mathbf{A}^{(k-1)}\, \mathbf{R}^{(k-1)}}{[\boldsymbol{N}^{(k-1)} \quad -1\,]\, \mathbf{A}^{(k-1)} \begin{Bmatrix} \boldsymbol{N}^{(k-1)} \\ -1 \end{Bmatrix}} \tag{10.130}$$

and

$$\begin{Bmatrix} \delta\boldsymbol{\sigma}^{(k)} \\ \delta\kappa^{(k)} \end{Bmatrix} = -\mathbf{A}^{(k-1)} \left[\mathbf{R}^{(k-1)} + \delta\Delta\gamma^{(k)} \begin{Bmatrix} \boldsymbol{N}^{(k-1)} \\ -1 \end{Bmatrix} \right], \tag{10.131}$$

where we have defined

$$\mathbf{A} \equiv \begin{bmatrix} \{[\mathsf{D}^e]^{-1} + \Delta\gamma\, \partial\boldsymbol{N}/\partial\boldsymbol{\sigma}\}^{-1} & 0 \\ 0 & H \end{bmatrix}. \tag{10.132}$$

With the above solution at hand, we update the Newton guess for the unknowns according to

$$\boldsymbol{\Sigma}^{(k)}_{n+1} = \boldsymbol{\Sigma}^{(k-1)}_{n+1} + \delta\boldsymbol{\Sigma}^{(k)}, \tag{10.133}$$

where we have used the notation

$$\boldsymbol{\Sigma}^{(k)}_{n+1} \equiv \begin{Bmatrix} \boldsymbol{\sigma}^{(k)}_{n+1} \\ \kappa^{(k)}_{n+1} \\ \Delta\gamma^{(k)} \end{Bmatrix}; \quad \delta\boldsymbol{\Sigma}^{(k)} \equiv \begin{Bmatrix} \delta\boldsymbol{\sigma}^{(k)} \\ \delta\kappa^{(k)} \\ \delta\Delta\gamma^{(k)} \end{Bmatrix}. \tag{10.134}$$

Line-search procedure

Dutko *et al.* (1993) observed that the convergence bowl of the above Newton–Raphson scheme can be dramatically reduced for larger values of material constant M. Such an ill-conditioned behaviour stems from the high curvature present in the Barlat–Lian model for large M in the neighbourhood of points in stress space corresponding to the corners of the Tresca yield surface. Under such conditions, small changes in stress on the yield surface result in large changes in flow vector direction, characterising a set of *stiff* evolution equations. To tackle the problem, Dutko *et al.* (1993) adopted a line-search procedure which has effectively stabilised the Newton–Raphson scheme for the range of strain increments expected to be present in practical finite element computations with values of constant M as high as 40. Line-search procedures are discussed in detail by Fletcher (1980) (see also Matthies and Strang 1979, and Crisfield 1991).

The line-search in the present case is activated whenever the rate of convergence of the Newton–Raphson iterations falls below a prescribed minimum; that is, if for a Newton iteration $k \geq 2$ and a prescribed tolerance ϵ, the relation

$$|\delta\boldsymbol{\Sigma}^{(k)} \cdot \bar{\mathbf{R}}^{(k)}| \geq \epsilon\, |\delta\boldsymbol{\Sigma}^{(k)} \cdot \bar{\mathbf{R}}^{(k-1)}| \tag{10.135}$$

is satisfied, then the line-search algorithm is carried out. The interested reader is referred to Dutko *et al.* (1993) for a complete description of the algorithm.

Consistent tangent operator

Again, an expression for the elastoplastic consistent tangent can be easily obtained by means of a specialisation of the general procedure of Section 7.4.4. Then, by differentiating the return-mapping system (10.126) with respect to its unknowns $\boldsymbol{\sigma}_{n+1}$, κ_{n+1} and $\Delta\gamma$ as well as to its input, $\boldsymbol{\varepsilon}^{e\ \mathrm{trial}} = [\mathsf{D}^e]^{-1} : \boldsymbol{\sigma}_{n+1}^{\mathrm{trial}}$, we obtain

$$\begin{bmatrix} [\mathsf{D}^e]^{-1} + \Delta\gamma\, \partial\boldsymbol{N}/\partial\boldsymbol{\sigma} & 0 & \boldsymbol{N} \\ 0 & 1/H & -1 \\ \boldsymbol{N} & -1 & 0 \end{bmatrix} \begin{Bmatrix} \mathrm{d}\boldsymbol{\sigma}_{n+1} \\ \mathrm{d}\kappa_{n+1} \\ \mathrm{d}\Delta\gamma \end{Bmatrix} = \begin{Bmatrix} \mathrm{d}\boldsymbol{\varepsilon}^{e\ \mathrm{trial}} \\ 0 \\ 0 \end{Bmatrix}. \tag{10.136}$$

Note that this differential relation has the same symbolic format as expression (7.141), obtained for the von Mises isotropically hardening model. Accordingly, the elastoplastic tangent for the Barlat–Lian model with implicit return mapping is a complete analogy to (7.148) and can be expressed as

$$\mathsf{D}^{ep} \equiv \frac{\partial\boldsymbol{\sigma}_{n+1}}{\partial\boldsymbol{\varepsilon}^{e\ \mathrm{trial}}} = \mathsf{P} - \frac{1}{\boldsymbol{N} : \mathsf{P} : \boldsymbol{N} + H} (\mathsf{P} : \boldsymbol{N}) \otimes (\mathsf{P} : \boldsymbol{N}), \tag{10.137}$$

where

$$\mathsf{P} \equiv \left(\mathsf{I}_S + \Delta\gamma\, \mathsf{D}^e : \frac{\partial\boldsymbol{N}}{\partial\boldsymbol{\sigma}} \right)^{-1} : \mathsf{D}^e. \tag{10.138}$$

As a result of the associative nature of the plastic flow rule and hardening law (refer to Section 7.4.6, page 243), the above operator is symmetric.

11 VISCOPLASTICITY

THE elastoplastic constitutive theories presented so far in Part Two of this book are classed as *rate independent* or *time independent*; that is, the material response is regarded as independent of the rate of application of loads and/or the timescale of the problems considered. Time (or, more precisely, *pseudo-time*) is used merely to describe the sequence of events that defines the history of the loading process. For such theories, identical solutions are produced when a given load (or sequence of loads) is applied at different rates.

However, the observed behaviour of real materials is generally time dependent; that is, the stress response always depends on the rate of loading and/or the timescale considered. The extent of such dependence may or may not be significant according to the physical conditions of the problem. In situations where the rates of loading and/or the timescale of the analysis remain within a range where the time-dependent phenomena can be neglected, rate-independent elastoplasticity models can provide good descriptions of the material behaviour (Lemaitre and Chaboche, 1990; Lubliner, 1990; Skrzypek, 1993). If such conditions are not met, then accurate predictions can only be obtained if rate dependence is adequately accounted for by the constitutive model. Rate-dependence effects are described by means of so-called *viscoplasticity* (or *rate-dependent plasticity*) models, to which the present chapter is devoted.

This chapter is organised as follows. Section 11.1 presents a brief introduction to phenomenological aspects of viscoplasticity. It motivates the establishment of a one-dimensional mathematical model of viscoplasticity in Section 11.2. Here, some simple analytical solutions are presented to demonstrate the ability of the one-dimensional model in capturing the fundamental phenomenological features of viscoplastic behaviour. In Section 11.3 the one-dimensional viscoplastic theory is generalised to the multidimensional case within the context of von Mises plasticity. A more general multidimensional model is presented in Section 11.4. The general model can be rigorously described within the constitutive framework of internal variable theories initially referred to in Chapter 3 (Section 3.5.2, from page 71). Rate-independent plasticity is shown to be, under some circumstances, a limit case of the general viscoplasticity model. This establishes a formal link between rate-independent plasticity and the general constitutive framework of Chapter 3. Section 11.5 proceeds to introduce a numerical framework to treat the general viscoplasticity model within the finite element environment of Chapter 4. This includes the numerical integration algorithm for the general viscoplastic constitutive equations as well as a symbolic form of the associated consistent tangent modulus. Then, in Section 11.6, the general numerical framework is specialised to a von Mises-based model presented in Section 11.3. The integration algorithm and the associated consistent tangent operator are derived step by step. In addition, an error assessment of the numerical integration procedure is presented by means of iso-error maps.

We remark that Section 11.6 is essential for the reader interested in the computational implementation of viscoplasticity. The chapter ends with finite element examples being shown in Section 11.7. In the reported examples, the procedures of Section 11.6 are used.

11.1. Viscoplasticity: phenomenological aspects

Many of the microscopic phenomena underlying the inelastic deformation of solids depend on time. Materials such as metals, rubbers, geomaterials in general, concrete and composites may all present substantially time-dependent mechanical behaviour under many practical circumstances. In metals, for instance, the phenomenological effects of time-dependent mechanisms become apparent typically at absolute temperatures higher than around one third of the melting point and can be clearly identified by a number of experimental procedures. To illustrate this fact, typical results of simple uniaxial tension tests with metallic bars at higher temperatures are schematically represented in Figure 11.1. Figure 11.1(a) shows stress–strain curves obtained in uniaxial tensile tests carried out under different prescribed strain rates. In general, the elasticity modulus is largely independent of the rate of loading but, clearly, the initial yield limit as well as the hardening curve depend strongly on the rate of straining. This rate-dependence is also observed at low temperatures, but usually becomes significant only at higher temperatures. Strain-rate dependence may be of crucial importance, for instance, in metal-forming operations such as hot forging and may have to be taken into consideration in the design of the process. It is also important to emphasise that, although not shown in Figure 11.1(a), the rupture limit, that is, the strain at which the specimen breaks, can also be strongly dependent on the rate of straining.

Another aspect of time dependence is the phenomenon of *creep*. This is illustrated in Figure 11.1(b). The curves of Figure 11.1(b) show the evolution of plastic strains over time in experiments where tensile specimens have been loaded to different stress levels and left at constant stress during long periods of time. The material experiences a continuous plastic flow that is accelerated for higher stress levels. The high strain rates shown towards the end of the schematic curves for high and moderate stresses is the phenomenon known as *tertiary creep*. Tertiary creep leads to the final rupture of the material and is associated with the evolution of internal damage. Internal damage will be discussed in Chapter 12. Prediction of creep behaviour is important, for instance, in situations where load-carrying metallic components are subjected to long duration loads at high temperatures. The need for consideration of creep response arises typically in the design and analysis of nuclear reactor and jet turbine engine components.

The third aspect of rate dependence, illustrated in Figure 11.1, is the phenomenon of stress *relaxation*. The graph of Figure 11.1(c) shows the typical evolution of stress in a relaxation test. The relaxation test consists of stretching the specimen (virtually instantaneously) to a prescribed axial strain and maintaining it strained (at constant strain) over a long period of time. The time-dependent response in this case is characterised by the continuous decay of stress in time. The prediction of stress relaxation can be vital, for instance, in the design of pre-stressed load-carrying components. We remark that the strain rate dependence of the stress response as well as the phenomena of creep and stress relaxation illustrated above for metals can also be observed for other materials by means of appropriate experiments.

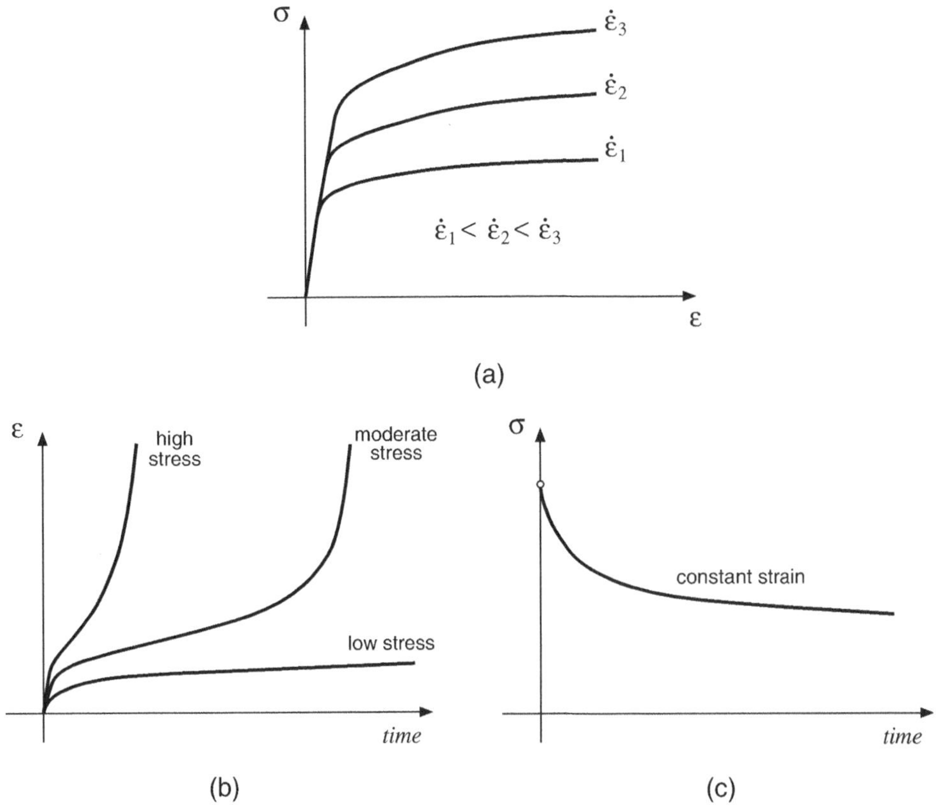

Figure 11.1. Viscoplasticity. Phenomenological aspects: uniaxial tensile tests at high temperature. (a) *Strain rate dependence*. Uniaxial tests at different strain rates. (b) *Creep*. Plastic flow at constant stress. (c) *Relaxation*. Stress decay at constant strain.

11.2. One-dimensional viscoplasticity model

Similarly to what has been done in Chapter 6 for rate-independent plasticity (refer to Section 6.2, from page 141), we find it convenient to introduce viscoplasticity by focusing first on a simple one-dimensional theory. Thus, we devote this section to the description of a simple uniaxial viscoplastic constitutive model. As we shall see, in spite of its simplicity, the uniaxial model possesses all the basic ingredients of the multidimensional models discussed in the remainder of this chapter. In particular, the model is able to capture many of the main features of the viscoplastic behaviour depicted in Figure 11.1.

11.2.1. ELASTOPLASTIC DECOMPOSITION OF THE AXIAL STRAIN

As for the rate-independent case, the decomposition of the total axial strain into a sum of an elastic (recoverable) and a plastic (permanent) component is introduced,

$$\varepsilon = \varepsilon^e + \varepsilon^p. \tag{11.1}$$

11.2.2. THE ELASTIC LAW

The axial stress is again assumed to be related to the elastic component of the axial strain by means of the standard linear elastic constitutive relation

$$\sigma = E\,\varepsilon^e. \tag{11.2}$$

11.2.3. THE YIELD FUNCTION AND THE ELASTIC DOMAIN

Here, the existence of an elastic domain for the stress within which the material behaviour is purely elastic is also experimentally observed in many cases.† Thus, analogously to the rate-independent model of Section 6.2, the elastic domain can be conveniently defined by means of a *yield function*

$$\Phi(\sigma, \sigma_y) = |\sigma| - \sigma_y, \tag{11.3}$$

where σ_y is the yield stress. The elastic domain is defined as the set

$$\mathscr{E} = \{\sigma \mid \Phi(\sigma, \sigma_y) < 0\}, \tag{11.4}$$

so that the behaviour is purely elastic whenever $|\sigma| < \sigma_y$.

11.2.4. VISCOPLASTIC FLOW RULE

The crucial difference between the uniaxial elastoplastic model of Section 6.2 and the viscoplasticity model introduced here lies in the definition of the flow rule, which describes the evolution of ε^p. The viscoplastic flow rule can be postulated with a format similar to that of the rate-independent case (see equation (6.10), page 144)

$$\dot{\varepsilon}^p = \dot{\gamma}(\sigma, \sigma_y)\,\mathrm{sign}(\sigma), \tag{11.5}$$

where sign is the *signum* function defined by (6.11).

In spite of its similarity to the rate-independent flow rule, the above constitutive equation for ε^p differs fundamentally from (6.10). Firstly, it needs to be emphasised that in the rate-independent model the plastic strain rate is in fact a *pseudo*-time rate; that is, $\dot{\varepsilon}^p$ in the rate-independent theory is the derivative of the plastic strain in respect to a *pseudo*-time parameter used solely to describe the sequence of events. In that case, the actual timescale is irrelevant. In contrast, the plastic strain rate in (11.5) is the actual time derivative of ε^p. In addition to this conceptual difference, $\dot{\gamma}$ – named the plastic multiplier and determined by the procedure of Section 6.2.7 (page 146) in the rate-independent theory – is here a *given* explicit function of σ and σ_y. Essentially, the explicit function for $\dot{\gamma}$ should model how the rate of plastic straining varies with the level of stress. Many forms are possible for $\dot{\gamma}$ and a discussion on this issue will be left for Section 11.3. Here, we will define the one-dimensional viscoplasticity model by adopting the following particular definition

$$\dot{\gamma}(\sigma, \sigma_y) = \begin{cases} \dfrac{1}{\mu}\left[\left(\dfrac{|\sigma|}{\sigma_y}\right)^{1/\epsilon} - 1\right] & \text{if } \Phi(\sigma, \sigma_y) \geq 0 \\[2ex] 0 & \text{if } \Phi(\sigma, \sigma_y) < 0, \end{cases} \tag{11.6}$$

† As we shall see later in this chapter, some models of viscoplasticity do not have an elastic domain. Such models do not require the definition of a yield function.

where the material constants are the *viscosity-related* parameter μ, whose dimension is time, and the non-dimensional *rate-sensitivity* parameter, ϵ. Both parameters are strictly positive. This particular form has been introduced by Perić (1993) similarly to the power law form of the viscoplastic potential proposed by Perzyna (1963). It is important to emphasise that the material parameters μ and ϵ are *temperature dependent*. As a general rule, as temperature increases (decreases) μ and ϵ increase (decrease). For many metals, $\mu, \epsilon \rightarrow 0$ for sufficiently low temperatures, when the material behaviour may be assumed rate-independent.

11.2.5. HARDENING LAW

In the rate-independent case, the phenomenon of hardening describes the changes in yield stress that result from plastic straining. In the viscoplastic model, hardening can be incorporated in the same manner as in the elastoplastic case by letting the yield stress, σ_y, be a given (experimentally determined) function

$$\sigma_y = \sigma_y(\bar{\varepsilon}^p) \tag{11.7}$$

of the accumulated plastic strain

$$\bar{\varepsilon}^p = \int_0^t |\dot{\varepsilon}^p| \, dt. \tag{11.8}$$

Note that (11.6) implies that at a given constant applied stress σ, an increase (decrease) in σ_y will produce a decrease (increase) in the plastic strain rate $\dot{\varepsilon}^p$. As in the elastoplastic case, an increase of σ_y will be referred to as *hardening* whereas its reduction will be described as *softening*. If σ_y is a constant, the model is referred to as *perfectly* viscoplastic.

11.2.6. SUMMARY OF THE MODEL

The overall one-dimensional viscoplasticity model is defined by the constinutive equations (11.1)–(11.8). For convenience we summarise the model in Box 11.1.

11.2.7. SOME SIMPLE ANALYTICAL SOLUTIONS

Section 11.1 discussed some of the phenomenological aspects of viscoplastic behaviour. One important aspect to be emphasised here is that, in spite of its simplicity, the above-defined one-dimensional model can capture the key phenomenological features of rate-dependent plasticity shown in Figure 11.1. To illustrate this and give the reader a better insight into the theory, we derive in this section analytical solutions for three simple problems where the basic properties of *creep* at constant stress, *strain rate dependence* of the stress response and *stress relaxation* under constant strain are reproduced by the one-dimensional model.

Creeping at constant stress

Let us consider the case of a bar subjected to an axial load that produces a uniform stress $\sigma > \sigma_y$ over its cross-section. The load is applied instantaneously and, after being applied, remains constant in time.

With the instantaneous loading (at time $t = 0$), the bar will initially deform (also instantaneously) in a purely elastic manner. The fact that the model behaviour is elastic under

Box 11.1. One-dimensional viscoplastic constitutive model.

1. Elastoplastic split of the axial strain

$$\varepsilon = \varepsilon^e + \varepsilon^p$$

2. Uniaxial elastic law

$$\sigma = E\,\varepsilon^e$$

3. Yield function

$$\Phi(\sigma, \sigma_y) = |\sigma| - \sigma_y$$

4. Plastic flow rule

$$\dot{\varepsilon}^p = \dot{\gamma}\ \mathrm{sign}(\sigma)$$

$$\dot{\gamma} = \begin{cases} \dfrac{1}{\mu}\left[\left(\dfrac{|\sigma|}{\sigma_y}\right)^{1/\epsilon} - 1\right] & \text{if } \Phi(\sigma, \sigma_y) \geq 0 \\ 0 & \text{if } \Phi(\sigma, \sigma_y) < 0 \end{cases}$$

5. Hardening law

$$\sigma_y = \sigma_y(\bar{\varepsilon}^p)$$

$$\dot{\bar{\varepsilon}}^p = \dot{\gamma}$$

instantaneous loading is formally demonstrated in the next example when the strain rate dependence of the stress response is discussed. However, even without a formal proof, it makes sense to accept that, as there is no time for plastic strains to develop over an (idealised) instantaneous loading event, the behaviour must be purely elastic under such a condition. Assuming zero initial plastic strain, it follows from the elastoplastic split of the total strain together with the elastic law that the total strain in the bar at $t = 0$, immediately after the instantaneous application of load, will be

$$\varepsilon_0 = \varepsilon_0^e = \frac{\sigma}{E}, \tag{11.9}$$

where the zero subscript denotes quantities at $t = 0$. From this moment on, the bar is kept under a constant stress above the yield limit. Under constant stress, the elastic law implies that the elastic strain will also remain constant. Thus, the straining of the bar after the instantaneous loading will be due purely to viscoplastic flow and will be modelled by constitutive equations (11.5), (11.6). Assuming that σ is positive (tensile), we then have

$$\dot{\varepsilon}^p = \frac{1}{\mu}\left[\left(\frac{\sigma}{\sigma_y}\right)^{1/\epsilon} - 1\right]. \tag{11.10}$$

For a perfectly viscoplastic material (constant σ_y), the integration of the above equation, in conjunction with the elastoplastic decomposition of the total strain and the initial condition (11.9), gives the following solution for the straining of the bar

$$\varepsilon(t) = \frac{\sigma}{E} + \frac{1}{\mu}\left[\left(\frac{\sigma}{\sigma_y}\right)^{1/\epsilon} - 1\right] t. \tag{11.11}$$

The creep rate in this case is constant and proportional to $(\sigma/\sigma_y)^{1/\epsilon} - 1$. The material constants (μ and ϵ) could be calibrated, for instance, so as to capture the initial branches of the creep curves of the material (refer to Figure 11.1(b)). The initial branches describe the phenomenon of *primary creep*. A hardening law could be incorporated to include the follow-up of the curves to their second straight branch that describes secondary creep.

Strain-rate dependence of the stress response

Here we analyse the monotonic stretching of an initially unstrained (and unstressed) bar at constant strain rates. These are the conditions typically encountered in conventional tensile tests. The objective here is to show that the one-dimensional model is capable of predicting the strain-rate dependence of the stress response as generally illustrated in Figure 11.1(a).

Before stating the associated initial value problem, let us first recall that the material response within the elastic domain ($\sigma < \sigma_y$) is purely elastic. Thus, at the initial stages of the monotonic stretching process, the stress response does not depend on the rate of stretching. The stress–strain response during this phase is expressed simply as

$$\sigma = E\,\varepsilon \quad \text{if} \quad \sigma < \sigma_y, \tag{11.12}$$

or, equivalently,

$$\sigma = E\,\varepsilon \quad \text{if} \quad \varepsilon < \varepsilon^*, \tag{11.13}$$

where ε^* is the strain at which the yield stress is reached

$$\varepsilon^* = \frac{\sigma_y}{E}. \tag{11.14}$$

Viscoplastic flow (and rate-dependent behaviour) may only take place when $\sigma \geq \sigma_y$ or, in terms of the applied strain, when $\varepsilon \geq \varepsilon^*$. Then, as our purpose is to illustrate the rate dependence predicted by the model, our initial value problem will be defined only over the portion of the loading process where $\sigma \geq \sigma_y$ (or $\varepsilon \geq \varepsilon^*$). To simplify the problem, we will assume that the material is perfectly viscoplastic (constant σ_y) and, in addition, the rate sensitivity parameter will be set to

$$\epsilon = 1,$$

so that an analytical solution to the initial value problem can be easily found.

The initial value problem.

The evolution of the plastic strain for the present model is defined by equation (11.10). Under the above assumptions, the associated initial value problem consists of finding a function $\varepsilon^p(t)$ such that

$$\dot{\varepsilon}^p(t) = \frac{1}{\mu}\left[\frac{\sigma(t)}{\sigma_y} - 1\right] = \frac{1}{\mu}\left\{\frac{E\,[\varepsilon(t) - \varepsilon^p(t)]}{\sigma_y} - 1\right\} = \frac{1}{\mu}\left[\frac{\varepsilon(t) - \varepsilon^p(t)}{\varepsilon^*} - 1\right], \tag{11.15}$$

where the total strain is the prescribed function

$$\varepsilon(t) = \alpha\, t + \varepsilon^* = \alpha\, t + \frac{\sigma_y}{E}, \tag{11.16}$$

corresponding to monotonic stretching with arbitrary (constant) strain rate $\alpha \geq 0$. Note that $t=0$ corresponds to the onset of viscoplastic flow ($\sigma = \sigma_y \Leftrightarrow \varepsilon = \varepsilon^*$). The initial condition for ε^p (the plastic strain at the onset of viscoplastic flow) is obviously

$$\varepsilon^p(0) = 0. \tag{11.17}$$

The analytical solution.

By substituting (11.16) into the differential equation (11.15), the initial value problem can be written as

$$\dot{\varepsilon}^p(t) = \frac{1}{\mu\, \varepsilon^*}[\alpha\, t - \varepsilon^p(t)]; \quad \varepsilon^p(0) = 0. \tag{11.18}$$

The analytical solution can then be promptly obtained by standard methods for first-order ordinary linear differential equations as

$$\varepsilon^p(t) = \alpha\big[t - \mu\, \varepsilon^*\big(1 - \mathrm{e}^{\frac{-t}{\mu\, \varepsilon^*}}\big)\big]. \tag{11.19}$$

By placing the above solution together with (11.16) into the elastic law, $\sigma = E(\varepsilon - \varepsilon^p)$, we obtain, after a straightforward manipulation, the following solution for the stress as a function of time

$$\sigma(t) = \sigma_y\big[1 + \alpha\, \mu\big(1 - \mathrm{e}^{\frac{-t}{\mu\, \varepsilon^*}}\big)\big]. \tag{11.20}$$

The stress–strain response.

The strain rate dependence of the stress response can be more clearly illustrated by expressing the stress as a function of strain for an arbitrary (but constant in time) strain rate, so that curves such as those depicted in Figure 11.1(a) can be produced. In the present case, such a function can be obtained by means of a simple variable transformation in (11.20). Indeed, note that by inverting the function defined by (11.16), we have

$$t = t(\varepsilon) = \frac{\varepsilon - \varepsilon^*}{\alpha}. \tag{11.21}$$

With the introduction of this relation into (11.20), we obtain

$$\tilde{\sigma}(\varepsilon) = \sigma(t(\varepsilon)) = \sigma_y\big\{1 + \alpha\, \mu\big[1 - \mathrm{e}^{\frac{1}{\mu\, \alpha}(1 - \frac{\varepsilon}{\varepsilon^*})}\big]\big\}, \tag{11.22}$$

which gives the stress–strain curve for an arbitrary strain rate α. Insight into the problem can be gained by looking into the limit stress–strain curves, i.e. the curves obtained at infinitely slow rates ($\alpha \to 0$) and infinitely fast processes ($\alpha \to \infty$). In order to obtain the limit for infinitely slow rates, we first observe that, in the present monotonic loading process, the term

$$1 - \varepsilon/\varepsilon^*$$

in (11.22) is always negative. In addition, μ and α are positive so that the last term on the right-hand side of (11.22) is the exponential of a negative number. The limit is then easily obtained as

$$\lim_{\alpha \to 0} \tilde{\sigma}(\varepsilon) = \sigma_y; \tag{11.23}$$

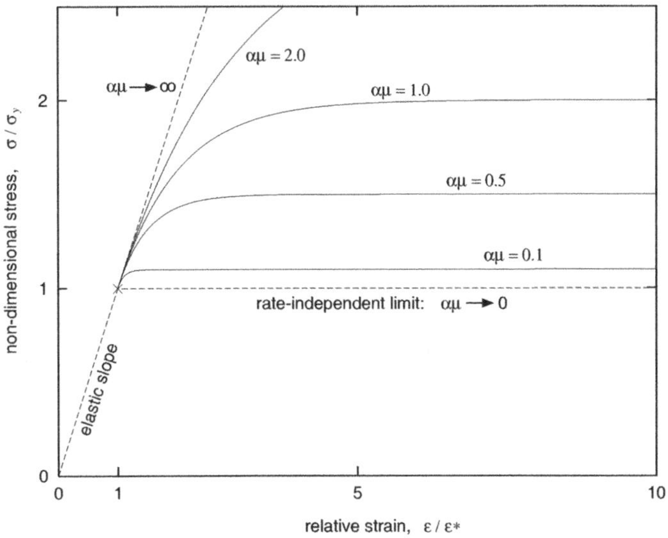

Figure 11.2. One-dimensional viscoplasticity model. Analytical solution showing the dependence of the stress response on the applied strain rate/viscosity parameter.

that is, at infinitely slow rates the perfectly viscoplastic model rigorously recovers the behaviour of the rate-independent plasticity model with yield stress σ_y. The rate-independent model was described in Section 6.2 (page 141). It is also very important to note that the same limit is obtained for the vanishing viscosity parameter, i.e. when $\mu \to 0$. At infinitely fast rates, the limit is derived by a standard limiting procedure which gives

$$\lim_{\alpha \to \infty} \tilde{\sigma}(\varepsilon) = E\,\varepsilon, \tag{11.24}$$

i.e. the process is purely elastic and the stress–strain curve after the yield limit is the continuation (with the same slope, E) of the elastic curve. Also note that the identical limit is found for $\mu \to \infty$ (infinitely viscous material). For any other rate (or viscosity parameter), the corresponding stress–strain curve will lie between these two limits with higher stress obtained at higher strain rates (or higher viscosity). To illustrate better the behaviour of the model under the present conditions, the analytical solution (11.22) is shown in the graph of Figure 11.2, where the non-dimensional stress, σ/σ_y, is plotted against the relative strain, $\varepsilon/\varepsilon^*$, for various normalised strain rates $\mu\alpha$. The limits $\mu\alpha \to 0$ (infinitely slow rates or non-viscous material) and $\mu\alpha \to \infty$ (infinitely fast rates or infinitely viscous material) are also included. Clearly, the model is able to capture the experimentally observed rate-dependence phenomenon illustrated in Figure 11.1(a).

Remark 11.1. In fact, even though it is not formally shown here, the above limits remain valid for any hardening curve and any rate sensitivity parameter ϵ; that is, at infinitely slow strain rates, the model recovers the rate-independent behaviour of the plasticity model of Section 6.2 (this limit is also obtained for $\mu \to 0$) and, at infinitely fast rates (or when $\mu \to \infty$), the model behaves in a purely elastic manner, regardless of the given hardening curve and rate-sensitivity parameter. In addition (again not formally shown in this section), the rate-independent behaviour is also recovered with vanishing rate-sensitivity, i.e. when $\epsilon \to 0$. This last property will be demonstrated in Section 11.4.3 in the context of the general multidimensional theory.

Stress relaxation at constant strain

In this final example, we consider the case of a bar which is instantaneously stretched (stretched at an infinitely fast strain rate) to a total strain ε and then kept stretched indefinitely at that constant strain. The instantaneous stretching to ε is assumed to produce a stress above the yield limit of the material. Here the model should be able to capture the phenomenon of stress relaxation alluded to in Figure 11.1(c).

Over the instantaneous stretching (at time $t = 0$), the bar will deform purely elastically (refer to the limit expression (11.24) and the text surrounding it). Thus, assuming that the plastic strain is zero at $t = 0$ (immediately after the instantaneous stretching), we have

$$\varepsilon_0^e = \varepsilon, \tag{11.25}$$

and, in view of the elastic law, the corresponding stress is given by

$$\sigma_0 = E\,\varepsilon_0^e = E\,\varepsilon. \tag{11.26}$$

From this point on, the stress in the bar will be governed by the law

$$\sigma = E(\varepsilon - \varepsilon^p) = \sigma_0 - E\,\varepsilon^p, \tag{11.27}$$

where ε^p evolves in time according to the differential equation (11.10) which, in view of the above expression can be equivalently written as

$$\dot{\varepsilon}^p = \frac{1}{\mu}\left[\left(\frac{\sigma_0 - E\,\varepsilon^p}{\sigma_y}\right)^{1/\epsilon} - 1\right]. \tag{11.28}$$

To simplify the problem, we will assume, as in the previous example, that the material is perfectly viscoplastic (constant σ_y) and $\epsilon = 1$. In this case, the initial value problem is to find a function $\varepsilon^p(t)$ such that

$$\dot{\varepsilon}^p(t) = c_1 - c_2\,\varepsilon^p(t), \tag{11.29}$$

with initial condition

$$\varepsilon^p(0) = 0, \tag{11.30}$$

where the constants c_1 and c_2 are defined as

$$c_1 = \frac{1}{\mu}\left(\frac{\sigma_0}{\sigma_y} - 1\right), \quad c_2 = \frac{E}{\mu\,\sigma_y}. \tag{11.31}$$

The analytical solution to (11.29–11.30) can be trivially obtained as

$$\varepsilon^p(t) = \frac{c_1}{c_2}(1 - \mathrm{e}^{-c_2\,t}). \tag{11.32}$$

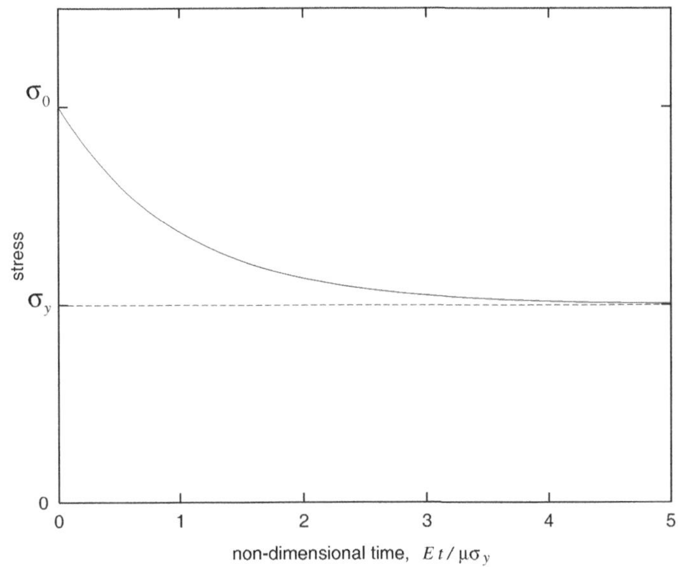

Figure 11.3. One-dimensional viscoplasticity model. Analytical solution to the stress relaxation problem.

Finally, by placing the above solution into (11.27), and taking into account the definition of c_1 and c_2, we obtain the stress as a function of time

$$\sigma(t) = \sigma_0 - (\sigma_0 - \sigma_y)(1 - \mathrm{e}^{-\frac{E}{\mu\,\sigma_y}t}). \tag{11.33}$$

Clearly, the above function describes the stress relaxation process of the bar, with the stress taking the value $\sigma = \sigma_0 > \sigma_y$ at $t = 0$ and subsequently relaxing asymptotically to σ_y as $t \to \infty$. This is illustrated in Figure 11.3 where a graph of the analytical function $\sigma(t)$ (with $\sigma_0 = 2\,\sigma_y$) is plotted. The analytical solution with the present one-dimensional model clearly captures the experimentally observed behaviour referred to in Figure 11.1(c).

11.3. A von Mises-based multidimensional model

This section introduces a multidimensional extension of the one-dimensional model discussed above (and summarised in Box 11.1). Rather than define a generic extension at the outset (as in Section 6.3 (page 148) where a general multidimensional extension of the one-dimensional rate-independent plasticity model of Section 6.2 (page 141) was presented), we chose here to focus first on a von Mises-based generalisation of the uniaxial theory. A discussion of a more generic model, including its potential structure and its relation to time-independent plasticity as a limit case, will be left for Section 11.4.

11.3.1. A VON MISES-TYPE VISCOPLASTIC MODEL WITH ISOTROPIC STRAIN HARDENING

The multidimensional generalisation of the uniaxial viscoplastic model follows the same basic steps as the generalisation presented in Section 6.3 for the rate-independent theory;

Box 11.2. A von Mises-type viscoplastic constitutive model.

1. Elastoplastic split of the strain tensor

$$\boldsymbol{\varepsilon} = \boldsymbol{\varepsilon}^e + \boldsymbol{\varepsilon}^p$$

2. Linear elastic law

$$\boldsymbol{\sigma} = \mathsf{D}^e : \boldsymbol{\varepsilon}^e$$

3. Yield function

$$\Phi(\boldsymbol{\sigma}, \sigma_y) = q(\boldsymbol{s}(\boldsymbol{\sigma})) - \sigma_y; \quad q = \sqrt{\tfrac{3}{2}\, \boldsymbol{s} : \boldsymbol{s}}$$

and elastic domain

$$\mathscr{E} = \{\boldsymbol{\sigma} \mid \Phi(\boldsymbol{\sigma}, \sigma_y) < 0\}$$

4. Plastic flow rule

$$\dot{\boldsymbol{\varepsilon}}^p = \dot{\gamma}\, \frac{\partial \Phi}{\partial \boldsymbol{\sigma}} = \dot{\gamma} \sqrt{\frac{3}{2}}\, \frac{\boldsymbol{s}}{\|\boldsymbol{s}\|}$$

$$\dot{\gamma} = \begin{cases} \dfrac{1}{\mu}\left[\left(\dfrac{q}{\sigma_y}\right)^{1/\epsilon} - 1\right] & \text{if } \Phi(\boldsymbol{\sigma}, \sigma_y) \geq 0 \\ 0 & \text{if } \Phi(\boldsymbol{\sigma}, \sigma_y) < 0 \end{cases}$$

5. Isotropic strain hardening law

$$\sigma_y = \sigma_y(\bar{\varepsilon}^p); \quad \bar{\varepsilon}^p = \int_0^t \|\dot{\boldsymbol{\varepsilon}}^p\| \, \mathrm{d}t$$

6. Evolution of accumulated plastic strain

$$\dot{\bar{\varepsilon}}^p = \dot{\gamma}$$

that is, the elastoplastic split of the total strain, the linear elasticity law, the flow rule and yield function are recast in terms of the corresponding *tensor* quantities (total, elastic and plastic strain tensors, stress tensor and flow vector). The yield function is also redefined as a function of variables of appropriate tensorial order. The usual concept of an elastic domain bounded by a yield surface in the rate-independent theory will remain valid in the viscoplastic case. Here, as our multidimensional extension is von Mises-based, the yield function and the plastic flow rule (including the hardening internal variable) will have the same format as that of the standard rate-independent von Mises model with (associative) Prandtl–Reuss flow vector. The resulting model is a viscoplastic version of the rate-independent isotropically hardening von Mises model summarised in Section 7.3.1 (from page 216). The constitutive equations of the von Mises-based extension are listed in Box 11.2.

Remark 11.2. Analogously to its rate-independent counterpart, under uniaxial stress conditions, the model of Box 11.2 reduces exactly to the one-dimensional theory of Box 11.1. It is also important to emphasise that the basic properties of creep, stress relaxation and strain-rate dependence of the stress response (including the behaviour at limits) as demonstrated for the uniaxial model in Section 11.2.7 are reproduced by the multidimensional theory under any state of stress.

11.3.2. ALTERNATIVE PLASTIC STRAIN RATE DEFINITIONS

So far, the explicit function for $\dot{\gamma}$ that takes part in the definition of the plastic flow equation (see expression (11.6) and item 4 of Boxes 11.1 and 11.2) has been assumed to be of the form proposed by Perić (1993). As mentioned in Section 11.2.4, many forms for $\dot{\gamma}$ have been proposed and, in practice, a particular choice should be dictated by its ability to model the dependence of the plastic strain rate on the state of stress for the material under consideration. In this section, we list some of the most widely used forms. Clearly, each form of $\dot{\gamma}$ defines a different model of viscoplasticity. However, within the framework of von Mises viscoplasticity, the format of the flow rule

$$\dot{\varepsilon}^p = \dot{\gamma}\, \boldsymbol{N}, \tag{11.34}$$

with associative flow vector $\boldsymbol{N}$ defined by

$$\boldsymbol{N} = \frac{\partial \Phi}{\partial \boldsymbol{\sigma}} = \sqrt{\frac{3}{2}}\, \frac{\boldsymbol{s}}{\|\boldsymbol{s}\|}, \tag{11.35}$$

and yield function

$$\Phi(\boldsymbol{\sigma}, \sigma_y) = q(\boldsymbol{s}(\boldsymbol{\sigma})) - \sigma_y, \tag{11.36}$$

will remain unchanged for any definition of $\dot{\gamma}$.

Bingham model

The Bingham model is the simplest model of viscoplasticity. The multiplier $\dot{\gamma}$ in this case is defined as

$$\dot{\gamma}(\boldsymbol{\sigma}, \sigma_y) = \begin{cases} \dfrac{1}{\eta}\, \Phi(\boldsymbol{\sigma}, \sigma_y) = \dfrac{q(\boldsymbol{\sigma}) - \sigma_y}{\eta} & \text{if } \Phi(\boldsymbol{\sigma}, \sigma_y) \geq 0 \\ 0 & \text{if } \Phi(\boldsymbol{\sigma}, \sigma_y) < 0. \end{cases} \tag{11.37}$$

The only material constant in this case is the (temperature-dependent) viscosity parameter η and the strain rate is modelled as a *linear* function of the von Mises effective stress. Note that this law is obtained from Perić's model given in item 4 of Box 11.2 (and also from the Perzyna model described below) by setting

$$\epsilon = 1; \quad \mu = \frac{\eta}{\sigma_y}. \tag{11.38}$$

In the uniaxial case, the plastic strain rate for the Bingham model is a linear function of the axial stress:

$$\dot{\varepsilon}^p = \frac{1}{\eta}\, (|\sigma| - \sigma_y)\, \mathrm{sign}(\sigma). \tag{11.39}$$

This may severely limit the ability of the model to fit experimental data as, in many cases, the observed strain rate may be a markedly nonlinear function of the stress. However, over a relatively narrow range of stresses, the linear approximation may give good results. Other models, with more material constants, have, in general, better flexibility to allow a wider range of experimental data to be fitted.

Perzyna model

This model was introduced by Perzyna (1966, 1971) and is widely used in computational applications of viscoplasticity. It is defined by

$$\dot{\gamma}(\boldsymbol{\sigma}, \sigma_y) = \begin{cases} \dfrac{1}{\mu}\left[\dfrac{q(\boldsymbol{\sigma})}{\sigma_y} - 1\right]^{1/\epsilon} & \text{if } \Phi(\boldsymbol{\sigma}, \sigma_y) \geq 0 \\ 0 & \text{if } \Phi(\boldsymbol{\sigma}, \sigma_y) < 0. \end{cases} \tag{11.40}$$

As in Perić's model, the material constants are the viscosity-related parameter, μ, and the rate sensitivity, ϵ. We remark here that, in spite of its similarity to Perić's definition, as the rate-independent limit is approached with vanishing rate-sensitivity $\epsilon \to 0$ (refer to Remark 11.1 on page 443), the Perzyna model does not reproduce the uniaxial stress–strain curve of the corresponding rate-independent model with yield stress σ_y. As shown by Perić (1993), in this limit, the Perzyna model produces a curve with $\sigma = 2\,\sigma_y$ instead. However, for vanishing viscosity ($\mu \to 0$) or vanishing strain rates, the response of both Perzyna and Perić models coincide with the standard rate-independent model with yield stress σ_y.

11.3.3. OTHER ISOTROPIC AND KINEMATIC HARDENING LAWS

In the viscoplasticity model of Box 11.2, only isotropic strain hardening has been taken into account. Other hardening laws, such as isotropic work hardening (where the plastic work is taken as the internal variable) as well as kinematic hardening and more general mixed isotropic/kinematic hardening rules can be considered in a manner completely analogous to that of the rate-independent theory as described in Section 6.6 (page 177); that is, isotropic work hardening is obtained by having σ_y as a given function of the plastic work, w^p, defined by expression (6.177)

$$\sigma_y = \sigma_y(w^p). \tag{11.41}$$

Kinematic hardening is introduced by simply replacing the von Mises effective stress, q, with the *relative* effective stress

$$\bar{q} = \sqrt{\tfrac{3}{2}\,\boldsymbol{\eta} : \boldsymbol{\eta}}, \tag{11.42}$$

where $\boldsymbol{\eta}$ is the relative stress

$$\boldsymbol{\eta} = \boldsymbol{s} - \boldsymbol{\beta}, \tag{11.43}$$

and $\boldsymbol{\beta}$ is the backstress tensor. Evolution laws for $\boldsymbol{\beta}$, such as Prager's rule and the Armstrong–Frederick kinematic hardening law, can be defined as in Section 6.6.

11.3.4. VISCOPLASTIC MODELS WITHOUT A YIELD SURFACE

The assumption of the existence of an elastic domain bounded by a yield surface is essential in the formulation of rate-independent plasticity models. For viscoplasticity models, however, such an assumption is by no means required. In fact, particularly at higher temperatures, many materials can be modelled as flowing whenever under stress; that is, the yield stress is effectively zero. For example, many metals at high temperatures will flow at virtually any stress state with a non-zero deviatoric component. In such cases, a yield surface and a corresponding elastic domain do not need to be introduced in the formulation of the theory. Viscoplasticity models without a yield surface have been used widely, especially in the analysis of creep and hot metal forming operations. Within the present framework, such models can be defined simply by postulating the explicit function for $\dot{\gamma}$ accordingly.

Norton's creep law

The classical Norton creep law has been employed extensively in the analysis of creep of metals. It is used mainly in the description of secondary creep. In its original (uniaxial) version, the flow rule is given by

$$\dot{\varepsilon}^p = \left(\frac{|\sigma|}{\lambda}\right)^N \text{sign}(\sigma), \tag{11.44}$$

where N and λ are temperature-dependent material constants. Clearly, plastic flow is assumed to occur whenever $\sigma \neq 0$. Its multidimensional generalisation, sometimes referred to as Odqvist's law, is obtained by simply replacing the definition of the function for $\dot{\gamma}$ in item 4 of Box 11.2 with the following

$$\dot{\gamma}(\boldsymbol{\sigma}) = \left[\frac{q(\boldsymbol{\sigma})}{\lambda}\right]^N. \tag{11.45}$$

Here, plastic flow takes place for any stress with non-zero deviator. Note that, by setting $\sigma_y = 0$ in (11.37) the Bingham model recovers the Norton law with $N = 1$ and $\lambda = \eta$.

Lemaitre–Chaboche law

A modification of Norton's law in order to improve its ability to model secondary creep over a wider range of stresses and strain rates is provided by the Lemaitre–Chaboche law (Lemaitre and Chaboche, 1990). The function $\dot{\gamma}$ in this case reads

$$\dot{\gamma}(\boldsymbol{\sigma}) = \left[\frac{q(\boldsymbol{\sigma})}{\lambda}\right]^N \exp[\alpha\, q(\boldsymbol{\sigma})^{N+1}]. \tag{11.46}$$

In addition to the material parameters N and λ required by Norton's law, the present model has a third (also temperature-dependent) parameter α.

Other creep laws

A rather general class of viscoplastic laws can be obtained by assuming that $\dot{\gamma}$ is a function of the stress, time and temperature, with the following multiplicative format

$$\dot{\gamma} = \dot{\gamma}(\boldsymbol{\sigma}, t, T) = f_\sigma(\boldsymbol{\sigma})\, f_t(t)\, f_T(T), \tag{11.47}$$

where t and T denote, respectively, the time and absolute temperature and f_σ, f_t and f_T are experimentally defined functions. A comprehensive list of proposed empirical functions is given by Skrzypek (1993), to which the interested reader is referred. For instance, f_σ could be Norton's law or the Lemaitre–Chaboche relation above. The temperature function f_T is normally defined by the Arrhenius law

$$f_T(T) = C \exp\left[\frac{-Q}{RT}\right] \tag{11.48}$$

where C is a constant, Q is the *activation energy* usually independent of the temperature, R is the gas constant 8.31 J mol^{-1} K^{-1}. A typical example of an empirical relation of the above format is given by the law (Boyle and Spence, 1983)

$$\dot{\gamma} = C \exp\left[\frac{-Q}{RT}\right] t^M\, q^N, \tag{11.49}$$

with M and N being material parameters.

Another interesting viscoplastic model used primarily in the description of the behaviour of metallic alloys at high temperatures is the *Bodner–Partom* model (Bodner and Partom, 1975). An implicit computational implementation of the Bodner–Partom model has been recently described by Anderson (2003).

11.4. General viscoplastic constitutive model

Having described in the previous section some of the most commonly used viscoplasticity models, we proceed here to formulate a more general constitutive theory of viscoplasticity. The theory presented here is a viscoplastic version of the general rate-independent model described in Section 6.3 (from page 148) and summarised in Box 6.2 (page 151). At this point, note that we will use here the notation of Section 6.3. The reader who is not familiar with that notation, or concepts used in that section, is advised to review them before proceeding further. The formulation of the viscoplastic model is analogous to that of its rate-independent counterpart. It follows the same considerations as Sections 6.3.1 to 6.3.4, except that the flow rule and hardening law are defined as

$$\begin{aligned} \dot{\boldsymbol{\varepsilon}}^p &= \boldsymbol{G}(\boldsymbol{\sigma}, \boldsymbol{A}) \\ \dot{\boldsymbol{\alpha}} &= \boldsymbol{J}(\boldsymbol{\sigma}, \boldsymbol{A}); \end{aligned} \tag{11.50}$$

that is, the plastic strain rate and the evolution law for the set $\boldsymbol{\alpha}$ of hardening internal variables are defined by means of the *explicit* constitutive functions $\boldsymbol{G}$ and $\boldsymbol{J}$ of $\boldsymbol{\sigma}$ and the set $\boldsymbol{A}$ of hardening thermodynamic forces. In addition, as we have seen above, an elastic domain may not exist. Thus, a yield function is not necessarily present in the viscoplastic formulation. The constitutive equations of the general viscoplasticity model are listed in Box 11.3.

Note that the von Mises-based model of Box 11.2 (which incorporates an elastic domain) is trivially recovered by defining the functions $\boldsymbol{G}$ and $\boldsymbol{J}$ as well as the free-energy potential ψ and the internal variable set $\boldsymbol{\alpha}$ accordingly. The same applies to all other models (with or without an elastic domain) described in Section 11.3.

11.4.1. RELATION TO THE GENERAL CONTINUUM CONSTITUTIVE THEORY

The above viscoplasticity model fits within the generic internal variable-based constitutive framework discussed in Section 3.5.2 (from page 71). Indeed, it can be trivially established

Box 11.3. A general viscoplastic constitutive model.

1. Additive decomposition of the strain tensor

$$\boldsymbol{\varepsilon} = \boldsymbol{\varepsilon}^e + \boldsymbol{\varepsilon}^p$$

2. Free-energy function

$$\psi = \psi(\boldsymbol{\varepsilon}^e, \boldsymbol{\alpha})$$

where $\boldsymbol{\alpha}$ is a set of hardening internal variables

3. Constitutive equation for $\boldsymbol{\sigma}$ and hardening thermodynamic forces $\boldsymbol{A}$

$$\boldsymbol{\sigma} = \bar{\rho}\,\frac{\partial \psi}{\partial \boldsymbol{\varepsilon}^e} \qquad \boldsymbol{A} = \bar{\rho}\,\frac{\partial \psi}{\partial \boldsymbol{\alpha}}$$

4. Plastic flow rule and hardening law

$$\dot{\boldsymbol{\varepsilon}}^p = \boldsymbol{G}(\boldsymbol{\sigma}, \boldsymbol{A})$$

$$\dot{\boldsymbol{\alpha}} = \boldsymbol{J}(\boldsymbol{\sigma}, \boldsymbol{A})$$

that the model of Box 11.3 is a particular case of the general purely mechanical infinitesimal constitutive law defined by (3.165) on page 76. The general viscoplasticity model is obtained by simply defining the set $\boldsymbol{\alpha}$ of (3.165) as composed of the plastic strain tensor and the set of hardening internal variables (as described in Section 6.3.2) and then introducing the explicit constitutive functions for the rates of plastic strain and hardening variables listed in item 4 of Box 11.3.

11.4.2. POTENTIAL STRUCTURE AND DISSIPATION INEQUALITY

A specialisation of the general theory of Box 11.3 can be obtained by endowing the model with a potential structure (refer to the discussion surrounding expression (3.162) on page 74). In this case, we define a *dissipation potential*

$$\Xi = \Xi(\boldsymbol{\sigma}, \boldsymbol{A}), \tag{11.51}$$

from which, through the hypothesis of *normal dissipativity*, the evolution of the internal variables of the problem are derived as

$$\begin{aligned} \dot{\boldsymbol{\varepsilon}}^p &= \frac{\partial \Xi}{\partial \boldsymbol{\sigma}} \\ \dot{\boldsymbol{\alpha}} &= -\frac{\partial \Xi}{\partial \boldsymbol{A}}. \end{aligned} \tag{11.52}$$

At this point, it is important to recall that the *plastic dissipation* in the present case is given by (again, refer to Section 6.3.2)

$$\boldsymbol{\sigma} : \dot{\boldsymbol{\varepsilon}}^p - \boldsymbol{A} * \dot{\boldsymbol{\alpha}},$$

so that the dissipation inequality reads

$$\Upsilon^p(\boldsymbol{\sigma}, \boldsymbol{A}; \dot{\boldsymbol{\varepsilon}}^p, \dot{\boldsymbol{\alpha}}) \geq 0, \tag{11.53}$$

where

$$\Upsilon^p(\boldsymbol{\sigma}, \boldsymbol{A}; \dot{\boldsymbol{\varepsilon}}^p, \dot{\boldsymbol{\alpha}}) \equiv \boldsymbol{\sigma} : \dot{\boldsymbol{\varepsilon}}^p - \boldsymbol{A} * \dot{\boldsymbol{\alpha}} \tag{11.54}$$

is the *dissipation function*.

By defining Ξ such that it is convex with respect to both variables, non-negative and zero-valued at $\{\boldsymbol{\sigma}, \boldsymbol{A}\} = \{\mathbf{0}, \mathbf{0}\}$ it is ensured that the dissipation inequality is satisfied *a priori* by the model.

11.4.3. RATE-INDEPENDENT PLASTICITY AS A LIMIT CASE

In this section we show that rate-independent plasticity can be recovered as a limit case of the above general viscoplastic theory with a potential structure. As emphasised above, the general viscoplastic model is a particular instance of the internal variable-based constitutive framework of Section 3.5.2. Thus, the demonstration that follows here shows effectively that, as anticipated in Section 6.3.7, the elastoplastic model of Box 6.3 (page 151) can be rigorously described, under some circumstances, as a particular case of the general continuum constitutive theory of Section 3.5.2.

The indicator function of a convex set

The demonstration presented here is based on arguments of convex analysis.[‡] Crucial to the proof to be shown is the concept of *indicator function* of a convex set. In this context, let us consider the closure, $\mathscr{A}$, of the elastic domain defined by means of a yield function Φ:

$$\mathscr{A} = \{(\boldsymbol{\sigma}, \boldsymbol{A}) \mid \Phi(\boldsymbol{\sigma}, \boldsymbol{A}) \leq 0\}. \tag{11.55}$$

In rate-independent plasticity, $\mathscr{A}$ is the set of all *admissible* states $(\boldsymbol{\sigma}, \boldsymbol{A})$ of stress and hardening thermodynamical forces. The set $\mathscr{A}$ is *convex*, i.e. it defines a convex region in the space of stresses and hardening forces. Following the above considerations, we now introduce the *indicator function*, $\Psi_{\mathscr{A}}$, of the convex set $\mathscr{A}$ as the scalar-valued function defined by

$$\Psi_{\mathscr{A}}(\boldsymbol{\sigma}, \boldsymbol{A}) = \begin{cases} 0 & \text{if } (\boldsymbol{\sigma}, \boldsymbol{A}) \in \mathscr{A} \\ \infty & \text{if } (\boldsymbol{\sigma}, \boldsymbol{A}) \notin \mathscr{A}. \end{cases} \tag{11.56}$$

The indicator function is clearly *non-differentiable*.

The rate-independent limit

In what follows, we shall see that an associative rate-independent plasticity model is obtained by adopting $\Psi_{\mathscr{A}}$ as the dissipation potential in the general viscoplastic theory; that is, we choose

$$\Xi(\boldsymbol{\sigma}, \boldsymbol{A}) \equiv \Psi_{\mathscr{A}}(\boldsymbol{\sigma}, \boldsymbol{A}). \tag{11.57}$$

[‡]Readers not familiar with convex analysis are referred to Rockafellar (1970).

At this point we need to make use of the concept of *subdifferential*.[§] In view of the non-differentiability of the indicator function, the constitutive equations (11.52), which follow from normal dissipativity, are replaced with the *subdifferential* relations

$$\begin{aligned} \dot{\varepsilon}^p &\in \partial_{\boldsymbol{\sigma}} \Psi_{\mathscr{A}} \\ \dot{\boldsymbol{\alpha}} &\in -\partial_{\boldsymbol{A}} \Psi_{\mathscr{A}}, \end{aligned} \tag{11.58}$$

where $\partial_{\boldsymbol{\sigma}} \Psi_{\mathscr{A}}$ and $\partial_{\boldsymbol{A}} \Psi_{\mathscr{A}}$ are the subdifferentials of $\Psi_{\mathscr{A}}$ with respect to $\boldsymbol{\sigma}$ and $\boldsymbol{A}$, respectively. From the subdifferential definition (6.69) together with (11.56) it can easily be established that (11.58) is equivalent to the inequality

$$\dot{\varepsilon}^p : (\boldsymbol{\sigma} - \boldsymbol{\sigma}^*) + \dot{\boldsymbol{\alpha}} * (\boldsymbol{A} - \boldsymbol{A}^*) \geq 0, \quad \forall\, (\boldsymbol{\sigma}^*, \boldsymbol{A}^*) \in \Psi_{\mathscr{A}}, \tag{11.59}$$

or, equivalently, in terms of the dissipation function (11.54)

$$\Upsilon^p(\boldsymbol{\sigma}, \boldsymbol{A}; \dot{\varepsilon}^p, \dot{\boldsymbol{\alpha}}) \geq \Upsilon^p(\boldsymbol{\sigma}^*, \boldsymbol{A}^*; \dot{\varepsilon}^p, \dot{\boldsymbol{\alpha}}), \quad \forall\, (\boldsymbol{\sigma}^*, \boldsymbol{A}^*) \in \mathscr{A}. \tag{11.60}$$

This last inequality states that, among all states $(\boldsymbol{\sigma}^*, \boldsymbol{A}^*) \in \mathscr{A}$, the actual stress and hardening force $(\boldsymbol{\sigma}, \boldsymbol{A})$ maximise the dissipation function. This is known as the *principle of maximum plastic dissipation*, discussed in Section 6.5.2 (page 170) to which the reader is referred for details. The solution to the maximisation problem associated with the principle of maximum plastic dissipation is the classical associative laws

$$\begin{aligned} \dot{\varepsilon}^p &= \dot{\gamma}\, \frac{\partial \Phi}{\partial \boldsymbol{\sigma}} \\ \dot{\boldsymbol{\alpha}} &= -\dot{\gamma}\, \frac{\partial \Phi}{\partial \boldsymbol{A}}, \end{aligned} \tag{11.61}$$

together with the loading/unloading conditions of rate-independent plasticity

$$\Phi(\boldsymbol{\sigma}, \boldsymbol{A}) \leq 0, \quad \dot{\gamma} \geq 0, \quad \Phi(\boldsymbol{\sigma}, \boldsymbol{A})\dot{\gamma} = 0. \tag{11.62}$$

In summary, it has been shown above that the classical rate-independent associative plasticity equations are rigorously recovered from the general viscoplasticity model when the indicator function of the set $\mathscr{A}$ is taken as the dissipation potential.

Example: von Mises-based model

Let us now consider the von Mises-based model of Box 11.2 and, for simplicity, assume that the model is perfectly viscoplastic (constant σ_y). Our purpose here is to illustrate the above ideas by demonstrating that the viscoplastic model can be defined in terms of a dissipation potential whose limit when $\epsilon \rightarrow 0$ or $\mu \rightarrow 0$ is the indicator function of the set of admissible stresses of the perfectly plastic von Mises model. Thus, in such limits, the perfectly viscoplastic model rigorously recovers the classical perfectly elastoplastic von Mises model.

[§] Refer to Section 6.3.9 (from page 153) for the definition of subdifferential.

We start by defining the dissipation potential as

$$\Xi(\boldsymbol{\sigma}) = \begin{cases} \dfrac{\sigma_y}{\mu}\left\{\dfrac{1}{1+\epsilon} + \dfrac{\epsilon}{1+\epsilon}\left[\dfrac{q(\boldsymbol{\sigma})}{\sigma_y}\right]^{\frac{1+\epsilon}{\epsilon}} - \dfrac{q(\boldsymbol{\sigma})}{\sigma_y}\right\} & \text{if } q/\sigma_y \geq 1 \\ 0 & \text{if } q/\sigma_y < 1, \end{cases} \tag{11.63}$$

where $\boldsymbol{\sigma}$ is the only variable. With the above definition, the flow rule

$$\dot{\boldsymbol{\varepsilon}}^p = \frac{\partial \Xi}{\partial \boldsymbol{\sigma}}, \tag{11.64}$$

is found through a straightforward differentiation to be that of item 4 of Box 11.2; that is, the above potential indeed defines the von Mises-based viscoplasticity model of Box 11.2 when hardening is not considered.

Finally, by simple inspection, we can easily see that, when $\epsilon \to 0$ or $\mu \to 0$, the limit of the potential Ξ of (11.63) is the indicator function of the set of admissible stresses defined by the von Mises yield function:

$$\lim_{\epsilon \to 0} \Xi(\boldsymbol{\sigma}) = \lim_{\mu \to 0} \Xi(\boldsymbol{\sigma}) = \Psi_{\mathscr{A}}(\boldsymbol{\sigma}), \tag{11.65}$$

where

$$\mathscr{A} = \{\boldsymbol{\sigma} \mid q(\boldsymbol{\sigma}) - \sigma_y \leq 0\}. \tag{11.66}$$

This completes the demonstration. The schematic illustration of Figure 11.4 shows the potential Ξ for various choices of the rate-sensitivity parameter ϵ. Clearly, as $\epsilon \to 0$, Ξ tends to the indicator function of $\mathscr{A}$.

11.5. General numerical framework

This section describes the basic ingredients needed to incorporate the general viscoplasticity model of Box 11.3 into the finite element framework of Chapter 4. The basic requirements are:

(i) an algorithm for numerical integration of the viscoplastic constitutive equations, to be used to update stresses and other state variables of the model;

(ii) the associated consistent tangent modulus, to be used in the assemblage of the finite element stiffness matrix.

For further discussions and analysis of various aspects of the numerical treatment of viscoplasticity, we refer to Simo and Govindjee (1991), Perić (1993), Chaboche and Cailletaud (1996), Simo and Hughes (1998), Simo (1998), Runesson and Mahler (1999), and Alfano and Rosati (2001).

11.5.1. A GENERAL IMPLICIT INTEGRATION ALGORITHM

Before proceeding to the derivation of an integration algorithm for the general viscoplastic model, it seems convenient, for the sake of clarity, to start by stating the underlying initial value problem we wish to solve. The problem here is analogous to its rate-independent counterpart, Problem 7.1, stated on page 193.

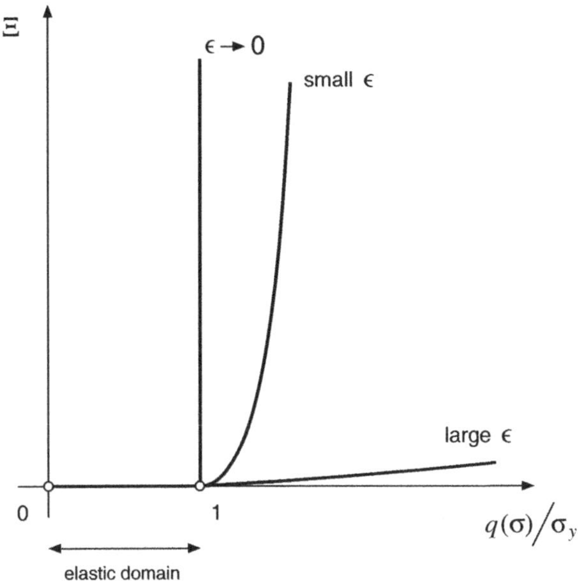

Figure 11.4. Viscoplastic potential Ξ for various rate-sensitivity parameters ϵ.

Problem 11.1 (The viscoplastic constitutive initial value problem). *Given the initial values $\boldsymbol{\varepsilon}^e(t_0)$ and $\boldsymbol{\alpha}(t_0)$ and given the history of the strain tensor, $\boldsymbol{\varepsilon}(t)$, $t \in [t_0, T]$, find the functions $\boldsymbol{\varepsilon}^e(t)$ and $\boldsymbol{\alpha}(t)$, for the elastic strain tensor and hardening internal variable set that satisfy the reduced general viscoplastic constitutive equations*

$$\dot{\boldsymbol{\varepsilon}}^e(t) = \dot{\boldsymbol{\varepsilon}}(t) - \boldsymbol{G}(\boldsymbol{\sigma}(t), \boldsymbol{A}(t)), \quad \dot{\boldsymbol{\alpha}}(t) = \boldsymbol{J}(\boldsymbol{\sigma}(t), \boldsymbol{A}(t)) \tag{11.67}$$

for each instant $t \in [t_0, T]$, with

$$\boldsymbol{\sigma}(t) = \bar{\rho}\left.\frac{\partial \psi}{\partial \boldsymbol{\varepsilon}^e}\right|_t, \quad \boldsymbol{A}(t) = \bar{\rho}\left.\frac{\partial \psi}{\partial \boldsymbol{\alpha}}\right|_t. \tag{11.68}$$

As in the definition of the rate-independent problem, the reduced system (11.67) of ordinary differential equations has been obtained by incorporating the viscoplastic flow equation $(11.50)_1$ into the elastoplastic split of the total strain rate so that the plastic strain does not appear explicitly in the initial value problem. Clearly, once the history of elastic strain is determined in the solution to the above problem, the history of the plastic strain is promptly obtained as

$$\boldsymbol{\varepsilon}^p(t) = \boldsymbol{\varepsilon}(t) - \boldsymbol{\varepsilon}^e(t). \tag{11.69}$$

The fully implicit algorithm for the numerical solution of the above problem is derived by simply applying a standard backward Euler time discretisation of the rate equations. The resulting incremental problem is presented in Box 11.4, where a typical step over the time interval $[t_n, t_{n+1}]$ is considered. The time and strain increments are defined in the usual way as

$$\Delta t = t_{n+1} - t_n, \quad \Delta \boldsymbol{\varepsilon} = \boldsymbol{\varepsilon}_{n+1} - \boldsymbol{\varepsilon}_n. \tag{11.70}$$

Box 11.4. Fully implicit algorithm for numerical integration of general viscoplastic constitutive equations.

Given the strain and time increment, $\Delta\boldsymbol{\varepsilon}$ and Δt, over $[t_n, t_{n+1}]$ and the state variables at t_n, compute the updated state by solving the nonlinear system of equations

$$\left\{\begin{matrix} \boldsymbol{\varepsilon}^e_{n+1} - \boldsymbol{\varepsilon}^e_n - \Delta\boldsymbol{\varepsilon} + \Delta t\, \boldsymbol{G}(\boldsymbol{\sigma}_{n+1}, \boldsymbol{A}_{n+1}) \\ \boldsymbol{\alpha}_{n+1} - \boldsymbol{\alpha}_n - \Delta t\, \boldsymbol{J}(\boldsymbol{\sigma}_{n+1}, \boldsymbol{A}_{n+1}) \end{matrix}\right\} = \left\{\begin{matrix} \boldsymbol{0} \\ \boldsymbol{0} \end{matrix}\right\}$$

for $\boldsymbol{\varepsilon}^e_{n+1}$ and $\boldsymbol{\alpha}_{n+1}$, with

$$\boldsymbol{\sigma}_{n+1} = \bar{\rho}\, \frac{\partial\psi}{\partial\boldsymbol{\varepsilon}^e}\bigg|_{n+1}, \qquad \boldsymbol{A}_{n+1} = \bar{\rho}\, \frac{\partial\psi}{\partial\boldsymbol{\alpha}}\bigg|_{n+1}$$

Models with a yield surface

Note that in the algorithm of Box 11.4 no assumption is made on the existence of an elastic domain. The algorithm is valid for models with or without a yield surface. If a yield surface is present, however, the specialisation of the algorithm of Box 11.4 takes a two-stage format completely analogous to the elastic predictor/return-mapping procedure of the rate-independent case. To see this, let us first consider that for a general model with a yield surface, the constitutive functions $\boldsymbol{G}$ and $\boldsymbol{J}$ can be defined with the following form

$$\begin{aligned} \boldsymbol{G}(\boldsymbol{\sigma}, \boldsymbol{A}) &= \dot{\gamma}(\boldsymbol{\sigma}, \boldsymbol{A})\, \boldsymbol{N}(\boldsymbol{\sigma}, \boldsymbol{A}) \\ \boldsymbol{J}(\boldsymbol{\sigma}, \boldsymbol{A}) &= \dot{\gamma}(\boldsymbol{\sigma}, \boldsymbol{A})\, \boldsymbol{H}(\boldsymbol{\sigma}, \boldsymbol{A}), \end{aligned} \tag{11.71}$$

where, following the terminology of the rate-independent theory, $\boldsymbol{N}$ is the flow vector and $\boldsymbol{H}$ is the generalised hardening modulus. The scalar $\dot{\gamma}$ is zero within the elastic domain or on the yield surface and may only be non-zero outside the elastic domain. Clearly, evolution of $\boldsymbol{\varepsilon}^p$ and $\boldsymbol{\alpha}$ may only occur here at states with $\Phi(\boldsymbol{\sigma}, \boldsymbol{A}) > 0$, i.e. states lying neither in the elastic domain nor on the yield surface. Then, as in the rate-independent case, it makes sense to first compute an *elastic trial state* by assuming that the material behaviour is purely elastic within the interval $[t_n, t_{n+1}]$. If the trial state is within the elastic domain or on the yield surface, then no viscoplastic flow takes place within the considered time step and the trial state is the actual state at the end of the step. Otherwise, the evolution of $\boldsymbol{\varepsilon}^p$ and $\boldsymbol{\alpha}$ is computed by means of the standard backward Euler method. The resulting algorithm, which we shall refer to as the *elastic predictor/viscoplastic corrector* or *elastic predictor/viscoplastic return mapping* algorithm, is listed in Box 11.5.

Remark 11.3. The viscoplastic return mapping differs from its elastoplastic (rate-independent) counterpart (refer to Box 7.1, page 199) in that, here, the updated stress state at t_{n+1} generally lies on the outside of the yield surface, i.e.

$$\Phi(\boldsymbol{\sigma}_{n+1}, \boldsymbol{A}_{n+1}) > 0.$$

This is in contrast with the rate-independent case in which the consistency equation, $\Phi_{n+1} = 0$, forces the updated state to be on the yield surface when there is plastic flow over

Box 11.5. Fully implicit elastic predictor/viscoplastic return-mapping algorithm for numerical integration of general viscoplastic constitutive equations with a yield surface over a generic time interval $[t_n, t_{n+1}]$ with $\Delta t = t_{n+1} - t_n$.

(i) Elastic predictor. Given $\Delta\boldsymbol{\varepsilon}$ and the state variables at t_n, evaluate the *elastic trial state*

$$\boldsymbol{\varepsilon}^{e\,\text{trial}}_{n+1} = \boldsymbol{\varepsilon}^e_n + \Delta\boldsymbol{\varepsilon}$$

$$\boldsymbol{\alpha}^{\text{trial}}_{n+1} = \boldsymbol{\alpha}_n$$

$$\boldsymbol{\sigma}^{\text{trial}}_{n+1} = \bar{\rho}\,\frac{\partial\psi}{\partial\boldsymbol{\varepsilon}^e}\bigg|^{\text{trial}}_{n+1}, \qquad \boldsymbol{A}^{\text{trial}}_{n+1} = \bar{\rho}\,\frac{\partial\psi}{\partial\boldsymbol{\alpha}}\bigg|^{\text{trial}}_{n+1}$$

(ii) Check for viscoplastic flow

IF $\Phi(\boldsymbol{\sigma}^{\text{trial}}_{n+1}, \boldsymbol{A}^{\text{trial}}_{n+1}) \le 0$

THEN set $(\cdot)_{n+1} = (\cdot)^{\text{trial}}_{n+1}$ and EXIT

(iii) Viscoplastic return mapping. Solve the system

$$\begin{Bmatrix} \boldsymbol{\varepsilon}^e_{n+1} - \boldsymbol{\varepsilon}^{e\,\text{trial}}_{n+1} + \Delta\gamma\,\boldsymbol{N}(\boldsymbol{\sigma}_{n+1}, \boldsymbol{A}_{n+1}) \\ \boldsymbol{\alpha}_{n+1} - \boldsymbol{\alpha}^{\text{trial}}_{n+1} - \Delta\gamma\,\boldsymbol{H}(\boldsymbol{\sigma}_{n+1}, \boldsymbol{A}_{n+1}) \end{Bmatrix} = \begin{Bmatrix} \boldsymbol{0} \\ \boldsymbol{0} \end{Bmatrix}$$

for $\boldsymbol{\varepsilon}^e_{n+1}$, and $\boldsymbol{\alpha}_{n+1}$ with

$$\Delta\gamma = \Delta\gamma(\boldsymbol{\sigma}_{n+1}, \boldsymbol{A}_{n+1}) = \Delta t\,\dot{\gamma}(\boldsymbol{\sigma}_{n+1}, \boldsymbol{A}_{n+1})$$

and

$$\boldsymbol{\sigma}_{n+1} = \bar{\rho}\,\frac{\partial\psi}{\partial\boldsymbol{\varepsilon}^e}\bigg|_{n+1}, \qquad \boldsymbol{A}_{n+1} = \bar{\rho}\,\frac{\partial\psi}{\partial\boldsymbol{\alpha}}\bigg|_{n+1}$$

(iv) EXIT

the considered interval. Nevertheless, the terminology viscoplastic *return mapping* remains justifiable in the present case since, upon application of the procedure, the updated stress is obtained by moving (or *returning*) the trial stress towards the yield surface.

11.5.2. ALTERNATIVE EULER-BASED ALGORITHMS

Similarly to the rate-independent case (refer to Section 7.2.7, page 201), different numerical integration algorithms can be employed in the stress updating procedure. In what follows we list the basic equations of the generalised trapezoidal and midpoint algorithms. For further details on alternative integration algorithms we refer to Cormeau (1975), Zienkiewicz and Cormeau (1974), Hughes and Taylor (1978), Marques and Owen (1983), Peirce *et al.* (1984) and Kojić and Bathe (1987).

The generalised trapezoidal algorithm

Here, the backward Euler discrete equations of Box 11.4 are replaced with the following system

$$
\begin{aligned}
\boldsymbol{\varepsilon}^e_{n+1} &= \boldsymbol{\varepsilon}^e_n + \Delta\boldsymbol{\varepsilon} - \Delta t\,[(1-\theta)\,\boldsymbol{G}_n + \theta\,\boldsymbol{G}_{n+1}] \\
\boldsymbol{\alpha}_{n+1} &= \boldsymbol{\alpha}_n + \Delta t\,[(1-\theta)\,\boldsymbol{J}_n + \theta\,\boldsymbol{J}_{n+1}],
\end{aligned}
\tag{11.72}
$$

where θ is a prescribed parameter

$$
0 \leq \theta \leq 1. \tag{11.73}
$$

For the choice $\theta = 1$, the implicit algorithm of Box 11.4 is recovered and $\theta = 0$ corresponds to the explicit algorithm.

The generalised midpoint algorithm

For the generalised midpoint rule, the discrete system of equations reads

$$
\begin{aligned}
\boldsymbol{\varepsilon}^e_{n+1} &= \boldsymbol{\varepsilon}^e_n + \Delta\boldsymbol{\varepsilon} - \Delta t\,\boldsymbol{G}_{n+\theta} \\
\boldsymbol{\alpha}_{n+1} &= \boldsymbol{\alpha}_n + \Delta t\,\boldsymbol{J}_{n+\theta},
\end{aligned}
\tag{11.74}
$$

where the prescribed parameter, θ, also lies within the interval $[0, 1]$ and

$$
\begin{aligned}
\boldsymbol{G}_{n+\theta} &= \boldsymbol{G}((1-\theta)\boldsymbol{\sigma}_{n+1} + \theta\,\boldsymbol{\sigma}_n, (1-\theta)\boldsymbol{A}_n + \theta\,\boldsymbol{A}_{n+1}) \\
\boldsymbol{J}_{n+\theta} &= \boldsymbol{J}((1-\theta)\boldsymbol{\sigma}_{n+1} + \theta\,\boldsymbol{\sigma}_n, (1-\theta)\boldsymbol{A}_n + \theta\,\boldsymbol{A}_{n+1}).
\end{aligned}
\tag{11.75}
$$

Again, for $\theta = 1$, the implicit algorithm of Box 11.4 is recovered and $\theta = 0$ defines the explicit algorithm.

11.5.3. GENERAL CONSISTENT TANGENT OPERATOR

To complete the requirements for the implementation of the model within an implicit finite element environment, the tangent modulus consistent with the general algorithm is needed. Let us then consider the algorithm of Box 11.4. Given all variables of the problem at t_n and a prescribed time increment Δt, the task here is to find the exact tangent operator

$$
\mathsf{D} \equiv \frac{\mathrm{d}\boldsymbol{\sigma}_{n+1}}{\mathrm{d}\boldsymbol{\varepsilon}_{n+1}} = \frac{\mathrm{d}\boldsymbol{\sigma}_{n+1}}{\mathrm{d}\Delta\boldsymbol{\varepsilon}}, \tag{11.76}
$$

consistent with the stress updating procedure defined by the backward Euler algorithm of Box 11.4.

Analogously to the general procedure for the rate-independent case (refer to Section 7.4.4, from page 238), we start by linearising the system of time-discrete equations of Box 11.4. The linearised system reads

$$
\left\{
\begin{array}{c}
\mathrm{d}\boldsymbol{\varepsilon}^e + \Delta t\,\dfrac{\partial \boldsymbol{G}}{\partial \boldsymbol{\sigma}} : \mathrm{d}\boldsymbol{\sigma} + \Delta t\,\dfrac{\partial \boldsymbol{G}}{\partial \boldsymbol{A}} * \mathrm{d}\boldsymbol{A} \\[2em]
\mathrm{d}\boldsymbol{\alpha} - \Delta t\,\dfrac{\partial \boldsymbol{J}}{\partial \boldsymbol{\sigma}} * \mathrm{d}\boldsymbol{\sigma} - \Delta t\,\dfrac{\partial \boldsymbol{J}}{\partial \boldsymbol{A}} * \mathrm{d}\boldsymbol{A}
\end{array}
\right\}
=
\left\{
\begin{array}{c}
\mathrm{d}\Delta\boldsymbol{\varepsilon} \\[2em]
\boldsymbol{0}
\end{array}
\right\}, \tag{11.77}
$$

where the symbol * denotes the product of the appropriate type and the subscripts $n+1$ have been omitted for notational convenience. With the introduction of the differential relations (7.129) (page 239), the linearised system is equivalently written as

$$\begin{bmatrix} \mathsf{C} + \Delta t\,\dfrac{\partial \boldsymbol{G}}{\partial \boldsymbol{\sigma}} & \mathsf{B} + \Delta t\,\dfrac{\partial \boldsymbol{G}}{\partial \boldsymbol{A}} \\ & \\ \mathsf{A} - \Delta t\,\dfrac{\partial \boldsymbol{J}}{\partial \boldsymbol{\sigma}} & \mathsf{J} - \Delta t\,\dfrac{\partial \boldsymbol{J}}{\partial \boldsymbol{A}} \end{bmatrix} \begin{bmatrix} \mathrm{d}\boldsymbol{\sigma} \\ \\ \mathrm{d}\boldsymbol{A} \end{bmatrix} = \begin{bmatrix} \mathrm{d}\Delta\boldsymbol{\varepsilon} \\ \\ \mathbf{0} \end{bmatrix}. \tag{11.78}$$

By inverting the linearised system above, we finally obtain a tangent relation which can be written symbolically as

$$\begin{bmatrix} \mathrm{d}\boldsymbol{\sigma} \\ \\ \mathrm{d}\boldsymbol{A} \end{bmatrix} = \begin{bmatrix} \mathsf{D}_{11} & \mathsf{D}_{12} \\ & \\ \mathsf{D}_{21} & \mathsf{D}_{22} \end{bmatrix} \begin{bmatrix} \mathrm{d}\Delta\boldsymbol{\varepsilon} \\ \\ \mathbf{0} \end{bmatrix}, \tag{11.79}$$

where D_{ij} are tensors of appropriate order resulting from the inversion of (11.78). The consistent tangent operator we are looking for is the fourth-order tensor

$$\mathsf{D} \equiv \frac{\mathrm{d}\boldsymbol{\sigma}_{n+1}}{\mathrm{d}\Delta\boldsymbol{\varepsilon}} = \mathsf{D}_{11}. \tag{11.80}$$

Models with a yield surface

For models with a yield surface, the tangent modulus is elastic if the state is within the elastic domain; that is, as in rate-independent plasticity, when $\Phi(\boldsymbol{\sigma}_{n+1}, \boldsymbol{A}_{n+1}) \leq 0$, we have

$$\mathsf{D} = \mathsf{D}^e = \bar{\rho}\frac{\partial^2 \psi}{\partial \boldsymbol{\varepsilon}^{e\,2}}. \tag{11.81}$$

Under viscoplastic flow, i.e. when $\Phi(\boldsymbol{\sigma}_{n+1}, \boldsymbol{A}_{n+1}) > 0$, the stress is the result from the solution of the equation system of item (iii) of Box 11.5. In this case, the tangent operator is a specialisation of the general tangent modulus (11.80) where the functions $\boldsymbol{G}$ and $\boldsymbol{J}$ taking part in the symbolic matrix (11.78) are defined by (11.71). The derivatives of $\boldsymbol{G}$ then specialise as

$$\begin{aligned} \frac{\partial \boldsymbol{G}}{\partial \boldsymbol{\sigma}} &= \dot{\gamma}\,\frac{\partial \boldsymbol{N}}{\partial \boldsymbol{\sigma}} + \boldsymbol{N} \otimes \frac{\partial \dot{\gamma}}{\partial \boldsymbol{\sigma}} \\ \frac{\partial \boldsymbol{G}}{\partial \boldsymbol{A}} &= \dot{\gamma}\,\frac{\partial \boldsymbol{N}}{\partial \boldsymbol{A}} + \boldsymbol{N} * \frac{\partial \dot{\gamma}}{\partial \boldsymbol{A}}. \end{aligned} \tag{11.82}$$

Similarly, the derivatives of $\boldsymbol{J}$ specialise as

$$\begin{aligned} \frac{\partial \boldsymbol{J}}{\partial \boldsymbol{\sigma}} &= \dot{\gamma}\,\frac{\partial \boldsymbol{H}}{\partial \boldsymbol{\sigma}} + \boldsymbol{H} * \frac{\partial \dot{\gamma}}{\partial \boldsymbol{\sigma}} \\ \frac{\partial \boldsymbol{J}}{\partial \boldsymbol{A}} &= \dot{\gamma}\,\frac{\partial \boldsymbol{H}}{\partial \boldsymbol{A}} + \boldsymbol{H} * \frac{\partial \dot{\gamma}}{\partial \boldsymbol{A}}. \end{aligned} \tag{11.83}$$

11.6. Application: computational implementation of a von Mises-based model

To illustrate the application of the above numerical framework, this section describes in detail the basic ingredients of the computational implementation of the von Mises-based model with isotropic strain hardening given in Box 11.2. In addition to the detailed description of the associated integration algorithm and consistent tangent operator, we present an accuracy analysis of the algorithm based on iso-error maps. We remark that the procedures presented here are *not* incorporated in the standard version of program HYPLAS that accompanies this book.

11.6.1. INTEGRATION ALGORITHM

The integration algorithm described here is a specialisation of the generic algorithm described in Section 11.5 to the model whose constitutive equations are summarised in Box 11.2. The algorithm comprises the standard *elastic predictor* and the *viscoplastic return mapping* which, for the present model, has the following format.

1. *Elastic predictor*. The material is assumed to behave purely elastically within the time interval $[t_n, t_{n+1}]$. The *elastic trial* state is then computed as

$$\begin{aligned} \boldsymbol{\varepsilon}^{e\ \mathrm{trial}} &= \boldsymbol{\varepsilon}^e_n + \Delta\boldsymbol{\varepsilon} \\ \boldsymbol{\varepsilon}^{p\ \mathrm{trial}} &= \boldsymbol{\varepsilon}^p_n \\ \bar{\varepsilon}^{p\ \mathrm{trial}} &= \bar{\varepsilon}^p_n \\ \boldsymbol{\sigma}^{\mathrm{trial}} &= \mathsf{D}^e : \boldsymbol{\varepsilon}^{e\ \mathrm{trial}}. \end{aligned} \tag{11.84}$$

   If $\Phi(\boldsymbol{\sigma}^{\mathrm{trial}}, \sigma_y(\bar{\varepsilon}^{p\ \mathrm{trial}})) \leq 0$, then the process is indeed elastic within the interval and the variables at t_{n+1} are assigned the values of the trial variables. Otherwise, we apply the viscoplastic return-mapping algorithm described in the following.

2. *Viscoplastic return mapping*. At this stage, we solve the system of discretised equations of item (iii) of Box 11.5 which, for the present model, by taking the linear elastic law into consideration, are specialised as

$$\begin{aligned} \boldsymbol{\sigma}_{n+1} &= \boldsymbol{\sigma}^{\mathrm{trial}} - \Delta\gamma\, \mathsf{D}^e : \left.\frac{\partial\Phi}{\partial\boldsymbol{\sigma}}\right|_{n+1} \\ \bar{\varepsilon}^p_{n+1} &= \bar{\varepsilon}^p_n + \Delta\gamma, \end{aligned} \tag{11.85}$$

   where the *incremental multiplier*, $\Delta\gamma$, is given by

$$\Delta\gamma = \frac{\Delta t}{\mu}\left[\left(\frac{q(\boldsymbol{\sigma}_{n+1})}{\sigma_y(\bar{\varepsilon}^p_{n+1})}\right)^{1/\epsilon} - 1\right], \tag{11.86}$$

   with Δt denoting the time increment within the considered interval. After solving (11.85), we can update

$$\boldsymbol{\varepsilon}^p_{n+1} = \boldsymbol{\varepsilon}^p_n + \Delta\gamma \left.\frac{\partial\Phi}{\partial\boldsymbol{\sigma}}\right|_{n+1}, \quad \boldsymbol{\varepsilon}^e_{n+1} = \boldsymbol{\varepsilon}^{e\ \mathrm{trial}} - \Delta\gamma \left.\frac{\partial\Phi}{\partial\boldsymbol{\sigma}}\right|_{n+1}. \tag{11.87}$$

Single-equation corrector

The viscoplastic corrector can be more efficiently implemented by reducing (11.85) to a single scalar equation. The situation here is completely analogous to that of the implementation of the elastoplastic (rate-independent) von Mises model described in Section 7.3.2 (page 217). For convenience, the steps leading to the system reduction are repeated here. Firstly, we observe that the plastic flow vector

$$\frac{\partial \Phi}{\partial \boldsymbol{\sigma}} = \sqrt{\frac{3}{2}}\, \frac{\boldsymbol{s}}{\|\boldsymbol{s}\|} \tag{11.88}$$

is deviatoric so that the hydrostatic stress is independent of the viscoplastic flow. The stress update equation $(11.85)_1$ can then be split as

$$\begin{aligned} \boldsymbol{s}_{n+1} &= \boldsymbol{s}^{\text{trial}} - \Delta\gamma\, 2G \sqrt{\frac{3}{2}}\, \frac{\boldsymbol{s}_{n+1}}{\|\boldsymbol{s}_{n+1}\|} \\ p_{n+1} &= p^{\text{trial}}. \end{aligned} \tag{11.89}$$

Further, simple inspection of $(11.89)_1$ shows that $\boldsymbol{s}_{n+1}$ is a scalar multiple of $\boldsymbol{s}^{\text{trial}}$ so that, trivially, we have the identity

$$\frac{\boldsymbol{s}_{n+1}}{\|\boldsymbol{s}_{n+1}\|} = \frac{\boldsymbol{s}^{\text{trial}}}{\|\boldsymbol{s}^{\text{trial}}\|}, \tag{11.90}$$

which allows us to rewrite $(11.89)_1$ as

$$\boldsymbol{s}_{n+1} = \left(1 - \sqrt{\frac{3}{2}}\, \frac{\Delta\gamma\, 2G}{\|\boldsymbol{s}^{\text{trial}}\|}\right) \boldsymbol{s}^{\text{trial}} = \left(1 - \frac{\Delta\gamma\, 3G}{q^{\text{trial}}}\right) \boldsymbol{s}^{\text{trial}} \tag{11.91}$$

where q^{trial} is the elastic trial von Mises equivalent stress. Application of the definition of the von Mises equivalent stress to the above equation gives the update formula

$$q_{n+1} = q^{\text{trial}} - 3G\, \Delta\gamma. \tag{11.92}$$

Finally, with the substitution of the above formula together with $(11.85)_2$ into (11.86) we obtain the following scalar algebraic equation for the multiplier $\Delta\gamma$

$$\Delta\gamma - \frac{\Delta t}{\mu}\left[\left(\frac{q^{\text{trial}} - 3G\,\Delta\gamma}{\sigma_y(\bar{\varepsilon}^p_n + \Delta\gamma)}\right)^{1/\epsilon} - 1\right] = 0, \tag{11.93}$$

or, equivalently, after a straightforward rearrangement,

$$(q^{\text{trial}} - 3G\,\Delta\gamma)\left(\frac{\Delta t}{\mu\,\Delta\gamma + \Delta t}\right)^{\epsilon} - \sigma_y(\bar{\varepsilon}^p_n + \Delta\gamma) = 0. \tag{11.94}$$

The single-equation viscoplastic corrector comprises the solution of (11.93) or (11.94) for the unknown $\Delta\gamma$ followed by the straightforward update of $\boldsymbol{\sigma}$, $\bar{\varepsilon}^p$, $\boldsymbol{\varepsilon}^p$, $\boldsymbol{\varepsilon}^e$ according to the relevant formulae. The solution of the equation for $\Delta\gamma$ is, as usual, undertaken by the Newton–Raphson iterative scheme. The overall algorithm is summarised in Box 11.6 in pseudo-code format.

Box 11.6. Integration algorithm for von Mises-type viscoplastic model (over a generic time interval $[t_n, t_{n+1}]$ with $\Delta t = t_{n+1} - t_n$).

(i) Elastic predictor. Given $\Delta\boldsymbol{\varepsilon}$, and the state variables at t_n, evaluate the *elastic trial state*

$$\boldsymbol{\varepsilon}^{e\ \mathrm{trial}} := \boldsymbol{\varepsilon}^e_n + \Delta\boldsymbol{\varepsilon}$$

$$\bar{\varepsilon}^{p\ \mathrm{trial}}_{n+1} := \bar{\varepsilon}^p_n$$

$$p^{\mathrm{trial}}_{n+1} := K\, \varepsilon^{e\ \mathrm{trial}}_{\mathrm{v}\ n+1}; \qquad \boldsymbol{s}^{\mathrm{trial}} := 2G\, \boldsymbol{\varepsilon}^{e\ \mathrm{trial}}_{\mathrm{d}\ n+1}$$

$$q^{\mathrm{trial}}_{n+1} := \sqrt{\tfrac{3}{2}}\, \|\boldsymbol{s}^{\mathrm{trial}}_{n+1}\|$$

(ii) Check for viscoplastic flow

$$\text{IF} \quad q^{\mathrm{trial}}_{n+1} - \sigma_y(\bar{\varepsilon}^{p\ \mathrm{trial}}_{n+1}) \leq 0 \quad \text{(elastic step)}$$

$$\text{THEN set} \quad (\cdot)_{n+1} := (\cdot)^{\mathrm{trial}}_{n+1} \quad \text{and EXIT}$$

(iii) Viscoplastic flow. Solve the return-mapping equation

$$R(\Delta\gamma) \equiv (q^{\mathrm{trial}}_{n+1} - 3G\,\Delta\gamma)\left(\frac{\Delta t}{\mu\,\Delta\gamma + \Delta t}\right)^{\epsilon} - \sigma_y(\bar{\varepsilon}^p_n + \Delta\gamma) = 0$$

for $\Delta\gamma$ using the Newton–Raphson scheme. Then update

$$p_{n+1} := p^{\mathrm{trial}}_{n+1}; \qquad \boldsymbol{s}_{n+1} := \left(1 - \frac{\Delta\gamma\, 3G}{q^{\mathrm{trial}}_{n+1}}\right)\boldsymbol{s}^{\mathrm{trial}}_{n+1}$$

$$\boldsymbol{\sigma}_{n+1} := \boldsymbol{s}_{n+1} + p_{n+1}\,\boldsymbol{I}$$

$$\boldsymbol{\varepsilon}^e_{n+1} := \frac{1}{2G}\,\boldsymbol{s}_{n+1} + \frac{1}{3}\,\varepsilon^{e\ \mathrm{trial}}_{\mathrm{v}\ n+1}\,\boldsymbol{I}$$

$$\bar{\varepsilon}^p_{n+1} := \bar{\varepsilon}^p_n + \Delta\gamma$$

(iv) EXIT

Remark 11.4 (Rate-independent limit). Note that, as expected, equation (11.94) rigorously recovers its elastoplastic (rate-independent) counterpart (7.91) (refer to page 219) when $\mu \to 0$ (no viscosity), $\epsilon \to 0$ (no rate-sensitivity) or $\Delta t \to \infty$ (infinitely slow straining). Clearly, in such cases, the algorithm of Box 11.6 reproduces the rate-independent elastoplastic numerical solution.

Remark 11.5 (Computational implementation aspects). In the computer implementation of the model (as shown in Box 11.6), it is more convenient to solve (11.94) rather than (11.93) in the viscoplastic corrector stage of the algorithm. The reason for this lies in the fact that, for low rate-sensitivity, i.e. small values of ϵ, the Newton–Raphson scheme for solution of (11.93) becomes unstable as its convergence bowl is sharply reduced with decreasing ϵ. The reduction of the convergence bowl stems from the fact that large exponents $1/\epsilon$ can easily produce numbers which are computationally intractable. This fact has been recognised by Perić (1993) in the context of a more general viscoplastic algorithm. In equation (11.94), on the other hand, the term to the power ϵ on the left-hand side can only assume values within the interval $[0, 1]$ and causes no numerical problems within practical ranges of material constants.

Remark 11.6 (Solution existence and uniqueness). Within a viscoplastic step, we have

$$q^{\text{trial}} > \sigma_y(\bar{\varepsilon}^{p\,\text{trial}}) = \sigma_y(\bar{\varepsilon}^p_n).$$

Let $R(\Delta\gamma)$ be the function defined by the right-hand side of (11.94). The above inequality clearly implies that $R(0) > 0$. In addition, taking into account the strict positiveness of the hardening function σ_y, we can easily verify that $R(q^{\text{trial}}/3G) < 0$. The continuity of R then implies that (11.94) has a root within the interval $(0, q^{\text{trial}}/3G)$. Let us now consider the derivative of R,

$$R'(\Delta\gamma) = -\left(3G + \epsilon\mu\,\frac{q^{\text{trial}} - 3G\Delta\gamma}{\mu\,\Delta\gamma + \Delta t}\right)\left(\frac{\Delta t}{\mu\,\Delta\gamma + \Delta t}\right)^{\epsilon} - H(\bar{\varepsilon}^p_n + \Delta\gamma),$$

where H is the derivative of the isotropic hardening function σ_y. Upon simple inspection, we can easily establish that the derivative R' is strictly negative for $\Delta\gamma \in (0, q^{\text{trial}}/3G)$ if the viscoplastic model is *non-softening*, i.e. if H is non-negative for any value of accumulated plastic strain. The strict negativeness of R' in conjunction with the existence of a root for R established in the above implies that the root of R (the solution of the viscoplastic corrector equation) within the interval $(0, q^{\text{trial}}/3G)$ is unique for non-softening materials.

11.6.2. ISO-ERROR MAPS

To illustrate the accuracy of the above integration algorithm in practical situations, this section presents some iso-error maps, produced with material constants covering a range of high rate-sensitivity to rate-independence. The material is assumed perfectly viscoplastic (no hardening). The maps have been generated in the standard fashion as described in Section 7.2.10 (refer to Figure 7.7, page 215). Using the three-dimensional implementation of the model, we start from a stress point at time t_n, with $\boldsymbol{\sigma}_n$ lying on the yield surface, and apply a sequence of strain increments (at *constant strain rate* within the interval $[t_n, t_{n+1}]$), corresponding to linear combinations of trial stress increments in the direction normal and tangential (directions of the unit tensors $\boldsymbol{N}$ and $\boldsymbol{T}$ of Figure 7.7, respectively) to the von Mises circle in the deviatoric plane. Figures 11.5 and 11.6 show iso-error maps obtained at low and high strain rates with the non-dimensional rate

$$\mu\,\|\dot{\varepsilon}\|$$

set respectively to 1 and 1000. For each non-dimensional rate, three values of rate-sensitivity parameter, ϵ, have been used: 10^0, 10^{-1} and 0. Recall that for $\epsilon = 0$ the algorithm reproduces the rate-independent solution. The resulting map in this case is obviously identical to the rate-independent map of Figure 7.7(b) and is shown here only to emphasise the effect of rate-dependence on the integration error. The main conclusion drawn from the iso-error maps is that, in general, increasing (decreasing) rate-sensitivity and/or increasing (decreasing) strain rates tend to produce decreasing (increasing) integration errors. The largest errors are expected in the rate-independent limit.

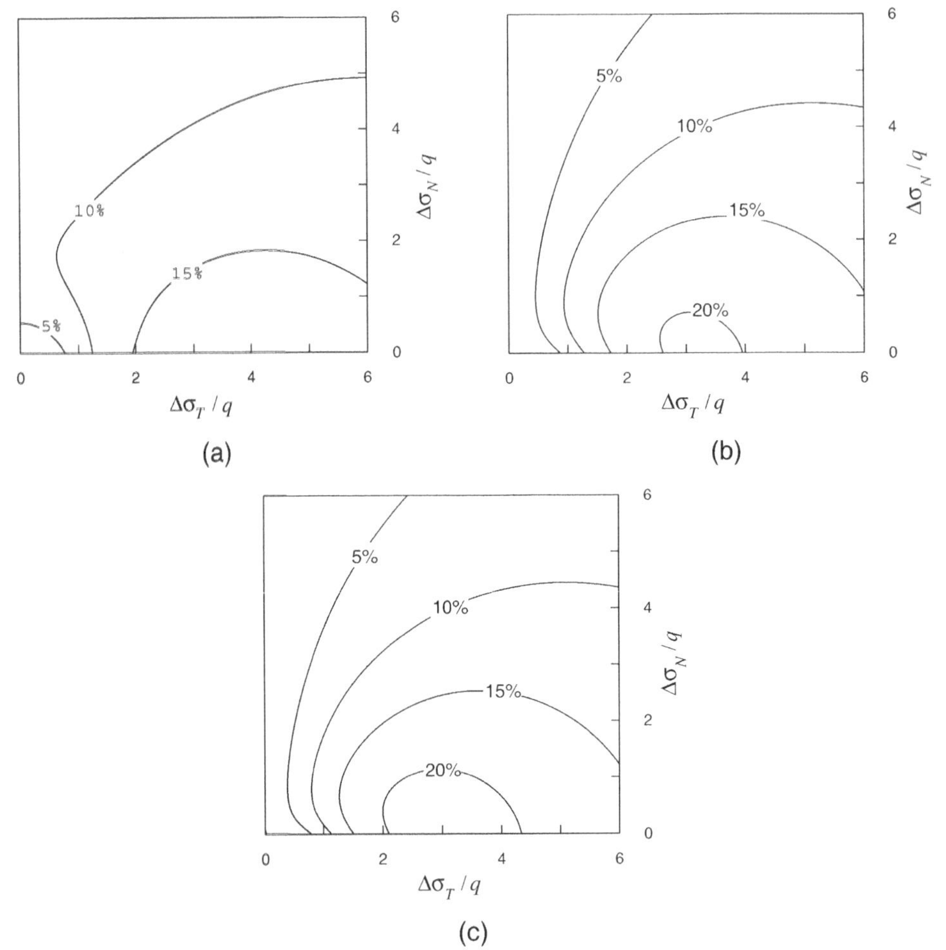

Figure 11.5. Iso-error maps with $\mu\,\|\dot{\varepsilon}\| = 1$; (a) $\epsilon = 10^0$; (b) $\epsilon = 10^{-1}$; (c) $\epsilon = 0$ (rate-independent).

11.6.3. CONSISTENT TANGENT OPERATOR

The consistent tangent operator here is a particular case of the general tangent operator derived in Section 11.5.3. Clearly, when the stress state lies within the elastic domain and no viscoplastic flow is possible, the tangent operator is the elastic tangent, $\mathbf{D}^e$. Under viscoplastic flow, the tangent operator which (as in the rate-independent case) will be denoted $\mathbf{D}^{ep}$, is derived by consistently linearising the viscoplastic return-mapping algorithm referred to in item (iii) of Box 11.6. Its closed-form expression can be obtained by following the same steps of the derivation of the elastoplastic (rate-independent) tangent presented in Section 7.4.2 (from page 232). The incremental constitutive function for the stress tensor in the present case has identical format to that of the rate-independent implementation given by (7.93) – which reduces to (7.113) under plastic flow – but the incremental plastic multiplier $\Delta\gamma$ here is the solution of viscoplastic return-mapping equation (11.94). Thus, to obtain the viscoplastic consistent tangent, we simply replace the derivative of the incremental plastic

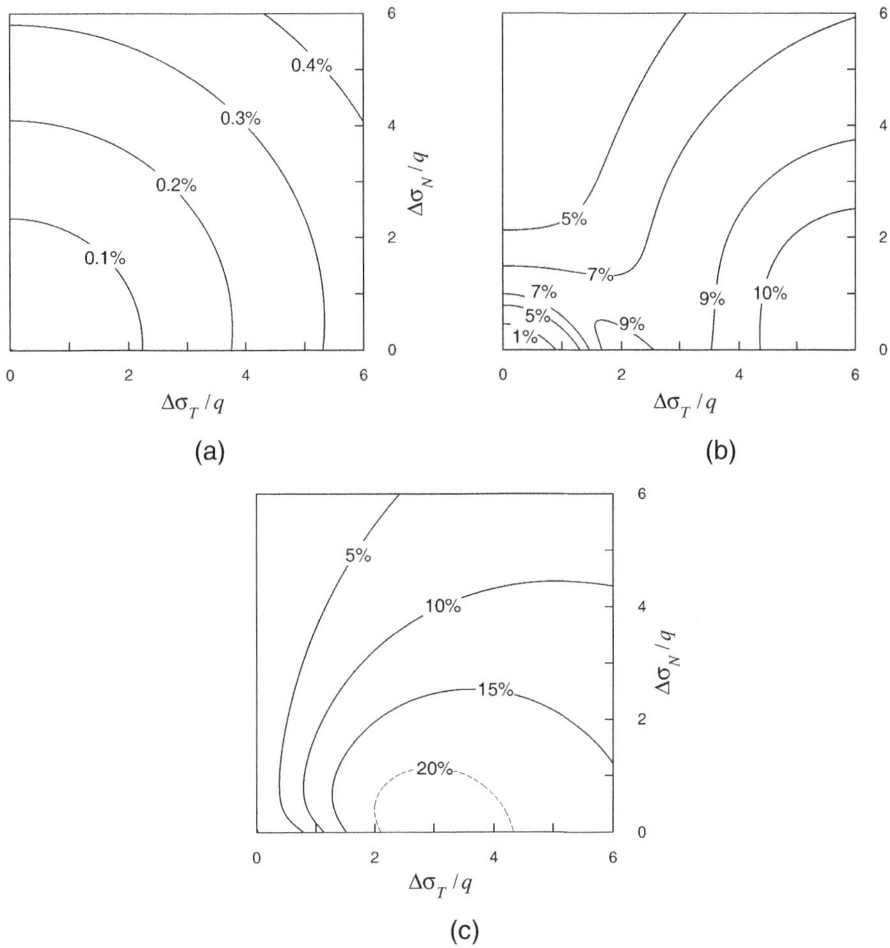

Figure 11.6. Iso-error maps with $\mu\,\|\dot{\varepsilon}\| = 10^3$; (a) $\epsilon = 10^0$; (b) $\epsilon = 10^{-1}$; (c) $\epsilon = 0$ (rate-independent).

multiplier (7.118) with the expression

$$\frac{\partial \Delta\gamma}{\partial \varepsilon^{e\ \mathrm{trial}}_{n+1}} = \frac{2G\,\sqrt{\frac{3}{2}}}{3G + \left(\frac{\Delta t}{\mu\,\Delta\gamma + \Delta t}\right)^{-\epsilon} H + \frac{\epsilon\mu\, q_{n+1}}{\mu\,\Delta\gamma + \Delta t}}\,\bar{\boldsymbol{N}}_{n+1}, \tag{11.95}$$

which is consistent with (11.94). Analogously to the elastoplastic case, this expression is obtained by taking the differential of the viscoplastic corrector equation (11.94), having $\Delta\gamma$ and q^{trial}_{n+1} as variables, and equating it to zero. With the above differential relation, the final elasto-viscoplastic consistent tangent operator is obtained in closed form as

$$\begin{aligned}\mathsf{D}^{ep} &= 2G\left(1-\frac{\Delta\gamma\, 3G}{q_{n+1}^{\text{trial}}}\right)\mathsf{I}_{\text{d}} \\ &\quad + 6G^2\left[\frac{\Delta\gamma}{q_{n+1}^{\text{trial}}} - \frac{1}{3G+\left(\frac{\Delta t}{\mu\,\Delta\gamma+\Delta t}\right)^{-\epsilon}H + \frac{\epsilon\mu\, q_{n+1}}{\mu\,\Delta\gamma+\Delta t}}\right]\bar{\boldsymbol{N}}_{n+1}\otimes\bar{\boldsymbol{N}}_{n+1} + K\,\boldsymbol{I}\otimes\boldsymbol{I}.\end{aligned} \tag{11.96}$$

Note that the tangent operator is symmetric.

Remark 11.7 (Rate-independent limit). By simple inspection we find that in the limits $\epsilon \to 0$ (vanishing rate-sensitivity parameter), $\mu \to 0$ (vanishing viscosity) or $\Delta t \to \infty$ (infinitely slow straining), expression (11.96) rigorously recovers the elastoplastic consistent tangent operator of the isotropically hardening rate-independent von Mises model with implicit return mapping given by expression (7.120).

11.6.4. PERZYNA-TYPE MODEL IMPLEMENTATION

The implementation of the von Mises-based model with Perzyna's viscoplastic law (11.40) follows exactly the same procedure as described in the above except that, consistently with the backward Euler time discretization of (11.40), the return-mapping equation (11.94) (or item (iii) of Box 11.6) is replaced with

$$q^{\text{trial}} - 3G\,\Delta\gamma - \left[1+\left(\frac{\mu\,\Delta\gamma}{\Delta t}\right)^{\epsilon}\right]\sigma_y(\bar{\varepsilon}_n^p+\Delta\gamma) = 0. \tag{11.97}$$

Here, we have assumed isotropic strain hardening. Note that, as $\mu \to \infty$ (vanishing viscosity) or $\Delta t \to \infty$ (infinitely slow process) equation (11.97) reduces to that of the elastoplastic rate-independent von Mises model with yield stress σ_y. For vanishing rate sensitivity parameter, $\epsilon \to 0$, (11.97) reduces to a von Mises elastoplastic return-mapping equation with yield stress $2\sigma_y$. This is, as one should expect, in agreement with the theoretical limits of the Perzyna model discussed in the text immediately following equation (11.40).

Elasto-viscoplastic consistent tangent operator

The differential relation between the incremental plastic multiplier and $\varepsilon_{n+1}^{e\ \text{trial}}$ consistent with the return-mapping equation (11.97) reads

$$\frac{\partial\Delta\gamma}{\partial\varepsilon_{n+1}^{e\ \text{trial}}} = \frac{2G\sqrt{\frac{3}{2}}}{3G+\left[1+\left(\frac{\mu\,\Delta\gamma}{\Delta t}\right)^{\epsilon}\right]H + \frac{\epsilon\mu}{\Delta t}\left(\frac{\mu\,\Delta\gamma}{\Delta t}\right)^{\epsilon-1}\sigma_y}\,\bar{\boldsymbol{N}}_{n+1}, \tag{11.98}$$

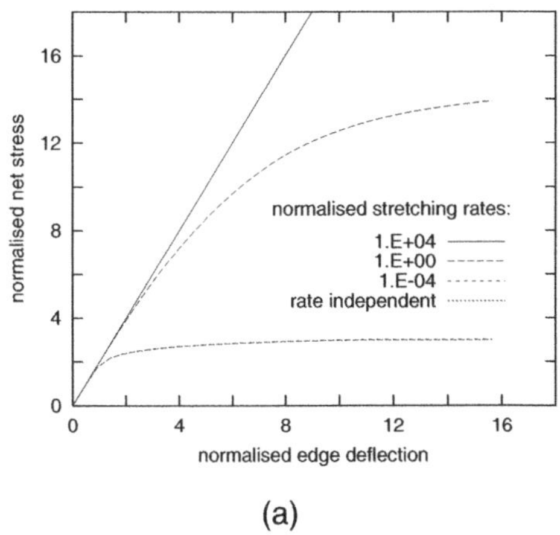

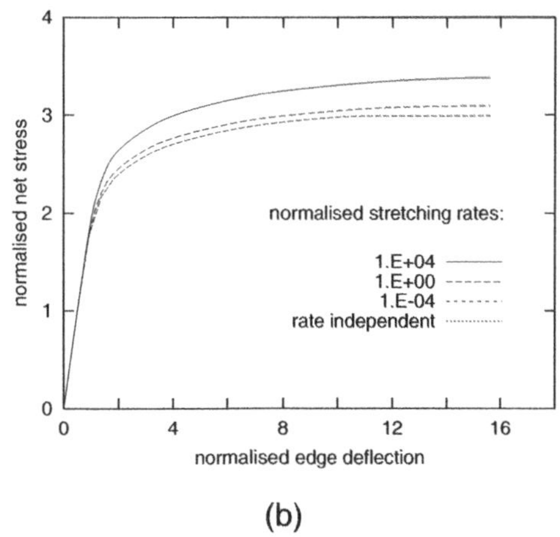

Figure 11.7. Double-notched specimen. Reaction–deflection diagrams: (a) $\epsilon = 10^0$; (b) $\epsilon = 10^{-2}$.

where σ_y is evaluated at $\bar{\varepsilon}^p_{n+1} = \bar{\varepsilon}^p_n + \Delta\gamma$. This expression is the counterpart of (11.95) for the present implementation of Perzyna's viscoplasticity law. The corresponding elasto-viscoplastic consistent tangent operator is obtained following the usual procedure as

$$\begin{aligned} \mathsf{D}^{ep} &= 2G\left(1 - \frac{\Delta\gamma\, 3G}{q^{\text{trial}}_{n+1}}\right)\mathsf{I}_{\text{d}} \\ &\quad + 6G^2\left[\frac{\Delta\gamma}{q^{\text{trial}}_{n+1}} - \frac{1}{3G + \left[1 + \left(\frac{\mu\,\Delta\gamma}{\Delta t}\right)^{\epsilon}\right]H + \frac{\epsilon\mu}{\Delta t}\left(\frac{\mu\,\Delta\gamma}{\Delta t}\right)^{\epsilon-1}\sigma_y}\right]\bar{\boldsymbol{N}}_{n+1}\otimes\bar{\boldsymbol{N}}_{n+1} \\ &\quad + K\,\boldsymbol{I}\otimes\boldsymbol{I}. \end{aligned} \tag{11.99}$$

Its format is completely analogous to that of (11.96).

11.7. Examples

The finite element examples presented in this section illustrate applications of the computational treatment of viscoplasticity described above. The underlying viscoplastic material model is the one shown in Box 11.2, which includes isotropic strain hardening.

11.7.1. DOUBLE-NOTCHED TENSILE SPECIMEN

The rate-independent version of this problem has been studied in Section 7.5.5 (from page 255). The problem consists of the plane strain analysis of a deep double-notched tensile specimen. The geometry of the specimen and the finite element mesh used are shown in Figure 7.29 (page 256). Analogously to the prescription of edge displacement u (refer to Figure 7.29), the simulation consists of stretching the specimen by prescribing a constant (in time) vertical velocity v on the top nodes of the mesh. For convenience, we define the normalised stretching rate

$$v^* = \frac{\mu\, v}{l/2}$$

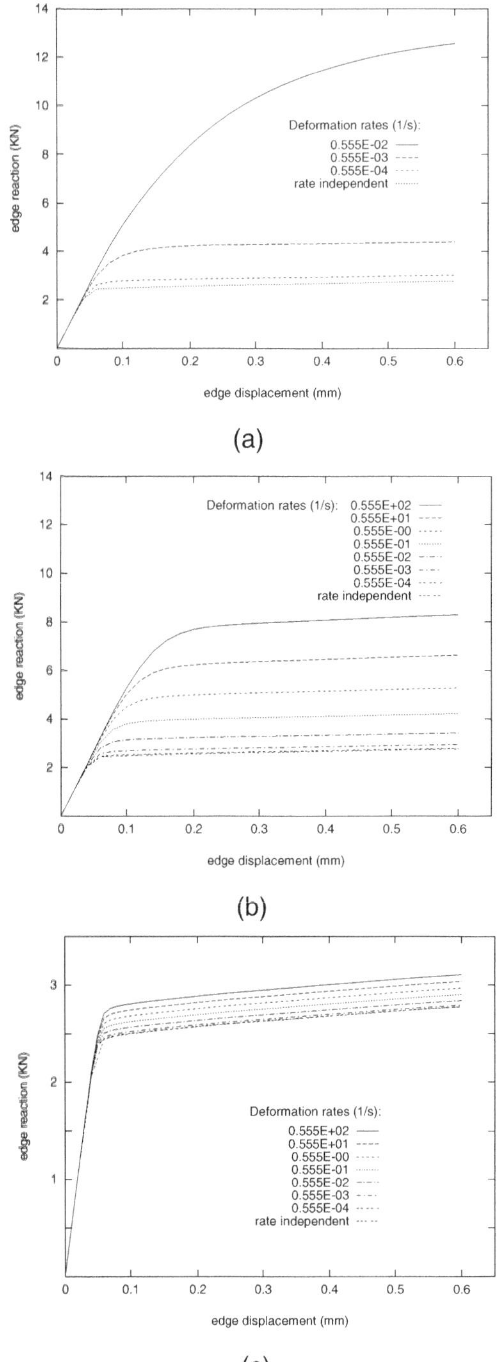

Figure 11.8. Stretching of a perforated plate. Displacement-reaction diagrams. (a) $\epsilon = 10^0$; (b) $\epsilon = 10^{-1}$; (c) $\epsilon = 10^{-2}$.

and the simulation is carried out for three different values of v^*

$$v^* = 10^{-4},\ 10^0,\ 10^4.$$

This choice covers very slow to very fast strain rates and is meant to demonstrate the robustness of the integration algorithm over a wide range of strain rates. The following material parameters are adopted

$$E = 206.9\,\text{GPa};\quad \nu = 0.29;\quad \sigma_y = 0.45\,\text{GPa (constant)}.$$

The linearly hardening case listed in Figure 7.29 is not considered here. In order to show the effect of the rate-sensitivity parameter on the behaviour of the model, two values of ϵ are considered

$$\epsilon = 10^0 \quad \text{and} \quad 10^{-2}.$$

The results of the finite element simulations are presented in Figure 11.7 whose diagrams show the evolution of the reaction forces on the constrained edge against the corresponding edge deflection. As in the rate-independent case, the results are plotted in terms of the *normalised net stress* and the *normalised edge deflection* defined in Section 7.5.5. The results for $\epsilon = 10^0$ and 10^{-2} are shown, respectively, in Figures 11.7(a) and (b). They illustrate the expected higher reactions and limit loads for higher rates of stretching. For the lowest non-dimensional rate of 10^{-4}, the rate-independent solution is recovered for any rate-sensitivity parameter. We remark that the rate-independent solution shown in the graphs for comparison can be obtained with the present model/algorithm simply by setting $\epsilon = 0$ or $\mu = 0$.

11.7.2. PLANE STRESS: STRETCHING OF A PERFORATED PLATE

This section describes the viscoplastic version of the plane stress problem of Section 9.5.3 (from page 390). This example has been analysed by Perić (1993). Here, a plane stress version of the numerical integration algorithm discussed in Section 11.6.1 is employed. The plane stress implementation adopted follows the nested iteration approach described in Section 9.2.2 (page 362) in the context of rate-independent plasticity. The problem consists of the axial stretching at constant rate of a perforated rectangular plate whose geometry is shown in Figure 9.7 (page 392). The mesh, boundary conditions and the material parameters that are common to both plastic and viscoplastic models are also shown in Figure 9.7. Note that linear strain hardening is assumed. The viscosity parameter adopted (required for the viscoplastic model) is

$$\mu = 500\ \text{s}.$$

Similarly to the previous example, in order to illustrate the response predicted by the viscoplastic model over a wide range of conditions, several simulations are carried out with various stretching rates, with three values of rate sensitivity coefficients ($\epsilon = 1, 10^{-1}, 10^{-2}$) being considered. The results obtained in the simulations are shown in Figure 11.8 where the reaction on the constrained edge of the plate is plotted against the prescribed edge displacement for the various conditions considered. The stretching rate in the present case is defined as

$$2v/l,$$

where v is the stretching velocity imposed on the nodes of the upper edge. The three graphs of Figure 11.8 show the effect of stretching rates on the response of the plate, with higher reactions obtained at high rates and the rate-independent solution being approached as the stretching rate vanishes. The effects of the rate sensitivity parameter are also clearly illustrated. At higher (lower) values of ϵ, a greater (smaller) variation of reaction as a function of the stretching rate is produced.

12 DAMAGE MECHANICS

INTERNAL damage can be defined as the presence and evolution of cracks and cavities at the microscopic level which may, eventually, lead to failure – a complete loss of load-carrying capability of the material. In many engineering applications, particularly those where mechanical/structural components are subjected to severe service conditions, the useful life of components is a crucial item of information which has to be carefully considered during the design process. In such cases, the ability of the designer to predict mechanical failure becomes an important factor. In some applications, such as in certain types of industrial machinery, non-scheduled stops for maintenance owing to unpredicted failure may incur serious economic consequences. In the design of manufacturing processes, such as metal-forming operations, prediction of failure is also a crucial issue. In safety-critical applications, frequently encountered in the aeronautical and nuclear industries, unpredicted failure may have catastrophic effects with consequences far beyond purely economical issues.

Traditionally, the prediction of useful life/failure of materials is based on mostly empirical experience accumulated over long periods of time. In some cases, failure prediction is achieved by the systematic (and expensive) testing of real models under laboratory-reproduced service conditions. However, with the growing knowledge of the mechanisms of progressive internal damage that cause failure in a wide range of materials, it is becoming possible to formulate continuum constitutive models capable of accounting for the evolution of internal deterioration. This relatively new branch of continuum solid mechanics is known as *Continuum Damage Mechanics* (*CDM*). This fact, allied to the fast development of computational mechanics techniques, has made the use of computational tools to carry out life/failure prediction a realistic alternative that can be successfully adopted in many design and damage assessment situations.

The present chapter is devoted to computational continuum damage mechanics. Our intention here is to provide the reader with an introduction to this new and promising ramification of computational solid mechanics that has been gaining widespread acceptance over the last two decades. The material presented in this chapter is summarised as follows. After providing a brief review of some basic mechanisms that characterise the presence and evolution of damage in Section 12.1, we give in Section 12.2 a brief historical account of CDM together with a discussion on the continuum modelling of damage phenomena. Sections 12.3, 12.4 and 12.5, describe, respectively, Lemaitre's ductile damage model (Lemaitre, 1985b), a simplified version of Lemaitre's model where kinematic hardening is not considered and Gurson's void growth model (Gurson, 1977). In each of these sections, the computational implementation of the corresponding constitutive models within an implicit finite element environment is described in detail. Note that the simplified version of Lemaitre's model discussed in Section 12.4 is fully incorporated into program HYPLAS. Further issues, including crack closure effects and damage anisotropy are addressed in Section 12.6.

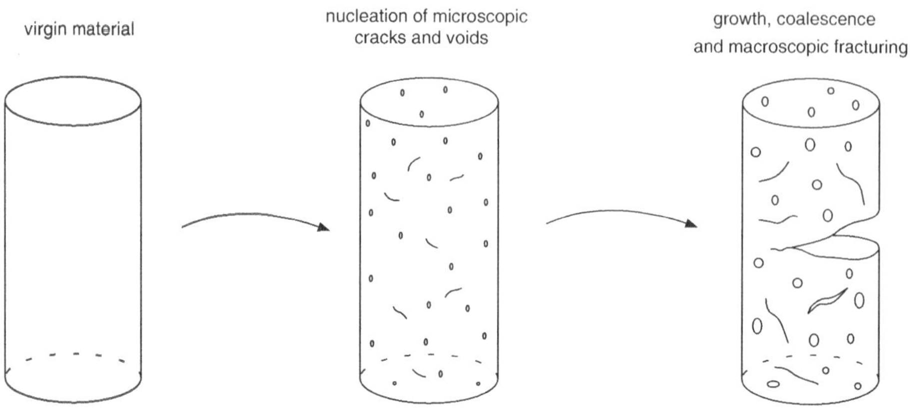

Figure 12.1. Ductile damage in metals. Schematic illustration.

12.1. Physical aspects of internal damage in solids

The characterisation of internal damage as well as the scale at which it occurs in common engineering materials depend crucially upon the specific type of material considered. In addition, for the same material, damage evolution may take place triggered by very different physical mechanisms which depend fundamentally on the type, rate of loading, temperature as well as environmental factors such as exposure to corrosive substances or radiation. Therefore, rather than the material alone, the material-process-environment triad must be considered in the study of internal damage. To illustrate the diversity of phenomena that may be involved in the process of internal degradation of solids, some basic physical mechanisms underlying damage evolution in metals and rubbery polymers are outlined below.

12.1.1. METALS

In metals, the primary mechanisms that characterise the phenomenon of mechanical degradation may be divided into two distinct classes: *brittle* and *ductile* damage. Brittle damaging occurs mainly in the form of cleavage of crystallographic planes in the presence of negligible inelastic deformations. This behaviour is observed for many polycrystalline metals, usually at low temperatures. At high temperatures, brittle damage can also be observed associated with creep processes. In this case, the decohesion of interatomic bonds is concentrated at grain boundaries. At low stresses they are accompanied by relatively small strains. Ductile damage, on the other hand, is normally associated with the presence of large plastic deformations in the neighbourhood of crystalline defects. The decohesion of interatomic bonds is initiated at the boundary interface of inclusions, precipitates and particles of alloy elements leading to the formation of microscopic cracks and cavities. Further evolution of local plastic deformation may cause the cavities to coalesce, resulting in final rupture. This mechanism is schematically illustrated in Figure 12.1. For most metallic materials, the damage behaviour is a combination of brittle and ductile response and the contribution of each mode is, to a significant extent, dependent on the temperature, loading rate, etc.

Another important mode of material deterioration in metals is *fatigue damage*. It is normally observed in mechanical components subjected to a large number of load and/or

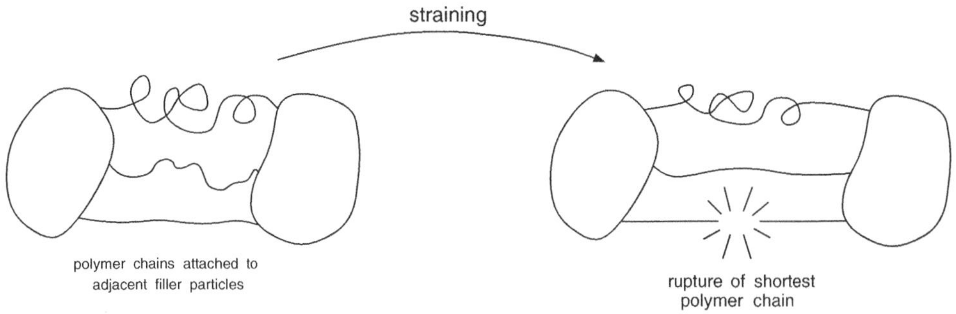

Figure 12.2. Damage in filled rubbery polymers. Schematic representation.

temperature cycles. Although fatigue damage occurs at overall stress levels below the macroscopic plastic yield limit, the nucleation of microcracks is attributed to the accumulation of dislocations observed in connection with cyclic plastic deformation due to stress concentration near microscopic defects. A large number of complex interactive physical mechanisms take place from the nucleation of cracks to the complete failure of the material, and the understanding of fatigue degradation processes in metals remains a challenging issue in the field of materials science. Some of the most important mechanisms of material damage are described by Engel and Klingele (1981).

12.1.2. RUBBERY POLYMERS

Rubbery polymers are widely employed in engineering applications. Essentially, these materials are made of long cross-linked molecular chains which differ radically from the structure of crystalline metals (refer to Arridge (1985), for further details). Although rubbery polymers exhibit a behaviour which, under a variety of circumstances, may be regarded as purely elastic, damaging does take place due to straining and/or thermal activation. The internal degradation in this case is mainly characterised by the rupture of molecular bonds concentrated in regions containing impurities and defects. In general, the damage response of such materials is predominantly *brittle* (in the sense that permanent deformations are small).

Filled rubbers are particularly susceptible to internal damaging. Those materials are obtained by addition of a filler in order to enhance the strength properties of the original rubber. In this case, even at very small overall straining, damage can occur in the form of progressive breakage of shorter polymer chains attached between filler particles. This phenomenon, as described by Bueche (1960), is schematically illustrated in Figure 12.2.

12.2. Continuum damage mechanics

Since the pioneering work by Kachanov (1958), a considerable body of the literature on applied mechanics has been devoted to the formulation of constitutive models to describe internal degradation of solids within the framework of continuum mechanics. After over two decades of uninterrupted development, significant progress has been achieved and such theories have merged into what is currently known as Continuum Damage Mechanics (Krajčinović, 1996; Lemaitre and Chaboche, 1990). The concepts underlying the development of

CDM models along with a brief historical review of this new branch of continuum mechanics are presented below.

In the previous section, some basic microscopic mechanisms associated with internal damage evolution in solids have been reviewed. It is clear that the underlying phenomena which characterise damage are essentially different from those characterising deformation. While damage manifests itself in the form of the irreversible rupture of atomic bonds, deformation can be associated with reversible variations of interatomic spacing (in purely elastic processes) and movement and accumulation of dislocations (in permanent deformations of metals). Therefore, it should be expected that in order to describe the internal degradation of solids within the framework of the continuum mechanics theory, new variables intrinsically connected with the internal damage process will have to be introduced in addition to the standard variables (such as the strain tensor, plastic strain, etc.) employed in the description of deformation. In this context, we shall refer to *Continuum Damage Mechanics Model* as any continuum constitutive model which features special internal variables representing, directly or indirectly, the density and/or distribution of the microscopic defects that characterise damage.

12.2.1. ORIGINAL DEVELOPMENT: CREEP-DAMAGE

The first continuum damage mechanics model was proposed by Kachanov (1958). Without a clear physical meaning for damage, Kachanov introduced a scalar internal variable to model the creep failure of metals under *uniaxial* loads. A physical significance for the damage variable was given later by Rabotnov (1963) who proposed the reduction of the cross-sectional area due to microcracking as a suitable measure of the state of internal damage. In this context, denoting respectively by A and A_0 the effective load bearing areas of the virgin and damaged materials, the damage variable D was introduced as

$$D = \frac{A - A_0}{A}, \tag{12.1}$$

with $D = 0$ corresponding to the virgin material and $D = 1$ representing a total loss of load-bearing capacity.[†] In order to describe the strain rate increase which characterises tertiary creep, Kachanov replaced the observed uniaxial stress σ with the *effective stress*

$$\sigma_{\text{eff}} = \frac{\sigma}{1 - D} \tag{12.2}$$

in the standard Norton's Law for creep. In this case, the uniaxial constitutive equation (11.44) (page 449) for the plastic strain rate is replaced with

$$\dot{\varepsilon}^p = \left(\frac{|\sigma|}{\lambda\,(1 - D)} \right)^N \text{sign}(\sigma), \tag{12.3}$$

so that the phenomenon of damaging (increase in D) may produce the marked acceleration of the plastic strain rate observed during tertiary creep (refer to Figure 11.1(b) and the discussion in Section 11.1).

[†]Kachanov in fact used the *material continuity* or *integrity*, $\omega = 1 - D$, as the variable associated with the internal deterioration process.

Since Kachanov–Rabotnov's original developments, it did not take long before the concept of internal damage variable was generalised to three-dimensional situations by a number of authors. Leckie and Hayhurst (1974) exploited the idea of the effective load-bearing area reduction as a scalar measure of material deterioration to define a model for creep-rupture under multiaxial stresses. The theories derived later by Chaboche (1978, 1981, 1984) and Murakami and Ohno (1981) deserve special mention. Chaboche proposed a phenomenological theory for creep-damage based on rigorous thermodynamic foundations, in which, as a consequence of the hypothesis of *strain equivalence*, the damage variable appears as a *fourth-order* non-symmetric tensor in the most general anisotropic case. In the theory derived by Murakami and Ohno, the anisotropic damage variable is represented by a *second-rank* symmetric tensor. In this case, the definition of the damage variable follows from the extension of the effective stress concept to three dimensions by means of the hypothesis of the existence of a mechanically equivalent *fictitious undamaged configuration*. Murakami's fictitious undamaged configuration concept was later extended to describe general anisotropic states of internal damage in solids with particular reference to the analysis of elastic-brittle materials (Murakami, 1988). Still within the context of creep-rupture, Saanouni *et al.* (1989) used a non-local formulation to predict the nucleation and growth of cracks.

12.2.2. OTHER THEORIES

Despite its origin in the description of creep rupture, Continuum Damage Mechanics has been shown to provide an effective tool to describe the phenomenon of internal degradation in other areas of solid mechanics.

Within the theory of elastoplasticity, Gurson (1977) proposed a model for ductile damage where the (scalar) damage variable is obtained from the consideration of microscopic spherical voids embedded in an elastoplastic matrix. Gurson's void growth theory has been shown to be particularly suitable for the representation of the behaviour of porous metals. A scalar damage variable was also considered by Lemaitre (1983) in the definition of a purely phenomenological model for ductile isotropic damage in metals. By appealing to the hypothesis of strain equivalence, which states that '*the deformation behaviour of the damaged material is represented by the constitutive laws of the virgin material with the true stress replaced by the effective stress*', this author postulates the following elastic constitutive law for a damaged material:

$$\sigma_{\text{eff}} = E_0\, \varepsilon, \tag{12.4}$$

or, equivalently,

$$\sigma = E\, \varepsilon, \tag{12.5}$$

where E_0 and

$$E = (1 - D)E_0 \tag{12.6}$$

are the Young's moduli of the virgin (undamaged) and damaged materials, respectively. As a consequence, the standard definition of damage in terms of reduction of the (neither well-defined nor easily measurable) load-carrying area is replaced in Lemaitre's model by the reduction of the Young's modulus in the ideally isotropic case. The damage variable (12.1) is then redefined as

$$D = \frac{E - E_0}{E}. \tag{12.7}$$

This theory was further elaborated on by Lemaitre (1985a,b) and ageing effects were later incorporated by Marquis and Lemaitre (1988). More recently, the original isotropic model has been extended by Lemaitre *et al.* (2000) to account for the anisotropy of damage as well as for partial closure of microcracks under compressive stresses. The damage variable in this case is a second-order tensor whose evolution is linked to the principal directions of the plastic strain rate. Based on the concept of *energy* equivalence (as opposed to Lemaitre's strain equivalence) another model for elastoplastic damage worth mentioning was proposed by Cordebois and Sidoroff (1982). The damage variable in this case takes the form of a second-order tensor under general anisotropy. Also within the theory of elastoplasticity, Simo and Ju (1987) proposed a framework for the development of (generally anisotropic) strain- and stress-based damage models. In this case, Lemaitre's hypothesis of strain equivalence and its dual *hypothesis of stress equivalence* are used, respectively, in the formulation of models in stress and strain spaces. Application of the proposed framework was made in the description of brittle damage in concrete. The more recent developments by Armero and Oller (2000) are also worth mentioning. These authors propose a framework whereby the strain tensor is decomposed additively into elastic, plastic and damage parts. Damage component of ε in this case is then split as a sum of individual contributions from various damage mechanisms.

A somewhat different approach was followed by Krajčinović and Fonseka (1981) (see also Fonseka and Krajčinović 1981), in the derivation of a continuum damage theory for brittle materials. Assuming that damage in this case is characterised mainly by planar penny-shaped microcracks, a *vectorial* variable was proposed as the local measure of internal deterioration. Later, in (Krajčinović, 1983), the model was endowed with a thermodynamical structure and extended to account for ductile damage. Further developments were introduced by Krajčinović (1985) with the distinction between active and passive systems of microcracks. Other vectorial models are described by Kachanov (1977) and Mitchell (1990).

Continuum damage mechanics has also been applied to the description of fatigue processes. Janson (1978) developed a continuum theory to model fatigue crack propagation which showed good agreement with simple uniaxial experiments. A general formulation incorporating low and high-cycle fatigue as well as creep-fatigue interaction at arbitrary stress states has been presented by Lemaitre (1987). Further discussion on these models is provided by Chaboche (1988) and Lemaitre and Chaboche (1990). In order to model the effects of fatigue, the evolution law for the damage variable is usually formulated in terms of a differential equation which relates damage growth with the mean stress, maximum stress and number of cycles.

12.2.3. REMARKS ON THE NATURE OF THE DAMAGE VARIABLE

As pointed out in Section 3.5.3 (page 74), the appropriate definition of internal variables associated with a specific phenomenon is one of the most important factors determining the success or failure of the continuum model intended for its description.

Due to the diversity of forms in which internal damage manifests itself at the microscopic level, the definition of adequate damage variables is certainly not an easy task. During the development of *CDM*, briefly reviewed above, variables of different mathematical nature (scalars, vectors, tensors) possessing different physical meaning (reduction of load bearing area, loss of stiffness, distribution of voids) have been employed in the description of damage under various circumstances.

Physical significance

With regard to the physical significance of damage variables, it is convenient to separate the *CDM* theories into two main categories: *micromechanical* and *phenomenological* models.

In micromechanical models, the damage internal variable must represent some average of the microscopic defects that characterise the state of internal deterioration. Despite the physical appeal of internal variables such as the reduction of load-bearing area, as suggested by Rabotnov (1963), or distribution of microcracks, as adopted by Krajčinović (1983, 1985) in his vectorial model, the enormous amount of bookkeeping required in conjunction with the serious difficulties involved in the experimental identification of damaged states and evolution laws preclude most micromechanical theories from practical applications at present. This is especially true if the final objective is the analysis of large-scale problems for engineering design purposes.

Phenomenological damage variables, on the other hand, can be defined on the basis of the influence that internal degradation exerts on the macroscopic properties of the material. In particular, properties such as the elastic moduli (Cordebois and Sidoroff, 1979; Horii and Nemat-Nasser, 1983), yield stress, density and electric resistance can be strongly affected by the presence of damage in the form of microscopic cavities. Needless to say, the measurement of such quantities is, in general, far easier than the determination of the geometry or distribution of micro-defects. Based on such concepts, the class of models presented by Lemaitre and Chaboche (1990) rely mostly on the use of the degradation of the elastic moduli as the macroscopic measure of damage. In its simplest form, i.e. under ideally isotropic conditions, the damage variable is the scalar defined by expression (12.7). Under anisotropy, the damage variable is a second-order tensor (Lemaitre *et al.*, 2000). A similar definition for the isotropic damage variable is employed by Cordebois and Sidoroff (1982). A model relying on the volume changes due to void growth as a measure of internal degradation is described by Gelin and Mrichcha (1992).

Current methods of experimental identification of damage, comprising direct as well as indirect techniques, are described in detail by Lemaitre and Dufailly (1987). Such techniques range from the direct observation of microscopic pictures to the measurement of the degradation of the elastic moduli by means of ultrasonic emissions and micro-hardness tests. The potential as well as the limitations of both micromechanical and phenomenological approaches to damage mechanics are discussed by Basista *et al.* (1992). In the present state of development of *CDM* it has been verified that, in general, the loss of microscopic information resulting from a phenomenological approach is compensated for by the gain in analytical, experimental and computational tractability of the model.

Mathematical representation

In view of the many possibilities regarding the choice of the damage internal variable, Leckie and Onat (1981) showed that the distribution of voids on the grain boundaries can be mathematically represented by a sequence of *even* rank irreducible tensors. Although this result was obtained in the context of creep-damage theories, Onat (1986) showed that the same phenomenological representation for the damage variable applies to general micro-cracked continua, regardless of the underlying deformation processes.

The conclusions drawn by Onat were based on the use of averaging techniques to transform the distribution of micro-defects into a mathematically well-defined continuum measure of damage. In spite of the micromechanical nature of Onat's argument, it is desirable that, in purely phenomenological theories, such restriction on the mathematical representation of the internal variables related to damage be satisfied also. This is obviously an expression of the requirement, stated in Section 3.5.3, that 'good' phenomenological internal variables be somehow connected to the underlying physical mechanisms they are intended to represent.

12.3. Lemaitre's elastoplastic damage theory

The constitutive equations for elastoplasticity coupled with damage described in this section have been proposed by Lemaitre (1985a,b). Based on the concept of *effective stress* and the *hypothesis of strain equivalence,* Lemaitre's model includes evolution of internal damage as well as nonlinear isotropic and kinematic hardening in the description of the behaviour of ductile metals. The description of Lemaitre's model together with the main components necessary for its computational implementation – the integration algorithm and associated tangent operators – are presented in the following.

12.3.1. THE MODEL

State variables

The starting point of the theory is the assumption that the free energy is a function of the set $\{\varepsilon^e, R, \boldsymbol{X}, D\}$ of state variables, i.e.

$$\psi = \psi(\varepsilon^e, R, \boldsymbol{X}, D), \tag{12.8}$$

where ε^e is the elastic strain tensor and R and D are the *scalar* internal variables associated, respectively, with *isotropic hardening* and *isotropic damage*. The second-order tensor $\boldsymbol{X}$ is the internal variable related to *kinematic hardening*.

Thermodynamical aspects of isotropic and kinematic hardening and the physical meaning of the related internal variables have been discussed in Sections 6.6.2 and 6.6.4, respectively. The continuum damage variable D, as discussed in Section 12.2, can be interpreted as an indirect measure of density of microvoids and microcracks (Leckie and Onat, 1981). In the present theory, such microscopic defects are assumed isotropically distributed and, as we shall see below, will be phenomenologically reflected in the degradation of the elastic modulus. A critical value D_c, for the damage variable, is an experimentally determined parameter that defines the initiation of a macrocrack (Lemaitre and Chaboche, 1990).

Under the hypothesis of decoupling between elasticity–damage and plastic hardening, the specific free energy is assumed to be given by the sum

$$\psi = \psi^{ed}(\varepsilon^e, D) + \psi^p(R, \boldsymbol{X}), \tag{12.9}$$

where ψ^{ed} and ψ^p are, respectively, the elastic-damage and plastic contribution to the free energy.

The elastic-damage potential: elasticity–damage coupling

In the present theory the following form is postulated for the elastic-damage potential:

$$\bar{\rho}\, \psi^{ed}(\varepsilon^e, D) = \tfrac{1}{2}\, \varepsilon^e : (1 - D)\mathsf{D}^e : \varepsilon^e, \tag{12.10}$$

where $\mathbf{D}^e$ is the standard isotropic elasticity tensor. For this particular potential, the elasticity law is given by

$$\boldsymbol{\sigma} = \bar{\rho}\,\frac{\partial \psi}{\partial \boldsymbol{\varepsilon}^e} = (1 - D)\mathbf{D}^e : \boldsymbol{\varepsilon}^e. \tag{12.11}$$

Equivalently, the above damaged elastic law can be written as

$$\boldsymbol{\sigma}_{\text{eff}} = \mathbf{D}^e : \boldsymbol{\varepsilon}^e, \tag{12.12}$$

where $\boldsymbol{\sigma}_{\text{eff}}$ is the *effective stress tensor* that generalises the uniaxial effective stress of (12.2):

$$\boldsymbol{\sigma}_{\text{eff}} \equiv \frac{1}{1 - D}\,\boldsymbol{\sigma}. \tag{12.13}$$

The thermodynamical force conjugate to the damage internal variable is given by

$$Y \equiv \bar{\rho}\,\frac{\partial \psi}{\partial D} = -\tfrac{1}{2}\,\boldsymbol{\varepsilon}^e : \mathbf{D}^e : \boldsymbol{\varepsilon}^e, \tag{12.14}$$

or, using the inverse of the elastic stress/strain law,

$$\begin{aligned} Y &= \frac{-1}{2(1 - D)^2}\,\boldsymbol{\sigma} : [\mathbf{D}^e]^{-1} : \boldsymbol{\sigma} \\ &= \frac{-1}{2E(1 - D)^2}[(1 + \nu)\,\boldsymbol{\sigma} : \boldsymbol{\sigma} - \nu\,(\operatorname{tr}\boldsymbol{\sigma})^2] \\ &= \frac{-q^2}{2E(1 - D)^2}\left[\frac{2}{3}(1 + \nu) + 3(1 - 2\nu)\left(\frac{p}{q}\right)^2\right] \\ &= \frac{-q^2}{6G\,(1 - D)^2} - \frac{p^2}{2K\,(1 - D)^2}, \end{aligned} \tag{12.15}$$

where E and ν are, respectively, the Young's modulus and Poisson ratio associated with G and K. In the above, p is the hydrostatic stress and q is the von Mises effective stress. Commonly known as the *damage energy release rate*, $-Y$ corresponds to the variation of internal energy density due to damage growth at constant stress. It is the continuum damage analogue of the J-integral used in fracture mechanics (Rice, 1968). The product $-Y\dot{D}$ represents the power dissipated by the process of internal deterioration (mainly as decohesion of interatomic bonds).

Remark 12.1. Stress–strain rule (12.11) has an important experimental consequence. With the elasticity–damage coupling introduced via the hypothesis of strain equivalence (stated in Section 12.2.2), the *effective* elastic modulus of the material, which can be measured from experiments, is given by

$$\mathbf{D}_{\text{eff}} = (1 - D)\mathbf{D}^e, \tag{12.16}$$

where the damage variable assumes values within the interval $[0, 1]$. In the absence of damage ($D = 0$), the effective modulus equals the modulus $\mathbf{D}^e$ of the virgin material. For a completely damaged state ($D = 1$), $\mathbf{D}_{\text{eff}} = \mathbf{0}$, corresponding to a total loss of stiffness and load bearing capacity of the material. The identification of a generic damaged state, with $D \in [0, 1]$, is then restricted to the measurement of the degradation of the current effective elastic modulus with respect to the virgin state ($D = 0$) as described by Lemaitre (1985a) and Lemaitre and Chaboche (1990).

Isotropic and kinematic hardening forces

The plastic contribution $\psi^p(R, \boldsymbol{X})$ to the free energy is chosen as a sum of an isotropic and a kinematic hardening-related term:‡

$$\bar{\rho}\,\psi^p(R, \boldsymbol{X}) = \bar{\rho}\,\psi^I(R) + \frac{a}{2}\,\boldsymbol{X} : \boldsymbol{X}, \tag{12.17}$$

where a is a material constant and the isotropic hardening contribution, $\psi^I(R)$, is an arbitrary function of the single argument R. The thermodynamical force associated with isotropic hardening is, then, defined as

$$\kappa \equiv \bar{\rho}\,\frac{\partial \psi^p(R, \boldsymbol{X})}{\partial R} = \bar{\rho}\,\frac{\partial \psi^I(R)}{\partial R} = \kappa(R). \tag{12.18}$$

From (12.17), it follows that the thermodynamic force associated with kinematic hardening – the *back-stress tensor*, $\boldsymbol{\beta}$ – is given by

$$\boldsymbol{\beta} \equiv \bar{\rho}\,\frac{\partial \psi}{\partial \boldsymbol{X}} = a\,\boldsymbol{X}. \tag{12.19}$$

The yield function

For the *yield function* Φ the following von Mises-type form is adopted:

$$\Phi(\boldsymbol{\sigma}, \kappa, \boldsymbol{\beta}, D) = \frac{\sqrt{3\,J_2(\boldsymbol{s} - \boldsymbol{\beta})}}{1 - D} - \sigma_{y_0} - \kappa, \tag{12.20}$$

where the material parameter σ_{y_0} is the uniaxial yield stress of the virgin material.

The flow potential: internal variables evolution

The *flow potential* is assumed to be given by

$$\Psi = \Phi + \frac{b}{2a}\boldsymbol{\beta} : \boldsymbol{\beta} + \frac{r}{(1 - D)(s + 1)}\left(\frac{-Y}{r}\right)^{s+1}, \tag{12.21}$$

where a, b, r and s are material constants. The damage evolution constants r and s can be identified by integrating the damage evolution law for particular cases of (constant) stress triaxiality rate as described in Section 7.4 of Lemaitre and Chaboche (1990). The constants a and b, associated with the resulting Armstrong–Frederick kinematic hardening law (refer to the text surrounding expression (6.223), page 188, for further details) and can be obtained from cyclic loading experiments (Lemaitre and Chaboche, 1990). The convexity of the flow potential Ψ with respect to the thermodynamical forces for positive constants a, b, r and s ensures that the dissipation inequality is satisfied *a priori* by the present constitutive model. The constitutive equations of Lemaitre's ductile damage model are conveniently grouped in Box 12.1.

‡Refer to Subsections 6.6.3 and 6.6.4, from pages 182 and 185, for the definitions of individual contributions associated, respectively, with isotropic and kinematic hardening.

Box 12.1. Lemaitre's ductile damage model.

(i) Elastoplastic split of the strain tensor

$$\boldsymbol{\varepsilon} = \boldsymbol{\varepsilon}^e + \boldsymbol{\varepsilon}^p$$

(ii) Coupled elastic-damage law

$$\boldsymbol{\sigma} = (1 - D)\mathbf{D}^e : \boldsymbol{\varepsilon}^e$$

(iii) Yield function

$$\Phi = \frac{\sqrt{3\, J_2(\boldsymbol{s} - \boldsymbol{\beta})}}{1 - D} - \sigma_y$$

where $\sigma_y = \sigma_y(R)$

(iv) Plastic flow and evolution equations for R, $\boldsymbol{\beta}$ and D

$$\dot{\boldsymbol{\varepsilon}}^p = \dot{\gamma}\, \boldsymbol{N}$$

$$\dot{R} = \dot{\gamma}$$

$$\dot{\boldsymbol{\beta}} = \dot{\gamma}\,(a\, \boldsymbol{N} - b\, \boldsymbol{\beta})$$

$$\dot{D} = \dot{\gamma}\, \frac{1}{1 - D} \left(\frac{-Y}{r} \right)^s$$

with Y given by (12.15) and the flow vector

$$\boldsymbol{N} \equiv \sqrt{\frac{3}{2}}\, \frac{\boldsymbol{s} - \boldsymbol{\beta}}{(1 - D)\|\boldsymbol{s} - \boldsymbol{\beta}\|}$$

(v) Loading/unloading criterion

$$\Phi \le 0, \quad \dot{\gamma} \ge 0, \quad \Phi\, \dot{\gamma} = 0$$

Damage threshold

At low values of accumulated plastic strain, $\bar{\varepsilon}^p$, elastic modulus degradation can be hardly detected in experiments. Thus, we can assume that damage growth starts only at a critical value, denoted $\bar{\varepsilon}^p_D$. This critical value will be called the *damage threshold* and can be included in the model by redefining the damage evolution law of item (iv) of Box 12.1 as

$$\dot{D} = \dot{\gamma}\, \frac{\hat{H}(\bar{\varepsilon}^p - \bar{\varepsilon}^p_D)}{1 - D} \left(\frac{-Y}{r} \right)^s, \tag{12.22}$$

where $\hat{H}$ here denotes the Heaviside step function defined by (7.94) on page 221.

If such a threshold is adopted, then the evolution law for $\bar{\varepsilon}^p$ has to be defined for the model to be complete. From its definition (we recall expression (6.168)) we have

$$\dot{\bar{\varepsilon}}^p = \sqrt{\tfrac{2}{3}}\, \|\dot{\boldsymbol{\varepsilon}}^p\|. \tag{12.23}$$

Box 12.2. Integration algorithm for Lemaitre's ductile damage model.

(i) Elastic predictor

- Given the elastic trial strain, $\boldsymbol{\varepsilon}^{e\ \mathrm{trial}}_{n+1} := \boldsymbol{\varepsilon}_n + \Delta\boldsymbol{\varepsilon}$, evaluate the elastic trial stress

$$\boldsymbol{\sigma}^{\mathrm{trial}}_{n+1} := (1 - D_n)\mathbf{D}^e : \boldsymbol{\varepsilon}^{e\ \mathrm{trial}}_{n+1}$$

- Check plastic consistency

IF $\Phi^{\mathrm{trial}} := \frac{[3\, J_2(\boldsymbol{s}^{\mathrm{trial}}_{n+1} - \boldsymbol{\beta}_n)]^{1/2}}{1 - D_n} - \kappa(R_n) - \sigma_{y0} \leq 0$ THEN

Set $(\cdot)_{n+1} := (\cdot)^{\mathrm{trial}}_{n+1}$ and RETURN

ELSE go to (ii)

(ii) Return mapping (solve the system for $\boldsymbol{\sigma}_{n+1}$, $\boldsymbol{\beta}_{n+1}$, D_{n+1} and $\Delta\gamma$)

$$\left\{\begin{array}{l} \dfrac{[3\, J_2(\boldsymbol{s}_{n+1} - \boldsymbol{\beta}_{n+1})]^{1/2}}{1 - D_{n+1}} - \kappa(R_n + \Delta\gamma) - \sigma_{y0} \\ \boldsymbol{\sigma}_{n+1} - (1 - D_{n+1})\, \mathbf{D}^e : (\boldsymbol{\varepsilon}^{e\ \mathrm{trial}}_{n+1} - \Delta\gamma\, \boldsymbol{N}_{n+1}) \\ \boldsymbol{\beta}_{n+1} - \boldsymbol{\beta}_n - \Delta\gamma\, (a\, \boldsymbol{N}_{n+1} - b\, \boldsymbol{\beta}_{n+1}) \\ D_{n+1} - D_n - \dfrac{1}{1 - D_{n+1}} \left(\dfrac{-Y_{n+1}}{r}\right)^s \Delta\gamma \end{array}\right\} = \left\{\begin{array}{c} 0 \\ \mathbf{0} \\ \mathbf{0} \\ 0 \end{array}\right\}$$

where $\boldsymbol{N}_{n+1} = \frac{3}{2} \frac{\boldsymbol{s}_{n+1} - \boldsymbol{\beta}_{n+1}}{(1 - D_{n+1})\, \sqrt{3\, J_2(\boldsymbol{s}_{n+1} - \boldsymbol{\beta}_{n+1})}}$

(iii) Update R and $\boldsymbol{\varepsilon}^e$

$$R_{n+1} := R_n + \Delta\gamma\,, \quad \boldsymbol{\varepsilon}^e_{n+1} := \boldsymbol{\varepsilon}^{e\ \mathrm{trial}}_{n+1} - \Delta\gamma\, \boldsymbol{N}_{n+1}$$

(iv) EXIT

By taking the plastic flow rule for the present model into consideration, the above equation results in the following evolution law for the accumulated plastic strain:

$$\dot{\bar{\varepsilon}}^p = \frac{\dot{\gamma}}{1 - D}. \tag{12.24}$$

12.3.2. INTEGRATION ALGORITHM

The implicit elastic predictor/return-mapping scheme for integration of Lemaitre's coupled elastoplastic damage equations is just another particularisation of the general algorithm discussed in Chapter 7. Its pseudo-code is listed in Box 12.2. This particular algorithm was originally proposed by Benallal *et al.* (1988) and later adopted in the finite strain context by the authors (de Souza Neto and Perić, 1996; de Souza Neto *et al.*, 1993, 1994a, 1998).

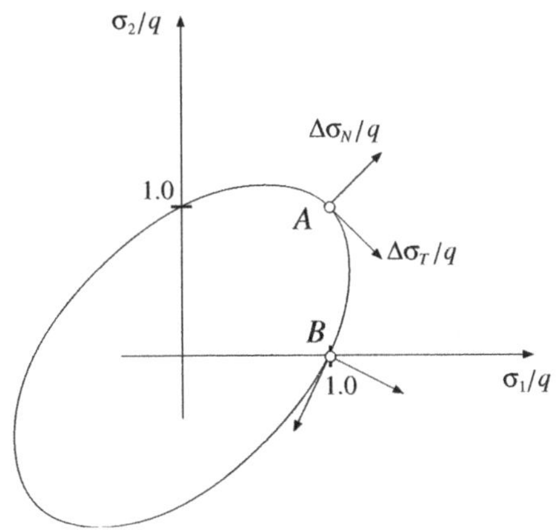

Figure 12.3. Iso-error maps for Lemaitre's damage model in plane stress. Departure points.

Including the damage threshold

Note that the damage threshold discussed above is not included in the numerical scheme shown in Box 12.2. Its inclusion, however, is trivial. The corresponding algorithm will be as follows. Two situations are possible:

1. If $\bar{\varepsilon}^p_n < \bar{\varepsilon}^p_D$, perform the calculations initially *without damage evolution*; that is, set the trial updated damage

$$D_{n+1} := D_n = 0,$$

and solve the equation system of item (ii) of Box 12.2 with the damage equation (the last equation of the system) 'switched off'. Then compute the trial updated accumulated plastic strain

$$\bar{\varepsilon}^p_{n+1} := \bar{\varepsilon}^p_n + \Delta\gamma.$$

Now check the validity of the trial updated values:

   (a) If $\bar{\varepsilon}^p_{n+1} < \bar{\varepsilon}^p_D$, the trial values are accepted as the updated values.

   (b) Otherwise, solve the return-mapping equations as in Box 12.2.

2. If $\bar{\varepsilon}^p_n \geq \bar{\varepsilon}^p_D$, then the calculations follow the procedure of Box 12.2.

Finite step accuracy: iso-error maps

A study of the finite step accuracy properties of the integration algorithm of Box 12.2 has been carried out by the authors (de Souza Neto *et al.*, 1994a) for a *plane stress-projected* implementation of Lemaitre's model. The main conclusion is that the accuracy of the algorithm deteriorates as damage increases. To illustrate this fact, we show here the error maps obtained at different stages of damage evolution for two departure points on the plane stress von Mises surface: biaxial and uniaxial stress states (Figure 12.3). The error maps are plotted in Figure 12.4.

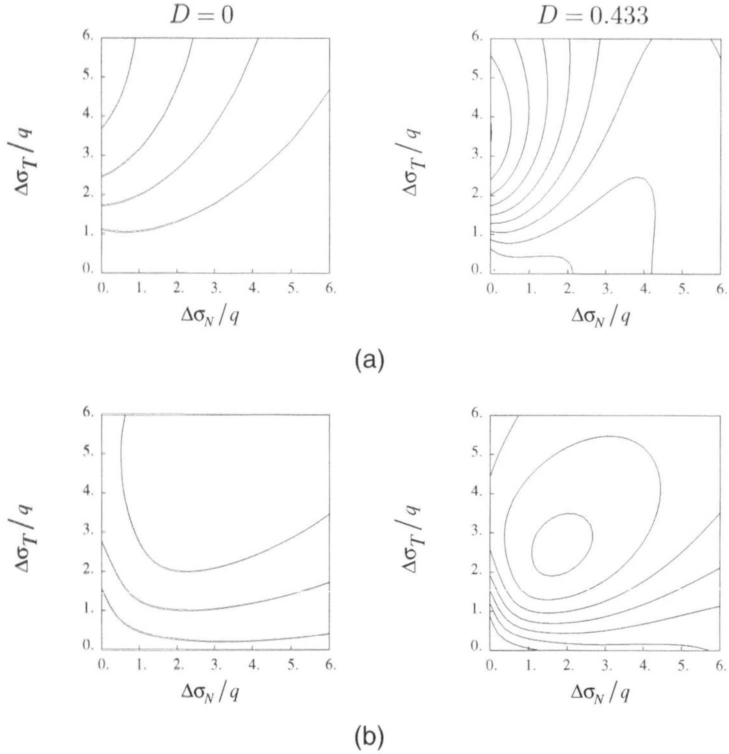

Figure 12.4. Lemaitre's model in plane stress. Iso-error maps at $D = 0$ and $D = 43.3\%$. (a) Biaxial state – point A; (b) Uniaxial state – point B. Contour intervals $= 4\%$.

Stabilisation of the local Newton–Raphson scheme

Unfortunately, the accuracy of the algorithm is not the only property that deteriorates with damage evolution. The convergence of the Newton–Raphson scheme used to solve the return-mapping system of nonlinear equations (item (ii) of Box 12.2) is also affected. This issue was briefly addressed in Remark 7.3 (page 199). As damage grows, the bowl of convergence of the local Newton algorithm shrinks and an appropriate initial guess needs to be supplied to ensure convergence. At highly damaged states the choice of initial guess becomes crucial. Within the finite element context, failure of the return mapping to converge for a single Gauss point requires that the global incrementation procedure be restarted from the beginning of the current increment with a reduced load step. This may incur a dramatic increase in computational costs, especially for large problems. To overcome this problem, the following strategy was found by the authors to provide an effective solution to the problem in the present case.

- Firstly, the Newton–Raphson scheme is applied taking $\boldsymbol{\sigma}_n$, $\boldsymbol{\beta}_n$, D_n and $\Delta\gamma = 0$ as initial guesses for the system variables.
- If convergence is not achieved, the Newton–Raphson scheme is restarted. The initial guess now is $\boldsymbol{\sigma}^{\text{proj}}$, $\boldsymbol{\beta}_n$, D_n and $\Delta\gamma = 0$. The *projected stress*, $\boldsymbol{\sigma}^{\text{proj}}$, is obtained by

solving, for $\Delta\gamma$, the *scalar* equation

$$\Phi(\text{dev}[\boldsymbol{\sigma}^{\text{proj}}], R_n, \boldsymbol{\beta}_n, D_n) = 0,$$

with

$$\text{dev}[\boldsymbol{\sigma}^{\text{proj}}] = \frac{\text{dev}[\boldsymbol{\sigma}^{\text{trial}}_{n+1} - \boldsymbol{\beta}_n]}{1 + \sqrt{\frac{3}{2}}\,\Delta\gamma} + \boldsymbol{\beta}_n,$$

and corresponds to the return of $\boldsymbol{\sigma}^{\text{trial}}_{n+1}$ to the *frozen* yield surface of time t_n *as if the material were perfectly plastic* with no damage or hardening evolution.

This procedure stabilises the local Newton–Raphson scheme and assures convergence at any stage of damage evolution for relatively large increments of elastic trial stress.

12.3.3. THE TANGENT OPERATORS

If the outcome $\{\boldsymbol{\sigma}_{n+1}, R_{n+1}, \boldsymbol{\beta}_{n+1}, D_{n+1}\}$ of the integration algorithm of Box 12.2 lies inside the elastic domain ($\Phi_{n+1} < 0$) then the corresponding algorithmic constitutive functional for stress,

$$\bar{\boldsymbol{\sigma}}\,(R_n, \boldsymbol{\beta}_n, D_n, \boldsymbol{\varepsilon}^{e\ \text{trial}}_{n+1}), \tag{12.25}$$

is the (damaged) elastic law. The consistent tangent operator is simply given by

$$\mathbf{D} = (1 - D_{n+1})\mathbf{D}^e. \tag{12.26}$$

The above tangent also applies when the current state is on the yield surface and *elastic unloading* is assumed to occur. Under plastic straining, $\boldsymbol{\sigma}_{n+1}$ is delivered as the solution of the nonlinear system of the plastic corrector stage (item (ii)). In this case, the nonlinear system is differentiated leading to the linear form

$$\begin{bmatrix} A_{1,\sigma} & A_{1,D} & A_{1,\Delta\gamma} & A_{1,\beta} \\ \boldsymbol{A}_{2,\sigma} & \boldsymbol{A}_{2,D} & \boldsymbol{A}_{2,\Delta\gamma} & \boldsymbol{A}_{2,\beta} \\ \boldsymbol{A}_{3,\sigma} & \boldsymbol{A}_{3,D} & \boldsymbol{A}_{3,\Delta\gamma} & \boldsymbol{A}_{3,\beta} \\ A_{4,\sigma} & A_{4,D} & A_{4,\Delta\gamma} & \mathbf{0} \end{bmatrix} \begin{bmatrix} \mathrm{d}\boldsymbol{\sigma}_{n+1} \\ \mathrm{d}D_{n+1} \\ \mathrm{d}\Delta\gamma \\ \mathrm{d}\boldsymbol{\beta}_{n+1} \end{bmatrix} = \begin{bmatrix} 0 \\ (1 - D_{n+1})\mathbf{D}^e : \mathrm{d}\boldsymbol{\varepsilon}^{e\ \text{trial}}_{n+1} \\ \mathbf{0} \\ 0 \end{bmatrix} \tag{12.27}$$

where the coefficients $\boldsymbol{A}_{1,\sigma}$, $A_{1,D}, \ldots,$ are the partial derivatives of the left-hand sides of item (ii) with respect to the system variables computed at the converged solution of the nonlinear system of equations of the plastic corrector procedure. Note that the same coefficients matrix is computed for each trial solution obtained during the Newton–Raphson iterations of the plastic corrector stage. Inversion of (12.27) gives the tangent relations between the system variables ($\boldsymbol{\sigma}_{n+1}$, D_{n+1}, $\Delta\gamma$ and $\boldsymbol{\beta}_{n+1}$) and $\boldsymbol{\varepsilon}^{e\ \text{trial}}_{n+1}$

$$\begin{bmatrix} \mathrm{d}\boldsymbol{\sigma}_{n+1} \\ \mathrm{d}D_{n+1} \\ \mathrm{d}\Delta\gamma \\ \mathrm{d}\boldsymbol{\beta}_{n+1} \end{bmatrix} = \begin{bmatrix} C_{11} & \mathsf{C}_{12} & \mathsf{C}_{13} & C_{14} \\ C_{21} & C_{22} & C_{23} & C_{24} \\ C_{31} & C_{32} & C_{33} & C_{34} \\ C_{41} & \mathsf{C}_{42} & \mathsf{C}_{43} & C_{44} \end{bmatrix} \begin{bmatrix} 0 \\ (1 - D_{n+1})\mathbf{D}^e : \mathrm{d}\boldsymbol{\varepsilon}^{e\ \text{trial}}_{n+1} \\ \mathbf{0} \\ 0 \end{bmatrix}. \tag{12.28}$$

In particular, for the elastoplastic consistent tangent operator, we have

$$\mathsf{D} \equiv \frac{\mathrm{d}\boldsymbol{\sigma}_{n+1}}{\mathrm{d}\boldsymbol{\varepsilon}^{e\ \mathrm{trial}}_{n+1}} = (1 - D_{n+1})\, \mathsf{C}_{12} : \mathsf{D}^e. \tag{12.29}$$

A closed formula for the small-strain consistent tangent operator, which does not require inversion of the linear system, was derived by Doghri (1995) for a variant of the present version of Lemaitre's ductile damage model.

Remark 12.2. The above tangent operator D is generally *unsymmetric*.

12.4. A simplified version of Lemaitre's model

By removing kinematic hardening from Lemaitre's original ductile damage theory, we obtain a simplified version of the model whose numerical implementation assumes a remarkably simple form. The use of such a simplified theory can be justified whenever the effects of kinematic hardening are not relevant, i.e. in any process where reverse plastic loading does not occur or has little influence on the overall evolution of damage and plastic flow.

In this section we describe the numerical implementation of the simplified Lemaitre model. We note that the simplified model is fully incorporated into the finite element program `HYPLAS` that accompanies this book. The FORTRAN subroutines of `HYPLAS` associated with this model are: `SUDAMA`, `CTDAMA`, `RDDAMA`, `SWDAMA` and `ORDAMA`.

Let us start by stating the constitutive equations of the model in Box 12.3. The equations are obtained from those of the original model (Box 12.1) by simply setting $\boldsymbol{\beta} = \mathbf{0}$ and $a = b = 0$.

12.4.1. THE SINGLE-EQUATION INTEGRATION ALGORITHM

One particularly remarkable aspect of the present model is that, as in the standard von Mises theory, the return-mapping stage of the constitutive integration algorithm can be reduced to the solution of a *single* nonlinear equation (see Section 7.3.2, from page 217). This results in an extremely fast algorithm, in sharp contrast to the original model. Note that for the original model (refer to Box 12.2) the independent variables of the return-mapping equations are the stress and back-stress components, the damage internal variable and the plastic multiplier, leading to a far more costly stress updating procedure. The integration algorithm for the simplified damage model is derived in this section. Readers who wish to skip the details of derivation are referred directly to Box 12.4 where the algorithm is listed in the standard pseudo-code format. The procedure is coded in subroutine `SUDAMA` of program `HYPLAS` for the plane strain and axisymmetric cases.

The *elastic trial* stage of the algorithm proceeds exactly in the same way as the one for the original model. Here, we conveniently perform the deviatoric/hydrostatic split of the stress tensor and evaluate

$$\begin{aligned} \boldsymbol{s}^{\mathrm{trial}}_{n+1} &:= (1 - D_n)\, 2G\, \boldsymbol{\varepsilon}^{e\ \mathrm{trial}}_{\mathrm{d}\ n+1} \\ p^{\mathrm{trial}}_{n+1} &:= (1 - D_n)\, K\, \varepsilon^{e\ \mathrm{trial}}_{\mathrm{v}\ n+1}. \end{aligned} \tag{12.30}$$

The elastic trial value of the yield function in the present case is then evaluated as

$$\Phi^{\mathrm{trial}} := \tilde{q}^{\mathrm{trial}}_{n+1} - \sigma_y(R_n), \tag{12.31}$$

Box 12.3. Simplified Lemaitre model (isotropic hardening only).

(i) Elastoplastic split of the strain tensor

$$\boldsymbol{\varepsilon} = \boldsymbol{\varepsilon}^e + \boldsymbol{\varepsilon}^p$$

(ii) Coupled elastic-damage law

$$\boldsymbol{\sigma} = (1 - D)\mathbf{D} : \boldsymbol{\varepsilon}^e$$

(iii) Yield function

$$\Phi = \frac{\sqrt{3\, J_2(\boldsymbol{s})}}{1 - D} - \sigma_y$$

(iv) Plastic flow and evolution equations for R and D

$$\dot{\boldsymbol{\varepsilon}}^p = \dot{\gamma}\, \boldsymbol{N}$$

$$\dot{R} = \dot{\gamma}$$

$$\dot{D} = \dot{\gamma}\, \frac{1}{1 - D}\left(\frac{-Y}{r}\right)^s$$

with Y given by (12.15) and flow vector

$$\boldsymbol{N} \equiv \sqrt{\frac{3}{2}}\, \frac{\boldsymbol{s}}{(1 - D)\|\boldsymbol{s}\|}$$

(v) Loading/unloading criterion

$$\Phi \leq 0, \quad \dot{\gamma} \geq 0, \quad \Phi\, \dot{\gamma} = 0$$

where we have defined the *effective* elastic trial von Mises equivalent stress,

$$\tilde{q}_{n+1}^{\text{trial}} \equiv \frac{q_{n+1}^{\text{trial}}}{1 - D_n} = \frac{\sqrt{3J_2(\boldsymbol{s}_{n+1}^{\text{trial}})}}{1 - D_n} = \sqrt{\tfrac{3}{2}}\, \frac{\|\boldsymbol{s}_{n+1}^{\text{trial}}\|}{1 - D_n}. \tag{12.32}$$

As usual, if $\Phi^{\text{trial}} \leq 0$ the process is elastic within the step and the elastic trial state coincides with the updated state at t_{n+1}. Otherwise, we apply the single equation return-mapping procedure derived in the following.

The single-equation return mapping

The implicit return-mapping equations for a generic plasticity model are given by (7.25), page 196. In the present case, with the model defined in Box 12.3, the set of return-mapping equations assumes the particular form

$$\begin{aligned}
&\boldsymbol{\varepsilon}^e_{n+1} = \boldsymbol{\varepsilon}^{e\ \mathrm{trial}}_{n+1} - \Delta\gamma \sqrt{\frac{3}{2}} \frac{\boldsymbol{s}_{n+1}}{(1 - D_{n+1})\|\boldsymbol{s}_{n+1}\|} \\
&R_{n+1} = R_n + \Delta\gamma \\
&D_{n+1} = D_n + \frac{\Delta\gamma}{1 - D_{n+1}} \left(\frac{-Y_{n+1}}{r}\right)^s \\
&\frac{q_{n+1}}{1 - D_{n+1}} - \sigma_y(R_{n+1}) = 0.
\end{aligned} \tag{12.33}$$

As we shall see, by performing relatively straightforward algebraic manipulations, the above system can be reduced to a single nonlinear algebraic equation for the plastic multiplier $\Delta\gamma$.

To start with, let us consider the deviatoric/volumetric split of $(12.33)_1$. This gives

$$\begin{aligned}
&\boldsymbol{\varepsilon}^e_{\mathrm{d}\ n+1} = \boldsymbol{\varepsilon}^{e\ \mathrm{trial}}_{\mathrm{d}\ n+1} - \Delta\gamma \sqrt{\frac{3}{2}} \frac{\boldsymbol{s}_{n+1}}{(1 - D_{n+1})\|\boldsymbol{s}_{n+1}\|} \\
&\varepsilon^e_{\mathrm{v}\ n+1} = \varepsilon^{e\ \mathrm{trial}}_{\mathrm{v}\ n+1}.
\end{aligned} \tag{12.34}$$

Expression $(12.34)_2$, together with the elastic law, gives the following updating relation for the hydrostatic pressure:

$$p_{n+1} = (1 - D_{n+1})\, \tilde{p}_{n+1}, \tag{12.35}$$

where we have defined

$$\tilde{p}_{n+1} = K\, \varepsilon^e_{\mathrm{v}\ n+1} = K\, \varepsilon^{e\ \mathrm{trial}}_{\mathrm{v}\ n+1}. \tag{12.36}$$

Let us now consider the stress deviator updating formula. With the introduction of the elastic law into $(12.34)_1$, it follows that

$$\boldsymbol{s}_{n+1} = (1 - D_{n+1})\, 2G\, \boldsymbol{\varepsilon}^e_{\mathrm{d}\ n+1} = (1 - D_{n+1})\, 2G\, \boldsymbol{\varepsilon}^{e\ \mathrm{trial}}_{\mathrm{d}\ n+1} - 2G\, \Delta\gamma \sqrt{\frac{3}{2}} \frac{\boldsymbol{s}_{n+1}}{\|\boldsymbol{s}_{n+1}\|}$$

and we obtain the update equation for the stress deviator:

$$\boldsymbol{s}_{n+1} = (1 - D_{n+1})\, \tilde{\boldsymbol{s}}^{\mathrm{trial}}_{n+1} - 2G\, \Delta\gamma \sqrt{\frac{3}{2}} \frac{\boldsymbol{s}_{n+1}}{\|\boldsymbol{s}_{n+1}\|}, \tag{12.37}$$

where we have defined

$$\tilde{\boldsymbol{s}}^{\mathrm{trial}}_{n+1} \equiv 2G\, \boldsymbol{\varepsilon}^{e\ \mathrm{trial}}_{\mathrm{d}\ n+1}. \tag{12.38}$$

From (12.37), it is clear that $\tilde{\boldsymbol{s}}^{\mathrm{trial}}_{n+1}$ is proportional to $\boldsymbol{s}_{n+1}$ so that we may equivalently write

$$\boldsymbol{s}_{n+1} = (1 - D_{n+1})\, \tilde{\boldsymbol{s}}^{\mathrm{trial}}_{n+1} - 2G\, \Delta\gamma \sqrt{\frac{3}{2}} \frac{\tilde{\boldsymbol{s}}^{\mathrm{trial}}_{n+1}}{\|\tilde{\boldsymbol{s}}^{\mathrm{trial}}_{n+1}\|}. \tag{12.39}$$

After a straightforward manipulation, the above equation yields the following simpler update formula for $\boldsymbol{s}_{n+1}$:

$$\boldsymbol{s}_{n+1} = \left(1 - D_{n+1} - \frac{3G\, \Delta\gamma}{\tilde{q}^{\mathrm{trial}}}\right) \tilde{\boldsymbol{s}}^{\mathrm{trial}}_{n+1}. \tag{12.40}$$

From this last expression and the definition of the von Mises equivalent stress, we obtain

$$q_{n+1} = (1 - D_{n+1})\tilde{q}^{\mathrm{trial}} - 3G\, \Delta\gamma. \tag{12.41}$$

Two-equation system.

By introducing (12.41) and $(12.33)_2$ into $(12.33)_4$, we obtain the consistency equation

$$\tilde{\Phi}(\Delta\gamma, D_{n+1}) \equiv \tilde{q}^{\text{trial}} - \frac{3G\,\Delta\gamma}{(1 - D_{n+1})} - \sigma_y(R_n + \Delta\gamma) = 0. \tag{12.42}$$

With (12.41) and (12.35) introduced into the definition of the damage energy release rate, $(12.33)_3$ can be written as

$$D_{n+1} - D_n - \frac{\Delta\gamma}{1 - D_{n+1}}\left(\frac{-Y(\Delta\gamma, D_{n+1})}{r}\right)^s = 0, \tag{12.43}$$

where

$$-Y(\Delta\gamma, D_{n+1}) \equiv \frac{[(1 - D_{n+1})\tilde{q}^{\text{trial}} - 3G\,\Delta\gamma]^2}{6G\,(1 - D_{n+1})^2} + \frac{\tilde{p}_{n+1}^2}{2K}. \tag{12.44}$$

In summary, the return mapping has been reduced to the set of *two* scalar equations (12.42) and (12.43). The unknowns of this system of equations are $\Delta\gamma$ and D_{n+1}. After solution, with $\Delta\gamma$ and D_{n+1} at hand, s_{n+1} and p_{n+1} are trivially updated respectively by (12.40) and (12.35). This two-equation return mapping was proposed by Vaz Jr. (1998) in the context of fracture prediction for metal-cutting applications. A similar two-equation algorithm was proposed by Steinmann *et al.* (1994) for a variation of the simplified Lemaitre model, where the damage energy release rate depends only on the deviatoric part of the strain energy function. The work of Steinmann *et al.* (1994) was carried out within the framework of multiplicative finite strain plasticity. Return-mapping convergence difficulties similar to those discussed in Section 12.3.2 have been identified by these authors. To tackle the problem, they introduced a *line-search* procedure within the Newton–Raphson scheme for solution of the return-mapping equations.

One-equation return mapping.

Further reduction in the above system, leading to a computationally more efficient single-equation return-mapping algorithm, is possible for the present model. The single-equation algorithm is derived as follows. Firstly, for convenience, we define the material *integrity* as

$$\omega \equiv 1 - D. \tag{12.45}$$

With the above definition and (12.42) we may write

$$\omega_{n+1} \equiv 1 - D_{n+1} = w(\Delta\gamma) \equiv \frac{3G\,\Delta\gamma}{\tilde{q}^{\text{trial}} - \sigma_y(R_n + \Delta\gamma)}. \tag{12.46}$$

In addition, by combining (12.42) and (12.44), the (updated) damage energy release rate may be expressed as a function of $\Delta\gamma$ *only*, i.e. we may redefine

$$-Y(\Delta\gamma) \equiv \frac{[\sigma_y(R_n + \Delta\gamma)]^2}{6G} + \frac{\tilde{p}_{n+1}^2}{2K}. \tag{12.47}$$

Finally, by combining (12.46) and (12.47) with (12.43), our return-mapping procedure is reduced to the solution of the following scalar equation for $\Delta\gamma$:

$$F(\Delta\gamma) \equiv \omega(\Delta\gamma) - \omega_n + \frac{\Delta\gamma}{\omega(\Delta\gamma)}\left(\frac{-Y(\Delta\gamma)}{r}\right)^s = 0. \tag{12.48}$$

Once a solution $\Delta\gamma$ has been found, we update the hardening and damage variables, the hydrostatic stress and the stress deviator using the relevant equations listed above. The overall algorithm has been implemented in subroutine SUDAMA (**S**tate **U**pdate procedure for the simplified Lemaitre **DAMA**ge model) of program HYPLAS. The corresponding pseudo-code is conveniently listed in Box 12.4. As for the other models discussed in this book, the return-mapping equation is solved by the standard Newton–Raphson iterative scheme.

Remark 12.3. In the Newton–Raphson scheme for the iterative solution of the return mapping equation (12.48) we adopt the following initial guess for $\Delta\gamma$:

$$\Delta\gamma^{(0)} = \frac{[\tilde{q}^{\text{trial}} - \sigma_y(R_n)]\,\omega_n}{3G}, \tag{12.49}$$

which corresponds to the perfectly plastic solution for the increment with *frozen* (damaged) yield surface at t_n. The above initial guess (used in subroutine SUDAMA) reduces the total number of iterations required for convergence, as compared to the usual choice $\Delta\gamma^{(0)} = 0$. The overall integration algorithm has been found to be extremely stable. No need for line-searches appears necessary since convergence is attained for large strain increments even at highly damaged states with D approaching unity.

Inclusion of the damage threshold

As for the original model, we have presented the numerical integration of the simplified version without considering the damage threshold discussed on page 481. The incorporation of the damage threshold here follows the same steps as in the original model (refer to page 483). Here, when damage evolution is 'switched off' we replace the solution of (12.48) with the solution of the standard return-mapping equation of the isotropically hardening von Mises model; that is, we solve (12.46) for $\Delta\gamma$ with $w(\Delta\gamma) \equiv 1$. The incorporation of the threshold is straightforward and will be left here as an exercise for the interested reader. Note that within the structure of program HYPLAS, a new state variable, the accumulated plastic strain, may be conveniently stored and updated accordingly. This will affect the corresponding material-specific state update and switching routines. Also, a new material constant, the damage threshold $\bar{\varepsilon}^p_D$ is required and needs to be read in the appropriate material-specific input data reading routine.

12.4.2. THE TANGENT OPERATOR

We now turn our attention to the tangent operators required for the complete finite element implementation of the present damage model. Because the elastic law in the present case is that of the original Lemaitre model, the corresponding (damaged) elastic operator is given by (12.26) and requires no further consideration. The elastoplastic consistent operator is a

Box 12.4. Fully implicit elastic predictor/return-mapping algorithm for the simplified version of Lemaitre's damage model.

HYPLAS procedure:	SUDAMA

(i) Elastic predictor. Given $\Delta\varepsilon$ and the state variables at t_n, evaluate the *elastic trial state*

$$\varepsilon^{e\,\text{trial}}_{n+1} := \varepsilon^e_n + \Delta\varepsilon$$

$$R^{\text{trial}}_{n+1} := R_n$$

$$\tilde{p}_{n+1} := K\,\varepsilon^{e\,\text{trial}}_{v\,n+1}; \quad \tilde{\boldsymbol{s}}^{\text{trial}}_{n+1} := 2G\,\boldsymbol{\varepsilon}^{e\,\text{trial}}_{d\,n+1}$$

$$\tilde{q}^{\text{trial}}_{n+1} := \sqrt{\tfrac{3}{2}}\,\|\tilde{\boldsymbol{s}}^{\text{trial}}_{n+1}\|/(1-D_n)$$

(ii) Check plastic admissibility

IF $\tilde{q}^{\text{trial}}_{n+1} - \sigma_y(R^{\text{trial}}_{n+1}) \leq 0$

THEN set $(\cdot)_{n+1} := (\cdot)^{\text{trial}}_{n+1}$ and EXIT

(iii) Return mapping. Solve the equation

$$F(\Delta\gamma) \equiv \omega(\Delta\gamma) - \omega_n + \frac{\Delta\gamma}{\omega(\Delta\gamma)}\left(\frac{-Y(\Delta\gamma)}{r}\right)^s = 0$$

for $\Delta\gamma$ using the Newton–Raphson method. Start iterations with the initial guess (12.49). The functions $\omega(\Delta\gamma)$ and $Y(\Delta\gamma)$ are defined, respectively, by (12.46) and (12.47). Then update

$$R_{n+1} := R_n + \Delta\gamma$$

$$p_{n+1} := \omega(\Delta\gamma)\,\tilde{p}_{n+1}; \quad q_{n+1} := \omega(\Delta\gamma)\,\sigma_y(R_{n+1})$$

$$\boldsymbol{s}_{n+1} := \frac{q_{n+1}}{\tilde{q}^{\text{trial}}_{n+1}}\,\tilde{\boldsymbol{s}}^{\text{trial}}_{n+1}; \quad \boldsymbol{\sigma}_{n+1} := \boldsymbol{s}_{n+1} + p_{n+1}\,\boldsymbol{I}$$

$$\boldsymbol{\varepsilon}^e_{n+1} = \frac{1}{2G}\,\boldsymbol{s}_{n+1} + \tfrac{1}{3}\,\varepsilon^{e\,\text{trial}}_{v\,n+1}\,\boldsymbol{I}$$

(iv) EXIT

particular case of the original one given by (12.29) and could be obtained by setting $\boldsymbol{\beta} = \boldsymbol{0}$ and $a = b = 0$ in the procedure described in Section 12.3.3. However, due to the relative simplicity of the one-equation return mapping, it is possible to derive a closed-form expression for the elastoplastic tangent in a straightforward manner. The evaluation of the elastoplastic tangent by means of its closed form is far simpler (and computationally more efficient) than its calculation as a particular case of the original model. The closed-form expression is implemented in program HYPLAS (subroutine CTDAMA). Its derivation follows the standard application of consistent linearisation concepts to the one-equation return-mapping scheme of Box 12.4 and we shall leave it as an exercise for the interested reader. The final closed form expression is the following:

$$\mathbf{D}^{ep} = a\,\mathbf{I}_d + b\,\bar{\boldsymbol{s}}_{n+1} \otimes \bar{\boldsymbol{s}}_{n+1} + c\,\bar{\boldsymbol{s}}_{n+1} \otimes \boldsymbol{I} + d\,\boldsymbol{I} \otimes \bar{\boldsymbol{s}}_{n+1} + e\,\boldsymbol{I} \otimes \boldsymbol{I}, \tag{12.50}$$

where $\bar{\boldsymbol{s}}_{n+1}$ is the normalised stress deviator,

$$\bar{\boldsymbol{s}}_{n+1} = \frac{\boldsymbol{s}_{n+1}}{\|\boldsymbol{s}_{n+1}\|}, \tag{12.51}$$

and the scalars a, b, c, d and e are given by

$$\begin{aligned}
a &= \frac{2G\,\omega_{n+1}\,\sigma_y(R_{n+1})}{\tilde{q}^{\,\text{trial}}_{n+1}} \\
b &= 2G\left(a_1\,H\,\omega_{n+1} + a_4\,\sigma_y(R_{n+1}) - \frac{\omega_{n+1}\,\sigma_y(R_{n+1})}{\tilde{q}^{\,\text{trial}}_{n+1}}\right) \\
c &= K\,\sqrt{\tfrac{2}{3}}\,[a_2\,H\,\omega_{n+1} + a_3\,\sigma_y(R_{n+1})] \\
d &= 2G\,\sqrt{\tfrac{3}{2}}\,\tilde{p}_{n+1}\,a_4 \\
e &= K\,(\omega_{n+1} + a_3\,\tilde{p}_{n+1}).
\end{aligned} \tag{12.52}$$

In the above, H denotes the updated slope of the hardening curve at t_{n+1}:

$$H = \left.\frac{\mathrm{d}\sigma_y}{\mathrm{d}R}\right|_{R_{n+1}}, \tag{12.53}$$

and a_1, a_2, a_3 and a_4 are defined as

$$\begin{aligned}
a_1 &= \frac{1}{F'}\left[\frac{\omega_{n+1}}{\tilde{q}^{\,\text{trial}}_{n+1} - \sigma_y(R_{n+1})} - \frac{1}{3G}\left(\frac{-Y_{n+1}}{r}\right)^s\right] \\
a_2 &= -\frac{s\,\tilde{p}_{n+1}[\tilde{q}^{\,\text{trial}}_{n+1} - \sigma_y(R_{n+1})]}{3G\,r\,K\,F'}\left(\frac{-Y_{n+1}}{r}\right)^{s-1} \\
a_3 &= a_2\,\omega' \\
a_4 &= a_1\,\omega' - \frac{\omega_{n+1}}{\tilde{q}^{\,\text{trial}}_{n+1} - \sigma_y(R_{n+1})}.
\end{aligned} \tag{12.54}$$

The scalar F' is the derivative (at the solution $\Delta\gamma$) of the return-mapping residual function defined by (12.48)

$$F' = \omega' - \frac{H}{3G}\left(\frac{-Y_{n+1}}{r}\right)^s + \frac{s\,H\,\sigma_y(R_{n+1})[\tilde{q}^{\,\text{trial}}_{n+1} - \sigma_y(R_{n+1})]}{9G^2 r}\left(\frac{-Y_{n+1}}{r}\right)^{s-1} \tag{12.55}$$

and ω' denotes the derivative (also at the solution $\Delta\gamma$) of the function defined by (12.46)

$$\omega' = \frac{3G + \omega_{n+1}H}{\tilde{q}^{\,\text{trial}}_{n+1} - \sigma_y(R_{n+1})}. \tag{12.56}$$

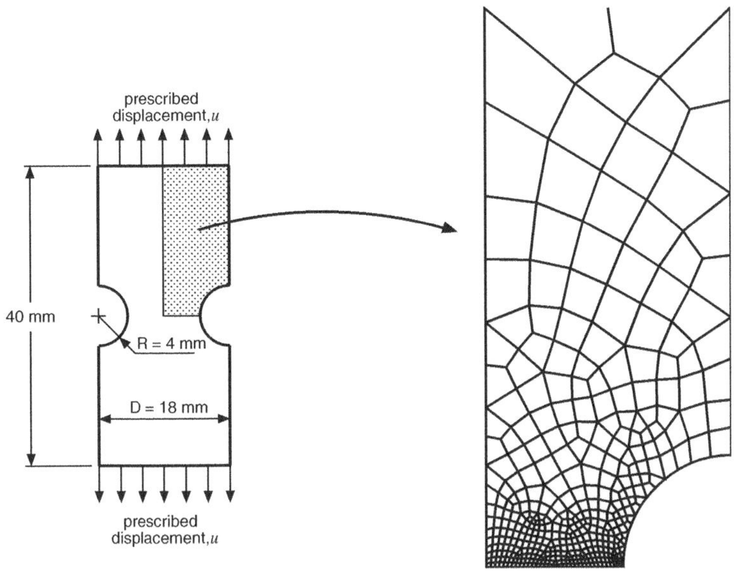

Figure 12.5. Cylindrical notched bar. Geometry, boundary conditions and finite element mesh.

Remark 12.4. In the first equilibrium iteration of any load increment, corresponding to $\Delta\gamma = 0$ and $\tilde{q}^{\text{trial}} = \sigma_y$, the above expression for $\mathbf{D}^{ep}$ is used in subroutine CTDAMA by setting a small perturbation

$$\Delta\gamma := 10^{-8}.$$

Remark 12.5. It should also be noted that as c is generally different from d in (12.50), the elastoplastic tangent operator for the simplified model is also *unsymmetric*.

12.4.3. EXAMPLE. FRACTURING OF A CYLINDRICAL NOTCHED SPECIMEN

In this example, Lemaitre's coupled plasticity-damage model is used to simulate fracture initiation in a cylindrical pre-notched bar subjected to monotonic axial stretching. As the loading is monotonic, kinematic hardening is not considered and the above described implementation of the simplified version of Lemaitre's model with one-equation return mapping is used in the present simulation. Similar analyses have been carried out by Benallal *et al.* (1991) using a coupled viscoplasticity-damage model, by Cescotto and Zhu (1995) with an elastoplastic damage model accounting for separate deviatoric and bulk damage, together with the distinct effects of tensile and compressive hydrostatic pressures, and by Vaz Jr. and Owen (2001). The geometry of the problem, boundary conditions and the finite element mesh adopted are given in Figure 12.5. The mesh discretises one symmetric quarter of the bar with appropriate symmetry boundary conditions imposed along the relevant edges. The loading consists of a prescribed monotonically increasing vertical displacement (with free horizontal displacement) of the nodes on the top edge of the mesh. A total number of 399 eight-noded axisymmetric quadrilaterals (with reduced 2×2 Gauss integration rule) was used in the discretisation amounting to a total of 1276 nodes. A relatively fine discretisation

Table 12.1. Cylindrical notched bar. Lemaitre's model parameters.

E	210 GPa
ν	0.3
Hardening curve	$\sigma_y(R) = 620 + 3300[1 - \exp(-0.4\,R)]$ (MPa)
S	1.0
r	3.5 MPa

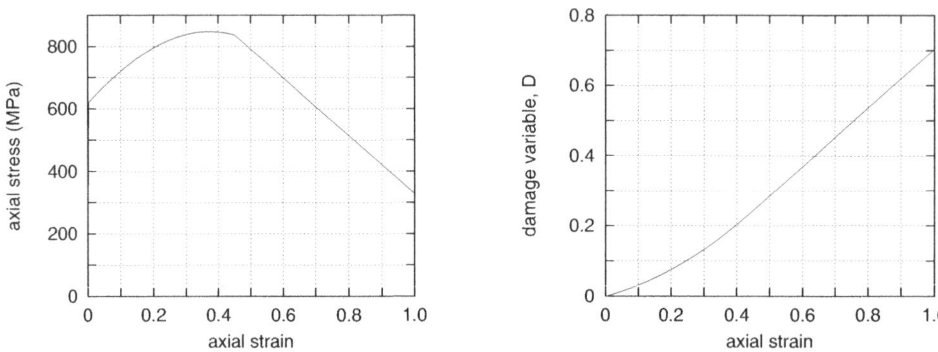

Figure 12.6. Cylindrical notched bar. Lemaitre's model uniaxial behaviour with the present material constants.

is used in the region surrounding the smallest cross-section. The material parameters adopted in the analysis are listed in Table 12.1. These parameters have been calibrated by Benallal *et al.* (1987) from uniaxial experiments with AISI 1010 low carbon steel in a rolled state. The corresponding uniaxial stress–strain and damage–strain curves predicted by Lemaitre's model with the material constants of Table 12.1 are shown in Figure 12.6.

The analysis was carried out by applying the vertical displacement of the constrained edge incrementally. The simulation was stopped when the first Gauss point in the structure reached a damaged state with $D > 0.999$. This state was attained in 62 increments with a prescribed edge displacement $u = 0.576$ mm. The evolution of the damage variable field obtained in the finite element analysis is illustrated in the contour plots shown in Figure 12.7. It can be seen that during the early stages of the loading process, maximum damage is detected near the root of the notch. As the specimen is progressively stretched, the maximum damage area moves gradually towards the centre of the specimen and localises there. At the final stage with $u = 0.576$ mm, damage is highly localised around the centre. It indicates, therefore, that fracture initiation should be expected in that area. This prediction is in agreement with experimental observations by Hancock and Mackenzie (1976) and Cescotto and Zhu (1995) which show that for certain notched specimen configurations, fracturing initiates at the centre of the specimen and propagates radially towards the notch. The reason for faster damaging at the centre lies in the fact that damage growth in ductile metals is strongly dependent on the stress triaxiality ratio

$$p/q,$$

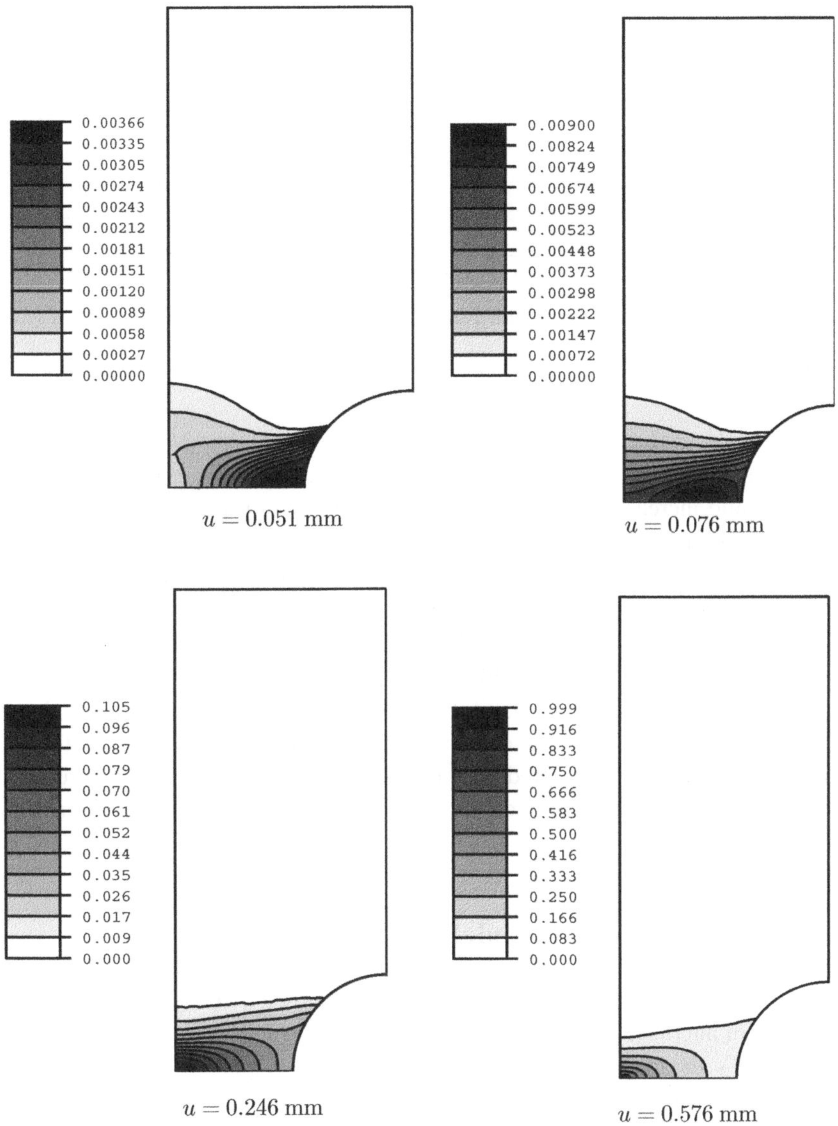

Figure 12.7. Cylindrical notched bar. Damage contour plots.

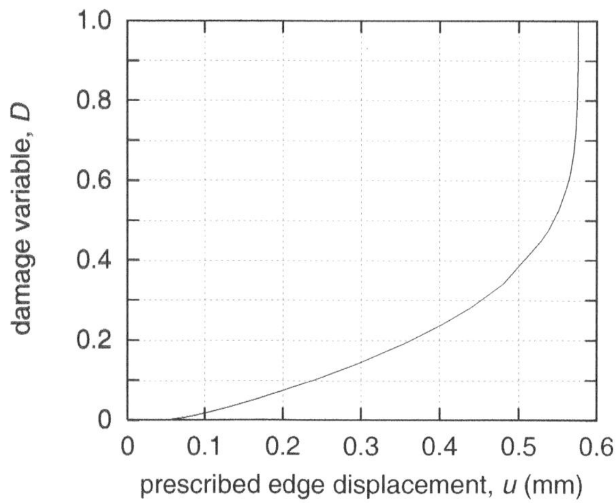

Figure 12.8. Cylindrical notched bar. Damage variable evolution at the centre of the specimen.

which is highest at the centre of the specimen. A marked decrease in ductility occurs as the triaxiality ratio increases. This phenomenon is captured by Lemaitre's ductile damage model. The evolution of the damage variable at the centre of the specimen is depicted in Figure 12.8 where the value of D computed at that point is plotted against the prescribed edge deflection. The corresponding reaction forces on the constrained edge of the specimen are shown in Figure 12.9. The diamond marks along the reaction–deflection curve correspond to the converged equilibrium states obtained at the end of every other step.

It is important to emphasise here that if the purely elastoplastic von Mises model (without damage) is used and a critical value of *accumulated plastic strain* is taken as the fracture criterion, crack initiation will be predicted at the notch root. It is, therefore, crucial that a damage model taking into consideration the effect of stress triaxiality be used in analyses of this type.

12.5. Gurson's void growth model

The constitutive model addressed in this section was proposed by Gurson (1977) to describe the mechanism of internal damaging in the form of void growth in porous metals. The starting point of Gurson's theory is the microscopic idealisation of porous metals as aggregates containing voids of simple geometric shapes embedded in a metallic matrix whose behaviour is governed by a rigid-plastic von Mises constitutive law. Approximate functional forms for the corresponding macroscopic yield functions are derived based on the analysis of single void cells and use of the upper bound plasticity theorem. In contrast to Lemaitre's damage model, the evolution of the damage variable of Gurson's model is *not* directly associated with a dissipative mechanism. The damage variable D in this case is the *void volume fraction*, i.e. the local fraction of volume occupied by voids and its evolution law follows as a direct consequence of mass conservation.

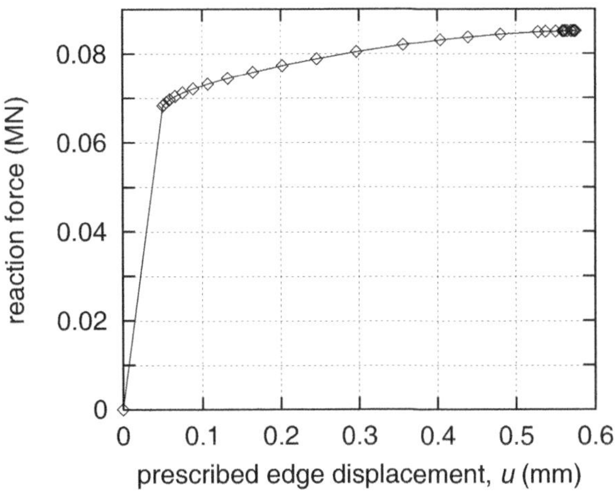

Figure 12.9. Cylindrical notched bar. Reaction–deflection diagram.

12.5.1. THE MODEL

The free energy potential

In the original version of Gurson's ductile damage model (Gurson, 1977), the matrix material was assumed to be incompressible rigid-perfectly plastic and the resulting macroscopic model was compressible rigid-plastic with hardening and softening associated, respectively, with healing and growth of voids. Here, elasticity as well as the possibility of additional isotropic hardening/softening due to straining of the matrix material are introduced and the free energy potential is assumed to be given by

$$\psi = \psi(\varepsilon^e, R) = \psi^e(\varepsilon^e) + \psi^p(R). \tag{12.57}$$

The elastic contribution ψ^e is taken as the standard quadratic strain-energy function

$$\bar{\rho}\, \psi^e(\varepsilon^e) = \tfrac{1}{2}\, \varepsilon^e : \mathsf{D}^e : \varepsilon^e, \tag{12.58}$$

giving the standard linear elastic relation

$$\sigma = \bar{\rho}\, \frac{\partial \psi}{\partial \varepsilon^e} = \mathsf{D}^e : \varepsilon^e. \tag{12.59}$$

As in Lemaitre's model, the isotropic hardening contribution is left as an arbitrary function of a single argument, so that the thermodynamic force κ associated with R is given by

$$\kappa = \bar{\rho}\, \frac{\partial \psi}{\partial R} = \bar{\rho}\, \frac{\partial \psi^p}{\partial R} = \kappa(R). \tag{12.60}$$

Remark 12.6. Note that, in contrast to Lemaitre's ductile damage model, the effect of internal damage on the elastic behaviour of the material is ignored in the present model; that is, the elasticity tensor is not a function of the damage variable.

The yield function and the damage variable

The *pressure-sensitive* yield function derived by Gurson (1977) for a void-matrix aggregate is expressed by

$$\Phi(\boldsymbol{\sigma}, \kappa, D) \equiv J_2(\boldsymbol{s}) - \frac{1}{3}\left\{1 + D^2 - 2\,D\cosh\left[\frac{3\,p}{2\,(\kappa + \sigma_{y_0})}\right]\right\}(\kappa + \sigma_{y_0})^2, \tag{12.61}$$

where σ_{y_0} and $\kappa + \sigma_{y_0}$ are, respectively, the initial and current uniaxial yield stress of the *matrix* material, p is the hydrostatic pressure, and D is the *damage variable*. The damage variable in Gurson's theory is the *void volume fraction*, i.e. the void volume per unit aggregate volume.

Remark 12.7. As for Lemaitre's model, the damage variable in the present case ranges between 0 and 1, with $D = 0$ corresponding to the virgin (undamaged) material and $D = 1$ to the fully damaged state with complete loss of load-carrying capacity. Note that damage growth induces softening, i.e. shrinkage of the yield surface defined by

$$\Phi = 0.$$

For $D = 0$ Gurson's yield surface reduces to that of the (pressure-insensitive) von Mises model. In the presence of damage, i.e. for $D \neq 0$, the yield surface becomes pressure sensitive with its size reducing to zero for $D = 1$.

Plastic flow and hardening evolution

In the present theory, the yield function is taken as the flow potential, $\Psi \equiv \Phi$. The resulting plastic rule then reads

$$\dot{\boldsymbol{\varepsilon}}^p = \dot{\gamma}\,\frac{\partial \Phi}{\partial \boldsymbol{\sigma}} = \dot{\gamma}\left\{\boldsymbol{s} + \frac{1}{3}\,D\,(\kappa + \sigma_{y_0})\sinh\left[\frac{3\,p}{2\,(\kappa + \sigma_{y_0})}\right]\boldsymbol{I}\right\}. \tag{12.62}$$

The evolution law for the hardening variable follows as

$$\begin{aligned}\dot{R} &= -\dot{\gamma}\,\frac{\partial \Phi}{\partial \kappa}\\ &= \dot{\gamma}\,\frac{\frac{2}{3}\{1 + D^2 - 2\,D\cosh\left[\frac{3\,p}{2\,(\kappa+\sigma_{y_0})}\right](\kappa + \sigma_{y_0})\} + p\,D\sinh\left[\frac{3\,p}{2\,(\kappa+\sigma_{y_0})}\right]}{1 - D}.\end{aligned} \tag{12.63}$$

Remark 12.8. Recall that the *volumetric* plastic strain rate is given by

$$\dot{\varepsilon}^p_{\mathrm{v}} = \mathrm{tr}\,\dot{\boldsymbol{\varepsilon}}^p = \dot{\boldsymbol{\varepsilon}}^p : \boldsymbol{I}.$$

From the flow rule defined by (12.62) it follows that the volumetric plastic strain rate in Gurson's damage model is given by

$$\dot{\varepsilon}^p_{\mathrm{v}} = \dot{\gamma}\,D\,(\kappa + \sigma_{y_0})\sinh\left[\frac{3\,p}{2\,(\kappa + \sigma_{y_0})}\right]. \tag{12.64}$$

This expression implies that Gurson's model predicts *plastic compressibility* in the presence of voids with dilatant (compressive) plastic flow under tensile (compressive) hydrostatic pressures. This phenomenon cannot be captured by Lemaitre's theory in which damage evolution can cause softening but does not change the original (pressure insensitive) von Mises shape of the yield surface.

Conservation of mass. Damage evolution

As the present material is assumed to be an aggregate of voids embedded in a solid matrix, its density can be expressed as

$$\rho = \rho_m \, v_m, \tag{12.65}$$

where ρ_m and v_m are, respectively, the density of the matrix material and the volume of matrix material per unit aggregate volume. With the variable D defined as the voids fraction, i.e. the volume of voids per unit aggregate volume, we have the identity

$$D + v_m = 1, \tag{12.66}$$

so that the density of the aggregate can be written as

$$\rho = \rho_m \, (1 - D). \tag{12.67}$$

Time differentiation of this formula gives

$$(1 - D) \, \dot{\rho}_m = \dot{\rho} + \rho_m \, \dot{D}. \tag{12.68}$$

The matrix material is assumed to be plastically incompressible (von Mises type). In addition, it is assumed that elastic volumetric strains are negligible. Under such hypotheses, mass conservation requires that

$$\dot{\rho}_m = 0. \tag{12.69}$$

Note that in the original derivation of Gurson's model, the matrix material is assumed to be rigid-plastic and the above identity holds exactly. Substitution of this equation into (12.68) gives

$$\dot{D} = -\frac{\dot{\rho}}{\rho_m} = -\frac{\dot{\rho}}{\rho}(1 - D). \tag{12.70}$$

Now let us recall that the axiom of mass conservation requires that

$$\dot{\varepsilon}_{\rm v} = -\frac{\dot{\rho}}{\rho}. \tag{12.71}$$

Again, by disregarding the elastic volumetric strains, we assume

$$\dot{\varepsilon}_{\rm v} = \dot{\varepsilon}^p_{\rm v}, \tag{12.72}$$

and, by combining the last three equations, we obtain the following evolution law for D:

$$\dot{D} = (1 - D) \, \dot{\varepsilon}^p_{\rm v}. \tag{12.73}$$

Finally, in view of the constitutive equation (12.64) for the volumetric plastic flow, the evolution law for the damage variable is obtained as

$$\dot{D} = \dot{\gamma} \, (D - D^2) \, (\kappa + \sigma_{y0}) \sinh\left[\frac{3\,p}{2\,(\kappa + \sigma_{y0})}\right]. \tag{12.74}$$

The overall model is summarised in Box 12.5.

Box 12.5. Gurson's void growth model.

(i) Elastoplastic split of the strain tensor

$$\boldsymbol{\varepsilon} = \boldsymbol{\varepsilon}^e + \boldsymbol{\varepsilon}^p$$

(ii) Elastic law (uncoupled from damage)

$$\boldsymbol{\sigma} = \mathsf{D}^e : \boldsymbol{\varepsilon}^e$$

(iii) Gurson's yield function

$$\Phi = J_2(\boldsymbol{s}) - \frac{1}{3}\left\{1 + D^2 - 2\,D\cosh\left[\frac{3\,p}{2\,\sigma_y}\right]\right\}\sigma_y^2$$

where $\sigma_y = \sigma_y(R)$

(iv) Plastic flow and evolution equations for R and D

$$\dot{\boldsymbol{\varepsilon}}^p = \dot{\gamma}\left\{\boldsymbol{s} + \frac{1}{3}\,D\,\sigma_y\sinh\left[\frac{3\,p}{2\,\sigma_y}\right]\boldsymbol{I}\right\}$$

$$\dot{R} = \dot{\gamma}\,\frac{\frac{2}{3}\{1 + D^2 - 2\,D\cosh[\frac{3\,p}{2\,\sigma_y}]\sigma_y\} + p\,D\sinh[\frac{3\,p}{2\,\sigma_y}]}{1 - D}$$

$$\dot{D} = \dot{\gamma}\,(D - D^2)\,\sigma_y\sinh\left[\frac{3\,p}{2\,\sigma_y}\right]$$

(v) Loading/unloading criterion

$$\Phi \le 0, \quad \dot{\gamma} \ge 0, \quad \Phi\,\dot{\gamma} = 0$$

Remark 12.9. Note from (12.74) that if the initial voids ratio is zero ($D = 0$), no damage evolution will be predicted by the standard Gurson model, whatever strain history the material might be subjected to. A non-zero initial voids ratio is required to produce damage growth. To improve the model, a mechanism of *damage nucleation*, whereby voids are nucleated depending on the strain history, should be incorporated. In this context, some nucleation laws have been proposed by Tvergaard (1982a) and Tvergaard and Needleman (1984) whereby voids may nucleate in the absence of damage. Further acceleration in voids growth, intended to produce a more realistic response, can be relatively easily incorporated by introducing a modification into Gurson's macroscopic yield criterion as suggested by Tvergaard (1981, 1982a,b).

Remark 12.10. Again from (12.74), it follows that the voids ratio growth rate is positive (negative) under tensile (compressive) hydrostatic stress. Positive and negative voids ratio growth rate correspond, respectively, to *damaging* and *healing* of the material. Under cyclic loading (in the plastic range), Gurson's model will not predict the experimentally observed continuous increase in damage, as the increase in voids ratio predicted in the tensile part of the cycle is reversed by the healing process predicted over the compressive part of the cycle (de Souza Neto *et al.*, 1998). This feature precludes the use of Gurson's model for damage prediction in situations where cyclic loads are important. On the other hand, the healing behaviour can be useful, for instance, in the prediction of voids distribution in powder compaction processes (Gethin *et al.*, 1998; Ransing *et al.*, 1998). In this case, the initial voids ratio is relatively high and is gradually reduced by application of hydrostatic pressure under highly confining boundary conditions. This is in sharp contrast to Lemaitre's model where only damaging (no healing) can be predicted.

12.5.2. INTEGRATION ALGORITHM

With the particular definition (12.62) for the plastic flow rule, if plastic yielding occurs within the time interval of interest, the general implicit elastoplastic integration algorithm discussed in Chapter 7 leads to the following update formulae:

$$\begin{aligned} \varepsilon^e_{\mathrm{d}\,n+1} &= \varepsilon^{e\ \mathrm{trial}}_{\mathrm{d}\,n+1} - \Delta\gamma\, \boldsymbol{s}_{\mathrm{d}\,n+1}, \\ \varepsilon^e_{\mathrm{v}\,n+1} &= \varepsilon^{e\ \mathrm{trial}}_{\mathrm{v}\,n+1} - \Delta\gamma \left\{ D_{n+1}\,[\sigma_y(R_{n+1})] \sinh\left[\frac{3\,p_{n+1}}{2\,\sigma_y(R_{n+1})} \right] \right\}, \end{aligned} \tag{12.75}$$

where we have performed the deviatoric/volumetric decomposition of the elastic strain. Use of the elastic law in the above expression gives

$$\begin{aligned} \boldsymbol{s}_{n+1} &= \frac{2\,G}{1+\Delta\gamma\,2\,G}\, \varepsilon^{e\ \mathrm{trial}}_{\mathrm{d}\,n+1}, \\ p_{n+1} &= \kappa\, \varepsilon^{e\ \mathrm{trial}}_{\mathrm{v}\,n+1} - \Delta\gamma\,\kappa \left\{ D_{n+1}\,\sigma_y(R_{n+1}) \sinh\left[\frac{3\,p_{n+1}}{2\,\sigma_y(R_{n+1})} \right] \right\}. \end{aligned} \tag{12.76}$$

With introduction of the above update formula for the stress deviator into definition (12.61) of Gurson's model yield function, the following algorithmic counterpart of Φ is obtained:

$$\tilde{\Phi}(p_{n+1}, R_{n+1}, D_{n+1}, \Delta\gamma) = \left(\frac{2\,G}{1+2\,G\,\Delta\gamma} \right)^2 J_2(\varepsilon^{e\ \mathrm{trial}}_{\mathrm{d}\,n+1}) - \tfrac{1}{3}\, a\, [\sigma_y(R_{n+1})]^2, \tag{12.77}$$

with a defined as

$$a \equiv 1 + D^2_{n+1} - 2\,D_{n+1} \cosh\left[\frac{3\,p_{n+1}}{2\,\sigma_y(R_{n+1})} \right].$$

Thus, for the present material model, the plastic corrector stage comprises the requirement of plastic consistency by means of the algorithmic yield function above, the pressure update $(12.76)_2$ and the backward Euler discrete counterparts of the evolution equations (12.63) and (12.74). The resulting implicit elastic predictor/return-mapping algorithm is listed in Box 12.6. Note that, here, a set of *four* coupled nonlinear equations has to be solved in the plastic corrector phase for any stress state.

Remark 12.11. As in Lemaitre's model, convergence difficulties have been detected in the Newton–Raphson scheme adopted to solve the equations of the plastic corrector phase of Box 12.6. To expand the convergence bowl of the Newton algorithm, the authors have tested a line-search procedure (as suggested by Steinmann *et al.* 1994 for a variant of the present model in the finite strain context). Some improvement was observed. Nevertheless, in some circumstances, convergence can only be attained for relatively small strain increments.

Box 12.6. State-update procedure for Gurson's void growth model.

(i) Elastic predictor

– Evaluate trial elastic stress

$$\boldsymbol{\sigma}_{n+1}^{\text{trial}} := \mathbf{D}^e : \boldsymbol{\varepsilon}_{n+1}^{e\ \text{trial}}$$

– Check plastic consistency

$$\Phi^{\text{trial}} := J_2(s_{n+1}^{\text{trial}}) - \frac{1}{3}\left\{1 + D_n^2 - 2D_n \cosh\left[\frac{3\,p_{n+1}^{\text{trial}}}{2\,\sigma_y(R_n)}\right]\right\}[\sigma_y(R_n)]^2$$

IF $\Phi^{\text{trial}} \leq 0$ THEN

Set $(\cdot)_{n+1} = (\cdot)_{n+1}^{\text{trial}}$ and EXIT

ELSE go to (ii)

(ii) Plastic corrector (solve the system for the unknowns p_{n+1}, R_{n+1}, D_{n+1} and $\Delta\gamma$)

$$\left\{\begin{array}{l} \left(\dfrac{2\,G}{1+2\,G\,\Delta\gamma}\right)^2 J_2(\varepsilon_{\mathrm{d}_{n+1}}^{e\ \text{trial}}) - \dfrac{1}{3}\,a\,[\sigma_y(R_{n+1})]^2 \\ p_{n+1} - \kappa\,\varepsilon_{\mathrm{v}_{n+1}}^{e\ \text{trial}} + \Delta\gamma\,\kappa\,b\,\sigma_y(R_{n+1}) \\ D_{n+1} - D_n - \Delta\gamma\,b\,(D_{n+1} - D_{n+1}^2)\,\sigma_y(R_{n+1}) \\ R_{n+1} - R_n - \dfrac{\Delta\gamma}{1-D_{n+1}}\left[\dfrac{2}{3}a\,\sigma_y(R_{n+1}) + b\,p_{n+1}\,D_{n+1}\right] \end{array}\right\} = \left\{\begin{array}{c}0\\0\\0\\0\end{array}\right\}$$

where

$$a = 1 + D_{n+1}^2 - 2D_{n+1}\cosh\left[\frac{3\,p_{n+1}}{2\,\sigma_y(R_{n+1})}\right], \quad b = \sinh\left[\frac{3\,p_{n+1}}{2\,\sigma_y(R_{n+1})}\right]$$

(iii) Update $\boldsymbol{\varepsilon}^e$ and $\boldsymbol{\sigma}$

$$\boldsymbol{\varepsilon}_{n+1}^e := \frac{1}{1+2G\,\Delta\gamma}\,\boldsymbol{\varepsilon}_{\mathrm{d}_{n+1}}^{e\ \text{trial}} + \frac{p_{n+1}}{\kappa}\,\boldsymbol{I}$$

$$\boldsymbol{\sigma}_{n+1} := \frac{2G}{1+2G\,\Delta\gamma}\,\boldsymbol{\varepsilon}_{\mathrm{d}_{n+1}}^{e\ \text{trial}} + p_{n+1}\,\boldsymbol{I}$$

(iv) RETURN

12.5.3. THE TANGENT OPERATOR

The elastic tangent for the present model is simply the standard linear elasticity tensor. In this case we have

$$\mathbf{D} = \mathbf{D}^e. \tag{12.78}$$

Under plastic loading, the elastoplastic operator is obtained by the standard linearisation procedure. The system of equations of the plastic corrector phase is differentiated at the converged state resulting in the identity

$$\begin{bmatrix} A_{1,\Delta\gamma} & A_{1,p} & A_{1,D} & A_{1,R} \\ A_{2,\Delta\gamma} & A_{2,p} & A_{2,D} & A_{2,R} \\ A_{3,\Delta\gamma} & A_{3,p} & A_{3,D} & A_{3,R} \\ A_{4,\Delta\gamma} & A_{4,p} & A_{4,D} & A_{4,R} \end{bmatrix} \begin{bmatrix} \mathrm{d}\Delta\gamma \\ \mathrm{d}p_{n+1} \\ \mathrm{d}D_{n+1} \\ \mathrm{d}R_{n+1} \end{bmatrix} = \begin{bmatrix} -A_{1,\varepsilon^{e\,\mathrm{trial}}_{\mathrm{d}_{n+1}}} : \mathrm{d}\varepsilon^{e\,\mathrm{trial}}_{\mathrm{d}_{n+1}} \\ -A_{2,\varepsilon^{e\,\mathrm{trial}}_{\mathrm{v}_{n+1}}}\, \mathrm{d}\varepsilon^{e\,\mathrm{trial}}_{\mathrm{v}_{n+1}} \\ 0 \\ 0 \end{bmatrix}, \tag{12.79}$$

where $A_{1,\Delta\gamma}$, $A_{1,p}, \ldots$ denote the derivatives of the plastic corrector system components. Inversion of the above expression then leads to

$$\begin{bmatrix} \mathrm{d}\Delta\gamma \\ \mathrm{d}p_{n+1} \\ \mathrm{d}D_{n+1} \\ \mathrm{d}R_{n+1} \end{bmatrix} = \begin{bmatrix} C_{11} & C_{12} & C_{13} & C_{14} \\ C_{21} & C_{22} & C_{23} & C_{24} \\ C_{31} & C_{32} & C_{33} & C_{34} \\ C_{41} & C_{42} & C_{43} & C_{44} \end{bmatrix} \begin{bmatrix} -A_{1,\varepsilon^{e\,\mathrm{trial}}_{\mathrm{d}_{n+1}}} : \mathrm{d}\varepsilon^{e\,\mathrm{trial}}_{\mathrm{d}_{n+1}} \\ -A_{2,\varepsilon^{e\,\mathrm{trial}}_{\mathrm{v}_{n+1}}}\, \mathrm{d}\varepsilon^{e\,\mathrm{trial}}_{\mathrm{v}_{n+1}} \\ 0 \\ 0 \end{bmatrix}, \tag{12.80}$$

which provides the tangent relations between the system variables ($\Delta\gamma$, p, D and R) and the system input $\varepsilon^{e\,\mathrm{trial}}_{n+1}$. Note that, as the stress tensor is one of the system variables in Lemaitre's original model, the tangent operator D^{ep} in that case is obtained directly from the inversion of the system derivative. Here, the consistent tangent operator can be obtained by differentiating the stress update formula of item (iii) of Box 12.6 which gives

$$\mathrm{d}\boldsymbol{\sigma}_{n+1} = \frac{2G}{1+2G\,\Delta\gamma}\,\mathrm{d}\varepsilon^{e\,\mathrm{trial}}_{\mathrm{d}_{n+1}} - \left(\frac{2G}{1+2G\,\Delta\gamma}\right)^2 \mathrm{d}\Delta\gamma\,\varepsilon^{e\,\mathrm{trial}}_{\mathrm{d}_{n+1}} + \mathrm{d}p_{n+1}\,\boldsymbol{I}. \tag{12.81}$$

Then, substitution of $\mathrm{d}\Delta\gamma$ and $\mathrm{d}p_{n+1}$ by the relations given in (12.80) and use of the identities

$$A_{1,\varepsilon^{e\,\mathrm{trial}}_{\mathrm{d}_{n+1}}} = \left(\frac{2\,G}{1+2\,G\,\Delta\gamma}\right)^2 \varepsilon^{e\,\mathrm{trial}}_{\mathrm{d}_{n+1}}; \quad A_{2,\varepsilon^{e\,\mathrm{trial}}_{\mathrm{v}_{n+1}}} = -\kappa,$$

result, after some straightforward manipulations, in the following expression for the elastoplastic consistent tangent operator:

$$\begin{aligned} \mathsf{D}^{ep} &= \frac{\mathrm{d}\boldsymbol{\sigma}_{n+1}}{\mathrm{d}\varepsilon^{e\,\mathrm{trial}}_{n+1}} \\ &= g\,\mathsf{I}_{\mathrm{d}} + g^2\,\varepsilon^{e\,\mathrm{trial}}_{\mathrm{d}_{n+1}} \otimes [C_{11}\,g^2\,\varepsilon^{e\,\mathrm{trial}}_{\mathrm{d}_{n+1}} - C_{12}\,\kappa\,\boldsymbol{I}] \\ &\quad - \boldsymbol{I} \otimes [C_{21}\,g^2\,\varepsilon^{e\,\mathrm{trial}}_{\mathrm{d}_{n+1}} - C_{22}\,\kappa\,\boldsymbol{I}], \end{aligned} \tag{12.82}$$

where

$$g \equiv \frac{2G}{1+2G\,\Delta\gamma}.$$

Remark 12.12. It is important to note that, as in Lemaitre's model, the resulting elastoplastic tangent operator D^{ep} is generally *unsymmetric*.

12.6. Further issues in damage modelling

In spite of the progress achieved over the last two decades in modelling material damage, further refinements are undoubtedly required to produce models with greater predictive capability. Such a requirement becomes particularly apparent in situations involving complex strain paths. Even though many models – such as the ones presented in the preceding sections of this chapter – are able to predict damage growth with reasonable accuracy over simple strain paths, increasing deviations from experimental observations should be expected as strain paths become more complex. In fact, this is true not only for damage models but for inelastic models of continua in general and, at present, it can be said that constitutive refinement in inelasticity remains largely an open issue. To give the reader an idea of possible improvements to the damage theories described in the above, we include in this section two topics that have been receiving attention recently. Namely, we address briefly the issues of *crack closure effects* and *anisotropic damage*.

12.6.1. CRACK CLOSURE EFFECTS IN DAMAGED ELASTIC MATERIALS

In the models discussed in Sections 12.3 and 12.4 the damage internal variable, D, was defined as the degradation of the elastic modulus as a result of evolution of voids and micro-cracks (refer to Remark 12.1, from page 479). For those models, the elastic law was assumed to remain linear in the presence of damage with equal response in tension and compression. However, it is frequently observed in simple uniaxial tests that the elasticity modulus degradation due to the presence of voids and micro-cracks is much more noticeable in tension than in compression. This is due to the fact that cracks that open in tension resulting in loss of load-carrying area and stiffness may partially close and increase the load-bearing area and stiffness under compression. A possible alternative to account for the effects of partial crack closure in isotropically damaged elastic materials is the model described by Ladèveze (1983), Ladevèze and Lemaitre (1984) and Lemaitre (1996) (anisotropic extensions of this model are discussed by Desmorat 2000, and Lemaitre *et al.* 2000). The model, described in the following, is incorporated into program HYPLAS.

Uniaxial crack closure model

Let us start by considering the uniaxial stress state case. For the standard Lemaitre model discussed in previous sections, a material with damage state D, has an effective Young's modulus

$$E = (1 - D)\, E_0, \tag{12.83}$$

where E_0 is the Young's modulus of the virgin (undamaged) material. The uniaxial stress–strain constitutive equation for the damaged material is given by

$$\sigma = (1 - D)\, E_0\, \varepsilon \quad \text{or} \quad \varepsilon = \frac{\sigma}{(1 - D)\, E_0}. \tag{12.84}$$

The crucial point in the definition of the crack closure model is the assumption that the above relationship is valid only under *tensile* stresses ($\sigma \geq 0$). Under *compressive* stresses ($\sigma < 0$), the uniaxial stress–strain relation is assumed to take the form

$$\sigma = (1 - h\, D)\, E_0\, \varepsilon \quad \text{or} \quad \varepsilon = \frac{\sigma}{(1 - h\, D)\, E_0}, \tag{12.85}$$

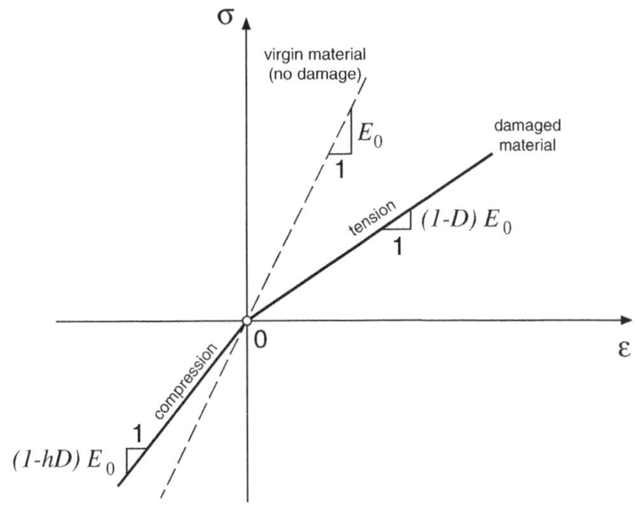

Figure 12.10. Uniaxial elastic model with damage and partial crack closure effect.

where h is an experimentally determined constant which satisfies

$$0 \leq h \leq 1. \tag{12.86}$$

This constant describes the effect of partial microcrack/void closure. A value $h \approx 0.2$ is typically observed in many experiments (Lemaitre, 1996). Note that for $h = 1$, the behaviour of the original damage model, without crack closure effects, is recovered, whereas the other extreme value, $h = 0$, represents full crack closure with $E = E_0$ under compression. Any other value of h describes a partial crack closure effect. The uniaxial stress–strain diagram of Figure 12.10 illustrates the behaviour of the model.

Tensile/compressive split of the uniaxial stress

The constitutive equation of the above uniaxial model can be more elegantly expressed by introducing the following *tensile/compressive split* of the uniaxial stress:

$$\sigma = \sigma_+ + \sigma_-, \tag{12.87}$$

where

$$\sigma_+ = \langle \sigma \rangle \quad \text{and} \quad \sigma_- = -\langle -\sigma \rangle, \tag{12.88}$$

are, respectively, the *tensile* and *compressive components* of σ and $\langle \; \rangle$ is the *Macauley bracket*, that is, for any scalar, a,

$$\langle a \rangle = \begin{cases} a & \text{if } a \geq 0 \\ 0 & \text{if } a < 0. \end{cases} \tag{12.89}$$

With the above notation, the uniaxial stress–strain relation for the damaged elastic material crack closure effect can be written simply as

$$\varepsilon = \frac{1}{E_0}\left(\frac{\sigma_+}{1-D} + \frac{\sigma_-}{1-h\,D}\right). \tag{12.90}$$

Tensile/compressive split of the stress tensor

As we have seen in the above, it is relatively easy to establish a piecewise linear damaged elastic model capable of accounting for crack closure effects in the uniaxial case. The extension of such a simple model to the general three-dimensional situation, however, is not trivial. The main problem here is how to distinguish between tensile and compressive stresses in a multiaxially stressed state. In the present model, such a distinction is made on the basis of a *tensile/compressive split* of the stress tensor. To introduce this concept, let us first recall (refer to Section 3.3.3, from page 62) that any stress tensor may be represented as a diagonal matrix with the diagonal terms – the principal stresses – representing normal stresses along three mutually perpendicular directions – the principal directions; that is, any stress tensor $\boldsymbol{\sigma}$ can be written as

$$\boldsymbol{\sigma} = \sum_{i=1}^{3} \sigma_i\, \boldsymbol{e}_i \otimes \boldsymbol{e}_i, \tag{12.91}$$

where σ_i are the principal stresses and $\{\boldsymbol{e}_1, \boldsymbol{e}_2, \boldsymbol{e}_3\}$ is an orthonormal basis of vectors along the principal directions. The matrix representation of $\boldsymbol{\sigma}$ in this basis, reads

$$[\boldsymbol{\sigma}] = \begin{bmatrix} \sigma_1 & 0 & 0 \\ 0 & \sigma_2 & 0 \\ 0 & 0 & \sigma_3 \end{bmatrix}. \tag{12.92}$$

The tensile/compressive split.

The tensile/compressive split of the stress tensor consists in splitting $\boldsymbol{\sigma}$ additively as

$$\boldsymbol{\sigma} = \boldsymbol{\sigma}_+ + \boldsymbol{\sigma}_-, \tag{12.93}$$

where $\boldsymbol{\sigma}_+$ and $\boldsymbol{\sigma}_-$ are, respectively, the *tensile* and *compressive components* of $\boldsymbol{\sigma}$ defined as

$$\boldsymbol{\sigma}_+ = \sum_{i=1}^{3} \langle \sigma_i \rangle\, \boldsymbol{e}_i \otimes \boldsymbol{e}_i \tag{12.94}$$

and

$$\boldsymbol{\sigma}_- = -\sum_{i=1}^{3} \langle -\sigma_i \rangle\, \boldsymbol{e}_i \otimes \boldsymbol{e}_i. \tag{12.95}$$

The matrix representation of $\boldsymbol{\sigma}_+$ and $\boldsymbol{\sigma}_-$ in principal stress basis reads

$$[\boldsymbol{\sigma}_+] \equiv \begin{bmatrix} \langle \sigma_1 \rangle & 0 & 0 \\ 0 & \langle \sigma_2 \rangle & 0 \\ 0 & 0 & \langle \sigma_3 \rangle \end{bmatrix} \tag{12.96}$$

and

$$[\boldsymbol{\sigma}_-] \equiv -\begin{bmatrix} \langle -\sigma_1 \rangle & 0 & 0 \\ 0 & \langle -\sigma_2 \rangle & 0 \\ 0 & 0 & \langle -\sigma_3 \rangle \end{bmatrix}. \tag{12.97}$$

Note that, under uniaxial stresses, i.e. when only one of the principal stresses is non-zero, the above tensile/compressive split reduces to that of expression (12.87).

The three-dimensional damaged elasticity law

The last step to complete the definition of the multidimensional damaged elastic model with crack closure effect is to generalise the uniaxial stress–strain law (12.90) to three dimensions. To do this, let us first recall the standard three-dimensional linear elastic stress–strain law (12.11) (page 479) of Lemaitre's damage model. Its inverse relation reads

$$\begin{aligned} \boldsymbol{\varepsilon} &= \frac{1}{1-D}\,\mathsf{D}^{-1} : \boldsymbol{\sigma} \\ &= \frac{1}{1-D}\left[\frac{1+\nu}{2E_0}\,\boldsymbol{\sigma} - \frac{\nu}{2E_0}\,(\operatorname{tr}\boldsymbol{\sigma})\,\boldsymbol{I}\right]. \end{aligned} \tag{12.98}$$

The three-dimensional generalisation of (12.90) is obtained by modifying the above rule with the inclusion of the tensile/compressive split of the stress tensor as follows

$$\boldsymbol{\varepsilon} = \frac{1+\nu}{2E_0}\left(\frac{\boldsymbol{\sigma}_+}{1-D} + \frac{\boldsymbol{\sigma}_-}{1-hD}\right) - \frac{\nu}{2E_0}\left(\frac{\langle \operatorname{tr}\boldsymbol{\sigma}\rangle}{1-D} - \frac{\langle -\operatorname{tr}\boldsymbol{\sigma}\rangle}{1-hD}\right)\boldsymbol{I}. \tag{12.99}$$

Under uniaxial stress, the above damaged elasticity law with crack closure effect reduces to that of (12.90). It is important to note that the stress–strain relation (12.99) remains isotropic ($\boldsymbol{\varepsilon}$ and $\boldsymbol{\sigma}$ share the same principal directions) but, differently from (12.98), is no longer linear. More precisely, the stress–strain relation is piecewise linear with derivative discontinuities occurring when any of the principal stresses and/or the hydrostatic pressure vanishes.

Computational implementation

Now, we turn our attention to the computational implementation of the above model within an implicit finite element environment. As usual, the basic ingredients of the computer implementation are: (a) the stress updating procedure which, given a strain tensor, should compute the stress tensor that satisfies the constitutive law (12.99); and (b) the computation of the tangent operator associated with (12.99). These will be described in the following.

At the outset, it should be noted that the stress tensor in (12.99) cannot be written explicitly as a function of the strain tensor. Thus, given $\boldsymbol{\varepsilon}$, the corresponding stress tensor will be obtained as the solution of the following nonlinear tensorial equation for $\boldsymbol{\sigma}$:

$$\boldsymbol{R}(\boldsymbol{\sigma}) \equiv \boldsymbol{\varepsilon} - \frac{1+\nu}{2E_0}\left(\frac{\boldsymbol{\sigma}_+}{1-D} + \frac{\boldsymbol{\sigma}_-}{1-hD}\right) - \frac{\nu}{2E_0}\left(\frac{\langle \operatorname{tr}\boldsymbol{\sigma}\rangle}{1-D} - \frac{\langle -\operatorname{tr}\boldsymbol{\sigma}\rangle}{1-hD}\right) = \boldsymbol{0}. \tag{12.100}$$

This solution can be obtained by means of the Newton–Raphson algorithm as described in what follows. Before proceeding, however, we note that as $\boldsymbol{\sigma}_+$ and $\boldsymbol{\sigma}_-$ are defined on the basis of principal stresses, it is convenient to rewrite (12.100) in terms of principal stresses and strains. The approach here is analogous to that adopted in the numerical treatment of the Tresca and Mohr–Coulomb elastoplastic models discussed in Chapter 8 and, due to the isotropy of the stress–strain relation, incurs no loss of generality.

Array/matrix definitions

For convenience, we define the following arrays of principal stresses and strains

$$\boldsymbol{\sigma}^* \equiv [\,\sigma_1\;\sigma_2\;\sigma_3\,]^T, \qquad \boldsymbol{\varepsilon}^* \equiv [\,\varepsilon_1\;\varepsilon_2\;\varepsilon_3\,]^T, \tag{12.101}$$

and the tensile/compressive split of the principal stresses

$$\boldsymbol{\sigma}^*_+ \equiv \begin{bmatrix} \langle\sigma_1\rangle \\ \langle\sigma_2\rangle \\ \langle\sigma_3\rangle \end{bmatrix}, \qquad \boldsymbol{\sigma}^*_- \equiv -\begin{bmatrix} \langle-\sigma_1\rangle \\ \langle-\sigma_2\rangle \\ \langle-\sigma_3\rangle \end{bmatrix}. \tag{12.102}$$

Further, we define the operators

$$\mathbf{P}^D_+ \equiv \frac{1}{1-D}\begin{bmatrix} \hat{H}(\sigma_1) & 0 & 0 \\ 0 & \hat{H}(\sigma_2) & 0 \\ 0 & 0 & \hat{H}(\sigma_3) \end{bmatrix}, \tag{12.103}$$

and

$$\mathbf{P}^D_- \equiv \frac{1}{1-hD}\begin{bmatrix} \bar{H}(\sigma_1) & 0 & 0 \\ 0 & \bar{H}(\sigma_2) & 0 \\ 0 & 0 & \bar{H}(\sigma_3) \end{bmatrix} \tag{12.104}$$

where $\hat{H}$ is the Heaviside step function given in (7.94), page 221, and $\bar{H}$ is defined such that, for any scalar a,

$$\bar{H}(a) \equiv \begin{cases} 0 & \text{if } a > 0 \\ 1 & \text{if } a \leq 0. \end{cases} \tag{12.105}$$

We also define the following:

$$\mathbf{I}^D_+ \equiv \begin{cases} \dfrac{1}{1-D}\,\mathbf{I} & \text{if } \operatorname{tr}\boldsymbol{\sigma} > 0 \\ \mathbf{0} & \text{if } \operatorname{tr}\boldsymbol{\sigma} \leq 0 \end{cases} \tag{12.106}$$

and

$$\mathbf{I}^D_- \equiv \begin{cases} \mathbf{0} & \text{if } \operatorname{tr}\boldsymbol{\sigma} > 0 \\ \dfrac{1}{1-hD}\,\mathbf{I} & \text{if } \operatorname{tr}\boldsymbol{\sigma} \leq 0, \end{cases} \tag{12.107}$$

where we have used the definitions

$$\mathbf{I} \equiv [\,1\,1\,1\,]^T \quad \mathbf{0} \equiv [\,0\,0\,0\,]^T. \tag{12.108}$$

Note that

$$\operatorname{tr} \boldsymbol{\sigma} = \mathbf{I}^T \boldsymbol{\sigma}^*. \tag{12.109}$$

The principal stress/strain-based constitutive equation

With the above notation, the three-dimensional damaged elasticity law (12.99) can be reformulated in terms of principal stresses and strains as

$$\boldsymbol{\varepsilon}^* = [\mathbf{D}^*]^{-1}\, \boldsymbol{\sigma}^*, \tag{12.110}$$

or, equivalently,

$$\boldsymbol{\sigma}^* = \mathbf{D}^*\, \boldsymbol{\varepsilon}^*, \tag{12.111}$$

where $\mathbf{D}^*$ – a function of the principal stresses – is defined as

$$\mathbf{D}^* \equiv \left\{ \frac{1+\nu}{2E_0} [\, \mathbf{P}_+^D + \mathbf{P}_-^D\,] - \frac{\nu}{E} [(\mathbf{I}_+^D + \mathbf{I}_-^D)\, \mathbf{I}^T] \right\}^{-1}. \tag{12.112}$$

At points where the stress–strain relation is differentiable, $\mathbf{D}^*$ is the matrix of principal stress derivatives

$$\mathbf{D}^* = \frac{\mathrm{d}\boldsymbol{\sigma}^*}{\mathrm{d}\boldsymbol{\varepsilon}^*}, \tag{12.113}$$

with elements

$$\mathrm{D}^*_{ij} = \frac{\mathrm{d}\sigma_i}{\mathrm{d}\varepsilon_j}. \tag{12.114}$$

The loci of the principal stress space where the tangent principal stress–principal strain relation is discontinuous are

- the *deviatoric plane*, where $\sigma_1 + \sigma_2 + \sigma_3 = 0$;
- the three planes defined respectively by $\sigma_1 = 0$, $\sigma_2 = 0$ and $\sigma_3 = 0$.

Stress updating with the Newton–Raphson algorithm

The principal stress/strain-based analogy of nonlinear equation (12.100), which will be solved for $\boldsymbol{\sigma}^*$ in our stress-updating procedure, is given by

$$\mathbf{R}(\boldsymbol{\sigma}^*) \equiv \boldsymbol{\varepsilon}^* - [\mathbf{D}^*(\boldsymbol{\sigma}^*)]^{-1}\, \boldsymbol{\sigma}^* = \mathbf{0}. \tag{12.115}$$

The Newton update formula for the typical iteration k, reads

$$\boldsymbol{\sigma}^*_{(k)} := \boldsymbol{\sigma}^*_{(k-1)} + \mathbf{D}^*(\boldsymbol{\sigma}^*_{(k-1)})\, \mathbf{R}(\boldsymbol{\sigma}^*_{(k-1)}). \tag{12.116}$$

The stress updating procedure is summarised below in Box 12.7 in pseudo-code format. The scheme is coded in subroutine `SUDMEL` of program `HYPLAS` (not listed here) to which the reader is referred for further details. At points where the derivative of the principal stress–principal strain relation is discontinuous, the tangent on one side of the discontinuity is used in the Newton–Raphson iteration (see definition of $\mathbf{P}_+^D$, $\mathbf{P}_-^D$, $\mathbf{I}_+^D$ and $\mathbf{I}_-^D$).

Box 12.7. Stress-updating algorithm for damaged elastic material model with crack closure effects.

HYPLAS procedure:	SUDMEL

(i) Given ε, compute initial guess for stress, $\boldsymbol{\sigma}_{(0)}$ (use damaged elastic constitutive equation without crack closure effect)

$$\boldsymbol{\sigma}_{(0)} := (1 - D)\, \mathbf{D} : \boldsymbol{\varepsilon}$$

(ii) Spectral decomposition of ε (use routine SPDEC2) and $\boldsymbol{\sigma}_{(0)}$.
Compute:

$$\boldsymbol{\varepsilon}^*, \quad \boldsymbol{\sigma}^*_{(0)} \quad \text{and} \quad \boldsymbol{e}_i \otimes \boldsymbol{e}_i \quad (i = 1, 2, 3)$$

(iii) Solve the algebraic system

$$\boldsymbol{\varepsilon}^* - [\mathbf{D}(\boldsymbol{\sigma}^*)]^{-1}\, \boldsymbol{\sigma}^* = \mathbf{0}$$

for $\boldsymbol{\sigma}^* = [\sigma_1\, \sigma_2\, \sigma_3]^T$ using the Newton–Raphson algorithm. The Newton update formula is given by (12.116)

(iv) Assemble updated stress tensor

$$\boldsymbol{\sigma} := \sum_{i=1}^{3} \sigma_i\, \boldsymbol{e}_i \otimes \boldsymbol{e}_i$$

(vi) EXIT

The consistent tangent operator

The computation of the consistent tangent operator,

$$\mathbf{D} \equiv \frac{\mathrm{d}\boldsymbol{\sigma}}{\mathrm{d}\boldsymbol{\varepsilon}}, \tag{12.117}$$

is carried out in the present case in the same fashion as the elastoplastic tangent operator for the Tresca and Mohr–Coulomb models thoroughly discussed in Chapter 8 (refer to Sections 8.1.4 and 8.2.4). As for the present case, the stress-updating procedures for those models is carried out in principal stress space. With the principal stress derivative (12.113) at hand, the consistent tangent operator (12.117) can be computed as the derivative of a general isotropic tensor-valued function of a single tensor as described in Sections A.3 and A.4 of Appendix A. The tangent computation for the present model is performed in subroutine CTDMEL of program HYPLAS (not listed here).

12.6.2. CRACK CLOSURE EFFECTS IN DAMAGE EVOLUTION

The above model describes the effect of crack closure in damaged elastic materials that remain elastic, without evolution of damage. Crack closure effects can also have a strong influence on damage evolution. It is a well-known fact that many materials have considerably higher strength in compression than in tension. The consideration of such an effect may be crucial under complex loading conditions and is relevant to many practical applications. An alternative to describe this phenomenon is provided by the damage evolution law proposed by Ladevèze and Lemaitre (1984) (described by Lemaitre 1996 in more detail). Their approach consists in modifying the damage energy release rate (12.15) of the original Lemaitre ductile damage model by including the tensile/compressive split of the stress tensor. Thus, the original expression (12.15) is replaced with

$$\begin{aligned} Y = & \frac{-1}{2E(1-D)^2}[(1+\nu)\,\boldsymbol{\sigma}_+ : \boldsymbol{\sigma}_+ - \nu\,\langle \operatorname{tr}\boldsymbol{\sigma}\rangle^2] \\ & - \frac{h}{2E(1-hD)^2}[(1+\nu)\,\boldsymbol{\sigma}_- : \boldsymbol{\sigma}_- - \nu\,\langle -\operatorname{tr}\boldsymbol{\sigma}\rangle^2], \end{aligned} \tag{12.118}$$

and the elastoplastic damage evolution equation keeps the same format as in the original model:

$$\dot{D} = \dot{\gamma}\,\frac{1}{1-D}\left(\frac{-Y}{r}\right)^s. \tag{12.119}$$

The complete fully coupled elastoplastic model for ductile damage is obtained by considering the above damage evolution equation together with the damaged isotropic elasticity law with crack closure effects (discussed in the previous section) and the standard plasticity equations of the original Lemaitre model.

Note on computer implementation

An extra degree of complexity will be introduced in the computer implementation if the fully coupled model is adopted. A possible alternative to avoid such a complexity and have a fast algorithm that estimates damage evolution with crack closure effects is to *decouple* the above damage evolution equation from the plasticity equations. In this case, the *decoupled* state update algorithm could follow the steps:

1. Given the incremental strain, $\Delta\boldsymbol{\varepsilon}$, compute the updated stress, $\boldsymbol{\sigma}_{n+1}$, and the incremental plastic multiplier, $\Delta\gamma$, by using the conventional von Mises model (possibly with kinematic hardening). The computer implementation of the von Mises model was described in Chapter 7.

2. With the updated stress and incremental multiplier at hand, compute the new damage state, D_{n+1}, '*a posteriori*' by solving (for D_{n+1}) the backward Euler discrete version of (12.119) which reads

$$D_{n+1} = D_n + \frac{\Delta\gamma}{1-D_{n+1}}\left(\frac{-Y_{n+1}}{r}\right)^s, \tag{12.120}$$

where Y_{n+1} is computed according to definition (12.118) as a function of the unknown updated damage variable, D_{n+1}, and the known updated stress tensor, $\boldsymbol{\sigma}_{n+1}$. Clearly, the damage variable does not affect the plasticity law in the above decoupled implementation. Damage growth does not cause plastic softening. Thus, if the decoupled approach is adopted, the hardening material parameters of the plasticity model have to be calibrated accordingly. For further details on implementation refer to Pires *et al.* (2004).

12.6.3. ANISOTROPIC DUCTILE DAMAGE

Another important issue in damage modelling is the anisotropy of micro-crack and void distribution. This can be of particular importance, for instance, in situations where two or more highly directional loads are applied sequentially. In such cases, each load will cause micro-cracks to grow in one preferential direction, affecting the material response to subsequent loads in different directions. Consideration of damage anisotropy can be relevant under such conditions and, even though the usual isotropy hypothesis may offer a good first approximation, it may lead to substantial errors in many practical applications.

Modelling anisotropic damage evolution is not an easy task. Even the experimental characterisation of anisotropically damaged states – an essential requirement for the determination of damage evolution laws – is complex. To date, many continuum theories have been proposed to describe anisotropic damage phenomena (Chaboche, 1984; Cordebois and Sidoroff, 1982; Krajčinović and Fonseka, 1981; Lemaitre, 1996; Lemaitre and Chaboche, 1990; Lemaitre *et al.*, 2000; Murakami and Ohno, 1981; Simo and Ju, 1987). Nevertheless, at present, this issue remains largely open and at the leading edge of material modelling science.

To give the reader an idea of more advanced topics in damage mechanics, we find it convenient to close this chapter with an outline of the anisotropic ductile damage elastoplastic model proposed by Lemaitre and co-workers (2000). This theory is an extension of the isotropic ductile damage model discussed in Sections 12.3 and 12.4.

The damage tensor

The scalar damage measure of the original isotropic model is replaced here with *two* measures of damage:

1. a *second-order symmetric tensor*, $\boldsymbol{D}$, which is associated with the variation of the deviatoric elastic constitutive behaviour due to damage;
2. a *scalar*, d_H, defined as
$$d_H = \eta\, D_H, \tag{12.121}$$
where η is a material parameter associated with the variation of Poisson's ratio due to damage, and
$$D_H \equiv \tfrac{1}{3}\,\mathrm{tr}\,\boldsymbol{D}. \tag{12.122}$$

The second-order damage tensor, $\boldsymbol{D}$, is the actual damage internal variable.

The orthotropic damaged elasticity law

Following the principle of strain equivalence, the damaged elasticity law for this model is obtained by replacing the stress tensor with the effective stress in the standard linear elastic law. The effective stress is here defined as

$$\boldsymbol{\sigma}_{\rm eff} = \boldsymbol{s}_{\rm eff} + p_{\rm eff}\, \boldsymbol{I}, \tag{12.123}$$

where $\boldsymbol{s}_{\rm eff}$ and $p_{\rm eff}$ are, respectively, the effective stress deviator and effective hydrostatic pressure defined as

$$\boldsymbol{s}_{\rm eff} \equiv \mathrm{dev}[\boldsymbol{H}\, \boldsymbol{s}\, \boldsymbol{H}], \quad p_{\rm eff} \equiv \frac{p}{1-d_H}\, \boldsymbol{I}. \tag{12.124}$$

In the above expression, $\boldsymbol{H}$ is the second-order tensor defined as

$$\boldsymbol{H} \equiv (\boldsymbol{I} - \boldsymbol{D})^{-\frac{1}{2}}. \tag{12.125}$$

The damaged elastic law is then given by

$$\boldsymbol{\sigma}_{\text{eff}} = \mathsf{D}^e : \boldsymbol{\varepsilon}^e. \tag{12.126}$$

In the present theory, an isotropically damaged state is characterised by

$$\boldsymbol{D} = D\,\boldsymbol{I}, \tag{12.127}$$

and

$$\eta = 1, \tag{12.128}$$

which imply

$$d_H = D. \tag{12.129}$$

Under such states, straightforward algebra shows that (12.126) reduces to the damaged elastic law (12.11) of the original isotropic model. For a more general non-isotropic damage state $\boldsymbol{D}$, the linear elastic constitutive equation (12.126) is *orthotropic*. The directions of orthotropy are the principal directions of the damage tensor. In an orthonormal basis aligned with its principal directions the damage tensor is represented as

$$[\boldsymbol{D}] = \begin{bmatrix} D_1 & 0 & 0 \\ 0 & D_2 & 0 \\ 0 & 0 & D_3 \end{bmatrix}. \tag{12.130}$$

The eigenvalues values D_1, D_2 and D_3 are associated to the degradation of the elasticity modulus along the corresponding directions of orthotropy. Their experimental identification is discussed by Lemaitre *et al.* (2000).

Damage evolution

To complete the definition of the model, an evolution law for the damage tensor is required. In the present theory, the rate of damage tensor is assumed to follow the directions of plastic straining. The evolution law for $\boldsymbol{D}$ is defined by

$$\dot{\boldsymbol{D}} = \left(\frac{-\tilde{Y}}{r}\right)^s \dot{\boldsymbol{\varepsilon}}^p, \tag{12.131}$$

where

$$\tilde{Y} \equiv \frac{-q_{\text{eff}}^2}{2E}\left[\frac{2}{3}(1+\nu) + 3(1-2\nu)\left(\frac{p_{\text{eff}}}{q_{\text{eff}}}\right)^2\right], \tag{12.132}$$

with q_{eff} defined as

$$q_{\text{eff}} \equiv \sqrt{3\,J_2(\boldsymbol{s}_{\text{eff}})}. \tag{12.133}$$

The tensor $\dot{\bar{\varepsilon}}^p$ is the *absolute plastic strain rate* defined as

$$\dot{\bar{\varepsilon}}^p \equiv \sum_{i=1}^{3} |\dot{\varepsilon}_i^p| \, \boldsymbol{e}_i^p \otimes \boldsymbol{e}_i^p, \tag{12.134}$$

where $\dot{\varepsilon}_i^p$ are the eigenvalues of the plastic strain rate tensor, $\dot{\varepsilon}^p$, and $\{\boldsymbol{e}_i^p\}$ is an orthonormal basis of eigenvectors of $\dot{\varepsilon}^p$. In the basis $\{\boldsymbol{e}_i^p\}$, the absolute plastic strain rate has the matrix representation

$$[\dot{\bar{\varepsilon}}^p] = \begin{bmatrix} |\dot{\varepsilon}_1^p| & 0 & 0 \\ 0 & |\dot{\varepsilon}_2^p| & 0 \\ 0 & 0 & |\dot{\varepsilon}_3^p| \end{bmatrix}. \tag{12.135}$$

The plastic flow equation

Analogously to the original Lemaitre ductile damage model, the present theory also accounts for the softening effect of damage. Here, however, the corresponding yield function is a generally anisotropic function of the stress tensor. In the present model, the von Mises-type isotropically hardening yield function (iii) of Box 12.3 (page 487) is replaced by the following:

$$\Phi = q_{\text{eff}} - \sigma_y, \tag{12.136}$$

where q_{eff} is the generally anisotropic function of $\boldsymbol{\sigma}$ defined by (12.133) and, as usual,

$$\sigma_y = \sigma_y(R), \tag{12.137}$$

where R is the strain-hardening variable. Clearly, with this yield function, the softening experienced by the material as a result of damage is generally anisotropic. The yield von Mises equivalent stress depends on the orientation of the applied stress with respect to the principal directions of damage. For an isotropically damaged state, where $\boldsymbol{D} = D\boldsymbol{I}$, the above yield function reduces to the isotropic function of Box 12.3.

The flow rule is defined by assuming associativity of the plastic flow. We then have the equation

$$\dot{\varepsilon}^p = \dot{\gamma}\, \boldsymbol{N}, \tag{12.138}$$

where the associative flow vector,

$$\boldsymbol{N} \equiv \frac{\partial \Phi}{\partial \boldsymbol{\sigma}}, \tag{12.139}$$

is found, after some straightforward algebra, to be given by

$$\boldsymbol{N} = \sqrt{\frac{3}{2}}\, \frac{\text{dev}[\boldsymbol{H} \boldsymbol{s}_{\text{eff}} \boldsymbol{H}]}{\|\boldsymbol{s}_{\text{eff}}\|}. \tag{12.140}$$

The evolution of the hardening variable has the usual format

$$\dot{R} = \dot{\gamma}. \tag{12.141}$$

Kinematic hardening

A kinematic hardening-type variable can also be introduced in the model. In this case, we replace $q_{\rm eff}$ with

$$\bar{q}_{\rm eff} \equiv \sqrt{3\,J_2(\boldsymbol{s}_{\rm eff} - \boldsymbol{\beta})} \tag{12.142}$$

in (12.136) and redefine the flow vector as

$$\boldsymbol{N} \equiv \sqrt{\frac{3}{2}}\,\frac{\text{dev}[\boldsymbol{H}(\boldsymbol{s}_{\rm eff} - \boldsymbol{\beta})\boldsymbol{H}\,]}{\|\boldsymbol{s}_{\rm eff} - \boldsymbol{\beta}\|}, \tag{12.143}$$

where $\boldsymbol{\beta}$ is the backstress tensor. The evolution of the backstress can be defined analogously to the original Lemaitre model:

$$\dot{\boldsymbol{\beta}} = \dot{\gamma}\,(a\,\boldsymbol{N} - b\,\boldsymbol{\beta}). \tag{12.144}$$

Remarks on computer implementation

As for the elastoplasticity models discussed in the preceding chapters and in the previous sections of this chapter, the derivation of an elastic predictor/plastic corrector algorithm in the present case follows the general methodology described in Section 7.2.3 (from page 196). However, for the present model, the computational implementation is far more complex than that of the original Lemaitre model discussed in Section 12.3. Here, even in the absence of kinematic hardening, the damage-induced anisotropy of elastic and plastic behaviours does not allow the use of the simplifying relations that led to the reduction in the number of plastic corrector equations in Section 12.4.

Part Three

Large strains

13 FINITE STRAIN HYPERELASTICITY

PART III of this book is devoted to the formulation and numerical treatment of finite strain solid mechanics problems. We begin by describing in this chapter some useful finite hyperelasticity constitutive theories together with their implementation within the implicit finite element environment of Chapter 4. Two of the models addressed here are implemented in program HYPLAS: the Hencky (logarithmic strain-based) and the Ogden hyperelastic models. Recall (refer to Chapter 4) that the most relevant components of the finite element implementation of a material model in HYPLAS are:

1. the *state-update procedure*, where the stresses are computed for a given state of deformation. The stresses computed by such a procedure are used to assemble the *element internal force vector*;
2. the computation of the corresponding spatial *tangent modulus*, which is used in the assembly of the *element tangent stiffness matrix*.

To provide the reader with a clear idea of the complete finite element implementation of finite hyperelastic theories, the FORTRAN source code of these procedures is provided and explained in detail for the particular case of the Ogden material model.

Hyperelastic constitutive models are suitable to describe the behaviour of a large number of engineering materials. In the finite strain range, such theories are particularly appropriate for the analysis of rubber-like solids and in many cases show excellent agreement with experiments involving strains over 700% (Ogden, 1984).

This chapter is divided as follows: after reviewing some basic concepts of finite hyperelasticity in Section 13.1, some popular models are described in Section 13.2. The formulation of plane stress theories is addressed in Section 13.3, with particular reference to the Ogden and Hencky models. The explicit formulae for the associated tangent moduli are derived in Section 13.4. Section 13.5 describes in detail the computational implementation of the Ogden model, including the listing of the relevant FORTRAN source code. A comprehensive set of benchmarking finite element examples of hyperelasticity applications is provided in Section 13.6. Finally, Section 13.7 discusses a relatively new topic: the modelling of internal damage coupled with hyperelasticity. Hyperelastic damage theories are useful in modelling the so-called *Mullins effect* – the loss of stiffness frequently observed in filled polymers at strain levels below the maximum previously attained strain.

We remark that much of the notation adopted in this chapter as well as throughout the remainder of Part Three of this book has been set up in Chapter 3. Readers who are unfamiliar with it are advised to review that chapter before proceeding further.

13.1. Hyperelasticity: basic concepts

In contrast to the elastoplastic and viscoplastic theories described in Part Two of this book, hyperelastic models are *non-dissipative* and, therefore, do not require the consideration of internal variables. A generic finite hyperelasticity model is characterised by the existence of a specific free-energy function of the form

$$\psi = \psi(\boldsymbol{F}). \tag{13.1}$$

Functions of this form (without dependence on internal variables) are particular cases of the general potential (3.150) (page 72) which include the description of dissipative materials (with internal variables). In this case (refer to (3.155)), the first Piola–Kirchhoff stress tensor is given by the constitutive relation

$$\boldsymbol{P} = \boldsymbol{P}(\boldsymbol{F}) = \bar{\rho}\,\frac{\partial\psi(\boldsymbol{F})}{\partial\boldsymbol{F}}, \tag{13.2}$$

where $\bar{\rho}$ is the reference density. The stress here depends solely on the *current* deformation gradient and is not affected by the past deformation history.† The free-energy function, ψ, completely defines a hyperelastic model.

Accordingly, the constitutive equation for the Kirchoff stress tensor,

$$\boldsymbol{\tau} \equiv \boldsymbol{P}\,\boldsymbol{F}^T, \tag{13.3}$$

for a hyperelastic material is given by

$$\boldsymbol{\tau}(\boldsymbol{F}) = \bar{\rho}\,\frac{\partial\psi(\boldsymbol{F})}{\partial\boldsymbol{F}}\,\boldsymbol{F}^T \tag{13.4}$$

and the Cauchy stress tensor, $\boldsymbol{\sigma} = \boldsymbol{\tau}/J$, has the following constitutive equation

$$\boldsymbol{\sigma}(\boldsymbol{F}) = \frac{\bar{\rho}}{J}\,\frac{\partial\psi(\boldsymbol{F})}{\partial\boldsymbol{F}}\,\boldsymbol{F}^T, \tag{13.5}$$

where

$$J \equiv \det \boldsymbol{F}. \tag{13.6}$$

In Sections 13.1.1 and 13.1.2 below, we consider some useful constraints imposed by the general constitutive axioms of Section 3.5.1 (from page 69) upon the constitutive equations for hyperelastic materials.

13.1.1. MATERIAL OBJECTIVITY: REDUCED FORM OF THE FREE-ENERGY FUNCTION

Let us start by recalling the axiom of material objectivity. This axiom requires the free energy (a *scalar* function) to be invariant under changes in the observer. This is represented mathematically in the general case by equation $(3.147)_2$ (page 71). The specialisation of this

†This assertion applies also to the so-called *Cauchy elastic* models (not discussed in this book) for which the first Piola–Kirchhoff stress is a function of the deformation gradient only, but is not necessarily the derivative of a free-energy function.

equation to the present case, where the free energy depends exclusively on the current value of $\boldsymbol{F}$, requires that the function ψ satisfies

$$\psi(\boldsymbol{QF}) = \psi(\boldsymbol{F}) \tag{13.7}$$

for any proper orthogonal (rotation) tensor $\boldsymbol{Q}$. In particular, if we choose $\boldsymbol{Q} = \boldsymbol{R}^T$ where $\boldsymbol{R}$ is the rotation obtained from the polar decomposition $\boldsymbol{F} = \boldsymbol{RU}$, we obtain the identity

$$\psi(\boldsymbol{F}) = \psi(\boldsymbol{U}). \tag{13.8}$$

Thus, material objectivity implies that ψ depends on $\boldsymbol{F}$ solely through the right stretch tensor, $\boldsymbol{U}$.

Equivalently, the free energy may be expressed as a function of the right Cauchy–Green strain tensor, $\boldsymbol{C} \equiv \boldsymbol{F}^T\boldsymbol{F} = \boldsymbol{U}^2$; that is, the free energy may be expressed by a function $\tilde{\psi}$ defined as

$$\psi(\boldsymbol{F}) = \tilde{\psi}(\boldsymbol{C}) \equiv \psi(\sqrt{\boldsymbol{C}}). \tag{13.9}$$

In terms of the reduced form, $\tilde{\psi}$, the first Piola–Kirchhoff stress tensor given by (13.2) can be equivalently expressed as

$$\boldsymbol{P} = \bar{\rho}\,\frac{\partial\tilde{\psi}}{\partial\boldsymbol{C}} : \frac{\partial\boldsymbol{C}}{\partial\boldsymbol{F}} = 2\bar{\rho}\boldsymbol{F}\,\frac{\partial\tilde{\psi}}{\partial\boldsymbol{C}}. \tag{13.10}$$

In deriving the rightmost part of the above formula, use has been made of the component form of the derivative $\partial\boldsymbol{C}/\partial\boldsymbol{F}$, obtained by simply applying the product rule to $C_{ij} = F_{mi}F_{mj}$, which gives

$$\frac{\partial C_{ij}}{\partial F_{kl}} = \delta_{il}F_{kj} + \delta_{jl}F_{ki}. \tag{13.11}$$

From (13.10) and (13.3), it then follows that the corresponding constitutive equation for the Kirchhoff stress is

$$\boldsymbol{\tau} = 2\bar{\rho}\boldsymbol{F}\,\frac{\partial\tilde{\psi}}{\partial\boldsymbol{C}}\,\boldsymbol{F}^T. \tag{13.12}$$

Similarly, for the Cauchy stress, we have

$$\boldsymbol{\sigma} = \frac{2\bar{\rho}}{J}\,\boldsymbol{F}\,\frac{\partial\tilde{\psi}}{\partial\boldsymbol{C}}\,\boldsymbol{F}^T. \tag{13.13}$$

13.1.2. ISOTROPIC HYPERELASTICITY

We now consider constraints posed by material symmetry on the constitutive equations for hyperelastic materials. More specifically, we will focus on material *isotropy* – a symmetry property that requires the general equations (3.148) (page 71) to hold for any rotation $\boldsymbol{Q}$. The specialisation of $(3.148)_2$ to the present case implies that the free-energy function of an isotropic hyperelastic material must satisfy

$$\psi(\boldsymbol{F}\,\boldsymbol{Q}) = \psi(\boldsymbol{F}), \tag{13.14}$$

for all rotations $\boldsymbol{Q}$. This equation introduces a further restriction on the possible representations for the free-energy function. Indeed, by choosing $\boldsymbol{Q} = \boldsymbol{R}^T$ again, we promptly establish that

$$\psi(\boldsymbol{F}) = \psi(\boldsymbol{V}); \tag{13.15}$$

that is, the free energy of an isotropic hyperelastic material must depend on $\boldsymbol{F}$ only through the left stretch tensor, $\boldsymbol{V}$.

The free energy in this case may be equivalently expressed as a function of the left Cauchy–Green strain tensor, $\boldsymbol{B} \equiv \boldsymbol{F}\,\boldsymbol{F}^T = \boldsymbol{V}^2$, as follows:

$$\tilde{\psi}(\boldsymbol{B}) = \psi(\sqrt{\boldsymbol{B}}). \tag{13.16}$$

Equation (13.16), together with (13.8), (13.9) and (13.15), imply that

$$\psi(\boldsymbol{F}) = \psi(\boldsymbol{U}\,) = \psi(\boldsymbol{V}) = \tilde{\psi}(\boldsymbol{C}) = \tilde{\psi}(\boldsymbol{B}) \tag{13.17}$$

for an isotropic hyperelastic material.

Note that from the definition of the right and left stretch tensors, we have

$$\boldsymbol{U} = \boldsymbol{R}^T \boldsymbol{V}\boldsymbol{R}, \tag{13.18}$$

which, together with (13.17), yields

$$\psi(\boldsymbol{U}\,) = \psi(\boldsymbol{R}\,\boldsymbol{U}\boldsymbol{R}^T). \tag{13.19}$$

As the above holds for any deformation (and, hence, any rotation $\boldsymbol{R}$), it follows from definition (A.2) (page 731) that ψ is an *isotropic* function of $\boldsymbol{U}$ (or $\boldsymbol{V}$). Following a completely analogous argument in terms of the right and left Cauchy–Green tensors, we establish that

$$\tilde{\psi}(\boldsymbol{C}) = \tilde{\psi}(\boldsymbol{R}\,\boldsymbol{B}\boldsymbol{R}^T), \tag{13.20}$$

for any rotation $\boldsymbol{R}$, and conclude that $\tilde{\psi}$ is an isotropic function of $\boldsymbol{C}$ (or $\boldsymbol{B}$).

Isotropic stress constitutive equations

With a straightforward application of the chain rule, we find that the first Piola–Kirchhoff stress constitutive equation (13.2) can be rewritten in terms of the function $\tilde{\psi}$ as

$$\boldsymbol{P} = \bar{\rho}\,\frac{\partial\tilde{\psi}}{\partial\boldsymbol{B}} : \frac{\partial\boldsymbol{B}}{\partial\boldsymbol{F}} = 2\bar{\rho}\,\frac{\partial\tilde{\psi}}{\partial\boldsymbol{B}}\,\boldsymbol{F}. \tag{13.21}$$

Analogously to the derivation of (13.10), in obtaining the rightmost part of (13.21) we have made use of the component form of $\partial\boldsymbol{B}/\partial\boldsymbol{F}$, which can be derived by differentiating $B_{ij} = F_{im}F_{jm}$. The corresponding constitutive equation for the Kirchhoff stress is obtained by combining (13.21) and (13.3) as

$$\boldsymbol{\tau} = \tilde{\boldsymbol{\tau}}(\boldsymbol{B}) \equiv 2\,\bar{\rho}\,\frac{\partial\tilde{\psi}}{\partial\boldsymbol{B}}\,\boldsymbol{B}. \tag{13.22}$$

Further, the isotropy of $\tilde{\psi}$ implies that $\boldsymbol{B}$ and $\partial\tilde{\psi}/\partial\boldsymbol{B}$ share the same principal axes and, thus, commute (refer to property (A.6), page 732). Hence, the Kirchhoff stress constitutive equation can be equivalently expressed as

$$\tilde{\tau}(\boldsymbol{B}) \equiv 2\,\bar{\rho}\,\boldsymbol{B}\,\frac{\partial\tilde{\psi}}{\partial\boldsymbol{B}}. \tag{13.23}$$

Accordingly, for the Cauchy stress, we have

$$\boldsymbol{\sigma} = \tilde{\boldsymbol{\sigma}}(\boldsymbol{B}) \equiv \frac{2\bar{\rho}}{J}\frac{\partial\tilde{\psi}}{\partial\boldsymbol{B}}\,\boldsymbol{B} = \frac{2\bar{\rho}}{J}\,\boldsymbol{B}\,\frac{\partial\tilde{\psi}}{\partial\boldsymbol{B}}. \tag{13.24}$$

The stress constitutive functions (13.22)–(13.24) are themselves isotropic functions of $\boldsymbol{B}$ (see Section A.1.2); that is,

$$\boldsymbol{Q}\tilde{\boldsymbol{\sigma}}(\boldsymbol{B})\boldsymbol{Q}^T = \tilde{\boldsymbol{\sigma}}(\boldsymbol{Q}\boldsymbol{B}\boldsymbol{Q}^T) \quad \boldsymbol{Q}\tilde{\tau}(\boldsymbol{B})\boldsymbol{Q}^T = \tilde{\tau}(\boldsymbol{Q}\boldsymbol{B}\boldsymbol{Q}^T) \tag{13.25}$$

for all rotations $\boldsymbol{Q}$.

Principal invariants representation

The isotropy of $\tilde{\psi}$ implies that (refer to the general representation formula (A.12), page 733) the free energy can be equivalently expressed as a function of the principal invariants, $I_1(\boldsymbol{B})$, $I_2(\boldsymbol{B})$ and $I_3(\boldsymbol{B})$, of the left Cauchy–Green strain tensor (see definition (2.72), page 27, of the principal invariants of a tensor); that is, there exists a function $\bar{\psi}$ such that

$$\tilde{\psi}(\boldsymbol{B}) = \bar{\psi}(I_1(\boldsymbol{B}), I_2(\boldsymbol{B}), I_3(\boldsymbol{B})). \tag{13.26}$$

As (13.22) defines the Kirchhoff stress (what follows is equally true for equation (13.24) for the Cauchy stress) as an isotropic symmetric tensor-valued function of $\boldsymbol{B}$ – an *invertible* symmetric tensor, then it follows from the general representation (A.14) (page 733) that the stress response of an isotropic hyperelastic solid may be cast in the form

$$\boldsymbol{\tau} = J\,(\beta_0\,\boldsymbol{I} + \beta_1\,\boldsymbol{B} + \beta_{-1}\boldsymbol{B}^{-1}), \tag{13.27}$$

or, equivalently,

$$\boldsymbol{\sigma} = \beta_0\,\boldsymbol{I} + \beta_1\,\boldsymbol{B} + \beta_{-1}\boldsymbol{B}^{-1}, \tag{13.28}$$

where the scalar coefficients $\beta_\Gamma = \beta_\Gamma(I_1, I_2, I_3)$, $(\Gamma = 0, 1, -1)$, are the *elastic response functions* which, in terms of the strain-energy function (13.26) are expressed as

$$\begin{aligned}\beta_0 &= \frac{2}{\sqrt{I_3}}\left[I_2\,\bar{\rho}\,\frac{\partial\bar{\psi}}{\partial I_2} + I_3\,\bar{\rho}\,\frac{\partial\bar{\psi}}{\partial I_3}\right]\\ \beta_1 &= \frac{2}{\sqrt{I_3}}\bar{\rho}\,\frac{\partial\bar{\psi}}{\partial I_1}\\ \beta_{-1} &= -2\sqrt{I_3}\,\bar{\rho}\,\frac{\partial\bar{\psi}}{\partial I_2}.\end{aligned} \tag{13.29}$$

Principal stretches representation

Other alternative representations of the free-energy function for isotropic hyperelastic materials are possible and the particular choice is normally a matter of mathematical and experimental convenience. A particularly appealing approach, which has gained popularity over the last two decades, is the explicit representation of ψ in terms of the *principal stretches*, λ_i. As ψ is an isotropic function of $\boldsymbol{V}$ (or $\boldsymbol{U}$), the general representation (A.4) implies that there exists a function $\hat{\psi}$ such that

$$\psi(\boldsymbol{V}) = \hat{\psi}(\lambda_1, \lambda_2, \lambda_3) \tag{13.30}$$

with the symmetries

$$\hat{\psi}(\lambda_1, \lambda_2, \lambda_3) = \hat{\psi}(\lambda_2, \lambda_1, \lambda_3) = \hat{\psi}(\lambda_1, \lambda_3, \lambda_2). \tag{13.31}$$

Likewise, with $b_i = \lambda_i^2$ denoting the principal values of $\boldsymbol{B}$ (or $\boldsymbol{C}$), there exists a function $\check{\psi}$ such that

$$\tilde{\psi}(\boldsymbol{B}) = \check{\psi}(b_1, b_2, b_3) = \hat{\psi}(b_1^{\frac{1}{2}}, b_2^{\frac{1}{2}}, b_3^{\frac{1}{2}}). \tag{13.32}$$

By combining (13.22), property (A.8), (13.32) and the fact that $\partial\boldsymbol{\tau}/\partial\boldsymbol{B}$ and $\boldsymbol{B}$ have the same principal axes, we find that the constitutive equation for the eigenvalues τ_i of the Kirchhoff stresses tensor is given by

$$\tau_i = 2\bar{\rho}\,\frac{\partial\check{\psi}}{\partial b_i}\,b_i = \bar{\rho}\,\frac{\partial\hat{\psi}}{\partial\lambda_i}\,\lambda_i, \tag{13.33}$$

with no summation on repeated indices. Accordingly, the constitutive function for the Kirchhoff stress tensor in terms of principal stretches can be obtained via its spectral representation

$$\boldsymbol{\tau} = \sum_i \bar{\rho}\,\frac{\partial\hat{\psi}}{\partial\lambda_i}\lambda_i\,\boldsymbol{e}_i \otimes \boldsymbol{e}_i, \tag{13.34}$$

where $\{\boldsymbol{e}_i\}$ is an orthonormal basis of eigenvectors of $\boldsymbol{V}$ (or $\boldsymbol{B}$).

13.1.3. INCOMPRESSIBLE HYPERELASTICITY

For many rubbery materials, very little volumetric deformation is observed even at highly strained states. For this reason, some hyperelastic models of rubber assume perfect incompressibility. In such cases, the invariant I_3 is constant, i.e.

$$I_3 \equiv \det \boldsymbol{B} = 1,$$

and the dependence of the strain-energy function upon I_3 is removed. We then have strain-energy functions of the type

$$\psi = \bar{\psi}(I_1, I_2). \tag{13.35}$$

Under perfect incompressibility, the stress tensor is determined from the potential relation (13.5) or (13.4) up to an arbitrary hydrostatic stress; that is, only the stress deviator,

$$\boldsymbol{s} \equiv \operatorname{dev}[\boldsymbol{\sigma}] \equiv \boldsymbol{\sigma} - \tfrac{1}{3}(\operatorname{tr}\boldsymbol{\sigma})\,\boldsymbol{I}, \tag{13.36}$$

is obtained from the hyperelastic constitutive law. In the solution of incompressible hyperelasticity boundary value problems, the hydrostatic pressure can only be determined from equilibrium and the boundary conditions.

13.1.4. COMPRESSIBLE REGULARISATION

The numerical treatment of perfect incompressibility is not trivial, except in the plane stress case addressed in Section 13.3. In the context of finite element solutions, it requires the use of *mixed* finite element techniques where the hydrostatic pressure (and not only displacements as in virtual work-based finite elements) is one of the structural variables. The problem, however, can be dealt with in the present context of virtual work-based finite elements by introducing compressibility in originally incompressible models. This *compressible regularisation* will be described in the next section for the Mooney–Rivlin/neo-Hookean and the Ogden material models. In fact, we should also bear in mind that real materials are never perfectly incompressible and also that the introduction of compressibility allows the extension of the original theories to treat materials whose compressibility cannot be disregarded.

13.2. Some particular models

Let us now review the basic expressions of some important hyperelastic theories applicable to the modelling of the behaviour of many materials in situations of practical interest. Here, we present only their free-energy functions and the resulting stress constitutive functions.

13.2.1. THE MOONEY–RIVLIN AND THE NEO-HOOKEAN MODELS

Two very important constitutive models of rubber-like materials are the *Mooney–Rivlin* and the *neo-Hookean* models (Green and Zerna, 1954). Due to their mathematical simplicity, as well as their predictive accuracy with a range of moderately large strains, these models have been widely employed in the description of the behaviour of rubbery materials. In their original versions, these models assume *perfect incompressibility*. The *Mooney–Rivlin* strain-energy function is expressed by

$$\bar{\rho}\,\bar{\psi}(I_1, I_2) = C_1\,(I_1 - 3) + C_2\,(I_2 - 3), \tag{13.37}$$

where C_1 and C_2 are material constants.

The *neo-Hookean* material is *the simplest model of rubber-like response*. It is obtained as a particular case of the Mooney–Rivlin model by setting $C_2 = 0$, i.e.

$$\bar{\rho}\,\bar{\psi}(I_1) = C_1\,(I_1 - 3). \tag{13.38}$$

The regularised (compressible) version

In order to derive *regularised* (or *compressible*) versions of the Mooney–Rivlin and neo-Hookean models, we resort to the concept of *isochoric/volumetric split* of the deformation gradient introduced in Section 3.1.5 (page 49), whose main equations are repeated below for convenience:

$$\boldsymbol{F} = \boldsymbol{F}_{\text{iso}}\,\boldsymbol{F}_{\text{v}}, \tag{13.39}$$

where

$$\boldsymbol{F}_{\text{v}} \equiv (\det \boldsymbol{F})^{\frac{1}{3}}\,\boldsymbol{I} \quad \text{and} \quad \boldsymbol{F}_{\text{iso}} \equiv (\det \boldsymbol{F})^{-\frac{1}{3}}\,\boldsymbol{F} \tag{13.40}$$

are, respectively, the *volumetric* and *isochoric* components of $\boldsymbol{F}$.

With the above decomposition at hand, we define the *isochoric left Cauchy–Green strain tensor* as

$$\boldsymbol{B}_{\text{iso}} \equiv \boldsymbol{F}_{\text{iso}}\, \boldsymbol{F}_{\text{iso}}^T = (\det \boldsymbol{F})^{-\frac{2}{3}}\, \boldsymbol{F}\, \boldsymbol{F}^T, \tag{13.41}$$

and the principal invariants

$$I_1^* \equiv \text{tr}\boldsymbol{B}_{\text{iso}} \quad I_2^* \equiv \tfrac{1}{2}\{(I_1^*)^2 - \text{tr}[\boldsymbol{B}_{\text{iso}}^2]\}. \tag{13.42}$$

With the above definitions, a regularised version of the Mooney–Rivlin strain-energy function can be postulated as

$$\bar{\rho}\,\tilde{\psi}^*(I_1^*, I_2^*, J) = C_1\,(I_1^* - 3) + C_2\,(I_2^* - 3) + \tfrac{1}{2}K\,(\ln J)^2, \tag{13.43}$$

where the material constant K is the *logarithmic bulk modulus* and relates the hydrostatic pressure to the purely volumetric component of the deformation gradient. The corresponding neo-Hookean function reads

$$\bar{\rho}\,\tilde{\psi}^*(I_1^*, J) = \tfrac{1}{2}G\,(I_1^* - 3) + \tfrac{1}{2}K\,(\ln J)^2, \tag{13.44}$$

where $G \equiv 2C_1$ is the *shear modulus*.

The stress constitutive function

The constitutive function for the Kirchhoff stress for the regularised Mooney–Rivlin material is obtained by using the above strain-energy function in the potential relation (13.22). As the Mooney–Rivlin function has been defined in terms of principal invariants, we apply the chain rule and obtain

$$\frac{\partial \tilde{\psi}}{\partial \boldsymbol{B}} = \left(\frac{\partial \tilde{\psi}}{\partial I_1^*}\,\frac{\partial I_1^*}{\partial \boldsymbol{B}_{\text{iso}}} + \frac{\partial \tilde{\psi}}{\partial I_2^*}\,\frac{\partial I_2^*}{\partial \boldsymbol{B}_{\text{iso}}}\right) : \frac{\partial \boldsymbol{B}_{\text{iso}}}{\partial \boldsymbol{B}} + \frac{\partial \tilde{\psi}}{\partial J}\,\frac{\partial J}{\partial \boldsymbol{B}}. \tag{13.45}$$

The final equation for the Kirchhoff stress is obtained, after some straightforward algebra, in compact form as

$$\boldsymbol{\tau} = 2(C_1 + C_2\,I_1^*)\,\text{dev}[\boldsymbol{B}_{\text{iso}}] - 2C_2\,\text{dev}[\boldsymbol{B}_{\text{iso}}^2] + K\,(\ln J)\,\boldsymbol{I}. \tag{13.46}$$

For the compressible neo-Hookean material, we have

$$\boldsymbol{\tau} = G\,\text{dev}[\boldsymbol{B}_{\text{iso}}] + K\,(\ln J)\,\boldsymbol{I}. \tag{13.47}$$

Remark 13.1. The incompressible limit (where $J = 1$) is approached as $K \rightarrow \infty$. Regularised models of the above type are frequently used to emulate incompressibility in finite element analyses. In this case, the bulk modulus K may be seen as a penalty factor that penalises volumetric deformations. This approach is adopted in the plane strain/axisymmetric implementation of the Ogden material in HYPLAS. Note also that other formats of volumetric (penalty) contribution to the strain-energy function may be adopted in this context. The function

$$\tfrac{1}{2}\,K\,(J-1)^2,$$

for instance, is also frequently adopted (see, among others, Crisfield 1997) in the finite element analysis of nearly incompressible materials.

Remark 13.2. In modelling materials whose compressibility is an important feature of the overall response, the functional form of the volumetric contribution to the strain-energy function should be determined on the basis of experimental evidence rather than just postulated as above.

13.2.2. THE OGDEN MATERIAL MODEL

At very large strains, it is a well-known fact that the neo-Hookean and Mooney–Rivlin models fail to represent the behaviour of rubbery materials (Ogden, 1984). To overcome this problem, a particularly useful form of strain-energy function – based on principal stretches representation (13.30) – has been proposed by Ogden (1972, 1984). Its general form is given by

$$\bar{\rho}\,\hat{\psi}(\lambda_1, \lambda_2, \lambda_3) = \sum_{p=1}^{N} \frac{\mu_p}{\alpha_p}\,(\lambda_1^{\alpha_p} + \lambda_2^{\alpha_p} + \lambda_3^{\alpha_p} - 3), \tag{13.48}$$

where N is the total number of terms in the series and μ_p and α_p $(p = 1, \ldots, N)$ are the corresponding material constants. Perfect incompressibility is also assumed so that $\lambda_1\lambda_2\lambda_3 = 1$. Thus, equivalently, the strain-energy can be expressed as a function of λ_1 and λ_2 only

$$\bar{\rho}\,\tilde{\psi}(\lambda_1, \lambda_2) = \sum_{p=1}^{N} \frac{\mu_p}{\alpha_p}\left(\lambda_1^{\alpha_p} + \lambda_2^{\alpha_p} + \frac{1}{\lambda_1^{\alpha_p}\lambda_2^{\alpha_p}} - 3\right). \tag{13.49}$$

As a particular case of an Ogden material, the Mooney–Rivlin model is obtained by setting $N = 2$, $\mu_1 = 2C_1$, $\mu_2 = -2C_2$ and $\alpha_1 = 2$, $\alpha_2 = -2$. The standard neo-Hookean material is recovered with $N = 1$, $\mu_1 = 2C_1$ and $\alpha_1 = 2$.

The regularised Ogden model

Let $\{\lambda_1^*, \lambda_2^*, \lambda_3^*\}$ be the principal *isochoric* stretches, i.e. the eigenvalues of the isochoric stretch tensor

$$\boldsymbol{V}_{\text{iso}} \equiv \sqrt{\boldsymbol{B}_{\text{iso}}}. \tag{13.50}$$

The principal isochoric stretches λ_i^* are related to the principal stretches λ_i by

$$\lambda_i^* = \frac{\lambda_i}{J^{\frac{1}{3}}} = \frac{\lambda_i}{(\lambda_1\lambda_2\lambda_3)^{\frac{1}{3}}} = \lambda_i^{\frac{2}{3}}(\lambda_j\lambda_k)^{-\frac{1}{3}}, \tag{13.51}$$

where (i, j, k) are cyclic permutations of $(1, 2, 3)$. The strain-energy function for a regularised compressible version of the Ogden model can be defined as

$$\bar{\rho}\,\hat{\psi}^*(\lambda_1^*, \lambda_2^*, \lambda_3^*, J) = \sum_{p=1}^{N} \frac{\mu_p}{\alpha_p}[(\lambda_1^*)^{\alpha_p} + (\lambda_2^*)^{\alpha_p} + (\lambda_3^*)^{\alpha_p} - 3] + \tfrac{1}{2}K\,(\ln J)^2. \tag{13.52}$$

The corresponding constitutive function for the principal Kirchhoff stresses is obtained from the potential relation (13.33). With the above regularised strain-energy function, we have

$$\tau_i = \lambda_i\,\bar{\rho}\,\frac{\partial\hat{\psi}}{\partial\lambda_i} = \lambda_i\,\bar{\rho}\left(\frac{\partial\hat{\psi}^*}{\partial\lambda_i^*}\frac{\partial\lambda_i^*}{\partial\lambda_i} + \frac{\partial\hat{\psi}^*}{\partial\lambda_j^*}\frac{\partial\lambda_j^*}{\partial\lambda_i} + \frac{\partial\hat{\psi}^*}{\partial\lambda_k^*}\frac{\partial\lambda_k^*}{\partial\lambda_i} + \frac{\partial\hat{\psi}^*}{\partial J}\frac{\partial J}{\partial\lambda_i}\right) \tag{13.53}$$

where no summation is implied on repeated indices and (i, j, k) are permutations of $(1, 2, 3)$. The final expression for the principal Kirchhoff stresses reads

$$\tau_i = \sum_{p=1}^{N} \mu_p \, J^{-\alpha_p/3}[\lambda_i^{\alpha_p} - \tfrac{1}{3}(\lambda_1^{\alpha_p} + \lambda_2^{\alpha_p} + \lambda_3^{\alpha_p})] + K \ln J. \tag{13.54}$$

This formula is implemented in subroutine SUOGD of HYPLAS.

13.2.3. THE HENCKY MATERIAL

The Hencky model is the finite logarithmic strain-based extension of the standard linear elastic material. It was proposed by Hencky (1933) to model the behaviour of vulcanised rubbers. Let $\boldsymbol{\varepsilon}$ be the *Eulerian logarithmic strain* tensor

$$\boldsymbol{\varepsilon} \equiv \ln \boldsymbol{V} = \tfrac{1}{2} \ln \boldsymbol{B}. \tag{13.55}$$

The Hencky strain-energy function is defined in compact form as

$$\bar{\rho}\, \psi(\boldsymbol{\varepsilon}) = \tfrac{1}{2}\, \boldsymbol{\varepsilon} : \mathsf{D} : \boldsymbol{\varepsilon}, \tag{13.56}$$

where D has the format of the infinitesimal isotropic elasticity tensor

$$\mathsf{D} \equiv 2G\, \mathsf{I}_S + (K - \tfrac{2}{3}G)\boldsymbol{I} \otimes \boldsymbol{I}. \tag{13.57}$$

Note that the functional format of the Hencky strain-energy function is identical to that of infinitesimal elasticity. Equivalently, in terms of principal stretches, the Hencky strain-energy function is given by

$$\begin{aligned} \bar{\rho}\, \hat{\psi}(\lambda_1, \lambda_2, \lambda_3) = {} & G\, [(\ln \lambda_1)^2 + (\ln \lambda_2)^2 + (\ln \lambda_3)^2] \\ & + \tfrac{1}{2}(K - \tfrac{2}{3}G)[\ln(\lambda_1 \lambda_2 \lambda_3)]^2. \end{aligned} \tag{13.58}$$

The above strain-energy renders the following *linear* relationship between the Kirchhoff stress and the Eulerian logarithmic strain:

$$\boldsymbol{\tau} = \bar{\rho} \frac{\partial \psi}{\partial \boldsymbol{\varepsilon}} = \mathsf{D} : \boldsymbol{\varepsilon}, \tag{13.59}$$

which has the same functional format as the infinitesimal linear elastic stress–strain relation.

That (13.59) follows from definition (13.56) (or (13.58)) can be demonstrated as follows. Recall equation (13.22), which is valid for any isotropic hyperlastic material. By taking definition (13.55) into account and applying the chain rule to differentiate the free-energy function (13.56), we obtain the expression

$$\frac{\partial \psi}{\partial \boldsymbol{B}} = \frac{1}{2} \frac{\partial \psi}{\partial \boldsymbol{\varepsilon}} : \frac{\partial (\ln \boldsymbol{B})}{\partial \boldsymbol{B}}, \tag{13.60}$$

which, substituted into (13.22), gives

$$\boldsymbol{\tau} = \bar{\rho} \left[\frac{\partial \psi}{\partial \boldsymbol{\varepsilon}} : \frac{\partial (\ln \boldsymbol{B})}{\partial \boldsymbol{B}} \right] \boldsymbol{B}. \tag{13.61}$$

Now, note that by definition of ε, the tensors ε and $\boldsymbol{B}$ share the same principal axes (have the same eigenvectors/eigenprojection tensors). Also, due to the isotropy of ψ, the tensor $\partial\psi/\partial\varepsilon$ has the same principal axes. In addition, it should be noted that the tensor logarithm is a member of the class of isotropic tensor functions discussed in Section A.5 (from page 740). These observations in conjunction with the specialisation of the formulae (A.52) (page 744) for the derivative $\partial(\ln \boldsymbol{B})/\partial\boldsymbol{B}$ lead, after some algebra, to the identity

$$\left[\frac{\partial\psi}{\partial\varepsilon} : \frac{\partial(\ln \boldsymbol{B})}{\partial\boldsymbol{B}}\right]\boldsymbol{B} = \frac{\partial\psi}{\partial\varepsilon}, \tag{13.62}$$

which, replaced in (13.61), yields the constitutive equation (13.59) for the Hencky material.

Deviatoric and volumetric logarithmic strains

By splitting the logarithmic strain into deviatoric and volumetric components, we have

$$\varepsilon = \varepsilon_{\text{d}} + \tfrac{1}{3}\,\varepsilon_{\text{v}}\boldsymbol{I}, \tag{13.63}$$

where

$$\varepsilon_{\text{d}} \equiv \varepsilon - \tfrac{1}{3}\,\varepsilon_{\text{v}}\boldsymbol{I} \tag{13.64}$$

is the deviatoric logarithmic strain and

$$\varepsilon_{\text{v}} \equiv \operatorname{tr} \varepsilon \tag{13.65}$$

is the volumetric logarithmic strain.

As in the infinitesimal strain theory, a traceless logarithmic strain tensor (ε with $\varepsilon_{\text{v}} = 0$) corresponds to a finite volume-preserving deformation (a deformation with $\det \boldsymbol{F} = 1$). To see this, without loss of generality, we represent ε in its diagonalised form, i.e. we use an orthonormal basis formed by unit eigenvectors of ε:

$$[\varepsilon] = \begin{bmatrix} \ln\,\lambda_1 & 0 & 0 \\ 0 & \ln\,\lambda_2 & 0 \\ 0 & 0 & \ln\,\lambda_3 \end{bmatrix}, \tag{13.66}$$

where λ_i are the eigenvalues of $\boldsymbol{V}$. On the same basis, $\boldsymbol{V}$ is represented as

$$[\boldsymbol{V}] = \begin{bmatrix} \lambda_1 & 0 & 0 \\ 0 & \lambda_2 & 0 \\ 0 & 0 & \lambda_3 \end{bmatrix}. \tag{13.67}$$

It is then clear that

$$\operatorname{tr} \varepsilon = \ln(\det \boldsymbol{V}). \tag{13.68}$$

By recalling the standard identity $\det \boldsymbol{V} = \det \boldsymbol{F}$, it then follows that

$$\varepsilon_{\text{v}} \equiv \operatorname{tr} \varepsilon = 0 \Longleftrightarrow \det \boldsymbol{F} = 1. \tag{13.69}$$

Let us now look at the properties of the logarithmic strain deviator. Using the definition of $\varepsilon_{\rm d}$, we obtain after a straightforward manipulation the following diagonal representation

$$[\varepsilon_{\rm d}] = \begin{bmatrix} \ln \lambda_1^{\frac{2}{3}} - \ln \lambda_2^{\frac{1}{3}} - \ln \lambda_3^{\frac{1}{3}} & 0 & 0 \\ 0 & \ln \lambda_2^{\frac{2}{3}} - \ln \lambda_3^{\frac{1}{3}} - \ln \lambda_1^{\frac{1}{3}} & 0 \\ 0 & 0 & \ln \lambda_3^{\frac{2}{3}} - \ln \lambda_1^{\frac{1}{3}} - \ln \lambda_2^{\frac{1}{3}} \end{bmatrix}. \tag{13.70}$$

From the above and the definition (13.50) of the isochoric left stretch tensor, we then find that

$$\varepsilon_{\rm d} = \ln \boldsymbol{V}_{\rm iso}; \tag{13.71}$$

that is, as for the infinitesimal strain tensor, the logarithmic strain deviator is a measure of the purely isochoric component of the strain.

13.2.4. THE BLATZ–KO MATERIAL

Another interesting isotropic hyperelastic solid model is the so called *Blatz–Ko* material. This model was originally proposed by Blatz and Ko (1962) to describe the behaviour of foamed (compressible) elastomers. Using the general representation (13.27), the Blatz–Ko stress constitutive function is expressed as

$$\boldsymbol{\tau} = J\left[\, g(J)\, \boldsymbol{I} + \frac{\mu_0\, f}{2}\, \boldsymbol{B} - \frac{\mu_0(1-f)}{2}\, \boldsymbol{B}^{-1}\right], \tag{13.72}$$

where μ_0 and f are material constants and $g(J)$ is the derivative of the strain-energy function with respect to J. This derivative is a function of J only. Note that for the Blatz–Ko material the corresponding elastic response functions β_Γ (refer to representation (13.27)) depend exclusively on J.

To validate the model, these authors have conducted a series of experiments, comprising uniaxial and biaxial tension tests, on polyurethane rubbers. The explicit expression adopted for the function g has been chosen by setting

$$J\, g(J) = -\mu_0[f J^{-2\nu/(1-2\nu)} - (1-f) J^{2\nu/(1-2\nu)}] \tag{13.73}$$

where the constant ν is a finite strain extension to the conventional Poisson's ratio of the infinitesimal theory. The model recovers the incompressible limit when $\nu \rightarrow 0.5$.

13.3. Isotropic finite hyperelasticity in plane stress

The treatment of plane stress problems has been discussed in detail in Chapter 9 in the context of infinitesimal elastoplasticity. There, it was shown that in spite of its triviality in the case of linear elasticity, the plane stress constraint generally requires further consideration in the elastoplastic range. This is also generally true for finite hyperelasticity and concepts such as the use of plane stress-projected constitutive equations (discussed in Section 9.4, page 370) and nested iterations for enforcement of the plane stress constraint (see Section 9.2.2, from page 362) can be applied to the hyperelastic case. As in infinitesimal elastoplasticity, the derivation of plane stress-projected equations here is feasible when the relevant constitutive model is sufficiently simple to allow the out-of-plane strain to be eliminated from the formulation, becoming a *dependent* variable that can be expressed in closed form as an explicit function of the in-plane strains. This approach is precisely that adopted in plane stress linear elasticity reviewed in Section 9.1.1 (from page 358). As we shall see, the same concept can be trivially applied to derive plane stress constitutive equations for the *incompressible* Ogden model and the logarithmic strain-based Hencky model (both implemented in program HYPLAS). These will be discussed, respectively, in Sections 13.3.1 and 13.3.2. Our discussion on plane stress ends in Section 13.3.3 where we briefly outline the use of a nested iteration approach in the computational treatment of finite hyperelasticity.

13.3.1. THE PLANE STRESS INCOMPRESSIBLE OGDEN MODEL

Let the direction 3 – associated with the principal stretch λ_3 – be the *out-of-plane* direction. The plane stress constraint requires that the principal stress in that direction vanishes:

$$\tau_3 = 0, \tag{13.74}$$

and only the in-plane principal stresses, τ_1 and τ_2, may be non-zero. For a general hyperelastic *incompressible* material, the stress tensor can be determined from the constitutive law up to an arbitrary hydrostatic pressure. Thus, considering the particular case of the incompressible Ogden material with strain-energy function defined by (13.48), the principal Kirchhoff stresses are given by

$$\tau_i = \lambda_i\, \bar{\rho}\, \frac{\partial \hat{\psi}}{\partial \lambda_i} + Jp\, \boldsymbol{I} = \left\{ \sum_{p=1}^{N} \mu_p\, \lambda_i^{\alpha_p} \right\} + Jp, \tag{13.75}$$

where Jp is the Kirchhoff pressure. The pressure Jp can be determined by introducing the plane stress state constraint. Indeed, (13.74) together with (13.75) gives

$$\tau_3 = \left\{ \sum_{p=1}^{N} \mu_p\, \lambda_3^{\alpha_p} \right\} + J\,p = 0, \tag{13.76}$$

which implies

$$J\,p = -\sum_{p=1}^{N} \mu_p\, \lambda_3^{\alpha_p}, \tag{13.77}$$

or, equivalently, since $\lambda_1\lambda_2\lambda_3 = 1$ (incompressibility),

$$J\,p = -\sum_{p=1}^{N} \mu_p\, (\lambda_1\lambda_2)^{-\alpha_p}. \tag{13.78}$$

Finally, the substitution of (13.78) into (13.75) results in the following *plane stress-projected* constitutive relation for the incompressible Ogden model;

$$\tau_\beta = \sum_{p=1}^{N} \mu_p\, [\lambda_\beta^{\alpha_p} - (\lambda_1\lambda_2)^{-\alpha_p}] \quad \text{for } \beta = 1, 2. \tag{13.79}$$

This formula is implemented in subroutine SUOGD of HYPLAS.

Thickness update

At finite strains, the thickness at any point depends on the out-of-plane stretch λ_3. In spatial finite element formulations, where integrations are performed over the deformed configuration, the current thickness (thickness in the deformed configuration) is required at each Gauss point during the computation of the internal force vector and stiffness matrix. With t_0 denoting the thickness of the Gauss point of interest in its reference configuration, the current thickness can be computed as

$$t = \lambda_3 \, t_0. \tag{13.80}$$

In view of the incompressibility of the model, the above update formula is equivalent to

$$t = \frac{t_0}{\lambda_1 \lambda_2}. \tag{13.81}$$

Remark 13.3. One important point to be considered is that for isotropic elastic materials, the Kirchhoff stress tensor is coaxial with the left Cauchy–Green strain tensor $\boldsymbol{B}$. Consequently, since the hypothesis of plane stress implies that the non-zero stress components are in-plane, the deformation pattern must be such that transverse shear strains vanish. Clearly, that can only be a good approximation as long as the thickness deformation does not vary abruptly within the plane. If abrupt thickness variations do occur, then the plane stress assumption no longer makes sense.

13.3.2. THE PLANE STRESS HENCKY MODEL

For the Hencky model, the plane stress constraint can also be enforced in a trivial manner. As its standard general version, the stress–strain relation of the plane stress counterpart has also a functional format identical to that of plane stress linear elasticity. That is, its stress–strain relation has the same format as (13.59) with $\boldsymbol{\tau}$ and $\boldsymbol{\varepsilon}$ having only in-plane components and

$$\mathsf{D} \equiv 2G \, \mathsf{I}_S + (K - \tfrac{2}{3}G) \frac{6G}{3K + 4G} \, \boldsymbol{I} \otimes \boldsymbol{I} \tag{13.82}$$

– now a plane tensor – defined as the standard plane stress linear elasticity tensor.

Thickness update

Analogously to the linear elastic case, we have

$$\varepsilon_{33} = -\frac{3K - 2G}{3K + 4G} \, \mathrm{tr}[\boldsymbol{\varepsilon}] = -\frac{\nu}{1 - \nu} \, \mathrm{tr}[\boldsymbol{\varepsilon}], \tag{13.83}$$

where, here,

$$\mathrm{tr}[\boldsymbol{\varepsilon}] = \varepsilon_{11} + \varepsilon_{22} \tag{13.84}$$

is the trace of the *plane* tensor $\boldsymbol{\varepsilon}$. In terms of logarithmic stretches, expression (13.83) is equivalent to

$$\ln \lambda_3 = -\frac{3K - 2G}{3K + 4G} (\ln \lambda_1 + \ln \lambda_2) = -\frac{\nu}{1 - \nu} (\ln \lambda_1 + \ln \lambda_2). \tag{13.85}$$

The thickness stretch is then trivially obtained as

$$\lambda_3 = \exp\left[-\frac{3K-2G}{3K+4G}(\ln\lambda_1 + \ln\lambda_2)\right] = \exp\left[-\frac{\nu}{1-\nu}(\ln\lambda_1 + \ln\lambda_2)\right]. \tag{13.86}$$

With the thickness stretch at hand, the updated thickness, t, can be computed according to (13.80).

13.3.3. PLANE STRESS WITH NESTED ITERATIONS

In the above we have shown two examples where plane stress constitutive equations can be easily derived in closed form from the general three-dimensional law. This, however, is not generally feasible, especially when the nonlinear stress–strain relation has a more complex format. In such cases, one possible alternative in the finite element implementation is to adopt the nested iteration approach to enforce the plane stress constraint upon the three-dimensional law. This methodology has been thoroughly discussed in Section 9.2.2, from page 362, in the infinitesimal elastoplasticity context. As we shall see in what follows, its application to finite hyperelasticity is straightforward.

Let us consider the case where the constitutive law is formulated in terms of principal stretches/stresses, such as the Ogden model. The three-dimensional constitutive equation for the principal Kirchhoff stresses reads

$$\tau_i = \tau(\lambda_i, \lambda_j, \lambda_k) \quad \text{for } i = 1, 2, 3, \tag{13.87}$$

where (i, j, k) are cyclic permutations of $(1, 2, 3)$ and the function τ is nonlinear. For the regularised Ogden model, for instance, the function τ is defined by (13.54). Now let direction 3 be the out-of-plane principal direction and assume that the in-plane stretches $\{\lambda_1, \lambda_2\}$ are *given*. The enforcement of the plane stress constraint,

$$\tau_3 = 0,$$

then reduces to the solution of the following nonlinear equation for the *unknown* thickness stretch λ_3:

$$\bar{\tau}_3(\lambda_3) \equiv \tau(\lambda_3, \lambda_1, \lambda_2) = 0. \tag{13.88}$$

The nested iteration stress update

A possible stress-updating procedure in this case can be summarised in the following steps:

1. Given the *in-plane* principal stretches $\{\lambda_1, \lambda_2\}$, use the Newton–Raphson method (any other root-finding technique may be used) to solve the nonlinear equation

   $$\bar{\tau}(\lambda_3) = 0,$$

   for λ_3. For the typical k^{th} Newton iteration, we update

   $$\lambda_3^{(k+1)} = \lambda_3^{(k)} - \frac{\tau_3^{(k)}}{\bar{\tau}'},$$

   where $\bar{\tau}'$ is the derivative of τ with respect to its first argument evaluated at $(\lambda_3^{(k)}, \lambda_1, \lambda_2)$. The initial guess, $\lambda_3^{(0)}$, may be taken as the thickness stretch obtained in the last converged equilibrium configuration.

2. With the solution λ_3 at hand, use the three-dimensional constitutive function to update the in-plane Kirchhoff stresses τ_1 and τ_2:

$$\tau_1 = \tau(\lambda_1, \lambda_2, \lambda_3)$$
$$\tau_2 = \tau(\lambda_2, \lambda_3, \lambda_1).$$

Clearly, τ_3 vanishes within the prescribed tolerance for solution of the nonlinear plane stress enforcement equation.

3. The current thickness is updated by the formula

$$t = \lambda_3 \, t_0.$$

•

The above procedure can be applied to any model. The remarks made in Section 9.2.2 concerning the computational implications of this approach, i.e. the introduction of an iteration loop for plane stress enforcement at the Gauss point level – nested within an outer equilibrium loop – applies equally to the present case.

13.4. Tangent moduli: the elasticity tensors

Firstly, let us recall the general component formula (C.31), given on page 757, for the tangent modulus in the *spatial* frame – the *spatial elasticity tensor*. Under material *isotropy*, where $\boldsymbol{\tau}$ can be expressed as a function of $\boldsymbol{B}$ only, the spatial elasticity tensor can be equivalently represented as

$$\mathsf{a}_{ijkl} = \frac{2}{J}\frac{\partial \tau_{ij}}{\partial B_{km}} \, B_{ml} - \sigma_{il} \, \delta_{jk}. \tag{13.89}$$

The above formula is obtained from (C.31) with the application of the chain rule considering the functional dependence of $\boldsymbol{\tau}$ on $\boldsymbol{B}$ and the dependence of $\boldsymbol{B}$ on $\boldsymbol{F}$. Our objective in this section is to derive *exact* expressions for the spatial tangent moduli consistent with some of the above described hyperelastic constitutive models. In the finite element context, the expressions derived can be employed in the assemblage of the element *tangent stiffness*, needed by the Newton–Raphson scheme for solution of the (equilibrium) discrete boundary value problem. We remark that for hyperelastic models (whether isotropic or not), the fourth-order tensor a is symmetric, i.e. its Cartesian components have the symmetry

$$\mathsf{a}_{ijkl} = \mathsf{a}_{klij}, \tag{13.90}$$

which results in symmetric stiffness matrices in the spatial formulation.

It should be noted that the only term in (13.89) that depends on the particular hyperelastic model is the components of the derivative (fourth-order tensor)

$$\frac{\partial \boldsymbol{\tau}}{\partial \boldsymbol{B}}.$$

Thus, in what follows, for each of the models considered, we derive an expression for the derivative of the Kirchhoff stress which, when substituted in (13.89), yields a final model-specific expression for a. Details of the computational implementation of formula (13.89) for the particular case of the Ogden material model is described in Box 13.2, page 543.

13.4.1. REGULARISED NEO-HOOKEAN MODEL

To obtain a compact representation of the spatial elasticity tensor for the regularised neo-Hookean model, we start by differentiating expression (13.47), which gives

$$\frac{\partial \boldsymbol{\tau}}{\partial \boldsymbol{B}} = G\,\mathsf{I}_{\mathrm{d}} : \frac{\partial \boldsymbol{B}_{\mathrm{iso}}}{\partial \boldsymbol{B}} + \frac{K}{J}\,\boldsymbol{I} \otimes \frac{\partial J}{\partial \boldsymbol{B}} \tag{13.91}$$

where the fourth-order tensor I_{d} is the deviatoric projection operator in the space of symmetric tensors, defined by expression (3.94), page 59. Next, a straightforward differentiation of (13.41) leads to

$$\frac{\partial \boldsymbol{B}_{\mathrm{iso}}}{\partial \boldsymbol{B}} = -\tfrac{2}{3}\,J^{-\frac{5}{3}}\,\boldsymbol{B} \otimes \frac{\partial J}{\partial \boldsymbol{B}} + J^{-\frac{2}{3}}\,\mathsf{I}_S. \tag{13.92}$$

Substitution of this formula into (13.91), together with the expression

$$\frac{\partial J}{\partial \boldsymbol{B}} = \tfrac{1}{2}\,J\,\boldsymbol{B}^{-1}, \tag{13.93}$$

gives

$$\begin{aligned}\frac{\partial \boldsymbol{\tau}}{\partial \boldsymbol{B}} &= G\,\mathsf{I}_{\mathrm{d}} : [J^{-\frac{2}{3}}\,\mathsf{I}_S - \tfrac{1}{3}\,J^{-\frac{2}{3}}\,\boldsymbol{B} \otimes \boldsymbol{B}^{-1}] + \tfrac{1}{2}\,K\,\boldsymbol{I} \otimes \boldsymbol{B}^{-1} \\ &= G\,J^{-\frac{2}{3}}\,\mathsf{I}_{\mathrm{d}} - \tfrac{1}{3}\,\boldsymbol{\tau}_{\mathrm{d}} \otimes \boldsymbol{B}^{-1} + \tfrac{1}{2}\,K\,\boldsymbol{I} \otimes \boldsymbol{B}^{-1},\end{aligned} \tag{13.94}$$

where $\boldsymbol{\tau}_{\mathrm{d}}$ is the deviatoric component of the Kirchhoff stress tensor. Expression (13.93) is obtained by direct application of the formula for the derivative of the determinant

$$\frac{\mathrm{d}(\det \boldsymbol{A})}{\mathrm{d}\boldsymbol{A}} = (\det \boldsymbol{A})\,\boldsymbol{A}^{-T}, \tag{13.95}$$

valid for any invertible $\boldsymbol{A}$, to the definition $J \equiv \sqrt{\det \boldsymbol{B}}$. Finally, by replacing the component form of (13.94) into (13.89), we arrive, after some algebra, at the following expression for a:

$$\mathsf{a}_{ijkl} = \mathsf{c}_{ijkl} + \delta_{ik}\sigma_{jl} \tag{13.96}$$

where

$$\mathsf{c} = \frac{2G}{3J}\,\mathrm{tr}[\boldsymbol{B}_{\mathrm{iso}}]\,\mathsf{I}_{\mathrm{d}} - 2p\,\mathsf{I}_S + \frac{K}{J}\,\boldsymbol{I} \otimes \boldsymbol{I} - \tfrac{2}{3}\,[\boldsymbol{s} \otimes \boldsymbol{I} + \boldsymbol{I} \otimes \boldsymbol{s}]. \tag{13.97}$$

In the above expression, $\boldsymbol{s}$ is the deviatoric *Cauchy* stress and p denotes the *Cauchy* hydrostatic pressure, i.e. $p \equiv \mathrm{tr}[\boldsymbol{\sigma}]/3$.

13.4.2. PRINCIPAL STRETCHES REPRESENTATION: OGDEN MODEL

Clearly, for all models discussed in this chapter, the Kirchhoff stress is an *isotropic* tensor function of a single tensor, $\boldsymbol{B}$. For the Ogden material model, however, the function $\boldsymbol{\tau}(\boldsymbol{B})$ is defined in such a way that the eigenvalues of the function, τ_i, can be expressed explicitly in simple format as functions of the eigenvalues, b_j, of the tensor argument $\boldsymbol{B}$; that is, we have

$$\tau_i = \hat{\tau}(\lambda_i, \lambda_j, \lambda_k) = \tilde{\tau}(b_i, b_j, b_k), \tag{13.98}$$

where (i, j, k) are cyclic permutations of $(1, 2, 3)$ and

$$\boldsymbol{\tau}(\boldsymbol{B}) = \sum_{i=1}^{3} \tilde{\tau}(b_i, b_j, b_k)\, \boldsymbol{e}_i \otimes \boldsymbol{e}_i, \tag{13.99}$$

where $\boldsymbol{e}_i$ are unit eigenvectors of both $\boldsymbol{\tau}$ and $\boldsymbol{B}$. Note that

$$b_i = \lambda_i^2, \quad \text{for } i = 1, 2, 3. \tag{13.100}$$

In the present case, it is crucial to observe that in (13.99) $\boldsymbol{\tau}(\boldsymbol{B})$ is represented explicitly in the same format as the general isotropic function defined by (A.15) in Appendix A (page 734). Thus, the derivative $\partial\boldsymbol{\tau}/\partial\boldsymbol{B}$ can be conveniently evaluated simply by setting $\boldsymbol{Y} = \boldsymbol{\tau}$, $y_i = \tau_i$, $\boldsymbol{X} = \boldsymbol{B}$ and $x_i = b_i$ in the general expressions for the derivative of isotropic tensor functions summarised in Boxes A.3 and A.6, respectively, for the two- and three-dimensional cases. In the computational implementation of the model (described in Section 13.5) we first evaluate the components of $\partial\boldsymbol{\tau}/\partial\boldsymbol{B}$ and then use (13.89) to compute the components of the spatial elasticity tensor. Note (see Boxes A.3 and A.6) that the eigenvalue derivatives

$$\frac{\partial \tau_i}{\partial b_j}, \quad i, j = 1, 2, 3$$

are required to compute $\partial\boldsymbol{\tau}/\partial\boldsymbol{B}$. The explicit form of the eigenvalue derivatives is given in what follows for the regularised version of the Ogden model as well as for the incompressible model under plane stress. These expressions are implemented in subroutine `CSTOGD` of program `HYPLAS`.

Regularised Ogden material

Straightforward differentiation of (13.54) leads to the following principal Kirchhoff stress derivatives for the regularised Ogden material:

$$\begin{aligned}\frac{\partial \tau_i}{\partial b_j} &= \frac{\partial \tau_i}{\partial \lambda_j}\frac{\partial \lambda_j}{\partial b_j}\\ &= \sum_{p=1}^{N} \frac{\mu_p \alpha_p\, J^{-\alpha_p/3}}{6\lambda_j^2}\left[\frac{1}{3}(\lambda_1^{\alpha_p} + \lambda_2^{\alpha_p} + \lambda_3^{\alpha_p}) - \lambda_i^{\alpha_p} - \lambda_j^{\alpha_p} + 3\lambda_i^{\alpha_p}\delta_{ij}\right] + \frac{K}{2\lambda_j^2}\end{aligned} \tag{13.101}$$

where no summation is implied on repeated indices and δ_{ij} is the Krönecker delta.

Incompressible Ogden material in plane stress

For the incompressible version under plane stress, we differentiate (13.79) to obtain

$$\frac{\partial \tau_i}{\partial b_j} = \sum_{p=1}^{N} \frac{\mu_p \alpha_p}{2\lambda_j^2}[\lambda_j^{\alpha_p}\delta_{ij} + (\lambda_1\lambda_2)^{-\alpha_p}] \tag{13.102}$$

with no summation on repeated indices.

Remark 13.4. As any isotropic model admits a principal stretch-based representation of the function $\boldsymbol{\tau}(\boldsymbol{B})$, we can, in principle, apply the methodology described above for the Ogden material to compute the elasticity tensor for any isotropic elastic model. For the regularised neo-Hookean model, whose expression for a is given by (13.96, 13.97), the principal stretch-based computation of a can be carried out by setting $N=1$, $\mu_1 = G$ and $\alpha_1 = 2$ in (13.101). For this particular model, however, it is computationally more efficient to use the more compact form (13.96, 13.97).

13.4.3. HENCKY MODEL

As the Kirchhoff stress for the Hencky material is a linear function of the logarithmic Eulerian strain which, through its definition (13.55), is a function of $\boldsymbol{B}$, the derivative $\partial\boldsymbol{\tau}/\partial\boldsymbol{B}$ can be derived by a straightforward application of the chain rule to (13.59). This gives

$$\frac{\partial \boldsymbol{\tau}}{\partial \boldsymbol{B}} = \frac{\partial \boldsymbol{\tau}}{\partial \boldsymbol{\varepsilon}} : \frac{\partial \boldsymbol{\varepsilon}}{\partial \boldsymbol{B}} = \frac{1}{2}\,\mathsf{D} : \mathsf{L}, \tag{13.103}$$

where

$$\mathsf{L} \equiv \frac{\partial(\ln \boldsymbol{B})}{\partial \boldsymbol{B}} \tag{13.104}$$

is the derivative of the *tensor logarithm function* evaluated at $\boldsymbol{B}$.

At this point, we should observe that the tensor logarithm is a member of the class of isotropic tensor functions discussed in Section A.5 of Appendix A. Thus, the fourth-order tensor L can be evaluated according to the formulae shown in that section. The computation of such derivatives in program `HYPLAS` is implemented in subroutine `DISO2`.

In order to derive a compact expression for the Hencky elasticity tensor, we note that (13.89) is equivalent to

$$\mathsf{a}_{ijkl} = \frac{1}{J}\left[\frac{\partial \boldsymbol{\tau}}{\partial \boldsymbol{B}} : \mathsf{B}\right]_{ijkl} - \sigma_{il}\delta_{jk}, \tag{13.105}$$

where we have defined the fourth-order tensor B by

$$\mathsf{B}_{ijkl} \equiv \delta_{ik}B_{jl} + \delta_{jk}B_{il}. \tag{13.106}$$

Finally, formula (13.105) together with (13.103) yields the following expression for the Hencky spatial elasticity tensor

$$\mathsf{a}_{ijkl} = \frac{1}{2J}\,[\mathsf{D} : \mathsf{L} : \mathsf{B}]_{ijkl} - \sigma_{il}\delta_{jk}, \tag{13.107}$$

which has a particularly simple format. We remark that the above formula is equally valid for the plane stress case discussed in Section 13.3.2.

13.4.4. BLATZ–KO MATERIAL

Let us now consider the Blatz–Ko material with the stress constitutive function given in (13.72). To derive an explicit formula for $\partial\boldsymbol{\tau}/\partial\boldsymbol{B}$, we differentiate expression (13.72) and obtain

$$\frac{\partial \boldsymbol{\tau}}{\partial \boldsymbol{B}} = g'(J)\,\boldsymbol{I} \otimes \frac{\partial J}{\partial \boldsymbol{B}} + f\,\mu_0\,\mathsf{I}_S - (1-f)\mu_0\,\frac{\partial \boldsymbol{B}^{-1}}{\partial \boldsymbol{B}}. \tag{13.108}$$

By differentiation of the identity[‡] $B_{ik}B^{-1}_{kj} = \delta_{ij}$ the following component expression for the last derivative on the right-hand side of the above formula is obtained:

$$\frac{\partial B^{-1}_{ij}}{\partial B_{kl}} = -\frac{1}{2}(B^{-1}_{ik}B^{-1}_{lj} + B^{-1}_{il}B^{-1}_{jk}). \tag{13.109}$$

Substitution of this expression together with (13.93) into (13.108) then gives

$$\frac{\partial \tau_{ij}}{\partial B_{kl}} = \frac{J}{2}\, g'(J)\, \delta_{ij} B^{-1}_{kl} + \frac{f\,\mu_0}{2}\, (\delta_{ik}\delta_{jl} + \delta_{il}\delta_{jk}) + \frac{(1-f)\mu_0}{2}\, (B^{-1}_{ik}B^{-1}_{jl} + B^{-1}_{il}B^{-1}_{jk}). \tag{13.110}$$

Finally, by substituting the above expression into (13.89), the following explicit formula for the Cartesian components of a is obtained:

$$\begin{aligned}\mathsf{a}_{ijkl} = {} & g'(J)\, \delta_{ij}\delta_{kl} + \frac{f\,\mu_0}{J}\, (\delta_{ik}B_{jl} + \delta_{jk}B_{il}) \\ & + \frac{(1-f)\mu_0}{J}\, (\delta_{il}B^{-1}_{jk} + \delta_{jl}B^{-1}_{ik}) - \sigma_{il}\delta_{jk}.\end{aligned} \tag{13.111}$$

The computational implementation of this formula is a relatively straightforward programming exercise.

13.5. Application: Ogden material implementation

Practical use of the concepts introduced in the previous sections of this chapter is made in this section where the computational implementation of the Ogden material model is described in detail. The fundamental routines of HYPLAS shown here are: SUOGD (**S**tate **U**pdate procedure for the **OGD**en material model), and CSTOGD (**C**onsistent **S**patial **T**angent modulus for the **OGD**en material model). These have been coded for the plane stress, plane strain and axisymmetric cases and use the two versions of the model (incompressible and regularised) as follows:

1. plane stress – incompressible version only (described in Section 13.3.1);
2. plane strain – regularised version only;
3. axisymmetric – regularised version only.

13.5.1. SUBROUTINE SUOGD

The pseudo-code of this procedure is provided in Box 13.1 which shows the steps required (with the corresponding expressions) in the computation of the stress for the Ogden material model. For a given deformation gradient $\boldsymbol{F}$, SUOGD returns the corresponding *Cauchy* stress components. In program HYPLAS, the stresses computed in SUOGD are used in subroutine IFSTD2 (or IFFB2) to assemble the element internal force vector. The FORTRAN source code of SUOGD in listed below.

[‡] B^{-1}_{kj} denotes the component kj of the tensor $\boldsymbol{B}^{-1}$.

Box 13.1. Stress-updating procedure for the Ogden material model.

HYPLAS procedure:	SUOGD

(i) Given the deformation gradient, $\boldsymbol{F}$, compute $\boldsymbol{B} := \boldsymbol{F}\,\boldsymbol{F}^T$

(ii) Perform spectral decomposition of $\boldsymbol{B}$ (routine SPDEC2) and compute principal stretches

$$\boldsymbol{B} = \sum_i b_i\,\boldsymbol{M}_i, \qquad \lambda_i := \sqrt{b_i}$$

(iii) Compute principal Kirchhoff stresses

(a) Plane stress. Use *incompressible* model

$$\tau_\beta := \sum_{p=1}^{N} \mu_p[\lambda_\beta^{\alpha_p} - (\lambda_1\lambda_2)^{-\alpha_p}]$$

(b) Plane strain and axisymmetric. Use *regularised* model

$$\tau_i := \sum_{p=1}^{N} \mu_p J^{-\alpha_p/3}[\lambda_i^{\alpha_p} - \tfrac{1}{3}(\lambda_1^{\alpha_p} + \lambda_2^{\alpha_p} + \lambda_3^{\alpha_p})] + K \ln J$$

(iv) Assemble Cauchy stress

$$\boldsymbol{\sigma} := J^{-1} \sum_i \tau_i\,\boldsymbol{M}_i$$

```
      SUBROUTINE SUOGD
     1(   B           ,IPROPS      ,NTYPE       ,RPROPS      ,RSTAVA      ,
     2    STRES       ,THICK       )
      IMPLICIT DOUBLE PRECISION (A-H,O-Z)
      PARAMETER(IPOGDC=2)
      LOGICAL DUMMY
      PARAMETER
     1(   MCOMP=4     ,MSTRE=4     ,NDIM=2      )
      DIMENSION
     1    B(MCOMP)            ,IPROPS(*)           ,RPROPS(*)           ,
     2    RSTAVA(MSTRE)       ,STRES(MSTRE)
      DIMENSION
     1    EIGPRJ(MCOMP,NDIM)  ,EIGB(NDIM)          ,PSTRES(3)           ,
     2    PSTRTC(3)
      DATA  R1    ,R3    /
     1      1.0D0,3.0D0/
C***********************************************************************
C STRESS UPDATE PROCEDURE FOR OGDEN TYPE HYPERELASTIC MATERIAL MODEL.
C PLANE STRESS, PLANE STRAIN AND AXISYMMETRIC IMPLEMENTATIONS.
C***********************************************************************
C Retrieve Ogden material constants
C =================================
C Number of terms in Ogden's strain-energy function
      NOGTRM=IPROPS(3)
```

```
C Bulk modulus
      BULK=RPROPS(IPOGDC+NOGTRM*2)
C Compute principal stretches
C ===========================
C Perform spectral decomposition of the left Cauchy-Green tensor B
      CALL SPDEC2
     1(   EIGPRJ     ,EIGB       ,DUMMY      ,B          )
C Compute in-plane principal stretches
      PSTRTC(1)=SQRT(EIGB(1))
      PSTRTC(2)=SQRT(EIGB(2))
C...and out-of-plane stretches
      IF(NTYPE.EQ.1)THEN
        PSTRTC(3)=R1/(PSTRTC(1)*PSTRTC(2))
      ELSEIF(NTYPE.EQ.2)THEN
        PSTRTC(3)=R1
      ELSEIF(NTYPE.EQ.3)THEN
        PSTRTC(3)=SQRT(B(4))
      ENDIF
C Compute principal Kirchhoff stresses
C ====================================
      CALL RVZERO(PSTRES,3)
      IF(NTYPE.EQ.1) THEN
C Plane stress: Exact incompressibility assumed
C --------------------------------------------
        DO 10 I=1,NOGTRM
          CMU=RPROPS(IPOGDC-1+I*2-1)
          ALPHA=RPROPS(IPOGDC-1+I*2)
          PSTRES(1)=PSTRES(1)+CMU*(PSTRTC(1)**ALPHA-
     1               (PSTRTC(1)*PSTRTC(2))**(-ALPHA))
          PSTRES(2)=PSTRES(2)+CMU*(PSTRTC(2)**ALPHA-
     1               (PSTRTC(1)*PSTRTC(2))**(-ALPHA))
   10   CONTINUE
        DETF=R1
      ELSE IF(NTYPE.EQ.2.OR.NTYPE.EQ.3)THEN
C Plane strain and axisymmetric: Regularised Ogden constitutive law
C -----------------------------------------------------------------
C Compute principal deviatoric Kirchhoff stresses
        R1D3=R1/R3
        DETF=PSTRTC(1)*PSTRTC(2)
        IF(NTYPE.EQ.3)DETF=DETF*PSTRTC(3)
        DO 20 I=1,NOGTRM
          CMU=RPROPS(IPOGDC-1+I*2-1)
          ALPHA=RPROPS(IPOGDC-1+I*2)
          FACTOR=R1D3*(PSTRTC(1)**ALPHA+PSTRTC(2)**ALPHA+
     1            PSTRTC(3)**ALPHA)
          FACVOL=DETF**(-ALPHA*R1D3)
          PSTRES(1)=PSTRES(1)+CMU*FACVOL*(PSTRTC(1)**ALPHA-FACTOR)
          PSTRES(2)=PSTRES(2)+CMU*FACVOL*(PSTRTC(2)**ALPHA-FACTOR)
          PSTRES(3)=PSTRES(3)+CMU*FACVOL*(PSTRTC(3)**ALPHA-FACTOR)
   20   CONTINUE
C Add hydrostatic Kirchhoff pressure (incompressibility penalty term)
        PRESS=BULK*LOG(DETF)
        DO 30 I=1,3
          PSTRES(I)=PSTRES(I)+PRESS
   30   CONTINUE
      ENDIF
C Assemble array of Cauchy stress tensor components
C =================================================
      CALL RVZERO(STRES,3)
      R1DDET=R1/DETF
```

```
      PSTRES(1)=PSTRES(1)*R1DDET
      PSTRES(2)=PSTRES(2)*R1DDET
      DO 50 ICOMP=1,3
        DO 40 IDIR=1,2
          STRES(ICOMP)=STRES(ICOMP)+PSTRES(IDIR)*EIGPRJ(ICOMP,IDIR)
   40   CONTINUE
   50 CONTINUE
      IF(NTYPE.EQ.2.OR.NTYPE.EQ.3)STRES(4)=PSTRES(3)*R1DDET
C Update thickness (plane stress only) and store left Cauchy-Green
C tensor components in state variables vector RSTAVA
C =========================================================================
      RSTAVA(1)=B(1)
      RSTAVA(2)=B(2)
      RSTAVA(3)=B(3)
      IF(NTYPE.EQ.1)THEN
        THICK=THICK*PSTRTC(3)
        RSTAVA(4)=PSTRTC(3)*PSTRTC(3)
      ELSEIF(NTYPE.EQ.2)THEN
        RSTAVA(4)=R1
      ELSEIF(NTYPE.EQ.3)THEN
        RSTAVA(4)=B(4)
      ENDIF
C
      RETURN
      END
```

The arguments of SUOGD

$\rightarrow$ B $[\boldsymbol{B}]$. Array containing the components of the current left Cauchy–Green strain tensor.

$\rightarrow$ IPROPS. Array of integer material properties. IPROPS(3) contains the number of terms in the Ogden strain-energy function, NOGTRM. This is the only integer material property required by SUOGD. IPROPS(3) is set in subroutine RDOGD during the input phase of HYPLAS.

$\rightarrow$ NTYPE. Stress state-type flag. Its value is: 1 for plane stress, 2 for plane strain and 3 for axisymmetric case.

$\rightarrow$ RPROPS. Array of real material properties. It contains the Ogden constants $\{\mu_p, \alpha_p\}$ for $p = 1, \ldots,$ NOGTRM and the logarithmic bulk modulus, K. The bulk modulus is used only in the compressible version of the Ogden model, adopted in the plane strain and axisymmetric cases only. It is not used in plane stress where the incompressible version of the model is adopted. Array RPROPS is set in subroutine RDOGD during the input phase of HYPLAS.

$\leftarrow$ RSTAVA $[\boldsymbol{B}]$. Array of state variables other than the stress components. For the present implementation of the Ogden material, this array stores the current left Cauchy–Green tensor, $\boldsymbol{B}$.

$\leftarrow$ STRES $[\boldsymbol{\sigma}]$. Array containing the updated Cauchy stress tensor components.

$\leftrightarrow$ THICK $[t_0, t]$. Initial Gauss point thickness, t_0, on entry. Returns as the updated thickness, t. Used only in the plane stress case.

Some local variables and arrays of SUOGD

- EIGB $[b_i]$. Array of eigenvalues of $\boldsymbol{B}$.
- EIGPRJ $[\boldsymbol{M}_i, i = 1, 2]$. Matrix containing the components of the in-plane eigenprojection tensors of $\boldsymbol{B}$.
- PSTRES $[\tau_i]$. Array of principal Kirchhoff stresses.
- PSTRTC $[\lambda_i]$. Array of principal stretches.

Function calls from SUOGD

- RVZERO. Initialises relevant arrays with zeros.
- SPDEC2. Called to perform the spectral decomposition of $\boldsymbol{B}$.

13.5.2. SUBROUTINE CSTOGD

This routine computes the consistent (exact) spatial tangent modulus, $\mathbf{a}$, for the Ogden material model whose stress update is implemented in SUOGD. The main steps and expressions (previously derived) involved in the computation of $\mathbf{a}$ are shown in Box 13.2 in pseudo-code format. The FORTRAN source code of CSTOGD is given below.

```
      SUBROUTINE CSTOGD
     1(   AMATX      ,B           ,IPROPS      ,NTYPE       ,RPROPS      ,
     2    STRES      )
      IMPLICIT DOUBLE PRECISION (A-H,O-Z)
      LOGICAL OUTOFP ,REPEAT
      PARAMETER
     1(   MADIM=5    ,MSTRE=4     ,NDIM=2      )
      PARAMETER(IPOGDC=2)
      DIMENSION
     1    AMATX(MADIM,MADIM) ,B(MSTRE)             ,IPROPS(*)         ,
     2    RPROPS(*)          ,STRES(MSTRE)
      DIMENSION
     1    DELTA(3,3)         ,DPSTRE(3,3)          ,DTAUDB(MSTRE,MSTRE),
     2    EIGPRJ(MSTRE,NDIM) ,EIGB(NDIM)           ,PSTALP(3)         ,
     3    PSTRES(3)          ,PSTRTC(3)
      DATA
     1    DELTA(1,1)    ,DELTA(1,2)    ,DELTA(1,3)    /
     2    1.0D0         ,0.0D0         ,0.0D0         /
     3    DELTA(2,1)    ,DELTA(2,2)    ,DELTA(2,3)    /
     4    0.0D0         ,1.0D0         ,0.0D0         /
     5    DELTA(3,1)    ,DELTA(3,2)    ,DELTA(3,3)    /
     6    0.0D0         ,0.0D0         ,1.0D0         /
      DATA
     1    R1    ,R2    ,R3    ,R6    /
     2    1.0D0,2.0D0,3.0D0,6.0D0/
C***********************************************************************
C COMPUTATION OF THE CONSISTENT SPATIAL TANGENT MODULUS 'a' FOR
C OGDEN TYPE HYPERELASTIC MATERIAL MODEL.
C PLANE STRESS, PLANE STRAIN AND AXISYMMETRIC IMPLEMENTATIONS.
C***********************************************************************
```

Box 13.2. Computation of the spatial tangent modulus for the Ogden material model.

HYPLAS procedure:	CSTOGD

(i) Given the deformation gradient, $\boldsymbol{F}$, compute $\boldsymbol{B} := \boldsymbol{F}\,\boldsymbol{F}^T$

(ii) Perform spectral decomposition of $\boldsymbol{B}$ (routine SPDEC2) and compute principal stretches

$$\boldsymbol{B} = \sum_i b_i\,\boldsymbol{M}_i, \qquad \lambda_i := \sqrt{b_i}$$

(iii) Recover principal Kirchhoff stresses (from given $\boldsymbol{\sigma}$)

$$\tau_i := J\,\boldsymbol{\sigma} : \boldsymbol{M}_i$$

(iv) Compute principal Kirchhoff stress derivatives

(a) Plane stress. Use *incompressible* model

$$\frac{\partial \tau_i}{\partial b_j} := \sum_{p=1}^{N} \frac{\mu_p \alpha_p}{2\lambda_j^2}\,[\lambda_j^{\alpha_p}\delta_{ij} + (\lambda_1\lambda_2)^{-\alpha_p}]$$

(b) Plane strain and axisymmetric. Use *regularised* model

$$f := \tfrac{1}{3}(\lambda_1^{\alpha_p} + \lambda_2^{\alpha_p} + \lambda_3^{\alpha_p})$$

$$\frac{\partial \tau_i}{\partial b_j} := \sum_{p=1}^{N} \frac{\mu_p \alpha_p\, J^{-\alpha_p/3}}{6\lambda_j^2}[f - \lambda_i^{\alpha_p} - \lambda_j^{\alpha_p} + 3\lambda_i^{\alpha_p}\delta_{ij}] + \frac{K}{2\lambda_j^2}$$

(v) Assemble $\partial\boldsymbol{\tau}/\partial\boldsymbol{B}$ (use routine DGISO2)

(vi) Compute spatial tangent modulus

$$\mathsf{a}_{ijkl} = \frac{2}{J}\frac{\partial \tau_{ij}}{\partial B_{km}}\,B_{ml} - \sigma_{il}\,\delta_{jk}$$

```
C Set Ogden material constants
C ============================
C Number of terms in Ogden's strain-energy function
      NOGTRM=IPROPS(3)
C Bulk modulus (incompressibility penalty parameter)
      BULK=RPROPS(IPOGDC+NOGTRM*2)
C Compute principal stretches
C ===========================
C Perform spectral decomposition of the left Cauchy-Green tensor B
      CALL SPDEC2
     1(   EIGPRJ     ,EIGB       ,REPEAT     ,B          )
C Compute in-plane principal stretches
      PSTRTC(1)=SQRT(EIGB(1))
      PSTRTC(2)=SQRT(EIGB(2))
C and out-of-plane stretches
      IF(NTYPE.EQ.1)THEN
```

```
          PSTRTC(3)=R1/(PSTRTC(1)*PSTRTC(2))
          DETF=R1
        ELSEIF(NTYPE.EQ.2)THEN
          PSTRTC(3)=R1
          DETF=PSTRTC(1)*PSTRTC(2)
        ELSEIF(NTYPE.EQ.3)THEN
          PSTRTC(3)=SQRT(B(4))
          DETF=PSTRTC(1)*PSTRTC(2)*PSTRTC(3)
        ENDIF
C Recover principal Kirchhoff stresses (from the given Cauchy stress)
        PSTRES(1)=(STRES(1)*EIGPRJ(1,1)+STRES(2)*EIGPRJ(2,1)+
     1                                R2*STRES(3)*EIGPRJ(3,1))*DETF
        PSTRES(2)=(STRES(1)*EIGPRJ(1,2)+STRES(2)*EIGPRJ(2,2)+
     1                                R2*STRES(3)*EIGPRJ(3,2))*DETF
        IF(NTYPE.EQ.2.OR.NTYPE.EQ.3)PSTRES(3)=STRES(4)*DETF
C Compute derivatives of principal Kirchhoff stresses
C ===================================================
        CALL RVZERO(DPSTRE,9)
        IF(NTYPE.EQ.1) THEN
C Plane stress: Perfectly incompressibility assumed
C -------------------------------------------------
          NSTRA=2
          DO 10 IP=1,NOGTRM
            ALPHA=RPROPS(IPOGDC+IP*2-1)
            ALPHMU=ALPHA*RPROPS(IPOGDC+IP*2-2)
            PSTALP(1)=PSTRTC(1)**ALPHA
            PSTALP(2)=PSTRTC(2)**ALPHA
            FACTOR=R1/(PSTALP(1)*PSTALP(2))
            DO I=1,NSTRA
              DO J=1,NSTRA
                DPSTRE(I,J)=DPSTRE(I,J)+ALPHMU/(R2*PSTRTC(J)**2)*
     1                      (PSTALP(J)*DELTA(I,J)+FACTOR)
              END DO
            END DO
   10     CONTINUE
        ELSE IF(NTYPE.EQ.2.OR.NTYPE.EQ.3)THEN
C Plane strain and axisymmetric: Regularised Ogden model
C ------------------------------------------------------
C compute principal Kirchhoff stresses derivatives
          R1D3=R1/R3
          IF(NTYPE.EQ.2)THEN
            NSTRA=2
          ELSEIF(NTYPE.EQ.3)THEN
            NSTRA=3
          ENDIF
          DO 40 IP=1,NOGTRM
            CMU=RPROPS(IPOGDC-1+IP*2-1)
            ALPHA=RPROPS(IPOGDC-1+IP*2)
            PSTALP(1)=PSTRTC(1)**ALPHA
            PSTALP(2)=PSTRTC(2)**ALPHA
            PSTALP(3)=PSTRTC(3)**ALPHA
            ALPHMU=ALPHA*CMU
            FACTOR=R1D3*(PSTALP(1)+PSTALP(2)+PSTALP(3))
            FACVOL=DETF**(-ALPHA*R1D3)
            DO 30 I=1,NSTRA
              DO 20 J=1,NSTRA
                DPSTRE(I,J)=DPSTRE(I,J)+ALPHMU*FACVOL/(R6*PSTRTC(J)**2)*
     1                      (FACTOR-PSTALP(I)-PSTALP(J)+R3*PSTALP(I)*
     2                      DELTA(I,J))
   20         CONTINUE
   30       CONTINUE
```

```
   40 CONTINUE
      DO 60 I=1,NSTRA
        DO 50 J=1,NSTRA
          DPSTRE(I,J)=DPSTRE(I,J)+BULK/(R2*PSTRTC(J)**2)
   50   CONTINUE
   60 CONTINUE
      ENDIF
C Compute the derivative of the Kirchhoff stress with respect to B
C (use routine for computation of derivative of general isotropic
C tensor functions of one tensor)
C =================================================================
      IF(NTYPE.EQ.3)THEN
        OUTOFP=.TRUE.
        NADIM=5
      ELSE
        OUTOFP=.FALSE.
        NADIM=4
      ENDIF
      CALL DGISO2
     1(   DPSTRE     ,DTAUDB     ,EIGPRJ     ,EIGB       ,PSTRES     ,
     2    OUTOFP     ,REPEAT     )
C Assemble the spatial tangent modulus 'a'
C ========================================
      R2DDET=R2/DETF
C upper triangle and diagonal terms
      AMATX(1,1)=R2DDET*(DTAUDB(1,1)*B(1)+DTAUDB(1,3)*B(3))-STRES(1)
      AMATX(1,2)=R2DDET*(DTAUDB(1,3)*B(1)+DTAUDB(1,2)*B(3))
      AMATX(1,3)=R2DDET*(DTAUDB(1,1)*B(3)+DTAUDB(1,3)*B(2))-STRES(3)
      AMATX(1,4)=R2DDET*(DTAUDB(1,3)*B(3)+DTAUDB(1,2)*B(2))
      AMATX(2,2)=R2DDET*(DTAUDB(3,3)*B(1)+DTAUDB(3,2)*B(3))
      AMATX(2,3)=R2DDET*(DTAUDB(3,1)*B(3)+DTAUDB(3,3)*B(2))-STRES(2)
      AMATX(2,4)=R2DDET*(DTAUDB(3,3)*B(3)+DTAUDB(3,2)*B(2))
      AMATX(3,3)=R2DDET*(DTAUDB(3,1)*B(3)+DTAUDB(3,3)*B(2))
      AMATX(3,4)=R2DDET*(DTAUDB(3,3)*B(3)+DTAUDB(3,2)*B(2))-STRES(3)
      AMATX(4,4)=R2DDET*(DTAUDB(2,3)*B(3)+DTAUDB(2,2)*B(2))-STRES(2)
      IF(NTYPE.EQ.3) THEN
        AMATX(1,5)=R2DDET*DTAUDB(1,4)*B(4)
        AMATX(2,5)=R2DDET*DTAUDB(3,4)*B(4)
        AMATX(3,5)=R2DDET*DTAUDB(3,4)*B(4)
        AMATX(4,5)=R2DDET*DTAUDB(2,4)*B(4)
        AMATX(5,5)=R2DDET*DTAUDB(4,4)*B(4)-STRES(4)
      ENDIF
C lower triangle
      DO 80 J=1,NADIM
        DO 70 I=J+1,NADIM
          AMATX(I,J)=AMATX(J,I)
   70   CONTINUE
   80 CONTINUE
C
      RETURN
      END
```

The arguments of CSTOGD

← AMATX [a]. Spatial tangent modulus arranged in matrix form.

→ B [B]. Array containing the components of the current left Cauchy–Green strain tensor.

→ IPROPS. Array of integer material properties (refer to page 541).

→ NTYPE. Stress state type flag (refer to page 541).

→ RPROPS. Array of real material properties (refer to page 541).

→ STRES $[\boldsymbol{\sigma}]$. Array of current Cauchy stress tensor components.

Some local variables and arrays of CSTOGD

- DELTA $[\delta_{ij}]$. Krönecker delta.
- DPSTRE $[\partial\tau_i/\partial b_j]$. Matrix containing the derivatives of the principal Kirchhoff stresses.
- EIGB $[b_i]$. Array of eigenvalues of $\boldsymbol{B}$.
- EIGPRJ $[\boldsymbol{M}_i, i = 1, 2]$. Matrix containing the components of the in-plane eigenprojection tensors of $\boldsymbol{B}$.
- PSTRES $[\tau_i]$. Array of principal Kirchhoff stresses.
- PSTRTC $[\lambda_i]$. Array of principal stretches.

Function calls from CSTOGD

- DGISO2. Called to assemble the isotropic tensor function derivative $\partial\boldsymbol{\tau}/\partial\boldsymbol{B}$.
- RVZERO. Initialises relevant arrays with zeros.
- SPDEC2. Called to perform the spectral decomposition of $\boldsymbol{B}$.

13.6. Numerical examples

A set of seven finite element examples of finite hyperelasticity applications is presented this section. The examples presented here illustrate the application of the framework discussed above in the numerical simulation of solids modelled by means of hyperelastic theories. The numerical results of the first two examples have been obtained with program HYPLAS. The remaining examples include three-dimensional membrane applications as well as a plane and an axisymmetric problem involving contact constraints. We recall that such features – three-dimensional membranes and contact constraints – are *not* available in the standard version of HYPLAS that accompanies this book. Different finite element programs have been used in such examples. It should be emphasised, however, that the basic hyperelastic algorithm – the stress-updating procedure and the computation of the elasticity tensor – employed in such programs are the ones described in the previous sections of this chapter. These are fully incorporated into HYPLAS.

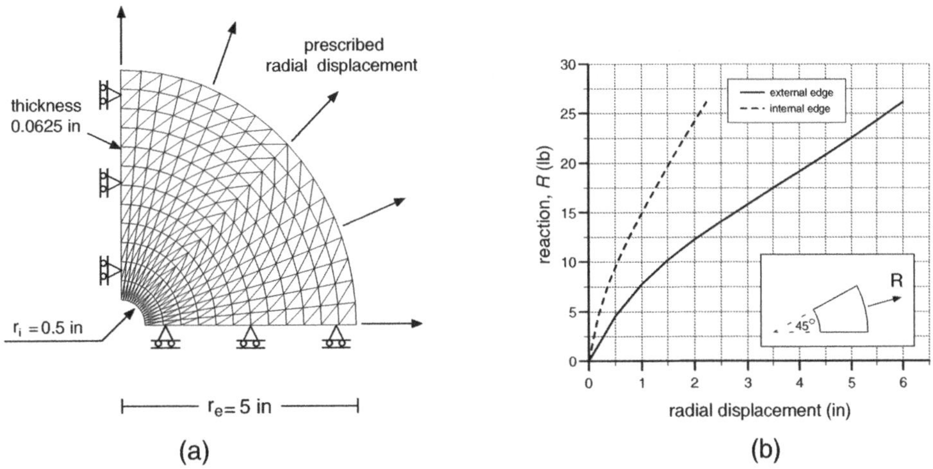

Figure 13.1. Annular plate: (a) geometry and boundary conditions; (b) reaction-displacement diagram. (Reproduced with permission from Finite elasticity in spatial description: Linearization aspects with membrane applications, EA de Souza Neto, D Perić and DRJ Owen, *International Journal for Numerical Methods in Engineering*, Vol 38 © 1995 John Wiley & Sons, Ltd.)

13.6.1. AXISYMMETRIC EXTENSION OF AN ANNULAR PLATE

This example studies the axisymmetric stretching of an annular rubber sheet. This problem has been considered by Parisch (1986). The sheet, with initial geometry illustrated in Figure 13.1(a), is assumed to be made of an *incompressible Mooney–Rivlin material* with constants

$$C_1 = 18.35\,\text{psi} \quad \text{and} \quad C_2 = 1.468\,\text{psi}.$$

In program HYPLAS, the Ogden model in plane stress is used with the following set of material parameters:

$$N = 2, \quad \mu_1 = 2C_1, \quad \mu_2 = -2C_2, \quad \alpha_1 = 2, \quad \alpha_2 = -2,$$

which correspond to the above Mooney–Rivlin material. Recall that in the HYPLAS implementation of the Ogden model, perfect incompressibility is assumed under plane stress. Due to symmetry, only one quadrant of the sheet is considered. A mesh containing 432 three-noded constant strain triangular elements is used to discretize the sheet with a total number of 247 nodes. A radial displacement of 6 in is prescribed on the external edge as indicated in Figure 13.1(a). The reaction force obtained on the external edge, corresponding to a 45° slice of the plate, is plotted in Figure 13.1(b) against the radial displacement of the external and internal edges. The results are in very close agreement with the shell simulation reported by Parisch (1986). We emphasise that, due to the use of the full Newton–Raphson algorithm, the final configuration (6 in radial displacement of the external edge) can be reached in one single step with the quadratic rate of convergence depicted in Table 13.1.

13.6.2. STRETCHING OF A SQUARE PERFORATED RUBBER SHEET

The problem here consists of the axial stretching of a square rubber sheet containing a circular hole. Again, the incompressible Mooney–Rivlin strain-energy function is adopted.

Table 13.1. Annular plate. Newton–Raphson convergence.

Iteration	Residual force norm
1	$0.545187\ E+01$
2	$0.134653\ E+01$
3	$0.148804\ E+00$
4	$0.148151\ E-02$
5	$0.198144\ E-06$
6	$0.974865\ E-13$

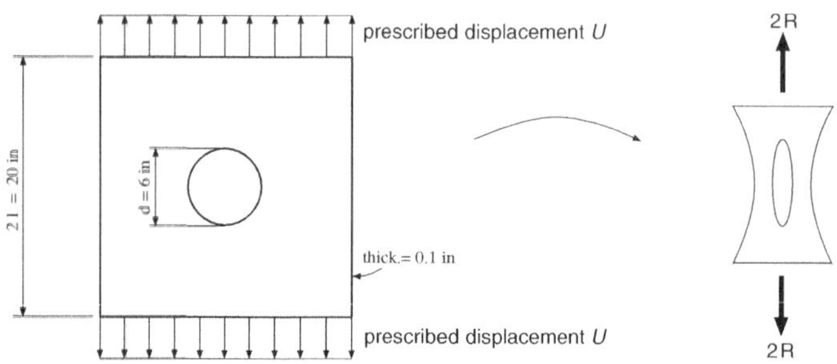

Figure 13.2. Square perforated sheet. Initial geometry and boundary conditions. (Reproduced with permission from Finite elasticity in spatial description: Linearization aspects with membrane applications, EA de Souza Neto, D Perić and DRJ Owen, *International Journal for Numerical Methods in Engineering*, Vol 38 © 1995 John Wiley & Sons, Ltd.)

The material constants are

$$C_1 = 25 \quad \text{and} \quad C_2 = 7.$$

As in the previous example, the corresponding incompressible Ogden material in plane stress is adopted in HYPLAS. The initial geometry and the boundary conditions are schematically illustrated in Figure 13.2. We define the non-dimensional factor

$$\gamma \equiv \frac{U}{l},$$

and the simulation is carried out up to $\gamma = 1$. A mesh containing 528 three-noded triangles discretises the symmetric quarter of the sheet. It corresponds to a total 549 degrees of freedom in the structure. Figure 13.3 shows the finite element mesh at the initial configuration ($\gamma = 0$) alongside the final deformed mesh ($\gamma = 1$). The final configuration is reached in two equal-size steps requiring five Newton–Raphson iterations per step for convergence with a tolerance of 10^{-12} in the Euclidean norm of the out-of-balance forces. For both increments, the residual norm of the kth iteration, $||G_k||$ is plotted in Figure 13.4 against the residual of the subsequent iteration $k+1$. Note that, as a logarithmic scale is used in both axes, the slope 2:1 indicated corresponds to quadratic convergence. In Figure 13.5, we plot the results obtained for the reaction force R on the restrained edge versus the load parameter γ. It is in good agreement with the numerical solutions presented by Parisch (1986) and Oden (1972).

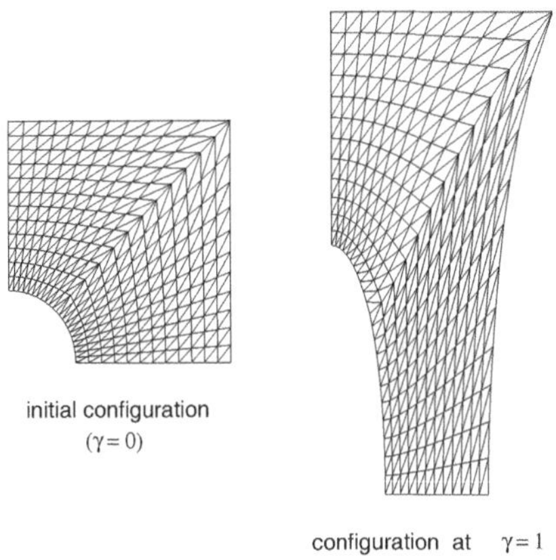

Figure 13.3. Square perforated sheet. Finite element meshes. (Reproduced with permission from Finite elasticity in spatial description: Linearization aspects with membrane applications, EA de Souza Neto, D Perić and DRJ Owen, *International Journal for Numerical Methods in Engineering*, Vol 38 © 1995 John Wiley & Sons, Ltd.)

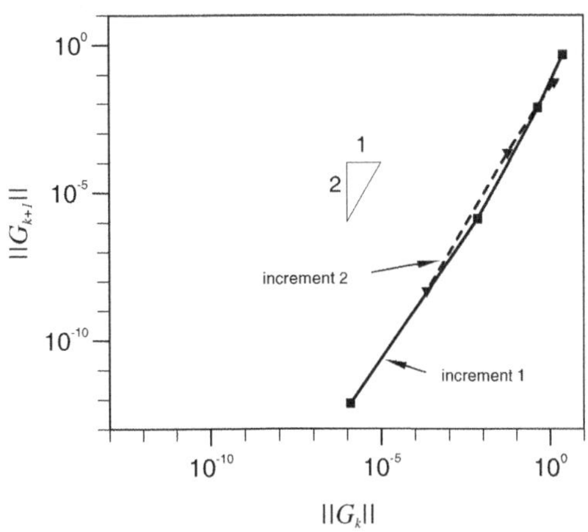

Figure 13.4. Square perforated sheet. Convergence behaviour. (Reproduced with permission from Finite elasticity in spatial description: Linearization aspects with membrane applications, EA de Souza Neto, D Perić and DRJ Owen, *International Journal for Numerical Methods in Engineering*, Vol 38 © 1995 John Wiley & Sons, Ltd.)

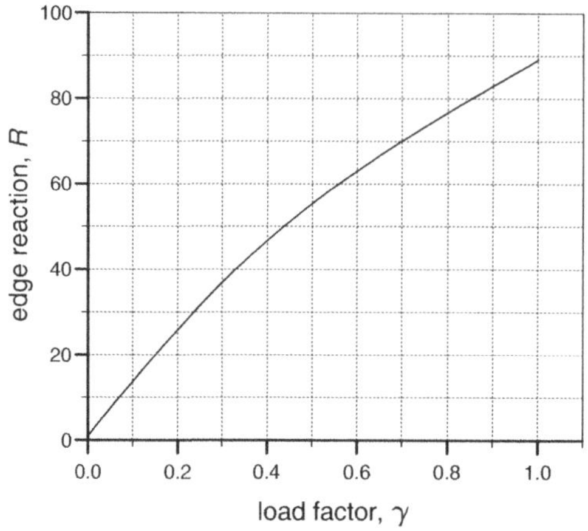

Figure 13.5. Square perforated sheet. Edge reactions. (Reproduced with permission from Finite elasticity in spatial description: Linearization aspects with membrane applications, EA de Souza Neto, D Perić and DRJ Owen, *International Journal for Numerical Methods in Engineering*, Vol 38 © 1995 John Wiley & Sons, Ltd.)

13.6.3. INFLATION OF A SPHERICAL RUBBER BALLOON

In this problem, we consider the simulation of a spherical rubber membrane inflated under internal pressure. A mesh of 675 isoparametric three-noded plane stress *three-dimensional membrane* elements, shown in Figure 13.6, discretises one octant of the sphere with symmetry boundary conditions imposed along the edges. Recall that three-dimensional membranes are not available in program HYPLAS. The present simulation was carried out with a different finite element program. Also note that, here, the pressure load is *non-conservative*. In order to achieve quadratic rates of convergence to equilibrium under such a configuration-dependent load, appropriate linearisation of the external finite element force vector is needed which results in an extra term – the *load-stiffness matrix* – to be added to the element stiffness matrix.[§] This facility is not available in HYPLAS, but has been considered in the simulation reported here. A three-term incompressible Ogden strain energy function is assumed. The material constants are

$$\alpha_1 = 1.3, \quad \alpha_2 = 5.0, \quad \alpha_3 = -2.0,$$
$$\mu_1 = 6.3, \quad \mu_2 = 0.012, \quad \mu_3 = -0.1 \text{ kg/cm}^2.$$

These parameters have been determined by Ogden (1972) in order to fit the experiments carried out by Treloar (1944) on natural vulcanised rubber. The analytical solution for a general hyperelastic spherical balloon has been presented by Ogden (1972, 1984). Needleman (1977) has used the Ritz–Galerkin method to investigate aspherical equilibrium configurations of initially imperfect balloons. The elastic stability of this problem is known to be crucially dependent on the specific strain-energy function adopted (Beatty, 1987; Ogden, 1972). Pressure instability is detected, in particular, for the three-term Ogden function with

[§] The derivation of load-stiffness matrices is discussed in Section 4.3.6, page 106.

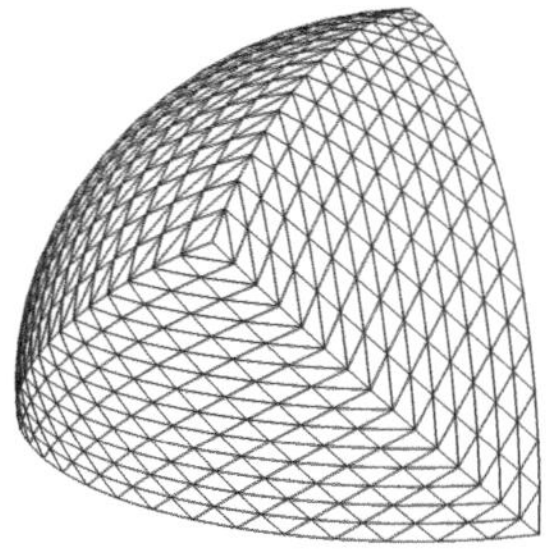

Figure 13.6. Spherical balloon. Finite element mesh. (Reproduced with permission from Finite elasticity in spatial description: Linearization aspects with membrane applications, EA de Souza Neto, D Perić and DRJ Owen, *International Journal for Numerical Methods in Engineering*, Vol 38 © 1995 John Wiley & Sons, Ltd.)

the constants chosen above. For this reason, the *Arc-Length Method* was employed here in conjunction with the Newton–Raphson algorithm to allow equilibrium to be found beyond the instability point. The Arc-Length Method has been described in Chapter 4. For convenience, we define the normalised internal pressure

$$p^* \equiv \frac{p\, r_0}{2\, t_0},$$

and the expansion ratio of the balloon

$$\lambda \equiv \frac{r}{r_0},$$

where r and r_0 are, respectively, the current and initial radii of the balloon, t_0 is the initial thickness of the rubber membrane and p is the current internal pressure. Ten increments were employed to reach the final configuration with $\lambda = 747\%$. In Figure 13.7, the expansion ratio obtained in the finite element simulation is plotted versus p^*. For comparison, Ogden's membrane analytical solution is also plotted. The numerical result matches the exact pressure-expansion curve very closely.

13.6.4. RUGBY BALL

The simulation of the inflation of an axisymmetric ellipsoidal balloon is carried out in this example. The initial geometry of the balloon corresponds to the shape of an official rugby ball: the axial and the transverse radii, r_1 and r_2, measure respectively 14.5 and 9.5 cm. The ball is assumed to be made of the three-term Ogden material of the previous example with 3 mm uniform initial thickness. The topology of the finite element mesh as well as the boundary conditions are the same as in the spherical balloon problem and the arc-length method is also used in conjunction with the Newton–Raphson algorithm. The evolution of the radii r_1 and r_2 obtained in the simulation is plotted in the diagram of Figure 13.8. The *snap-through* behaviour observed in the example of Section 13.6.3 is also detected for the present problem. As shown in the graph, starting from its initial configuration (Figure 13.9(a)), the ball inflates under internal pressure maintaining its ellipsoidal shape ($r_2 < r_1$) until it reaches a critical point with internal pressure $p = 0.149$ kg/cm^2. Then, a sharp decrease of

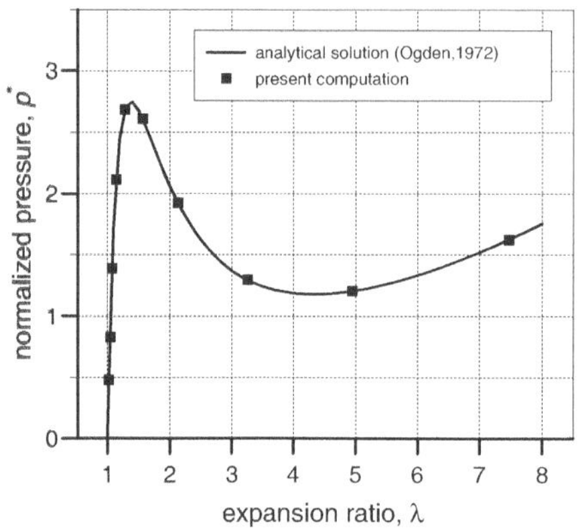

Figure 13.7. Spherical balloon. Pressure-expansion diagram. (Reproduced with permission from Finite elasticity in spatial description: Linearization aspects with membrane applications, EA de Souza Neto, D Perić and DRJ Owen, *International Journal for Numerical Methods in Engineering*, Vol 38 © 1995 John Wiley & Sons, Ltd.)

internal pressure occurs followed by a monotonic increase. Interestingly, in the descending branch of the pressure-expansion diagram, the radii r_1 and r_2 are nearly equal, i.e. the ball has an approximately spherical shape. The configuration corresponding to point A of the diagram, with $r_1 = r_2 = 22.6$ cm and $p \approx 0.109$ kg/cm^2, is depicted in Figure 13.9(b). After the minimum pressure ($p = 0.065$ kg/cm^2) is reached, the ball recovers its ellipsoidal shape with $r_2 < r_1$.

13.6.5. INFLATION OF INITIALLY FLAT MEMBRANES

In this example, we simulate the inflation of a circular and a square initially flat thin rubber sheet. The simulation of the circular membrane has been carried out by Wriggers and Taylor (1990) employing an axisymmetric membrane formulation. The inflation of the square membrane illustrates a truly three-dimensional application. The geometry of the problems and the finite element discretisations are shown in Figure 13.10. In both cases, a mesh containing 512 three-noded three-dimensional membrane triangular elements was employed to discretise the corresponding symmetric quadrant of the problem. The material parameters are the same as in the previous two examples. Again, the Arc-Length Method is used in order to capture the snap-through behaviour. To avoid numerical problems arising from the zero tangent stiffness of the flat membranes in the normal direction, both sheets are slightly prestretched before the pressure loading is applied. The results for the normal deflections of the mid points A and B (indicated in Figure 13.10) are plotted in the diagram of Figure 13.11 versus the applied pressure p. The pressure–deflection diagram obtained for both membranes are similar. After reaching a critical point, a considerable drop is followed by a monotonic increase of the pressure load. This behaviour has also been observed in the spherical balloon example. Note that instability occurs at lower pressure (and larger mid-deflection) for the square sheet. Figure 13.12 depicts the evolution of the deformed geometry of the square

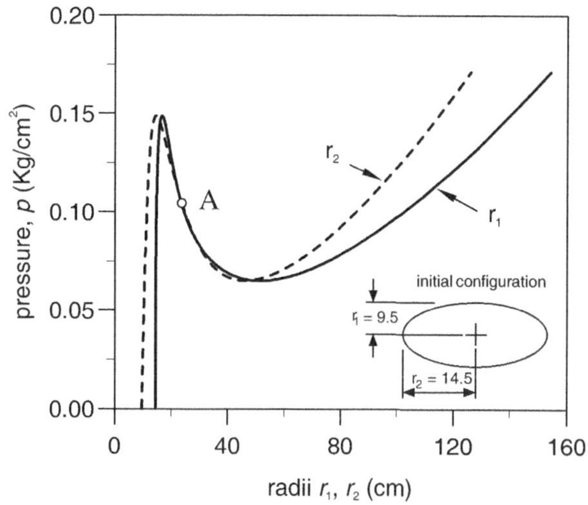

Figure 13.8. Rugby ball. Pressure-expansion diagram. (Reproduced with permission from Finite elasticity in spatial description: Linearization aspects with membrane applications, EA de Souza Neto, D Perić and DRJ Owen, *International Journal for Numerical Methods in Engineering*, Vol 38 © 1995 John Wiley & Sons, Ltd.)

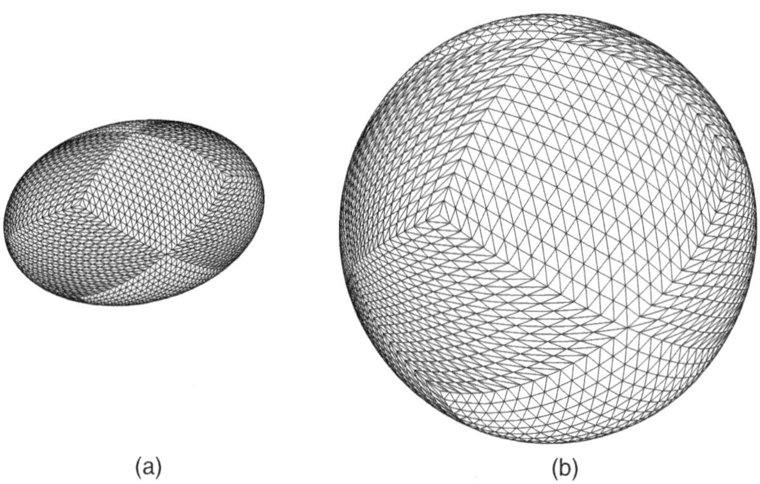

Figure 13.9. Rugby ball. (a) Initial configuration, and (b) spherical configuration corresponding to point A of the pressure–expansion diagram. (Reproduced with permission from Finite elasticity in spatial description: Linearization aspects with membrane applications, EA de Souza Neto, D Perić and DRJ Owen, *International Journal for Numerical Methods in Engineering*, Vol 38 © 1995 John Wiley & Sons, Ltd.)

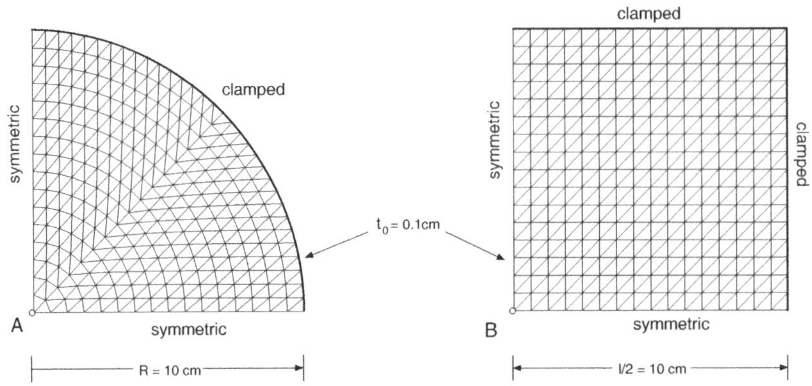

Figure 13.10. Flat membranes. Initial geometry and boundary conditions. (Reproduced with permission from Finite elasticity in spatial description: Linearization aspects with membrane applications, EA de Souza Neto, D Perić and DRJ Owen, *International Journal for Numerical Methods in Engineering*, Vol 38 © 1995 John Wiley & Sons, Ltd.)

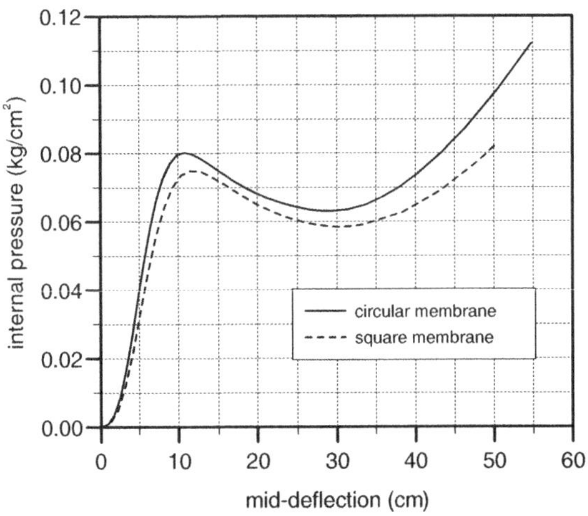

Figure 13.11. Flat membranes. Pressure-deflection diagram. (Reproduced with permission from Finite elasticity in spatial description: Linearization aspects with membrane applications, EA de Souza Neto, D Perić and DRJ Owen, *International Journal for Numerical Methods in Engineering*, Vol 38 © 1995 John Wiley & Sons, Ltd.)

membrane. The initially flat sheet evolves into a spherical-type shape. Figures 13.12(a), (b), (c) and (d) correspond, respectively, to the mid-deflections $d_B = 4.94$, $d_B = 12.01$, $d_B = 17.49$ and $d_B = 27.34$ cm. The deformed state shown in Figure 13.12(b) is very close to the instability point observed in the pressure–deflection curve.

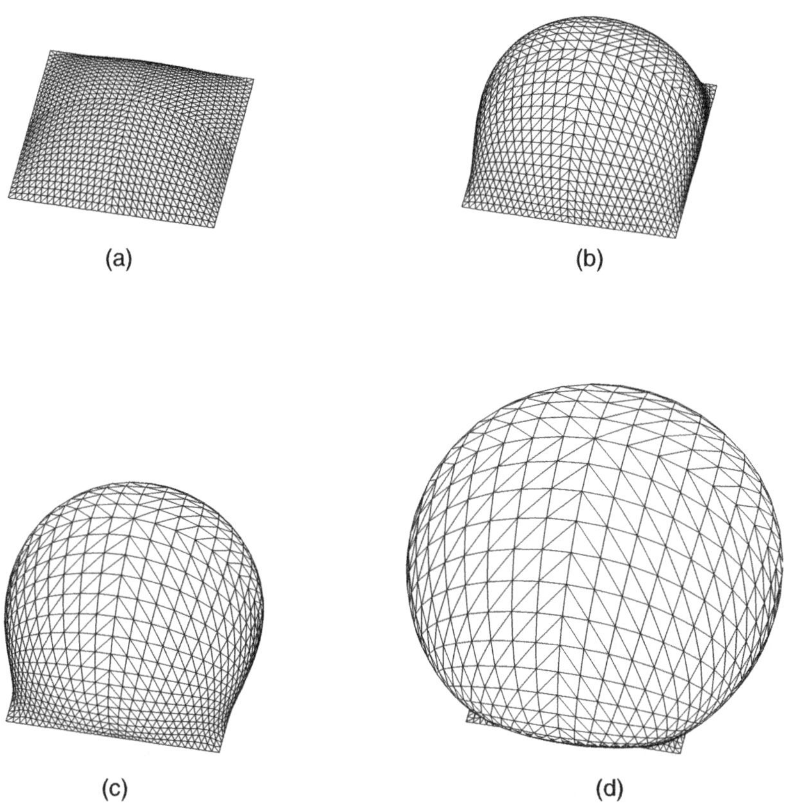

Figure 13.12. Square membrane. Deformed configurations; (a) $d_B = 4.94$ cm; (b) $d_B = 12.01$ cm; (c) $d_B = 17.49$ cm, and (d) $d_B = 27.34$ cm. (Reproduced with permission from Finite elasticity in spatial description: Linearization aspects with membrane applications, EA de Souza Neto, D Perić and DRJ Owen, *International Journal for Numerical Methods in Engineering*, Vol 38 © 1995 John Wiley & Sons, Ltd.)

13.6.6. RUBBER CYLINDER PRESSED BETWEEN TWO PLATES

The plane strain simulation of the compression of a long rubber cylinder between two frictionless rigid plates is carried out in this example. Two different material models which fit the same experimental data are used:

1. the *regularised Mooney–Rivlin* material, with constants

$$C_1 = 0.293\,\text{MPa}, \quad C_2 = 0.177\,\text{MPa}, \quad K = 1410\,\text{MPa};$$

2. a three-term *regularised Ogden model* with coefficients

$$\mu_1 = 0.746\,\text{MPa}, \quad \mu_2 = -0.306\,\text{MPa}, \quad \mu_3 = 6.609 \times 10^{-5}\,\text{MPa},$$
$$\alpha_1 = 1.748, \qquad \alpha_2 = -1.656, \qquad \alpha_3 = 7.671$$

and the same bulk modulus as for the Mooney–Rivlin simulation.

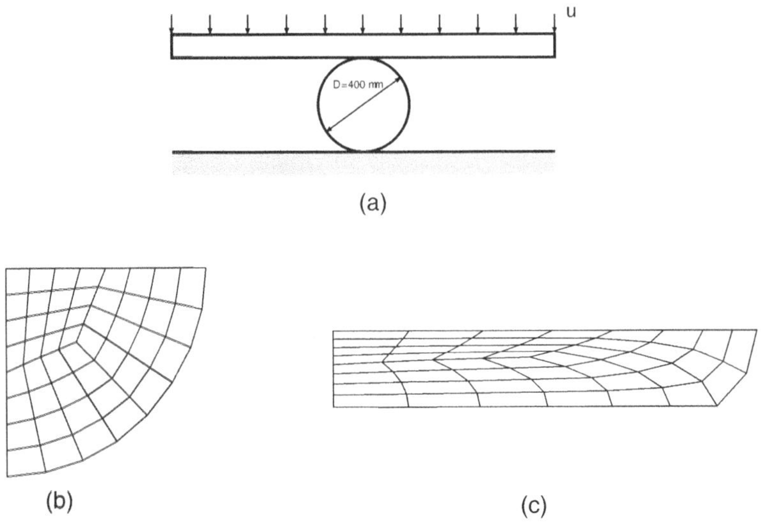

Figure 13.13. Rubber cylinder pressed between two plates: (a) geometry and boundary conditions; (b) initial mesh; and (c) deformed mesh at $u = 250$ mm. (Reproduced with permission from Design of simple low-order finite elements for large-strain analysis of nearly incompressible solids, EA de Souza Neto, D Perić, M Dutko and DRJ Owen, *International Journal of Solids and Structures*, Vol 33, Issue 20–22 © 1996 Elsevier Science Ltd.)

Note that the high bulk modulus adopted here makes the material virtually incompressible, so that appropriate treatment of quasi-incompressibility is required for a successful finite element simulation. Here, the *F-bar* methodology thoroughly discussed in Chapter 15 is adopted to overcome the problem. This problem has been considered originally by Sussman and Bathe (1987) in the context of hybrid *u/p* finite elements and by Simo and Taylor (1991), who employed a mixed formulation in conjunction with an augmented Lagrangian procedure to handle near-incompressibility. The geometry of the problem and the boundary conditions are illustrated in Figure 13.13(a). For symmetry reasons, only one quarter of the cylinder cross-section is considered in the simulation. It is discretised with the mesh of 48 four-noded F-bar elements shown in Figure 13.13(b). We remark that this type of finite element is available in program HYPLAS. The deformed mesh obtained with a prescribed displacement $u = 250$ mm of the plate is depicted in Figure 13.13(c). It is in good qualitative agreement with the deformed mesh shown by Simo and Taylor (1991). The reaction forces per unit thickness of the plate, obtained for the Mooney–Rivlin and Ogden models, are plotted in Figure 13.14 against the plate deflection u. The curves are plotted up to $u = 200$ mm. For both materials the results shown here agree with those obtained by Sussman and Bathe (1987).

13.6.7. ELASTOMERIC BEAD COMPRESSION

The numerical simulation of the compression of an elastomeric axisymmetric bead is carried out in this example. The bead – a circular ring with trapezoidal cross-section – is schematically represented in Figure 13.15. Its function is to provide sealing when the plate, which contacts its top edge, is pressed downwards. Extremely high compressive strains occur in this problem. In the finite element simulation, the bottom edge of the ring seal is

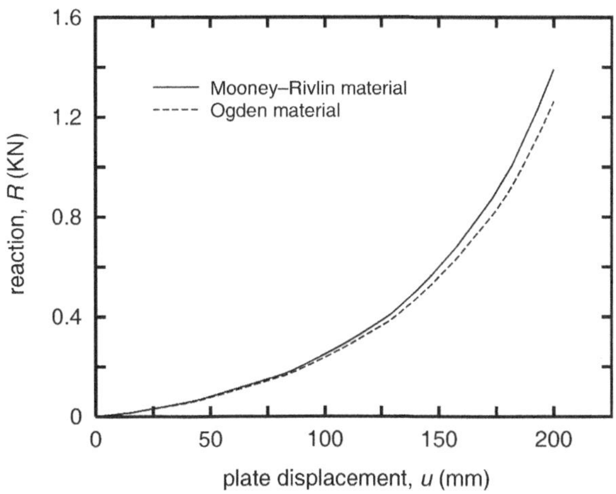

Figure 13.14. Rubber cylinder pressed between two plates. Load-deflection curves. (Reproduced with permission from Design of simple low-order finite elements for large-strain analysis of nearly incompressible solids, EA de Souza Neto, D Perić, M Dutko and DRJ Owen, *International Journal of Solids and Structures*, Vol 33, Issue 20–22 © 1996 Elsevier Science Ltd.)

assumed clamped to a flat rigid base and both plate and base are idealised as rigid bodies with frictionless contact condition on the boundaries. The bead is modelled as a regularised neo-Hookean material with constant $C = 2.5$ and bulk modulus $k = 1000$. A mesh of 520 four-noded *F-bar* axisymmetric elements is used to discretise the bead. Figure 13.16(a) shows the mesh in its initial configuration. A total vertical displacement $u = 0.25$ is applied to the plate in 25 increments. Deformed meshes obtained at different stages of the compression process are depicted in Figures 13.16(b), (c) and (d). They correspond, respectively, to $u = 0.09$, 0.17 and the final configuration with $u = 0.25$. At the early stage shown in Figure 13.16(b), the lateral surfaces of the seal make contact only with the top plate. At the later stages of Figures 13.16(c) and (d), contact also occurs between the bead and the rigid base. The reaction force obtained in the plate is plotted in the diagram of Figure 13.17 versus the imposed displacement u.

13.7. Hyperelasticity with damage: the Mullins effect

It has been mentioned in Section 12.1.2 (page 473) that some rubbery materials are particularly susceptible to internal damaging. The underlying microscopic mechanism associated with internal damaging of *filled polymers* has been briefly discussed in the text surrounding Figure 12.2.

The phenomenological manifestation of the damage mechanism in filled rubbers is the so-called *Mullins effect*. In a uniaxial cyclic extension experiment, the Mullins effect is characterised by the degradation of the elastic stiffness at strain levels below the maximum strain attained in the history of deformation (Mullins, 1969; Stern, 1967). A simplified schematic illustration of the Mullins effect is given in Figure 13.18. During a typical (quasi-static) uniaxial experiment with a filled polymer, the initial stretching up to ε_1 follows the stress–strain path A with unloading from ε_1 via curve B. A subsequent stretching up to ε_2

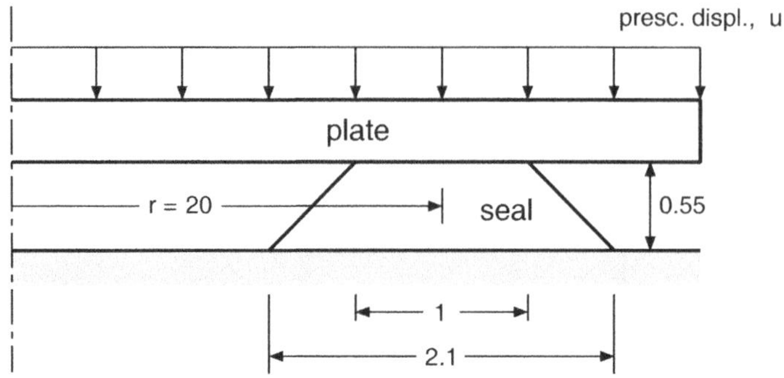

Figure 13.15. Elastomeric bead compression. Initial geometry and boundary conditions. (Reproduced with permission from Design of simple low-order finite elements for large-strain analysis of nearly incompressible solids, EA de Souza Neto, D Perić, M Dutko and DRJ Owen, *International Journal of Solids and Structures*, Vol 33, Issue 20–22 © 1996 Elsevier Science Ltd.)

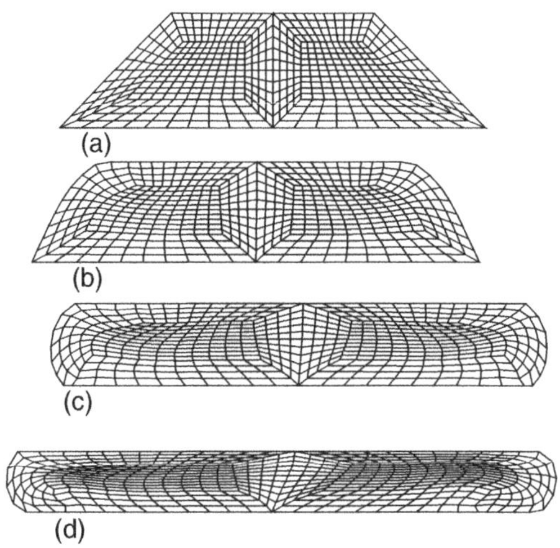

Figure 13.16. Elastomeric bead compression. Finite element discretisation on the initial configuration and deformed meshes obtained with the present element: (a) initial configuration; (b) $u = 0.09$; (c) $u = 0.17$; and (d) final configuration, $u = 0.25$. (Reproduced with permission from Design of simple low-order finite elements for large-strain analysis of nearly incompressible solids, EA de Souza Neto, D Perić, M Dutko and DRJ Owen, *International Journal of Solids and Structures*, Vol 33, Issue 20–22 © 1996 Elsevier Science Ltd.)

will follow path BC. Then, unloading will follow curve D with a third stretch occurring via DE and so on. The above description shows that the Mullins effect is a clearly *dissipative* phenomenon. Therefore, it cannot be modelled by means of purely hyperelastic theories.

The first attempt to model the Mullins effect through a phenomenological constitutive theory appears to have been made by Gurtin and Francis (1981) in the one-dimensional context. The Gurtin–Francis uniaxial model is very simple in concept. It is described in the following.

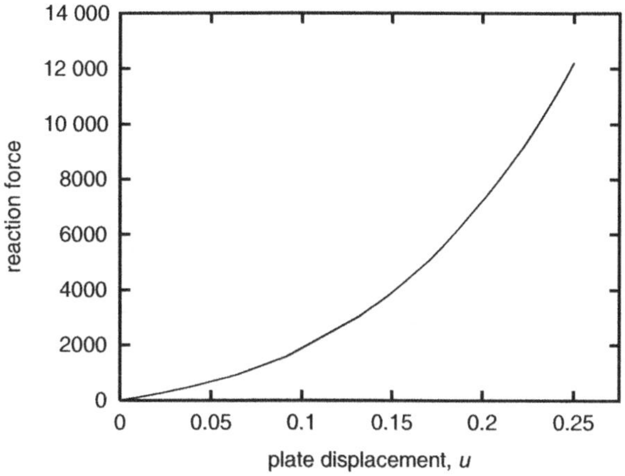

Figure 13.17. Elastomeric bead compression. Displacement–reaction diagram. (Reproduced with permission from Design of simple low-order finite elements for large-strain analysis of nearly incompressible solids, EA de Souza Neto, D Perić, M Dutko and DRJ Owen, *International Journal of Solids and Structures*, Vol 33, Issue 20–22 © 1996 Elsevier Science Ltd.)

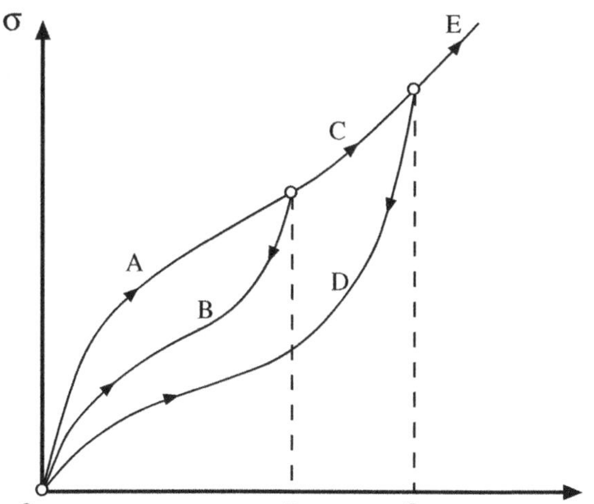

Figure 13.18. Mullins effect. Schematic representation.

13.7.1. THE GURTIN–FRANCIS UNIAXIAL MODEL

The starting point of this model is the choice of the *maximum axial strain* attained in the history of deformation

$$\varepsilon^m(t) \equiv \max_{0\leq s\leq t} \{\varepsilon(s)\}, \tag{13.112}$$

as a measure of damage. Note that this phenomenological measure of damage is in fact closely related to the microscopic mechanism of internal deterioration in filled polymers discussed in Section 12.1.2. The number of broken polymer chains between filler particles (effectively defining the state of internal damage) increases as the material is stretched and will remain constant at strains below the maximum attained strain.¶

Having defined the damage measure, a constitutive equation expressing the uniaxial stress, σ, as a function of the current axial strain, ε, and the state of damage (characterised by ε^m) is then postulated as

$$\sigma(\varepsilon, \varepsilon^m) = \bar{f}(\zeta)\, \sigma^0(\varepsilon^m), \tag{13.113}$$

where ζ is the *relative strain*, defined as

$$\zeta \equiv \frac{\varepsilon}{\varepsilon^m}. \tag{13.114}$$

From the definition of ε^m, it follows that

$$\zeta \leq 1 \quad \forall\, \varepsilon. \tag{13.115}$$

The function $\bar{f}(\zeta)$, named the *master damage curve*, satisfies for any ζ

$$0 \leq \bar{f}(\zeta) \leq 1, \tag{13.116}$$

and the factor $\bar{f}$ defines the loss of stiffness experienced by the polymer at strain levels below the maximum previously attained strain, ε^m. The following additional condition is imposed on $\bar{f}(\zeta)$:

$$\bar{f}(1) = 1, \tag{13.117}$$

so that when the maximum strain occurs at the current time, i.e. when

$$\varepsilon^m = \varepsilon \Leftrightarrow \zeta = 1,$$

the axial stress is given by

$$\sigma = \sigma^0(\varepsilon^m). \tag{13.118}$$

The stress–strain curve defined by the function σ^0 is called the *virgin curve*. It is the stress–strain curve obtained in a simple uniaxial experiment with monotonically increasing strain. To characterise completely the material parameters for this model, we need, in addition to the virgin curve, to determine the master damage curve $\bar{f}(\zeta)$. This curve is obtained from unloading experiments.

¶Here, we disregard other potentially important phenomena such as thermally activated damage, strain rate effects, etc.

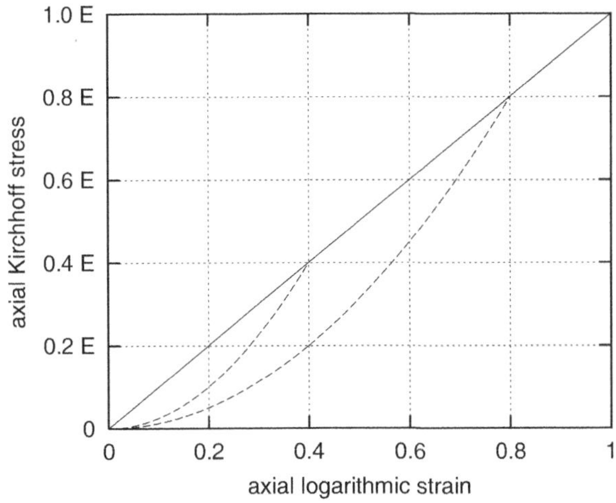

Figure 13.19. Hencky-based uniaxial hyperelastic damage model. Stress–strain curves.

Example

The above concepts can be illustrated in the following simple example. Let us consider the uniaxial version of the *Hencky* hyperelastic model of Section 13.2.3. For this model, the axial Kirchhoff stress is given by

$$\tau = E\,\varepsilon, \tag{13.119}$$

where E is the logarithmic Young's modulus and ε is the axial logarithmic strain. A Gurtin–Francis-type model with underlying Hencky hyperelastic behaviour can be obtained by taking the above logarithmic-based stress–strain relation as the *virgin curve*:

$$\tau^0(\varepsilon^m) \equiv E\,\varepsilon^m, \tag{13.120}$$

where ε^m is the maximum logarithmic strain, and postulating a *master damage curve* defined by, say,[||]

$$\bar{f}(\zeta) \equiv \zeta^2 = (\varepsilon/\varepsilon^m)^2. \tag{13.121}$$

The final hyperelastic damage stress–strain relation will then be given by

$$\tau = \bar{f}(\zeta)\,\tau^0(\varepsilon^m) = E\,\varepsilon^2/\varepsilon^m. \tag{13.122}$$

The resulting behaviour is illustrated in the graph of Figure 13.19 where the solid straight line corresponds to monotonic loading and the parabolas plotted with dashed lines correspond to the stress–strain relation upon unloading/reloading below the maximum attained strains: $\varepsilon^m = 0.4$ and $\varepsilon^m = 0.8$.

[||] We remark that the master damage curve adopted in the present example has been chosen only for the purpose of illustrating the damage model. It is not based on experimental data.

Maximum strain-dependent damage curve

A constitutive refinement of the above model can be introduced by making $\bar{f}$ dependent also on the maximum strain, ε^m; that is, we now have

$$\bar{f} = \bar{f}(\zeta, \varepsilon^m), \tag{13.123}$$

and the stress constitutive equation reads

$$\sigma(\varepsilon, \varepsilon^m) = \bar{f}(\zeta, \varepsilon^m)\, \sigma^0(\varepsilon^m). \tag{13.124}$$

The above refinement allows different experimental unloading curves, obtained by unloading from different maximum strain levels, to be accurately reproduced by the model.

13.7.2. THREE-DIMENSIONAL MODELLING. A BRIEF REVIEW

A pioneering step in the development of three-dimensional phenomenological constitutive models capable of capturing the Mullins effect in finite element simulations was provided by Simo (1987). This author proposed a continuum damage theory based on the *principle of strain equivalence* (refer to Chapter 12) within a viscoelastic constitutive framework. Subsequently, Govindjee and Simo (1991) developed a rate-independent theory meant to model the Mullins effect in carbon black-filled rubbers. Their model, which was based on micromechanical considerations, was further extended to account for viscous effects (Govindjee and Simo, 1992). In the context of the rate-independent theory, the authors (de Souza Neto *et al.*, 1994b) introduced a purely phenomenological model whose finite element implementation assumes a particularly simple format. This model will be described in detail in Section 13.7.3. Further work in the field has been developed by Miehe (1995) and Miehe and Keck (2000). In particular, the model proposed by Miehe and Keck (2000), which incorporates viscous effects and plastic deformations, is reported to model accurately the behaviour of certain filled rubbers subjected to relatively complex strain paths. Further reference to recent developments in the constitutive description of damage in rubbers is provided by Holzapfel (2000).

13.7.3. A SIMPLE RATE-INDEPENDENT THREE-DIMENSIONAL MODEL

In this section we describe the three-dimensional rate-independent hyperelastic damage model proposed by the authors (de Souza Neto *et al.*, 1994b). Its computational implementation into an implicit finite element framework will be addressed in Section 13.7.5. The objective here is mainly to outline the basic concepts associated with hyperelastic damage and its finite element implementation in the multidimensional case. The inherent simplicity of this particular model, we believe, makes it especially suitable for practical applications.

The model is based on concepts analogous to those underlying the Gurtin–Francis uniaxial theory. Firstly, let us consider a conventional isotropic hyperelastic model governed by a free-energy function ψ^0. The derivation that follows will be based on the *principal stretch* representation of the free-energy function. Note, however, that the particular choice for the representation of ψ^0 is immaterial and all arguments of the following discussion remain valid if any other equivalent representation is adopted. According to expression (13.33), the principal Kirchhoff stresses for such a material are given by

$$\tau_i = \bar{\rho}\,\frac{\partial \psi^0}{\partial \lambda_i}\lambda_i, \tag{13.125}$$

with no summation implied on repeated indices.

The crucial step in the definition of the hyperelastic damage model is the assumption that the above constitutive equation for stress is valid *only* upon monotonic loading and reloading with damage evolution. In addition, similarly to (13.113), we postulate a general stress constitutive function given by

$$\tau_i = f(\xi)\,\tau_i^0(\lambda_1, \lambda_2, \lambda_3), \tag{13.126}$$

where

$$\tau_i^0(\lambda_1, \lambda_2, \lambda_3) \equiv \bar{\rho}\,\frac{\partial \psi^0}{\partial \lambda_i}\lambda_i, \tag{13.127}$$

again with no summation implied on repeated indices. Analogously to its one-dimensional counterpart, $\bar{f}$, the function $f : [0, 1] \rightarrow [0, 1]$ is expressed in terms of some, as yet not defined, three-dimensional measure of relative strain ξ. It also satisfies

$$f(1) = 1. \tag{13.128}$$

The damage variable

Generalising the idea of maximum strain of the Gurtin–Francis model (see expression (13.112)), the new variable, D, is defined as a history recording parameter for the phenomenon of internal damage in general three-dimensional situations:

$$D(t) \equiv \max_{0\leq s\leq t}\{\bar{\rho}\,\psi^0(s)\}. \tag{13.129}$$

Remark 13.5. Definition (13.129) implies that the damage variable grows equally in tension and compression so that the model cannot distinguish between these two situations. As filled rubbers typically suffer substantially more damage in tension than in compression, the use of the present model in situations where the damageable body is subjected to both tension and compression is expected to produce unrealistic predictions. One possible alternative to overcome this deficiency could be to link the damage variable to the *maximum principal stretch* instead. This approach was adopted, for instance, by Govindjee and Simo (1992).

The relative strain measure

With the above-defined damage variable, the uniaxial *relative strain*, ζ, defined in (13.114) may then be immediately generalised to the three-dimensional case as

$$\xi \equiv \frac{\bar{\rho}\,\psi^0}{D}. \tag{13.130}$$

This completes the characterisation of the model.

Box 13.3. Three-dimensional constitutive model for damage in highly filled polymers.

(i) Damage variable

$$D(t) \equiv \max_{0 \le s \le t} \{\bar{\rho}\, \psi^0(s)\}$$

(ii) Stress constitutive relation

$$\tau_i = f(\xi)\, \bar{\rho}\, \frac{\partial \psi^0}{\partial \lambda_i} \lambda_i \quad \text{(no summation)}$$

$$\xi \equiv \frac{\bar{\rho}\, \psi^0}{D}$$

Material properties identification

For convenience, we summarise the overall model in Box 13.3. The identification of material properties for the present model consists solely of the determination of the functions ψ^0 and f.

Elastoplasticity analogy

Following the definition (13.129) for the damage internal variable, a straightforward analogy between classical elastoplasticity (refer to Chapter 6) and the present model for elastic damage may be established by introducing a *damage surface* (*cf.* yield surface) in the space of principal stretches, defined by the equation

$$\Phi(\lambda_1, \lambda_2, \lambda_3, D) \equiv \bar{\rho}\, \psi^0(\lambda_1, \lambda_2, \lambda_3) - D = 0. \tag{13.131}$$

For fixed D, the damage surface delimits the region of the space of principal stretches in which the behaviour of the material is purely hyperelastic without evolution of damage. The above definitions also imply the following *complementarity law* associated with the evolution of damage or damage loading/unloading (*cf.* plastic loading/unloading criterion):

$$\Phi \le 0, \quad \dot{D} \ge 0, \quad \dot{D}\Phi = 0. \tag{13.132}$$

Thermodynamical aspects

Alternatively to the above arguments, the present constitutive equations for damage in filled polymers can be obtained by postulating the existence of a free-energy function of the form

$$\bar{\rho}\, \psi(\lambda_1, \lambda_2, \lambda_3, D) \equiv \bar{\rho}\, \psi(\bar{\rho}\, \psi^0(\lambda_1, \lambda_2, \lambda_3), D) = \int_0^{\bar{\rho}\psi^0} f\left(\frac{\kappa}{D}\right) \mathrm{d}\kappa, \tag{13.133}$$

recalling that ψ^0 is the free-energy of the hypothetical hyperelastic (non-damageable) rubber that governs the behaviour of the material upon continuous loading and f is the damage master curve. Indeed, with the free-energy defined by (13.133), the principal Kirchhoff stresses are given by

$$\tau_i = \bar{\rho}\,\frac{\partial\psi}{\partial\lambda_i}\lambda_i = f\left(\frac{\bar{\rho}\,\psi^0}{D}\right)\bar{\rho}\,\frac{\partial\psi^0}{\partial\lambda_i}\lambda_i, \tag{13.134}$$

which is precisely the constitutive function (13.126). Note that within the elastic domain, i.e. in the region of the principal stretches space delimited by the damage surface (with fixed D), the expression above defines an essentially *hyperelastic* behaviour characterised by the strain–energy function ψ.

Second law of thermodynamics

With the specific free-energy defined by (13.133), the second law of thermodynamics, in the form of the *Clausius–Duhem inequality* (refer to Sections 3.4 and 3.5) requires that

$$-\bar{\rho}\,\frac{\partial\psi}{\partial D}\,\dot{D} \geq 0. \tag{13.135}$$

It has been shown by the authors de Souza Neto *et al.* (1994b) that a sufficient condition for the Clausius–Duhem inequality to be satisfied is that f be a *non-decreasing* function. This, together with the fact that f is a non-negative function, also guarantees that a ψ^0 convex in λ_i produces a potential ψ also convex in λ_i.

13.7.4. EXAMPLE: THE MODEL PROBLEM

To give the reader a better picture of the behaviour of the above model, this example utilises a uniaxial stress–strain curve obtained under cyclic loading in the above model. The objective is to show the capability of the theory in describing the Mullins effect at the local level. The particular form of the function ψ^0 employed to describe the stress–strain behaviour during loading corresponds to an incompressible *neo-Hookean* strain-energy function (see expression (13.38)) with the constant C_1 chosen as

$$C_1 = 135 \text{ psi}.$$

The master damage curve adopted is plotted in Figure 13.20 ($\bar{\rho}$ taken as unity). In a uniaxial test, when subjected to the cyclic axial strain path shown in Figure 13.21(a), the material described by the parameters above produces the uniaxial stress–strain curve plotted in Figure 13.21(b). The axial strain measure referred to in the graph of Figure 13.21(b) has been defined as $\lambda - 1$ where λ is the principal stretch in the axial direction. The stress plotted in Figure 13.21(b) is the corresponding Cauchy principal stress. The solid line is the *virgin curve* followed during monotonic loading. The dotted lines correspond to the material behaviour with $\bar{\rho}\psi^0 < D$ and illustrate the ability of the model to capture the Mullins effect.

13.7.5. COMPUTATIONAL IMPLEMENTATION

The computational implementation of the above model within the implicit finite element framework of Chapter 4 requires, as usual, the two essential ingredients: (1) a procedure to update the stress consistently with the constitutive equation of the model, and (2) the computation of the corresponding tangent modulus. These are discussed in the following.

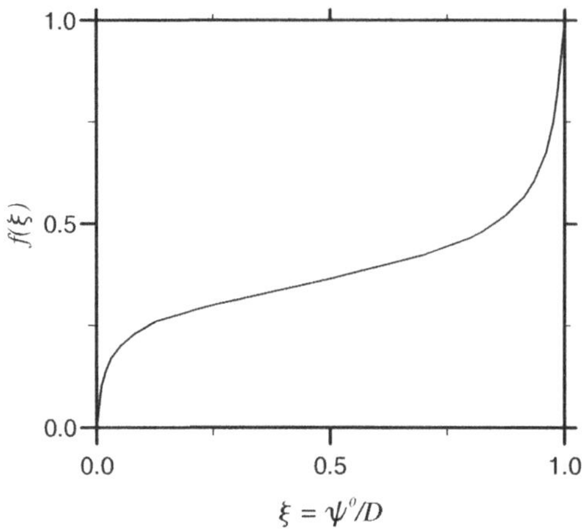

Figure 13.20. Master damage curve. (Reproduced with permission from A phenomenological three-dimensional rate-independent continuum damage model for highly filled polymers: Formulation and computational aspects, EA de Souza Neto, D Perić and DRJ Owen, *Journal of the Mechanics and Physics of Solids*, Vol 42, Issue 10 © 1994 Elsevier Science Ltd.)

We remark that the hyperelastic-damage model is *not* incorporated into the standard version of program HYPLAS that accompanies this book. However, as we shall see below, its incorporation is a straightforward programming exercise that we will leave for the interested reader.

Stress updating

Due to the path-dependent nature of the above elastic damage constitutive equations, a numerical algorithm is required for the evaluation of stress and damage evolution throughout the deformation history. The derivation of such an algorithm for the present model is straightforward. The underlying idea is the following. Considering the typical (pseudo-) time interval $[t_n, t_{n+1}]$, if the current value of $\bar{\rho}\psi^0$ at t_{n+1} is greater than the previous maximum value (D_n), then damage evolution necessarily occurs within the interval. If the current value of $\bar{\rho}\psi^0$ is smaller than or equal to the previous maximum, then it will be assumed that no damage evolution has occurred throughout the interval. The algorithm is listed in Box 13.4 in pseudo-code format.

Remark 13.6. In contrast to numerical integration procedures employed in classical elasto-plasticity (these are thoroughly discussed in Chapter 7), the algorithm of Box 13.4 is *exact*, independently of the increment size, provided that in the actual deformation path between times t_n and t_{n+1}, the material is being loaded monotonically (ψ^0 increasing monotonically) or unloaded monotonically with possible reloading below the maximum attained value of $\bar{\rho}\psi^0$.

Remark 13.7. The implementation of the algorithm of Box 13.4 in program HYPLAS with, for instance, the Ogden material as the underlying hyperelastic law, is a straightforward exercise.

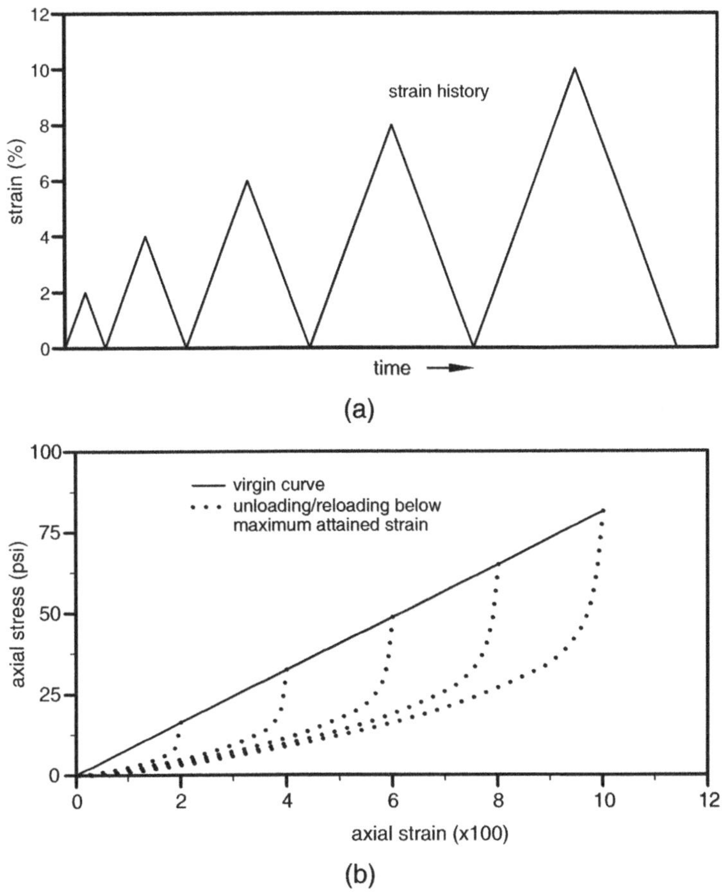

Figure 13.21. The model problem: (a) strain history, and (b) uniaxial stress–strain curve. (Reproduced with permission from A phenomenological three-dimensional rate-independent continuum damage model for highly filled polymers: Formulation and computational aspects, EA de Souza Neto, D Perić and DRJ Owen, *Journal of the Mechanics and Physics of Solids*, Vol 42, Issue 10 © 1994 Elsevier Science Ltd.)

The essential modification required to the existing Ogden stress updating implementation (refer to subroutine SUOGD) in order to accommodate damage is the incorporation of the operations carried out in items (ii)–(iv). Also, the storage of one extra state variable – the damage internal variable, D – is required. This variable is to be handled completely analogously to other internal state variables such as, for instance, the accumulated plastic strain in conventional elastoplasticity models (refer, for example, to subroutine SUVM described on page 224 for details). A logical flag similar to IFPLAS of SUVM is also needed. This is addressed in Remark 13.8 below.

The tangent modulus

The second crucial component of the implicit finite element implementation of the model is the calculation of its associated tangent moduli.

Box 13.4. Algorithm for integration of hyperelastic damage constitutive equations.

(i) Given the deformation gradient $\boldsymbol{F}_{n+1}$, compute $\boldsymbol{B}_{n+1} := \boldsymbol{F}\,\boldsymbol{F}^T$ and perform the spectral decomposition of $\boldsymbol{B}_{n+1}$

$$\boldsymbol{B}_{n+1} = \sum_{i=1}^{3} \lambda_i^2\, \boldsymbol{M}_i$$

(ii) Evaluate current value of ψ^0

$$\psi^0_{n+1} := \psi^0(\lambda_1, \lambda_2, \lambda_3)$$

(iii) Update damage variable

IF $\Phi^{\text{trial}}_{n+1} := \bar{\rho}\psi^0_{n+1} - D_n \leq 0$ THEN

no damage evolution $\Rightarrow D_{n+1} := D_n$

ELSE

damage evolution $\Rightarrow D_{n+1} := \bar{\rho}\psi^0_{n+1}$

ENDIF

(iv) Update principal Kirchhoff stresses

$$\xi := \frac{\bar{\rho}\psi^0_{n+1}}{D_{n+1}}$$

$$\tau_i := f(\xi)\, \bar{\rho}\, \frac{\partial \psi^0}{\partial \lambda_i} \lambda_i \bigg|_{n+1}$$

(v) Assemble Kirchhoff stress tensor

$$\boldsymbol{\tau}_{n+1} := \sum_{i=1}^{3} \tau_i\, \boldsymbol{M}_i$$

In contrast to purely hyperelastic theories, the situation here is similar to that of classical elastoplasticity (with elastic and elastoplastic tangents) in that two possible tangents exist at points on the virgin curve: one tangent to the loading path and another tangent to the unloading path (the situation is clearly illustrated in Figure 13.21(b)). Below the maximum attained strain level of ψ^0 (the dotted lines of Figure 13.21(b), excluding their intersection with the virgin curve), the tangent is unique.

Now let us recall (refer to Section 13.4.2) that, when the principal stretch representation is adopted, the essential model-related derivatives required to assemble the spatial tangent modulus, a, are the principal Kirchhoff stress derivatives: $\partial\tau_i/\partial b_j$, where $b_j = \lambda_j^2$ are the eigenvalues of $\boldsymbol{B}$. The only difference between the present model and a standard hyperelastic one in the calculation of the tangent modulus is that, following what was said in the previous paragraph, two distinct situations have to be considered:

1. *unloading* or *reloading with* $\bar{\rho}\psi^0 < D$ is assumed to occur, i.e. the process is purely elastic. The tangent is unique and the corresponding derivatives are obtained by differentiating the principal stress constitutive equation (ii) of Box 13.3 with constant D:

$$\frac{\partial \tau_i}{\partial b_j} = \frac{1}{2\lambda_j}\left[f(\xi)\,\frac{\partial \tau_i^0}{\partial \lambda_j} + \frac{f'(\xi)\,\tau_i^0 \tau_j^0}{D\,\lambda_j}\right] \tag{13.136}$$

   with no summation implied on repeated indices and with f' denoting the derivative of f;

2. otherwise damage evolution occurs (loading). In this case, $\xi = \text{constant} = 1$, $f(\xi) = 1$ and $f' = 0$ so that

$$\frac{\partial \tau_i}{\partial b_j} = \frac{\partial \tau_i^0}{\partial b_j} = \frac{1}{2\lambda_j}\,\frac{\partial \tau_i^0}{\partial \lambda_j}, \tag{13.137}$$

   with no summation implied.

Remark 13.8. The incorporation of the above into an existing hyperelastic computational implementation is also straightforward. One important practical aspect that needs to be taken into consideration in the implementation of this hyperelastic damage model is that the decision on whether to use the expression of item 1 or item 2 above has to be made in the computation of the tangent modulus. This can be done in the same manner as in the computation of the tangent modulus in elastoplasticity where a logical flag, set before the tangent modulus computation subroutine is called, indicates which tangent (either elastic or elastoplastic) is to be computed. The interested reader is referred to subroutine `CTVM` (Consistent Tangent computation for the Von Mises model), described in Section 7.4.3 (page 235), where the logical argument `EPFLAG` (set in subroutine `MATICT` before `CTVM` is called) defines which tangent will be computed. Damage *loading* (and use of item 2 above) and *unloading/reloading without damage evolution* (with use of item 1 above) should be assumed under the same conditions that set `EPFLAG`, respectively, to `.TRUE.` and `.FALSE.` in subroutine `MATICT` of program `HYPLAS`). Note that a logical algorithmic variable (analogous to `IFPLAS` in subroutine `SUVM`), indicating whether the step is purely elastic (no damage evolution) or whether there is damage evolution, must be set in the corresponding stress-updating procedure. As in elastoplasticity, this flag can be transferred between the stress-updating routine and the tangent modulus computation routine stored in the logical global array `LALGVA`.

13.7.6. EXAMPLE: INFLATION/DEFLATION OF A DAMAGEABLE RUBBER BALLOON

In this example, we use the above hyperelastic damage model to simulate one inflation/deflation cycle of a rubber balloon. The geometry of the problem, loading and finite element mesh are identical to those of the purely hyperelastic simulation described in Section 13.6.5. The underlying function ψ^0 is assumed to be a three-term Ogden function with the same constants as the strain–energy function of the example of Section 13.6.5. The master damage curve adopted is the straight line shown in Figure 13.22 (again, $\bar{\rho}$ is taken as unity).

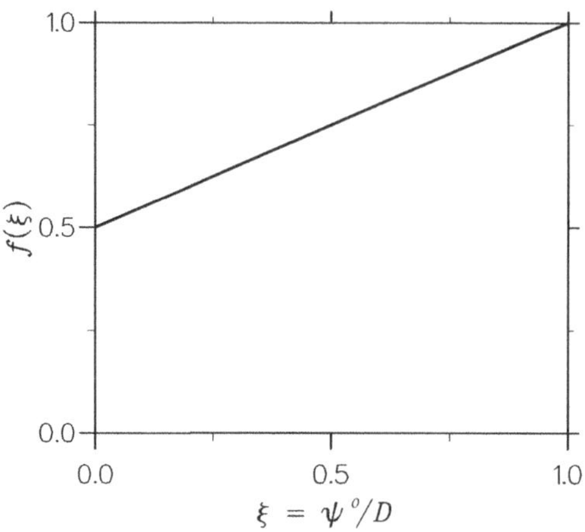

Figure 13.22. Damageable rubber balloon. Master damage curve. (Reproduced with permission from A phenomenological three-dimensional rate-independent continuum damage model for highly filled polymers: Formulation and computational aspects, EA de Souza Neto, D Perić and DRJ Owen, *Journal of the Mechanics and Physics of Solids*, Vol 42, Issue 10 © 1994 Elsevier Science Ltd.)

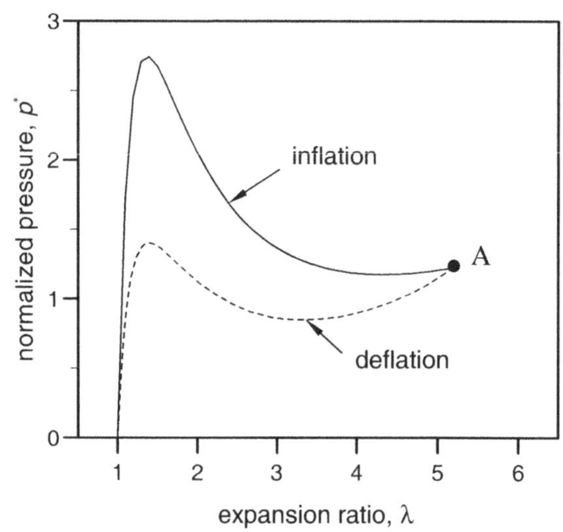

Figure 13.23. Damageable rubber balloon. Pressure–expansion diagram. (Reproduced with permission from A phenomenological three-dimensional rate-independent continuum damage model for highly filled polymers: Formulation and computational aspects, EA de Souza Neto, D Perić and DRJ Owen, *Journal of the Mechanics and Physics of Solids*, Vol 42, Issue 10 © 1994 Elsevier Science Ltd.)

By means of arc-length control, starting from the initial configuration ($\lambda = 1$), the internal pressure is applied gradually and the membrane is inflated until the configuration defined by $\lambda = 5.182$ (point A of Figure 13.23) is reached. At this stage, the load is reversed and the balloon is deflated returning to its initial configuration. The pressure–expansion curve obtained is presented in Figure 13.23. As inflation occurs under monotonically increasing circumferential stretching, the inflation branch of the pressure–expansion diagram corresponds to the behaviour governed by the strain-energy function ψ^0. Indeed, it matches exactly the hyperelastic solution obtained in the example of Section 13.6.5. The deflation branch of the curve shows clearly the softening effect of damage at the global level. The pressure–expansion curve shown here has a good qualitative agreement with the balloon inflation experiment discussed by Beatty (1987). However, in the experiment studied by this author, a residual circumferential strain was observed after complete deflation of the balloon ($p = 0$). Incorporation of this effect would require the consideration of additional internal variables leading to a theory that allows for description of inelastic deformations with possible inclusion of viscous effects.

14 FINITE STRAIN ELASTOPLASTICITY

THE present chapter is devoted to the theory and finite element implementation of *finite strain* elastoplasticity. The theory and numerical algorithms discussed here are an extension of those seen in Part Two of this book where *infinitesimal* elastoplasticity has been thoroughly discussed.

As we have seen in Part Two, infinitesimal theories can be adopted to model the behaviour of inelastic solids so long as the strains and rotations remain sufficiently small. In many industrial applications of the theory of plasticity, however, the infinitesimal deformation hypothesis cannot be introduced without significant loss of accuracy. Typical examples include the analysis of metal-forming operations and crashworthiness problems where the underlying deformation processes are dominated by very large strains and rotations. In fact, over the last two decades, the use of finite element models to simulate large deformations of inelastic solids has become an essential part of the design process, particularly in highly competitive sectors such as the automotive manufacturing industry. The use of finite element models in such cases is gradually replacing classical predictive tools (which have been largely based on empirical knowledge developed on a trial-and-error basis) with a more rational and scientifically based approach. Such a move towards continuum mechanics/finite element-based simulation tools follows the ever-increasing need for reduction of design-to-production time and costs.

The outline of this chapter is as follows. We start in Section 14.1 with a brief review of the development of finite strain computational plasticity. In Section 14.2 we introduce a one-dimensional finite strain plasticity model. We then move on, in Section 14.3, to the description of a general multiaxial hyperelastic-based multiplicative plasticity model. The description of the general model is followed, in Section 14.4, by the detailed derivation of an elastic predictor/return-mapping-type algorithm for integration of the finite plasticity constitutive equations. Section 14.5 focuses on the derivation of the associated general consistent spatial tangent modulus, which is crucial to the implicit finite element implementation of program HYPLAS. The corresponding principal stress-based implementation is addressed in Section 14.6. The issue of treatment of the plane stress constraint within the multiplicative finite plasticity framework is discussed in Section 14.7 and the viscoplastic formulation is briefly discussed in Section 14.8. This is followed, in Section 14.9, by a set of finite element examples illustrating applications of the multiplicative plasticity framework. Section 14.10 reviews alternative hypoelastic-based approaches to the modelling of finite inelastic deformations. Finally, Section 14.11 addresses the issue of kinematic hardening modelling under finite strains.

Computational Methods for Plasticity: Theory and Applications EA de Souza Neto, D Perić and DRJ Owen

14.1. Finite strain elastoplasticity: a brief review

The formulation of elastoplastic constitutive models in the geometrically nonlinear range can be traced back to the late 1950s. Early references to this fascinating subject are provided by the pioneering works of Hill (1958) and Green and Naghdi (1965). In view of its potential application in many areas of practical engineering interest, it did not take long before large strain elastoplastic formulations were brought within the finite element context. The first papers on the subject were published in the 1970s. Initial developments (Argyris and Kleiber, 1977; Argyris *et al.*, 1978; Hibbitt *et al.*, 1970; McMeeking and Rice, 1975; Nagtegaal and de Jong, 1981) have relied exclusively on the use of *hypoelastic*-based constitutive formulations whereby standard infinitesimal elastoplasticity models are extended to the finite strain range by recasting the original evolution equations in terms of suitably chosen objective (or frame-invariant) stress rates. During early stages of development, hypoelastic-based descriptions have been the subject of intense debate within the finite element community. Many controversial issues have arisen, ranging from the use of different objective stress rates in the formulation of the constitutive equations (Atluri, 1984; Nemat-Nasser, 1982; Perić, 1992) to fundamental drawbacks such as the possible lack of objectivity of (algorithmic) incremental constitutive laws (Hughes, 1984; Hughes and Winget, 1980; Rubinstein and Atluri, 1983) as well as observed oscillatory stress response under monotonic loading (Nagtegaal and de Jong, 1982) and dissipative behaviour within the 'elastic' range (Kojić and Bathe, 1987; Simo and Pister, 1984). In this context, *hyperelastic*-based formulations of finite plasticity have emerged (Simo, 1985; Simo and Ortiz, 1985). Based on the hyperelastic description of the reversible behaviour (as in the general constitutive theory alluded to in Chapter 3) in conjunction with the multiplicative elastoplastic split of the deformation gradient (Lee and Liu, 1967), such theories naturally by-pass the inherent drawbacks of hypoelastic-based approaches. In particular, dissipative response becomes impossible within the elastic range and the requirement of *incremental objectivity* (frame invariance of the algorithmic constitutive rule) is trivially satisfied. Moreover, when the Hencky strain energy function (described in Section 13.2.3, page 528) is adopted to model the elastic behaviour, a small strain format return mapping-based stress-updating procedure can be elegantly recovered with the use of a suitable exponential map-based integrator for the plastic flow equation (Cuitiño and Ortiz, 1992a; Eterovic and Bathe, 1990; Perić and Owen, 1991; Perić *et al.*, 1992; Simo, 1992; Simo and Miehe, 1992; Weber and Anand, 1990). Throughout the last decade, the hyperelastic-based multiplicative approach has gained widespread acceptance (see Simo (1998), for an extensive review of both theory and related numerical methods) and is currently adopted in many commercial finite element codes for large-strain elastoplastic analysis. Hyperelastic-based formulations have proven successful not only in the context of conventional isotropic J_2 metal plasticity but also with other underlying theories such as soil plasticity (Meschke and Liu, 1999), damage mechanics (de Souza Neto and Perić, 1996; de Souza Neto *et al.*, 1992, 1994a, 1998; Li, 1995; Steinmann *et al.*, 1994) and anisotropic single crystal plasticity (discussed in detail in Chapter 16) as well as in the modelling of various structural elements such as membranes (Ibrahimbegović, 1994) and shells (Miehe, 1998b).

14.2. One-dimensional finite plasticity model

Following a format similar to that of Chapter 6, we choose here to introduce the basic concepts of finite elastoplasticity by firstly formulating a one-dimensional constitutive model. The model presented here extends the uniaxial infinitesimal theory of Section 6.2 (from page 141) to the finite strain regime.

Before proceeding, we remark that the generalisation of the infinitesimal one-dimensional model to the finite strain range is by no means unique; that is, an infinite number of large strain models may be defined such that their small strain limit is the infinitesimal theory of Section 6.2. The particular finite strain generalisation presented here is one that contains all the basic ingredients of the large strain isotropic multiaxial model to be introduced in Section 14.3. This should assist the reader in becoming familiarised with the essential concepts involved in the derivation of the multidimensional finite plasticity model.

At the outset, we introduce the *axial stretch* (in analogy with the infinitesimal axial strain of the small strain theory) as the main kinematic variable of the model. Referring back to Section 6.2, if we assume constant strain over the uniaxially loaded bar of Figure 6.2 and with l_0 and l denoting, respectively, the reference length and the length of the bar at an arbitrarily deformed configuration, the axial stretch is given by

$$\lambda = \frac{l}{l_0}. \tag{14.1}$$

14.2.1. THE MULTIPLICATIVE SPLIT OF THE AXIAL STRETCH

Let us now assume that, at the deformed configuration defined by l, the bar has already been loaded past the elastic limit. Upon complete unloading from this configuration, the plastified bar will have a (permanently deformed) length l_p. The length l_p, thus, defines an *unstressed configuration* of the bar. This configuration will also be referred to as the *intermediate* or *plastic* configuration. The *axial plastic stretch*, that is, the stretch associated with the unstressed configuration is defined, similarly to (14.1), as

$$\lambda^p = \frac{l_p}{l_0}. \tag{14.2}$$

The unloading process, that is, the deformation from the configuration defined by l to the unstressed configuration defined by l_p, is elastic. This motivates the definition of the *elastic axial stretch*,

$$\lambda^e = \frac{l}{l_p}. \tag{14.3}$$

From the above, we establish that any deformed state of the bar is characterised by the *multiplicative split* of the axial stretch (14.1) into an elastic stretch, λ^e, and a plastic stretch, λ^p, i.e.

$$\lambda = \lambda^e \lambda^p. \tag{14.4}$$

This expression is a generalisation of the additive split (6.2) of the infinitesimal strain to the finite strain case.

14.2.2. LOGARITHMIC STRETCHES AND THE HENCKY HYPERELASTIC LAW

The Hencky hyperelastic model has been presented in Section 13.2.3 (from page 528) as an immediate generalisation of the conventional linear elastic law to the finite strain range. Under uniaxial stress conditions, Hencky's hyperelasticity law (13.59) reads

$$\tau = E \ln \lambda, \tag{14.5}$$

where τ denotes the Kirchhoff axial stress. Here we shall adopt an analogous law in terms of the *elastic* stretch, i.e. we will use the following finite extension of the one-dimensional elastic constitutive equation (6.4)

$$\tau = E\, \varepsilon^e, \tag{14.6}$$

where τ is the axial Kirchhoff stress and ε^e is the *elastic logarithmic strain*

$$\varepsilon^e = \ln \lambda^e. \tag{14.7}$$

14.2.3. THE YIELD FUNCTION

The *yield function*, Φ, is defined in complete analogy with (6.5) as

$$\Phi(\tau, \tau_y) = |\tau| - \tau_y, \tag{14.8}$$

where $|\tau_y|$ is the yield Kirchhoff axial stress. The corresponding elastic domain is given by

$$\mathscr{E} = \{\tau \mid \Phi(\tau, \tau_y) < 0\}. \tag{14.9}$$

14.2.4. THE PLASTIC FLOW RULE

In addition to the above, a plastic flow rule is required to define the evolution of the plastic stretch, λ^p. Here we shall postulate the following law

$$\dot{\lambda}^p (\lambda^p)^{-1} = \dot{\gamma}\, \text{sign}(\tau), \tag{14.10}$$

which is complemented by the usual condition

$$\Phi \leq 0, \quad \dot{\gamma} \geq 0, \quad \dot{\gamma}\Phi = 0. \tag{14.11}$$

Note that, from the definition of λ^p, we have the identity

$$\dot{\lambda}^p (\lambda^p)^{-1} = \frac{\dot{l}_p}{l_p}, \tag{14.12}$$

that is, the right-hand side of (14.10) is the instantaneous rate of plastic straining measured with respect to the plastic configuration defined by λ^p. Also note, with the *logarithmic axial plastic strain* defined by

$$\varepsilon^p = \ln \lambda^p, \tag{14.13}$$

that (14.10) can be equivalently expressed as

$$\dot{\varepsilon}^p = \dot{\gamma}\, \text{sign}(\tau), \tag{14.14}$$

which has the same functional format as its infinitesimal counterpart (6.10). We remark, however, that the above law in terms of the logarithmic axial plastic strain does not generalise to the multiaxial case treated in Section 14.3.

Box 14.1. One-dimensional finite strain elastoplastic constitutive model.

(i) Multiplicative split of the axial stretch

$$\lambda = \lambda^e \lambda^p$$

(ii) Uniaxial Hencky elastic law

$$\tau = E \ln \lambda^e$$

(iii) Yield function

$$\Phi(\tau, \tau_y) = |\tau| - \tau_y$$

(iv) Plastic flow rule

$$\dot{\lambda}^p (\lambda^p)^{-1} = \dot{\gamma} \, \text{sign}(\tau)$$

(v) Hardening law

$$\tau_y = \tau_y(\bar{\varepsilon}^p)$$

$$\dot{\bar{\varepsilon}}^p = \dot{\gamma}$$

(vi) Loading/unloading criterion

$$\Phi \le 0, \quad \dot{\gamma} \ge 0, \quad \dot{\gamma}\Phi = 0$$

14.2.5. THE HARDENING LAW

To define completely the one-dimensional constitutive model, a hardening law is required. Following the same arguments of Section 6.2.5, we introduce hardening by letting τ_y be a given function

$$\tau_y = \tau_y(\bar{\varepsilon}^p), \tag{14.15}$$

of the *accumulated logarithmic axial plastic strain*, $\bar{\varepsilon}^p$, defined by the evolution equation

$$\dot{\bar{\varepsilon}}^p = |\dot{\varepsilon}^p| = \dot{\gamma}. \tag{14.16}$$

The resulting one-dimensional finite strain elastoplasticity constitutive model is summarised in Box 14.1.

14.2.6. THE PLASTIC MULTIPLIER

The determination of the plastic multiplier $\dot{\gamma}$ under plastic flow follows the same steps as in Section 6.2.7. The final expression reads

$$\dot{\gamma} = \frac{E}{H + E} \, \text{sign}(\tau) \, \frac{\dot{\lambda}}{\lambda}. \tag{14.17}$$

Equivalently, in terms of the total logarithmic axial strain,

$$\varepsilon = \ln \lambda, \tag{14.18}$$

expression (14.17) can be rewritten as

$$\dot{\gamma} = \frac{E}{H + E}\, |\dot{\varepsilon}|, \tag{14.19}$$

which has the identical functional format as its infinitesimal counterpart given by equation (6.30).

14.3. General hyperelastic-based multiplicative plasticity model

A generic finite strain hyperelastic-based multiplicative plasticity model is described in detail in this section. The theory introduced here forms the basis of the isotropic large strain plasticity framework whose finite element implementation (incorporated in program HYPLAS) is addressed in Sections 14.4 and 14.5. We remark that some of the crucial features of the model, namely the multiplicative elastoplastic kinematics and the use of a Hencky (logarithmic strain-based) hyperelasticity law, have already been introduced in the simpler context of the one-dimensional finite plasticity model discussed in the preceding section.

14.3.1. MULTIPLICATIVE ELASTOPLASTICITY KINEMATICS

The main hypothesis underlying the finite strain elastoplasticity constitutive framework described here is the *multiplicative decomposition* of the deformation gradient, $\boldsymbol{F}$, into elastic and plastic contributions; that is, it is assumed that the deformation gradient can be decomposed as the product

$$\boldsymbol{F} = \boldsymbol{F}^{e} \boldsymbol{F}^{p}, \tag{14.20}$$

where $\boldsymbol{F}^e$ and $\boldsymbol{F}^p$ are named, respectively, the *elastic* and *plastic deformation gradients*. The multiplicative split of $\boldsymbol{F}$, introduced by Lee and Liu (1967) and Lee (1969), embodies the assumption of the existence of a local *unstressed intermediate configuration* defined by the plastic deformation gradient, $\boldsymbol{F}^p$. At each material point, the local intermediate configuration is obtained from the fully deformed configuration by a purely elastic unloading (associated with the inverse of $\boldsymbol{F}^e$) of its neighbourhood. The concept is schematically illustrated in Figure 14.1. It is a multiaxial generalisation of the one-dimensional multiplicative split (14.4). It must be emphasised that, unlike in the one-dimensional case, the unstressed intermediate configuration concept in the multiaxial case is valid only in the *local* (pointwise) sense. Compatible unstressed configurations of multiaxially stretched elastoplastic *bodies* do not exist in general; that is, it is generally not possible to find a deformation φ^p such that

$$\boldsymbol{F}^p = \frac{\mathrm{d}\varphi^p}{\mathrm{d}\boldsymbol{p}}.$$

For a number of years since its introduction, the multiplicative kinematic description has been the subject of intense investigation (Dafalias, 1984, 1985; Dashner, 1986; Lubarda and Lee, 1981; Mandel, 1973; Nemat-Nasser, 1979, 1982) and much light has been shed on its constitutive implications. Some aspects and kinematic consequences of the multiplicative decomposition assumption are addressed in the following.

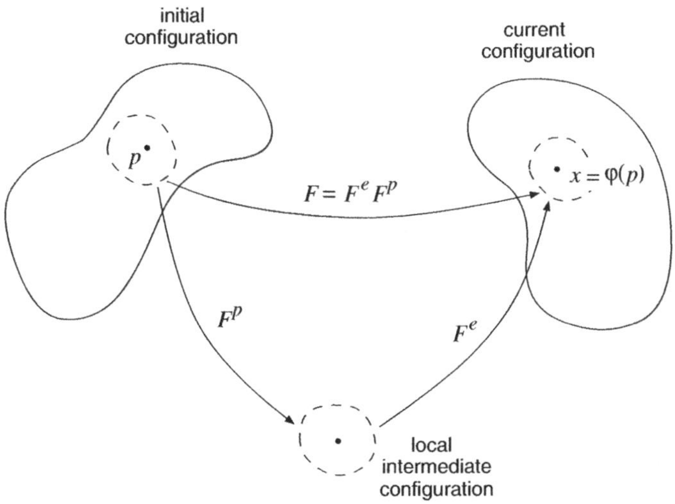

Figure 14.1. Multiplicative decomposition of the deformation gradient.

Remark 14.1. Rather than being just a convenient mathematical extension of the standard additive decomposition of the strain tensor adopted in the infinitesimal theory, the multiplicative decomposition of $\boldsymbol{F}$ finds a solid and consistent physical justification in the slip theory of crystals (Asaro, 1983; Peirce *et al.*, 1982, 1983; Rice, 1971). Essentially, the macroscopic plastic deformation of metallic crystals is the result of microscopic sliding between blocks of crystals along certain crystallographic planes (for more details on the plastic deformation of crystals, we refer to Chapter 16, where the computational treatment of an elastoplastic single crystal model is addressed). This mechanism is schematically illustrated in Figure 14.2. Under generic finite straining, crystal blocks slide over each other under the simultaneous action of lattice rotation and distortion. The plastic deformation gradient, $\boldsymbol{F}^p$, is associated with the pure sliding between crystal blocks. The elastic contribution, $\boldsymbol{F}^e$, on the other hand, represents the crystal lattice rotation and distortion.

Elastic and plastic stretch and rotation tensors

Following the above multiplicative split of $\boldsymbol{F}$, the stretches and rotations associated with the elastic and plastic parts of the deformation gradient are obtained by performing the polar decompositions

$$\boldsymbol{F}^e = \boldsymbol{R}^e\,\boldsymbol{U}^e = \boldsymbol{V}^e\boldsymbol{R}^e \tag{14.21}$$

and

$$\boldsymbol{F}^p = \boldsymbol{R}^p\,\boldsymbol{U}^p = \boldsymbol{V}^p\boldsymbol{R}^p. \tag{14.22}$$

The resulting tensors $\boldsymbol{U}^e$ ($\boldsymbol{U}^p$), $\boldsymbol{V}^e$ ($\boldsymbol{V}^p$) and $\boldsymbol{R}^e$ ($\boldsymbol{R}^p$), are named, respectively, the *elastic* (*plastic*) right stretch tensor, the *elastic* (*plastic*) left stretch tensor and the *elastic* (*plastic*) rotation tensor.

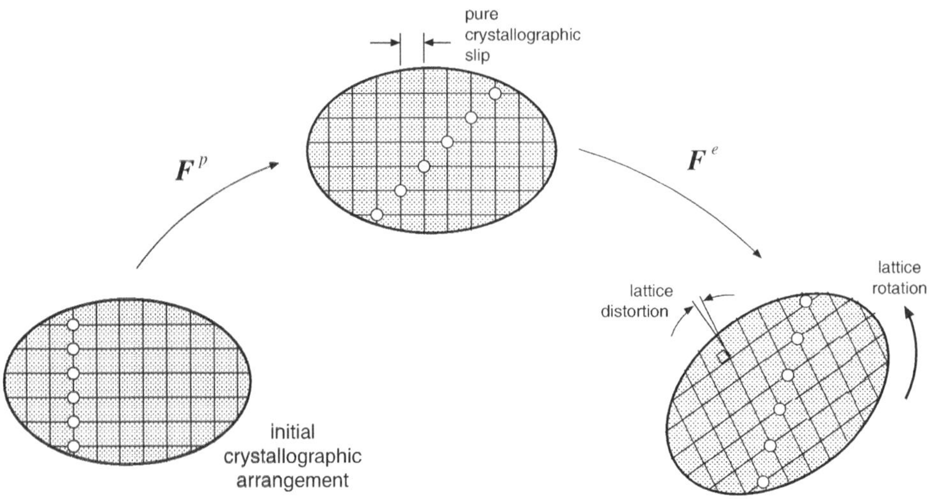

Figure 14.2. Multiplicative decomposition of the deformation gradient. Micromechanical basis.

The meaning of the above tensors is analogous to that of the tensors obtained from the polar decomposition (3.40) (page 49) of the total deformation gradient; that is, $\boldsymbol{U}^p$ represents a pure plastic stretch along three mutually orthogonal axes starting from the reference configuration, $\boldsymbol{U}^e$ represents a pure elastic stretch along three mutually orthogonal axes starting from the intermediate configuration, $\boldsymbol{R}^p$ is the pure rotation between the reference and intermediate configurations, and so forth.

The velocity gradient: elastic and plastic contributions

Let us now look at the consequences of the assumed multiplicative split of $\boldsymbol{F}$ on the *velocity gradient* (refer to Section 3.1.8, from page 55) defined by

$$\boldsymbol{L} \equiv \nabla_x \boldsymbol{v} = \dot{\boldsymbol{F}}\, \boldsymbol{F}^{-1}. \tag{14.23}$$

With the introduction of (14.20) into the above formula, a straightforward differentiation using the product rule gives the following additive decomposition of $\boldsymbol{L}$,

$$\boldsymbol{L} = \boldsymbol{L}^e + \boldsymbol{F}^e \boldsymbol{L}^p (\boldsymbol{F}^e)^{-1}, \tag{14.24}$$

where $\boldsymbol{L}^e$ and $\boldsymbol{L}^p$ are defined as

$$\boldsymbol{L}^e \equiv \dot{\boldsymbol{F}}^e (\boldsymbol{F}^e)^{-1}, \quad \boldsymbol{L}^p \equiv \dot{\boldsymbol{F}}^p (\boldsymbol{F}^p)^{-1}. \tag{14.25}$$

We shall refer to $\boldsymbol{L}^e$ and $\boldsymbol{L}^p$, respectively, as the *elastic* and *plastic velocity gradients*.

Analogously to the velocity gradient discussed in Section 3.1.8, the plastic velocity gradient as defined in the above is a quantity of the intermediate configuration. The velocity gradient, $\boldsymbol{L}$, itself is a quantity of the spatial (or deformed) configuration. Note that, accordingly, in the plastic contribution

$$F^e L^p (F^e)^{-1},$$

to L, the plastic velocity gradient L^p is pre- and post-multiplied, respectively, by F^e (which maps the local intermediate configuration into the deformed configuration) and its inverse. This renders the above plastic contribution to L a spatial quantity.

The plastic stretching and spin tensors

In order to formulate a plastic flow rule within the multiplicative plasticity framework, it is convenient here to define the *plastic stretching* (or *rate of plastic deformation*) and the *plastic spin* tensors as

$$D^p \equiv \text{sym}[L^p], \quad W^p \equiv \text{skew}[L^p]. \tag{14.26}$$

The above definitions are analogous to the decomposition of the velocity gradient in Section 3.1.8 into the stretching and spin tensors.

The meaning of the plastic stretching and spin

It has been shown in Section 3.1.8 that the stretching tensor D (defined by expression (3.65)) is associated with a purely straining velocity field that stretches the body locally along the three mutually orthogonal directions of its current deformed configuration defined by the eigenvectors of D (refer to expression (3.69) and the surrounding text). An analogous interpretation can be given to D^p. As D^p is associated only with F^p, which defines the intermediate configuration, D^p measures the instantaneous rate of plastic straining that stretches the intermediate configuration along the three mutually orthogonal directions defined by the eigenvectors of D^p. With $\{e_i\}$ denoting an orthonormal basis of eigenvectors of D^p, we have

$$D^p = \sum_{i=1}^{3} d_i^p \, e_i \otimes e_i, \tag{14.27}$$

where d_i^p is the eigenvalue of D^p associated with the eigenvector e_i. Each d_i^p gives the instantaneous rate of stretching of the intermediate configuration along the direction of e_i.

The plastic spin tensor, W^p, defined by $(14.26)_2$, on the other hand, represents an instantaneous rate of rigid spinning of the intermediate configuration.

Spatially rotated plastic stretching

In the following sections, the *spatial* configuration will be used to formulate the general finite isotropic multiplicative plasticity constitutive model. Accordingly, all kinematic quantities involved in the constitutive formulation will be expressed as spatial quantities (i.e. in the deformed configuration). In the context of the present constitutive framework, we shall work with the following rotation of D^p,

$$\tilde{D}^p \equiv R^e \, D^p \, R^{eT} = R^e \, \text{sym}[\dot{F}^p F^{p^{-1}}] \, R^{eT} \tag{14.28}$$

in the definition of the plastic flow rule. The above tensor represents the rate of plastic stretching rotated (by the elastic rotation) to the spatial (or deformed) configuration and will be called the *spatially rotated* (or simply *spatial*) plastic stretching (or rate of plastic deformation). It can be represented as

$$\tilde{\boldsymbol{D}}^p = \sum_{i=1}^{3} d_i^p \, \tilde{\boldsymbol{e}}_i \otimes \tilde{\boldsymbol{e}}_i, \tag{14.29}$$

where $\tilde{\boldsymbol{e}}_i \equiv \boldsymbol{R}^e \, \boldsymbol{e}_i$.

14.3.2. THE LOGARITHMIC ELASTIC STRAIN MEASURE

Recall that in Chapter 3, Eulerian (or spatial) strain measures have been defined as functions of the left stretch tensor $\boldsymbol{V}$. Likewise, spatial *elastic* strain measures can be defined as functions of the *elastic* left stretch tensor, $\boldsymbol{V}^e$. In the present general model of finite strain elastoplasticity, the *logarithmic* strain will be adopted to measure elastic deformations. As we shall see in what follows, the use of the *logarithmic* (or *natural*) strain measure to describe the elastic behaviour is particularly convenient. In addition to its physical appeal as an extension of the infinitesimal elastic strain measure, its use, allied to suitable approximations to the plastic flow rule, results in substantial simplifications in the stress integration algorithm and allows a natural extension, to the finite strain range, of the elastic predictor/return-mapping algorithms of infinitesimal elastoplasticity described in Chapter 7. The Eulerian logarithmic elastic strain is defined by

$$\boldsymbol{\varepsilon}^e \equiv \ln \boldsymbol{V}^e = \tfrac{1}{2} \ln \boldsymbol{B}^e, \tag{14.30}$$

where $\ln(\cdot)$ above denotes the *tensor logarithm* of $(\cdot)$ and

$$\boldsymbol{B}^e = \boldsymbol{F}^e (\boldsymbol{F}^e)^T = (\boldsymbol{V}^e)^2$$

is the left elastic Cauchy–Green strain tensor.

Deviatoric and volumetric logarithmic strains

By performing the deviatoric/volumetric split of the elastic logarithmic strain, we obtain

$$\boldsymbol{\varepsilon}^e = \boldsymbol{\varepsilon}_{\mathrm{d}}^e + \tfrac{1}{3} \varepsilon_{\mathrm{v}}^e \, \boldsymbol{I}, \tag{14.31}$$

where

$$\boldsymbol{\varepsilon}_{\mathrm{d}}^e \equiv \boldsymbol{\varepsilon}^e - \tfrac{1}{3} \operatorname{tr}[\boldsymbol{\varepsilon}^e] \, \boldsymbol{I},$$

and the volumetric logarithmic elastic strain is given by

$$\varepsilon_{\mathrm{v}}^e \equiv \operatorname{tr}[\boldsymbol{\varepsilon}^e] = \ln J^e,$$

with

$$J^e \equiv \det \boldsymbol{F}^e.$$

Due to the properties of the logarithmic strain measure (refer to the discussion starting on page 529), analogously to the infinitesimal theory, a traceless $\boldsymbol{\varepsilon}^e$ ($\varepsilon_{\mathrm{v}}^e = 0$) corresponds to a finite volume-preserving elastic deformation, i.e. a deformation with $\det \boldsymbol{F}^e = 1$.

14.3.3. A GENERAL ISOTROPIC LARGE-STRAIN PLASTICITY MODEL

A rather general class of isotropic hyperelastic-based finite strain elastoplastic constitutive models, formulated in the spatial configuration, can be defined by following the formalism of thermodynamics with internal variables described in Section 3.4. Essentially, the model is defined by postulating

1. a *free-energy potential*, from which the hyperelastic law is derived;
2. a *yield function* to define the onset of plastic yielding;
3. a *dissipation potential*, from which the plastic flow rule and the evolution laws for the internal variables are derived.

It is emphasised that the model described here is restricted to the isothermal case. Its general constitutive equations are presented in the following.

The free-energy potential: hyperelastic law

At the outset, the free-energy potential is assumed to have the general form

$$\psi(\boldsymbol{\varepsilon}^e, \boldsymbol{\alpha}); \tag{14.32}$$

that is, ψ is here expressed as a function of the elastic *logarithmic* strain and a generic set $\boldsymbol{\alpha} \equiv \{\alpha_1, \alpha_2, \ldots, \alpha_k\}$ of k internal variables associated with dissipative mechanisms.

It will be shown in Section 14.3.4 that the above defined potential yields the following constitutive law for the *Kirchhoff* stress tensor

$$\boldsymbol{\tau} = \bar{\rho}\, \frac{\partial \psi}{\partial \boldsymbol{\varepsilon}^e}. \tag{14.33}$$

The set $\boldsymbol{A} \equiv \{A_1, A_2, \ldots, A_k\}$ of *thermodynamical forces* conjugate to the internal variables is defined as usual by

$$A_i = \bar{\rho}\, \frac{\partial \psi}{\partial \alpha_i} \quad (i = 1, \ldots, k). \tag{14.34}$$

The yield criterion

A general yield function, $\Phi(\boldsymbol{\tau}, \boldsymbol{A})$, expressed in terms of the *Kirchhoff* stress and the set $\boldsymbol{A}$ of conjugate thermodynamical forces is adopted to define the onset of plastic flow. Accordingly, the *elastic domain* where only reversible phenomena take place is defined as the set

$$\mathscr{E} = \{\boldsymbol{\tau} \mid \Phi(\boldsymbol{\tau}, \boldsymbol{A}) < 0\},$$

of the space of stresses. Plastic yielding may take place only if $\boldsymbol{\tau}$ lies on the *yield surface* – the boundary of the elastic domain – defined by

$$\mathscr{Y} = \{\boldsymbol{\tau} \mid \Phi(\boldsymbol{\tau}, \boldsymbol{A}) = 0\}. \tag{14.35}$$

Admissible Kirchhoff stresses must lie either in the elastic domain or on the yield surface. For a given $\boldsymbol{A}$, the *set of admissible Kirchhoff stresses* is defined as

$$\bar{\mathscr{E}} = \{\boldsymbol{\tau} \mid \Phi(\boldsymbol{\tau}, \boldsymbol{A}) \le 0\}. \tag{14.36}$$

Finite strain plastic flow rule

A general *plastic flow potential*, $\Psi(\boldsymbol{\tau}, \boldsymbol{A})$, also expressed as a function of the *Kirchhoff* stress and the thermodynamical force set, $\boldsymbol{A}$, is adopted. With the dissipation potential, the evolution of the plastic deformation gradient – the plastic flow – is defined by the following constitutive equation for the rotated plastic stretching

$$\tilde{\boldsymbol{D}}^p = \dot{\gamma}\,\frac{\partial \Psi}{\partial \boldsymbol{\tau}}, \tag{14.37}$$

complemented by postulating a zero plastic spin†

$$\boldsymbol{W}^p = \boldsymbol{0}. \tag{14.38}$$

The two equations above completely define the evolution of $\boldsymbol{F}^p$. Indeed, the definition (14.28) of $\tilde{\boldsymbol{D}}^p$ together with the constitutive laws (14.37) and (14.38) are equivalent to the following evolution law for the plastic deformation gradient

$$\boldsymbol{L}^p \equiv \dot{\boldsymbol{F}}^p(\boldsymbol{F}^p)^{-1} = \dot{\gamma}\,(\boldsymbol{R}^e)^T\,\frac{\partial \Psi}{\partial \boldsymbol{\tau}}\,\boldsymbol{R}^e, \tag{14.39}$$

which is a multidimensional extension of the one-dimensional plastic flow equation (14.14).

Remark 14.2. Plastic isotropy implies that $\boldsymbol{\tau}$ and $\partial\Psi/\partial\boldsymbol{\tau}$ are coaxial (i.e. they share the same principal axes). Thus, in the above constitutive equation, the (instantaneous) principal directions of stretching of the intermediate configuration are the directions obtained by rotating the principal axes of the Kirchhoff stress back to the intermediate configuration. In the absence of elastic rotations, the principal directions of plastic stretching coincide with those of the Kirchhoff stress.

Under elastoplastic isotropy (a condition to which the present model is restricted), the above plastic flow rule is equivalent to the following constitutive equation for the total plastic contribution, $\boldsymbol{F}^e\boldsymbol{L}^p(\boldsymbol{F}^e)^{-1}$, to the velocity gradient (14.39),

$$\boldsymbol{F}^e\boldsymbol{L}^p(\boldsymbol{F}^e)^{-1} = \dot{\gamma}\,\frac{\partial \Psi}{\partial \boldsymbol{\tau}}. \tag{14.40}$$

The equivalence between the two equations can be established by first making use of the left polar decomposition of $\boldsymbol{F}^e$, which after a straightforward rearrangement gives

$$\boldsymbol{L}^p = \dot{\gamma}\,(\boldsymbol{R}^e)^T(\boldsymbol{V}^e)^{-1}\,\frac{\partial \Psi}{\partial \boldsymbol{\tau}}\,\boldsymbol{V}^e\boldsymbol{R}^e. \tag{14.41}$$

Due to the assumed elastoplastic isotropy, $\boldsymbol{V}^e$ and $\partial\Psi/\partial\boldsymbol{\tau}$ commute so that the terms $\boldsymbol{V}^e$ and $(\boldsymbol{V}^e)^{-1}$ are cancelled out of the above equation, leading to (14.39).

†The assumption of zero plastic spin is compatible with plastic *isotropy* – the condition to which the present model is restricted. Introduction of plastic anisotropy requires the definition of an appropriate constitutive equation for the plastic spin. In the anisotropic single-crystal model described in Chapter 16, a micromechanically-based constitutive law which defines a generally non-vanishing plastic spin is adopted. The reader is referred to Dafalias (1984, 1985) for an in-depth discussion on this topic.

Equivalent elastic rate-based equations

Recall that in the infinitesimal theory, the additive split of the strain tensor allows, in a rather trivial manner, the elastoplasticity model to be equivalently cast in terms of either the plastic strain rate, as in (6.49), page 150, or the elastic strain rate, as in $(7.6)_1$, page 193. The definition of equivalent elastic rate-based rate equations is also possible within the present framework for finite plasticity. However, due to the multiplicative nature of the elastoplastic kinematic split, the relationship between the elastic and plastic-based rates is more complex than that of the infinitesimal theory. The elastic deformation gradient evolution equation corresponding to the plastic flow equation (14.39) within the multiplicative kinematics framework is given by

$$\dot{\boldsymbol{F}}^e\,(\boldsymbol{F}^e)^{-1} = \dot{\boldsymbol{F}}\,\boldsymbol{F}^{-1} - \dot{\gamma}\,\frac{\partial\Psi}{\partial\boldsymbol{\tau}} \tag{14.42}$$

or, with a more compact notation,

$$\boldsymbol{L}^e = \boldsymbol{L} - \dot{\gamma}\,\frac{\partial\Psi}{\partial\boldsymbol{\tau}}. \tag{14.43}$$

The above equation is a finite strain counterpart of the elastic strain evolution law $(7.6)_1$ and is obtained simply by introducing the plastic flow equation (14.40) into the velocity gradient expression (14.24)–(14.25).

Another alternative definition of the present model in terms of an elastic rate can be made by exploiting the so-called *Lie derivative* concept (the reader is referred to Marsden and Hughes 1983, for a precise mathematical definition of the Lie derivative.) Let

$$\pounds_v\boldsymbol{B}^e \equiv \dot{\boldsymbol{B}}^e - \boldsymbol{L}\boldsymbol{B}^e - \boldsymbol{B}^e\boldsymbol{L}^T \tag{14.44}$$

denote the Lie derivative of $\boldsymbol{B}^e$ with respect to the velocity field $\boldsymbol{v}$. The above derivative is identified with the *Oldroyd objective rate*‡ introduced in (14.130)–(14.131). The plastic flow rule (14.39) (or (14.40)) is equivalent to the following constitutive equation for the Lie derivative of the elastic left Cauchy–Green tensor

$$\frac{1}{2}\pounds_v\boldsymbol{B}^e = -\dot{\gamma}\,\frac{\partial\Psi}{\partial\boldsymbol{\tau}}\,\boldsymbol{B}^e. \tag{14.45}$$

The equivalence between the two constitutive equations can be established as follows. We start by post-multiplying both sides of (14.40) by $\boldsymbol{B}^e$ and expanding the resulting expression with use of (14.24). This gives

$$\begin{aligned}\dot{\gamma}\,\frac{\partial\Psi}{\partial\boldsymbol{\tau}}\,\boldsymbol{B}^e &= \boldsymbol{F}^e\boldsymbol{L}^p(\boldsymbol{F}^e)^{-1}\,\boldsymbol{B}^e\\ &= [\boldsymbol{L} - \boldsymbol{L}^e]\boldsymbol{B}^e\\ &= \boldsymbol{L}\,\boldsymbol{B}^e - \dot{\boldsymbol{F}}^e(\boldsymbol{F}^e)^T.\end{aligned} \tag{14.46}$$

‡The concept of objective rates will be discussed only in Section 14.10.1 in connection with the definition of *hypoelastic*-based elastoplasticity models. The reader who is unfamiliar with the concept of objective rates of a tensor may prefer to read that section before proceeding further. We remark, however, that the alternative formulation presented here in terms of the Lie derivative (or Oldroyd objective rate) of the elastic left Cauchy–Green tensor is not required for the computational treatment of the model discussed in Section 14.4.

Due to the isotropy of Ψ and the elastic response, the tensors $\boldsymbol{B}^e$ and $\partial\Psi/\partial\boldsymbol{\tau}$ commute (refer to Section A.1.2, page 732) so that the left-hand side of the above equation is symmetric and the right-hand side must equal its symmetrised counterpart; that is, we have

$$\dot{\gamma}\,\frac{\partial\Psi}{\partial\boldsymbol{\tau}}\,\boldsymbol{B}^e = \frac{1}{2}\,[\boldsymbol{L}\,\boldsymbol{B}^e + \boldsymbol{B}^e\,\boldsymbol{L}^T - \dot{\boldsymbol{F}}^e(\boldsymbol{F}^e)^T - (\boldsymbol{F}^e)^T\dot{\boldsymbol{F}}^e]. \tag{14.47}$$

Now, note that the definition of $\boldsymbol{B}^e$ gives

$$\dot{\boldsymbol{B}}^e = [\boldsymbol{F}^e(\boldsymbol{F}^e)^T]^{\cdot} = \dot{\boldsymbol{F}}^e(\boldsymbol{F}^e)^T + (\boldsymbol{F}^e)^T\dot{\boldsymbol{F}}^e. \tag{14.48}$$

Equation (14.45) is recovered by substituting this formula into (14.47) and taking the definition (14.44) into account.

Evolution of the internal variables

A general evolution equation for the internal variables set is postulated as having the same functional format as the infinitesimal law

$$\dot{\boldsymbol{\alpha}} = \dot{\gamma}\,\boldsymbol{H}(\boldsymbol{\tau}, \boldsymbol{A}). \tag{14.49}$$

If the potential formulation of the internal variable evolution is adopted, then the constitutive function $\boldsymbol{H}$ is given by

$$\boldsymbol{H}(\boldsymbol{\tau}, \boldsymbol{A}) = -\frac{\partial\Psi}{\partial\boldsymbol{A}}. \tag{14.50}$$

In the above, the *plastic multiplier*, $\dot{\gamma}$, is required to satisfy the standard complementarity relation

$$\Phi \le 0, \quad \dot{\gamma} \ge 0, \quad \dot{\gamma}\Phi = 0. \tag{14.51}$$

Summary of the model

The overall finite strain elastoplastic constitutive model is completely defined by equations (14.20), (14.32)–(14.38) and (14.49)–(14.51). The model is summarised in Box 14.2.

14.3.4. THE DISSIPATION INEQUALITY

Following assumption (14.32), the time derivative of the free energy reads

$$\dot{\psi} = \frac{\partial\psi}{\partial\boldsymbol{\varepsilon}^e} : \dot{\boldsymbol{\varepsilon}}^e + \frac{1}{\bar{\rho}}\,\boldsymbol{A} * \dot{\boldsymbol{\alpha}}, \tag{14.52}$$

where the appropriate product (denoted by *) is assumed between the thermodynamical forces $\boldsymbol{A}$ and internal variables rate $\dot{\boldsymbol{\alpha}}$. Equivalently, by applying the chain rule to the definition (14.30) of $\boldsymbol{\varepsilon}^e$,

$$\begin{aligned}\dot{\psi} &= \frac{1}{2}\frac{\partial\psi}{\partial\boldsymbol{\varepsilon}^e} : \frac{\partial(\ln\boldsymbol{B}^e)}{\partial\boldsymbol{B}^e} : \dot{\boldsymbol{B}}^e + \frac{1}{\bar{\rho}}\,\boldsymbol{A} * \dot{\boldsymbol{\alpha}}\\ &= \frac{1}{2}\left[\frac{\partial\psi}{\partial\boldsymbol{\varepsilon}^e} : \frac{\partial(\ln\boldsymbol{B}^e)}{\partial\boldsymbol{B}^e}\right]\boldsymbol{B}^e : \dot{\boldsymbol{B}}^e\,\boldsymbol{B}^{e\,-1} + \frac{1}{\bar{\rho}}\,\boldsymbol{A} * \dot{\boldsymbol{\alpha}}.\end{aligned} \tag{14.53}$$

Box 14.2. General isotropic finite strain multiplicative elastoplastic model.

(i) Multiplicative decomposition of the deformation gradient

$$\boldsymbol{F} = \boldsymbol{F}^{e}\boldsymbol{F}^{p}$$

(ii) Isotropic hyperelastic law

$$\boldsymbol{\tau} = \bar{\rho}\,\frac{\partial\psi(\boldsymbol{\varepsilon}^{e},\boldsymbol{\alpha})}{\partial\boldsymbol{\varepsilon}^{e}}$$

(iii) Evolution equations for $\boldsymbol{F}^{p}$ and internal variable set $\boldsymbol{\alpha}$

$$\dot{\boldsymbol{F}}^{p}\boldsymbol{F}^{p^{-1}} = \dot{\gamma}\,\boldsymbol{R}^{eT}\frac{\partial\Psi}{\partial\boldsymbol{\tau}}\boldsymbol{R}^{e}$$

$$\dot{\boldsymbol{\alpha}} = \dot{\gamma}\,\boldsymbol{H}$$

(iv) Loading/unloading criterion

$$\Phi \le 0, \quad \dot{\gamma} \ge 0, \quad \dot{\gamma}\Phi = 0$$

To obtain the first summand of the last expression in the above, we have made use of identity (ii) of page 22. Note that the tensors $\boldsymbol{\varepsilon}^{e}$, $\boldsymbol{B}^{e}$ and $\partial\psi/\partial\boldsymbol{\varepsilon}^{e}$ in (14.53) share the same principal axes. The situation here is completely analogous to that leading to equation (13.62) (page 529) for the Hencky hyperelastic model. Accordingly, the following identity holds in the present case

$$\frac{\partial\psi}{\partial\boldsymbol{\varepsilon}^{e}} : \frac{\partial(\ln\boldsymbol{B}^{e})}{\partial\boldsymbol{B}^{e}}\,\boldsymbol{B}^{e} = \frac{\partial\psi}{\partial\boldsymbol{\varepsilon}^{e}}, \tag{14.54}$$

and then, (14.53) can be rewritten as

$$\dot{\psi} = \frac{1}{2}\,\frac{\partial\psi}{\partial\boldsymbol{\varepsilon}^{e}} : \dot{\boldsymbol{B}}^{e}\,\boldsymbol{B}^{e^{-1}} + \frac{1}{\bar{\rho}}\,\boldsymbol{A} * \dot{\boldsymbol{\alpha}}. \tag{14.55}$$

By definition, $\boldsymbol{B}^{e} = \boldsymbol{F}^{e}\boldsymbol{F}^{e^{T}}$, or, in view of the multiplicative elastoplastic decomposition assumption, $\boldsymbol{B}^{e} = \boldsymbol{F}(\boldsymbol{F}^{p})^{-1}(\boldsymbol{F}^{p})^{-T}\boldsymbol{F}^{T}$. Time differentiation of this last expression and substitution into (14.55) result, after some straightforward manipulations, in

$$\begin{aligned}\dot{\psi} &= \frac{\partial\psi}{\partial\boldsymbol{\varepsilon}^{e}} : \left\{\boldsymbol{D} + \frac{1}{2}[\boldsymbol{F}^{e}\boldsymbol{F}^{p}(\boldsymbol{F}^{p^{-1}})^{\cdot}\,\boldsymbol{F}^{e^{T}} + \boldsymbol{F}^{e}(\boldsymbol{F}^{p^{-T}})^{\cdot}\,\boldsymbol{F}^{p^{T}}\boldsymbol{F}^{e^{-1}}]\right\} + \frac{1}{\bar{\rho}}\,\boldsymbol{A} * \dot{\boldsymbol{\alpha}} \\ &= \frac{\partial\psi}{\partial\boldsymbol{\varepsilon}^{e}} : \left\{\boldsymbol{D} - \frac{1}{2}\boldsymbol{V}^{e}\boldsymbol{R}^{e}[\bar{\boldsymbol{L}}^{p} + \bar{\boldsymbol{L}}^{p^{T}}]\boldsymbol{R}^{e^{T}}\boldsymbol{V}^{e^{-1}}\right\} + \frac{1}{\bar{\rho}}\,\boldsymbol{A} * \dot{\boldsymbol{\alpha}},\end{aligned} \tag{14.56}$$

where use has been made of the relations

$$\boldsymbol{F}^{p}(\boldsymbol{F}^{p^{-1}})^{\cdot} = -\dot{\boldsymbol{F}}^{p}\boldsymbol{F}^{p^{-1}} = -\bar{\boldsymbol{L}}^{p} \tag{14.57}$$

and

$$(\boldsymbol{F}^{p^{-T}})^{\cdot}\,\boldsymbol{F}^{p^{T}} = -\boldsymbol{F}^{p^{-T}}(\boldsymbol{F}^{p^{T}})^{\cdot} = -\bar{\boldsymbol{L}}^{p^{T}}, \tag{14.58}$$

obtained, respectively, from the time differentiation of the identities

$$\boldsymbol{F}^p \boldsymbol{F}^{p-1} = \boldsymbol{I} \quad \text{and} \quad \boldsymbol{F}^{p-T} \boldsymbol{F}^{pT} = \boldsymbol{I}. \tag{14.59}$$

Finally, with the introduction of definition (14.28) of the spatially rotated plastic stretching tensor, $\tilde{\boldsymbol{D}}^p$, and by taking into account the elastic isotropy, the rate of change of free energy can be expressed as

$$\dot{\psi} = \frac{\partial \psi}{\partial \boldsymbol{\varepsilon}^e} : (\boldsymbol{D} - \tilde{\boldsymbol{D}}^p) + \frac{1}{\bar{\rho}} \boldsymbol{A} * \dot{\boldsymbol{\alpha}}. \tag{14.60}$$

When thermal effects are ignored, the *Clausius–Duhem* inequality (3.142) can be expressed as

$$\boldsymbol{\tau} : \boldsymbol{D} - \bar{\rho}\, \dot{\psi} \geq 0, \tag{14.61}$$

so that by introducing (14.60), we obtain

$$\left(\boldsymbol{\tau} - \bar{\rho} \frac{\partial \psi}{\partial \boldsymbol{\varepsilon}^e} \right) : \boldsymbol{D} + \bar{\rho} \frac{\partial \psi}{\partial \boldsymbol{\varepsilon}^e} : \tilde{\boldsymbol{D}}^p - \boldsymbol{A} * \dot{\boldsymbol{\alpha}} \geq 0. \tag{14.62}$$

As the above dissipation inequality must hold for any motion (and, consequently, any $\boldsymbol{D}$), it implies the constitutive equation (14.33) for the Kirchhoff stress. The non-negative dissipation requirement is then reduced to

$$\boldsymbol{\tau} : \tilde{\boldsymbol{D}}^p - \boldsymbol{A} * \dot{\boldsymbol{\alpha}} \geq 0. \tag{14.63}$$

14.3.5. FINITE STRAIN EXTENSION TO INFINITESIMAL THEORIES

Expressions (14.33) and (14.63) as well as the adopted plastic flow rule (14.37) are completely analogous to their small strain counterparts. In the small strain limit, $\boldsymbol{\varepsilon}^e$ and $\tilde{\boldsymbol{D}}^p$ reduce, respectively, to the standard infinitesimal elastic strain tensor and plastic strain rate. In fact, all equations of the finite strain model reduce to those of small strain plasticity as the small strain limit is approached. Thus, the present approach allows a natural extension, to the finite strain range, of general isotropic infinitesimal elastoplastic models. A generic small strain model defined by an elastic potential ψ_s, a yield function Φ_s and a flow potential Ψ_s can be extended to finite strains by adopting, in the constitutive equations above, ψ, Φ and Ψ with the same functional format as the respective small strain counterparts. Finite strain extensions obtained as such preserve some very important properties of the original small strain model. These are briefly discussed below.

Volume-preserving plastic deformations

Firstly, a large strain measure of volumetric plastic strain is defined. With $J^p \equiv \det \boldsymbol{F}^p$ the *volumetric plastic strain* is defined as

$$\begin{aligned} \varepsilon^p_v \equiv \ln J^p &= \ln[\lambda^p_{(1)}\, \lambda^p_{(2)}\, \lambda^p_{(3)}] \\ &= \ln \lambda^p_{(1)} + \ln \lambda^p_{(2)} + \ln \lambda^p_{(3)} = \operatorname{tr}[\ln \boldsymbol{V}^p] \end{aligned} \tag{14.64}$$

where $\lambda^p_{(i)}$ are the principal *plastic* stretches, i.e. the eigenvalues of the plastic stretch tensor

$$\boldsymbol{V}^p \equiv \sqrt{\boldsymbol{F}^p \boldsymbol{F}^{pT}}.$$

A *volume-preserving* plastic deformation gradient satisfies

$$\det \boldsymbol{F}^p = 1,$$

which in view of definition (14.64) holds if and only if

$$\varepsilon^p_{\mathrm{v}} = 0.$$

In other words, the (local) plastic deformation is volume-preserving if and only if the above defined volumetric plastic strain vanishes.

Finite plastic incompressibility

Attention is now focused on the condition under which *isochoric plastic flow* occurs within the present finite strain plasticity theory. As J^p is *constant* under volume-preserving plastic flow, it follows that the plastic flow is isochoric if and only if the *rate* of the above-defined volumetric plastic strain vanishes,

$$\dot{\varepsilon}^p_{\mathrm{v}} = 0.$$

By using the chain rule together with the standard rule for differentiation of the determinant of an invertible tensor, we obtain the identity

$$\begin{aligned}\dot{\varepsilon}^p_{\mathrm{v}} &\equiv (\ln[\det \boldsymbol{F}^p])^{\cdot} = \frac{1}{\det \boldsymbol{F}^p}\,(\det \boldsymbol{F}^p)^{\cdot}\\ &= \mathrm{tr}[\dot{\boldsymbol{F}}^p (\boldsymbol{F}^p)^{-1}].\end{aligned}$$

From the evolution law (14.39) for $\boldsymbol{F}^p$, it follows that the rate of the above defined volumetric plastic strain is given by

$$\dot{\varepsilon}^p_{\mathrm{v}} = \dot{\gamma}\ \mathrm{tr}\left[\frac{\partial \Psi}{\partial \boldsymbol{\tau}}\right], \tag{14.65}$$

so that, clearly, as in the infinitesimal theory, flow potentials whose derivative with respect to stress – the *flow vector* – is traceless (such as the classical von Mises and Tresca functions) produce *isochoric* plastic flow. In summary, within the present framework, the crucial infinitesimal plasticity property of plastic incompressibility under a traceless flow vector is naturally extended to the finite strain range.

Associativity and maximum plastic dissipation at large strains

Analogously to the small strain theory, if Φ is taken as the flow potential, then the *principle of maximum plastic dissipation* discussed in Chapter 6 (see page 170) is extended to the finite strain range. In that case, the loading/unloading criterion (14.51) is the Kuhn–Tucker optimality condition for the left-hand side of (14.63) to reach a maximum subjected to the plastic admissibility constraint, $\Phi \leq 0$.

14.4. The general elastic predictor/return-mapping algorithm

The use of elastic predictor/return-mapping algorithms in the integration of elastoplastic constitutive equations has been thoroughly discussed in Chapter 7, in the context of the infinitesimal theory. In this section, we show that an integration algorithm can be derived for the general finite multiplicative plasticity model based on completely analogous concepts. As we will see, the finite strain algorithm is particularly appealing from the computational point of view. Its structure is such that it allows an elegant extension of existing infinitesimal plasticity codes with reuse of the basic routines for stress updating and computation of associated tangent operators. We remark that the procedure is fully implemented in program HYPLAS.

14.4.1. THE BASIC CONSTITUTIVE INITIAL VALUE PROBLEM

Let us start by stating in the following the finite strain elastoplastic constitutive initial value problem associated with the general isotropic model of Box 14.2.

Problem 14.1 (The finite elastoplasticity initial value problem). *Given the initial values $\boldsymbol{F}^p(t_0)$ and $\boldsymbol{\alpha}(t_0)$ and given the history of the deformation gradient $\boldsymbol{F}(t)$, $t \in [t_0, T]$, find the functions $\boldsymbol{F}^p(t)$, $\boldsymbol{\alpha}(t)$ and $\dot{\gamma}(t)$ that satisfy*

$$\begin{aligned} \dot{\boldsymbol{F}}^p(t)\,[\boldsymbol{F}^p(t)]^{-1} &= \dot{\gamma}(t)\,\boldsymbol{R}^e(t)^T \left.\frac{\partial \Psi}{\partial \boldsymbol{\tau}}\right|_t \boldsymbol{R}^e(t) \\ \dot{\boldsymbol{\alpha}}(t) &= \dot{\gamma}(t)\,\boldsymbol{H}(\boldsymbol{\tau}(t), \boldsymbol{A}(t)) \end{aligned} \tag{14.66}$$

and

$$\dot{\gamma}(t) \geq 0, \quad \Phi(\boldsymbol{\tau}(t), \boldsymbol{A}(t)) \leq 0, \quad \dot{\gamma}(t)\Phi(\boldsymbol{\tau}(t), \boldsymbol{A}(t)) = 0 \tag{14.67}$$

for each instant $t \in [t_0, T]$, with

$$\boldsymbol{\tau}(t) = \bar{\rho}\left.\frac{\partial \psi}{\partial \boldsymbol{\varepsilon}^e}\right|_t, \quad \boldsymbol{A}(t) = \bar{\rho}\left.\frac{\partial \psi}{\partial \boldsymbol{\alpha}}\right|_t \tag{14.68}$$

and the kinematic relations

$$\begin{aligned} \boldsymbol{\varepsilon}^e(t) &= \ln \boldsymbol{V}^e(t) \\ \boldsymbol{V}^e(t) &= [\boldsymbol{F}^e(t)\,\boldsymbol{F}^e(t)^T]^{\frac{1}{2}} \\ \boldsymbol{R}^e(t) &= [\boldsymbol{V}^e(t)]^{-1}\,\boldsymbol{F}^e(t) \\ \boldsymbol{F}^e(t) &= \boldsymbol{F}(t)\,[\boldsymbol{F}^p(t)]^{-1}. \end{aligned} \tag{14.69}$$

Remark 14.3. In the above definition of the constitutive initial value problem, we have used the *plastic* deformation gradient as the primary kinematic unknown in the system of differential equations of evolution. This is conceptually at variance with Problem 7.1 (page 193) for the infinitesimal strain case, where the *elastic* strain has been chosen as the primary kinematic variable. The choice of the plastic deformation gradient in Problem 14.1 is motivated only by the fact that it allows, in the finite strain context, a more straightforward derivation of the integration algorithm. We stress, however, that in both cases (small and large strains) the basic initial value problem (and the corresponding numerical integration algorithm) may be equivalently defined in terms of evolution of either the elastic or the plastic kinematic variable. In Problem 14.1, the equivalent formulation having $\boldsymbol{F}^e$ as the primary kinematic unknown is obtained by replacing $(14.66)_1$ with the differential equation (14.42) for the elastic deformation gradient.

14.4.2. EXPONENTIAL MAP BACKWARD DISCRETISATION

In Section 7.2.2 (page 194), the small-strain elastoplastic constitutive initial value problem has been discretised by means of a standard backward Euler difference scheme, which has led to the incremental version – Problem 7.2. Then, it has been shown that the discrete evolution problem gives rise to the infinitesimal elastic predictor/plastic corrector scheme of Box 7.1 (page 199). As we shall see in the following, the numerical solution of Problem 14.1 can also be undertaken by an elastic predictor/return-mapping scheme. The procedure is derived here following the same basic steps as in the derivation of its infinitesimal counterpart.

The starting point is the use of a numerical scheme for integration of the differential equations (14.66) and constraints (14.67). As in the infinitesimal case, the evolution of the hardening internal variables will be discretised here by a conventional backward Euler scheme. Then, $(14.66)_2$ is transformed into the following incremental counterpart

$$\boldsymbol{\alpha}_{n+1} = \boldsymbol{\alpha}_n + \Delta\gamma\, \boldsymbol{H}_{n+1}. \tag{14.70}$$

Similarly, the discrete version of the constraints (14.67) is obtained as

$$\Delta\gamma \geq 0, \quad \Phi(\boldsymbol{\tau}_{n+1}, \boldsymbol{A}_{n+1}) \leq 0, \quad \Delta\gamma\, \Phi(\boldsymbol{\tau}_{n+1}, \boldsymbol{A}_{n+1}) = 0. \tag{14.71}$$

The above expressions have the same functional format as their infinitesimal counterparts.

The exponential integrator for the plastic flow equation

The crucial difference between the discretisation of the large strain problem and the infinitesimal one lies in the numerical approximation of the plastic flow equation. The structure of the plastic flow equation $(14.66)_1$ makes algorithms based on *exponential map integrators* ideal for numerical approximation. At this point, the reader who is not familiar with the exponential map integrator concept is referred to Section B.3 (from page 751) for further details. Here we shall choose to adopt the *backward* exponential integrator, whose general formula is obtained from expression (B.24) by setting $\theta = 1$. The discretisation of $(14.66)_1$ by means of the resulting scheme, leads to the update formula for the plastic deformation gradient

$$\boldsymbol{F}^p_{n+1} = \exp\left[\Delta\gamma\, \boldsymbol{R}^{e\,T}_{n+1} \left.\frac{\partial\Psi}{\partial\boldsymbol{\tau}}\right|_{n+1} \boldsymbol{R}^e_{n+1}\right] \boldsymbol{F}^p_n, \tag{14.72}$$

or, equivalently, in view of the isotropy of the tensor exponential function (refer to (B.8)),

$$\boldsymbol{F}^p_{n+1} = \boldsymbol{R}^{e\,T}_{n+1} \exp\left[\Delta\gamma \left.\frac{\partial\Psi}{\partial\boldsymbol{\tau}}\right|_{n+1}\right] \boldsymbol{R}^e_{n+1}\, \boldsymbol{F}^p_n. \tag{14.73}$$

Remark 14.4. The incompressibility of the plastic flow for pressure insensitive flow potentials is carried over *exactly* to the incremental rule (14.73). Indeed, for a traceless flow vector, $\partial\Psi/\partial\boldsymbol{\tau}$, $\det[\exp[\Delta\gamma\,\partial\Psi/\partial\boldsymbol{\tau}]] = 1$ which ensures that the updating formula (14.73) is volume preserving. This basic property of the exponential map is listed in Section B.1.1 of Appendix B. Had we adopted, for instance, a standard backward Euler difference scheme to discretise the plastic flow equation $(14.66)_1$, the resulting updating formula for the plastic deformation gradient would be

$$\boldsymbol{F}^p_{n+1} = \left(\boldsymbol{I} - \Delta\gamma\,\boldsymbol{R}^{e\,T}_{n+1}\frac{\partial\Psi}{\partial\boldsymbol{\tau}}\bigg|_{n+1}\boldsymbol{R}^e_{n+1}\right)^{-1}\boldsymbol{F}^p_n. \tag{14.74}$$

By simple inspection, we can easily establish that the above update formula for $\boldsymbol{F}^p$ is not volume-preserving in general; that is, given $\boldsymbol{F}^p_n$ such that $\det\boldsymbol{F}^p_n = 1$ and a traceless $\partial\Psi/\partial\boldsymbol{\tau}$, the resulting $\boldsymbol{F}^p_{n+1}$ is such that, in general, $\det\boldsymbol{F}^p_{n+1} \neq 1$. This approach would result in substantial accuracy loss in the numerical integration of elastoplastic constitutive equations of plastically incompressible models.

Elastic deformation gradient update

By combining the above formula for $\boldsymbol{F}^p_{n+1}$ with the multiplicative elastoplastic split and making use of property (B.9), we obtain the following equivalent kinematic update expression in terms of the elastic deformation gradient

$$\boldsymbol{F}^e_{n+1} = \boldsymbol{F}_\Delta\boldsymbol{F}^e_n\,\boldsymbol{R}^{e\,T}_{n+1}\exp\left[-\Delta\gamma\,\frac{\partial\Psi}{\partial\boldsymbol{\tau}}\bigg|_{n+1}\right]\boldsymbol{R}^e_{n+1}, \tag{14.75}$$

where $\boldsymbol{F}_\Delta$ is the *incremental* deformation gradient,

$$\boldsymbol{F}_\Delta \equiv \boldsymbol{F}_{n+1}(\boldsymbol{F}_n)^{-1} = \boldsymbol{I} + \nabla_{x_n}[\Delta\boldsymbol{u}], \tag{14.76}$$

i.e. the gradient of the deformation that maps the configuration of time t_n onto the configuration of t_{n+1}, with $\Delta\boldsymbol{u}$ denoting the corresponding incremental displacement field. Note that (14.75) is a finite strain version of the infinitesimal elastic strain incremental equation $(7.10)_1$ (page 194).

With the above discretised evolution equations at hand, we now proceed to state the incremental version of Problem 14.1. In order to remain consistent with the infinitesimal counterpart (Problem 7.2, where the *elastic* strain is the primary kinematic unknown), rather than using (14.73), we shall here adopt the equivalent formula (14.75) in the definition of the incremental problem. This will result in a reduced system having the *elastic* deformation gradient as the primary kinematic unknown.

Problem 14.2 (The incremental finite plasticity problem). *Given $\boldsymbol{F}^e_n$ and $\boldsymbol{\alpha}_n$ at the beginning of the interval $[t_n, t_{n+1}]$ and given the prescribed incremental deformation gradient, $\boldsymbol{F}_\Delta$, for this interval, solve the following system of algebraic equations*

$$\begin{aligned}\boldsymbol{F}^e_{n+1} &= \boldsymbol{F}_\Delta\boldsymbol{F}^e_n\,\boldsymbol{R}^{e\,T}_{n+1}\exp\left[-\Delta\gamma\,\frac{\partial\Psi}{\partial\boldsymbol{\tau}}\bigg|_{n+1}\right]\boldsymbol{R}^e_{n+1}\\ \boldsymbol{\alpha}_{n+1} &= \boldsymbol{\alpha}_n + \Delta\gamma\,\boldsymbol{H}_{n+1}\end{aligned} \tag{14.77}$$

for the unknowns $\boldsymbol{F}^e_{n+1}$, $\boldsymbol{\alpha}_{n+1}$ *and* $\Delta\gamma$, *subjected to the constraints*

$$\Delta\gamma \geq 0, \quad \Phi(\boldsymbol{\tau}_{n+1}, \boldsymbol{A}_{n+1}) \leq 0, \quad \Delta\gamma\, \Phi(\boldsymbol{\tau}_{n+1}, \boldsymbol{A}_{n+1}) = 0, \tag{14.78}$$

with

$$\boldsymbol{\tau}_{n+1} = \bar{\rho}\frac{\partial\psi}{\partial\boldsymbol{\varepsilon}^e}\bigg|_{n+1}, \qquad \boldsymbol{A}_{n+1} = \bar{\rho}\frac{\partial\psi}{\partial\boldsymbol{\alpha}}\bigg|_{n+1} \tag{14.79}$$

and the kinematic relations

$$\begin{aligned}\boldsymbol{\varepsilon}^e_{n+1} &= \ln \boldsymbol{V}^e_{n+1}\\ \boldsymbol{V}^e_{n+1} &= [\boldsymbol{F}^e_{n+1}\,(\boldsymbol{F}^e_{n+1})^T]^{\frac{1}{2}}\\ \boldsymbol{R}^e_{n+1} &= [\boldsymbol{V}^e_{n+1}]^{-1}\,\boldsymbol{F}^e_{n+1}.\end{aligned} \tag{14.80}$$

After the solution of the above problem, the updated plastic strain can be determined promptly from the multiplicative elastoplastic split as

$$\boldsymbol{F}^p_{n+1} = (\boldsymbol{F}^e_{n+1})^{-1}\boldsymbol{F}_{n+1}. \tag{14.81}$$

Clearly, Problem 14.2 has the same structure as its infinitesimal counterpart (Problem 7.2) and all arguments leading to the establishment of the elastic predictor/return-mapping scheme in Section 7.2.3 remain valid in the present case. The resulting large-strain algorithm is described next.

The elastic predictor/return-mapping scheme

The *elastic predictor* step, obtained by solving (14.77) with $\Delta\gamma = 0$, gives the *elastic trial state*

$$\begin{aligned}\boldsymbol{F}^{e\,\text{trial}}_{n+1} &= \boldsymbol{F}_{\Delta}\boldsymbol{F}^e_n\\ \boldsymbol{\alpha}^{\text{trial}}_{n+1} &= \boldsymbol{\alpha}_n.\end{aligned} \tag{14.82}$$

If the pair $\{\boldsymbol{\tau}^{\text{trial}}_{n+1}, \boldsymbol{A}^{\text{trial}}_{n+1}\}$, obtained from the potential constitutive relations at the elastic trial state, is plastically admissible, i.e. if

$$\Phi(\boldsymbol{\tau}^{\text{trial}}_{n+1}, \boldsymbol{A}^{\text{trial}}_{n+1}) \leq 0, \tag{14.83}$$

then the elastic trial state is accepted as the actual state at t_{n+1}. Otherwise, we solve the *return-mapping* equations

$$\begin{aligned}\boldsymbol{F}^e_{n+1} &= \boldsymbol{F}^{e\,\text{trial}}_{n+1}\,\boldsymbol{R}^{e\,T}_{n+1}\,\exp\left[-\Delta\gamma\,\frac{\partial\Psi}{\partial\boldsymbol{\tau}}\bigg|_{n+1}\right]\boldsymbol{R}^e_{n+1}\\ \boldsymbol{\alpha}_{n+1} &= \boldsymbol{\alpha}^{\text{trial}}_{n+1} + \Delta\gamma\,\boldsymbol{H}_{n+1}\\ \Phi(\boldsymbol{\tau}_{n+1}, \boldsymbol{A}_{n+1}) &= 0,\end{aligned} \tag{14.84}$$

for $\boldsymbol{F}^e_{n+1}$, $\boldsymbol{\alpha}_{n+1}$ and $\Delta\gamma$, with $\boldsymbol{\tau}_{n+1}$ and $\boldsymbol{A}_{n+1}$ obtained from their potential constitutive relations.

Logarithmic strains and infinitesimal format return mapping

A substantial simplification of the return-mapping equation $(14.84)_1$ can be achieved by rewriting it equivalently in terms of the *logarithmic elastic strain measure* (rather than the elastic deformation gradient). This is shown in the following.

Firstly, note that the post-multiplication of both sides of (14.75) by $\boldsymbol{R}^{e\,T}_{n+1}$ results in

$$\boldsymbol{V}^e_{n+1} = \boldsymbol{F}^{e\ \text{trial}}_{n+1}\ \boldsymbol{R}^{e\,T}_{n+1} \exp\left[-\Delta\gamma \left.\frac{\partial\Psi}{\partial\boldsymbol{\tau}}\right|_{n+1}\right], \tag{14.85}$$

or, equivalently, by making use of (B.9) in the inversion of the exponential term,

$$\boldsymbol{V}^e_{n+1} \exp\left[\Delta\gamma \left.\frac{\partial\Psi}{\partial\boldsymbol{\tau}}\right|_{n+1}\right] = \boldsymbol{F}^{e\ \text{trial}}_{n+1}\ \boldsymbol{R}^{e\,T}_{n+1}. \tag{14.86}$$

Then, a further post-multiplication of each side by its transpose, together with the use of property (B.10), gives

$$\boldsymbol{V}^e_{n+1} \exp\left[2\,\Delta\gamma \left.\frac{\partial\Psi}{\partial\boldsymbol{\tau}}\right|_{n+1}\right] \boldsymbol{V}^e_{n+1} = (\boldsymbol{V}^{e\ \text{trial}}_{n+1})^2. \tag{14.87}$$

Recall that, due to the assumed elastic and plastic isotropy, $\boldsymbol{V}^e$ and $\partial\Psi/\partial\boldsymbol{\tau}$ commute. Then, by rearranging the terms and taking the square root of both sides of (14.87), we obtain

$$\boldsymbol{V}^e_{n+1} = \boldsymbol{V}^{e\ \text{trial}}_{n+1}\ \exp\left[-\Delta\gamma \left.\frac{\partial\Psi}{\partial\boldsymbol{\tau}}\right|_{n+1}\right]. \tag{14.88}$$

Further, by taking the tensor logarithm of both sides of (14.88), we obtain the much simpler formula in terms of logarithmic Eulerian strain tensors

$$\boldsymbol{\varepsilon}^e_{n+1} = \boldsymbol{\varepsilon}^{e\ \text{trial}}_{n+1} - \Delta\gamma \left.\frac{\partial\Psi}{\partial\boldsymbol{\tau}}\right|_{n+1}. \tag{14.89}$$

Remarkably, the above expression has the *same format* as the elastic strain update formula of the *backward* return-mapping algorithms of the infinitesimal theory derived in Chapter 7 (compare (14.89) with expression $(7.25)_1$ of page 196).

Finally, by replacing $(14.84)_1$ with its equivalent (14.89) in terms of logarithmic strains, the return-mapping equation system of the finite strain incremental problem is reduced to

$$\begin{aligned}
&\boldsymbol{\varepsilon}^e_{n+1} = \boldsymbol{\varepsilon}^{e\ \text{trial}}_{n+1} - \Delta\gamma \left.\frac{\partial\Psi}{\partial\boldsymbol{\tau}}\right|_{n+1}\\
&\boldsymbol{\alpha}_{n+1} = \boldsymbol{\alpha}_n + \Delta\gamma\,\boldsymbol{H}_{n+1}\\
&\Phi(\boldsymbol{\tau}_{n+1}, \boldsymbol{A}_{n+1}) = 0,
\end{aligned} \tag{14.90}$$

which has the same functional format as the infinitesimal counterpart (7.25).

Elastic rotation update

For the elastic rotation, the above algorithm results in the following update formula

$$\boldsymbol{R}^e_{n+1} = \boldsymbol{R}^{e\ \text{trial}}_{n+1}. \tag{14.91}$$

To see this, note that by comparing (14.85) to (14.88), we find

$$\boldsymbol{V}^{e\ \text{trial}}_{n+1} = \boldsymbol{F}^{e\ \text{trial}}_{n+1}\, \boldsymbol{R}^{e\,T}_{n+1} = \boldsymbol{V}^{e\ \text{trial}}_{n+1}\, \boldsymbol{R}^{e\ \text{trial}}_{n+1}\, \boldsymbol{R}^{e\,T}_{n+1}, \tag{14.92}$$

which leads to (14.91).

14.4.3. COMPUTATIONAL IMPLEMENTATION OF THE GENERAL ALGORITHM

As the elastic law is defined here in terms of the spatial elastic logarithmic strain, in the actual computational implementation of the above elastic predictor/return-mapping procedure we shall take $\boldsymbol{\varepsilon}^e$ as the kinematic variable to be stored in the computer memory. Accordingly, the computational procedure follows the steps:

- firstly, after retrieving the elastic logarithmic strain, $\boldsymbol{\varepsilon}^e_n$, of t_n, we recover the corresponding elastic left Cauchy–Green tensor according to

$$\boldsymbol{B}^e_n = \exp[2\, \boldsymbol{\varepsilon}^e_n]; \tag{14.93}$$

- the elastic trial left Cauchy–Green tensor is computed next by means of the formula

$$\boldsymbol{B}^{e\ \text{trial}}_{n+1} \equiv \boldsymbol{F}^{e\ \text{trial}}_{n+1} (\boldsymbol{F}^{e\ \text{trial}}_{n+1})^T = \boldsymbol{F}_\Delta \boldsymbol{B}^e_n (\boldsymbol{F}_\Delta)^T; \tag{14.94}$$

- with $\boldsymbol{B}^{e\ \text{trial}}_{n+1}$ at hand, the elastic trial logarithmic strain tensor is computed as

$$\boldsymbol{\varepsilon}^{e\ \text{trial}} = \tfrac{1}{2} \ln[\boldsymbol{B}^{e\ \text{trial}}_{n+1}].$$

  In program HYPLAS, this calculation is carried out in subroutine LOGSTR. The actual computation of $\frac{1}{2}\ln[\boldsymbol{B}^{e\text{trial}}_{n+1}]$ is performed in the general subroutine ISO2, called by LOGSTR. Subroutine ISO2 computes general isotropic tensor functions of a single tensor belonging to the family described in Section A.5 of Appendix A. The above function, $\frac{1}{2}\ln[\cdot]$, is a particular member of this class of isotropic tensor functions;

- the elastic trial internal variable set is simply given by

$$\boldsymbol{\alpha}^{\text{trial}}_{n+1} = \boldsymbol{\alpha}_n;$$

- having computed the elastic trial state, we proceed now to the standard *small strain* integration algorithm to update $\boldsymbol{\varepsilon}^e$, α and the corresponding forces $\boldsymbol{\tau}$ and $\boldsymbol{A}$. This procedure comprises the standard admissibility check followed by the application of the return-mapping procedure – solution of the system (14.90) – when the trial state is not plastically admissible;

Box 14.3. General integration algorithm for isotropic multiplicative finite strain elastoplasticity.

HYPLAS procedure:	MATISU

(i) Given incremental displacement $\Delta \boldsymbol{u}$, update the deformation gradient

$$\boldsymbol{F}_{\Delta} := \boldsymbol{I} + \nabla_n[\Delta \boldsymbol{u}], \quad \boldsymbol{F}_{n+1} := \boldsymbol{F}_{\Delta}\, \boldsymbol{F}_n$$

(ii) Compute elastic trial state

$$\begin{aligned} \boldsymbol{B}_n^e &:= \exp[2\,\boldsymbol{\varepsilon}_n^e] \\ \boldsymbol{B}_{n+1}^{e\ \text{trial}} &:= \boldsymbol{F}_{\Delta}\, \boldsymbol{B}_n^e\, (\boldsymbol{F}_{\Delta})^T \\ \boldsymbol{\varepsilon}_{n+1}^{e\ \text{trial}} &:= \ln[\boldsymbol{V}_{n+1}^{e\ \text{trial}}] = \tfrac{1}{2}\ln[\boldsymbol{B}_{n+1}^{e\ \text{trial}}] \\ \boldsymbol{\alpha}_{n+1}^{\text{trial}} &:= \boldsymbol{\alpha}_n \end{aligned}$$

(iii) GOTO BOX 14.4 – small-strain algorithm (update $\boldsymbol{\tau}$, $\boldsymbol{\varepsilon}^e$ and $\boldsymbol{\alpha}$)

(iv) Update the Cauchy stress

$$\boldsymbol{\sigma}_{n+1} := \det[\boldsymbol{F}_{n+1}]^{-1}\, \boldsymbol{\tau}_{n+1}$$

- finally, the updated *Cauchy* stress tensor, required to assemble the internal finite element force vector $\mathbf{f}_{(e)}^{\text{int}}$, is evaluated

$$\boldsymbol{\sigma}_{n+1} = \frac{1}{\det\ \boldsymbol{F}_{n+1}}\, \boldsymbol{\tau}_{n+1}.$$

The overall algorithm for integration of the large-strain elastoplastic constitutive equations is listed in Boxes 14.3 and 14.4.

Remark 14.5. The operations carried out in items (i), (ii) and (iv) of Box 14.3 are related exclusively to the kinematics of finite strains. Due to the use of the logarithmic elastic strain measure in conjunction with the backward exponential approximation (14.73) to the plastic flow rule, the essential material-related stress-updating procedure, shown in Box 14.4, preserves the format of the general elastic predictor/return-mapping algorithm for infinitesimal plasticity described in Chapter 7 and summarised in Box 7.1 (page 199). Further, if the Hencky model (linear relation between the Kirchoff stress and the elastic logarithmic strain) is adopted to describe the elastic behaviour, then the resulting algorithms will have the same functional format as the specific procedures seen in Part Two of this book, for which the return mapping is effectively carried out in the space of stresses. In computational terms, it means that all small-strain elastoplastic integration subroutines described in Part Two can be reused in the finite strain range *without modification*. For instance, in the finite strain extension to the von Mises model, subroutine SUVM (Boxes 7.3 and 7.4 and Section 7.3.5) is used in program HYPLAS to carry out the procedure of Box 14.4. Clearly, the kinematic operations of Box 14.3 have to be performed before and after these routines are called.

Box 14.4. General integration procedure – small strains.

(i) Given $\varepsilon^{e\ \mathrm{trial}}_{n+1}$ and $\boldsymbol{\alpha}^{\mathrm{trial}}_{n+1}$, compute

$$\boldsymbol{\tau}^{\mathrm{trial}}_{n+1} = \bar{\rho}\, \frac{\partial \psi}{\partial \boldsymbol{\varepsilon}^e}\bigg|^{\mathrm{trial}}_{n+1}, \qquad \boldsymbol{A}^{\mathrm{trial}}_{n+1} = \bar{\rho}\, \frac{\partial \psi}{\partial \boldsymbol{\alpha}}\bigg|^{\mathrm{trial}}_{n+1}$$

(ii) Check plastic admissibility

$$\text{IF } \Phi(\boldsymbol{\tau}^{\mathrm{trial}}_{n+1}, \boldsymbol{A}^{\mathrm{trial}}_{n+1}) \le 0$$

$$\text{THEN set } \ (\cdot)_{n+1} := (\cdot)^{\mathrm{trial}}_{n+1} \ \text{ and EXIT}$$

(iii) Return mapping. Solve the algebraic system

$$\begin{Bmatrix} \boldsymbol{\varepsilon}^e_{n+1} - \boldsymbol{\varepsilon}^{e\ \mathrm{trial}}_{n+1} + \Delta\gamma\, \dfrac{\partial \Psi}{\partial \boldsymbol{\tau}}\bigg|_{n+1} \\ \boldsymbol{\alpha}_{n+1} - \boldsymbol{\alpha}_n - \Delta\gamma\, \boldsymbol{H}(\boldsymbol{\tau}_{n+1}, \boldsymbol{A}_{n+1}) \\ \Phi(\boldsymbol{\tau}_{n+1}, \boldsymbol{A}_{n+1}) \end{Bmatrix} = \begin{Bmatrix} \mathbf{0} \\ \mathbf{0} \\ 0 \end{Bmatrix}$$

for $\boldsymbol{\varepsilon}^e_{n+1}$, $\boldsymbol{\alpha}_{n+1}$ and $\Delta\gamma$, with

$$\boldsymbol{\tau}_{n+1} = \bar{\rho}\, \frac{\partial \psi}{\partial \boldsymbol{\varepsilon}^e}\bigg|_{n+1}, \qquad \boldsymbol{A}_{n+1} = \bar{\rho}\, \frac{\partial \psi}{\partial \boldsymbol{\alpha}}\bigg|_{n+1}$$

(iv) EXIT

Remark 14.6. It is important to emphasise that the simplicity of the integration algorithm of Boxes 14.3 and 14.4 came as a result of the assumptions of *elastoplastic isotropy* and the particular implicit exponential approximation adopted to discretise the plastic flow rule. As already mentioned, other schemes such as the standard backward Euler method of equation (14.74) or the fully explicit (forward) Euler method may be used to discretise the plastic flow law. However, the adoption of schemes other than the exponential approximation will generally result in quite complex integration algorithms in which kinematic and material-related operations cannot be carried out separately.

14.5. The consistent spatial tangent modulus

The next step towards the complete incorporation of the general finite strain plasticity model into the implicit finite element framework of Chapter 4 is the derivation of a closed formula for the spatial tangent modulus consistent with the integration algorithm of Boxes 14.3 and 14.4. In Appendix C, the Cartesian components of the spatial tangent modulus have been shown to be given by the following expression

$$\mathsf{a}_{ijkl} = \frac{1}{J}\, \frac{\partial \tau_{ij}}{\partial F_{kq}}\, F_{lq} - \sigma_{il}\, \delta_{jk}, \tag{14.95}$$

where $J = \det \boldsymbol{F}$. The derivation of a for the finite strain elastoplasticity case is described in the following.

14.5.1. DERIVATION OF THE SPATIAL TANGENT MODULUS

Firstly, note that in the small strain integration algorithm of Box 14.4, the updated stress Kirchhoff stress, $\boldsymbol{\tau}_{n+1}$, is obtained as a function of the internal variable set $\boldsymbol{\alpha}_n$, at t_n, and the elastic trial logarithmic strain. This procedure can be regarded as an incremental constitutive function of the form

$$\boldsymbol{\tau}_{n+1} = \tilde{\boldsymbol{\tau}}(\boldsymbol{\alpha}_n, \boldsymbol{\varepsilon}^{e\ \mathrm{trial}}_{n+1}). \tag{14.96}$$

This algorithmic function has the same functional format and properties as the function $\bar{\boldsymbol{\sigma}}$ referred to in expression (7.110), page 230, in the small-strain context.

In the general procedure of Box 14.3, $\boldsymbol{\varepsilon}^{e\ \mathrm{trial}}_{n+1}$ is computed as a function of $\boldsymbol{B}^{e\ \mathrm{trial}}_{n+1}$ which, in turn, is a function of $\boldsymbol{F}^p_n$ and $\boldsymbol{F}_{n+1}$. With $\boldsymbol{\varepsilon}^{e\ \mathrm{trial}}_{n+1}$ at hand, the Kirchhoff stress is then updated by means of the incremental constitutive function $\tilde{\boldsymbol{\tau}}$ (small strain algorithm – Box 14.4). Thus, the overall operator split algorithm, comprising the operations carried out in Boxes 14.3 and 14.4, defines an implicit function $\hat{\boldsymbol{\tau}}$ (the incremental constitutive function), for the updated Kirchhoff stress, that can be generally expressed as

$$\hat{\boldsymbol{\tau}}(\boldsymbol{\alpha}_n, \boldsymbol{F}_{n+1}) = \tilde{\boldsymbol{\tau}}(\boldsymbol{\alpha}_n, \boldsymbol{\varepsilon}^{e\ \mathrm{trial}}_{n+1}(\boldsymbol{B}^{e\ \mathrm{trial}}_{n+1}(\boldsymbol{F}^p_n, \boldsymbol{F}_{n+1}))). \tag{14.97}$$

The term $\partial \tau_{ij}/\partial F_{kq}$ taking part in (14.95) is the derivative of this implicit function. To obtain this derivative, we apply the chain rule to (14.97) and obtain

$$\frac{\partial \hat{\boldsymbol{\tau}}}{\partial \boldsymbol{F}_{n+1}} = \frac{\partial \tilde{\boldsymbol{\tau}}}{\partial \boldsymbol{\varepsilon}^{e\ \mathrm{trial}}_{n+1}} : \frac{\partial \boldsymbol{\varepsilon}^{e\ \mathrm{trial}}_{n+1}}{\partial \boldsymbol{B}^{e\ \mathrm{trial}}_{n+1}} : \frac{\partial \boldsymbol{B}^{e\ \mathrm{trial}}_{n+1}}{\partial \boldsymbol{F}_{n+1}}. \tag{14.98}$$

Substitution of this expression into (14.95) results, after straightforward manipulations, in the following closed formula for the components of the spatial tangent modulus consistent with the present operator split algorithm

$$\mathsf{a}_{ijkl} = \frac{1}{2J}\,[\,\mathsf{D} : \mathsf{L} : \mathsf{B}\,]_{ijkl} - \sigma_{il}\,\delta_{jk}, \tag{14.99}$$

where D is the *small-strain elastic or elastoplastic consistent tangent operator*, associated exclusively with the integration algorithm of Box 14.4

$$\mathsf{D} = \frac{\partial \tilde{\boldsymbol{\tau}}}{\partial \boldsymbol{\varepsilon}^{e\ \mathrm{trial}}_{n+1}}. \tag{14.100}$$

The fourth-order tensor L is defined as

$$\mathsf{L} = \frac{\partial \ln[\boldsymbol{B}^{e\ \mathrm{trial}}_{n+1}]}{\partial \boldsymbol{B}^{e\ \mathrm{trial}}_{n+1}}, \tag{14.101}$$

i.e. it is the derivative of the tensor logarithm function at $\boldsymbol{B}^{e\ \mathrm{trial}}_{n+1}$. The fourth-order tensor B is defined by the Cartesian components

$$\mathsf{B}_{ijkl} = \delta_{ik}\,(\boldsymbol{B}^{e\ \mathrm{trial}}_{n+1})_{jl} + \delta_{jk}\,(\boldsymbol{B}^{e\ \mathrm{trial}}_{n+1})_{il}. \tag{14.102}$$

The structure of the tangent modulus

Note that D is the *only* material-related contribution to the spatial modulus a; that is, D is the only term of (14.99) that depends on the elastic law and yield surface. All other terms taking part in its assemblage in (14.99) are related purely to the geometry of finite deformations and are completely independent of the particular material model adopted. The operator D has exactly the same functional format as the tangent consistent with the corresponding implicit integration algorithms for *infinitesimal* plasticity.

Thus, the complete separation between large-strain kinematics and material-related operations observed in the general integration algorithm of Boxes 14.3 and 14.4 is also possible in the assemblage of the spatial consistent tangent modulus.

Elastic and elastoplastic moduli

It is important here to recall the remarks made in Chapter 7 on the differentiability of the algorithmic stress constitutive function associated with the small-strain elastic predictor/return-mapping algorithm (refer to Section 7.4.1, from page 229). When the stress is in the elastic domain, the function $\tilde{\boldsymbol{\tau}}$ is differentiable and its derivative D is the standard small-strain elasticity modulus. However, when the Kirchhoff stress lies on the yield surface, $\tilde{\boldsymbol{\tau}}$ is non-differentiable. In this case, if elastic unloading is assumed to occur, then D is taken as the elastic modulus – the infinitesimal elasticity tensor – in the assemblage of a. If plastic loading is assumed to occur, then D is taken as the infinitesimal elastoplastic consistent tangent modulus, D^{ep}, whose derivation in the general case is described in Section 7.4.4 (from page 238).

14.5.2. COMPUTATIONAL IMPLEMENTATION

From the above comments on the structure of a, it is clear that the same subroutines used to compute the consistent tangent operator for infinitesimal plasticity models can be used *without modification* to compute D in the assemblage of the spatial tangent modulus; that is, for the finite strain extension to the von Mises model, D will be computed in subroutine `CTVM` (Section 7.4.3, page 235). For the finite strain extension to the Tresca model, subroutine `CTTR` (Section 8.1.5, page 291) will be used, and so on. Tangent operators for other infinitesimal plasticity models have been derived in Chapters 8 and 10.

Another important observation regarding the computer implementation of a is the fact that the tensor logarithm is a member of the class of isotropic tensor functions described in Section A.5. In expression (A.48), the tensor logarithm corresponds to $y(\cdot) \equiv \ln(\cdot)$ and $\boldsymbol{X} \equiv \boldsymbol{B}_{n+1}^{e\ \mathrm{trial}}$. In program `HYPLAS`, L is computed in subroutine `DISO2`. `DISO2` is a general routine for computation of the derivative of any isotropic tensor function of the class discussed in Section A.5.

14.6. Principal stress space-based implementation

In the context of the *infinitesimal* theory, it was seen in Chapter 8 that the computational implementation of some isotropic plasticity models may be more efficient and clear if a principal stress-based representation is adopted. This was the case, for instance, for the Tresca and Mohr–Coulomb isotropically hardening models whose principal stress-based implementations are described in Sections 8.1 and 8.2, respectively. Even though *any* isotropic model may be formulated and implemented in terms of principal stresses and strains, it has been emphasised in Chapter 8 that this approach may not be optimal in certain cases. For example, the computer implementation of the infinitesimal isotropically hardening von Mises model turns out to be more efficient and clear when the standard component representation is adopted.

Equivalent principal stress representations are also possible in *finite* isotropic plasticity. The general computational schemes for stress updating and tangent modulus evaluation for isotropic finite strain plasticity described in Sections 14.4 and 14.5 can be alternatively implemented in principal stress space-based format. This approach has been extensively exploited by Simo (1992). We also refer to Rosati and Valoroso (2004) for a recently proposed general framework, also based on the principal space description. An outline of the principal stress space-based treatment of finite plasticity is presented in the following.

14.6.1. STRESS-UPDATING ALGORITHM

In this section, we reformulate the general numerical integration algorithm of Boxes 14.3 and 14.4 in an equivalent principal stress-based form. We begin the derivation of the fully principal stress-based algorithm by recalling that, for *any* isotropic model, $\boldsymbol{\varepsilon}^{e\ \mathrm{trial}}_{n+1}$ and $\boldsymbol{\tau}_{n+1}$ have the same principal axes. This means that the small-strain part of the algorithm (Box 14.4) can be implemented in terms of the principal values of $\boldsymbol{\tau}_{n+1}$ and $\boldsymbol{\varepsilon}^{e\ \mathrm{trial}}_{n+1}$. By observing that $\boldsymbol{B}^{e\ \mathrm{trial}}_{n+1}$ and $\boldsymbol{\varepsilon}^{e\ \mathrm{trial}}_{n+1}$ also share the same principal directions, we may reformulate the algorithm of Boxes 14.3 and 14.4 in the following equivalent form

1. Given $\Delta \boldsymbol{u}$, compute $\boldsymbol{F}_\Delta$, $\boldsymbol{F}_{n+1}$ and $\boldsymbol{B}^{e\ \mathrm{trial}}_{n+1}$ (as in Box 14.3).

2. Perform spectral decomposition of $\boldsymbol{B}^{e\ \mathrm{trial}}_{n+1}$

$$\boldsymbol{B}^{e\ \mathrm{trial}}_{n+1} = \sum_{i=1}^{3} b^{e\ \mathrm{trial}}_i \, \boldsymbol{e}_i \otimes \boldsymbol{e}_i, \tag{14.103}$$

   and compute the principal elastic trial strains

$$\varepsilon^{e\ \mathrm{trial}}_i := \tfrac{1}{2} \ln b^{e\ \mathrm{trial}}_i. \tag{14.104}$$

3. With the principal elastic trial strains and the set $\boldsymbol{\alpha}_n$ of internal variables at t_n at hand, use a *principal stress-based* integration algorithm to obtain the eigenvalues τ_i, of $\boldsymbol{\tau}_{n+1}$, and ε^e_i, of $\boldsymbol{\varepsilon}^e_{n+1}$.

4. Assemble the updated Cauchy stress tensor

$$\boldsymbol{\sigma}_{n+1} := \frac{1}{J} \sum_{i=1}^{3} \tau_i \, \boldsymbol{e}_i \otimes \boldsymbol{e}_i. \tag{14.105}$$

14.6.2. TANGENT MODULUS COMPUTATION

Instead of computing the tensors D and L of expression (14.99) individually and then calculating the product $\mathsf{D} : \mathsf{L}$ (as we do in HYPLAS – refer to subroutines MATICT and CSTEP2), the computation of the spatial tangent modulus can be made more efficient by exploiting a fully principal stress-based approach. This can be achieved by recalling that $\boldsymbol{\tau}_{n+1}$, $\boldsymbol{\varepsilon}^{e\ \text{trial}}_{n+1}$ and $\boldsymbol{B}^{e\ \text{trial}}_{n+1}$ all share the same principal directions. Thus, the product

$$\frac{1}{2}\mathsf{D} : \mathsf{L} = \frac{\partial \boldsymbol{\tau}_{n+1}}{\partial \boldsymbol{B}^{e\ \text{trial}}_{n+1}} = \frac{\partial \boldsymbol{\tau}_{n+1}}{\partial \boldsymbol{\varepsilon}^{e\ \text{trial}}_{n+1}} : \frac{\partial \boldsymbol{\varepsilon}^{e\ \text{trial}}_{n+1}}{\partial \boldsymbol{B}^{e\ \text{trial}}_{n+1}}, \tag{14.106}$$

that takes part in (14.99) can be evaluated by directly computing the principal stress derivative

$$\frac{\partial \tau_i}{\partial b^{e\ \text{trial}}_j} = \frac{1}{2\, b^{e\ \text{trial}}_j} \frac{\partial \tau_i}{\partial \varepsilon^{e\ \text{trial}}_j} \quad \text{(no summation implied)} \tag{14.107}$$

and then assembling the corresponding fourth-order tensor (the product (14.106)) as an isotropic tensor-valued function of one tensor of the type alluded to in Section 14.5.2.

Remark 14.7. Further efficiency can be gained if the eigenprojection tensors $\boldsymbol{M}_i \equiv \boldsymbol{e}_i \otimes \boldsymbol{e}_i$ (no summation over i) obtained in the spectral decomposition carried out in the stress-updating stage (item 2 of Section 14.6.1) is stored in memory and then retrieved later for use in the computation of the above fourth-order tensor. This approach (not adopted in HYPLAS) avoids the re-computation of the eigenprojections of $\boldsymbol{B}^{e\ \text{trial}}_{n+1}$ during the evaluation of the tangent modulus.

14.7. Finite plasticity in plane stress

The treatment of plane stress problems has been discussed in Chapter 9 and Section 13.3, respectively, in the context of infinitesimal plasticity and finite hyperelasticity. The finite element implementation of hyperelastic-based multiplicative plasticity models in plane stress follows completely analogous concepts; that is, procedures such as the nested iteration methodology for enforcement of the plane stress constraint or plane stress-projected constitutive models can be equally derived for finite strain elastoplasticity. Again, as in infinitesimal plasticity or finite elasticity, plane stress-projected constitutive models can be practically derived only when the equations of the model are sufficiently simple to allow the elimination of out-of-plane terms from the formulation. This is the case, for instance, of the finite strain extension of the von Mises model as we shall see below. The (generally computationally more expensive) nested iteration method, on the other hand, can be relatively easily implemented for any model whose axisymmetric implementation already exists. The two approaches are described in this section. We start by presenting in the sequel the plane-stress projected implementation of the finite strain extension of the von Mises model. The application of the nested iteration methodology is then described for a generic underlying plasticity model.

14.7.1. THE PLANE STRESS-PROJECTED FINITE VON MISES MODEL

Let us consider the plane stress-projected version of the finite strain extension of the isotropically hardening von Mises model. Before proceeding further, the reader is advised to review Section 9.4 where the *infinitesimal* plane stress-projected von Mises model is described in detail. Of particular relevance to our discussion are the constitutive model and the corresponding integration algorithm summarised, respectively, in Boxes 9.3 (page 374) and 9.4 (page 377).

The extension of the infinitesimal plane stress-projected model (Box 9.3) to the finite strain range within the present logarithmic strain-based multiplicative framework is given by the following set of constitutive equations:

1. the standard multiplicative split (item (i) of Box 14.2) where, now, only the *in-plane* components of $\boldsymbol{F}$, $\boldsymbol{F}^e$ and $\boldsymbol{F}^p$ are required;

2. the Hencky hyperelastic relationship

$$\boldsymbol{\tau} = \mathbf{D}^e \, \boldsymbol{\varepsilon}^e,$$

   where $\boldsymbol{\tau}$ and $\boldsymbol{\varepsilon}^e$ are the arrays of in-plane components of the *Kirchhoff* stress and *logarithmic* elastic strain

$$\boldsymbol{\tau} = [\tau_{11}, \tau_{22}, \tau_{12}]^T, \quad \boldsymbol{\varepsilon}^e = [\varepsilon^e_{11}, \varepsilon^e_{22}, 2\varepsilon^e_{12}]^T;$$

3. a yield function defined in terms of in-plane stress components

$$\Phi = \tfrac{1}{2}\,\boldsymbol{\tau}^T \mathbf{P}\,\boldsymbol{\tau} - \tfrac{1}{3}\sigma_y^2(\bar{\varepsilon}^p)$$

   where matrix $\mathbf{P}$ is defined in Box 9.3 (page 374);

4. the plastic flow rule

$$\tilde{\mathbf{d}}^p = \dot{\gamma}\,\mathbf{P}\,\boldsymbol{\tau}$$

   where $\tilde{\mathbf{d}}^p$ is the array of in-plane components of $\tilde{\boldsymbol{D}}^p$;

5. hardening variable evolution

$$\dot{\bar{\varepsilon}}^p = \dot{\gamma}\,\sqrt{\tfrac{2}{3}\,\boldsymbol{\tau}^T \mathbf{P}\,\boldsymbol{\tau}};$$

6. together with the standard loading/unloading criterion

$$\dot{\gamma} \geq 0, \quad \Phi \leq 0, \quad \dot{\gamma}\,\Phi = 0.$$

The integration algorithm

The operator split integration algorithm in the present case has essentially the same format as the general finite strain procedure listed in Box 14.3. It comprises the following steps:

(a) Perform the calculations of items (i) and (ii) of Box 14.3 taking into consideration *only* the *in-plane* components of the tensor variables $\boldsymbol{F}_\Delta$, $\boldsymbol{F}_{n+1}$, etc.

(b) Having computed the elastic trial state, apply the integration algorithm of the small strain plane stress-projected model (Box 9.4) to update the in-plane components of $\boldsymbol{\tau}$ and $\boldsymbol{\varepsilon}^e$ as well as the hardening variable. From the Hencky hyperelastic law, the updated thickness logarithmic elastic strain

$$\varepsilon_{33}^e \equiv \ln \lambda_3^e \tag{14.108}$$

can also be computed according to the trivial relation

$$\varepsilon_{33}^e = -\frac{\nu}{1-\nu}\,(\varepsilon_{11}^e + \varepsilon_{22}^e). \tag{14.109}$$

In program HYPLAS, these calculations are carried out in subroutine SUVMPS.

(c) Perform the calculations of item (iv) of Box 14.3, including the computation of the out-of-plane component of the elastic left Cauchy–Green tensor. Note that before computing the updated Cauchy stress components, it is necessary to compute $\det \boldsymbol{F}_{n+1}$; that is, the determinant of the full deformation gradient which includes the contribution from the thickness (or out-of-plane) strain. At this point an extra operation is required. The situation here is completely analogous to that of the plane stress Hencky hyperelastic model (see Section 13.3.2, from page 532). Note that, due to the plastic incompressibility of the von Mises model, the determinants of the total and elastic deformation gradient coincide

$$\det \boldsymbol{F} = \det \boldsymbol{F}^e. \tag{14.110}$$

This, together with the use of the logarithmic elastic strain measure, allows $\det \boldsymbol{F}_{n+1}$ to be computed from the relation

$$\det \boldsymbol{F} = \mathrm{tr}[\ln \boldsymbol{U}^e] = \exp[\varepsilon_{11}^e + \varepsilon_{22}^e + \varepsilon_{33}^e]. \tag{14.111}$$

The above computation is carried out in subroutine TUVM of HYPLAS). Once $\det \boldsymbol{F}_{n+1}$ is computed, we then update the Cauchy stress

$$\boldsymbol{\sigma}_{n+1} := (\det \boldsymbol{F}_{n+1}^{-1})\,\boldsymbol{\tau}_{n+1}. \tag{14.112}$$

Thickness update

Within the finite element environment, the current thickness at Gauss points is required in the computation of the internal force vector and tangent stiffness. With t_0 denoting the initial thickness (in the reference configuration) at a generic Gauss point, the current thickness t is given by

$$t = \lambda_3\, t_0. \tag{14.113}$$

Within the structure of program HYPLAS, where the determinant of the current total deformation gradient (in-plane deformation gradient in the plane stress case) is readily available during the stress-update phase (see subroutine MATISU), the thickness stretch, λ_3, for the von Mises model can be conveniently computed as follows. With $\tilde{\boldsymbol{F}}_{n+1}$ denoting the current *in-plane* deformation gradient, we have the trivial relation

$$\det \boldsymbol{F}_{n+1} = (\det \tilde{\boldsymbol{F}}_{n+1})\,\lambda_3. \tag{14.114}$$

Thus, after the computation of $\det \boldsymbol{F}_{n+1}$ by (14.111), the thickness stretch is promptly evaluated as

$$\lambda_3 := \frac{\det \boldsymbol{F}_{n+1}}{\det \tilde{\boldsymbol{F}}_{n+1}}. \tag{14.115}$$

The thickness is then updated according to (14.113). These operations are performed in subroutine TUVM of HYPLAS.

The tangent modulus

The consistent tangent modulus in the present case is computed quite trivially. Basically, the spatial tangent operator is a *plane* tensor defined by expression (14.99) where D is either the standard plane stress infinitesimal elasticity tensor or, under plastic flow, the elastoplastic tangent tensor consistent with the small strain plane stress return-mapping algorithm whose expression is given in Box 9.6 (page 384). The elastoplastic infinitesimal tangent operator for the plane stress-projected von Mises model is computed in subroutine CTVMPS of program HYPLAS.

14.7.2. NESTED ITERATION FOR PLANE STRESS ENFORCEMENT

Readers not already familiar with the nested iteration procedure for plane stress enforcement are advised to review Section 9.2 with particular attention to Section 9.2.2 where the methodology is described in the context of infinitesimal elastoplasticity. The generic infinitesimal plasticity nested iteration algorithm is presented in pseudo-code format in Box 9.1 (page 364). It is also worth revisiting Section 13.3.3 which describes the application of this method to finite hyperelasticity models.

In the present context, the nested iteration algorithm may be cast in the following form:

1. Perform the calculations of items (i) and (ii) of Box 14.3 taking into consideration *only* the *in-plane* components of the tensor variables $\boldsymbol{F}_\Delta$, $\boldsymbol{F}_{n+1}$, etc.

2. With the elastic trial state at hand, apply the small-strain nested iteration procedure (refer to Box 9.1). The procedure will update $\boldsymbol{\tau}$, the in-plane components of $\boldsymbol{\varepsilon}^e$ and the internal variables of the model.

3. Perform the calculations of item (iv) of Box 14.3 again taking into consideration only in-plane components. Again (see item (c), page 603), the determinant of $\boldsymbol{F}_{n+1}$ (which includes the thickness stretching) needs to be computed before the Cauchy stress can be updated. For a plastically incompressible model, $\det \boldsymbol{F}_{n+1}$ can be computed as described in item (c) of the plane stress-projected finite von Mises algorithm. For more general models, the following procedure may be adopted. Firstly, let us recall that at a (plane stress) converged state of the plane stress enforcement loop, both the elastic trial and updated elastic thickness logarithmic strains, $\varepsilon_{33}^{e\ \mathrm{trial}}$ and ε_{33}^{e} (with subscripts $n+1$ dropped for convenience), are available as a by-product of the algorithm. Now let us define the plastic logarithmic thickness strain as

$$\varepsilon_{33}^{p} \equiv \ln \lambda_3^{p}.$$

Trivially, the total logarithmic thickness strain satisfies

$$\varepsilon_{33} \equiv \ln \lambda_3 = \varepsilon^e_{33} + \varepsilon^p_{33} = \ln(\lambda^e_3 \lambda^p_3) \tag{14.116}$$

and, as in the infinitesimal case,

$$\varepsilon^p_{33} = (\varepsilon^p_{33})_n + \varepsilon^{e\ \mathrm{trial}}_{33} - \varepsilon^e_{33}, \tag{14.117}$$

where the subscript n denotes the (equilibrium) converged value at the end of the previous load step. To compute $\det \boldsymbol{F}_{n+1}$, we proceed as follows:

(a) update the plastic thickness strain using (14.117);

(b) then compute the total logarithmic thickness strain using (14.116) and obtain the total thickness stretch

$$\lambda_3 := \exp(\varepsilon_{33});$$

(c) finally compute

$$\det \boldsymbol{F}_{n+1} := (\det \tilde{\boldsymbol{F}}_{n+1})\, \lambda_3,$$

and update the Cauchy stress according to (14.112).

Thickness update

With λ_3 already computed, the current Gauss point thickness is promptly updated with expression (14.113).

The tangent modulus

The comments made on page 604 regarding the spatial tangent modulus for the plane stress-projected model apply equally to the present case. Here, however, the infinitesimal elastoplastic tangent operator is generally obtained from formula (9.24) (page 366).

14.8. Finite viscoplasticity

Isotropic finite strain viscoplasticity can be equally accommodated within the hyperelastic-based multiplicative plasticity framework discussed in the previous sections of this chapter. Here, we will assume that the reader is already familiar with the theory of viscoplasticity discussed in Chapter 11 in the infinitesimal strain context. Analogously to the infinitesimal case, the general finite viscoplasticity model is defined by simply replacing, in Box 14.2, the large-strain plastic flow rule (evolution equation for $\boldsymbol{F}^p$) and the evolution equation for the set $\boldsymbol{\alpha}$ of internal variables with the following expressions:

$$\begin{aligned} \tilde{\boldsymbol{D}}^p &= \boldsymbol{G}(\boldsymbol{\tau}, \boldsymbol{A}) \\ \dot{\boldsymbol{\alpha}} &= \boldsymbol{J}(\boldsymbol{\tau}, \boldsymbol{A}), \end{aligned} \tag{14.118}$$

where $\boldsymbol{G}$ and $\boldsymbol{J}$ are given explicit functions of $\boldsymbol{\tau}$ and the set $\boldsymbol{A}$ of thermodynamical forces. The resulting model is a finite strain extension (under isotropic conditions) of the infinitesimal viscoplasticity model of Box 11.3 (page 451). It incorporates, for instance, the finite strain extension of all von Mises-based infinitesimal viscoplasticity models discussed in Section 11.3 (from page 445).

For viscoplastic models with a yield surface, the above evolution equations can be written in the more specialised format (refer to equations (11.71) on page 456)

$$\tilde{\boldsymbol{D}}^p = \dot{\gamma}(\boldsymbol{\tau}, \boldsymbol{A})\, \boldsymbol{N}(\boldsymbol{\sigma}, \boldsymbol{A})$$
$$\dot{\boldsymbol{\alpha}} = \dot{\gamma}(\boldsymbol{\tau}, \boldsymbol{A})\, \boldsymbol{H}(\boldsymbol{\sigma}, \boldsymbol{A}), \tag{14.119}$$

where $\boldsymbol{N}$ is the flow vector and $\boldsymbol{H}$ represents the generalised hardening moduli. The scalar function $\dot{\gamma}$ is zero within the elastic domain or on the yield surface and positive outside the elastic domain as, for instance, in the definitions given in Sections 11.3.1 and 11.3.2.

14.8.1. NUMERICAL TREATMENT

The numerical treatment of hyperelastic-based multiplicative viscoplasticity models is completely analogous to the rate-independent case thoroughly discussed earlier in this chapter; that is, in the constitutive integration algorithm, the effective material-related integration scheme has the same format as the small strain counterpart, with the finite strain-related operations confined to the level of kinematics. Thus, the integration can be performed by means of the procedure of Box 14.3, with the small-strain algorithm referred to in item (iii) replaced with a small-strain viscoplasticity algorithm described in Chapter 11 (refer to Boxes 11.4 and 11.5 on pages 456 and 457, respectively). The same applies to the computation of the consistent tangent matrix. In the viscoplastic case, the small-strain tangent modulus D on the right-hand side of (14.99) becomes the consistent tangent operator of the corresponding infinitesimal viscoplasticity model/algorithm. Infinitesimal viscoplastic consistent tangent operators were discussed in Sections 11.5.3 and 11.6.3.

We note that this framework can be readily extended to include the combined effects of hyperelasticity, viscoplasticity, viscoelasticity and damage as described, for example, by Reese and Govindjee (1998), Miehe and Keck (2000) and Perić and Dettmer (2003).

14.9. Examples

In this section we present a set of finite element examples where the hyperelastic-based multiplicative plasticity framework described in the preceding sections has been used. Unless otherwise stated, the results presented have been obtained with program HYPLAS. Again, in all examples, the full Newton–Raphson algorithm is selected to solve the incremental equilibrium problem.

14.9.1. FINITE STRAIN BENDING OF A V-NOTCHED TRESCA BAR

This is the finite strain version of the numerical example presented is Section 8.4.1 (from page 343). The problem consists of a wide rectangular metal bar containing a deep 90° V-shaped notch and subjected to bending. The material model adopted for the bar is the hyperelastic-based multiplicative finite strain extension of the Tresca elasto-perfectly plastic model. Plane strain state is assumed. The initial geometry and material parameters are identical to those of the infinitesimal analysis and are shown in Figure 8.17 (page 344).

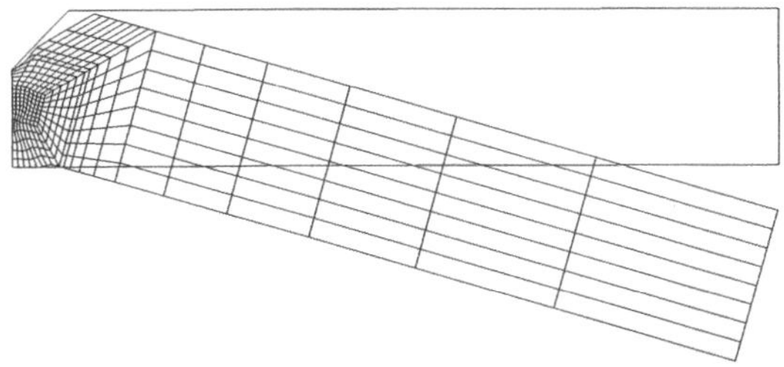

Figure 14.3. Large strain bending of a V-notched bar. Deformed configuration with eight-noded quadrilaterals mesh. (Reproduced with permission from A new computational model for Tresca plasticity at finite strains with an optimal parametrization in the principal space, D Perić and EA de Souza Neto, *Computer Methods in Applied Mechanics and Engineering*, Vol 171 © 1999 Elsevier Science S.A.)

Finite element solutions are obtained here with two different meshes. The first mesh, of eight-noded quadrilaterals with four-point reduced integration, is identical to that employed in the infinitesimal analysis. The second solution is obtained with a mesh of 312 *F-bar* four-noded quadrilaterals. The *F-bar* four-noded quadrilateral element is described in detail in Chapter 15. This element is especially suited for large strain analysis of nearly incompressible materials. The adopted discretisation coincides with that shown in Figure 8.18 (page 344). In this case, however, the mesh contains only 345 nodes, as opposed to 1001 nodes used in the corresponding discretisation with eight-noded quadrilaterals. Due to the occurrence of geometric softening of the bar under large strains, the *arc-length* procedure described in Chapter 4 (see Section 4.4, from page 107) is used in conjunction with the Newton–Raphson algorithm in the equilibrium iterations. A deformed configuration obtained with the eight-noded elements is depicted in Figure 14.3. The corresponding vertical displacement of the mid-node of the edge where the forces are applied is 10.288 mm. This configuration was reached in 30 steps. The results obtained with the *F-bar* element mesh are shown in Figure 14.4. Figures 14.4(a) and (b) show the deformed configurations of the bar, with the corresponding contour plot of accumulated plastic strains, respectively with 2.50 mm and 12.77 mm of displacement of the mid-node. These configurations have been reached, respectively, at load steps 10 and 30. The corresponding moment–deflection diagram obtained in the simulation is shown in Figure 14.5. The normalised applied moment plotted in the diagram is measured in the current configuration of the specimen. It should be noted that after the limit load is reached, global softening is observed. This phenomenon – not captured in the corresponding small strain analysis – is due to the reduction of cross-sectional area near the notch. The behaviour of the Newton–Raphson algorithm for equilibrium iterations is illustrated in Table 14.1, where the evolution of the relative residual norm is shown during a typical load increment. The quadratic rates of convergence of the Newton algorithm, which results from the consistent linearisation of the algorithms involved, are evident.

14.9.2. NECKING OF A CYLINDRICAL BAR

This example consists of the simulation of the necking of a cylindrical bar, with radius 6.413 mm and length 53.334 mm, in a tensile test. Two material models are adopted, the

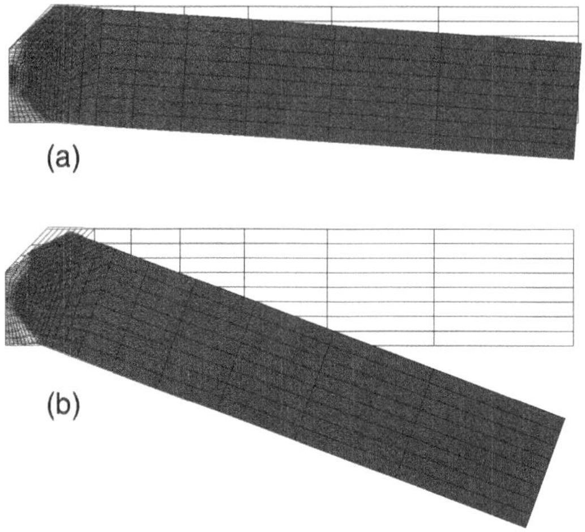

Figure 14.4. Large strain bending of a V-notched bar. Deformed configuration with four-noded *F-bar* quadrilaterals mesh. Mid-node displacements: (a) 2.503 mm; and (b) 12.836 mm. (Reproduced with permission from A new computational model for Tresca plasticity at finite strains with an optimal parametrization in the principal space, D Perić and EA de Souza Neto, *Computer Methods in Applied Mechanics and Engineering*, Vol 171 © 1999 Elsevier Science S.A.)

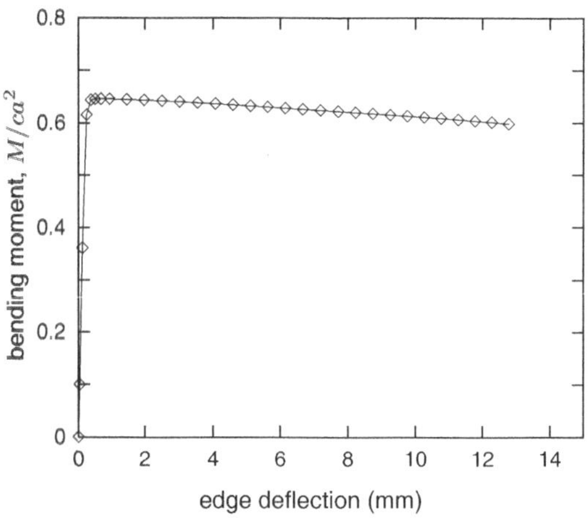

Figure 14.5. Finite strain bending of a V-notched bar. Moment–deflection diagram.

Tresca and the von Mises model with elastic constants

$$E = 206.9 \text{ GPa}, \quad \nu = 0.29,$$

and the following hardening curve

$$\sigma_y(\bar{\varepsilon}^p) \equiv (\sigma_\infty - \sigma_0)[1 - \exp(-\delta\, \bar{\varepsilon}^p)] + H\, \bar{\varepsilon}^p,$$

where

$$\sigma_0 = 0.45 \text{ GPa}, \quad \sigma_\infty = 0.715 \text{ GPa}, \quad \delta = 16.93, \quad H = 0.12924 \text{ GPa}.$$

Table 14.1. V-notched Tresca bar. Typical convergence table.

Iteration	Relative residual norm (%)
1	$0.167404\,E{+}03$
2	$0.777646\,E{+}01$
3	$0.368556\,E{+}00$
4	$0.650100\,E{-}01$
5	$0.606889\,E{-}04$
6	$0.385324\,E{-}09$

Simo and Armero (1992) used this problem, with the von Mises model, to study the performance of enhanced assumed strain elements in large strain localisation problems.

We first carry out an axisymmetric analysis of the problem. Due to obvious symmetry, only one quarter of the bar is discretised with the appropriate boundary conditions being imposed on the symmetry lines. The mesh of 200 four-node axisymmetric *F-bar* elements (refer to Chapter 15) shown in Figure 14.6(a) is used. The mesh contains 231 nodal points. A geometric imperfection of 1.8% of the radius is introduced at the centre of the bar to trigger the necking. A vertical displacement $u = 7.0$ mm is imposed incrementally at the top of the bar. The final deformed meshes, in which the development of necking in the central zone can be clearly seen, are plotted in Figure 14.6(b) and (c) and correspond, respectively, to the Tresca and von Mises material models. The final configuration was reached in 23 steps for the Tresca model and 17 steps for the von Mises model. Figure 14.7 shows the reaction–displacement curves obtained in the present computation. It is noted that up to approximately $u = 3.0$ mm – where the limit reaction is attained – the reactions obtained with the Tresca model are virtually identical to those obtained with the von Mises material. This is an obvious consequence of the fact that the dominant stresses up to that stage are uniaxial – a condition under which the behaviour of both material models coincide in the present set-up. Some discrepancy in reactions is observed after strain localisation is triggered, in the descent branch of the displacement–reaction curve. This is attributed to the stronger influence of shear under localisation. It should be noted that the final deformed configurations of Figures 14.6(b) and (c) differ quite considerably in the necking zone.

Three-dimensional analysis

We now carry out a *three-dimensional* analysis of the cylindrical bar necking problem. Only the von Mises model is used. The analysis reported here was *not* carried out in program HYPLAS. Two different meshes – one with 120 and another with 960 eight-noded *F-bar* hexahedra – are used to discretise the symmetric octant of the specimen, with the appropriate boundary conditions imposed on the symmetry planes. The initial meshes are shown in Figure 14.8. The final deformed configurations with imposed edge displacement $u = 7.0$ mm obtained for both meshes are shown in Figure 14.9 along with the von Mises effective stress contour plot obtained for the finer mesh. For ease of visualisation, the deformed

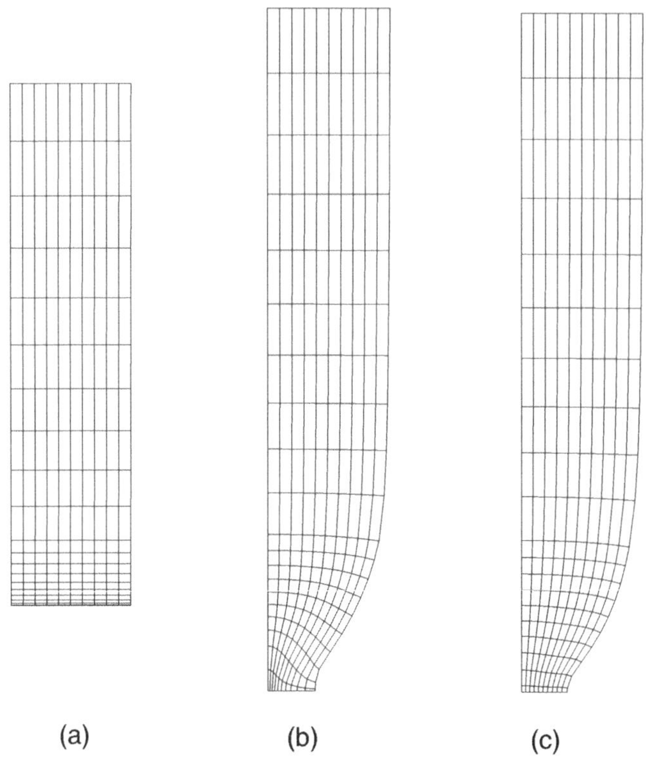

Figure 14.6. Necking of a cylindrical bar. Axisymmetric analysis: (a) initial mesh. Deformed meshes with imposed displacement $u = 7.0$ mm; (b) Tresca model; (c) von Mises model. (Reproduced with permission from A new computational model for Tresca plasticity at finite strains with an optimal parametrization in the principal space, D Perić and EA de Souza Neto, *Computer Methods in Applied Mechanics and Engineering*, Vol 171 © 1999 Elsevier Science S.A.)

symmetric octant has been mirrored to produce a full deformed specimen in Figures 14.9(a) and (b). The upper half of Figure 14.9(c) shows the distribution of von Mises effective stress within the longitudinal cross-section of the specimen whereas its lower half shows the distribution on the boundary of the specimen. Both fine and coarse meshes are able to predict correctly the necking phenomenon. The final shape resulting from the simulation with 960 eight-node hexahedra is virtually identical to that predicted by the above axisymmetric simulation. The deformed transversal cross-section obtained with the fine mesh simulation is depicted in Figure 14.10 where, again, the deformed symmetric octant has been mirrored for ease of visualisation. The reaction–displacement curves obtained for both three-dimensional simulations are plotted in Figure 14.11 along with results of the axisymmetric analysis. The results for the finer three-dimensional mesh are almost indistinguishable from those obtained in the axisymmetric simulation.

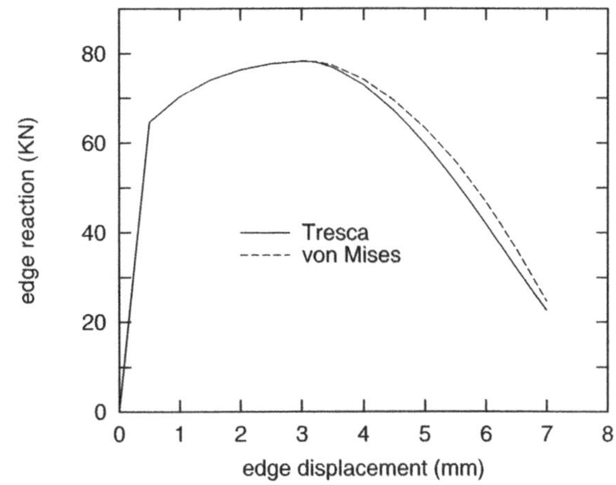

Figure 14.7. Necking of a cylindrical bar. Axisymmetric analysis. Force–displacement diagrams. (Reproduced with permission from A new computational model for Tresca plasticity at finite strains with an optimal parametrization in the principal space, D Perić and EA de Souza Neto, *Computer Methods in Applied Mechanics and Engineering*, Vol 171 © 1999 Elsevier Science S.A.)

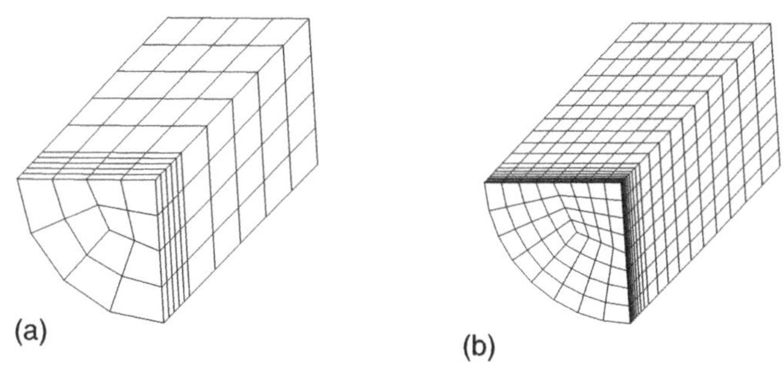

Figure 14.8. Necking of a cylindrical bar. Initial meshes for three-dimensional analysis: (a) 120 eight-noded *F-bar* hexahedra; (b) 960 eight-noded *F-bar* hexahedra. (Reproduced with permission from Design of simple low-order finite elements for large-strain analysis of nearly incompressible solids, EA de Souza Neto, D Perić, M Dutko and DRJ Owen, *International Journal of Solids and Structures*, Vol 33, Issue 20–22 © 1996 Elsevier Science Ltd.)

14.9.3. PLANE STRAIN LOCALISATION

In this example we simulate the occurrence of shear bands during the finite stretching of an elastoplastic rectangular bar. This problem was considered by Simo and Armero (1992) to test the ability of assumed enhanced strain element formulations in capturing strain localisation. Here, we use a mesh of *F-bar* four-noded quadrilaterals under plane strain condition. Due to symmetry, only one quarter of the bar is discretised. A total number of 200 elements is used. Figure 14.12(a) shows the initial geometry and the finite element mesh. The von Mises model is adopted. The material constants are the same as in the previous example except that the hardening modulus here is

$$H = -0.012924\,\text{GPa}.$$

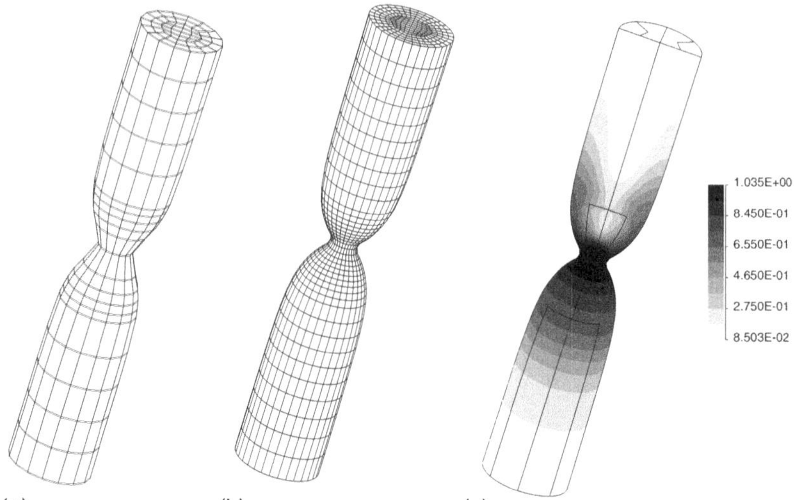

(a) (b) (c)

Figure 14.9. Necking of a cylindrical bar. Three-dimensional analysis. Results at $u = 7.0$ mm. (a) Deformed configuration with coarse mesh; (b) deformed configuration with fine mesh; (c) von Mises effective stress contour obtained with the fine mesh. (Reproduced with permission from Design of simple low-order finite elements for large-strain analysis of nearly incompressible solids, EA de Souza Neto, D Perić, M Dutko and DRJ Owen, *International Journal of Solids and Structures*, Vol 33, Issue 20–22 © 1996 Elsevier Science Ltd.)

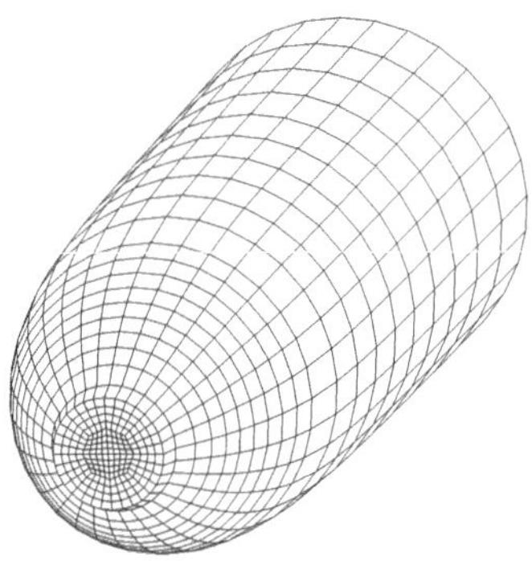

Figure 14.10. Necking of a cylindrical bar. Three-dimensional analysis. Deformed transversal cross-section obtained with fine mesh at $u = 7.0$ mm.

To trigger strain localisation, a width reduction of 1.8% is introduced in the centre of the bar. The displacement u on the constrained (top) edge is prescribed incrementally until the final deformed configuration with $u = 5.0$ mm is reached. The final deformed mesh obtained in the simulation is shown in Figure 14.12(b). At that stage, a very localised shear band can be observed. Figure 14.13 shows the displacement–reaction diagram obtained. The substantial decrease in reaction forces occurring near $u = 3.0$ mm corresponds to the development of the localised shear band. To emphasise the need for an appropriate element formulation in

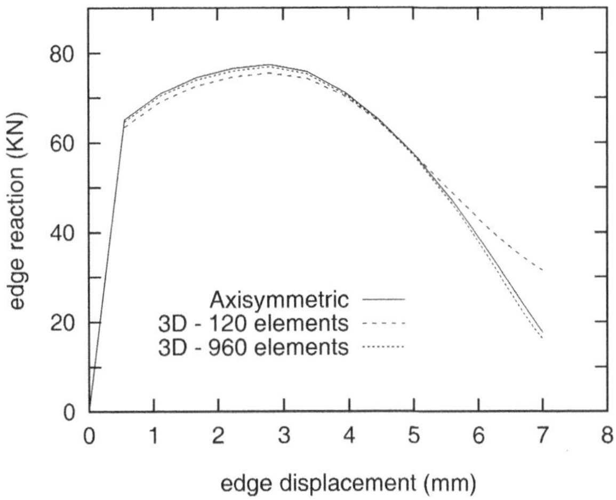

Figure 14.11. Necking of a cylindrical bar. Force–displacement diagrams for axisymmetric and three-dimensional analysis with the von Mises model. (Reproduced with permission from Design of simple low-order finite elements for large-strain analysis of nearly incompressible solids, EA de Souza Neto, D Perić, M Dutko and DRJ Owen, *International Journal of Solids and Structures*, Vol 33, Issue 20–22 © 1996 Elsevier Science Ltd.)

capturing strain localisation phenomena,[§] we also plot the results obtained with the *Q1/E4* enhanced strain element of Simo and Armero (1992) as well as the results obtained with the standard four-noded quadrilateral. Both *F-bar* and *Q1/E4* elements are able to capture localisation (producing substantial global softening) whereas the conventional four-noded element produces an over-stiff solution with no localisation.

14.9.4. STRETCHING OF A PERFORATED PLATE

This example consists of the plane stress analysis of the stretching of a rectangular perforated plate and has been studied in Sections 9.5.3 (page 390) and 11.7.2 (page 469), respectively, in the context of infinitesimal (rate-independent) elastoplasticity and viscoplasticity. Here we consider the large strain version of this problem and analyse both elastoplastic and viscoplastic cases. Note that the viscoplastic case results presented here were not obtained with program HYPLAS. The reader is referred to Sections 9.5.3 and 11.7.2 for the details of geometry, material parameters and finite element mesh used. The results obtained for the constrained edge reaction as a function of the edge prescribed displacement are shown in Figure 14.14. The rate-independent solution is plotted together with the viscoplastic solution obtained for different values of rate-sensitivity parameter, ϵ. Figure 14.14 is to be compared with the infinitesimal strain solutions shown in Figure 11.8 (page 468). All comments made in Section 11.7.2 regarding the behaviour of the viscoplastic model with various rate-sensitivity parameters and stretching rates apply to the finite strain results shown here. The major difference between the small and large strain results lies in the fact that, in spite of the linear hardening law considered, the finite strain analysis show a marked softening in the reaction–displacement curves in all cases considered. This global *geometric softening* is caused by the

[§]Refer to Chapter 15 for further discussions on element technology issues.

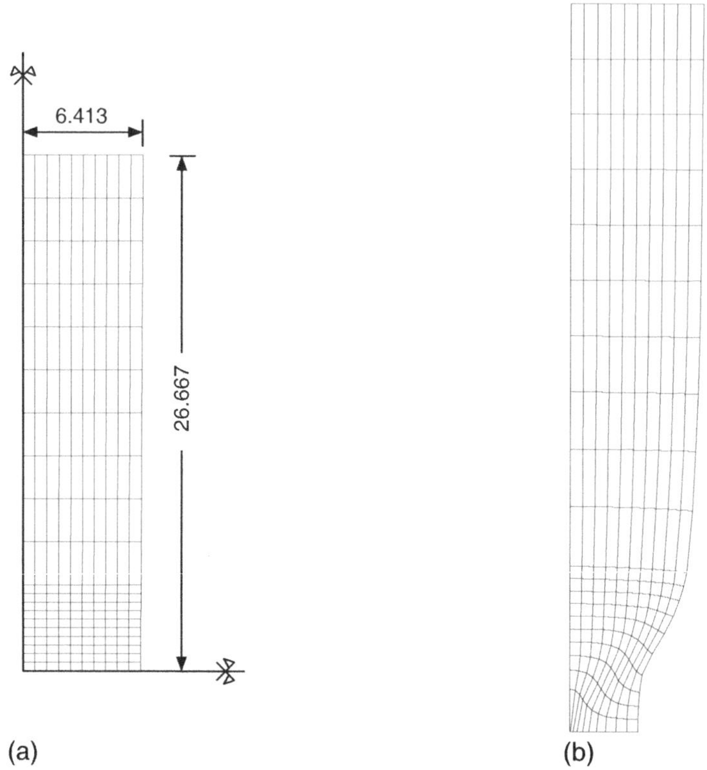

Figure 14.12. Plane strain localisation: (a) initial geometry (dimensions in mm) and finite element mesh; (b) final deformed mesh with $u = 5.0$ mm. (Reproduced with permission from Design of simple low-order finite elements for large-strain analysis of nearly incompressible solids, EA de Souza Neto, D Perić, M Dutko and DRJ Owen, *International Journal of Solids and Structures*, Vol 33, Issue 20–22 © 1996 Elsevier Science Ltd.)

change in geometry (reduction of cross-section) that occurs as the plate is stretched. In the infinitesimal theory, the change in geometry is neglected as all calculations are based on the initial geometry. Thus, this phenomenon cannot be captured by the infinitesimal strain theory that predicts global hardening in all cases.

14.9.5. THIN SHEET METAL-FORMING APPLICATION

This example considers the simulation of a thin sheet metal-forming operation. This example has *not* been solved with program HYPLAS. The problem consists of a thin circular sheet stretched by a rigid spherical punch and is frequently taken as a benchmark example (Oñate *et al.*, 1995). The sheet lies on a rigid cylindrical die and its edge is assumed to be clamped. The geometry and material parameters are shown in Figure 14.15. The material is modelled as a hyperelastic-based extension of the von Mises model with isotropic strain hardening. Due to the symmetry of the problem, only one quarter of the domain is considered in the finite element simulation. A mesh with 736 three-noded membrane elements is used in the

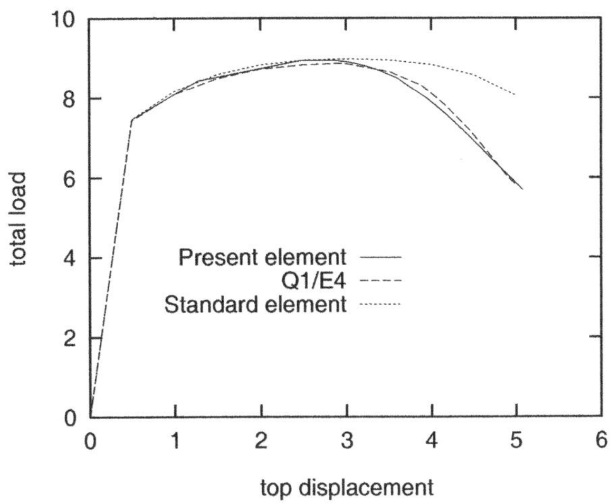

Figure 14.13. Plane strain localisation. Force–displacement diagrams. (Reproduced with permission from Design of simple low-order finite elements for large-strain analysis of nearly incompressible solids, EA de Souza Neto, D Perić, M Dutko and DRJ Owen, *International Journal of Solids and Structures*, Vol 33, Issue 20–22 © 1996 Elsevier Science Ltd.)

discretisation of the sheet. The penalty based approach described by Perić and Owen (1992) is employed in the treatment of the frictional contact between the sheet, punch and die. Coulomb dry friction with friction coefficient $\mu = 0.3$ is assumed between the sheet, punch and die. The surfaces of the punch and the die are discretised, respectively, by 2145 and 612 flat triangular elements. Figure 14.16 shows the finite element meshes used. Note that within membrane elements, a *plane stress* state is assumed. In the present case, the plane stress projected approach to the finite strain von Mises model (described in Section 14.7.1) is adopted. The reaction force obtained on the punch is plotted in Figure 14.17 against the punch travel. Figure 14.18 shows the corresponding distributions of radial strain and thickness along the sheet radius obtained in the simulation. The results are plotted for 10, 20, 30 and 40 mm of punch displacement (d_p). The peak in thickness reduction and radial strain observed at $d_p = 40$ mm corresponds to a strain localisation phenomenon leading to the rupture of the workpiece. Strain localisation initiates when the peak reaction is reached at around $d_p = 34$ mm. The ability accurately to predict rupture in industrial processes of this nature is of crucial importance in the design of tools. It can bring substantial savings in the overall design process. For this reason, the Finite Element Method is currently widely employed in industry for the simulation of thin sheet-forming operations.

14.10. Rate forms: hypoelastic-based plasticity models

The inelastic response of finitely deformed solids can also be modelled by means of so-called *hypoelastic-based* constitutive theories. In contrast to the above described family of (hyperelastic-based) finite plasticity models, hypoelastic-based theories do not rely on the assumption of the existence of a free-energy potential to model the reversible behaviour. The starting point of hypoelastic-based models is the formulation of the constitutive equations for stress in terms of objective (frame invariant) stress rates. As already mentioned in

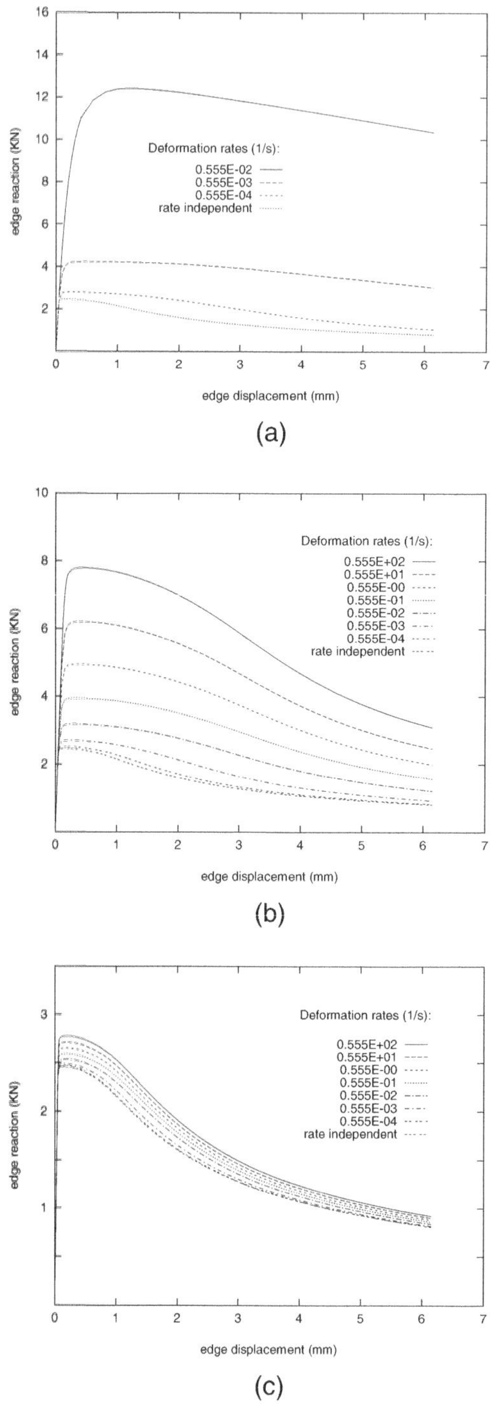

Figure 14.14. Stretching of a perforated plate. Finite strain solution. Displacement–reaction diagrams. (a) $\epsilon = 10^0$; (b) $\epsilon = 10^{-1}$; (c) $\epsilon = 10^{-2}$.

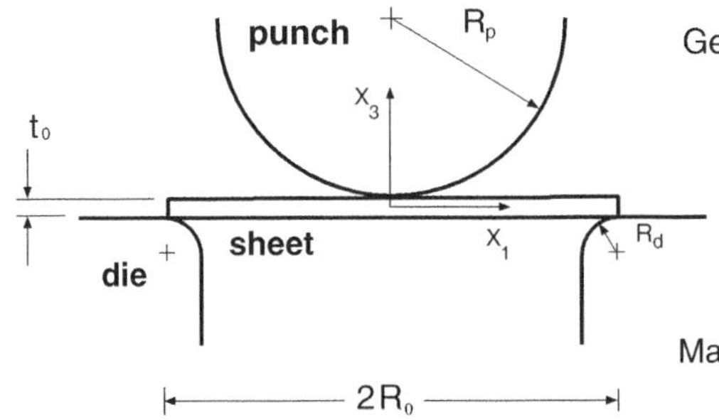

Geometry:

$R_p = 50.8$ mm
$R_d = 6.35$ mm
$R_0 = 59.18$ mm
$t_0 = 1.0$ mm

Material properties (von Mises):

$E = 69004$ N/mm^2
$\nu = 0.3$

isotropic hardening curve:

$\sigma(\varepsilon^p) = 80.559 + [589\,(10^{-4} + \varepsilon^p)^{0.216} - 80.559]$ N/mm^2

Figure 14.15. Thin sheet metal forming. Tool/workpiece configuration.

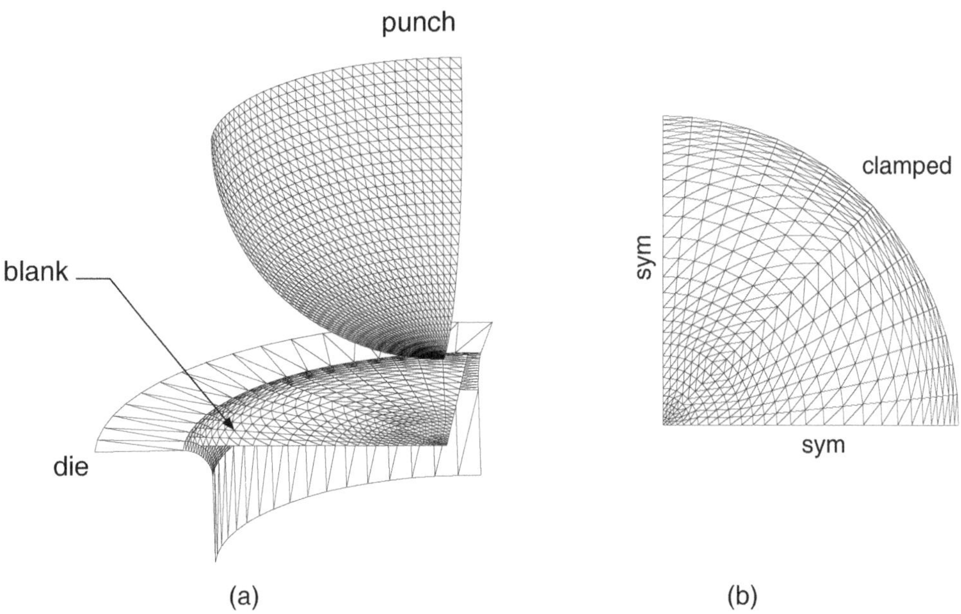

Figure 14.16. Thin sheet metal forming. (a) Finite element discretization of the sheet, die and punch; (b) finite element mesh and boundary conditions for the sheet.

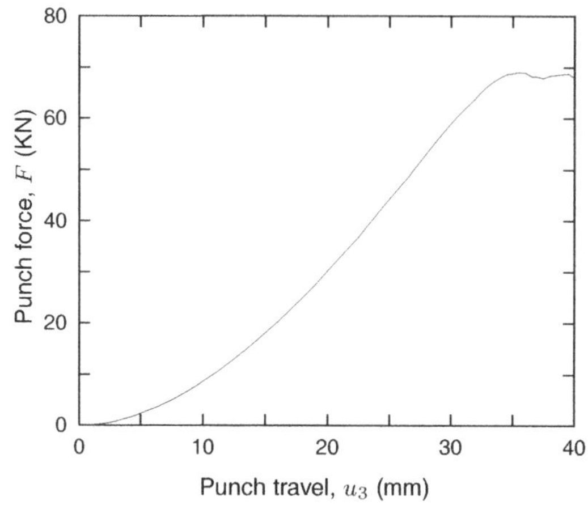

Figure 14.17. Thin sheet metal forming. Reaction forces on punch.

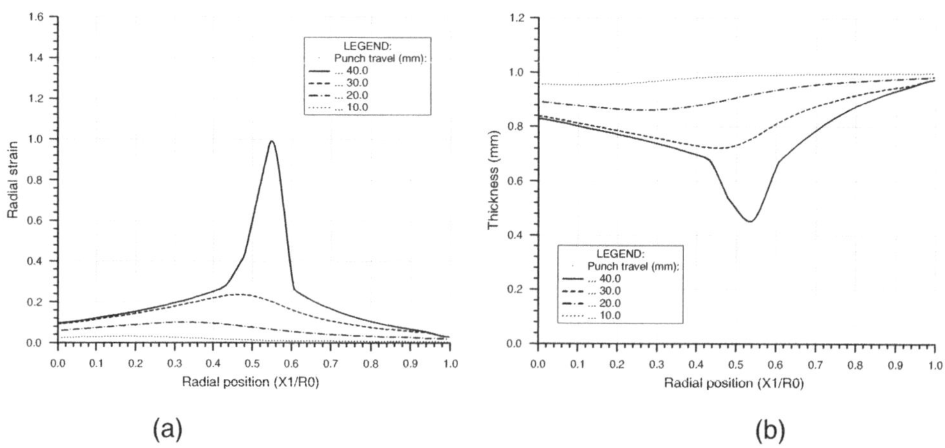

Figure 14.18. Thin sheet metal forming: (a) radial strain distributions plotted over the initial configuration; (b) thickness distributions.

Section 14.1, early formulations of finite strain plasticity have relied exclusively on the hypoelastic-based approach to provide ad hoc finite strain extensions to existing infinitesimal models. In spite of their relative simplicity, hypoelastic-based plasticity models do not usually account for truly reversible behaviour, even in the absence of plastic flow. In other words, dissipative behaviour may be predicted even within what is meant to be an 'elastic' (reversible) domain. This is certainly an undesirable feature from the theoretical point of view. In addition, the formulation of incremental constitutive equations which preserve the objectivity (frame-invariance) of the rate (time-continuum) forms is not trivial and may, in some circumstances, result in rather cumbersome numerical procedures. Despite such disadvantages, hypoelastic-based models are used extensively by many researchers and are currently available (in some cases as the only option) in many commercial finite element packages. We, therefore, find it appropriate to devote this section to this class of constitutive models of plastic material.

14.10.1. OBJECTIVE STRESS RATES

Crucial to the formulation of hypoelastic-based models of plasticity is the definition of *objective stress rates*. To introduce this concept, let us first recall from Chapter 3 (refer to the text surrounding expression (3.146), page 70) that the principle of material objectivity requires that under a change in observer, the Cauchy stress tensor, $\boldsymbol{\sigma}$, transforms according to the rule

$$\boldsymbol{\sigma}(t) \longrightarrow \boldsymbol{Q}(t)\,\boldsymbol{\sigma}(t)\,\boldsymbol{Q}^T(t). \tag{14.120}$$

A given *stress rate*, which will be here denoted generically $\overset{\bullet}{\boldsymbol{\sigma}}$ (note the bold dot), is said to be *objective* if, under a change of observer, it transforms according to the same rule; that is, a stress rate $\overset{\bullet}{\boldsymbol{\sigma}}$ is objective only if it transforms as

$$\overset{\bullet}{\boldsymbol{\sigma}} \longrightarrow \boldsymbol{Q}\,\overset{\bullet}{\boldsymbol{\sigma}}\,\boldsymbol{Q}^T \tag{14.121}$$

for any change in observer. Note, for instance, that the material time derivative $\dot{\boldsymbol{\sigma}}$ of the Cauchy stress is *not* an objective stress rate. In this case, the transformation reads

$$\dot{\boldsymbol{\sigma}} \longrightarrow \boldsymbol{Q}\,\dot{\boldsymbol{\sigma}}\,\boldsymbol{Q}^T + \dot{\boldsymbol{Q}}\,\boldsymbol{\sigma}\,\boldsymbol{Q}^T + \boldsymbol{Q}\,\boldsymbol{\sigma}\,\dot{\boldsymbol{Q}}^T, \tag{14.122}$$

and satisfies (14.121) only for changes in observer with time-independent rotation ($\dot{\boldsymbol{Q}} = \boldsymbol{0}$).

In order to ensure material objectivity in the formulation of finite strain constitutive laws directly in terms of *stress rates*, it is essential that the constitutive equation for the stress tensor be defined in terms of objective stress rates. Objective stress rates are usually defined by suitably modifying the material time derivative of the stress tensor to ensure that (14.121) is satisfied. Their definition is somewhat arbitrary and many different objective rates of stress have been proposed in the formulation of hypoelastic-based constitutive theories. The definition of some of the most commonly used objective rates is reviewed in the following. We remark that proof of objectivity is provided here only for the Jaumann rate of Cauchy stress described below. The objectivity of the remaining stress rates can be proved in a completely analogous way. This will be left as an exercise for the interested reader.

The Jaumann rate

The *Jaumann rate* of Cauchy stress, denoted $\overset{\nabla}{\boldsymbol{\sigma}}$, is defined as

$$\overset{\nabla}{\boldsymbol{\sigma}} \equiv \dot{\boldsymbol{\sigma}} - \boldsymbol{W}\,\boldsymbol{\sigma} + \boldsymbol{\sigma}\,\boldsymbol{W}, \tag{14.123}$$

where $\boldsymbol{W}$ is the *spin* tensor

$$\boldsymbol{W} \equiv \text{skew}[\boldsymbol{L}], \quad \boldsymbol{L} \equiv \nabla_x \dot{\boldsymbol{v}} = \dot{\boldsymbol{F}}\,\boldsymbol{F}^{-1}. \tag{14.124}$$

This rate is also frequently referred to as the *Jaumann–Zaremba* rate.

To show that $\overset{\triangledown}{\boldsymbol{\sigma}}$ is indeed objective, let us recall (again, refer to page 70) that under a change in observer, the deformation gradient transforms as

$$\boldsymbol{F} \longrightarrow \boldsymbol{Q}\,\boldsymbol{F}.$$

By introducing the above into the definition of $\boldsymbol{L}$ and then into the definition of $\boldsymbol{W}$, we obtain the following transformation for the spin tensor

$$\boldsymbol{W} \longrightarrow \dot{\boldsymbol{Q}}\,\boldsymbol{Q}^T + \boldsymbol{Q}\,\boldsymbol{W}\,\boldsymbol{Q}^T. \tag{14.125}$$

With the substitution of the above transformation for $\boldsymbol{W}$, together with (14.120) and (14.122), into the definition of the Jaumann rate, we find, after some straightforward tensor algebra, that the Jaumann rate transforms objectively

$$\overset{\triangledown}{\boldsymbol{\sigma}} \longrightarrow \boldsymbol{Q}\,\dot{\boldsymbol{\sigma}}\,\boldsymbol{Q}^T - \boldsymbol{Q}\,\boldsymbol{W}\,\boldsymbol{\sigma}\,\boldsymbol{Q}^T + \boldsymbol{Q}\,\boldsymbol{\sigma}\,\boldsymbol{W}\,\boldsymbol{Q}^T = \boldsymbol{Q}\,\overset{\triangledown}{\boldsymbol{\sigma}}\,\boldsymbol{Q}^T. \tag{14.126}$$

Generalising the above definition, let $\boldsymbol{T}$ be a generic spatial tensor; that is, a tensor that transforms according to (14.120) under a change in observer. The Jaumann rate of $\boldsymbol{T}$ is defined as

$$\overset{\triangledown}{\boldsymbol{T}} \equiv \dot{\boldsymbol{T}} - \boldsymbol{W}\,\boldsymbol{T} + \boldsymbol{T}\boldsymbol{W}. \tag{14.127}$$

In particular, the Jaumann rate of *Kirchhoff* stress

$$\overset{\triangledown}{\boldsymbol{\tau}} \equiv \dot{\boldsymbol{\tau}} - \boldsymbol{W}\,\boldsymbol{\tau} + \boldsymbol{\tau}\,\boldsymbol{W} \tag{14.128}$$

is often employed in the formulation of hypoelastic-based finite strain plasticity models.

The Truesdell rate

The *Truesdell* rate of $\boldsymbol{T}$ is defined as

$$\overset{\circ}{\boldsymbol{T}} \equiv \dot{\boldsymbol{T}} - \boldsymbol{L}\,\boldsymbol{T} - \boldsymbol{T}\,\boldsymbol{L}^T + (\mathrm{tr}\boldsymbol{L})\,\boldsymbol{T}. \tag{14.129}$$

The Oldroyd rate

The *Oldroyd* rate is associated with the concept of Lie derivative, already exploited in the description of the hyperelastic-based isotropic plasticity model addressed earlier in this chapter. The Oldroyd rate is defined as

$$\overset{\diamond}{\boldsymbol{T}} \equiv \dot{\boldsymbol{T}} - \boldsymbol{L}\,\boldsymbol{T} - \boldsymbol{T}\,\boldsymbol{L}^T, \tag{14.130}$$

or, equivalently,

$$\overset{\diamond}{\boldsymbol{T}} \equiv \boldsymbol{F}\left[\frac{\mathrm{d}}{\mathrm{d}t}(\boldsymbol{F}^{-1}\boldsymbol{T}\boldsymbol{F}^{-T})\right]\boldsymbol{F}^T. \tag{14.131}$$

Note in the above expression that $\boldsymbol{T}$ is pulled back to the reference configuration, the time derivative of the reference quantity is taken and, subsequently, the resulting time derivative is pushed forward to the spatial configuration. The equivalence between (14.130) and (14.131) can be established by expanding the right-hand side of (14.131) and making use of the definition of $\boldsymbol{L}$, together with the identity $(\dot{\boldsymbol{F}})^{-1} = -\boldsymbol{F}^{-1}\dot{\boldsymbol{F}}\,\boldsymbol{F}^{-1}$, which can be obtained with the help of the differential relation (vii), page 36, for the tensor inverse.

The convected rate

The *convected* rate is defined by

$$\overset{\triangle}{T} \equiv F^{-T}\left[\frac{\mathrm{d}}{\mathrm{d}t}(F^{T} T F)\right] F^{-1} = \dot{T} + L^{T}\, T + T L. \tag{14.132}$$

The Green–Naghdi rate

The *Green–Naghdi* rate of the generic spatial tensor T, here denoted $\overset{\square}{T}$, is obtained by rotating T back to the reference configuration, taking the time material derivative of the rotated quantity and then rotating the resulting derivative forward to the deformed configuration; that is,

$$\overset{\square}{T} \equiv R\left[\frac{\mathrm{d}}{\mathrm{d}t}(R^{T}\, T R)\right] R^{T} = \dot{T} - \Omega\, T + T\, \Omega, \tag{14.133}$$

where the skew-symmetric tensor

$$\Omega \equiv \dot{R}\, R^{T}, \tag{14.134}$$

is the spin of the Eulerian triad relative to the Lagrangian triad. The use of the corresponding Green–Naghdi rate of Kirchhoff stress

$$\overset{\square}{\tau} = \dot{\tau} - \Omega\, \tau + \tau\, \Omega, \tag{14.135}$$

in the formulation of hypoelastic-based finite plasticity models will be discussed later in this chapter.

14.10.2. HYPOELASTIC-BASED PLASTICITY MODELS

As the starting point in the formulation of hypoelastic-based plasticity models, it is postulated that the evolution of the stress tensor is governed by a constitutive law of the form

$$\overset{\bullet}{\Sigma} = \mathsf{D}^{e} : (D - D^{p}), \tag{14.136}$$

where $\overset{\bullet}{\Sigma}$ denotes some objective rate of some stress measure Σ, D^{e} is some suitably defined tangential elasticity operator, D is the usual stretching tensor and D^{p} is some measure of rate of plastic deformation (not necessarily related to that defined by $(14.26)_1$ which arises naturally in the multiplicative theory). The above rate equation is complemented by a constitutive law for D^{p} (a *plastic flow rule*) usually stated as

$$D^{p} = \dot{\gamma}\, \frac{\partial \Psi}{\partial \Sigma},$$

where $\Psi(\Sigma, \alpha)$ is a flow potential, with α denoting a set of internal variables whose evolution is governed by

$$\dot{\alpha} = \dot{\gamma}\, H(\Sigma, \alpha),$$

where H is a given evolution function and the plastic multiplier, $\dot{\gamma}$, defines the loading/unloading criterion through the usual complementarity condition

$$\Phi \leq 0, \quad \dot{\gamma} \geq 0, \quad \dot{\gamma}\Phi = 0, \tag{14.137}$$

with $\Phi(\Sigma, \alpha)$ denoting a yield function.

14.10.3. THE JAUMANN RATE-BASED MODEL

Of particular relevance is the model defined in terms of the Jaumann rate of Kirchhoff stress. This model has been widely used to extend conventional isotropic infinitesimal plasticity models to the finite strain range. Its basic rate evolution law for stress is the following

$$\overset{\nabla}{\boldsymbol{\tau}} = \mathsf{D}^e : (\boldsymbol{D} - \boldsymbol{D}^p), \tag{14.138}$$

where D^e is taken simply as the standard *infinitesimal* elasticity tensor

$$\mathsf{D}^e \equiv 2G\,\mathsf{I}_S + (K - \tfrac{2}{3}G)\boldsymbol{I} \otimes \boldsymbol{I},$$

with constant coefficients G and K.

The 'elastic' range

In the 'elastic' range (where $\boldsymbol{D}^p = \mathbf{0}$) the stress rate equation reads simply

$$\overset{\nabla}{\boldsymbol{\tau}} = \mathsf{D}^e : \boldsymbol{D}. \tag{14.139}$$

Plasticity equations

The model is completed with the definition of a plastic flow law

$$\boldsymbol{D}^p = \dot{\gamma}\,\frac{\partial \Psi}{\partial \boldsymbol{\tau}}, \tag{14.140}$$

together with the general evolution equation for the internal variables

$$\dot{\boldsymbol{\alpha}} = \dot{\gamma}\,\boldsymbol{H}(\boldsymbol{\tau}, \boldsymbol{\alpha}), \tag{14.141}$$

and the standard load/unload criterion (14.137). The extension of a given infinitesimal model to the finite strain range is obtained by adopting in the above equations Ψ, Φ and $\boldsymbol{H}$ with the same functional format as those of the corresponding infinitesimal model. Analogously to the rate form of infinitesimal plasticity, the stress rate equation (14.138) is reduced under plastic flow to the following form

$$\overset{\nabla}{\boldsymbol{\tau}} = \mathsf{D}^{ep} : \boldsymbol{D}, \tag{14.142}$$

where the elastoplastic tangent operator D^{ep} is obtained from (14.138), (14.140) and (14.141) together with the consistency condition

$$\dot{\Phi} = 0.$$

In fact, the operator D^{ep} above has exactly the same format as the elastoplastic (continuum) tangent operator derived in Section 6.3.8 (page 153) for the generic infinitesimal plasticity model. This is demonstrated in the following.

The elastoplastic (continuum) tangent operator

The derivation of the continuum operator D^{ep} shown here is completely analogous to that shown in Section 6.3.8. We start by taking the time derivative of the yield function Φ which, in the present case, gives

$$\begin{aligned}\dot{\Phi} &= \frac{\partial \Phi}{\partial \boldsymbol{\tau}} : \dot{\boldsymbol{\tau}} + \frac{\partial \Phi}{\partial \boldsymbol{\alpha}} * \dot{\boldsymbol{\alpha}} \\ &= \frac{\partial \Phi}{\partial \boldsymbol{\tau}} : (\overset{\nabla}{\boldsymbol{\tau}} + \boldsymbol{W}\,\boldsymbol{\tau} - \boldsymbol{\tau}\,\boldsymbol{W}) + \frac{\partial \Phi}{\partial \boldsymbol{\alpha}} * \dot{\boldsymbol{\alpha}} \\ &= \frac{\partial \Phi}{\partial \boldsymbol{\tau}} : \left[\mathsf{D}^e : \left(\boldsymbol{D} - \dot{\gamma}\frac{\partial \Psi}{\partial \boldsymbol{\tau}}\right) + \boldsymbol{W}\,\boldsymbol{\tau} - \boldsymbol{\tau}\,\boldsymbol{W}\right] + \dot{\gamma}\,\frac{\partial \Phi}{\partial \boldsymbol{\alpha}} * \boldsymbol{H}(\boldsymbol{\tau}, \boldsymbol{\alpha}).\end{aligned} \tag{14.143}$$

The above expression can be simplified initially by observing that, in view of the symmetry of $\boldsymbol{\tau}$ and $\partial\Phi/\partial\boldsymbol{\tau}$ and skew-symmetry of $\boldsymbol{W}$, we have

$$\begin{aligned}\frac{\partial \Phi}{\partial \boldsymbol{\tau}} : (\boldsymbol{W}\,\boldsymbol{\tau} - \boldsymbol{\tau}\,\boldsymbol{W}) &= \frac{\partial \Phi}{\partial \boldsymbol{\tau}} : [(\boldsymbol{W}\,\boldsymbol{\tau})^T - \boldsymbol{\tau}\,\boldsymbol{W}] \\ &= -2\,\frac{\partial \Phi}{\partial \boldsymbol{\tau}} : \boldsymbol{\tau}\,\boldsymbol{W} \\ &= -2\,\boldsymbol{\tau}\frac{\partial \Phi}{\partial \boldsymbol{\tau}} : \boldsymbol{W}.\end{aligned}$$

In addition, due to the isotropy of the model, $\boldsymbol{\tau}$ and $\partial\Phi/\partial\boldsymbol{\tau}$ commute and their product remains symmetric, leading to the identity

$$\frac{\partial \Phi}{\partial \boldsymbol{\tau}} : (\boldsymbol{W}\,\boldsymbol{\tau} - \boldsymbol{\tau}\,\boldsymbol{W}) = 0, \tag{14.144}$$

which substituted into (14.143) results in the simpler formula for $\dot{\Phi}$

$$\dot{\Phi} = \frac{\partial \Phi}{\partial \boldsymbol{\tau}} : \mathsf{D}^e : \left(\boldsymbol{D} - \dot{\gamma}\frac{\partial \Psi}{\partial \boldsymbol{\tau}}\right) + \dot{\gamma}\,\frac{\partial \Phi}{\partial \boldsymbol{\alpha}} * \boldsymbol{H}(\boldsymbol{\tau}, \boldsymbol{\alpha}).$$

The above expression has identical format to that of the infinitesimal deformation theory. Finally, by equating the above formula to zero (the consistency condition), we obtain the general expression for the plastic multiplier

$$\dot{\gamma} = \frac{\partial\Phi/\partial\boldsymbol{\tau} : \mathsf{D}^e : \boldsymbol{D}}{\partial\Phi/\partial\boldsymbol{\tau} : \mathsf{D}^e : \partial\Psi/\partial\boldsymbol{\tau} - \partial\Phi/\partial\boldsymbol{\alpha} * \boldsymbol{H}}.$$

Substitution of this expression into (14.140) and the resulting formula into (14.138) followed by comparison with (14.142) leads to

$$\mathsf{D}^{ep} = \mathsf{D}^e - \frac{(\mathsf{D}^e : \partial\Psi/\partial\boldsymbol{\tau}) \otimes (\mathsf{D}^e : \partial\Phi/\partial\boldsymbol{\tau})}{\partial\Phi/\partial\boldsymbol{\tau} : \mathsf{D}^e : \partial\Psi/\partial\boldsymbol{\tau} - \partial\Phi/\partial\boldsymbol{\alpha} * \boldsymbol{H}}, \tag{14.145}$$

which has the format of the small strain counterpart (6.67), shown on page 153.

Remark 14.8. Note that the small strain format of D^{ep} in (14.145) is a direct consequence of the identity (14.144). It will hold whenever the difference between the material time derivative of the stress and the corresponding objective rate comprises only terms with products between the stress and a *skew-symmetric* tensor. It holds, for instance, for the *Green–Naghdi* rate-based model defined by

$$\overset{\square}{\boldsymbol{\tau}} = \mathsf{D}^e : (\boldsymbol{D} - \boldsymbol{D}^p),$$

with $\overset{\square}{\boldsymbol{\tau}}$ given in (14.135). For similar models defined in terms of other objective stress rates, such as the Truesdell or the convected rate, different elastoplastic operators have to be derived.

14.10.4. HYPERELASTIC-BASED MODELS AND EQUIVALENT RATE FORMS

The choice of 'appropriate' objective stress rates in the definition of finite plasticity models analogous to the one described above has, for a considerable time, been the subject of intense debate (Atluri, 1984; Dienes, 1979; Johnson and Bammann, 1984; Moss, 1984; Nemat-Nasser, 1982; Sowerby and Chu, 1984). It has been observed, for instance, that spurious stress oscillations occur under pure large elastic shearing deformations with the Jaumann rate-based model. This problem can be eliminated by replacing the Jaumann rate with the Green–Naghdi or Truedell stress rates in the original model. However, no matter what objective rate is chosen, fundamental drawbacks will be present in models defined on a purely ad hoc basis simply by choosing one particular objective stress rate and postulating an evolution law of the type (14.136) with constant isotropic elasticity tensor D^e. It has been shown by Simo and Pister (1984) that models postulated as such fail to define an elastic material (even in the absence of plastic flow); that is, dissipation of energy will occur within closed (supposedly elastic) cycles. It should be emphasised, however, that such inconsistencies remain negligible in metal plasticity (where elastic strains are small) but pose serious limitations in the use of hypoelastic-based theories in situations where elastic deformations become considerable. Equivalent rate descriptions of hyperelastic models defined in terms of different strain measures have been thoroughly discussed by Perić (1992). In particular, it has been shown that the Jaumann rate- and the Green–Naghdi rate-based models above (with constant and isotropic elasticity tensor) provide different levels of approximation to problems governed by the logarithmic strain-based Hencky hyperelastic law, justifying the use of hypoelastic-based theories in metal plasticity.

Remark 14.9. The issue of choice of 'proper' rates in hypoelastic formulations has often been treated in the computational literature as a matter of numerical convenience alone. Rather than being just a numerical issue, the reader should bear in mind that different stress rates (with a constant isotropic elasticity tensor) define in fact *different material models* with *different responses*. To emphasise this point, we plot in Figure 14.19 the stress response under monotonic finite shear deformation for the finite (hypo-) elasticity models defined by the Jaumann, Green–Naghdi and Truesdell rates as well as for the Hencky (logarithmic strain-based) hyperelastic model used in the formulation of multiplicative plasticity in Section 14.3. The results shown are the elastic analytical solutions to the monotonic finite shearing problem. These solutions are given by Moss (1984) for the Jaumann, Green–Naghdi and Truesdell rate models and by Perić *et al.* (1992) for the Hencky model. The response also differs in the elastoplastic case, becoming apparent in the presence of moderate to large elastic strains (Moss, 1984; Perić *et al.*, 1992).

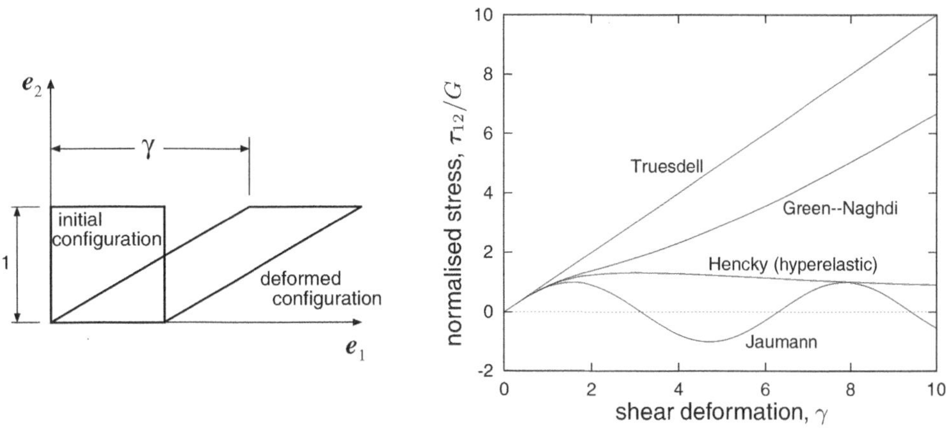

Figure 14.19. Elastic response for different models.

14.10.5. INTEGRATION ALGORITHMS AND INCREMENTAL OBJECTIVITY

An essential step to incorporate a hypoelastic-based plasticity model within a finite element framework is the definition of an integration algorithm to transform the original time-continuum constitutive equations into an incremental law. As we shall see in this section, in contrast to the case of hyperelastic-based theories, it is by no means easy to formulate integration algorithms for rate-based finite plasticity models that preserve the crucial property of material objectivity. The objectivity of incremental constitutive laws is usually referred to as *incremental objectivity*. If incremental objectivity is not satisfied, then spurious stress changes may be produced within increments with rigid incremental displacement fields. Such a feature is clearly undesirable and will, unfortunately, be present if not carefully considered in the derivation of incremental constitutive laws for hypoelastic-based models. This is illustrated in the following example.

Example: a non-objective algorithm for the Jaumann rate model

For simplicity let us start by considering the Jaumann rate formulation in the 'elastic' regime only. Our basic objective is to devise a numerical algorithm to integrate the stress rate equation (14.139). Given the Kirchhoff stress $\boldsymbol{\tau}_n$ at the beginning of the standard interval $[t_n, t_{n+1}]$ and given the incremental displacement field $\Delta \boldsymbol{u}$, we want to find an updated stress $\boldsymbol{\tau}_{n+1}$ consistently with (14.139). In view of definition (14.128) of the Jaumann rate, (14.139) is equivalent to

$$\dot{\boldsymbol{\tau}} = \mathsf{D}^e : \boldsymbol{D} + \boldsymbol{W}\boldsymbol{\tau} - \boldsymbol{\tau}\boldsymbol{W}. \tag{14.146}$$

A trivial choice to integrate numerically the above equation consists in making use of a midpoint formula as follows. Firstly, we approximate the *velocity* field at the mid-point

$$t_{n+\frac{1}{2}} = t_n + \tfrac{1}{2}\,\Delta t$$

as

$$\boldsymbol{v}_{n+\frac{1}{2}} = \frac{\Delta \boldsymbol{u}}{\Delta t}. \tag{14.147}$$

The velocity gradient is then computed as

$$\boldsymbol{L}_{n+\frac{1}{2}} = \nabla_{n+\frac{1}{2}} \boldsymbol{v}_{n+\frac{1}{2}}, \tag{14.148}$$

where $\nabla_{n+\frac{1}{2}}$ denotes the spatial gradient taken at the midpoint configuration defined by the displacement field

$$\boldsymbol{u}_{n+\frac{1}{2}} = \boldsymbol{u}_n + \tfrac{1}{2}\,\Delta \boldsymbol{u}. \tag{14.149}$$

With the above at hand, the rate of deformation and spin tensors at $t_{n+\frac{1}{2}}$ are computed as

$$\boldsymbol{D}_{n+\frac{1}{2}} = \mathrm{sym}[\boldsymbol{L}_{n+\frac{1}{2}}]; \quad \boldsymbol{W}_{n+\frac{1}{2}} = \mathrm{skew}[\boldsymbol{L}_{n+\frac{1}{2}}]. \tag{14.150}$$

Finally, the Kirchhoff stress tensor can be updated according to the formula

$$\boldsymbol{\tau}_{n+1} = \boldsymbol{\tau}_n + \Delta t\, (\mathsf{D}^e : \boldsymbol{D}_{n+\frac{1}{2}} + \boldsymbol{W}_{n+\frac{1}{2}}\, \boldsymbol{\tau}_n - \boldsymbol{\tau}_n\, \boldsymbol{W}_{n+\frac{1}{2}}), \tag{14.151}$$

or, equivalently,

$$\begin{aligned}\boldsymbol{\tau}_{n+1} = \boldsymbol{\tau}_n &+ \mathsf{D}^e : \mathrm{sym}[\nabla_{n+\frac{1}{2}} \Delta \boldsymbol{u}] \\ &+ \mathrm{skew}[\nabla_{n+\frac{1}{2}} \Delta \boldsymbol{u}]\, \boldsymbol{\tau}_n - \boldsymbol{\tau}_n\, \mathrm{skew}[\nabla_{n+\frac{1}{2}} \Delta \boldsymbol{u}].\end{aligned} \tag{14.152}$$

Lack of incremental objectivity.

To show that the above incremental constitutive law for $\boldsymbol{\tau}$ is not objective, let us assume that $\Delta \boldsymbol{u}$ is a finite *rigid rotation*. Accordingly, the *incremental deformation gradient*, $\boldsymbol{F}_\Delta$, is a rotation (here denoted $\boldsymbol{Q}_\Delta$):

$$\boldsymbol{F}_\Delta \equiv \boldsymbol{I} + \nabla_n (\Delta \boldsymbol{u}) = \boldsymbol{Q}_\Delta, \tag{14.153}$$

or, equivalently,

$$\boldsymbol{F}_\Delta = [\boldsymbol{I} + \tfrac{1}{2} \nabla_{n+\frac{1}{2}} (\Delta \boldsymbol{u})]\, \boldsymbol{F}_{\frac{1}{2}\Delta} = \boldsymbol{Q}_\Delta, \tag{14.154}$$

with

$$\boldsymbol{F}_{\frac{1}{2}\Delta} \equiv \boldsymbol{I} + \tfrac{1}{2} \nabla_n (\Delta \boldsymbol{u}). \tag{14.155}$$

From the above, we can easily obtain

$$\nabla_{n+\frac{1}{2}} (\Delta \boldsymbol{u}) = 2(\boldsymbol{Q}_\Delta \boldsymbol{F}_{\frac{1}{2}\Delta}^{-1} - \boldsymbol{I}). \tag{14.156}$$

The principle of material objectivity requires that the following identity be satisfied

$$\boldsymbol{\tau}_{n+1} = \boldsymbol{Q}_\Delta\, \boldsymbol{\tau}_n\, \boldsymbol{Q}_\Delta^T. \tag{14.157}$$

However, by replacing (14.156) into (14.152) we find that the present algorithm updates $\boldsymbol{\tau}$ as

$$\begin{aligned}\boldsymbol{\tau}_{n+1} = \boldsymbol{\tau}_n &+ 2(\mathsf{D}^e : \mathrm{sym}[\boldsymbol{Q}_\Delta \boldsymbol{F}_{\frac{1}{2}\Delta}^{-1} - \boldsymbol{I}] \\ &+ \mathrm{skew}[\boldsymbol{Q}_\Delta \boldsymbol{F}_{\frac{1}{2}\Delta}^{-1} - \boldsymbol{I}]\, \boldsymbol{\tau}_n - \boldsymbol{\tau}_n\, \mathrm{skew}[\boldsymbol{Q}_\Delta \boldsymbol{F}_{\frac{1}{2}\Delta}^{-1} - \boldsymbol{I}]),\end{aligned} \tag{14.158}$$

which, in general, does not satisfy (14.157). To see this, note that, in particular, if $\boldsymbol{\tau}_n = \boldsymbol{0}$ then under an incremental pure rotation ($\boldsymbol{Q}_\Delta \neq \boldsymbol{I}$) we should have $\boldsymbol{\tau}_{n+1} = \boldsymbol{0}$. The algorithm instead gives

$$\boldsymbol{\tau}_{n+1} = 2\,\mathsf{D}^e : \text{sym}[\boldsymbol{Q}_\Delta \boldsymbol{F}_{\frac{1}{2}\Delta}^{-1} - \boldsymbol{I}] \neq \boldsymbol{0}. \tag{14.159}$$

It is worth remarking that, in spite of its lack of objectivity, the use of the update formula (14.152) may be justified if displacement increments are sufficiently small. This is the case, for instance, with explicit transient dynamic finite element schemes, where global stability criteria impose stringent restrictions on increment sizes. The use of (14.152) within implicit schemes, where steps can be large, will, however, produce unacceptable results.

The elastoplastic case.

Under plastic flow, the elastoplastic tangent operator D^{ep} replaces the elasticity tensor in (14.152) and the stress update formula reads

$$\boldsymbol{\tau}_{n+1} = \boldsymbol{\tau}_n + \mathsf{D}^{ep} : \text{sym}[\nabla_{n+\frac{1}{2}} \Delta\boldsymbol{u}] + \text{skew}[\nabla_{n+\frac{1}{2}} \Delta\boldsymbol{u}]\,\boldsymbol{\tau}_n - \boldsymbol{\tau}_n\,\text{skew}[\nabla_{n+\frac{1}{2}} \Delta\boldsymbol{u}]. \tag{14.160}$$

Alternatively, a *return-mapping-type* algorithm can be used as follows. Firstly, we assume that the increment is elastic and compute the elastic trial stress as

$$\boldsymbol{\tau}^{\text{trial}}_{n+1} = \boldsymbol{\tau}_n + \mathsf{D}^{e} : \text{sym}[\nabla_{n+\frac{1}{2}} \Delta\boldsymbol{u}] + \text{skew}[\nabla_{n+\frac{1}{2}} \Delta\boldsymbol{u}]\,\boldsymbol{\tau}_n - \boldsymbol{\tau}_n\,\text{skew}[\nabla_{n+\frac{1}{2}} \Delta\boldsymbol{u}]. \tag{14.161}$$

If $\Phi(\boldsymbol{\tau}^{\text{trial}}_{n+1}, \boldsymbol{\alpha}_n) \leq 0$, then we accept $\boldsymbol{\tau}_{n+1} := \boldsymbol{\tau}^{\text{trial}}_{n+1}$. Otherwise, we obtain $\boldsymbol{\tau}_{n+1}$ by applying a return-mapping algorithm of the type discussed in Section 14.4.

Incrementally objective algorithms for rate-based models

The issue of incremental objectivity of integration algorithms for hypoelastic-based constitutive models has been thoroughly investigated by many researchers (Flanagan and Taylor, 1987; Hughes and Winget, 1980; Pinsky *et al.*, 1983; Rubinstein and Atluri, 1983; Simo and Hughes, 1998). There are many possible alternatives in the formulation of incrementally objective algorithms. The underlying idea, however, is the same and comprises the following steps:

(i) Firstly, the original rate constitutive equation is mapped into a *rigid motion-insensitive* (or *rotation-neutralised*) local configuration.

(ii) Time discretisation is then performed over the resulting equation which involves only rigid motion-insensitive quantities.

(iii) Finally, the discretised equation is mapped back to the spatial configuration.

The reader is referred to Chapter 8 of Simo and Hughes (1998) for a detailed description of the general methodology. Here we shall limit our discussion to a midpoint rule-based algorithm that is closely related to the popular *Hughes–Winget* algorithm originally proposed by Hughes and Winget (1980).

14.10.6. OBJECTIVE ALGORITHM FOR JAUMANN RATE-BASED MODELS

Readers who wish to skip the details of derivation presented below are referred directly to Box 14.5 (page 631) where the incrementally objective integration algorithm for the Jaumann rate-based model is summarised in pseudo-code format. At the outset, we define the *rotated Kirchhoff stress tensor* and the *rotated stretching tensor*

$$\bar{\boldsymbol{\tau}} \equiv \boldsymbol{\Lambda}^T \, \boldsymbol{\tau} \, \boldsymbol{\Lambda}, \qquad \bar{\boldsymbol{D}} \equiv \boldsymbol{\Lambda}^T \, \boldsymbol{D} \, \boldsymbol{\Lambda}, \tag{14.162}$$

where $\boldsymbol{\Lambda}$ is the rotation tensor that solves the initial value problem

$$\begin{cases} \dot{\boldsymbol{\Lambda}} = \boldsymbol{W} \, \boldsymbol{\Lambda} \\ \boldsymbol{\Lambda}|_{t=0} = \boldsymbol{I}. \end{cases} \tag{14.163}$$

From the above definitions, it follows after straightforward tensor algebra that the material time derivative of the rotated Kirchhoff stress is given by

$$\dot{\bar{\boldsymbol{\tau}}} \equiv \boldsymbol{\Lambda}^T \, \overset{\nabla}{\boldsymbol{\tau}} \, \boldsymbol{\Lambda}. \tag{14.164}$$

Again, considering the 'elastic' regime, the first step (item (i) above) in the formulation of the present algorithm is to rotate both sides of the original Jaumann rate equation (14.139) with the rotation $\boldsymbol{\Lambda}^T$. This results in the following equivalent rate form defined in the local rotated configuration

$$\dot{\bar{\boldsymbol{\tau}}} = \mathsf{D}^e : \bar{\boldsymbol{D}}. \tag{14.165}$$

In rotating the right-hand side of (14.139) we have made use of the fact that D^e is isotropic. The above rate form of constitutive equation for the stress is *rigid motion-insensitive*, in the sense that it involves only quantities which are not affected by superimposed rigid-body motions. This is proved in the following.

Rigid motion-insensitivity of the rotated rate form

Let us consider a superimposed rigid-body motion under which the deformation gradient transforms according to the standard relation

$$\boldsymbol{F} \longrightarrow \boldsymbol{F}^* = \boldsymbol{Q} \, \boldsymbol{F}.$$

Under the same superimposed motion, it can be easily established that the stretching tensor transforms as

$$\boldsymbol{D} \longrightarrow \boldsymbol{D}^* = \boldsymbol{Q} \, \boldsymbol{D} \, \boldsymbol{Q}^T.$$

To see this, we firstly note that the spin tensor follows the transformation rule (14.125). This expression for the spin together with (14.163) yields

$$\dot{\boldsymbol{\Lambda}} \, \boldsymbol{\Lambda}^{-1} \longrightarrow \dot{\boldsymbol{\Lambda}}^* \, (\boldsymbol{\Lambda}^*)^{-1} = \boldsymbol{Q} \, \dot{\boldsymbol{\Lambda}} \, \boldsymbol{\Lambda}^{-1} \, \boldsymbol{Q}^T + \dot{\boldsymbol{Q}} \, \boldsymbol{Q}^T. \tag{14.166}$$

The expression for the transformed rotation $\boldsymbol{\Lambda}^*$ is obtained simply by integrating the differential equation on the right-hand side of (14.166). This gives the following transformation rule for $\boldsymbol{\Lambda}$

$$\boldsymbol{\Lambda} \longrightarrow \boldsymbol{\Lambda}^* = \boldsymbol{Q} \, \boldsymbol{\Lambda}. \tag{14.167}$$

From the above and definition $(14.162)_2$ of the rotated stretching, we then find

$$\bar{\boldsymbol{D}} \longrightarrow \boldsymbol{\Lambda}^T \boldsymbol{Q}^T (\boldsymbol{Q}\, \boldsymbol{D}\, \boldsymbol{Q}^T) \boldsymbol{Q}\, \boldsymbol{\Lambda} = \boldsymbol{\Lambda}^T\, \boldsymbol{D}\, \boldsymbol{\Lambda} = \bar{\boldsymbol{D}},$$

so that the right-hand side of (14.165) is indeed rigid motion-insensitive.

To complete the proof, let us now concentrate on the time derivative of the rotated stress. By taking the time derivative of $(14.162)_1$ we obtain

$$\dot{\bar{\boldsymbol{\tau}}} = \dot{\boldsymbol{\Lambda}}^T\, \boldsymbol{\tau}\, \boldsymbol{\Lambda} + \boldsymbol{\Lambda}^T\, \dot{\boldsymbol{\tau}}\, \boldsymbol{\Lambda} + \boldsymbol{\Lambda}^T\, \boldsymbol{\tau}\, \dot{\boldsymbol{\Lambda}}^T.$$

This, together with (14.167) and the transformation rule for $\boldsymbol{\tau}$

$$\boldsymbol{\tau} \longrightarrow \boldsymbol{Q}\, \boldsymbol{\tau}\, \boldsymbol{Q}^T,$$

yields, after some straightforward algebra,

$$\begin{aligned}\dot{\bar{\boldsymbol{\tau}}} \longrightarrow\, & \boldsymbol{\Lambda}^T\, \dot{\boldsymbol{Q}}^T\, \boldsymbol{Q}\, \boldsymbol{\tau}\, \boldsymbol{\Lambda} + \dot{\boldsymbol{\Lambda}}^T\, \boldsymbol{\tau}\, \boldsymbol{\Lambda} \\ & - \boldsymbol{\Lambda}^T\, \dot{\boldsymbol{Q}}^T\, \boldsymbol{Q}\, \boldsymbol{\tau}\, \boldsymbol{\Lambda} + \boldsymbol{\Lambda}^T\, \dot{\boldsymbol{\tau}}\, \boldsymbol{\Lambda} \\ & + \boldsymbol{\Lambda}^T\, \boldsymbol{\tau} \dot{\boldsymbol{Q}}^T\, \boldsymbol{Q}\, \boldsymbol{\Lambda} - \boldsymbol{\Lambda}^T\, \boldsymbol{\tau} \dot{\boldsymbol{Q}}^T\, \boldsymbol{Q}\, \boldsymbol{\Lambda} + \boldsymbol{\Lambda}^T\, \boldsymbol{\tau}\, \dot{\boldsymbol{\Lambda}} = \dot{\bar{\boldsymbol{\tau}}}.\end{aligned}$$

This concludes the proof that all quantities taking part in (14.165) are rigid motion-insensitive.

Incrementally objective stress integration

We now proceed to discretise the rotation-insensitive rate equation (14.165). This corresponds to item (ii) listed on page 627. Again, we consider the standard interval $[t_n, t_{n+1}]$. We start by approximating the midpoint velocity gradient, rate of stretching and spin tensors according to (14.148) and (14.150). Then, by applying the midpoint rule to (14.165) the following update formula for the rotated stress $\bar{\boldsymbol{\tau}}$ is obtained

$$\bar{\boldsymbol{\tau}}_{n+1} = \bar{\boldsymbol{\tau}}_n + \Delta t\, \mathsf{D}^e : \bar{\boldsymbol{D}}_{n+\frac{1}{2}}, \tag{14.168}$$

where the midpoint rotated stretching tensor, $\bar{\boldsymbol{D}}_{n+\frac{1}{2}}$, is computed as

$$\bar{\boldsymbol{D}}_{n+\frac{1}{2}} = \boldsymbol{\Lambda}^T_{n+\frac{1}{2}}\, \boldsymbol{D}_{n+\frac{1}{2}}\, \boldsymbol{\Lambda}_{n+\frac{1}{2}}. \tag{14.169}$$

The midpoint rotation tensor, $\boldsymbol{\Lambda}_{n+\frac{1}{2}}$, is obtained through the numerical integration of the initial value problem (14.163), which we will discuss later.

Finally, with the updated rotated Kirchhoff stress computed by (14.168), the Kirchhoff stress tensor at t_{n+1} is then obtained simply by rotating $\bar{\boldsymbol{\tau}}_{n+1}$ back to the spatial configuration (item (iii) listed on page 627)

$$\boldsymbol{\tau}_{n+1} = \boldsymbol{\Lambda}_{n+1}\, \bar{\boldsymbol{\tau}}_{n+1}\, \boldsymbol{\Lambda}^T_{n+1}, \tag{14.170}$$

where the rotation $\boldsymbol{\Lambda}_{n+1}$ is obtained by solving problem (14.163) numerically. For implementation purposes it is convenient to recast (14.170) in the equivalent form

$$\begin{aligned}\boldsymbol{\tau}_{n+1} &= \boldsymbol{\Lambda}_\Delta\, \boldsymbol{\tau}_n\, \boldsymbol{\Lambda}^T_\Delta + \Delta t\, \mathsf{D}^e : \boldsymbol{\Lambda}_\delta\, \boldsymbol{D}_{n+\frac{1}{2}}\, \boldsymbol{\Lambda}^T_\delta \\ &= \boldsymbol{\Lambda}_\Delta\, \boldsymbol{\tau}_n\, \boldsymbol{\Lambda}^T_\Delta + \mathsf{D}^e : \boldsymbol{\Lambda}_\delta\, \mathrm{sym}[\nabla_{n+\frac{1}{2}} \Delta \boldsymbol{u}]\, \boldsymbol{\Lambda}^T_\delta,\end{aligned} \tag{14.171}$$

which is obtained by combining (14.168–14.170) together with the obvious identity

$$\bar{\boldsymbol{\tau}}_n = \boldsymbol{\Lambda}_n^T\, \boldsymbol{\tau}_n\, \boldsymbol{\Lambda}_n \tag{14.172}$$

and using the incremental rotation definitions

$$\boldsymbol{\Lambda}_\Delta \equiv \boldsymbol{\Lambda}_{n+1}\, \boldsymbol{\Lambda}_n^T, \quad \boldsymbol{\Lambda}_\delta \equiv \boldsymbol{\Lambda}_{n+1}\, \boldsymbol{\Lambda}_{n+\frac{1}{2}}^T. \tag{14.173}$$

Computation of the incremental rotation tensors.

As in the numerical integration of the plastic flow equation of the hyperelastic-based formulation (see expression (14.73), page 591), *exponential map integrators* (refer to Section B.3, from page 751) are particularly suitable for the numerical solution of the initial value problem defined by (14.163). By adopting the *midpoint* exponential map integrator, $\boldsymbol{\Lambda}_{n+1}$ is computed as

$$\boldsymbol{\Lambda}_{n+1} = \exp[\Delta t\, \boldsymbol{W}_{n+\frac{1}{2}}]\, \boldsymbol{\Lambda}_n. \tag{14.174}$$

Note that since the tensor exponential function maps skew-symmetric tensors onto rotations (refer to item (e) listed in Section B.1.1, page 748), the above formula guarantees that the updated value of $\boldsymbol{\Lambda}$ is always a rotation. Similarly to (14.174) the midpoint rotation at $t_n + \frac{1}{2}\Delta t$ can be computed by

$$\boldsymbol{\Lambda}_{n+\frac{1}{2}} = \exp\left[\frac{\Delta t}{2}\, \boldsymbol{W}_{n+\frac{1}{2}}\right] \boldsymbol{\Lambda}_n. \tag{14.175}$$

From $(14.173)_1$ and (14.174), it follows that

$$\boldsymbol{\Lambda}_\Delta = \exp[\Delta t\, \boldsymbol{W}_{n+\frac{1}{2}}], \tag{14.176}$$

or, for the purpose of computer implementation,

$$\boldsymbol{\Lambda}_\Delta = \exp[\mathrm{skew}[\nabla_{n+\frac{1}{2}} \Delta \boldsymbol{u}]]. \tag{14.177}$$

Finally, by combining (14.176), (14.175) and (14.173) and making use of property (B.11) of the exponential map (see page 748), we find

$$\boldsymbol{\Lambda}_\delta = \sqrt{\boldsymbol{\Lambda}_\Delta} = \exp\left[\frac{\Delta t}{2}\, \boldsymbol{W}_{n+\frac{1}{2}}\right], \tag{14.178}$$

which, for implementation, is better expressed as

$$\boldsymbol{\Lambda}_\delta = \exp[\tfrac{1}{2}\, \mathrm{skew}[\nabla_{n+\frac{1}{2}} \Delta \boldsymbol{u}]]. \tag{14.179}$$

Remark 14.10. In practical computations, the tensor exponential above can be evaluated according to expressions (B.13) or (B.14), noting that the use of (B.14) should be avoided near its singularities. An alternative way to compute $\boldsymbol{\Lambda}_\delta$ and $\boldsymbol{\Lambda}_\Delta$ is by using a *quaternion* parametrisation as suggested by Crisfield (1997) and Simo and Hughes (1998).

Box 14.5. Incrementally objective integration algorithm for a general Jaumann rate-based finite plasticity model.

(i) Given the incremental displacement, $\Delta \boldsymbol{u}$, compute

$$\boldsymbol{G}_{n+\frac{1}{2}} := \nabla_{n+\frac{1}{2}} \Delta \boldsymbol{u};$$

$$\boldsymbol{\Lambda}_\delta := \exp[\tfrac{1}{2}\, \mathrm{skew}[\boldsymbol{G}_{n+\frac{1}{2}}]]; \quad \boldsymbol{\Lambda}_\Delta := \boldsymbol{\Lambda}_\delta^2$$

(ii) Compute the elastic trial state

$$\boldsymbol{\tau}_{n+1}^{\mathrm{trial}} := \boldsymbol{\Lambda}_\Delta\, \boldsymbol{\tau}_n\, \boldsymbol{\Lambda}_\Delta^T + \mathsf{D}^e : \boldsymbol{\Lambda}_\delta\, \mathrm{sym}[\boldsymbol{G}_{n+\frac{1}{2}}]\, \boldsymbol{\Lambda}_\delta^T$$

$$\boldsymbol{\alpha}_{n+1}^{\mathrm{trial}} := \boldsymbol{\alpha}_n$$

(iii) Check for plastic admissibility

$$\text{IF } \Phi(\boldsymbol{\tau}_{n+1}^{\mathrm{trial}}, \boldsymbol{\alpha}_{n+1}^{\mathrm{trial}}) \leq 0$$

$$\text{THEN set } \quad (\cdot)_{n+1} := (\cdot)_{n+1}^{\mathrm{trial}} \quad \text{and EXIT}$$

(iv) Use small-strain format return-mapping algorithm to compute $\boldsymbol{\tau}_{n+1}$ and $\boldsymbol{\alpha}_{n+1}$

The elastoplastic case

Application of the above algorithm to the elastoplastic case is as follows. Firstly, the elastic update formulae (14.168)–(14.170) are used to compute the *elastic trial* Kirchhoff stress

$$\boldsymbol{\tau}_{n+1}^{\mathrm{trial}}.$$

Next, the plastic consistency check is carried out. If the elastic trial stress lies within the trial elastic domain, then $\boldsymbol{\tau}_{n+1} = \boldsymbol{\tau}_{n+1}^{\mathrm{trial}}$. Otherwise, a small-strain format return-mapping algorithm is used to compute $\boldsymbol{\tau}_{n+1}$ and the updated values of the internal variables of the model. The overall incrementally objective algorithm for the Jaumann rate-based finite plasticity model is summarised in Box 14.5.

The Hughes–Winget algorithm

The original Hughes–Winget algorithm differs from the above scheme in that the rotation that appears in the second term on the right-hand side of (14.171) is not present in the Hughes–Winget update formula and, in addition, a further approximation is introduced in the computation of $\boldsymbol{\Lambda}_\Delta$. The Hughes–Winget formula reads

$$\boldsymbol{\tau}_{n+1} = \tilde{\boldsymbol{\Lambda}}_\Delta\, \boldsymbol{\tau}_n\, \tilde{\boldsymbol{\Lambda}}_\Delta^T + \Delta t\, \mathsf{D}^e : \boldsymbol{D}_{n+\frac{1}{2}}, \tag{14.180}$$

where $\tilde{\boldsymbol{\Lambda}}_\Delta$ is the following approximation to $\boldsymbol{\Lambda}_\Delta$

$$\tilde{\boldsymbol{\Lambda}}_\Delta = [\boldsymbol{I} - \tfrac{1}{2}\, \Delta t\, \boldsymbol{W}_{n+\frac{1}{2}}]^{-1}\, [\boldsymbol{I} + \tfrac{1}{2}\, \Delta t\, \boldsymbol{W}_{n+\frac{1}{2}}]. \tag{14.181}$$

The above approximation is valid if $\Delta t\, \boldsymbol{W}_{n+\frac{1}{2}}$ is sufficiently small (sufficiently small incremental rotations). To see this, we appeal to the series representation of the tensor exponential (expression (B.4), page 747) and assume that second order and higher terms in $\Delta t\, \boldsymbol{W}_{n+\frac{1}{2}}$ can be neglected. This gives

$$\exp[\tfrac{1}{2}\Delta t\, \boldsymbol{W}_{n+\frac{1}{2}}] \approx [\boldsymbol{I} + \tfrac{1}{2}\, \Delta t\, \boldsymbol{W}_{n+\frac{1}{2}}]. \tag{14.182}$$

In view of property (B.9) of the tensor exponential, we have the alternative approximation of the same order

$$\exp[\tfrac{1}{2}\Delta t\, \boldsymbol{W}_{n+\frac{1}{2}}] = (\exp[-\tfrac{1}{2}\Delta t\, \boldsymbol{W}_{n+\frac{1}{2}}])^{-1} \approx [\boldsymbol{I} - \tfrac{1}{2}\, \Delta t\, \boldsymbol{W}_{n+\frac{1}{2}}]^{-1}. \tag{14.183}$$

The multiplication of these two expressions yields the Hughes–Winget approximation (14.181) to the exponential map integrator (14.176).

14.10.7. INTEGRATION OF GREEN–NAGHDI RATE-BASED MODELS

An algorithm very similar to that of Box 14.5 can be derived for Green–Naghdi rate-based plasticity models by following the methodology employed in the Jaumann rate case above. The basic difference is that for Green–Naghdi rate models the evolution problem (14.163) is rephrased as

$$\begin{cases} \dot{\boldsymbol{\Lambda}} = \boldsymbol{\Omega}\, \boldsymbol{\Lambda} \\ \boldsymbol{\Lambda}|_{t=0} = \boldsymbol{I}, \end{cases} \tag{14.184}$$

where the skew-symmetric tensor $\boldsymbol{\Omega}$ is defined as

$$\boldsymbol{\Omega} \equiv \dot{\boldsymbol{R}}\, \boldsymbol{R}^T,$$

with $\boldsymbol{R}$ being the rotation tensor resulting from the polar decomposition of the deformation gradient

$$\boldsymbol{F} = \boldsymbol{R}\boldsymbol{U} = \boldsymbol{V}\boldsymbol{R}.$$

Trivially, the exact solution to problem (14.184) is

$$\boldsymbol{\Lambda} = \boldsymbol{R}. \tag{14.185}$$

As a result, the incrementally objective algorithm for this model follows the same steps of Box 14.5 except that the incremental rotation tensors $\boldsymbol{\Lambda}_\Delta$ and $\boldsymbol{\Lambda}_\delta$ are redefined as

$$\begin{aligned} \boldsymbol{\Lambda}_\Delta &= \boldsymbol{R}_\Delta = \boldsymbol{R}_{n+1}\, \boldsymbol{R}_n^T \\ \boldsymbol{\Lambda}_\delta &= \boldsymbol{R}_\delta = \boldsymbol{R}_{n+1}\, \boldsymbol{R}_{n+\frac{1}{2}}^T. \end{aligned} \tag{14.186}$$

The rotations $\boldsymbol{R}_n$, $\boldsymbol{R}_{n+1}$ and $\boldsymbol{R}_{n+\frac{1}{2}}$ are obtained, respectively from the polar decomposition of $\boldsymbol{F}_n$, $\boldsymbol{F}_{n+1}$ and the deformation gradient at the midpoint configuration defined by the displacement field $\boldsymbol{u}_{n+\frac{1}{2}} = \boldsymbol{u}_n + \frac{1}{2}\, \Delta\boldsymbol{u}$

$$\begin{aligned} \boldsymbol{F}_{n+\frac{1}{2}} &= \boldsymbol{F}_n + \tfrac{1}{2}\, \nabla_0(\Delta\boldsymbol{u}) \\ &= \tfrac{1}{2}(\boldsymbol{F}_n + \boldsymbol{F}_{n+1}). \end{aligned} \tag{14.187}$$

Note that $\boldsymbol{R}_{\Delta}$ is the actual local rigid rotation associated with the incremental displacement $\Delta\boldsymbol{u}$, which can be equivalently obtained from the polar decomposition of the incremental deformation gradient

$$\boldsymbol{F}_{\Delta} = \boldsymbol{I} + \nabla_n(\Delta\boldsymbol{u}). \tag{14.188}$$

$\boldsymbol{R}_{\delta}$, in turn, is the rigid rotation associated with the deformation mapping between the midpoint configuration and the configuration at the end of the increment. In can be computed directly from the polar decomposition of

$$\boldsymbol{F}_{\delta} = \boldsymbol{I} + \tfrac{1}{2}\,\nabla_{n+\frac{1}{2}}(\Delta\boldsymbol{u}). \tag{14.189}$$

14.11. Finite plasticity with kinematic hardening

This section addresses the treatment of kinematic hardening within the framework of hyperelastic-based multiplicative finite strain plasticity of Sections 14.3 and 14.4. Readers who are not familiar with kinematic hardening are advised to review Sections 6.6 and 7.6 (starting, respectively, on pages 177 and 257) before proceeding. We remark that the introduction of kinematic hardening into the framework of multiplicative plasticity has been postponed until now simply because, as we shall see, evolution equations for the back-stress tensor can be conveniently formulated in the present context in terms of objective rates of back-stress. Objective rates, with the corresponding numerical integration issues, have just been explored in the previous section.

The first model of kinematic hardening behaviour embedded within a hyperelastic-based multiplicative description of elastoplasticity appears to have been proposed by Eterovic and Bathe (1990). The model was originally formulated in terms of a *rotated* back-stress tensor measure defined on the intermediate (plastic) configuration. This approach was later adopted by the authors (de Souza Neto and Perić, 1996; de Souza Neto *et al.*, 1994a) in the finite strain extension of the fully coupled Lemaitre ductile damage model of Section 12.3, which incorporates the Armstrong–Frederick kinematic hardening law. One important aspect of this formulation is the fact that, to within an error of second order in the elastic strains, the essential stress-updating procedure resulting from the exponential map-based integration of the plastic flow equation retains the small-strain format of the fully isotropic case discussed in Section 14.4. This allows a straightforward implementation of the model within the algorithmic framework of Section 14.4, but limits its applicability to situations where the elastic strains are small (under arbitrary plastic strains). This restriction is clearly immaterial in the solution of metal plasticity problems.

14.11.1. A MODEL OF FINITE STRAIN KINEMATIC HARDENING

We introduce here a generic model of finite strain kinematic hardening based on the formulation originally proposed by Eterovic and Bathe (1990). Before proceeding, however, we remark that to maintain consistency with Section 14.3, where the generic isotropic plasticity model was expressed in terms of *spatial* quantities, what we shall present in the following is based on an *equivalent spatial description* of the formulation of Eterovic and Bathe (originally devised, as briefly mentioned earlier, in terms of quantities defined upon the *intermediate* configuration).

At the outset, in order to describe kinematic hardening, we introduce a deviatoric symmetric back-stress tensor, $\boldsymbol{\beta}$, defined on the *spatial* configuration. Accordingly, a von Mises-type yield function generalising (6.208) (page 185) to the finite strain regime can be defined as

$$\Phi(\boldsymbol{\tau}, \boldsymbol{\beta}) \equiv \sqrt{3\, J_2(\boldsymbol{s}(\boldsymbol{\tau}) - \boldsymbol{\beta})} - \sigma_y, \tag{14.190}$$

where

$$\boldsymbol{s} \equiv \boldsymbol{\tau} - \tfrac{1}{3}\mathrm{tr}[\boldsymbol{\tau}]\, \boldsymbol{I} \tag{14.191}$$

is now the Kirchhoff stress deviator and σ_y is the usual isotropic hardening stress defining the size of the yield surface. Again, mixed hardening can be accounted for by having σ_y as a function of, for example, the accumulated plastic strain

$$\sigma_y = \sigma_y(\bar{\varepsilon}^p). \tag{14.192}$$

A class of von Mises-based finite strain kinematic hardening models can be defined within the multiplicative plasticity framework by adopting $\Psi \equiv \Phi$ in Box 14.2, where Φ denotes the above yield function, together with the standard Hencky hyperelastic law and an appropriate (frame-invariant) evolution equation for the back-stress tensor.

Green–Naghdi rate of back-stress tensor

Frame-invariant kinematic hardening laws can be postulated in terms of *objective rates* of the spatial back-stress tensor, $\boldsymbol{\beta}$, that takes part in definition (14.190). In particular, we shall describe here a formulation based on the *Green–Naghdi rate* (refer to expression (14.184)) of the back-stress tensor with respect to the *elastic* rotation, defined as

$$\overset{\square}{\boldsymbol{\beta}} \equiv \boldsymbol{R}^e \left[\frac{\mathrm{d}}{\mathrm{d}t}(\boldsymbol{R}^{e\,T}\, \boldsymbol{\beta}\, \boldsymbol{R}^e)\right] \boldsymbol{R}^{e\,T} = \dot{\boldsymbol{\beta}} - \boldsymbol{\Omega}^e\, \boldsymbol{\beta} + \boldsymbol{\beta}\, \boldsymbol{\Omega}^e \tag{14.193}$$

where the skew-symmetric tensor

$$\boldsymbol{\Omega}^e \equiv \dot{\boldsymbol{R}}^e\, \boldsymbol{R}^{e\,T}, \tag{14.194}$$

is the *elastic spin*.

Objectivity of the back-stress evolution law

The Green–Naghdi rate (14.193) of the back-stress tensor with respect to the *elastic* rotation (rather than to the *total* rotation of the original Green–Naghdi rate definition of expression (14.133)) is *objective*; that is, under any superimposed rigid-body motion (change in observer) associated with a time-dependent rotation $\boldsymbol{Q}$, it transforms according to

$$\overset{\square}{\boldsymbol{\beta}} \longrightarrow \boldsymbol{Q}\, \overset{\square}{\boldsymbol{\beta}} \boldsymbol{Q}^T. \tag{14.195}$$

To see this, we must first recall (refer to expression (3.145), page 70), that, under the superimposed rigid motions, the (total) deformation gradient transforms as

$$\boldsymbol{F} \longrightarrow \boldsymbol{Q}\boldsymbol{F}. \tag{14.196}$$

In the context of multiplicative plasticity kinematics, this transformation is equivalent to

$$\boldsymbol{F}^e \boldsymbol{F}^p \longrightarrow \boldsymbol{Q}\boldsymbol{F}^e \boldsymbol{F}^p, \tag{14.197}$$

so that $\boldsymbol{F}^e$ and $\boldsymbol{F}^p$ transform individually according to the rules

$$\boldsymbol{F}^e \longrightarrow \boldsymbol{Q}\boldsymbol{F}^e; \quad \boldsymbol{F}^p \longrightarrow \boldsymbol{F}^p. \tag{14.198}$$

Expression $(14.198)_1$ implies that the elastic rotation obtained from the polar decomposition $\boldsymbol{F}^e = \boldsymbol{R}^e \boldsymbol{U}^e$ transforms as

$$\boldsymbol{R}^e \longrightarrow \boldsymbol{Q}\boldsymbol{R}^e. \tag{14.199}$$

The proof that (14.195) holds follows analogous steps to those leading to (14.126); that is, we simply introduce (14.199) and the obvious transformation rule for the spatial tensor $\boldsymbol{\beta}$,

$$\boldsymbol{\beta} \longrightarrow \boldsymbol{Q}\boldsymbol{\beta}\boldsymbol{Q}^T, \tag{14.200}$$

together with their corresponding material time derivatives, into the right-hand side of (14.193).

Finite strain extensions of infinitesimal evolution laws

With the above definition at hand, objective finite strain extensions of the infinitesimal models discussed in Section 6.6.4 are obtained as particular cases of the equation

$$\overset{\square}{\boldsymbol{\beta}} = \dot{\gamma}\, \hat{\boldsymbol{\beta}}(\boldsymbol{\tau}, \boldsymbol{\beta}, \bar{\varepsilon}^p), \tag{14.201}$$

where $\hat{\boldsymbol{\beta}}$ is a given constitutive function of $\boldsymbol{\tau}$, $\boldsymbol{\beta}$ and the accumulated plastic strain, $\bar{\varepsilon}^p$, having the same functional format as in the back-stress evolution equation of the corresponding infinitesimal model. In this context, the finite strain extension of Prager's linear kinematic hardening rule (6.212) is defined as

$$\overset{\square}{\boldsymbol{\beta}} = \dot{\gamma}\, \frac{2}{3}\, H\, \frac{\partial \Phi}{\partial \boldsymbol{\tau}}, \tag{14.202}$$

with nonlinear hardening extension

$$\overset{\square}{\boldsymbol{\beta}} = \dot{\gamma}\, \frac{2}{3}\, H(\bar{\varepsilon}^p)\, \frac{\partial \Phi}{\partial \boldsymbol{\tau}}. \tag{14.203}$$

For the Armstrong–Frederick model (6.223), we have

$$\overset{\square}{\boldsymbol{\beta}} = \dot{\gamma}\left(\frac{2}{3}\, H\, \frac{\partial \Phi}{\partial \boldsymbol{\tau}} - b\, \boldsymbol{\beta}\right). \tag{14.204}$$

For completeness, we summarise in Box 14.6 a finite strain von Mises-type multiplicative plasticity model with mixed hardening, based on the Hencky hyperelastic description of the reversible behaviour and Green–Naghdi rate of back-stress tensor.

Box 14.6. Multiplicative finite strain extension of a general infinitesimal von Mises-type mixed hardening plasticity model based on the Hencky hyperelastic law and Green–Naghdi rate of back-stress tensor.

(i) Multiplicative split of $\boldsymbol{F}$

$$\boldsymbol{F} = \boldsymbol{F}^e \boldsymbol{F}^p$$

(ii) Hencky hyperelastic law

$$\boldsymbol{\tau} = \mathsf{D} : \boldsymbol{\varepsilon}^e; \quad \boldsymbol{\varepsilon}^e = \ln \boldsymbol{V}^e$$

(iii) Yield function

$$\Phi(\boldsymbol{\tau}, \boldsymbol{\beta}, \sigma_y) = \sqrt{3\, J_2(\boldsymbol{s}(\boldsymbol{\tau}) - \boldsymbol{\beta})} - \sigma_y$$

(iv) Plastic flow rule

$$\tilde{\boldsymbol{D}}^p = \dot{\gamma} \frac{\partial \Phi}{\partial \boldsymbol{\tau}} = \dot{\gamma} \sqrt{\frac{3}{2}} \frac{\boldsymbol{\eta}}{\|\boldsymbol{\eta}\|}; \quad \boldsymbol{\eta} = \boldsymbol{s} - \boldsymbol{\beta}$$

$$\boldsymbol{W}^p = \boldsymbol{0}$$

(v) Isotropic and kinematic hardening laws

$$\sigma_y = \sigma_y(\bar{\varepsilon}^p); \quad \dot{\bar{\varepsilon}}^p = \dot{\gamma}$$

$$\overset{\square}{\boldsymbol{\beta}} = \dot{\gamma}\, \hat{\boldsymbol{\beta}}(\boldsymbol{\tau}, \boldsymbol{\beta}, \bar{\varepsilon}^p)$$

(vi) Loading/unloading criterion

$$\Phi \leq 0, \quad \dot{\gamma} \geq 0, \quad \dot{\gamma}\Phi = 0$$

Model response

To illustrate the behaviour predicted by models of the present type, we plot in Figures 14.20 and 14.21 the response of the finite strain extension (14.202) of Prager's linear kinematic hardening model, respectively, over a cyclic strain-controlled uniaxial test and under monotonic shearing deformation (refer to Figure 14.19, page 625, for the definition of the shearing problem). For comparison, the response predicted by the finite strain linear isotropic hardening model with identical hardening modulus is also shown. The adopted material constants are

$$E = 210; \quad \nu = 0.3; \quad \sigma_y = 0.45; \quad H = 1.0.$$

In the uniaxial test, the behaviour predicted by the finite strain Prager linear law is in line with what is expected from an extension of its infinitesimal counterpart. In the shearing problem, however, the Green–Naghdi rate-based extension to Prager's rule produces unrealistic stress oscillations, which start at strains of around 150%. Such unexpected oscillations of shear stress in the finite strain formulation of Prager's kinematic hardening law were first reported by Nagtegaal and de Jong (1981) for a hypoelastic-based finite plasticity model with a Jaumann rate extension of the infinitesimal Prager kinematic hardening law. A more

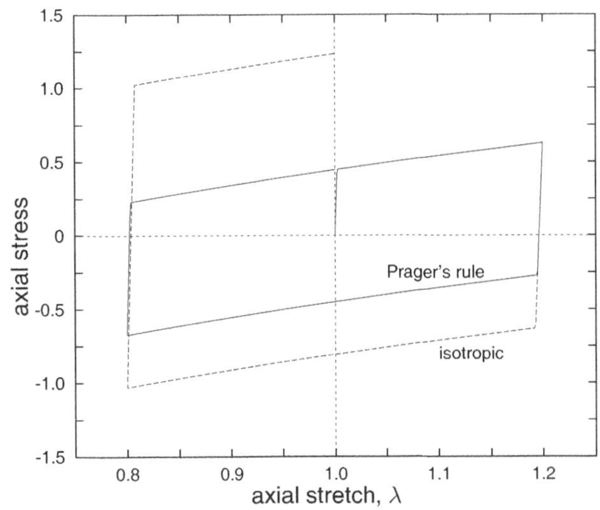

Figure 14.20. Cyclic strain-controlled uniaxial test. Finite strain extension of Prager's linear kinematic hardening model and linear isotropic hardening model.

recent study by Dettmer and Reese (2004) has also reported shear stress oscillations in a hyperelastic-based plasticity model with a Jaumann rate extension of the infinitesimal linear kinematic hardening rule. For the model discussed in this section, such oscillations can be eliminated, for instance, by adopting the Green–Naghdi rate-based extension (14.204) of the Armstrong–Frederick law (note that this law reduces to Prager's linear rule if $b = 0$). This is shown in Figure 14.21. The curve for the Armstrong–Frederick model was obtained with the same material parameters as above and the additional non-dimensional saturation hardening parameter b set to

$$b = 3.$$

With this choice of b in the present case, before the oscillations start, the stress–strain curve reaches a saturation state where the behaviour becomes perfectly plastic in the absence of load reversal, producing a more realistic behaviour under finite shearing. Dettmer and Reese (2004) showed that, in the presence of linear kinematic hardening only (Prager's rule), stress oscillations can be removed and monotonically increasing shear stress obtained with increasing shear deformation if suitable alternative finite strain extensions of the infinitesimal model are adopted (such alternatives are briefly described below in Section 14.11.5).

In summary, the example presented here serves to emphasise that care should be exercised when modelling kinematic hardening effects within the framework of finite plasticity, particularly if high strains are expected to occur. The use of the finite strain Prager rule (14.202), for instance, appears to be justifiable only in the presence of moderately large plastic strains.

14.11.2. INTEGRATION ALGORITHM

Crucial to the numerical integration algorithm for the finite strain model with kinematic hardening is the derivation of an incrementally objective update formula consistent with the basic back-stress evolution equation (14.201). The algorithm described in the following

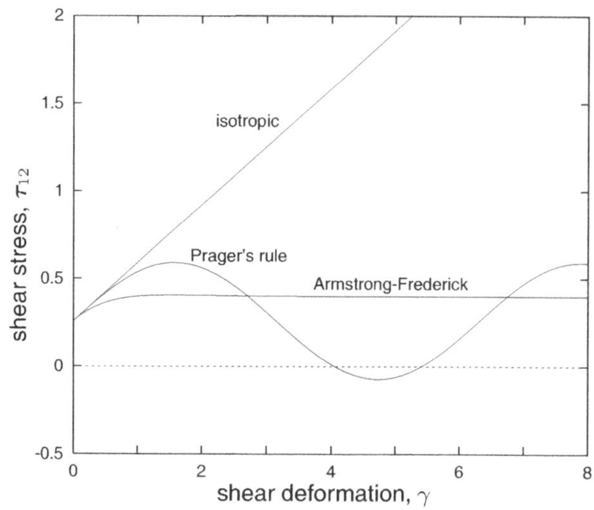

Figure 14.21. Behaviour under monotonic shearing (see problem definition in Figure 14.19). Finite strain extension of Prager's linear kinematic hardening model, Armstrong–Frederick kinematic hardening model and linear isotropic hardening model.

is analogous to that discussed in Section 14.10.7 in connection with the integration of the hypoelastic-based plasticity model defined in terms of the Green–Naghdi stress rate. Here, however, we shall adopt a *fully implicit* integration scheme. In this case, the incremental version of (14.201) reads

$$\boldsymbol{\beta}_{n+1} = \boldsymbol{\Lambda}_\Delta\, \boldsymbol{\beta}_n\, \boldsymbol{\Lambda}_\Delta^T + \Delta\gamma\, \hat{\boldsymbol{\beta}}(\boldsymbol{\tau}_{n+1}, \boldsymbol{\beta}_{n+1}, \bar{\varepsilon}^p_{n+1}), \tag{14.205}$$

where $\boldsymbol{\Lambda}_\Delta$ here denotes the incremental elastic rotation

$$\boldsymbol{\Lambda}_\Delta = \boldsymbol{R}^e_{n+1}\, (\boldsymbol{R}^e_n)^T. \tag{14.206}$$

Overall integration scheme

Similarly to the fully isotropic case, an exponential map-based numerical integration algorithm for finite strain plasticity including kinematic hardening can be devised such that the essential stress-update procedure maintains the small-strain format. However, as we shall demonstrate later, the infinitesimal format can only be justified here in the presence of small elastic strains (with arbitrarily large plastic strains). This is in contrast to the fully isotropic case where the small-strain format holds regardless of the magnitude of elastic strains. A summary of the algorithm, which is here applied to the generic model of Box 14.6, is presented in Box 14.7.

Remark 14.11 (The small-strain format algorithm). The algorithm called in item (iii) of Box 14.7 has exactly the same format as the return mapping-based scheme of the corresponding infinitesimal model. For instance, consider the nonlinear mixed hardening model defined by the laws (14.192) and (14.203). The implementation of the infinitesimal counterpart of this model was fully described in Section 7.6 (from page 257) with the associated elastic

Box 14.7. Numerical integration algorithm for the general multiplicative finite strain elastoplasticity model with kinematic hardening.

(i) Given incremental displacement $\Delta \boldsymbol{u}$, update the deformation gradient

$$\boldsymbol{F}_\Delta := \boldsymbol{I} + \nabla_n[\Delta \boldsymbol{u}], \quad \boldsymbol{F}_{n+1} := \boldsymbol{F}_\Delta \, \boldsymbol{F}_n$$

(ii) Compute elastic trial state

$$\begin{aligned}
\boldsymbol{F}^{e\ \mathrm{trial}}_{n+1} &:= \boldsymbol{F}_\Delta \, \boldsymbol{F}^e_n = \boldsymbol{F}_{n+1} (\boldsymbol{F}^p_n)^{-1} \\
\boldsymbol{B}^{e\ \mathrm{trial}}_{n+1} &:= \boldsymbol{F}^{e\ \mathrm{trial}}_{n+1} (\boldsymbol{F}^{e\ \mathrm{trial}}_{n+1})^T \\
\boldsymbol{\varepsilon}^{e\ \mathrm{trial}}_{n+1} &:= \ln[\boldsymbol{V}^{e\ \mathrm{trial}}_{n+1}] = \tfrac{1}{2} \ln[\boldsymbol{B}^{e\ \mathrm{trial}}_{n+1}] \\
\boldsymbol{\Lambda}_\Delta &:= \boldsymbol{R}^{e\ \mathrm{trial}}_{n+1} (\boldsymbol{R}^e_n)^T \\
\boldsymbol{\beta}^{\mathrm{trial}}_{n+1} &:= \boldsymbol{\Lambda}_\Delta \, \boldsymbol{\beta}_n \, \boldsymbol{\Lambda}^T_\Delta \\
\bar{\varepsilon}^{p\ \mathrm{trial}}_{n+1} &:= \bar{\varepsilon}^p_n
\end{aligned}$$

(iii) **USE SMALL STRAIN ALGORITHM** to update $\boldsymbol{\tau}$, $\boldsymbol{\varepsilon}^e$, $\boldsymbol{\beta}$ and $\bar{\varepsilon}^p$

(iv) Update Cauchy stress and $\boldsymbol{F}^e$

$$\begin{aligned}
\boldsymbol{\sigma}_{n+1} &:= \det[\boldsymbol{F}_{n+1}]^{-1} \, \boldsymbol{\tau}_{n+1} \\
\boldsymbol{V}^e_{n+1} &:= \exp[\boldsymbol{\varepsilon}^e_{n+1}] \\
\boldsymbol{R}^e_{n+1} &:= \boldsymbol{R}^{e\ \mathrm{trial}}_{n+1} \\
\boldsymbol{F}^e_{n+1} &:= \boldsymbol{V}^e_{n+1} \, \boldsymbol{R}^e_{n+1}
\end{aligned}$$

predictor/return-mapping algorithm summarised in Box 7.5, which corresponds to HYPLAS subroutine SUVMMX. In the computational implementation of the finite strain model, we would follow the steps of Box 14.7 with item (iii) being a call to the same subroutine.

Remark 14.12 (Further computer implementation issues). The overall algorithm of Box 14.7 is an extension of the isotropic-only procedure of Box 14.3. It should be noted that, under kinematic hardening, both elastic stretch *and* rotation tensors are needed in the kinematics related operations carried out before the infinitesimal integration algorithm is called. Thus, in this case we would store the elastic deformation gradient as a state variable, rather than just the logarithmic Eulerian elastic strain stored in the isotropic case (the interested reader should refer to the source code of HYPLAS). In addition, note that one of the state variables – the back-stress tensor – has to be rotated in item (ii) prior to the infinitesimal algorithm being applied and a few extra operations have to be performed in item (iv) to update the elastic deformation gradient. For those interested in implementing the finite strain kinematic hardening model in HYPLAS, we would suggest its inclusion as a member of a *new* class of material models. This would keep this implementation clearly distinct from the isotropic class of models and, in our view, would contribute to maintenance of code clarity.

Validity under small elastic strains

To demonstrate that the above algorithm is valid under sufficiently small elastic strains, we start by referring back to the exponential map-based updating equation (14.87) for $\boldsymbol{V}^e_{n+1}$, obtained in the course of derivation of the small-strain format return mapping for the fully isotropic case. Note that, in obtaining (14.87), no assumptions of material isotropy have been made; that is, (14.87) is a direct consequence of the general exponential map-based formula (14.73) for the numerical integration of the plastic flow equation, valid regardless of material symmetry. Particularisation of (14.87) for the present model gives

$$\boldsymbol{V}^e_{n+1} \exp[2\,\Delta\gamma\,\boldsymbol{N}_{n+1}]\,\boldsymbol{V}^e_{n+1} = (\boldsymbol{V}^{e\,\mathrm{trial}}_{n+1})^2, \tag{14.207}$$

where the flow vector, $\boldsymbol{N}_{n+1}$, is here defined by

$$\boldsymbol{N}_{n+1} \equiv \left.\frac{\partial \Phi}{\partial \boldsymbol{\tau}}\right|_{n+1} = \sqrt{\frac{3}{2}}\,\frac{\boldsymbol{\eta}_{n+1}}{\|\boldsymbol{\eta}_{n+1}\|}, \tag{14.208}$$

with $\boldsymbol{\eta}$ denoting the relative stress tensor

$$\boldsymbol{\eta} \equiv \boldsymbol{s} - \boldsymbol{\beta}. \tag{14.209}$$

Before proceeding, we should note the crucial step in establishing the equivalence between (14.87) and the small-strain format update formula (14.89) was the introduction of the assumption of elastoplastic isotropy, under which, the flow vector, the updated elastic and the elastic trial left stretch tensor all commute. Here, due to the plastic anisotropy introduced by kinematic hardening, $\boldsymbol{N}_{n+1}$, $\boldsymbol{V}^e_{n+1}$ and $\boldsymbol{V}^{e\,\mathrm{trial}}_{n+1}$ do not commute in general (note that they do commute if $\boldsymbol{s}$ and $\boldsymbol{\beta}$ share the same principal axes). Consequently, equivalence between (14.207) and the infinitesimal format update formula for the logarithmic elastic strain does not hold in general. Fortunately, however, such an equivalence can be established *approximately* as we demonstrate in the following.

In terms of the Eulerian logarithmic strains, $\boldsymbol{\varepsilon}^e_{n+1}$ and $\boldsymbol{\varepsilon}^{e\,\mathrm{trial}}_{n+1}$, expression (14.207) can be written as

$$\exp[\boldsymbol{\varepsilon}^e_{n+1}]\exp[2\,\Delta\gamma\,\boldsymbol{N}_{n+1}]\exp[\boldsymbol{\varepsilon}^e_{n+1}] = \exp[2\,\boldsymbol{\varepsilon}^{e\,\mathrm{trial}}_{n+1}]. \tag{14.210}$$

From the series representation (B.4) for the tensor exponential, it follows that

$$\exp[\boldsymbol{\varepsilon}^e_{n+1}] = \boldsymbol{I} + \boldsymbol{\varepsilon}^e_{n+1} + \boldsymbol{o}(\boldsymbol{\varepsilon}^e_{n+1}), \tag{14.211}$$

where $\boldsymbol{o}(\boldsymbol{\varepsilon}^e)$ is a term of second order in the elastic logarithmic strain. This together with (14.210) gives

$$\begin{aligned}\exp[2\,\boldsymbol{\varepsilon}^{e\,\mathrm{trial}}_{n+1}] = \exp[2\,\Delta\gamma\,\boldsymbol{N}_{n+1}] &+ \boldsymbol{\varepsilon}^e_{n+1}\,\exp[2\,\Delta\gamma\,\boldsymbol{N}_{n+1}] \\ &+ \exp[2\,\Delta\gamma\,\boldsymbol{N}_{n+1}]\,\boldsymbol{\varepsilon}^e_{n+1} + \boldsymbol{o}(\boldsymbol{\varepsilon}^e_{n+1}).\end{aligned} \tag{14.212}$$

For convenience, we now define the strain-like tensor

$$\boldsymbol{\varepsilon}^p_{\Delta} \equiv \tfrac{1}{2}\,[\exp(2\Delta\gamma\,\boldsymbol{N}_{n+1}) - \boldsymbol{I}\,]. \tag{14.213}$$

This tensor is a Green–Lagrange-type measure (refer to Section 3.1.7, from page 52) of incremental plastic strain within the interval considered. Indeed, within the present exponential map-based discretisation of the plastic flow (14.73), the incremental plastic deformation gradient, $\boldsymbol{F}^p_\Delta$, defined by the relation

$$\boldsymbol{F}^p_{n+1} = \boldsymbol{F}^p_\Delta \, \boldsymbol{F}^p_n, \tag{14.214}$$

is the symmetric (pure stretch) tensor given by

$$\boldsymbol{F}^p_\Delta = \boldsymbol{U}^p_\Delta = \boldsymbol{V}^p_\Delta = \boldsymbol{R}^{e\,T}_{n+1} \, \exp[\Delta\gamma \, \boldsymbol{N}_{n+1}] \boldsymbol{R}^e_{n+1}. \tag{14.215}$$

The tensor defined by (14.213) is the Green–Lagrange strain measure associated with the above incremental plastic stretch, rotated to the spatial configuration (by $\boldsymbol{R}^e_{n+1}$).

With definition (14.213) at hand, we can rewrite (14.212) equivalently as

$$\exp[2\,\boldsymbol{\varepsilon}^{e\,\text{trial}}_{n+1}] = \boldsymbol{I} + 2\,\boldsymbol{\varepsilon}^p_\Delta + 2[\boldsymbol{\varepsilon}^e_{n+1} + \boldsymbol{\varepsilon}^e_{n+1}\,\boldsymbol{\varepsilon}^p_\Delta + \boldsymbol{\varepsilon}^p_\Delta\,\boldsymbol{\varepsilon}^e_{n+1}] + \boldsymbol{o}(\boldsymbol{\varepsilon}^e_{n+1}). \tag{14.216}$$

We then take the logarithm of both sides of the above equation, making use, on its right-hand side, of the series representation of the tensor logarithm

$$\ln(\boldsymbol{I} + \boldsymbol{X}) = \sum_{k=0}^{\infty} \frac{(-1)^k}{k+1} \, \boldsymbol{X}^{\,k+1}, \tag{14.217}$$

which is convergent for sufficiently small $\boldsymbol{X}$. This gives

$$\begin{aligned} 2\,\boldsymbol{\varepsilon}^{e\,\text{trial}}_{n+1} &= \sum_{k=0}^{\infty} \frac{(-1)^k}{k+1} \, \{2\,\boldsymbol{\varepsilon}^p_\Delta + 2[\boldsymbol{\varepsilon}^e_{n+1} + \boldsymbol{\varepsilon}^e_{n+1}\,\boldsymbol{\varepsilon}^p_\Delta + \boldsymbol{\varepsilon}^p_\Delta\,\boldsymbol{\varepsilon}^e_{n+1}] + \boldsymbol{o}(\boldsymbol{\varepsilon}^e_{n+1})\}^{k+1} \\ &= \sum_{k=0}^{\infty} \frac{(-1)^k}{k+1} \, (2\,\boldsymbol{\varepsilon}^p_\Delta)^{k+1} + \sum_{k=0}^{\infty} \frac{(-1)^k}{k+1} \, 2^{k+1} \sum_{i=0}^{\infty} [(\boldsymbol{\varepsilon}^p_\Delta)^i (\boldsymbol{\varepsilon}^e_{n+1} + \boldsymbol{\varepsilon}^e_{n+1}\,\boldsymbol{\varepsilon}^p_\Delta \\ &\quad + \boldsymbol{\varepsilon}^p_\Delta \boldsymbol{\varepsilon}^e_{n+1})(\boldsymbol{\varepsilon}^p_\Delta)^{k-i}] + \boldsymbol{o}(\boldsymbol{\varepsilon}^e_{n+1}). \end{aligned} \tag{14.218}$$

In addition, by noting that, in view of the definition of $\boldsymbol{\varepsilon}^p_\Delta$, the first term on the right-hand side of the above equals $2\Delta\gamma \boldsymbol{N}_{n+1}$, and expanding the second term, we obtain

$$\boldsymbol{\varepsilon}^{e\,\text{trial}}_{n+1} = \boldsymbol{\varepsilon}^e_{n+1} + \Delta\gamma \, \boldsymbol{N}_{n+1} + \boldsymbol{o}(\boldsymbol{\varepsilon}^p_\Delta) + \boldsymbol{o}(\boldsymbol{\varepsilon}^e_{n+1}), \tag{14.219}$$

where $\boldsymbol{o}(\boldsymbol{\varepsilon}^p_\Delta)$ is a term of second order in the incremental plastic strain.

We now recall that the implicit exponential map integrator leading to (14.207) is *first-order accurate* (refer to Section B.3.1); that is, the exponential map-based update formula is an approximation of first order in the incremental plastic strain. Thus, consistently with this order of accuracy (14.219) can be approximated as

$$\boldsymbol{\varepsilon}^{e\,\text{trial}}_{n+1} = \boldsymbol{\varepsilon}^e_{n+1} + \Delta\gamma \, \boldsymbol{N}_{n+1} + \boldsymbol{o}(\boldsymbol{\varepsilon}^e_{n+1}). \tag{14.220}$$

Finally, under infinitesimal elastic strains (typical of metal plasticity), the second-order term can be neglected in the above expression leading to the approximate update formula

$$\boldsymbol{\varepsilon}^e_{n+1} = \boldsymbol{\varepsilon}^{e\,\text{trial}}_{n+1} - \Delta\gamma \, \boldsymbol{N}_{n+1} \tag{14.221}$$

which, remarkably, has the same small-strain format as the fully isotropic formula discussed in earlier sections of this chapter.

14.11.3. SPATIAL TANGENT OPERATOR

The consistent spatial tangent modulus in the present case is slightly more complex than its isotropic counterpart (14.99). The extra complexity arises from the fact that, in the integration algorithm for the Green–Naghdi rate-based kinematic hardening model, the elastic trial state of one of the internal variables – the back-stress tensor – is not fixed as its value at the beginning of the time interval (refer to Box 14.7). Here, the elastic trial back-stress is obtained by rotating the back-stress tensor, $\boldsymbol{\beta}_n$, forward to the spatial configuration (refer to item (ii) of Box 14.7). Thus, the elastic trial back-stress, $\boldsymbol{\beta}^{\text{trial}}_{n+1}$, is a function of the (fixed) back-stress tensor, $\boldsymbol{\beta}_n$, at the beginning of the interval *and* the incremental elastic rotation, $\boldsymbol{\Lambda}_\Delta$. The incremental elastic rotation itself is a function of the elastic trial deformation gradient, $\boldsymbol{F}^{e\ \text{trial}}_{n+1}$, which, in turn, is a function of $\boldsymbol{F}^p_n$ and the deformation gradient $\boldsymbol{F}_{n+1}$ at the end of the integration interval. As a consequence, the general symbolic expression (14.97), valid for the Kirchhoff stress incremental constitutive function of the fully isotropic model, is redefined as

$$\hat{\boldsymbol{\tau}}(\boldsymbol{\alpha}_n, \boldsymbol{F}_{n+1}) = \tilde{\boldsymbol{\tau}}(\boldsymbol{\alpha}^*_n, \boldsymbol{\beta}^{\text{trial}}_{n+1}(\boldsymbol{\beta}_n, \boldsymbol{F}^{e\ \text{trial}}_{n+1}(\boldsymbol{F}^p_n, \boldsymbol{F}_{n+1})), \boldsymbol{\varepsilon}^{e\ \text{trial}}_{n+1}(\boldsymbol{B}^{e\ \text{trial}}_{n+1}(\boldsymbol{F}^p_n, \boldsymbol{F}_{n+1}))). \tag{14.222}$$

In the above, we have introduced the definition

$$\boldsymbol{\alpha}_n \equiv \{\boldsymbol{\alpha}^*_n, \boldsymbol{\beta}_n\}, \tag{14.223}$$

where α^*_n is the subset of state variables at t_n, excluding the back-stress tensor.

The key difference between the present and the isotropic case comes from the replacement of (14.97) with (14.222) in the derivation of the derivative of the Kirchhoff stress with respect to the deformation gradient that takes part in the general formula (14.95) for the spatial tangent modulus. Clearly, the elastic tangent here coincides with that of the isotropic model (as the same elastic law is adopted in both cases). In the elastoplastic case, as a result of (14.222), the derivative (14.98) is here given by

$$\frac{\partial \hat{\boldsymbol{\tau}}}{\partial \boldsymbol{F}_{n+1}} = \mathsf{D}^{ep} : \frac{\partial \boldsymbol{\varepsilon}^{e\ \text{trial}}_{n+1}}{\partial \boldsymbol{B}^{e\ \text{trial}}_{n+1}} : \frac{\partial \boldsymbol{B}^{e\ \text{trial}}_{n+1}}{\partial \boldsymbol{F}_{n+1}} + \mathsf{D}^{\beta} : \frac{\partial \boldsymbol{\beta}^{\text{trial}}_{n+1}}{\partial \boldsymbol{F}^{e\ \text{trial}}_{n+1}} : \frac{\partial \boldsymbol{F}^{e\ \text{trial}}_{n+1}}{\partial \boldsymbol{F}_{n+1}}, \tag{14.224}$$

where D^{ep} is the standard infinitesimal elastoplastic consistent tangent operator associated with the particular model/algorithm adopted and

$$\mathsf{D}^{\beta} \equiv \frac{\partial \tilde{\boldsymbol{\tau}}}{\partial \boldsymbol{\beta}^{\text{trial}}_{n+1}} \tag{14.225}$$

is a fourth-order tensor which represents the tangent relation between the updated Kirchhoff stress and the trial back-stress, also consistent with the infinitesimal return mapping. The tangential relation D^{β} is obtained in a manner analogous to the conventional infinitesimal consistent tangent operator D^{ep}, through appropriate differentiation of the corresponding return-mapping equations. Note that the conventional tangent measures the rate of change of $\boldsymbol{\tau}_{n+1}$ with changes in the input value of $\boldsymbol{\varepsilon}^{e\ \text{trial}}_{n+1}$ at a frozen value of $\boldsymbol{\beta}^{\text{trial}}_{n+1}$ and the other internal variables, whilst D^{β} measures the rate of change of $\boldsymbol{\tau}_{n+1}$ with changes in the input value of $\boldsymbol{\beta}^{\text{trial}}_{n+1}$ at a frozen value of $\boldsymbol{\varepsilon}^{e\ \text{trial}}_{n+1}$ and the other internal variables of the model.

The structure of the elastoplastic spatial tangent

The right-hand side of (14.224) comprises two summands. The first one has identical functional format as the right-hand side of (14.98) and, therefore, gives rise to a contribution to the spatial tangent a with identical format as its fully isotropic counterpart (14.99). The second summand, in turn, gives rise to a new contribution to a, specifically associated with the pre-rotation of $\boldsymbol{\beta}_n$ prior to the application of the infinitesimal return mapping. This allows the elastoplastic spatial tangent modulus to be expressed here as

$$\mathsf{a} = \mathsf{a}^{\tau} + \mathsf{a}^{\beta}, \tag{14.226}$$

where a^{τ} has the format of (14.99) and the new term a^{β} is defined by

$$\mathsf{a}^{\beta} \equiv \frac{2}{J}\,\mathsf{U} : \mathsf{V}, \tag{14.227}$$

where the fourth-order tensors U and V are defined by the Cartesian components

$$\mathsf{U}_{ijkl} \equiv [\mathsf{D}^{\beta}]_{ijkm}\;[\boldsymbol{\beta}^{\text{trial}}_{n+1}]_{ml} \tag{14.228}$$

and

$$\mathsf{V}_{ijkl} \equiv \left[\frac{\partial (\boldsymbol{B}^{e\ \text{trial}}_{n+1})^{-\frac{1}{2}}}{\partial (\boldsymbol{B}^{e\ \text{trial}}_{n+1})}\right]_{imkn} (\boldsymbol{V}^{e\ \text{trial}}_{n+1})_{mj}(\boldsymbol{B}^{e\ \text{trial}}_{n+1})_{ln} + (\boldsymbol{V}^{e\ \text{trial}}_{n+1})^{-1}_{ik}(\boldsymbol{V}^{e\ \text{trial}}_{n+1})_{jl}. \tag{14.229}$$

Expression (14.227) is obtained through a lengthy but straightforward manipulation with repeated application of the chain and product rules after the substitution of the second summand of the right-hand side of (14.224) into the first summand of the right-hand side of (14.95). Note that the term within square brackets on the right-hand side of (14.229) is the derivative of the inverse of the square root of a symmetric tensor. The inverse of the tensor square root is a member of the class of isotropic tensor-valued functions discussed in Section A.5 (from page 740), where closed formulae for the corresponding derivative are also given.

Implementation aspects

Those interested in the computational implementation of finite strain kinematic hardening models of the present type should note that the computation of a^{τ} follows the same steps as those already included in program HYPLAS for the isotropic plasticity models. The computation of tensors U and V and subsequent assemblage of a^{β} need to be added to the relevant part of the code. In program HYPLAS, this is probably best done in the material interface subroutine MATICT, having (as suggested in Remark 14.12, page 639) kinematic hardening models of this type as a new *class* of material models. In the computation of U, a new material model-specific quantity – D^{β} – is needed. In program HYPLAS, this would probably be best implemented in a separate subroutine specifically associated with the particular material model/algorithm *type* in question (just as with the other state-update procedures and tangent moduli computation routines of HYPLAS). In the computation of V, as given by expression (14.229), the derivative of the inverse tensor square root can be computed in practice by subroutine DISO2 of HYPLAS. Finally, note that for the particular nonlinear mixed hardening model alluded to in Remark 14.11, the tangent operator D^{ep} is the one coded in subroutine CTVMMX. This routine can be used in the extension of the mixed hardening model already implemented in HYPLAS to the finite strain range.

14.11.4. REMARKS ON PREDICTIVE CAPABILITY

At present, the issue of kinematic hardening modelling remains largely open. The actual microscopic mechanisms responsible for kinematic hardening are quite complex and their representation by means of purely phenomenological models, even in the infinitesimal strain regime, is expected to be accurate only for relatively simple strain paths. The situation is far more complex under finite plastic flow. In this case, given an accepted infinitesimal model, many different large-strain extensions can be postulated (mainly on a purely ad hoc basis), all having the same small-strain limit but with each one presenting a distinct stress response under finite straining. In the authors' opinion, a more accurate description of kinematic hardening effects, especially in the finite strain range, should be attainable by means of micromechanics-based models, possibly derived from multiscale representations relying on computational homogenisation. Such an approach to finite plasticity is discussed, for instance, by Miehe *et al.* (2002b) in the context of polycrystalline materials (without specific reference to kinematic hardening modelling).

In spite of the potentially limited predictive ability of phenomenological kinematic hardening models under complex strain paths, it is important to emphasise that models of this type can often provide useful estimates of material behaviour when load reversing-related phenomena are an important feature of the modelled process. They can be of particular relevance, for instance, in the simulation of multistage metal-forming operations, where substantial changes in loading direction occur between the different forming stages. Under such circumstances, isotropic models may fail to capture some essential characteristics of the process (Miehe *et al.*, 2002a; Schröder *et al.*, 2002).

14.11.5. ALTERNATIVE DESCRIPTIONS

Alternatives (also relying on hyperelastic-based multiplicative formulations) to the Green–Naghdi rate description of the kinematic hardening evolution law considered in the above are discussed, for instance, by Simo and Hughes (1998) and Dettmer and Reese (2004), who also address their relevant computational aspects.

One such alternative is the model introduced in Tsakmakis (1996a,b). For this model, the finite strain extension to the Armstrong–Frederick evolution law for the back-stress tensor can be cast in the following form

$$\dot{\boldsymbol{\beta}}^* = \tfrac{2}{3} H \, (\boldsymbol{F}^p)^{-1} \boldsymbol{D}^p (\boldsymbol{F}^p)^{-T} - \dot{\gamma} \, b \, \boldsymbol{\beta}^*, \tag{14.230}$$

where $\boldsymbol{\beta}^*$ is a measure of back-stress defined on the reference configuration and $\boldsymbol{D}^p$ is the plastic stretching $(14.26)_1$ having, in the present case, the constitutive equation

$$\boldsymbol{D}^p = \dot{\gamma} \, \frac{\boldsymbol{M}_{\rm d} - \boldsymbol{\kappa}_{\rm d}}{\|\boldsymbol{M}_{\rm d} - \boldsymbol{\kappa}_{\rm d}\|}. \tag{14.231}$$

In the above plastic flow rule, $\boldsymbol{M}_{\rm d}$ denotes the deviator of the *Mandel stress tensor*

$$\boldsymbol{M} \equiv (\boldsymbol{F}^e)^T \, \boldsymbol{\tau} \, (\boldsymbol{F}^e)^{-T}, \tag{14.232}$$

defined on the intermediate configuration, and $\boldsymbol{\kappa}_d$ is the deviatoric component of

$$\boldsymbol{\kappa} \equiv \boldsymbol{\beta} + \frac{3}{H} \boldsymbol{\beta} \, \boldsymbol{\beta}, \tag{14.233}$$

with $\boldsymbol{\beta}$ denoting the push-forward (to the intermediate configuration) of $\boldsymbol{\beta}^*$

$$\boldsymbol{\beta} \equiv \boldsymbol{F}^p\, \boldsymbol{\beta}^*\, (\boldsymbol{F}^p)^T. \tag{14.234}$$

Another possible representation is derived from a rheological model of kinematic hardening (Lion, 2000). The finite strain generalisation of the Armstrong–Frederick kinematic hardening law in this case is based on a further multiplicative split of the plastic deformation gradient into 'elastic' and 'inelastic' components (associated, respectively, with a rheological spring and a rate-independent dashpot)

$$\boldsymbol{F}^p = \boldsymbol{F}^p_e\, \boldsymbol{F}^p_i, \tag{14.235}$$

in conjunction with the plastic flow rule

$$\boldsymbol{D}^p = \dot{\gamma}\, \frac{\boldsymbol{M}_{\rm d} - \boldsymbol{\beta}_{\rm d}}{\|\boldsymbol{M}_{\rm d} - \boldsymbol{\beta}_{\rm d}\|} \tag{14.236}$$

and the following constitutive equation for the dashpot stretching

$$\boldsymbol{D}^p_i \equiv \text{sym}[\dot{\boldsymbol{F}}^p_i (\boldsymbol{F}^p_i)^{-1}] = \dot{\gamma}\, \frac{3b}{2H}\, \boldsymbol{M}_i. \tag{14.237}$$

In the above, $\boldsymbol{\beta}_{\rm d}$ is the deviator of

$$\boldsymbol{\beta} = \frac{H}{3}\, [\boldsymbol{F}^p_e (\boldsymbol{F}^p_e)^T - \boldsymbol{I}] \tag{14.238}$$

and

$$\boldsymbol{M}_i \equiv (\boldsymbol{F}^p_e)^T \boldsymbol{\beta}\, \boldsymbol{F}^p_e. \tag{14.239}$$

The reader is referred to Dettmer and Reese (2004) for further details on the two models above, including their numerical treatment by an exponential map-based integration algorithm. The computational implementation of these models is considerably more elaborate than that of the Green–Naghdi rate-based theory. However, under small elastic strains, the stress updating procedures for both models is reduced, essentially, to the solution of a single scalar nonlinear nonlinear equation and twelve linear equations in the three-dimensional case. It is interesting to note that Dettmer and Reese (2004) showed that neither of the two models exhibit the unrealistic shear stress oscillations reported earlier in this section for the Green–Naghdi rate-based model (refer to Figure 14.21) when the finite strain extension to Prager's (linear) kinematic hardening ($b = 0$) is adopted.

Yet another possibility, as suggested by Simo and Hughes (1998), is the definition of the evolution law for the spatial back-stress tensor, $\boldsymbol{\beta}$, in terms of its Oldroyd rate (refer to expressions (14.130) and (14.131) for the general definition of the Oldroyd rate of a spatial tensor),

$$\overset{\diamond}{\boldsymbol{\beta}} \equiv \dot{\boldsymbol{\beta}} - \boldsymbol{L}\, \boldsymbol{\beta} - \boldsymbol{\beta}\, \boldsymbol{L}^T. \tag{14.240}$$

These authors proposed a finite strain extension to Prager's linear kinematic hardening rule with the following format

$$\overset{\diamond}{\boldsymbol{\beta}} = \dot{\gamma}\, \tfrac{2}{3}\, H\, (\text{tr}\, \boldsymbol{B}^e)\, \bar{\boldsymbol{N}}, \tag{14.241}$$

where $\boldsymbol{B}^e = \boldsymbol{F}^e(\boldsymbol{F}^e)^T$ and $\bar{\boldsymbol{N}}$ is the unit flow vector

$$\bar{\boldsymbol{N}} = \frac{\boldsymbol{s} - \boldsymbol{\beta}}{\|\boldsymbol{s} - \boldsymbol{\beta}\|}, \tag{14.242}$$

with $\boldsymbol{s}$ the Kirchhoff stress deviator.

15 FINITE ELEMENTS FOR LARGE-STRAIN INCOMPRESSIBILITY

THE present chapter is devoted to special finite element techniques for the analysis of large deformations of nearly incompressible solids. It is a well-known fact that the performance of low-order virtual work-based finite elements becomes extremely poor as the incompressible limit is approached. Problems of practical engineering interest for which incompressibility plays a crucial role include the analysis of rubbery solids, typically modelled as incompressible hyperelastic materials (refer to Chapter 13), as well as elastoplastic simulations under plastic dominant deformations and the assumption of isochoric plastic flow, such as in metal plasticity models. In such situations, spurious *volumetric locking*, i.e. overstiff solutions, are frequently obtained as a consequence of the inability of low-order interpolation polynomials adequately to represent general volume-preserving displacement fields. The problem can be remedied simply by adopting elements of sufficiently high order. However, due to their simplicity, low-order elements are often preferred in large-scale computations. For this reason, several formulations have been proposed to allow the use of low-order elements near the incompressible limit.

Within the context of the geometrically linear theory, the class of enhanced assumed strain (*EAS*) methods described by Simo and Rifai (1990), which incorporates popular procedures such as the classical incompatible modes formulation proposed by Taylor *et al.* (1976) and *B-bar* methods (Hughes, 1980), is well established and is employed with success in a number of existing commercial finite element codes.

At this point, it is worth mentioning that such issues have not been raised by the authors in the infinitesimal strain context of Part Two of this book, the reason for this being that, under infinitesimal kinematics, the use of *underintegrated* standard finite elements (such as the eight-noded quadrilateral with four Gauss point quadrature) can provide simple and satisfactory solutions near the incompressible limit. However, in the geometrically nonlinear regime, particularly in the simulation of more realistic industrially relevant problems, more complex phenomena such as frictional contact with large sliding distances under extreme strains are often present. In such cases, lower-order elements are usually preferred due to their robustness and simplicity.

To tackle the problem, different approaches have been proposed. Among others, the class of mixed variational methods developed by Simo *et al.* (1985), the mixed *u/p* (displacement-pressure) formulation proposed by Sussman and Bathe (1987), the geometrically nonlinear extension of the *B-bar* methodology devised by Hughes *et al.* (1975) and Moran *et al.* (1990), the family of enhanced assumed strain elements of Simo and Armero (1992), the *F-bar* technique proposed by the authors and co-workers (de Souza Neto *et al.*, 1996), the co-rotational incompatible modes approach of Crisfield and Moita (1996), the geometrically nonlinear selective reduced integration scheme of Doll *et al.* (2000), the more recent versions of the enhanced assumed strain method proposed by Reese and Wriggers (2000) and Wall *et al.* (2000) are possible alternatives that allow the use of low-order elements near the incompressibility limit.

In the context of explicit transient dynamic analysis, methods based on so-called *hourglass control techniques* are almost invariably employed. Essentially, this class of methods relies on the use of reduced numerical integration quadrature within the element. For linear elements this implies single point integration which, however, can result in spurious zero energy (hourglass) modes of deformation (Belytschko *et al.*, 2000, 1984; Hughes, 1987). In order to obtain reliable results, various control methods have been proposed to eliminate hourglassing by providing the restraint that the element lacks under single point integration, but without stiffening the element's adequate response to other modes (Belytschko and Bindeman, 1991, 1993; Belytschko *et al.*, 2000, 1984). Two principal ways of resisting hourglassing are with viscous damping and by introduction of artificial stiffness, both of which are capable of eliminating the spurious singular modes but have a negligible effect on the stable global modes. However, it should be stressed that hourglass control does not fully remove the kinematic modes and, in particular, coarse meshes and meshes loaded with large nodal forces, resulting either from boundary conditions or from contact, are susceptible to hourglassing despite the use of control techniques.

One aspect that should be pointed out here is that, in addition to handling near incompressibility, robust formulations should also be able to cope with the extra requirements that different problems may present. For instance, in applications such as the prediction of failure in metal-forming processes, the ability to capture strain localisation phenomena becomes crucial; in problems involving extremely large strains, frequently encountered in the analysis of rubbery materials and metal-forming problems, it is not unusual that a solution can be obtained only if adaptive mesh refinement is employed. Thus, since a single formulation is normally not sufficiently robust to produce an optimal performance under a very wide range of conditions, the design of low-order finite elements for large-strain analysis of quasi-incompressible materials remains an open issue.

Rather than providing a general overview of a large number of different approaches, we have opted in this chapter to provide the reader with a reasonably detailed description (to a level of detail sufficient to allow the reader to produce the necessary computer codes) of only three methods. These methods, which we consider to be of sufficient generality, are, namely, the *F-bar* methodology (including its more recent *F-bar-Patch* variant for simplex elements), the geometrically nonlinear *EAS* method and the mixed u/p formulation. These are described, respectively, in Sections 15.1, 15.2 and 15.3. We remark that the *F-bar* method is incorporated into the standard version of program HYPLAS that accompanies this book.

15.1. The *F-bar* methodology

The basic idea behind the *F-bar* (de Souza Neto *et al.*, 1996) procedure is simple: an *F-bar* element is obtained from the corresponding standard (displacement-based) finite element simply by adopting a suitably modified deformation gradient in the computation of the stress tensor. To avoid volumetric *locking*, the modified deformation gradient is constructed such that the incompressibility constraint can be enforced in an approximate average (not pointwise) sense throughout the element. The method is in fact closely related to the classical *B-bar* procedures (Hughes, 1980). However, as we shall see, it cannot be regarded as a direct geometrically nonlinear extension of the *B-bar* method. Geometrically nonlinear *B-bar* elements (Hughes *et al.*, 1975; Moran *et al.*, 1990) possess a more complex structure, particularly with respect to the corresponding consistent tangent stiffness matrices. Some basic features of the *F-bar* approach are summarised below:

1. The methodology is *simple* in concept.

2. The computer implementation of *F-bar* elements is also *simple*. In fact, *F-bar* element routines can be obtained by introducing straightforward modifications into existing standard element routines. Interested readers are referred to the source code of program HYPLAS where both standard and *F-bar* four-noded quadrilaterals are implemented.

3. As for *B-bar* methods and assumed enhanced strain approaches, *F-bar* elements can be used with *any* material model, elastic or inelastic, regardless of deviatoric/volumetric coupling of the constitutive equations. This is not true if *selective reduced integration* is used.

4. Numerical experience shows that *F-bar* elements produce good solutions with reasonably sized meshes for a wide range of industrially relevant problems. The methodology avoids volumetric locking and, in addition, is particularly suitable to capture strain localisation phenomena.

5. In view of the absence of non-physical internal element parameters (as opposed to enhanced assumed strain and incompatible modes methods in general) *F-bar* elements can be easily used in conjunction with *adaptive* remeshing techniques. This is also true for geometrically nonlinear *B-bar* methods and *u/p* mixed formulations. This property becomes essential in the realistic simulation of finite strain industrial problems where excessively large strains occur.

We start below by focusing on the formulation of four-noded quadrilaterals (for plane strain and axisymmetric analysis) and a three-dimensional eight-noded hexahedron. The formulation of these relatively simple elements is then followed by an outline of a general procedure for the formulation of elements for the analysis of near-incompressible solids.

15.1.1. STRESS COMPUTATION: THE *F-BAR* DEFORMATION GRADIENT

Consider an ordinary displacement-based four-node quadrilateral and an eight-node hexahedron, with local coordinates denoted $\boldsymbol{\xi}$, as illustrated in Figure 15.1. The numerical integration of the element internal force vector requires the computation of the Cauchy stress tensor at a prescribed number of Gauss points (refer to Chapter 4). Let us consider the typical interval $[t_n, t_{n+1}]$ and let $\boldsymbol{\sigma}_{n+1}$ and $\boldsymbol{F}$ be, respectively, the Cauchy stress tensor and deformation gradient[†] for Gauss point i (with coordinate $\boldsymbol{\xi}_i$, indicated in Figure 15.1) at t_{n+1}. For finite strain problems with a generic dissipative underlying material model, the Cauchy stress

[†] The deformation gradient at t_{n+1} is usually denoted $\boldsymbol{F}_{n+1}$ in this book. Here, for notational convenience, we shall exceptionally use $\boldsymbol{F}$ to denote the deformation gradient at the end of the typical time (or pseudo-time) interval.

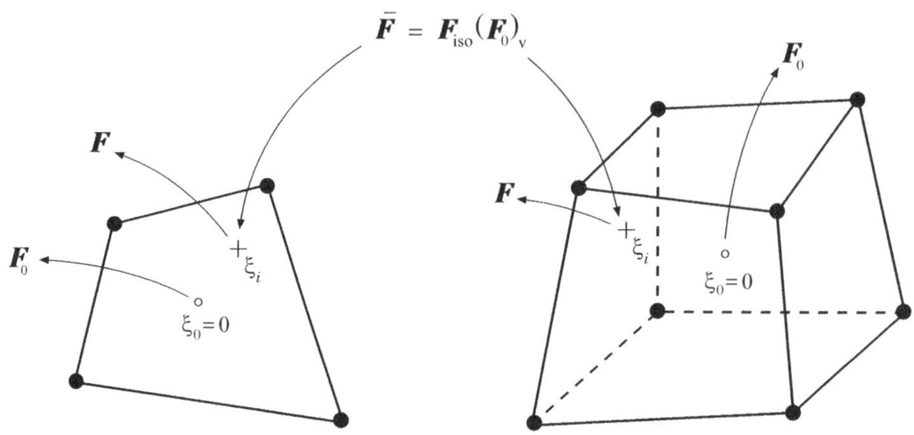

Figure 15.1. The *F-bar* four-node quadrilateral and eight-node hexahedron. (Reproduced with permission from Design of simple low-order finite elements for large-strain analysis of nearly incompressible solids, EA de Souza Neto, D Perić, M Dutko and DRJ Owen, *International Journal of Solids and Structures*, Vol 33, Issue 20–22 © 1996 Elsevier Science Ltd.)

tensor is obtained from the deformation gradient at the end of the time interval by means of incremental constitutive functions of the general form[‡]

$$\boldsymbol{\sigma}_{n+1} = \hat{\boldsymbol{\sigma}}(\boldsymbol{\alpha}_n, \boldsymbol{F}). \tag{15.1}$$

where $\boldsymbol{\alpha}_n$ denotes the set of internal variables of the model at t_n. For conventional (or standard) finite elements, the deformation gradient $\boldsymbol{F}$ is computed directly from the standard interpolation (bi-linear for the quadrilateral and tri-linear for the hexahedron) of the displacement field at the generic integration point i.

The isochoric/volumetric split of the deformation gradient

Central to the formulation of the present method is the concept of multiplicative split of the deformation gradient, $\boldsymbol{F}$, into an isochoric (volume-preserving) and a volumetric (purely dilatational) contribution. This concept was introduced in Section 3.1.5 (page 49). It has been exploited by many authors (Hughes *et al.* 1975, Simo *et al.* 1985, Moran *et al.* 1990, and Simo and Taylor 1991) in the treatment of the incompressibility constraint in finite deformation problems. Essentially, $\boldsymbol{F}$ is split according to

$$\boldsymbol{F} = \boldsymbol{F}_{\rm iso}\, \boldsymbol{F}_{\rm v}, \tag{15.2}$$

where

$$\boldsymbol{F}_{\rm iso} = (\det \boldsymbol{F})^{-1/3}\, \boldsymbol{F} \quad \text{and} \quad \boldsymbol{F}_{\rm v} = (\det \boldsymbol{F})^{1/3}\, \boldsymbol{I} \tag{15.3}$$

denote, respectively, the *isochoric* and *volumetric* components of $\boldsymbol{F}$.

[‡]Recall that for hyperelastic materials, no internal variables are present in the definition of $\hat{\boldsymbol{\sigma}}$ and the computation of the stress tensor usually involves only straightforward function evaluations. For dissipative materials, $\hat{\boldsymbol{\sigma}}$ is in general associated with numerical algorithms for integration of the constitutive equations of the underlying material model. Such algorithms have been thoroughly discussed in the preceding chapters of this book.

The F-bar deformation gradient

The key idea of the *F-bar* method is simply to replace the conventional $\boldsymbol{F}$ with a modified counterpart, which we will denote $\bar{\boldsymbol{F}}$ (the *F-bar* deformation gradient), in (15.1) in order to circumvent the spurious volumetric *locking* exhibited by standard low-order elements near the incompressible limit. To construct the *F-bar* deformation gradient the isochoric/volumetric split (15.3) is firstly applied to the conventional deformation gradient $\boldsymbol{F}$ at the Gauss point of interest as well as to the deformation gradient $\boldsymbol{F}_0$ that results from the conventional displacement interpolation at the *centroid* of the element, $\boldsymbol{\xi} = \boldsymbol{\xi}_0$ (Figure 15.1 illustrates the procedure for the four-noded quadrilateral and eight-noded hexahedron):

$$\begin{aligned} \boldsymbol{F} &= \boldsymbol{F}_{\text{iso}}\, \boldsymbol{F}_{\text{v}}, \\ \boldsymbol{F}_0 &= (\boldsymbol{F}_0)_{\text{iso}}\, (\boldsymbol{F}_0)_{\text{v}}. \end{aligned} \tag{15.4}$$

The *F-bar* deformation gradient is then defined as the product of the isochoric component of $\boldsymbol{F}$ with the volumetric component of $\boldsymbol{F}_0$, i.e. we compute

$$\bar{\boldsymbol{F}} = \boldsymbol{F}_{\text{iso}}\, (\boldsymbol{F}_0)_{\text{v}} = \left(\frac{\det \boldsymbol{F}_0}{\det \boldsymbol{F}} \right)^{1/3} \boldsymbol{F}. \tag{15.5}$$

Having defined the modified deformation gradient, the *F-bar* four- and eight-noded elements are obtained simply by replacing $\boldsymbol{F}$ with $\bar{\boldsymbol{F}}$ in (15.1); that is, for the present elements, the Cauchy stress at each Gauss point is computed as

$$\boldsymbol{\sigma}_{n+1} = \hat{\boldsymbol{\sigma}}(\boldsymbol{\alpha}_n, \bar{\boldsymbol{F}}). \tag{15.6}$$

Remark 15.1. By construction of $\bar{\boldsymbol{F}}$, the isochoric/volumetric split of the modified deformation gradient gives

$$\begin{cases} \bar{\boldsymbol{F}}_{\text{iso}} = (\det \boldsymbol{F})^{-1/3}\, \boldsymbol{F} = \boldsymbol{F}_{\text{iso}} \\ \bar{\boldsymbol{F}}_{\text{v}} = (\det \boldsymbol{F}_0)^{1/3}\, \boldsymbol{I} = (\boldsymbol{F}_0)_{\text{v}} \end{cases} \tag{15.7}$$

i.e. the isochoric component of $\bar{\boldsymbol{F}}$ coincides with the current (integration point) isochoric deformation gradient (as obtained from the conventional interpolation functions) while its volumetric part corresponds to the dilatation at the centroid of the element. In view of (15.6), this implies that, for materials whose deviatoric and volumetric constitutive responses are *decoupled*, the present formulation results in constant pressure throughout one element.

15.1.2. THE INTERNAL FORCE VECTOR

With the stress tensor computed according to (15.6), we define the modified internal virtual work functional for an *F-bar* element as

$$G^{\text{int}}_{(e)}(\boldsymbol{u}, \boldsymbol{\eta}) \equiv \int_{\varphi(\Omega^{(e)})} \hat{\boldsymbol{\sigma}}(\boldsymbol{\alpha}_n, \bar{\boldsymbol{F}}) : \nabla_x \boldsymbol{\eta}\, \mathrm{d}v. \tag{15.8}$$

The corresponding *F-bar* element internal force vector is then given simply by

$$\mathbf{f}^{\text{int}}_{(e)} = \int_{\varphi(\Omega^{(e)})} \mathbf{B}^T\, \hat{\sigma}(\boldsymbol{\alpha}_n, \bar{\boldsymbol{F}})\, \mathrm{d}v, \tag{15.9}$$

where $\mathbf{B}$ denotes the *standard* discrete spatial symmetric gradient operator (the same as that used to assemble the internal force vector of the corresponding standard element – see expression (4.92), page 103).

Remark 15.2. As far as the computation of the internal force vector is concerned, the *only* difference between *F-bar* elements and their conventional displacement-based counterparts is the replacement of $\boldsymbol{F}$ with $\bar{\boldsymbol{F}}$ in the computation of the Cauchy stress by the incremental constitutive function $\hat{\boldsymbol{\sigma}}$ (the stress-updating procedure). Thus, in addition to the evaluation of $\boldsymbol{F}$ and $\boldsymbol{\sigma}$ at each Gauss point – procedures carried out normally for the standard elements – the *F-bar* elements require only the evaluation of det $\boldsymbol{F}_0$, at their centroids, and the computation of $\bar{\boldsymbol{F}}$ which, from its definition (15.5), is obtained simply by multiplying $\boldsymbol{F}$ by a *scalar* factor. The added computational effort is, therefore, minimal (refer to Box 15.1). We emphasise that, as for the conventional elements, four and eight-point Gauss quadratures are adopted for the *F-bar* four-node quadrilateral and eight-node hexahedron respectively.

Remark 15.3. Geometrically nonlinear extensions of the so-called *B-bar* methods, such as the class of elements proposed by Hughes *et al.* (1975), Moran *et al.* (1990) and particular cases of the general methodology described by Simo *et al.* (1985), are also based on the replacement of the compatible deformation gradient with an assumed modified counterpart defined similarly to (15.5). However, in contrast to such procedures, modified gradient operators (or *B-bar* matrices) do not take part in the present formulation. Note that the *standard* B-matrix appears in (15.9). This is due to the fact that, in the *F-bar* procedure, the assumed deformation gradient has been introduced in the stress constitutive function rather than in the corresponding strain energy function. As we shall see in what follows, this crucial difference allows a relatively straightforward adaptation of existing finite strain displacement-based element routines to incorporate *F-bar* elements.

15.1.3. CONSISTENT LINEARISATION: THE TANGENT STIFFNESS

Let us now turn our attention to another crucial point in the computational implementation of the above *F-bar* elements – the computation of their tangent stiffness matrices. This is addressed in the following proposition.

Proposition 15.1. *Let the internal force vector* $\mathbf{f}^{\,\text{int}}$ *be defined as the assembly of the element vectors (15.9). Then, the corresponding tangent stiffness,* $\mathbf{K}_T$*, is obtained by the assemblage of the element stiffness matrices*

$$\mathbf{K}_T^{(e)} = \underbrace{\int_{\varphi(\Omega^{(e)})} \mathbf{G}^T\, \mathbf{a}|_{F=\bar{F}}\, \mathbf{G}\, \mathrm{d}v}_{\substack{\text{standard element}\\ \text{stiffness at } \boldsymbol{F}=\bar{\boldsymbol{F}}}} + \underbrace{\int_{\varphi(\Omega^{(e)})} \mathbf{G}^T\, \mathbf{q}\, (\mathbf{G}_0 - \mathbf{G})\, \mathrm{d}v}_{\substack{\text{additional}\\ \text{stiffness}}}, \tag{15.10}$$

where $\mathbf{G}$ *is the* standard *discrete spatial gradient operator,* $\mathbf{G}_0$ *is the gradient operator at the element centroid,* $\mathbf{a}|_{F=\bar{F}}$ *denotes the matrix form of the consistent spatial tangent modulus evaluated at* $\boldsymbol{F} = \bar{\boldsymbol{F}}$*, and* $\mathbf{q}$ *is the matrix form of the fourth-order tensor defined by*

$$\mathsf{q} = \tfrac{1}{3}\, \mathsf{a} : (\boldsymbol{I} \otimes \boldsymbol{I}) - \tfrac{2}{3}(\boldsymbol{\sigma} \otimes \boldsymbol{I}), \tag{15.11}$$

also computed at $\boldsymbol{F} = \bar{\boldsymbol{F}}$.

Proof. This follows from the linearisation of the element internal force vector (15.9). The main steps are the following. Firstly, we map (15.9) to the material configuration. Then we linearise the material form and map the linearised expression back to the spatial configuration. Let us start by observing the following relation for the first Piola–Kirchhoff stress:

$$\boldsymbol{P} \equiv (\det \boldsymbol{F})\, \boldsymbol{\sigma}\, \boldsymbol{F}^{-T} = (\det \boldsymbol{F})\, \hat{\boldsymbol{\sigma}}(\bar{\boldsymbol{F}})\, \boldsymbol{F}^{-T} = \left(\frac{\det \boldsymbol{F}_0}{\det \boldsymbol{F}}\right)^{-\frac{2}{3}} \hat{\boldsymbol{P}}(\bar{\boldsymbol{F}}) \tag{15.12}$$

where we have defined

$$\hat{\boldsymbol{P}}(\bar{\boldsymbol{F}}) \equiv (\det \bar{\boldsymbol{F}})\, \hat{\boldsymbol{\sigma}}(\bar{\boldsymbol{F}})\, \bar{\boldsymbol{F}}^{-T}. \tag{15.13}$$

For notational convenience we have omitted $\boldsymbol{\alpha}_n$ from the arguments of $\hat{\boldsymbol{\sigma}}$ (and $\hat{\boldsymbol{P}}$). With the above definition, we write the following material version of the *F-bar* internal virtual work functional (15.8):

$$G^{\text{int}}_{(e)}(\boldsymbol{u}, \boldsymbol{\eta}) = \int_{\Omega^{(e)}} \left(\frac{\det \boldsymbol{F}_0}{\det \boldsymbol{F}}\right)^{-\frac{2}{3}} \hat{\boldsymbol{P}}(\bar{\boldsymbol{F}}) : \nabla_p \boldsymbol{\eta}\, \mathrm{d}v, \tag{15.14}$$

where dependence of G^{int} on $\boldsymbol{u}$ comes from the dependence of $\bar{\boldsymbol{F}}$ on $\boldsymbol{u}$. The directional derivative of G^{int} in a direction $\boldsymbol{d}$ is given by

$$\mathrm{D}G^{\text{int}}_{(e)}(\boldsymbol{u}, \boldsymbol{\eta})\,[\boldsymbol{d}] = \int_{\Omega^{(e)}} \frac{\mathrm{d}}{\mathrm{d}\epsilon}\bigg|_{\epsilon=0} \left[\left(\frac{\det (\boldsymbol{F}_0)_\epsilon}{\det \boldsymbol{F}_\epsilon}\right)^{-\frac{2}{3}} \hat{\boldsymbol{P}}(\bar{\boldsymbol{F}}_\epsilon)\right] : \nabla_p \boldsymbol{\eta}\, \mathrm{d}v, \tag{15.15}$$

where

$$\begin{aligned} \boldsymbol{F}_\epsilon &\equiv \boldsymbol{I} + \nabla_p(\boldsymbol{u} + \epsilon\, \boldsymbol{d}) \\ (\boldsymbol{F}_0)_\epsilon &\equiv \boldsymbol{I} + \nabla_p(\boldsymbol{u} + \epsilon\, \boldsymbol{d})|_{\xi=0} \\ \bar{\boldsymbol{F}}_\epsilon &\equiv \left(\frac{\det (\boldsymbol{F}_0)_\epsilon}{\det \boldsymbol{F}_\epsilon}\right)^{\frac{1}{3}} \boldsymbol{F}_\epsilon. \end{aligned} \tag{15.16}$$

After a lengthy but straightforward application of the chain and product rules to (15.15) together with the use of basic tensor algebra relations, we obtain

$$\begin{aligned} \mathrm{D}G^{\text{int}}_{(e)}(\boldsymbol{u}, \boldsymbol{\eta})\,[\boldsymbol{d}] = \int_{\varphi(\Omega^{(e)})} &\{\mathsf{a}(\bar{\boldsymbol{F}}) : \nabla_x \boldsymbol{d} + [\tfrac{1}{3}\, \mathsf{a}(\bar{\boldsymbol{F}}) : (\boldsymbol{I} \otimes \boldsymbol{I}) \\ &- \tfrac{2}{3}\, \hat{\boldsymbol{\sigma}}(\bar{\boldsymbol{F}}) \otimes \boldsymbol{I}] : (\nabla_0 \boldsymbol{d} - \nabla_x \boldsymbol{d})\} : \nabla_x \boldsymbol{\eta}\, \mathrm{d}v, \end{aligned} \tag{15.17}$$

where $\nabla_0 \boldsymbol{d}$ denotes the spatial gradient of $\boldsymbol{d}$ evaluated at the centroid of the element. Finally, the replacement of the gradient operators, tangent modulus and other relevant tensors with the corresponding finite element matrices in (15.17) leads to the tangent stiffness formula (15.10). □

Remark 15.4. (The structure of the *F-bar* tangent stiffness) The first term on the right-hand side of (15.10) is identical to the tangent stiffness of conventional displacement elements evaluated at $\boldsymbol{F} = \bar{\boldsymbol{F}}$. The computation of the extra term indicated in (15.10) is relatively simple. In addition to the spatial tangent modulus, stress and the discrete gradient at the Gauss point of interest – already used in the calculation of the stiffness of standard elements – only the discrete gradient at the element centroid, $\mathbf{G}_0$, and the matrix $\mathbf{q}$ are required. The evaluation of $\mathbf{G}_0$ is straightforward and, from its definition, the computation of $\mathbf{q}$ also requires little computational effort. Thus, the c.p.u. time needed to compute the tangent stiffnesses of the proposed elements is just slightly longer than that consumed by their conventional counterparts. However, note that the additional stiffness term appearing in (15.10) is generally *unsymmetric* and, therefore, requires an unsymmetric solver in the finite element computations. Note that this fact is immaterial whenever the problem to be solved is characterised by an unsymmetric tangent modulus in the continuum setting, i.e. before finite element discretisation. This condition is encountered in a wide variety of practical engineering applications, such as, for instance, the simulation of metal-forming operations in which frictional contact almost invariably plays an essential role and the use of an unsymmetric solver is inevitable. More sophisticated constitutive laws, such as the coupled elastoplastic damage models discussed in Chapter 12, single crystal plasticity laws (Chapter 16), and non-associative plasticity models in general, also lead to unsymmetric tangent moduli.

Axisymmetric F-bar elements

To illustrate the simplicity of the extra stiffness term of expression (15.10), the explicit form of the matrices involved in its computation is shown here for the axisymmetric case. These are implemented in HYPLAS for the *F-bar* four-node axisymmetric quadrilateral. As the evaluation of the discrete gradient is standard, only the terms required for computation of $\mathbf{q}$ are presented below. Adopting the usual finite element convention, in which the matrix format indices $\{1,2,3,4,5\}$ correspond to the fourth-order counterparts $\{11,21,12,22,33\}$, the matrix form of the term $\mathbf{a}:(\boldsymbol{I}\otimes\boldsymbol{I})$ that takes part in the definition (15.11) of $\mathbf{q}$ is given by

$$[\mathbf{a}:(\boldsymbol{I}\otimes\boldsymbol{I})] = \begin{bmatrix} a_{11}+a_{14}+a_{15} & 0 & 0 & a_{11}+a_{14}+a_{15} & a_{11}+a_{14}+a_{15} \\ a_{21}+a_{24}+a_{25} & 0 & 0 & a_{21}+a_{24}+a_{25} & a_{21}+a_{24}+a_{25} \\ a_{31}+a_{34}+a_{35} & 0 & 0 & a_{31}+a_{34}+a_{35} & a_{31}+a_{34}+a_{35} \\ a_{41}+a_{44}+a_{45} & 0 & 0 & a_{41}+a_{44}+a_{45} & a_{41}+a_{44}+a_{45} \\ a_{51}+a_{54}+a_{55} & 0 & 0 & a_{51}+a_{54}+a_{55} & a_{51}+a_{54}+a_{55} \end{bmatrix} \tag{15.18}$$

where a_{ij} are the components of the spatial tangent modulus matrix $\mathbf{a}$. The matrix form of the remaining term $\boldsymbol{\sigma}\otimes\boldsymbol{I}$ appearing in (15.11) is simply given by

$$[\boldsymbol{\sigma}\otimes\boldsymbol{I}] = \begin{bmatrix} \sigma_{11} & 0 & 0 & \sigma_{11} & \sigma_{11} \\ \sigma_{12} & 0 & 0 & \sigma_{12} & \sigma_{12} \\ \sigma_{12} & 0 & 0 & \sigma_{12} & \sigma_{12} \\ \sigma_{22} & 0 & 0 & \sigma_{22} & \sigma_{22} \\ \sigma_{33} & 0 & 0 & \sigma_{33} & \sigma_{33} \end{bmatrix}. \tag{15.19}$$

In the expressions above, the matrix index 5 (or tensorial index 33) represents the circumferential directions, other indices correspond to the in-plane components.

15.1.4. PLANE STRAIN IMPLEMENTATION

We now turn our attention to the plane strain implementation of *F-bar* elements. To ensure that the modified deformation gradient corresponds to a plane strain deformation, expression (15.5) is replaced in plane strain *F-bar* elements with

$$\bar{\boldsymbol{F}} = \left[\begin{array}{cc|c} \multicolumn{2}{c|}{\multirow{2}{*}{$\bar{\boldsymbol{F}}_{\rm p}$}} & 0 \\ & & 0 \\ \hline 0 & 0 & 1 \end{array}\right], \tag{15.20}$$

where $\bar{\boldsymbol{F}}_{\rm p}$ is the assumed modified counterpart of the *in-plane component*, $\boldsymbol{F}_{\rm p}$, of the deformation gradient, defined by

$$\bar{\boldsymbol{F}}_{\rm p} = \left(\frac{\det \boldsymbol{F}_{\rm p0}}{\det \boldsymbol{F}_{\rm p}}\right)^{1/2} \boldsymbol{F}_{\rm p}. \tag{15.21}$$

Remark 15.5. Note from (15.20) that the redefined modified deformation gradient represents indeed a plane strain state. This would not be the case if the general definition (15.5) were applied under plane strain conditions. Note that, under the present definition, $(15.7)_2$ remains valid and pressure distributions within elements will be constant for materials with decoupled volumetric/deviatoric response. Expression $(15.7)_1$, however, is no longer valid, i.e. the deviatoric component of the modified deformation gradient does not coincide in general with that of the deviatoric deformation gradient that results from the standard bilinear interpolation at the Gauss points.

The algorithmic implementation of the plane strain version of the proposed four-node quadrilateral element follows that described in Boxes 15.1 and 15.2, except that the exponent $1/3$ must be replaced by $1/2$ in the computation of the modified deformation gradient. Also, as a consequence of definition (15.20), expression (15.11) for the tensor q, which arises from the linearisation of the internal force vector, must be replaced by

$$\mathsf{q} = \tfrac{1}{2}\, \mathsf{a} : (\boldsymbol{I} \otimes \boldsymbol{I}) - \tfrac{1}{2}(\boldsymbol{\sigma} \otimes \boldsymbol{I}), \tag{15.22}$$

in the computation of the element tangent stiffness matrix (Box 15.2). The explicit form of the matrices $[\mathsf{a} : (\boldsymbol{I} \otimes \boldsymbol{I})]$ and $[\boldsymbol{\sigma} \otimes \boldsymbol{I}]$, in this case, is obtained from expressions (15.18) and (15.19) simply by deleting all components related to the circumferential direction, that is, we have

$$[\mathsf{a} : (\boldsymbol{I} \otimes \boldsymbol{I})] = \begin{bmatrix} a_{11} + a_{14} & 0 & 0 & a_{11} + a_{14} \\ a_{21} + a_{24} & 0 & 0 & a_{21} + a_{24} \\ a_{31} + a_{34} & 0 & 0 & a_{31} + a_{34} \\ a_{41} + a_{44} & 0 & 0 & a_{41} + a_{44} \end{bmatrix} \tag{15.23}$$

and

$$[\boldsymbol{\sigma} \otimes \boldsymbol{I}] = \begin{bmatrix} \sigma_{11} & 0 & 0 & \sigma_{11} \\ \sigma_{12} & 0 & 0 & \sigma_{12} \\ \sigma_{12} & 0 & 0 & \sigma_{12} \\ \sigma_{22} & 0 & 0 & \sigma_{22} \end{bmatrix}. \tag{15.24}$$

15.1.5. COMPUTATIONAL IMPLEMENTATION ASPECTS

As pointed out in Remarks 15.2 and 15.4, in addition to the computations carried out normally for the conventional elements, only a few extra operations are required for the *F-bar* four-node quadrilateral and eight-node hexahedron in the evaluation of the internal force vector and the tangent stiffness matrix. In a code which already supports the standard four-node quadrilateral (or eight-node brick, if three-dimensional analysis is sought) and allows access to an unsymmetric solver, the necessary modifications are relatively simple. The routines for calculation of the internal force and tangent stiffness can be reused and adapted to accommodate the new elements. The computational procedures for internal force and stiffness evaluation of the *F-bar* elements are schematically illustrated in Boxes 15.1 and 15.2, where the corresponding conventional element routines have been extended with the additional operations (framed) required. The procedures of Boxes 15.1 and 15.2 are implemented, respectively, in subroutines `IFFBA2` (**I**nternal **F**orce calculation for **F**-**BA**r element in **2**-D) and `STFBA2` (**ST**iffness matrix calculation for **F**-**BA**r element in **2**-D). The reader will note that these routines have been obtained by introducing relatively small modifications to the corresponding standard element routines `IFSTD2` and `STSTD2`.

15.1.6. NUMERICAL TESTS

This section presents a set of numerical examples whose purpose is to assess some basic performance indicators of the four-noded quadrilateral *F-bar* element and to comment on its suitability for analysis in various situations of interest. Some of the results are compared with similar computations carried out with enhanced assumed strain elements for near incompressible analysis. Further examples of finite element analysis with *F-bar* elements (including the eight-noded *F-bar* brick) under near incompressibility are provided in Sections 13.6, 14.9 and 16.8.

Cook's membrane.

This example is frequently used in the literature to assess the convergence properties of finite elements under a mixture of shear and bending strains (Glaser and Armero, 1997; Korelc and Wriggers, 1996; Simo and Armero, 1992; Simo and Rifai, 1990). The purpose here is only to illustrate the convergence properties of the four-noded *F-bar* element under near-incompressibility. It is a well-known fact that, under near-incompressibility, conventional low-order elements require excessively fine discretisations to produce solutions sufficiently close to the converged one. 'Adequate' elements should converge with more sensible coarser meshes. The problem consists of a tapered and swept panel of unit thickness, illustrated in

Box 15.1. Internal force vector computation for *F-bar* element.

HYPLAS procedure:	IFFBA2

(i)
- Compute $\mathbf{G}_0$ (standard $\mathbf{G}$ matrix at $\boldsymbol{\xi} = \boldsymbol{\xi}_0$)
- Evaluate $\det \boldsymbol{F}_0$, with $\boldsymbol{F}_0 := \boldsymbol{I} + \mathbf{G}_0\, \mathbf{u}$

(ii) Set Gauss point coordinates $\boldsymbol{\xi}_i$, weights w_i and Jacobian determinants j_i

(iii) Do $i = 1,\ n_{\text{gausp}}$ (loop over Gauss points)

- compute standard $\mathbf{G}$ matrix at $\boldsymbol{\xi}_i$
- $\boldsymbol{F} := \boldsymbol{I} + \mathbf{G}\,\mathbf{u}$ (conventional deformation gradient)
- $\bar{\boldsymbol{F}} := \left(\dfrac{\det \boldsymbol{F}_0}{\det \boldsymbol{F}}\right)^{\frac{1}{3}} \boldsymbol{F}$ (modified deformation gradient)
- $\boldsymbol{\sigma} := \hat{\boldsymbol{\sigma}}(\boldsymbol{\alpha}_n, \bar{\boldsymbol{F}})$ (call stress update routine)
- compute standard $\mathbf{B}$ matrix at $\boldsymbol{\xi}_i$
- $\mathbf{f}^{\text{int}}_{(e)} := \mathbf{f}^{\text{int}}_{(e)} + w_i\, j_i\, \mathbf{B}^T \boldsymbol{\sigma}$ (add i^{th} Gauss point contribution)

End do

Figure 15.2(a), with one edge clamped and a uniformly distributed vertical load of linear density $f = 6.25$ applied to the opposite edge. The corresponding resultant applied load in the initial configuration is $F = 100$. Throughout the loading programme, the applied load remains vertical (it does not follow the direction tangent to the faces of the loaded elements), but the resultant load will change according to the variation of length of the loaded faces so as to preserve the prescribed constant load density. Plane strain condition is assumed and a regularised neo-Hookean material with shear modulus $\mu = 80.1938$ and bulk modulus $k = 40.0942 \times 10^4$ is adopted. Note that near incompressibility is achieved with the high ratio k/μ, of order 10^4. Several meshes are considered, so that the convergence of the solution with mesh refinement can be assessed. A mesh of 4×4 elements is depicted in Figure 15.2(a). For all meshes considered, the total load is applied in five increments. The results are shown in the graph of Figure 15.2(b) where the final vertical displacement obtained at the upper right corner of the panel (point A of Figure 15.2(a)) is plotted against the number of elements per side. Results obtained with the *Q1/E4* element of Simo and Armero (1992) are also shown for comparison. It can be seen that the convergence of the *F-bar* quadrilateral in this case is almost identical to the behaviour of the element *Q1/E4* with five-point integration rule.[§]

[§]The five-point integration rule for quadrilateral domains (whose corresponding weights and integration point positions are found in subroutine GAUS2D of program HYPLAS) has been introduced (Simo *et al.*, 1993) as an improvement to the earlier version of the *Q1/E4* enhanced assumed strain element.

Box 15.2. Stiffness matrix computation for *F-bar* elements.

HYPLAS procedure:	STFBA2

(i)
- Compute $\mathbf{G}_0$ (standard $\mathbf{G}$ matrix at $\boldsymbol{\xi} = \boldsymbol{\xi}_0$)
- Evaluate $\det \boldsymbol{F}_0$, with $\boldsymbol{F}_0 := \boldsymbol{I} + \mathbf{G}_0\,\mathbf{u}$

(ii) Set Gauss point coordinates $\boldsymbol{\xi}_i$, weights w_i and Jacobian determinants j_i

(iii) Do $i = 1,\ n_{\text{gausp}}$ (loop over Gauss points)

- compute standard $\mathbf{G}$ matrix at $\boldsymbol{\xi}_i$
- $\boldsymbol{F} := \boldsymbol{I} + \mathbf{G}\,\mathbf{u}$
- $\bar{\boldsymbol{F}} := \left(\dfrac{\det \boldsymbol{F}_0}{\det \boldsymbol{F}}\right)^{1/3} \boldsymbol{F}$ (modified deformation gradient)
- $\mathbf{a} := \hat{\mathbf{a}}(\bar{\boldsymbol{F}})$ (tangent modulus computation routine)
- $\mathbf{K}_T^{(e)} := \mathbf{K}_T^{(e)} + w_i\, j_i\, \mathbf{G}^T \mathbf{a}\, \mathbf{G}$
- – compute matrix $\mathbf{q}$ defined by (15.11)
  – $\mathbf{K}_T^{(e)} := \mathbf{K}_T^{(e)} + w_i\, j_i\, \mathbf{G}^T \mathbf{q}\, (\mathbf{G}_0 - \mathbf{G})$

End do

Stretching of a double-notched specimen.

This example, originally considered by Nagtegaal *et al.* (1974) in the geometrically linear regime, illustrates the need for appropriate treatment of the incompressibility constraint to allow for accurate prediction of limit loads. It consists of the plane strain simulation of the stretching of a deep double-notched elastoplastic specimen. The geometry of the problem is illustrated in Figure 15.3(a). Only one symmetric quarter of the specimen, discretised by the 5×15 element mesh shown in Figure 15.3(a), is used in the simulation. The material is assumed to be elastic-perfectly plastic with J_2 flow rule and yield stress:

$$\sigma_y = 0.45.$$

The elastic behaviour is defined by the Hencky material model with Young's modulus $E = 206.9$ and Poisson ratio $\nu = 0.29$. This corresponds to the shear and bulk moduli

$$\mu = 80.1938, \quad K = 164.21.$$

Vertical displacement is applied to the top nodes of the mesh up to a total displacement $u = 0.3$. The simulation is carried out here using the four-node element presented in this section as well as the *Q1/E4* element and the standard displacement-based four-node quadrilateral.

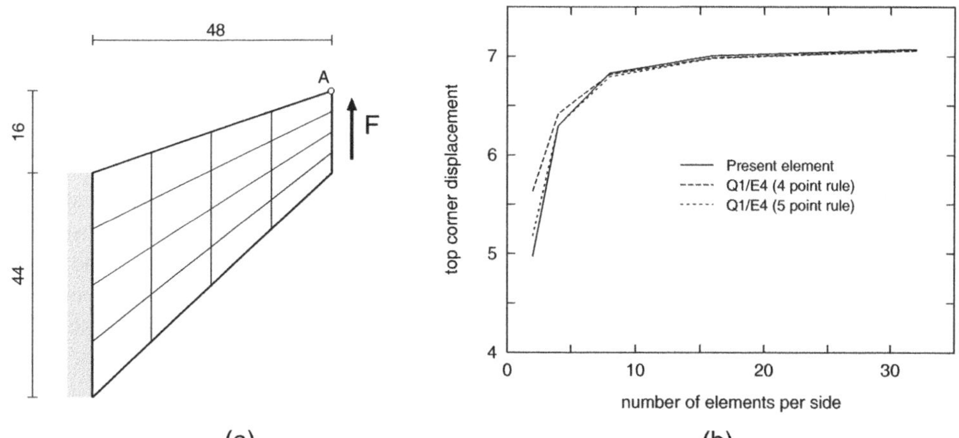

Figure 15.2. Cook's membrane: (a) geometry and boundary conditions; (b) convergence of the solution with mesh refinement. (Reproduced with permission from Design of simple low-order finite elements for large-strain analysis of nearly incompressible solids, EA de Souza Neto, D Perić, M Dutko and DRJ Owen, *International Journal of Solids and Structures*, Vol 33, Issue 20–22 © 1996 Elsevier Science Ltd.)

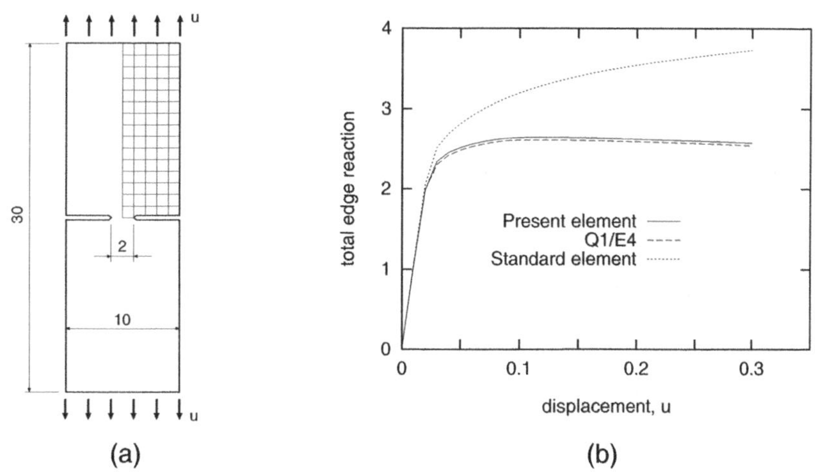

Figure 15.3. Stretching of a double notched elastoplastic specimen: (a) geometry and finite element discretisation; (b) reaction–displacement diagram. (Reproduced with permission from Design of simple low-order finite elements for large-strain analysis of nearly incompressible solids, EA de Souza Neto, D Perić, M Dutko and DRJ Owen, *International Journal of Solids and Structures*, Vol 33, Issue 20–22 © 1996 Elsevier Science Ltd.)

The total edge reactions per unit thickness obtained for each computation is plotted versus the prescribed displacement in Figure 15.3(b). It can be seen that both the present element and *Q1/E4* are able to predict the existence of a limit load – a phenomenon that is not captured by the standard four-node quadrilateral. The force–displacement curve predicted by the present formulation is very close to that obtained with the enhanced element *Q1/E4*.

Unconstrained single elastoplastic element test.

The purpose of this test is to give a practical insight into the adequacy of a finite element to capture strain localisation phenomena. It has been used by Simo and Armero (1992) to assess the performance of assumed enhanced strain elements. Here, the unconstrained single element test is carried out for the four-node *F-bar* quadrilateral. The results are compared to similar computations with the enhanced element *Q1/E4*. The problem is schematically represented in Figure 15.4(a). The element, with the dimensions and kinematic constraints indicated, is subjected to prescribed horizontal displacement, u, at its top node. Plane strain condition is assumed. The material model used is the same as in the notched specimen example except that the assumption of perfect plasticity is replaced by the following isotropic hardening/softening law:

$$\sigma_y(\bar{\varepsilon}^p) = (\sigma_\infty - \sigma_0)\,[1 - \exp(-\delta\,\bar{\varepsilon}^p)] + H\bar{\varepsilon}^p,$$

with constants

$$\sigma_0 = 0.45, \quad \sigma_\infty = 0.715, \quad \delta = 16.93, \quad H = -0.012924.$$

The corresponding force per unit thickness, P, obtained during the tests is plotted in Figure 15.4(b). It can be seen that both the *F-bar* element and, as verified by Simo and Armero (1992), the *Q1/E4* element with four-point quadrature are able to capture the global softening that characterises strain localisation. This indicates a possible suitability of these elements for localisation problems. Interestingly, if the five-point integration rule suggested by Simo *et al.* (1993) is employed, the forces obtained for the *Q1/E4* element increase dramatically after following closely the results for the four-Gauss point rule up to approximately $u = 3.0$. This phenomenon might be attributed to a bifurcation of the solution and was noticed by Schönauer *et al.* (1995). The results obtained with the standard four-node quadrilateral are also plotted and show that, for this element, no softening occurs. The deformed configurations, at $u = 10.0$, of each of the elements discussed are depicted in Figure 15.5. The final geometry of the proposed element lies between the deformed geometries of the *Q1/E4* with four and five integration points. We point out the substantial difference that results from the application of distinct quadratures in the *Q1/E4* element. Severe locking is exhibited by the standard quadrilateral as a consequence of the pointwise enforcement of the plastic incompressibility constraint.

Suitability for adaptive analysis.

Due to the large element distortions involved, as well as accuracy considerations, many finite strain problems of industrial interest can only be solved if the finite element mesh is redefined at some stage of the solution procedure. Whenever a new mesh is chosen, all relevant variables that define the problem at hand must be appropriately transferred from the old mesh to the new one so that the solution process can continue. Formulations such as enhanced assumed strain methods (Glaser and Armero, 1997; Korelc and Wriggers, 1996; Simo and Armero, 1992; Simo *et al.*, 1993) and incompatible modes elements (Crisfield *et al.*, 1995) are based on the enrichment of the element strain field with internal deformation modes defined by a set of local internal element parameters. For such methodologies, the appropriate transfer of internal parameters between meshes is crucial and seems not to be

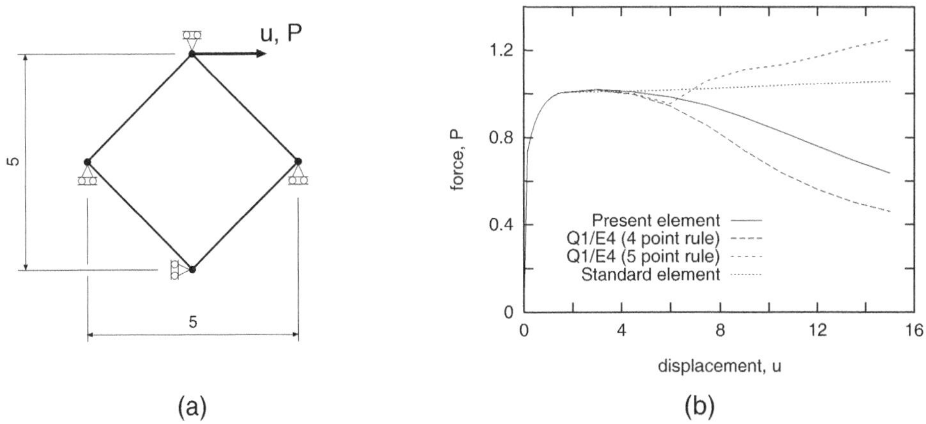

Figure 15.4. Single element test: (a) geometry and load; (b) force-displacement curve. (Reproduced with permission from Design of simple low-order finite elements for large-strain analysis of nearly incompressible solids, EA de Souza Neto, D Perić, M Dutko and DRJ Owen, *International Journal of Solids and Structures*, Vol 33, Issue 20–22 © 1996 Elsevier Science Ltd.)

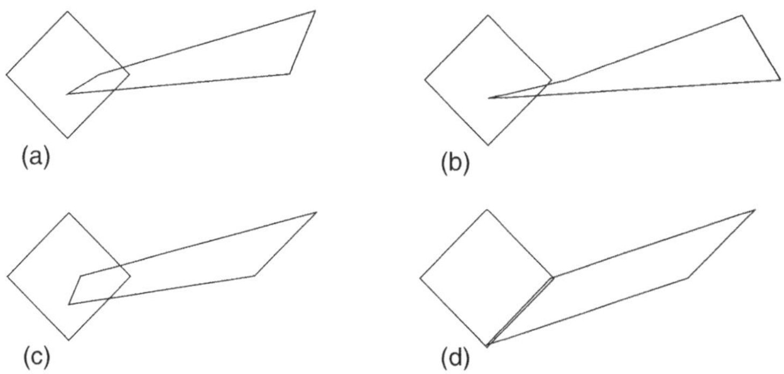

Figure 15.5. Single element test. Deformed configurations at $u = 10.0$: (a) present element; (b) *Q1/E4* with four-point rule; (c) *Q1/E4* with five-point rule; (d) standard four-node quadrilateral. (Reproduced with permission from Design of simple low-order finite elements for large-strain analysis of nearly incompressible solids, EA de Souza Neto, D Perić, M Dutko and DRJ Owen, *International Journal of Solids and Structures*, Vol 33, Issue 20–22 © 1996 Elsevier Science Ltd.)

clear at present. In contrast, such an issue does not arise in methodologies such as the *F-bar* procedure described above and geometrically nonlinear extensions of the classical *B-bar* method (Moran *et al.*, 1990). The absence of element internal parameters makes the incorporation of such methods into adaptive remeshing environments rather straightforward. Exactly as for standard displacement-based formulations, such methods require only nodal displacements and *physical* variables (e.g. plastic, elastic strains, hardening parameters, etc.) to be transferred between meshes. The suitability of the four-noded *F-bar* quadrilateral in particular for adaptive analysis is illustrated here in the simulation of the upsetting of an elastoplastic cylindrical billet. The billet, with radius $r = 9$ mm and height $h = 30$ mm is compressed between two flat tools (assumed rigid) subjected to sticking contact condition on the interface. The initial tool/workpiece configuration is schematically illustrated in

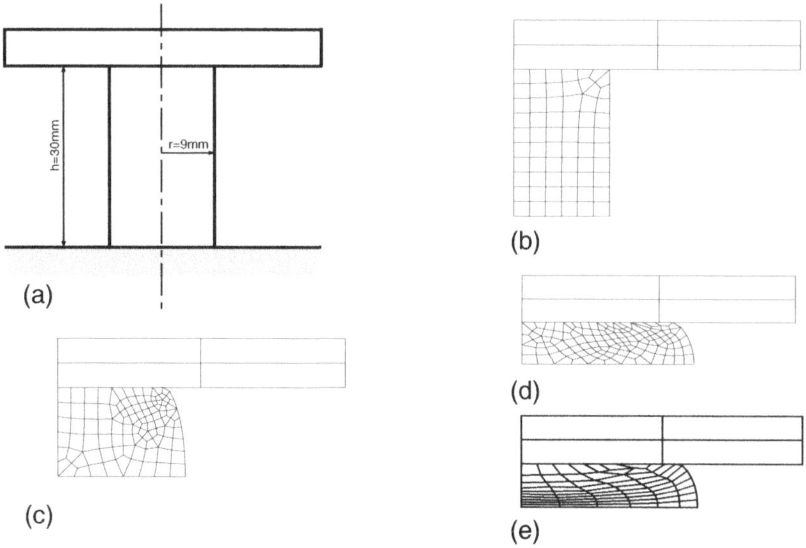

Figure 15.6. Adaptive analysis: (a) initial tool/workpiece configuration; (b) initial mesh; (c) deformed mesh at 40% compression; (d) deformed mesh at 70% compression; (e) deformed mesh at 70% compression *without remeshing*. (Reproduced with permission from Design of simple low-order finite elements for large-strain analysis of nearly incompressible solids, EA de Souza Neto, D Perić, M Dutko and DRJ Owen, *International Journal of Solids and Structures*, Vol 33, Issue 20–22 © 1996 Elsevier Science Ltd.)

Figure 15.6(a). The Hencky elastic model and the von Mises elastoplastic law are used to model the billet. Linear isotropic hardening is assumed:

$$\sigma_y(\bar{\varepsilon}^p) = \sigma_{y_0} + H\bar{\varepsilon}^p,$$

with hardening modulus $H = 0.7$ GPa and initial yield stress $\sigma_{y_0} = 0.45$ GPa. The Young's modulus and Poisson ratio are, respectively, $E = 200$ GPa and $\nu = 0.3$, corresponding to

$$\mu = 76.92 \text{ GPa}, \quad K = 166.67 \text{ GPa}.$$

The initial mesh employed to discretise the symmetric quarter of the cylinder is plotted in Figure 15.6(b). The criterion for mesh refinement/de-refinement is based on the *incremental plastic work* (refer to Perić *et al.* (1994) for details). Two other meshes, generated during the adaptive analysis, are shown in Figures 15.6(c) and 15.6(d). The configurations illustrated correspond to 40% and 70% compression of the billet respectively. Due to the concentration of the plastic process near the corner of the billet, substantial refinement is detected in that region of the mesh of Figure 15.6(c). At the later stage of 70% compression, shown in Figure 15.6(d), the rate of plastic dissipation and, therefore, the elements, are more uniformly distributed. Note that, at any instant, all elements have a very good aspect ratio. The result of a similar computation without adaptivity, i.e. maintaining the initial discretisation throughout the entire process, is shown in Figure 15.6(e) where the excessive distortion of the mesh is clear.

15.1.7. OTHER CENTROID SAMPLING-BASED *F-BAR* ELEMENTS

For the *F-bar* elements described so far, we have assumed that the volumetric component of the deformation gradient is sampled only at the centroid of the element. Within this framework, finite element codes incorporating popular standard elements such as the six-noded triangle, the eight-noded bilinear quadrilateral and the ten-noded tetrahedron can be easily extended to include *F-bar* versions of such elements. The required modifications are exactly the same as those described above for the four-noded quadrilateral and the eight-noded brick.

15.1.8. A MORE GENERAL *F-BAR* METHODOLOGY

More general *F-bar* elements can be devised by replacing the centroid-only volumetric sampling with a suitable approximation of the volumetric deformation gradient throughout the element generated by volumetric sampling in more points. The general procedure is outlined in the following. Let $n_{\rm vol}$ be the prescribed number of volumetric sampling points and let

$$\bar{N}_j(\boldsymbol{\xi}), \quad j = 1, 2, \ldots, n_{\rm vol}$$

be the prescribed functions which generate the approximate volumetric deformation gradient field within the element. With $\bar{\boldsymbol{\xi}}_i$ $(i = 1, \ldots, n_{\rm vol})$ denoting the prescribed local coordinates of the volumetric sampling points, we assume that, analogously to the shape functions for displacement interpolation, the functions $\bar{N}_j$ satisfy

$$\bar{N}_j(\bar{\boldsymbol{\xi}}_i) = \delta_{ij}. \tag{15.25}$$

The approximation to the volumetric deformation gradient field, denoted $\bar{J}$, is constructed as

$$\bar{J}(\boldsymbol{\xi}) = \sum_{j=1}^{n_{\rm vol}} J(\bar{\boldsymbol{\xi}}_j)\, \bar{N}_j(\boldsymbol{\xi}), \tag{15.26}$$

where $J(\bar{\boldsymbol{\xi}}_j) \equiv \det \boldsymbol{F}(\bar{\boldsymbol{\xi}}_j)$ is the standard deformation gradient obtained as usual from the interpolated displacement field of the element, evaluated at the volumetric sampling point with local coordinate $\bar{\boldsymbol{\xi}}_j$. The modified deformation gradient, $\bar{\boldsymbol{F}}$, at each Gauss point with coordinates $\boldsymbol{\xi}_i$ is obtained as the composition of the isochoric deformation gradient at $\boldsymbol{\xi}_i$ with the approximate volumetric deformation gradient (15.26) at $\boldsymbol{\xi}_i$; that is, we compute

$$\begin{aligned} \bar{\boldsymbol{F}}(\boldsymbol{\xi}_i) &= \left(\frac{\bar{J}(\boldsymbol{\xi}_i)}{J(\boldsymbol{\xi}_i)}\right)^{1/3} \boldsymbol{F}(\boldsymbol{\xi}_i) \\ &= \left(\frac{\sum_{j=1}^{n_{\rm vol}} J(\bar{\boldsymbol{\xi}}_j)\, \bar{N}_j(\boldsymbol{\xi}_i)}{J(\boldsymbol{\xi}_i)}\right)^{1/3} \boldsymbol{F}(\boldsymbol{\xi}_i). \end{aligned} \tag{15.27}$$

Clearly, (15.27) reduces to the centroid-only formula (15.5) if $n_{\rm vol} = 1$ with $\bar{\boldsymbol{\xi}}_1$ being the coordinate of the centroid. Following (15.21), in plane strain, J in the above is replaced with the determinant of the in-plane deformation gradient $J_{\rm p} = \det \boldsymbol{F}_{\rm p}$ and the exponent $1/3$ is replaced with $1/2$. Having computed $\bar{\boldsymbol{F}}$ according to (15.27), the remainder of the procedure is identical to that of the centroid-only sampling *F-bar* elements, except for the tangent stiffness matrix whose explicit expression is given below.

The tangent stiffness

By means of a linearisation procedure completely analogous to that outlined in Section 15.1.3 we can obtain the formula for the element tangent stiffness for general *F-bar* elements. For a generic element based on the approximation (15.26) to J we have

$$\mathbf{K}_T^{(e)} = \underbrace{\int_{\varphi(\Omega^{(e)})} \mathbf{G}^T \, \mathbf{a}|_{F=\bar{F}} \mathbf{G} \, dv}_{\substack{\text{standard element} \\ \text{stiffness at } F=\bar{F}}} + \underbrace{\int_{\varphi(\Omega^{(e)})} \mathbf{G}^T \, \mathbf{q} \, (\bar{\mathbf{G}} - \mathbf{G}) \, dv}_{\substack{\text{additional} \\ \text{stiffness}}}, \tag{15.28}$$

where

$$\bar{\mathbf{G}}(\boldsymbol{\xi}) \equiv \frac{1}{\bar{J}(\boldsymbol{\xi})} \sum_{j=1}^{n_{\text{vol}}} J(\bar{\boldsymbol{\xi}}_j) \, \bar{N}_j(\boldsymbol{\xi}) \, \mathbf{G}(\bar{\boldsymbol{\xi}}_j). \tag{15.29}$$

The element stiffness in this case has the same structure as its counterpart (15.10) for centroid sampling-only *F-bar* elements. The only difference lies in the additional stiffness term where the original gradient operator $\mathbf{G}_0$ (evaluated at the centroid) used in (15.10) is replaced here with the operator $\bar{\mathbf{G}}$ defined by (15.29).

Logarithm-based volumetric approximation

Alternatively to (15.26) we may adopt, as suggested by Moran *et al.* (1990), a similar approximation to the *logarithm* of J. In this case (15.26) is replaced with

$$\log \bar{J}(\boldsymbol{\xi}) = \sum_{j=1}^{n_{\text{vol}}} \log[J(\bar{\boldsymbol{\xi}}_j)] \, \bar{N}_j(\boldsymbol{\xi}), \tag{15.30}$$

and the modified deformation gradient is computed as

$$\bar{\boldsymbol{F}}(\boldsymbol{\xi}_i) = \left(\frac{\exp[\sum_{j=1}^{n_{\text{vol}}} \log J(\bar{\boldsymbol{\xi}}_j) \, \bar{N}_j(\boldsymbol{\xi}_i)]}{J(\boldsymbol{\xi}_i)} \right)^{1/3} \boldsymbol{F}(\boldsymbol{\xi}_i). \tag{15.31}$$

Number of nodes and volumetric sampling points

Recall that the basic reason why formulations such as the *F-bar* method are used is to avoid the overstiff solutions (*locking*) produced by low-order conventional displacement-based finite elements near the incompressibility limit. With conventional elements, locking can be overcome by using a sufficiently large number of nodes per element or by increasing sufficiently the number of elements in the mesh. In other words, we can say that an increase in the number of degrees of freedom reduces the stiffness of the element. This is also true for *F-bar* elements; that is, an increase in the number of nodes reduces the stiffness of *F-bar* elements. For such elements, however, if the number of volumetric sampling points is increased, the element stiffens and may become prone to locking behaviour. Note that the corresponding standard element is recovered if the same number of volumetric sampling points and Gauss integration points are adopted in *F-bar* elements. For instance, a four-noded *F-bar* quadrilateral with four volumetric sampling points and four Gauss integration points has identical behaviour as the standard four-noded element with four Gauss points. Successful

F-bar elements will have fewer volumetric sampling points than Gauss integration points. The arguments here are similar to those used in mixed u/p formulations described in Section 15.3 (refer to the comments made at the end of Section 15.3.2).

Combined standard/F-bar elements

Further refinement upon the *F-bar* formulation can be introduced by letting the deformation gradient be a linear combination of the $\bar{\boldsymbol{F}}$ field constructed within the element and the field $\boldsymbol{F}$ obtained as in the standard displacement element. In this case, as suggested by Moran *et al.* (1990) in the context of geometrically nonlinear *B-bar* extensions, we replace the $\bar{\boldsymbol{F}}$ gradient of the above formulation with the linear combination

$$\hat{\boldsymbol{F}} \equiv \alpha\, \boldsymbol{F} + (1 - \alpha)\, \bar{\boldsymbol{F}}, \tag{15.32}$$

where α is a prescribed parameter. If $\alpha = 0$, the *F-bar* formulation described above is recovered. The standard displacement-based element is obtained if $\alpha = 1$. For $0 < \alpha < 1$, the element behaves as a combination of standard and *F-bar* procedures. For instance, if the stiffness of a specific *F-bar* element ($\alpha = 0$) is too low and makes the element prone to instabilities, the problem can be corrected suitably by increasing α (stiffening the element). However, it is the view of the authors that artificial *stabilisation* parameters such as α in the combined *standard/F-bar* approach should be avoided as much as possible. The optimal value of parameters of this type is usually problem-dependent and its determination requires a relatively good experience of the finite element analyst.

15.1.9. THE *F-BAR*-PATCH METHOD FOR SIMPLEX ELEMENTS

The *F-bar* procedure described above cannot be applied directly, without modification, to simplex elements (linear triangles in two-dimensional/axisymmetric problems and linear tetrahedra in three dimensions). Simplex elements produce a uniform deformation gradient so that, at any point within the element domain,

$$\boldsymbol{F} = \boldsymbol{F}_0. \tag{15.33}$$

The use of an *F-bar* procedure based on the original definition (15.5) of the modified deformation gradient would not make sense in this case.

However, the use of simplex finite elements is very desirable, particularly in the solution of three-dimensional problems with complex geometry. The generation of hexahedral element meshes for complex three-dimensional geometries remains largely an open issue. Many difficulties exist and this poses serious limitations on the use of such elements in the solution of many problems of interest.

An effective alternative proposed in de Souza Neto *et al.* (2005) is to redefine the *F-bar* deformation gradient so as to produce the required relaxation of the pointwise incompressibility constraint in the context of simplex elements. Then, rather than working separately within individual elements, we will consider a *patch* of simplex elements. A typical patch is illustrated in Figure 15.7 (drawn in solid lines). With $\mathscr{P}$ denoting the set of elements forming a predefined patch, for each element $e \in \mathscr{P}$, the modified deformation gradient, $\bar{\boldsymbol{F}}_e$,

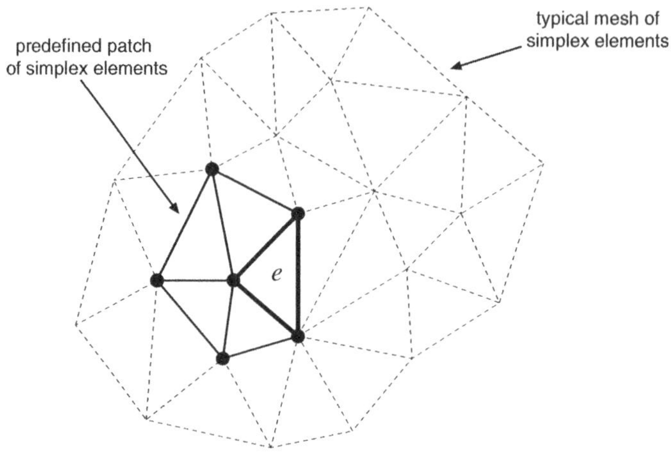

Figure 15.7. The *F-bar*-Patch Method. Patch of simplex elements. (Reproduced with permission from F-bar-based linear triangles and tetrahedra for finite strain analysis of nearly incompressible solids. Part I: Formulation and benchmarking, EA de Souza Neto, FM Andrade Pires and DRJ Owen, *International Journal for Numerical Methods in Engineering*, Vol 62 © 2005 John Wiley & Sons, Ltd.)

is defined as:

$$\bar{\boldsymbol{F}}_e = \left[\frac{v_{\text{patch}}}{V_{\text{patch}}\,(\det \boldsymbol{F}_e)} \right]^{\frac{1}{3}} \boldsymbol{F}_e, \tag{15.34}$$

where $\boldsymbol{F}_e$ is the deformation gradient obtained from the standard linear displacement interpolation within element e and v_{patch} and V_{patch} are, respectively, the deformed and reference (undeformed) volume of the *patch* $\mathscr{P}$:

$$v_{\text{patch}} = \sum_{i \in \mathscr{P}} v_i, \quad V_{\text{patch}} = \sum_{i \in \mathscr{P}} V_i. \tag{15.35}$$

In the above, v_i and V_i denote, respectively, the deformed and reference (undeformed) volume of the generic element i of patch $\mathscr{P}$.

The above definition implies that the determinant of the modified deformation gradient is identical for all elements e of patch $\mathscr{P}$ and is given by:

$$\bar{J}_e \equiv \det \bar{\boldsymbol{F}}_e = \frac{v_{\text{patch}}}{V_{\text{patch}}}, \tag{15.36}$$

that is, $\det \bar{\boldsymbol{F}}_e$ is the ratio between the deformed and undeformed volume of the predefined *patch* $\mathscr{P}$ of elements. Thus, the use of (15.34) under the incompressibility constraint requires that the *patch* $\mathscr{P}$ of elements preserves its volume, *even though individual elements of the patch may suffer volume changes* during deformation.

Definition of element patches

The application of the *F-bar*-Patch method to the solution of a problem discretised by an arbitrary mesh of simplex elements requires the mesh to be split into a number of

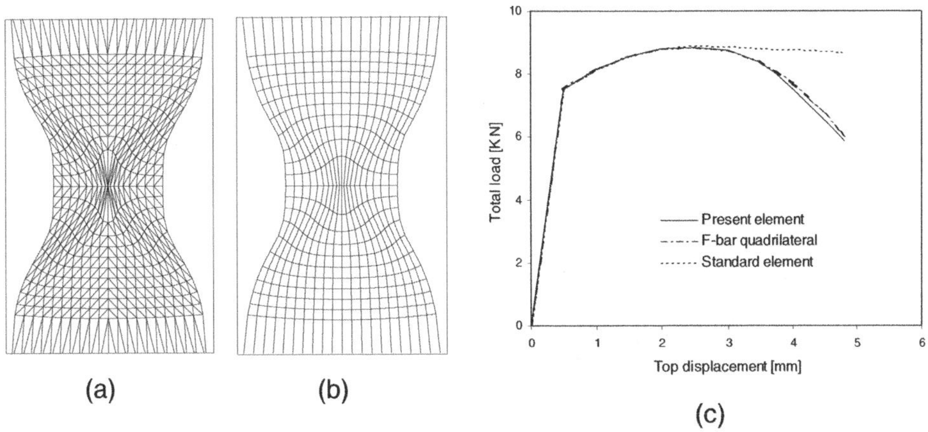

Figure 15.8. Plane strain localisation: (a) *F-bar*-Patch solution with patches of two linear triangles; (b) four-noded quadrilateral *F-bar* solution, and; (c) force–displacement diagrams. (Reproduced with permission from F-bar-based linear triangles and tetrahedra for finite strain analysis of nearly incompressible solids. Part I: Formulation and benchmarking, EA de Souza Neto, FM Andrade Pires and DRJ Owen, *International Journal for Numerical Methods in Engineering*, Vol 62 © 2005 John Wiley & Sons, Ltd.)

non-overlapping patches of elements. The more elements in each patch, the greater the constraint relaxation achieved. Then, in defining such patches, it is crucial to have in mind that excessive constraint relaxation (too many elements in the patch) may lead to spurious zero-energy mechanisms. On the other hand, insufficient constraint relaxation (too few elements in the patch) may result in incompressibility locking. Numerical experience shows that patches of two elements in two-dimensional/axisymmetric problems and eight elements in three-dimensional analyses produce sufficient constraint relaxation to avoid incompressibility locking without showing spurious mechanisms. Figure 15.8(a) shows the detail of the deformed mesh obtained with patches of two triangles in the solution of the plane strain localisation problem whose four-noded *F-bar* solution is plotted in Figure 14.12(b) (page 614). The detail of the localisation zone for the *F-bar* quadrilateral is shown for comparison alongside the *F-bar*-Patch result, in Figure 15.8(b). The deformation pattern obtained with the *F-bar*-Patch element is virtually identical to that obtained with the four-noded *F-bar* quadrilateral. The force–displacement curves for both elements is shown in Figure 15.8(c). The curves for both elements virtually coincide. An *F-bar*-Patch solution of the three-dimensional necking problem of Section 14.9.2 (from page 607) is shown in Figure 15.9(a), where patches of eight linear tetrahedra have been used. The deformed mesh is in close agreement with that produced with the *F-bar* hexahedron (shown in Figure 14.9). The force-displacement diagrams of Figure 15.9(b) show that the results obtained with the *F-bar*-Patch method using patches of eight tetrahedra virtually coincide with analogous results obtained in an axisymmetric analysis using patches of two linear triangles. These results are in very close agreement with those obtained with the *F-bar* hexahedral and axisymmetric quadrilateral elements (shown in Figure 14.11).

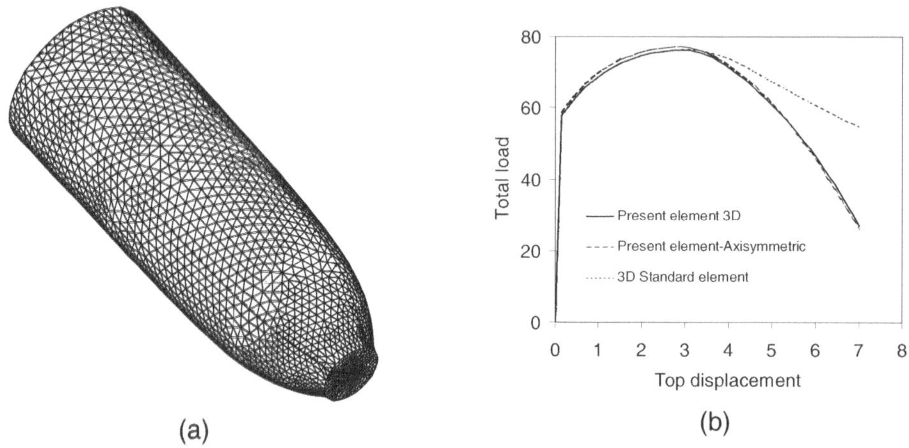

Figure 15.9. Necking of a cylindrical bar. *F-bar*-Patch solution with patches of eight linear tetrahedra: (a) deformed configuration at $u = 7.0$ mm; (b) force–displacement diagrams (forces in KN, displacements in mm). (Reproduced with permission from F-bar-based linear triangles and tetrahedra for finite strain analysis of nearly incompressible solids. Part I: Formulation and benchmarking, EA de Souza Neto, FM Andrade Pires and DRJ Owen, *International Journal for Numerical Methods in Engineering*, Vol 62 © 2005 John Wiley & Sons, Ltd.)

Element tangent stiffness

In contrast to conventional (and *F-bar* elements), the internal force vector of an *F-bar*-Patch element depends not only on the degrees of freedom of the element itself, but also on the degrees of freedom of all other elements belonging to its patch. Such a dependency stems from the use of the modified deformation gradient defined by (15.34) in the computation of the stress tensor components used to assemble the internal force vector. Note in (15.34) that volumetric changes in any element of a patch (produced by variations of any nodal displacements in a patch of elements) affects the *F-bar* gradient for all elements of the patch which, in turn, changes the stress state and the internal force vector of all elements of the patch. As a result, the tangent stiffness in the present case has a non-conventional structure. Consider an *F-bar*-Patch element e, belonging to a patch $\mathscr{P}$. The internal force vector of element e depends on the vector $\mathbf{u}_e$ of nodal displacements of element e, as well as on the vectors:

$$\mathbf{u}_s, \quad s \in \mathscr{P};\ s \neq e,$$

of nodal displacements of the other elements of the patch. The application of a lengthy but standard exact linearisation procedure (details are given in (de Souza Neto *et al.*, 2005)) to the internal force vector of element e gives rise to the following element tangent stiffnesses:

$$\mathbf{K}_{ee}^{(e)} = \int_{\varphi(\Omega_e)} \mathbf{G}_e^T\, \mathbf{a}\, \mathbf{G}_e \,\mathrm{d}v + \left(\frac{v_e}{v_{\text{patch}}} - 1\right) \int_{\varphi(\Omega_e)} \mathbf{G}_e^T\, \mathbf{q}\, \mathbf{G}_e \,\mathrm{d}v \tag{15.37}$$

$$\mathbf{K}_{es}^{(e)} = \frac{v_e}{v_{\text{patch}}} \int_{\varphi(\Omega_e)} \mathbf{G}_e^T\, \mathbf{q}\, \mathbf{G}_s \,\mathrm{d}v, \quad s \in \mathscr{P};\ s \neq e. \tag{15.38}$$

Both $\mathbf{K}_{ee}$ and $\mathbf{K}_{es}$ are generally *unsymmetric*, regardless of the material model adopted.

Remark 15.6 (The structure of the tangent stiffness). The stiffness contribution given in (15.37) gives the tangent relation between the internal force vector components and nodal displacements of element e. Its format is similar to that of the stiffness matrix (15.10) for the *F-bar* elements. Its rows and columns are associated with the degrees of freedom of element e only. The additional contributions (15.38), on the other hand, are the tangent relations between the internal force components of element e and nodal displacements of the *other* elements s of the patch. Their rows are associated with the degrees of freedom of element e and their columns are associated with the degrees of freedom of the other elements s of the patch. The global tangent stiffness required in the solution of the equilibrium problem is obtained by adding, for each element e of the mesh, the contributions of $\mathbf{K}_{ee}$ and $\mathbf{K}_{es}$ $(s \in \mathscr{P};\ s \neq e)$ to the appropriate global stiffness matrix position.

15.2. Enhanced assumed strain methods

This section describes the class of geometrically nonlinear *enhanced assumed strain* (*EAS*) methods proposed by Simo and Armero (1992), which extends to the finite strain range the original ideas proposed by Simo and Hughes (1986) and Simo and Rifai (1990) in the geometrically linear context. This class of methods finds a variational justification on a re-parametrisation of the geometrically nonlinear version of the three-field *Hu–Washizu* variational principle (Washizu, 1968) and can be used to tackle incompressibility locking as well as *bending* locking of low-order elements.

15.2.1. ENHANCED THREE-FIELD VARIATIONAL PRINCIPLE

Let $\boldsymbol{u}$, $\boldsymbol{P}$ and $\boldsymbol{H}$ be, respectively, the displacement, first Piola–Kirchhoff stress and displacement gradient fields. The geometrically nonlinear version of the three-field Hu–Washizu variational principle is defined for a generic hyperelastic material by means of the functional

$$\Pi(\boldsymbol{u}, \boldsymbol{H}, \boldsymbol{P}) \equiv \int_{\Omega} [\bar{\rho}\psi(\tilde{\boldsymbol{C}}) + \boldsymbol{P} : (\nabla_p \boldsymbol{u} - \boldsymbol{H})]\, \mathrm{d}v + \Pi^{\mathrm{ext}}(\boldsymbol{u}), \tag{15.39}$$

where ψ is the specific free-energy function characterising the hyperelastic material, $\Pi^{\mathrm{ext}}(\boldsymbol{u})$ is the potential energy functional of the external loading at the configuration defined by the displacement field $\boldsymbol{u}$ and

$$\tilde{\boldsymbol{C}} = \tilde{\boldsymbol{C}}(\boldsymbol{H}) = [\boldsymbol{I} + \boldsymbol{H}]^T\, [\boldsymbol{I} + \boldsymbol{H}]$$

is the (generally incompatible) right Cauchy–Green strain tensor.

The crucial idea underlying the finite strain *EAS* method is to let the independent displacement gradient field be constructed as

$$\boldsymbol{H} = \nabla_p \boldsymbol{u} + \tilde{\boldsymbol{H}}, \tag{15.40}$$

where $\tilde{\boldsymbol{H}}$ is the *enhanced displacement gradient*, and then re-parametrise the Hu–Washizu functional so as to have displacement, first Piola–Kirchhoff stress and *enhanced* displacement gradient, $\tilde{\boldsymbol{H}}$, as the independent variables; that is, the *three-field enhanced functional* is defined as

$$\begin{aligned}\tilde{\Pi}(\boldsymbol{u}, \tilde{\boldsymbol{H}}, \boldsymbol{P}) &\equiv \Pi(\boldsymbol{u}, \nabla_p \boldsymbol{u} + \tilde{\boldsymbol{H}}, \boldsymbol{P}) \\ &= \int_\Omega [\bar{\rho}\psi(\tilde{\boldsymbol{C}}) + \boldsymbol{P} : \tilde{\boldsymbol{H}}]\, \mathrm{d}v + \Pi^{\mathrm{ext}}(\boldsymbol{u}).\end{aligned} \tag{15.41}$$

With the above decomposition of $\boldsymbol{H}$, we define the (generally) *incompatible* deformation gradient

$$\tilde{\boldsymbol{F}} = \boldsymbol{F} + \tilde{\boldsymbol{H}}, \tag{15.42}$$

where $\boldsymbol{F}$ is the standard *compatible* deformation gradient

$$\boldsymbol{F} = \boldsymbol{I} + \nabla_p \boldsymbol{u}.$$

In (15.41) we have

$$\tilde{\boldsymbol{C}} = \tilde{\boldsymbol{C}}(\tilde{\boldsymbol{H}}) = \tilde{\boldsymbol{F}}^T \tilde{\boldsymbol{F}} = [\boldsymbol{I} + \nabla_p \boldsymbol{u} + \tilde{\boldsymbol{H}}]^T\, [\boldsymbol{I} + \nabla_p \boldsymbol{u} + \tilde{\boldsymbol{H}}]. \tag{15.43}$$

Stationary condition

The enhanced three-field functional $\tilde{\Pi}$ is stationary if and only if the following integral equation holds

$$\int_\Omega \left\{ 2\bar{\rho}\, \tilde{\boldsymbol{F}} \frac{\partial \psi}{\partial \tilde{\boldsymbol{C}}} : \nabla_p \boldsymbol{\eta} - \tilde{\boldsymbol{H}} : \delta \boldsymbol{P} - \left[\boldsymbol{P} - 2\bar{\rho}\, \tilde{\boldsymbol{F}} \frac{\partial \psi}{\partial \tilde{\boldsymbol{C}}} \right] : \delta \tilde{\boldsymbol{H}} \right\} \mathrm{d}v - G^{\mathrm{ext}}(\boldsymbol{\eta}) = 0 \tag{15.44}$$

for all variations $\boldsymbol{\eta}$, $\delta\tilde{\boldsymbol{H}}$ and $\delta\boldsymbol{P}$ of displacement, enhanced displacement gradient and first Piola–Kirchhoff stress, respectively. In the above, G^{ext} is the functional of virtual work of the external forces:

$$G^{\mathrm{ext}}(\boldsymbol{\eta}) \equiv \int_\Omega \bar{\boldsymbol{b}} \cdot \boldsymbol{\eta}\, \mathrm{d}v + \int_{\partial\Omega} \bar{\boldsymbol{t}} \cdot \boldsymbol{\eta}\, \mathrm{d}a, \tag{15.45}$$

where $\bar{\boldsymbol{b}}$ and $\bar{\boldsymbol{t}}$ are, respectively, the material description of the body force and boundary surface traction fields.

Strong form

By means of arguments analogous to those leading to the equivalence between the weak and strong form of the equilibrium in Section 3.6.1, it can be shown that the above integral form leads to the strong (pointwise) equations

$$\left.\begin{aligned} &\mathrm{div}_p\left(2\bar{\rho}\, \tilde{\boldsymbol{F}} \frac{\partial \psi}{\partial \tilde{\boldsymbol{C}}}\right) + \bar{\boldsymbol{b}} = 0 \\ &\tilde{\boldsymbol{H}} = \boldsymbol{0} \\ &\boldsymbol{P} = 2\bar{\rho}\, \tilde{\boldsymbol{F}} \frac{\partial \psi}{\partial \tilde{\boldsymbol{C}}} \end{aligned}\right\} \text{ in } \Omega \tag{15.46}$$
$$\boldsymbol{P}\boldsymbol{m} = \bar{\boldsymbol{t}} \qquad \text{in } \partial\Omega,$$

where $\boldsymbol{m}$ denotes the outward unit normal to the reference boundary $\partial\Omega$. Note that the above strong form has recovered in $(15.46)_{1,3,4}$ the momentum balance equation (3.133) (see page 68) together with the standard hyperelasticity law (13.10) (page 521). Equation $(15.46)_2$ enforces the compatibility of the displacement gradient field, i.e. $\boldsymbol{H} = \nabla_p \boldsymbol{u}$, and implies $\tilde{\boldsymbol{F}} = \boldsymbol{F}$.

15.2.2. EAS FINITE ELEMENTS

The procedure for derivation of EAS elements for finite strain analysis can be summarised as follows:

1. Firstly, a finite element discretisation is applied to the three-field integral equation (15.44). The original infinite-dimensional functional spaces of independent fields $\boldsymbol{u}$, $\boldsymbol{P}$ and $\tilde{\boldsymbol{H}}$ and corresponding variations are approximated by finite-dimensional counterparts spanned by finite element-based interpolation functions.

2. The interpolation of displacement fields, $\boldsymbol{u}$, and virtual displacements, $\boldsymbol{\eta}$, follows exactly the standard procedure employed for virtual work-based finite elements discussed in Chapter 4.

3. The interpolation functions for the remaining independent fields, $\boldsymbol{P}$ and $\tilde{\boldsymbol{H}}$, and their variations are chosen in such a way that the stress is eliminated from the formulation and, in addition, the degrees of freedom associated with the enhanced gradient field, $\tilde{\boldsymbol{H}}$, can be determined via static condensation at the element level.

4. The resulting finite element formulation has effectively the same format as virtual work-based methods with the displacement degrees of freedom being the only unknowns in the global solution procedure.

To formulate the finite element procedure in the *spatial* configuration (as we do throughout this book in the treatment of large-strain problems) we perform the finite element discretisation over the equivalent *spatial* version of the three-field equation (15.44), expressed as

$$\int_{\varphi(\Omega)} \frac{1}{J}\bigg\{ 2\bar{\rho}\, \tilde{\boldsymbol{F}}\, \frac{\partial \psi}{\partial \tilde{\boldsymbol{C}}}\, \tilde{\boldsymbol{F}}^T : \tilde{\nabla}_x \boldsymbol{\eta} - \tilde{\boldsymbol{E}} : \delta\boldsymbol{\tau} \\ - \bigg[\boldsymbol{\tau} - 2\bar{\rho}\, \tilde{\boldsymbol{F}}\, \frac{\partial \psi}{\partial \tilde{\boldsymbol{C}}}\, \tilde{\boldsymbol{F}}^T \bigg] : \delta\tilde{\boldsymbol{E}} \bigg\}\, \mathrm{d}v - G^{\mathrm{ext}}(\boldsymbol{\eta}) = 0, \tag{15.47}$$

where the following definitions are used:

$$\begin{aligned} \boldsymbol{\tau} &= \boldsymbol{P}\, \tilde{\boldsymbol{F}}^T \\ \delta\boldsymbol{\tau} &= \delta\boldsymbol{P}\, \tilde{\boldsymbol{F}}^T \\ \tilde{\boldsymbol{E}} &= \tilde{\boldsymbol{H}}\, \tilde{\boldsymbol{F}}^{-1} \\ \delta\tilde{\boldsymbol{E}} &= \delta\tilde{\boldsymbol{H}}\, \tilde{\boldsymbol{F}}^{-1} \\ \tilde{\nabla}_x^s \boldsymbol{\eta} &= \mathrm{sym}[\nabla_p \boldsymbol{\eta}\, \tilde{\boldsymbol{F}}^{-1}], \end{aligned} \tag{15.48}$$

and

$$J \equiv \det \boldsymbol{F} \tag{15.49}$$

is the determinant of the *compatible* deformation gradient. Note that in view of the definition (15.42) of the enhanced deformation gradient, the *modified* spatial gradient, $\tilde{\nabla}_x \boldsymbol{\eta}$, does not coincide in general with the true spatial gradient $\nabla_x \boldsymbol{\eta}$. However, they do coincide, i.e.

$$\tilde{\nabla}_x \boldsymbol{\eta} = \nabla_x \boldsymbol{\eta}$$

if the strong (pointwise) compatibility equation $(15.46)_2$ holds. As we shall see below, the modified spatial gradient gives rise to *modified discrete gradient operators* in the finite element formulation.

Interpolation of the enhanced displacement gradient field

At the outset, we assume that the interpolated enhanced displacement gradient is *discontinuous* across element boundaries. Within a typical element e, with local coordinates $\boldsymbol{\xi}$, let ${}^h\tilde{\boldsymbol{H}}_{(e)}(\boldsymbol{\xi})$ denote the interpolated displacement gradient tensor field and let ${}^h\tilde{\mathbf{H}}_{(e)}(\boldsymbol{\xi})$ (with upright $\mathbf{H}$) denote the corresponding array of components of ${}^h\tilde{\boldsymbol{H}}_{(e)}(\boldsymbol{\xi})$ arranged in the usual finite element convention. In two dimensions, for instance, we have

$$ {}^h\tilde{\boldsymbol{H}}_{(e)}(\boldsymbol{\xi}) = \begin{bmatrix} {}^hH_{11}(\boldsymbol{\xi}) & {}^hH_{12}(\boldsymbol{\xi}) \\ {}^hH_{21}(\boldsymbol{\xi}) & {}^hH_{22}(\boldsymbol{\xi}) \end{bmatrix}, \quad {}^h\tilde{\mathbf{H}}_{(e)}(\boldsymbol{\xi}) = \begin{bmatrix} {}^hH_{11}(\boldsymbol{\xi}) \\ {}^hH_{12}(\boldsymbol{\xi}) \\ {}^hH_{21}(\boldsymbol{\xi}) \\ {}^hH_{22}(\boldsymbol{\xi}) \end{bmatrix}, \tag{15.50}$$

where ${}^hH_{ij}(\boldsymbol{\xi})$ are the corresponding interpolated components. The array ${}^h\tilde{\mathbf{H}}$ is assumed to be constructed as

$$ {}^h\tilde{\mathbf{H}}_{(e)}(\boldsymbol{\xi}) = \mathbf{H}(\boldsymbol{\xi})\, \boldsymbol{\beta}_{(e)}, \tag{15.51}$$

where each $\mathbf{H}$ is the interpolation matrix for the enhanced displacement gradient and $\boldsymbol{\beta}_{(e)}$ is the vector of element internal degrees of freedom. In two dimensions, the matrix $\mathbf{H}$ has the format

$$\mathbf{H}(\boldsymbol{\xi}) = \begin{bmatrix} H_1^{11}(\boldsymbol{\xi}) & H_2^{11}(\boldsymbol{\xi}) & \cdots & H_{n_{\rm enh}}^{11}(\boldsymbol{\xi}) \\ H_1^{12}(\boldsymbol{\xi}) & H_2^{12}(\boldsymbol{\xi}) & \cdots & H_{n_{\rm enh}}^{12}(\boldsymbol{\xi}) \\ H_1^{21}(\boldsymbol{\xi}) & H_2^{21}(\boldsymbol{\xi}) & \cdots & H_{n_{\rm enh}}^{21}(\boldsymbol{\xi}) \\ H_1^{22}(\boldsymbol{\xi}) & H_2^{22}(\boldsymbol{\xi}) & \cdots & H_{n_{\rm enh}}^{22}(\boldsymbol{\xi}) \end{bmatrix}, \tag{15.52}$$

where $H_i^{jk}(\boldsymbol{\xi})$ denotes the component jk of the $i^{\rm th}$ prescribed *enhanced mode*:

$$\boldsymbol{H}_i(\boldsymbol{\xi}) = \begin{bmatrix} H_i^{11}(\boldsymbol{\xi}) & H_i^{12}(\boldsymbol{\xi}) \\ H_i^{21}(\boldsymbol{\xi}) & H_i^{22}(\boldsymbol{\xi}) \end{bmatrix}, \tag{15.53}$$

and $n_{\rm enh}$ is the total number of enhanced modes or *element internal degrees of freedom* adopted. The vector $\boldsymbol{\beta}_{(e)}$ is generally given by

$$\boldsymbol{\beta}_{(e)} = [\, \beta_1\, \beta_2\, \cdots\, \beta_{n_{\rm enh}}]^T. \tag{15.54}$$

The interpolated tensor field ${}^h\tilde{\boldsymbol{H}}_{(e)}(\boldsymbol{\xi})$ is obviously a linear combination of the $n_{\rm enh}$ modes:

$$ {}^h\tilde{\boldsymbol{H}}_{(e)}(\boldsymbol{\xi}) = \sum_{i=1}^{n_{\rm enh}} \beta_i \, \boldsymbol{H}_i(\boldsymbol{\xi}). \tag{15.55} $$

Variations of the enhanced displacement gradient are interpolated in the same way, i.e.

$$ \delta\, {}^h\tilde{\mathbf{H}}_{(e)}(\boldsymbol{\xi}) = \mathbf{H}\, \delta\boldsymbol{\beta}_{(e)}, \quad \delta {}^h\tilde{\boldsymbol{H}}_{(e)}(\boldsymbol{\xi}) = \sum_{i=1}^{n_{\rm enh}} \delta\beta_i \, \boldsymbol{H}_i(\boldsymbol{\xi}), \tag{15.56} $$

where $\delta\boldsymbol{\beta}_{(e)}$ (with components $\delta\beta_i$) is the array of variations of the element internal degrees of freedom.

Stress interpolation and orthogonality

From standard arguments of functional analysis it follows that (15.44) implies the following L_2-orthogonality condition:¶

$$ \int_\Omega \tilde{\boldsymbol{H}} : \delta\boldsymbol{P} \, \mathrm{d}v = 0 \tag{15.57} $$

for all variations $\delta\boldsymbol{P}$. The above condition will be identically satisfied in the discretised version of (15.44) if the space of interpolated first Piola–Kirchhoff stress fields is L_2-orthogonal to the space of interpolated enhanced displacement gradients; that is, let

$$ {}^h\boldsymbol{P}_{(e)}(\boldsymbol{\xi}) $$

be the generic interpolation function for the nominal stress tensor within an element e. At the outset, we assume that, as for the interpolated enhanced displacement gradients, the global interpolated first Piola–Kirchhoff stress field may be discontinuous across element boundaries. In the discrete setting, the L_2-orthogonality is expressed as

$$ \int_{\Omega^{(e)}} {}^h\tilde{\boldsymbol{H}}_{(e)} : {}^h\boldsymbol{P}_{(e)} \, \mathrm{d}v = 0 $$

and, in view of the assumed enhanced displacement gradient interpolation (15.51), requires that

$$ \int_{\Omega^{(e)}} {}^h\boldsymbol{P}_{(e)}(\boldsymbol{\xi}) : \boldsymbol{H}_i(\boldsymbol{\xi}) \, \mathrm{d}v = 0 \tag{15.58} $$

for all functions ${}^h\boldsymbol{P}_{(e)}$ and $\boldsymbol{H}_i^{(e)}$. With the interpolation functions ${}^h\boldsymbol{P}_{(e)}$ satisfying the above condition, equation (15.57) is automatically satisfied. In addition, by observing that (15.58) also implies that the product $\boldsymbol{P} : \delta\tilde{\boldsymbol{H}}$ in (15.44) vanishes in the discretised version of the variational principle, we find that the nominal stress field $\boldsymbol{P}$ is completely eliminated from the original three-field formulation.

¶Readers not familiar with this nomenclature are referred to standard texts on functional analysis (Oden, 1979) for a definition of L_2-orthogonality.

The reduced finite element equations

In summary, in the discrete setting, the three-field principle has been reduced to a finite element formulation where only displacements, defined as usual by nodal degrees of freedom, and enhanced displacement gradients, defined by element internal degrees of freedom, remain as independent variables. Based on the (computationally more convenient) equivalent form (15.47), the *EAS* finite element equations are obtained from the discretisation of the reduced integral equation

$$\int_{\varphi(\Omega)} \frac{1}{J}\left\{2\bar{\rho}\,\tilde{\boldsymbol{F}}\,\frac{\partial\psi}{\partial\tilde{\boldsymbol{C}}}\,\tilde{\boldsymbol{F}}^T{:}\nabla^s_x\boldsymbol{\eta} + 2\bar{\rho}\,\tilde{\boldsymbol{F}}\,\frac{\partial\psi}{\partial\tilde{\boldsymbol{C}}}\,\tilde{\boldsymbol{F}}^T{:}\,\mathrm{sym}[\delta\tilde{\boldsymbol{E}}]\right\}\,\mathrm{d}v - G^{\mathrm{ext}}(\boldsymbol{\eta}) = 0. \tag{15.59}$$

The corresponding finite element equations read

$$\begin{cases} \tilde{\mathbf{f}}^{\mathrm{int}}(\mathbf{u},\boldsymbol{\beta}) - \mathbf{f}^{\mathrm{ext}} = \mathbf{0} \\ \mathbf{s}_{(e)}(\mathbf{u}_{(e)},\boldsymbol{\beta}_{(e)}) = \mathbf{0} \quad (e = 1, 2, \ldots, n_{\mathrm{el}}), \end{cases} \tag{15.60}$$

where $\tilde{\mathbf{f}}^{\mathrm{int}}$ is a *modified internal force vector* obtained by assembling the element vectors

$$\tilde{\mathbf{f}}^{\mathrm{int}}_{(e)}(\mathbf{u}_{(e)},\boldsymbol{\beta}_{(e)}) = \int_{\varphi(\Omega^{(e)})} \frac{1}{J}\,\tilde{\mathbf{B}}^T\,\boldsymbol{\tau}(\tilde{\boldsymbol{F}})\,\mathrm{d}v, \tag{15.61}$$

where $\mathbf{f}^{\mathrm{ext}}$ is the standard external load vector – the same as in the virtual work-based formulation – and the n_{enh}-dimensional element vectors $\mathbf{s}_{(e)}$ are defined by

$$\mathbf{s}_{(e)}(\mathbf{u}_{(e)},\boldsymbol{\beta}_{(e)}) = \int_{\varphi(\Omega^{(e)})} \frac{1}{J}\,\tilde{\mathbf{Q}}^T\,\boldsymbol{\tau}(\tilde{\boldsymbol{F}})\,\mathrm{d}v. \tag{15.62}$$

In the above, the matrix $\tilde{\mathbf{B}}$ is a *modified* discrete symmetric gradient operator and $\tilde{\mathbf{Q}}$ is the discrete operator associated with variations of interpolated enhanced displacement gradients. The structure of $\tilde{\mathbf{B}}$ and $\tilde{\mathbf{Q}}$ will be discussed below. The function $\boldsymbol{\tau}$ gives the finite element array of Kirchhoff stress tensor components (arranged in the same order as in the arrays defined in (4.20–4.22)) obtained from the standard hyperelastic law

$$\boldsymbol{\tau}(\tilde{\boldsymbol{F}}) = 2\bar{\rho}\,\tilde{\boldsymbol{F}}\frac{\partial\psi}{\partial\tilde{\boldsymbol{C}}}\,\tilde{\boldsymbol{F}}^T. \tag{15.63}$$

Dissipative material models.

For general dissipative materials, $\boldsymbol{\tau}$ is replaced by an incremental (algorithmic) constitutive function, $\hat{\boldsymbol{\tau}}$, whereby for an interval $[t_n, t_{n+1}]$ the Kirchhoff stress is computed as

$$\boldsymbol{\tau}_{n+1} = \hat{\boldsymbol{\tau}}(\boldsymbol{\alpha}_n, \tilde{\boldsymbol{F}}_{n+1}),$$

where $\boldsymbol{\alpha}_n$ denotes the set of internal variables of the material model in question at t_n. In such cases, if there exists an incremental potential $\psi_{n+1}(\boldsymbol{C})$ such that

$$\boldsymbol{\tau}_{n+1} = 2\bar{\rho}\,\tilde{\boldsymbol{F}}\left.\frac{\partial\psi_{n+1}}{\partial\boldsymbol{C}}\right|_{\boldsymbol{C}=\tilde{\boldsymbol{C}}_{n+1}}\tilde{\boldsymbol{F}}^T, \tag{15.64}$$

then the variational structure of the *EAS* method is preserved. Otherwise, the variational structure of the method is lost.

The operator $\tilde{\mathbf{B}}$.

Analogously to the standard symmetric discrete gradient operator (or strain-displacement matrix) $\mathbf{B}$ of the virtual work-based formulation, the matrix $\tilde{\mathbf{B}}$ that appears in the definition of the *EAS* element internal force vector is defined such that, for a generic element vector $\boldsymbol{\eta}^{(e)}$ of nodal virtual displacements generating the interpolated field

$$^{h}\boldsymbol{\eta}^{(e)}(\boldsymbol{\xi}) = \mathbf{N}(\boldsymbol{\xi})\,\boldsymbol{\eta}^{(e)},$$

we have

$$\tilde{\nabla}^{s}_{x}[^{h}\boldsymbol{\eta}^{(e)}] = \tilde{\mathbf{B}}\,\boldsymbol{\eta}^{(e)}. \tag{15.65}$$

It can be easily established that for a generic *EAS* element with n_{node} nodal points, the modified discrete symmetric gradient operator, $\tilde{\mathbf{B}}$, is given by

$$\tilde{\mathbf{B}} = \begin{bmatrix} N^{(e)}_{1;1} & 0 & N^{(e)}_{2;1} & 0 & \cdots & N^{(e)}_{n_{\text{node}};1} & 0 \\ 0 & N^{(e)}_{1;2} & 0 & N^{(e)}_{2;2} & \cdots & 0 & N^{(e)}_{n_{\text{node}};2} \\ N^{(e)}_{1;2} & N^{(e)}_{1;1} & N^{(e)}_{2;2} & N^{(e)}_{2;1} & \cdots & N^{(e)}_{n_{\text{node}};2} & N^{(e)}_{n_{\text{node}};1} \end{bmatrix}, \tag{15.66}$$

for plane stress and plane strain problems. Similar expressions are obtained for axisymmetric and three-dimensional elements. The structure of $\tilde{\mathbf{B}}$ is completely analogous to that of the standard $\mathbf{B}$ matrix (refer to expression (4.30), page 89). The difference between $\tilde{\mathbf{B}}$ and $\mathbf{B}$ lies in the fact that, as $\tilde{\mathbf{B}}$ stems from the *modified* spatial gradient $(15.48)_5$, the standard spatial derivatives,

$$N_{i,j} \equiv \frac{\partial N_i}{\partial x_j},$$

that appear in the definition of the standard $\mathbf{B}$ matrix are replaced in $\tilde{\mathbf{B}}$ with the *modified shape function derivatives*, defined as

$$N_{i;j} \equiv \frac{\partial N_i}{\partial p_k}\,\tilde{F}^{-1}_{kj}, \tag{15.67}$$

where $\partial N_i/\partial p_k$ are the components of the *material* derivatives of the shape functions and attention should be paid to the fact that $\tilde{F}^{-1}_{kj}$ are the components of the inverse of the *incompatible* deformation gradient. Note that if the enhanced displacement gradient vanishes, then $N_{i,j} = N_{i;j}$.

The operator $\tilde{\mathbf{Q}}$.

The matrix $\tilde{\mathbf{Q}}$ taking part in the definition of $\mathbf{s}_{(e)}$ in (15.62) originates from the variation $\text{sym}[\delta\tilde{\boldsymbol{E}}]$ in the second term within the curly brackets of (15.59). It is defined such that

$$\boldsymbol{\tau}:\text{sym}[\delta\,{}^{h}\tilde{\boldsymbol{E}}_{(e)}] = \boldsymbol{\tau}:\text{sym}[\delta\,{}^{h}\tilde{\boldsymbol{H}}_{(e)}\,\tilde{\boldsymbol{F}}^{-1}] = \boldsymbol{\tau}\cdot\tilde{\mathbf{Q}}\,\delta\boldsymbol{\beta}_{(e)}.$$

Again considering the two-dimensional case (plane strain/plane stress), we can easily find that, in view of (15.56) and the definition of the Kirchhoff stress array $\boldsymbol{\tau}$, $\tilde{\mathbf{Q}}$ must be given by

$$\tilde{\mathbf{Q}} = \begin{bmatrix} h^{11}_{1} & h^{11}_{2} & \cdots & h^{11}_{n_{\text{enh}}} \\ h^{22}_{1} & h^{22}_{2} & \cdots & h^{22}_{n_{\text{enh}}} \\ h^{12}_{1}+h^{21}_{1} & h^{12}_{2}+h^{21}_{2} & \cdots & h^{12}_{n_{\text{enh}}}+h^{21}_{n_{\text{enh}}} \end{bmatrix}, \tag{15.68}$$

where h_i^{jk} is the component jk of the tensor

$$\boldsymbol{h}_i^{(e)}(\boldsymbol{\xi}) \equiv \boldsymbol{H}_i^{(e)}(\boldsymbol{\xi})\, \tilde{\boldsymbol{F}}^{-1}. \tag{15.69}$$

15.2.3. FINITE ELEMENT EQUATIONS: STATIC CONDENSATION

Crucial to the derivation of the final virtual work-like format of the method is the observation that, due to the assumed interelement discontinuity of the interpolation (15.51) of $\tilde{\boldsymbol{H}}$ (and the variations $\delta\tilde{\boldsymbol{H}}$ and $\delta\tilde{\boldsymbol{E}}$), equation $(15.60)_2$ holds *individually* for each element of the mesh. Within a Newton–Raphson iterative scheme for solution of the nonlinear system (15.60), this property gives rise to a *static condensation* procedure whereby the iterative increment of enhanced parameters is determined at the element level for each iteration and the main system of linearised equations to be solved at the global level is reduced to a displacement-like format. The procedure is derived as follows. For a typical Newton iteration (k), the linearised version of (15.60) reads[||]

$$\begin{cases} \displaystyle\mathop{\mathsf{A}}_{e=1}^{n_{\rm elem}} [\mathbf{r}_{(e)}^{(k-1)} + \mathbf{K}_{uu}^{(e)}\, \delta\mathbf{u}_{(e)}^{(k)} + \mathbf{K}_{u\beta}^{(e)}\, \delta\boldsymbol{\beta}_{(e)}^{(k)}] = \mathbf{0} \\ \mathbf{s}_{(e)}^{(k-1)} + \mathbf{K}_{\beta u}^{(e)}\, \delta\mathbf{u}_{(e)}^{(k)} + \mathbf{K}_{\beta\beta}^{(e)}\, \delta\boldsymbol{\beta}_{(e)}^{(k)} = \mathbf{0} \quad (e = 1, 2, \ldots, n_{\rm el}), \end{cases} \tag{15.70}$$

where

$$\mathbf{r}_{(e)}^{(k-1)} \equiv \tilde{\mathbf{f}}_{(e)}^{\rm int}(\mathbf{u}_{(e)}^{(k-1)}, \boldsymbol{\beta}_{(e)}^{(k-1)}) - \mathbf{f}_{(e)}^{\rm ext} \tag{15.71}$$

and $(\mathbf{u}^{(k-1)}, \boldsymbol{\beta}^{(k-1)})$, $\mathbf{K}_{uu}$ and $\mathbf{K}_{u\beta}$ are, respectively, the derivatives of the global residual $\mathbf{r}$ with respect to the global vectors $\mathbf{u}$ and $\boldsymbol{\beta}$. The element tangent stiffnesses $\mathbf{K}_{\beta u}^{(e)}$ and $\mathbf{K}_{\beta\beta}^{(e)}$ are, respectively, the derivatives of the left-hand side of $(15.60)_2$ with respect to the element vectors $\delta\mathbf{u}_{(e)}$ and $\delta\boldsymbol{\beta}_{(e)}$. All stiffnesses are evaluated at $(\mathbf{u}^{(k-1)}, \boldsymbol{\beta}^{(k-1)})$. Their expressions are given below. Solution of $(15.70)_2$ for $\delta\boldsymbol{\beta}_{(e)}^{(k)}$ gives

$$\delta\boldsymbol{\beta}_{(e)}^{(k)} = -[\mathbf{K}_{\beta\beta}^{(e)}]^{-1}\, (\mathbf{s}_{(e)}^{(k-1)} + \mathbf{K}_{\beta u}^{(e)}\, \delta\mathbf{u}_{(e)}^{(k)}). \tag{15.72}$$

The tangent matrices

Let $\tilde{\mathsf{a}}$ be the matrix form of the *modified spatial tangent modulus* defined by the components

$$\tilde{\mathsf{a}}_{ijkl} \equiv \frac{1}{J}\, \mathsf{A}_{imkn}\, \tilde{F}_{jm}\, \tilde{F}_{ln}, \tag{15.73}$$

where

$$\mathsf{A}_{imkn} = \frac{\partial P_{im}}{\partial \tilde{F}_{kn}}, \tag{15.74}$$

and it should be noted that J is the determinant of the *compatible* deformation gradient. A standard linearisation procedure applied to the functional (15.59) produces the linearised

[||] Following the notation introduced in Chapter 4 (refer to Section 4.2.4, from page 96), we use the notation $\delta(\cdot)$ in the remainder of this section to denote the *iterative increment* of $(\cdot)$ within the Newton–Raphson scheme and *not* the variation of $(\cdot)$ as we did previously in this chapter.

version of (15.60) where the element matrices $\mathbf{K}_{uu}^{(e)}$, $\mathbf{K}_{u\beta}^{(e)}$, $\mathbf{K}_{\beta u}^{(e)}$ and $\mathbf{K}_{\beta\beta}^{(e)}$ are found to be given by

$$\begin{aligned}
\mathbf{K}_{uu}^{(e)} &= \int_{\varphi(\Omega^{(e)})} \tilde{\mathbf{G}}^T \, \tilde{\mathbf{a}} \, \tilde{\mathbf{G}} \, dv \\
\mathbf{K}_{u\beta}^{(e)} &= \int_{\varphi(\Omega^{(e)})} \tilde{\mathbf{G}}^T \, \tilde{\mathbf{a}} \, \tilde{\mathbf{R}} \, dv \\
\mathbf{K}_{\beta u}^{(e)} &= \int_{\varphi(\Omega^{(e)})} \tilde{\mathbf{R}}^T \, \tilde{\mathbf{a}} \, \tilde{\mathbf{G}} \, dv \\
\mathbf{K}_{\beta\beta}^{(e)} &= \int_{\varphi(\Omega^{(e)})} \tilde{\mathbf{R}}^T \, \tilde{\mathbf{a}} \, \tilde{\mathbf{R}} \, dv,
\end{aligned} \tag{15.75}$$

where $\tilde{\mathbf{G}}$ is the *modified full discrete gradient operator* which, in the two-dimensional case, is defined by

$$\tilde{\mathbf{G}} = \begin{bmatrix} N_{1;1}^{(e)} & 0 & N_{2;1}^{(e)} & 0 & \cdots & N_{n_{\text{node}};1}^{(e)} & 0 \\ 0 & N_{1;1}^{(e)} & 0 & N_{2;1}^{(e)} & \cdots & 0 & N_{n_{\text{node}};1}^{(e)} \\ N_{1;2}^{(e)} & 0 & N_{2;2}^{(e)} & 0 & \cdots & N_{n_{\text{node}};2}^{(e)} & 0 \\ 0 & N_{1;2}^{(e)} & 0 & N_{2;2}^{(e)} & \cdots & 0 & N_{n_{\text{node}};2}^{(e)} \end{bmatrix}, \tag{15.76}$$

and $\tilde{\mathbf{R}}$ is defined as

$$\tilde{\mathbf{R}} = \begin{bmatrix} h_1^{11} & h_2^{11} & \cdots & h_{n_{\text{enh}}}^{11} \\ h_1^{12} & h_2^{12} & \cdots & h_{n_{\text{enh}}}^{12} \\ h_1^{21} & h_2^{21} & \cdots & h_{n_{\text{enh}}}^{21} \\ h_1^{22} & h_2^{22} & \cdots & h_{n_{\text{enh}}}^{22} \end{bmatrix}. \tag{15.77}$$

The reduced displacement-like linearised equation

By introducing (15.72) into $(15.70)_1$ we finally obtain the following reduced (virtual work-like) linearised system whose only unknown is the iterative increment of displacement $\delta\mathbf{u}^{(k)}$:

$$\tilde{\mathbf{K}}_T \, \delta\mathbf{u}^{(k)} = -\tilde{\mathbf{r}}^{(k-1)}, \tag{15.78}$$

where $\tilde{\mathbf{K}}_T$ is the *modified tangent stiffness* obtained by assembling the element matrices

$$\tilde{\mathbf{K}}_T^{(e)} \equiv \mathbf{K}_{uu}^{(e)} - \mathbf{K}_{u\beta}^{(e)} \, [\mathbf{K}_{\beta\beta}^{(e)}]^{-1} \, \mathbf{K}_{\beta u}^{(e)}, \tag{15.79}$$

and $\tilde{\mathbf{r}}^{(k-1)}$ is the global *modified residual vector* obtained from the element vectors

$$\tilde{\mathbf{r}}_{(e)}^{(k-1)} \equiv \mathbf{r}_{(e)}^{(k-1)} - \mathbf{K}_{u\beta}^{(e)} \, [\mathbf{K}_{\beta\beta}^{(e)}]^{-1} \, \mathbf{s}_{(e)}^{(k-1)}. \tag{15.80}$$

The overall Newton–Raphson/static condensation procedure for solution of (15.70) based on the above expressions is summarised in Box 15.3.

Remark 15.7. Recall that for hyperelastic models (or dissipative models whose stress is derived from an incremental potential), $\mathbf{a}$ is symmetric. In this case, the modified tangent stiffness $\tilde{\mathbf{K}}_T$ is also symmetric. Such symmetry is an obvious consequence of the variational structure of the *EAS* method and is in contrast with the *F-bar* approach described earlier in this chapter. For materials with unsymmetric spatial tangent modulus (such as non-associative plasticity models in general) the *EAS* approach does not possess a variational structure (refer to the text surrounding equation (15.64)) and the tangent stiffness $\tilde{\mathbf{K}}_T$ is unsymmetric.

15.2.4. IMPLEMENTATION ASPECTS

As can be observed from Box 15.3 the computational implementation of an EAS element involves far more operations than those of the conventional virtual work-based element of the same order. The extra computational effort concentrates around the evaluation of the matrices $\mathbf{K}_{u\beta}$ and $\mathbf{K}_{\beta\beta}$ (and inversion of $\mathbf{K}_{\beta\beta}$), computation of the iterative internal parameters and the element vectors $\mathbf{s}_{(e)}$. The additional matrices increase in size with the number of internal degrees of freedom and their computation cost may become large particularly for three-dimensional elements (see Simo *et al.* 1993). Also, note that the extra matrices are initially computed during the assembly of the modified tangent stiffness $\tilde{\mathbf{K}}_T$ in the solution of the condensed linearised system. The same matrices are reused later in the computation of the internal forces and residuals. To avoid the costly recomputation of those matrices at this stage, they can be stored in memory immediately after their initial computation and retrieved later. This approach, however, causes a considerable increase in memory storage and may become critical for three-dimensional elements. To circumvent this problem, Simo *et al.* (1993) have proposed an alternative static condensation procedure whereby the relevant matrices can be recomputed at a smaller additional cost.

15.2.5. THE STABILITY OF EAS ELEMENTS

The stability of geometrically nonlinear *EAS* elements has been for some time the subject of research by many authors. In spite of the substantial reduction of incompressibility *locking*, the ability to capture strain localisation phenomena and the high coarse mesh accuracy achieved by many *EAS* formulations, analytical and numerical studies (Crisfield *et al.*, 1995; de Souza Neto *et al.*, 1996; Glaser and Armero, 1997; Korelc and Wriggers, 1996; Wriggers and Reese, 1996) have shown that severe instabilities in the form of spurious *hourglass* deformation modes may occur in some *EAS* elements under large strains. This is particularly true for early formulations such as the original *Q1/E4* (plane strain), the *Q1/E5* (axisymmetric) and *Q1/E9* (three-dimensional) elements proposed by Simo and Armero (1992). The stability of such elements is briefly discussed below.**

**A mathematical analysis of the stability and convergence properties of the EAS method, restricted to the *infinitesimal* strain theory, is provided by Reddy and Simo (1995). For the large strain formulation, we refer to a recent contribution by Auricchio *et al.* (2005).

Box 15.3. *EAS* elements. Newton–Raphson/static condensation loop.

(i) Compute the element matrices $\mathbf{K}_{uu}^{(e)}$, $\mathbf{K}_{\beta\beta}^{(e)}$, $\mathbf{K}_{u\beta}^{(e)}$ and $\mathbf{K}_{\beta u}^{(e)}$. Then compute the modified tangent stiffness

$$\tilde{\mathbf{K}}_T^{(e)} := \mathbf{K}_{uu}^{(e)} - \mathbf{K}_{u\beta}^{(e)}\,[\mathbf{K}_{\beta\beta}^{(e)}]^{-1}\,\mathbf{K}_{\beta u}^{(e)}$$

(ii) Solve the condensed linearised equation for $\delta\mathbf{u}^{(k)}$

$$\tilde{\mathbf{K}}_T\,\delta\mathbf{u}^{(k)} = \tilde{\mathbf{r}}^{(k-1)}$$

(iii) Apply Newton–Raphson correction to the nodal displacement vector

$$\mathbf{u}^{(k)} := \mathbf{u}^{(k-1)} + \delta\mathbf{u}^{(k)}$$

(iv) Compute the vector of iterative internal parameters $\delta\boldsymbol{\beta}_{(e)}^{(k)}$

$$\delta\boldsymbol{\beta}_{(e)}^{(k)} = -[\mathbf{K}_{\beta\beta}^{(e)}]^{-1}\,(\mathbf{s}_{(e)}^{(k-1)} + \mathbf{K}_{\beta u}^{(e)}\,\delta\mathbf{u}_{(e)}^{(k)})$$

(v) Apply Newton–Raphson correction to internal element parameters

$$\boldsymbol{\beta}_{(e)}^{(k)} := \boldsymbol{\beta}_{(e)}^{(k-1)} + \delta\boldsymbol{\beta}_{(e)}^{(k)}$$

(vi) Update deformation gradient (for each Gauss point of each element)

$$\tilde{\boldsymbol{F}}^{(k)} := \boldsymbol{I} + \nabla_p \boldsymbol{u}^{(k)} + \sum_{i=1}^{n_{\text{enh}}} \beta_i^{(k)}\,\boldsymbol{H}_i$$

(vii) Compute Kirchhoff stresses (for each Gauss point of each element)

$$\boldsymbol{\tau}^{(k)} := \hat{\boldsymbol{\tau}}(\tilde{\boldsymbol{F}}^{(k)})$$

(viii) Compute the modified element internal force and s vector

$$(\tilde{\mathbf{f}}_{(e)}^{\text{int}})^{(k)} := \sum_{i=1}^{n_{\text{gausp}}} (\tilde{\mathbf{B}}^{(k)})^T\,w_i\,j_i\,\boldsymbol{\tau}_i^{(k)}$$

$$\mathbf{s}_{(e)}^{(k)} := \sum_{i=1}^{n_{\text{gausp}}} w_i\,j_i\,(\tilde{\mathbf{Q}}_i^{(k)})^T\,\boldsymbol{\tau}_i^{(k)}$$

(ix) Compute the modified residual

$$\tilde{\mathbf{r}}_{(e)}^{(k)} := (\tilde{\mathbf{f}}_{(e)}^{\text{int}})^{(k)} - \mathbf{f}_{(e)}^{\text{ext}} - \mathbf{K}_{u\beta}^{(e)}\,[\mathbf{K}_{\beta\beta}^{(e)}]^{-1}\,\mathbf{s}_{(e)}^{(k)}$$

(x) Check convergence. If $\|\tilde{\mathbf{r}}^{(k)}\| \leq \epsilon_{\text{tol}}$ then accept $(\cdot)^{(k)}$ as the solution to (15.70) and EXIT. Else, set $k := k + 1$ and GOTO (i)

The Q1/E4 and Q1/E9 elements

The so-called *Q1/E4* element is a four-noded *EAS* quadrilateral with four enhanced modes obtained by setting

$$\boldsymbol{E}_1 = \begin{bmatrix} \xi & 0 \\ 0 & 0 \end{bmatrix}, \quad \boldsymbol{E}_2 = \begin{bmatrix} 0 & 0 \\ \xi & 0 \end{bmatrix}, \quad \boldsymbol{E}_3 = \begin{bmatrix} 0 & \eta \\ 0 & 0 \end{bmatrix}, \quad \boldsymbol{E}_4 = \begin{bmatrix} 0 & 0 \\ 0 & \eta \end{bmatrix} \tag{15.81}$$

and defining the following interpolation functions for the enhanced displacement gradient field:

$$\boldsymbol{H}_i(\boldsymbol{\xi}) \equiv \frac{j_0}{j(\boldsymbol{\xi})}\, \mathbf{J}_0^{-T}\, \boldsymbol{E}_i(\boldsymbol{\xi}), \tag{15.82}$$

where $\mathbf{J}_0$ denotes the Jacobian of the isoparametric mapping of the element in the reference configuration evaluated at the centroid ($\boldsymbol{\xi} = \mathbf{0}$),

$$j_0 = \det \mathbf{J}_0, \tag{15.83}$$

and

$$j(\boldsymbol{\xi}) = \det[\mathbf{J}(\boldsymbol{\xi})], \tag{15.84}$$

where $\mathbf{J}(\boldsymbol{\xi})$ is the Jacobian at the generic point with local coordinates $\boldsymbol{\xi}$.

The direct extension of the above element to three-dimensions yields the so-called *Q1/E9* eight-noded brick which contains a total number of nine enhanced modes. The *Q1/E9* element is obtained as above by setting

$$\begin{gathered}
\boldsymbol{E}_1 = \begin{bmatrix} \xi & 0 & 0 \\ 0 & 0 & 0 \\ 0 & 0 & 0 \end{bmatrix}, \quad \boldsymbol{E}_2 = \begin{bmatrix} 0 & 0 & 0 \\ \xi & 0 & 0 \\ 0 & 0 & 0 \end{bmatrix}, \quad \boldsymbol{E}_3 = \begin{bmatrix} 0 & 0 & 0 \\ 0 & 0 & 0 \\ \xi & 0 & 0 \end{bmatrix}, \\
\boldsymbol{E}_4 = \begin{bmatrix} 0 & \eta & 0 \\ 0 & 0 & 0 \\ 0 & 0 & 0 \end{bmatrix}, \quad \boldsymbol{E}_5 = \begin{bmatrix} 0 & 0 & 0 \\ 0 & \eta & 0 \\ 0 & 0 & 0 \end{bmatrix}, \quad \boldsymbol{E}_6 = \begin{bmatrix} 0 & 0 & 0 \\ 0 & 0 & 0 \\ 0 & \eta & 0 \end{bmatrix}, \\
\boldsymbol{E}_7 = \begin{bmatrix} 0 & 0 & \zeta \\ 0 & 0 & 0 \\ 0 & 0 & 0 \end{bmatrix}, \quad \boldsymbol{E}_8 = \begin{bmatrix} 0 & 0 & 0 \\ 0 & 0 & \zeta \\ 0 & 0 & 0 \end{bmatrix}, \quad \boldsymbol{E}_9 = \begin{bmatrix} 0 & 0 & 0 \\ 0 & 0 & 0 \\ 0 & 0 & \zeta \end{bmatrix}.
\end{gathered} \tag{15.85}$$

The stability issue

As compared to the standard virtual work-based four-noded quadrilateral and eight-noded hexahedron, the above elements, as well as the axisymmetric version named *Q1/E5*, have superior performance in situations such as bending dominated problems, being able to provide accurate solutions with relatively coarse meshes. Their tendency to lock under nearly incompressible conditions is also substantially reduced and, in addition, they are able to capture strain localisation phenomena. Unfortunately, however, their instability in finitely

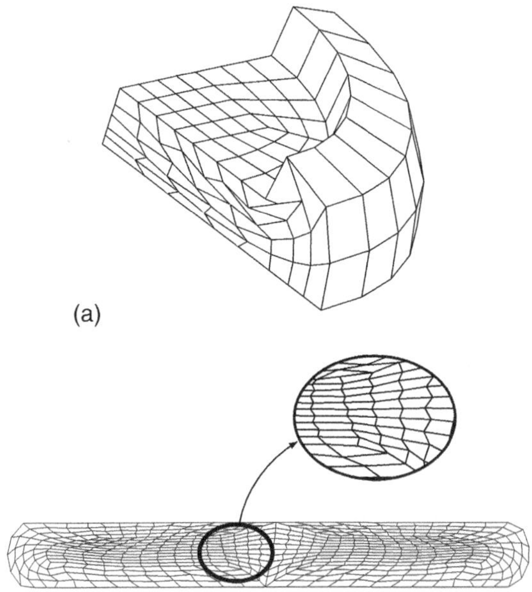

(a) (b)

Figure 15.10. Spurious hourglassing in some *EAS* elements: (a) *Q1/E9* three-dimensional element; (b) *Q1/E5* axisymmetric element. ((a) Reproduced with permission from Remarks on the stability of enhanced assumed strain elements in finite elasticity and elastoplasticity, EA de Souza Neto, D Perić, GC Huang and DRJ Owen, *Communications in Numerical Methods in Engineering*, Vol 11 © 1995 John Wiley & Sons, Ltd. (b) Reproduced with permission from Design of simple low-order finite elements for large-strain analysis of nearly incompressible solids, EA de Souza Neto, D Perić, M Dutko and DRJ Owen, *International Journal of Solids and Structures*, Vol 33, Issue 20–22 © 1996 Elsevier Science Ltd.)

strained configurations can be a source of major problems. This problem seems to have been first systematically investigated in the analytical study of Wriggers and Reese (1996) who found that severe hourglass instabilities can occur under moderate-to-large compressive strains for hyperelastic material models. Numerical studies carried out by the authors and co-workers de Souza Neto *et al.* (1996) found in addition that, for elastoplastic models, hourglass instabilities can be activated not only in compressive regimes but also (although less severe) under finite stretching. The unstable behaviour is illustrated in Figure 15.10. Figure 15.10(a) shows the results of the analysis of the compression of a cylindrical block made of compressible rubber (modelled as a regularised Ogden material). The initially cylindrical block is clamped at both ends and the analysis considers only the symmetric octant of the actual geometry. The *Q1/E9* eight-noded brick (with $3 \times 3 \times 3$ Gauss quadrature) was used and the configuration shown corresponds to 70% compression. Figure 15.10(b) depicts a deformed configuration obtained with the axisymmetric *EAS* element *Q1/E5* in the analysis of the elastomeric bead compression problem described in Section 13.6.7 (page 556). The configuration shown was obtained for a prescribed top plate displacement $u = 0.235$. In both examples, severe spurious hourglassing invalidates the numerical results.

Other two-dimensional enhanced elements

To overcome such instabilities, a number of alternative formulations of low-order *EAS* elements have been proposed. To avoid the spurious hourglass modes of the original *Q1/E4* quadrilateral, Korelc and Wriggers (1996), for instance, proposed an element, named *CG4*, obtained by replacing the primary modes (15.81) with their transpose:

$$\boldsymbol{E}_1 = \begin{bmatrix} \xi & 0 \\ 0 & 0 \end{bmatrix}, \quad \boldsymbol{E}_2 = \begin{bmatrix} 0 & \xi \\ 0 & 0 \end{bmatrix}, \quad \boldsymbol{E}_3 = \begin{bmatrix} 0 & 0 \\ \eta & 0 \end{bmatrix}, \quad \boldsymbol{E}_4 = \begin{bmatrix} 0 & 0 \\ 0 & \eta \end{bmatrix}, \tag{15.86}$$

in conjunction with the following alternative definition for the enhanced displacement gradient interpolation functions:

$$\boldsymbol{H}_i(\boldsymbol{\xi}) \equiv \frac{j_0}{j(\boldsymbol{\xi})}\, \mathbf{J}_0^{-T}\, \boldsymbol{E}_i(\boldsymbol{\xi})\, \mathbf{J}_0^{-1}. \tag{15.87}$$

Glaser and Armero (1997) proposed the elements named *Q1/ET4* and *Q1/ES4* obtained by replacing the primary modes $\boldsymbol{E}_i$ of (15.81) respectively with their transpose (as for the *GC4* element above) and their symmetrised counterparts

$$\boldsymbol{E}_1 = \begin{bmatrix} \xi & 0 \\ 0 & 0 \end{bmatrix}, \quad \boldsymbol{E}_2 = \begin{bmatrix} 0 & \xi \\ \xi & 0 \end{bmatrix}, \quad \boldsymbol{E}_3 = \begin{bmatrix} 0 & \eta \\ \eta & 0 \end{bmatrix}, \quad \boldsymbol{E}_4 = \begin{bmatrix} 0 & 0 \\ 0 & \eta \end{bmatrix}. \tag{15.88}$$

These authors also replace (15.82) with

$$\boldsymbol{H}_i(\boldsymbol{\xi}) \equiv \frac{j_0}{j(\boldsymbol{\xi})}\, \boldsymbol{F}_0\, \mathbf{J}_0^{-T}\, \boldsymbol{E}_i(\boldsymbol{\xi})\, \mathbf{J}_0^{-1} \tag{15.89}$$

and

$$\boldsymbol{H}_i(\boldsymbol{\xi}) \equiv \frac{j_0}{j(\boldsymbol{\xi})}\, \boldsymbol{F}_0\, \mathbf{J}_0\, \boldsymbol{E}_i(\boldsymbol{\xi})\, \mathbf{J}_0^{-1}, \tag{15.90}$$

where $\boldsymbol{F}_0$ denotes the (compatible) deformation gradient at the element centroid. The inclusion of $\boldsymbol{F}_0$ (absent in the formulation of the *CG4* element and the original *Q1/E4* and *Q1/E9* elements) guarantees the frame invariance of the interpolated enhanced displacement gradient in (15.89) and (15.90); that is, under an arbitrary superimposed rigid-body deformation with rotation $\boldsymbol{Q}$, we have the transformations

$$\begin{aligned} {}^h\tilde{\boldsymbol{H}} &\longrightarrow \boldsymbol{Q}\, {}^h\tilde{\boldsymbol{H}} \\ \tilde{\boldsymbol{F}} \equiv \boldsymbol{F} + {}^h\tilde{\boldsymbol{H}} &\longrightarrow \boldsymbol{Q}\,\boldsymbol{F} + \boldsymbol{Q}\,{}^h\tilde{\boldsymbol{H}} = \boldsymbol{Q}\,\tilde{\boldsymbol{F}}. \end{aligned} \tag{15.91}$$

The interpolation form (15.90) was introduced by Simo *et al.* (1993) to ensure objectivity of improved versions of the original eight-noded *EAS* hexahedron.

The comparative study carried out by Glaser and Armero (1997) shows that the elements *Q1/ET4*, *Q1/ES4* and *CG4* overcome the hourglass instability under compressive strains. However, the problem persists in highly stretched configurations. These authors avoid this instability by introducing an *artificial hourglass control* term in the formulation of the elements *Q1/ET4* and *Q1/ES4*.

15.3. Mixed *u/p* formulations

Another attractive alternative in the treatment of nearly incompressible models is the class of finite element formulations for geometrically nonlinear problems originally introduced by Sussman and Bathe (1987). This class of methods extends to the geometrically nonlinear range of the classical mixed *u/p* methods for infinitesimal problems (Bathe, 1996). Such methods are based on a form of the Hu–Washizu variational principle where the independent variables are the *displacement* and *hydrostatic pressure* fields. One important feature of this mixed method is that the adoption of pressure interpolation functions discontinuous across element boundaries allows the use of a static condensation procedure (analogous to that described above for *EAS* elements) that reduces the formulation to a displacement-like format, where only nodal displacements remain as global variables. However, in contrast to *EAS* and *F-bar* elements, the use of u/p formulations is limited to a class of material models whose constitutive equations satisfy a certain restriction, as will be briefly discussed in the following.

15.3.1. THE TWO-FIELD VARIATIONAL PRINCIPLE

The starting point of the geometrically nonlinear *u/p* method is the variational principle derived for a hyperelastic material with free-energy potential ψ, from the two-field functional

$$\Pi(\boldsymbol{u}, p) \equiv \int_{\Omega} [\psi(\boldsymbol{F}) + \pi(\bar{p}, p)] \, \mathrm{d}v + \Pi^{\text{ext}}(\boldsymbol{u}), \tag{15.92}$$

where $\boldsymbol{u}$ and p are, respectively, the displacement and the *independent* Cauchy hydrostatic pressure fields and $\bar{p}$ is the (dependent) Cauchy pressure obtained from the pointwise deformation gradient through the hyperelastic constitutive law

$$\bar{p} = \bar{p}(\boldsymbol{F}) \equiv \tfrac{1}{3}\, \bar{\boldsymbol{\sigma}}(\boldsymbol{F}) : \boldsymbol{I}, \tag{15.93}$$

where $\bar{\boldsymbol{\sigma}}$ is the hyperelastic constitutive function for the Cauchy stress tensor

$$\bar{\boldsymbol{\sigma}}(\boldsymbol{F}) \equiv \frac{\bar{\rho}}{\det \boldsymbol{F}} \frac{\partial \psi}{\partial \boldsymbol{F}} \boldsymbol{F}^T. \tag{15.94}$$

Note that ψ depends on $\boldsymbol{u}$ owing to the dependence of the pointwise value of the deformation gradient, $\boldsymbol{F}$, on the field $\boldsymbol{u}$. The dependence of π on $\boldsymbol{u}$ follows from the dependence of $\bar{p}$ on $\boldsymbol{F}$. Π^{ext} is the potential energy functional of the external load. The potential π is assumed to have the general form

$$\pi(\boldsymbol{u}, p) \equiv \frac{(\bar{p} - p)^2}{2\, P(\bar{p})}, \tag{15.95}$$

where P is the *functional* defined by

$$P(\bar{p}) \equiv \frac{1}{3 \det \boldsymbol{F}} \frac{\partial \bar{p}}{\partial \boldsymbol{F}} \boldsymbol{F}^T : \boldsymbol{I}. \tag{15.96}$$

The functional P is assumed to satisfy the condition

$$P(P(\bar{p})) = 0, \tag{15.97}$$

which is equivalent to

$$\frac{\partial P(\bar{p})}{\partial \boldsymbol{F}}\,\boldsymbol{F}^T : \boldsymbol{I} = 0. \tag{15.98}$$

Expression (15.98) represents a constraint on the possible constitutive models that can be used in conjunction with the u/P formulation. In other words, only constitutive models whose constitutive function $\bar{p}$ for the hydrostatic pressure satisfy (15.98) are amenable to treatment via the present u/p formulation. The need for the above constraint is discussed below.

Stationary condition

Mixed u/p finite element formulations are derived simply by applying standard finite element discretisations to the integral equations expressing the stationary condition of Π. It can be easily established that the stationary condition of the functional defined by (15.92) is given by

$$\begin{cases} \displaystyle\int_\Omega \left\{ \bar{\boldsymbol{P}} - \frac{\bar{p}-p}{P(\bar{p})}\frac{\partial \bar{p}}{\partial \boldsymbol{F}} + \frac{(\bar{p}-p)^2}{2\,[P(\bar{p})]^2}\frac{\partial P(\bar{p})}{\partial \boldsymbol{F}} \right\} : \nabla_p \boldsymbol{\eta}\, \mathrm{d}v - G^{\mathrm{ext}}(\boldsymbol{\eta}) = 0 \\[2ex] \displaystyle\int_\Omega \frac{\bar{p}-p}{P(\bar{p})}\,\delta p\, \mathrm{d}v = 0 \end{cases} \tag{15.99}$$

for all variations $\boldsymbol{\eta}$ and $\delta\bar{p}$ of displacement and independent pressure fields. In the above,

$$\bar{\boldsymbol{P}} = \bar{\boldsymbol{P}}(\boldsymbol{F}) \equiv \frac{1}{\det \boldsymbol{F}}\,\bar{\boldsymbol{\sigma}}(\boldsymbol{F})\,\boldsymbol{F}^T \tag{15.100}$$

is the first Piola–Kirchhoff stress obtained from the constitutive law.

For the discussion that follows, and also to outline the computer implementation based on the spatial configuration, it is convenient to recast (15.99) in its spatial form

$$\begin{cases} \displaystyle\int_{\varphi(\Omega)} \boldsymbol{\sigma}(\boldsymbol{F}, p) : \nabla_x \boldsymbol{\eta}\, \mathrm{d}v - G^{\mathrm{ext}}(\boldsymbol{\eta}) = 0 \\[2ex] \displaystyle\int_{\varphi(\Omega)} \frac{\bar{p}-p}{J\,P(\bar{p})}\,\delta p\, \mathrm{d}v = 0, \end{cases} \tag{15.101}$$

where we have used the standard definition $J \equiv \det \boldsymbol{F}$ and

$$\boldsymbol{\sigma}(\boldsymbol{F}, \bar{p}) \equiv \bar{\boldsymbol{\sigma}}(\boldsymbol{F}) - \frac{\bar{p}(\boldsymbol{F}) - p}{\det \boldsymbol{F}\, P(\bar{p})}\frac{\partial \bar{p}}{\partial \boldsymbol{F}}\,\boldsymbol{F}^T + \frac{[\bar{p}(\boldsymbol{F}) - p]^2}{2\,\det \boldsymbol{F}\,[P(\bar{p})]^2}\frac{\partial P(\bar{p})}{\partial \boldsymbol{F}}\,\boldsymbol{F}^T. \tag{15.102}$$

Now let us focus on the meaning of (15.101). Equation $(15.101)_1$ is analogous to the standard virtual work equation. It expresses the weak form of equilibrium for the above-defined modified Cauchy stress field $\boldsymbol{\sigma}$. By introducing definition (15.96) together with the assumed constitutive constraint (15.98) we find, after straightforward manipulations, that

$$\tfrac{1}{3}\,\boldsymbol{\sigma}(\boldsymbol{F}, p) : \boldsymbol{I} = p; \tag{15.103}$$

that is, the hydrostatic component of the modified stress defined by (15.102) coincides with the interpolated pressure and is not obtained from the deformation map through the constitutive equations of the model. The need for the constraint (15.98) stems from the fact that if it is not satisfied by the constitutive model in question, then (15.103) does not hold in general. We finish our discussion on the meaning of the stationary conditions of the u/p formulation, by considering equation $(15.101)_2$. This equation implies that the independent pressure field satisfies the constitutive equation of the material only in an averaged (weak or integral) sense. Its corresponding strong (pointwise) form is simply

$$p = \bar{p}. \tag{15.104}$$

Remark 15.8. To see how (15.98) limits the possible constitutive models for which the above interpretation of the stationary conditions remains valid, let us consider the class of hyperelastic models whose free-energy function can be split into a sum of a purely isochoric and a purely volumetric contribution; that is, we consider materials with free-energy function of the type

$$\psi(\boldsymbol{F}) \equiv \psi_{\text{iso}}(\boldsymbol{F}_{\text{iso}}) + \psi_{\text{v}}(J). \tag{15.105}$$

The Hencky model as well as the regularised versions of the neo-Hookean, Mooney–Rivlin and Ogden models discussed in Chapter 13 fall into this category. For such models, it can be shown that (15.98) is equivalent to the restriction

$$\frac{\partial^2 \psi_{\text{v}}}{\partial J^2} = 0, \tag{15.106}$$

or, equivalently,

$$\frac{\partial \bar{p}}{\partial J} = 0. \tag{15.107}$$

Free-energy functions based on the *logarithmic* bulk term, such as those defined by (13.43) (13.44), (13.52) and (13.56), do *not* satisfy the above constraint in general. However, modified versions of such models, obtained by replacing the volumetric contribution to ψ with the alternative function referred to in Remark 13.1 (page 526), do satisfy (15.105) and can be used in conjunction with u/p formulations.

15.3.2. FINITE ELEMENT EQUATIONS

In the discretisation of (15.101) standard finite element shape functions are used to approximate the displacement and virtual displacement fields. Within each u/p finite element, pressure fields are interpolated according to

$$^{h}p(\boldsymbol{\xi}) = \sum_{i=1}^{n_{\text{pres}}} p_i \, \hat{N}_i(\boldsymbol{\xi}), \tag{15.108}$$

where p_i and $\hat{N}_i$ $(i = 1, 2, \ldots, n_{\text{pres}})$ are, respectively, the element pressure degrees of freedom and the prescribed pressure interpolation functions and n_{pres} is the prescribed total number of pressure degrees of freedom of the element. The interpolated pressure within the element can be equivalently expressed as

$$^{h}p(\boldsymbol{\xi}) = \mathbf{p}_{(e)}^{T} \, \hat{\mathbf{N}}(\boldsymbol{\xi}), \tag{15.109}$$

Table 15.1. Pressure interpolation functions for u/p quadrilaterals.

n_{pres}	$\hat{N}_1$	$\hat{N}_2$	$\hat{N}_3$	$\hat{N}_4$	$\hat{N}_5$	$\hat{N}_6$	$\hat{N}_7$	$\hat{N}_8$	$\hat{N}_9$
1	1								
3	1	ξ	η						
4	1	ξ	η	$\xi\eta$					
6	1	ξ	η	$\xi\eta$	ξ^2	η^2			
8	1	ξ	η	$\xi\eta$	ξ^2	η^2	$\xi^2\eta$	$\xi\eta^2$	
9	1	ξ	η	$\xi\eta$	ξ^2	η^2	$\xi^2\eta$	$\xi\eta^2$	$\xi^2\eta^2$

where we have introduced the finite element array definitions

$$\begin{aligned} \mathbf{N}(\boldsymbol{\xi}) &\equiv [\hat{N}_1(\boldsymbol{\xi})\ \hat{N}_2(\boldsymbol{\xi})\ \cdots\ \hat{N}_{n_{\text{pres}}}(\boldsymbol{\xi})]^T \\ \mathbf{p}_{(e)} &\equiv [p_1\ p_2\ \cdots\ p_{n_{\text{pres}}}]^T. \end{aligned} \tag{15.110}$$

Variations of pressure are interpolated in the same way. The interpolation functions are assumed discontinuous across element boundaries. Typical pressure interpolation functions for two-dimensional quadrilateral u/p elements are shown in Table 15.1

The above interpolations result in the following discrete version of (15.101):

$$\begin{cases} \mathbf{r}(\mathbf{u}, \mathbf{p}) \equiv \mathbf{f}^{\text{int}}(\mathbf{u}, \mathbf{p}) - \mathbf{f}^{\text{ext}} = \mathbf{0} \\ \mathbf{s}_{(e)}(\mathbf{u}_{(e)}, \mathbf{p}_{(e)}) = \mathbf{0} \quad (e = 1, 2, \ldots, n_{\text{el}}), \end{cases} \tag{15.111}$$

where $\mathbf{f}^{\text{ext}}$ is the global standard external load vector and $\mathbf{f}^{\text{int}}$ is the assembly of element internal force vectors

$$\mathbf{f}^{\text{int}}_{(e)}(\boldsymbol{u}_{(e)}, \mathbf{p}_{(e)} \equiv \int_{\varphi(\Omega^{(e)})} \mathbf{B}^T\, \boldsymbol{\sigma}(\boldsymbol{F}, \mathbf{p}_{(\mathbf{e})})\, \mathrm{d}v, \tag{15.112}$$

where the array $\boldsymbol{\sigma}$ contains the components of the modified stress defined by (15.102). The arrays $\mathbf{p}$ and $\mathbf{p}_{(e)}$ are, respectively, the global and element vectors of pressure degrees of freedom and $\mathbf{B}$ is the *standard* discrete symmetric gradient operator (the same as in virtual work-based elements). The element residual vector $\mathbf{s}_{(e)}$ – associated with the weak enforcement of the pressure constitutive relation – is given by

$$\mathbf{s}_{(e)}(\mathbf{u}_{(e)}, \mathbf{p}_{(e)}) \equiv \int_{\varphi(\Omega^{(e)})} \frac{\bar{p} - p}{J\, P(\bar{p})}\, \hat{\mathbf{N}}\, \mathrm{d}v. \tag{15.113}$$

Dissipative material models

For dissipative materials in general, the constitutive-related component $\bar{\boldsymbol{\sigma}}$ of the Cauchy stress taking part in the assemblage of the internal force vector is computed by an incremental (algorithmic) constitutive function, $\hat{\boldsymbol{\sigma}}$, such that

$$\bar{\boldsymbol{\sigma}}_{n+1} = \hat{\boldsymbol{\sigma}}(\boldsymbol{\alpha}_n, \tilde{\boldsymbol{F}}_{n+1}).$$

If the algorithmic constitutive function does not possess a potential structure (refer to the text surrounding expression (15.64) on page 674) the variational basis of the u/p formulation is lost.

Pressure interpolation order

The choice of a right balance between the number of displacement and pressure degrees of freedom is crucial for the success of u/p finite elements. Analogously to *F-bar* elements, an increase in the number of displacement degrees of freedom reduces the stiffness of u/p elements and an increase in the number of *pressure* degrees of freedom stiffens the element. In fact, u/p elements with equal number of displacement nodes and pressure degrees of freedom behave in the same way as their purely displacement-based counterparts. Thus, successful u/p elements must have fewer pressure degrees of freedom than displacement nodes. In the infinitesimal theory, the tendency of u/p formulations to lock is formally established by the so-called *Babuška–Brezzi, LBB*[††] or *inf-sup* condition (Bathe, 1996; Brezzi and Fortin, 1991; Hughes, 1987), which guarantees that an assembly of elements over a domain Ω will not lock if there exists a constant $\beta > 0$ such that

$$\inf_{{}^h p \,\in\, {}^h\mathscr{P}} \; \sup_{{}^h \boldsymbol{u} \,\in\, {}^h\mathscr{U}} \frac{\int_\Omega {}^h p \,\mathrm{div}[{}^h \boldsymbol{u}]\, \mathrm{d}v}{\|{}^h p\| \, \|{}^h \boldsymbol{u}\|} \geq \beta, \tag{15.114}$$

where ${}^h\mathscr{U}$ and ${}^h\mathscr{P}$ are, respectively, the spaces of interpolated displacements and interpolated pressures generated by the finite element discretisation. In practice, we are looking for elements that have the least possible number of displacement degrees of freedom combined with the greatest number of pressure degrees of freedom that do not lock. In two dimensions, the nine-noded quadrilateral with three pressure degrees of freedom is known to satisfy the *LBB* condition. This element, named *9/3* is recommended by Sussman and Bathe (1987) for finite strain analysis. For three-dimensional analysis, these authors suggest the use of the so-called *27/4* elements – 27-noded element with four pressure degrees of freedom.

15.3.3. SOLUTION: STATIC CONDENSATION

As, in view of the interelement discontinuity of the interpolated pressure, $(15.111)_2$ holds for each element separately, it is possible to derive a *static condensation* procedure for the Newton–Raphson solution of (15.111). Within each iteration of the resulting Newton–Raphson/static condensation scheme, iterative increments of nodal displacements are the only unknowns at the global level. Iterative increments of the pressure degrees of freedom are determined at the element level. The procedure is completely analogous to that devised in Section 15.2.3 (page 676) and summarised in Box 15.3 (page 679) for *EAS* elements. Thus, in what follows, we skip the derivation of the static condensation procedure and summarise only the resulting formulae relevant to the computer implementation of the method.

[††] Short for Ladyzhenskaya–Babuška–Brezzi.

Summary of the iteration loop

The Newton–Raphson/static condensation loop for mixed u/p elements starts with the computation of the element tangent matrices

$$\begin{aligned}
\mathbf{K}_{uu}^{(e)} &\equiv \frac{\partial \mathbf{r}_{(e)}}{\partial \mathbf{u}_{(e)}} = \int_{\varphi(\Omega^{(e)})} \mathbf{G}^T\,(\mathbf{a}+\mathbf{b})\,\mathbf{G}\,dv \\
\mathbf{K}_{up}^{(e)} &\equiv \frac{\partial \mathbf{r}_{(e)}}{\partial \mathbf{p}_{(e)}} = \int_{\varphi(\Omega^{(e)})} \mathbf{G}^T\,\mathbf{q}\,\hat{\mathbf{N}}^T\,dv \\
\mathbf{K}_{pu}^{(e)} &\equiv \frac{\partial \mathbf{s}_{(e)}}{\partial \mathbf{u}_{(e)}} = \int_{\varphi(\Omega^{(e)})} \hat{\mathbf{N}}\,\mathbf{q}^T\,\mathbf{G}\,dv \;=\; [\mathbf{K}_{up}^{(e)}]^T \\
\mathbf{K}_{pp}^{(e)} &\equiv \frac{\partial \mathbf{s}_{(e)}}{\partial \mathbf{p}_{(e)}} = \int_{\varphi(\Omega^{(e)})} \frac{-1}{J\,P(\bar{p})}\,\hat{\mathbf{N}}\,\hat{\mathbf{N}}^T\,dv,
\end{aligned} \tag{15.115}$$

where $\mathbf{r}_{(e)}$ is the element residual vector associated with the global vector $\mathbf{r}$. The matrix $\mathbf{a}$ is the matrix form of the spatial tangent modulus (the same as that computed for the corresponding standard displacement-based element) and $\mathbf{b}$ is the matrix form of the fourth-order tensor defined as

$$\begin{aligned}
\mathsf{b} \equiv\; & \frac{-1}{P(\bar{p})}\,\boldsymbol{M}\otimes\boldsymbol{M} + \frac{\bar{p}-p}{P(\bar{p})}(\boldsymbol{M}\otimes\boldsymbol{N}+\boldsymbol{N}\otimes\boldsymbol{M}) \\
& - \frac{(\bar{p}-p)^2}{[P(\bar{p})]^3}\,\boldsymbol{N}\otimes\boldsymbol{N} - \frac{\bar{p}-p}{P(\bar{p})}\,\mathsf{d} + \frac{(\bar{p}-p)^2}{2\,[P(\bar{p})]^2}\,\mathsf{e},
\end{aligned} \tag{15.116}$$

where

$$\boldsymbol{M} \equiv \frac{1}{J}\,\frac{\partial \bar{p}}{\partial \boldsymbol{F}}\,\boldsymbol{F}^T, \qquad \boldsymbol{N} \equiv \frac{1}{J}\,\frac{\partial P(\bar{p})}{\partial \boldsymbol{F}}\,\boldsymbol{F}^T \tag{15.117}$$

and d and e are the fourth-order tensors defined by the components

$$\mathsf{d} \equiv \frac{1}{J}\left[\frac{\partial^2 \bar{p}}{\partial \boldsymbol{F}\,\partial \boldsymbol{F}}\right]_{imkn} F_{jm}\,F_{ln}, \qquad \mathsf{e} \equiv \frac{1}{J}\left[\frac{\partial^2 P(\bar{p})}{\partial \boldsymbol{F}\,\partial \boldsymbol{F}}\right]_{imkn} F_{jm}\,F_{ln}. \tag{15.118}$$

The array $\mathbf{q}$ is the vector form of the second-order tensor

$$\boldsymbol{Q} \equiv \frac{1}{J}\left(\frac{1}{P(\bar{p})}\,\frac{\partial \bar{p}}{\partial \boldsymbol{F}} - \frac{\bar{p}-p}{[P(\bar{p})]^2}\,\frac{\partial P(\bar{p})}{\partial \boldsymbol{F}}\right)\boldsymbol{F}^T. \tag{15.119}$$

Following the standard finite element convention, $\mathbf{q}$ is given explicitly by

$$\begin{aligned}
\mathbf{q} &\equiv [Q_{11}\;\; Q_{12}\;\; Q_{21}\;\; Q_{22}]^T \\
\mathbf{q} &\equiv [Q_{11}\;\; Q_{12}\;\; Q_{21}\;\; Q_{22}\;\; Q_{33}]^T \\
\mathbf{q} &\equiv [Q_{11}\;\; Q_{12}\;\; Q_{13}\;\; Q_{21}\;\; Q_{22}\;\; Q_{23}\;\; Q_{31}\;\; Q_{32}\;\; Q_{33}]^T,
\end{aligned} \tag{15.120}$$

respectively in the two-dimensional, axisymmetric and three-dimensional cases.

With $\mathbf{K}_{uu}^{(e)}$, $\mathbf{K}_{up}^{(e)}$, $\mathbf{K}_{pu}^{(e)}$ and $\mathbf{K}_{pp}^{(e)}$ at hand, we compute the *modified element tangent stiffnesses*

$$\hat{\mathbf{K}}_T^{(e)} \equiv \mathbf{K}_{uu}^{(e)} - \mathbf{K}_{up}^{(e)}\,[\mathbf{K}_{pp}^{(e)}]^{-1}\,\mathbf{K}_{pu}^{(e)}, \tag{15.121}$$

and solve during a typical iteration (k) the condensed global system[‡‡]

$$\hat{\mathbf{K}}_T\,\delta\mathbf{u}^{(k)} = \hat{\mathbf{r}}^{(k-1)} \tag{15.122}$$

for the iterative displacement $\delta\mathbf{u}^{(k)}$, where $\hat{\mathbf{K}}_T$ is the assembly of the element matrices $\hat{\mathbf{K}}_T^{(e)}$ and $\hat{\mathbf{r}}^{(k-1)}$ is the *modified residual* obtained from the element vectors

$$\hat{\mathbf{r}}_{(e)}^{(k-1)} \equiv \mathbf{r}_{(e)}^{(k-1)} - \mathbf{K}_{up}^{(e)}\,[\mathbf{K}_{pp}^{(e)}]^{-1}\,\mathbf{s}_{(e)}^{(k-1)}. \tag{15.123}$$

With $\delta\mathbf{u}^{(k)}$ we apply the Newton–Raphson correction to the vector $\mathbf{u}$ of nodal displacements:

$$\mathbf{u}^{(k)} = \mathbf{u}^{(k-1)} + \delta\mathbf{u}^{(k)}, \tag{15.124}$$

and to the element vectors $\mathbf{p}^{(e)}$ of pressure degrees of freedom:

$$\begin{aligned} \delta\mathbf{p}_{(e)}^{(k)} &= -[\mathbf{K}_{pp}^{(e)}]^{-1}\,(\mathbf{s}_{(e)}^{(k-1)} + \mathbf{K}_{pu}^{(e)}\,\delta\mathbf{u}_{(e)}^{(k)}) \\ \mathbf{p}_{(e)}^{(k)} &= \mathbf{p}_{(e)}^{(k-1)} + \delta\mathbf{p}_{(e)}^{(k)}. \end{aligned} \tag{15.125}$$

Having computed the new displacement and pressure degrees of freedom, we compute, for each element, the modified Cauchy stress according to (15.102) and then assemble the internal force vector defined by (15.112) and the element vector defined by (15.113). The iterative process is stopped when the norm of the global vector $\mathbf{r}$ and the element vectors $\mathbf{s}_{(e)}$ fall below a prescribed convergence tolerance.

Remark 15.9. As for *EAS* elements (refer to Remark 15.7, page 678) the variational basis of the u/p formulation ensures that the stiffness matrix $\hat{\mathbf{K}}_T$ is symmetric for materials that possess a potential structure (materials with symmetric $\mathbf{a}$). For materials with unsymmetric tangent modulus, the method loses its variational justification and $\hat{\mathbf{K}}_T$ is unsymmetric.

15.3.4. IMPLEMENTATION ASPECTS

The comments made in Section 15.2.4 (page 678) regarding the extra memory requirements associated with the use of the static condensation procedure for *EAS* elements apply equally to u/p mixed formulations. One important aspect of the u/p method that needs to be observed is the fact that extra material model-related quantities, such as

$$\frac{\partial \bar{p}}{\partial \boldsymbol{F}}, \quad P(\bar{p}), \quad \frac{\partial P(\bar{p})}{\partial \boldsymbol{F}}, \quad \text{etc.}$$

need to be computed during the evaluation of the residual vector and stiffness matrix within the Newton–Raphson loop. Thus, the implementation of u/p methods requires, in addition to the standard procedures for stress-updating and tangent modulus computation (refer to Chapter 4), the provision of new material-related routines for computation of the extra terms.

[‡‡] In the remainder of this section the symbol $\delta(\cdot)$ will be used to denote the *iterative increment of* $(\cdot)$ and *not* the variation of $(\cdot)$.

16 ANISOTROPIC FINITE PLASTICITY: SINGLE CRYSTALS

THE formulation and the numerical treatment of *isotropic* finite elastoplasticity, with a possible simpler form of plastic anisotropy modelled by kinematic hardening, have been thoroughly discussed in Chapter 14. Despite the widespread use of techniques such as those discussed in that chapter in the industrial simulation of finite deformations of polycrystalline metals, we should bear in mind that the underlying hypothesis of elastoplastic isotropy is only an approximation of the actual phenomena. In practice, when subjected to finite inelastic straining, polycrystalline metals rarely remain truly isotropic but, under a wide range of conditions, the isotropy hypothesis provides a very good approximation. For some materials, however, the hypothesis of isotropy cannot be introduced without substantial loss of accuracy of the resulting constitutive model. This is particularly true for *single crystal* alloys. Mechanical components made from single crystals show a strongly *anisotropic* response and any attempt to simulate their behaviour under the assumption of isotropy will almost certainly result in unacceptably erroneous predictions. In such cases, the formulation of an appropriate anisotropic plasticity model is required together with a suitable numerical scheme.

This chapter introduces a framework for the formulation and numerical simulation of anisotropic finite strain single crystal (rate-independent and rate-dependent) plasticity models. On the modelling side, the micro-mechanically-based constitutive theory introduced here is relatively well established. Based on the slip theory of crystals (Asaro, 1983; Asaro and Needleman, 1985; Havner, 1992; Peirce *et al.*, 1982, 1983; Rice, 1971), the constitutive model extends the hyperelastic-based multiplicative plasticity framework described in Chapter 14. On the computational side, the elastic predictor/return-mapping methodology adopted in Chapter 14, with its elastic predictor/return-mapping format, is naturally extended to cope with the inherent anisotropy of the single crystal model. As we shall see, the resulting integration algorithm is far more complex than its isotropic counterpart. In particular, the simple structure where the finite strain related operations are confined to the kinematical level is lost. It should be emphasised, however, that in spite of the complexity of the integration algorithm, quadratic rates of convergence at both local (return-mapping) and global (equilibrium) levels are preserved. Of course, this is a mere consequence of the fully consistent linearisation of the integration algorithm that is described here in detail. The framework presented here is particularised and applied to a planar double-slip model whose computational implementation is fully incorporated into HYPLAS.

Computational Methods for Plasticity: Theory and Applications EA de Souza Neto, D Perić and DRJ Owen

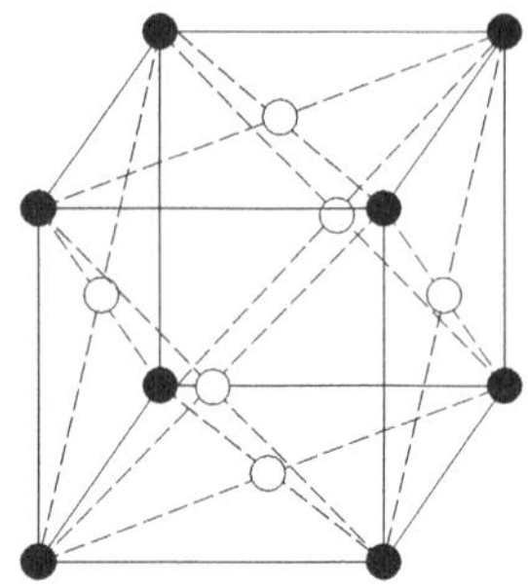

Figure 16.1. The fcc cell structure. Schematic illustration.

16.1. Physical aspects

A summary of the physical concepts that underly the plastic deformation of crystals is presented in this section. The material reviewed here is rather basic and is provided only to familiarise the reader with some important concepts that will be needed throughout this chapter. The interested reader is referred to Dieter (1986) for a more comprehensive account of the subject.

The atomic structure of a metallic crystal is generally characterised by the arrangement of atoms in regular, repeated three-dimensional geometric patterns. A common type of atomic arrangement found in many metal crystals is the so-called *face-centred cubic* (fcc) structure. The fcc structure is schematically illustrated in Figure 16.1. The atoms are arranged in cubic cells with one atom at each vertex (the black circles) and one atom at the centre of each face (the white circles). The fcc crystal consists of the repeated assembly of basic fcc cells.

16.1.1. PLASTIC DEFORMATION BY SLIP: SLIP-SYSTEMS

The plastic deformation of crystals is mainly the result of sliding between crystal blocks along well-defined crystallographic planes. A simple analogy can be made between such planes and playing cards which slide relative to each other in a pack of cards. As atomic bonds strengthen with the decrease in atomic distance, crystallographic sliding (which requires atomic bonds to be cyclically broken and re-established) tends to occur between preferential planes of greatest atomic density. These planes are called *slip planes*. This principle is schematically illustrated in Figure 16.2(a). The bonds between atoms lying within slip planes are the strongest, whilst bonds between atoms of neighbouring slip planes are weaker. By the same token, the direction of sliding follows one of the closest-packed directions within the slip plane. The closest-packed directions are called *slip directions*. A pair comprising a slip plane and a slip direction defines a *slip-system*. In an fcc crystal, for instance, the closest-packed slip planes are the four possible planes of the type illustrated in Figure 16.2(b). Each slip plane contains three slip directions (indicated by the arrows in Figure 16.2(b)). The total number of slip systems in this case is twelve and crystallographic sliding may happen in any such system. Other crystallographic arrangements found in metal crystals are the *body-centred cubic* (bcc) and the *hexagonal close-packed* (hcp) structures. These are illustrated in Figure 16.3.

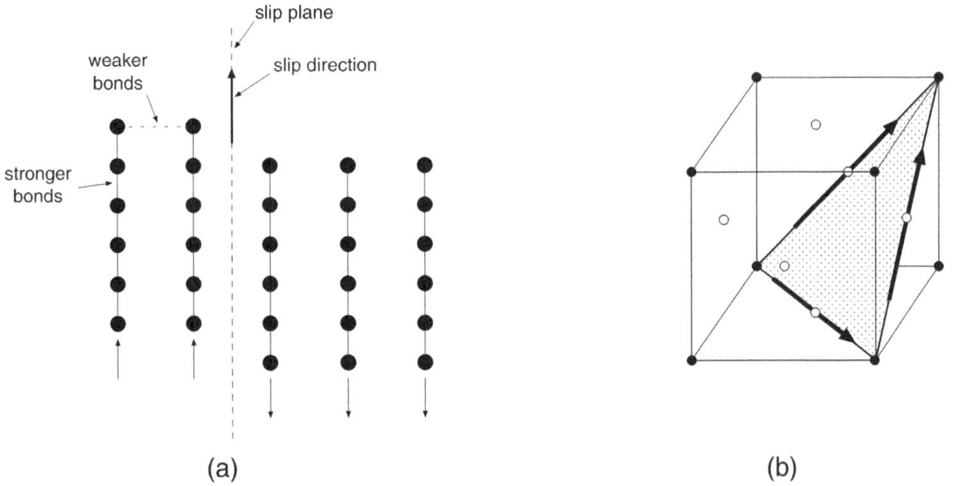

Figure 16.2. Slip planes and slip directions: (a) the concept; (b) slip planes and slip directions in an fcc crystal.

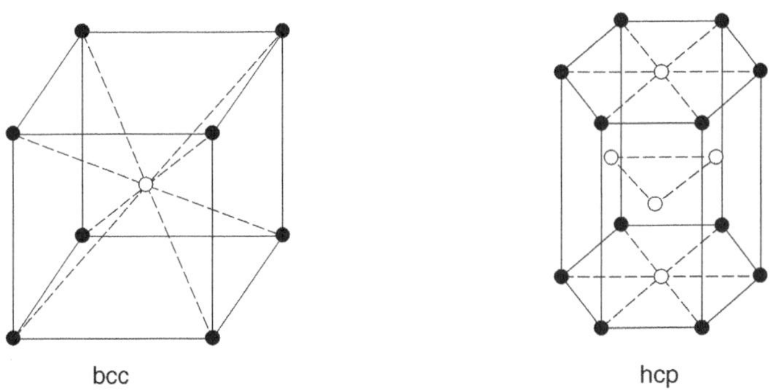

Figure 16.3. The bcc and hcp cell structures. Schematic illustration.

16.2. Plastic slip and the Schmid resolved shear stress

A simple micromechanically-based continuum model of inelastic behaviour of single crystals can be devised by assuming that plastic slip on a particular slip system takes place when the component parallel to the slip direction of the stress vector acting upon the corresponding slip plane reaches a critical value. The continuum description of this principle can be formulated as follows. Let $\boldsymbol{m}$ and $\boldsymbol{s}$ be, respectively, the unit normal vector to the slip plane in question and the unit vector in the slip direction. The pair $\{\boldsymbol{s}, \boldsymbol{m}\}$ defines the slip system. With $\boldsymbol{\sigma}$ denoting the stress tensor, we define the so-called *Schmid resolved shear stress* as

$$\tau \equiv (\boldsymbol{\sigma}\, \boldsymbol{m}) \cdot \boldsymbol{s} = \boldsymbol{\sigma} : (\boldsymbol{s} \otimes \boldsymbol{m}). \tag{16.1}$$

The resolved stress τ is the shear force transferred across the slip plane in the direction of $\boldsymbol{s}$. Thus, plastic slip occurs in system $\{\boldsymbol{s}, \boldsymbol{m}\}$ when τ reaches a critical level. This critical value is analogous to the uniaxial yield stress of the von Mises theory of Chapter 6 (refer to Subsection 6.4.2, page 162) and, in general, will be made dependent on temperature as well as on the history of sliding. The dependence of the critical resolved Schmid stress on the history of sliding is the crystal analogue of the phenomenon of *hardening* discussed in Section 6.6 (from page 177) in the context of purely phenomenological plasticity.

16.3. Single crystal simulation: a brief review

Over the last fifteen years or so, considerable research has been focused on constitutive and numerical aspects of single crystal modelling. Due to certain intrinsic characteristics, such as strong anisotropy, latent hardening, etc., the issue of modelling and numerical simulation of this class of materials remains open. This is particularly true within the context of finite deformations where an extra degree of complexity is added by the nonlinearity of the underlying kinematics. Some substantial early work on the subject has been undertaken by Peirce *et al.* (1982, 1983), Peirce (1983), Asaro and Needleman (1985) and Needleman *et al.* (1985) who have formulated constitutive models for single crystals within the context of both rate-independent and rate-dependent plasticity theories. More recently, the implementation of a rate-dependent model within an explicit finite element scheme has been addressed by Rashid and Nemat-Nasser (1995). Still within the context of rate-dependent formulations, the implicit approach has been reviewed by Cuitiño and Ortiz (1992b). In line with a more modern treatment of nonlinear computational solid mechanics, these last authors discuss issues such as the consistent linearisation of the associated numerical algorithms. Restricted to the infinitesimal theory, the rigorous derivation of a general return-mapping scheme for rate-independent single crystal models has been presented by Borja and Wren (1993). In their paper, Borja and Wren have addressed in detail the important issue of possible non-unique combinations of active slip systems in the rate-independent theory. Based on their analysis, these authors have proposed an algorithm for selection of active systems within the stress-update procedure. Their selection algorithm was later adopted by Miehe (1996a,b) in the finite strain context. Miehe (1996a) brought the concept of exponential map integrators into the realm of single crystal simulation. The use of the exponential map in the integration of the plastic flow equation has been described in the framework for isotropic finite plasticity of Chapter 14. Recall that, among other advantages, the exponential map integrator naturally preserves crucial properties such as the incompressibility of the plastic flow usually associated with pressure-independent metal plasticity models. As we shall see, this property becomes crucially important in the treatment of finite crystal plasticity and the exponential integrator will therefore be adopted in the framework presented in this chapter.

16.4. A general continuum model of single crystals

This section describes in detail a general finite strain continuum elastoplastic model of single crystals. The model is formulated within the framework of hyperelastic-based multiplicative plasticity.

In Chapter 14 (see Remark 14.1, page 579), the physical mechanisms of plastic deformation of metal crystals have been pointed out as the main reason for postulating the multiplicative split of the deformation gradient into elastic and plastic components:

$$\boldsymbol{F} = \boldsymbol{F}^e\, \boldsymbol{F}^p. \tag{16.2}$$

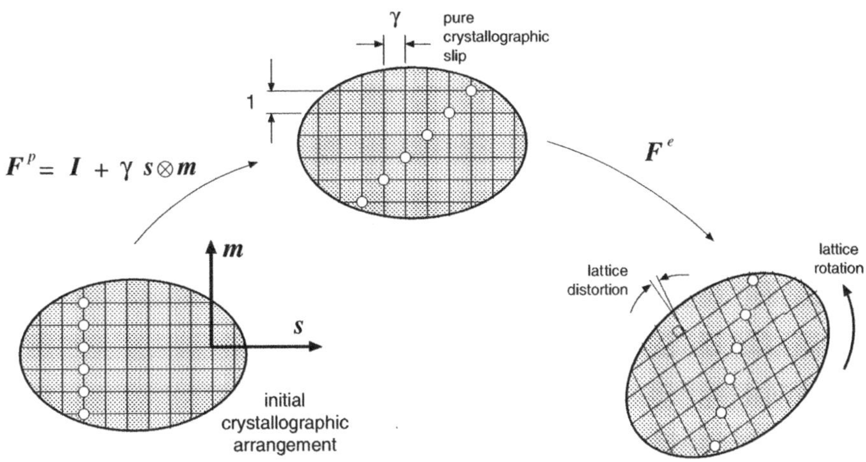

Figure 16.4. The multiplicative decomposition of the deformation gradient. Schematic illustration.

The plastic deformation gradient represents a continuum measure of the microscopic sliding along crystallographic planes. The elastic deformation gradient, on the other hand, provides a continuum measure of crystal lattice distortion, which can be recovered upon complete unloading of the material, as well as the rigid rotation. Consider a crystal containing a single slip system $\{\boldsymbol{s}, \boldsymbol{m}\}$ (see schematic illustration of Figure 16.4). A plastic slip of magnitude γ along the system $\{\boldsymbol{s}, \boldsymbol{m}\}$ corresponds to the plastic deformation gradient

$$\boldsymbol{F}^p = \boldsymbol{I} + \gamma\, \boldsymbol{s} \otimes \boldsymbol{m}. \tag{16.3}$$

It is worth remarking here that whilst rigid rotations contained in $\boldsymbol{F}^e$ are generally large, the actual elastic distortion of the lattice observed in metal crystals is usually small (infinitesimal).

16.4.1. THE PLASTIC FLOW EQUATION

Let us recall now the plastic rate of deformation defined in Chapter 14 (see page 580)

$$\boldsymbol{L}^p \equiv \dot{\boldsymbol{F}}^p\, (\boldsymbol{F}^p)^{-1}. \tag{16.4}$$

Single slip

For a crystal yielding along a single slip-system $\{\boldsymbol{s}, \boldsymbol{m}\}$ the evolution of the plastic deformation gradient is defined by the equation

$$\boldsymbol{L}^p = \dot{\gamma}\, \boldsymbol{s} \otimes \boldsymbol{m}. \tag{16.5}$$

To show that this equation leads to (16.3), note that it can be rephrased as

$$\dot{\boldsymbol{F}}^p = [\dot{\gamma}\, \boldsymbol{s} \otimes \boldsymbol{m}]\, \boldsymbol{F}^p. \tag{16.6}$$

Without loss of generality, $\dot{\gamma}$ can be assumed constant within any interval where the process is monotonic. Then, the bracketed term on the right-hand side of (16.6) becomes constant and the following exact solution (see Section B.1, page 747) is obtained if $\boldsymbol{F}^p|_{\gamma=0} = \boldsymbol{I}$ (the virgin state) is taken as the initial condition:

$$\boldsymbol{F}^p = \exp[\gamma\, \boldsymbol{s} \otimes \boldsymbol{m}]. \tag{16.7}$$

By using the series representation (B.4) (page 747) of the tensor exponential function and taking into account that the orthogonality between $\boldsymbol{s}$ and $\boldsymbol{m}$ implies

$$(\boldsymbol{s} \otimes \boldsymbol{m})(\boldsymbol{s} \otimes \boldsymbol{m}) = \boldsymbol{0}, \tag{16.8}$$

we conclude that (16.7) yields (16.3).

Multislip

In the presence of n_{act} *active slip systems* (systems undergoing slip), the above plastic flow equation is generalised as

$$\boldsymbol{L}^p = \sum_{\alpha=1}^{n_{\text{act}}} \dot{\gamma}^\alpha\, \boldsymbol{s}^\alpha \otimes \boldsymbol{m}^\alpha, \tag{16.9}$$

where $\{\boldsymbol{s}^\alpha, \boldsymbol{m}^\alpha\}$ are the unit vectors that define slip system α and the multiplier $\dot{\gamma}^\alpha$ is the plastic shear rate in system α.

16.4.2. THE RESOLVED SCHMID SHEAR STRESS

Plastic slip may occur on a slip system only if the resolved shear stress on that system reaches a critical level. Here, this principle will be expressed in terms of the *Kirchhoff* stress tensor, $\boldsymbol{\tau}$. Accordingly, definition (16.1) is generalised and the *resolved shear stress* on a system α is defined as

$$\tau^\alpha \equiv \boldsymbol{R}^{e\,T}\, \boldsymbol{\tau}\, \boldsymbol{R}^e : \boldsymbol{s}^\alpha \otimes \boldsymbol{m}^\alpha. \tag{16.10}$$

Note that, as vectors $\boldsymbol{s}^\alpha$ and $\boldsymbol{m}^\alpha$ are defined on the *intermediate* configuration, the Kirchhoff stress tensor (which is defined on the *deformed* configuration) has been appropriately rotated back (with $\boldsymbol{R}^e$) to the intermediate configuration before being resolved along slip system α.

16.4.3. MULTISURFACE FORMULATION OF THE FLOW RULE

Yield criterion

To formulate a yield criterion for single crystals within the framework of multisurface plasticity, it is convenient to split each physical slip system of the crystal into two mirrored systems

$$\{\boldsymbol{s}^\alpha, \boldsymbol{m}^\alpha\} \quad \text{and} \quad \{\boldsymbol{s}^\beta, \boldsymbol{m}^\beta\} \equiv \{-\boldsymbol{s}^\alpha, \boldsymbol{m}^\alpha\}. \tag{16.11}$$

Now we have a total number of $2 \times n_{\text{syst}}$ systems.

Yield functions: the yield surface

In order to define the onset of plastic slip, we introduce $2 \times n_{\rm syst}$ yield functions of the form

$$\Phi^\alpha(\tau^\alpha, \tau_y^\alpha) \equiv \tau^\alpha - \tau_y^\alpha, \quad \alpha = 1, \ldots, 2n_{\rm syst}, \tag{16.12}$$

where τ_y^α is the critical resolved shear stress for system α. In the presence of hardening, the value of τ_y^α depends on the history of the deformation process. It is important to note that the functions Φ^α are *anisotropic* functions of $\boldsymbol{\tau}$. For a given τ_y^α a hypersurface in the space of Kirchhoff stresses is defined by

$$\Phi^\alpha(\tau^\alpha(\boldsymbol{\tau}), \tau_y^\alpha) = 0, \tag{16.13}$$

for each system α. Plastic slip in system α may occur only if the stress state is on the above defined surface. The $2 \times n_{\rm syst}$ surfaces of this type form the envelope of the elastic domain where the mechanical response of the crystal is fully reversible. The elastic domain is defined by

$$\mathscr{E} \equiv \{\boldsymbol{\tau} \,|\, \Phi^\alpha(\tau^\alpha(\boldsymbol{\tau}), \tau_y^\alpha) < 0, \ \alpha = 1, \ldots, 2n_{\rm syst}\}. \tag{16.14}$$

The set of *plastically admissible stresses* is the closure of $\mathscr{E}$, defined as

$$\bar{\mathscr{E}} \equiv \{\boldsymbol{\tau} \,|\, \Phi^\alpha(\tau^\alpha(\boldsymbol{\tau}), \tau_y^\alpha) \leq 0, \ \alpha = 1, \ldots, 2n_{\rm syst}\}. \tag{16.15}$$

The *boundary* of the elastic domain is the *yield surface*, where plastic slip may occur. The yield surface is defined by

$$\mathscr{Y} \equiv \{\boldsymbol{\tau} \in \bar{\mathscr{E}} \mid \Phi^\alpha(\tau^\alpha(\boldsymbol{\tau}), \tau_y^\alpha) = 0 \quad \text{for at least one } \alpha,\ \Phi^\beta(\tau^\beta(\boldsymbol{\tau}), \tau_y^\beta) \leq 0 \quad \forall\, \beta \neq \alpha\}. \tag{16.16}$$

Unlike isotropic criteria such as the von Mises, Tresca and Mohr–Coulomb (refer to Chapter 6) the visualisation of the above anisotropic yield surface is extremely difficult and will not be attempted here. A graphical representation of yield surfaces for fcc and bcc crystals is presented by Kocks (1970).

The multisurface plastic flow rule

Following the standard multisurface plasticity format (see page 156) the evolution law for the plastic deformation gradient is then formulated as

$$\boldsymbol{L}^p = \sum_{\alpha=1}^{2n_{\rm syst}} \dot{\gamma}^\alpha \, \boldsymbol{s}^\alpha \otimes \boldsymbol{m}^\alpha, \tag{16.17}$$

or, equivalently,

$$\dot{\boldsymbol{F}}^p = \left[\sum_{\alpha=1}^{2n_{\rm syst}} \dot{\gamma}^\alpha \, \boldsymbol{s}^\alpha \otimes \boldsymbol{m}^\alpha \right] \boldsymbol{F}^p, \tag{16.18}$$

where the plastic multipliers $\dot{\gamma}^\alpha$ satisfy the complementarity condition

$$\Phi^\alpha \leq 0, \quad \dot{\gamma}^\alpha \geq 0, \quad \Phi^\alpha \, \dot{\gamma}^\alpha = 0, \tag{16.19}$$

for $\alpha = 1, \ldots, 2n_{\rm syst}$. Summation on the repeated index is *not* implied in $(16.19)_3$.

16.4.4. ISOTROPIC TAYLOR HARDENING

A hardening law must be defined in order to describe the evolution of the resolved yield stresses, τ_y^α, in each system. The physical phenomena behind hardening of crystals are quite complex and their appropriate representation within the continuum setting remains an open issue. A popular approach consists in postulating the following evolution equation for the resolved yield stresses (Asaro and Needleman, 1985):

$$\dot{\tau}_y^\alpha = \sum_{\beta=1}^{2n_{\rm syst}} h_{\alpha\beta}\,\dot{\gamma}^\beta, \qquad \alpha = 1, \ldots, 2n_{\rm syst},$$

where $h_{\alpha\beta}$ defines a matrix of hardening coefficients, whose components may depend on the history of deformation.

In what follows, we shall adopt a much simpler law, known as *Taylor hardening*, in which the resolved yield stress is the same for all systems. With τ_y denoting the common critical value, the yield function for the generic system α is redefined as

$$\Phi^\alpha(\tau^\alpha, \tau_y) = \tau^\alpha - \tau_y. \tag{16.20}$$

In addition, the resolved critical stress is assumed to be a function of a single internal variable, γ:

$$\tau_y = \tau_y(\gamma). \tag{16.21}$$

The hardening variable is the *accumulated slip* which, analogously to the accumulated (or effective) plastic strain of the isotropic theory (see expression (6.167), page 179), is defined as

$$\gamma \equiv \int_0^t \sum_{\alpha=1}^{2n_{\rm syst}} \dot{\gamma}^\alpha\,{\rm d}t. \tag{16.22}$$

Its evolution is then given by

$$\dot{\gamma} \equiv \sum_{\alpha=1}^{2n_{\rm syst}} \dot{\gamma}^\alpha. \tag{16.23}$$

16.4.5. THE HYPERELASTIC LAW

Finally, a hyperelastic law is incorporated to model the reversible behaviour. The elastic contribution to the specific free energy

$$\psi^e(\boldsymbol{F}^e) \tag{16.24}$$

is introduced, and the standard constitutive law for the Kirchhoff stress tensor follows as

$$\boldsymbol{\tau} = \boldsymbol{\tau}(\boldsymbol{F}^e) = \bar{\rho}\,\frac{\partial\psi^e}{\partial\boldsymbol{F}^e}\,\boldsymbol{F}^{eT}. \tag{16.25}$$

This completes the definition of the continuum constitutive model of single crystals. The model is summarised in Box 16.1.

Box 16.1. General finite strain anisotropic single crystal plasticity model.

(i) Multiplicative decomposition of the deformation gradient

$$\boldsymbol{F} = \boldsymbol{F}^{e}\boldsymbol{F}^{p}$$

(ii) Hyperelastic law

$$\boldsymbol{\tau} = \bar{\rho}\,\frac{\partial \psi^e}{\partial \boldsymbol{F}^e}\,\boldsymbol{F}^{eT}$$

(iii) Resolved shear stresses

$$\tau^\alpha = \boldsymbol{R}^{eT}\,\boldsymbol{\tau}\,\boldsymbol{R}^{e} : (\boldsymbol{s}^\alpha \otimes \boldsymbol{m}^\alpha)$$

(iv) Yield functions

$$\Phi^\alpha = \tau^\alpha - \tau_y(\gamma)$$

(v) Evolution equations for $\boldsymbol{F}^p$ and Taylor hardening internal variable γ

$$\dot{\boldsymbol{F}}^p = \left[\sum_{\alpha=1}^{2n_{\rm syst}} \dot{\gamma}^\alpha\, \boldsymbol{s}^\alpha \otimes \boldsymbol{m}^\alpha\right] \boldsymbol{F}^p$$

$$\dot{\gamma} = \sum_{\alpha=1}^{2n_{\rm syst}} \dot{\gamma}^\alpha$$

(vi) Loading/unloading criterion

$$\Phi^\alpha \leq 0, \quad \dot{\gamma}^\alpha \geq 0, \quad \dot{\gamma}^\alpha \Phi^\alpha = 0$$

for $\alpha = 1, \ldots, 2n_{\rm syst}$

16.5. A general integration algorithm

The elastic predictor/return-mapping algorithm for integration of the general constitutive equations of Box 16.1 is a natural extension of the scheme thoroughly described in Chapter 14 in the isotropic context. As we shall see, the anisotropic extension is considerably more complex than its isotropic counterpart. In particular, it does not preserve the simple and convenient structure where the essential stress integration has the small strain format and the finite strain nonlinearity is confined to the kinematic level. Here, the geometric nonlinearity cannot be decoupled from the stress-updating procedure.

The elastic trial step

Let us once more consider the typical (pseudo-) time interval $[t_n, t_{n+1}]$. With the known values of $\boldsymbol{F}^e_n$, $\boldsymbol{F}^p_n$ and γ_n at hand, and given an incremental deformation gradient

$$\boldsymbol{F}_\Delta \equiv \boldsymbol{I} + \nabla_{x_n}[\Delta \boldsymbol{u}], \tag{16.26}$$

we firstly compute the elastic trial deformation gradient

$$\boldsymbol{F}^{e\ \text{trial}}_{n+1} = \boldsymbol{F}_{\Delta}\, \boldsymbol{F}^{e}_{n}, \tag{16.27}$$

and the corresponding elastic trial Kirchhoff stress

$$\boldsymbol{\tau}^{\text{trial}}_{n+1} = \bar{\rho}\, \frac{\partial \psi^{e}}{\partial \boldsymbol{F}^{e}}\bigg|^{\text{trial}}_{n+1} \boldsymbol{F}^{e\ \text{trial}\,T}. \tag{16.28}$$

The *elastic trial resolved Schmid stresses* are then computed as

$$\tau^{\alpha\ \text{trial}}_{n+1} = (\boldsymbol{R}^{e\ \text{trial}}_{n+1})^{T}\, \boldsymbol{\tau}^{\text{trial}}_{n+1}\, \boldsymbol{R}^{e\ \text{trial}}_{n+1} : (\boldsymbol{s}^{\alpha} \otimes \boldsymbol{m}^{\alpha}), \tag{16.29}$$

for $\alpha = 1, \ldots, 2n_{\text{syst}}$. The elastic trial rotation, $\boldsymbol{R}^{e\ \text{trial}}_{n+1}$, is obtained from the polar decomposition of the elastic trial deformation gradient. The elastic trial Taylor hardening variable is set to

$$\gamma^{\text{trial}}_{n+1} = \gamma_{n}. \tag{16.30}$$

Consistency check

The next step is to check whether the elastic trial state violates plastic admissibility. If the trial state lies in the elastic domain or on the yield surface, i.e. if

$$\Phi^{\alpha\ \text{trial}} \equiv \tau^{\alpha\ \text{trial}}_{n+1} - \tau_{y}(\gamma^{\text{trial}}_{n+1}) \leq 0, \tag{16.31}$$

for $\alpha = 1, \ldots, 2n_{\text{syst}}$, then we simply set

$$(\cdot)_{n+1} := (\cdot)^{\text{trial}}_{n+1} \tag{16.32}$$

and the state update is complete. Otherwise, the multisurface return mapping described below is applied.

The multivector return mapping

In deriving the multisurface return mapping for the present general model, we shall assume that the slip systems undergoing plastic slip over the current increment are known in advance. The search for active systems – a crucial component of the overall scheme – is discussed below in Section 16.5.1. Let us then assume that $\mathscr{A}$ is the set of n_{act} active slip systems within $[t_n, t_{n+1}]$. As in the isotropic case (cf. expression (14.73), page 591), the implicit *exponential integrator* is used to discretise the plastic flow equation. In the present case, the discretised plastic flow equation is given by

$$\boldsymbol{F}^{p}_{n+1} = \exp\left[\sum_{\alpha\in\mathscr{A}} \Delta\gamma^{\alpha}\, \boldsymbol{s}^{\alpha} \otimes \boldsymbol{m}^{\alpha}\right] \boldsymbol{F}^{p}_{n}. \tag{16.33}$$

The corresponding elastic deformation gradient update formula reads

$$\boldsymbol{F}^{e}_{n+1} = \boldsymbol{F}^{e\ \text{trial}}_{n+1}\, \exp\left[-\sum_{\alpha\in\mathscr{A}} \Delta\gamma^{\alpha}\, \boldsymbol{s}^{\alpha} \otimes \boldsymbol{m}^{\alpha}\right]. \tag{16.34}$$

The standard backward Euler discretisation of the Taylor hardening variable evolution equation gives

$$\gamma_{n+1} = \gamma_n + \sum_{\alpha \in \mathscr{A}} \Delta\gamma^\alpha. \tag{16.35}$$

The corresponding discrete consistency condition reads

$$\Phi^\alpha_{n+1} \equiv \boldsymbol{\tau}(\boldsymbol{F}^e_{n+1}) - \tau_y(\gamma_{n+1}) = 0, \quad \forall\, \alpha \in \mathscr{A}. \tag{16.36}$$

Remark 16.1. By construction, the *incremental flow vector*,

$$\sum_{\alpha \in \mathscr{A}} \Delta\gamma^\alpha \, \boldsymbol{s} \otimes \boldsymbol{m},$$

is *traceless*. As a result of the use of the exponential integrator, the corresponding incremental plastic deformation gradient,

$$\boldsymbol{F}^p_\Delta \equiv \exp\left[\sum_{\alpha \in \mathscr{A}} \Delta\gamma^\alpha \, \boldsymbol{s}^\alpha \otimes \boldsymbol{m}^\alpha\right], \tag{16.37}$$

is *isochoric* (the basic properties of the tensor exponential are listed in Section B.1.1, page 748), i.e.

$$\det[\boldsymbol{F}^p_\Delta] = 1. \tag{16.38}$$

Thus, the above update formula for $\boldsymbol{F}^p$ preserves *exactly* the plastic incompressibility of the continuum model. This property is pointed out by Steinmann and Stein (1996) as a vital source of accuracy of the numerical integration scheme.

Remark 16.2. Recall that in the isotropic context (see derivation in Section 14.4.2, page 591) the tensor exponential function does not need to be explicitly evaluated in the return-mapping procedure. In that case, when the logarithmic strain-based model is adopted, the tensor exponential function is conveniently cancelled out of the relevant update formulae. In contrast, in the present anisotropic case, the tensor exponential function cannot be eliminated from the return-mapping equations and will have to be evaluated (expressions (16.33) and (16.34)) in the computational procedure.

Remark 16.3. Note that the incremental flow vector is generally *unsymmetric*. Thus, the evaluation of the exponential of an unsymmetric tensor is required. The evaluation will be carried out here by means of the procedure described in Section B.1 (page 747) which makes use of the series representation of the tensor exponential function.

The return-mapping system of equations

In the above update formulae, $\boldsymbol{F}^e_{n+1}$ and γ_{n+1} are functions of the multipliers $\Delta\gamma^\alpha$ only. As the resolved stresses τ^α_{n+1} are functions of $\boldsymbol{R}^e$ and the Kirchhoff stress tensor which, in turn, are functions of $\boldsymbol{F}^e$, the above discrete consistency condition is equivalent to the following system of n_{act} nonlinear algebraic equations for the plastic multipliers

$$\tilde{\Phi}^\alpha(\Delta\boldsymbol{\gamma}) \equiv \tau^\alpha(\boldsymbol{F}^e_{n+1}(\Delta\boldsymbol{\gamma})) - \tau_y(\gamma_{n+1}(\Delta\boldsymbol{\gamma})) = 0, \quad \forall\, \alpha \in \mathscr{A}, \tag{16.39}$$

where we have defined

$$\Delta\gamma \equiv \{\Delta\gamma^{\alpha} \mid \alpha \in \mathscr{A}\}. \tag{16.40}$$

Note that if the given set $\mathscr{A}$ of active slip systems is correct, then the resulting updated variables, $\boldsymbol{F}^{e}_{n+1}$ and γ_{n+1}, must satisfy

$$\Phi^{\beta} \equiv \tau^{\beta}(\boldsymbol{F}^{e}_{n+1}) - \tau_{y}(\gamma_{n+1}) \leq 0, \quad \beta = 1, \ldots, 2n_{\text{syst}}. \tag{16.41}$$

Also, the active plastic multipliers must all be non-negative:

$$\Delta\gamma^{\alpha} \geq 0, \quad \forall\, \alpha \in \mathscr{A}. \tag{16.42}$$

The above requirement is the discrete counterpart of the complementarity condition (16.19).

The Newton–Raphson scheme

As for the other models discussed in this book, the standard Newton–Raphson scheme can be adopted for solution of the above algebraic system. Here, in the kth Newton iteration we update

$$\Delta\gamma^{\alpha}_{(k)} := \Delta\gamma^{\alpha}_{(k-1)} + \delta\gamma^{\alpha}, \tag{16.43}$$

where the iterative corrections $\delta\gamma^{\alpha}$ are obtained by solving the linearised counterpart of (16.39)

$$\sum_{\beta \in \mathscr{A}} \mathsf{J}^{\alpha\beta}\, \delta\gamma^{\beta} = -\tilde{\Phi}^{\alpha}. \tag{16.44}$$

The components

$$\mathsf{J}^{\alpha\beta} \equiv \frac{\mathrm{d}\tilde{\phi}^{\alpha}}{\mathrm{d}\Delta\gamma^{\beta}} = \frac{\mathrm{d}\tau^{\alpha}}{\mathrm{d}\boldsymbol{F}^{e}} : \frac{\mathrm{d}\boldsymbol{F}^{e}}{\mathrm{d}\Delta\gamma^{\beta}} - H, \tag{16.45}$$

where the subscripts $n+1$ have been omitted for notational convenience, define the Jacobian matrix $\mathbf{J}$ of the return-mapping system. The tensor $\mathrm{d}\tau^{\alpha}/\mathrm{d}\boldsymbol{F}^{e}$ on the right-hand side of the above expression depends on the particular hyperelastic model adopted. The derivative $\mathrm{d}\boldsymbol{F}^{e}/\mathrm{d}\Delta\gamma^{\beta}$ is defined by the Cartesian components

$$\left[\frac{\mathrm{d}\boldsymbol{F}^{e}}{\mathrm{d}\Delta\gamma^{\beta}}\right]_{ij} = -F^{e\,\text{trial}}_{im}\, \mathsf{E}_{mjkl}\, [\boldsymbol{s}^{\beta} \otimes \boldsymbol{m}^{\beta}]_{kl}, \tag{16.46}$$

where E_{mjkl} denote the components of the derivative of the tensor exponential function evaluated at the (generally unsymmetric) argument

$$-\sum_{\alpha \in \mathscr{A}} \Delta\gamma^{\alpha}\, \boldsymbol{s}^{\alpha} \otimes \boldsymbol{m}^{\alpha}.$$

The exact computation of the derivative of the exponential of a generally unsymmetric tensor is discussed in detail in Section B.2 of Appendix B and is implemented in `HYPLAS` subroutine `DEXPMP`. The term H is the *Taylor hardening modulus*, defined as

$$H \equiv \frac{\mathrm{d}\tau_{y}}{\mathrm{d}\gamma}. \tag{16.47}$$

In deriving the above Jacobian, we have made use of the fact that

$$\frac{\mathrm{d}\gamma}{\mathrm{d}\Delta\gamma^{\beta}} = 1, \tag{16.48}$$

which follows by differentiation of (16.35).

16.5.1. THE SEARCH FOR AN ACTIVE SET OF SLIP SYSTEMS

In the above return mapping, it has been assumed that the set $\mathscr{A}$ of active slip systems, i.e. systems with $\Delta\gamma^\alpha \neq 0$ within the interval $[t_n, t_{n+1}]$ is known in advance. However, it is not possible in general to know which systems will be active in the interval of interest. Recall that this was also the case in the integration of isotropic multisurface models (such as Tresca and Mohr–Coulomb) discussed in Chapter 8 (refer to Sections 8.1 and 8.2). There, robust algorithms have been developed to search for the active set of yield surfaces and to ensure that, under any circumstance, the final converged state rigorously satisfies the discrete consistency condition. The search algorithms discussed in Chapter 8 have been devised largely based on the geometric representation (in principal stress space) of the underlying criteria. Unfortunately, the proper geometric visualisation of single crystal yield surfaces is very difficult and the derivation of a geometrically-based search algorithm is virtually impossible. The issue is here made more complex due to the fact that the active set that satisfies the discrete plastic consistency may not be unique. Also, for a given active set $\mathscr{A}$, many combinations of plastic multipliers (i.e. solution vectors $\Delta\boldsymbol{\gamma}$) may exist which produce the same incremental plastic deformation gradient. Non-uniqueness is possible whenever the set of all slip-system tensors,

$$\{\boldsymbol{s}^\alpha \otimes \boldsymbol{m}^\alpha \mid \alpha = 1, \ldots, n_{\text{syst}}\}$$

is *linearly dependent* in the space of traceless tensors to which the incremental flow vector belongs. Note that in two dimensions, for instance, the space of traceless tensors is spanned by *three* linearly independent base tensors. In three dimensions *eight* linearly independent tensors generate the traceless tensors space. Thus, if the total number of active slip systems is greater than three in two dimensions (greater than eight in three dimensions), then the set $\mathscr{A}$ is necessarily linearly dependent and multiple solutions $\Delta\boldsymbol{\gamma}$ exist for the return-mapping equations. In this case, the Jacobian of the return-mapping system of equations becomes singular.

To overcome the problems associated with non-uniqueness, many authors have resorted to the viscoplastic (rate-dependent) regularisation of the original elastoplastic (rate-independent) model (Asaro and Needleman, 1985; Needleman *et al.*, 1985; Peirce, 1983; Peirce *et al.*, 1982, 1983). Non-uniqueness is not an issue in the viscoplastic context and the rate-independent limit can be recovered by setting suitably low values for the rate-sensitivity constant. However, the resulting set of equations to be integrated becomes extremely stiff as the rate-independent limit is approached and the formulation of robust algorithms under such conditions is not an easy task.

A completely different approach has been proposed by Borja and Wren (1993). Restricted to the infinitesimal theory, they have developed a fairly robust algorithm for selection of active systems. However, in spite of its relative robustness, the active set determination algorithm of Borja and Wren (1993) is reported to fail under certain circumstances. This algorithm is described next.

The active set determination algorithm

The basic idea underlying the active set determination algorithm is to start the application of the above described return-mapping scheme with some trial active set $\mathscr{A}^{(0)}$, and then, to generate a sequence of sets $\{\mathscr{A}^{(1)}, \mathscr{A}^{(2)}, \ldots\}$ as the Newton iterations for solution of (16.39) are applied. It is expected that, at the end, the algorithm will converge to a state (and an active set) where the discrete complementarity condition expressed by (16.41, 16.42) is satisfied.

We may then proceed as follows. During a generic Newton iteration k, just after the new guesses $\Delta\gamma^{\alpha}_{(k)}$ have been obtained according to (16.43), we reset the incremental multipliers, which are currently negative, to zero:

$$\Delta\gamma^{\alpha}_{(k)} := \max[\, 0, \Delta\gamma^{\alpha}_{(k)}\,]. \tag{16.49}$$

With the above reset plastic multipliers, we compute the corresponding yield function values, $\Phi^{\alpha}_{(k)}$, and define the following sets:

$$\begin{aligned} \mathscr{A}^{(k)}_{\Phi} &\equiv \{\alpha \in \{1, 2, \ldots, 2n_{\text{syst}}\} \mid \Phi^{\alpha}_{(k)} > 0\} \\ \mathscr{A}^{(k)}_{\Delta\gamma} &\equiv \{\alpha \in \{1, 2, \ldots, 2n_{\text{syst}}\} \mid \Delta\gamma^{\alpha}_{(k)} > 0\}. \end{aligned} \tag{16.50}$$

Having defined the above sets, a set $\mathscr{A}^{(k+1)}$ is assembled as

$$\mathscr{A}^{(k+1)} := \mathscr{A}^{(k)}_{\Phi} \cup \mathscr{A}^{(k)}_{\Delta\gamma}. \tag{16.51}$$

The above set is then used as the new trial active set in the following Newton iteration $k+1$. A solution to (16.39) that satisfies the discrete complementarity will have been obtained if, after an iteration k, we have

$$\mathscr{A}^{(k+1)}_{\Phi} = \emptyset.$$

To start up the algorithm, Borja and Wren (1993) suggest the following trial active set

$$\mathscr{A}^{(0)} := \mathscr{A}_n, \tag{16.52}$$

where $\mathscr{A}_n$ is the converged active set obtained for the Gauss point in question at the end of the previous global load increment. Miehe (1996a,b), on the other hand, adopts

$$\mathscr{A}^{(0)} := \mathscr{A}^{\,\text{trial}} \equiv \{\alpha \in \{1, \ldots, 2n_{\text{syst}}\} \mid \Phi^{\alpha\ \text{trial}} > 0\}. \tag{16.53}$$

Dealing with a linearly dependent active set

If a set $\mathscr{A}^{(k)}$ contains linearly dependent slip-system tensors, the standard Newton iteration cannot be performed, for the Jacobian matrix $\mathbf{J}$ in (16.44) is singular. To overcome this problem, Borja and Wren (1993) suggested the triangular factorisation of $\mathbf{J}$ in the solution of (16.44) followed by the elimination of the redundant equations (corresponding to zero diagonal terms) in the backsubstitution phase. In this way, one of the possible solutions to the return-mapping system will be picked up. It should be noted that, with isotropic Taylor hardening, even when multiple solutions exist for the plastic multipliers, the final stress tensor is unique. This may not be true, however, for more general crystal hardening laws where plastic slip in one system affects hardening on distinct latent systems in different ways. In such cases, the basic evolution problem may be ill-posed.

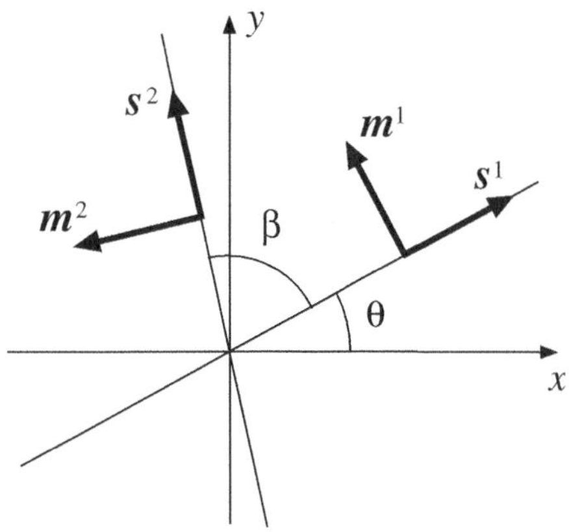

Figure 16.5. Planar double-slip single crystal model.

16.6. An algorithm for a planar double-slip model

To illustrate the concepts conveyed in Section 16.5, we describe in this section the particularisation of the above discussed general integration algorithm for a planar double-slip single crystal model. The choice of this model is motivated by the simplicity of its implementation. In particular, the slip-system tensors of the model are linearly independent and the return-mapping equations have unique solution. Thus, the need for special treatment of redundant equations addressed above is completely by-passed. Here, the active slip systems can be determined by a trial-and-error procedure conceptually similar to those employed in the implementation of the Tresca and Mohr–Coulomb models described in Chapter 8. The procedure described in this section has been fully incorporated into program `HYPLAS`.

16.6.1. A PLANAR DOUBLE-SLIP MODEL

The planar double-slip single crystal model is suitable to describe the plane deformation of fcc crystals under certain crystal orientations and load conditions. In spite of the twelve slip systems that characterise fcc crystals, their behaviour under such conditions can be modelled by means of only two 'effective' slip systems ($n_{\rm syst} = 2$) whose slip and normal direction vectors lie on the plane of loading (see schematic illustration of Figure 16.5). Plastic deformation in this case may occur only in that plane. In Figure 16.5, θ denotes the initial orientation of slip system 1 with respect to the x-axis of the plane and β is the relative orientation of slip system 2 with respect to system 1. In the multisurface plasticity description of the model, we define the mirrored systems

$$\{\boldsymbol{s}^3, \boldsymbol{m}^3\} \equiv \{-\boldsymbol{s}^1, \boldsymbol{m}^1\}, \quad \{\boldsymbol{s}^4, \boldsymbol{m}^4\} \equiv \{-\boldsymbol{s}^2, \boldsymbol{m}^2\}. \tag{16.54}$$

The hyperelastic model: compressible neo-Hookean material

As elastic distortions of metal crystal lattices are typically infinitesimal, the choice of a particular hyperelastic law to describe the reversible behaviour is likely to have little or no impact on the results of numerical simulations. Thus, the definition of the elastic law in the present context will be regarded as a matter of numerical convenience rather than a physically-based choice. In the isotropic context of Chapter 14, the logarithmic strain-based Hencky model was found to be particularly convenient. Here a *compressible neo-Hookean model* will be adopted instead. Originally proposed by Miehe (1996a,b), the use of the compressible neo-Hookean material in the single crystal plasticity context was shown to lead to a relatively simple format of return-mapping equations.

The compressible neo-Hookean model was described in detail in Chapter 13. In the present elastoplastic context, the regularised neo-Hookean potential (13.44) (page 526) is redefined in terms of the *elastic* left Cauchy–Green tensor as

$$\bar{\rho}\,\psi^e(\boldsymbol{B}^e_{\rm iso}) \equiv \tfrac{1}{2}\,G\,({\rm tr}[\boldsymbol{B}^e_{\rm iso}] - 3) + \tfrac{1}{2}\,K\,(\ln J^e)^2, \tag{16.55}$$

where G and K denote, respectively the shear and bulk moduli and

$$J^e \equiv \det[\boldsymbol{F}^e]. \tag{16.56}$$

$\boldsymbol{B}^e_{\rm iso}$ is the isochoric (volume-preserving) component of the elastic left Cauchy–Green strain tensor, $\boldsymbol{B}^e$,

$$\boldsymbol{B}^e_{\rm iso} \equiv \boldsymbol{F}^e_{\rm iso}(\boldsymbol{F}^e_{\rm iso})^T = (J^e)^{-\frac{2}{3}}\,\boldsymbol{B}^e, \tag{16.57}$$

with $\boldsymbol{F}^e_{\rm iso}$ denoting the isochoric component of $\boldsymbol{F}^e$,

$$\boldsymbol{F}^e_{\rm iso} \equiv (J^e)^{-\frac{1}{3}}\,\boldsymbol{F}^e. \tag{16.58}$$

The constitutive law for the Kirchhoff stress is obtained directly from the above free-energy potential as

$$\boldsymbol{\tau} = 2\,\bar{\rho}\,\frac{\partial\psi}{\partial\boldsymbol{B}^e}\,\boldsymbol{B}^e = G\,{\rm dev}[\boldsymbol{B}^e_{\rm iso}] + K\,(\ln J^e)\,\boldsymbol{I}. \tag{16.59}$$

The Schmid-resolved stresses

One crucial aspect to be emphasised is the fact that as a result of the use of the compressible neo-Hookean material, the resolved Schmid stresses admit a strikingly simple representation. Because slip-system tensors are deviatoric by construction, their internal products by the hydrostatic Kirchhoff stress components vanish and (16.59) together with (16.10) give

$$\begin{aligned}\tau^\alpha &= G\,\boldsymbol{R}^{e\,T}\,{\rm dev}[\boldsymbol{B}^e_{\rm iso}]\,\boldsymbol{R}^e : \boldsymbol{s}^\alpha \otimes \boldsymbol{m}^\alpha \\ &= G\,\boldsymbol{R}^{e\,T}\,\boldsymbol{B}^e_{\rm iso}\,\boldsymbol{R}^e : \boldsymbol{s}^\alpha \otimes \boldsymbol{m}^\alpha \\ &= G\,\boldsymbol{C}^e_{\rm iso} : \boldsymbol{s}^\alpha \otimes \boldsymbol{m}^\alpha\end{aligned} \tag{16.60}$$

where $\boldsymbol{C}^e_{\rm iso} \equiv (\boldsymbol{F}^e_{\rm iso})^T\,\boldsymbol{F}^e_{\rm iso} = \boldsymbol{R}^{e\,T}\,\boldsymbol{B}^e_{\rm iso}\,\boldsymbol{R}^e$ is the isochoric right elastic Cauchy–Green strain tensor. Then, after a straightforward manipulation, we end up with the simpler final formula

$$\tau^\alpha = G\,\bar{\boldsymbol{s}}^\alpha \cdot \bar{\boldsymbol{m}}^\alpha, \tag{16.61}$$

where $\bar{\boldsymbol{s}}^\alpha$ ($\bar{\boldsymbol{m}}^\alpha$) is the isochoric elastic push-forward of $\boldsymbol{s}^\alpha$ ($\boldsymbol{m}^\alpha$),

$$\begin{aligned}\bar{\boldsymbol{s}}^\alpha &\equiv \boldsymbol{F}^e_{\rm iso}\,\boldsymbol{s}^\alpha \\ \bar{\boldsymbol{m}}^\alpha &\equiv \boldsymbol{F}^e_{\rm iso}\,\boldsymbol{m}^\alpha.\end{aligned} \tag{16.62}$$

16.6.2. THE INTEGRATION ALGORITHM

The essential algorithm described here is a specialisation of the general integration procedure of Section 16.5 for the present model. Again within the interval $[t_n, t_{n+1}]$, we start by computing the elastic trial state. Firstly, we evaluate the elastic trial deformation gradient using the usual expression (16.27) and compute the corresponding isochoric component $\boldsymbol{F}_{\text{iso}}^{e\ \text{trial}}$. We then evaluate the elastic trial Schmid-resolved shear stresses, which for the present model have the simple format

$$\tau_{n+1}^{\alpha\ \text{trial}} = G\,\bar{\boldsymbol{s}}^{\alpha\ \text{trial}} \cdot \bar{\boldsymbol{m}}^{\alpha\ \text{trial}}, \tag{16.63}$$

where

$$\bar{\boldsymbol{s}}^{\alpha\ \text{trial}} = \boldsymbol{F}_{\text{iso}}^{e\ \text{trial}}\,\boldsymbol{s}^{\alpha}, \quad \bar{\boldsymbol{m}}^{\alpha\ \text{trial}} = \boldsymbol{F}_{\text{iso}}^{e\ \text{trial}}\,\boldsymbol{m}^{\alpha}. \tag{16.64}$$

Next, we proceed to the standard consistency check of page 700. If the process is elastic, the stresses are updated according to the compressible neo-Hookean relation (16.59). Otherwise we apply the return-mapping algorithm whose equations are described in the following.

The multivector return-mapping equations

In view of the representation (16.61) for the resolved Schmid shear stress, the return-mapping equation for $\Delta\gamma$ in the present model reads

$$\tilde{\Phi}^{\alpha}(\Delta\gamma) \equiv G\,\bar{\boldsymbol{s}}^{\alpha}(\Delta\gamma) \cdot \bar{\boldsymbol{m}}^{\alpha}(\Delta\gamma) - \tau_y(\gamma_{n+1}(\Delta\gamma)) = 0, \quad \alpha \in \mathscr{A}, \tag{16.65}$$

where we have defined

$$\bar{\boldsymbol{s}}^{\alpha}(\Delta\gamma) \equiv \boldsymbol{F}_{\text{iso}}^{e}(\Delta\gamma)\,\boldsymbol{s}^{\alpha}, \quad \bar{\boldsymbol{m}}^{\alpha}(\Delta\gamma) \equiv \boldsymbol{F}_{\text{iso}}^{e}(\Delta\gamma)\,\boldsymbol{m}^{\alpha}, \tag{16.66}$$

and $\boldsymbol{F}_{\text{iso}}^{e}$ is the function of $\Delta\gamma$ defined through the exponential map-based update formula

$$\boldsymbol{F}_{\text{iso}}^{e}(\Delta\gamma) \equiv \boldsymbol{F}_{\text{iso}}^{e\ \text{trial}} \exp\Big[-\sum_{\alpha\in\mathscr{A}} \Delta\gamma^{\alpha}\,\boldsymbol{s}^{\alpha} \otimes \boldsymbol{m}^{\alpha}\Big]. \tag{16.67}$$

The determination of the above active set $\mathscr{A}$ will be discussed later. For the time being, we assume $\mathscr{A}$ to be known *a priori.*

Due to the isochoric/volumetric split of the hyperelastic response of the present model, the return mapping, which effectively involves only the isochoric component of the elastic deformation gradient, affects only the deviatoric component of the Kirchhoff stress tensor. The updated hydrostatic Kirchhoff stress is the one at the elastic trial state. Note that this is not necessarily true in the general case described in Section 16.5 where the underlying hyperelastic model may have coupling between isochoric and volumetric responses. Here, after solution of (16.65), we update

$$\boldsymbol{\tau}_{n+1} := G\,\text{dev}[\boldsymbol{F}_{\text{iso}}^{e}(\boldsymbol{F}_{\text{iso}}^{e})^{T}] + K\,(\ln J_{n+1}^{e})\,\boldsymbol{I}, \tag{16.68}$$

where

$$J_{n+1}^{e} = J_{n+1} \equiv \det[\boldsymbol{F}_{n+1}]. \tag{16.69}$$

The Newton–Raphson iterations

By linearising (16.65) we obtain the following expression for the Jacobian matrix components to be used in the Newton iterations for solution of the return-mapping equation:

$$\mathrm{J}^{\alpha\beta} = G\,(\bar{\boldsymbol{s}}^{\alpha}\otimes \boldsymbol{m}^{\alpha} + \bar{\boldsymbol{m}}^{\alpha}\otimes \boldsymbol{s}^{\alpha}) : [\boldsymbol{F}^{e\ \mathrm{trial}}_{\mathrm{iso}} \cdot \mathsf{E} : (\boldsymbol{s}^{\alpha}\otimes \boldsymbol{m}^{\alpha})] - H, \tag{16.70}$$

where H is the Taylor hardening modulus and E denotes the derivative of the exponential map at $-\sum_{\alpha\in\mathscr{A}} \Delta\gamma^{\alpha}\, \boldsymbol{s}^{\alpha} \otimes \boldsymbol{m}^{\alpha}$.

Search for active systems

As in the present case the crystal is defined by two physical slip systems (four systems in the multisurface description), only two possibilities exist for plastic slip:

1. the set $\mathscr{A}$ contains only one system; or,
2. the set $\mathscr{A}$ contains two systems.

Due to the reduced number of possible combinations of active systems, the active set $\mathscr{A}$ can be determined here simply by solving the return-mapping equations for each possible combination of active systems and then selecting the one whose results satisfy the discrete consistency condition. Note that because the two physical slip systems are linearly independent, the active set that produces plastic consistency is unique and no special treatment of redundant equations is required.

In the actual algorithm we proceed as follows. Firstly, we solve the system with the tentative active set

$$\mathscr{A}^{(1)} := \mathscr{A}^{\,\mathrm{trial}}, \tag{16.71}$$

where $\mathscr{A}^{\,\mathrm{trial}}$ is the set defined in (16.53). If the solution satisfies consistency, then it is the one we are looking for: we stop the process and exit the state-update procedure. Otherwise, we define a new tentative set $\mathscr{A}^{(2)}$ and solve the return-mapping equations again. We check for consistency of the solution and redefine the tentative active set if necessary. The process is repeated until the plastically consistent active set is found.

As one system cannot be active at the same time as its mirrored counterpart, the possible active sets are (see definition (16.54) of mirrored systems for the present model)

$$\{1\}, \{2\}, \{3\}, \{4\}, \{1, 2\}, \{2, 3\}, \{3, 4\}, \{1, 4\}.$$

In practice, however, we do not need to try all the above combinations of active systems. The combinations we try will depend on the set $\mathscr{A}^{\,\mathrm{trial}}$. The tentative active sets shown in Table 16.1 are sufficient.

Remark 16.4. For large numbers of slip systems, the number of possible combinations of active systems increases dramatically and the present active set search procedure becomes computationally prohibitive.

The overall elastic predictor/return-mapping integration algorithm for the planar double-slip single crystal model is summarised in Boxes 16.2 and 16.3 in standard pseudo-code format. The procedure is implemented in HYPLAS subroutine SUPDSC (**S**tate **U**pdate for the **P**lanar **D**ouble-slip **S**ingle **C**rystal model).

Table 16.1. Planar double-slip model. Tentative active sets.

$\mathscr{A}^{(1)} \equiv \mathscr{A}^{\,\mathrm{trial}}$	1	2	3	4	1, 2	2, 3	3, 4	1, 4
$\mathscr{A}^{(2)}$	1, 4	1, 2	2, 3	3, 4	1	2	3	1
$\mathscr{A}^{(3)}$	1, 2	2, 3	3, 4	1, 4	2	3	4	4
$\mathscr{A}^{(4)}$	–	–	–	–	1, 4	1, 2	2, 3	1, 2
$\mathscr{A}^{(5)}$	–	–	–	–	2, 3	3, 4	1, 4	3, 4

Box 16.2. Elastic predictor/return-mapping algorithm for the planar double-slip single crystal model.

HYPLAS procedure: SUPDSC

(i) Given the deformation gradient $\boldsymbol{F}_\Delta := \boldsymbol{I} + \nabla_n[\Delta \boldsymbol{u}]$, evaluate the elastic trial state

$$\boldsymbol{F}^{e\,\mathrm{trial}}_{n+1} := \boldsymbol{F}_\Delta\, \boldsymbol{F}^e_n$$
$$\gamma^{\mathrm{trial}}_{n+1} := \gamma_n$$

(ii) Isochoric/volumetric decomposition

$$J := \det[\boldsymbol{F}_{n+1}], \qquad \boldsymbol{F}^{e\,\mathrm{trial}}_{\mathrm{iso}} := J^{-\frac{1}{3}}\, \boldsymbol{F}^{e\,\mathrm{trial}}_{n+1}$$

(iii) Elastic trial resolved Schmid stresses

$$\bar{\boldsymbol{s}}^{\alpha\,\mathrm{trial}} := \boldsymbol{F}^{e\,\mathrm{trial}}_{\mathrm{iso}}\, \boldsymbol{s}^\alpha, \qquad \bar{\boldsymbol{m}}^{\alpha\,\mathrm{trial}} := \boldsymbol{F}^{e\,\mathrm{trial}}_{\mathrm{iso}}\, \boldsymbol{m}^\alpha$$

$$\tau^{\alpha\,\mathrm{trial}}_{n+1} := G\, \bar{\boldsymbol{s}}^{\alpha\,\mathrm{trial}} \cdot \bar{\boldsymbol{m}}^{\alpha\,\mathrm{trial}}$$

for $\alpha = 1, \ldots, 2n_{\mathrm{syst}}$

(iv) Consistency check

IF $\Phi^{\alpha\,\mathrm{trial}} \equiv \tau^{\alpha\,\mathrm{trial}}_{n+1} - \tau_y(\gamma^{\mathrm{trial}}_{n+1}) \le \epsilon_{\mathrm{tol}}$ THEN

set $(\cdot)_{n+1} := (\cdot)^{\mathrm{trial}}_{n+1}$ and GOTO (v)

ELSE GOTO BOX 16.3 – return map (update $\boldsymbol{F}^e_{\mathrm{iso}}$ and γ_{n+1})

(v) Update $\boldsymbol{F}^e_{n+1}$ and the Cauchy stress tensor

$$\boldsymbol{F}^e_{n+1} := J^{\frac{1}{3}}\, \boldsymbol{F}^e_{\mathrm{iso}}$$
$$\boldsymbol{B}^e_{\mathrm{iso}} := \boldsymbol{F}^e_{\mathrm{iso}}\, (\boldsymbol{F}^e_{\mathrm{iso}})^T$$
$$\boldsymbol{\tau}_{n+1} := G\, \mathrm{dev}[\boldsymbol{B}^e_{\mathrm{iso}}] + K\, (\ln J_{n+1})\, \boldsymbol{I}$$
$$\boldsymbol{\sigma}_{n+1} := J^{-1}\, \boldsymbol{\tau}_{n+1}$$

Remark 16.5. The algorithm of Boxes 16.2 and 16.3 is equally applicable to three-dimensional cases as well as to other two-dimensional models, provided that the slip systems

Box 16.3. Return-mapping algorithm for the planar double-slip single crystal model.

HYPLAS procedure:	SUPDSC

(i) Set tentative active sets according to Table 16.1 and initialise tentative set counter: $i_{\rm tent} := 1$

(ii) Initial guess for plastic multiplier(s)

$$\Delta\gamma^{\alpha} := 0, \quad \alpha \in \mathscr{A}^{(i_{\rm tent})}$$

(iii) Initialise Newton iteration counter: $i_{\rm iter} := 1$

(iv) Compute Jacobian matrix $\mathbf{J}$ according to expression (16.70)

(v) Apply Newton–Raphson correction to multiplier(s)

$$\Delta\gamma^{\alpha} := \Delta\gamma^{\alpha} - \sum_{\beta\in\mathscr{A}^{(i_{\rm tent})}} [\mathbf{J}^{-1}]^{\alpha\beta}\, \tilde{\Phi}^{\beta}, \quad \alpha \in \mathscr{A}^{(i_{\rm tent})}$$

(vi) Update $\boldsymbol{F}^{e}_{\rm iso}$ (expression (16.67)) and γ_{n+1} (expression (16.35)) and compute yield function values, $\tilde{\Phi}^{\alpha}$, for *all* systems $\alpha = 1, \ldots, 2n_{\rm syst}$ (expression (16.65))

(vii) Check convergence

IF $|\Phi^{\alpha}| \leq \epsilon_{\rm tol},\ \forall\, \alpha \in \mathscr{A}^{(i_{\rm tent})}$ THEN

system has converged – **GOTO (viii)**

ELSE set $i_{\rm iter} := i_{\rm iter} + 1$ and **GOTO (iv)**

(viii) Check validity of converged solution

IF $\Delta\gamma^{\alpha} < 0$ for any $\alpha \in \mathscr{A}^{(i_{\rm tent})}$

OR $|\Phi^{\alpha}| > \epsilon_{\rm tol}$ for any $\alpha \in \{1, \ldots, 2n_{\rm syst}\}$ THEN

invalid solution – set $i_{\rm tent} := i_{\rm tent} + 1$ and **GOTO (ii)**

ELSE converged solution is valid – EXIT

are linearly independent. The only modification required is the redefinition of the possible combinations of active systems described in Table 16.1. However, as pointed out in Remark 16.4, the use of the present trial-and-error search procedure with a larger number of slip systems may become impractical.

16.6.3. EXAMPLE: THE MODEL PROBLEM

A simple numerical example is described in this section in order to illustrate the constitutive response of the planar double-slip model as well as to test the accuracy of the above described integration algorithm. The model problem consists of the deformation-controlled uniform shearing of a block whose material is modelled as a planar double-slip single crystal (Figure 16.6). This problem was used by Miehe (1996a,b) to test integration algorithms for the present planar model. Starting from a virgin state, the block is deformed monotonically

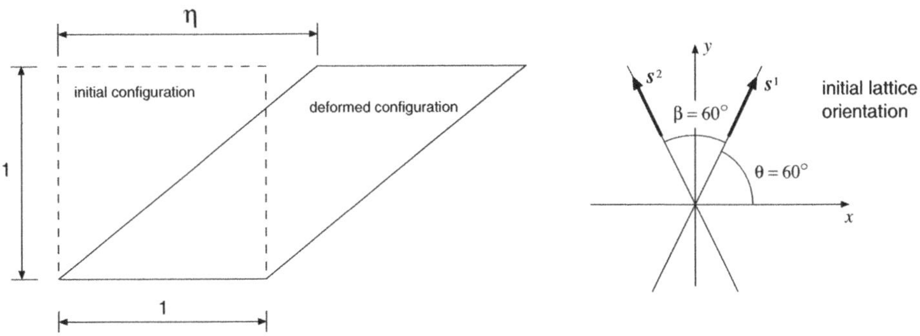

Figure 16.6. The model problem. Simple monotonic shear.

Table 16.2. The model problem. Material parameters.

Shear modulus: $G = 49.98$ kN/mm^2
Bulk modulus: $K = 21.10$ kN/mm^2
Hardening curve: $\tau_y(\gamma) = 0.06 + 0.048(1 - \exp(-\gamma/0.929)) + 0.001\gamma$ kN/mm^2
Initial lattice orientation: $\theta = 60^o$, $\beta = 60^o$

until the final state with a shear deformation parameter $\eta = 3$ is reached. During the process, the state variables (stress, elastic deformation gradient and hardening variable) are obtained by means of the exponential map-based integration algorithm. In order to assess the finite step accuracy of the algorithm under the present circumstances, the problem is solved with the final shear parameter $\eta = 3$ attained in 10, 30 and 300 equally sized increments. The material constants are the same as those adopted by Miehe (1996a,b) and are listed in Table 16.2.

The values of the Taylor hardening variable, γ, obtained along the process are plotted in Figure 16.7. The rotation of the crystal lattice is also shown. The crystal lattice rotation is the (in-plane) rotation angle (here assumed counterclockwise-positive) corresponding to the *elastic rotation tensor*, $\boldsymbol{R}^e$. The elastic rotation tensor is obtained from the polar decomposition of the elastic deformation gradient, $\boldsymbol{F}^e$. The lattice rotation tends asymptotically to -60° – orientation where slip-system 1 is aligned with the x-axis. Note that in all cases (10, 30 and 300 increments) virtually identical results are obtained, which emphasises the high accuracy of the exponential map-based integration algorithm. This is also observed in the graphs of Figure 16.8 where the in-plane Kirchhoff stress components are plotted. The high accuracy of the exponential map-based algorithm is due partly to its natural plastic volume conservation property. As shown by Miehe (1996b), the stress components τ_{xx} and τ_{yy} are particularly sensitive to this property in the present problem. When using an integration algorithm where the plastic flow rule is discretised by a standard backward Euler scheme together with *a posteriori* plastic volume corrections, Miehe (1996b) showed that a large number of increments are required to produce reasonable accuracy for stress components τ_{xx} and τ_{yy}.

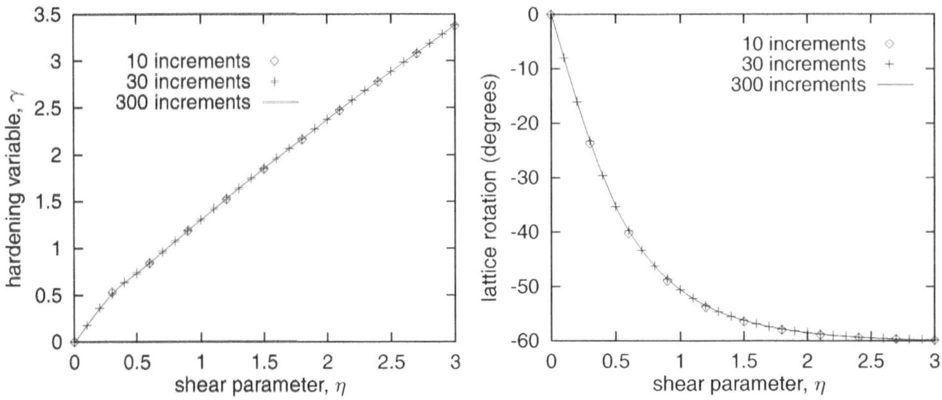

Figure 16.7. The model problem. Results for hardening internal variable and lattice rotation.

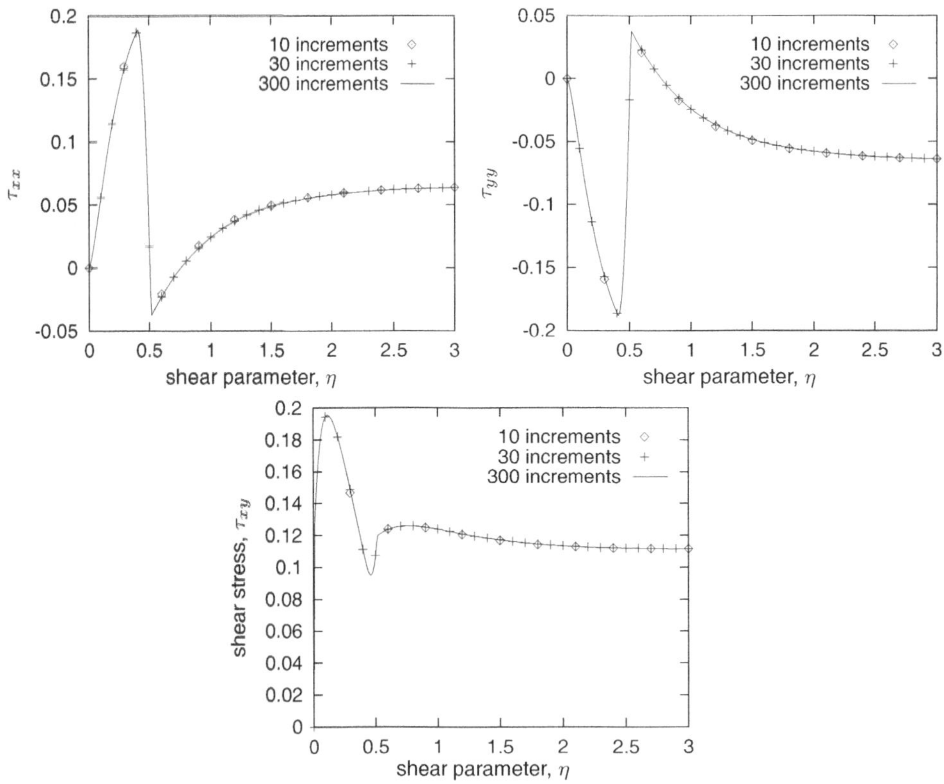

Figure 16.8. The model problem. In-plane Kirchhoff stress components (kN/mm^2).

Table 16.3. Return-mapping residuals. Table of convergence.

Iteration number	Increment 2	Increment 5	Increment 8
1	$6.38339109\ E+00$	$4.84948965\ E+00$	$6.21917562\ E+00$
2	$3.19909200\ E-01$	$1.84531024\ E-01$	$3.09099084\ E-01$
3	$5.17419932\ E-05$	$4.73003185\ E-06$	$4.90717916\ E-05$
4	$2.04754400\ E-11$	$1.95572560\ E-13$	$1.89060095\ E-11$

As the exactly linearised Newton–Raphson algorithm is adopted in the present implementation, quadratic rates of convergence are attained in the iterative solution of the return-mapping equations. To illustrate this important aspect, the evolution of the relative residual norm of the return-mapping equations,

$$r \equiv \sum_{\alpha=1}^{n_{\rm act}} |\tilde{\Phi}^{\alpha}/\tau_y|,$$

is shown in Table 16.3 for typical iterations. The results of Table 16.3 were obtained in the 10 increment case.

16.7. The consistent spatial tangent modulus

In this section, we focus our attention on the derivation of the spatial tangent modulus consistent with the implicit integration scheme for the rate-independent planar double-slip single crystal model described in Section 16.6. The tangent moduli for the planar model has been fully incorporated into program HYPLAS. The corresponding subroutine is named CSTPDS (**C**onsistent **S**patial **T**angent modulus for the **P**lanar **D**ouble-slip **S**ingle crystal model). We anticipate that, due to the inherent complexities of the model and the corresponding integration algorithm, the derivation of the spatial tangent moduli is a tedious exercise of consistent linearisation. For the sake of clarity, some of the longest tensor algebra manipulations involved have been omitted in the derivation shown in this section. The reader who is not interested in the details of derivation is referred directly to the final relevant expressions (16.72) and (16.85).

Remark 16.6. Before proceeding, it is worth remarking that the tangent moduli derived in this section are valid for any (two- or three-dimensional) single crystal models based on the neo-Hookean hyperelastic description and the present exponential-map based integration algorithm. Only the active set search procedure has to be changed to accommodate other models of this class and this procedure does not affect in any way the tangent moduli.

16.7.1. THE ELASTIC MODULUS: COMPRESSIBLE NEO-HOOKEAN MODEL

The spatial elasticity tensor for the compressible neo-Hookean model has been derived in Chapter 13 (see final expressions (13.96, 13.97), page 535). In the present elastoplastic context, the corresponding *elastic* spatial tangent modulus, a^e, is given by the completely analogous expression

$$\mathsf{a}^e_{ijkl} = \mathsf{c}^e_{ijkl} + \delta_{ik}\sigma_{jl}, \tag{16.72}$$

where c^e has the compact representation

$$\mathsf{c}^e = \frac{2G}{3J}\,\mathrm{tr}[\boldsymbol{B}^e_{\mathrm{iso}}]\,\mathsf{I}_{\mathrm{d}} - 2p\,\mathsf{I}_S + \frac{K}{J}\,\boldsymbol{I}\otimes\boldsymbol{I} - \frac{2}{3}\,[\boldsymbol{\sigma}_{\mathrm{d}}\otimes\boldsymbol{I} + \boldsymbol{I}\otimes\boldsymbol{\sigma}_{\mathrm{d}}], \tag{16.73}$$

with $\boldsymbol{\sigma}_{\mathrm{d}}$ being the Cauchy stress deviator and p denoting the Cauchy hydrostatic pressure, $p \equiv \mathrm{tr}[\boldsymbol{\sigma}]/3$.

16.7.2. THE ELASTOPLASTIC CONSISTENT TANGENT MODULUS

Let us now turn our attention to the *elastoplastic* tangent modulus consistent with the return-mapping scheme of Box 16.3. The derivation of the elastoplastic tangent modulus for the present model/algorithm is conceptually identical to the simpler (isotropic) cases addressed in previous chapters. Here, the return-mapping algorithm of Box 16.3 defines an implicit incremental constitutive function for the isochoric elastic deformation gradient of the form

$$\boldsymbol{F}^e_{\mathrm{iso}} = \boldsymbol{F}^e_{\mathrm{iso}}(\boldsymbol{F}^{e\ \mathrm{trial}}_{\mathrm{iso}}), \tag{16.74}$$

where the isochoric elastic trial deformation gradient is a function of the total deformation gradient $\boldsymbol{F}_{n+1}$. In view of the hyperelastic law, the incremental constitutive function for the Kirchhoff stress, $\hat{\boldsymbol{\tau}}$, consistent with the return-mapping algorithm, is defined as

$$\begin{aligned}\boldsymbol{\tau}_{n+1} &= \hat{\boldsymbol{\tau}}(\boldsymbol{F}_{n+1})\\ &\equiv \boldsymbol{\tau}_{\mathrm{d}}(\boldsymbol{F}^e_{\mathrm{iso}}(\boldsymbol{F}^{e\ \mathrm{trial}}_{\mathrm{iso}}(\boldsymbol{F}_{n+1}))) + K\,\ln(\det[\boldsymbol{F}_{n+1}])\,\boldsymbol{I}\end{aligned} \tag{16.75}$$

where

$$\boldsymbol{\tau}_{\mathrm{d}}(\boldsymbol{F}^e_{\mathrm{iso}}) \equiv G\,\mathrm{dev}[\boldsymbol{F}^e_{\mathrm{iso}}\boldsymbol{F}^{e\ T}_{\mathrm{iso}}]. \tag{16.76}$$

Our task now is to derive the elastoplastic spatial tangent modulus, a^{ep}, defined as

$$\mathsf{a}^{ep}_{ijkl} = \frac{1}{J_{n+1}}\left[\frac{\partial\hat{\boldsymbol{\tau}}}{\partial\boldsymbol{F}_{n+1}}\right]_{ijkm}(\boldsymbol{F}_{n+1})_{lm} - \sigma_{il}\,\delta_{jk}. \tag{16.77}$$

We then start by concentrating on the derivation of an explicit formula for the derivative

$$\frac{\partial\hat{\boldsymbol{\tau}}}{\partial\boldsymbol{F}_{n+1}},$$

which is the only term in (16.77) that depends on the algorithmic constitutive function. By straightforward differentiation of (16.75) we obtain

$$\frac{\partial\hat{\boldsymbol{\tau}}}{\partial\boldsymbol{F}} = \frac{\partial\boldsymbol{\tau}_{\mathrm{d}}}{\partial\boldsymbol{B}^e_{\mathrm{iso}}} : \frac{\partial\boldsymbol{B}^e_{\mathrm{iso}}}{\partial\boldsymbol{F}^e_{\mathrm{iso}}} : \mathsf{P} : \frac{\partial\boldsymbol{F}_{\mathrm{iso}}}{\partial\boldsymbol{F}} + \frac{K}{J}\,\boldsymbol{I}\otimes\frac{\partial}{\partial\boldsymbol{F}}\det[\boldsymbol{F}],$$

where the subscript $n+1$ has been omitted for notational convenience and the fourth-order tensor

$$\mathsf{P} \equiv \frac{\partial\boldsymbol{F}^e_{\mathrm{iso}}}{\partial\boldsymbol{F}_{\mathrm{iso}}}, \tag{16.78}$$

is the only contribution to the derivative of $\hat{\boldsymbol{\tau}}$ that depends on the particular return-mapping algorithm adopted. After some straightforward derivations considering the elastic law and the definitions of $\boldsymbol{B}^e_{\rm iso}$ and $\boldsymbol{F}_{\rm iso}$ we obtain the components

$$\left[\frac{\partial \hat{\boldsymbol{\tau}}}{\partial \boldsymbol{F}}\right]_{ijkl} = \frac{G}{J^{\frac{1}{3}}}\,\mathsf{d}:\mathsf{b}:\mathsf{P}:\mathsf{f} + K\,\boldsymbol{I}\otimes\boldsymbol{F}^T, \tag{16.79}$$

where

$$\mathsf{b}_{ijkl} \equiv [\delta_{ik}\,(\boldsymbol{F}^e_{\rm iso})_{jl} + \delta_{jk}\,(\boldsymbol{F}^e_{\rm iso})_{il}]$$

$$\mathsf{f}_{ijkl} \equiv \delta_{ik}\,\delta_{jl} - \tfrac{1}{3}\,(\boldsymbol{F}^e_{\rm iso})_{ij}(\boldsymbol{F}^e_{\rm iso})_{lk}.$$

The crucial point now is the derivation of an explicit expression for the term P which depends on the return-mapping scheme. This is addressed in the following.

The algorithm-consistent derivative P

From the update formula (16.67) we obtain the differential relation

$$\mathrm{d}\boldsymbol{F}^e_{\rm iso} = \mathrm{d}\boldsymbol{F}^{e\ \rm trial}_{\rm iso}\,(\boldsymbol{F}^p_\Delta)^{-1} + \boldsymbol{F}^{e\ \rm trial}_{\rm iso}\,\mathrm{d}(\boldsymbol{F}^p_\Delta)^{-1}, \tag{16.80}$$

where

$$(\boldsymbol{F}^p_\Delta)^{-1} = \exp\Big[-\sum_{\alpha\in\mathscr{A}} \Delta\gamma^\alpha\,\boldsymbol{s}^\alpha\otimes\boldsymbol{m}^\alpha\Big]$$

and

$$\mathrm{d}(\boldsymbol{F}^p_\Delta)^{-1} = -\mathsf{E}:\sum_{\alpha\in\mathscr{A}} \mathrm{d}\Delta\gamma^\alpha\,\boldsymbol{s}^\alpha\otimes\boldsymbol{m}^\alpha.$$

Recall that E denotes the derivative of the exponential map evaluated at the argument $-\sum_{\alpha\in\mathscr{A}}\Delta\gamma^\alpha\,\boldsymbol{s}^\alpha\otimes\boldsymbol{m}^\alpha$. The differentials of the plastic multiplier, $\mathrm{d}\Delta\gamma^\alpha$, which take part in the above expression are obtained by the standard procedure that consists in differentiating the discrete consistency condition (16.65), having $\boldsymbol{F}^e_{\rm iso}$ also as a variable, to obtain

$$\mathrm{d}\Phi^\alpha = G\,[\bar{\boldsymbol{m}}^\alpha\otimes\boldsymbol{s}^\alpha + \bar{\boldsymbol{s}}^\alpha\otimes\boldsymbol{m}^\alpha]:\mathrm{d}\boldsymbol{F}^e_{\rm iso} - H\sum_{\beta\in\mathscr{A}}\mathrm{d}\Delta\gamma^\beta, \quad \forall\,\alpha\in\mathscr{A},$$

and then enforcing the condition

$$\mathrm{d}\Phi^\alpha = 0, \quad \forall\,\alpha\in\mathscr{A}.$$

These last two expressions together with (16.80) yield, after some tensor algebra, the following differential relation

$$\mathrm{d}\Delta\gamma^\alpha = \boldsymbol{S}^\alpha : \mathrm{d}\boldsymbol{F}_{\rm iso}, \quad \alpha\in\mathscr{A}, \tag{16.81}$$

where the second-order tensors $\boldsymbol{S}^\alpha$ are defined by the components

$$[\boldsymbol{S}^\alpha]_{ij} \equiv G\sum_{\beta\in\mathscr{A}}[\boldsymbol{J}^{-1}]^{\alpha\beta}[\bar{\boldsymbol{m}}^\alpha\otimes\boldsymbol{s}^\alpha + \bar{\boldsymbol{s}}^\alpha\otimes\boldsymbol{m}^\alpha]_{ik}[\boldsymbol{F}^{p-1}]_{jk}.$$

With the substitution of (16.81) into (16.80), we obtain, again after some manipulations, the following differential relation

$$\mathrm{d}\boldsymbol{F}^e_{\text{iso}} = \mathsf{P} : \mathrm{d}\boldsymbol{F}_{\text{iso}}, \tag{16.82}$$

where the tangent operator P is defined by the components

$$\mathsf{P}_{ijkl} \equiv \delta_{ik}\, F^p_{lj} + (\boldsymbol{F}^{e\,\text{trial}}_{\text{iso}})_{im}\, \mathsf{E}_{mjpq}\, \mathsf{Q}_{pqkl}, \tag{16.83}$$

and Q_{pqkl} are the components of the fourth-order tensor

$$\mathsf{Q} \equiv -\sum_{\alpha\in\mathscr{A}} \boldsymbol{s}^\alpha \otimes \boldsymbol{m}^\alpha \otimes \boldsymbol{S}^\alpha. \tag{16.84}$$

The final expression for a^{ep}

Finally, with the substitution of (16.83) into (16.79) followed by the introduction of the resulting expression into (16.77) we arrive, after some tedious but straightforward tensor algebra, at a relatively simple expression for the consistent spatial tangent modulus given by

$$\mathsf{a}^{ep} = \mathsf{a}^e + \mathsf{a}^p, \tag{16.85}$$

where the elastic contribution a^e is that defined by (16.72) and a^p is the plastic contribution to the tangent modulus defined as

$$\mathsf{a}^p \equiv -\frac{2G^2}{J}\, \mathsf{I}_{\mathrm{d}} : \mathsf{e} : \mathsf{v}, \tag{16.86}$$

with the fourth-order tensors e and v given by the components

$$\begin{aligned} \mathsf{e}_{ijkl} &\equiv (\boldsymbol{F}^e_{\text{iso}})_{il}\, (\boldsymbol{F}^{e\,\text{trial}}_{\text{iso}})_{jk} \\ \mathsf{v}_{ijkl} &\equiv [\mathsf{E} : \mathsf{U}]_{ijkm}\, (\boldsymbol{F}^e_{\text{iso}})_{lm} - \tfrac{1}{3}[\mathsf{E} : \mathsf{U} : \boldsymbol{F}^e_{\text{iso}}]_{ij}\, \delta_{kl}. \end{aligned} \tag{16.87}$$

In the above, the fourth-order tensor U is defined as

$$\mathsf{U} \equiv \sum_{\alpha\in\mathscr{A}} \sum_{\beta\in\mathscr{A}} [\mathbf{J}^{-1}]^{\alpha\beta}\, \boldsymbol{s}^\alpha \otimes \boldsymbol{m}^\alpha \otimes [\bar{\boldsymbol{s}}^\beta \otimes \boldsymbol{m}^\beta + \bar{\boldsymbol{m}}^\beta \otimes \boldsymbol{s}^\beta]. \tag{16.88}$$

An approximate formula for the elastoplastic tangent modulus tensor has been employed by Miehe (1996a) which can be obtained by setting $\mathsf{E} := \mathsf{I}_S$ and $\boldsymbol{F}^{e\,\text{trial}}_{\text{iso}} := \boldsymbol{F}^e_{\text{iso}}$ in the above exact expression. At the beginning of the first global equilibrium iteration of any load increment – when all plastic multipliers have been reset to zero – we have $\mathsf{E} = \mathsf{I}_S$ and $\boldsymbol{F}^{e\,\text{trial}}_{\text{iso}} = \boldsymbol{F}^e_{\text{iso}}$ and the approximate operator coincides with the exact one.

16.8. Numerical examples

To illustrate the application of the numerical framework described in the previous sections of this chapter, two numerical examples involving the planar double-slip single crystal implementation are shown here. The examples comprise a classical problem of symmetric strain localisation on a rectangular strip as well as an unsymmetric localisation problem. Both examples have been run with the standard version of program HYPLAS that accompanies this book. The full Newton–Raphson scheme, which relies on the above described spatial tangent moduli, has been adopted in the global equilibrium iterations.

16.8.1. SYMMETRIC STRAIN LOCALISATION ON A RECTANGULAR STRIP

In the present example we simulate the phenomenon of strain localisation on a single-crystal rectangular strip subjected to finite axial stretching. This problem has been thoroughly analysed by Miehe (1996a,b). Similar problems have been considered by Peirce *et al.* (1982), Needleman *et al.* (1985), Rashid and Nemat-Nasser (1995) and Steinmann and Stein (1996) in the rate-dependent context. The geometry, boundary conditions and finite element meshes adopted are illustrated in Figure 16.9. The material parameters are the same as in the model problem of Section 16.6. They are listed in Table 16.2 (page 713). Note that in the present case, where $\theta = \beta = 60^o$, the orientation of the crystal lattice is perfectly symmetric with respect to the vertical axis. Due to this symmetry, only one quarter of the strip with appropriate boundary conditions imposed on the symmetry edges is used in the finite element discretisation. In view of the plastic incompressibility of single crystals, the use of an appropriate finite element is absolutely essential in order to avoid volumetric locking. Here we adopt the four-noded *F-bar* quadrilateral described in Chapter 15. As shown in Chapter 15, this element is particularly suitable to treat finite isochoric deformations as well as to capture strain localisation phenomena. Two different meshes are used to discretise the symmetric quarter: a 10×30 and a 20×60 element mesh. In order to trigger strain localisation, for both meshes the element nearest the centre of the strip is made weaker than the others. For the weakened element, all material properties, except the hardening curve, are the same as those of Table 16.2. In the hardening curve for the weakened element the resolved yield stress, τ_y, for any value of γ is 90% of that given by the hardening curve of Table 16.2.

A final vertical displacement $u = 5$ mm is imposed in 13 and 23 increments, respectively, for the 10×30 and the 20×60 element mesh. The reactions on the constrained edge obtained in the numerical simulation are plotted in Figure 16.10 against the prescribed vertical deflection. The force–deflection curve obtained with a larger number of smaller displacement increments is also plotted (the solid line) in Figure 16.10. It gives a better picture of the complete equilibrium path. In spite of the size of the incremental displacements used in the calculations with 13 and 23 increments (in both cases the first two increments are: $\Delta u_1 = 2.0$ mm and $\Delta u_2 = 1.5$ mm), the corresponding reactions are virtually identical to the reactions shown on the solid line. This is an obvious consequence of the high accuracy of the exponential map-based numerical integration scheme. The deformed meshes at $u = 3.5$ mm, $u = 4.5$ mm and $u = 5.0$ mm are depicted in Figures 16.11 and 16.12. For better visualisation, the deformed discretised symmetric quarter is mirrored into the other three quadrants. The localised shear bands are clearly visible in both meshes at later stages of deformation.

For both meshes, the convergence tolerance for the relative residual norm in the equilibrium iterations has been set to 10^{-7}%. Such a tolerance, which is far smaller than that required for engineering purposes, has been selected only to emphasise the quadratic rates of asymptotic convergence achieved as a result of the exact linearisation of the relevant equations. Table 16.4 shows the evolution of the relative residual norm during Newton–Raphson iterations for typical increments with the 20×60 element mesh. The relatively higher number of iterations needed for convergence in increment 6 (ending at $u_6 = 3.875$ mm) is due to the fact that the shear band is being formed at that stage. Under such conditions, large rotations of the crystal lattice occur within the increment in elements along the shear band. This may cause the set of active systems at the corresponding Gauss points to change a few times during the global iterations before the converged equilibrium configuration is found. Similar phenomena can be observed with other multisurface plasticity models such as the Tresca

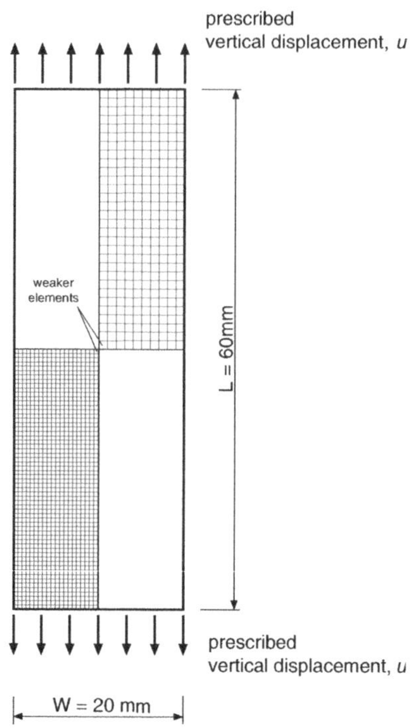

Figure 16.9. Rectangular strip. Geometry and finite element mesh.

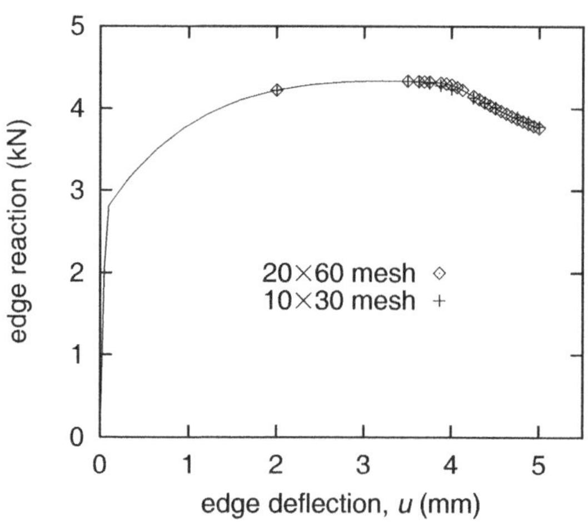

Figure 16.10. Rectangular strip. Reaction–deflection diagram.

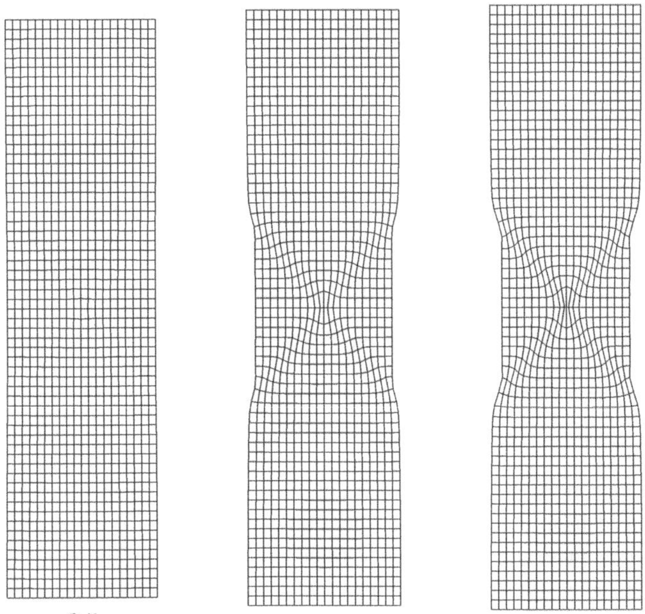

$u = 3.5$ mm $u = 4.5$ mm $u = 5.0$ mm

Figure 16.11. Rectangular strip. Deformed configurations for 10×30 mesh.

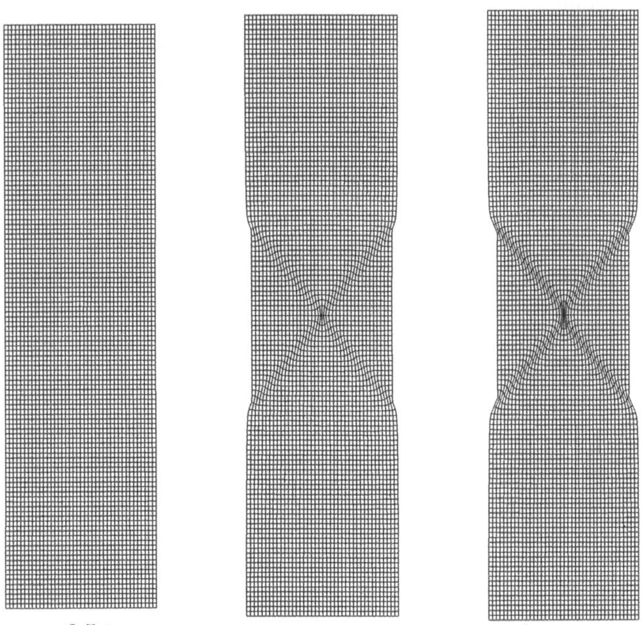

$u = 3.5$ mm $u = 4.5$ mm $u = 5.0$ mm

Figure 16.12. Rectangular strip. Deformed configurations for 20×60 mesh.

Table 16.4. Rectangular strip. Convergence table.

	Relative residual norm (%) – 20 × 60 mesh		
Iteration number	Increment 6	Increment 13	Increment 20
1	$0.201249\ E + 01$	$0.210379\ E + 02$	$0.187277\ E + 02$
2	$0.543621\ E + 01$	$0.138596\ E + 01$	$0.889781\ E + 01$
3	$0.129385\ E + 01$	$0.171953\ E - 00$	$0.291015\ E - 01$
4	$0.106128\ E + 00$	$0.142736\ E - 01$	$0.171479\ E - 05$
5	$0.592669\ E - 01$	$0.645938\ E - 04$	$0.444649\ E - 10$
6	$0.603329\ E - 01$	$0.212211\ E - 09$	
7	$0.693394\ E - 02$		
8	$0.367075\ E - 03$		
9	$0.916724\ E - 09$		

and Mohr–Coulomb models whose implementation is discussed in Chapter 8. This is also frequently observed in contact problems near sharp contacting corners. It is important to note, however, that quadratic convergence rates are restored as the converged solution is approached. A possible alternative to improving convergence under such conditions could be the use of line-searches (Crisfield, 1991) in conjunction with the global Newton iterations.

16.8.2. UNSYMMETRIC LOCALISATION

We now solve a problem where strain localisation on the rectangular strip occurs in an unsymmetric mode. The geometry of the problem is the same as in the previous example (Figure 16.9, page 718). With the exception of the lattice orientation angle θ, the material parameters are also the same as in the previous example (Table 16.2, page 713). Here the crystal lattice orientation angle θ is chosen as

$$\theta = 75^o.$$

The angle between the slip systems remains $\beta = 60^o$. The slip systems are then no longer symmetrically oriented with respect to the vertical axis and the whole strip (rather than the symmetric quarter adopted in the previous example) has to be discretised. A mesh of 20 × 60 four-noded *F-bar* elements is used to discretise the strip (Figure 16.13). The vertical displacement is prescribed for all nodes at the top and bottom edges. Here, we consider two situations:

(a) clamped condition – constrained horizontal displacement of the top and bottom edges;

(b) free edges – only the mid-side nodes A and B are fixed in the horizontal direction.

The strip is stretched up to $u = 9$ mm. A total of 77 and 56 increments are used, respectively, for the clamped and free edge conditions. The reaction–deflection diagrams obtained are shown in Figure 16.14. A peak reaction of approximately 3.8 kN is observed for both cases. The corresponding configurations are $u = 3.25$ mm for the clamped condition and $u = 4.25$ mm for the free edge. These configurations are reached, respectively, in ten and eight displacement increments. The sharp decrease in reactions observed in the graphs

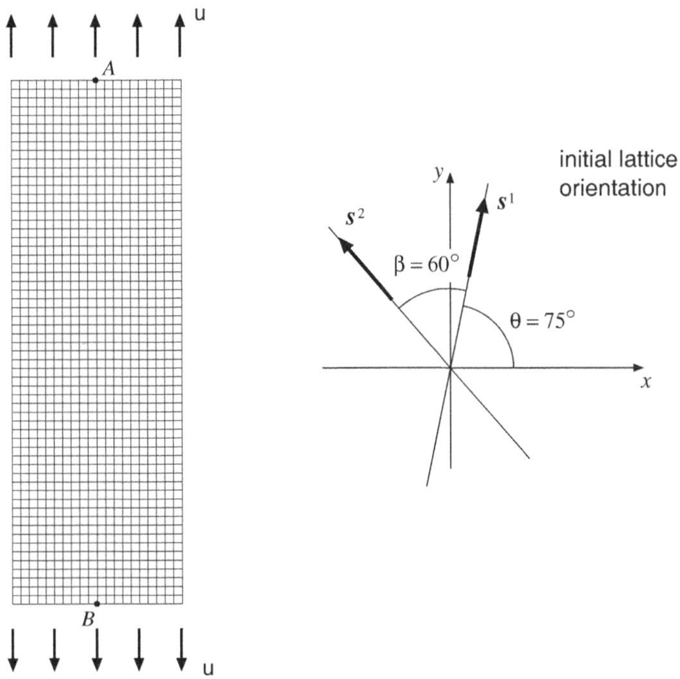

Figure 16.13. Unsymmetric localisation. Finite element model.

corresponds to the formation of shear bands which, in contrast to the previous example, are unsymmetrically oriented with respect to the axis of stretching. The unsymmetric shear bands are depicted in Figures 16.15 and 16.16, which show the deformed configurations at $u = 4.0$ mm, $u = 6.5$ mm and $u = 9.0$ mm. Unlike in the previous example, no material imperfection (weakened element) is required here to trigger localisation. The non-uniform state of stress generated due to the unsymmetric lattice orientation is sufficient. In fact, if the four elements that share the central node are weakened (with the same properties as the weakened elements of the previous example), the results obtained are virtually identical to those shown here.

16.9. Viscoplastic single crystals

One of the important physical properties of single crystal alloys is their distinctive resistance to creep failure at high service temperatures when the crystallographic microstructure is favourably oriented with respect to the direction of loading. This makes such materials particularly suitable for applications such as, for instance, jet aero-engine turbine blades which may operate at temperatures as high as 700–1000 °C for long periods of time. Under such high temperatures the time-dependence of the plastic flow becomes an important property that cannot be disregarded if numerical simulations are to produce realistic results.

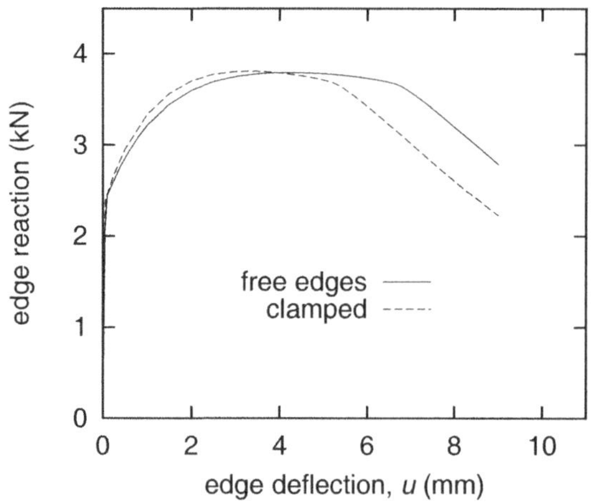

Figure 16.14. Unsymmetric localisation. Reaction–deflection diagram.

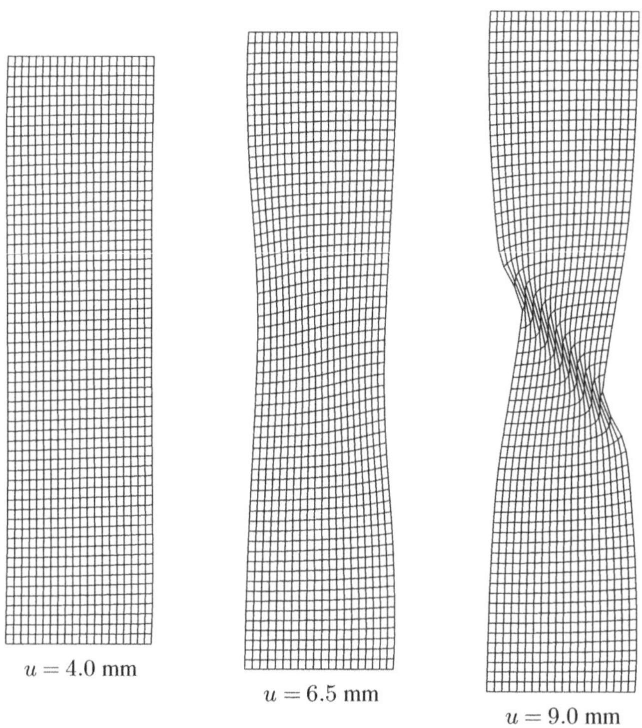

Figure 16.15. Unsymmetric localisation. Deformed configurations (free edges).

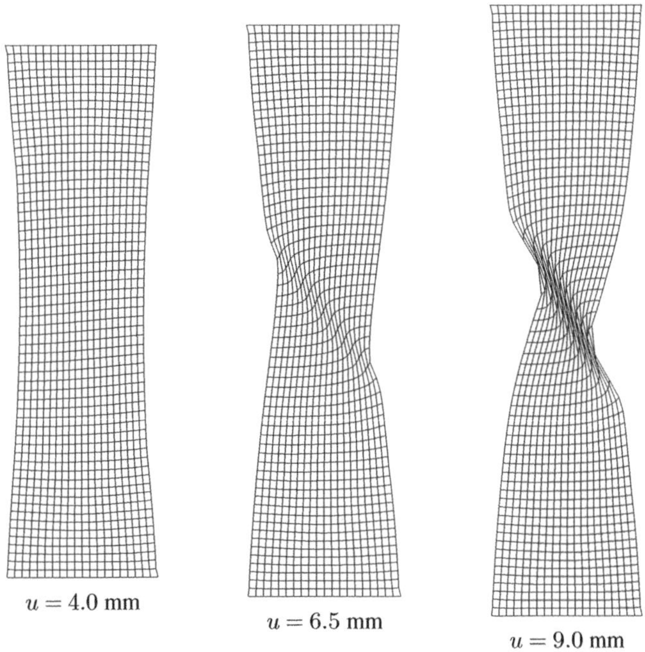

Figure 16.16. Unsymmetric localisation. Deformed configurations (clamped).

16.9.1. RATE-DEPENDENT FORMULATION

A rate-dependent theory to model creep in single crystals can be obtained by introducing, for instance, the following power law-type constitutive equation for the rates $\dot{\gamma}^\alpha$ of inelastic deformation in each slip system:

$$\dot{\gamma}^\alpha = \dot{\gamma}_0 \left(\frac{\langle \tau^\alpha \rangle}{g} \right)^{1/M}, \quad \alpha = 1, \ldots, 2n_{\text{syst}}, \tag{16.89}$$

where, to maintain consistency with the material presented in the previous sections of this chapter, we have adopted the mirrored slip-system convention (16.11). In the above, $\langle \cdot \rangle$ denotes the Macauley bracket (refer to definition (12.89), page 505). The constants $\dot{\gamma}_0$ and M are, respectively, the *reference shear rate* and the *rate-sensitivity parameter*. In the absence of hardening, the *reference shear stress*, g, is also a constant. All material parameters are generally temperature-dependent. Note that the reference shear rate is the slip rate in system α when $\tau^\alpha = g$.

The model defined by the above equation does not possess yield surfaces and inelastic flow occurs for all slip systems with $\tau^\alpha > 0$. Also note that, in terms of the n_{syst} *physical* slip systems (without the mirrored system convention), equation (16.89) can be equivalently expressed as

$$\dot{\gamma}^\alpha = \dot{\gamma}_0 \left(\frac{|\tau^\alpha|}{g} \right)^{1/M} \text{sign}(\tau^\alpha), \quad \alpha = 1, \ldots, n_{\text{syst}}. \tag{16.90}$$

Clearly, this law is a generalisation of Norton's uniaxial creep equation (11.44) to single crystal plasticity.

Alternative slip-rate laws

Slip-rate laws alternative to the above definition can be adopted. The suitability of any law for a particular single-crystal material should be determined by experimental evidence. In defining alternative slip-rate equations, we could adopt, for instance, models which include the use of yield surfaces (in analogy with the elastoplastic model discussed in previous sections of this chapter). In such a case, one possibility could be, for instance, a multisurface single-crystal generalisation of equation (11.6):

$$\dot{\gamma}^\alpha = \begin{cases} \dfrac{1}{\mu}\left[\left(\dfrac{|\tau^\alpha|}{\tau_y}\right)^{1/\epsilon} - 1\right] & \text{if } \Phi^\alpha(\tau^\alpha, \tau_y) \geq 0 \\[2ex] 0 & \text{if } \Phi^\alpha(\tau^\alpha, \tau_y) < 0, \end{cases} \tag{16.91}$$

or a generalisation of Perzyna's model:

$$\dot{\gamma}^\alpha = \begin{cases} \dfrac{1}{\mu}\left[\dfrac{|\tau^\alpha|}{\tau_y} - 1\right]^{1/\epsilon} & \text{if } \Phi^\alpha(\tau^\alpha, \tau_y) \geq 0 \\[2ex] 0 & \text{if } \Phi^\alpha(\tau^\alpha, \tau_y) < 0, \end{cases} \tag{16.92}$$

where the material constants are analogous to those of the corresponding original models and the yield functions Φ^α are defined according to (16.20).

16.9.2. THE EXPONENTIAL MAP-BASED INTEGRATION ALGORITHM

In view of the hyperelastic constitutive law, which defines $\boldsymbol{\tau}$ as a function of $\boldsymbol{F}^e$, and the viscoplastic slip rate equation (any of the above) that defines the slip rate $\dot{\gamma}^\alpha$ as a function of $\boldsymbol{\tau}$, we may write

$$\tau^\alpha = \tau^\alpha(\boldsymbol{F}^e) \quad \text{and} \quad \dot{\gamma}^\alpha = \dot{\gamma}^\alpha(\boldsymbol{F}^e). \tag{16.93}$$

Then, following a standard backward Euler discretisation of any of the above slip-rate laws, we have

$$\Delta\gamma^\alpha = \Delta t\, \dot{\gamma}^\alpha(\boldsymbol{F}^e_{n+1}). \tag{16.94}$$

By substituting this last expression into the update formula (16.34) the integration of the constitutive equations of the general viscoplastic single crystal model† within the interval $[t_n, t_{n+1}]$ is reduced to the solution of the following residual equation for $\boldsymbol{F}^e_{n+1}$:

$$\boldsymbol{S}(\boldsymbol{F}^e) \equiv \boldsymbol{F}^e - \boldsymbol{F}^{e\ \text{trial}} \exp\left[-\Delta t \sum_{\alpha=1}^{n_{\text{syst}}} \dot{\gamma}^\alpha(\boldsymbol{F}^e)\, \boldsymbol{s}^\alpha \otimes \boldsymbol{m}^\alpha\right] = \boldsymbol{0}, \tag{16.95}$$

where the subscript $n+1$ has been omitted for notational convenience.

†Here, we assume that the models are perfectly viscoplastic. If hardening is considered, then an additional hardening equation(s) needs to be added to the system.

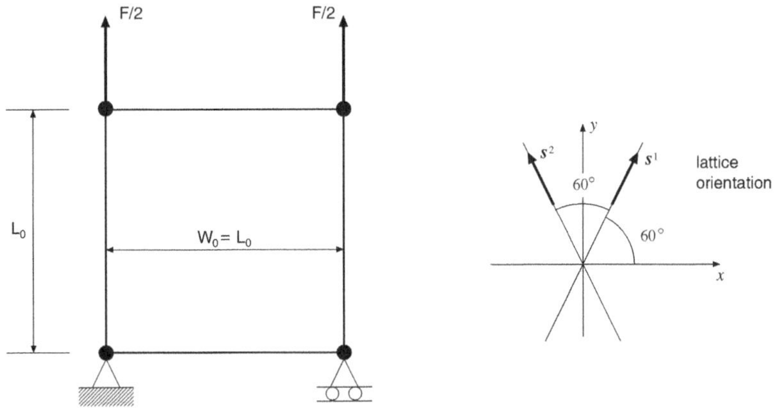

Figure 16.17. Rate-dependent model. Problem definition.

Table 16.5. Rate-dependent model problem. Material parameters.

Shear modulus:	$G = 80\ 769$ MPa
Bulk modulus:	$K = 175\ 000$ MPa
Reference shear stress:	$g = 180$ MPa
Reference shear strain rate:	$\dot{\gamma}_0 = 0.002\ \mathrm{h}^{-1}$
Rate sensitivity parameter:	$M = 0.1961$

The local Newton–Raphson algorithm

The exact Jacobian used in the Newton–Raphson scheme for the solution of residual equation (16.95) is obtained simply by differentiating $\boldsymbol{S}$ with respect to the equation variable $\boldsymbol{F}^e_{n+1}$. Again omitting the $n+1$ subscript, we have in Cartesian components

$$\mathsf{J}_{ijkl} \equiv \left[\frac{\mathrm{d}\boldsymbol{S}}{\mathrm{d}\boldsymbol{F}^e}\right]_{ijkl} = \delta_{ik}\delta_{jl} + \Delta t\ F^{e\ \mathrm{trial}}_{im}\ \mathsf{E}_{mjpq}\left[\sum_{\alpha=1}^{n_{\mathrm{syst}}} \boldsymbol{s}_0^\alpha \otimes \boldsymbol{m}_0^\alpha \otimes \frac{\mathrm{d}\dot{\gamma}^\alpha}{\mathrm{d}\boldsymbol{F}^e}\right]_{pqkl}, \tag{16.96}$$

where only the derivative $\mathrm{d}\dot{\gamma}^\alpha/\mathrm{d}\boldsymbol{F}^e$ depends on the particular slip-rate model adopted. Clearly, the use of the above exact Jacobian results in quadratic rates of asymptotic convergence in the iterative solution of the nonlinear algebraic residual system.

16.9.3. THE SPATIAL TANGENT MODULUS: NEO-HOOKEAN-BASED MODEL

Here we shall assume, as in the rate-independent case discussed earlier, the hyperelastic law to be given by the regularised neo-Hookean model (16.59). Again, the derivation of the tangent modulus associated with the above integration algorithm is relatively lengthy. Only the final

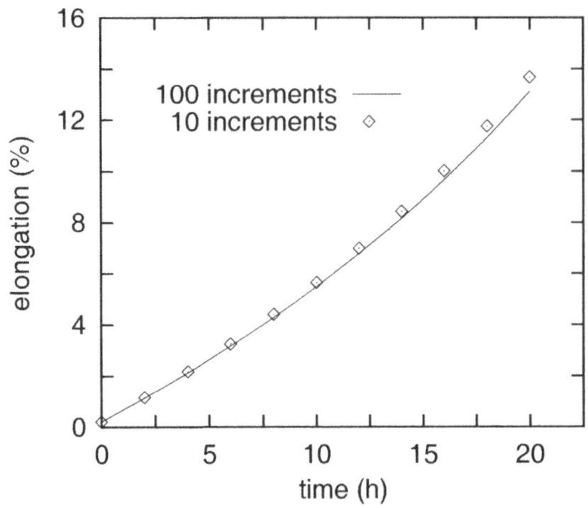

Figure 16.18. Rate-dependent model problem. Time versus elongation.

expression is shown here. In the present case, the spatial tangent modulus is given by

$$\mathsf{a}_{ijkl} = \left[\frac{G}{J^{4/3}} \mathsf{F} : \mathsf{A} \; : \mathsf{H} + \frac{K}{J} \boldsymbol{I} \otimes \boldsymbol{I} \right]_{ijkl} - \sigma_{il} \delta_{jk}, \tag{16.97}$$

where

$$\mathsf{F}_{ijkl} \equiv \delta_{ik} (\boldsymbol{F}^e_{\text{iso}})_{jl} + \delta_{jk} (\boldsymbol{F}^e_{\text{iso}})_{il} - \tfrac{1}{3} \delta_{ij} (\boldsymbol{F}^e_{\text{iso}})_{kl} \tag{16.98}$$

and

$$\mathsf{H}_{ijkl} \equiv \delta_{ik} F_{lj} - \tfrac{1}{3} F_{ij} \delta_{kl}. \tag{16.99}$$

The tensor A is the only algorithm-related term taking part in the assemblage of a. It is defined as

$$\mathsf{A} \equiv \frac{\partial \boldsymbol{F}^e_{\text{iso}}}{\partial \boldsymbol{F}_{\text{iso}}}. \tag{16.100}$$

The final expression for A reads

$$\mathsf{A} = \mathsf{J}^{-1} : \mathsf{C}, \tag{16.101}$$

where C is defined by the components

$$\mathsf{C}_{ijkl} \equiv -\delta_{ik} \, (\boldsymbol{F}^{p\,-1}_{n+1})_{lj}. \tag{16.102}$$

The above expression for A is obtained by linearising (16.95), also taking into account the relationship between $\boldsymbol{F}^e$ and $\boldsymbol{F}$ and the respective isochoric components.

16.9.4. RATE-DEPENDENT CRYSTAL: MODEL PROBLEM

Here, we consider a simple model problem where a single finite element is subjected to uniform creep stretching under the action of a constant load. A planar double-slip-rate-dependent single crystal model with slip-rate equation (16.89) is adopted. The problem is defined in Figure 16.17. The dimensions of the element are $L_0 = W_0 = 1$ mm. Its thickness is also 1 mm. The material properties of the rate-dependent model are listed in Table 16.3. Initially, a total load $F = 500$ N is applied instantaneously (with $\Delta t = 0$), subjecting the element to a uniform axial stress of 500 MPa. The element is subsequently allowed to creep for 20 hours under constant load (stresses will increase due to cross-section reduction). The creeping phase is carried out in 10 and 100 equal-size time increments $\Delta t = 2$h and $\Delta t = 0.2$h, respectively. For both cases, the elongation of the element, L/L_0 where L is the deformed length, obtained in the simulation is plotted against time in Figure 16.18. The results show that, in the present algorithmic set-up, large time steps are possible without significant loss of accuracy. In addition, we remark that, as in the rate-independent problems reported in Section 16.8, the rates of convergence to equilibrium obtained with the rate-dependent model are also quadratic.

Appendices

A ISOTROPIC FUNCTIONS OF A SYMMETRIC TENSOR

This appendix presents some important definitions, properties and expressions involving isotropic scalar- and symmetric tensor-valued functions of one symmetric tensor. Such functions and their properties are exploited in various parts of this book in connection with the definition of elastic and elastoplastic constitutive models as well as with their computer implementation. Much of the material presented here is quite standard, so the proof to most relations is omitted.

We remark that some of the formulae presented here for the tensor-valued functions have been coded in program `HYPLAS` and are used in practice for the evaluation of functions as well as function derivatives. The corresponding subroutines of `HYPLAS` are the following:

- `DGISO2` – **D**erivative of a **G**eneral **ISO**tropic symmetric tensor-valued function of a symmetric tensor in **2**-D (plane strain and axisymmetric conditions).
- `DISO2` – **D**erivative of a particular class of **ISO**tropic symmetric tensor-valued function of a symmetric tensor in **2**-D/plane strain/axisymmetric conditions.
- `ISO2` – evaluation of a particular class of **ISO**tropic symmetric tensor-valued functions of a symmetric tensor in **2**-D/plane strain/axisymmetric conditions.
- `SPDEC2` – closed form **SP**ectral **DEC**omposition of symmetric tensors in **2**-D.

In line with the general layout adopted throughout this book, the corresponding formulae have been conveniently grouped in Boxes A.1–A.6 in the standard pseudo-code format. Readers who are interested only in the computational implementation of the above-mentioned formulae are referred directly to those boxes.

A.1. Isotropic scalar-valued functions

Let us start with the definition of the isotropic scalar-valued function of a symmetric tensor. In what follows, $\mathscr{S}$ will denote the space of symmetric tensors in an n-dimensional space. A function

$$\phi(\boldsymbol{X}) : \mathscr{L} \subset \mathscr{S} \rightarrow \mathscr{R} \tag{A.1}$$

is said to be isotropic if

$$\phi(\boldsymbol{X}) = \phi(\boldsymbol{Q}\boldsymbol{X}\boldsymbol{Q}^T) \tag{A.2}$$

for all rotations $\boldsymbol{Q}$.

A.1.1. REPRESENTATION

In three-dimensional space, a scalar function of a symmetric tensor is isotropic if and only if it admits the representation

$$\phi(\boldsymbol{X}) = \bar{\phi}(I_1(\boldsymbol{X}), I_2(\boldsymbol{X}), I_3(\boldsymbol{X})), \quad \forall\, \boldsymbol{X} \in \mathscr{L}, \tag{A.3}$$

where $I_i(\boldsymbol{X})$ $(i = 1, 2, 3)$ are the principal invariants of $\boldsymbol{X}$ (refer to definition (2.72), page 27). Note that the principal invariants themselves are isotropic functions.

Another useful representation of isotropic scalar functions of a symmetric tensor is in terms of principal values (eigenvalues) of the argument. A scalar function of a symmetric tensor (in three-dimensional space) is isotropic if and only if it has the representation

$$\phi(\boldsymbol{X}) = \hat{\phi}(x_1, x_2, x_3), \quad \forall\, \boldsymbol{X} \in \mathscr{L}, \tag{A.4}$$

where x_i $(i = 1, 2, 3)$ are the eigenvalues of $\boldsymbol{X}$ and $\hat{\phi} : \mathscr{R}^3 \to \mathscr{R}$ has the symmetries

$$\hat{\phi}(x_1, x_2, x_3) = \hat{\phi}(x_2, x_1, x_3) = \hat{\phi}(x_1, x_3, x_2). \tag{A.5}$$

A.1.2. THE DERIVATIVE OF AN ISOTROPIC SCALAR FUNCTION

The derivative of an isotropic function ϕ of the above type is the symmetric second-order tensor

$$\frac{\partial \phi}{\partial \boldsymbol{X}}.$$

This tensor is *coaxial* with $\boldsymbol{X}$, i.e. any eigenvector of $\boldsymbol{X}$ is an eigenvector of $\partial\phi/\partial\boldsymbol{X}$, so that these two tensors commute

$$\frac{\partial \phi}{\partial \boldsymbol{X}}\,\boldsymbol{X} = \boldsymbol{X}\,\frac{\partial \phi}{\partial \boldsymbol{X}}. \tag{A.6}$$

With $\{\boldsymbol{e}_i\}$ denoting an orthonormal basis of eigenvectors of $\boldsymbol{X}$, the spectral representation of $\partial\phi/\partial\boldsymbol{X}$ reads

$$\frac{\partial \phi}{\partial \boldsymbol{X}} = \sum_i \frac{\partial \hat{\phi}}{\partial x_i}\,\boldsymbol{e}_i \otimes \boldsymbol{e}_i; \tag{A.7}$$

that is, the principal values $(\partial\phi/\partial\boldsymbol{X})_i$ of $\partial\phi/\partial\boldsymbol{X}$ are given by

$$\left(\frac{\partial \phi}{\partial \boldsymbol{X}}\right)_i = \frac{\partial \hat{\phi}}{\partial x_i}. \tag{A.8}$$

This relation is proved by Ogden (1984).

It is important to note that the derivative $\partial\phi/\partial\boldsymbol{X}$, as well as the product (A.6), are themselves isotropic symmetric tensor-valued functions of a symmetric tensor. That this statement is true becomes clear in the next section, which discusses isotropic tensor functions of a tensor.

A.2. Isotropic tensor-valued functions

A symmetric tensor-valued function of a symmetric tensor

$$\boldsymbol{Y}(\boldsymbol{X}) : \mathscr{L} \subset \mathscr{S} \rightarrow \mathscr{S} \tag{A.9}$$

is said to be isotropic if

$$\boldsymbol{Q}\,\boldsymbol{Y}(\boldsymbol{X})\,\boldsymbol{Q}^T = \boldsymbol{Y}(\boldsymbol{Q}\boldsymbol{X}\boldsymbol{Q}^T) \tag{A.10}$$

for all rotations $\boldsymbol{Q}$.

The following important property holds for any isotropic function defined as above. If $\boldsymbol{Y}$ is isotropic, then $\boldsymbol{Y}(\boldsymbol{X})$ and $\boldsymbol{X}$ are coaxial and $\boldsymbol{Y}(\boldsymbol{X})$ and $\boldsymbol{X}$ commute

$$\boldsymbol{Y}(\boldsymbol{X})\,\boldsymbol{X} = \boldsymbol{X}\,\boldsymbol{Y}(\boldsymbol{X}). \tag{A.11}$$

A.2.1. REPRESENTATION

Let us now focus on the representation of isotropic tensor-valued functions of a tensor in three-dimensional space. A tensor function is isotropic if and only if it admits the representation

$$\boldsymbol{Y}(\boldsymbol{X}) = \alpha_0\,\boldsymbol{I} + \alpha_1\,\boldsymbol{X} + \alpha_2\,\boldsymbol{X}^2, \tag{A.12}$$

where

$$\begin{aligned}\alpha_0 &= \alpha_0(I_1(\boldsymbol{X}), I_2(\boldsymbol{X}), I_3(\boldsymbol{X}));\\ \alpha_1 &= \alpha_1(I_1(\boldsymbol{X}), I_2(\boldsymbol{X}), I_3(\boldsymbol{X}));\\ \alpha_2 &= \alpha_2(I_1(\boldsymbol{X}), I_2(\boldsymbol{X}), I_3(\boldsymbol{X})),\end{aligned} \tag{A.13}$$

are scalar-valued functions of the principal invariants of $\boldsymbol{X}$. Note that the functions α_0, α_1 and α_2 are themselves isotropic scalar functions of a symmetric tensor.

Further, if the domain of $\boldsymbol{Y}$ is a set of *invertible* symmetric tensors, then $\boldsymbol{Y}$ is isotropic if and only if it has the representation

$$\boldsymbol{Y}(\boldsymbol{X}) = \beta_0\,\boldsymbol{I} + \beta_1\,\boldsymbol{X} + \beta_{-1}\,\boldsymbol{X}^{-1}, \tag{A.14}$$

where the scalars β_Γ ($\Gamma = 0, 1, -1$) are functions of the principal invariants (2.72) of $\boldsymbol{X}$.

Assertions (A.12) and (A.14) are well-known representation theorems for isotropic tensor functions. Their formal proof can be found, for instance, in Gurtin (1981) and Ogden (1984).

Principal values representation

An alternative to (A.12) and (A.14), which is of particular relevance in the context of the present book, is the *principal values* representation of isotropic tensor functions. We remark that this representation is extensively exploited in the computational implementation of hyperelastic and elastoplastic solids in the program HYPLAS. The remainder of this appendix focuses exclusively on the principal values representation of isotropic tensor functions and many of the formulae presented are used in the computation of function values as well as function derivatives in HYPLAS.

The principal values representation is established in the following. A symmetric tensor-valued function of a symmetric tensor is isotropic if and only if there exists a scalar function $y : \mathscr{R}^n \to \mathscr{R}$ such that

$$\boldsymbol{Y}(\boldsymbol{X}) = \sum_{i=1}^{p} y_i \, \boldsymbol{E}_i, \tag{A.15}$$

where p is the number of distinct eigenvalues of $\boldsymbol{Y}$ (and $\boldsymbol{X}$) and $\boldsymbol{E}_i$ denotes the eigenprojections of $\boldsymbol{X}$ (and $\boldsymbol{Y}$) (refer to Section 2.2.8, from page 25). In two-dimensional space ($n = 2$), the eigenvalues y_i of $\boldsymbol{Y}$ are obtained from the eigenvalues x_i of $\boldsymbol{X}$ as

$$\begin{aligned} y_1 &= y(x_1, x_2) \\ y_2 &= y(x_2, x_1). \end{aligned} \tag{A.16}$$

In three-dimensional space ($n = 3$), the eigenvalues y_i are given by

$$\begin{aligned} y_1 &= y(x_1, x_2, x_3) \\ y_2 &= y(x_2, x_3, x_1) \\ y_3 &= y(x_3, x_1, x_2), \end{aligned} \tag{A.17}$$

and the function y has the symmetry property

$$y(a, b, c) = y(a, c, b) \tag{A.18}$$

for arbitrary a, b and c.

A.2.2. THE DERIVATIVE OF AN ISOTROPIC TENSOR FUNCTION

Assume that the isotropic tensor function $\boldsymbol{Y}$ is *differentiable* and let the fourth-order tensor D be its derivative

$$\mathsf{D}(\boldsymbol{X}) \equiv \frac{\mathrm{d}\boldsymbol{Y}(\boldsymbol{X})}{\mathrm{d}\boldsymbol{X}}. \tag{A.19}$$

In an n-dimensional space, if $\boldsymbol{X}$ has n distinct eigenvalues ($p = n$), a straightforward application of the product rule to the definition (A.15) leads to the following expression

$$\mathsf{D}(\boldsymbol{X}) = \sum_{i=1}^{n} \left\{ \boldsymbol{E}_i \otimes \frac{\mathrm{d}y_i}{\mathrm{d}\boldsymbol{X}} + y_i \, \frac{\mathrm{d}\boldsymbol{E}_i}{\mathrm{d}\boldsymbol{X}} \right\}, \tag{A.20}$$

or, after applying the chain rule,

$$\mathsf{D}(\boldsymbol{X}) = \sum_{i=1}^{n} \left\{ y_i \, \frac{\mathrm{d}\boldsymbol{E}_i}{\mathrm{d}\boldsymbol{X}} + \sum_{j=1}^{n} \frac{\partial y_i}{\partial x_j} \boldsymbol{E}_i \otimes \frac{\mathrm{d}x_j}{\mathrm{d}\boldsymbol{X}} \right\}. \tag{A.21}$$

Remark A.1. If there are no repeated eigenvalues, closed formulae for $\mathsf{D}(\boldsymbol{X})$ in the two- and three-dimensional cases can be obtained simply by taking the derivatives of explicit expressions for x_i and $\boldsymbol{E}_i$. However, in the presence of repeated eigenvalues, the eigenprojections (see expression (2.65)) as well as the eigenvalues are not differentiable – in spite of the differentiability of $\boldsymbol{Y}$. The final explicit formulae for the derivative $\mathsf{D}(\boldsymbol{X})$ at arguments with repeated eigenvalues are obtained as limits of the above explicit expression.

Box A.1. Computation of general isotropic tensor functions of a tensor in two dimensions.

(i) Given $\boldsymbol{X}$, compute its eigenvalues, x_α, and eigenprojections, $\boldsymbol{E}_\alpha$ ($\alpha = 1, 2$) – GOTO Box A.2

(ii) Compute the eigenvalues of $\boldsymbol{Y}$ as:

$$y_1 := y(x_1, x_2)$$
$$y_2 := y(x_2, x_1)$$

(iii) Assemble $\boldsymbol{Y}$:

$$\boldsymbol{Y}(\boldsymbol{X}) := \sum_{\alpha=1}^{p} y_\alpha \, \boldsymbol{E}_\alpha$$

where p is the number of distinct eigenvalues

A.3. The two-dimensional case

In the two-dimensional space, the characteristic equation (2.69) (page 27) yields the following quadratic equation for the eigenvalues x_α of $\boldsymbol{X}$:

$$x_\alpha^2 - I_1 \, x_\alpha + I_2 = 0, \quad \alpha = 1, 2, \tag{A.22}$$

where

$$\begin{aligned} I_1 &= \operatorname{tr}[\boldsymbol{X}] = X_{11} + X_{22} \\ I_2 &= \det[\boldsymbol{X}] = X_{11}X_{22} - X_{12}X_{21}. \end{aligned} \tag{A.23}$$

The solution of the quadratic equation (A.22) provides the exact formula for the eigenvalues of $\boldsymbol{X}$:

$$x_1 = \frac{I_1 + \sqrt{I_1^2 - 4I_2}}{2}; \quad x_2 = \frac{I_1 - \sqrt{I_1^2 - 4I_2}}{2}. \tag{A.24}$$

If $x_1 \neq x_2$, $(2.65)_1$ results in the following closed formula for computation of the eigenprojections of $\boldsymbol{X}$ in two dimensions:

$$\boldsymbol{E}_\alpha = \frac{1}{2\, x_\alpha - I_1} \, [\boldsymbol{X} + (x_\alpha - I_1)\boldsymbol{I}]. \tag{A.25}$$

On the other hand, if $x_1 = x_2$, then $(2.65)_2$ applies.

With the above closed expressions for eigenvalues and eigenprojections at hand, the computation of general isotropic tensor functions in two dimensions is carried out according to the pseudo-code summarised in Boxes A.1 and A.2. The algorithm of Box A.2 carries out only the spectral decomposition required in the function evaluation. This procedure has been implemented in subroutine `SPDEC2` (**SPEC**tral **DEC**omposition of tensors in **2**-D) of program `HYPLAS`.

Box A.2. Computation of eigenvalues and eigenprojections in two dimensions.

HYPLAS procedure:	SPDEC2

(i) Given $\boldsymbol{X}$, compute its principal invariants

$$I_1 = \text{tr}[\boldsymbol{X}] = X_{11} + X_{22}$$
$$I_2 = \det[\boldsymbol{X}] = X_{11}X_{22} - X_{12}X_{21}$$

(ii) Compute the eigenvalues of $\boldsymbol{X}$

$$x_1 = \frac{I_1 + \sqrt{I_1^2 - 4I_2}}{2}$$
$$x_2 = \frac{I_1 - \sqrt{I_1^2 - 4I_2}}{2}$$

(iii) Compute eigenprojections of $\boldsymbol{X}$

– If $x_i \neq x_2$, then for $\alpha = 1, 2$,

$$\boldsymbol{E}_\alpha = \frac{1}{2\,x_\alpha - I_1}\,[\boldsymbol{X} + (x_\alpha - I_1)\boldsymbol{I}]$$

– Else, if $x_1 = x_2$ $(p = 1)$, then

$$\boldsymbol{E}_1 = \boldsymbol{I}$$

A.3.1. TENSOR FUNCTION DERIVATIVE

In the two-dimensional case, we want to find closed-form expressions for the derivative

$$\mathsf{D}(\boldsymbol{X}) = \sum_{\alpha=1}^{2}\left\{y_\alpha \frac{\mathrm{d}\boldsymbol{E}_\alpha}{\mathrm{d}\boldsymbol{X}} + \sum_{\beta=1}^{2}\frac{\partial y_\alpha}{\partial x_\beta}\boldsymbol{E}_\alpha \otimes \frac{\mathrm{d}x_\beta}{\mathrm{d}\boldsymbol{X}}\right\}. \tag{A.26}$$

If the eigenvalues of $\boldsymbol{X}$ are *distinct*, we have (Carlson and Hoger, 1986)

$$\frac{\mathrm{d}x_i}{\mathrm{d}\boldsymbol{X}} = \boldsymbol{E}_i. \tag{A.27}$$

A straightforward differentiation of (A.24) and (A.25), followed by the substitution of the result together with (A.27) into (A.26), gives

$$\mathsf{D}(\boldsymbol{X}) = \frac{y_1 - y_2}{x_1 - x_2}\,[\mathsf{I}_S - \boldsymbol{E}_1 \otimes \boldsymbol{E}_1 - \boldsymbol{E}_2 \otimes \boldsymbol{E}_2] + \sum_{\alpha=1}^{2}\sum_{\beta=1}^{2}\frac{\partial y_\alpha}{\partial x_\beta}\boldsymbol{E}_\alpha \otimes \boldsymbol{E}_\beta, \tag{A.28}$$

where I_S is the fourth-order tensor defined by (2.110) (page 31). As pointed out in Remark A.2, if the eigenvalues are *repeated*, (A.25) is singular and cannot be differentiated.

Box A.3. Computation of the derivative of a general isotropic tensor function in two dimensions.

HYPLAS procedure:	DGISO2

(i) Given $\boldsymbol{X}$, compute its eigenvalues, x_α, and eigenprojections, $\boldsymbol{E}_\alpha$ ($\alpha = 1, 2$) – GOTO Box A.2

(ii) Compute the eigenvalues y_α of $\boldsymbol{Y}$ and their derivatives $\partial y_\alpha / \partial x_\beta$ for $\alpha = 1, 2$ and $\beta = 1, 2$

(iii) Assemble the derivative

$$\mathsf{D}(\boldsymbol{X}) := \begin{cases} \dfrac{y_1 - y_2}{x_1 - x_2}\,[\mathsf{I}_S - \boldsymbol{E}_1 \otimes \boldsymbol{E}_1 - \boldsymbol{E}_2 \otimes \boldsymbol{E}_2] \\ \quad + \displaystyle\sum_{\alpha=1}^{2}\sum_{\beta=1}^{2} \frac{\partial y_\alpha}{\partial x_\beta}\, \boldsymbol{E}_\alpha \otimes \boldsymbol{E}_\beta & \text{if } x_1 \neq x_2 \\ \left(\dfrac{\partial y_1}{\partial x_1} - \dfrac{\partial y_1}{\partial x_2}\right) \mathsf{I}_S + \dfrac{\partial y_1}{\partial x_2}\, \boldsymbol{I} \otimes \boldsymbol{I} & \text{if } x_1 = x_2 \end{cases}$$

Also, the right-hand side of the above formula for D becomes indeterminate. In this case, D is the limit

$$\lim_{x_1 \to x_2} \mathsf{D}(\boldsymbol{X}), \tag{A.29}$$

which is obtained by repeated applications of l'Hôspital's rule to (A.28). This tedious but straightforward derivation of the limit is omitted here. The final expression for the derivative at arguments with $x_1 = x_2$, which exists for sufficiently smooth functions y, is

$$\mathsf{D}(\boldsymbol{X}) = \left(\frac{\partial y_1}{\partial x_1} - \frac{\partial y_1}{\partial x_2}\right) \mathsf{I}_S + \frac{\partial y_1}{\partial x_2}\, \boldsymbol{I} \otimes \boldsymbol{I}. \tag{A.30}$$

The algorithm for computation of the derivative of $\boldsymbol{Y}(\boldsymbol{X})$ based on the above closed formulae is described in Box A.3. The procedure is implemented in subroutine `DGISO2` (**D**erivative of a **G**eneral **ISO**tropic tensor function of one tensor in **2**-D) of program `HYPLAS`.

Remark A.2. In practical computations, the signs $=$ and $\neq$, that decide which formula is to be used in the Boxes A.2 and A.3, are replaced by a check that takes the numerical precision of the machine used into account. For generic eigenvalues x_i and x_j we proceed as follows:

$$\text{If} \left| \frac{x_i - x_j}{x_i} \right| < \epsilon_{\text{tol}}, \quad \text{then assume } x_i = x_j.$$

Otherwise, assume $x_i \neq x_j$, where ϵ_{tol} is a small tolerance whose value depends on the machine precision.

A.3.2. PLANE STRAIN AND AXISYMMETRIC PROBLEMS

In plane strain and axisymmetric solid mechanics problems, isotropic tensor functions of a tensor are frequently defined as

$$\boldsymbol{Y}(\boldsymbol{X}) = \sum_{\alpha=1}^{p} y_\alpha \, \boldsymbol{E}_\alpha + y_3 \, \boldsymbol{E}_3, \tag{A.31}$$

where $p \leq 2$ is the number of *in-plane* distinct eigenvalues, $\boldsymbol{E}_3$ is the *out-of-plane* eigenprojection and y_3 is the eigenvalue associated with the out-of-plane direction. In this case, the eigenvalues $\{y_1, y_2, y_3\}$ satisfy the general three-dimensional definition (A.17) and the out-of-plane eigenprojection is *fixed*. Accordingly, the derivative of $\boldsymbol{Y}$ when the in-plane eigenvalues are distinct, is given by:

$$\begin{aligned}\mathsf{D}(\boldsymbol{X}) = {} & \sum_{\alpha=1}^{2} \left\{ y_\alpha \frac{\mathrm{d}\boldsymbol{E}_\alpha}{\mathrm{d}\boldsymbol{X}} + \sum_{\beta=1}^{2} \frac{\partial y_\alpha}{\partial x_\beta} \boldsymbol{E}_\alpha \otimes \frac{\mathrm{d}x_\beta}{\mathrm{d}\boldsymbol{X}} \right\} \\ & + \sum_{\alpha=1}^{2} \frac{\partial y_\alpha}{\partial x_3} \boldsymbol{E}_\alpha \otimes \frac{\mathrm{d}x_3}{\mathrm{d}\boldsymbol{X}} + \sum_{j=1}^{3} \frac{\partial y_3}{\partial x_j} \boldsymbol{E}_3 \otimes \frac{\mathrm{d}x_j}{\mathrm{d}\boldsymbol{X}}.\end{aligned} \tag{A.32}$$

The first summation on the right-hand side of the above equation is identical to the corresponding two-dimensional expression and is computed by the closed formula listed in Box A.3 for $x_1 \neq x_2$. After straightforward manipulations, the contribution of the remaining terms is found to be

$$\mathsf{D}_3(\boldsymbol{X}) \equiv \sum_{\alpha=1}^{2} \left\{ \frac{\partial y_\alpha}{\partial x_3} \boldsymbol{E}_\alpha \otimes \boldsymbol{E}_3 + \frac{\partial y_3}{\partial x_\alpha} \boldsymbol{E}_3 \otimes \boldsymbol{E}_\alpha \right\} + \frac{\partial y_3}{\partial x_3} \boldsymbol{E}_3 \otimes \boldsymbol{E}_3. \tag{A.33}$$

The assemblage of the complete derivative is attained by adding $\mathsf{D}_3(\boldsymbol{X})$ to the corresponding equation of Box A.3. This operation is included in subroutine `DGISO2`.

With repeated in-plane eigenvalues, $x_1 = x_2$, the above formula is replaced with

$$\mathsf{D}_3(\boldsymbol{X}) = \frac{\partial y_1}{\partial x_3} \boldsymbol{I}_p \otimes \boldsymbol{E}_3 + \frac{\partial y_3}{\partial x_1} \boldsymbol{E}_3 \otimes \boldsymbol{I}_p + \frac{\partial y_3}{\partial x_3} \boldsymbol{E}_3 \otimes \boldsymbol{E}_3, \tag{A.34}$$

where $\boldsymbol{I}_p$ is the second-order plane orthogonal projection tensor. It is important to note that in Box A.3, I_S and $\boldsymbol{I}$ are tensors on the *plane*. When applying the formulae of Box A.3 in plane strain/axisymmetric situations, I_S and $\boldsymbol{I}$ must be replaced, respectively, with the plane projection operators I_p and $\boldsymbol{I}_p$. The tensor $\boldsymbol{I}_p$ has the following matrix representation:

$$\boldsymbol{I}_p = \begin{bmatrix} 1 & 0 & 0 \\ 0 & 1 & 0 \\ 0 & 0 & 0 \end{bmatrix} \tag{A.35}$$

in an orthonormal system where the out-of-plane direction is associated with the third line and column. Analogously, the Cartesian components of I_p are given by

$$[\mathsf{I}_p]_{ijkl} = \begin{cases} \frac{1}{2}(\delta_{ik}\delta_{jl} + \delta_{il}\delta_{jk}) & i, j, k, l \in \{1, 2\} \\ 0 & \text{otherwise.} \end{cases} \tag{A.36}$$

A.4. The three-dimensional case

A.4.1. FUNCTION COMPUTATION

In the three-dimensional space, the characteristic equation (2.73) (page 27) yields the cubic equation

$$x_i^3 - I_1\, x_i^2 + I_2\, x_i - I_3 = 0, \tag{A.37}$$

for the eigenvalues of $\boldsymbol{X}$, where

$$\begin{aligned} I_1 &= \mathrm{tr}[\boldsymbol{X}] \\ I_2 &= \tfrac{1}{2}\{(\mathrm{tr}\boldsymbol{X})^2 - \mathrm{tr}[\boldsymbol{X}^2]\} \\ I_3 &= \det[\boldsymbol{X}]. \end{aligned} \tag{A.38}$$

As the $\boldsymbol{X}$ is real by assumption, its eigenvalues x_i are real and the solution of (A.37) can be computed exactly by the following expressions:

$$\begin{aligned} x_1 &= -2\sqrt{Q}\,\cos\left[\frac{\theta}{3}\right] + \frac{I_1}{3}; \\ x_2 &= -2\sqrt{Q}\,\cos\left[\frac{\theta + 2\pi}{3}\right] + \frac{I_1}{3}; \\ x_3 &= -2\sqrt{Q}\,\cos\left[\frac{\theta - 2\pi}{3}\right] + \frac{I_1}{3}, \end{aligned} \tag{A.39}$$

where Q and θ are defined as

$$Q = \frac{I_1^2 - 3I_2}{9} \tag{A.40}$$

and

$$\theta = \cos^{-1}\left[\frac{R}{\sqrt{Q^3}}\right], \tag{A.41}$$

with

$$R = \frac{-2I_1^3 + 9I_1I_2 - 27I_3}{54}. \tag{A.42}$$

If x_i is not repeated, the following closed formula for the eigenprojection $\boldsymbol{E}_i$ in three dimensions is obtained from (2.65):

$$\boldsymbol{E}_i = \frac{x_i}{2\,x_i^3 - I_1\,x_i^2 + I_3}\left[\boldsymbol{X}^2 - (I_1 - x_i)\,\boldsymbol{X} + \frac{I_3}{x_i}\,\boldsymbol{I}\right]. \tag{A.43}$$

If x_j is repeated (with multiplicity 2) and $x_i \neq x_j$, then $\boldsymbol{E}_i$ can be computed by the expression above and $\boldsymbol{E}_j$ is given simply by

$$\boldsymbol{E}_j = \boldsymbol{I} - \boldsymbol{E}_i, \tag{A.44}$$

and, if $x_1 = x_2 = x_3$, then $(2.65)_2$ applies.

Given $\boldsymbol{X}$, the tensor function $\boldsymbol{Y}(\boldsymbol{X})$ is computed as described in Boxes A.4 and A.5, where the above closed formulae for eigenvalues and eigenprojections are used.

A.4.2. COMPUTATION OF THE FUNCTION DERIVATIVE

In three dimensions the derivation of the closed-form expressions for the derivative is far more laborious than the derivation of the two-dimensional counterpart. The path for derivation, however, is the same. Firstly, we write the symbolic expression for $\mathsf{D}(\boldsymbol{X})$:

$$\mathsf{D}(\boldsymbol{X}) = \sum_{i=1}^{3} \left\{ y_i \frac{\mathrm{d}\boldsymbol{E}_i}{\mathrm{d}\boldsymbol{X}} + \sum_{j=1}^{3} \frac{\partial y_i}{\partial x_j} \boldsymbol{E}_i \otimes \frac{\mathrm{d}x_j}{\mathrm{d}\boldsymbol{X}} \right\}. \tag{A.45}$$

If there are three distinct eigenvalues, the explicit formula for D is obtained simply by taking the derivatives of the closed expressions for x_i and $\boldsymbol{E}_i$, given in Box A.5. Again, if there are two or three identical eigenvalues, the expression for $\boldsymbol{E}_i$ becomes indefinite and the final formula for the derivative is obtained by applying a limiting procedure to the general formula valid for three distinct eigenvalues. Note that the limiting procedure in the three-dimensional case is quite lengthy but, as in the two-dimensional case, involves only the repetitive application of l'Hôspital's rule. The final expressions are gathered in Box A.6.

In Box A.6, $\mathrm{d}\boldsymbol{X}^2/\mathrm{d}\boldsymbol{X}$ denotes the derivative of the square of a tensor with Cartesian components given by

$$\left[\frac{\mathrm{d}\boldsymbol{X}^2}{\mathrm{d}\boldsymbol{X}}\right]_{ijkl} = \frac{1}{2}\left(\delta_{ik}X_{lj} + \delta_{il}X_{kj} + \delta_{jl}X_{ik} + \delta_{kj}X_{il}\right) \tag{A.46}$$

and the scalars $s_1, s_2, \ldots, s_6$ have been defined as

$$\begin{aligned}
s_1 &= \frac{y_a - y_c}{(x_a - x_c)^2} + \frac{1}{x_a - x_c}\left(\frac{\partial y_c}{\partial x_b} - \frac{\partial y_c}{\partial x_c}\right) \\
s_2 &= 2x_c\frac{y_a - y_c}{(x_a - x_c)^2} + \frac{x_a + x_c}{x_a - x_c}\left(\frac{\partial y_c}{\partial x_b} - \frac{\partial y_c}{\partial x_c}\right) \\
s_3 &= 2\frac{y_a - y_c}{(x_a - x_c)^3} + \frac{1}{(x_a - x_c)^2}\left(\frac{\partial y_a}{\partial x_c} + \frac{\partial y_c}{\partial x_a} - \frac{\partial y_a}{\partial x_a} - \frac{\partial y_c}{\partial x_c}\right) \\
s_4 &= 2x_c\frac{y_a - y_c}{(x_a - x_c)^3} + \frac{1}{x_a - x_c}\left(\frac{\partial y_a}{\partial x_c} - \frac{\partial y_c}{\partial x_b}\right) + \frac{x_c}{(x_a - x_c)^2}\left(\frac{\partial y_a}{\partial x_c} + \frac{\partial y_c}{\partial x_a} - \frac{\partial y_a}{\partial x_a} - \frac{\partial y_c}{\partial x_c}\right) \\
s_5 &= 2x_c\frac{y_a - y_c}{(x_a - x_c)^3} + \frac{1}{x_a - x_c}\left(\frac{\partial y_c}{\partial x_a} - \frac{\partial y_c}{\partial x_b}\right) + \frac{x_c}{(x_a - x_c)^2}\left(\frac{\partial y_a}{\partial x_c} + \frac{\partial y_c}{\partial x_a} - \frac{\partial y_a}{\partial x_a} - \frac{\partial y_c}{\partial x_c}\right) \\
s_6 &= 2x_c^2\frac{y_a - y_c}{(x_a - x_c)^3} + \frac{x_a x_c}{(x_a - x_c)^2}\left(\frac{\partial y_a}{\partial x_c} + \frac{\partial y_c}{\partial x_a}\right) \\
&\quad - \frac{x_c^2}{(x_a - x_c)^2}\left(\frac{\partial y_a}{\partial x_a} + \frac{\partial y_c}{\partial x_c}\right) - \frac{x_a + x_c}{x_a - x_c}\frac{\partial y_c}{\partial x_b}.
\end{aligned} \tag{A.47}$$

Here and in the closed expression for two distinct eigenvalues, (a, b, c) are cyclic permutations of (1,2,3).

A.5. A particular class of isotropic tensor functions

Let us now consider the subset of the general class of functions discussed above for which y is a function of a *single* argument. Given $y : \mathscr{R} \rightarrow \mathscr{R}$, a class of isotropic tensor function of a

Box A.4. Computation of general isotropic tensor functions of a tensor in three dimensions.

(i) Given $\boldsymbol{X}$, compute its eigenvalues, x_i, and eigenprojections, $\boldsymbol{E}_i$ ($i = 1, 2, 3$) – GOTO Box A.5

(ii) Compute the eigenvalues of $\boldsymbol{Y}$ as

$$
\begin{aligned}
y_1 &:= y(x_1, x_2, x_3) \\
y_2 &:= y(x_2, x_3, x_1) \\
y_3 &:= y(x_3, x_1, x_2)
\end{aligned}
$$

(iii) Assemble $\boldsymbol{Y}$

$$\boldsymbol{Y}(\boldsymbol{X}) := \sum_{i=1}^{p} y_i \, \boldsymbol{E}_i$$

where p is the number of distinct eigenvalues

symmetric tensor can be constructed as

$$\boldsymbol{Y}(\boldsymbol{X}) \equiv \sum_{i=1}^{p} y(x_i) \, \boldsymbol{E}_i. \tag{A.48}$$

Closed-form expressions for the derivative of functions of this class, employing the eigenprojection-based representation, were derived by Carlson and Hoger (1986). Complete algorithms using compact expressions for computation of functions of the above class (and their derivatives) at invertible arguments have been proposed by Miehe (1993).

Remark A.3. Functions expressed as such define an important class of isotropic tensor-valued functions of a tensor and are, clearly, particular cases of the general form (A.15). The families of strain measures defined by (3.56) (page 53) and (3.60), for instance, are members of this class. Other important functions, such as the tensor square root and the tensor exponential, can also be expressed in the format (A.48) by setting $y(x_i) \equiv \sqrt{x_i}$ and $y(x_i) \equiv \exp(x_i)$, respectively.

As (A.48) is a particular case of (A.15), the computation of $\boldsymbol{Y}(\boldsymbol{X})$ and its derivative can be carried out in a manner entirely analogous to the procedure described in the previous section. Only the main expressions are summarised below. They are promptly obtained from their general counterparts simply by eliminating the dependence of the eigenvalues y_i of $\boldsymbol{Y}$ on x_j for $i \neq j$.

Box A.5. Computation of eigenvalues and eigenprojections in three dimensions.

(i) Given $\boldsymbol{X}$, compute its principal invariants

$$I_1 = \mathrm{tr}[\boldsymbol{X}]$$
$$I_2 = \tfrac{1}{2}\{(\mathrm{tr}\,\boldsymbol{X})^2 - \mathrm{tr}[\boldsymbol{X}^2]\}$$
$$I_3 = \det[\boldsymbol{X}]$$

(ii) Compute the eigenvalues of $\boldsymbol{X}$

$$R = \frac{-2\,I_1^3 + 9\,I_1\,I_2 - 27\,I_3}{54}$$
$$\theta = \cos^{-1}\left[\frac{R}{\sqrt{Q^3}}\right]$$
$$Q = \frac{I_1^2 - 3\,I_2}{9}$$
$$x_1 = -2\,\sqrt{Q}\,\cos\left[\frac{\theta}{3}\right] + \frac{I_1}{3}$$
$$x_2 = -2\,\sqrt{Q}\,\cos\left[\frac{\theta + 2\pi}{3}\right] + \frac{I_1}{3}$$
$$x_3 = -2\,\sqrt{Q}\,\cos\left[\frac{\theta - 2\pi}{3}\right] + \frac{I_1}{3}$$

(iii) Compute the eigenprojections of $\boldsymbol{X}$

– If $x_i \neq x_2 \neq x_3$, then for $i = 1, 2, 3$,

$$\boldsymbol{E}_i = \frac{x_i}{2\,x_i^3 - I_1\,x_i^2 + I_3}\left[\boldsymbol{X}^2 - (I_1 - x_i)\,\boldsymbol{X} + \frac{I_3}{x_i}\,\boldsymbol{I}\right]$$

– Else, if $x_i \neq x_j = x_k$, then compute $\boldsymbol{E}_i$ using the expression above and

$$\boldsymbol{E}_j = \boldsymbol{I} - \boldsymbol{E}_i$$

– Else ($x_1 = x_2 = x_3$),

$$\boldsymbol{E}_1 = \boldsymbol{I}$$

A.5.1. TWO DIMENSIONS

After closed-form evaluation of x_i and $\boldsymbol{E}_i$ in Box A.2, the following expression is used to compute $\boldsymbol{Y}(\boldsymbol{X})$ in two dimensions

$$\boldsymbol{Y}(\boldsymbol{X}) = \begin{cases} \sum_{i=1}^{2} y(x_i)\,\boldsymbol{E}_i & \text{if } x_1 \neq x_2 \\ y(x_1)\,\boldsymbol{I} & \text{if } x_1 = x_2. \end{cases} \tag{A.49}$$

Box A.6. Computation of the derivative of a general isotropic tensor function in three dimensions.

(i) Given $\boldsymbol{X}$, compute its eigenvalues, x_i, and eigenprojections, $\boldsymbol{E}_i$ $(i = 1, 2, 3)$ – GOTO Box A.5

(ii) Compute the eigenvalues y_i of $\boldsymbol{Y}$ and their derivatives $\partial y_i/\partial x_j$ for $i, j = 1, 2, 3$

(iii) Assemble the derivative

$$\mathsf{D}(\boldsymbol{X}) = \begin{cases} \displaystyle\sum_{a=1}^{3} \frac{y_a}{(x_a - x_b)(x_a - x_c)} \Bigg\{ \frac{\mathrm{d}\boldsymbol{X}^2}{\mathrm{d}\boldsymbol{X}} - (x_b + x_c)\, \mathsf{I}_S \\ \quad - [(x_a - x_b) + (x_a - x_c)] \boldsymbol{E}_a \otimes \boldsymbol{E}_a \\ \quad\quad - (x_b - x_c)(\boldsymbol{E}_b \otimes \boldsymbol{E}_b - \boldsymbol{E}_c \otimes \boldsymbol{E}_c) \Bigg\} \\ \quad\quad\quad + \displaystyle\sum_{i=1}^{3}\sum_{j=1}^{3} \frac{\partial y_i}{\partial x_j} \boldsymbol{E}_i \otimes \boldsymbol{E}_j \\ \quad\quad\quad\quad \text{if } x_1 \neq x_2 \neq x_3 \\ s_1 \dfrac{\mathrm{d}\boldsymbol{X}^2}{\mathrm{d}\boldsymbol{X}} - s_2\, \mathsf{I}_S - s_3\, \boldsymbol{X} \otimes \boldsymbol{X} + s_4\, \boldsymbol{X} \otimes \boldsymbol{I} \\ \quad\quad + s_5\, \boldsymbol{I} \otimes \boldsymbol{X} - s_6\, \boldsymbol{I} \otimes \boldsymbol{I} \\ \quad\quad\quad \text{if } x_a \neq x_b = x_c \\ \left(\dfrac{\partial y_1}{\partial x_1} - \dfrac{\partial y_1}{\partial x_2} \right) \mathsf{I}_S + \dfrac{\partial y_1}{\partial x_2} \boldsymbol{I} \otimes \boldsymbol{I} \quad\quad \text{if } x_1 = x_2 = x_3 \end{cases}$$

The derivative of $\boldsymbol{Y}(\boldsymbol{X})$ is computed by the expressions

$$\mathsf{D}(\boldsymbol{X}) = \begin{cases} \dfrac{y(x_1) - y(x_2)}{x_1 - x_2} [\mathsf{I}_S - \boldsymbol{E}_1 \otimes \boldsymbol{E}_1 - \boldsymbol{E}_2 \otimes \boldsymbol{E}_2] \\ \quad + \displaystyle\sum_{\alpha=1}^{2} y'(x_\alpha)\, \boldsymbol{E}_\alpha \otimes \boldsymbol{E}_\alpha & \text{if } x_1 \neq x_2 \\ y'(x_1)\, \mathsf{I}_S & \text{if } x_1 = x_2. \end{cases} \tag{A.50}$$

A.5.2. THREE DIMENSIONS

In three dimensions, the eigenvalues and eigenprojection tensors of $\boldsymbol{X}$ are firstly evaluated following the algorithm of Box A.5. With those at hand, the function $\boldsymbol{Y}(\boldsymbol{X})$ is computed as

$$\boldsymbol{Y}(\boldsymbol{X}) = \begin{cases} \displaystyle\sum_{i=1}^{3} y(x_i)\, \boldsymbol{E}_i & \text{if } x_1 \neq x_2 \neq x_3 \\ y(x_a)\, \boldsymbol{E}_a + y(x_b)\, (\boldsymbol{I} - \boldsymbol{E}_a) & \text{if } x_a \neq x_b = x_c \\ y(x_1)\, \boldsymbol{I} & \text{if } x_1 = x_2 = x_3 \end{cases} \tag{A.51}$$

where the subscripts (a, b, c) are cyclic permutations of (1,2,3). The computation of the derivative, $\mathsf{D}(\boldsymbol{X})$, follows the expressions

$$\mathsf{D}(\boldsymbol{X}) = \begin{cases} \displaystyle\sum_{a=1}^{3} \left\{ \frac{y(x_a)}{(x_a - x_b)(x_a - x_c)} \left[\frac{\mathrm{d}\boldsymbol{X}^2}{\mathrm{d}\boldsymbol{X}} - (x_b + x_c)\,\mathsf{I}_S \right.\right. \\ \quad - [(x_a - x_b) + (x_a - x_c)]\boldsymbol{E}_a \otimes \boldsymbol{E}_a \\ \quad \left.\left. - (x_b - x_c)(\boldsymbol{E}_b \otimes \boldsymbol{E}_b - \boldsymbol{E}_c \otimes \boldsymbol{E}_c) \right] + y'(x_a)\,\boldsymbol{E}_a \otimes \boldsymbol{E}_a \right\} \\ \qquad\qquad\qquad\qquad\qquad\qquad \text{if } x_1 \neq x_2 \neq x_3 \\ s_1 \dfrac{\mathrm{d}\boldsymbol{X}^2}{\mathrm{d}\boldsymbol{X}} - s_2\,\mathsf{I}_S - s_3\,\boldsymbol{X} \otimes \boldsymbol{X} + s_4\,\boldsymbol{X} \otimes \boldsymbol{I} + s_5\,\boldsymbol{I} \otimes \boldsymbol{X} - s_6\,\boldsymbol{I} \otimes \boldsymbol{I} \\ \qquad\qquad\qquad\qquad\qquad\qquad \text{if } x_a \neq x_b = x_c \\ y'(x_1)\,\mathsf{I}_S \qquad\qquad\qquad\qquad\qquad \text{if } x_1 = x_2 = x_3 \end{cases} \tag{A.52}$$

with the scalars $s_1, \ldots, s_6$ now defined as

$$\begin{aligned} s_1 &= \frac{y(x_a) - y(x_c)}{(x_a - x_c)^2} - \frac{y'(x_c)}{x_a - x_c} \\ s_2 &= 2x_c \frac{y(x_a) - y(x_c)}{(x_a - x_c)^2} - \frac{x_a + x_c}{x_a - x_c}\, y'(x_c) \\ s_3 &= 2\,\frac{y(x_a) - y(x_c)}{(x_a - x_c)^3} - \frac{y'(x_a) + y'(x_c)}{(x_a - x_c)^2} \\ s_4 &= s_5 = x_c\, s_3 \\ s_6 &= x_c^2\, s_3. \end{aligned} \tag{A.53}$$

Remark A.4. The closed expressions (A.50) and (A.52) for the function derivative are equivalent to those derived by Carlson and Hoger (1986). Their equivalence can be established (after some algebra) by considering the standard identity for the directional derivative of the square of a tensor in a generic direction $\boldsymbol{T}$,

$$\frac{\mathrm{d}\boldsymbol{X}^2}{\mathrm{d}\boldsymbol{X}}[\boldsymbol{T}] = \boldsymbol{X}\boldsymbol{T} + \boldsymbol{T}\boldsymbol{X},$$

together with the identity

$$(\boldsymbol{E}_i \otimes \boldsymbol{E}_i)\,\boldsymbol{T} = \boldsymbol{E}_i\,\boldsymbol{T}\boldsymbol{E}_i,$$

demonstrated by Carlson and Hoger, and the general closed formulae (2.65) for $\boldsymbol{E}_i$.

A.6. Alternative procedures

We remark that the methodology described in the preceding sections is by no means the only possible way to compute isotropic tensor function/derivatives of the present type. The formulae adopted here are an alternative to the eigenvector-based representation obtained by Chadwick and Ogden (1971), which is widely used in the computational treatment of solid mechanics problems (see Crisfield (1997), Chapter 13, for instance). Also relying on the eigenprojection-based representation, a methodology similar to that adopted here was introduced by Miehe (1993, 1998a), where a particularly compact representation for the function derivative is used. However, the compact representation allows only the computation of the derivative at *invertible* arguments and cannot be used, for instance, to compute tangent operators consistent with principal stress-based algorithms described in Chapter 8 for multisurface plasticity models. In that case, evaluation of the derivative at a generally non-invertible argument (the elastic trial strain tensor) is necessary.

B THE TENSOR EXPONENTIAL

This appendix describes the computational procedures for evaluation of the tensor exponential function (or exponential map) and its derivative. It also describes the application of the exponential map in the numerical solution of a class of initial value problems of particular relevance in computational mechanics. Note that the exponential of a *symmetric* tensor is a particular member of the class of isotropic tensor-valued functions described in Section A.5 (page 740) and is obtained by setting $y(x_i) \equiv \exp(x_i)$ in expression (A.48). In this case, the function as well as its derivatives can be computed by the procedures already explained in that section. In the present appendix, however, we are concerned with the more general tensor exponential function whose domain is the entire space of (symmetric and *unsymmetric*) real tensors in $n_{\rm dim}$ dimensions. The computational procedures described here have been implemented in the following subroutines of the program HYPLAS:

- EXPMAP – **EXP**onential **MAP** computation.
- DEXPMP – Evaluation of the **D**erivative of the **EXP**onential **M**a**P**.

The corresponding pseudo-codes are provided, respectively, in Boxes B.1 and B.2.

B.1. The tensor exponential function

Consider the initial value problem defined by the tensor ordinary differential equation

$$\dot{\boldsymbol{Y}}(t) = \boldsymbol{A}\,\boldsymbol{Y}(t), \tag{B.1}$$

with initial condition

$$\boldsymbol{Y}(t_0) = \boldsymbol{Y}_0, \tag{B.2}$$

where the superimposed dot denotes differentiation with respect to t and $\boldsymbol{A}$ and $\boldsymbol{Y}_0$ are given (generally unsymmetric) constant tensors. The *tensor exponential function* (or *exponential map*), $\exp[\,\cdot\,]$, is the (unique) solution to the above problem

$$\boldsymbol{Y}(t) = \exp[(t - t_0)\boldsymbol{A}]\,\boldsymbol{Y}_0. \tag{B.3}$$

Explicitly, the tensor exponential function can be expressed by means of its series representation (Hirsch and Smale, 1974). For a generic tensor $\boldsymbol{X}$, we have

$$\exp[\boldsymbol{X}\,] = \sum_{n=0}^{\infty} \frac{1}{n!}\,\boldsymbol{X}^{\,n}. \tag{B.4}$$

The above series is absolutely convergent for any argument $\boldsymbol{X}$ and, as its scalar counterpart, can be used to evaluate the tensor exponential function to any prescribed degree of accuracy.

B.1.1. SOME PROPERTIES OF THE TENSOR EXPONENTIAL FUNCTION

The tensor exponential possesses some important properties (Gurtin, 1981; Hirsch and Smale, 1974) which are relevant in computational solid mechanics applications:

(a) The determinant of the exponential of a tensor satisfies

$$\det[\exp[\boldsymbol{X}]] = \exp[\operatorname{tr}[\boldsymbol{X}]]. \tag{B.5}$$

Thus, the exponential tensor function maps *traceless* tensors onto *unimodular* tensors, i.e.

$$\operatorname{tr}[\boldsymbol{X}] = 0 \Longleftrightarrow \det[\exp[\boldsymbol{X}]] = 1. \tag{B.6}$$

This property is crucially important in the numerical treatment of finite plasticity described in this book.

(b) By straightforward use of (B.4) we find that, for any invertible $\boldsymbol{D}$,

$$\exp[\boldsymbol{D}\,\boldsymbol{X}\,\boldsymbol{D}^{-1}] = \boldsymbol{D}\,\exp[\boldsymbol{X}]\,\boldsymbol{D}^{-1}. \tag{B.7}$$

In particular, for any *orthogonal* tensor $\boldsymbol{Q}$, we have

$$\exp[\boldsymbol{Q}\,\boldsymbol{X}\,\boldsymbol{Q}^T] = \boldsymbol{Q}\,\exp[\boldsymbol{X}]\,\boldsymbol{Q}^T, \tag{B.8}$$

so that the tensor exponential is an *isotropic* function.

(c) For a generic tensor $\boldsymbol{X}$, we have

$$\exp[-\boldsymbol{X}] = (\exp[\boldsymbol{X}])^{-1}. \tag{B.9}$$

(d) If $\boldsymbol{C}$ and $\boldsymbol{D}$ commute, i.e. if $\boldsymbol{C}\boldsymbol{D} = \boldsymbol{D}\boldsymbol{C}$, then

$$\exp[\boldsymbol{C} + \boldsymbol{D}] = \exp[\boldsymbol{C}]\,\exp[\boldsymbol{D}] = \exp[\boldsymbol{D}]\,\exp[\boldsymbol{C}]. \tag{B.10}$$

This property and (c) imply that for any $\boldsymbol{X}$ and integer n,

$$\exp[n\,\boldsymbol{X}] = (\exp[\boldsymbol{X}])^n. \tag{B.11}$$

(e) Let $\boldsymbol{X}$ be a skew-symmetric tensor ($\boldsymbol{X} = -\boldsymbol{X}^T$). Then,

$$\boldsymbol{Q} = \exp[\boldsymbol{X}], \tag{B.12}$$

is a proper orthogonal tensor (a *rotation*) and the exponential map has the following representation known as *Rodrigues' formula*

$$\exp[\boldsymbol{X}] = \boldsymbol{I} + \frac{\sin(\|\boldsymbol{x}\|)}{\|\boldsymbol{x}\|}\,\boldsymbol{X} + \frac{1}{2}\left[\frac{\sin(\|\boldsymbol{x}\|/2)}{\|\boldsymbol{x}\|/2}\right]^2 \boldsymbol{X}^2, \tag{B.13}$$

where $\boldsymbol{x}$ is the axial vector of $\boldsymbol{X}$. If $\|\boldsymbol{x}\| \neq n\,\pi$ for any odd n, then we have the additional equivalent representation

$$\exp[\boldsymbol{X}] = \boldsymbol{I} + \frac{2}{1 + \|\bar{\boldsymbol{x}}\|^2}[\bar{\boldsymbol{X}} + \bar{\boldsymbol{X}}^2] \tag{B.14}$$

where $\bar{\boldsymbol{X}}$ is the skew-symmetric tensor with axial vector

$$\bar{\boldsymbol{x}} = \frac{\tan(\|\boldsymbol{x}\|/2)}{\|\boldsymbol{x}\|}\,\boldsymbol{x}. \tag{B.15}$$

Box B.1. Computation of the exponential of a generally unsymmetric tensor.

HYPLAS procedure:	EXPMAP

(i) Given $\boldsymbol{X}$, initialise $n = 0$ and

$$\exp[\boldsymbol{X}] := \boldsymbol{I}$$

(ii) Increment counter

$$n := n + 1$$

(iii) Compute $n!$ and $\boldsymbol{X}^n$

(iv) Add new term to the series

$$\exp[\boldsymbol{X}] := \exp[\boldsymbol{X}] + \frac{1}{n!}\boldsymbol{X}^n$$

(v) Check convergence

IF $\|\boldsymbol{X}^n\|/n! < \epsilon_{\text{tol}}$ THEN EXIT

ELSE GOTO (ii)

B.1.2. COMPUTATION OF THE TENSOR EXPONENTIAL FUNCTION

For a generic unsymmetric argument† the computation of the tensor exponential function is carried out by simply truncating the infinite series with $n_{\max}$ terms, i.e. we compute

$$\exp[\boldsymbol{X}] = \sum_{n=0}^{n_{\max}} \frac{1}{n!}\boldsymbol{X}^n, \tag{B.16}$$

with $n_{\max}$ such that

$$\frac{1}{n_{\max}!}\|\boldsymbol{X}^{n_{\max}}\| < \epsilon_{\text{tol}}, \tag{B.17}$$

where ϵ_{tol} is a prescribed tolerance (typically of the order of the machine precision). Note that the above principle is invariably adopted in the computation of computer intrinsic mathematical functions such as the exponential of a *scalar*. The above formula is implemented in subroutine EXPMAP of HYPLAS. The corresponding pseudo-code is illustrated in Box B.1.

Remark B.1. It is important to emphasise that in the single-crystal plasticity application described in Section 16.5, where the argument of the tensor exponential is a deformation gradient (with components of order 1), only a few terms of the above series (about six or seven) are normally required for convergence compatible with the usual tolerances involved in nonlinear computational continuum mechanics. The practical use of the above formula in finite element computations is, therefore, absolutely feasible.

†Note that for *symmetric* arguments, the computation of the tensor exponential can be performed more efficiently by the procedure described in Section A.5. For *skew-symmetric* arguments, the closed formulae (B.13) or (B.14) can be used for a more efficient computation of the tensor exponential.

B.2. The tensor exponential derivative

A series representation of the derivative of the tensor exponential function, suitable for computer implementation, is derived in this section. Crucial to the proposed representation is the general expression for the derivative of positive integer powers of a tensor, derived in the following.

Proposition B.1. *Let n be a strictly positive integer and let $\boldsymbol{F}$ be the tensor-valued function of a tensor argument defined as*

$$\boldsymbol{F}(\boldsymbol{X}) = \boldsymbol{X}^{\,n}. \tag{B.18}$$

The derivative $\mathrm{D}\boldsymbol{F}(\boldsymbol{X})$ of the function $\boldsymbol{F}$ at $\boldsymbol{X}$ is the fourth-order tensor whose Cartesian components are given by

$$[\mathrm{D}\boldsymbol{F}(\boldsymbol{X})]_{ijkl} = \sum_{m=1}^{n} [\boldsymbol{X}^{\,m-1}]_{ik}\,[\boldsymbol{X}^{\,n-m}]_{lj}. \tag{B.19}$$

Proof. The Cartesian components of the derivative of $\boldsymbol{F}(\boldsymbol{X})$ are given by

$$[\mathrm{D}\boldsymbol{F}(\boldsymbol{X})]_{ijkl} = \frac{\mathrm{d}[\boldsymbol{X}^{\,n}]_{ij}}{\mathrm{d}X_{kl}}.$$

Let us consider the following general expression for the components of a positive integer power of a tensor

$$[\boldsymbol{X}^{\,n}]_{a_1 a_{n+1}} = \prod_{m=1}^{n} X_{a_m a_{m+1}}.$$

By straightforward differentiation of the above product, we obtain

$$\begin{aligned}
\frac{\mathrm{d}[\boldsymbol{X}^{\,n}]_{a_1 a_{n+1}}}{\mathrm{d}X_{kl}} &= \delta_{a_1 k}\,\delta_{a_2 l} \prod_{m=2}^{n} X_{a_m a_{m+1}} + X_{a_1 a_2}\,\delta_{a_2 k}\,\delta_{a_3 l} \prod_{m=3}^{n} X_{a_m a_{m+1}} \\
&\quad + \cdots + \left(\prod_{m=1}^{n-1} X_{a_m a_{m+1}}\right)\delta_{a_n k}\delta_{a_{n+1} l} \\
&= \delta_{a_1 k}\,[\boldsymbol{X}^{\,n-1}]_{l a_{n+1}} + X_{a_1 k}\,[\boldsymbol{X}^{\,n-2}]_{l a_{n+1}} + \cdots + [\boldsymbol{X}^{n-1}]_{a_1 k}\,\delta_{l a_{n+1}} \\
&= [\boldsymbol{X}^{\,0}]_{a_1 k}\,[\boldsymbol{X}^{\,n-1}]_{l a_{n+1}} + [\boldsymbol{X}^{1}]_{a_1 k}\,[\boldsymbol{X}^{\,n-2}]_{l a_{n+1}} \\
&\quad + \cdots + [\boldsymbol{X}^{\,n-1}]_{a_1 k}\,[\boldsymbol{X}^{\,0}]_{l a_{n+1}}
\end{aligned}$$

which corresponds exactly to the proposed general formula (B.19). □

With the above result at hand, a general formula for the derivative of the tensor exponential function is established in the following proposition.

Proposition B.2. *The Cartesian components of the derivative of the tensor exponential function at $\boldsymbol{X}$ have the following series representation*

$$[\mathrm{D}\exp(\boldsymbol{X})]_{ijkl} = \sum_{n=1}^{\infty} \frac{1}{n!} \sum_{m=1}^{n} [\boldsymbol{X}^{\,m-1}]_{ik}\,[\boldsymbol{X}^{\,n-m}]_{lj}. \tag{B.20}$$

Proof. Follows by direct differentiation of the series representation (B.4) of the tensor exponential and subsequent use of the general formula (B.19) for the Cartesian components of the derivative of a positive integer power of a tensor. □

Box B.2. Computation of the derivative of the exponential of a generally unsymmetric tensor in $n_{\rm dim}$ dimensions.

HYPLAS procedure:	DEXPMAP

(i) Given $\boldsymbol{X}$ and the numerical tolerance $\epsilon_{\rm tol}$, compute $n_{\rm max}$

(a) initialise $n := 0$

(b) increment counter $n := n + 1$

(c) compute $n!$ and $\boldsymbol{X}^{\,n}$

(d) If $\|\boldsymbol{X}^{\,n}\|/n! < \epsilon_{\rm tol}$ then

set $n_{\rm max} := n$ and GOTO (ii)

Else

GOTO (i.b)

(ii) Compute exponential map derivative

(a) initialise $[\mathrm{D}\exp(\boldsymbol{X})]_{ijkl} := 0, \quad i, j, k, l = 1, \ldots, n_{\rm dim}$

(b) initialise $n := 0$

(c) increment counter $n := n + 1$

(d) add new term to the series. For $i, j, k, l = 1, \ldots, n_{\rm dim}$ compute

$$[\mathrm{D}\exp(\boldsymbol{X})]_{ijkl} := [\mathrm{D}\exp(\boldsymbol{X})]_{ijkl} + \frac{1}{n!}\sum_{m=1}^{n} [\boldsymbol{X}^{\,m-1}]_{ik}\,[\boldsymbol{X}^{\,n-m}]_{lj}$$

(e) If $n = n_{\rm max}$ then EXIT

Else GOTO (ii.c)

B.2.1. COMPUTER IMPLEMENTATION

Exactly as in the computation of the tensor exponential itself, the actual evaluation of the tensor exponential derivative is carried out by means of the truncated series

$$[\mathrm{D}\exp(\boldsymbol{X})]_{ijkl} = \sum_{n=1}^{n_{\rm max}} \frac{1}{n!} \sum_{m=1}^{n} [\boldsymbol{X}^{\,m-1}]_{ik}\,[\boldsymbol{X}^{\,n-m}]_{lj} \tag{B.21}$$

where $n_{\rm max}$ satisfies the accuracy requirement (B.17). The computer implementation of the above expression follows the pseudo-code shown in Box B.2. The procedure has been implemented in HYPLAS subroutine DEXPMP.

B.3. Exponential map integrators

In this section we discuss the use of the exponential map in the numerical solution of the class of initial value problems defined by a tensor differential equation with general format

$$\dot{\boldsymbol{Y}}(t) = \boldsymbol{A}(t)\,\boldsymbol{Y}(t), \tag{B.22}$$

with given initial condition

$$\boldsymbol{Y}(t_0) = \boldsymbol{Y}_0. \tag{B.23}$$

The above class of initial value problems is of particular relevance to finite deformation computational mechanics and generalises the class of problems defined by (B.1) and (B.2) by letting the tensor $\boldsymbol{A}$ now be a function of t.

B.3.1. THE GENERALISED EXPONENTIAL MAP MIDPOINT RULE

The exact solution (B.3) to problem (B.1, B.2) can be used to generate approximate solutions to problem (B.22, B.23). Analogously to the generalised midpoint algorithm based on the standard Euler approximation, it is possible to approximate the solution of (B.22, B.23) by a similar algorithm based on the exponential map – the *generalised exponential map midpoint rule*. Let us consider a generic time interval $[t_n, t_{n+1}]$. Basically, the algorithm approximates $\boldsymbol{Y}_{n+1}$ as the exact solution that would be obtained at t_{n+1} if $\boldsymbol{A}$ were constant over $[t_n, t_{n+1}]$. The corresponding update formula is the following

$$\boldsymbol{Y}_{n+1} = \exp[\Delta t\; \boldsymbol{A}(t_{n+\theta})]\; \boldsymbol{Y}_n, \tag{B.24}$$

where

$$\Delta t \equiv t_{n+1} - t_n, \quad t_{n+\theta} \equiv t_n + \theta\, \Delta t. \tag{B.25}$$

The prescribed parameter θ satisfies

$$0 \leq \theta \leq 1. \tag{B.26}$$

The choice $\theta = 0$ yields the *explicit* exponential map integrator. With choices $\theta = \frac{1}{2}$ and $\theta = 1$ the above formula is called, respectively, the *midpoint* and the *backward* (or fully *implicit*) exponential map integrators. The algorithm is second-order accurate for $\theta = \frac{1}{2}$ and only first-order accurate otherwise.

C LINEARISATION OF THE VIRTUAL WORK

IN this appendix we present the derivation of the linearised version of the virtual work equations. Linearisation of the virtual work gives rise to the *tangent moduli* which take part in the assemblage of the tangent stiffness matrix – a crucial component of the implicit finite element solution procedure described in Chapter 4. Tangent moduli associated with specific material models and constitutive integration procedures are discussed in Parts Two and Three of this book.

C.1. Infinitesimal deformations

We start by considering the simplest case – the virtual work equation under infinitesimal deformations and strains. Consider Problem 3.5 (page 81). Here, we add the assumption that the stress tensor is a function of the current strain only:†

$$\boldsymbol{\sigma} = \boldsymbol{\sigma}(\boldsymbol{\varepsilon}), \tag{C.1}$$

so that the problem is reduced to finding a kinematically admissible field $\boldsymbol{u}$ that satisfies

$$G(\boldsymbol{u}, \boldsymbol{\eta}) = 0, \quad \forall\, \boldsymbol{\eta} \in \mathscr{V}, \tag{C.2}$$

where G is *the virtual work functional* defined as

$$G(\boldsymbol{u}, \boldsymbol{\eta}) = \int_{\Omega} (\boldsymbol{\sigma} : \nabla^s \boldsymbol{\eta} - \boldsymbol{b} \cdot \boldsymbol{\eta})\, \mathrm{d}v - \int_{\partial\Omega_t} \boldsymbol{t} \cdot \boldsymbol{\eta}\, \mathrm{d}a. \tag{C.3}$$

The dependence of G on the unknown function $\boldsymbol{u}$ follows from the constitutive dependence of the stress tensor on the strain tensor which, in turn depends on the *field* $\boldsymbol{u}$ (not only its pointwise value); that is, $\boldsymbol{\sigma}$ is a *functional* of the field $\boldsymbol{u}$:

$$\boldsymbol{\sigma} = \boldsymbol{\sigma}(\nabla^s \boldsymbol{u}). \tag{C.4}$$

Here we want to linearise equation (C.2) with respect to the unknown $\boldsymbol{u}$ about an arbitrary argument $\boldsymbol{u}^*$ (refer to Section 2.6, from page 38, where the concept of linearisation is discussed). In abstract notation, the linearised problem consists in finding the field $\delta\boldsymbol{u}$ such

† This assumption encompasses elastic constitutive models as well as incremental constitutive functions employed in the numerical implementation of path-dependent (dissipative) material models. In the incremental case, $\boldsymbol{\sigma}$ is a function of $\boldsymbol{\varepsilon}$ alone (analogous to an elastic constitutive function) within each time or load step.

that

$$L(\delta \boldsymbol{u}, \boldsymbol{\eta}) \equiv G(\boldsymbol{u}^*, \boldsymbol{\eta}) + \mathrm{D}G(\boldsymbol{u}^*, \boldsymbol{\eta})\,[\delta \boldsymbol{u}] = 0, \quad \forall\, \boldsymbol{\eta} \in \mathscr{V}, \tag{C.5}$$

where L is the linearised virtual work functional and

$$\mathrm{D}G(\boldsymbol{u}^*, \boldsymbol{\eta})\,[\delta \boldsymbol{u}] = \left.\frac{\mathrm{d}}{\mathrm{d}\epsilon}\right|_{\epsilon=0} G(\boldsymbol{u}^* + \epsilon\, \delta \boldsymbol{u}, \boldsymbol{\eta}) \tag{C.6}$$

is the directional derivative of G at $\boldsymbol{u}^*$ in the direction of $\delta \boldsymbol{u}$.

Explicitly, the directional derivative of G is given by

$$\begin{aligned}\mathrm{D}G(\boldsymbol{u}^*, \boldsymbol{\eta})\,[\delta \boldsymbol{u}] &= \left.\frac{\mathrm{d}}{\mathrm{d}\epsilon}\right|_{\epsilon=0} \int_{\Omega} [\boldsymbol{\sigma}(\boldsymbol{\varepsilon}(\epsilon)) : \nabla^s \boldsymbol{\eta} - \boldsymbol{b} \cdot \boldsymbol{\eta}]\, \mathrm{d}v - \int_{\partial\Omega_t} \boldsymbol{t} \cdot \boldsymbol{\eta}\, \mathrm{d}a \\ &= \left.\frac{\mathrm{d}}{\mathrm{d}\epsilon}\right|_{\epsilon=0} \int_{\Omega} \boldsymbol{\sigma}(\boldsymbol{\varepsilon}(\epsilon)) : \nabla^s \boldsymbol{\eta}\, \mathrm{d}v,\end{aligned} \tag{C.7}$$

where we have used the definition

$$\boldsymbol{\varepsilon}(\epsilon) = \nabla^s(\boldsymbol{u}^* + \epsilon\, \delta \boldsymbol{u}) = \boldsymbol{\varepsilon}^* + \epsilon\, \nabla^s \delta \boldsymbol{u}, \tag{C.8}$$

with

$$\boldsymbol{\varepsilon}^* = \nabla^s \boldsymbol{u}^* \tag{C.9}$$

denoting the strain tensor field at $\boldsymbol{u}^*$.

The infinitesimal tangent modulus

Straightforward application of the chain rule to the functional $\boldsymbol{\sigma}$ in (C.7) yields

$$\mathrm{D}G(\boldsymbol{u}^*, \boldsymbol{\eta})\,[\delta \boldsymbol{u}] = \int_{\Omega} \mathsf{D} : \nabla^s \delta \boldsymbol{u} : \nabla^s \boldsymbol{\eta}\, \mathrm{d}v. \tag{C.10}$$

The fourth-order tensor D is the *infinitesimal tangent modulus* defined as

$$\mathsf{D} \equiv \left.\frac{\partial \boldsymbol{\sigma}}{\partial \boldsymbol{\varepsilon}}\right|_{\boldsymbol{\varepsilon}^*}. \tag{C.11}$$

The linearised virtual work equation

With the above at hand, we arrive at the following final expression for the linearised virtual work equation at $\boldsymbol{u}^*$:

$$\int_{\Omega} \mathsf{D} : \nabla^s \delta \boldsymbol{u} : \nabla^s \boldsymbol{\eta}\, \mathrm{d}v = -\int_{\Omega} (\boldsymbol{\sigma} : \nabla^s \boldsymbol{\eta} - \boldsymbol{b} \cdot \boldsymbol{\eta})\, \mathrm{d}v + \int_{\partial\Omega_t} \boldsymbol{t} \cdot \boldsymbol{\eta}\, \mathrm{d}a, \quad \forall\, \boldsymbol{\eta} \in \mathscr{V}, \tag{C.12}$$

where $\boldsymbol{\sigma}$ is the stress corresponding to the field $\boldsymbol{u}^*$.

C.2. Finite strains and deformations

C.2.1. MATERIAL DESCRIPTION

Now let us consider the finite strain case. We assume that the first Piola–Kirchhoff stress tensor is a function of the deformation gradient:

$$\boldsymbol{P} = \boldsymbol{P}(\boldsymbol{F}). \tag{C.13}$$

The material version of the virtual work functional under large deformations and strains is given by

$$G(\boldsymbol{u}, \boldsymbol{\eta}) = \int_{\Omega} [\boldsymbol{P} : \nabla_p \boldsymbol{\eta} - \bar{\boldsymbol{b}} \cdot \boldsymbol{\eta}] \, \mathrm{d}v - \int_{\partial\Omega_t} \bar{\boldsymbol{t}} \cdot \boldsymbol{\eta} \, \mathrm{d}a, \tag{C.14}$$

where $\bar{\boldsymbol{b}}$ and $\bar{\boldsymbol{t}}$ denote, respectively, the reference body force and surface traction fields. In the above, $\boldsymbol{P}$ is a functional of the displacement field due to its constitutive dependence on the deformation gradient $\boldsymbol{F} = \boldsymbol{I} + \nabla_p \boldsymbol{u}$. The material description of the virtual work equation for the unknown displacement field $\boldsymbol{u}$ under finite deformations is given by (C.2) with the functional G defined by (C.14).

The linearisation of equation (C.2, C.14) at a state defined by the field $\boldsymbol{u}^*$ has the representation (C.5), with the directional derivative (C.6) in the present case being given by‡

$$\begin{aligned} \mathrm{D}G(\boldsymbol{u}^*, \boldsymbol{\eta}) \, [\delta \boldsymbol{u}] &= \left.\frac{\mathrm{d}}{\mathrm{d}\epsilon}\right|_{\epsilon=0} \int_{\Omega} [\boldsymbol{P}(\boldsymbol{F}\,(\epsilon)) : \nabla \boldsymbol{\eta} - \bar{\boldsymbol{b}} \cdot \boldsymbol{\eta}] \, \mathrm{d}v - \int_{\partial\Omega_t} \bar{\boldsymbol{t}} \cdot \boldsymbol{\eta} \, \mathrm{d}a \\ &= \left.\frac{\mathrm{d}}{\mathrm{d}\epsilon}\right|_{\epsilon=0} \int_{\Omega} \boldsymbol{P}(\boldsymbol{F}\,(\epsilon)) : \nabla \boldsymbol{\eta} \, \mathrm{d}v, \end{aligned} \tag{C.15}$$

where we have defined

$$\boldsymbol{F}\,(\epsilon) = \boldsymbol{I} + \nabla_p (\boldsymbol{u}^* + \epsilon \, \delta \boldsymbol{u}) = \boldsymbol{F}^{\,*} + \epsilon \, \nabla_p \delta \boldsymbol{u}, \tag{C.16}$$

with

$$\boldsymbol{F}^{\,*} = \boldsymbol{I} + \nabla_p \boldsymbol{u}^* \tag{C.17}$$

denoting the deformation gradient for the displacement field $\boldsymbol{u}^*$.

The material tangent modulus

With a straightforward application of the chain rule to the above expression for the directional derivative, we obtain

$$\mathrm{D}G(\boldsymbol{u}^*, \boldsymbol{\eta}) \, [\delta \boldsymbol{u}] = \int_{\Omega} \mathsf{A} : \nabla_p \delta \boldsymbol{u} : \nabla_p \boldsymbol{\eta} \, \mathrm{d}v \tag{C.18}$$

where

$$\mathsf{A} \equiv \left.\frac{\partial \boldsymbol{P}}{\partial \boldsymbol{F}}\right|_{\boldsymbol{F}^*} \tag{C.19}$$

is generally termed the *material tangent modulus*. For materials with an elastic constitutive law, this is also referred to as the *first elasticity tensor*.

‡We have implicitly assumed here that the prescribed body force and surface traction fields are independent of the displacement field. If the applied loads are *configuration-dependent*, i.e. functions of the displacement field, then extra contributions associated to the linearisation of such terms have to be added to the directional derivative.

The linearised virtual work equation in material description

Finally, expressions (C.5), (C.14) and (C.18) yield the material description of the linearised virtual work equation for large-strain problems:

$$\int_\Omega \mathsf{A} : \nabla_p \delta \boldsymbol{u} : \nabla_p \boldsymbol{\eta} \, \mathrm{d}v = -\int_\Omega (\boldsymbol{P} : \nabla_p \boldsymbol{\eta} - \bar{\boldsymbol{b}} \cdot \boldsymbol{\eta}) \, \mathrm{d}v + \int_{\partial\Omega_t} \bar{\boldsymbol{t}} \cdot \boldsymbol{\eta} \, \mathrm{d}a, \quad \forall\, \boldsymbol{\eta} \in \mathscr{V}. \tag{C.20}$$

C.2.2. SPATIAL DESCRIPTION

We consider now the spatial version of the virtual work equation. The spatial virtual work functional is defined by

$$G(\boldsymbol{u}, \boldsymbol{\eta}) = \int_{\varphi(\Omega)} [\boldsymbol{\sigma} : \nabla_x \boldsymbol{\eta} - \boldsymbol{b} \cdot \boldsymbol{\eta}] \, \mathrm{d}v - \int_{\varphi(\partial\Omega_t)} \boldsymbol{t} \cdot \boldsymbol{\eta} \, \mathrm{d}a, \tag{C.21}$$

where $\boldsymbol{b}$ and $\boldsymbol{t}$ are, respectively, forces per unit volume and area of the deformed configuration of the body in question and $\boldsymbol{\sigma}$ is the Cauchy stress tensor which, for an elastic constitutive model, is a function of the deformation gradient:

$$\boldsymbol{\sigma} = \boldsymbol{\sigma}(\boldsymbol{F}) = \boldsymbol{\sigma}(\boldsymbol{I} + \nabla_p \boldsymbol{u}). \tag{C.22}$$

To derive the corresponding linearised equilibrium equation, it is worth recalling that the material and spatial virtual work functionals are equivalent, with the transformation between (C.14) and (C.21) requiring only straightforward use of the standard identities

$$\nabla_x \boldsymbol{a} = \nabla_p \boldsymbol{a} \, \boldsymbol{F}^{-1}, \quad \int_{\varphi(\Omega)} a(\boldsymbol{x}) \, \mathrm{d}v = \int_\Omega J(\boldsymbol{p}) a(\boldsymbol{\varphi}(\boldsymbol{p})) \, \mathrm{d}v, \tag{C.23}$$

valid for any vector field $\boldsymbol{a}$ and scalar field a. Likewise, the spatial version of the virtual work directional derivative can be obtained from its material counterpart (C.18) simply by replacing the material description with the corresponding spatial description of the relevant fields.

The spatial tangent modulus

With use of the above identities in (C.18), we obtain

$$\mathrm{D}G(\boldsymbol{u}^*, \boldsymbol{\eta}) \, [\delta \boldsymbol{u}] = \int_{\varphi(\Omega)} \frac{1}{J} \mathsf{A} : (\nabla_x \delta \boldsymbol{u} \, \boldsymbol{F}) : (\nabla_x \boldsymbol{\eta} \, \boldsymbol{F}) \, \mathrm{d}v. \tag{C.24}$$

By regrouping the terms of the above integrand, the directional derivative may be equivalently written as

$$\mathrm{D}G(\boldsymbol{u}^*, \boldsymbol{\eta}) \, [\delta \boldsymbol{u}] = \int_{\varphi(\Omega)} \mathsf{a} : \nabla_x \delta \boldsymbol{u} : \nabla_x \boldsymbol{\eta} \, \mathrm{d}v, \tag{C.25}$$

where a is termed generally the *spatial tangent modulus*. Strictly under elasticity, it is often called the *spatial elasticity tensor*. It is defined by the Cartesian components

$$\mathsf{a}_{ijkl} = \frac{1}{J} \mathsf{A}_{imkn} \, F_{jm} \, F_{ln}. \tag{C.26}$$

In expression (C.26) we have

$$\mathsf{A}_{imkn} = \frac{\partial P_{im}}{\partial F_{kn}}, \tag{C.27}$$

or, in view of the definition $\boldsymbol{P} = J\boldsymbol{\sigma}\,\boldsymbol{F}^{-T} = \boldsymbol{\tau}\,\boldsymbol{F}^{-T}$,

$$\begin{aligned}\mathsf{A}_{imkn} &= \frac{\partial}{\partial F_{kn}}(\tau_{ip}F^{-1}_{mp}) \\ &= \frac{\partial \tau_{ip}}{\partial F_{kn}}\,F^{-1}_{mp} + \tau_{ip}\,\frac{\partial F^{-1}_{mp}}{\partial F_{kn}}.\end{aligned} \tag{C.28}$$

From relation (vii) of Section 2.5.4 (from page 35) for the derivative of the inverse of a tensor, the derivative in the last term of the right-hand side of the above equation is given by

$$\frac{\partial F^{-1}_{mp}}{\partial F_{kn}} = -F^{-1}_{mk}\,F^{-1}_{np}, \tag{C.29}$$

so that we obtain

$$\mathsf{A}_{imkn} = \frac{\partial \tau_{ip}}{\partial F_{kn}}\,F^{-1}_{mp} - \tau_{ip}\,F^{-1}_{mk}\,F^{-1}_{np}. \tag{C.30}$$

Finally, by substituting (C.30) into (C.26) we obtain, after straightforward manipulations, the following equivalent expression for the components of the spatial tangent modulus:

$$\mathsf{a}_{ijkl} = \frac{1}{J}\frac{\partial \tau_{ij}}{\partial F_{km}}\,F_{lm} - \sigma_{il}\,\delta_{jk}. \tag{C.31}$$

The linearised virtual work equation in spatial description

Finally, expressions (C.5), (C.21) and (C.25) lead to the spatial version of the linearised virtual work equation:

$$\int_{\varphi(\Omega)} \mathsf{a} : \nabla_x \delta\boldsymbol{u} : \nabla_x \boldsymbol{\eta}\,\mathrm{d}v = -\int_{\varphi(\Omega)} [\boldsymbol{\sigma} : \nabla_x \boldsymbol{\eta} - \boldsymbol{b}\cdot\boldsymbol{\eta}]\,\mathrm{d}v + \int_{\varphi(\partial\Omega_t)} \boldsymbol{t}\cdot\boldsymbol{\eta}\,\mathrm{d}a, \quad \forall\,\boldsymbol{\eta}\in\mathscr{V}. \tag{C.32}$$

D ARRAY NOTATION FOR COMPUTATIONS WITH TENSORS

This appendix describes the handling of second and fourth-order tensors in finite element computer programs. It should be particularly helpful to those wishing to follow, in program HYPLAS, the implementation of techniques discussed in this book that have been presented almost exclusively in compact tensorial notation.

In finite element computer programs, the components of a symmetric second-order tensor are usually stored as a single column array, whereas fourth-order tensor components are stored in two-dimensional arrays. By arranging the relevant components consistently, operations such as internal products between tensors and products between fourth and second-order tensors can be conveniently carried out in the computer program as matrix products.

The order in which components of a tensor can be stored in array format is not unique. In the following, we show the convention adopted in many finite element programs and, in particular, in the program HYPLAS.

D.1. Second-order tensors

Let us start with second-order tensors. Expression (2.27) (page 21) shows the matrix representation of a generic tensor in terms of its Cartesian components. Here we shall be concerned only with symmetric tensors (which are of relevance for finite element computations). Second-order symmetric tensors will be converted into single column arrays and their actual single array representation will depend on whether the tensor is a stress-like or strain-like quantity. Let us start by considering the stress tensor, $\boldsymbol{\sigma}$, in plane stress and plane strain problems. In this case, the in-plane components of the matrix $[\boldsymbol{\sigma}]$ will be converted into a single column array $\boldsymbol{\upsigma}$ (the computer array representation will be denoted here by *upright* bold-faced symbols) according to the rule

$$[\boldsymbol{\sigma}] = \begin{bmatrix} \sigma_{11} & \sigma_{12} \\ \sigma_{12} & \sigma_{22} \end{bmatrix} \longrightarrow \boldsymbol{\upsigma} = \begin{bmatrix} \sigma_{11} \\ \sigma_{22} \\ \sigma_{12} \end{bmatrix}. \tag{D.1}$$

Only the three relevant independent components are stored in $\boldsymbol{\sigma}$. In axisymmetric problems we store, in addition to the above, the component σ_{33}; that is

$$[\boldsymbol{\sigma}] = \begin{bmatrix} \sigma_{11} & \sigma_{12} & 0 \\ \sigma_{12} & \sigma_{22} & 0 \\ 0 & 0 & \sigma_{33} \end{bmatrix} \longrightarrow \boldsymbol{\sigma} = \begin{bmatrix} \sigma_{11} \\ \sigma_{22} \\ \sigma_{12} \\ \sigma_{33} \end{bmatrix}, \tag{D.2}$$

where the index 3 is associated with the circumferential direction. It should be noted here that the σ_{33} stress is generally non-zero also under plane strain conditions. In this case, the stress may be stored as in the above (i.e. including σ_{33}), but the last element of the stress array will be ignored in product operations such as (D.5).

In three-dimensions (this case is not implemented in HYPLAS), we have the conversion rule

$$[\boldsymbol{\sigma}] = \begin{bmatrix} \sigma_{11} & \sigma_{12} & \sigma_{13} \\ \sigma_{12} & \sigma_{22} & \sigma_{23} \\ \sigma_{13} & \sigma_{23} & \sigma_{33} \end{bmatrix} \longrightarrow \boldsymbol{\sigma} = \begin{bmatrix} \sigma_{11} \\ \sigma_{22} \\ \sigma_{33} \\ \sigma_{12} \\ \sigma_{23} \\ \sigma_{13} \end{bmatrix}. \tag{D.3}$$

Let us now consider the strain tensor, $\boldsymbol{\varepsilon}$. The rule for storage in this case is

$$[\boldsymbol{\varepsilon}] = \begin{bmatrix} \varepsilon_{11} & \varepsilon_{12} & \varepsilon_{13} \\ \varepsilon_{12} & \varepsilon_{22} & \varepsilon_{23} \\ \varepsilon_{13} & \varepsilon_{23} & \varepsilon_{33} \end{bmatrix} \longrightarrow \boldsymbol{\varepsilon} = \begin{bmatrix} \varepsilon_{11} \\ \varepsilon_{22} \\ \varepsilon_{33} \\ 2\varepsilon_{12} \\ 2\varepsilon_{23} \\ 2\varepsilon_{13} \end{bmatrix}. \tag{D.4}$$

Note that the shear components have been multiplied by a factor of two; that is, $\boldsymbol{\varepsilon}$ is the array of *engineering* strains. The reason for this is that, in this way, the internal product between a stress- and a strain-like tensor can be computed as a matrix product

$$\boldsymbol{\sigma} : \boldsymbol{\varepsilon} = \boldsymbol{\sigma}^T \boldsymbol{\varepsilon}. \tag{D.5}$$

If we denote by $\delta\boldsymbol{\varepsilon}$ a virtual strain tensor, i.e. the symmetric gradient virtual displacement field, by applying the above conversion rules the corresponding virtual work reads

$$\boldsymbol{\sigma} : \delta\boldsymbol{\varepsilon} = \boldsymbol{\sigma}^T \delta\boldsymbol{\varepsilon}. \tag{D.6}$$

In plane problems, the general conversion rule for strains is

$$[\boldsymbol{\varepsilon}] = \begin{bmatrix} \varepsilon_{11} & \varepsilon_{12} \\ \varepsilon_{12} & \varepsilon_{22} \end{bmatrix} \longrightarrow \boldsymbol{\varepsilon} = \begin{bmatrix} \varepsilon_{11} \\ \varepsilon_{22} \\ 2\varepsilon_{12} \end{bmatrix}. \tag{D.7}$$

However, note that in elastoplasticity under plane strain, even though the *total* strain component ε_{33} vanishes, the corresponding elastic and plastic components do not. Thus, we adopt the following storage rule

$$[\boldsymbol{\varepsilon}^e] = \begin{bmatrix} \varepsilon^e_{11} & \varepsilon^e_{12} & 0 \\ \varepsilon^e_{12} & \varepsilon^e_{22} & 0 \\ 0 & 0 & \varepsilon^e_{33} \end{bmatrix} \longrightarrow \boldsymbol{\varepsilon}^e = \begin{bmatrix} \varepsilon^e_{11} \\ \varepsilon^e_{22} \\ 2\varepsilon^e_{12} \\ \varepsilon^e_{33} \end{bmatrix} \tag{D.8}$$

for the elastic strain (and the plastic strain). Axisymmetric implementations follow the above rule also for the array conversion of the total strain tensor, $\boldsymbol{\varepsilon}$.

D.2. Fourth-order tensors

We now consider fourth-order tensors. Let D be a tangent modulus tensor with Cartesian components D_{ijkl} on the basis $\{\boldsymbol{e}_i\}$

$$\mathsf{D} = \mathsf{D}_{ijkl}\, \boldsymbol{e}_i \otimes \boldsymbol{e}_j \otimes \boldsymbol{e}_k \otimes \boldsymbol{e}_l. \tag{D.9}$$

In tensorial compact form, the tangent stress–strain relation reads

$$\mathrm{d}\boldsymbol{\sigma} = \mathsf{D} : \mathrm{d}\boldsymbol{\varepsilon}. \tag{D.10}$$

Fourth-order tensors will be stored in two-dimensional arrays. In its array form, the components of D in plane strain problems (i.e. with $i, j = 1, 2$) will be arranged as

$$\mathbf{D} = \begin{bmatrix} \mathsf{D}_{1111} & \mathsf{D}_{1122} & \mathsf{D}_{1112} \\ \mathsf{D}_{2211} & \mathsf{D}_{2222} & \mathsf{D}_{2212} \\ \mathsf{D}_{1211} & \mathsf{D}_{1222} & \mathsf{D}_{1212} \end{bmatrix}, \tag{D.11}$$

so that the tangential relation between the in-plane stress array can be represented as the matrix product

$$\mathrm{d}\mathbf{\sigma} = \mathbf{D}\, \mathrm{d}\mathbf{\varepsilon}. \tag{D.12}$$

That the above is equivalent to (D.10) is left as an exercise for the interested reader. Note that in elasticity and associative plasticity problems the tensor D has the symmetries

$$\mathsf{D}_{ijkl} = \mathsf{D}_{jikl} = \mathsf{D}_{jilk} = \mathsf{D}_{klij}. \tag{D.13}$$

In such cases, the conversion rule produces a *symmetric* two-dimensional matrix.

In axisymmetric problems we have

$$\mathbf{D} = \begin{bmatrix} \mathsf{D}_{1111} & \mathsf{D}_{1122} & \mathsf{D}_{1112} & \mathsf{D}_{1133} \\ \mathsf{D}_{2211} & \mathsf{D}_{2222} & \mathsf{D}_{2212} & \mathsf{D}_{2233} \\ \mathsf{D}_{1211} & \mathsf{D}_{1222} & \mathsf{D}_{1212} & \mathsf{D}_{1233} \\ \mathsf{D}_{3311} & \mathsf{D}_{3322} & \mathsf{D}_{3312} & \mathsf{D}_{3333} \end{bmatrix}, \tag{D.14}$$

and in three-dimensions,

$$
\mathbf{D} = \begin{bmatrix} \mathsf{D}_{1111} & \mathsf{D}_{1122} & \mathsf{D}_{1133} & \mathsf{D}_{1112} & \mathsf{D}_{1123} & \mathsf{D}_{1113} \\ \mathsf{D}_{2211} & \mathsf{D}_{2222} & \mathsf{D}_{2233} & \mathsf{D}_{2212} & \mathsf{D}_{2223} & \mathsf{D}_{2213} \\ \mathsf{D}_{3311} & \mathsf{D}_{3322} & \mathsf{D}_{3333} & \mathsf{D}_{3312} & \mathsf{D}_{3323} & \mathsf{D}_{3313} \\ \mathsf{D}_{1211} & \mathsf{D}_{1222} & \mathsf{D}_{1233} & \mathsf{D}_{1212} & \mathsf{D}_{1223} & \mathsf{D}_{1213} \\ \mathsf{D}_{2311} & \mathsf{D}_{2322} & \mathsf{D}_{2333} & \mathsf{D}_{2312} & \mathsf{D}_{2323} & \mathsf{D}_{2313} \\ \mathsf{D}_{1311} & \mathsf{D}_{1322} & \mathsf{D}_{1333} & \mathsf{D}_{1312} & \mathsf{D}_{1323} & \mathsf{D}_{1313} \end{bmatrix}. \tag{D.15}
$$

Note that, according to the above rule, the fourth-order symmetric identity tensor defined by (2.108) (page 31) is represented in plane problems as

$$
\mathbf{I}_S = \begin{bmatrix} 1 & 0 & 0 \\ & 1 & 0 \\ \text{sym} & & \frac{1}{2} \end{bmatrix}. \tag{D.16}
$$

In using the above representation in computations, account should be taken of the fact that

$$
\mathbf{I}_S \boldsymbol{\varepsilon} = \begin{bmatrix} \varepsilon_{11} \\ \varepsilon_{22} \\ \varepsilon_{12} \end{bmatrix} \neq \boldsymbol{\varepsilon}. \tag{D.17}
$$

For axisymmetric problems,

$$
\mathbf{I}_S = \begin{bmatrix} 1 & 0 & 0 & 0 \\ & 1 & 0 & 0 \\ \text{sym} & & \frac{1}{2} & 0 \\ & & 0 & 1 \end{bmatrix}. \tag{D.18}
$$

In three-dimensions, we have

$$
\mathbf{I}_S = \begin{bmatrix} 1 & 0 & 0 & 0 & 0 & 0 \\ & 1 & 0 & 0 & 0 & 0 \\ & & 1 & 0 & 0 & 0 \\ & & & \frac{1}{2} & 0 & 0 \\ & \text{sym} & & & \frac{1}{2} & 0 \\ & & & & & \frac{1}{2} \end{bmatrix}. \tag{D.19}
$$

D.2.1. OPERATIONS WITH NON-SYMMETRIC TENSORS

Note (refer to expressions (C.18), page 755, and (C.25)) that the product between a fourth-order tensor (the material or spatial modulus) and a generally *non-symmetric* second-order tensor (the *full* material or spatial gradient of a vector field) arises naturally in the linearisation of the virtual work equation under large deformations. In such cases, the above representation cannot be used. Modifications in the array conversion rules are needed to allow such products to be carried out in the computer program as matrix products. Let us use the spatial version (C.32) of the linearised virtual work as an example. We start by defining the second-order tensors

$$\boldsymbol{T} \equiv \nabla_x \delta \boldsymbol{u}, \quad \boldsymbol{U} \equiv \nabla_x \boldsymbol{\eta}, \tag{D.20}$$

respectively, as the (full) gradients of $\delta \boldsymbol{u}$ and the virtual displacement field $\boldsymbol{\eta}$. In plane problems, we will adopt the following computer array representation:

$$\mathbf{T} = \begin{bmatrix} T_{11} \\ T_{21} \\ T_{12} \\ T_{22} \end{bmatrix}, \tag{D.21}$$

with the same rule applying for $\boldsymbol{U}$, and

$$\mathbf{a} = \begin{bmatrix} \mathsf{a}_{1111} & \mathsf{a}_{1121} & \mathsf{a}_{1112} & \mathsf{a}_{1122} \\ \mathsf{a}_{2111} & \mathsf{a}_{2121} & \mathsf{a}_{2112} & \mathsf{a}_{2122} \\ \mathsf{a}_{1211} & \mathsf{a}_{1221} & \mathsf{a}_{1212} & \mathsf{a}_{1222} \\ \mathsf{a}_{2211} & \mathsf{a}_{2221} & \mathsf{a}_{2212} & \mathsf{a}_{2222} \end{bmatrix}. \tag{D.22}$$

With the above notation, the integrand on the right-hand side of (C.25) has the representation

$$\mathsf{a} : \boldsymbol{T} : \boldsymbol{U} = \mathbf{U}^T \mathbf{a}\, \mathbf{T}. \tag{D.23}$$

Again, note that when the fourth-order tensor has the major symmetries

$$\mathsf{a}_{ijkl} = \mathsf{a}_{klij}, \tag{D.24}$$

which occurs in hyperelasticity and hyperelastic-based associative plasticity, the conversion rule gives a symmetric two-dimensional matrix representation.

In axisymmetric problems, we adopt

$$\mathbf{T} = \begin{bmatrix} T_{11} \\ T_{21} \\ T_{12} \\ T_{22} \\ T_{33} \end{bmatrix} \tag{D.25}$$

and

$$\mathbf{a} = \begin{bmatrix} \mathsf{a}_{1111} & \mathsf{a}_{1121} & \mathsf{a}_{1112} & \mathsf{a}_{1122} & \mathsf{a}_{1133} \\ \mathsf{a}_{2111} & \mathsf{a}_{2121} & \mathsf{a}_{2112} & \mathsf{a}_{2122} & \mathsf{a}_{2133} \\ \mathsf{a}_{1211} & \mathsf{a}_{1221} & \mathsf{a}_{1212} & \mathsf{a}_{1222} & \mathsf{a}_{1233} \\ \mathsf{a}_{2211} & \mathsf{a}_{2221} & \mathsf{a}_{2212} & \mathsf{a}_{2222} & \mathsf{a}_{2233} \\ \mathsf{a}_{3311} & \mathsf{a}_{3321} & \mathsf{a}_{3312} & \mathsf{a}_{3322} & \mathsf{a}_{3333} \end{bmatrix}. \tag{D.26}$$

An analogous conversion rule can be defined for three-dimensional problems.

REFERENCES

Alfano, G., de Angelis F. and Rosati, L. 2001. General Solution Procedures in Elasto/Viscoplasticity. *Comp. Meth. Appl. Mech. Engng*, **190**, 5123–5147.

Anderson, H. 2003. An Implicit Formulation of the Bodner–Partom Constitutive Equations. *Comp. Struct*, **81**, 1405–1414.

ANSI. 1978. *American National Standard Programming Language FORTRAN*. New York.

Aravas, N. 1987. On the Numerical Integration of a Class of Pressure-Dependent Plasticity Models. *Int. J. Num. Meth. Engng*, **24**, 1395–1416.

Argyris, J.H. and Kleiber, M. 1977. Incremental Formulation in Nonlinear Mechanics and Large Strain Elasto-plasticity – Natural Approach. Part I. *Comp. Meth. Appl. Mech. Engng*, **11**, 215–247.

Argyris, J.H. and Symeonidis, Sp. 1981. Nonlinear Finite Element Analysis of Elastic Systems Under Nonconservative Loading – Natural Formulation. Part I. Quasistatic Problems. *Comp. Meth. Appl. Mech. Engng*, **26**, 75–123.

Argyris, J.H., Doltsinis, J.St. and Kleiber, M. 1978. Incremental Formulation in Nonlinear Mechanics and Large Strain Elasto-plasticity – Natural Approach. Part II. *Comp. Meth. Appl. Mech. Engng*, **14**, 259–294.

Armero, F. and Oller, S. 2000. A General Framework for Continuum Damage Models. I. Infinitesimal Plastic Damage Models in Stress Space. *Int. J. Solids Structs*, **37**, 7409–7436.

Armero, F. and Pérez-Foguet, A. 2002. On the Formulation of Closest-Point Projection Algorithms in Elastoplasticity – Part I: The Variational Structure. *Int. J. Numer. Meth. Engng*, **53**, 297–329.

Armstrong, P.J. and Frederick, C.O. 1966. *A Mathematical Representation of the Multiaxial Bauschinger Effect*. CEGB: Report RD/B/N 731.

Arridge, R.G.C. 1985. *An Introduction to Polymer Mechanics*. London: Taylor and Francis.

Asaro, R.J. 1983. Micromechanics of Crystals and Polycrystals. *Adv. Appl. Mech.*, **23**, 1–115.

Asaro, R.J. and Needleman, A. 1985. Texture Development and Strain Hardening in Rate Dependent Polycrystals. *Acta Metall.*, **33**(6), 923–953.

Ascher, U.M. and Petzold, L.R. 1998. *Computer Methods for Ordinary Differential Equations and Differential-Algebraic Equations*. Philadelphia: SIAM.

Atluri, S.N. 1984. On Constitutive Relations at Finite Strains: Hypo-Elasticity and Elasto-Plasticity with Isotropic or Kinematic Hardening. *Comp. Meth. Appl. Mech. Engng*, **43**, 137–171.

Auricchio, F., Beirão da Veiga, L., Lovadina, C. and Reali, A. 2005. A Stability Study of Some Mixed Finite Elements for Large Deformation Elasticity Problems. *Comp. Meth. Appl. Mech. Engng*, **194**, 1075–1092.

Barlat, F. and Lian, J. 1989. Plastic Behavior and Stretchability of Sheet Metals. Part I: A Yield Function for Orthotropic Sheets Under Plane Stress Condition. *Int. J. Plast.*, **5**, 51–66.

Computational Methods for Plasticity: Theory and Applications EA de Souza Neto, D Perić and DRJ Owen

Basista, M., Krajčinović, D. and Šumarac, D. 1992. Micromechanics, Phenomenology and Statistics of Brittle Deformation. *Pages 1479–1490 of:* Owen, D.R.J., Oñate, E. and Hinton, E. (eds), *Computational Plasticity: Fundamentals and Applications – Proceedings of the Third International Conference held in Barcelona, 6–10 April 1992*. Swansea: Pineridge Press.

Bataille, J. and Kestin, J. 1979. Irreversible Processes and Physical Interpretation of Rational Thermodynamics. *J. Non-Equilib. Thermodyn.*, **4**, 229–258.

Bathe, K.-J. 1996. *Finite Element Procedures*. Englewood Cliffs, New Jersey: Prentice-Hall.

Bazaraa, M. and Shetty, C. 1979. *Non-linear Programming: Theory and Applications*. New York: Wiley.

Beatty, M.F. 1987. Topics in Finite Elasticity: Hyperelasticity of Rubber, Elastomers, and Biological Tissues – with Examples. *Appl. Mech. Rev.*, **40**, 1699–1734.

Belytschko, T. and Bindeman, L.P. 1991. Assumed Strain Stabilization of the 4-Node Quadrilateral with 1-Point Quadrature for Nonlinear Problems. *Comp. Meth. Appl. Mech. Engng*, **88**, 311–340.

Belytschko, T. and Bindeman, L.P. 1993. Assumed Strain Stabilization of the Eight Node Hexahedral Element. *Comp. Meth. Appl. Mech. Engng*, **105**, 225–260.

Belytschko, T., Liu, W.-K. and Moran, B. 2000. *Nonlinear Finite Elements for Continua and Structures*. Chichester: Wiley.

Belytschko, T., Ong, J.S.-J., Liu, W.K. and Kennedy, J.M. 1984. Hourglass Control in Linear and Nonlinear Problems. *Comp. Meth. Appl. Mech. Engng*, **43**, 251–276.

Benallal, A., Billardon, R. and Doghri, I. 1988. An Integration Algorithm and the Corresponding Consistent Tangent Operator for Fully Coupled Elastoplastic and Damage Equations. *Comm. Appl. Num. Meth.*, **4**, 731–740.

Benallal, A., Billardon, R. and Lemaitre, J. 1991. Continuum Damage Mechanics and Local Approach to Fracture: Numerical Procedures. *Comp. Meth. Appl. Mech. Engng*, **92**, 141–155.

Benallal, A., Billardon, R., Doghri, I. and Moret-Bailly, L. 1987. Crack Initiation and Propagation Analyses Taking into Account Initial Strain Hardening and Damage Fields. *Pages 337–351 of:* Luxmore, A.L., Owen, D.R.J., Rajapakse, Y.P.S. and Kanninen, M.F. (eds), *Numerical Methods in Fracture Mechanics: Proceedings of the Fourth International Conference held in San Antonio, Texas, 23–27 March 1987*. Pineridge Press.

Billington, E.W. and Tate, A. 1981. *The Physics of Deformation and Flow*. New York: McGraw-Hill.

Blatz, P.J. and Ko, W.L. 1962. Application of Finite Elasticity to the Deformation of Rubbery Materials. *Trans. Soc. Rheol.*, **6**, 223–251.

Bodner, S.R. and Partom, Y. 1975. Constitutive Equations for Elastic Viscoplastic Strain Hardening. *J. Appl. Mech.*, **42**, 385–389.

Boisse, Ph., Bussy, P. and Ladevèze, P. 1990. A New Approach in Non-linear Mechanics: The Large Time Increment Method. *Int. J. Num. Meth. Engng*, **29**, 647–663.

Boisse, Ph., Ladevèze, P. and Rougee, P. 1989. A Large Time Increment Method for Elastoplastic Problems. *Eur. J. Mech. A/Solids*, **8**(4), 257–275.

Bonet, J. and Wood, R.D. 1997. *Nonlinear Continuum Mechanics*. Cambridge: Cambridge University Press.

Borja, R. and Wren, J.R. 1993. Discrete Micromechanics of Elastoplastic Crystals. *Int. J. Num. Meth. Engng*, **36**, 3815–3840.

Boucard, P.-A., Ladevèze, P., Poss, M. and Rougée, P. 1997. A Nonincremental Approach for Large Displacement Problems. *Comput. Struct.*, **64**(1–4), 499–508.

Boyle, J.T. and Spence, J. 1983. *Stress Analysis for Creep*. London: Butterworths.

Brezzi, F. and Fortin, M. 1991. *Mixed and Hybrid Finite Element Methods*. New York: Springer-Verlag.

Bueche, F. 1960. Molecular Basis for the Mullins Effect. *J. Appl. Poly. Sci.*, **4**(10), 107–114.

Caddemi, S. and Martin, J.B. 1991. Convergence of the Newton–Raphson Algorithm in Elastic-plastic Incremental Analysis. *Int. J. Num. Meth. Engng*, **31**, 177–191.

Carlson, D.E. and Hoger, A. 1986. The Derivative of a Tensor-valued Function of a Tensor. *Quart. Appl. Math.*, **44**(3), 409–423.

Cescotto, S. and Zhu, Y.Y. 1995. Modelling of Ductile Fracture Initiation During Bulk Forming. *Pages 987–998 of:* Owen, D.R.J., Oñate, E. and Hinton, E. (eds), *Computational Plasticity: Fundamentals and Applications – Proceedings of the Fourth International Conference held in Barcelona, 3–6 April 1995*. Swansea: Pineridge Press.

Chaboche, J.L. 1978. *Description thermodynamique et phénoménologique de la viscoplasticité cyclique avec endommagement*. Technical Report 1978–3. Office National d'Etudes et de Recherches Aérospatiales.

Chaboche, J.L. 1981. Continuous Damage Mechanics – A Tool to Describe Phenomena Before Crack Initiation. *Nucl. Engng. Des.*, **64**, 233–247.

Chaboche, J.L. 1984. Anisotropic Creep Damage in the Framework of Continuum Damage Mechanics. *Nucl. Engng. Des.*, **79**, 309–319.

Chaboche, J.L. 1988. Continuum Damage Mechanics: Part I – General Concepts and Part II – Damage Growth, Crack Initiation and Crack Growth. *J. Appl. Mech.*, **55**, 59–72.

Chaboche, J.L. and Cailletaud, G. 1996. Integration Methods for Complex Plastic Constitutive Equations. *Comp. Meth. Appl. Mech. Engng*, **133**, 125–155.

Chadwick, P. and Ogden, R.W. 1971. A Theorem in Tensor Calculus and its Application to Isotropic Elasticity. *Arch. Rat. Mech. Anal.*, **44**, 54–68.

Chakrabarty, J. 1987. *Theory of Plasticity*. London: McGraw-Hill.

Chen, W-F. 1975. *Limit Analysis and Soil Plasticity*. New York: Elsevier.

Chen, W-F. 1982. *Plasticity in Reinforced Concrete*. McGraw-Hill.

Chen, W-F. and Mizuno, E. 1990. *Nonlinear Analysis in Soil Mechanics. Theory and Implementation*. New York: Elsevier.

Chorin, A.J., Hughes, T.J.R., McCracken, M.F. and Marsden, J.E. 1978. Product Formulas and Numerical Algorithms. *Comm. Pure and Appl. Math.*, **31**, 205–256.

Ciarlet, P.G. 1988. *Mathematical Elasticity. Volume I: Three-dimensional Elasticity*. Amsterdam: North-Holland.

Cognard, J.Y. and Ladèveze, P. 1993. A Large Time Increment Approach for Cyclic Viscoplasticity. *Int. J. Plast.*, **9**(2), 141–157.

Cognard, J.Y., Ladèveze, P. and Talbot, P. 1999. A Large Time Increment Approach for Thermo-Mechanical Problems. *Adv. Engng. Software*, **30**(9–11), 583–593.

Coleman, B.D. and Gurtin, M.E. 1967. Thermodynamics with Internal State Variables. *J. Chem. Phys.*, **47**(2), 597–613.

Comi, C., Corigliano, A. and Maier, G. 1991. Extremum Properties of Finite-step Solutions in Elastoplasticity with Nonlinear Mixed Hardening. *Int. J. Solids Struct.*, **27**(8), 965–981.

Cordebois, J.P. and Sidoroff, F. 1979. Damage Induced Elastic Anisotropy. *Pages 761–774 of: Euromech 115 – Villard de Lans – Juin 1979*.

Cordebois, J.P. and Sidoroff, F. 1982. Endomagement Anisotrope en Élasticité et Plasticité. *Journal de Mécanique Théorique et Appliquée*, 45–60. Numéro spécial.

Cormeau, I.C. 1975. Numerical Stability in Quasi-Static Elasto/Visco-Plasticity. *Int. J. Num. Meth. Engng*, **9**, 109–127.

Cosserrat, E. and Cosserrat, F. 1909. *Théorie des corps déformables*. Paris: Hermann.

Cox, A.D., Eason, G. and Hopkins, H.G. 1961. Axially Symmetric Plastic Deformation in Soils. *Trans. R. Soc. Lond. A*, **254**, 1–45.

Crisfield, M.A. 1981. A Fast Incremental/Iterative Solution Procedure that Handles 'Snap-Through'. *Comp. Structs*, **13**, 55–62.

Crisfield, M.A. 1983. An Arc-Length Method Including Line Searches and Accelerations. *Int. J. Num. Meth. Engng*, **19**, 1269–1289.

Crisfield, M.A. 1987. Plasticity Computations Using the Mohr–Coulomb Yield Criterion. *Engng. Comp.*, **4**, 300–308.

Crisfield, M.A. 1991. *Non-linear Finite Element Analysis of Solids and Structures. Volume 1: Essentials*. Chichester: John Wiley & Sons.

Crisfield, M.A. 1997. *Non-linear Finite Element Analysis of Solids and Structures. Volume 2: Advanced Topics*. Chichester: John Wiley & Sons.

Crisfield, M.A. and Moita, G.F. 1996. A Co-rotational Formulation for 2-D Continua Including Incompatible Modes. *Int. J. Numer. Meth. Engng*, **39**, 2619–2633.

Crisfield, M.A., Moita, G.F., Jelenić, G. and Lyons, L.P.R. 1995. Enhanced Lower-Order Element Formulations for Large Strains. *Pages 293–320 of:* Owen, D.R.J., Oñate, E. and Hinton, E. (eds), *Computational Plasticity: Fundamentals and Applications – Proceedings of the Fourth International Conference held in Barcelona, 3–6 April 1995*. Swansea: Pineridge Press.

Cuitiño, A. and Ortiz, M. 1992a. A Material-independent Method for Extending Stress Update Algorithms from Small-strain Plasticity to Finite Plasticity with Multiplicative Kinematics. *Engng. Comp.*, **9**, 437–451.

Cuitiño, A.M. and Ortiz, M. 1992b. Computational Modelling of Single Crystals. *Modelling Simul. Mater. Sci. Engng*, **1**, 225–263.

Dafalias, Y.F. 1984. The Plastic Spin Concept and a Simple Illustration of its Role in Finite Plastic Transformations. *Mech. Mater.*, **3**, 223–233.

Dafalias, Y.F. 1985. The Plastic Spin. *J. Appl. Mech.*, **52**, 865–871.

Dashner, P.A. 1986. Invariance Considerations in Large Strain Elasto-Plasticity. *J. Appl. Mech.*, **53**, 55–60.

de Borst, R. 1987. Integration of Plasticity Equations for Singular Yield Functions. *Comp. Structs*, **26**, 823–829.

de Borst, R. 1991. The Zero-normal-stress Condition in Plane-stress and Shell Elastoplasticity. *Comm. Appl. Numer. Meth.*, **7**, 29–33.

de Borst, R. and Feenstra, P.H. 1990. Studies in Anisotropic Plasticity with Reference to the Hill Criterion. *Int. J. Numer. Meth. Engng*, **29**, 315–336.

de Souza Neto, E.A., Andrade Pires, F.M. and Owen, D.R.J. 2005. F-bar-based Linear Triangles and Tetrahedra for Finite Strain Analysis of Nearly Incompressible Solids. Part I: Formulation and Benchmarking. *Int. J. Numer. Meth. Engng*, **62**(3), 353–383.

de Souza Neto, E.A. and Feng, Y.T. 1999. On the Determination of the Path Direction for Arc-length Methods in the Presence of Bifurcations and 'Snap-Backs'. *Comp. Meth. Appl. Mech. Engng*, **179**, 81–89.

de Souza Neto, E.A. and Perić, D. 1996. A Computational Framework for a Class of Models for Fully Coupled Elastoplastic Damage at Finite Strains with Reference to the Linearization Aspects. *Comp. Meth. Appl. Mech. Engng*, **130**, 179–193.

de Souza Neto, E.A., Perić, D. and Owen, D.R.J. 1992. A Computational Model for Ductile Damage at Finite Strains. *Pages 1425–1441 of:* Owen, D.R.J., Oñate, E. and Hinton, E. (eds), *Computational Plasticity: Fundamentals and Applications – Proceedings of the Third International Conference held in Barcelona, 6–10 April 1992.* Swansea: Pineridge Press.

de Souza Neto, E.A., Perić, D. and Owen, D.R.J. 1993. Some Aspects of Formulation and Implementation of Ductile Damage at Finite Strains. *In: Proceedings of the ACME 93 Conference: Computational Mechanics in the UK.*

de Souza Neto, E.A., Perić, D. and Owen, D.R.J. 1994a. A Model for Elasto-plastic Damage at Finite Strains: Computational Issues and Applications. *Engng. Comp.*, **11**(3), 257–281.

de Souza Neto, E.A., Perić, D. and Owen, D.R.J. 1994b. A Phenomenological Three-dimensional Rate-independent Continuum Damage Model for Highly Filled Polymers: Formulation and Computational Aspects. *J. Mech. Phys. Solids*, **42**(10), 1533–1550.

de Souza Neto, E.A., Perić, D. and Owen, D.R.J. 1998. Continuum Modelling and Numerical Simulation of Material Damage at Finite Strains. *Arch. Comput. Meth. Engng*, **5**, 311–384.

de Souza Neto, E.A., Perić, D., Dutko, M. and Owen, D.R.J. 1996. Design of Simple Low Order Finite Elements for Large Strain Analysis of Nearly Incompressible Solids. *Int. J. Solids Structs*, **33**, 3277–3296.

Desmorat, R. 2000. Dissymétrie de comportement élastique anisotrope couplé ou non à l'endomagement. *C. R. Acad. Sci. Paris, Série II b*, **328**, 445–450.

Dettmer, W. and Reese, S. 2004. On the Theoretical and Numerical Modelling of Armstrong–Friederick Kinematic Hardening in the Finite Strain Regime. *Comp. Meth. Appl. Mech. Engng*, **192**, 87–116.

Dienes, J.K. 1979. On the Analysis of Rotation and Stress Rates in Deforming Bodies. *Acta Mech.*, **32**, 217–232.

Dieter, G. 1986. *Mechanical Metallurgy*. McGraw-Hill.

Dodds, Jr., R.H. 1987. Numerical Techniques for Plasticity Computations in Finite Element Analysis. *Comput. Structs*, **26**(5), 767–779.

Doghri, I. 1995. Numerical Implementation and Analysis of a Class of Metal Plasticity Models Coupled with Ductile Damage. *Int. J. Numer. Meth. Engng*, **38**, 3403–3431.

Doll, S., Hauptmann, R., Schweizerhof, K. and Freischlager, C. 2000. On Volumetric Locking of Low-order Solid and Solid-shell Elements for Finite Elastoviscoplastic Deformations and Selective Reduced Integration. *Engng. Comp.*, **17**, 874–902.

Drucker, D.C. and Prager, W. 1952. Soil Mechanics and Plasticity Analysis of Limit Design. *Quart. J. Appl. Math.*, **10**, 157–162.

Dureisseix, D., Ladevèze, P. and Schrefler, B.A. 2003. A LATIN Computational Strategy for Multi-physics Problems: Application to Poroelasticity. *Int. J. Numer. Meth. Engng*, **56**(10), 1489–1510.

Dutko, M., Perić, D. and Owen, D.R.J. 1993. Universal Anisotropic Yield Criterion Based on Superquadric Functional Representation. Part 1: Algorithmic Issues and Accuracy Analysis. *Comp. Meth. Appl. Mech. Engng*, **109**, 73–93.

Duvaut, G. and Lions, J.L. 1976. *Inequalities in Mechanics and Physics*. Springer-Verlag: Berlin.

Engel, L. and Klingele, H. 1981. *An Atlas of Metal Damage*. Wolfe Science Books.

Eterovic, A.L. and Bathe, K.-J. 1990. A Hyperelastic Based Large Strain Elasto-plastic Constitutive Formulation with Combined Isotropic-kinematic Hardening Using the Logarithmic Stress and Strain Measures. *Int. J. Numer. Meth. Engng*, **30**, 1099–1114.

Eve, R.A., Reddy, B.D. and Rockafellar, R.T. 1990. An Internal Variable Theory of Elastoplasticity based on the Maximum Plastic Work Inequality. *Quart. Appl. Math.*, **48**, 59–83.

Feijóo, R.A. and Zouain, N. 1988. Formulations in Rates and Increments for Elastic-plastic Analysis. *Int. J. Numer. Meth. Engng*, **26**, 2031–2048.

Feng, Y.T., Owen, D.R.J. and Perić, D. 1997. On the Sign of the Determinant of the Structural Stiffness Matrix for Determination of Loading Increment in Arc-length Methods. *Comm. Numer. Meth. Engng*, **13**, 47–49.

Feng, Y.T., Perić, D. and Owen, D.R.J. 1995. Determination of Travel Directions in Path-following Methods. *Math. and Comput. Modelling*, **21**, 43–59.

Feng, Y.T., Perić, D. and Owen, D.R.J. 1996. A New Criterion for Determination of Initial Loading Parameter in Arc-length Methods. *Comput. Structs*, **58**, 479–485.

Flanagan, D.P. and Taylor, L.M. 1987. An Accurate Numerical Algorithm for Stress Integration with Finite Rotations. *Comp. Meth. Appl. Mech. Engng*, **62**, 305–320.

Fletcher, R. 1980. *Practical Methods of Optimization. Volume 1: Unconstrained Problems, and Volume 2: Constrained Problems*. New York: Wiley.

Fonseka, G.U. and Krajčinović, D. 1981. The Continuous Damage Theory of Brittle Materials – Part 2: Uniaxial and Plane Response Modes. *J. Appl. Mech.*, **48**, 816–824.

Fuschi, P., Perić, D. and Owen, D.R.J. 1992. Studies on Generalized Midpoint Integration in Rate-Independent Plasticity with Reference to Plane Stress J_2-Flow Theory. *Comput. Structs*, **43**, 1117–1133.

Gear, C.W. 1971. *Numerical Initial Value Problems in Ordinary Differential Equations*. Englewood Cliffs, New Jersey: Prentice-Hall.

Gelin, J.C. and Mrichcha, A. 1992. Computational Procedures for Finite Strain Elasto Plasticity with Isotropic Damage. *Pages 1401–1412 of:* Owen, D.R.J., Oñate, E. and Hinton, E. (eds), *Computational Plasticity: Fundamentals and Applications – Proceedings of the Third International Conference held in Barcelona, 6–10 April 1992*. Swansea: Pineridge Press.

Gethin, D.T., Lewis, R.W. and Ransing, R.S. 1998. Compaction of Powder via a Deformable Discrete Element Approach. *Pages 45–50 of: Proceedings of the 1998 Powder Metallurgy World Congress and Exhibition*. Shrewsbury, UK: European Powder Metallurgy Association.

Glaser, S. and Armero, F. 1997. On the Formulation of Enhanced Strain Finite Element Methods in Finite Deformations. *Engng. Comp.*, **14**(7), 759–791.

Govindjee, S. and Simo, J. 1991. A Micro-mechanically Based Continuum Damage Model for Carbon Black-filled Rubbers Incorporating Mullins' Effect. *J. Mech. Phys. Solids*, **39**(1), 87–112.

Govindjee, S. and Simo, J.C. 1992. Mullins' Effect and the Strain Amplitude Dependence of the Storage Modulus. *Int. J. Solids Structs*, **20**, 1737–1751.

Green, A.E. and Naghdi, P.M. 1965. A General Theory of an Elasto-plastic Continuum. *Arch. Rat. Mech. Anal.*, **18**, 251–281.

Green, A.E. and Zerna, W. 1954. *Theoretical Elasticity*. Oxford University Press.

Green, A.P. 1953. The Plastic Yielding of Notched Bars Due to Bending. *Quart. J. Mech. Appl. Math.*, **6**, 223–239.

Green, A.P. 1954. A Theory of the Plastic Yielding Due to Bending of Cantilevers and Fixed-ended Beams. Part I. *J. Mech. Phys. Solids*, **3**, 1–15.

Gurson, A.L. 1977. Continuum Theory of Ductile Rupture by Void Nucleation and Growth – Part I: Yield Criteria and Flow Rule for Porous Media. *J. Engng. Mater. Tech.*, **99**, 2–15.

Gurtin, M.E. 1972. The Linear Theory of Elasticity. *Pages 1–295 of:* Flügge, S. and Truesdell, C. (eds), *Handbuch der Physik*, vol. VIa/2. Springer-Verlag.

Gurtin, M.E. 1981. *An Introduction to Continuum Mechanics*. New York: Academic Press.

Gurtin,.M.E. and Francis, E.C. 1981. Simple Rate-Independent Model for Damage. *J. Spacecraft*, **18**(3), 285–286.

Gurtin, M.E. and Martins, L.C. 1976. Cauchy's Theorem in Classical Physics. *Arch. Rat. Mech. Anal.*, **60**(4), 305–324.

Haftka, R.T., Gürdal, Z. and Kamat, M.P. 1990. *Elements of Structural Optimization*. 2nd edn. Kluwer Academic Publishers.

Halphen, B. and Nguyen, Q.S. 1975. Sue les matériaux Standards généralisés. *J. de Mécanique*, **14**, 39–63.

Han, W. and Reddy, B.D. 1999. *Plasticity: Mathematical Theory and Numerical Analysis*. Berlin: Springer-Verlag.

Hancock, J.W. and Mackenzie, A.C. 1976. On the Mechanism of Ductile Fracture in High-strength Steels Subjected to Multi-axial Stress-states. *J. Mech. Phys. Solids*, **24**, 147–169.

Havner, K.S. 1992. *Finite Plastic Deformation of Crystalline Solids*. Cambridge: Cambridge University Press.

Hencky, H. 1933. The Elastic Behavior of Vulcanized Rubber. *J. Appl. Mech.*, **1**, 45–53.

Hibbitt, H.D., Marcal, P.V. and Rice, J.R. 1970. A Finite Element Formulation for Problems of Large Strain and Large Displacement. *Int. J. Solids Structs*, **6**, 1069–1086.

Hill, R. 1948. A Theory of the Yielding and Plastic Flow in Anisotropic Metals. *Proc. Roy. Soc. A*, **193**, 281–297.

Hill, R. 1950. *The Mathematical Theory of Plasticity*. London: Oxford University Press.

Hill, R. 1958. A General Theory of Uniqueness and Stability in Elastic-plastic Solids. *J. Mech. Phys. Solids*, **6**, 236–249.

Hill, R. 1978. Aspects of Invariance in Solid Mechanics. *Adv. Appl. Mech.*, **18**, 1–75.

Hinton, E. and Owen, D.R.J. 1977. *Finite Element Programming*. New York: Academic Press.

Hirsch, M.W. and Smale, S. 1974. *Differential Equations, Dynamical Systems, and Linear Algebra*. New York: Academic Press.

Hoffman, O. 1967. The Brittle Strength of Orthotropic Materials. *J. Composite Mater.*, **1**, 200–206.

Holzapfel, G.A. 2000. *Nonlinear Solid Mechanics. A Continuum Approach for Engineering*. London: John Wiley & Sons.

Horii, H. and Nemat-Nasser, S. 1983. Overall Moduli of Solids with Microcracks: Load Induced Anisotropy. *J. Mech. Phys. Solids*, **31**(2), 155–171.

Huerta, A., Pérez-Foguet, A. and Rodrígues-Ferran, A. 1999. Consistent Tangent Matrices for Complex Problems: Substepping Schemes and Numerical Differentiation. *In:* Wunderlich, W. (ed), *Proceedings of the ECCM'99 – European Conference on Computational Mechanics, Munich 1999*.

Hughes, T.J., Taylor, R.L. and Sackman, J.L. 1975. *Finite Element Formulation and Solution of Contact-impact Problems in Continuum Mechanics – III*. Technical Report UC SESM 75-7. Department of Civil Engineering, University of California, Berkeley.

Hughes, T.J.R. 1980. Generalization of Selective Integration Procedures to Anisotropic and Nonlinear Media. *Int. J. Num. Meth. Engng*, **15**, 1413–1418.

Hughes, T.J.R. 1984. Numerical Implementation of Constitutive Models: Rate-independent Deviatoric Plasticity. *Pages 29–63 of:* Nemat-Nasser *et al.* (Ed.), *Theoretical Foundations for Large-scale Computations for Nonlinear Material Behavior*. Dordrecht, The Netherlands: Martinus Nijhoff.

Hughes, T.J.R. 1987. *The Finite Element Method. Linear Static and Dynamic Finite Element Analysis*. Englewood Cliffs, New Jersey: Prentice-Hall.

Hughes, T.J.R. and Pister, K. 1978. Consistent Linearization in Mechanics of Solids and Structures. *Comp. Structs*, **8**, 391–397.

Hughes, T.J.R. and Taylor, R.L. 1978. Unconditionally Stable Algorithms for Quasi-static Elasto/Viscoplastic Finite Element Analysis. *Comput. Structs*, **8**, 169–173.

Hughes, T.J.R. and Winget, J. 1980. Finite Rotation Effects in Numerical Integration of Rate Constitutive Equations Arising in Large-deformation Analysis. *Int. J. Numer. Meth. Engng*, **15**, 1862–1867.

Ibrahimbegović, A. 1994. Finite Elastoplastic Deformations of Space-Curved Membranes. *Comp. Meth. Appl. Mech. Engng*, **119**, 371–394.

Irons, B.M. 1970. A Frontal Solution Program for Finite Element Analysis. *Int. J. Numer. Meth. Engng*, **2**, 5–32.

Janson, J. 1978. A Continuous Damage Approach to the Fatigue Process. *Engng. Fract. Mech.*, **10**, 651–657.

Jetteur, Ph. 1986. Implicit Integration Algorithm for Elastoplasticity in Plane Stress Analysis. *Engng. Comp.*, **3**, 251–253.

Jirásek, M. and Bažant, Z.P. 2002. *Inelastic Analysis of Structures*. Chichester: Wiley.

Johnson, G. and Bammann, D.J. 1984. A Discussion of Stress Rates in Finite Deformation Problems. *Int. J. Solids Structs*, **20**, 725–737.

Kachanov, L.M. 1958. Time of the Rupture Process under Creep Condition. *Izv. Akad. Nauk. SSSR, Otd. Tekhn. Nauk.*, **8**, 26–31.

Kachanov, L.M. 1977. Creep and Rupture under Complex Loading. *Problemi Prochnosti*, **6**.

Kestin, J. and Bataille, J. 1977. Irreversible Thermodynamics of Continua and Internal Variables. *Pages 39–67 of: Proceedings of the International Symposium on Continuum Models of Discrete Systems*. University of Waterloo Press.

Kocks, U.F. 1970. The Relation Between Polycrystal Deformation and Single-crystal Deformation. *Metall. Trans.*, **1**, 1121–1143.

Koh, C.G., Owen, D.R.J. and Perić, D. 1995. Explicit Dynamic Analysis of Elasto-plastic Laminated Composite Shells: Implementation of Non-iterative Stress Update Schemes for the Hoffman Yield Criterion. *Comput. Mech.*, **16**, 307–314.

Koiter, W.T. 1953. Stress-strain Relations, Uniqueness and Variational Theorems for Elasto-plastic Materials with Singular Yield Surface. *Quart. Appl. Math.*, **11**, 350–354.

Kojić, M. and Bathe, K.-J. 1987. Studies of Finite Element Procedures – Stress Solution of a Closed Elastic Strain Path with Stretching and Shearing Using the Updated Lagrangian Jaumann Formulation. *Comp. Structs*, **26**, 175–179.

Kojić, M. and Bathe, K.J. 1987. The 'Effective-stress-function' for Thermo-elasto-plasticity and Creep. *Int. J. Numer. Meth. Engng*, **24**, 1509–1532.

Korelc, J. and Wriggers, P. 1996. Consistent Gradient Formulation for a Stable Enhanced Strain Method for Large Deformations. *Engng. Comp.*, **13**(1), 103–123.

Krajčinović, D. 1996. *Damage Mechanics*. Amsterdam: North-Holland.

Krajčinović, D. 1983. Constitutive Equations for Damaging Materials. *J. Appl. Mech.*, **50**, 355–360.

Krajčinović, D. 1985. Continuous Damage Mechanics Revisited: Basic Concepts and Definitions. *J. Appl. Mech.*, **52**, 829–834.

Krajčinović, D. and Fonseka, G.U. 1981. The Continuous Damage Theory of Brittle Materials – Part 1: General Theory. *J. Appl. Mech.*, **48**, 809–815.

Krieg, R.D. and Krieg, D.B. 1977. Accuracies of Numerical Solution Methods for the Elastic-perfectly Plastic Model. *J. Pressure Vessel Tech.*, **99**, 510–515.

Ladèveze, P. 1983. On an Anisotropic Damage Theory. *Pages 355–363 of:* Boehler, J.P. (ed), *Proc. CNRS Int. Coll. 351 – Failure Criteria of Structured Media*. Villars-de-Lans.

Ladevèze, P. 1984. Sur une famille d'algorithmes en méchanique des structures. *Comptes-Rendus de l'Academie des Sciences (Paris), Sér. II*, **300**(2), 41–44.

Ladevèze, P. 1989. La méthode à grand incrément de temps pour l'analyse de structures à comportement nonlinéaire décrit par variable internes. *Comptes Rendus de l'Academie des Sciences (Paris), Sér. II*, **309**(11), 1095–1099.

Ladevèze, P. 1999. *Nonlinear Computational Structural Mechanics – New Approaches and Non-Incremental Methods of Calculation*. Springer.

Ladevèze, P. and Lemaitre, J. 1984. Damage Effective Stress in Quasi Unilateral Conditions. *In: 16th Int. Congress Theor. Appl. Mech.*

Ladevèze, P. and Perego, U. 2000. Duality Preserving Discetization of the Large Time Increment Method. *Comp. Meth. Appl. Mech. Engng*, **189**(1), 205–232.

Leckie, F.A. and Hayhurst, D.R. 1974. Creep Rupture of Structures. *Proc. Roy. Soc. Lond. A*, **340**, 323–347.

Leckie, F.A. and Onat, E.T. 1981. Tensorial Nature of Damage Measuring Internal Variables. *Pages 140–155 of: Proceedings of the IUTAM Symposium on Physical Nonlinearities in Structures*. Springer.

Lee, E.H. 1969. Elastic-plastic Deformation at Finite Strains. *J. Appl. Mech.*, **36**, 1–6.

Lee, E.H. and Liu, D.T. 1967. Finite Strain Elastic-plastic Theory with Application to Plane-wave Analysis. *J. Appl. Phys.*, **38**, 19–27.

Lege, D.J., Barlat, F. and Brem, J.C. 1989. Characterization and Modeling of the Mechanical Behaviour and Formability of a 2008-T4 Sheet Sample. *Int. J. Mech. Sci.*, **31**(7), 549–563.

Lemaitre, J. 1983. A Three-dimensional Ductile Damage Model Applied to Deep-drawing Forming Limits. *Pages 1047–1053 of: ICM 4 Stockholm*, vol. 2.

Lemaitre, J. 1985a. A Continuous Damage Mechanics Model for Ductile Fracture. *J. Engng. Mat. Tech.*, **107**, 83–89.

Lemaitre, J. 1985b. Coupled Elasto-plasticity and Damage Constitutive Equations. *Comp. Meth. Appl. Mech. Engng*, **51**, 31–49.

Lemaitre, J. 1987. Formulation and Identification of Damage Kinetic Constitutive Equations. *Pages 37–89 of:* Krajčinović, D. and Lemaitre, J. (eds), *Continuum Damage Mechanics: Theory and Applications*. Springer-Verlag.

Lemaitre, J. 1996. *A Course on Damage Mechanics*. 2nd edn. Springer.

Lemaitre, J. and Chaboche, J.L. 1990. *Mechanics of Solid Materials*. Cambridge University Press.

Lemaitre, J. and Dufailly, J. 1987. Damage Measurements. *Engng. Fract. Mech.*, **28**(8), 643–661.

Lemaitre, J., Desmorat, R. and Sauzay, M. 2000. Anisotropic Damage Law of Evolution. *Eur. J. Mech. A/Solids*, **19**, 187–208.

Li, X. 1995. Large Strain Constitutive Modelling and Computation for Isotropic, Creep Elastoplastic Damage Solids. *Int. J. Numer. Meth. Engng*, **38**, 841–860.

Lion, A. 2000. Constitutive Modelling in Finite Thermoviscoplasticity: A Physical Approach Based on Nonlinear Rheological Models. *Int. J. Plast.*, **16**, 469–494.

Lubarda, V.A. and Lee, E.H. 1981. A Correct Definition of Elastic and Plastic Deformation and its Computational Significance. *J. Appl. Mech.*, **48**, 35–40.

Lubliner, J. 1990. *Plasticity Theory*. New York: Macmillan.

Luenberger, D.G. 1973. *Introduction to Linear and Nonlinear Programming*. Reading, Massachusetts: Addison-Wesley.

Maier, G. 1970. A Matrix Structural Theory of Piece-wise Linear Plasticity with Interacting Yield Planes. *Meccanica*, **6**, 55–66.

Mandel, J. 1973. Equations constitutives et directeurs dans les milieux plastiques et viscoplastiques. *Int. J. Solids Structs*, **9**, 725–740.

Marques, J.M.M.C. 1984. Stress Computation is Elastoplasticity. *Engng. Comp.*, **1**, 42–51.

Marques, J.M.M.C. and Owen, D.R.J. 1983. Strain Hardening Representation for Implicit Quasistatic Elasto-viscoplastic Algorithms. *Comput. Structs*, **17**, 301–304.

Marquis, D. and Lemaitre, J. 1988. Constitutive Equations for the Coupling Between Elasto-plasticity Damage and Ageing. *Rev. Phys. Applic*, **23**, 615–624.

Marsden, J.E. and Hughes, T.J.R. 1983. *Mathematical Foundations of Elasticity*. Englewood Cliffs, New Jersey: Prentice-Hall.

Martin, J.B. 1975. *Plasticity: Fundamentals and General Results*. Cambridge: MIT Press.

Martin, J.B., Reddy, B.D. and Griffin, T.B. 1987. Applications of Mathematical Programming Concepts to Incremental Elastic-plastic Analysis. *Engng. Structs*, **9**, 171–176.

Matthies, H. 1979. Existence Theorems in Thermo-plasticity. *J. Mécanique*, **18**(4), 695–712.

Matthies, H. and Strang, G. 1979. The Solution of Nonlinear Finite Element Equations. *Int. J. Numer. Meth. Engng*, **14**, 1613–1626.

McMeeking, R.M. and Rice, J.R. 1975. Finite-element Formulations for Problems of Large Elastic-plastic Deformation. *Int. J. Solids Structs*, **11**, 601–616.

Meschke, G. and Liu, W.N. 1999. A Re-formulation of the Exponential Algorithm for Finite Strain Plasticity in Terms of Cauchy Stresses. *Comp. Meth. Appl. Mech. Engng*, **173**, 167–187.

Miehe, C. 1993. Computation of Isotropic Tensor Functions. *Comm. Numer. Meth. Engng*, **9**, 889–896.

Miehe, C. 1995. Discontinuous and Continuous Damage Evolution in Ogden-type large-strain elastic materials. *Eur. J. Mech. A/Solids*, **14**(5), 697–720.

Miehe, C. 1996a. Exponential Map Algorithm for Stress Updates in Anisotropic Multiplicative Elastoplasticity for Single Crystals. *Int. J. Numer. Meth. Engng*, **39**, 3367–3390.

Miehe, C. 1996b. Multisurface Thermoplasticity for Single Crystals at Large Strains in Terms of Eulerian Vector Updates. *Int. J. Solids Structs*, **33**, 3103–3130.

Miehe, C. 1998a. Comparison of Two Algorithms for the Computation of Fourth-order Isotropic Tensor Functions. *Comp. Structs*, **66**, 37–43.

Miehe, C. 1998b. A Theoretical and Computational Model for Isotropic Elastoplastic Stress Analysis in Shells at Large Strains. *Comp. Meth. Appl. Mech. Engng*, **155**, 193–233.

Miehe, C. and Keck, J. 2000. Superimposed Finite Elastic-viscoelastic-plastoelastic Stress Response with Damage in Filled Rubbery Polymers. Experiments, Modelling and Algorithmic Implementation. *J. Mech. Phys. Solids*, **48**, 323–365.

Miehe, C., Apel, N. and Lambrecht, M. 2002a. Anisotropic Additive Plasticity in the Logarithmic Strain Space: Modular Kinematic Formulation and Implementation based on Incremental Minimization Principles for Standard Materials. *Comp. Meth. Appl. Mech. Engng*, **191**, 5383–5425.

Miehe, C., Shotte, J. and Lambrecht, M. 2002b. Homogenization of Inelastic Solid Materials at Finite Strains Based on Incremental Minimization Principles. Application to the Texture Analysis of Polycrystals. *J. Mech. Phys. Solids*, **50**(10), 2123–2167.

Millard, A. 1995. Numerical Algorithms for Plane Stress Elastoplasticity: Review and Recommendation. *Pages 237–247 of:* Owen, D.R.J., Oñate, E. and Hinton, E. (eds), *Computational Plasticity: Fundamentals and Applications – Proceedings of the Fourth International Conference held in Barcelona, 3–6 April 1995*. Swansea: Pineridge Press.

Mitchell, G.P. 1990. *Topics in the Numerical Analysis of Inelastic Solids*. PhD thesis, Department of Civil Engineering, University College of Swansea.

Moran, B., Ortiz, M. and Shih, F. 1990. Formulation of Implicit Finite Element Methods for Multiplicative Finite Deformation Plasticity. *Int. J. Numer. Meth. Engng*, **29**, 483–514.

Moss, W.C. 1984. On Instabilities in Large Deformation Simple Shear Loading. *Comp. Meth. Appl. Mech. Engng*, **46**, 329–338.

Mróz, Z. 1967. On the Description of Anisotropic Work Hardening. *J. Mech. Phys. Solids*, **15**, 163–175.

Muir Wood, D. 1990. *Soil Behaviour and Critical State Soil Mechanics*. Cambridge: Cambridge University Press.

Mullins, L. 1969. Softening of Rubber by Deformation. *Rubber Chemistry and Technology*, **42**, 339–351.

Murakami, S. 1988. Mechanical Modeling of Material Damage. *J. Appl. Mech.*, **55**, 280–286.

Murakami, S. and Ohno, N. 1981. A Continuum Theory of Creep and Creep Damage. *Pages 422–443 of:* Ponter, A.R.S. (ed), *Proceedings of the IUTAM Symposium on Creep in Structures, Leicester, 1980*. Berlin: Springer.

Nagtegaal, J.C. 1982. On the Implementation of Inelastic Constitutive Equations with Special Reference to Large Deformation Problems. *Comp. Meth. Appl. Mech. Engng*, **33**, 469–484.

Nagtegaal, J.C. and de Jong, J.E. 1981. Some Computational Aspects of Elastic-plastic Large Strain Analysis. *Int. J. Numer. Meth. Engng*, **17**, 15–41.

Nagtegaal, J.C. and de Jong, J.E. 1982. Some Aspects of Non-isotropic Work-hardening in Finite Strain Plasticity. *Pages 65–102 of:* Lee, E.H. and Mallet, R.L. (eds), *Plasticity of Metals at Finite Strain: Theory, Experiment and Computation*. Division of Applied Mechanics, Stanford University and Department of Engineering & Mechanics, Rensselaer Polytechnic Institute.

Nagtegaal, J.C., Parks, D.M. and Rice, J.M. 1974. On Numerically Accurate Finite Element Solutions in the Fully Plastic Range. *Comp. Meth. Appl. Mech. Engng*, **4**, 153–177.

Nayak, G.C. and Zienkiewicz, O.C. 1972. Elastoplastic Stress Analysis: A Generalisation for Various Constitutive Laws including Strain Softening. *Int. J. Numer. Meth. Engng*, **5**, 113–135.

Naylor, D.J., Pande, G.N., Simpson, B. and Tabb, R. 1981. *Finite Elements in Geotechnical Engineering*. Swansea: Pineridge Press.

Needleman, A. 1977. Inflation of Spherical Rubber Balloons. *Int. J. Solids Structs*, **13**, 409–421.

Needleman, A., Asaro, R.J., Lemonds, J. and Peirce, D. 1985. Finite Element Analysis of Crystalline Solids. *Comp. Meth. Appl. Mech. Engng*, **52**, 689–708.

Nemat-Nasser, S. 1979. Decomposition of Strain Measures and their Rates in Finite Deformation Elastoplasticity. *Int. J. Solids Structs*, **15**, 155–166.

Nemat-Nasser, S. 1982. On Finite Deformation Elasto-plasticity. *Int. J. Solids Structs*, **18**(10), 857–872.

Nemat-Nasser, S. 1999. Averaging Theorems in Finite Deformation Plasticity. *Mech. Mater.*, **31**, 493–523.

Oñate, E., Rojek, J. and García Garino, C. 1995. NUMISTAMP: A Research Project for Assessment of Finite-element Models for Stamping Processes. *J. Mater. Proc. Technol.*, **50**, 17–38.

Oden, J.T. 1972. *Finite Elements of Nonlinear Continua*. London: McGraw-Hill.

Oden, J.T. 1979. *Applied Functional Analysis*. Englewood Cliffs, New Jersey: Prentice-Hall.

Ogden, R.W. 1972. Large Deformation Isotropic Elasticity – On the Correlation of Theory and Experiment for Incompressible Rubberlike Solids. *Proc. R. Soc. Lond. A*, **326**, 565–584.

Ogden, R.W. 1984. *Non-Linear Elastic Deformations*. Chichester: Ellis Horwood.

Onat, E.T. 1986. Representation of Mechanical Behaviour in the Presence of Internal Damage. *Engng. Fract. Mech.*, **25**, 605–614.

Onat, E.T. and Leckie, F.A. 1988. Representation of Mechanical Behavior in the Presence of Changing Internal Structure. *J. Appl. Mech.*, **55**, 1–10.

Ortiz, M. and Martin, J.B. 1989. Symmetry-preserving Return Mapping Algorithms and Incrementally Extremal Paths: A Unification of Concepts. *Int. J. Numer. Meth. Engng*, **28**, 1839–1853.

Ortiz, M. and Popov, E.P. 1985. Accuracy and Stability of Integration Algorithms for Elastoplastic Constitutive Relations. *Int. J. Numer. Meth. Engng*, **21**, 1561–1576.

Ortiz, M. and Simo, J.C. 1986. An Analysis of a New Class of Integration Algorithms for Elastoplastic Constitutive Relations. *Int. J. Numer. Meth. Engng*, **23**, 353–366.

Owen, D.R.J. and Hinton, E. 1980. *Finite Elements in Plasticity: Theory and Practice*. Swansea: Pineridge Press.

Owen, D.R.J., Souza Neto, E.A. de, Zhao, S.Y., Perić, D., and Loughran, J.G. 1998. Finite Element Simulation of the Rolling and Extrusion of Multi-phase Materials: Application to the Rolling of Prepared Sugar Cane. *Comp. Meth. Appl. Mech. Engng*, **151**, 479–495.

Pankaj, and Bićanić, N. 1989. On Multivector Stress Returns in Mohr–Coulomb Plasticity. *Pages 421–436 of:* Owen, D.R.J., Hinton, E. and Oñate, E. (eds), *Proceedings of the Second International Conference on Computational Plasticity: Models, Software and Applications*. Swansea: Pineridge Press.

Papadopoulos, P. and Taylor, R.L. 1994. On the Application of Multi-step Integration Methods to Infinitesimal Elastoplasticity. *Int. J. Numer. Meth. Engng*, **37**, 3169–3184.

Parisch, H. 1986. Efficient Non-linear Finite Element Shell Formulation Involving Large Strains. *Engng. Comp.*, **3**, 121–128.

Peirce, D. 1983. Shear Band Bifurcation in Ductile Single Crystals. *J. Mech. Phys. Solids*, **31**, 133–153.

Peirce, D., Asaro, R.J. and Needleman, A. 1982. An Analysis of Nonuniform and Localized Deformation in Ductile Single Crystals. *Acta Metall.*, **30**, 1087–1119.

Peirce, D., Asaro, R.J. and Needleman, A. 1983. Material Rate Dependence and Localized Deformation in Crystalline Solids. *Acta Metall.*, **31**, 1951–1976.

Peirce, D., Shih, C.F. and Needleman, A. 1984. A Tangent Modulus Method for Rate Dependent Solids. *Comput. Sructs*, **18**, 875–887.

Pérez-Foguet, A. and Armero, F. 2002. On the Formulation of Closest-point Projection Algorithms in Elastoplasticity – Part II: Globally Convergent Schemes. *Int. J. Numer. Meth. Engng*, **53**, 331–374.

Perić, D. 1992. On Consistent Stress Rates in Solid Mechanics: Computational Implications. *Int. J. Numer. Meth. Engng*, **33**, 799–817.

Perić, D. 1993. On a Class of Constitutive Equations in Viscoplasticity: Formulation and Computational Issues. *Int. J. Numer. Meth. Engng*, **36**, 1365–1393.

Perić, D. and de Souza Neto, E.A. 1999. A New Computational Model for Tresca Plasticity at Finite Strains with an Optimal Parametrization in the Principal Space. *Comp. Meth. Appl. Mech. Engng*, **171**, 463–489.

Perić, D. and Dettmer, W. 2003. A Computational Model for Generalized Inelastic Materials at Finite Strains Combining Elastic, Viscoelastic and Plastic Material Behaviour. *Engng. Comput.*, **20**, 768–787.

Perić, D. and Owen, D.R.J. 1991. A Model for Large Deformation of Elasto-viscoplastic Solids at Finite Strains: Computational Issues. *Pages 299–312 of:* Besdo, D. and Stein, E. (eds), *Proceedings of the IUTAM Symposium on Finite Inelastic Deformations – Theory and Applications*. Berlin: Springer.

Perić, D. and Owen, D.R.J. 1992. Computational Model for 3-D Contact Problems with Friction Based on the Penalty Method. *Int. J. Numer. Meth. Engng*, **35**, 1289–1309.

Perić, D., Owen, D.R.J. and Honnor, M.E. 1992. A Model for Finite Strain Elasto-plasticity Based on Logarithmic Strains: Computational Issues. *Comp. Meth. Appl. Mech Engng*, **94**, 35–61.

Perić, D., Yu, J. and Owen, D.R.J. 1994. On Error Estimates and Adaptivity in Elastoplastic Solids: Application to the Numerical Simulation of Strain Localization in Classical and Cosserat Continua. *Int. J. Numer. Meth. Engng*, **37**, 1351–1379.

Perzyna, P. 1963. The Constitutive Equations for Rate Sensitive Plastic Materials. *Quart. Appl. Math.*, **20**, 321–332.

Perzyna, P. 1966. *Fundamental Problems in Viscoplasticity*. Advances in Applied Mechanics, vol. 9. New York: Academic Press. Pages 243–377.

Perzyna, P. 1971. *Thermodynamic Theory of Viscoplasticity*. Advances in Applied Mechanics, vol. 11. New York: Academic Press. Pages 313–354.

Pinsky, P.M., Ortiz, M. and Pister, K.S. 1983. Numerical Integration of Rate Constitutive Equations in Finite Deformation Analysis. *Comp. Meth. Appl. Mech. Engng*, **40**, 137–158.

Pires, F.M.A., de Souza Neto, E.A. and Owen, D.R.J. 2004. On the Finite Element Prediction of Damage Growth and Fracture Initiation in Finitely Deforming Ductile Materials. *Comp. Meth. Appl. Mech. Engng*, **193**, 5223–5256.

Prager, W. 1959. *An Introduction to Plasticity*. London: Addison-Wesley.

Pramono, E. and Willam, K. 1989. Implicit Integration of Composite Yield Surfaces with Corners. *Engng. Comp.*, **6**, 186–197.

Rabotnov, Y.N. 1963. On the Equations of State for Creep. *Page 307 of: Progress in Applied Mechanics, Prager Anniversary Volume*. New York: Macmillan.

Ramm, E. 1981. Strategies for Tracing the Nonlinear Response Near Limit Point. *Pages 63–89 of:* Wunderlich, W. Stein, E. and Bathe, K.-J. (eds), *Nonlinear Finite Element Analysis in Structural Mechanics*. Berlin: Springer-Verlag.

Ramm, E. and Matzenmiller, A. 1987. Computational Aspects of Elasto-plasticity in Shell Analysis. *Pages 711–734 of:* Owen, D.R.J., Hinton, E. and Oñate, E. (eds), *Computational Plasticity: Models, Software and Applications – Proceedings of the International Conference held in Barcelona, 6–10 April 1987*. Swansea: Pineridge Press.

Ransing, R.S., Lewis, R.W., Gethin, D.T., Crook, A.J.L. and Dutko, M. 1998. Compaction of Granular Materials using Deformable Discrete Elements. *Pages 143–147 of:* Huétink, J. and Baaijens, F.P.T. (eds), *Proceedings of the Sixth International Conference on Numerical Methods in Industrial Forming*

Processes – NUMIFORM 98 – Enschede, Netherlands, 22–25 June 1998 – Simulation of Materials Processing: Theory, Methods and Applications. Rotterdam: Balkema.

Rashid, M.M. and Nemat-Nasser, S. 1995. A Constitutive Algorithm for Rate-dependent Crystal Plasticity. *Comp. Meth. Appl. Mech. Engng*, **94**, 201–228.

Reddy, B.D. 1998. *Introductory Functional Analysis*. Springer.

Reddy, B.D. and Martin, J.B. 1991. Algorithms for the Solution of Internal Variable Problems in Plasticity. *Comp. Meth. Appl. Mech. Engng*, **93**, 253–273.

Reddy, B.D. and Simo, J.C. 1995. Stability and Convergence of a Class of Enhanced Strain Methods. *SIAM J. Num. Anal.*, **32**(6), 1705–1728.

Reese, S. and Govindjee, S. 1998. A Theory of Finite Viscoelasticity and Numerical Aspects. *Int. J. Solid Structs*, **35**, 3455–3482.

Reese, S. and Wriggers, P. 2000. A Stabilization Technique to Avoid Hourglassing in Finite Elasticity. *Int. J. Numer. Meth. Engng*, **48**, 79–109.

Rice, J.R. 1968. A Path Independent Integral and the Approximate Analysis of Strain Concentration by Notches and Cracks. *J. Appl. Mech.*, **35**, 379–386.

Rice, J.R. 1971. Inelastic Constitutive Relations for Solids: An Internal-Variable Theory and its Application to Metal Plasticity. *J. Mech. Phys. Solids*, **19**, 433–455.

Rice, J.R. 1976. The Localization of Plastic Deformation. *Pages 207–220 of:* Koiter, W.T. (ed), *Theoretical and Applied Mechanics*, vol. 1. Amsterdam: North-Holland.

Riks, E. 1972. The Application of Newton's Method to the Problem of Elastic Stability. *J. Appl. Mech.*, **39**, 1060–1066.

Riks, E. 1979. An Incremental Approach to the Solution of Snapping and Buckling Problems. *Int. J. Solids Structs*, **15**, 529–551.

Rockafellar, R.T. 1970. *Convex Analysis*. Princeton University Press.

Rockafellar, R.T. and Wets, R.J.-B. 1998. *Variational Analysis*. A Series of Comprehensive Studies in Mathematics, vol. 317. Berlin: Springer.

Romano, G., Rosati, L. and Marotti de Sciarra, M. 1993. Variational Principles for a Class of Finite Step Elastoplastic Problems with Non-linear Mixed Hardening. *Comp. Meth. Appl. Mech. Engng*, **109**, 293–314.

Rosati, L. and Valoroso, N. 2004. A Return Map Algorithm for General Isotropic Elasto/Visco-plastic Materials in Principal Space. *Int. J. Numer. Meth. Engng*, **60**(2), 461–498.

Roscoe, K.H. and Burland, J.B. 1968. On the Generalised Stress-strain Behaviour of 'Wet' Clay. *Pages 535–609 of:* Heyman, J. and Leckie, F.A. (eds), *Engineering Plasticity*. Cambridge: Cambridge University Press.

Rubinstein, R. and Atluri, S.N. 1983. Objectivity of Incremental Constitutive Relations Over Finite Time Steps in Computational Finite Deformation Analysis. *Comp. Meth. Appl. Mech. Engng*, **36**, 277–290.

Runesson, K., Ristinmaa, M. and Mahler, L. 1999. A Comparison of Viscoplasticity Formats and Algorithms. *Mech. Cohesive-Frict. Mater.*, **4**, 75–98.

Saanouni, K., Chaboche, J.L. and Lesne, P.M. 1989. Creep Crack-growth Prediction by a Non-local Damage Formulation. *Pages 404–414 of:* Mazars, J. and Bazant, Z.P. (eds), *Cracking and Damage, Strain Localization and Size Effect*. Amsterdam: Elsevier.

Schellekens, J.C.J. and de Borst, R. 1990. The Use of the Hoffman Yield Criterion in Finite-element Analysis of Anisotropic Composites. *Comput. Structs*, **37**, 1087–1096.

Schönauer, M., de Souza Neto, E.A. and Owen, D.R.J. 1995. Hencky Tensor Based Enhanced Large Strain Element for Elasto-plastic Analysis. *Pages 333–348 of:* Owen, D.R.J., Oñate, E. and Hinton, E. (eds), *Computational Plasticity: Fundamentals and Applications – Proceedings of the Fourth International Conference held in Barcelona, 3–6 April 1995*. Swansea: Pineridge Press.

Schröder, J., Gruttmann, F. and Löblein, J. 2002. A Simple Orthotropic Finite Elasto-plasticity Model based on Generalized Stress–strain Measures. *Comp. Mech.*, **30**, 48–64.

Schweizerhof, K. and Ramm, E. 1984. Displacement Dependent Pressure Loads in Nonlinear Finite Element Analyses. *Comput. Structs*, **18**, 1099–1114.

Seth, B.R. 1964. Generalized Strain Measure with Applications to Physical Problems. *Pages 162–172 of:* Reiner, M. and Abir, D. (eds), *Second-order Effects in Elasticity, Plasticity and Fluid Dynamics*. Oxford: Pergamon Press.

Simo, J.C. 1985. On the Computational Significance of the Intermediate Configuration and Hyperelastic Stress Relations in Finite Deformation Elastoplasticity. *Mech. Mater.*, **4**, 439–451.

Simo, J.C. 1987. On a Fully Three-dimensional Finite-strain Viscoelastic Damage Model: Formulation and Computational Aspects. *Comp. Meth. Appl. Mech. Engng*, **60**, 153–173.

Simo, J.C. 1992. Algorithms for Static and Dynamic Multiplicative Plasticity that Preserve the Classical Return Mapping Schemes of the Infinitesimal Theory. *Comp. Meth. Appl. Mech. Engng*, **99**, 61–112.

Simo, J.C. 1998. Numerical Analysis and Simulation of Plasticity. *Pages 183–499 of:* Ciarlet, P.G. and Lions, J.L. (eds), *Handbook of Numerical Analysis, Volume VI*. Amsterdam: Elsevier.

Simo, J.C. and Armero, F. 1992. Geometrically Non-linear Enhanced Strain Mixed Methods and the Method of Incompatible Modes. *Int. J. Numer. Meth. Engng*, **33**, 1413–1449.

Simo, J.C., Armero, F. and Taylor, R.L. 1993. Improved Versions of Assumed Enhanced Strain Tri-linear Elements for 3D Finite Deformation Problems. *Comp. Meth. Appl. Mech. Engng*, **110**, 359–386.

Simo, J.C. and Govindjee, S. 1991. Non-linear B-stability and Symmetry Preserving Return Mapping Algorithms for Plasticity and Viscoplasticity. *Int. J. Numer. Meth. Engng*, **31**, 151–176.

Simo, J.C. and Hughes, T.J.R. 1986. On the Variational Foundations of Assumed Strain Methods. *J. Appl. Mech.*, **53**, 1685–1695.

Simo, J.C. and Hughes, T.R.J. 1987. General Return Mapping Algorithms for Rate-independent Plasticity. *Pages 221–231 of:* Desai, C.S. *et al.* (Ed.), *Constitutive Laws for Engineering Materials: Theory and Applications*. Elsevier.

Simo, J.C. and Hughes, T.J.R. 1998. *Computational Inelasticity*. New York: Springer-Verlag.

Simo, J.C. and Ju, J.W. 1987. Strain- and Stress-based Continuum Damage Models – I. Formulation and II. Computational Aspects. *Int. J. Solids Structs*, **23**, 821–869.

Simo, J.C., Ju, J.-W., Pister, K.S. and Taylor, R.L. 1988a. Assessment of Cap Model: Consistent Return Algorithms and Rate-dependent Extension. *ASCE J. Engng. Mech.*, **114**(2), 191–218.

Simo, J.C., Kennedy, J.G. and Govindjee, S. 1988b. Non-smooth Multisurface Plasticity and Viscoplasticity. Loading/Unloading Conditions and Numerical Algorithms. *Int. J. Numer. Meth. Engng*, **26**, 2161–2185.

Simo, J.C. and Meschke, G. 1993. A New Class of Algorithms for Classical Plasticity Extended to Finite Strains. Application to Geomaterials. *Comp. Mech.*, **11**, 253–278.

Simo, J.C. and Miehe, C. 1992. Associative Coupled Thermoplasticity at Finite Strains: Formulation, Numerical Analysis and Implementation. *Comp. Meth. Appl. Mech. Engng*, **98**, 41–104.

Simo, J.C. and Ortiz, M. 1985. A Unified Approach to Finite Deformation Plasticity Based on the Use of Hyperelastic Constitutive Equations. *Comp. Meth. Appl. Mech. Engng*, **49**, 221–245.

Simo, J.C. and Pister, K.S. 1984. Remarks on the Rate Constitutive Equations for Finite Deformation Problems: Computational Implications. *Comp. Meth. Appl. Mech. Engng*, **46**, 201–215.

Simo, J.C. and Rifai, S. 1990. A Class of Mixed Assumed Strain Methods and the Method of Incompatible Modes. *Int. J. Numer. Meth. Engng*, **29**, 1595–1638.

Simo, J.C. and Taylor, R.L. 1985. Consistent Tangent Operators for Rate-Independent Elastoplasticity. *Comp. Meth. Appl. Mech. Engng*, **48**, 101–118.

Simo, J.C. and Taylor, R.L. 1986. A Return Mapping Algorithm for Plane Stress Elastoplasticity. *Int. J. Numer. Meth. Engng*, **22**, 649–670.

Simo, J.C. and Taylor, R.L. 1991. Quasi-incompressible Finite Elasticity in Principal Stretches. Continuum Basis and Numerical Algorithms. *Comp. Meth. Appl. Mech. Engng*, **85**, 273–310.

Simo, J.C., Taylor, R.L. and Pister, K.S. 1985. Variational and Projection Methods for the Volume Constraint in Finite Deformation Elasto-plasticity. *Comp. Meth. Appl. Mech. Engng*, **51**, 177–208.

Skrzypek, J.J. 1993. *Plasticity and Creep. Theory, Examples and Problems*. London: CRC Press.

Sloan, S.W. and Booker, J.R. 1992. Integration of Tresca and Mohr–Coulomb Constitutive Relations in Plane Strain Elastoplasticity. *Int. J. Numer. Meth. Engng*, **33**, 163–196.

Sowerby, R. and Chu, E. 1984. Rotations, Stress Rates and Strain Measures in Homogenous Deformation Processes. *Int. J. Solids Structs*, **20**, 1037–1048.

Spencer, A.J.M. 1980. *Continuum Mechanics*. London: Longman.

Steinmann, P. and Stein, E. 1996. On the Numerical Treatment and Analysis of Finite Deformation Ductile Single Crystal Plasticity. *Comp. Meth. Appl. Mech. Engng*, **129**, 235–254.

Steinmann, P., Miehe, C. and Stein, E. 1994. Comparison of Different Finite Deformation Inelastic Damage Models within Multiplicative Elastoplasticity for Ductile Metals. *Comput. Mech.*, **13**, 458–474.

Stern, H.J. 1967. *Rubber: Natural and Synthetic*. London: McLaren and Sons.

Suquet, P. 1981. Sur les équations de la plasticité: existence et régularité des solutions. *J. Mécanique*, **20**, 3–40.

Sussman, T. and Bathe, K.-J. 1987. A Finite Element Formulation for Nonlinear Incompressible Elastic and Inelastic Analysis. *Comp. Structs*, **26**, 357–409.

Taylor, R.L., Beresford, P.J. and Wilson, E.L. 1976. A Non-conforming Element for Stress Analysis. *Int. J. Numer. Meth. Engng*, **10**, 1211–1219.

Toupin, R.A. 1962. Elastic Materials with Couple-Stresses. *Arch. Rat. Mech. Anal.*, **11**, 385–414.

Treloar, L.R.G. 1944. Stress–strain Data for Vulcanized Rubber under Various Types of Deformation. *Trans. Faraday Soc.*, **40**, 59–70.

Tresca, H. 1868. Mémoire sur l'écoulement des corps solides. *Mém. pres. par div. sav.*, **18**, 733–799.

Truesdell, C. 1969. *Rational Thermodynamics*. New York: McGraw-Hill.

Truesdell, C. and Noll, W. 1965. The Non-linear Field Theories of Mechanics. *In:* Flügge, S. (ed), *Handbuch der Physik*, vol. III/3. Springer-Verlag.

Tsakmakis, C. 1996a. Kinematic Hardening Rules for Finite Plasticity. Part I: A Constitutive Approach. *Cont. Mech. Thermodyn.*, **8**, 215–231.

Tsakmakis, C. 1996b. Kinematic Hardening Rules for Finite Plasticity. Part II: Some Examples. *Cont. Mech. Thermodyn.*, **8**, 233–246.

Tvergaard, V. 1981. Influence of Voids on Shear Band Instabilities under Plane Strain Conditions. *Int. J. Fract.*, **17**, 389–407.

Tvergaard, V. 1982a. Material Failure by Void Coalescence in Localized Shear Bands. *Int. J. Solids Structs*, **18**, 659–672.

Tvergaard, V. 1982b. On Localization in Ductile Materials Containing Spherical Voids. *Int. J. Fract.*, **18**, 237–252.

Tvergaard, V. and Needleman, A. 1984. Analysis of the Cup-cone Fracture in a Round Tensile Bar. *Acta Metall.*, **32**, 157–169.

Vaz Jr., M. 1998. *Computational Approaches to Simulation of Metal Cutting Processes*. PhD thesis, Department of Civil Engineering, University of Wales Swansea.

Vaz Jr., M. and Owen, D.R.J. 2001. Aspects of Ductile Fracture and Adaptive mesh Refinement in Damaged Elasto-plastic Materials. *Int. J. Numer. Meth. Engng*, **50**, 29–54.

von Mises, R. 1913. Mechanik der Festen Körper im Plastische-deformablen Zustand. *Nachr. d. Gesellsch d. Wissensch. zu Göttingen, Math.-phys. Klasse*.

Wall, W.A., Bischoff, M. and Ramm, E. 2000. A Deformation Dependent Stabilization Technique Exemplified by EAS Elements at Large Strains. *Comp. Meth. Appl. Mech. Engng*, **188**, 859–871.

Wang, C.C. and Truesdell, C. 1973. *Introduction to Rational Elasticity*. Leyden: Noordhoff.

Washizu, K. 1968. *Variational Methods in Elasticity and Plasticity*. London: Pergamon Press.

Weber, G. and Anand, L. 1990. Finite Deformation Constitutive Equations and a Time Integration Procedure for Isotropic, Hyperelastic-Viscoplastic Solids. *Comp. Meth. Appl. Mech. Engng*, **79**, 173–202.

Wempner, G.A. 1971. Discrete Approximations Related to the Nonlinear Theory of Solids. *Int. J. Solids Structs*, **7**, 1581–1599.

Wilkins, M.L. 1964. Calculation of Elasto-plastic Flow. *In:* Alder, B. *et al.* (Ed.), *Methods of Computational Physics, Vol. 3*. New York: Academic Press.

Wolfram, S. 1991. *Mathematica: A System for Doing Mathematics by Computer*. New York: Addison-Wesley.

Wriggers, P. 2001. *Nichtlineare Finite-Element-Methoden*. Berlin: Springer-Verlag.

Wriggers, P. and Reese, S. 1996. A Note on Enhanced Strain Methods for Large Deformations. *Comp. Meth. Appl. Mech. Engng*, **135**, 201–209.

Wriggers, P. and Taylor, R.L. 1990. A Fully Non-linear Axisymmetrical Membrane Element for Rubber-like Materials. *Engng. Comp.*, **7**, 303–310.

Zienkiewicz, O.C. and Cormeau, I.C. 1974. Visco-plasticity – Plasticity and Creep in Elastic Solids - A Unified Numerical Solution Approach. *Int. J. Numer. Meth. Engng*, **8**, 821–845.

Zienkiewicz, O.C. and Taylor, R.L. 2000. *The Finite Element Method. Volume II: Solid Mechanics*. 5th edn. Butterworth–Heinemann.

Zienkiewicz, O.C., Norris, V.A., Winnicki, L.A., Naylor, D.J. and Lewis, R.W. 1978. A Unified Approach to the Soil Mechanics Problems of Offshore Foundations. *In:* Zienkiewicz, O.C., Lewis, R.W. and Stagg, K.G. (eds), *Numerical Methods in Offshore Engineering*. John Wiley and Sons.

Zouain, N., Borges, L.A. and Hecke, M.B. 1988. A Force Method for Elastic-plastic Analysis of Frames by Quadratic Optimization. *Int. J. Solids Structs*, **24**, 211–221.

Zouain, N., Hecke, M.B. and Feijóo, R.A. 1992. Elastic-plastic Analysis with Potential Functionals Using Internal Variables. *Pages 163–174 of:* Owen, D.R.J., Oñate, E. and Hinton, E. (eds), *Computational Plasticity: Fundamentals and Applications – Proceedings of the Third International Conference held in Barcelona, 6–10 April 1992*. Swansea: Pineridge Press.

INDEX

Printed and bound by CPI Group (UK) Ltd, Croydon, CR0 4YY

19/04/2026

14863920-0002